Complete Solutions Manual

Multivariable Calculus

TENTH EDITION

Ron Larson
The Pennsylvania State University,
The Behrend College

Bruce Edwards
University of Florida

BROOKS/COLE
CENGAGE Learning™

Australia • Brazil • Japan • Korea • Mexico • Singapore • Spain • United Kingdom • United States

BROOKS/COLE
CENGAGE Learning™

ISBN-13: 978-1-285-08580-7

ISBN-10: 1-285-08580-9

Brooks/Cole
20 Channel Center Street
Belmont, MA 02210
USA

Cengage Learning is a leading provider of customized learning solutions with office locations around the globe, including Singapore, the United Kingdom, Australia, Mexico, Brazil, and Japan. Locate your local office at: **www.cengage.com/global**

Cengage Learning products are represented in Canada by Nelson Education, Ltd.

To learn more about Brooks/Cole, visit **www.cengage.com/brookscole**

Purchase any of our products at your local college store or at our preferred online store **www.cengagebrain.com**

Printed in the United States of America
2 3 4 5 6 20 19 18 17 16

CONTENTS

CHAPTER 11
Vectors and the Geometry of Space

CHAPTER 11
Vectors and the Geometry of Space

Section 11.1 Vectors in the Plane

1. (a) $\mathbf{v} = \langle 5 - 1, 4 - 2 \rangle = \langle 4, 2 \rangle$

(b)

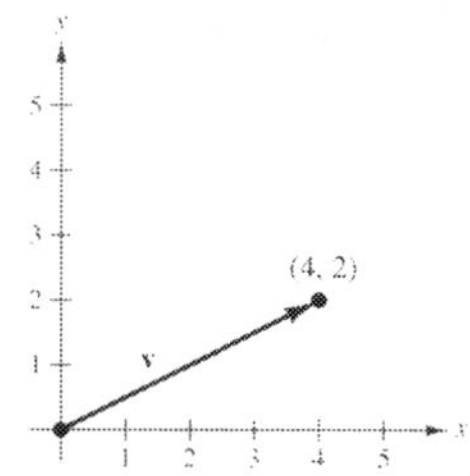

2. (a) $\mathbf{v} = \langle 3 - 3, -2 - 4 \rangle = \langle 0, -6 \rangle$

(b)

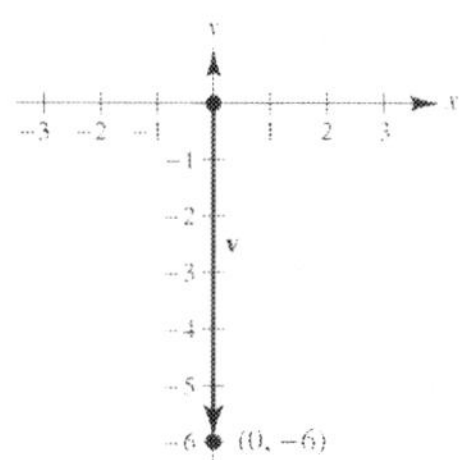

3. (a) $\mathbf{v} = \langle -4 - 2, -3 - (-3) \rangle = \langle -6, 0 \rangle$

(b)

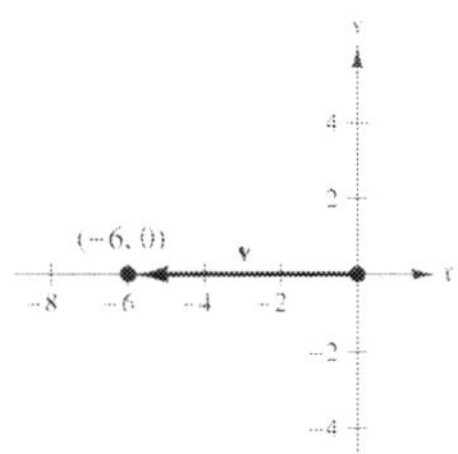

4. (a) $\mathbf{v} = \langle -1 - 2, 3 - 1 \rangle = \langle -3, 2 \rangle$

(b)

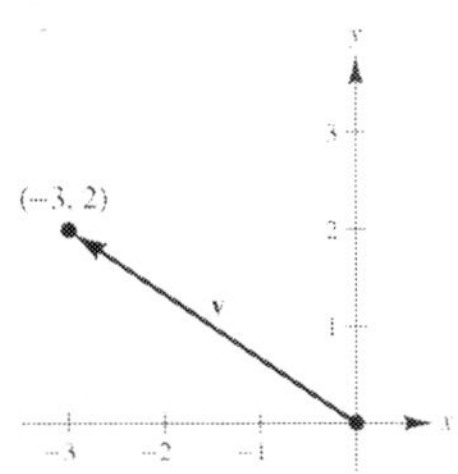

5. $\mathbf{u} = \langle 5 - 3, 6 - 2 \rangle = \langle 2, 4 \rangle$

$\mathbf{v} = \langle 3 - 1, 8 - 4 \rangle = \langle 2, 4 \rangle$

$\mathbf{u} = \mathbf{v}$

6. $\mathbf{u} = \langle 1 - (-4), 8 - 0 \rangle = \langle 5, 8 \rangle$

$\mathbf{v} = \langle 7 - 2, 7 - (-1) \rangle = \langle 5, 8 \rangle$

$\mathbf{u} = \mathbf{v}$

7. $\mathbf{u} = \langle 6 - 0, -2 - 3 \rangle = \langle 6, -5 \rangle$

$\mathbf{v} = \langle 9 - 3, 5 - 10 \rangle = \langle 6, -5 \rangle$

$\mathbf{u} = \mathbf{v}$

8. $\mathbf{u} = \langle 11 - (-4), -4 - (-1) \rangle = \langle 15, -3 \rangle$

$\mathbf{v} = \langle 25 - 0, 10 - 13 \rangle = \langle 15, -3 \rangle$

$\mathbf{u} = \mathbf{v}$

9. (b) $\mathbf{v} = \langle 5 - 2, 5 - 0 \rangle = \langle 3, 5 \rangle$

(c) $\mathbf{v} = 3\mathbf{i} + 5\mathbf{j}$

(a), (d)

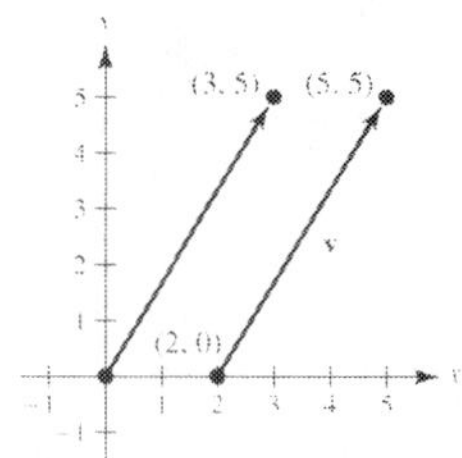

10. (b) $\mathbf{v} = \langle 3 - 4, 6 - (-6) \rangle = \langle -1, 12 \rangle$

(c) $\mathbf{v} = -\mathbf{i} + 12\mathbf{j}$

(a), (d)

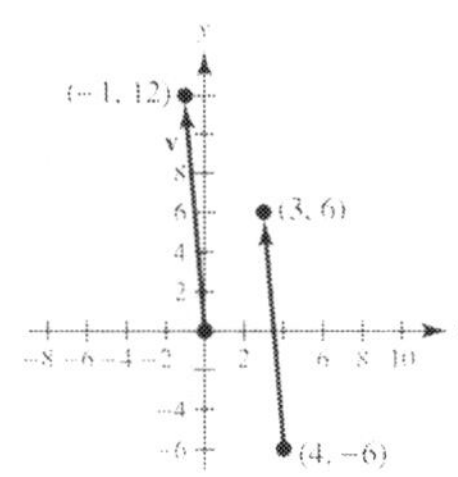

11. (b) $\mathbf{v} = \langle 6 - 8, -1 - 3 \rangle = \langle -2, -4 \rangle$

(c) $\mathbf{v} = -2\mathbf{i} - 4\mathbf{j}$

(a), (d)

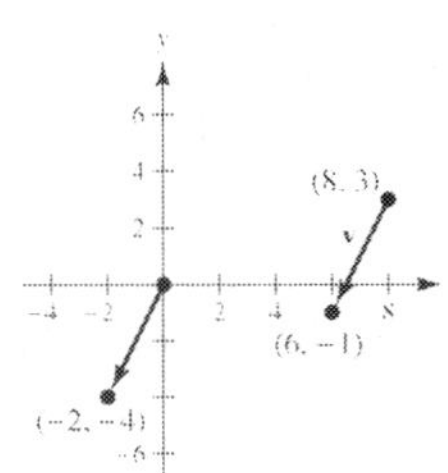

12. (b) $\mathbf{v} = \langle -5 - 0, -1 - (-4) \rangle = \langle -5, 3 \rangle$

(c) $\mathbf{v} = -5\mathbf{i} + 3\mathbf{j}$

(a) and (d).

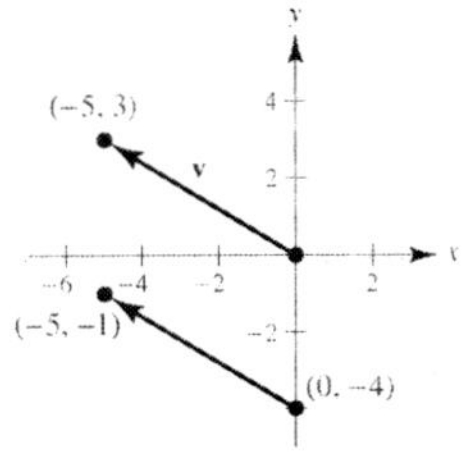

13. (b) $\mathbf{v} = \langle 6 - 6, 6 - 2 \rangle = \langle 0, 4 \rangle$

(c) $\mathbf{v} = 4\mathbf{j}$

(a) and (d).

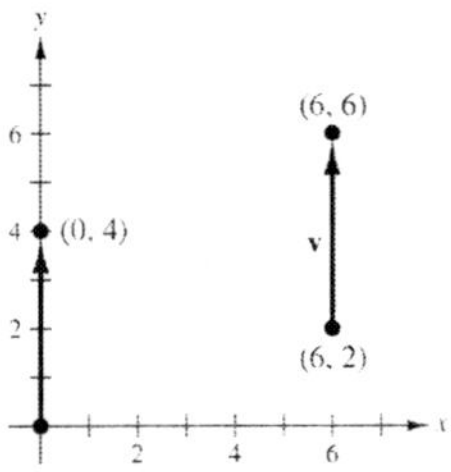

14. (b) $\mathbf{v} = \langle -3 - 7, -1 - (-1) \rangle = \langle -10, 0 \rangle$

(c) $\mathbf{v} = -10\mathbf{i}$

(a) and (d).

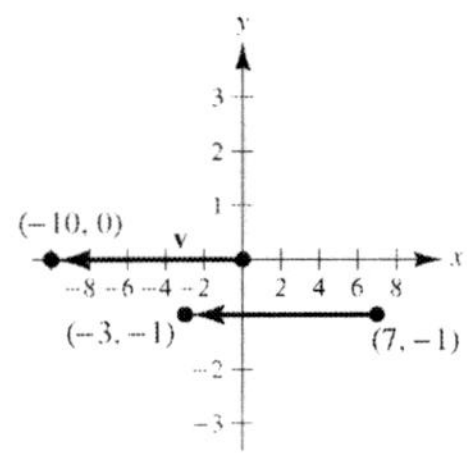

15. (b) $\mathbf{v} = \left\langle \frac{1}{2} - \frac{3}{2}, 3 - \frac{4}{3} \right\rangle = \left\langle -1, \frac{5}{3} \right\rangle$

(c) $\mathbf{v} = -\mathbf{i} + \frac{5}{3}\mathbf{j}$

(a) and (d)

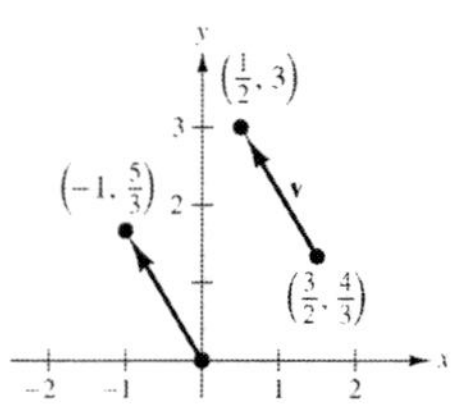

16. (b) $\mathbf{v} = \langle 0.84 - 0.12, 1.25 - 0.60 \rangle = \langle 0.72, 0.65 \rangle$

(c) $\mathbf{v} = 0.72\mathbf{i} + 0.65\mathbf{j}$

(a) and (d).

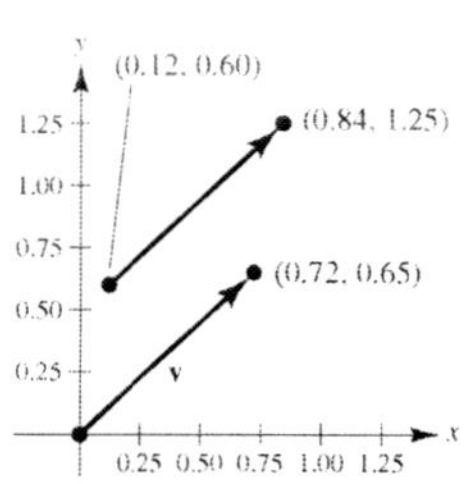

17. (a) $2\mathbf{v} = 2\langle 3, 5 \rangle = \langle 6, 10 \rangle$

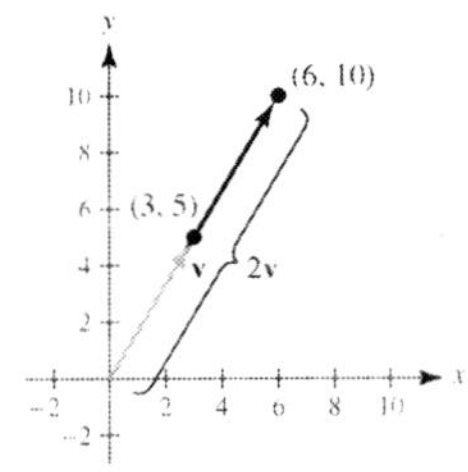

(b) $-3\mathbf{v} = \langle -9, -15 \rangle$

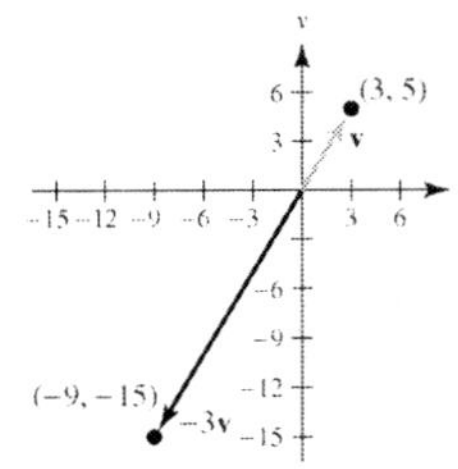

(c) $\frac{7}{2}\mathbf{v} = \left\langle \frac{21}{2}, \frac{35}{2} \right\rangle$

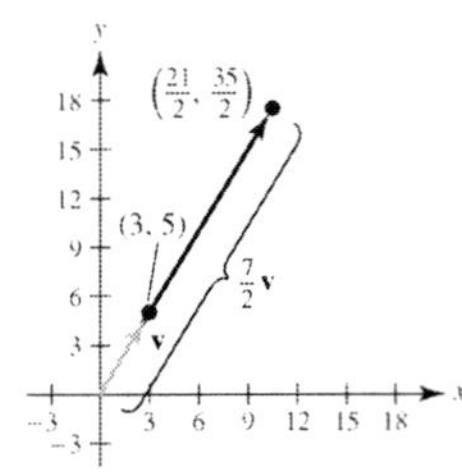

(d) $\frac{2}{3}\mathbf{v} = \left\langle 2, \frac{10}{3} \right\rangle$

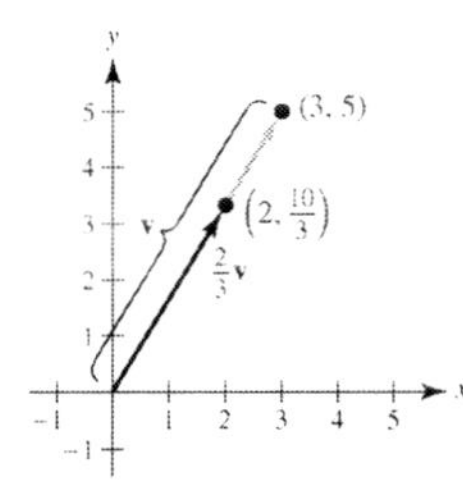

18. (a) $4\mathbf{v} = 4\langle -2, 3\rangle = \langle -8, 12\rangle$

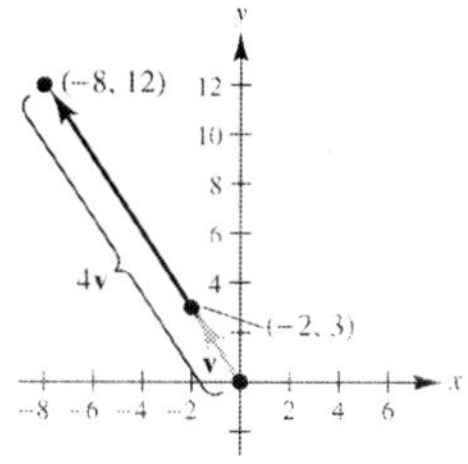

(b) $-\frac{1}{2}\mathbf{v} = \left\langle 1, -\frac{3}{2}\right\rangle$

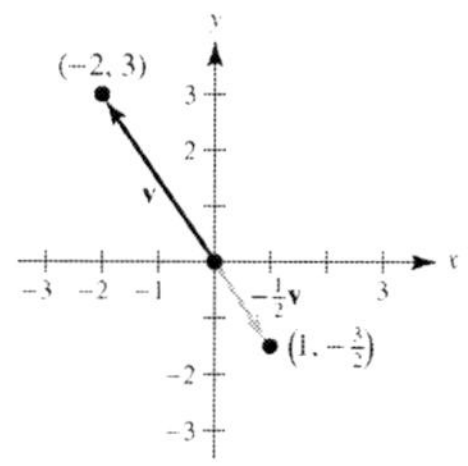

(c) $0\,\mathbf{v} = \langle 0, 0\rangle$

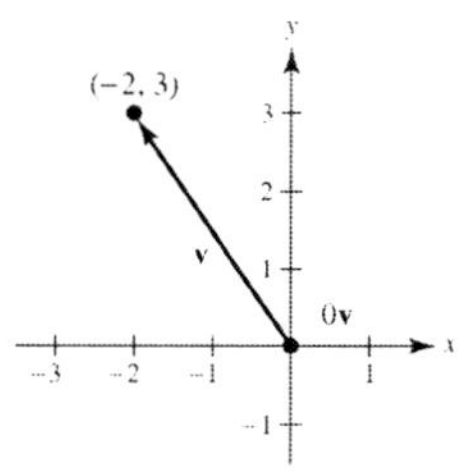

(d) $-6\mathbf{u} = \langle 12, -18\rangle$

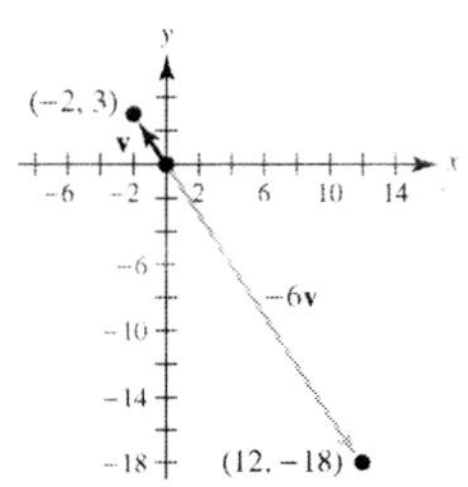

19. (a) $\frac{2}{3}\mathbf{u} = \frac{2}{3}\langle 4, 9\rangle = \left\langle \frac{8}{3}, 6\right\rangle$

(b) $3\mathbf{v} = 3\langle 2, -5\rangle = \langle 6, -15\rangle$

(c) $\mathbf{v} - \mathbf{u} = \langle 2, -5\rangle - \langle 4, 9\rangle = \langle -2, -14\rangle$

(d) $2\mathbf{u} + 5\mathbf{v} = 2\langle 4, 9\rangle + 5\langle 2, -5\rangle = \langle 18, -7\rangle$

20. (a) $\frac{2}{3}\mathbf{u} = \frac{2}{3}\langle -3, -8\rangle = \left\langle -2, -\frac{16}{3}\right\rangle$

(b) $3\mathbf{v} = 3\langle 8, 25\rangle = \langle 24, 75\rangle$

(c) $\mathbf{v} - \mathbf{u} = \langle 8, 25\rangle - \langle -3, -8\rangle = \langle 11, 33\rangle$

(d) $2\mathbf{u} + 5\mathbf{v} = 2\langle -3, -8\rangle + 5\langle 8, 25\rangle = \langle 34, 109\rangle$

21.

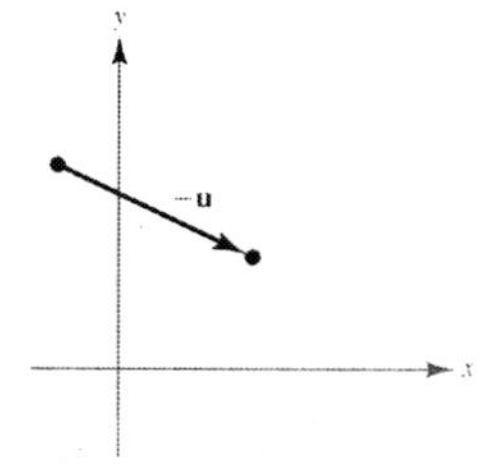

22. Twice as long as given vector **u**.

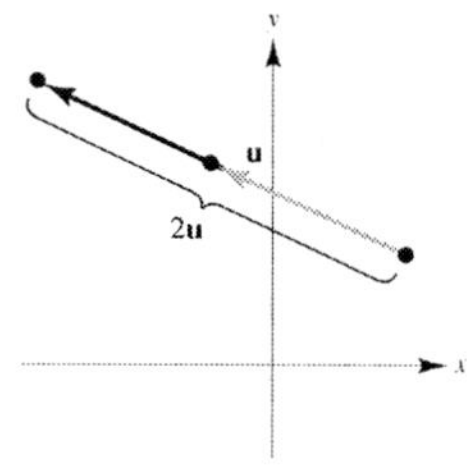

23.

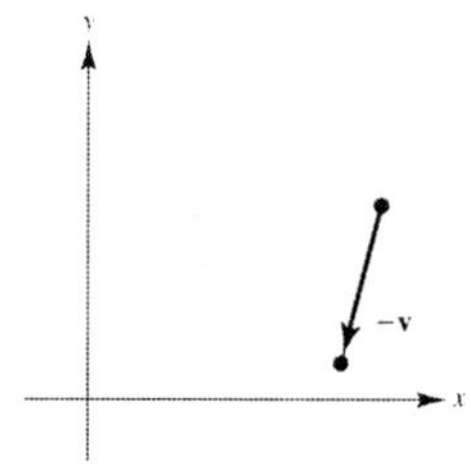

24.

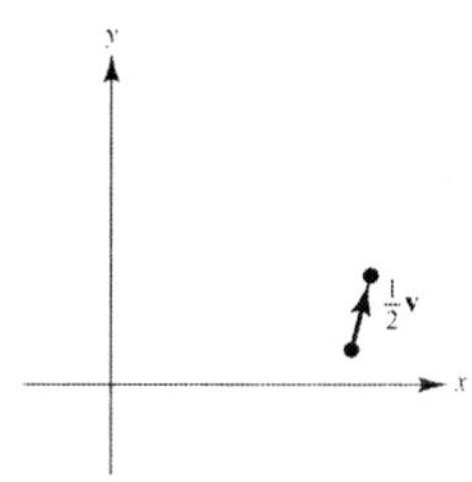

25.

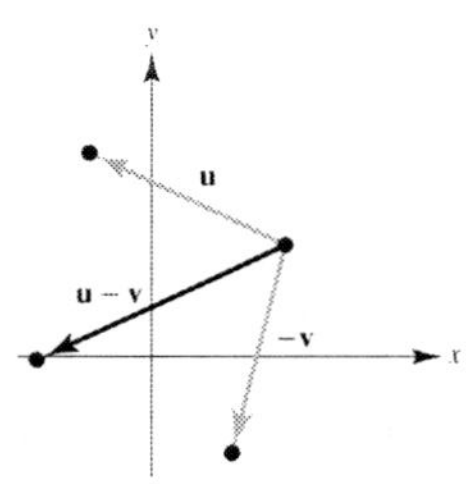

26.

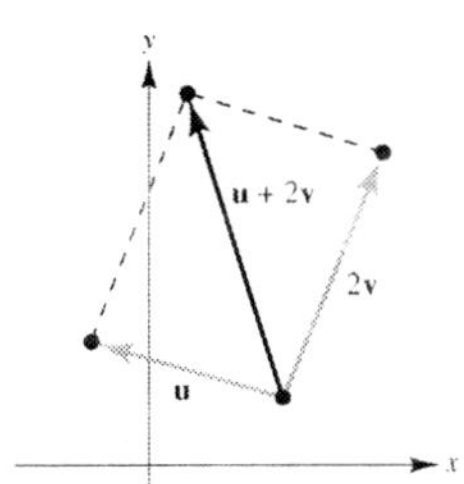

27. $u_1 - 4 = -1$ $\quad u_1 = 3$

$u_2 - 2 = 3$ $\quad u_2 = 5$

$Q = (3, 5)$

28. $u_1 - 5 = 4$ $\quad u_1 = 9$

$u_2 - 3 = -9$ $\quad u_2 = -6$

$Q = (9, -6)$ Terminal point

29. $\|\mathbf{v}\| = \sqrt{0 + 7^2} = 7$

30. $\|\mathbf{v}\| = \sqrt{(-3)^2 + 0} = 3$

31. $\|\mathbf{v}\| = \sqrt{4^2 + 3^2} = 5$

32. $\|\mathbf{v}\| = \sqrt{12^2 + (-5)^2} = 13$

33. $\|\mathbf{v}\| = \sqrt{6^2 + (-5)^2} = \sqrt{61}$

34. $\|\mathbf{v}\| = \sqrt{(-10)^2 + 3^2} = \sqrt{109}$

35. $\mathbf{v} = \langle 3, 12 \rangle$

$\|\mathbf{v}\| = \sqrt{3^2 + 12^2} = \sqrt{153}$

$\mathbf{u} = \dfrac{\mathbf{v}}{\|\mathbf{v}\|} = \dfrac{\langle 3, 12 \rangle}{\sqrt{153}} = \left\langle \dfrac{3}{\sqrt{153}}, \dfrac{12}{\sqrt{153}} \right\rangle$

$= \left\langle \dfrac{\sqrt{17}}{17}, \dfrac{4\sqrt{17}}{17} \right\rangle$ unit vector

36. $\mathbf{v} = \langle -5, 15 \rangle$

$\|\mathbf{v}\| = \sqrt{25 + 225} = \sqrt{250} = 5\sqrt{10}$

$\mathbf{u} = \dfrac{\mathbf{v}}{\|\mathbf{v}\|} = \dfrac{\langle -5, 15 \rangle}{5\sqrt{10}} = \left\langle -\dfrac{\sqrt{10}}{10}, \dfrac{3\sqrt{10}}{10} \right\rangle$ unit vector

37. $\mathbf{v} = \left\langle \dfrac{3}{2}, \dfrac{5}{2} \right\rangle$

$\|\mathbf{v}\| = \sqrt{\left(\dfrac{3}{2}\right)^2 + \left(\dfrac{5}{2}\right)^2} = \dfrac{\sqrt{34}}{2}$

$\mathbf{u} = \dfrac{\mathbf{v}}{\|\mathbf{v}\|} = \dfrac{\left\langle \left(\dfrac{3}{2}\right), \left(\dfrac{5}{2}\right) \right\rangle}{\dfrac{\sqrt{34}}{2}} = \left\langle \dfrac{3}{\sqrt{34}}, \dfrac{5}{\sqrt{34}} \right\rangle$

$= \left\langle \dfrac{3\sqrt{34}}{34}, \dfrac{5\sqrt{34}}{34} \right\rangle$ unit vector

38. $\mathbf{v} = \langle -6.2, 3.4 \rangle$

$\|\mathbf{v}\| = \sqrt{(-6.2)^2 + (3.4)^2} = \sqrt{50} = 5\sqrt{2}$

$\mathbf{u} = \dfrac{\mathbf{v}}{\|\mathbf{v}\|} = \dfrac{\langle -6.2, 3.4 \rangle}{5\sqrt{2}} = \left\langle -\dfrac{31\sqrt{2}}{50}, \dfrac{17\sqrt{2}}{50} \right\rangle$ unit vector

39. $\mathbf{u} = \langle 1, -1 \rangle, \mathbf{v} = \langle -1, 2 \rangle$

(a) $\|\mathbf{u}\| = \sqrt{1 + 1} = \sqrt{2}$

(b) $\|\mathbf{v}\| = \sqrt{1 + 4} = \sqrt{5}$

(c) $\mathbf{u} + \mathbf{v} = \langle 0, 1 \rangle$

$\|\mathbf{u} + \mathbf{v}\| = \sqrt{0 + 1} = 1$

(d) $\dfrac{\mathbf{u}}{\|\mathbf{u}\|} = \dfrac{1}{\sqrt{2}}\langle 1, -1 \rangle$

$\left\|\dfrac{\mathbf{u}}{\|\mathbf{u}\|}\right\| = 1$

(e) $\dfrac{\mathbf{v}}{\|\mathbf{v}\|} = \dfrac{1}{\sqrt{5}}\langle -1, 2 \rangle$

$\left\|\dfrac{\mathbf{v}}{\|\mathbf{v}\|}\right\| = 1$

(f) $\dfrac{\mathbf{u} + \mathbf{v}}{\|\mathbf{u} + \mathbf{v}\|} = \langle 0, 1 \rangle$

$\left\|\dfrac{\mathbf{u} + \mathbf{v}}{\|\mathbf{u} + \mathbf{v}\|}\right\| = 1$

40. $\mathbf{u} = \langle 0, 1 \rangle, \mathbf{v} = \langle 3, -3 \rangle$

(a) $\|\mathbf{u}\| = \sqrt{0 + 1} = 1$

(b) $\|\mathbf{v}\| = \sqrt{9 + 9} = 3\sqrt{2}$

(c) $\mathbf{u} + \mathbf{v} = \langle 3, -2 \rangle$

$\|\mathbf{u} + \mathbf{v}\| = \sqrt{9 + 4} = \sqrt{13}$

(d) $\dfrac{\mathbf{u}}{\|\mathbf{u}\|} = \langle 0, 1 \rangle$

$\left\|\dfrac{\mathbf{u}}{\|\mathbf{u}\|}\right\| = 1$

(e) $\dfrac{\mathbf{v}}{\|\mathbf{v}\|} = \dfrac{1}{3\sqrt{2}}\langle 3, -3 \rangle$

$\left\|\dfrac{\mathbf{v}}{\|\mathbf{v}\|}\right\| = 1$

(f) $\dfrac{\mathbf{u} + \mathbf{v}}{\|\mathbf{u} + \mathbf{v}\|} = \dfrac{1}{\sqrt{13}}\langle 3, -2 \rangle$

$\left\|\dfrac{\mathbf{u} + \mathbf{v}}{\|\mathbf{u} + \mathbf{v}\|}\right\| = 1$

41. $\mathbf{u} = \left\langle 1, \frac{1}{2} \right\rangle, \mathbf{v} = \langle 2, 3 \rangle$

(a) $\|\mathbf{u}\| = \sqrt{1 + \frac{1}{4}} = \frac{\sqrt{5}}{2}$

(b) $\|\mathbf{v}\| = \sqrt{4 + 9} = \sqrt{13}$

(c) $\mathbf{u} + \mathbf{v} = \left\langle 3, \frac{7}{2} \right\rangle$

$\|\mathbf{u} + \mathbf{v}\| = \sqrt{9 + \frac{49}{4}} = \frac{\sqrt{85}}{2}$

(d) $\frac{\mathbf{u}}{\|\mathbf{u}\|} = \frac{2}{\sqrt{5}}\left\langle 1, \frac{1}{2} \right\rangle$

$\left\| \frac{\mathbf{u}}{\|\mathbf{u}\|} \right\| = 1$

(e) $\frac{\mathbf{v}}{\|\mathbf{v}\|} = \frac{1}{\sqrt{13}}\langle 2, 3 \rangle$

$\left\| \frac{\mathbf{v}}{\|\mathbf{v}\|} \right\| = 1$

(f) $\frac{\mathbf{u} + \mathbf{v}}{\|\mathbf{u} + \mathbf{v}\|} = \frac{2}{\sqrt{85}}\left\langle 3, \frac{7}{2} \right\rangle$

$\left\| \frac{\mathbf{u} + \mathbf{v}}{\|\mathbf{u} + \mathbf{v}\|} \right\| = 1$

42. $\mathbf{u} = \langle 2, -4 \rangle, \mathbf{v} = \langle 5, 5 \rangle$

(a) $\|\mathbf{u}\| = \sqrt{4 + 16} = 2\sqrt{5}$

(b) $\|\mathbf{v}\| = \sqrt{25 + 25} = 5\sqrt{2}$

(c) $\mathbf{u} + \mathbf{v} = \langle 7, 1 \rangle$

$\|\mathbf{u} + \mathbf{v}\| = \sqrt{49 + 1} = 5\sqrt{2}$

(d) $\frac{\mathbf{u}}{\|\mathbf{u}\|} = \frac{1}{2\sqrt{5}}\langle 2, -4 \rangle$

$\left\| \frac{\mathbf{u}}{\|\mathbf{u}\|} \right\| = 1$

(e) $\frac{\mathbf{v}}{\|\mathbf{v}\|} = \frac{1}{5\sqrt{2}}\langle 5, 5 \rangle$

$\left\| \frac{\mathbf{v}}{\|\mathbf{v}\|} \right\| = 1$

(f) $\frac{\mathbf{u} + \mathbf{v}}{\|\mathbf{u} + \mathbf{v}\|} = \frac{1}{5\sqrt{2}}\langle 7, 1 \rangle$

$\left\| \frac{\mathbf{u} + \mathbf{v}}{\|\mathbf{u} + \mathbf{v}\|} \right\| = 1$

43. $\mathbf{u} = \langle 2, 1 \rangle$

$\|\mathbf{u}\| = \sqrt{5} \approx 2.236$

$\mathbf{v} = \langle 5, 4 \rangle$

$\|\mathbf{v}\| = \sqrt{41} \approx 6.403$

$\mathbf{u} + \mathbf{v} = \langle 7, 5 \rangle$

$\|\mathbf{u} + \mathbf{v}\| = \sqrt{74} \approx 8.602$

$\|\mathbf{u} + \mathbf{v}\| \le \|\mathbf{u}\| + \|\mathbf{v}\|$

$\sqrt{74} \le \sqrt{5} + \sqrt{41}$

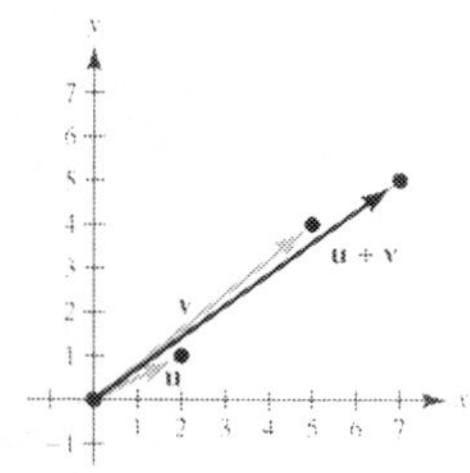

44. $\mathbf{u} = \langle -3, 2 \rangle$

$\|\mathbf{u}\| = \sqrt{13} \approx 3.606$

$\mathbf{v} = \langle 1, -2 \rangle$

$\|\mathbf{v}\| = \sqrt{5} \approx 2.236$

$\mathbf{u} + \mathbf{v} = \langle -2, 0 \rangle$

$\|\mathbf{u} + \mathbf{v}\| = 2$

$\|\mathbf{u} + \mathbf{v}\| \le \|\mathbf{u}\| + \|\mathbf{v}\|$

$2 \le \sqrt{13} + \sqrt{5}$

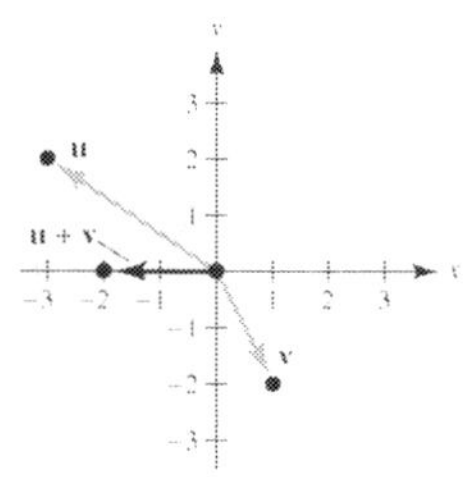

45. $\frac{\mathbf{u}}{\|\mathbf{u}\|} = \frac{1}{3}\langle 0, 3 \rangle = \langle 0, 1 \rangle$

$6\left(\frac{\mathbf{u}}{\|\mathbf{u}\|}\right) = 6\langle 0, 1 \rangle = \langle 0, 6 \rangle$

$\mathbf{v} = \langle 0, 6 \rangle$

46. $\frac{\mathbf{u}}{\|\mathbf{u}\|} = \frac{1}{\sqrt{2}}\langle 1, 1 \rangle$

$4\left(\frac{\mathbf{u}}{\|\mathbf{u}\|}\right) = 2\sqrt{2}\langle 1, 1 \rangle$

$\mathbf{v} = \langle 2\sqrt{2}, 2\sqrt{2} \rangle$

47. $\frac{\mathbf{u}}{\|\mathbf{u}\|} = \frac{1}{\sqrt{5}}\langle -1, 2 \rangle = \left\langle -\frac{1}{\sqrt{5}}, \frac{2}{\sqrt{5}} \right\rangle$

$5\left(\frac{\mathbf{u}}{\|\mathbf{u}\|}\right) = 5\left\langle -\frac{1}{\sqrt{5}}, \frac{2}{\sqrt{5}} \right\rangle = \langle -\sqrt{5}, 2\sqrt{5} \rangle$

$\mathbf{v} = \langle -\sqrt{5}, 2\sqrt{5} \rangle$

48. $\frac{\mathbf{u}}{\|\mathbf{u}\|} = \frac{1}{2\sqrt{3}}\langle \sqrt{3}, 3 \rangle$

$2\left(\frac{\mathbf{u}}{\|\mathbf{u}\|}\right) = \frac{1}{\sqrt{3}}\langle \sqrt{3}, 3 \rangle$

$\mathbf{v} = \langle 1, \sqrt{3} \rangle$

49. $\mathbf{v} = 3\left[(\cos 0^\circ)\mathbf{i} + (\sin 0^\circ)\mathbf{j}\right] = 3\mathbf{i} = \langle 3, 0 \rangle$

50. $\mathbf{v} = 5\left[(\cos 120°)\mathbf{i} + (\sin 120°)\mathbf{j}\right] = -\frac{5}{2}\mathbf{i} + \frac{5\sqrt{3}}{2}\mathbf{j} = \left\langle -\frac{5}{2}, \frac{5\sqrt{3}}{2} \right\rangle$

51. $\mathbf{v} = 2\left[(\cos 150°)\mathbf{i} + (\sin 150°)\mathbf{j}\right] = -\sqrt{3}\mathbf{i} + \mathbf{j} = \left\langle -\sqrt{3}, 1 \right\rangle$

52. $\mathbf{v} = 4\left[(\cos 3.5°)\mathbf{i} + (\sin 3.5°)\mathbf{j}\right] \approx 3.9925\mathbf{i} + 0.2442\mathbf{j} = \langle 3.9925, 0.2442 \rangle$

53.
$$\mathbf{u} = (\cos 0°)\mathbf{i} + (\sin 0°)\mathbf{j} = \mathbf{i}$$
$$\mathbf{v} = 3(\cos 45°)\mathbf{i} + 3(\sin 45°)\mathbf{j} = \frac{3\sqrt{2}}{2}\mathbf{i} + \frac{3\sqrt{2}}{2}\mathbf{j}$$
$$\mathbf{u} + \mathbf{v} = \left(\frac{2 + 3\sqrt{2}}{2}\right)\mathbf{i} + \frac{3\sqrt{2}}{2}\mathbf{j} = \left\langle \frac{2 + 3\sqrt{2}}{2}, \frac{3\sqrt{2}}{2} \right\rangle$$

54.
$$\mathbf{u} = 4(\cos 0°)\mathbf{i} + 4(\sin 0°)\mathbf{j} = 4\mathbf{i}$$
$$\mathbf{v} = 2(\cos 30°)\mathbf{i} + 2(\sin 30°)\mathbf{j} = \mathbf{i} + \sqrt{3}\mathbf{j}$$
$$\mathbf{u} + \mathbf{v} = 5\mathbf{i} + \sqrt{3}\mathbf{j} = \left\langle 5, \sqrt{3} \right\rangle$$

55.
$$\mathbf{u} = 2(\cos 4)\mathbf{i} + 2(\sin 4)\mathbf{j}$$
$$\mathbf{v} = (\cos 2)\mathbf{i} + (\sin 2)\mathbf{j}$$
$$\mathbf{u} + \mathbf{v} = (2\cos 4 + \cos 2)\mathbf{i} + (2\sin 4 + \sin 2)\mathbf{j} = \langle 2\cos 4 + \cos 2, 2\sin 4 + \sin 2 \rangle$$

56.
$$\mathbf{u} = 5\left[\cos(-0.5)\right]\mathbf{i} + 5\left[\sin(-0.5)\right]\mathbf{j} = 5(\cos 0.5)\mathbf{i} - 5(\sin 0.5)\mathbf{j}$$
$$\mathbf{v} = 5(\cos 0.5)\mathbf{i} + 5(\sin 0.5)\mathbf{j}$$
$$\mathbf{u} + \mathbf{v} = 10(\cos 0.5)\mathbf{i} = \langle 10\cos 0.5, 0 \rangle$$

57. Answers will vary. *Sample answer*: A scalar is a real number such as 2. A vector is represented by a directed line segment. A vector has both magnitude and direction. For example $\left\langle \sqrt{3}, 1 \right\rangle$ has direction $\frac{\pi}{6}$ and a magnitude of 2.

58. (a) Vector. The velocity has both magnitude and direction.

(b) Scalar. The price is a number.

(c) Scalar. The temperature is a number.

(d) Vector. The weight has magnitude and direction.

59. $(-4, -1), (6, 5), (10, 3)$

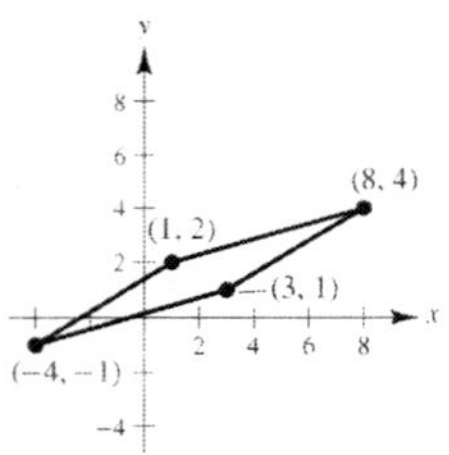

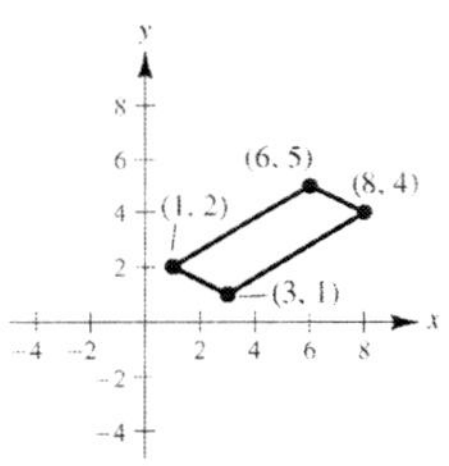

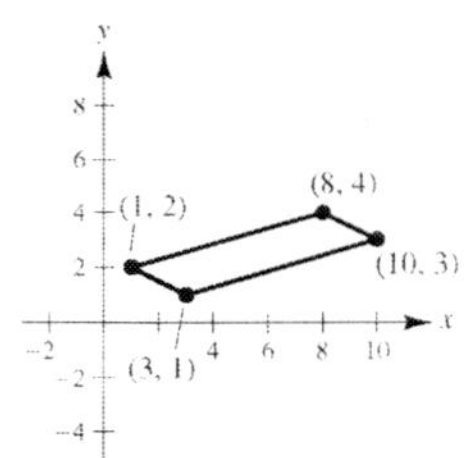

60. (a) True. **d** has the same magnitude as **a** but is in the opposite direction.

(b) True. **c** and **s** have the same length and direction.

(c) True. **a** and **u** are the adjacent sides of a parallelogram. So, the resultant vector, **a** + **u**, is the diagonal of the parallelogram, **c**.

(d) False. The negative of a vector has the opposite direction of the original vector.

(e) True. $\mathbf{a} + \mathbf{d} = \mathbf{a} + (-\mathbf{a}) = 0$

(f) False. $\mathbf{u} - \mathbf{v} = \mathbf{u} - (-\mathbf{u}) = 2\mathbf{u}$
$$-2(\mathbf{b} + \mathbf{t}) = -2(\mathbf{b} + \mathbf{b}) = -2(2\mathbf{b}) = -2\left[2(-\mathbf{u})\right] = -\mathbf{u}$$

For Exercises 61–66,
$\mathbf{au} + \mathbf{bw} = a(\mathbf{i} + 2\mathbf{j}) + b(\mathbf{i} - \mathbf{j}) = (a + b)\mathbf{i} + (2a - b)\mathbf{j}.$

61. $\mathbf{v} = 2\mathbf{i} + \mathbf{j}$. So, $a + b = 2, 2a - b = 1$. Solving simultaneously, you have $a = 1, b = 1$.

62. $\mathbf{v} = 3\mathbf{j}$. So, $a + b = 0, 2a - b = 3$. Solving simultaneously, you have $a = 1, b = -1$.

63. $\mathbf{v} = 3\mathbf{i}$. So, $a + b = 3, 2a - b = 0$. Solving simultaneously, you have $a = 1, b = 2$.

64. $\mathbf{v} = 3\mathbf{i} + 3\mathbf{j}$. So, $a + b = 3, 2a - b = 3$. Solving simultaneously, you have $a = 2, b = 1$.

65. $\mathbf{v} = \mathbf{i} + \mathbf{j}$. So, $a + b = 1, 2a - b = 1$. Solving simultaneously, you have $a = \frac{2}{3}, b = \frac{1}{3}$.

66. $\mathbf{v} = -\mathbf{i} + 7\mathbf{j}$. So, $a + b = -1, 2a - b = 7$. Solving simultaneously, you have $a = 2, b = -3$.

67. $f(x) = x^2, f'(x) = 2x, f'(3) = 6$

(a) $m = 6$. Let $\mathbf{w} = \langle 1, 6\rangle, \|\mathbf{w}\| = \sqrt{37}$, then $\pm\dfrac{\mathbf{w}}{\|\mathbf{w}\|} = \pm\dfrac{1}{\sqrt{37}}\langle 1, 6\rangle$.

(b) $m = -\frac{1}{6}$. Let $\mathbf{w} = \langle -6, 1\rangle, \|\mathbf{w}\| = \sqrt{37}$, then $\pm\dfrac{\mathbf{w}}{\|\mathbf{w}\|} = \pm\dfrac{1}{\sqrt{37}}\langle -6, 1\rangle$.

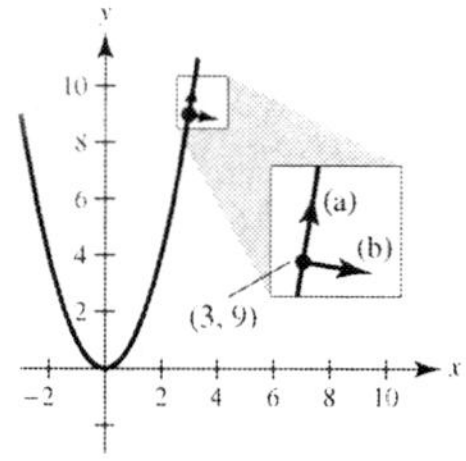

68. $f(x) = -x^2 + 5, f'(x) = -2x, f'(1) = -2$

(a) $m = -2$. Let $\mathbf{w} = \langle 1, -2\rangle, \|\mathbf{w}\| = \sqrt{5}$, then $\pm\dfrac{\mathbf{w}}{\|\mathbf{w}\|} = \pm\dfrac{1}{\sqrt{5}}\langle 1, -2\rangle$.

(b) $m = \dfrac{1}{2}$. Let $\mathbf{w} = \langle 2, 1\rangle, \|\mathbf{w}\| = \sqrt{5}$, then $\pm\dfrac{\mathbf{w}}{\|\mathbf{w}\|} = \pm\dfrac{1}{\sqrt{5}}\langle 2, 1\rangle$.

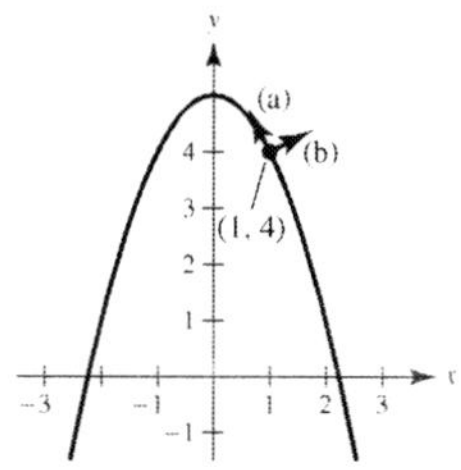

69. $f(x) = x^3, f'(x) = 3x^2 = 3$ at $x = 1$.

(a) $m = 3$. Let $\mathbf{w} = \langle 1, 3\rangle, \|\mathbf{w}\| = \sqrt{10}$, then $\dfrac{\mathbf{w}}{\|\mathbf{w}\|} = \pm\dfrac{1}{\sqrt{10}}\langle 1, 3\rangle$.

(b) $m = -\dfrac{1}{3}$. Let $\mathbf{w} = \langle 3, -1\rangle, \|\mathbf{w}\| = \sqrt{10}$, then $\dfrac{\mathbf{w}}{\|\mathbf{w}\|} = \pm\dfrac{1}{\sqrt{10}}\langle 3, -1\rangle$.

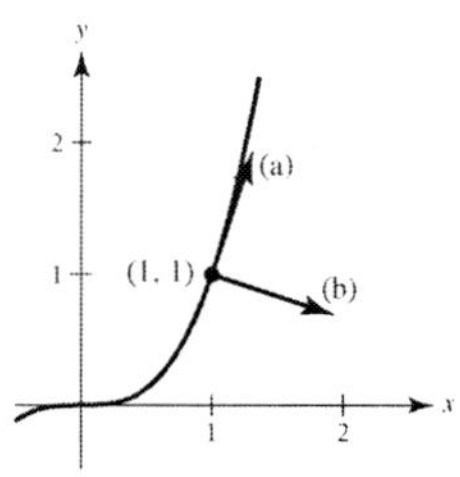

70. $f(x) = x^3, f'(x) = 3x^2 = 12$ at $x = -2$.

(a) $m = 12$. Let $\mathbf{w} = \langle 1, 12\rangle, \|\mathbf{w}\| = \sqrt{145}$, then $\dfrac{\mathbf{w}}{\|\mathbf{w}\|} = \pm\dfrac{1}{\sqrt{145}}\langle 1, 12\rangle$.

(b) $m = -\dfrac{1}{12}$. Let $\mathbf{w} = \langle 12, -1\rangle, \|\mathbf{w}\| = \sqrt{145}$, then $\dfrac{\mathbf{w}}{\|\mathbf{w}\|} = \pm\dfrac{1}{\sqrt{145}}\langle 12, -1\rangle$.

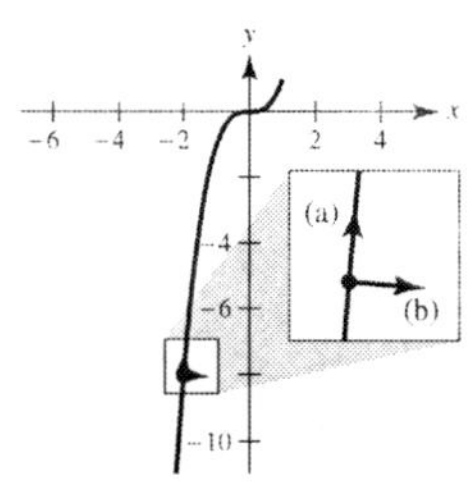

71. $f(x) = \sqrt{25 - x^2}$

$f'(x) = \dfrac{-x}{\sqrt{25 - x^2}} = \dfrac{-3}{4}$ at $x = 3$.

(a) $m = -\dfrac{3}{4}$. Let $\mathbf{w} = \langle -4, 3\rangle, \|\mathbf{w}\| = 5$, then $\dfrac{\mathbf{w}}{\|\mathbf{w}\|} = \pm\dfrac{1}{5}\langle -4, 3\rangle$.

(b) $m = \dfrac{4}{3}$. Let $\mathbf{w} = \langle 3, 4\rangle, \|\mathbf{w}\| = 5$, then $\dfrac{\mathbf{w}}{\|\mathbf{w}\|} = \pm\dfrac{1}{5}\langle 3, 4\rangle$.

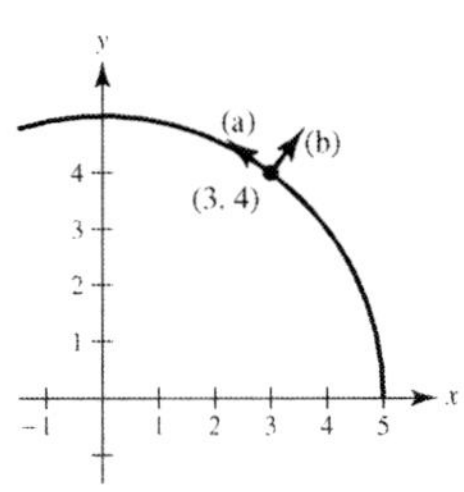

72. $f(x) = \tan x$

$f'(x) = \sec^2 x = 2$ at $x = \dfrac{\pi}{4}$

(a) $m = 2$. Let $\mathbf{w} = \langle 1, 2\rangle, \|\mathbf{w}\| = \sqrt{5}$, then $\dfrac{\mathbf{w}}{\|\mathbf{w}\|} = \pm\dfrac{1}{\sqrt{5}}\langle 1, 2\rangle$.

(b) $m = -\dfrac{1}{2}$. Let $\mathbf{w} = \langle -2, 1\rangle, \|\mathbf{w}\| = \sqrt{5}$, then $\dfrac{\mathbf{w}}{\|\mathbf{w}\|} = \pm\dfrac{1}{\sqrt{5}}\langle -2, 1\rangle$.

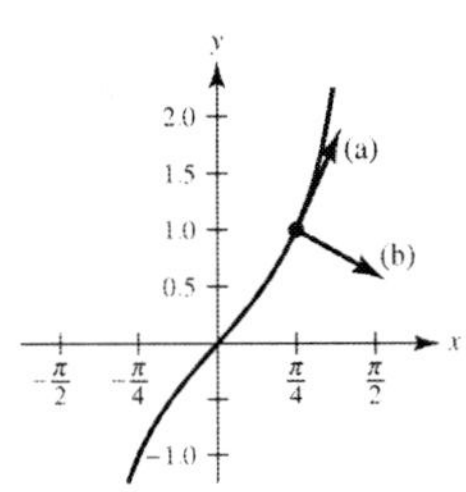

73. $\mathbf{u} = \frac{\sqrt{2}}{2}\mathbf{i} + \frac{\sqrt{2}}{2}\mathbf{j}$

$\mathbf{u} + \mathbf{v} = \sqrt{2}\mathbf{j}$

$\mathbf{v} = (\mathbf{u} + \mathbf{v}) - \mathbf{u} = -\frac{\sqrt{2}}{2}\mathbf{i} + \frac{\sqrt{2}}{2}\mathbf{j} = \left\langle -\frac{\sqrt{2}}{2}, \frac{\sqrt{2}}{2} \right\rangle$

74. $\mathbf{u} = 2\sqrt{3}\mathbf{i} + 2\mathbf{j}$

$\mathbf{u} + \mathbf{v} = -3\mathbf{i} + 3\sqrt{3}\mathbf{j}$

$\mathbf{v} = (\mathbf{u} + \mathbf{v}) - \mathbf{u} = \left(-3 - 2\sqrt{3}\right)\mathbf{i} + \left(3\sqrt{3} - 2\right)\mathbf{j}$

$= \left\langle -3 - 2\sqrt{3}, 3\sqrt{3} - 2 \right\rangle$

75. $\mathbf{F}_1 + \mathbf{F}_2 = (500 \cos 30°\mathbf{i} + 500 \sin 30°\mathbf{j}) + (200 \cos(-45°)\mathbf{i} + 200 \sin(-45°)\mathbf{j}) = \left(250\sqrt{3} + 100\sqrt{2}\right)\mathbf{i} + \left(250 - 100\sqrt{2}\right)\mathbf{j}$

$\|\mathbf{F}_1 + \mathbf{F}_2\| = \sqrt{\left(250\sqrt{3} + 100\sqrt{2}\right)^2 + \left(250 - 100\sqrt{2}\right)^2} \approx 584.6 \text{ lb}$

$\tan\theta = \frac{250 - 100\sqrt{2}}{250\sqrt{3} + 100\sqrt{2}} \Rightarrow \theta \approx 10.7°$

76. (a) $180(\cos 30°\mathbf{i} + \sin 30°\mathbf{j}) + 275\mathbf{i} \approx 430.88\mathbf{i} + 90\mathbf{j}$

Direction: $\alpha \approx \arctan\left(\frac{90}{430.88}\right) \approx 0.206 \ (\approx 11.8°)$

Magnitude: $\sqrt{430.88^2 + 90^2} \approx 440.18$ newtons

(b) $M = \sqrt{(275 + 180 \cos\theta)^2 + (180 \sin\theta)^2}$

$\alpha = \arctan\left[\frac{180 \sin\theta}{275 + 180 \cos\theta}\right]$

(c)

θ	0°	30°	60°	90°	120°	150°	180°
M	455	440.2	396.9	328.7	241.9	149.3	95
α	0°	11.8°	23.1°	33.2°	40.1°	37.1°	0

(d)

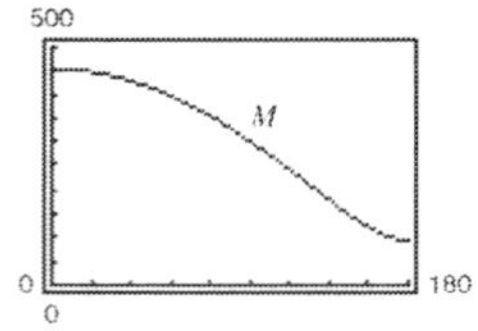

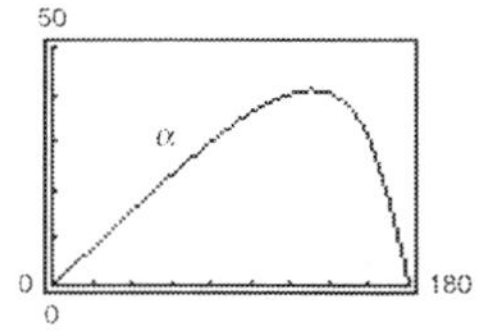

(e) M decreases because the forces change from acting in the same direction to acting in the opposite direction as θ increases from 0° to 180°.

77. $\mathbf{F}_1 + \mathbf{F}_2 + \mathbf{F}_3 = (75 \cos 30°\mathbf{i} + 75 \sin 30°\mathbf{j}) + (100 \cos 45°\mathbf{i} + 100 \sin 45°\mathbf{j}) + (125 \cos 120°\mathbf{i} + 125 \sin 120°\mathbf{j})$

$= \left(\frac{75}{2}\sqrt{3} + 50\sqrt{2} - \frac{125}{2}\right)\mathbf{i} + \left(\frac{75}{2} + 50\sqrt{2} + \frac{125}{2}\sqrt{3}\right)\mathbf{j}$

$\|\mathbf{R}\| = \|\mathbf{F}_1 + \mathbf{F}_2 + \mathbf{F}_3\| \approx 228.5 \text{ lb}$

$\theta_{\mathbf{R}} = \theta_{\mathbf{F}_1 + \mathbf{F}_2 + \mathbf{F}_3} \approx 71.3°$

78. $\mathbf{F}_1 + \mathbf{F}_2 + \mathbf{F}_3 = \left[400(\cos(-30°)\mathbf{i} + \sin(-30°)\mathbf{j})\right] + \left[280(\cos(45°)\mathbf{i} + \sin(45°)\mathbf{j})\right] + \left[350(\cos(135°)\mathbf{i} + \sin(135°)\mathbf{j})\right]$

$= \left[200\sqrt{3} + 140\sqrt{2} - 175\sqrt{2}\right]\mathbf{i} + \left[-200 + 140\sqrt{2} + 175\sqrt{2}\right]\mathbf{j}$

$\|\mathbf{R}\| = \sqrt{\left(200\sqrt{3} - 35\sqrt{2}\right)^2 + \left(-200 + 315\sqrt{2}\right)^2} \approx 385.2483$ newtons

$\theta_{\mathbf{R}} = \arctan\left(\dfrac{-200 + 315\sqrt{2}}{200\sqrt{3} - 35\sqrt{2}}\right) \approx 0.6908 \approx 39.6°$

79. (a) The forces act along the same direction. $\theta = 0°$.

(b) The forces cancel out each other. $\theta = 180°$.

(c) No, the magnitude of the resultant can not be greater than the sum.

80. (a) $\mathbf{u} = \overrightarrow{CB} = \|\mathbf{u}\|(\cos 30°\mathbf{i} + \sin 30°\mathbf{j})$

$\mathbf{v} = \overrightarrow{CA} = \|\mathbf{v}\|(\cos 130°\mathbf{i} + \sin 130°\mathbf{j})$

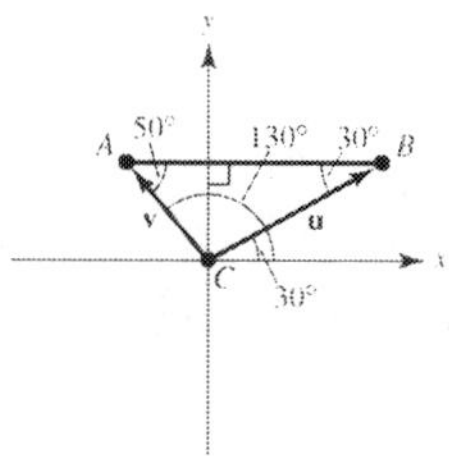

Vertical components:
$\|\mathbf{u}\|\sin 30° + \|\mathbf{v}\|\sin 130° = 3000$

Horizontal components:
$\|\mathbf{u}\|\cos 30° + \|\mathbf{v}\|\cos 130° = 0$

Solving this system, you obtain

$\|\mathbf{u}\| \approx 1958.1$ pounds

$\|\mathbf{v}\| \approx 2638.2$ pounds

(b) $\theta_1 = \arctan\left(\dfrac{24}{20}\right) \approx 0.8761$ or $50.2°$

$\theta_2 = \arctan\left(\dfrac{24}{-10}\right) + \pi \approx 1.9656$ or $112.6°$

$\mathbf{u} = \|\mathbf{u}\|(\cos\theta_1\mathbf{i} + \sin\theta_1\mathbf{j})$

$\mathbf{v} = \|\mathbf{v}\|(\cos\theta_2\mathbf{i} + \sin\theta_2\mathbf{j})$

Vertical components:
$\|\mathbf{u}\|\sin\theta_1 + \|\mathbf{v}\|\sin\theta_2 = 5000$

Horizontal components:
$\|\mathbf{u}\|\cos\theta_1 + \|\mathbf{v}\|\cos\theta_2 = 0$

Solving this system, you obtain

$\|\mathbf{u}\| \approx 2169.4$ and $\|\mathbf{v}\| \approx 3611.2$.

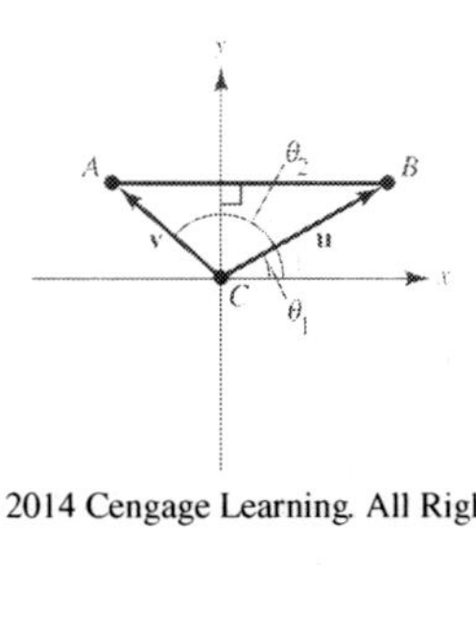

81. Horizontal component $= \|\mathbf{v}\|\cos\theta$

$= 1200\cos 6° \approx 1193.43$ ft/sec

Vertical component $= \|\mathbf{v}\|\sin\theta$

$= 1200\sin 6° \approx 125.43$ ft/sec

82. To lift the weight vertically, the sum of the vertical components of **u** and **v** must be 100 and the sum of the horizontal components must be 0.

$\mathbf{u} = \|\mathbf{u}\|(\cos 60°\mathbf{i} + \sin 60°\mathbf{j})$

$\mathbf{v} = \|\mathbf{v}\|(\cos 110°\mathbf{i} + \sin 110°\mathbf{j})$

So, $\|\mathbf{u}\|\sin 60° + \|\mathbf{v}\|\sin 110° = 100$, or

$\|\mathbf{u}\|\left(\dfrac{\sqrt{3}}{2}\right) + \|\mathbf{v}\|\sin 110° = 100.$

And $\|\mathbf{u}\|\cos 60° + \|\mathbf{v}\|\cos 110° = 0$ or

$\|\mathbf{u}\|\left(\dfrac{1}{2}\right) + \|\mathbf{v}\|\cos 110° = 0.$

Multiplying the last equation by $\left(\sqrt{3}\right)$ and adding to the first equation gives

$\|\mathbf{u}\|\left(\sin 110° - \sqrt{3}\cos 110°\right) = 100 \Rightarrow \|\mathbf{v}\| \approx 65.27$ lb.

Then, $\|\mathbf{u}\|\left(\dfrac{1}{2}\right) + 65.27\cos 110° = 0$ gives

$\|\mathbf{u}\| \approx 44.65$ lb.

(a) The tension in each rope: $\|\mathbf{u}\| = 44.65$ lb,

$\|\mathbf{v}\| = 65.27$ lb

(b) Vertical components: $\|\mathbf{u}\|\sin 60° \approx 38.67$ lb,

$\|\mathbf{v}\|\sin 110° \approx 61.33$ lb

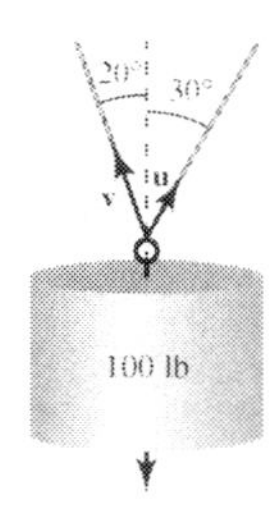

83. $\mathbf{u} = 900(\cos 148°\mathbf{i} + \sin 148°\mathbf{j})$

$\mathbf{v} = 100(\cos 45°\mathbf{i} + \sin 45°\mathbf{j})$

$$\mathbf{u} + \mathbf{v} = (900\cos 148° + 100\cos 45°)\mathbf{i} + (900\sin 148° + 100\sin 45°)\mathbf{j}$$
$$\approx -692.53\mathbf{i} + 547.64\mathbf{j}$$

$$\theta \approx \arctan\left(\frac{547.64}{-692.53}\right) \approx -38.34°;\ 38.34° \text{ North of West}$$

$$\|\mathbf{u} + \mathbf{v}\| \approx \sqrt{(-692.53)^2 + (547.64)^2} \approx 882.9 \text{ km/h}$$

84. $\mathbf{u} = 400\mathbf{i}$ (plane)

$\mathbf{v} = 50(\cos 135°\mathbf{i} + \sin 135°\mathbf{j}) = -25\sqrt{2}\mathbf{i} + 25\sqrt{2}\mathbf{j}$ (wind)

$$\mathbf{u} + \mathbf{v} = \left(400 - 25\sqrt{2}\right)\mathbf{i} + 25\sqrt{2}\mathbf{j} \approx 364.64\mathbf{i} + 35.36\mathbf{j}$$

$$\tan\theta = \frac{35.36}{364.64} \Rightarrow \theta \approx 5.54°$$

Direction North of East: $\approx$ N 84.46° E

Speed: $\approx$ 336.35 mi/h

85. True

86. True

87. True

88. False

$a = b = 0$

89. False

$\|a\mathbf{i} + b\mathbf{j}\| = \sqrt{2}|a|$

90. True

91. $\|\mathbf{u}\| = \sqrt{\cos^2\theta + \sin^2\theta} = 1,$

$\|\mathbf{v}\| = \sqrt{\sin^2\theta + \cos^2\theta} = 1$

92. Let the triangle have vertices at $(0, 0)$, $(a, 0)$, and (b, c). Let $\mathbf{u}$ be the vector joining $(0, 0)$ and (b, c), as indicated in the figure. Then $\mathbf{v}$, the vector joining the midpoints, is

$$\mathbf{v} = \left(\frac{a+b}{2} - \frac{a}{2}\right)\mathbf{i} + \frac{c}{2}\mathbf{j}$$
$$= \frac{b}{2}\mathbf{i} + \frac{c}{2}\mathbf{j}$$
$$= \frac{1}{2}(b\mathbf{i} + c\mathbf{j}) = \frac{1}{2}\mathbf{u}.$$

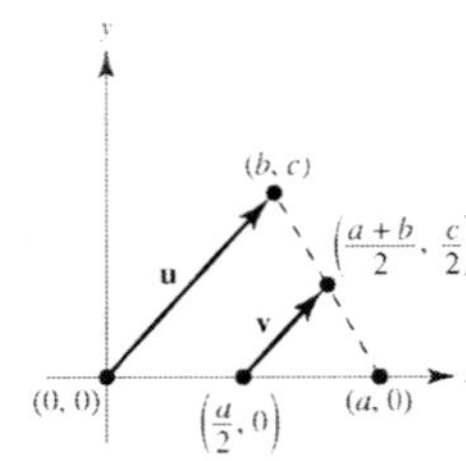

93. Let $\mathbf{u}$ and $\mathbf{v}$ be the vectors that determine the parallelogram, as indicated in the figure. The two diagonals are $\mathbf{u} + \mathbf{v}$ and $\mathbf{v} - \mathbf{u}$. So, $\mathbf{r} = x(\mathbf{u} + \mathbf{v})$, $\mathbf{s} = 4(\mathbf{v} - \mathbf{u})$. But,

$$\mathbf{u} = \mathbf{r} - \mathbf{s}$$
$$= x(\mathbf{u} + \mathbf{v}) - y(\mathbf{v} - \mathbf{u}) = (x + y)\mathbf{u} + (x - y)\mathbf{v}.$$

So, $x + y = 1$ and $x - y = 0$. Solving you have $x = y = \frac{1}{2}$.

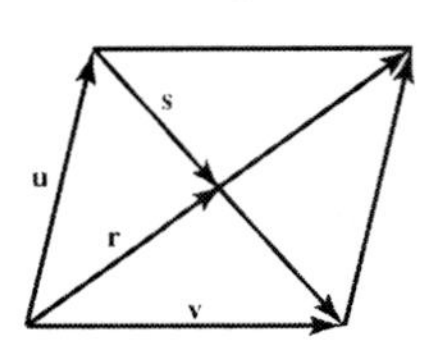

94. $\mathbf{w} = \|\mathbf{u}\|\mathbf{v} + \|\mathbf{v}\|\mathbf{u}$

$$= \|\mathbf{u}\|\left[\|\mathbf{v}\|\cos\theta_{\mathbf{v}}\mathbf{i} + \|\mathbf{v}\|\sin\theta_{\mathbf{v}}\mathbf{j}\right] + \|\mathbf{v}\|\left[\|\mathbf{u}\|\cos\theta_{\mathbf{u}}\mathbf{i} + \|\mathbf{u}\|\sin\theta_{\mathbf{u}}\mathbf{j}\right]$$

$$= \|\mathbf{u}\|\|\mathbf{v}\|\left[(\cos\theta_{\mathbf{u}} + \cos\theta_{\mathbf{v}})\mathbf{i} + (\sin\theta_{\mathbf{u}} + \sin\theta_{\mathbf{v}})\mathbf{j}\right]$$

$$= 2\|\mathbf{u}\|\|\mathbf{v}\|\left[\cos\left(\frac{\theta_{\mathbf{u}} + \theta_{\mathbf{v}}}{2}\right)\cos\left(\frac{\theta_{\mathbf{u}} - \theta_{\mathbf{v}}}{2}\right)\mathbf{i} + \sin\left(\frac{\theta_{\mathbf{u}} + \theta_{\mathbf{v}}}{2}\right)\cos\left(\frac{\theta_{\mathbf{u}} - \theta_{\mathbf{v}}}{2}\right)\mathbf{j}\right]$$

$$\tan\theta_{\mathbf{w}} = \frac{\sin\left(\frac{\theta_{\mathbf{u}} + \theta_{\mathbf{v}}}{2}\right)\cos\left(\frac{\theta_{\mathbf{u}} - \theta_{\mathbf{v}}}{2}\right)}{\cos\left(\frac{\theta_{\mathbf{u}} + \theta_{\mathbf{v}}}{2}\right)\cos\left(\frac{\theta_{\mathbf{u}} - \theta_{\mathbf{v}}}{2}\right)} = \tan\left(\frac{\theta_{\mathbf{u}} + \theta_{\mathbf{v}}}{2}\right)$$

So, $\theta_{\mathbf{w}} = (\theta_{\mathbf{u}} + \theta_{\mathbf{v}})/2$ and **w** bisects the angle between **u** and **v**.

95. The set is a circle of radius 5, centered at the origin.

$$\|\mathbf{u}\| = \|\langle x, y\rangle\| = \sqrt{x^2 + y^2} = 5 \Rightarrow x^2 + y^2 = 25$$

96. Let $x = v_0 t\cos\alpha$ and $y = v_0 t\sin\alpha - \frac{1}{2}gt^2$.

$$t = \frac{x}{v_0\cos\alpha} \Rightarrow y = v_0\sin\alpha\left(\frac{x}{v_0\cos\alpha}\right) - \frac{1}{2}g\left(\frac{x}{v_0\cos\alpha}\right)^2$$

$$= x\tan\alpha - \frac{g}{2v_0^2}x^2\sec^2\alpha$$

$$= x\tan\alpha - \frac{gx^2}{2v_0^2}\left(1 + \tan^2\alpha\right)$$

$$= \frac{v_0^2}{2g} - \frac{gx^2}{2v_0^2} - \frac{gx^2}{2v_0^2}\tan^2\alpha + x\tan\alpha - \frac{v_0^2}{2g}$$

$$= \frac{v_0^2}{2g} - \frac{gx^2}{2v_0^2} - \frac{gx^2}{2v_0^2}\left[\tan^2\alpha - 2\tan\alpha\left(\frac{v_0^2}{gx}\right) + \frac{v_0^4}{g^2x^2}\right]$$

$$= \frac{v_0^2}{2g} - \frac{gx^2}{2v_0^2} - \frac{gx^2}{2v_0^2}\left(\tan\alpha - \frac{v_0^2}{gx}\right)^2$$

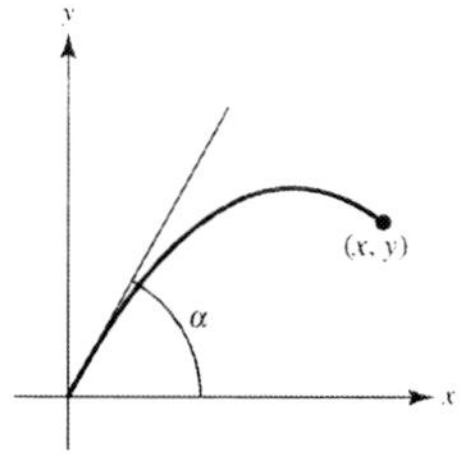

If $y \le \frac{v_0^2}{2g} - \frac{gx^2}{2v_0^2}$, then α can be chosen to hit the point (x, y). To hit $(0, y)$: Let $\alpha = 90°$. Then

$$y = v_0 t - \frac{1}{2}gt^2 = \frac{v_0^2}{2g} - \frac{v_0^2}{2g}\left(\frac{g}{v_0}t - 1\right)^2, \text{ and you need } y \le \frac{v_0^2}{2g}.$$

The set H is given by $0 \le x$, $0 < y$ and $y \le \frac{v_0^2}{2g} - \frac{gx^2}{2v_0^2}$

Note: The parabola $y = \frac{v_0^2}{2g} - \frac{gx^2}{2v_0^2}$ is called the "parabola of safety."

Section 11.2 Space Coordinates and Vectors in Space

1.

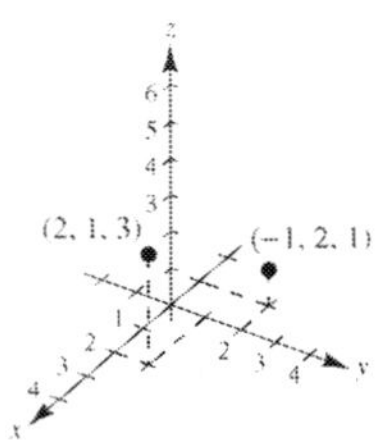

2.

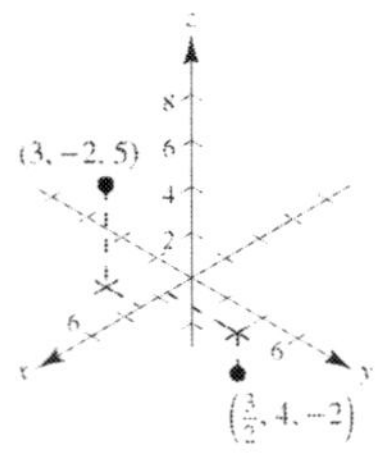

3.

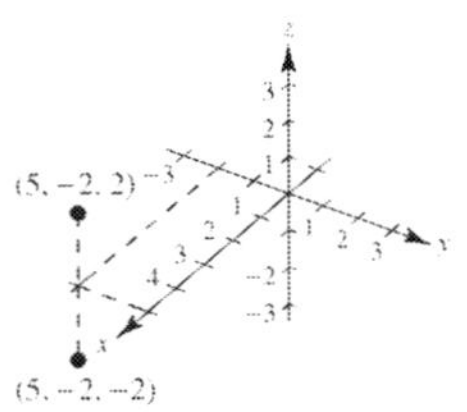

4.

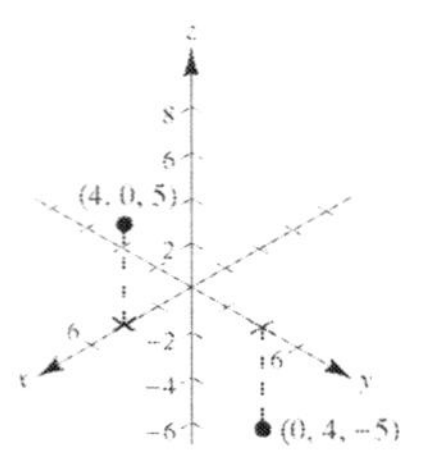

5. $x = -3,\ y = 4,\ z = 5: (-3, 4, 5)$

6. $x = 7,\ y = -2,\ z = -1:$

$(7, -2, -1)$

7. $y = z = 0, x = 12: (12, 0, 0)$

8. $x = 0, y = 3, z = 2: (0, 3, 2)$

9. The z-coordinate is 0.

10. The x-coordinate is 0.

11. The point is 6 units above the xy-plane.

12. The point is 2 units in front of the xz-plane.

13. The point is on the plane parallel to the yz-plane that passes through $x = -3$.

14. The point is on the plane parallel to the xy-plane that passes through $z = -5/2$.

15. The point is to the left of the xz-plane.

16. The point is in front of the yz-plane.

17. The point is on or between the planes $y = 3$ and $y = -3$.

18. The point is in front of the plane $x = 4$.

19. The point (x, y, z) is 3 units below the xy-plane, and below either quadrant I or III.

20. The point (x, y, z) is 4 units above the xy-plane, and above either quadrant II or IV.

21. The point could be above the xy-plane and so above quadrants II or IV, or below the xy-plane, and so below quadrants I or III.

22. The point could be above the xy-plane, and so above quadrants I and III, or below the xy-plane, and so below quadrants II or IV.

23. $d = \sqrt{(-4-0)^2 + (2-0)^2 + (7-0)^2}$

$= \sqrt{16 + 4 + 49} = \sqrt{69}$

24. $d = \sqrt{(2-(-2))^2 + (-5-3)^2 + (-2-2)^2}$

$= \sqrt{16 + 64 + 16} = \sqrt{96} = 4\sqrt{6}$

25. $d = \sqrt{(6-1)^2 + (-2-(-2))^2 + (-2-4)^2}$

$= \sqrt{25 + 0 + 36} = \sqrt{61}$

26. $d = \sqrt{(4-2)^2 + (-5-2)^2 + (6-3)^2}$

$= \sqrt{4 + 49 + 9} = \sqrt{62}$

27. $A(0, 0, 4), B(2, 6, 7), C(6, 4, -8)$

$|AB| = \sqrt{2^2 + 6^2 + 3^2} = \sqrt{49} = 7$

$|AC| = \sqrt{6^2 + 4^2 + (-12)^2} = \sqrt{196} = 14$

$|BC| = \sqrt{4^2 + (-2)^2 + (-15)^2} = \sqrt{245} = 7\sqrt{5}$

$|BC|^2 = 245 = 49 + 196 = |AB|^2 + |AC|^2$

Right triangle

28. $A(3, 4, 1), B(0, 6, 2), C(3, 5, 6)$

$|AB| = \sqrt{9 + 4 + 1} = \sqrt{14}$

$|AC| = \sqrt{0 + 1 + 25} = \sqrt{26}$

$|BC| = \sqrt{9 + 1 + 16} = \sqrt{26}$

Because $|AC| = |BC|$, the triangle is isosceles.

29. $A(-1, 0, -2), B(-1, 5, 2), C(-3, -1, 1)$

$|AB| = \sqrt{0 + 25 + 16} = \sqrt{41}$

$|AC| = \sqrt{4 + 1 + 9} = \sqrt{14}$

$|BC| = \sqrt{4 + 36 + 1} = \sqrt{41}$

Because $|AB| = |BC|$, the triangle is isosceles.

30. $A(4, -1, -1), B(2, 0, -4), C(3, 5, -1)$

$|AB| = \sqrt{4 + 1 + 9} = \sqrt{14}$

$|AC| = \sqrt{1 + 36 + 0} = \sqrt{37}$

$|BC| = \sqrt{1 + 25 + 9} = \sqrt{35}$

Neither

31. The z-coordinate is changed by 5 units:

$(0, 0, 9), (2, 6, 12), (6, 4, -3)$

32. The y-coordinate is changed by 3 units:

$(3, 7, 1), (0, 9, 2), (3, 8, 6)$

33. $\left(\frac{3 + 1}{2}, \frac{4 + 8}{2}, \frac{6 + 0}{2}\right) = (2, 6, 3)$

34. $\left(\frac{7 - 5}{2}, \frac{2 - 2}{2}, \frac{2 - 3}{2}\right) = \left(1, 0, -\frac{1}{2}\right)$

35. $\left(\frac{5 + (-2)}{2}, \frac{-9 + 3}{2}, \frac{7 + 3}{2}\right) = \left(\frac{3}{2}, -3, 5\right)$

36. $\left(\frac{4 + 8}{2}, \frac{0 + 8}{2}, \frac{-6 + 20}{2}\right) = (6, 4, 7)$

37. Center: $(0, 2, 5)$

Radius: 2

$(x - 0)^2 + (y - 2)^2 + (z - 5)^2 = 4$

38. Center: $(4, -1, 1)$

Radius: 5

$(x - 4)^2 + (y + 1)^2 + (z - 1)^2 = 25$

39. Center: $\frac{(2, 0, 0) + (0, 6, 0)}{2} = (1, 3, 0)$

Radius: $\sqrt{10}$

$(x - 1)^2 + (y - 3)^2 + (z - 0)^2 = 10$

40. Center: $(-3, 2, 4)$

Radius: 3

(tangent to yz-plane)

$(x + 3)^2 + (y - 2)^2 + (z - 4)^2 = 9$

41.

$$x^2 + y^2 + z^2 - 2x + 6y + 8z + 1 = 0$$
$$(x^2 - 2x + 1) + (y^2 + 6y + 9) + (z^2 + 8z + 16) = -1 + 1 + 9 + 16$$
$$(x - 1)^2 + (y + 3)^2 + (z + 4)^2 = 25$$

Center: $(1, -3, -4)$

Radius: 5

42.

$$x^2 + y^2 + z^2 + 9x - 2y + 10z + 19 = 0$$
$$\left(x^2 + 9x + \frac{81}{4}\right) + (y^2 - 2y + 1) + (z^2 + 10z + 25) = -19 + \frac{81}{4} + 1 + 25$$
$$\left(x + \frac{9}{2}\right)^2 + (y - 1)^2 + (z + 5)^2 = \frac{109}{4}$$

Center: $\left(-\frac{9}{2}, 1, -5\right)$

Radius: $\frac{\sqrt{109}}{2}$

43.
$$9x^2 + 9y^2 + 9z^2 - 6x + 18y + 1 = 0$$
$$x^2 + y^2 + z^2 - \tfrac{2}{3}x + 2y + \tfrac{1}{9} = 0$$
$$\left(x^2 - \tfrac{2}{3}x + \tfrac{1}{9}\right) + \left(y^2 + 2y + 1\right) + z^2 = -\tfrac{1}{9} + \tfrac{1}{9} + 1$$
$$\left(x - \tfrac{1}{3}\right)^2 + (y + 1)^2 + (z - 0)^2 = 1$$

Center: $\left(\frac{1}{3}, -1, 0\right)$

Radius: 1

44. $4x^2 + 4y^2 + 4z^2 - 24x - 4y + 8z - 23 = 0$
$$\left(x^2 - 6x + 9\right) + \left(y^2 - y + \tfrac{1}{4}\right) + \left(z^2 + 2z + 1\right) = \tfrac{23}{4} + 9 + \tfrac{1}{4} + 1$$
$$(x - 3)^2 + \left(y - \tfrac{1}{2}\right)^2 + (z + 1)^2 = 16$$

Center: $\left(3, \frac{1}{2}, -1\right)$

Radius: 4

45. (a) $\mathbf{v} = \langle 2 - 4, 4 - 2, 3 - 1 \rangle = \langle -2, 2, 2 \rangle$

(b) $\mathbf{v} = -2\mathbf{i} + 2\mathbf{j} + 2\mathbf{k}$

(c)

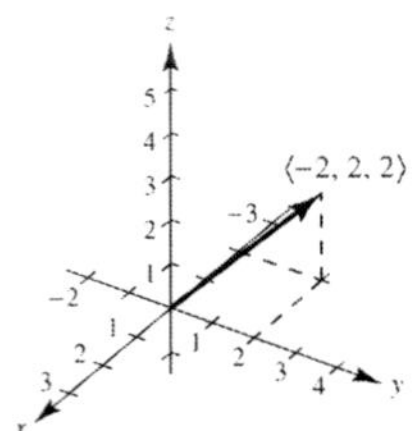

46. (a) $\mathbf{v} = \langle 4 - 0, 0 - 5, 3 - 1 \rangle = \langle 4, -5, 2 \rangle$

(b) $\mathbf{v} = 4\mathbf{i} - 5\mathbf{j} + 2\mathbf{h}$

(c)

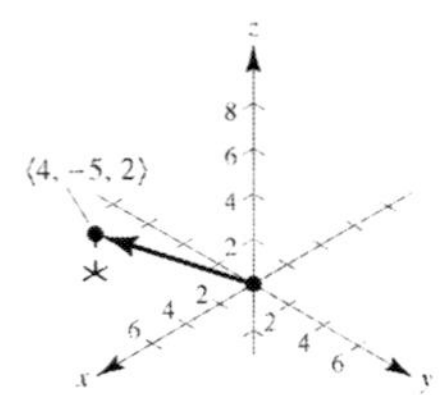

47. (a) $\mathbf{v} = \langle 0 - 3, 3 - 3, 3 - 0 \rangle = \langle -3, 0, 3 \rangle$

(b) $\mathbf{v} = -3\mathbf{i} + 3\mathbf{k}$

(c)

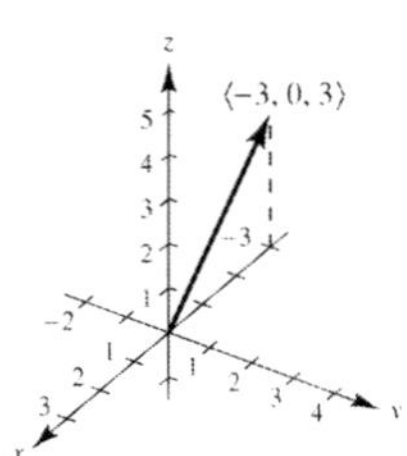

48. (a) $\mathbf{v} = \langle 2 - 2, 3 - 3, 4 - 0 \rangle = \langle 0, 0, 4 \rangle$

(b) $\mathbf{v} = 4\mathbf{k}$

(c)

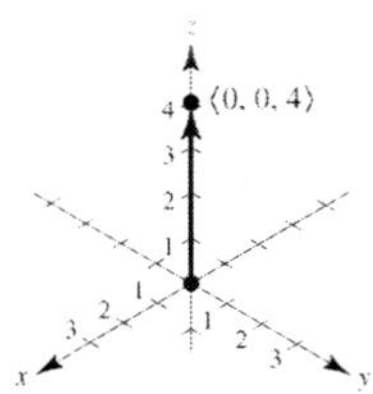

49. $\langle 4 - 3, 1 - 2, 6 - 0 \rangle = \langle 1, -1, 6 \rangle$

$\left\|\langle 1, -1, 6 \rangle\right\| = \sqrt{1 + 1 + 36} = \sqrt{38}$

Unit vector: $\dfrac{\langle 1, -1, 6 \rangle}{\sqrt{38}} = \left\langle \dfrac{1}{\sqrt{38}}, \dfrac{-1}{\sqrt{38}}, \dfrac{6}{\sqrt{38}} \right\rangle$

50. $\langle 2 - 1, 4 - (-2), -2 - 4 \rangle = \langle 1, 6, -6 \rangle$

$\left\|\langle 1, 6, -6 \rangle\right\| = \sqrt{1 + 36 + 36} = \sqrt{73}$

Unit vector: $\dfrac{\langle 1, 6, -6 \rangle}{\sqrt{73}} = \left\langle \dfrac{1}{\sqrt{73}}, \dfrac{6}{\sqrt{73}}, \dfrac{-6}{\sqrt{73}} \right\rangle$

51. (b) $\mathbf{v} = \langle 3 - (-1), 3 - 2, 4 - 3 \rangle = \langle 4, 1, 1 \rangle$

(c) $\mathbf{v} = 4\mathbf{i} + \mathbf{j} + \mathbf{k}$

(a), (d)

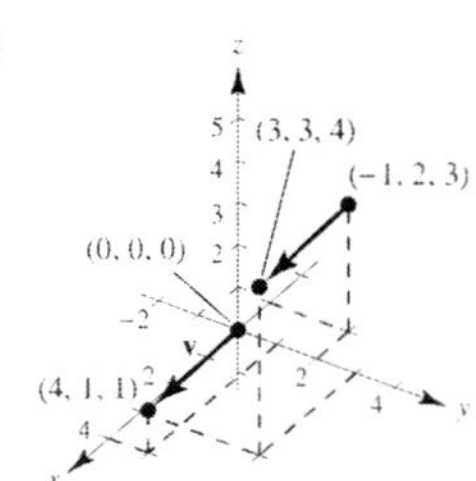

52. (b) $\mathbf{v} = \langle -4 - 2, 3 - (-1), 7 - (-2) \rangle = \langle -6, 4, 9 \rangle$

(c) $\mathbf{v} = 6\mathbf{i} + 4\mathbf{j} + 9\mathbf{k}$

(a), (d)

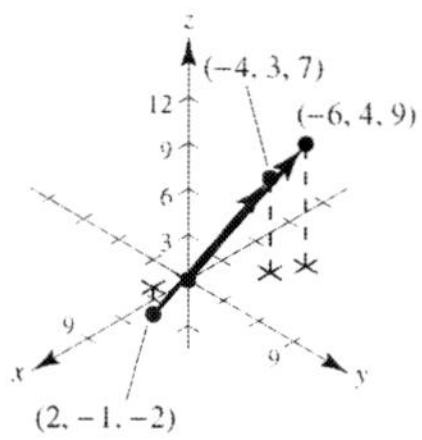

53. $(q_1, q_2, q_3) - (0, 6, 2) = (3, -5, 6)$

$Q = (3, 1, 8)$

54. $(q_1, q_2, q_3) - \left(0, 2, \frac{5}{2}\right) = \left(1, -\frac{2}{3}, \frac{1}{2}\right)$

$Q = \left(1, -\frac{4}{3}, 3\right)$

55. (a) $2\mathbf{v} = \langle 2, 4, 4 \rangle$

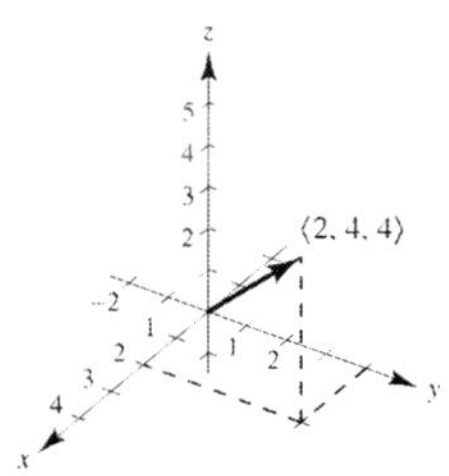

(b) $-\mathbf{v} = \langle -1, -2, -2 \rangle$

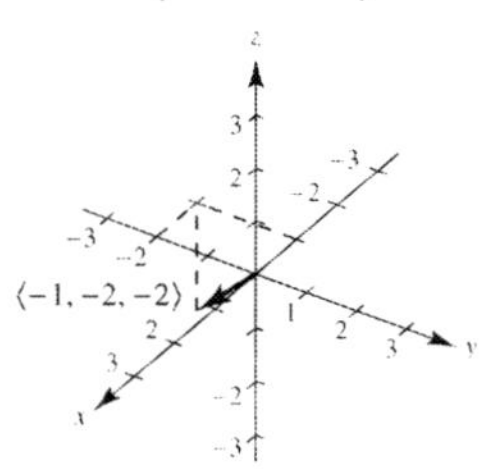

(c) $\frac{3}{2}\mathbf{v} = \left\langle \frac{3}{2}, 3, 3 \right\rangle$

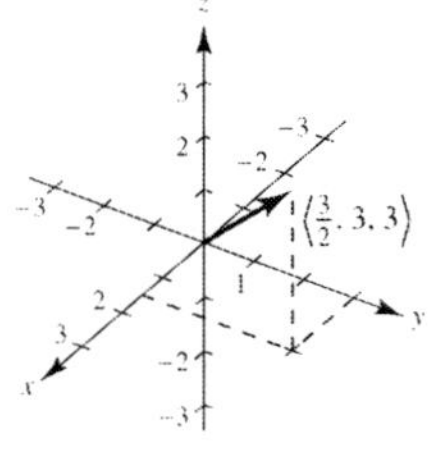

(d) $0\mathbf{v} = \langle 0, 0, 0 \rangle$

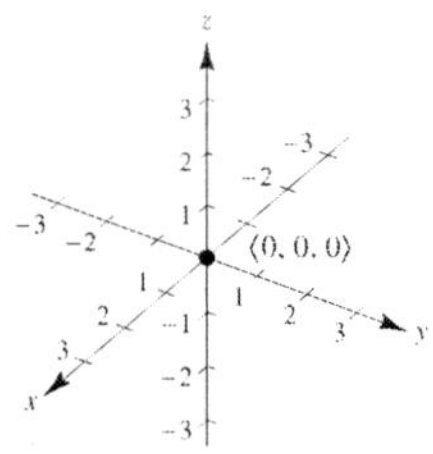

56. (a) $-\mathbf{v} = \langle -2, 2, -1 \rangle$

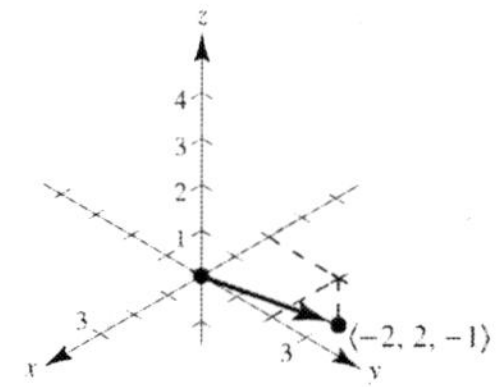

(b) $2\mathbf{v} = \langle 4, -4, 2 \rangle$

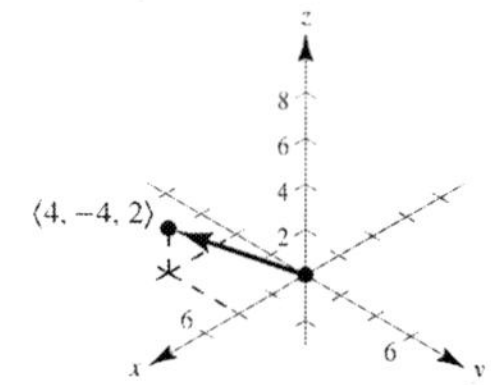

(c) $\frac{1}{2}\mathbf{v} = \left\langle 1, -1, \frac{1}{2} \right\rangle$

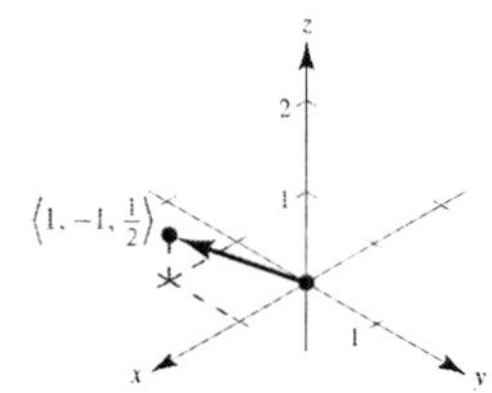

(d) $\frac{5}{2}\mathbf{v} = \left\langle 5, -5, \frac{5}{2} \right\rangle$

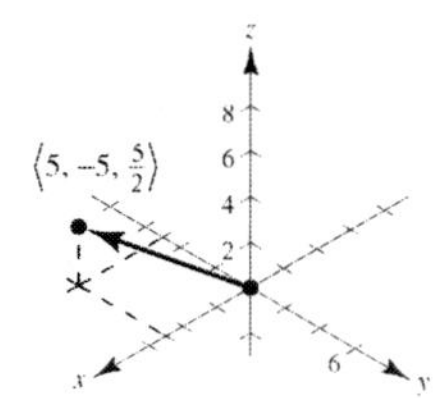

57. $\mathbf{z} = \mathbf{u} - \mathbf{v} + 2\mathbf{w}$

$= \langle 1, 2, 3 \rangle - \langle 2, 2, -1 \rangle + \langle 8, 0, -8 \rangle = \langle 7, 0, -4 \rangle$

58. $\mathbf{z} = 5\mathbf{u} - 3\mathbf{v} - \frac{1}{2}\mathbf{w}$

$= \langle 5, 10, 15 \rangle - \langle 6, 6, -3 \rangle - \langle 2, 0, -2 \rangle$

$= \langle -3, 4, 20 \rangle$

59. $2\mathbf{z} - 3\mathbf{u} = 2\langle z_1, z_2, z_3 \rangle - 3\langle 1, 2, 3 \rangle = \langle 4, 0, -4 \rangle$

$2z_1 - 3 = 4 \Rightarrow z_1 = \frac{7}{2}$

$2z_2 - 6 = 0 \Rightarrow z_2 = 3$

$2z_3 - 9 = -4 \Rightarrow z_3 = \frac{5}{2}$

$\mathbf{z} = \left\langle \frac{7}{2}, 3, \frac{5}{2} \right\rangle$

60. $2\mathbf{u} + \mathbf{v} - \mathbf{w} + 3\mathbf{z} = 2\langle 1, 2, 3\rangle + \langle 2, 2, -1\rangle - \langle 4, 0, -4\rangle + 3\langle z_1, z_2, z_3\rangle = \langle 0, 0, 0\rangle$

$\langle 0, 6, 9\rangle + \langle 3z_1, 3z_2, 3z_3\rangle = \langle 0, 0, 0\rangle$

$0 + 3z_1 = 0 \Rightarrow z_1 = 0$

$6 + 3z_2 = 0 \Rightarrow z_2 = -2$

$9 + 3z_3 = 0 \Rightarrow z_3 = -3$

$\mathbf{z} = \langle 0, -2, -3\rangle$

61. (a) and (b) are parallel because $\langle -6, -4, 10\rangle = -2\langle 3, 2, -5\rangle$ and $\left\langle 2, \frac{4}{3}, -\frac{10}{3}\right\rangle = \frac{2}{3}\langle 3, 2, -5\rangle$.

62. (b) and (d) are parallel because $-\mathbf{i} + \frac{4}{3}\mathbf{j} - \frac{3}{2}\mathbf{k} = -2\left(\frac{1}{2}\mathbf{i} - \frac{2}{3}\mathbf{j} + \frac{3}{4}\mathbf{k}\right)$ and $\frac{3}{4}\mathbf{i} - \mathbf{j} + \frac{9}{8}\mathbf{k} = \frac{3}{2}\left(\frac{1}{2}\mathbf{i} - \frac{2}{3}\mathbf{j} + \frac{3}{4}\mathbf{k}\right)$.

63. $\mathbf{z} = -3\mathbf{i} + 4\mathbf{j} + 2\mathbf{k}$

(a) is parallel because $-6\mathbf{i} + 8\mathbf{j} + 4\mathbf{k} = 2\mathbf{z}$.

64. $\mathbf{z} = \langle -7, -8, 3\rangle$

(b) is parallel because $(-z)\mathbf{z} = \langle 14, 16, -6\rangle$.

65. $P(0, -2, -5), Q(3, 4, 4), R(2, 2, 1)$

$\overrightarrow{PQ} = \langle 3, 6, 9\rangle$

$\overrightarrow{PR} = \langle 2, 4, 6\rangle$

$\langle 3, 6, 9\rangle = \frac{3}{2}\langle 2, 4, 6\rangle$

So, $\overrightarrow{PQ}$ and $\overrightarrow{PR}$ are parallel, the points are collinear.

66. $P(4, -2, 7), Q(-2, 0, 3), R(7, -3, 9)$

$\overrightarrow{PQ} = \langle -6, 2, -4\rangle$

$\overrightarrow{PR} = \langle 3, -1, 2\rangle$

$\langle 3, -1, 2\rangle = -\frac{1}{2}\langle -6, 2, -4\rangle$

So, $\overrightarrow{PQ}$ and $\overrightarrow{PR}$ are parallel. The points are collinear.

67. $P(1, 2, 4), Q(2, 5, 0), R(0, 1, 5)$

$\overrightarrow{PQ} = \langle 1, 3, -4\rangle$

$\overrightarrow{PR} = \langle -1, -1, 1\rangle$

Because $\overrightarrow{PQ}$ and $\overrightarrow{PR}$ are not parallel, the points are not collinear.

68. $P(0, 0, 0), Q(1, 3, -2), R(2, -6, 4)$

$\overrightarrow{PQ} = \langle 1, 3, -2\rangle$

$\overrightarrow{PR} = \langle 2, -6, 4\rangle$

Because $\overrightarrow{PQ}$ and $\overrightarrow{PR}$ are not parallel, the points are not collinear.

69. $A(2, 9, 1), B(3, 11, 4), C(0, 10, 2), D(1, 12, 5)$

$\overrightarrow{AB} = \langle 1, 2, 3\rangle$

$\overrightarrow{CD} = \langle 1, 2, 3\rangle$

$\overrightarrow{AC} = \langle -2, 1, 1\rangle$

$\overrightarrow{BD} = (-2, 1, 1)$

Because $\overrightarrow{AB} = \overrightarrow{CD}$ and $\overrightarrow{AC} = \overrightarrow{BD}$, the given points form the vertices of a parallelogram.

70. $A(1, 1, 3)$ $B(9, -1, -2), C(11, 2, -9), D(3, 4, -4)$

$\overrightarrow{AB} = \langle 8, -2, -5\rangle$

$\overrightarrow{DC} = \langle 8, -2, -5\rangle$

$\overrightarrow{AD} = \langle 2, 3, -7\rangle$

$\overrightarrow{BC} = \langle 2, 3, -7\rangle$

Because $\overrightarrow{AB} = \overrightarrow{DC}$ and $\overrightarrow{AD} = \overrightarrow{BC}$, the given points form the vertices of a parallelogram.

71. $\mathbf{v} = \langle 0, 0, 0\rangle$

$\|\mathbf{v}\| = 0$

72. $\mathbf{v} = \langle 1, 0, 3\rangle$

$\|\mathbf{v}\| = \sqrt{1 + 0 + 9} = \sqrt{10}$

73. $\mathbf{v} = 3\mathbf{j} - 5\mathbf{k} = \langle 0, 3, -5\rangle$

$\|\mathbf{v}\| = \sqrt{0 + 9 + 25} = \sqrt{34}$

74. $\mathbf{v} = 2\mathbf{i} + 5\mathbf{j} - \mathbf{k} = \langle 2, 5, -1\rangle$

$\|\mathbf{v}\| = \sqrt{4 + 25 + 1} = \sqrt{30}$

75. $\mathbf{v} = \mathbf{i} - 2\mathbf{j} - 3\mathbf{k} = \langle 1, -2, -3\rangle$

$\|\mathbf{v}\| = \sqrt{1 + 4 + 9} = \sqrt{14}$

76. $\mathbf{v} = -4\mathbf{i} + 3\mathbf{j} + 7\mathbf{k} = \langle -4, 3, 7\rangle$

$\|\mathbf{v}\| = \sqrt{16 + 9 + 49} = \sqrt{74}$

77. $\mathbf{v} = \langle 2, -1, 2\rangle$

$\|\mathbf{v}\| = \sqrt{4 + 1 + 4} = 3$

(a) $\dfrac{\mathbf{v}}{\|\mathbf{v}\|} = \dfrac{1}{3}\langle 2, -1, 2\rangle$

(b) $-\dfrac{\mathbf{v}}{\|\mathbf{v}\|} = -\dfrac{1}{3}\langle 2, -1, 2\rangle$

78. $\mathbf{v} = \langle 6, 0, 8\rangle$

$\|\mathbf{v}\| = \sqrt{36 + 0 + 64} = 10$

(a) $\dfrac{\mathbf{v}}{\|\mathbf{v}\|} = \dfrac{1}{10}\langle 6, 0, 8\rangle$

(b) $-\dfrac{\mathbf{v}}{\|\mathbf{v}\|} = -\dfrac{1}{10}\langle 6, 0, 8\rangle$

79. $\mathbf{v} = 4\mathbf{i} - 5\mathbf{j} + 3\mathbf{k}$

$\|\mathbf{v}\| = \sqrt{16 + 25 + 9} = \sqrt{50} = 5\sqrt{2}$

(a) $\dfrac{\mathbf{v}}{\|\mathbf{v}\|} = \dfrac{1}{5\sqrt{2}}(4\mathbf{i} - 5\mathbf{j} + 3\mathbf{k}) = \dfrac{2\sqrt{2}}{5}\mathbf{i} - \dfrac{\sqrt{2}}{2}\mathbf{j} + \dfrac{3\sqrt{2}}{10}\mathbf{k}$

(b) $-\dfrac{\mathbf{v}}{\|\mathbf{v}\|} = -\dfrac{1}{5\sqrt{2}}(4\mathbf{i} - 5\mathbf{j} + 3\mathbf{k}) = -\dfrac{2\sqrt{2}}{5}\mathbf{i} + \dfrac{\sqrt{2}}{2}\mathbf{j} - \dfrac{3\sqrt{2}}{10}\mathbf{k}$

80. $\mathbf{v} = 5\mathbf{i} + 3\mathbf{j} - \mathbf{k}$

$\|\mathbf{v}\| = \sqrt{25 + 9 + 1} = \sqrt{35}$

(a) $\dfrac{\mathbf{v}}{\|\mathbf{v}\|} = \dfrac{1}{\sqrt{35}}(5\mathbf{i} + 3\mathbf{j} - \mathbf{k}) = \dfrac{\sqrt{35}}{7}\mathbf{i} + \dfrac{3\sqrt{35}}{35}\mathbf{j} - \dfrac{\sqrt{35}}{35}\mathbf{k}$

(b) $-\dfrac{\mathbf{v}}{\|\mathbf{v}\|} = -\dfrac{1}{\sqrt{35}}(5\mathbf{i} + 3\mathbf{j} - \mathbf{k}) = -\dfrac{\sqrt{35}}{7}\mathbf{i} - \dfrac{3\sqrt{35}}{35}\mathbf{j} + \dfrac{\sqrt{35}}{35}\mathbf{k}$

81. The terminal points of the vectors $t\mathbf{u}$, $\mathbf{u} + t\mathbf{v}$ and $s\mathbf{u} + t\mathbf{v}$ are collinear.

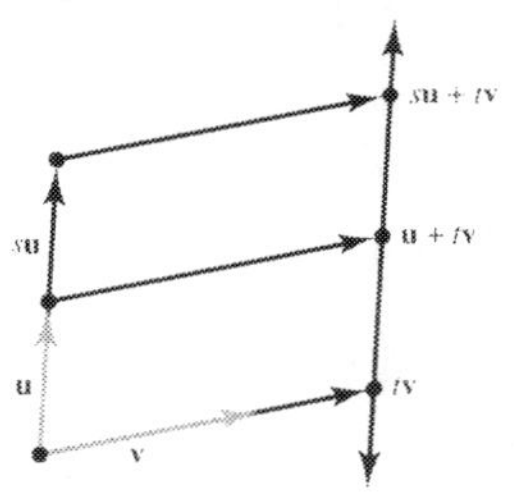

82. (a) $(x, y, z^2) = (3, 3, 3)$

$\mathbf{v} = \langle 3, 3, 3\rangle - \langle 3, 0, 0\rangle$

$= \langle 3 - 3, 3 - 0, 3 - 0\rangle = \langle 0, 3, 3\rangle$

(b) $(x, y, z) = (4, 4, 8)$

$\mathbf{v} = \langle 4, 4, 8\rangle - \langle 4, 0, 0\rangle$

$= \langle 4 - 4, 4 - 0, 8 - 0\rangle = \langle 0, 4, 8\rangle$

83. $\mathbf{v} = 10\dfrac{\mathbf{u}}{\|\mathbf{u}\|} = 10\dfrac{\langle 0, 3, 3\rangle}{3\sqrt{2}}$

$= 10\left\langle 0, \dfrac{1}{\sqrt{2}}, \dfrac{1}{\sqrt{2}}\right\rangle = \left\langle 0, \dfrac{10}{\sqrt{2}}, \dfrac{10}{\sqrt{2}}\right\rangle$

84. $\mathbf{v} = 3\dfrac{\mathbf{u}}{\|\mathbf{u}\|} = 3\dfrac{\langle 1, 1, 1\rangle}{\sqrt{3}}$

$= 3\left\langle \dfrac{1}{\sqrt{3}}, \dfrac{1}{\sqrt{3}}, \dfrac{1}{\sqrt{3}}\right\rangle = \left\langle \dfrac{3}{\sqrt{3}}, \dfrac{3}{\sqrt{3}}, \dfrac{3}{\sqrt{3}}\right\rangle$

85. $\mathbf{v} = \dfrac{3}{2}\dfrac{\mathbf{u}}{\|\mathbf{u}\|} = \dfrac{3}{2}\dfrac{(2, -2, 1)}{3} = \dfrac{3}{2}\left\langle \dfrac{2}{3}, \dfrac{-2}{3}, \dfrac{1}{3}\right\rangle = \left\langle 1, -1, \dfrac{1}{2}\right\rangle$

86. $\mathbf{v} = 7\dfrac{\mathbf{u}}{\|\mathbf{u}\|} = 7\dfrac{\langle -4, 6, 2\rangle}{2\sqrt{14}} = \left\langle \dfrac{-14}{\sqrt{14}}, \dfrac{21}{\sqrt{14}}, \dfrac{7}{14}\right\rangle$

87. $\mathbf{v} = 2\left[\cos(\pm 30°)\mathbf{j} + \sin(\pm 30°)\mathbf{k}\right]$

$= \sqrt{3}\mathbf{j} \pm \mathbf{k} = \langle 0, \sqrt{3}, \pm 1\rangle$

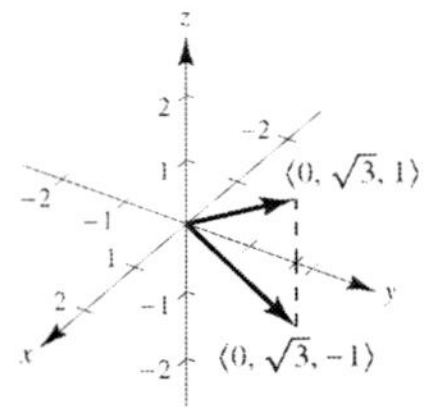

88. $\mathbf{v} = 5(\cos 45°\mathbf{i} + \sin 45°\mathbf{k}) = \dfrac{5\sqrt{2}}{2}(\mathbf{i} + \mathbf{k})$ or

$\mathbf{v} = 5(\cos 135°\mathbf{i} + \sin 135°\mathbf{k}) = \dfrac{5\sqrt{2}}{2}(-\mathbf{i} + \mathbf{k})$

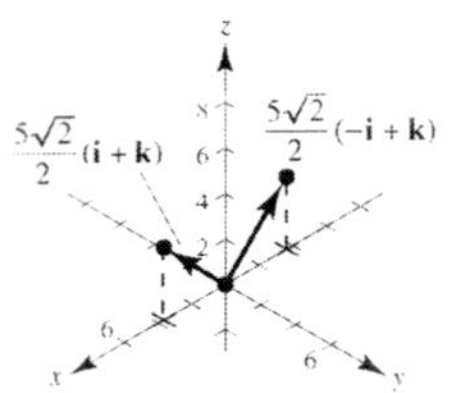

89.

$$\mathbf{v} = \langle -3, -6, 3\rangle$$
$$\tfrac{2}{3}\mathbf{v} = \langle -2, -4, 2\rangle$$
$$(4, 3, 0) + (-2, -4, 2) = (2, -1, 2)$$

90.

$$\mathbf{v} = \langle 5, 6, -3\rangle$$
$$\tfrac{2}{3}\mathbf{v} = \langle \tfrac{10}{3}, 4, -2\rangle$$
$$(1, 2, 5) + \left(\tfrac{10}{3}, 4, -2\right) = \left(\tfrac{13}{3}, 6, 3\right)$$

91. (a)

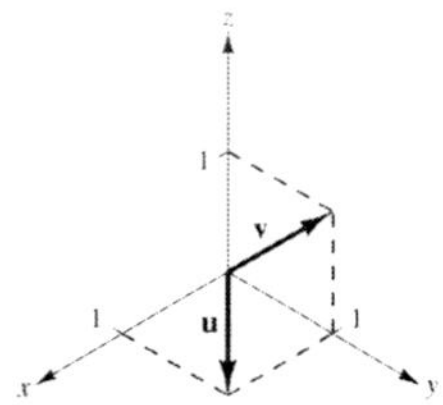

(b) $\mathbf{w} = a\mathbf{u} + b\mathbf{v} = a\mathbf{i} + (a + b)\mathbf{j} + b\mathbf{k} = 0$

$a = 0, a + b = 0, b = 0$

So, a and b are both zero.

(c) $a\mathbf{i} + (a + b)\mathbf{j} + b\mathbf{k} = \mathbf{i} + 2\mathbf{j} + \mathbf{k}$

$a = 1, a + b = 2, b = 1$

$\mathbf{w} = \mathbf{u} + \mathbf{v}$

(d) $a\mathbf{i} + (a + b)\mathbf{j} + b\mathbf{k} = \mathbf{i} + 2\mathbf{j} + 3\mathbf{k}$

$a = 1, a + b = 2, b = 3$

Not possible

92. A sphere of radius 4 centered at (x_1, y_1, z_1).

$\|\mathbf{v}\| = \|\langle x - x_2, y - y_1, z - z_1\rangle\|$

$= \sqrt{(x - x_1)^2 + (y - y_1)^2 + (z - z_1)^2} = 4$

$(x - x_1)^2 + (y - y_1)^2 + (z - z_1)^2 = 16$

93. x_0 is directed distance to yz-plane.

y_0 is directed distance to xz-plane.

z_0 is directed distance to xy-plane.

94. $d = \sqrt{(x_2 - x_1)^2 + (y_2 - y_1)^2 + (z_2 - z_1)^2}$

95. $(x - x_0)^2 + (y - y_0)^2 + (z - z_0)^2 = r^2$

96. Two nonzero vectors **u** and **v** are parallel if $\mathbf{u} = c\mathbf{v}$ for some scalar c.

97.

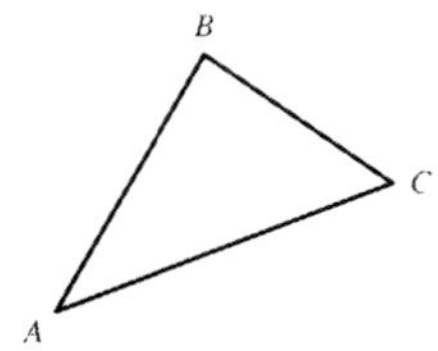

$\overrightarrow{AB} + \overrightarrow{BC} = \overrightarrow{AC}$

So, $\overrightarrow{AB} + \overrightarrow{BC} + \overrightarrow{CA} = \overrightarrow{AC} + \overrightarrow{CA} = \mathbf{0}$

98. $\|\mathbf{r} - \mathbf{r}_0\| = \sqrt{(x - 1)^2 + (y - 1)^2 + (z - 1)^2} = 2$

$(x - 1)^2 + (y - 1)^2 + (z - 1)^2 = 4$

This is a sphere of radius 2 and center $(1, 1, 1)$.

99. Let α be the angle between **v** and the coordinate axes.

$\mathbf{v} = (\cos\alpha)\mathbf{i} + (\cos\alpha)\mathbf{j} + (\cos\alpha)\mathbf{k}$

$\|\mathbf{v}\| = \sqrt{3}\cos\alpha = 1$

$\cos\alpha = \dfrac{1}{\sqrt{3}} = \dfrac{\sqrt{3}}{3}$

$\mathbf{v} = \dfrac{\sqrt{3}}{3}(\mathbf{i} + \mathbf{j} + \mathbf{k}) = \dfrac{\sqrt{3}}{3}\langle 1, 1, 1\rangle$

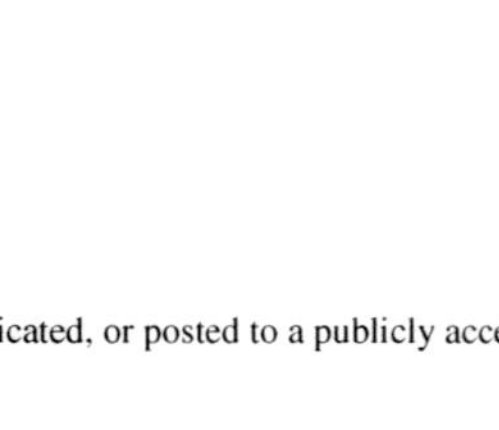

100.

$$\begin{aligned} 550 &= \|c(75\mathbf{i} - 50\mathbf{j} - 100\mathbf{k})\| \\ 302{,}500 &= 18{,}125c^2 \\ c^2 &= 16.689655 \\ c &\approx 4.085 \\ \mathbf{F} &\approx 4.085(75\mathbf{i} - 50\mathbf{j} - 100\mathbf{k}) \\ &\approx 306\mathbf{i} - 204\mathbf{j} - 409\mathbf{k} \end{aligned}$$

101. (a) The height of the right triangle is $h = \sqrt{L^2 - 18^2}$. The vector $\overrightarrow{PQ}$ is given by

$\overrightarrow{PQ} = \langle 0, -18, h\rangle$.

The tension vector **T** in each wire is

$\mathbf{T} = c\langle 0, -18, h\rangle$ where $ch = \dfrac{24}{3} = 8$.

So, $\mathbf{T} = \dfrac{8}{h}\langle 0, -18, h\rangle$ and

$$\begin{aligned} T = \|\mathbf{T}\| &= \frac{8}{h}\sqrt{18^2 + h^2} \\ &= \frac{8}{\sqrt{L^2 - 18^2}}\sqrt{18^2 + (L^2 - 18^2)} \\ &= \frac{8L}{\sqrt{L^2 - 18^2}}, \; L > 18. \end{aligned}$$

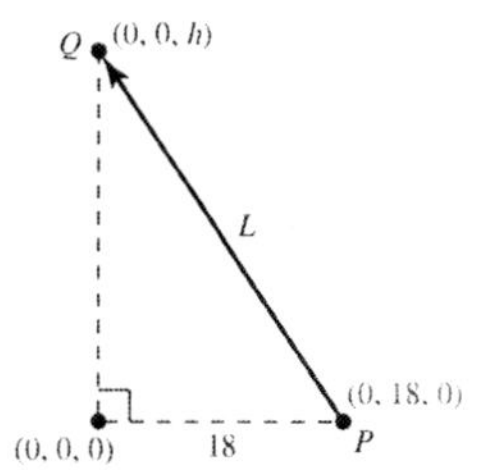

(b)

L	20	25	30	35	40	45	50
T	18.4	11.5	10	9.3	9.0	8.7	8.6

(c)

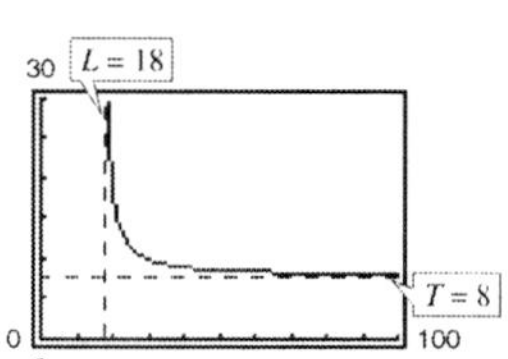

$x = 18$ is a vertical asymptote and $y = 8$ is a horizontal asymptote.

(d) $\displaystyle\lim_{L\to 18^+} \frac{8L}{\sqrt{L^2 - 18^2}} = \infty$

$$\lim_{L\to\infty} \frac{8L}{\sqrt{L^2 - 18^2}} = \lim_{L\to\infty} \frac{8}{\sqrt{1 - (18/L)^2}} = 8$$

(e) From the table, $T = 10$ implies $L = 30$ inches.

102. As in Exercise 109(c), $x = a$ will be a vertical asymptote. So, $\displaystyle\lim_{r_0\to a^-} T = \infty$.

103.

$$\begin{aligned} \overrightarrow{AB} &= \langle 0, 70, 115\rangle, \mathbf{F}_1 = C_1\langle 0, 70, 115\rangle \\ \overrightarrow{AC} &= \langle -60, 0, 115\rangle, \mathbf{F}_2 = C_2\langle -60, 0, 115\rangle \\ \overrightarrow{AD} &= \langle 45, -65, 115\rangle, \mathbf{F}_3 = C_3\langle 45, -65, 115\rangle \\ \mathbf{F} &= \mathbf{F}_1 + \mathbf{F}_2 + \mathbf{F}_3 = \langle 0, 0, 500\rangle \end{aligned}$$

So:

$$\begin{aligned} -60C_2 + 45C_3 &= 0 \\ 70C_1 \quad - 65C_3 &= 0 \\ 115(C_1 + C_2 + C_3) &= 500 \end{aligned}$$

Solving this system yields $C_1 = \frac{104}{69}$, $C_2 = \frac{28}{23}$, and $C_3 = \frac{112}{69}$. So:

$\|\mathbf{F}_1\| \approx 202.919\text{N}$

$\|\mathbf{F}_2\| \approx 157.909\text{N}$

$\|\mathbf{F}_3\| \approx 226.521\text{N}$

104. Let A lie on the y-axis and the wall on the x-axis. Then $A = (0, 10, 0)$, $B = (8, 0, 6)$, $C = (-10, 0, 6)$ and

$\overrightarrow{AB} = \langle 8, -10, 6\rangle$, $\overrightarrow{AC} = \langle -10, -10, 6\rangle$.

$\|AB\| = 10\sqrt{2}, \|AC\| = 2\sqrt{59}$

Thus, $\mathbf{F}_1 = 420\dfrac{\overrightarrow{AB}}{\|AB\|}$, $\mathbf{F}_2 = 650\dfrac{\overrightarrow{AC}}{\|AC\|}$

$\mathbf{F} = \mathbf{F}_1 + \mathbf{F}_2 \approx \langle 237.6, -297.0, 178.2\rangle + \langle -423.1, -423.1, 253.9\rangle \approx \langle -185.5, -720.1, 432.1\rangle$

$\|\mathbf{F}\| \approx 860.0$ lb

105. $d(AP) = 2d(BP)$

$$\sqrt{x^2 + (y+1)^2 + (z-1)^2} = 2\sqrt{(x-1)^2 + (y-2)^2 + z^2}$$

$$x^2 + y^2 + z^2 + 2y - 2z + 2 = 4(x^2 + y^2 + z^2 - 2x - 4y + 5)$$

$$0 = 3x^2 + 3y^2 + 3z^2 - 8x - 18y + 2z + 18$$

$$-6 + \frac{16}{9} + 9 + \frac{1}{9} = \left(x^2 - \frac{8}{3}x + \frac{16}{9}\right) + \left(y^2 - 6y + 9\right) + \left(z^2 + \frac{2}{3}z + \frac{1}{9}\right)$$

$$\frac{44}{9} = \left(x - \frac{4}{3}\right)^2 + (y-3)^2 + \left(z + \frac{1}{3}\right)^2$$

Sphere; center: $\left(\dfrac{4}{3}, 3, -\dfrac{1}{3}\right)$, radius: $\dfrac{2\sqrt{11}}{3}$

Section 11.3 The Dot Product of Two Vectors

1. $\mathbf{u} = \langle 3, 4\rangle$, $\mathbf{v} = \langle -1, 5\rangle$

(a) $\mathbf{u} \cdot \mathbf{v} = 3(-1) + 4(5) = 17$

(b) $\mathbf{u} \cdot \mathbf{u} = 3(3) + 4(4) = 25$

(c) $\|\mathbf{u}\|^2 = 3^2 + 4^2 = 25$

(d) $(\mathbf{u} \cdot \mathbf{v})\mathbf{v} = 17\langle -1, 5\rangle = \langle -17, 85\rangle$

(e) $\mathbf{u} \cdot (2\mathbf{v}) = 2(\mathbf{u} \cdot \mathbf{v}) = 2(17) = 34$

2. $\mathbf{u} = \langle 4, 10\rangle$, $\mathbf{v} = \langle -2, 3\rangle$

(a) $\mathbf{u} \cdot \mathbf{v} = 4(-2) + 10(3) = 22$

(b) $\mathbf{u} \cdot \mathbf{u} = 4(4) + 10(10) = 116$

(c) $\|\mathbf{u}\|^2 = 4^2 + 10^2 = 116$

(d) $(\mathbf{u} \cdot \mathbf{v})\mathbf{v} = 22\langle -2, 3\rangle = \langle -44, 66\rangle$

(e) $\mathbf{u} \cdot (2\mathbf{v}) = 2(\mathbf{u} \cdot \mathbf{v}) = 2(22) = 44$

3. $\mathbf{u} = \langle 6, -4\rangle$, $\mathbf{v} = \langle -3, 2\rangle$

(a) $\mathbf{u} \cdot \mathbf{v} = 6(-3) + (-4)(2) = -26$

(b) $\mathbf{u} \cdot \mathbf{u} = 6(6) + (-4)(-4) = 52$

(c) $\|\mathbf{u}\|^2 = 6^2 + (-4)^2 = 52$

(d) $(\mathbf{u} \cdot \mathbf{v})\mathbf{v} = -26\langle -3, 2\rangle = \langle 78, -52\rangle$

(e) $\mathbf{u} \cdot (2\mathbf{v}) = 2(\mathbf{u} \cdot \mathbf{v}) = 2(-26) = -52$

4. $\mathbf{u} = \langle -4, 8\rangle$, $\mathbf{v} = \langle 7, 5\rangle$

(a) $\mathbf{u} \cdot \mathbf{v} = -4(7) + 8(5) = 12$

(b) $\mathbf{u} \cdot \mathbf{u} = (-4)(-4) + 8(8) = 80$

(c) $\|\mathbf{u}\|^2 = (-4)^2 + 8^2 = 80$

(d) $(\mathbf{u} \cdot \mathbf{v})\mathbf{v} = 12\langle 7, 5\rangle = \langle 84, 60\rangle$

(e) $\mathbf{u} \cdot (2\mathbf{v}) = 2(\mathbf{u} \cdot \mathbf{v}) = 2(12) = 24$

5. $\mathbf{u} = \langle 2, -3, 4\rangle, \mathbf{v} = \langle 0, 6, 5\rangle$

(a) $\mathbf{u} \cdot \mathbf{v} = 2(0) + (-3)(6) + (4)(5) = 2$

(b) $\mathbf{u} \cdot \mathbf{u} = 2(2) + (-3)(-3) + 4(4) = 29$

(c) $\|\mathbf{u}\|^2 = 2^2 + (-3)^2 + 4^2 = 29$

(d) $(\mathbf{u} \cdot \mathbf{v})\mathbf{v} = 2\langle 0, 6, 5\rangle = \langle 0, 12, 10\rangle$

(e) $\mathbf{u} \cdot (2\mathbf{v}) = 2(\mathbf{u} \cdot \mathbf{v}) = 2(2) = 4$

6. $\mathbf{u} = \mathbf{i}, \mathbf{v} = \mathbf{i}$

(a) $\mathbf{u} \cdot \mathbf{v} = 1$

(b) $\mathbf{u} \cdot \mathbf{u} = 1$

(c) $\|\mathbf{u}\|^2 = 1$

(d) $(\mathbf{u} \cdot \mathbf{v})\mathbf{v} = \mathbf{i}$

(e) $\mathbf{u} \cdot (2\mathbf{v}) = 2(\mathbf{u} \cdot \mathbf{v}) = 2$

7. $\mathbf{u} = 2\mathbf{i} - \mathbf{j} + \mathbf{k}, \mathbf{v} = \mathbf{i} - \mathbf{k}$

(a) $\mathbf{u} \cdot \mathbf{v} = 2(1) + (-1)(0) + 1(-1) = 1$

(b) $\mathbf{u} \cdot \mathbf{u} = 2(2) + (-1)(-1) + (1)(1) = 6$

(c) $\|\mathbf{u}\|^2 = 2^2 + (-1)^2 + 1^2 = 6$

(d) $(\mathbf{u} \cdot \mathbf{v})\mathbf{v} = \mathbf{v} = \mathbf{i} - \mathbf{k}$

(e) $\mathbf{u} \cdot (2\mathbf{v}) = 2(\mathbf{u} \cdot \mathbf{v}) = 2$

8. $\mathbf{u} = 2\mathbf{i} + \mathbf{j} - 2\mathbf{k}, \mathbf{v} = \mathbf{i} - 3\mathbf{j} + 2\mathbf{k}$

(a) $\mathbf{u} \cdot \mathbf{v} = 2(1) + 1(-3) + (-2)(2) = -5$

(b) $\mathbf{u} \cdot \mathbf{u} = 2(2) + 1(1) + (-2)(-2) = 9$

(c) $\|\mathbf{u}\|^2 = 2^2 + 1^2 + (-2)^2 = 9$

(d) $(\mathbf{u} \cdot \mathbf{v})\mathbf{v} = -5(\mathbf{i} - 3\mathbf{j} + 2\mathbf{k}) = -5\mathbf{i} + 15\mathbf{j} - 10\mathbf{k}$

(e) $\mathbf{u} \cdot (2\mathbf{v}) = 2(\mathbf{u} \cdot \mathbf{v}) = 2(-5) = -10$

9. $\mathbf{u} = \langle 1, 1\rangle, \mathbf{v} = \langle 2, -2\rangle$

$$\cos\theta = \frac{\mathbf{u} \cdot \mathbf{v}}{\|\mathbf{u}\|\|\mathbf{v}\|} = \frac{0}{\sqrt{2}\sqrt{8}} = 0$$

(a) $\theta = \dfrac{\pi}{2}$ (b) $\theta = 90°$

10. $\mathbf{u} = \langle 3, 1\rangle, \mathbf{v} = \langle 2, -1\rangle$

$$\cos\theta = \frac{\mathbf{u} \cdot \mathbf{v}}{\|\mathbf{u}\|\|\mathbf{v}\|} = \frac{5}{\sqrt{10}\sqrt{5}} = \frac{1}{\sqrt{2}}$$

(a) $\theta = \dfrac{\pi}{4}$ (b) $\theta = 45°$

11. $\mathbf{u} = 3\mathbf{i} + \mathbf{j}, \mathbf{v} = -2\mathbf{i} + 4\mathbf{j}$

$$\cos\theta = \frac{\mathbf{u} \cdot \mathbf{v}}{\|\mathbf{u}\|\|\mathbf{v}\|} = \frac{-2}{\sqrt{10}\sqrt{20}} = \frac{-1}{5\sqrt{2}}$$

(a) $\theta = \arccos\left(-\dfrac{1}{5\sqrt{2}}\right) \approx 1.713$

(b) $\theta \approx 98.1°$

12.

$$\mathbf{u} = \cos\left(\frac{\pi}{6}\right)\mathbf{i} + \sin\left(\frac{\pi}{6}\right)\mathbf{j} = \frac{\sqrt{3}}{2}\mathbf{i} + \frac{1}{2}\mathbf{j}$$

$$\mathbf{v} = \cos\left(\frac{3\pi}{4}\right)\mathbf{i} + \sin\left(\frac{3\pi}{4}\right)\mathbf{j} = -\frac{\sqrt{2}}{2}\mathbf{i} + \frac{\sqrt{2}}{2}\mathbf{j}$$

$$\cos\theta = \frac{\mathbf{u} \cdot \mathbf{v}}{\|\mathbf{u}\|\|\mathbf{v}\|} = \frac{\sqrt{3}}{2}\left(-\frac{\sqrt{2}}{2}\right) + \frac{1}{2}\left(\frac{\sqrt{2}}{2}\right) = \frac{\sqrt{2}}{4}\left(1 - \sqrt{3}\right)$$

(a) $\theta = \arccos\left[\dfrac{\sqrt{2}}{4}\left(1 - \sqrt{3}\right)\right] = \dfrac{7\pi}{2}$

(b) $\theta = 105°$

13. $\mathbf{u} = \langle 1, 1, 1\rangle, \mathbf{v} = \langle 2, 1, -1\rangle$

$$\cos\theta = \frac{\mathbf{u} \cdot \mathbf{v}}{\|\mathbf{u}\|\|\mathbf{v}\|} = \frac{2}{\sqrt{3}\sqrt{6}} = \frac{\sqrt{2}}{3}$$

(a) $\theta = \arccos\dfrac{\sqrt{2}}{3} \approx 1.080$

(b) $\theta \approx 61.9°$

14. $\mathbf{u} = 3\mathbf{i} + 2\mathbf{j} + \mathbf{k}, \mathbf{v} = 2\mathbf{i} - 3\mathbf{j}$

$$\cos\theta = \frac{\mathbf{u} \cdot \mathbf{v}}{\|\mathbf{u}\|\|\mathbf{v}\|} = \frac{3(2) + 2(-3) + 0}{\|\mathbf{u}\|\|\mathbf{v}\|} = 0$$

(a) $\theta = \dfrac{\pi}{2}$

(b) $\theta = 90°$

15. $\mathbf{u} = 3\mathbf{i} + 4\mathbf{j}, \mathbf{v} = -2\mathbf{j} + 3\mathbf{k}$

$$\cos\theta = \frac{\mathbf{u} \cdot \mathbf{v}}{\|\mathbf{u}\|\|\mathbf{v}\|} = \frac{-8}{5\sqrt{13}} = \frac{-8\sqrt{13}}{65}$$

(a) $\theta = \arccos\left(-\dfrac{8\sqrt{13}}{65}\right) \approx 2.031$

(b) $\theta \approx 116.3°$

16. $\mathbf{u} = 2\mathbf{i} - 3\mathbf{j} + \mathbf{k}, \mathbf{v} = \mathbf{i} - 2\mathbf{j} + \mathbf{k}$

$$\cos\theta = \frac{\mathbf{u}\cdot\mathbf{v}}{\|\mathbf{u}\|\|\mathbf{v}\|} = \frac{9}{\sqrt{14}\sqrt{6}} = \frac{9}{2\sqrt{21}} = \frac{3\sqrt{21}}{14}$$

(a) $\theta = \arccos\left(\frac{3\sqrt{21}}{14}\right) \approx 0.190$

(b) $\theta \approx 10.9°$

17. $\frac{\mathbf{u}\cdot\mathbf{v}}{\|\mathbf{u}\|\|\mathbf{v}\|} = \cos\theta$

$\mathbf{u}\cdot\mathbf{v} = (8)(5)\cos\frac{\pi}{3} = 20$

18. $\frac{\mathbf{u}\cdot\mathbf{v}}{\|\mathbf{u}\|\|\mathbf{v}\|} = \cos\theta$

$\mathbf{u}\cdot\mathbf{v} = (40)(25)\cos\frac{5\pi}{6} = -500\sqrt{3}$

19. $\mathbf{u} = \langle 4, 3\rangle, \mathbf{v} = \left\langle \frac{1}{2}, -\frac{2}{3}\right\rangle$

$\mathbf{u} \neq c\mathbf{v} \Rightarrow$ not parallel

$\mathbf{u}\cdot\mathbf{v} = 0 \Rightarrow$ orthogonal

20. $\mathbf{u} = -\frac{1}{3}(\mathbf{i} - 2\mathbf{j}), \mathbf{v} = 2\mathbf{i} - 4\mathbf{j}$

$\mathbf{u} = -\frac{1}{6}\mathbf{v} \Rightarrow$ parallel

21. $\mathbf{u} = \mathbf{j} + 6\mathbf{k}, \mathbf{v} = \mathbf{i} - 2\mathbf{j} - \mathbf{k}$

$\mathbf{u} \neq c\mathbf{v} \Rightarrow$ not parallel

$\mathbf{u}\cdot\mathbf{v} = -8 \neq 0 \Rightarrow$ not orthogonal

Neither

22. $\mathbf{u} = -2\mathbf{i} + 3\mathbf{j} - \mathbf{k}, \mathbf{v} = 2\mathbf{i} + \mathbf{j} - \mathbf{k}$

$\mathbf{u} \neq c\mathbf{v} \Rightarrow$ not parallel

$\mathbf{u}\cdot\mathbf{v} = 0 \Rightarrow$ orthogonal

23. $\mathbf{u} = \langle 2, -3, 1\rangle, \mathbf{v} = \langle -1, -1, -1\rangle$

$\mathbf{u} \neq c\mathbf{v} \Rightarrow$ not parallel

$\mathbf{u}\cdot\mathbf{v} = 0 \Rightarrow$ orthogonal

24. $\mathbf{u} = \langle \cos\theta, \sin\theta, -1\rangle$,

$\mathbf{v} = \langle \sin\theta, -\cos\theta, 0\rangle$

$\mathbf{u} \neq c\mathbf{v} \Rightarrow$ not parallel

$\mathbf{u}\cdot\mathbf{v} = 0 \Rightarrow$ orthogonal

25. The vector $\langle 1, 2, 0\rangle$ joining $(1, 2, 0)$ and $(0, 0, 0)$ is perpendicular to the vector $\langle -2, 1, 0\rangle$ joining $(-2, 1, 0)$ and $(0, 0, 0)$: $\langle 1, 2, 0\rangle \cdot \langle -2, 1, 0\rangle = 0$

The triangle has a right angle, so it is a right triangle.

26. Consider the vector $\langle -3, 0, 0\rangle$ joining $(0, 0, 0)$ and $(-3, 0, 0)$, and the vector $\langle 1, 2, 3\rangle$ joining $(0, 0, 0)$ and $(1, 2, 3)$: $\langle -3, 0, 0\rangle \cdot \langle 1, 2, 3\rangle = -3 < 0$

The triangle has an obtuse angle, so it is an obtuse triangle.

27. $A(2, 0, 1), B(0, 1, 2), C\left(-\frac{1}{2}, \frac{3}{2}, 0\right)$

$\overrightarrow{AB} = \langle -2, 1, 1\rangle \qquad \overrightarrow{BA} = \langle 2, -1, -1\rangle$

$\overrightarrow{AC} = \left\langle -\frac{5}{2}, \frac{3}{2}, -1\right\rangle \qquad \overrightarrow{CA} = \left\langle \frac{5}{2}, -\frac{3}{2}, 1\right\rangle$

$\overrightarrow{BC} = \left\langle -\frac{1}{2}, \frac{1}{2}, -2\right\rangle \qquad \overrightarrow{CB} = \left\langle \frac{1}{2}, -\frac{1}{2}, 2\right\rangle$

$\overrightarrow{AB}\cdot\overrightarrow{AC} = 5 + \frac{3}{2} - 1 > 0$

$\overrightarrow{BA}\cdot\overrightarrow{BC} = -1 - \frac{1}{2} + 2 > 0$

$\overrightarrow{CA}\cdot\overrightarrow{CB} = \frac{5}{4} + \frac{3}{4} + 2 > 0$

The triangle has three acute angles, so it is an acute triangle.

28. $A(2, -7, 3), B(-1, 5, 8), C(4, 6, -1)$

$\overrightarrow{AB} = \langle -3, 12, 5\rangle \qquad \overrightarrow{BA} = \langle 3, -12, -5\rangle$

$\overrightarrow{AC} = \langle 2, 13, -4\rangle \qquad \overrightarrow{CA} = \langle -2, -13, 4\rangle$

$\overrightarrow{BC} = \langle 5, 1, -9\rangle \qquad \overrightarrow{CB} = \langle -5, -1, 9\rangle$

$\overrightarrow{AB}\cdot\overrightarrow{AC} = -6 + 156 - 20 > 0$

$\overrightarrow{BA}\cdot\overrightarrow{BC} = 15 - 12 + 45 > 0$

$\overrightarrow{CA}\cdot\overrightarrow{CB} = 10 + 13 + 36 > 0$

The triangle has three acute angles, so it is an acute triangle.

29. $\mathbf{u} = \mathbf{i} + 2\mathbf{j} + 2\mathbf{k}, \|\mathbf{u}\| = \sqrt{1 + 4 + 4} = 3$

$\cos\alpha = \frac{1}{3} \Rightarrow \alpha \approx 1.2310$ or $70.5°$

$\cos\beta = \frac{2}{3} \Rightarrow \beta \approx 0.8411$ or $48.2°$

$\cos\gamma = \frac{2}{3} \Rightarrow \gamma \approx 0.8411$ or $48.2°$

$\cos^2\alpha + \cos^2\beta + \cos^2\gamma = \frac{1}{9} + \frac{4}{9} + \frac{4}{9} = 1$

30. $\mathbf{u} = 5\mathbf{i} + 3\mathbf{j} - \mathbf{k} \quad \|\mathbf{u}\| = \sqrt{25 + 9 + 1} = \sqrt{35}$

$\cos\alpha = \frac{5}{\sqrt{35}} \Rightarrow \alpha \approx 0.5639$ or $32.3°$

$\cos\beta = \frac{3}{\sqrt{35}} \Rightarrow \beta \approx 1.0390$ or $59.5°$

$\cos\gamma = \frac{-1}{\sqrt{35}} \Rightarrow \gamma \approx 1.7406$ or $99.7°$

$\cos^2\alpha + \cos^2\beta + \cos^2\gamma = \frac{25}{35} + \frac{9}{35} + \frac{1}{35} = 1$

31. $\mathbf{u} = 3\mathbf{i} + 2\mathbf{j} - 2\mathbf{k}$ $\|\mathbf{u}\| = \sqrt{9 + 4 + 4} = \sqrt{17}$

$\cos\alpha = \dfrac{3}{\sqrt{17}} \Rightarrow \alpha \approx 0.7560 \text{ or } 43.3°$

$\cos\beta = \dfrac{2}{\sqrt{17}} \Rightarrow \beta \approx 1.0644 \text{ or } 61.0°$

$\cos\gamma = \dfrac{-2}{\sqrt{17}} \Rightarrow \gamma \approx 2.0772 \text{ or } 119.0°$

$\cos^2\alpha + \cos^2\beta + \cos^2\gamma = \dfrac{9}{17} + \dfrac{4}{17} + \dfrac{4}{17} = 1$

32. $\mathbf{u} = -4\mathbf{i} + 3\mathbf{j} + 5\mathbf{k}$ $\|\mathbf{u}\| = \sqrt{16 + 9 + 25} = \sqrt{50} = 5\sqrt{2}$

$\cos\alpha = \dfrac{-4}{5\sqrt{2}} \Rightarrow \alpha \approx 2.1721 \text{ or } 124.4°$

$\cos\beta = \dfrac{3}{5\sqrt{2}} \Rightarrow \beta \approx 1.1326 \text{ or } 64.9°$

$\cos\gamma = \dfrac{5}{5\sqrt{2}} = \dfrac{1}{\sqrt{2}} \Rightarrow \gamma \approx \dfrac{\pi}{4} \text{ or } 45°$

$\cos^2\alpha + \cos^2\beta + \cos^2\gamma = \dfrac{16}{50} + \dfrac{9}{50} + \dfrac{25}{50} = 1$

33. $\mathbf{u} = \langle 0, 6, -4\rangle, \|\mathbf{u}\| = \sqrt{0 + 36 + 16} = \sqrt{52} = 2\sqrt{13}$

$\cos\alpha = 0 \Rightarrow \alpha = \dfrac{\pi}{2} \text{ or } 90°$

$\cos\beta = \dfrac{3}{\sqrt{13}} \Rightarrow \beta \approx 0.5880 \text{ or } 33.7°$

$\cos\gamma = -\dfrac{2}{\sqrt{13}} \Rightarrow \gamma \approx 2.1588 \text{ or } 123.7°$

$\cos^2\alpha + \cos^2\beta + \cos^2\gamma = 0 + \dfrac{9}{13} + \dfrac{4}{13} = 1$

34. $\mathbf{u} = \langle -1, 5, 2\rangle$ $\|\mathbf{u}\| = \sqrt{1 + 25 + 4} = \sqrt{30}$

$\cos\alpha = \dfrac{-1}{\sqrt{30}} \Rightarrow \alpha \approx 1.7544 \text{ or } 100.5°$

$\cos\beta = \dfrac{5}{\sqrt{30}} \Rightarrow \beta \approx 0.4205 \text{ or } 24.1°$

$\cos\gamma = \dfrac{2}{\sqrt{30}} \Rightarrow \gamma \approx 1.1970 \text{ or } 68.6°$

$\cos^2\alpha + \cos^2\beta + \cos^2\gamma = \dfrac{1}{30} + \dfrac{25}{30} + \dfrac{4}{30} = 1$

35. $\mathbf{u} = \langle 6, 7\rangle, \mathbf{v} = \langle 1, 4\rangle$

(a) $\mathbf{w}_1 = \text{proj}_{\mathbf{v}}\mathbf{u} = \left(\dfrac{\mathbf{u}\cdot\mathbf{v}}{\|\mathbf{v}\|^2}\right)\mathbf{v}$

$= \dfrac{6(1) + 7(4)}{1^2 + 4^2}\langle 1, 4\rangle$

$= \dfrac{34}{17}\langle 1, 4\rangle = \langle 2, 8\rangle$

(b) $\mathbf{w}_2 = \mathbf{u} - \mathbf{w}_1 = \langle 6, 7\rangle - \langle 2, 8\rangle = \langle 4, -1\rangle$

36. $\mathbf{u} = \langle 9, 7\rangle, \mathbf{v} = \langle 1, 3\rangle$

(a) $\mathbf{w}_1 = \text{proj}_{\mathbf{v}}\mathbf{u} = \left(\dfrac{\mathbf{u}\cdot\mathbf{v}}{\|\mathbf{v}\|^2}\right)\mathbf{v}$

$= \dfrac{9(1) + 7(3)}{1 + 3^2}\langle 1, 3\rangle$

$= \dfrac{30}{10}\langle 1, 3\rangle = \langle 3, 9\rangle$

(b) $\mathbf{w}_2 = \mathbf{u} - \mathbf{w}_1 = \langle 9, 7\rangle = \langle 3, 9\rangle = \langle 6, -2\rangle$

37. $\mathbf{u} = 2\mathbf{i} + 3\mathbf{j} = \langle 2, 3\rangle, \mathbf{v} = 5\mathbf{i} + \mathbf{j} = \langle 5, 1\rangle$

(a) $\mathbf{w}_1 = \text{proj}_{\mathbf{v}}\mathbf{u} = \left(\dfrac{\mathbf{u}\cdot\mathbf{v}}{\|\mathbf{v}\|^2}\right)\mathbf{v}$

$= \dfrac{2(5) + 3(1)}{5^2 + 1}\langle 5, 1\rangle$

$= \dfrac{13}{26}\langle 5, 1\rangle = \left\langle \dfrac{5}{2}, \dfrac{1}{2}\right\rangle$

(b) $\mathbf{w}_2 = \mathbf{u} - \mathbf{w}_1 = \langle 2, 3\rangle - \left\langle \dfrac{5}{2}, \dfrac{1}{2}\right\rangle = \left\langle -\dfrac{1}{2}, \dfrac{5}{2}\right\rangle$

38. $\mathbf{u} = 2\mathbf{i} - 3\mathbf{j} = \langle 2, -3\rangle, \mathbf{v} = 3\mathbf{i} + 2\mathbf{j} = \langle 3, 2\rangle$

(a) $\mathbf{w}_1 = \text{proj}_{\mathbf{v}}\mathbf{u} = \left(\dfrac{\mathbf{u}\cdot\mathbf{v}}{\|\mathbf{v}\|^2}\right)\mathbf{v}$

$= \dfrac{2(3) + (-3)(2)}{3^2 + 2^2}\langle 3, 2\rangle$

$= 0\langle 3, 2\rangle = \langle 0, 0\rangle$

(b) $\mathbf{w}_2 = \mathbf{u} - \mathbf{w}_1 = \langle 2, -3\rangle$

39. $\mathbf{u} = \langle 0, 3, 3\rangle, \mathbf{v} = \langle -1, 1, 1\rangle$

(a) $\mathbf{w}_1 = \text{proj}_{\mathbf{v}}\mathbf{u} = \left(\dfrac{\mathbf{u}\cdot\mathbf{v}}{\|\mathbf{v}\|^2}\right)\mathbf{v}$

$= \dfrac{0(-1) + 3(1) + 3(1)}{1 + 1 + 1}\langle -1, 1, 1\rangle$

$= \dfrac{6}{3}\langle -1, 1, 1\rangle = \langle -2, 2, 2\rangle$

(b) $\mathbf{w}_2 = \mathbf{u} - \mathbf{w}_1 = \langle 0, 3, 3\rangle - \langle -2, 2, 2\rangle = \langle 2, 1, 1\rangle$

40. $\mathbf{u} = \langle 8, 2, 0\rangle, \mathbf{v} = (2, 1, -1)$

(a) $\mathbf{w}_1 = \text{proj}_{\mathbf{v}}\mathbf{u} = \left(\dfrac{\mathbf{u}\cdot\mathbf{v}}{\|\mathbf{v}\|^2}\right)\mathbf{v}$

$= \dfrac{8(2) + 2(1) + 0(-1)}{2^2 + 1 + 1}\langle 2, 1, -1\rangle$

$= \dfrac{18}{6}\langle 2, 1, -1\rangle = \langle 6, 3, -3\rangle$

(b) $\mathbf{w}_2 = \mathbf{u} - \mathbf{w}_1 = \langle 8, 2, 0\rangle - \langle 6, 3, -3\rangle = \langle 2, -1, 3\rangle$

41. $\mathbf{u} = 2\mathbf{i} + \mathbf{j} + 2\mathbf{k} = \langle 2, 1, 2\rangle$

$\mathbf{v} = 3\mathbf{j} + 4\mathbf{k} = \langle 0, 3, 4\rangle$

(a) $\mathbf{w}_1 = \text{proj}_{\mathbf{v}}\mathbf{u} = \left(\dfrac{\mathbf{u}\cdot\mathbf{v}}{\|\mathbf{v}\|^2}\right)\mathbf{v}$

$= \dfrac{2(0) + 1(3) + 2(4)}{3^2 + 4^2}\langle 0, 3, 4\rangle$

$= \dfrac{11}{25}\langle 0, 3, 4\rangle = \left\langle 0, \dfrac{33}{25}, \dfrac{44}{25}\right\rangle$

(b) $\mathbf{w}_2 = \mathbf{u} - \mathbf{w}_1 = \langle 2, 1, 2\rangle - \left\langle 0, \dfrac{33}{25}, \dfrac{44}{25}\right\rangle = \left\langle 2, -\dfrac{8}{25}, \dfrac{6}{25}\right\rangle$

42. $\mathbf{u} = \mathbf{i} + 4\mathbf{k} = \langle 1, 0, 4\rangle$

$\mathbf{v} = 3\mathbf{i} + 2\mathbf{k} = \langle 3, 0, 2\rangle$

(a) $\mathbf{w}_1 = \text{proj}_{\mathbf{v}}\mathbf{u} = \left(\dfrac{\mathbf{u}\cdot\mathbf{v}}{\|\mathbf{v}\|^2}\right)\mathbf{v}$

$= \dfrac{1(3) + 4(2)}{3^2 + 2^2}\langle 3, 0, 2\rangle$

$= \dfrac{11}{13}\langle 3, 0, 2\rangle = \left\langle \dfrac{33}{13}, 0, \dfrac{22}{13}\right\rangle$

(b) $\mathbf{w}_2 = \mathbf{u} - \mathbf{w}_1 = \langle 1, 0, 4\rangle - \left\langle \dfrac{33}{13}, 0, \dfrac{22}{13}\right\rangle$

$= \left\langle -\dfrac{20}{13}, 0, \dfrac{30}{13}\right\rangle$

43. $\mathbf{u}\cdot\mathbf{v} = \langle u_1, u_2, u_3\rangle \cdot \langle v_1, v_2, v_3\rangle = u_1v_1 + u_2v_2 + u_3v_3$

44. The vectors **u** and **v** are orthogonal if $\mathbf{u}\cdot\mathbf{v} = 0$. The angle θ between **u** and **v** is given by $\cos\theta = \dfrac{\mathbf{u}\cdot\mathbf{v}}{\|\mathbf{u}\|\|\mathbf{v}\|}$.

45. (a) and (b) are defined. (c) and (d) are not defined because it is not possible to find the dot product of a scalar and a vector or to add a scalar to a vector.

46. See page 769. Direction cosines of $\mathbf{v} = \langle v_1, v_2, v_3\rangle$ are

$\cos\alpha = \dfrac{v_1}{\|\mathbf{v}\|}, \cos\beta = \dfrac{v_2}{\|\mathbf{v}\|}, \cos\gamma = \dfrac{v_3}{\|\mathbf{v}\|}$. α, β, and γ are the direction angles. See Figure 11.26.

47. See figure 11.29, page 770.

48. (a) $\left(\dfrac{\mathbf{u}\cdot\mathbf{v}}{\|\mathbf{v}\|^2}\right)\mathbf{v} = \mathbf{u} \Rightarrow \mathbf{u} = c\mathbf{v} \Rightarrow$ **u** and **v** are parallel.

(b) $\left(\dfrac{\mathbf{u}\cdot\mathbf{v}}{\|\mathbf{v}\|^2}\right)\mathbf{v} = \mathbf{0} \Rightarrow \mathbf{u}\cdot\mathbf{v} = 0 \Rightarrow$ **u** and **v** are orthogonal.

49. Yes, $\left\|\dfrac{\mathbf{u}\cdot\mathbf{v}}{\|\mathbf{v}\|^2}\mathbf{v}\right\| = \left\|\dfrac{\mathbf{v}\cdot\mathbf{u}}{\|\mathbf{u}\|^2}\mathbf{u}\right\|$

$|\mathbf{u}\cdot\mathbf{v}|\dfrac{\|\mathbf{v}\|}{\|\mathbf{v}\|^2} = |\mathbf{v}\cdot\mathbf{u}|\dfrac{\|\mathbf{u}\|}{\|\mathbf{u}\|^2}$

$\dfrac{1}{\|\mathbf{v}\|} = \dfrac{1}{\|\mathbf{u}\|}$

$\|\mathbf{u}\| = \|\mathbf{v}\|$

50. (a) Orthogonal, $\theta = \dfrac{\pi}{2}$

(b) Acute, $0 < \theta < \dfrac{\pi}{2}$

(c) Obtuse, $\dfrac{\pi}{2} < \theta < \pi$

51. $\mathbf{u} = \langle 3240, 1450, 2235\rangle$

$\mathbf{v} = \langle 2.25, 2.95, 2.65\rangle$

$\mathbf{u}\cdot\mathbf{v} = 3240(2.25) + 1450(2.95) + 2235(2.65)$

$= \$17{,}490.25$

This represents the total revenue the restaurant earned on its three products.

52. $\mathbf{u} = \langle 3240, 1450, 2235\rangle$

$\mathbf{v} = \langle 2.25, 2.95, 2.65\rangle$

Increase prices by $4\% = 1.04\mathbf{v}$

New total amount: $1.04(\mathbf{u}\cdot\mathbf{v}) = 1.04(17{,}490.25)$

$= \$18{,}189.86$

53. Answers will vary. *Sample answer:*

$\mathbf{u} = -\dfrac{1}{4}\mathbf{i} + \dfrac{3}{2}\mathbf{j}$. Want $\mathbf{u}\cdot\mathbf{v} = 0$.

$\mathbf{v} = 12\mathbf{i} + 2\mathbf{j}$ and $-\mathbf{v} = -12\mathbf{i} - 2\mathbf{j}$ are orthogonal to **u**.

54. Answers will vary. *Sample answer:*

$\mathbf{u} = 9\mathbf{i} - 4\mathbf{j}$. Want $\mathbf{u} \cdot \mathbf{v} = 0$.

$\mathbf{v} = 4\mathbf{i} + 9\mathbf{j}$ and $-\mathbf{v} - 4\mathbf{i} - 9\mathbf{j}$

are orthogonal to $\mathbf{u}$.

55. Answers will vary. *Sample answer:*

$\mathbf{u} = \langle 3, 1, -2\rangle$. Want $\mathbf{u} \cdot \mathbf{v} = 0$.

$\mathbf{v} = \langle 0, 2, 1\rangle$ and $-\mathbf{v} = \langle 0, -2, -1\rangle$ are orthogonal to $\mathbf{u}$.

56. Answers will vary. *Sample answer:*

$\mathbf{u} = \langle 4, -3, 6\rangle$. Want $\mathbf{u} \cdot \mathbf{v} = 0$

$\mathbf{v} = \langle 0, 6, 3\rangle$ and $-\mathbf{v} = \langle 0, -6, -3\rangle$

are orthogonal to $\mathbf{u}$.

57. Let s = length of a side.

$\mathbf{v} = \langle s, s, s\rangle$

$\|\mathbf{v}\| = s\sqrt{3}$

$$\cos\alpha = \cos\beta = \cos\gamma = \frac{s}{s\sqrt{3}} = \frac{1}{\sqrt{3}}$$

$$\alpha = \beta = \gamma = \arccos\left(\frac{1}{\sqrt{3}}\right) \approx 54.7°$$

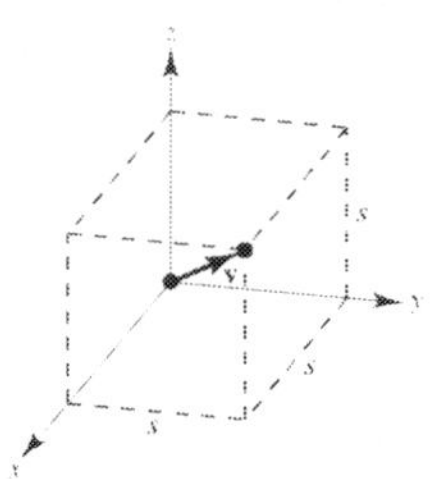

58.

$\mathbf{v}_1 = \langle s, s, s\rangle$

$\|\mathbf{v}_1\| = s\sqrt{3}$

$\mathbf{v}_2 = \langle s, s, 0\rangle$

$\|\mathbf{v}_2\| = s\sqrt{2}$

$$\cos\theta = \frac{2\sqrt{2}}{2\sqrt{3}} = \frac{\sqrt{6}}{3}$$

$$\theta = \arccos\frac{\sqrt{6}}{3} \approx 35.26°$$

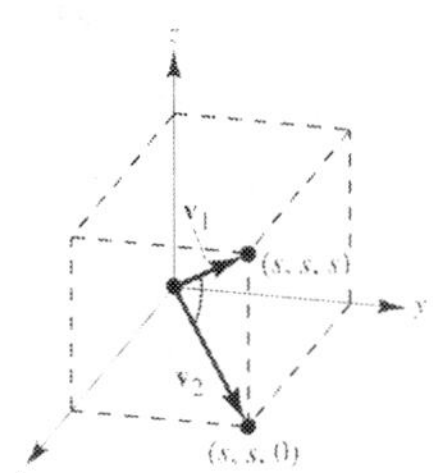

59. (a) Gravitational Force $\mathbf{F} = -48{,}000\mathbf{j}$

$\mathbf{v} = \cos 10°\mathbf{i} + \sin 10°\mathbf{j}$

$$\mathbf{w}_1 = \frac{\mathbf{F}\cdot\mathbf{v}}{\|\mathbf{v}\|^2}\mathbf{v} = (\mathbf{F}\cdot\mathbf{v})\mathbf{v}$$

$$= (-48{,}000)(\sin 10°)\mathbf{v}$$

$$\approx -8335.1(\cos 10°\mathbf{i} + \sin 10°\mathbf{j})$$

$\|\mathbf{w}_1\| \approx 8335.1$ lb

(b) $\mathbf{w}_2 = \mathbf{F} - \mathbf{w}_1$

$= -48{,}000\mathbf{j} + 8335.1(\cos 10°\mathbf{i} + \sin 10°\mathbf{j})$

$= 8208.5\mathbf{i} - 46{,}552.6\mathbf{j}$

$\|\mathbf{w}_2\| \approx 47{,}270.8$ lb

60. (a) Gravitational Force $\mathbf{F} = -5400\mathbf{j}$

$\mathbf{v} = \cos 18°\mathbf{i} + \sin 18°\mathbf{j}$

$$\mathbf{w}_1 = \frac{\mathbf{F}\cdot\mathbf{v}}{\|\mathbf{v}\|^2}\mathbf{v} = (\mathbf{F}\cdot\mathbf{v})\mathbf{v}$$

$$= (-5400)(\sin 18°)\mathbf{v}$$

$$\approx -1668.7(\cos 18°\mathbf{i} + \sin 18°\mathbf{j})$$

$\|\mathbf{w}_1\| = 1668.7$ lb

(b) $\mathbf{w}_2 = \mathbf{F} - \mathbf{w}_1$

$= -5400\mathbf{j} + 1668.7(\cos 18°\mathbf{i} + \sin 18°\mathbf{j})$

$\approx 1587.0\mathbf{i} - 4884.3\mathbf{j}$

$\|\mathbf{w}_2\| \approx 5135.7$ lb

61. $\mathbf{F} = 85\left(\frac{1}{2}\mathbf{i} + \frac{\sqrt{3}}{2}\mathbf{j}\right)$

$\mathbf{v} = 10\mathbf{i}$

$W = \mathbf{F}\cdot\mathbf{v} = 425$ ft-lb

62. $\mathbf{F} = 25(\cos 20°\mathbf{i} + \sin 20°\mathbf{j})$

$\mathbf{v} = 50\mathbf{i}$

$W = \mathbf{F}\cdot\mathbf{v} = 1250\cos 20° \approx 1174.6$ ft-lb

63. $\mathbf{F} = 1600(\cos 25°\,\mathbf{i} + \sin 25°\,\mathbf{j})$

$\mathbf{v} = 2000\mathbf{i}$

$W = \mathbf{F}\cdot\mathbf{v} = 1600(2000)\cos 25°$

$\approx 2{,}900{,}184.9$ Newton meters (Joules)

≈ 2900.2 km-N

64. $\overrightarrow{PQ} = 40\mathbf{i}$

$\mathbf{F} = 100\cos 25°\mathbf{i}$

$W = \mathbf{F}\cdot\overrightarrow{PQ} = 4000\cos 25° \approx 3625.2$ Joules

65. False.

For example, let $\mathbf{u} = \langle 1, 1\rangle$, $\mathbf{v} = \langle 2, 3\rangle$ and $\mathbf{w} = \langle 1, 4\rangle$. Then $\mathbf{u}\cdot\mathbf{v} = 2 + 3 = 5$ and $\mathbf{u}\cdot\mathbf{w} = 1 + 4 = 5$.

66. True

$\mathbf{w}\cdot(\mathbf{u} + \mathbf{v}) = \mathbf{w}\cdot\mathbf{u} + \mathbf{w}\cdot\mathbf{v} = 0 + 0 = 0$ so, $\mathbf{w}$ and $\mathbf{u} + \mathbf{v}$ are orthogonal.

67. (a) The graphs $y_1 = x^2$ and $y_2 = x^{1/3}$ intersect at $(0, 0)$ and $(1, 1)$.

(b) $y_1' = 2x$ and $y_2' = \dfrac{1}{3x^{2/3}}$.

At $(0, 0)$, $\pm\langle 1, 0\rangle$ is tangent to y_1 and $\pm\langle 0, 1\rangle$ is tangent to y_2.

At $(1, 1)$, $y_1' = 2$ and $y_2' = \dfrac{1}{3}$.

$\pm\dfrac{1}{\sqrt{5}}\langle 1, 2\rangle$ is tangent to y_1, $\pm\dfrac{1}{\sqrt{10}}\langle 3, 1\rangle$ is tangent to y_2.

(c) At $(0, 0)$, the vectors are perpendicular $(90°)$.

At $(1, 1)$,

$$\cos\theta = \frac{\dfrac{1}{\sqrt{5}}\langle 1, 2\rangle\cdot\dfrac{1}{\sqrt{10}}\langle 3, 1\rangle}{(1)(1)} = \frac{5}{\sqrt{50}} = \frac{1}{\sqrt{2}}$$

$\theta = 45°$

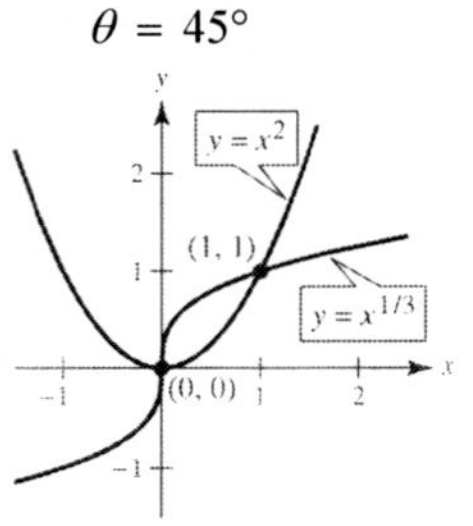

68. (a) The graphs $y_1 = x^3$ and $y_2 = x^{1/3}$ intersect at $(-1, -1)$, $(0, 0)$ and $(1, 1)$.

(b) $y_1' = 3x^2$ and $y_2' = \dfrac{1}{3x^{2/3}}$.

At $(0, 0)$, $\pm\langle 1, 0\rangle$ is tangent to y_1 and $\pm\langle 0, 1\rangle$ is tangent to y_2.

At $(1, 1)$, $y_1' = 3$ and $y_2' = \dfrac{1}{3}$.

$\pm\dfrac{1}{\sqrt{10}}\langle 1, 3\rangle$ is tangent to y_1, $\pm\dfrac{1}{\sqrt{10}}\langle 3, 1\rangle$ is tangent to y_2.

At $(-1, -1)$, $y_1' = 3$ and $y_2' = \dfrac{1}{3}$.

$\pm\dfrac{1}{\sqrt{10}}\langle 1, 3\rangle$ is tangent to y_1, $\pm\dfrac{1}{\sqrt{10}}\langle 3, 1\rangle$ is tangent to y_2.

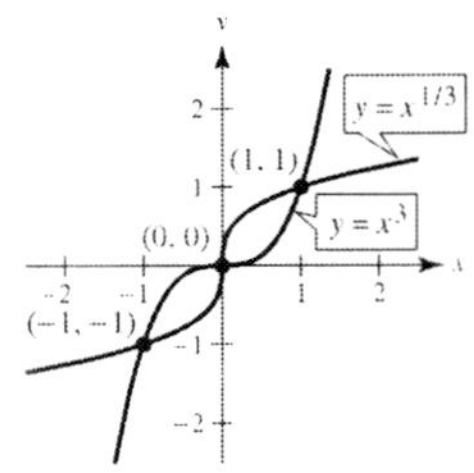

(c) At $(0, 0)$, the vectors are perpendicular $(90°)$.

At $(1, 1)$,

$$\cos\theta = \frac{\dfrac{1}{\sqrt{10}}\langle 1, 3\rangle\cdot\dfrac{1}{\sqrt{10}}\langle 3, 1\rangle}{(1)(1)} = \frac{6}{10} = \frac{3}{5}.$$

$\theta \approx 0.9273$ or $53.13°$

By symmetry, the angle is the same at $(-1, -1)$.

69. (a) The graphs of $y_1 = 1 - x^2$ and $y^2 = x^2 - 1$ intersect at $(1, 0)$ and $(-1, 0)$.

(b) $y_1' = -2x$ and $y_2' = 2x$.

At $(1, 0)$, $y_1' = -2$ and $y_2' = 2$. $\pm\dfrac{1}{\sqrt{5}}\langle 1, -2\rangle$ is tangent to y_1, $\pm\dfrac{1}{\sqrt{5}}\langle 1, 2\rangle$ is tangent to y_2.

At $(-1, 0)$, $y_1' = 2$ and $y_2' = -2$. $\pm\dfrac{1}{\sqrt{5}}\langle 1, 2\rangle$ is tangent to y_1, $\pm\dfrac{1}{\sqrt{5}}\langle 1, -2\rangle$ is tangent to y_2.

(c) At $(1, 0)$, $\cos\theta = \dfrac{1}{\sqrt{5}}\langle 1, -2\rangle\cdot\dfrac{-1}{\sqrt{5}}\langle 1, -2\rangle = \dfrac{3}{5}$.

$\theta \approx 0.9273$ or $53.13°$

By symmetry, the angle is the same at $(-1, 0)$.

70. (a) To find the intersection points, rewrite the second equation as $y + 1 = x^3$. Substituting into the first equation

$(y + 1)^2 = x \Rightarrow x^6 = x \Rightarrow x = 0, 1.$

There are two points of intersection, $(0, -1)$ and $(1, 0)$, as indicated in the figure.

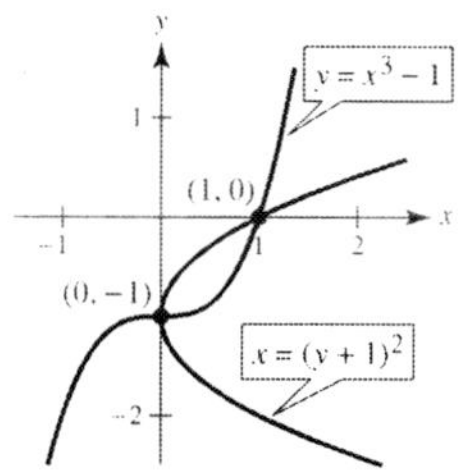

(b) First equation:

$$(y + 1)^2 = x \Rightarrow 2(y + 1)y' = 1 \Rightarrow y' = \frac{1}{2(y + 1)}$$

At $(1, 0)$, $y' = \frac{1}{2}$.

Second equation: $y = x^3 - 1 \Rightarrow y' = 3x^2$. At $(1, 0)$, $y' = 3$.

$\pm\frac{1}{\sqrt{5}}\langle 2, 1\rangle$ unit tangent vectors to first curve,

$\pm\frac{1}{\sqrt{10}}\langle 1, 3\rangle$ unit tangent vectors to second curve

At $(0, 1)$, the unit tangent vectors to the first curve are $\pm\langle 0, 1\rangle$, and the unit tangent vectors to the second curve are $\pm\langle 1, 0\rangle$.

(c) At $(1, 0)$,

$$\cos\theta = \frac{1}{\sqrt{5}}\langle 2, 1\rangle \cdot \frac{1}{\sqrt{10}}\langle 1, 3\rangle = \frac{5}{\sqrt{50}} = \frac{1}{\sqrt{2}}.$$

$$\theta \approx \frac{\pi}{4} \text{ or } 45°$$

At $(0, -1)$ the vectors are perpendicular, $\theta = 90°$.

71. In a rhombus, $\|\mathbf{u}\| = \|\mathbf{v}\|$. The diagonals are $\mathbf{u} + \mathbf{v}$ and $\mathbf{u} - \mathbf{v}$.

$$\begin{aligned}(\mathbf{u} + \mathbf{v}) \cdot (\mathbf{u} - \mathbf{v}) &= (\mathbf{u} + \mathbf{v}) \cdot \mathbf{u} - (\mathbf{u} + \mathbf{v}) \cdot \mathbf{v} \\ &= \mathbf{u} \cdot \mathbf{u} + \mathbf{v} \cdot \mathbf{u} - \mathbf{u} \cdot \mathbf{v} - \mathbf{v} \cdot \mathbf{v} \\ &= \|\mathbf{u}\|^2 - \|\mathbf{v}\|^2 = 0\end{aligned}$$

So, the diagonals are orthogonal.

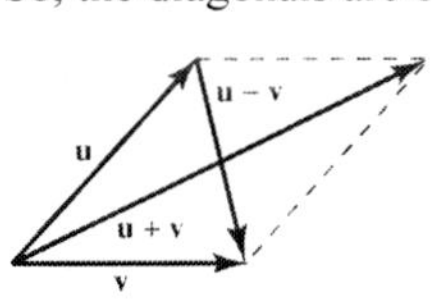

72. If $\mathbf{u}$ and $\mathbf{v}$ are the sides of the parallelogram, then the diagonals are $\mathbf{u} + \mathbf{v}$ and $\mathbf{u} - \mathbf{v}$, as indicated in the figure.

the parallelogram is a rectangle.

$\Leftrightarrow \mathbf{u} \cdot \mathbf{v} = 0$

$\Leftrightarrow 2\mathbf{u} \cdot \mathbf{v} = -2\mathbf{u} \cdot \mathbf{v}$

$\Leftrightarrow (\mathbf{u} + \mathbf{v}) \cdot (\mathbf{u} + \mathbf{v}) = (\mathbf{u} - \mathbf{v}) \cdot (\mathbf{u} - \mathbf{v})$

$\Leftrightarrow \|\mathbf{u} + \mathbf{v}\|^2 = \|\mathbf{u} - \mathbf{v}\|^2$

$\Leftrightarrow$ The diagonals are equal in length.

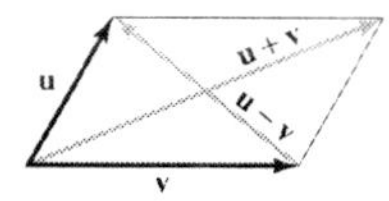

73. (a)

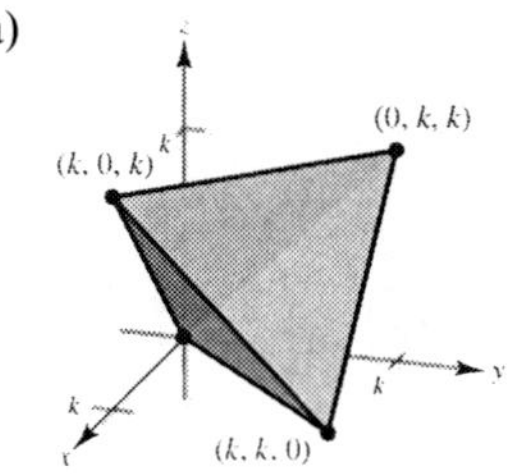

(b) Length of each edge: $\sqrt{k^2 + k^2 + 0^2} = k\sqrt{2}$

(c) $\cos\theta = \dfrac{k^2}{(k\sqrt{2})(k\sqrt{2})} = \dfrac{1}{2}$

$\theta = \arccos\left(\frac{1}{2}\right) = 60°$

(d) $\vec{r_1} = \langle k, k, 0\rangle - \left\langle \frac{k}{2}, \frac{k}{2}, \frac{k}{2}\right\rangle = \left\langle \frac{k}{2}, \frac{k}{2}, -\frac{k}{2}\right\rangle$

$\vec{r_2} = \langle 0, 0, 0\rangle - \left\langle \frac{k}{2}, \frac{k}{2}, \frac{k}{2}\right\rangle = \left\langle -\frac{k}{2}, -\frac{k}{2}, -\frac{k}{2}\right\rangle$

$$\cos\theta = \frac{-\frac{k^2}{4}}{\left(\frac{k}{2}\right)^2 \cdot 3} = -\frac{1}{3}$$

$\theta = 109.5°$

74. $\mathbf{u} = \langle \cos\alpha, \sin\alpha, 0\rangle$, $\mathbf{v} = \langle \cos\beta, \sin\beta, 0\rangle$

The angle between $\mathbf{u}$ and $\mathbf{v}$ is $\alpha - \beta$. (Assuming that $\alpha > \beta$). Also,

$$\begin{aligned}\cos(\alpha - \beta) &= \frac{\mathbf{u} \cdot \mathbf{v}}{\|\mathbf{u}\|\|\mathbf{v}\|} \\ &= \frac{\cos\alpha\cos\beta + \sin\alpha\sin\beta}{(1)(1)} \\ &= \cos\alpha\cos\beta + \sin\alpha\sin\beta.\end{aligned}$$

75. $\|\mathbf{u} - \mathbf{v}\|^2 = (\mathbf{u} - \mathbf{v}) \cdot (\mathbf{u} - \mathbf{v})$

$= (\mathbf{u} - \mathbf{v}) \cdot \mathbf{u} - (\mathbf{u} - \mathbf{v}) \cdot \mathbf{v}$

$= \mathbf{u} \cdot \mathbf{u} - \mathbf{v} \cdot \mathbf{u} - \mathbf{u} \cdot \mathbf{v} + \mathbf{v} \cdot \mathbf{v}$

$= \|\mathbf{u}\|^2 - \mathbf{u} \cdot \mathbf{v} - \mathbf{u} \cdot \mathbf{v} + \|\mathbf{v}\|^2$

$= \|\mathbf{u}\|^2 + \|\mathbf{v}\|^2 - 2\mathbf{u} \cdot \mathbf{v}$

76. $\mathbf{u} \cdot \mathbf{v} = \|\mathbf{u}\|\|\mathbf{v}\| \cos\theta$

$|\mathbf{u} \cdot \mathbf{v}| = \big|\|\mathbf{u}\|\|\mathbf{v}\| \cos\theta\big|$

$= \|\mathbf{u}\|\|\mathbf{v}\||\cos\theta|$

$\le \|\mathbf{u}\|\|\mathbf{v}\|$ because $|\cos\theta| \le 1$.

77. $\|\mathbf{u} + \mathbf{v}\|^2 = (\mathbf{u} + \mathbf{v}) \cdot (\mathbf{u} + \mathbf{v})$

$= (\mathbf{u} + \mathbf{v}) \cdot \mathbf{u} + (\mathbf{u} + \mathbf{v}) \cdot \mathbf{v}$

$= \mathbf{u} \cdot \mathbf{u} + \mathbf{v} \cdot \mathbf{u} + \mathbf{u} \cdot \mathbf{v} + \mathbf{v} \cdot \mathbf{v}$

$= \|\mathbf{u}\|^2 + 2\mathbf{u} \cdot \mathbf{v} + \|\mathbf{v}\|^2$

$\le \|\mathbf{u}\|^2 + 2\|\mathbf{u}\|\|\mathbf{v}\| + \|\mathbf{v}\|^2 \le (\|\mathbf{u}\| + \|\mathbf{v}\|)^2$

So, $\|\mathbf{u} + \mathbf{v}\| \le \|\mathbf{u}\| + \|\mathbf{v}\|$.

78. Let $\mathbf{w}_1 = \text{proj}_{\mathbf{v}}\mathbf{u}$, as indicated in the figure. Because $\mathbf{w}_1$ is a scalar multiple of $\mathbf{v}$, you can write

$\mathbf{u} = \mathbf{w}_1 + \mathbf{w}_2 = c\mathbf{v} + \mathbf{w}_2.$

Taking the dot product of both sides with $\mathbf{v}$ produces

$\mathbf{u} \cdot \mathbf{v} = (c\mathbf{v} + \mathbf{w}_2) \cdot \mathbf{v} = c\mathbf{v} \cdot \mathbf{v} + \mathbf{w}_2 \cdot \mathbf{v}$

$= c\|\mathbf{v}\|^2$, because $\mathbf{w}_2$ and $\mathbf{v}$ are orthogonol.

So, $\mathbf{u} \cdot \mathbf{v} = c\|\mathbf{v}\|^2 \Rightarrow c = \dfrac{\mathbf{u} \cdot \mathbf{v}}{\|\mathbf{v}\|^2}$ and

$\mathbf{w}_1 = \text{proj}_{\mathbf{v}}\mathbf{u} = c\mathbf{v} = \dfrac{\mathbf{u} \cdot \mathbf{v}}{\|\mathbf{v}\|^2}\mathbf{v}.$

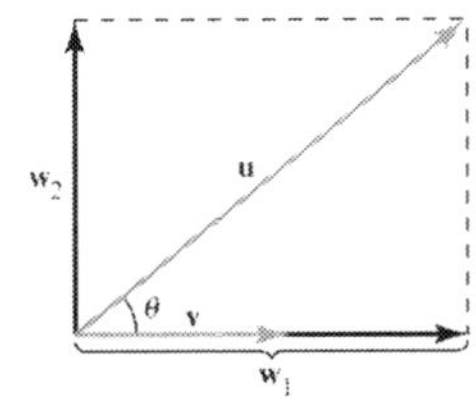

Section 11.4 The Cross Product of Two Vectors in Space

1. $\mathbf{j} \times \mathbf{i} = \begin{vmatrix} \mathbf{i} & \mathbf{j} & \mathbf{k} \\ 0 & 1 & 0 \\ 1 & 0 & 0 \end{vmatrix} = -\mathbf{k}$

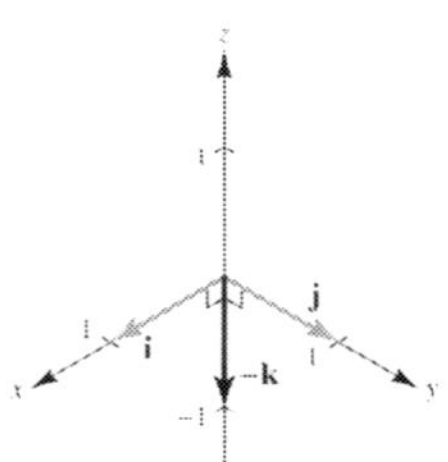

2. $\mathbf{i} \times \mathbf{j} = \begin{vmatrix} \mathbf{i} & \mathbf{j} & \mathbf{k} \\ 1 & 0 & 0 \\ 0 & 1 & 0 \end{vmatrix} = \mathbf{k}$

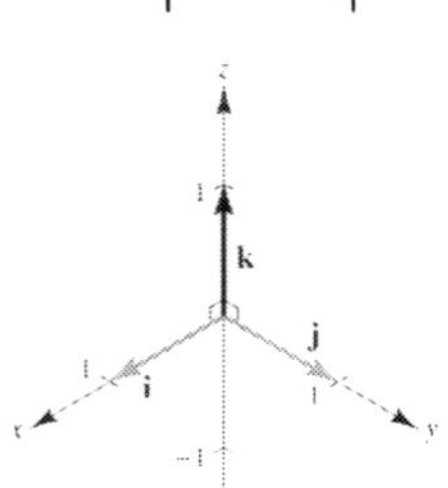

3. $\mathbf{j} \times \mathbf{k} = \begin{vmatrix} \mathbf{i} & \mathbf{j} & \mathbf{k} \\ 0 & 1 & 0 \\ 0 & 0 & 1 \end{vmatrix} = \mathbf{i}$

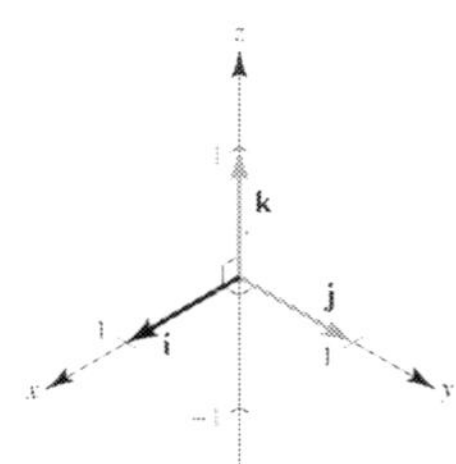

4. $\mathbf{k} \times \mathbf{j} = \begin{vmatrix} \mathbf{i} & \mathbf{j} & \mathbf{k} \\ 0 & 0 & 1 \\ 0 & 1 & 0 \end{vmatrix} = -\mathbf{i}$

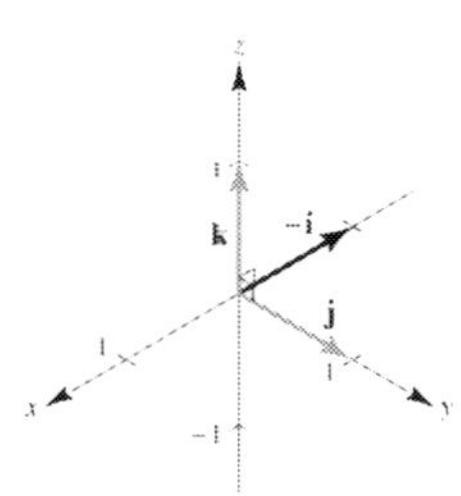

5. $\mathbf{i} \times \mathbf{k} = \begin{vmatrix} \mathbf{i} & \mathbf{j} & \mathbf{k} \\ 1 & 0 & 0 \\ 0 & 0 & 1 \end{vmatrix} = -\mathbf{j}$

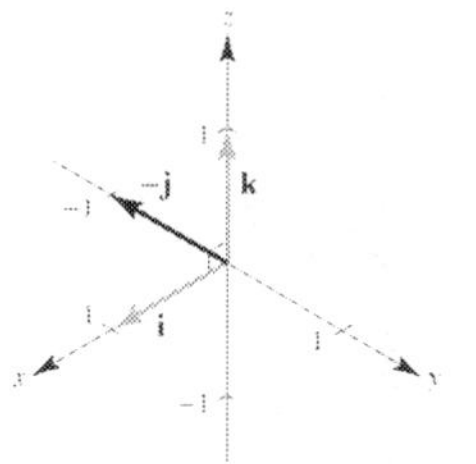

6. $\mathbf{k} \times \mathbf{i} = \begin{vmatrix} \mathbf{i} & \mathbf{j} & \mathbf{k} \\ 0 & 0 & 1 \\ 1 & 0 & 0 \end{vmatrix} = \mathbf{j}$

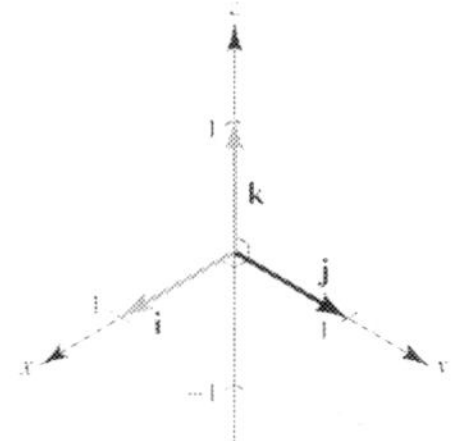

7. (a) $\mathbf{u} \times \mathbf{v} = \begin{vmatrix} \mathbf{i} & \mathbf{j} & \mathbf{k} \\ -2 & 4 & 0 \\ 3 & 2 & 5 \end{vmatrix} = 20\mathbf{i} + 10\mathbf{j} - 16\mathbf{k}$

(b) $\mathbf{v} \times \mathbf{u} = -(\mathbf{u} \times \mathbf{v}) = -20\mathbf{i} - 10\mathbf{j} + 16\mathbf{k}$

(c) $\mathbf{v} \times \mathbf{v} = \mathbf{0}$

8. (a) $\mathbf{u} \times \mathbf{v} = \begin{vmatrix} \mathbf{i} & \mathbf{j} & \mathbf{k} \\ 3 & 0 & 5 \\ 2 & 3 & -2 \end{vmatrix} = -15\mathbf{i} + 16\mathbf{j} + 9\mathbf{k}$

(b) $\mathbf{v} \times \mathbf{u} = -(\mathbf{u} \times \mathbf{v}) = 15\mathbf{i} - 16\mathbf{j} - 9\mathbf{k}$

(c) $\mathbf{v} \times \mathbf{v} = \mathbf{0}$

9. (a) $\mathbf{u} \times \mathbf{v} = \begin{vmatrix} \mathbf{i} & \mathbf{j} & \mathbf{k} \\ 7 & 3 & 2 \\ 1 & -1 & 5 \end{vmatrix} = 17\mathbf{i} - 33\mathbf{j} - 10\mathbf{k}$

(b) $\mathbf{v} \times \mathbf{u} = -(\mathbf{u} \times \mathbf{v}) = -17\mathbf{i} + 33\mathbf{j} + 10\mathbf{k}$

(c) $\mathbf{v} \times \mathbf{v} = \mathbf{0}$

10. (a) $\mathbf{u} \times \mathbf{v} = \begin{vmatrix} \mathbf{i} & \mathbf{j} & \mathbf{k} \\ 3 & -2 & -2 \\ 1 & 5 & 1 \end{vmatrix} = 8\mathbf{i} - 5\mathbf{j} + 17\mathbf{k}$

(b) $\mathbf{v} \times \mathbf{u} = -(\mathbf{u} \times \mathbf{v}) = -8\mathbf{i} + 5\mathbf{j} + 17\mathbf{k}$

(c) $\mathbf{v} \times \mathbf{v} = \mathbf{0}$

11. $\mathbf{u} = \langle 12, -3, 0\rangle$, $\mathbf{v} = \langle -2, 5, 0\rangle$

$$\mathbf{u} \times \mathbf{v} = \begin{vmatrix} \mathbf{i} & \mathbf{j} & \mathbf{k} \\ 12 & -3 & 0 \\ -2 & 5 & 0 \end{vmatrix} = 54\mathbf{k} = \langle 0, 0, 54\rangle$$

$$\mathbf{u} \cdot (\mathbf{u} \times \mathbf{v}) = 12(0) + (-3)(0) + 0(54) = 0 \Rightarrow \mathbf{u} \perp \mathbf{u} \times \mathbf{v}$$

$$\mathbf{v} \cdot (\mathbf{u} \times \mathbf{v}) = -2(0) + 5(0) + 0(54) = 0 \Rightarrow \mathbf{v} \perp \mathbf{u} \times \mathbf{v}$$

12. $\mathbf{u} = \langle -1, 1, 2\rangle$, $\mathbf{v} = \langle 0, 1, 0\rangle$

$$\mathbf{u} \times \mathbf{v} = \begin{vmatrix} \mathbf{i} & \mathbf{j} & \mathbf{k} \\ -1 & 1 & 2 \\ 0 & 1 & 0 \end{vmatrix} = -2\mathbf{i} - \mathbf{k} = \langle -2, 0, -1\rangle$$

$$\mathbf{u} \cdot (\mathbf{u} \times \mathbf{v}) = (-1)(-2) + (1)(0) + (2)(-1) = 0 \Rightarrow \mathbf{u} \perp \mathbf{u} \times \mathbf{v}$$

$$\mathbf{v} \cdot (\mathbf{u} \times \mathbf{v}) = (0)(-2) + (1)(0) + (0)(-1) = 0 \Rightarrow \mathbf{v} \perp \mathbf{u} \times \mathbf{v}$$

13. $\mathbf{u} = \langle 2, -3, 1\rangle$, $\mathbf{v} = \langle 1, -2, 1\rangle$

$$\mathbf{u} \times \mathbf{v} = \begin{vmatrix} \mathbf{i} & \mathbf{j} & \mathbf{k} \\ 2 & -3 & 1 \\ 1 & -2 & 1 \end{vmatrix} = -\mathbf{i} - \mathbf{j} - \mathbf{k} = \langle -1, -1, -1\rangle$$

$$\mathbf{u} \cdot (\mathbf{u} \times \mathbf{v}) = 2(-1) + (-3)(-1) + (1)(-1) = 0 \Rightarrow \mathbf{u} \perp \mathbf{u} \times \mathbf{v}$$

$$\mathbf{v} \cdot (\mathbf{u} \times \mathbf{v}) = 1(-1) + (-2)(-1) + (1)(-1) = 0 \Rightarrow \mathbf{v} \perp \mathbf{u} \times \mathbf{v}$$

14. $\mathbf{u} = \langle -10, 0, 6\rangle$, $\mathbf{v} = \langle 5, -3, 0\rangle$

$$\mathbf{u} \times \mathbf{v} = \begin{vmatrix} \mathbf{i} & \mathbf{j} & \mathbf{k} \\ -10 & 0 & 6 \\ 5 & -3 & 0 \end{vmatrix} = 18\mathbf{i} + 30\mathbf{j} + 30\mathbf{k} = \langle 18, 30, 30\rangle$$

$$\mathbf{u} \cdot (\mathbf{u} \times \mathbf{v}) = -10(18) + 0(30) + 6(30) = 0 \Rightarrow \mathbf{u} \perp \mathbf{u} \times \mathbf{v}$$

$$\mathbf{v} \cdot (\mathbf{u} \times \mathbf{v}) = 5(18) - 3(30) + 0(30) = 0 \Rightarrow \mathbf{v} \perp \mathbf{u} \times \mathbf{v}$$

15. $\mathbf{u} = \mathbf{i} + \mathbf{j} + \mathbf{k}$, $\mathbf{v} = 2\mathbf{i} + \mathbf{j} - \mathbf{k}$

$$\mathbf{u} \times \mathbf{v} = \begin{vmatrix} \mathbf{i} & \mathbf{j} & \mathbf{k} \\ 1 & 1 & 1 \\ 2 & 1 & -1 \end{vmatrix} = -2\mathbf{i} + 3\mathbf{j} - \mathbf{k} = \langle -2, 3, -1\rangle$$

$$\mathbf{u} \cdot (\mathbf{u} \times \mathbf{v}) = 1(-2) + 1(3) + 1(-1) = 0 \Rightarrow \mathbf{u} \perp \mathbf{u} \times \mathbf{v}$$

$$\mathbf{v} \cdot (\mathbf{u} \times \mathbf{v}) = 2(-2) + 1(3) + (-1)(-1) = 0 \Rightarrow \mathbf{v} \perp \mathbf{u} \times \mathbf{v}$$

$$(-\mathbf{v}) \times \mathbf{u} = -(\mathbf{v} \times \mathbf{u}) = \mathbf{u} \times \mathbf{v}$$

16. $\mathbf{u} = \mathbf{i} + 6\mathbf{j},\ \mathbf{v} = -2\mathbf{i} + \mathbf{j} + \mathbf{k}$

$$\mathbf{u} \times \mathbf{v} = \begin{vmatrix} \mathbf{i} & \mathbf{j} & \mathbf{k} \\ 1 & 6 & 0 \\ -2 & 1 & 1 \end{vmatrix} = 6\mathbf{i} - \mathbf{j} + 13\mathbf{k}$$

$$\mathbf{u} \cdot (\mathbf{u} \times \mathbf{v}) = 1(6) + 6(-1) = 0 \Rightarrow \mathbf{u} \perp (\mathbf{u} \times \mathbf{v})$$

$$\mathbf{v} \cdot (\mathbf{u} \times \mathbf{v}) = -2(6) + 1(-1) + 1(13) = 0 \Rightarrow \mathbf{v} \perp (\mathbf{u} \times \mathbf{v})$$

17. $\mathbf{u} = \langle 4, -3, 1 \rangle$

$\mathbf{v} = \langle 2, 5, 3 \rangle$

$$\mathbf{u} \times \mathbf{v} = \begin{vmatrix} \mathbf{i} & \mathbf{j} & \mathbf{k} \\ 4 & -3 & 1 \\ 2 & 5 & 3 \end{vmatrix} = -14\mathbf{i} - 10\mathbf{j} + 26\mathbf{k}$$

$$\frac{\mathbf{u} \times \mathbf{v}}{\|\mathbf{u} \times \mathbf{v}\|} = \frac{1}{\sqrt{972}} \langle -14, -10, 26 \rangle = \frac{1}{18\sqrt{3}} \langle -14, -10, 26 \rangle = \left\langle -\frac{7}{9\sqrt{3}}, -\frac{5}{9\sqrt{3}}, \frac{13}{9\sqrt{3}} \right\rangle$$

18. $\mathbf{u} = \langle -8, -6, 4 \rangle$

$\mathbf{v} = \langle 10, -12, -2 \rangle$

$$\mathbf{u} \times \mathbf{v} = \begin{vmatrix} \mathbf{i} & \mathbf{j} & \mathbf{k} \\ -8 & -6 & 4 \\ 10 & -12 & -2 \end{vmatrix} = 60\mathbf{i} + 24\mathbf{j} + 156\mathbf{k}$$

$$\frac{\mathbf{u} \times \mathbf{v}}{\|\mathbf{u} \times \mathbf{v}\|} = \frac{1}{36\sqrt{22}} \langle 60, 24, 156 \rangle = \left\langle \frac{5}{3\sqrt{22}}, \frac{2}{3\sqrt{22}}, \frac{13}{3\sqrt{22}} \right\rangle$$

19. $\mathbf{u} = -3\mathbf{i} + 2\mathbf{j} - 5\mathbf{k}$

$\mathbf{v} = \mathbf{i} - \mathbf{j} + 4\mathbf{k}$

$$\mathbf{u} \times \mathbf{v} = \begin{vmatrix} \mathbf{i} & \mathbf{j} & \mathbf{k} \\ -3 & 2 & -5 \\ 1 & -1 & -4 \end{vmatrix} = 3\mathbf{i} + 7\mathbf{j} + \mathbf{k}$$

$$\frac{\mathbf{u} \times \mathbf{v}}{\|\mathbf{u} \times \mathbf{v}\|} = \frac{1}{\sqrt{59}} \langle 3, 7, 1 \rangle = \left\langle \frac{3}{\sqrt{59}}, \frac{7}{\sqrt{59}}, \frac{1}{\sqrt{59}} \right\rangle$$

20. $\mathbf{u} = 2\mathbf{k}$

$\mathbf{v} = 4\mathbf{i} + 6\mathbf{k}$

$$\mathbf{u} \times \mathbf{v} = \begin{vmatrix} \mathbf{i} & \mathbf{j} & \mathbf{k} \\ 0 & 0 & 2 \\ 4 & 0 & 6 \end{vmatrix} = 8\mathbf{j}$$

$$\frac{\mathbf{u} \times \mathbf{v}}{\|\mathbf{u} \times \mathbf{v}\|} = \frac{1}{8}(8\mathbf{j}) = \mathbf{j} = \langle 0, 1, 0 \rangle$$

21. $\mathbf{u} = \mathbf{j}$

$\mathbf{v} = \mathbf{j} + \mathbf{k}$

$$\mathbf{u} \times \mathbf{v} = \begin{vmatrix} \mathbf{i} & \mathbf{j} & \mathbf{k} \\ 0 & 1 & 0 \\ 0 & 1 & 1 \end{vmatrix} = \mathbf{i}$$

$$A = \|\mathbf{u} \times \mathbf{v}\| = \|\mathbf{i}\| = 1$$

22. $\mathbf{u} = \mathbf{i} + \mathbf{j} + \mathbf{k}$

$\mathbf{v} = \mathbf{j} + \mathbf{k}$

$$\mathbf{u} \times \mathbf{v} = \begin{vmatrix} \mathbf{i} & \mathbf{j} & \mathbf{k} \\ 1 & 1 & 1 \\ 0 & 1 & 1 \end{vmatrix} = -\mathbf{j} + \mathbf{k}$$

$$A = \|\mathbf{u} \times \mathbf{v}\| = \|-\mathbf{j} + \mathbf{k}\| = \sqrt{2}$$

23. $\mathbf{u} = \langle 3, 2, -1 \rangle$

$\mathbf{v} = \langle 1, 2, 3 \rangle$

$$\mathbf{u} \times \mathbf{v} = \begin{vmatrix} \mathbf{i} & \mathbf{j} & \mathbf{k} \\ 3 & 2 & -1 \\ 1 & 2 & 3 \end{vmatrix} = \langle 8, -10, 4 \rangle$$

$$A = \|\mathbf{u} \times \mathbf{v}\| = \|\langle 8, -10, 4 \rangle\| = \sqrt{180} = 6\sqrt{5}$$

24. $\mathbf{u} = \langle 2, -1, 0 \rangle$

$\mathbf{v} = \langle -1, 2, 0 \rangle$

$$\mathbf{u} \times \mathbf{v} = \begin{vmatrix} \mathbf{i} & \mathbf{j} & \mathbf{k} \\ 2 & -1 & 0 \\ -1 & 2 & 0 \end{vmatrix} = \langle 0, 0, 3 \rangle$$

$$A = \|\mathbf{u} \times \mathbf{v}\| = \|\langle 0, 0, 3 \rangle\| = 3$$

25. $A(0, 3, 2), B(1, 5, 5), C(6, 9, 5), D(5, 7, 2)$

$$\overrightarrow{AB} = \langle 1, 2, 3\rangle$$
$$\overrightarrow{DC} = \langle 1, 2, 3\rangle$$
$$\overrightarrow{BC} = \langle 5, 4, 0\rangle$$
$$\overrightarrow{AD} = \langle 5, 4, 0\rangle$$

Because $\overrightarrow{AB} = \overrightarrow{DC}$ and $\overrightarrow{BC} = \overrightarrow{AD}$, the figure $ABCD$ is a parallelogram.

$\overrightarrow{AB}$ and $\overrightarrow{AD}$ are adjacent sides

$$\overrightarrow{AB} \times \overrightarrow{AD} = \begin{vmatrix} \mathbf{i} & \mathbf{j} & \mathbf{k} \\ 1 & 2 & 3 \\ 5 & 4 & 0 \end{vmatrix} = \langle -12, 15, -6\rangle$$

$$A = \left\|\overrightarrow{AB} \times \overrightarrow{AD}\right\| = \sqrt{144 + 225 + 36} = 9\sqrt{5}$$

26. $A(2, -3, 1), B(6, 5, -1), C(7, 2, 2), D(3, -6, 4)$

$$\overrightarrow{AB} = \langle 4, 8, -2\rangle$$
$$\overrightarrow{DC} = \langle 4, 8, -2\rangle$$
$$\overrightarrow{BC} = \langle 1, -3, 3\rangle$$
$$\overrightarrow{AD} = \langle 1, -3, 3\rangle$$

Because $\overrightarrow{AB} = \overrightarrow{DC}$ and $\overrightarrow{BC} = \overrightarrow{AD}$, the figure $ABCD$ is a parallelogram.

$\overrightarrow{AB}$ and $\overrightarrow{AD}$ are adjacent sides

$$\overrightarrow{AB} \times \overrightarrow{AD} = \begin{vmatrix} \mathbf{i} & \mathbf{j} & \mathbf{k} \\ 4 & 8 & -2 \\ 1 & -3 & 3 \end{vmatrix} = \langle 18, -14, -20\rangle$$

$$A = \left\|\overrightarrow{AB} \times \overrightarrow{AD}\right\| = \sqrt{324 + 196 + 400} = 2\sqrt{230}$$

27. $A(0, 0, 0), B(1, 0, 3), C(-3, 2, 0)$

$$\overrightarrow{AB} = \langle 1, 0, 3\rangle, \overrightarrow{AC} = \langle -3, 2, 0\rangle$$

$$\overrightarrow{AB} \times \overrightarrow{AC} = \begin{vmatrix} \mathbf{i} & \mathbf{j} & \mathbf{k} \\ 1 & 0 & 3 \\ -3 & 2 & 0 \end{vmatrix} = \langle -6, -9, 2\rangle$$

$$A = \frac{1}{2}\left\|\overrightarrow{AB} \times \overrightarrow{AC}\right\| = \frac{1}{2}\sqrt{36 + 81 + 4} = \frac{11}{2}$$

28. $A(2, -3, 4), B(0, 1, 2), C(-1, 2, 0)$

$$\overrightarrow{AB} = \langle -2, 4, -2\rangle, \overrightarrow{AC} = \langle -3, 5, -4\rangle$$

$$\overrightarrow{AB} \times \overrightarrow{AC} = \begin{vmatrix} \mathbf{i} & \mathbf{j} & \mathbf{k} \\ -2 & 4 & -2 \\ -3 & 5 & -4 \end{vmatrix} = -6\mathbf{i} - 2\mathbf{j} + 2\mathbf{k}$$

$$A = \frac{1}{2}\left\|\overrightarrow{AB} \times \overrightarrow{AC}\right\| = \frac{1}{2}\sqrt{44} = \sqrt{11}$$

29. $\mathbf{F} = -20\mathbf{k}$

$$\overrightarrow{PQ} = \frac{1}{2}(\cos 40°\mathbf{j} + \sin 40°\mathbf{k})$$

$$\overrightarrow{PQ} \times \mathbf{F} = \begin{vmatrix} \mathbf{i} & \mathbf{j} & \mathbf{k} \\ 0 & \cos 40°/2 & \sin 40°/2 \\ 0 & 0 & -20 \end{vmatrix} = -10\cos 40°\mathbf{i}$$

$$\left\|\overrightarrow{PQ} \times \mathbf{F}\right\| = 10\cos 40° \approx 7.66 \text{ ft-lb}$$

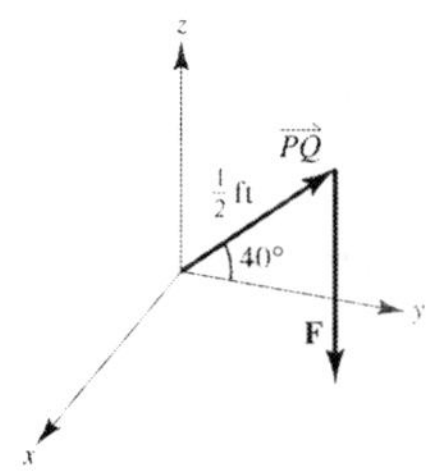

30. $\mathbf{F} = -2000(\cos 30°\mathbf{j} + \sin 30°\mathbf{k}) = -1000\sqrt{3}\mathbf{j} - 1000\mathbf{k}$

$$\overrightarrow{PQ} = 0.16\mathbf{k}$$

$$\overrightarrow{PQ} \times \mathbf{F} = \begin{vmatrix} \mathbf{i} & \mathbf{j} & \mathbf{k} \\ 0 & 0 & 0.16 \\ 0 & -1000\sqrt{3} & -1000 \end{vmatrix}$$
$$= 160\sqrt{3}\mathbf{i}$$

$$\left\|\overrightarrow{PQ} \times \mathbf{F}\right\| = 160\sqrt{3} \text{ ft-lb}$$

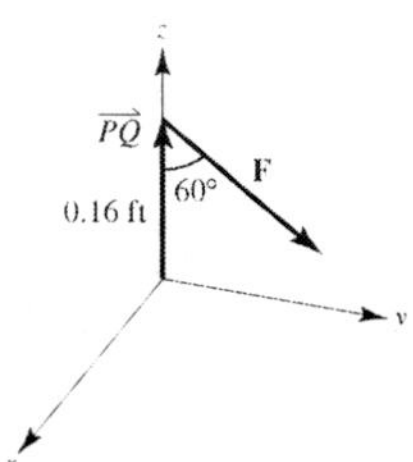

31. (a) $AC = 15 \text{ inches} = \frac{5}{4} \text{ feet}$

$$BC = 12 \text{ inches} = 1 \text{ foot}$$

$$\overrightarrow{AB} = -\frac{5}{4}\mathbf{j} + \mathbf{k}$$

$$\mathbf{F} = -180(\cos\theta\,\mathbf{j} + \sin\theta\,\mathbf{k})$$

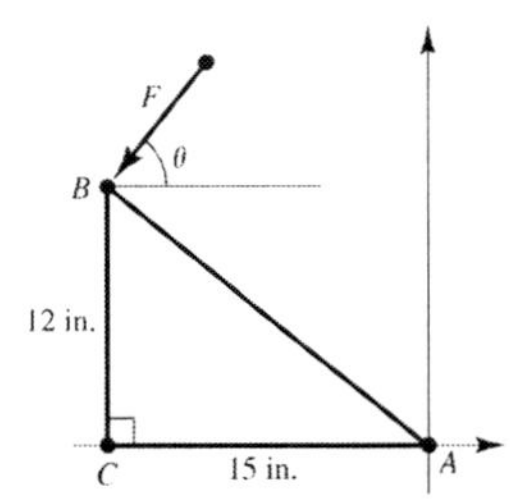

(b) $$\overrightarrow{AB} \times \mathbf{F} = \begin{vmatrix} \mathbf{i} & \mathbf{j} & \mathbf{k} \\ 0 & -\frac{5}{4} & 1 \\ 0 & -180\cos\theta & -180\sin\theta \end{vmatrix}$$

$$= (225\sin\theta + 180\cos\theta)i$$

$$\left\|\overrightarrow{AB} \times \mathbf{F}\right\| = |225\sin\theta + 180\cos\theta|$$

(c) When $\theta = 30°$, $\left\|\overrightarrow{AB} \times \mathbf{F}\right\| = 225\left(\frac{1}{2}\right) + 180\left(\frac{\sqrt{3}}{2}\right) \approx 268.38$

(d) If $T = |225\sin\theta + 180\cos\theta|$, $T = 0$ for $225\sin\theta = -180\cos\theta \Rightarrow \tan\theta = -\frac{4}{5} \Rightarrow \theta \approx 141.34°$.

For $0 < \theta < 141.34$, $T'(\theta) = 225\cos\theta - 180\sin\theta = 0 \Rightarrow \tan\theta = \frac{5}{4} \Rightarrow \theta \approx 51.34°$. $\overrightarrow{AB}$ and $\mathbf{F}$ are perpendicular.

(e)

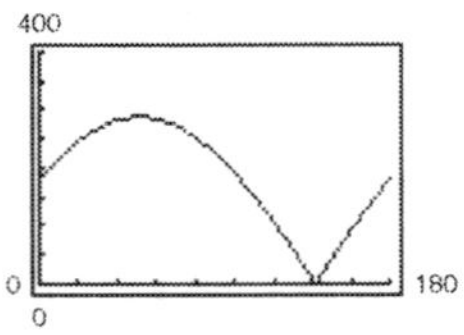

From part (d), the zero is $\theta \approx 141.34°$, when the vectors are parallel.

32. (a) Place the wrench in the xy-plane, as indicated in the figure. The angle from $\overrightarrow{AB}$ to $\mathbf{F}$ is $30° + 180° + \theta = 210° + \theta$

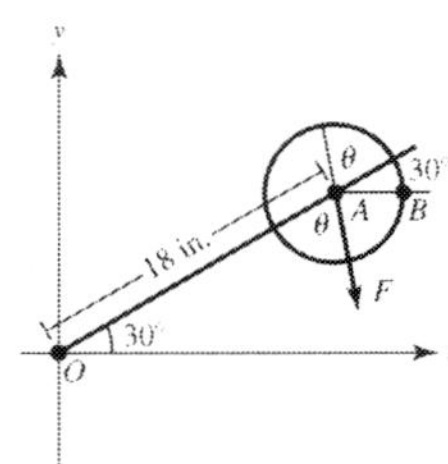

$$\left\|\overrightarrow{OA}\right\| = 18 \text{ inches} = 1.5 \text{ feet}$$

$$\overrightarrow{OA} = 1.5\left[\cos(30°)\mathbf{i} + \sin(30°)\mathbf{j}\right] = \frac{3\sqrt{3}}{4}\mathbf{i} + \frac{3}{4}\mathbf{j}$$

$$\mathbf{F} = 56\left[\cos(210° + \theta)\mathbf{i} + \sin(210° + \theta)\mathbf{j}\right]$$

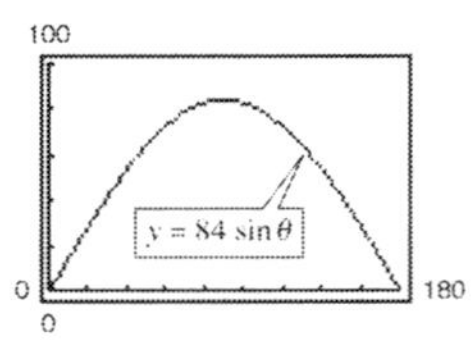

$$\overrightarrow{OA} \times \mathbf{F} = \begin{vmatrix} \mathbf{i} & \mathbf{j} & \mathbf{k} \\ \frac{3\sqrt{3}}{4} & \frac{3}{4} & 0 \\ 56\cos(210° + \theta) & 56\sin(210° + \theta) & 0 \end{vmatrix}$$

$$= \left[42\sqrt{3}\sin(210° + \theta) - 42\cos(210° + \theta)\right]\mathbf{k}$$

$$= \left[42\sqrt{3}(\sin 210°\cos\theta + \cos 210°\sin\theta) - 42(\cos 210°\cos\theta - \sin 210°\sin\theta)\right]\mathbf{k}$$

$$= \left[42\sqrt{3}\left(-\frac{1}{2}\cos\theta - \frac{\sqrt{3}}{2}\sin\theta\right) - 42\left(-\frac{\sqrt{3}}{2}\cos\theta + \frac{1}{2}\sin\theta\right)\right]\mathbf{k} = (-84\sin\theta)\mathbf{k}$$

$$\left\|\overrightarrow{OA} \times \mathbf{F}\right\| = 84\sin\theta,\ 0 \le \theta \le 180°$$

(b) When $\theta = 45°$, $\left\|\overrightarrow{OA} \times \mathbf{F}\right\| = 84\frac{\sqrt{2}}{2} = 42\sqrt{2} \approx 59.40$

(c) Let $T = 84\sin\theta$

$$\frac{dT}{d\theta} = 84\cos\theta = 0 \text{ when } \theta = 90°.$$

This is reasonable. When $\theta = 90°$, the force is perpendicular to the wrench.

33. $$\mathbf{u} \cdot (\mathbf{v} \times \mathbf{w}) = \begin{vmatrix} 1 & 0 & 0 \\ 0 & 1 & 0 \\ 0 & 0 & 1 \end{vmatrix} = 1$$

34. $$\mathbf{u} \cdot (\mathbf{v} \times \mathbf{w}) = \begin{vmatrix} 1 & 1 & 1 \\ 2 & 1 & 0 \\ 0 & 0 & 1 \end{vmatrix} = -1$$

35. $\mathbf{u}\cdot(\mathbf{v}\times\mathbf{w}) = \begin{vmatrix} 2 & 0 & 1 \\ 0 & 3 & 0 \\ 0 & 0 & 1 \end{vmatrix} = 6$

36. $\mathbf{u}\cdot(\mathbf{v}\times\mathbf{w}) = \begin{vmatrix} 2 & 0 & 0 \\ 1 & 1 & 1 \\ 0 & 2 & 2 \end{vmatrix} = 0$

37. $\mathbf{u}\cdot(\mathbf{v}\times\mathbf{w}) = \begin{vmatrix} 1 & 1 & 0 \\ 0 & 1 & 1 \\ 1 & 0 & 1 \end{vmatrix} = 2$

$V = |\mathbf{u}\cdot(\mathbf{v}\times\mathbf{w})| = 2$

38. $\mathbf{u}\cdot(\mathbf{v}\times\mathbf{w}) = \begin{vmatrix} 1 & 3 & 1 \\ 0 & 6 & 6 \\ -4 & 0 & -4 \end{vmatrix} = -72$

$V = |\mathbf{u}\cdot(\mathbf{v}\times\mathbf{w})| = 72$

39. $\mathbf{u} = \langle 3, 0, 0\rangle$

$\mathbf{v} = \langle 0, 5, 1\rangle$

$\mathbf{w} = \langle 2, 0, 5\rangle$

$\mathbf{u}\cdot(\mathbf{v}\times\mathbf{w}) = \begin{vmatrix} 3 & 0 & 0 \\ 0 & 5 & 1 \\ 2 & 0 & 5 \end{vmatrix} = 75$

$V = |\mathbf{u}\cdot(\mathbf{v}\times\mathbf{w})| = 75$

40. $\mathbf{u} = \langle 0, 4, 0\rangle$

$\mathbf{v} = \langle -3, 0, 0\rangle$

$\mathbf{w} = \langle -1, 1, 5\rangle$

$\mathbf{u}\cdot(\mathbf{v}\times\mathbf{w}) = \begin{vmatrix} 0 & 4 & 0 \\ -3 & 0 & 0 \\ -1 & 1 & 5 \end{vmatrix} = -4(-15) = 60$

$V = |\mathbf{u}\cdot(\mathbf{v}\times\mathbf{w})| = 60$

41. (a) $\mathbf{u}\cdot(\mathbf{v}\times\mathbf{w}) = (\mathbf{v}\times\mathbf{w})\cdot\mathbf{u}$ (b)

$= \mathbf{w}\cdot(\mathbf{u}\times\mathbf{v}) = (\mathbf{u}\times\mathbf{v})\cdot\mathbf{w}$ (c)

$= \mathbf{v}\cdot(\mathbf{w}\times\mathbf{u}) = (\mathbf{u}x - \mathbf{w})\cdot\mathbf{v}$ (d)

$= \mathbf{v}\cdot(\mathbf{w}\times\mathbf{u}) = (\mathbf{w}\times\mathbf{u})\cdot\mathbf{v}$ (h)

(e) $\mathbf{u}\cdot(\mathbf{w}\times\mathbf{v}) = \mathbf{w}\cdot(\mathbf{v}\times\mathbf{u})$ (f)

$= \mathbf{w}\cdot(\mathbf{v}\times\mathbf{u}) = (-\mathbf{u}\times\mathbf{v})\cdot\mathbf{w}$ (g)

So, $a = b = c = d = h$ and $e = f = g$

42. $\mathbf{u}\times\mathbf{v} = \mathbf{0} \Rightarrow \mathbf{u}$ and $\mathbf{v}$ are parallel.

$\mathbf{u}\cdot\mathbf{v} = \mathbf{0} \Rightarrow \mathbf{u}$ and $\mathbf{v}$ are orthogonal.

So, $\mathbf{u}$ or $\mathbf{v}$ (or both) is the zero vector.

43. $\mathbf{u}\times\mathbf{v} = \langle u_1, u_2, u_3\rangle \cdot \langle v_1, v_2, v_3\rangle$

$= (u_2v_3 - u_3v_2)\mathbf{i} - (u_1v_3 - u_3v_1)\mathbf{j} + (u_1v_2 - u_2v_1)\mathbf{k}$

44. See Theorem 11.8, page 377.

45. The magnitude of the cross product will increase by a factor of 4.

46. Form the vectors for two sides of the triangle, and compute their cross product.

$\langle x_2 - x_1, y_2 - y_1, z_2 - z_1\rangle \times \langle x_3 - x_1, y_3 - y_1, z_3 - z_1\rangle$

47. False. If the vectors are ordered pairs, then the cross product does not exist.

48. False. In general, $\mathbf{u}\times\mathbf{v} = -(\mathbf{v}\times\mathbf{u})$

49. False. Let $\mathbf{u} = \langle 1, 0, 0\rangle$, $\mathbf{v} = \langle 1, 0, 0\rangle$, $\mathbf{w} = \langle -1, 0, 0\rangle$.

Then, $\mathbf{u}\times\mathbf{v} = \mathbf{u}\times\mathbf{w} = \mathbf{0}$, but $\mathbf{v} \neq \mathbf{w}$.

50. True

51. $\mathbf{u} = \langle u_1, u_2, u_3\rangle$, $\mathbf{v} = \langle v_1, v_2, v_3\rangle$, $\mathbf{w} = \langle w_1, w_2, w_3\rangle$

$$\mathbf{u}\times(\mathbf{v}+\mathbf{w}) = \begin{vmatrix} \mathbf{i} & \mathbf{j} & \mathbf{k} \\ u_1 & u_2 & u_3 \\ v_1 + w_1 & v_2 + w_2 & v_3 + w_3 \end{vmatrix}$$

$$= [u_2(v_3 + w_3) - u_3(v_2 + w_2)]\mathbf{i} - [u_1(v_3 + w_3) - u_3(v_1 + w_1)]\mathbf{j} + [u_1(v_2 + w_2) - u_2(v_1 + w_1)]\mathbf{k}$$

$$= (u_2v_3 - u_3v_2)\mathbf{i} - (u_1v_3 - u_3v_1)\mathbf{j} + (u_1v_2 - u_2v_1)\mathbf{k} + (u_2w_3 - u_3w_2)\mathbf{i} - (u_1w_3 - u_3w_1)\mathbf{j} + (u_1w_2 - u_2w_1)\mathbf{k}$$

$$= (\mathbf{u}\times\mathbf{v}) + (\mathbf{u}\times\mathbf{w})$$

52. $\mathbf{u} = \langle u_1, u_2, u_3 \rangle$, $\mathbf{v} = \langle v_1, v_2, v_3 \rangle$, c is a scalar:

$$(c\mathbf{u}) \times \mathbf{v} = \begin{vmatrix} \mathbf{i} & \mathbf{j} & \mathbf{k} \\ cu_1 & cu_2 & cu_3 \\ v_1 & v_2 & v_3 \end{vmatrix}$$

$$= (cu_2v_3 - cu_3v_2)\mathbf{i} - (cu_1v_3 - cu_3v_1)\mathbf{j} + (cu_1v_2 - cu_2v_1)\mathbf{k}$$

$$= c\left[(u_2v_3 - u_3v_2)\mathbf{i} - (u_1v_3 - u_3v_1)\mathbf{j} + (u_1v_2 - u_2v_1)\mathbf{k}\right] = c(\mathbf{u} \times \mathbf{v})$$

53. $\mathbf{u} = \langle u_1, u_2, u_3 \rangle$

$$\mathbf{u} \times \mathbf{u} = \begin{vmatrix} \mathbf{i} & \mathbf{j} & \mathbf{k} \\ u_1 & u_2 & u_3 \\ u_1 & u_2 & u_3 \end{vmatrix} = (u_2u_3 - u_3u_2)\mathbf{i} - (u_1u_3 - u_3u_1)\mathbf{j} + (u_1u_2 - u_2u_1)\mathbf{k} = \mathbf{0}$$

54. $$\mathbf{u} \cdot (\mathbf{v} \times \mathbf{w}) = \begin{vmatrix} u_1 & u_2 & u_3 \\ v_1 & v_2 & v_3 \\ w_1 & w_2 & w_3 \end{vmatrix}$$

$$(\mathbf{u} \times \mathbf{v}) \cdot \mathbf{w} = \mathbf{w} \cdot (\mathbf{u} \times \mathbf{v}) = \begin{vmatrix} w_1 & w_2 & w_3 \\ u_1 & u_2 & u_3 \\ v_1 & v_2 & v_3 \end{vmatrix}$$

$$= w_1(u_2v_3 - v_2u_3) - w_2(u_1v_3 - v_1u_3) + w_3(u_1v_2 - v_1u_2)$$

$$= u_1(v_2w_3 - w_2v_3) - u_2(v_1w_3 - w_1v_3) + u_3(v_1w_2 - w_1v_2) = \mathbf{u} \cdot (\mathbf{v} \times \mathbf{w})$$

55. $$\mathbf{u} \times \mathbf{v} = (u_2v_3 - u_3v_2)\mathbf{i} - (u_1v_3 - u_3v_1)\mathbf{j} + (u_1v_2 - u_2v_1)\mathbf{k}$$

$$(\mathbf{u} \times \mathbf{v}) \cdot \mathbf{u} = (u_2v_3 - u_3v_2)u_1 + (u_3v_1 - u_1v_3)u_2 + (u_1v_2 - u_2v_1)u_3 = \mathbf{0}$$

$$(\mathbf{u} \times \mathbf{v}) \cdot \mathbf{v} = (u_2v_3 - u_3v_2)v_1 + (u_3v_1 - u_1v_3)v_2 + (u_1v_2 - u_2v_1)v_3 = \mathbf{0}$$

So, $\mathbf{u} \times \mathbf{v} \perp \mathbf{u}$ and $\mathbf{u} \times \mathbf{v} \perp \mathbf{v}$.

56. If **u** and **v** are scalar multiples of each other, $\mathbf{u} = c\mathbf{v}$ for some scalar c.

$$\mathbf{u} \times \mathbf{v} = (c\mathbf{v}) \times \mathbf{v} = c(\mathbf{v} \times \mathbf{v}) = c(\mathbf{0}) = \mathbf{0}$$

If $\mathbf{u} \times \mathbf{v} = \mathbf{0}$, then $\|\mathbf{u}\|\|\mathbf{v}\| \sin\theta = 0$. (Assume $\mathbf{u} \neq \mathbf{0}, \mathbf{v} \neq \mathbf{0}$.) So, $\sin\theta = 0, \theta = 0$, and **u** and **v** are parallel. So, $\mathbf{u} = c\mathbf{v}$ for some scalar c.

57. $\|\mathbf{u} \times \mathbf{v}\| = \|\mathbf{u}\|\|\mathbf{v}\| \sin\theta$

If **u** and **v** are orthogonal, $\theta = \pi/2$ and $\sin\theta = 1$. So, $\|\mathbf{u} \times \mathbf{v}\| = \|\mathbf{u}\|\|\mathbf{v}\|$.

58. $\mathbf{u} = \langle a_1, b_1, c_1 \rangle$, $\mathbf{v} = \langle a_2, b_2, c_2 \rangle$, $\mathbf{w} = \langle a_3, b_3, c_3 \rangle$

$$\mathbf{v} \times \mathbf{w} = \begin{vmatrix} \mathbf{i} & \mathbf{j} & \mathbf{k} \\ a_2 & b_2 & c_2 \\ a_3 & b_3 & c_3 \end{vmatrix} = (b_2c_3 - b_3c_2)\mathbf{i} - (a_2c_3 - a_3c_2)\mathbf{j} + (a_2b_3 - a_3b_2)\mathbf{k}$$

$$\mathbf{u} \times (\mathbf{v} \times \mathbf{w}) = \begin{vmatrix} \mathbf{i} & \mathbf{j} & \mathbf{k} \\ a_1 & b_1 & c_1 \\ (b_2c_3 - b_3c_2) & (a_3c_2 - a_2c_3) & (a_2b_3 - a_3b_2) \end{vmatrix}$$

$$\mathbf{u} \times (\mathbf{v} \times \mathbf{w}) = \left[b_1(a_2b_3 - a_3b_2) - c_1(a_3c_2 - a_2c_3)\right]\mathbf{i} - \left[a_1(a_2b_3 - a_3b_2) - c_1(b_2c_3 - b_3c_2)\right]\mathbf{j}$$
$$+ \left[a_1(a_3c_2 - a_2c_3) - b_1(b_2c_3 - b_3c_2)\right]\mathbf{k}$$
$$= \left[a_2(a_1a_3 + b_1b_3 + c_1c_3) - a_3(a_1a_2 + b_1b_2 + c_1c_2)\right]\mathbf{i} + \left[b_2(a_1b_3 + b_1b_3 + c_1c_3) - b_3(a_1a_2 + b_1b_2 + c_1c_2)\right]\mathbf{j}$$
$$+ \left[c_2(a_1a_3 + b_1b_3 + c_1c_3) - c_3(a_1a_2 + b_1b_2 + c_1c_2)\right]\mathbf{k}$$
$$= (a_1a_3 + b_1b_3 + c_1c_3)\langle a_2, b_2, c_2 \rangle - (a_1a_2 + b_1b_2 + c_1c_2)\langle a_3, b_3, c_3 \rangle = (\mathbf{u} \cdot \mathbf{w})\mathbf{v} - (\mathbf{u} \cdot \mathbf{v})\mathbf{w}$$

59. $\mathbf{u} = \mathbf{u}_1\mathbf{i} + \mathbf{u}_2\mathbf{j} + \mathbf{u}_3\mathbf{k}$, $\mathbf{v} = \mathbf{v}_1\mathbf{i} + \mathbf{v}_2\mathbf{j} + \mathbf{v}_3\mathbf{k}$, $\mathbf{w} = \mathbf{w}_1\mathbf{i} + \mathbf{w}_2\mathbf{j} + \mathbf{w}_3\mathbf{k}$

$$\mathbf{v} \times \mathbf{w} = \begin{vmatrix} \mathbf{i} & \mathbf{j} & \mathbf{k} \\ v_1 & v_2 & v_3 \\ w_1 & w_2 & w_3 \end{vmatrix} = \langle v_2w_3 - w_2v_3, -(v_1w_3 - w_1v_3), v_1w_2 - w_1v_2 \rangle$$

$$\mathbf{u} \times (\mathbf{v} \times \mathbf{w}) = \langle u_1, u_2, u_3 \rangle \cdot \langle v_2w_3 - w_2v_3, -(v_1w_3 - w_1v_3), v_1w_2 - w_1v_2 \rangle$$

$$= u_1v_2w_3 - u_1v_3w_2 - u_2v_1w_3 + u_2v_3w_1 + u_3v_1w_2 - u_3v_2w_1$$

$$= \begin{vmatrix} u_1 & u_2 & u_3 \\ v_1 & v_2 & v_3 \\ w_1 & w_2 & w_3 \end{vmatrix}$$

Section 11.5 Lines and Planes in Space

1. $x = -2 + t, y = 3t, z = 4 + t$

(a) $(0, 6, 6)$: For $x = 0 = -2 + t$, you have $t = 2$. Then $y = 3(2) = 6$ and $z = 4 + 2 = 6$. Yes, $(0, 6, 6)$ lies on the line.

(b) $(2, 3, 5)$: For $x = 2 = -2 + t$, you have $t = 4$. Then $y = 3(4) = 12 \neq 3$. No, $(2, 3, 5)$ does not lie on the line.

2. $\dfrac{x - 3}{2} = \dfrac{y - 7}{8} = z + 2$

(a) $(7, 23, 0)$: Substituting, you have

$$\frac{7 - 3}{2} = \frac{23 - 7}{8} = 0 + 2$$

$$2 = 2 = 2$$

Yes, $(7, 23, 0)$ lies on the line.

(b) $(1, -1, -3)$: Substituting, you have

$$\frac{1 - 3}{2} = \frac{-1 - 7}{8} = -3 + 2$$

$$-1 = -1 = -1$$

Yes, $(1, -1, -3)$ lies on the line.

3. Point: $(0, 0, 0)$

Direction vector: $\langle 3, 1, 5 \rangle$

Direction numbers: 3, 1, 5

(a) Parametric: $x = 3t, y = t, z = 5t$

(b) Symmetric: $\dfrac{x}{3} = y = \dfrac{z}{5}$

4. Point: $(0, 0, 0)$

Direction vector: $\mathbf{v} = \left\langle -2, \dfrac{5}{2}, 1 \right\rangle$

Direction numbers: $-4, 5, 2$

(a) Parametric: $x = -4t, y = 5t, z = 2t$

(b) Symmetric: $\dfrac{x}{-4} = \dfrac{y}{5} = \dfrac{z}{2}$

5. Point: $(-2, 0, 3)$

Direction vector: $\mathbf{v} = \langle 2, 4, -2 \rangle$

Direction numbers: $2, 4, -2$

(a) Parametric: $x = -2 + 2t, y = 4t, z = 3 - 2t$

(b) Symmetric: $\dfrac{x + 2}{2} = \dfrac{y}{4} = \dfrac{z - 3}{-2}$

6. Point: $(-3, 0, 2)$

Direction vector: $\mathbf{v} = \langle 0, 6, 3 \rangle$

Direction numbers: 0, 2, 1

(a) Parametric: $x = -3, y = 2t, z = 2 + t$

(b) Symmetric: $\dfrac{y}{2} = z - 2, x = -3$

7. Point: $(1, 0, 1)$

Direction vector: $\mathbf{v} = 3\mathbf{i} - 2\mathbf{j} + \mathbf{k}$

Direction numbers: $3, -2, 1$

(a) Parametric: $x = 1 + 3t, y = -2t, z = 1 + t$

(b) Symmetric: $\dfrac{x - 1}{3} = \dfrac{y}{-2} = \dfrac{z - 1}{1}$

8. Point: $(-3, 5, 4)$

Directions numbers: $3, -2, 1$

(a) Parametric: $x = -3 + 3t, y = 5 - 2t, z = 4 + t$

(b) Symmetric: $\frac{x+3}{3} = \frac{y-5}{-2} = z - 4$

9. Points: $(5, -3, -2), \left(-\frac{2}{3}, \frac{2}{3}, 1\right)$

Direction vector: $\mathbf{v} = \frac{17}{3}\mathbf{i} - \frac{11}{3}\mathbf{j} - 3\mathbf{k}$

Direction numbers: $17, -11, -9$

(a) Parametric:
$x = 5 + 17t, y = -3 - 11t, z = -2 - 9t$

(b) Symmetric: $\frac{x-5}{17} = \frac{y+3}{-11} = \frac{z+2}{-9}$

10. Points: $(0, 4, 3), (-1, 2, 5)$

Direction vector: $\langle 1, 2, -2 \rangle$

Direction numbers: $1, 2, -2$

(a) Parametric: $x = t, y = 4 + 2t, z = 3 - 2t$

(b) Symmetric: $x = \frac{y-4}{2} = \frac{z-3}{-2}$

11. Points: $(7, -2, 6), (-3, 0, 6)$

Direction vector: $\langle -10, 2, 0 \rangle$

Direction numbers: $-10, 2, 0$

(a) Parametric: $x = 7 - 10t, y = -2 + 2t, z = 6$

(b) Symmetric: Not possible because the direction number for z is 0. But, you could describe the line as $\frac{x-7}{10} = \frac{y+2}{-2}, z = 6.$

12. Points: $(0, 0, 25), (10, 10, 0)$

Direction vector: $\langle 10, 10, -25 \rangle$

Direction numbers: $2, 2, -5$

(a) Parametric: $x = 2t, y = 2t, z = 25 - 5t$

(b) Symmetric: $\frac{x}{2} = \frac{y}{2} = \frac{z-25}{-5}$

13. Point: $(2, 3, 4)$

Direction vector: $\mathbf{v} = \mathbf{k}$

Direction numbers: $0, 0, 1$

Parametric: $x = 2, y = 3, z = 4 + t$

14. Point: $(-4, 5, 2)$

Direction vector: $\mathbf{v} = \mathbf{j}$

Direction numbers: $0, 1, 0$

Parametric: $x = -4, y = 5 + t, z = 2$

15. Point: $(2, 3, 4)$

Direction vector: $\mathbf{v} = 3\mathbf{i} + 2\mathbf{j} - \mathbf{k}$

Direction numbers: $3, 2, -1$

Parametric: $x = 2 + 3t, y = 3 + 2t, z = 4 - t$

16. Point $(-4, 5, 2)$

Direction vector: $\mathbf{v} = -\mathbf{i} + 2\mathbf{j} + \mathbf{k}$

Direction numbers: $-1, 2, 1$

Parametric: $x = -4 - t, y = 5 + 2t, z = 2 + t$

17. Point: $(5, -3, -4)$

Direction vector: $\mathbf{v} = \langle 2, -1, 3 \rangle$

Direction numbers: $2, -1, 3$

Parametric: $x = 5 + 2t, y = -3 - t, z = -4 + 3t$

18. Point: $(-1, 4, -3)$

Direction vector: $\mathbf{v} = 5\mathbf{i} - \mathbf{j}$

Direction numbers: $5, -1, 0$

Parametric: $x = -1 + 5t, y = 4 - t, z = -3$

19. Point: $(2, 1, 2)$

Direction vector: $\langle -1, 1, 1 \rangle$

Direction numbers: $-1, 1, 1$

Parametric: $x = 2 - t, y = 1 + t, z = 2 + t$

20. Point: $(-6, 0, 8)$

Direction vector: $\langle -2, 2, 0 \rangle$

Direction numbers: $-2, 2, 0$

Parametric: $x = -6 - 2t, y = 2t, z = 8$

21. Let $t = 0$: $P = (3, -1, -2)$ (other answers possible)

$\mathbf{v} = \langle -1, 2, 0 \rangle$ (any nonzero multiple of $\mathbf{v}$ is correct)

22. Let $t = 0$: $P = (0, 5, 4)$ (other answers possible)

$\mathbf{v} = \langle 4, -1, 3 \rangle$ (any nonzero multiple of $\mathbf{v}$ is correct)

23. Let each quantity equal 0:

$P = (7, -6, -2)$ (other answers possible)

$\mathbf{v} = \langle 4, 2, 1 \rangle$ (any nonzero multiple of $\mathbf{v}$ is correct)

24. Let each quantity equal 0:

$P = (-3, 0, 3)$ (other answers possible)

$\mathbf{v} = \langle 5, 8, 6\rangle$ (any nonzero multiple of $\mathbf{v}$ is correct)

25. L_1: $\mathbf{v} = \langle -3, 2, 4\rangle$ $(6, -2, 5)$ on line

L_2: $\mathbf{v} = \langle 6, -4, -8\rangle$ $(6, -2, 5)$ on line

L_3: $\mathbf{v} = \langle -6, 4, 8\rangle$ $(6, -2, 5)$ not online

L_4: $\mathbf{v} = \langle 6, 4, -6\rangle$ not parallel to L_1, L_2, nor L_3

L_1 and L_2 are identical. $L_1 = L_2$ and is parallel to L_3.

26. L_1: $\mathbf{v} = \langle 2, -6, -2\rangle$ $(3, 0, 1)$ on line

L_2: $\mathbf{v} = \langle 2, -1, 3\rangle$ $(1, -1, 0)$ on line

L_3: $\mathbf{v} = \langle 2, -10, -4\rangle$ $(-1, 3, 1)$ on line

L_4: $\mathbf{v} = \langle 2, -1, 3\rangle$ $(5, 1, 8)$ on line

L_2 and L_4 are parallel, not identical, because $(1, -1, 0)$ is not on L_4.

27. L_1: $\mathbf{v} = \langle 4, -2, 3\rangle$ $(8, -5, -9)$ on line

L_2: $\mathbf{v} = \langle 2, 1, 5\rangle$

L_3: $\mathbf{v} = \langle -8, 4, -6\rangle$ $(8, -5, -9)$ on line

L_4: $\mathbf{v} = \langle -2, 1, 1.5\rangle$

L_1 and L_3 are identical.

28. L_1: $\mathbf{v} = \langle 2, 1, 2\rangle$ $(3, 2, -2)$ on line

L_2: $\mathbf{v} = \langle 4, 2, 4\rangle$ $(1, 1, -3)$ on line

L_3: $\mathbf{v} = \langle 1, \frac{1}{2}, 1\rangle$ $(-2, 1, 3)$ on line

L_4: $\mathbf{v} = \langle 2, 4, -1\rangle$ $(3, -1, 2)$ on line

L_1, L_2 and L_3 have same direction.

$(3, 2, -2)$ is not on L_2 nor L_3

$(1, 1, -3)$ is not on L_3

So, the three lines are parallel, not identical.

29. At the point of intersection, the coordinates for one line equal the corresponding coordinates for the other line. So,

(i) $4t + 2 = 2s + 2$, (ii) $3 = 2s + 3$, and

(iii) $-t + 1 = s + 1$.

From (ii), you find that $s = 0$ and consequently, from (iii), $t = 0$. Letting $s = t = 0$, you see that equation (i) is satisfied and so the two lines intersect. Substituting zero for s or for t, you obtain the point $(2, 3, 1)$.

$\mathbf{u} = 4\mathbf{i} - \mathbf{k}$ (First line)

$\mathbf{v} = 2\mathbf{i} + 2\mathbf{j} + \mathbf{k}$ (Second line)

$$\cos\theta = \frac{|\mathbf{u}\cdot\mathbf{v}|}{\|\mathbf{u}\|\|\mathbf{v}\|} = \frac{8-1}{\sqrt{17}\sqrt{9}} = \frac{7}{3\sqrt{17}} = \frac{7\sqrt{17}}{51}$$

30. By equating like variables, you have

(i) $-3t + 1 = 3s + 1$, (ii) $4t + 1 = 2s + 4$, and

(iii) $2t + 4 = -s + 1$.

From (i) you have $s = -t$, and consequently from (ii), $t = \frac{1}{2}$ and from (iii), $t = -3$. The lines do not intersect.

31. Writing the equations of the lines in parametric form you have

$x = 3t$ $y = 2 - t$ $z = -1 + t$

$x = 1 + 4s$ $y = -2 + s$ $z = -3 - 3s$.

For the coordinates to be equal, $3t = 1 + 4s$ and $2 - t = -2 + s$. Solving this system yields $t = \frac{17}{7}$ and $s = \frac{11}{7}$. When using these values for s and t, the z coordinates are not equal. The lines do not intersect.

32. Writing the equations of the lines in parametric form you have

$x = 2 - 3t$ $y = 2 + 6t$ $z = 3 + t$

$x = 3 + 2s$ $y = -5 + s$ $z = -2 + 4s$.

By equating like variables, you have
$2 - 3t = 3 + 2s$, $2 + 6t = -5 + s$, $3 + t = -2 + 4s$.
So, $t = -1$, $s = 1$ and the point of intersection is $(5, -4, 2)$.

$\mathbf{u} = \langle -3, 6, 1\rangle$ (First line)

$\mathbf{v} = \langle 2, 1, 4\rangle$ (Second line)

$$\cos\theta = \frac{|\mathbf{u}\cdot\mathbf{v}|}{\|\mathbf{u}\|\|\mathbf{v}\|} = \frac{4}{\sqrt{46}\sqrt{21}} = \frac{4}{\sqrt{966}} = \frac{2\sqrt{966}}{483}$$

33. $x + 2y - 4z - 1 = 0$

(a) $(-7, 2, -1)$: $(-7) + 2(2) - 4(-1) - 1 = 0$

Point is in plane

(b) $(5, 2, 2)$: $5 + 2(2) - 4(2) - 1 = 0$

Point is in plane

34. $2x + y + 3z - 6 = 0$

(a) $(3, 6, -2)$: $2(3) + 6 + 3(-2) - 6 = 0$

Point is in plane

(b) $(-1, 5, -1)$: $2(-1) + 5 + 3(-1) - 6 = -6 \neq 0$

Point is not in plane

35. Point: $(1, 3, -7)$

Normal vector: $\mathbf{n} = \mathbf{j} = \langle 0, 1, 0\rangle$

$$0(x - 1) + 1(y - 3) + 0(z - (-7)) = 0$$
$$y - 3 = 0$$

36. Point: $(0, -1, 4)$

Normal vector: $\mathbf{n} = \mathbf{k} = \langle 0, 0, 1\rangle$

$$0(x - 0) + 0(y + 1) + 1(z - 4) = 0$$
$$z - 4 = 0$$

37. Point: $(3, 2, 2)$

Normal vector: $\mathbf{n} = 2\mathbf{i} + 3\mathbf{j} - \mathbf{k}$

$$2(x - 3) + 3(y - 2) - 1(z - 2) = 0$$
$$2x + 3y - z = 10$$

38. Point: $(0, 0, 0)$

Normal vector: $\mathbf{n} = -3\mathbf{i} + 2\mathbf{k}$

$$-3(x - 0) + 0(y - 0) + 2(z - 0) = 0$$
$$-3x + 2z = 0$$

39. Point: $(-1, 4, 0)$

Normal vector: $\mathbf{v} = \langle 2, -1, -2\rangle$

$$2(x + 1) - 1(y - 4) - 2(z - 0) = 0$$
$$2x - y - 2z + 6 = 0$$

40. Point: $(3, 2, 2)$

Normal vector: $\mathbf{v} = 4\mathbf{i} + \mathbf{j} - 3\mathbf{k}$

$$4(x - 3) + (y - 2) - 3(z - 2) = 0$$
$$4x + y - 3z = 8$$

41. Let $\mathbf{u}$ be the vector from $(0, 0, 0)$ to $(2, 0, 3)$: $\mathbf{u} = \langle 2, 0, 3\rangle$

Let $\mathbf{u}$ be the vector from $(0, 0, 0)$ to $(-3, -1, 5)$: $\mathbf{v} = \langle -3, -1, 5\rangle$

Normal vectors: $\mathbf{u} \times \mathbf{v} = \begin{vmatrix} \mathbf{i} & \mathbf{j} & \mathbf{k} \\ 2 & 0 & 3 \\ -3 & -1 & 5 \end{vmatrix} = \langle 3, -19, -2\rangle$

$$3(x - 0) - 19(y - 0) - 2(z - 0) = 0$$
$$3x - 19y - 2z = 0$$

42. Let $\mathbf{u}$ be the vector from $(3, -1, 2)$ to $(2, 1, 5)$:

$\mathbf{u} = \langle -1, 2, 3\rangle$

Let $\mathbf{u}$ be the vector from $(3, -1, 2)$ to $(1, -2, -2)$:

$\mathbf{v} = \langle -2, -1, -4\rangle$

Normal vector:

$$\mathbf{u} \times \mathbf{v} = \begin{vmatrix} \mathbf{i} & \mathbf{j} & \mathbf{k} \\ -1 & 2 & 3 \\ -2 & -1 & -4 \end{vmatrix} = \langle -5, -10, 5\rangle = -5\langle 1, 2, -1\rangle$$

$$1(x - 3) + 2(y + 1) - (z - 2) = 0$$
$$x + 2y - z + 1 = 0$$

43. Let $\mathbf{u}$ be the vector from $(1, 2, 3)$ to $(3, 2, 1)$: $\mathbf{u} = 2\mathbf{i} - 2\mathbf{k}$

Let $\mathbf{v}$ be the vector from $(1, 2, 3)$ to $(-1, -2, 2)$: $\mathbf{v} = -2\mathbf{i} - 4\mathbf{j} - \mathbf{k}$

Normal vector:

$$\left(\tfrac{1}{2}\mathbf{u}\right) \times (-\mathbf{v}) = \begin{vmatrix} \mathbf{i} & \mathbf{j} & \mathbf{k} \\ 1 & 0 & -1 \\ 2 & 4 & 1 \end{vmatrix} = 4\mathbf{i} - 3\mathbf{j} + 4\mathbf{k}$$

$$4(x - 1) - 3(y - 2) + 4(z - 3) = 0$$
$$4x - 3y + 4z = 10$$

44. $(1, 2, 3)$, Normal vector: $\mathbf{v} = \mathbf{i}$, $1(x - 1) = 0$, $x = 1$

45. $(1, 2, 3)$, Normal vector: $\mathbf{v} = \mathbf{k}$, $1(z - 3) = 0$, $z = 3$

46. The plane passes through the three points $(0, 0, 0)$, $(0, 1, 0)$, $(\sqrt{3}, 0, 1)$.

The vector from $(0, 0, 0)$ to $(0, 1, 0)$: $\mathbf{u} = \mathbf{j}$

The vector from $(0, 0, 0)$ to $(\sqrt{3}, 0, 1)$: $\mathbf{v} = \sqrt{3}\mathbf{i} + \mathbf{k}$

Normal vector: $\mathbf{u} \times \mathbf{v} = \begin{vmatrix} \mathbf{i} & \mathbf{j} & \mathbf{k} \\ 0 & 1 & 0 \\ \sqrt{3} & 0 & 1 \end{vmatrix} = \mathbf{i} - \sqrt{3}\mathbf{k}$

$$x = \sqrt{3}z = 0$$

47. The direction vectors for the lines are $\mathbf{u} = -2\mathbf{i} + \mathbf{j} + \mathbf{k}$, $\mathbf{v} = -3\mathbf{i} + 4\mathbf{j} - \mathbf{k}$.

Normal vector: $\mathbf{u} \times \mathbf{v} = \begin{vmatrix} \mathbf{i} & \mathbf{j} & \mathbf{k} \\ -2 & 1 & 1 \\ -3 & 4 & -1 \end{vmatrix} = -5(\mathbf{i} + \mathbf{j} + \mathbf{k})$

Point of intersection of the lines: $(-1, 5, 1)$

$$(x + 1) + (y - 5) + (z - 1) = 0$$
$$x + y + z = 5$$

48. The direction of the line is $\mathbf{u} = 2\mathbf{i} - \mathbf{j} + \mathbf{k}$. Choose any point on the line, $[(0, 4, 0)$, for example$]$, and let $\mathbf{v}$ be the vector from $(0, 4, 0)$ to the given point $(2, 2, 1)$:

$$\mathbf{v} = 2\mathbf{i} - 2\mathbf{j} + \mathbf{k}$$

$$\text{Normal vector: } \mathbf{u} \times \mathbf{v} = \begin{vmatrix} \mathbf{i} & \mathbf{j} & \mathbf{k} \\ 2 & -1 & 1 \\ 2 & -2 & 1 \end{vmatrix} = \mathbf{i} - 2\mathbf{k}$$

$$(x - 2) - 2(z - 1) = 0$$
$$x - 2z = 0$$

49. Let $\mathbf{v}$ be the vector from $(-1, 1, -1)$ to $(2, 2, 1)$: $\mathbf{v} = 3\mathbf{i} + \mathbf{j} + 2\mathbf{k}$

Let $\mathbf{n}$ be a vector normal to the plane $2x - 3y + z = 3$: $\mathbf{n} = 2\mathbf{i} - 3\mathbf{j} + \mathbf{k}$

Because $\mathbf{v}$ and $\mathbf{n}$ both lie in the plane P, the normal vector to P is

$$\mathbf{v} \times \mathbf{n} = \begin{vmatrix} \mathbf{i} & \mathbf{j} & \mathbf{k} \\ 3 & 1 & 2 \\ 2 & -3 & 1 \end{vmatrix} = 7\mathbf{i} - \mathbf{j} - 11\mathbf{k}$$

$$7(x - 2) + 1(y - 2) - 11(z - 1) = 0$$
$$7x + y - 11z = 5$$

50. Let $\mathbf{v}$ be the vector from $(3, 2, 1)$ to $(3, 1, -5)$: $\mathbf{v} = -\mathbf{j} - 6\mathbf{k}$

Let $\mathbf{n}$ be the normal to the given plane: $\mathbf{n} = 6\mathbf{i} + 7\mathbf{j} + 2\mathbf{k}$

Because $\mathbf{v}$ and $\mathbf{n}$ both lie in the plane P, the normal vector to P is:

$$\mathbf{v} \times \mathbf{n} = \begin{vmatrix} \mathbf{i} & \mathbf{j} & \mathbf{k} \\ 0 & -1 & -6 \\ 6 & 7 & 2 \end{vmatrix} = 40\mathbf{i} - 36\mathbf{j} + 6\mathbf{k}$$
$$= 2(20\mathbf{i} - 18\mathbf{j} + 3\mathbf{k})$$

$$20(x - 3) - 18(y - 2) + 3(z - 1) = 0$$
$$20x - 18y + 3z = 27$$

51. Let $\mathbf{u} = \mathbf{i}$ and let $\mathbf{v}$ be the vector from $(1, -2, -1)$ to $(2, 5, 6)$: $\mathbf{v} = \mathbf{i} + 7\mathbf{j} + 7\mathbf{k}$

Because $\mathbf{u}$ and $\mathbf{v}$ both lie in the plane P, the normal vector to P is:

$$\mathbf{u} \times \mathbf{v} = \begin{vmatrix} \mathbf{i} & \mathbf{j} & \mathbf{k} \\ 1 & 0 & 0 \\ 1 & 7 & 7 \end{vmatrix} = -7\mathbf{j} + 7\mathbf{k} = -7(\mathbf{j} - \mathbf{k})$$

$$[y - (-2)] - [z - (-1)] = 0$$
$$y - z = -1$$

52. Let $\mathbf{u} = \mathbf{k}$ and let $\mathbf{v}$ be the vector from $(4, 2, 1)$ to $(-3, 5, 7)$: $\mathbf{v} = -7\mathbf{i} + 3\mathbf{j} + 6\mathbf{k}$

Because $\mathbf{u}$ and $\mathbf{v}$ both lie in the plane P, the normal vector to P is:

$$\mathbf{u} \times \mathbf{v} = \begin{vmatrix} \mathbf{i} & \mathbf{j} & \mathbf{k} \\ 0 & 0 & 1 \\ -7 & 3 & 6 \end{vmatrix} = -3\mathbf{i} - 7\mathbf{j} = -(3\mathbf{i} + 7\mathbf{j})$$

$$3(x - 4) + 7(y - 2) = 0$$
$$3x + 7y = 26$$

53. Let (x, y, z) be equidistant from $(2, 2, 0)$ and $(0, 2, 2)$.

$$\sqrt{(x - 2)^2 + (y - 2)^2 + (z - 0)^2} = \sqrt{(x - 0)^2 + (y - 2)^2 + (z - 2)^2}$$
$$x^2 - 4x + 4 + y^2 - 4y + 4 + z^2 = x^2 + y^2 - 4y + 4 + z^2 - 4z + 4$$
$$-4x + 8 = -4z + 8$$
$$x - z = 0 \text{ Plane}$$

54. Let (x, y, z) be equidistant from $(1, 0, 2)$ and $(2, 0, 1)$.

$$\sqrt{(x - 1)^2 + (y - 0)^2 + (z - 2)^2} = \sqrt{(x - 2)^2 + (y - 0)^2 + (z - 1)^2}$$
$$x^2 - 2x + 1 + y^2 + z^2 - 4z + 4 = x^2 - 4x + 4 + y^2 + z^2 - 2z + 1$$
$$-2x - 4z + 5 = -4x - 2z + 5$$
$$2x - 2z = 0$$
$$x - z = 0 \text{ Plane}$$

55. Let (x, y, z) be equidistant from $(-3, 1, 2)$ and $(6, -2, 4)$.

$$\begin{aligned}\sqrt{(x+3)^2+(y-1)^2+(z-2)^2} &= \sqrt{(x-6)^2+(y+2)^2+(z-4)^2}\\ x^2+6x+9+y^2-2y+1+z^2-4z+4 &= x^2-12x+36+y^2+4y+4+z^2-8z+16\\ 6x-2y-4z+14 &= -12x+4y-8z+56\\ 18x-6y+4z-42 &= 0\\ 9x-3y+2z-21 &= 0 \text{ Plane}\end{aligned}$$

56. Let (x, y, z) be equidistant from $(-5, 1, -3)$ and $(2, -1, 6)$

$$\begin{aligned}\sqrt{(x+5)^2+(y-1)^2+(z+3)^2} &= \sqrt{(x-2)^2+(y+1)^2+(z-6)^2}\\ x^2+10x+25+y^2-2y+1+z^2+6z+9 &= x^2-4x+4+y^2+2y+1+z^2-12z+36\\ 10x-2y+6z+35 &= -4x+2y-12z+41\\ 14x-4y+18z-6 &= 0\\ 7x-2y+9z-3 &= 0 \text{ Plane}\end{aligned}$$

57. The normal vectors to the planes are

$\mathbf{n}_1 = \langle 5, -3, 1\rangle$, $\mathbf{n}_2 = \langle 1, 4, 7\rangle$, $\cos\theta = \dfrac{|\mathbf{n}_1 \cdot \mathbf{n}_2|}{\|\mathbf{n}_1\|\|\mathbf{n}_2\|} = 0.$

So, $\theta = \pi/2$ and the planes are orthogonal.

58. The normal vectors to the planes are

$\mathbf{n}_1 = \langle 3, 1, -4\rangle$, $\mathbf{n}_2 = \langle -9, -3, 12\rangle$.

Because $\mathbf{n}_2 = -3\mathbf{n}_1$, the planes are parallel, but not equal.

59. The normal vectors to the planes are

$\mathbf{n}_1 = \mathbf{i} - 3\mathbf{j} + 6\mathbf{k}$, $\mathbf{n}_2 = 5\mathbf{i} + \mathbf{j} - \mathbf{k}$,

$$\cos\theta = \frac{|\mathbf{n}_1 \cdot \mathbf{n}_2|}{\|\mathbf{n}_1\|\|\mathbf{n}_2\|} = \frac{|5-3-6|}{\sqrt{46}\sqrt{27}} = \frac{4\sqrt{138}}{414} = \frac{2\sqrt{138}}{207}.$$

So, $\theta = \arccos\left(\dfrac{2\sqrt{138}}{207}\right) \approx 83.5^\circ$.

60. The normal vectors to the planes are

$\mathbf{n}_1 = 3\mathbf{i} + 2\mathbf{j} - \mathbf{k}$, $\mathbf{n}_2 = \mathbf{i} - 4\mathbf{j} + 2\mathbf{k}$,

$$\cos\theta = \frac{|\mathbf{n}_1 \cdot \mathbf{n}_2|}{\|\mathbf{n}_1\|\|\mathbf{n}_2\|} = \frac{|3-8-2|}{\sqrt{14}\sqrt{21}} = \frac{7\sqrt{6}}{42} = \frac{\sqrt{6}}{6}.$$

So, $\theta = \arccos\left(\dfrac{\sqrt{6}}{6}\right) \approx 65.9^\circ$.

61. The normal vectors to the planes are $\mathbf{n}_1 = \langle 1, -5, -1\rangle$ and $\mathbf{n}_2 = \langle 5, -25, -5\rangle$. Because $\mathbf{n}_2 = 5\mathbf{n}_1$, the planes are parallel, but not equal.

62. The normal vectors to the planes are

$\mathbf{n}_1 = \langle 2, 0, -1\rangle$, $\mathbf{n}_2 = \langle 4, 1, 8\rangle$,

$$\cos\theta = \frac{|\mathbf{n}_1 \cdot \mathbf{n}_2|}{\|\mathbf{n}_1\|\|\mathbf{n}_2\|} = 0$$

So, $\theta = \dfrac{\pi}{2}$ and the planes are orthogonal.

63. $4x + 2y + 6z = 12$

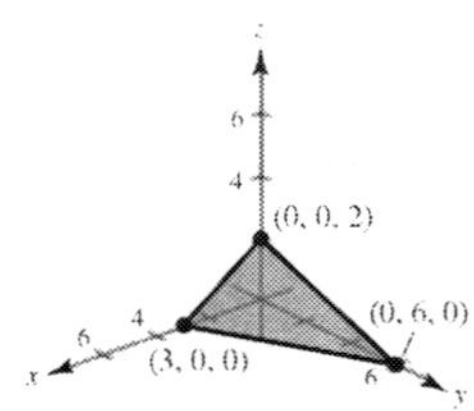

64. $3x + 6y + 2z = 6$

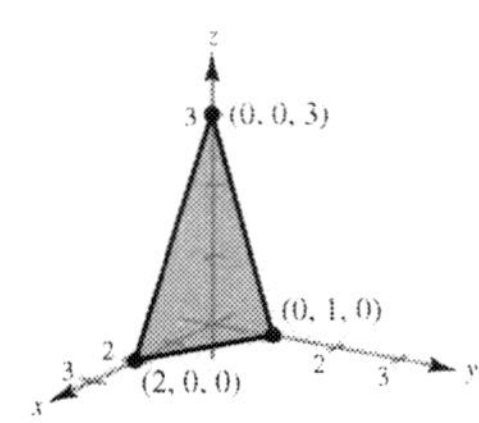

65. $2x - y + 3z = 4$

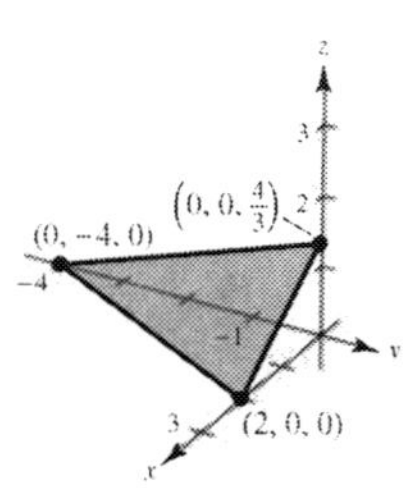

66. $2x - y + z = 4$

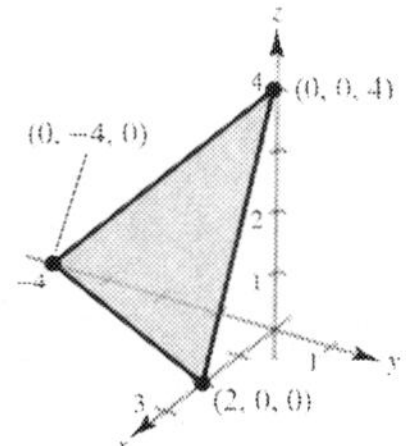

67. $x + z = 6$

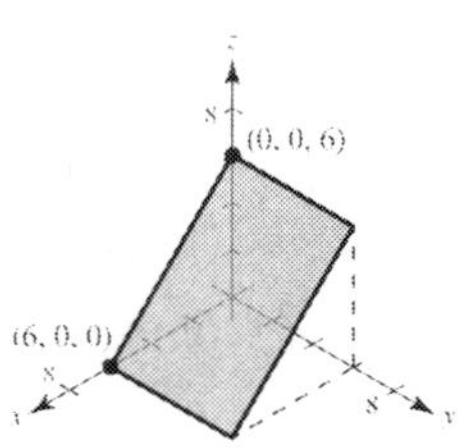

68. $2x + y = 8$

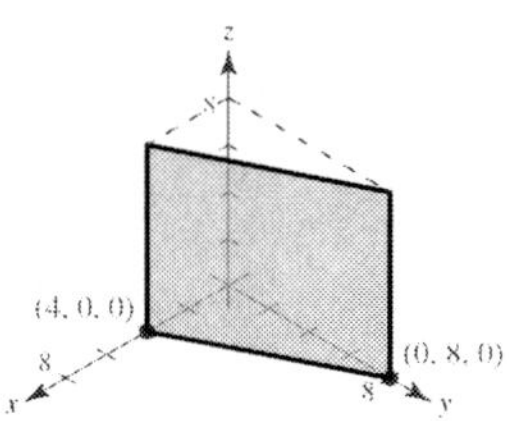

69. $x = 5$

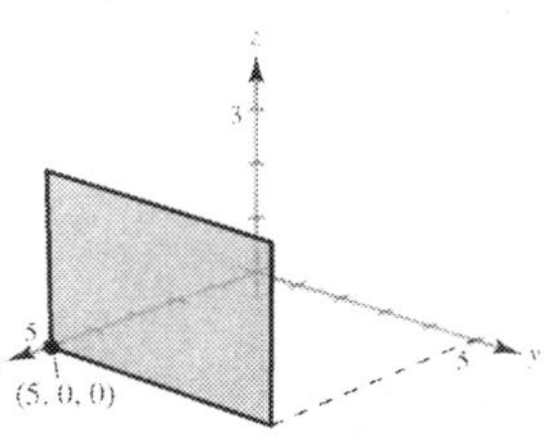

70. $z = 8$

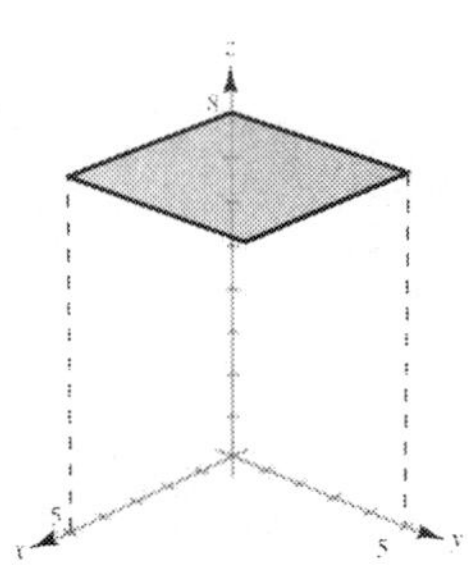

71. P_1: $\mathbf{n} = \langle -5, 2, -8\rangle$ $\quad (0, -1, -1)$ on plane

P_2: $\mathbf{n} = \langle 15, -6, 24\rangle$ $\quad (0, -1, -1)$ not on plane

P_3: $\mathbf{n} = \langle 6, -4, 4\rangle$

P_4: $\mathbf{n} = \langle 3, -2, -2\rangle$

Planes P_1 and P_2 are parallel.

72. P_1: $\mathbf{n} = \langle 2, -1, 3\rangle$ $\quad (4, 0, 0)$ on plane

P_2: $\mathbf{n} = \langle 3, -5, -2\rangle$

P_3: $\mathbf{n} = \langle 8, -4, 12\rangle$ $\quad (4, 0, 0)$ not on plane

P_4: $\mathbf{n} = \langle -4, -2, 6\rangle$

P_1 and P_3 are parallel.

73. P_1: $\mathbf{n} = \langle 3, -2, 5\rangle$ $\quad (1, -1, 1)$ on plane

P_2: $\mathbf{n} = \langle -6, 4, -10\rangle$ $\quad (1, -1, 1)$ not on plane

P_3: $\mathbf{n} = \langle -3, 2, 5\rangle$

P_4: $\mathbf{n} = \langle 75, -50, 125\rangle$ $\quad (1, -1, 1)$ on plane

P_1 and P_4 are identical.

$P_1 = P_4$ and is parallel to P_2.

74. P_1: $\mathbf{n} = \langle -60, 90, 30\rangle$ or $\langle -2, 3, 1\rangle$ $\quad \left(0, 0, \frac{9}{10}\right)$ on plane

P_2: $\mathbf{n} = \langle 6, -9, -3\rangle$ or $\langle -2, 3, 1\rangle$ $\quad \left(0, 0, -\frac{2}{3}\right)$ on plane

P_3: $\mathbf{n} = \langle -20, 30, 10\rangle$ or $\langle -2, 3, 1\rangle$ $\quad \left(0, 0, \frac{5}{6}\right)$ on plane

P_4: $\mathbf{n} = \langle 12, -18, 6\rangle$ or $\langle -2, 3, -1\rangle$

P_1, P_2, and P_3 are parallel.

75. (a) $\mathbf{n}_1 = 3\mathbf{i} + 2\mathbf{j} - \mathbf{k}$ and $\mathbf{n}_2 = \mathbf{i} - 4\mathbf{j} + 2\mathbf{k}$

$$\cos\theta = \frac{|\mathbf{n}_1 \cdot \mathbf{n}_2|}{\|\mathbf{n}_1\|\|\mathbf{n}_2\|} = \frac{|-7|}{\sqrt{14}\sqrt{21}} = \frac{\sqrt{6}}{6}$$

$\Rightarrow \theta \approx 1.1503 \approx 65.91°$

(b) The direction vector for the line is

$$\mathbf{n}_2 \times \mathbf{n}_1 = \begin{vmatrix} \mathbf{i} & \mathbf{j} & \mathbf{k} \\ 1 & -4 & 2 \\ 3 & 2 & -1 \end{vmatrix} = 7(\mathbf{j} + 2\mathbf{k}).$$

Find a point of intersection of the planes.

$$\begin{aligned} 6x + 4y - 2z &= 14 \\ x - 4y + 2z &= 0 \\ 7x &= 14 \\ x &= 2 \end{aligned}$$

Substituting 2 for x in the second equation, you have $-4y + 2z = -2$ or $z = 2y - 1$. Letting $y = 1$, a point of intersection is $(2, 1, 1)$.

$x = 2, y = 1 + t, z = 1 + 2t$

76. (a) $\mathbf{n}_1 = 6\mathbf{i} - 3\mathbf{j} + \mathbf{k},\ \mathbf{n}_2 = -\mathbf{i} + \mathbf{j} + 5\mathbf{k}$

$$\cos\theta = \frac{|\mathbf{n}_1 \cdot \mathbf{n}_2|}{\|\mathbf{n}_1\|\|\mathbf{n}_2\|} = \frac{-4}{\sqrt{46}\sqrt{27}} = \frac{-2\sqrt{138}}{207}$$

$$\theta \approx 1.6845 \approx 96.52°$$

(b) The direction vector for the line is

$$\mathbf{n}_1 \times \mathbf{n}_2 = \begin{vmatrix} \mathbf{i} & \mathbf{j} & \mathbf{k} \\ 6 & -3 & 1 \\ -1 & 1 & 5 \end{vmatrix} = \langle -16, -31, 3 \rangle.$$

Find a point of intersection of the planes.

$$\begin{aligned} 6x - 3y + z &= 5 \Rightarrow & 6x - 3y + z &= 5 \\ -x + y + 5z &= 5 \Rightarrow & -6x + 6y + 30z &= 30 \\ & & 3y + 31z &= 35 \end{aligned}$$

Let $y = -9, z = 2 \Rightarrow x = -4 \Rightarrow (-4, -9, 2)$.

$$x = -4 - 16t,\ y = -9 - 31t,\ z = 2 + 3t$$

77. Writing the equation of the line in parametric form and substituting into the equation of the plane you have:

$$x = \frac{1}{2} + t,\ y = \frac{-3}{2} - t,\ z = -1 + 2t$$

$$2\left(\frac{1}{2} + t\right) - 2\left(\frac{-3}{2} - t\right) + (-1 + 2t) = 12,\ t = \frac{3}{2}$$

Substituting $t = 3/2$ into the parametric equations for the line you have the point of intersection $(2, -3, 2)$. The line does not lie in the plane.

78. Writing the equation of the line in parametric form and substituting into the equation of the plane you have:

$$x = 1 + 4t,\ y = 2t,\ z = 3 + 6t$$

$$2(1 + 4t) + 3(2t) = -5,\ t = \frac{-1}{2}$$

Substituting $t = -\frac{1}{2}$ into the parametric equations for the line you have the point of intersection $(-1, -1, 0)$. The line does not lie in the plane.

79. Writing the equation of the line in parametric form and substituting into the equation of the plane you have:

$$x = 1 + 3t,\ y = -1 - 2t,\ z = 3 + t$$

$$2(1 + 3t) + 3(-1 - 2t) = 10,\ -1 = 10, \text{ contradiction}$$

So, the line does not intersect the plane.

80. Writing the equation of the line in parametric form and substituting into the equation of the plane you have:

$$x = 4 + 2t,\ y = -1 - 3t,\ z = -2 + 5t$$

$$5(4 + 2t) + 3(-1 - 3t) = 17,\ t = 0$$

Substituting $t = 0$ into the parametric equations for the line you have the point of intersection $(4, -1, -2)$. The line does not lie in the plane.

81. Point: $Q(0, 0, 0)$

Plane: $2x + 3y + z - 12 = 0$

Normal to plane: $\mathbf{n} = \langle 2, 3, 1 \rangle$

Point in plane: $P(6, 0, 0)$

Vector $\overrightarrow{PQ} = \langle -6, 0, 0 \rangle$

$$D = \frac{|\overrightarrow{PQ} \cdot \mathbf{n}|}{\|\mathbf{n}\|} = \frac{|-12|}{\sqrt{14}} = \frac{6\sqrt{14}}{7}$$

82. Point: $Q(0, 0, 0)$

Plane: $5x + y - z - 9 = 0$

Normal to plane: $\mathbf{n} = \langle 5, 1, -1 \rangle$

Point in plane: $P(0, 9, 0)$

Vector $\overrightarrow{PQ} = \langle 0, -9, 0 \rangle$

$$D = \frac{|\overrightarrow{PQ} \cdot \mathbf{n}|}{\|\mathbf{n}\|} = \frac{|-9|}{\sqrt{27}} = \sqrt{3}$$

83. Point: $Q(2, 8, 4)$

Plane: $2x + y + z = 5$

Normal to plane: $\mathbf{n} = \langle 2, 1, 1 \rangle$

Point in plane: $P\langle 0, 0, 5 \rangle$

Vector: $\overrightarrow{PQ} = \langle 2, 8, -1 \rangle$

$$D = \frac{|\overrightarrow{PQ} \cdot \mathbf{n}|}{\|\mathbf{n}\|} = \frac{11}{\sqrt{6}} = \frac{11\sqrt{6}}{6}$$

84. Point: $Q(1, 3, -1)$

Plane: $3x - 4y + 5z - 6 = 0$

Normal to plane: $\mathbf{n} = \langle 3, -4, 5 \rangle$

Point in plane: $P(2, 0, 0)$

Vector $\overrightarrow{PQ}$: $\langle -1, 3, -1 \rangle$

$$D = \frac{|\overrightarrow{PQ} \cdot \mathbf{n}|}{\|\mathbf{n}\|} = \frac{|-20|}{\sqrt{50}} = 2\sqrt{2}$$

85. The normal vectors to the planes are $\mathbf{n}_1 = \langle 1, -3, 4 \rangle$ and $\mathbf{n}_2 = \langle 1, -3, 4 \rangle$. Because $\mathbf{n}_1 = \mathbf{n}_2$, the planes are parallel. Choose a point in each plane.

$P(10, 0, 0)$ is a point in $x - 3y + 4z = 10$.

$Q(6, 0, 0)$ is a point in $x - 3y + 4z = 6$.

$$\overrightarrow{PQ} = \langle -4, 0, 0 \rangle,\ D = \frac{|\overrightarrow{PQ} \cdot \mathbf{n}_1|}{\|\mathbf{n}_1\|} = \frac{4}{\sqrt{26}} = \frac{2\sqrt{26}}{13}$$

86. The normal vectors to the planes are $\mathbf{n}_1 = \langle 4, -4, 9\rangle$ and $\mathbf{n}_2 = \langle 4, -4, 9\rangle$. Because $\mathbf{n}_1 = \mathbf{n}_2$, the planes are parallel. Choose a point in each plane.

$P(-5, 0, 3)$ is a point in $4x - 4y + 9z = 7$.

$Q(0, 0, 2)$ is a point in $4x - 4y + 9z = 18$.

$\overrightarrow{PQ} = \langle 5, 0, -1\rangle$

$$D = \frac{\left|\overrightarrow{PQ}\cdot\mathbf{n}_1\right|}{\|\mathbf{n}_1\|} = \frac{11}{\sqrt{113}} = \frac{11\sqrt{113}}{113}$$

87. The normal vectors to the planes are $\mathbf{n}_1 = \langle -3, 6, 7\rangle$ and $\mathbf{n}_2 = \langle 6, -12, -14\rangle$. Because $\mathbf{n}_2 = -2\mathbf{n}_1$, the planes are parallel. Choose a point in each plane.

$P(0, -1, 1)$ is a point in $-3x + 6y + 7z = 1$.

$Q\left(\frac{25}{6}, 0, 0\right)$ is a point in $6x - 12y - 14z = 25$.

$$\overrightarrow{PQ} = \left\langle\frac{25}{6}, 1, -1\right\rangle$$

$$D = \frac{\left|\overrightarrow{PQ}\cdot\mathbf{n}_1\right|}{\|\mathbf{n}_1\|} = \frac{|-27/2|}{\sqrt{94}} = \frac{27}{2\sqrt{94}} = \frac{27\sqrt{94}}{188}$$

88. The normal vectors to the planes are $\mathbf{n}_1 = \langle 2, 0, -4\rangle$ and $\mathbf{n}_2 = \langle 2, 0, -4\rangle$. Because $\mathbf{n}_1 = \mathbf{n}_2$, the planes are parallel. Choose a point in each plane.

$P(2, 0, 0)$ is a point in $2x - 4z = 4$.

$Q(5, 0, 0)$ is a point in $2x - 4z = 10$.

$$\overrightarrow{PQ} = \langle 3, 0, 0\rangle,\ D = \frac{\left|\overrightarrow{PQ}\cdot\mathbf{n}_1\right|}{\|\mathbf{n}_1\|} = \frac{6}{\sqrt{20}} = \frac{3\sqrt{5}}{5}$$

89. $\mathbf{u} = \langle 4, 0, -1\rangle$ is the direction vector for the line.

$Q(1, 5, -2)$ is the given point, and $P(-2, 3, 1)$ is on the line.

$\overrightarrow{PQ} = \langle 3, 2, -3\rangle$

$$\overrightarrow{PQ}\times\mathbf{u} = \begin{vmatrix}\mathbf{i} & \mathbf{j} & \mathbf{k}\\ 3 & 2 & -3\\ 4 & 0 & -1\end{vmatrix} = \langle -2, -9, -8\rangle$$

$$D = \frac{\left\|\overrightarrow{PQ}\times\mathbf{u}\right\|}{\|\mathbf{u}\|} = \frac{\sqrt{149}}{\sqrt{17}} = \frac{\sqrt{2533}}{17}$$

90. $\mathbf{u} = \langle 2, 1, 2\rangle$ is the direction vector for the line.

$Q(1, -2, 4)$ is the given point, and $P(0, -3, 2)$ is a point on the line (let $t = 0$).

$\overrightarrow{PQ} = \langle 1, 1, 2\rangle$

$$\overrightarrow{PQ}\times\mathbf{u} = \begin{vmatrix}\mathbf{i} & \mathbf{j} & \mathbf{k}\\ 1 & 1 & 2\\ 2 & 1 & 2\end{vmatrix} = \langle 0, 2, -1\rangle$$

$$D = \frac{\left\|\overrightarrow{PQ}\times\mathbf{u}\right\|}{\|\mathbf{u}\|} = \frac{\sqrt{5}}{\sqrt{9}} = \frac{\sqrt{5}}{3}$$

91. $\mathbf{u} = \langle -1, 1, -2\rangle$ is the direction vector for the line.

$Q(-2, 1, 3)$ is the given point, and $P(1, 2, 0)$ is on the line (let $t = 0$ in the parametric equations for the line).

$\overrightarrow{PQ} = \langle -3, -1, 3\rangle$

$$\overrightarrow{PQ}\times\mathbf{u} = \begin{vmatrix}\mathbf{i} & \mathbf{j} & \mathbf{k}\\ -3 & -1 & 3\\ -1 & 1 & -2\end{vmatrix} = \langle -1, -9, -4\rangle$$

$$D = \frac{\left\|\overrightarrow{PQ}\times\mathbf{u}\right\|}{\|\mathbf{u}\|} = \frac{\sqrt{1+81+16}}{\sqrt{1+1+4}} = \frac{\sqrt{98}}{6} = \frac{7}{\sqrt{3}} = \frac{7\sqrt{3}}{3}$$

92. $\mathbf{u} = \langle 0, 3, 1\rangle$ is the direction vector for the line.

$Q(4, -1, 5)$ is the given point, and $P(3, 1, 1)$ is on the line.

$\overrightarrow{PQ} = \langle 1, -2, 4\rangle$

$$\overrightarrow{PQ}\times\mathbf{u} = \begin{vmatrix}\mathbf{i} & \mathbf{j} & \mathbf{k}\\ 1 & -2 & 4\\ 0 & 3 & 1\end{vmatrix} = \langle -14, -1, 3\rangle$$

$$D = \frac{\left\|\overrightarrow{PQ}\times\mathbf{u}\right\|}{\|\mathbf{u}\|}$$

$$= \frac{\sqrt{14^2+1+9}}{\sqrt{9+1}} = \sqrt{\frac{206}{10}} = \sqrt{\frac{103}{5}} = \frac{\sqrt{515}}{5}$$

93. The direction vector for L_1 is $\mathbf{v}_1 = \langle -1, 2, 1\rangle$.

The direction vector for L_2 is $\mathbf{v}_2 = \langle 3, -6, -3\rangle$.

Because $\mathbf{v}_2 = -3\mathbf{v}_1$, the lines are parallel.

Let $Q(2, 3, 4)$ to be a point on L_1 and $P(0, 1, 4)$ a point on L_2. $\overrightarrow{PQ} = \langle 2, 2, 0\rangle$.

$\mathbf{u} = \mathbf{v}_2$ is the direction vector for L_2.

$$\overrightarrow{PQ} \times \mathbf{v}_2 = \begin{vmatrix} \mathbf{i} & \mathbf{j} & \mathbf{k} \\ 2 & 2 & 0 \\ 3 & -6 & -3 \end{vmatrix} = \langle -6, 6, -18\rangle$$

$$D = \frac{\left\|\overrightarrow{PQ} \times \mathbf{v}_2\right\|}{\left\|\mathbf{v}_2\right\|}$$

$$= \frac{\sqrt{36 + 36 + 324}}{\sqrt{9 + 36 + 9}} = \sqrt{\frac{396}{54}} = \sqrt{\frac{22}{3}} = \frac{\sqrt{66}}{3}$$

94. The direction vector for L_1 is $\mathbf{v}_1 = \langle 6, 9, -12\rangle$.

The direction vector for L_2 is $\mathbf{v}_2 = \langle 4, 6, -8\rangle$.

Because $\mathbf{v}_1 = \frac{3}{2}\mathbf{v}_2$, the lines are parallel.

Let $Q(3, -2, 1)$ to be a point on L_1 and $P(-1, 3, 0)$ a point on L_2. $\overrightarrow{PQ} = \langle 4, -5, 1\rangle$.

$\mathbf{u} = \mathbf{v}_2$ is the direction vector for L_2.

$$\overrightarrow{PQ} \times \mathbf{v}_2 = \begin{vmatrix} \mathbf{i} & \mathbf{j} & \mathbf{k} \\ 4 & -5 & 1 \\ 4 & 6 & -8 \end{vmatrix} = \langle 34, 36, 44\rangle$$

$$D = \frac{\left\|\overrightarrow{PQ} \times \mathbf{v}_2\right\|}{\left\|\mathbf{v}_2\right\|}$$

$$= \frac{\sqrt{34^2 + 36^2 + 44^2}}{\sqrt{16 + 36 + 64}}$$

$$= \frac{\sqrt{4388}}{\sqrt{116}} = \sqrt{\frac{1097}{29}} = \frac{\sqrt{31813}}{29}$$

95. The parametric equations of a line L parallel to $\mathbf{v} = \langle a, b, c,\rangle$ and passing through the point $P(x_1, y_1, z_1)$ are

$x = x_1 + at,\ y = y_1 + bt,\ z = z_1 + ct.$

The symmetric equations are

$$\frac{x - x_1}{a} = \frac{y - y_1}{b} = \frac{z - z_1}{c}.$$

96. The equation of the plane containing $P(x_1, y_1, z_1)$ and having normal vector $\mathbf{n} = \langle a, b, c\rangle$ is

$a(x - x_1) + b(y - y_1) + c(z - z_1) = 0.$

You need $\mathbf{n}$ and P to find the equation.

97. Simultaneously solve the two linear equations representing the planes and substitute the values back into one of the original equations. Then choose a value for t and form the corresponding parametric equations for the line of intersection.

98. (a) The planes are parallel if their normal vectors are parallel:

$\langle a_1, b_1, c_1\rangle = t\langle a_2, b_2, c_2\rangle,\ t \neq 0$

(b) The planes are perpendicular if their normal vectors are perpendicular:

$\langle a_1, b_1, c_1\rangle \cdot \langle a_2, b_2, c_2\rangle = 0$

99. Yes. If $\mathbf{v}_1$ and $\mathbf{v}_2$ are the direction vectors for the lines L_1 and L_2, then $\mathbf{v} = \mathbf{v}_1 \times \mathbf{v}_2$ is perpendicular to both L_1 and L_2.

100. (a) $ax + by + d = 0$ matches (iv). The plane is parallel to the z-axis.

(b) $ax + d = 0$ matches (i). The plane is parallel to the yz-plane.

(c) $cz + d = 0$ matches (ii). The plane is parallel to the xy-plane.

(d) $ax + cz + d = 0$ matches (iii). The plane is parallel to the y-axis.

101. (a)

Year	2005	2006	2007	2008	2009	2010
x	36.4	39.0	42.4	44.7	43.0	45.2
y	15.3	16.6	17.4	17.5	17.0	17.3
z	16.4	18.1	20.0	20.5	20.1	21.4
Model z	16.39	17.98	19.78	20.87	19.94	21.04

The approximations are close to the actual values.

(b) According to the model, if x and y increase, then so does z.

102. On one side you have the points $(0, 0, 0)$, $(6, 0, 0)$, and $(-1, -1, 8)$.

$$\mathbf{n}_1 = \begin{vmatrix} \mathbf{i} & \mathbf{j} & \mathbf{k} \\ 6 & 0 & 0 \\ -1 & -1 & 8 \end{vmatrix} = -48\mathbf{j} - 6\mathbf{k}$$

On the adjacent side you have the points $(0, 0, 0)$, $(0, 6, 0)$, and $(-1, -1, 8)$.

$$\mathbf{n}_2 = \begin{vmatrix} \mathbf{i} & \mathbf{j} & \mathbf{k} \\ 0 & 6 & 0 \\ -1 & -1 & 8 \end{vmatrix} = 48\mathbf{i} + 6\mathbf{k}$$

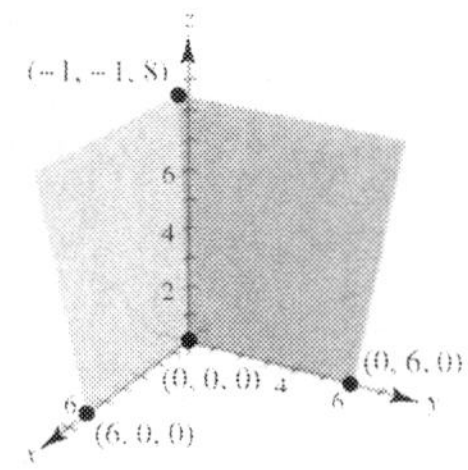

$$\cos\theta = \frac{|\mathbf{n}_1 \cdot \mathbf{n}_2|}{\|\mathbf{n}_1\|\|\mathbf{n}_2\|} = \frac{36}{2340} = \frac{1}{65}$$

$$\theta = \arccos\frac{1}{65} \approx 89.1°$$

103. L_1: $x_1 = 6 + t$; $y_1 = 8 - t$, $z_1 = 3 + t$

L_2: $x_2 = 1 + t$, $y_2 = 2 + t$, $z_2 = 2t$

(a) At $t = 0$, the first insect is at $P_1(6, 8, 3)$ and the second insect is at $P_2(1, 2, 0)$.

$$\text{Distance} = \sqrt{(6-1)^2 + (8-2)^2 + (3-0)^2} = \sqrt{70} \approx 8.37 \text{ inches}$$

(b) $\text{Distance} = \sqrt{(x_1 - x_2)^2 + (y_1 - y_2)^2 + (z_1 - z_2)^2} = \sqrt{5^2 + (6-2t)^2 + (3-t)^2} = \sqrt{5t^2 - 30t + 70}$, $0 \le t \le 10$

(c) The distance is never zero.

(d) Using a graphing utility, the minimum distance is 5 inches when $t = 3$ minutes.

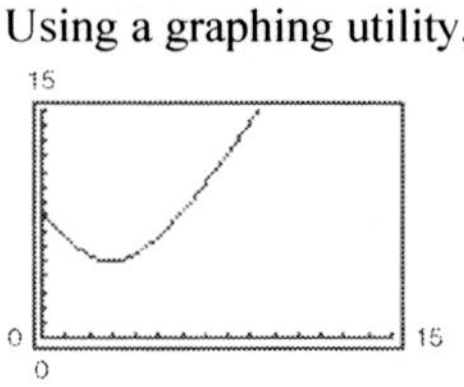

104. First find the distance D from the point $Q(-3, 2, 4)$ to the plane. Let $P(4, 0, 0)$ be on the plane.

$\mathbf{n} = \langle 2, 4, -3\rangle$ is the normal to the plane.

$$D = \frac{|\overrightarrow{PQ} \cdot \mathbf{n}|}{\|\mathbf{n}\|} = \frac{|\langle -7, 2, 4\rangle \cdot \langle 2, 4, -3\rangle|}{\sqrt{4 + 16 + 9}} = \frac{|-14 + 8 - 12|}{\sqrt{29}} = \frac{18}{\sqrt{29}} = \frac{18\sqrt{29}}{29}$$

The equation of the sphere with center $(-3, 2, 4)$ and radius $18\sqrt{29}/29$ is $(x + 3)^2 + (y - 2)^2 + (z - 4)^2 = \dfrac{324}{29}$.

105. The direction vector $\mathbf{v}$ of the line is the normal to the plane, $\mathbf{v} = \langle 3, -1, 4\rangle$.

The parametric equations of the line are $x = 5 + 3t$, $y = 4 - t$, $z = -3 + 4t$.

To find the point of intersection, solve for t in the following equation:

$$3(5 + 3t) - (4 - t) + 4(-3 + 4t) = 7$$
$$26t = 8$$
$$t = \tfrac{4}{13}$$

Point of intersection:

$$\left(5 + 3\left(\tfrac{4}{13}\right), 4 - \tfrac{4}{13}, -3 + 4\left(\tfrac{4}{13}\right)\right) = \left(\tfrac{77}{13}, \tfrac{48}{13}, -\tfrac{23}{13}\right)$$

106. The normal to the plane, $\mathbf{n} = \langle 2, -1, -3\rangle$ is perpendicular to the direction vector $\mathbf{v} = \langle 2, 4, 0\rangle$ of the line because $\langle 2, -1, -3\rangle \cdot \langle 2, 4, 0\rangle = 0$.

So, the plane is parallel to the line. To find the distance between them, let $Q(-2, -1, 4)$ be on the line and $P(2, 0, 0)$ on the plane. $\overrightarrow{PQ} = \langle -4, -1, 4\rangle$.

$$D = \frac{|\overrightarrow{PQ} \cdot \mathbf{n}|}{\|\mathbf{n}\|}$$
$$= \frac{|\langle -4, -1, 4\rangle \cdot \langle 2, -1, -3\rangle|}{\sqrt{4 + 1 + 9}} = \frac{19}{\sqrt{14}} = \frac{19\sqrt{14}}{14}$$

107. The direction vector of the line L through $(1, -3, 1)$ and $(3, -4, 2)$ is $\mathbf{v} = \langle 2, -1, 1 \rangle$.

The parametric equations for L are
$x = 1 + 2t,\ y = -3 - t,\ z = 1 + t.$

Substituting these equations into the equation of the plane gives

$$(1 + 2t) - (-3 - t) + (1 + t) = 2$$
$$4t = -3$$
$$t = -\tfrac{3}{4}.$$

Point of intersection:

$$\left(1 + 2\left(-\tfrac{3}{4}\right), -3 + \tfrac{3}{4}, 1 - \tfrac{3}{4}\right) = \left(-\tfrac{1}{2}, -\tfrac{9}{4}, \tfrac{1}{4}\right)$$

108. The unknown line L is perpendicular to the normal vector $\mathbf{n} = \langle 1, 1, 1 \rangle$ of the plane, and perpendicular to the direction vector $\mathbf{u} = \langle 1, 1, -1 \rangle$. So, the direction vector of L is

$$\mathbf{v} = \begin{vmatrix} \mathbf{i} & \mathbf{j} & \mathbf{k} \\ 1 & 1 & 1 \\ 1 & 1 & -1 \end{vmatrix} = \langle -2, 2, 0 \rangle.$$

The parametric equations for L are $x = 1 - 2t,\ y = 2t,\ z = 2.$

109. True

110. False. They may be skew lines. (See Section Project.)

111. True

112. False. The lines $x = t,\ y = 0,\ z = 1$ and $x = 0,\ y = t,\ z = 1$ are both parallel to the plane $z = 0$, but the lines are not parallel.

113. False. Planes $7x + y - 11z = 5$ and $5x + 2y - 4z = 1$ are both perpendicular to plane $2x - 3y + z = 3$, but are not parallel.

114. True

Section 11.6 Surfaces in Space

1. Ellipsoid
Matches graph (c)

2. Hyperboloid of two sheets
Matches graph (e)

3. Hyperboloid of one sheet
Matches graph (f)

4. Elliptic cone
Matches graph (b)

5. Elliptic paraboloid
Matches graph (d)

6. Hyperbolic paraboloid
Matches graph (a)

7. $y = 5$

Plane is parallel to the xz-plane.

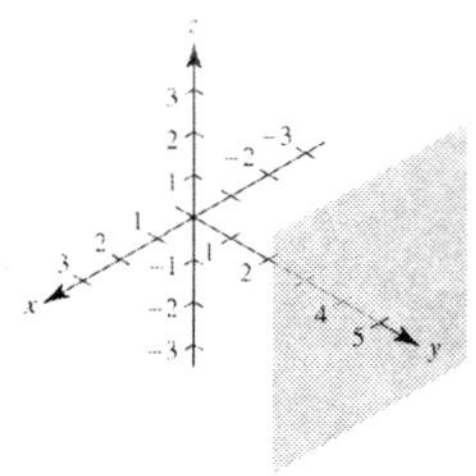

8. $z = 2$

Plane is parallel to the xy-plane.

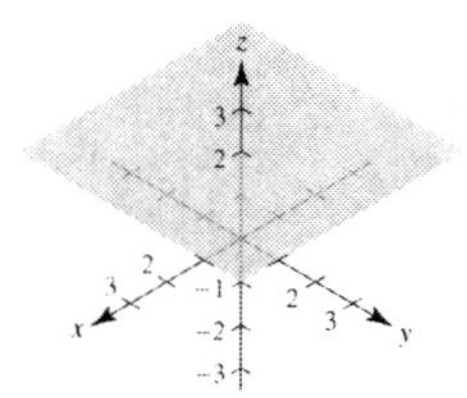

9. $y^2 + z^2 = 9$

The x-coordinate is missing so you have a right circular cylinder with rulings parallel to the x-axis. The generating curve is a circle.

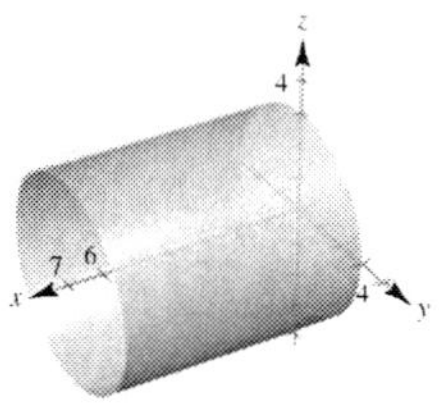

10. $y^2 + z = 6$

The x-coordinate is missing so you have a parabolic cylinder with the rulings parallel to the x-axis. The generating curve is a parabola.

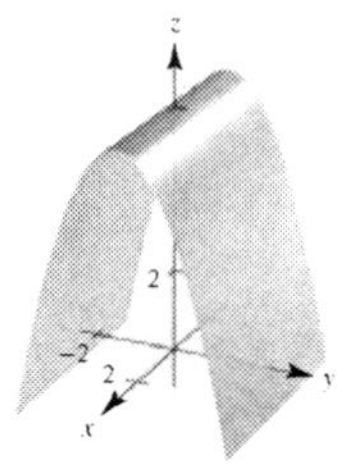

11. $4x^2 + y^2 = 4$

$$\frac{x^2}{1} + \frac{y^2}{4} = 1$$

The z-coordinate is missing so you have an elliptic cylinder with rulings parallel to the z-axis. The generating curve is an ellipse.

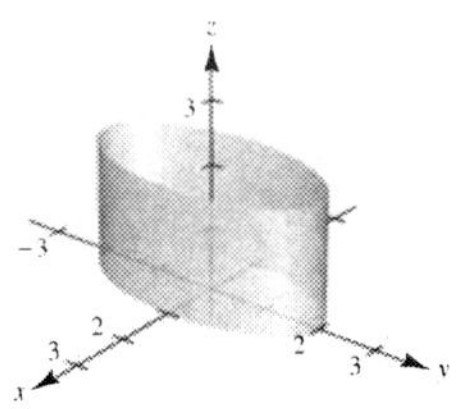

12. $y^2 - z^2 = 16$

$$\frac{y^2}{16} - \frac{z^2}{16} = 1$$

The x-coordinate is missing so you have a hyperbolic cylinder with rulings parallel to the x-axis. The generating curve is a hyperbola.

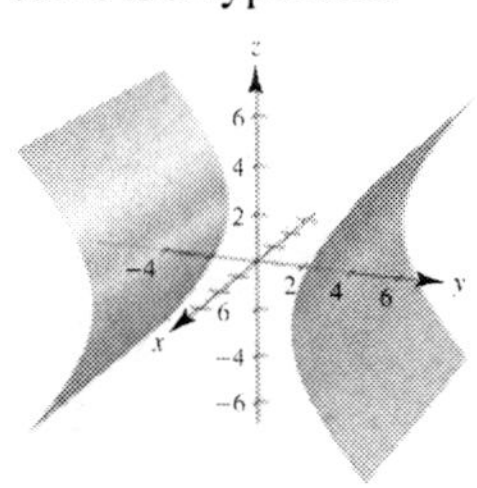

13. $\dfrac{x^2}{1} + \dfrac{y^2}{4} + \dfrac{z^2}{1} = 1$

Ellipsoid

xy-trace: $\dfrac{x^2}{1} + \dfrac{y^2}{4} = 1$ ellipse

xz-trace: $x^2 + z^2 = 1$ circle

yz-trace: $\dfrac{y^2}{4} + \dfrac{z^2}{1} = 1$ ellipse

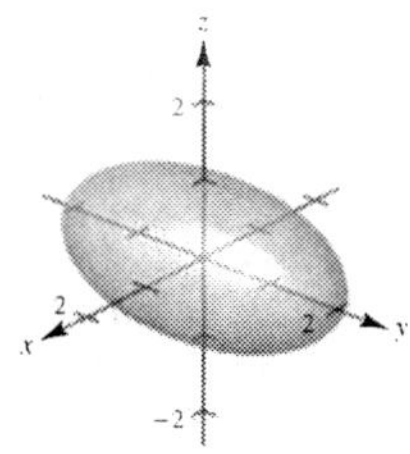

14. $\dfrac{x^2}{16} + \dfrac{y^2}{25} + \dfrac{z^2}{25} = 1$

Ellipsoid

xy-trace: $\dfrac{x^2}{16} + \dfrac{y^2}{25} = 1$ ellipse

xz-trace: $\dfrac{x^2}{16} + \dfrac{z^2}{25} = 1$ ellipse

yz-trace: $y^2 + z^2 = 25$ circle

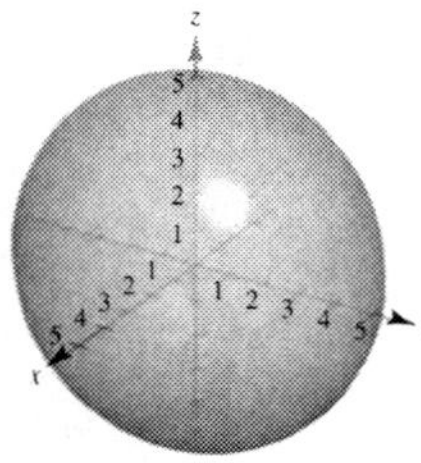

15. $16x^2 - y^2 + 16z^2 = 4$

$$4x^2 - \frac{y^2}{4} + 4z^2 = 1$$

Hyperboloid of one sheet

xy-trace: $4x^2 - \dfrac{y^2}{4} = 1$ hyperbola

xz-trace: $4\left(x^2 + z^2\right) = 1$ circle

yz-trace: $\dfrac{-y^2}{4} + 4z^2 = 1$ hyperbola

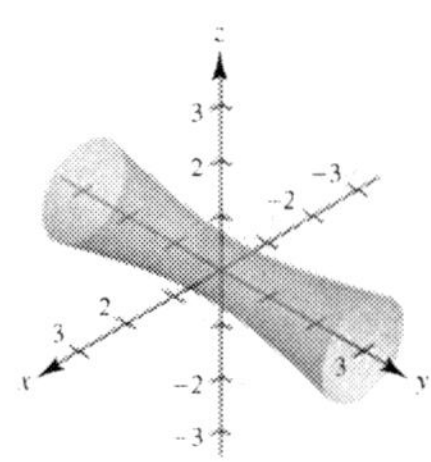

16. $-8x^2 + 18y^2 + 18z^2 = 2$

$9y^2 + 9z^2 - 4x^2 = 1$

Hyperboloid of one sheet

xy-trace: $9y^2 - 4x^2 = 1$ hyperbola

yz-trace: $9y^2 + 9z^2 = 1$ circle

xz-trace: $9z^2 - 4x^2 = 1$ hyperbola

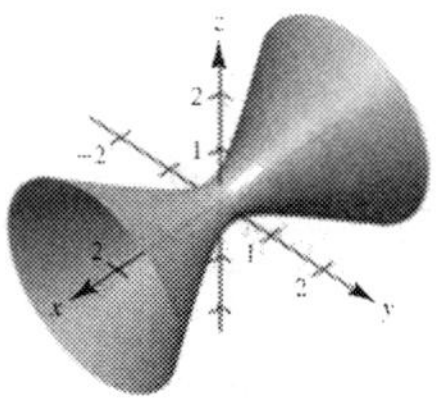

17. $4x^2 - y^2 - z^2 = 1$

Hyperboloid of two sheets

xy-trace: $4x^2 - y^2 = 1$ hyperbola

yz-trace: none

xz-trace: $4x^2 - z^2 = 1$ hyperbola

18. $z^2 - x^2 - \dfrac{y^2}{4} = 1$

Hyperboloid of two sheets

xy-trace: none

xz-trace: $z^2 - x^2 = 1$ hyperbola

yz-trace: $z^2 - \dfrac{y^2}{4} = 1$ hyperbola

$z = \pm\sqrt{10}$: $\dfrac{x^2}{9} + \dfrac{y^2}{36} = 1$ ellipse

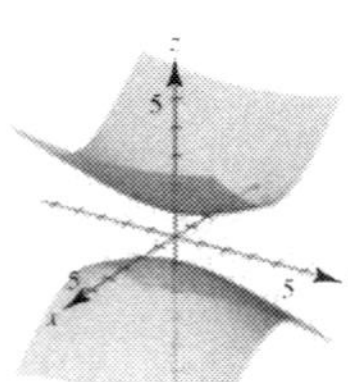

19. $x^2 - y + z^2 = 0$

Elliptic paraboloid

xy-trace: $y = x^2$

xz-trace: $x^2 + z^2 = 0$, point $(0, 0, 0)$

yz-trace: $y = z^2$

$y = 1$: $x^2 + z^2 = 1$

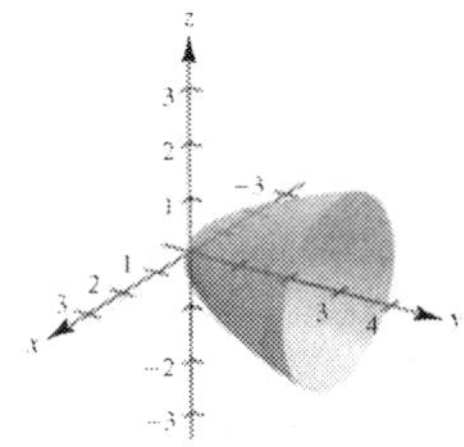

20. $z = x^2 + 4y^2$

Elliptic paraboloid

xy-trace: point $(0, 0, 0)$

xz-trace: $z = x^2$ parabola

yz-trace: $z = 4y^2$ parabola

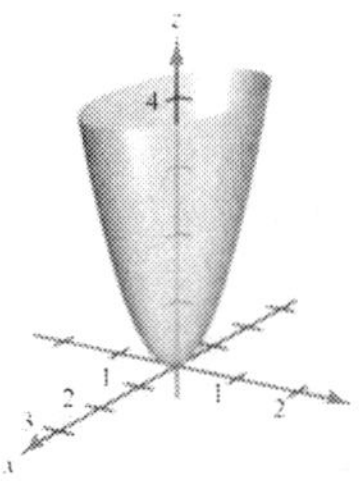

21. $x^2 - y^2 + z = 0$

Hyperbolic paraboloid

xy-trace: $y = \pm x$

xz-trace: $z = -x^2$

yz-trace: $z = y^2$

$y = \pm 1$: $z = 1 - x^2$

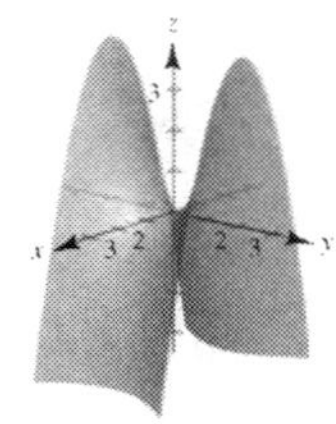

22. $3z = -y^2 + x^2$

Hyperbolic paraboloid

xy-trace: $y = \pm x$

xz-trace: $z = \frac{1}{3}x^2$

yz-trace: $z = -\frac{1}{3}y^2$

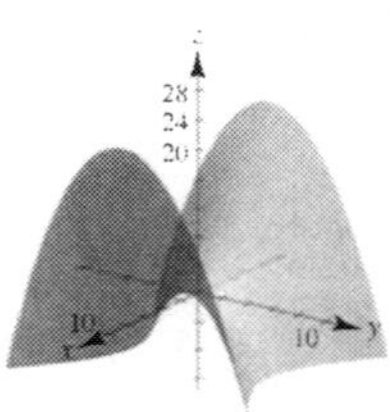

23. $z^2 = x^2 + \dfrac{y^2}{9}$

Elliptic cone

xy-trace: point $(0, 0, 0)$

xz-trace: $z = \pm x$

yz-trace: $z = \pm\dfrac{y}{3}$

When $z = \pm 1$, $x^2 + \dfrac{y^2}{9} = 1$ ellipse

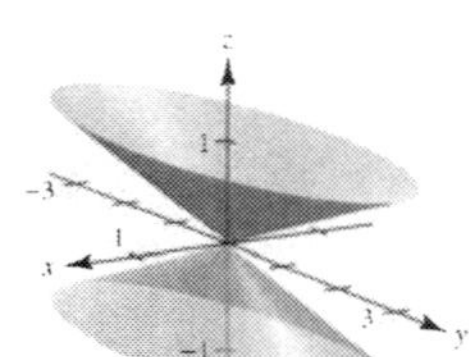

24. $x^2 = 2y^2 + 2z^2$

Elliptic Cone

xy-trace: $x = \pm\sqrt{2}y$

xz-trace: $x = \pm\sqrt{2}z$

yz-trace: point: $(0, 0, 0)$

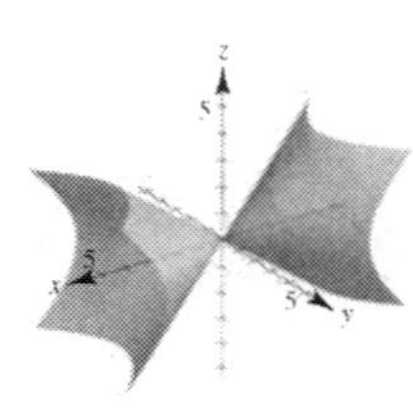

25. Let C be a curve in a plane and let L be a line not in a parallel plane. The set of all lines parallel to L and intersecting C is called a cylinder. C is called the generating curve of the cylinder, and the parallel lines are called rulings.

26. The trace of a surface is the intersection of the surface with a plane. You find a trace by setting one variable equal to a constant, such as $x = 0$ or $z = 2$.

27. See pages 796 and 797.

28. In the xz-plane, $z = x^2$ is a parabola.

In three-space, $z = x^2$ is a cylinder.

29. In the xy-plane, $4x^2 + 6y^2 - 3z^2 = 12$ is an ellipse.

In three-space, $4x^2 + 6y^2 - 3z^2 = 12$ is a hyperboloid of one sheet.

30. $z = x^2 + y^2$

(a) You are viewing the paraboloid from the x-axis: $(20, 0, 0)$

(b) You are viewing the paraboloid from above, but not on the z-axis: $(10, 10, 20)$

(c) You are viewing the paraboloid from the z-axis: $(0, 0, 20)$

(d) You are viewing the paraboloid from the y-axis: $(0, 20, 0)$

31. $x^2 + z^2 = \left[r(y)\right]^2$ and $z = r(y) = \pm 2\sqrt{y}$; so,

$x^2 + z^2 = 4y.$

32. $x^2 + z^2 = \left[r(y)\right]^2$ and $z = r(y) = 3y$; so,

$x^2 + z^2 = 9y^2.$

33. $x^2 + y^2 = \left[r(z)\right]^2$ and $y = r(z) = \dfrac{z}{2}$; so,

$x^2 + y^2 = \dfrac{z^2}{4}, 4x^2 + 4y^2 = z^2.$

34. $y^2 + z^2 = \left[r(x)\right]^2$ and $z = r(x) = \frac{1}{2}\sqrt{4 - x^2}$; so,

$y^2 + z^2 = \frac{1}{4}\left(4 - x^2\right), x^2 + 4y^2 + 4z^2 = 4.$

35. $y^2 + z^2 = \left[r(x)\right]^2$ and $y = r(x) = \dfrac{2}{x}$; so,

$y^2 + z^2 = \left(\dfrac{2}{x}\right)^2, y^2 + z^2 = \dfrac{4}{x^2}.$

36. $x^2 + y^2 = \left[r(z)\right]^2$ and $y = r(z) = e^z$; so,

$x^2 + y^2 = e^{2z}.$

37. $x^2 + y^2 - 2z = 0$

$x^2 + y^2 = \left(\sqrt{2z}\right)^2$

Equation of generating curve: $y = \sqrt{2z}$ or $x = \sqrt{2z}$

38. $x^2 + z^2 = \cos^2 y$

Equation of generating curve: $x = \cos y$ or $z = \cos y$

39. $V = 2\pi \int_0^4 x\left(4x - x^2\right) dx = 2\pi\left[\dfrac{4x^3}{3} - \dfrac{x^4}{4}\right]_0^4 = \dfrac{218\pi}{3}$

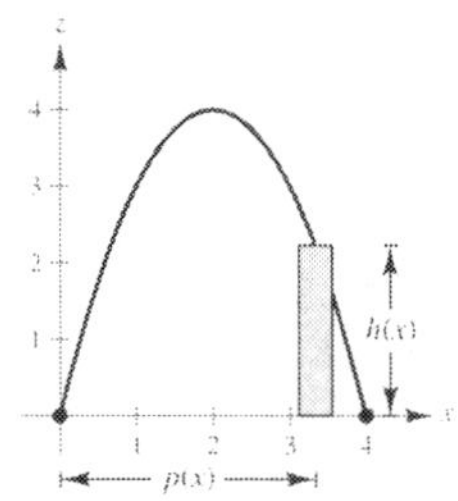

40. $V = 2\pi \int_0^{\pi} y \sin y \, dy$

$= 2\pi\left[\sin y - y \cos y\right]_0^{\pi} = 2\pi^2$

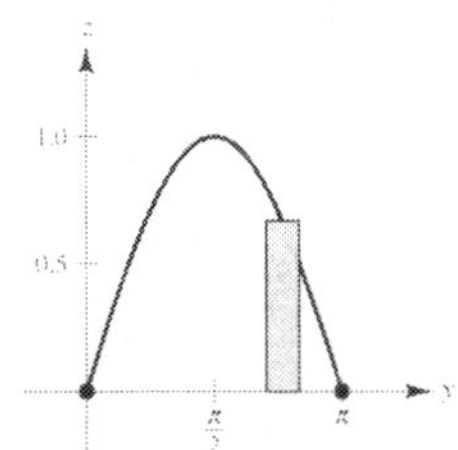

41. $z = \frac{x^2}{2} + \frac{y^2}{4}$

(a) When $z = 2$ we have $2 = \frac{x^2}{2} + \frac{y^2}{4}$, or

$1 = \frac{x^2}{4} + \frac{y^2}{8}$

Major axis: $2\sqrt{8} = 4\sqrt{2}$

Minor axis: $2\sqrt{4} = 4$

$c^2 = a^2 - b^2, c^2 = 4, c = 2$

Foci: $(0, \pm 2, 2)$

(b) When $z = 8$ we have $8 = \frac{x^2}{2} + \frac{y^2}{4}$, or

$1 = \frac{x^2}{16} + \frac{y^2}{32}$.

Major axis: $2\sqrt{32} = 8\sqrt{2}$

Minor axis: $2\sqrt{16} = 8$

$c^2 = 32 - 16 = 16, c = 4$

Foci: $(0, \pm 4, 8)$

42. $z = \frac{x^2}{2} + \frac{y^2}{4}$

(a) When $y = 4$ you have $z = \frac{x^2}{2} + 4$,

$4\left(\frac{1}{2}\right)(z - 4) = x^2$.

Focus: $\left(0, 4, \frac{9}{2}\right)$

(b) When $x = 2$ you have

$z = 2 + \frac{y^2}{4}, 4(z - 2) = y^2$.

Focus: $(2, 0, 3)$

43. If (x, y, z) is on the surface, then

$$(y + 2)^2 = x^2 + (y - 2)^2 + z^2$$
$$y^2 + 4y + 4 = x^2 + y^2 - 4y + 4 + z^2$$
$$x^2 + z^2 = 8y$$

Elliptic paraboloid

Traces parallel to xz-plane are circles.

44. If (x, y, z) is on the surface, then

$$z^2 = x^2 + y^2 + (z - 4)^2$$
$$z^2 = x^2 + y^2 + z^2 - 8z + 16$$
$$8z = x^2 + y^2 + 16 \Rightarrow z = \frac{x^2}{8} + \frac{y^2}{8} + 2$$

Elliptic paraboloid shifted up 2 units. Traces parallel to xy-plane are circles.

45. $\frac{x^2}{3963^2} + \frac{y^2}{3963^2} + \frac{z^2}{3950^2} = 1$

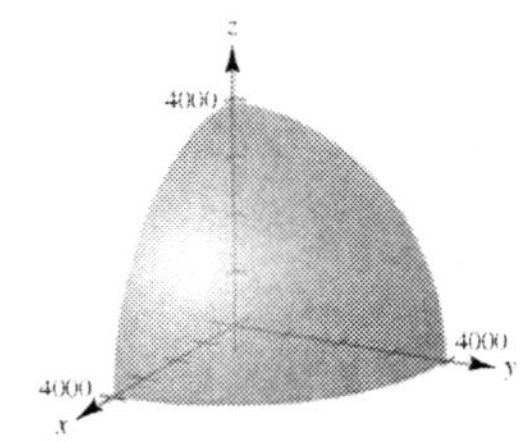

46. (a)
$$x^2 + y^2 = [r(z)]^2 = \left[\sqrt{2(z - 1)}\right]^2$$
$$x^2 + y^2 - 2z + 2 = 0$$

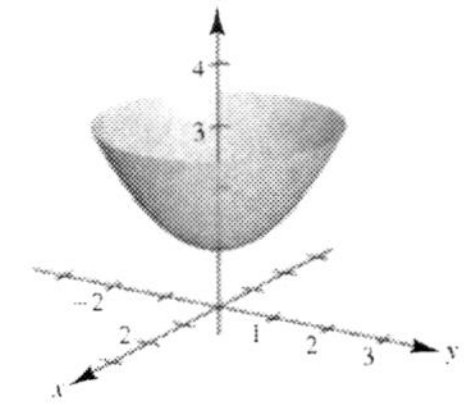

(b) $V = 2\pi \int_0^2 x\left[3 - \left(\frac{1}{2}x^2 + 1\right)\right] dx$

$= 2\pi \int_0^2 \left(2x - \frac{1}{2}x^3\right) dx$

$= 2\pi\left[x^2 - \frac{x^4}{8}\right]_0^2 = 4\pi \approx 12.6 \text{ cm}^3$

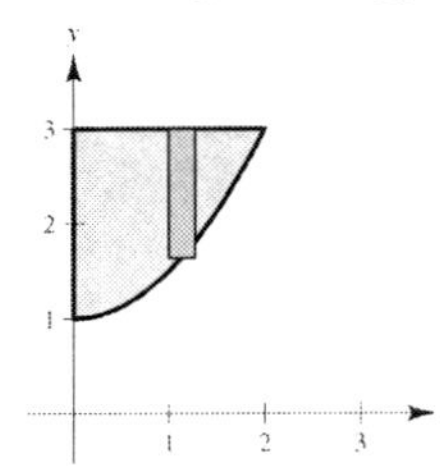

(c) $V = 2\pi \int_{1/2}^2 x\left[3 - \left(\frac{1}{2}x^2 + 1\right)\right] dx$

$= 2\pi \int_{1/2}^2 \left(2x - \frac{1}{2}x^3\right) dx$

$= 2\pi\left[x^2 - \frac{x^4}{8}\right]_{1/2}^2$

$= 4 - \frac{31\pi}{64} = \frac{225\pi}{64} \approx 11.04 \text{ cm}^3$

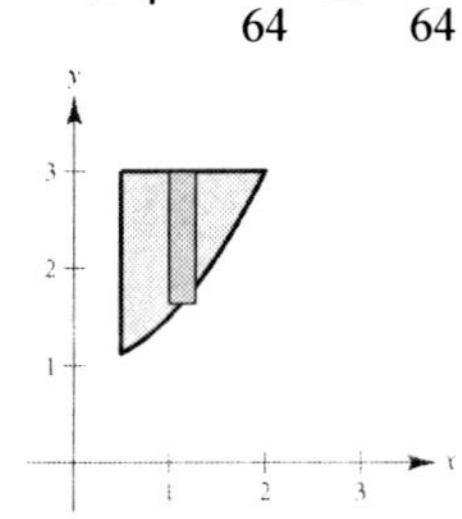

47. $z = \dfrac{y^2}{b^2} - \dfrac{x^2}{a^2}, z = bx + ay$

$$bx + ay = \frac{y^2}{b^2} - \frac{x^2}{a^2}$$

$$\frac{1}{a^2}\left(x^2 + a^2bx + \frac{a^4b^2}{4}\right) = \frac{1}{b^2}\left(y^2 - ab^2y + \frac{a^2b^4}{4}\right)$$

$$\frac{\left(x + \dfrac{a^2b}{2}\right)^2}{a^2} = \frac{\left(y - \dfrac{ab^2}{2}\right)^2}{b^2}$$

$$y = \pm\frac{b}{a}\left(x + \frac{a^2b}{2}\right) + \frac{ab^2}{2}$$

Letting $x = at$, you obtain the two intersecting lines $x = at, y = -bt, z = 0$ and $x = at,$ $y = bt + ab^2, z = 2abt + a^2b^2.$

48. Equating twice the first equation with the second equation:

$$2x^2 + 6y^2 - 4z^2 + 4y - 8 = 2x^2 + 6y^2 - 4z^2 - 3x - 2$$

$$4y - 8 = -3x - 2$$

$$3x + 4y = 6, \text{ a plane}$$

49. True. A sphere is a special case of an ellipsoid (centered at origin, for example)

$$\frac{x^2}{a^2} + \frac{y^2}{b^2} + \frac{z^2}{c^2} = 1$$

having $a = b = c$.

50. False. For example, the surface $x^2 + z^2 = e^{-2y}$ can be formed by revolving the graph of $x = e^{-y}$ about the y-axis, as the graph of $z = e^{-y}$ about the y-axis.

51. False. The trace $x = 2$ of the ellipsoid $\dfrac{x^2}{4} + \dfrac{y^2}{9} + z^2 = 1$ is the point $(2, 0, 0)$.

52. False. Traces perpendicular to the axis are ellipses.

53. The Klein bottle *does not* have both an "inside" and an "outside." It is formed by inserting the small open end through the side of the bottle and making it contiguous with the top of the bottle.

Section 11.7 Cylindrical and Spherical Coordinates

1. $(-7, 0, 5)$, cylindrical

$x = r\cos\theta = -7\cos 0 = -7$

$y = r\sin\theta = -7\sin 0 = 0$

$z = 5$

$(-7, 0, 5)$, rectangular

2. $(2, -\pi, -4)$, cylindrical

$x = r\cos\theta = 2\cos(-\pi) = -2$

$y = r\sin\theta = 2\sin(-\pi) = 0$

$z = -4$

$(-2, 0, -4)$, rectangular

3. $\left(3, \dfrac{\pi}{4}, 1\right)$, cylindrical

$x = 3\cos\dfrac{\pi}{4} = \dfrac{3\sqrt{2}}{2}$

$y = 3\sin\dfrac{\pi}{4} = \dfrac{3\sqrt{2}}{2}$

$z = 1$

$\left(\dfrac{3\sqrt{2}}{2}, \dfrac{3\sqrt{2}}{2}, 1\right)$, rectangular

4. $\left(6, -\dfrac{\pi}{4}, 2\right)$, cylindrical

$x = 6\cos\left(-\dfrac{\pi}{4}\right) = 3\sqrt{2}$

$y = 6\sin\left(-\dfrac{\pi}{4}\right) = -3\sqrt{2}$

$z = 2$

$\left(3\sqrt{2}, -3\sqrt{2}, 2\right)$, rectangular

5. $\left(4, \dfrac{7\pi}{6}, 3\right)$, cylindrical

$x = 4\cos\dfrac{7\pi}{6} = -2\sqrt{3}$

$y = 4\sin\dfrac{7\pi}{6} = -2$

$z = 3$

$\left(-2\sqrt{3}, -2, 3\right)$, rectangular

6. $\left(-0.5, \dfrac{4\pi}{3}, 8\right)$, cylindrical

$x = -\dfrac{1}{2}\cos\dfrac{4\pi}{3} = \dfrac{1}{4}$

$y = -\dfrac{1}{2}\sin\dfrac{4\pi}{3} = \dfrac{\sqrt{3}}{4}$

$z = 8$

$\left(\dfrac{1}{4}, \dfrac{\sqrt{3}}{4}, 8\right)$, rectangular

7. $(0, 5, 1)$, rectangular

$$r = \sqrt{(0)^2 + (5)^2} = 5$$
$$\theta = \arctan\frac{5}{0} = \frac{\pi}{2}$$
$$z = 1$$
$\left(5, \frac{\pi}{2}, 1\right)$, cylindrical

8. $\left(2\sqrt{2}, -2\sqrt{2}, 4\right)$, rectangular

$$r = \sqrt{\left(2\sqrt{2}\right)^2 + \left(-2\sqrt{2}\right)^2} = 4$$
$$\theta = \arctan(-1) = -\frac{\pi}{4}$$
$$z = -4$$
$\left(4, -\frac{\pi}{4}, 4\right)$, cylindrical

9. $(2, -2, -4)$, rectangular

$$r = \sqrt{2^2 + (-2)^2} = 2\sqrt{2}$$
$$\theta = \arctan(-1) = -\frac{\pi}{4}$$
$$z = -4$$
$\left(2\sqrt{2}, -\frac{\pi}{4}, -4\right)$, cylindrical

10. $(3, -3, 7)$, rectangular

$$r = \sqrt{3^2 + (-3)^2} = \sqrt{18} = 3\sqrt{2}$$
$$\theta = \arctan(-1) = -\frac{\pi}{4}$$
$$z = 7$$
$\left(3\sqrt{2}, -\frac{\pi}{4}, 7\right)$, cylindrical

11. $\left(1, \sqrt{3}, 4\right)$, rectangular

$$r = \sqrt{1^2 + \left(\sqrt{3}\right)^2} = 2$$
$$\theta = \arctan\sqrt{3} = \frac{\pi}{3}$$
$$z = 4$$
$\left(2, \frac{\pi}{3}, 4\right)$, cylindrical

12. $\left(2\sqrt{3}, -2, 6\right)$, rectangular

$$r = \sqrt{12 + 4} = 4$$
$$\theta = \arctan\left(-\frac{1}{\sqrt{3}}\right) = \frac{5\pi}{6}$$
$$z = 6$$
$\left(4, -\frac{\pi}{6}, 6\right)$, cylindrical

13. $z = 4$ is the equation in cylindrical coordinates. (plane)

14. $x = 9$, rectangular equation

$r\cos\theta = 9$

$r = 9\sec\theta$, cylindrical equation

15. $x^2 + y^2 + z^2 = 17$, rectangular equation

$r^2 + z^2 = 17$, cylindrical equation

16. $z = x^2 + y^2 - 11$, rectangular equation

$z = r^2 - 11$, cylindrical equation

17. $y = x^2$, rectangular equation

$r\sin\theta = (r\cos\theta)^2$

$\sin\theta = r\cos^2\theta$

$r = \sec\theta \cdot \tan\theta$, cylindrical equation

18. $x^2 + y^2 = 8x$, rectangular equation

$r^2 = 8r\cos\theta$

$r = 8\cos\theta$, cylindrical equation

19. $y^2 = 10 - z^2$, rectangular equation

$(r\sin\theta)^2 = 10 - z^2$

$r^2\sin^2\theta + z^2 = 10$, cylindrical equation

20. $x^2 + y^2 + z^2 - 3z = 0$, rectangular equation

$r^2 + z^2 - 3z = 0$, cylindrical equation

21. $r = 3$

$\sqrt{x^2 + y^2} = 3$

$x^2 + y^2 = 9$

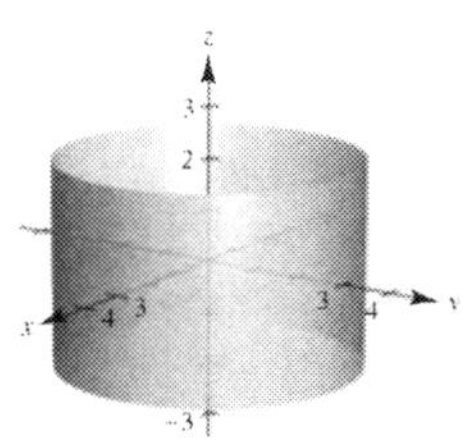

22. $z = 2$

Same

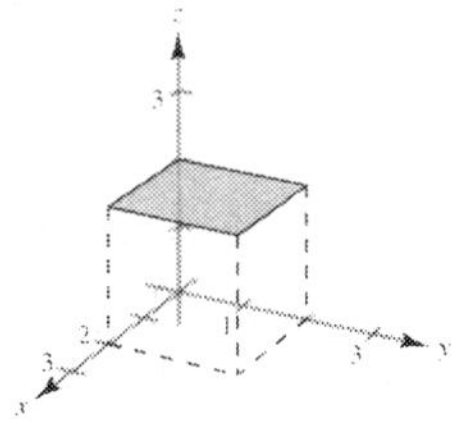

23.
$$\theta = \frac{\pi}{6}$$
$$\tan\frac{\pi}{6} = \frac{y}{x}$$
$$\frac{1}{\sqrt{3}} = \frac{y}{x}$$
$$x = \sqrt{3}y$$
$$x - \sqrt{3}y = 0$$

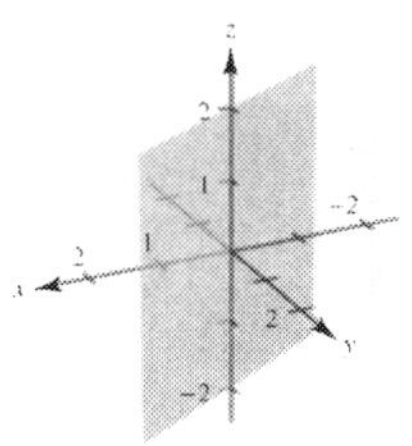

24.
$$r = \frac{z}{2}$$
$$\sqrt{x^2 + y^2} = \frac{z}{2}$$
$$x^2 + y^2 - \frac{z^2}{4} = 0$$

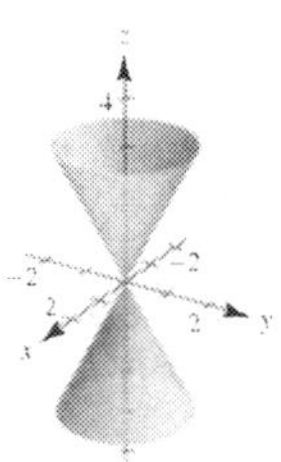

25.
$$r^2 + z^2 = 5$$
$$x^2 + y^2 + z^2 = 5$$

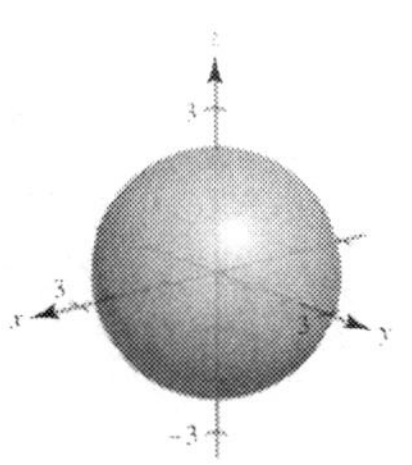

26. $z = r^2 \cos^2\theta$

$z = x^2$

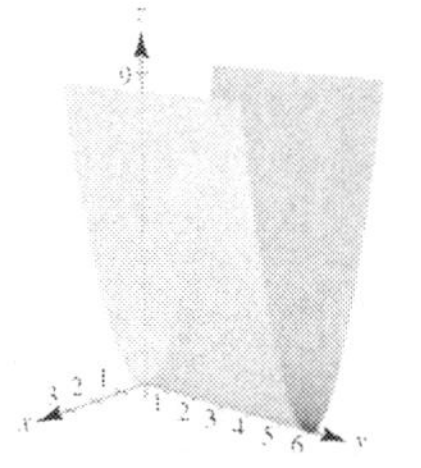

27.
$$r = 2\sin\theta$$
$$r^2 = 2r\sin\theta$$
$$x^2 + y^2 = 2y$$
$$x^2 + y^2 - 2y = 0$$
$$x^2 + (y - 1)^2 = 1$$

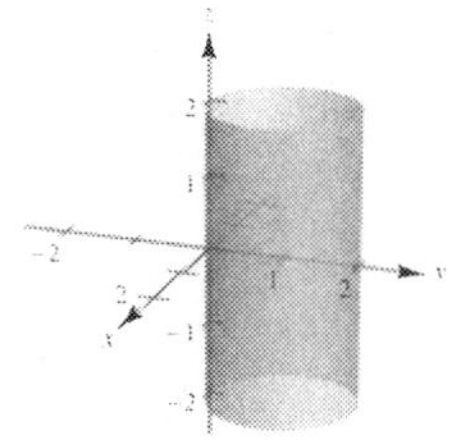

28.
$$r = 2\cos\theta$$
$$r^2 = 2r\cos\theta$$
$$x^2 + y^2 = 2x$$
$$x^2 + y^2 - 2x = 0$$
$$(x - 1)^2 + y^2 = 1$$

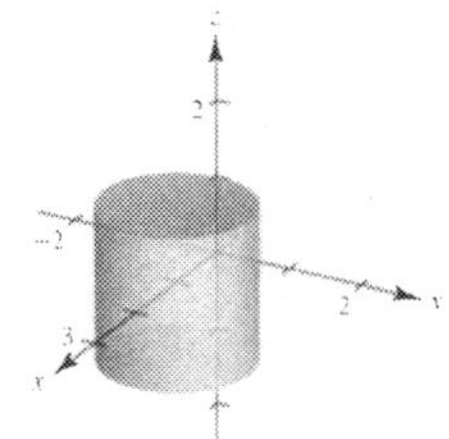

29. $(4, 0, 0)$, rectangular

$$\rho = \sqrt{4^2 + 0^2 + 0^2} = 4$$
$$\tan\theta = \frac{y}{x} = 0 \Rightarrow \theta = 0$$
$$\phi = \arccos 0 = \frac{\pi}{2}$$
$\left(4, 0, \frac{\pi}{2}\right)$, spherical

30. $(-4, 0, 0)$, rectangular

$$\rho = \sqrt{(-4)^2 + 0^2 + 0^2} = 4$$

$$\tan\theta = \frac{y}{x} = 0 \Rightarrow \theta = 0$$

$$\theta = \arccos\left(\frac{z}{e}\right) = \arccos(0) = \frac{\pi}{2}$$

$\left(4, 0, \frac{\pi}{2}\right)$, spherical

31. $\left(-2, 2\sqrt{3}, 4\right)$, rectangular

$$\rho = \sqrt{(-2)^2 + \left(2\sqrt{3}\right)^2 + 4^2} = 4\sqrt{2}$$

$$\tan\theta = \frac{y}{x} = \frac{2\sqrt{3}}{-2} = -\sqrt{3}$$

$$\theta = \frac{2\pi}{3}$$

$$\phi = \arccos\frac{1}{\sqrt{2}} = \frac{\pi}{4}$$

$\left(4\sqrt{2}, \frac{2\pi}{3}, \frac{\pi}{4}\right)$, spherical

32. $\left(2, 2, 4\sqrt{2}\right)$, rectangular

$$\rho = \sqrt{2^2 + 2^2 + \left(4\sqrt{2}\right)^2} = 2\sqrt{10}$$

$$\tan\theta = \frac{y}{x} = 1 \Rightarrow \theta = \frac{\pi}{4}$$

$$\phi = \arccos\frac{2}{\sqrt{5}}$$

$\left(2\sqrt{10}, \frac{\pi}{4}, \arccos\frac{2}{\sqrt{5}}\right)$, spherical

33. $\left(\sqrt{3}, 1, 2\sqrt{3}\right)$, rectangular

$$\rho = \sqrt{3 + 1 + 12} = 4$$

$$\tan\theta = \frac{y}{x} = \frac{1}{\sqrt{3}}$$

$$\theta = \frac{\pi}{6}$$

$$\phi = \arccos\frac{\sqrt{3}}{2} = \frac{\pi}{6}$$

$\left(4, \frac{\pi}{6}, \frac{\pi}{6}\right)$, spherical

34. $(-1, 2, 1)$, rectangular

$$\rho = \sqrt{(-1)^2 + 2^2 + 1^2} = \sqrt{6}$$

$$\tan\theta = \frac{y}{x} = -2 \Rightarrow \theta = \arctan(-2) + \pi$$

$$\phi = \arccos\left(\frac{1}{\sqrt{6}}\right)$$

$\left(\sqrt{6}, \arctan(-2) + \pi, \arccos\frac{1}{\sqrt{6}}\right)$, spherical

35. $\left(4, \frac{\pi}{6}, \frac{\pi}{4}\right)$, spherical

$$x = 4\sin\frac{\pi}{4}\cos\frac{\pi}{6} = \sqrt{6}$$

$$y = 4\sin\frac{\pi}{4}\sin\frac{\pi}{6} = \sqrt{2}$$

$$z = 4\cos\frac{\pi}{4} = 2\sqrt{2}$$

$\left(\sqrt{6}, \sqrt{2}, 2\sqrt{2}\right)$, rectangular

36. $\left(12, \frac{3\pi}{4}, \frac{\pi}{9}\right)$, spherical

$$x = 12\sin\frac{\pi}{9}\cos\frac{3\pi}{4} \approx -2.902$$

$$y = 12\sin\frac{\pi}{9}\sin\frac{3\pi}{4} \approx 2.902$$

$$z = 12\cos\frac{\pi}{9} \approx 11.276$$

$(-2.902, 2.902, 11.276)$, rectangular

37. $\left(12, -\frac{\pi}{4}, 0\right)$, spherical

$$x = 12\sin 0\cos\left(-\frac{\pi}{4}\right) = 0$$

$$y = 12\sin 0\sin\left(-\frac{\pi}{4}\right) = 0$$

$$z = 12\cos 0 = 12$$

$(0, 0, 12)$, rectangular

38. $\left(9, \frac{\pi}{4}, \pi\right)$, spherical

$$x = 9\sin\pi\cos\frac{\pi}{4} = 0$$

$$y = 9\sin\pi\sin\frac{\pi}{4} = 0$$

$$z = 9\cos\pi = -9$$

$(0, 0, -9)$, rectangular

39. $\left(5, \frac{\pi}{4}, \frac{3\pi}{4}\right)$, spherical

$$x = 5\sin\frac{3\pi}{4}\cos\frac{\pi}{4} = \frac{5}{2}$$
$$y = 5\sin\frac{3\pi}{4}\sin\frac{\pi}{4} = \frac{5}{2}$$
$$z = 5\cos\frac{3\pi}{4} = -\frac{5\sqrt{2}}{2}$$

$\left(\frac{5}{2}, \frac{5}{2}, -\frac{5\sqrt{2}}{2}\right)$, rectangular

40. $\left(6, \pi, \frac{\pi}{2}\right)$, spherical

$$x = 6\sin\frac{\pi}{2}\cos\pi = -6$$
$$y = 6\sin\frac{\pi}{2}\sin\pi = 0$$
$$z = 6\cos\frac{\pi}{2} = 0$$

$(-6, 0, 0)$, rectangular

41. $y = 2$, rectangular equation

$$\rho\sin\phi\sin\theta = 2$$
$$\rho = 2\csc\phi\csc\theta, \text{ spherical equation}$$

42. $z = 6$, rectangular equation

$$\rho\cos\phi = 6$$
$$\rho = 6\sec\phi, \text{ spherical equation}$$

43. $x^2 + y^2 + z^2 = 49$, rectangular equation

$$\rho^2 = 49$$
$$\rho = 7, \quad \text{spherical equation}$$

44. $x^2 + y^2 - 3z^2 = 0$, rectangular equation

$$x^2 + y^2 + z^2 = 4z^2$$
$$\rho^2 = 4\rho^2\cos^2\phi$$
$$1 = 4\cos^2\phi$$
$$\cos\phi = \frac{1}{2}$$
$$\phi = \frac{\pi}{3}, \text{ (cone) spherical equation}$$

45. $x^2 + y^2 = 16$, rectangular equation

$$\rho^2\sin^2\phi\sin^2\theta + \rho^2\sin^2\phi\cos^2\theta = 16$$
$$\rho^2\sin^2\phi\left(\sin^2\theta + \cos^2\theta\right) = 16$$
$$\rho^2\sin^2\phi = 16$$
$$\rho\sin\phi = 4$$
$$\rho = 4\csc\phi, \text{ spherical equation}$$

46. $x = 13$, rectangular equation

$$\rho\sin\phi\cos\theta = 13$$
$$\rho = 13\csc\phi\sec\theta, \text{ spherical equation}$$

47.
$$x^2 + y^2 = 2z^2, \text{ rectangular equation}$$
$$\rho^2\sin^2\phi\cos^2\theta + \rho^2\sin^2\phi\sin^2\theta = 2\rho^2\cos^2\phi$$
$$\rho^2\sin^2\phi\left[\cos^2\theta + \sin^2\theta\right] = 2\rho^2\cos^2\phi$$
$$\rho^2\sin^2\phi = 2\rho^2\cos^2\theta$$
$$\frac{\sin^2\phi}{\cos^2\phi} = 2$$
$$\tan^2\phi = 2$$
$$\tan\phi = \pm\sqrt{2}, \text{ spherical equation}$$

48. $x^2 + y^2 + z^2 - 9z = 0$, rectangular equation

$$\rho^2 - 9\rho\cos\phi = 0$$
$$\rho = 9\cos\phi, \text{ spherical equation}$$

49. $\rho = 5$

$x^2 + y^2 + z^2 = 25$

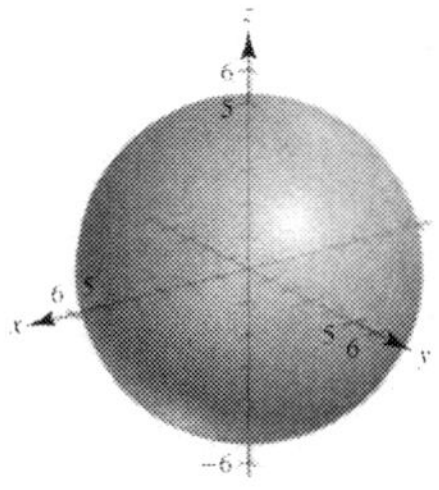

50. $\theta = \dfrac{3\pi}{4}$

$\tan\theta = \dfrac{y}{x}$

$-1 = \dfrac{y}{x}$

$x + y = 0$

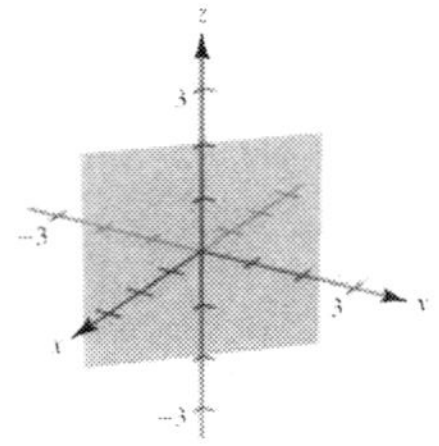

51. $\phi = \dfrac{\pi}{6}$

$\cos\phi = \dfrac{z}{\sqrt{x^2 + y^2 + z^2}}$

$\dfrac{\sqrt{3}}{2} = \dfrac{z}{\sqrt{x^2 + y^2 + z^2}}$

$\dfrac{3}{4} = \dfrac{z^2}{x^2 + y^2 + z^2}$

$3x^2 + 3y^2 - z^2 = 0, z \ge 0$

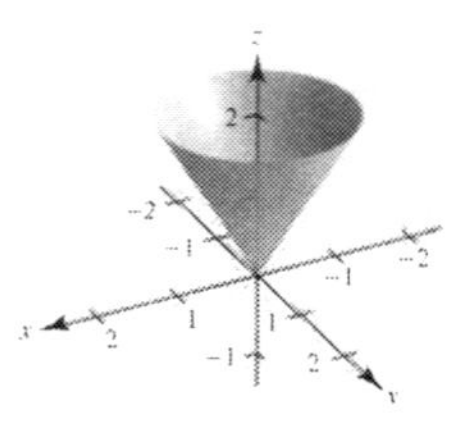

52. $\phi = \dfrac{\pi}{2}$

$\cos\phi = \dfrac{z}{\sqrt{x^2 + y^2 + z^2}}$

$0 = \dfrac{z}{\sqrt{x^2 + y^2 + z^2}}$

$z = 0$

xy-plane

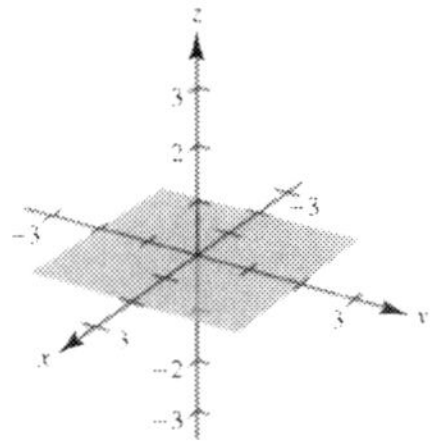

53. $\rho = 4\cos\phi$

$\sqrt{x^2 + y^2 + z^2} = \dfrac{4z}{\sqrt{x^2 + y^2 + z^2}}$

$x^2 + y^2 + z^2 - 4z = 0$

$x^2 + y^2 + (z - 2)^2 = 4, z \ge 0$

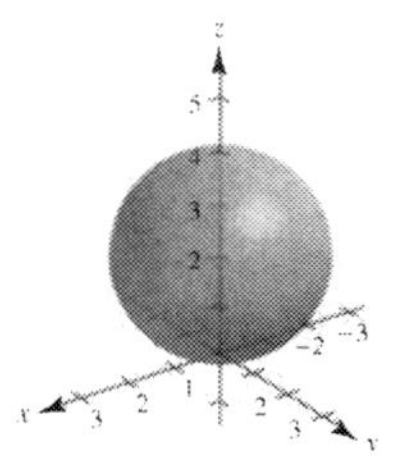

54. $\rho = 2\sec\phi$

$\rho\cos\phi = 2$

$z = 2$

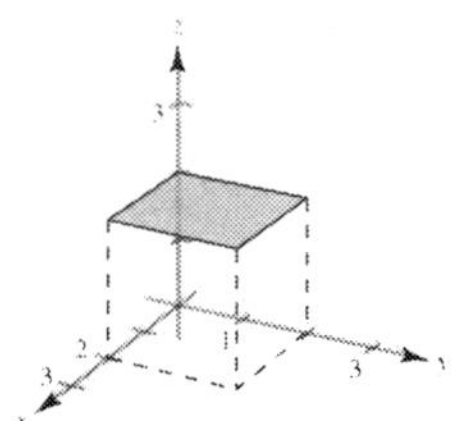

55. $\rho = \csc\phi$

$\rho\sin\phi = 1$

$\sqrt{x^2 + y^2} = 1$

$x^2 + y^2 = 1$

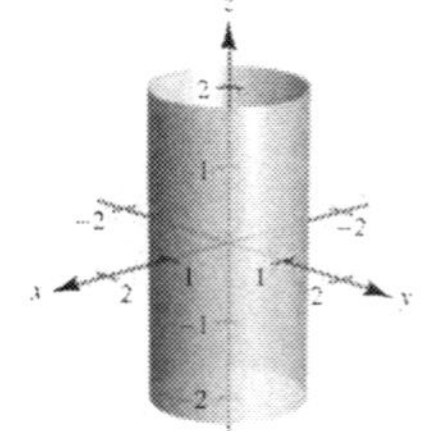

56. $\rho = 4 \csc \phi \sec \phi$

$= \dfrac{4}{\sin \phi \cos \theta}$

$\rho \sin \phi \cos \theta = 4$

$x = 4$

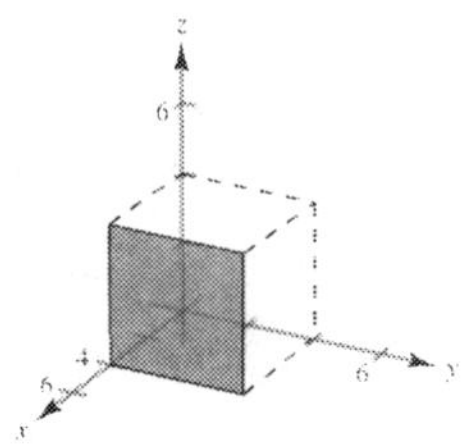

57. $r = 5$

Cylinder

Matches graph (d)

58. $\theta = \dfrac{\pi}{4}$

Plane

Matches graph (e)

59. $\rho = 5$

Sphere

Matches graph (c)

60. $\phi = \dfrac{\pi}{4}$

Cone

Matches graph (a)

61. $r^2 = z, x^2 + y^2 = z$

Paraboloid

Matches graph (f)

62. $\rho = 4 \sec \phi, z = \rho \cos \phi = 4$

Plane

Matches graph (b)

63. $\left(4, \dfrac{\pi}{4}, 0\right)$, cylindrical

$\rho = \sqrt{4^2 + 0^2} = 4$

$\theta = \dfrac{\pi}{4}$

$\phi = \arccos 0 = \dfrac{\pi}{2}$

$\left(4, \dfrac{\pi}{4}, \dfrac{\pi}{2}\right)$, spherical

64. $\left(3, -\dfrac{\pi}{4}, 0\right)$, cylindrical

$\rho = \sqrt{3^2 + 0^2} = 3$

$\theta = -\dfrac{\pi}{4}$

$\phi = \arccos\left(\dfrac{0}{9}\right) = \dfrac{\pi}{2}$

$\left(3, -\dfrac{\pi}{4}, \dfrac{\pi}{2}\right)$, spherical

65. $\left(4, \dfrac{\pi}{2}, 4\right)$, cylindrical

$\rho = \sqrt{4^2 + 4^2} = 4\sqrt{2}$

$\theta = \dfrac{\pi}{2}$

$\phi = \arccos\left(\dfrac{4}{4\sqrt{2}}\right) = \dfrac{\pi}{4}$

$\left(4\sqrt{2}, \dfrac{\pi}{2}, \dfrac{\pi}{4}\right)$, spherical

66. $\left(2, \dfrac{2\pi}{3}, -2\right)$, cylindrical

$\rho = \sqrt{2^2 + (-2)^2} = 2\sqrt{2}$

$\theta = \dfrac{2\pi}{3}$

$\phi = \arccos\left(\dfrac{-1}{\sqrt{2}}\right) = \dfrac{3\pi}{4}$

$\left(2\sqrt{2}, \dfrac{2\pi}{3}, \dfrac{3\pi}{4}\right)$, spherical

67. $\left(4, -\dfrac{\pi}{6} -\dfrac{\pi}{6}, 6\right)$, cylindrical

$\rho = \sqrt{4^2 + 6^2} = 2\sqrt{13}$

$\theta = -\dfrac{\pi}{6}$

$\phi = \arccos \dfrac{3}{\sqrt{13}}$

$\left(2\sqrt{13}, -\dfrac{\pi}{6}, \arccos \dfrac{3}{\sqrt{13}}\right)$, spherical

68. $\left(-4, \frac{\pi}{3}, 4\right)$, cylindrical

$$\rho = \sqrt{(-4)^2 + 4^2} = 4\sqrt{2}$$

$$\theta = \frac{\pi}{3}$$

$$\phi = \arccos\frac{1}{\sqrt{2}} = \frac{\pi}{4}$$

$\left(4\sqrt{2}, \frac{\pi}{3}, \frac{\pi}{4}\right)$, spherical

69. $(12, \pi, 5)$, cylindrical

$$\rho = \sqrt{12^2 + 5^2} = 13$$

$$\theta = \pi$$

$$\phi = \arccos\frac{5}{13}$$

$\left(13, \pi, \arccos\frac{5}{13}\right)$, spherical

70. $\left(4, \frac{\pi}{2}, 3\right)$, cylindrical

$$\rho = \sqrt{4^2 + 3^2} = 5$$

$$\theta = \frac{\pi}{2}$$

$$\phi = \arccos\frac{3}{5}$$

$\left(5, \frac{\pi}{2}, \arccos\frac{3}{5}\right)$, spherical

71. $\left(10, \frac{\pi}{6}, \frac{\pi}{2}\right)$, spherical

$$r = 10\sin\frac{\pi}{2} = 10$$

$$\theta = \frac{\pi}{6}$$

$$z = 10\cos\frac{\pi}{2} = 0$$

$\left(10, \frac{\pi}{6}, 0\right)$, cylindrical

72. $\left(4, \frac{\pi}{18}, \frac{\pi}{2}\right)$, spherical

$$r = 4\sin\frac{\pi}{2} = 4$$

$$\theta = \frac{\pi}{18}$$

$$z = 4\cos\frac{\pi}{2} = 0$$

$\left(4, \frac{\pi}{18}, 0\right)$, cylindrical

73. $\left(36, \pi, \frac{\pi}{2}\right)$, spherical

$$r = \rho\sin\phi = 36\sin\frac{\pi}{2} = 36$$

$$\theta = \pi$$

$$z = \rho\cos\phi = 36\cos\frac{\pi}{2} = 0$$

$(36, \pi, 0)$, cylindrical

74. $\left(18, \frac{\pi}{3}, \frac{\pi}{3}\right)$, spherical

$$r = \rho\sin\phi = 18\sin\frac{\pi}{3} = 9$$

$$\theta = \frac{\pi}{3}$$

$$z = \rho\cos\phi = 18\cos\frac{\pi}{3} = 9\sqrt{3}$$

$\left(9, \frac{\pi}{3}, 9\sqrt{3}\right)$, cylindrical

75. $\left(6, -\frac{\pi}{6}, \frac{\pi}{3}\right)$, spherical

$$r = 6\sin\frac{\pi}{3} = 3\sqrt{3}$$

$$\theta = -\frac{\pi}{6}$$

$$z = 6\cos\frac{\pi}{3} = 3$$

$\left(3\sqrt{3}, -\frac{\pi}{6}, 3\right)$, cylindrical

76. $\left(5, -\frac{5\pi}{6}, \pi\right)$, spherical

$$r = 5\sin\pi = 0$$

$$\theta = -\frac{5\pi}{6}$$

$$z = 5\cos\pi = -5$$

$\left(0, -\frac{5\pi}{6}, -5\right)$, cylindrical

77. $\left(8, \frac{7\pi}{6}, \frac{\pi}{6}\right)$, spherical

$$r = 8\sin\frac{\pi}{6} = 4$$

$$\theta = \frac{7\pi}{6}$$

$$z = 8\cos\frac{\pi}{6} = \frac{8\sqrt{3}}{2}$$

$\left(4, \frac{7\pi}{6}, 4\sqrt{3}\right)$, cylindrical

78. $\left(7, \frac{\pi}{4}, \frac{3\pi}{4}\right)$, spherical

$r = 7 \sin \frac{3\pi}{4} = \frac{7\sqrt{2}}{2}$

$\theta = \frac{\pi}{4}$

$z = 7 \cos \frac{3\pi}{4} = -\frac{7\sqrt{2}}{2}$

$\left(\frac{7\sqrt{2}}{2}, \frac{\pi}{4}, -\frac{7\sqrt{2}}{2}\right)$, cylindrical

79. Rectangular to cylindrical: $r^2 = x^2 + y^2$

$\tan \theta = \frac{y}{x}$

$z = z$

Cylindrical to rectangular: $x = r \cos \theta$

$y = r \sin \theta$

$z = z$

80. $\theta = c$ is a half-plane because of the restriction $r \geq 0$.

81. Rectangular to spherical: $\rho^2 = x^2 + y^2 + z^2$

$\tan \theta = \frac{y}{x}$

$\phi = \arccos\left(\frac{z}{\sqrt{x^2 + y^2 + z^2}}\right)$

Spherical to rectangular: $x = \rho \sin \phi \cos \theta$

$y = \rho \sin \phi \sin \theta$

$z = \rho \cos \phi$

82. (a) The surface is a cone. The equation is (i)

$x^2 + y^2 = \frac{4}{9}z^2$.

In cylindrical coordinates, the equation is

$x^2 + y^2 = \frac{4}{9}z^2$

$r^2 = \frac{4}{9}z^2$

$r = \frac{2}{3}z$.

(b) The surface is a hyperboloid of one sheet. The equation is (ii) $x^2 + y^2 - z^2 = 2$.

In cylindrical coordinates, the equation is

$x^2 + y^2 - z^2 = 2$

$r^2 - z^2 = 2$

$r^2 = z^2 + 2$.

83. $x^2 + y^2 + z^2 = 25$

(a) $r^2 + z^2 = 25$

(b) $\rho^2 = 25 \Rightarrow \rho = 5$

84. $4(x^2 + y^2) = z^2$

(a) $4r^2 = z^2 \Rightarrow 2r = z$

(b) $4(\rho^2 \sin^2 \phi \cos^2 \theta + \rho^2 \sin^2 \phi \sin^2 \theta) = \rho^2 \cos^2 \phi$

$4 \sin^2 \phi = \cos^2 \phi,$

$\tan^2 \phi = \frac{1}{4},$

$\tan \phi = \frac{1}{2} \Rightarrow \phi = \arctan \frac{1}{2}$

85. $x^2 + y^2 + z^2 - 2z = 0$

(a) $r^2 + z^2 - 2z = 0 \Rightarrow r^2 + (z - 1)^2 = 1$

(b) $\rho^2 - 2\rho \cos \phi = 0$

$\rho(\rho - 2 \cos \phi) = 0$

$\rho = 2 \cos \phi$

86. $x^2 + y^2 = z$

(a) $r^2 = z$

(b) $\rho^2 \sin^2 \phi = \rho \cos \phi$

$\rho \sin^2 \phi = \cos \phi$

$\rho = \frac{\cos \phi}{\sin^2 \phi}$

$\rho = \csc \phi \cot \phi$

87. $x^2 + y^2 = 4y$

(a) $r^2 = 4r \sin \theta, r = 4 \sin \theta$

(b) $\rho^2 \sin^2 \phi = 4\rho \sin \phi \sin \theta$

$\rho \sin \phi(\rho \sin \phi - 4 \sin \theta) = 0$

$\rho = \frac{4 \sin \theta}{\sin \phi}$

$\rho = 4 \sin \theta \csc \phi$

88. $x^2 + y^2 = 36$

(a) $r^2 = 36 \Rightarrow r = 6$

(b) $\rho^2 \sin^2 \phi \cos^2 \theta + \rho^2 \sin^2 \phi \sin^2 \theta = 36$

$\rho^2 \sin^2 \phi = 36$

$\rho = 6 \csc \phi$

89. $x^2 - y^2 = 9$

(a) $r^2 \cos^2 \theta - r^2 \sin^2 \theta = 9$

$r^2 = \frac{9}{\cos^2 \theta - \sin^2 \theta}$

(b) $\rho^2 \sin^2 \phi \cos^2 \theta - \rho^2 \sin^2 \phi \sin^2 \theta = 9$

$\rho^2 \sin^2 \phi = \frac{9}{\cos^2 \theta - \sin^2 \theta}$

$\rho^2 = \frac{9 \csc^2 \phi}{\cos^2 \theta - \sin^2 \theta}$

90. $y = 4$

(a) $r \sin\theta = 4 \Rightarrow r = 4\csc\theta$

(b) $\rho \sin\phi \sin\theta = 4,$

$\rho = 4\csc\phi\csc\theta$

91. $0 \le \theta \le \dfrac{\pi}{2}$

$0 \le r \le 2$

$0 \le z \le 4$

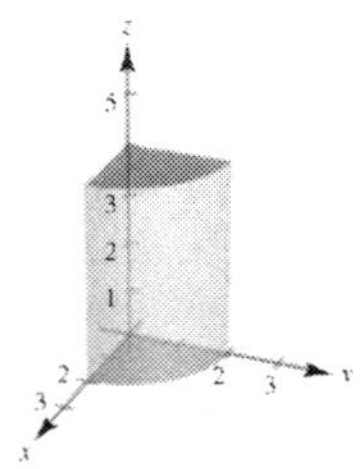

92. $-\dfrac{\pi}{2} \le \theta \le \dfrac{\pi}{2}$

$0 \le r \le 3$

$0 \le z \le r\cos\theta$

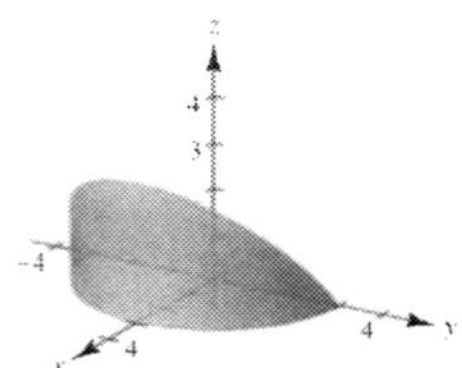

93. $0 \le \theta \le 2\pi$

$0 \le r \le a$

$r \le z \le a$

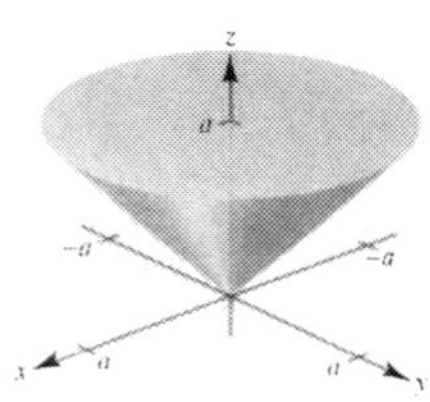

94. $0 \le \theta \le 2\pi$

$2 \le r \le 4$

$z^2 \le -r^2 + 6r - 8$

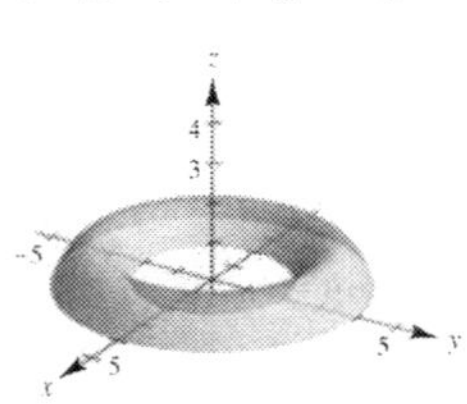

95. $0 \le \theta \le 2\pi$

$0 \le \phi \le \dfrac{\pi}{6}$

$0 \le \rho \le a\sec\phi$

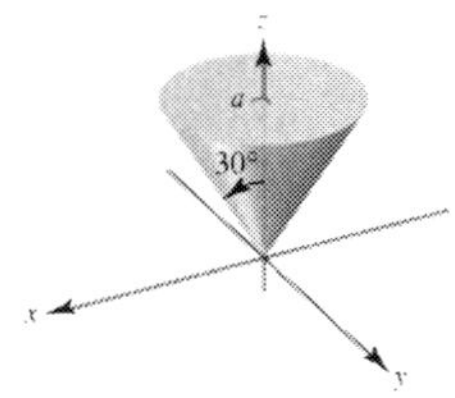

96. $0 \le \theta \le 2\pi$

$\dfrac{\pi}{4} \le \phi \le \dfrac{\pi}{2}$

$0 \le \rho \le 1$

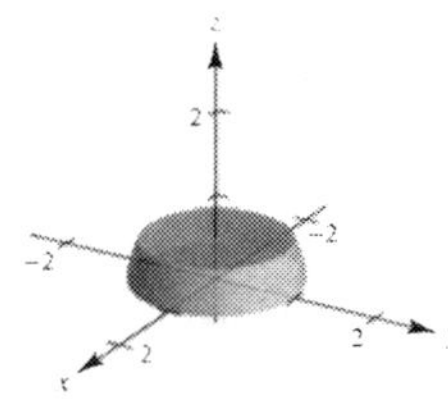

97. $0 \le \theta \le \dfrac{\pi}{2}$

$0 \le \phi \le \dfrac{\pi}{2}$

$0 \le \rho \le 2$

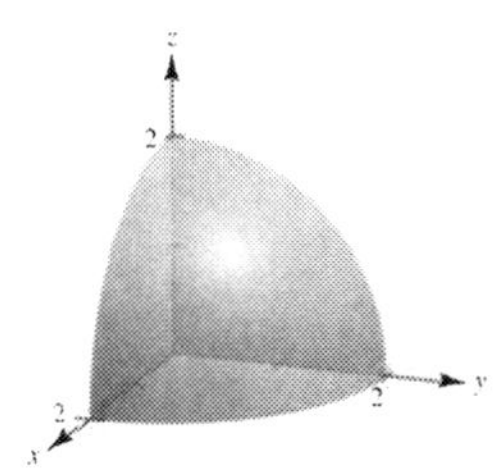

98. $0 \le \theta \le \pi$

$0 \le \phi \le \dfrac{\pi}{2}$

$1 \le \rho \le 3$

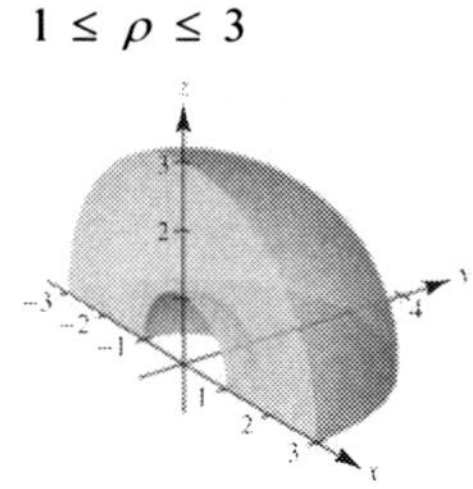

99. Rectangular

$0 \le x \le 10$

$0 \le y \le 10$

$0 \le z \le 10$

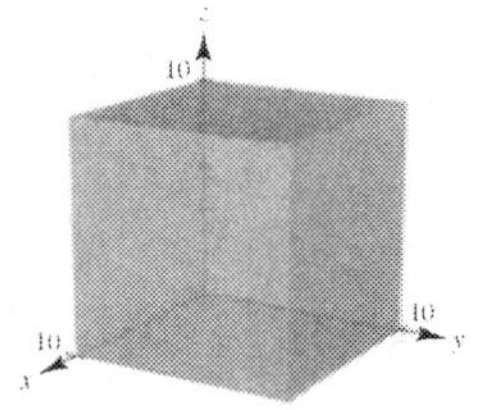

100. Cylindrical:

$0.75 \le r \le 1.25$

$0 \le z \le 8$

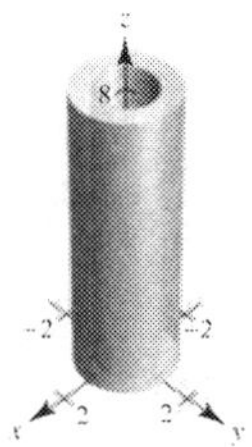

101. Spherical

$4 \le \rho \le 6$

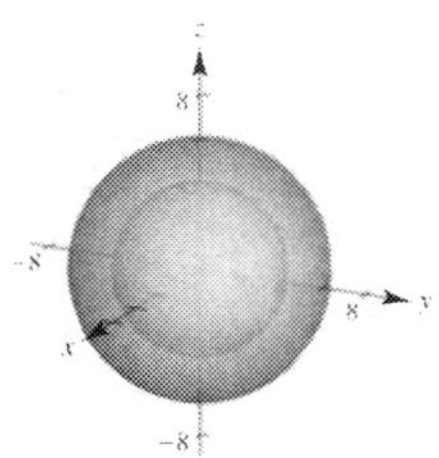

102. Cylindrical

$\frac{1}{2} \le r \le 3$

$0 \le \theta \le 2\pi$

$-\sqrt{9 - r^2} \le z \le \sqrt{9 - r^2}$

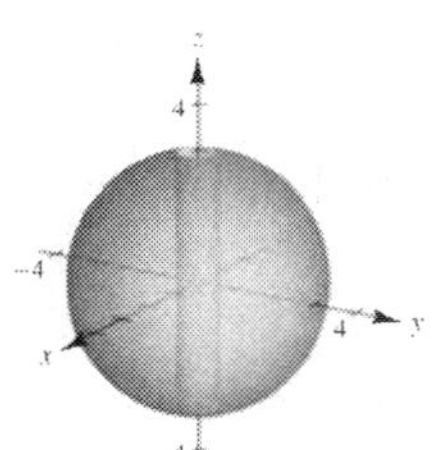

103. Cylindrical coordinates:

$r^2 + z^2 \le 9,$

$r \le 3\cos\theta, 0 \le \theta \le \pi$

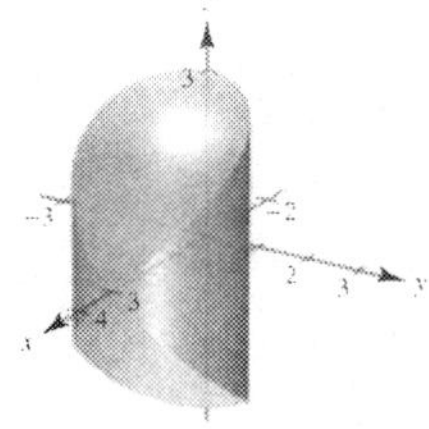

104. Spherical coordinates:

$\rho \ge 2$

$\rho \le 3$

$0 \le \phi \le \frac{\pi}{4}$

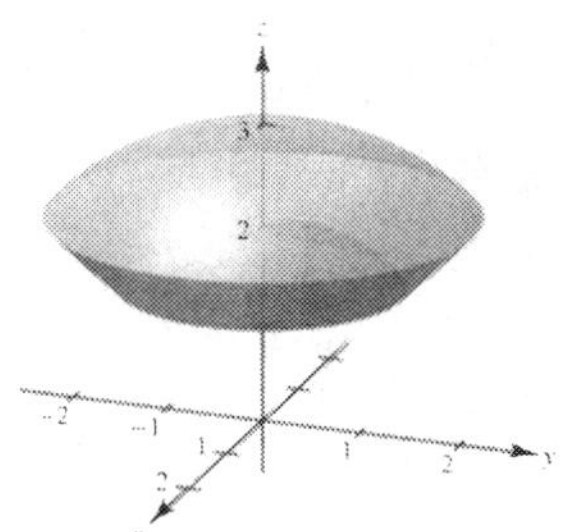

105. False. $r = z \Rightarrow x^2 + y^2 = z^2$ is a cone.

106. True. They both represent spheres of radius 2 centered at the origin.

107. False. $(r, \theta, z) = (0, 0, 1)$ and $(r, \theta, z) = (0, \pi, 1)$ represent the same point $(x, y, z) = (0, 0, 1)$.

108. True (except for the origin).

109. $z = \sin\theta, r = 1$

$$z = \sin\theta = \frac{y}{r} = \frac{y}{1} = y$$

The curve of intersection is the ellipse formed by the intersection of the plane $z = y$ and the cylinder $r = 1$.

110. $\rho = 2\sec\phi \Rightarrow \rho\cos\phi = 2 \Rightarrow z = 2$ plane

$\rho = 4$ sphere

The intersection of the plane and the sphere is a circle.

Review Exercises for Chapter 11

1. $P = (1, 2), Q = (4, 1), R = (5, 4)$

(a) $\mathbf{u} = \overline{PQ} = \langle 4 - 1, 1 - 2\rangle = \langle 3, -1\rangle$

$\mathbf{v} = \overline{PR} = \langle 5 - 1, 4 - 2\rangle = \langle 4, 2\rangle$

(b) $\mathbf{u} = 3\mathbf{i} - \mathbf{j}, \mathbf{v} = 4\mathbf{i} + 2\mathbf{j}$

(c) $\|\mathbf{u}\| = \sqrt{3^2 + (-1)^2} = \sqrt{10}$ $\|\mathbf{v}\| = \sqrt{4^2 + 2^2} = \sqrt{20} = 2\sqrt{5}$

(d) $2\mathbf{u} + \mathbf{v} = 2\langle 3, -1\rangle + \langle 4, 2\rangle = \langle 10, 0\rangle$

2. $P = (-2, -1), Q = (5, -1), R = (2, 4)$

(a) $\mathbf{u} = \overline{PQ} = \langle 5 - (-2), -1 - (-1)\rangle = \langle 7, 0\rangle$

$\mathbf{v} = \overline{PR} = \langle 2 - (-2), 4 - (-1)\rangle = \langle 4, 5\rangle$

(b) $\mathbf{u} = 7\mathbf{i}, \mathbf{v} = 4\mathbf{i} + 5\mathbf{j}$

(c) $\|\mathbf{u}\| = \sqrt{7^2 + 0^2} = \sqrt{49} = 7$ $\|\mathbf{v}\| = \sqrt{4^2 + 5^2} = \sqrt{41}$

(d) $2\mathbf{u} + \mathbf{v} = 2\langle 7, 0\rangle + \langle 4, 5\rangle = \langle 18, 5\rangle$

3. $\mathbf{v} = \|\mathbf{v}\|(\cos\theta\,\mathbf{i} + \sin\theta\,\mathbf{j})$

$= 8(\cos 60°\,\mathbf{i} + \sin 60°\,\mathbf{j})$

$= 8\left(\frac{1}{2}\mathbf{i} + \frac{\sqrt{3}}{2}\mathbf{j}\right) = 4\mathbf{i} + 4\sqrt{3}\,\mathbf{j} = \langle 4, 4\sqrt{3}\rangle$

4. $\mathbf{v} = \|\mathbf{v}\|\cos\theta\,\mathbf{i} + \|\mathbf{v}\|\sin\theta\,\mathbf{j}$

$= \frac{1}{2}\cos 225°\mathbf{i} + \frac{1}{2}\sin 225°\mathbf{j}$

$= -\frac{\sqrt{2}}{4}\mathbf{i} - \frac{\sqrt{2}}{4}\mathbf{j} = \left\langle -\frac{\sqrt{2}}{4}, -\frac{\sqrt{2}}{4}\right\rangle$

5. $z = 0, y = 4, x = -5: (-5, 4, 0)$

6. $x = z = 0,\ y = -7: (0, -7, 0)$

7. $d = \sqrt{(-2 - 1)^2 + (3 - 6)^2 + (5 - 3)^2}$

$= \sqrt{9 + 9 + 4} = \sqrt{22}$

8. $d = \sqrt{(4 - (-2))^2 + (-1 - 1)^2 + (-1 - (-5))^2}$

$= \sqrt{36 + 4 + 16} = \sqrt{56} = 2\sqrt{14}$

9. $(x - 3)^2 + (y + 2)^2 + (z - 6)^2 = \left(\frac{15}{2}\right)^2$

10. Center: $\left(\frac{0 + 4}{2}\right), \left(\frac{0 + 6}{2}\right), \left(\frac{4 + 0}{2}\right) = (2, 3, 2)$

Radius:

$\sqrt{(2 - 0)^2 + (3 - 0)^2 + (2 - 4)^2} = \sqrt{4 + 9 + 4} = \sqrt{17}$

$(x - 2)^2 + (y - 3)^2 + (z - 2)^2 = 17$

11. $(x^2 - 4x + 4) + (y^2 - 6y + 9) + z^2 = -4 + 4 + 9$

$(x - 2)^2 + (y - 3)^3 + z^2 = 9$

Center: $(2, 3, 0)$

Radius: 3

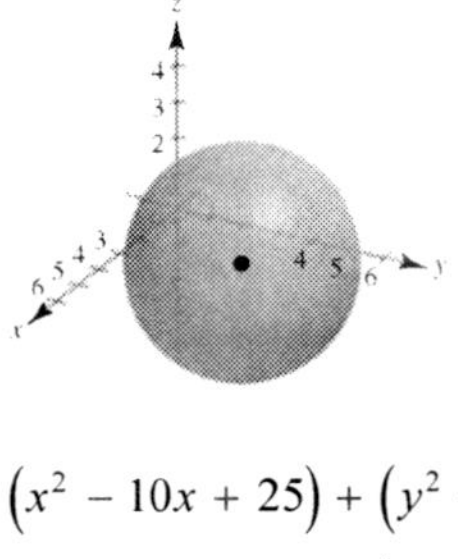

12. $(x^2 - 10x + 25) + (y^2 + 6y + 9)$

$+ (z^2 - 4z + 4) = -34 + 25 + 9 + 4$

$(x - 5)^2 + (y + 3)^2 + (z - 2)^2 = 4$

Center: $(5, -3, 2)$

Radius: 2

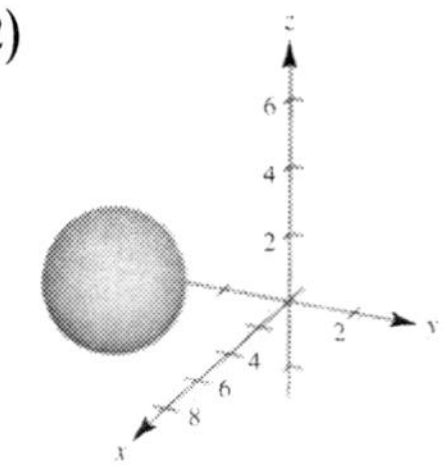

13. (a), (d)

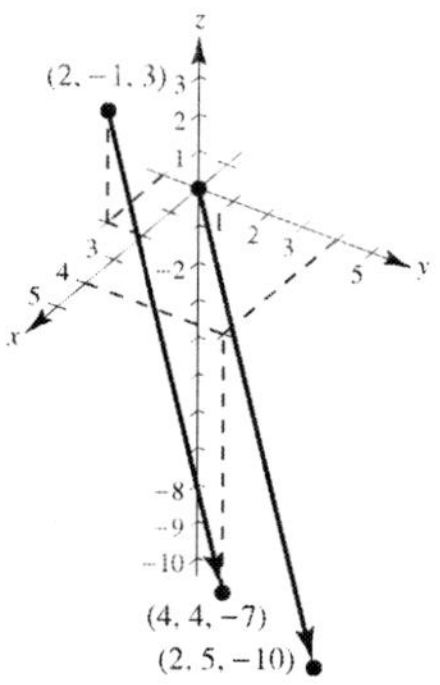

(b) $\mathbf{v} = \langle 4 - 2, 4 - (-1), -7 - 3 \rangle = \langle 2, 5, -10 \rangle$

(c) $\mathbf{v} = 2\mathbf{i} + 5\mathbf{j} - 10\mathbf{k}$

14. (a), (d)

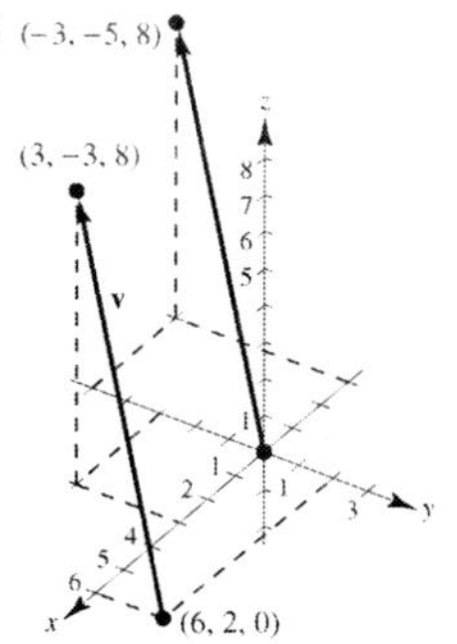

(b) $\mathbf{v} = \langle 3 - 6, -3 - 2, 8 - 0 \rangle = \langle -3, -5, 8 \rangle$

(c) $\mathbf{v} = -3\mathbf{i} - 5\mathbf{j} + 8\mathbf{k}$

15. $\mathbf{v} = \langle -1 - 3, 6 - 4, 9 + 1 \rangle = \langle -4, 2, 10 \rangle$

$\mathbf{w} = \langle 5 - 3, 3 - 4, -6 + 1 \rangle = \langle 2, -1, -5 \rangle$

Because $-2\mathbf{w} = \mathbf{v}$, the points lie in a straight line.

16. $\mathbf{v} = \langle 8 - 5, -5 + 4, 5 - 7 \rangle = \langle 3, -1, -2 \rangle$

$\mathbf{w} = \langle 11 - 5, 6 + 4, 3 - 7 \rangle = \langle 6, 10, -4 \rangle$

Because $\mathbf{v}$ and $\mathbf{w}$ are not parallel, the points do not lie in a straight line.

17. Unit vector: $\dfrac{\mathbf{u}}{\|\mathbf{u}\|} = \left\langle \dfrac{2, 3, 5}{\sqrt{38}} \right\rangle = \left\langle \dfrac{2}{\sqrt{38}}, \dfrac{3}{\sqrt{38}}, \dfrac{5}{\sqrt{38}} \right\rangle$

18. $8\dfrac{\langle 6, -3, 2 \rangle}{\sqrt{49}} = \dfrac{8}{7}\langle 6, -3, 2 \rangle = \left\langle \dfrac{48}{7}, -\dfrac{24}{7}, \dfrac{16}{7} \right\rangle$

19. $P = \langle 5, 0, 0 \rangle, Q = \langle 4, 4, 0 \rangle, R = \langle 2, 0, 6 \rangle$

(a) $\mathbf{u} = \overrightarrow{PQ} = \langle -1, 4, 0 \rangle$

$\mathbf{v} = \overrightarrow{PR} = \langle -3, 0, 6 \rangle$

(b) $\mathbf{u} \cdot \mathbf{v} = (-1)(-3) + 4(0) + 0(6) = 3$

(c) $\mathbf{v} \cdot \mathbf{v} = 9 + 36 = 45$

20. $P = \langle 2, -1, 3 \rangle, Q = \langle 0, 5, 1 \rangle, R = \langle 5, 5, 0 \rangle$

(a) $\mathbf{u} = \overrightarrow{PQ} = \langle -2, 6, -2 \rangle$

$\mathbf{v} = \overrightarrow{PR} = \langle 3, 6, -3 \rangle$

(b) $\mathbf{u} \cdot \mathbf{v} = (-2)(3) + (6)(6) + (-2)(-3) = 36$

(c) $\mathbf{v} \cdot \mathbf{v} = 9 + 36 + 9 = 54$

21. $\mathbf{u} = 5\left(\cos\dfrac{3\pi}{4}\mathbf{i} + \sin\dfrac{3\pi}{4}\mathbf{j}\right) = \dfrac{5\sqrt{2}}{2}[-\mathbf{i} + \mathbf{j}]$

$\mathbf{v} = 2\left(\cos\dfrac{2\pi}{3}\mathbf{i} + \sin\dfrac{2\pi}{3}\mathbf{j}\right) = -\mathbf{i} + \sqrt{3}\,\mathbf{j}$

$\mathbf{u} \cdot \mathbf{v} = \dfrac{5\sqrt{2}}{2}\left(1 + \sqrt{3}\right)$

$\|\mathbf{u}\| = \sqrt{\dfrac{25}{2} + \dfrac{25}{2}} = 5 \qquad \|\mathbf{v}\| = \sqrt{1 + 3} = 2$

$\cos\theta = \dfrac{|\mathbf{u} \cdot \mathbf{v}|}{\|\mathbf{u}\|\|\mathbf{v}\|} = \dfrac{\left(5\sqrt{2}/2\right)\left(1 + \sqrt{3}\right)}{5(2)} = \dfrac{\sqrt{2} + \sqrt{6}}{4}$

(a) $\theta = \arccos\dfrac{\sqrt{2} + \sqrt{6}}{4} = \dfrac{\pi}{12} \approx 0.262$

(b) $\theta \approx 15°$

22. $\mathbf{u} = 6\mathbf{i} + 2\mathbf{j} - 3\mathbf{k}$

$\mathbf{v} = -\mathbf{i} + 5\mathbf{j}$

$\mathbf{u} \cdot \mathbf{v} = 6(-1) + 2(5) + (-3)0 = 4$

$\|\mathbf{u}\| = \sqrt{36 + 4 + 9} = \sqrt{49} = 7$

$\|\mathbf{v}\| = \sqrt{1 + 25} = \sqrt{26}$

$$\cos\theta = \frac{|\mathbf{u} \cdot \mathbf{u}|}{\|\mathbf{u}\|\|\mathbf{v}\|} = \frac{4}{7\sqrt{26}}$$

(a) $\theta = \text{arcos}\left(\dfrac{4}{7\sqrt{26}}\right) \approx 1.458$

(b) $\theta \approx 83.6°$

23. $\mathbf{u} = \langle 10, -5, 15\rangle$, $\mathbf{v} = \langle -2, 1, -3\rangle$

$\mathbf{u} = -5\mathbf{v} \Rightarrow \mathbf{u}$ is parallel to $\mathbf{v}$ and in the opposite direction.

(a) $\theta = \pi$ (b) $\theta = 180°$

24. $\mathbf{u} = \langle 1, 0, -3\rangle$

$\mathbf{v} = (2, -2, 1)$

$\mathbf{u} \cdot \mathbf{v} = -1$

$\|\mathbf{u}\| = \sqrt{1 + 9} = \sqrt{10}$

$\|\mathbf{v}\| = \sqrt{4 + 4 + 1} = 3$

$$\cos\theta = \frac{|\mathbf{u} \cdot \mathbf{v}|}{\|\mathbf{u}\|\|\mathbf{v}\|} = \frac{1}{3\sqrt{10}}$$

(a) $\theta = \arccos\left(\dfrac{1}{3\sqrt{10}}\right) \approx 1.465$

(b) $\theta = 83.9°$

25. $\mathbf{u} = \langle 7, -2, 3\rangle$, $\mathbf{v} = \langle -1, 4, 5\rangle$

Because $\mathbf{u} \cdot \mathbf{v} = 0$, the vectors are orthogonal.

26. $\mathbf{u} = \langle -4, 3, -6\rangle$, $\mathbf{v} = \langle 16, -12, 24\rangle$

Because $\mathbf{v} = -4\mathbf{u}$, the vectors are parallel.

27. $\mathbf{u} = \langle 7, 9\rangle$, $\mathbf{v} = \langle 1, 5\rangle$

$$\text{proj}_{\mathbf{v}}\,\mathbf{u} = \frac{\mathbf{u} \cdot \mathbf{v}}{\|\mathbf{v}\|^2}\mathbf{v} = \frac{7 + 45}{\left(\sqrt{26}\right)^2}\langle 1, 5\rangle = \frac{52}{26}\langle 1, 5\rangle = \langle 2, 10\rangle$$

28. $\mathbf{u} = 4\mathbf{i} + 2\mathbf{j}$, $\mathbf{v} = 3\mathbf{i} + 4\mathbf{j}$

$$\text{proj}_{\mathbf{v}}\,\mathbf{u} = \frac{\mathbf{u} \cdot \mathbf{v}}{\|\mathbf{v}\|^2}\mathbf{v} = \frac{12 + 8}{\left(\sqrt{9 + 16}\right)^2}(3\mathbf{i} + 4\mathbf{j}) = \frac{20}{25}(3\mathbf{i} + 4\mathbf{j}) = \frac{12}{5}\mathbf{i} + \frac{16}{5}\mathbf{j}$$

29. $\mathbf{u} = \langle 1, -1, 1\rangle$, $\mathbf{v} = \langle 2, 0, 2\rangle$

$$\text{proj}_{\mathbf{v}}\,\mathbf{u} = \frac{\mathbf{u} \cdot \mathbf{v}}{\|\mathbf{v}\|^2}\mathbf{v} = \frac{2 + 2}{\left(\sqrt{4 + 4}\right)^2}\langle 2, 0, 2\rangle = \frac{1}{2}\langle 2, 0, 2\rangle = \langle 1, 0, 1\rangle$$

30. $\mathbf{u} = 5\mathbf{i} + \mathbf{j} + 3\mathbf{k}$, $\mathbf{v} = 2\mathbf{i} + 3\mathbf{j} + \mathbf{k}$

$$\text{proj}_{\mathbf{v}}\,\mathbf{u} = \frac{\mathbf{u} \cdot \mathbf{v}}{\|\mathbf{v}\|^2}\mathbf{v} = \frac{10 + 3 + 3}{\left(\sqrt{14}\right)^2}(2\mathbf{i} + 3\mathbf{j} + \mathbf{k}) = \frac{16}{14}(2\mathbf{i} + 3\mathbf{j} + \mathbf{k}) = \frac{16}{7}\mathbf{i} + \frac{24}{7}\mathbf{j} + \frac{8}{7}\mathbf{k}$$

31. There are many correct answers.

For example: $\mathbf{v} = \pm\langle 6, -5, 0\rangle$.

32. $W = \mathbf{F} \cdot \overline{PQ} = \|\mathbf{F}\|\|\overline{PQ}\|\cos\theta = (75)(8)\cos 30°$

$= 300\sqrt{3}$ ft-lb

33. (a) $\mathbf{u} \times \mathbf{v} = \begin{bmatrix} \mathbf{i} & \mathbf{j} & \mathbf{k} \\ 4 & 3 & 6 \\ 5 & 2 & 1 \end{bmatrix} = -9\mathbf{i} + 26\mathbf{j} - 7\mathbf{k}$

(b) $\mathbf{v} \times \mathbf{u} = -(\mathbf{u} \times \mathbf{v}) = 9\mathbf{i} - 26\mathbf{j} + 7\mathbf{k}$

(c) $\mathbf{v} \times \mathbf{v} = \mathbf{0}$

34. (a) $\mathbf{u} \times \mathbf{v} = \begin{bmatrix} \mathbf{i} & \mathbf{j} & \mathbf{k} \\ 6 & -5 & 2 \\ -4 & 2 & 3 \end{bmatrix} = -19\mathbf{i} - 26\mathbf{j} - 8\mathbf{k}$

(b) $\mathbf{v} \times \mathbf{u} = -(\mathbf{u} \times \mathbf{v}) = 19\mathbf{i} + 26\mathbf{j} + 8\mathbf{k}$

(c) $\mathbf{v} \times \mathbf{v} = \mathbf{0}$

35. (a) $\mathbf{u} \times \mathbf{v} = \begin{bmatrix} \mathbf{i} & \mathbf{j} & \mathbf{k} \\ 2 & -4 & -4 \\ 1 & 1 & 3 \end{bmatrix} = -8\mathbf{i} - 10\mathbf{j} + 6\mathbf{k}$

(b) $\mathbf{v} \times \mathbf{u} = -(\mathbf{u} \times \mathbf{v}) = 8\mathbf{i} + 10\mathbf{j} - 6\mathbf{k}$

(c) $\mathbf{v} \times \mathbf{v} = \mathbf{0}$

36. (a) $\mathbf{u} \times \mathbf{v} = \begin{bmatrix} \mathbf{i} & \mathbf{j} & \mathbf{k} \\ 0 & 2 & 1 \\ 1 & -3 & 4 \end{bmatrix} = 11\mathbf{i} + \mathbf{j} - 2\mathbf{k}$

(b) $\mathbf{v} \times \mathbf{u} = -(\mathbf{u} \times \mathbf{v}) = -11\mathbf{i} - \mathbf{j} + 2\mathbf{k}$

(c) $\mathbf{v} \times \mathbf{v} = \mathbf{0}$

37. $\mathbf{u} \times \mathbf{v} = \begin{bmatrix} \mathbf{i} & \mathbf{j} & \mathbf{k} \\ 2 & -10 & 8 \\ 4 & 6 & -8 \end{bmatrix} = 32\mathbf{i} + 48\mathbf{j} + 52\mathbf{k}$

$\|\mathbf{u} \times \mathbf{v}\| = \sqrt{6032} = 4\sqrt{377}$

Unit vector: $\dfrac{1}{\sqrt{377}}\langle 8, 12, 13\rangle$

38. $\mathbf{u} = \langle 3, -1, 5\rangle, \mathbf{v} = \langle 2, -4, 1\rangle$

$\mathbf{u} \times \mathbf{v} = \begin{bmatrix} \mathbf{i} & \mathbf{j} & \mathbf{k} \\ 3 & -1 & 5 \\ 2 & -4 & 1 \end{bmatrix} = 19\mathbf{i} + 7\mathbf{j} - 10\mathbf{k}$

$A = \|\mathbf{u} \times \mathbf{v}\| = \sqrt{19^2 + 7^2 + (-10)^2}$

$= \sqrt{510}$

39. $\mathbf{F} = c(\cos 20°\mathbf{j} + \sin 20°\mathbf{k})$

$\overrightarrow{PQ} = 2\mathbf{k}$

$\overrightarrow{PQ} \times \mathbf{F} = \begin{vmatrix} \mathbf{i} & \mathbf{j} & \mathbf{k} \\ 0 & 0 & 2 \\ 0 & c\cos 20° & c\sin 20° \end{vmatrix} = -2c\cos 20°\mathbf{i}$

$200 = \left\|\overrightarrow{PQ} \times \mathbf{F}\right\| = 2c\cos 20°$

$c = \dfrac{100}{\cos 20°}$

$\mathbf{F} = \dfrac{100}{\cos 20°}(\cos 20°\mathbf{j} + \sin 20°\mathbf{k}) = 100(\mathbf{j} + \tan 20°\mathbf{k})$

$\|\mathbf{F}\| = 100\sqrt{1 + \tan^2 20°} = 100\sec 20° \approx 106.4 \text{ lb}$

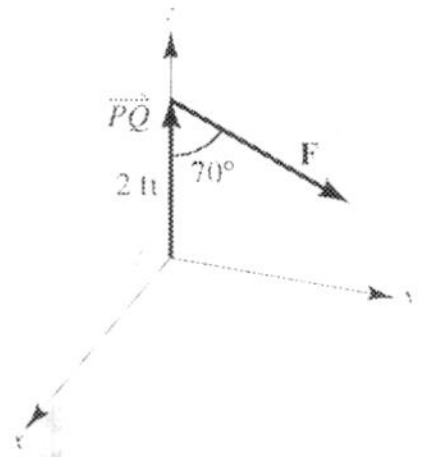

40. $V = |\mathbf{u} \cdot (\mathbf{v} \cdot \mathbf{w})| = \begin{bmatrix} 2 & 1 & 0 \\ 0 & 2 & 1 \\ 0 & -1 & 2 \end{bmatrix} = 2(5) = 10$

41. $\mathbf{v} = \langle 9 - 3, 11 - 0, 6 - 2\rangle = \langle 6, 11, 4\rangle$

(a) Parametric equations:
$x = 3 + 6t, y = 11t, z = 2 + 4t$

(b) Symmetric equations: $\dfrac{x - 3}{6} = \dfrac{y}{11} = \dfrac{z - 2}{4}$

42. $\mathbf{v} = \langle 8 + 1, 10 - 4, 5 - 3\rangle = \langle 9, 6, 2\rangle$

(a) Parametric equations:
$x = -1 + 9t, y = 4 + 6t, z = 3 + 2t$

(b) Symmetric equations: $\dfrac{x + 1}{9} = \dfrac{y - 4}{6} = \dfrac{z - 3}{2}$

43. $\mathbf{v} = \mathbf{j}, P(1, 2, 3)$

$x = 1, y = 2 + t, z = 3$

44. Direction numbers: $1, 1, 1, \mathbf{v} = \langle 1, 1, 1\rangle$

$P(1, 2, 3)$

$x = 1 + t, y = 2 + t, z = 3 + t$

45. $3x - 3y - 7z = -4, x - y + 2z = 3$

Solving simultaneously, you have $z = 1$. Substituting $z = 1$ into the second equation, you have $y = x - 1$. Substituting for x in this equation you obtain two points on the line of intersection, $(0, -1, 1)$, $(1, 0, 1)$. The direction vector of the line of intersection is $\mathbf{v} = \mathbf{i} + \mathbf{j}$.

$x = t, y = -1 + t, z = 1$

46. $$\mathbf{u} \times \mathbf{v} = \begin{vmatrix} \mathbf{i} & \mathbf{j} & \mathbf{k} \\ 2 & -5 & 1 \\ -3 & 1 & 4 \end{vmatrix} = -21\mathbf{i} - 11\mathbf{j} - 13\mathbf{k}$$

Direction numbers: 21, 11, 13

$x = 21t, y = 1 + 11t, z = 4 + 13t$

47. $P = (-3, -4, 2), Q = (-3, 4, 1), R = (1, 1, -2)$

$\overrightarrow{PQ} = \langle 0, 8, -1 \rangle, \overrightarrow{PR} = [4, 5, -4]$

$$\mathbf{n} = \overrightarrow{PQ} \times \overrightarrow{PR} = \begin{vmatrix} \mathbf{i} & \mathbf{j} & \mathbf{k} \\ 0 & 8 & -1 \\ 4 & 5 & -4 \end{vmatrix} = -27\mathbf{i} - 4\mathbf{j} - 32\mathbf{k}$$

$$-27(x + 3) - 4(y + 4) - 32(z - 2) = 0$$
$$27x + 4y + 32z = -33$$

48. $\mathbf{n} = 3\mathbf{i} - \mathbf{j} + \mathbf{k}$

$3(x + 2) - 1(y - 3) + 1(z - 1) = 0$

$3x - y + z + 8 = 0$

49. The two lines are parallel as they have the same direction numbers, $-2, 1, 1$. Therefore, a vector parallel to the plane is $\mathbf{v} = -2\mathbf{i} + \mathbf{j} + \mathbf{k}$. A point on the first line is $(1, 0, -1)$ and a point on the second line is $(-1, 1, 2)$. The vector $\mathbf{u} = 2\mathbf{i} - \mathbf{j} - 3\mathbf{k}$ connecting these two points is also parallel to the plane. Therefore, a normal to the plane is

$$\mathbf{v} \times \mathbf{u} = \begin{vmatrix} \mathbf{i} & \mathbf{j} & \mathbf{k} \\ -2 & 1 & 1 \\ 2 & -1 & -3 \end{vmatrix}$$
$$= -2\mathbf{i} - 4\mathbf{j} = -2(\mathbf{i} + 2\mathbf{j}).$$

Equation of the plane: $(x - 1) + 2y = 0$

$$x + 2y = 1$$

50. Let $\mathbf{v} = \langle 5 - 2, 1 + 2, 3 - 1 \rangle = \langle 3, 3, 2 \rangle$ be the direction vector for the line through the two points. Let $\mathbf{n} = \langle 2, 1, -1 \rangle$ be the normal vector to the plane. Then

$$\mathbf{v} \times \mathbf{n} = \begin{vmatrix} \mathbf{i} & \mathbf{j} & \mathbf{k} \\ 3 & 3 & 2 \\ 2 & 1 & -1 \end{vmatrix} = \langle -5, 7, -3 \rangle$$

is the normal to the unknown plane.

$-5(x - 5) + 7(y - 1) - 3(z - 3) = 0$

$-5x + 7y - 3z + 27 = 0$

51. $Q(1, 0, 2)$ point

$2x - 3y + 6z = 6$

A point P on the plane is $(3, 0, 0)$.

$\overrightarrow{PQ} = \langle -2, 0, 2 \rangle$

$\mathbf{n} = \langle 2, -3, 6 \rangle$ normal to plane

$$D = \frac{|\overrightarrow{PQ} \cdot \mathbf{n}|}{\|\mathbf{n}\|} = \frac{8}{7}$$

52. $Q(3, -2, 4)$ point

$2x - 5y + z = 10$

A point P on the plane is $(5, 0, 0)$.

$\overrightarrow{PQ} = \langle -2, -2, 4 \rangle$

$\mathbf{n} = \langle 2, -5, 1 \rangle$ normal to plane

$$D = \frac{|\overrightarrow{PQ} \cdot \mathbf{n}|}{\|\mathbf{n}\|} = \frac{10}{\sqrt{30}} = \frac{\sqrt{30}}{3}$$

53. The normal vectors to the planes are the same, $\mathbf{n} = (5, -3, 1)$.

Choose a point in the first plane $P(0, 0, 2)$. Choose a point in the second plane, $Q(0, 0, -3)$.

$\overrightarrow{PQ} = \langle 0, 0, -5 \rangle$

$$D = \frac{|\overrightarrow{PQ} \cdot \mathbf{n}|}{\|\mathbf{n}\|} = \frac{|-5|}{\sqrt{35}} = \frac{5}{\sqrt{35}} = \frac{\sqrt{35}}{7}$$

54. $Q(-5, 1, 3)$ point

$\mathbf{u} = \langle 1, -2, -1 \rangle$ direction vector

$P(1, 3, 5)$ point on line

$\overrightarrow{PQ} = \langle -6, -2, -2 \rangle$

$$\overrightarrow{PQ} \times \mathbf{u} = \begin{vmatrix} \mathbf{i} & \mathbf{j} & \mathbf{k} \\ -6 & -2 & -2 \\ 1 & -2 & -1 \end{vmatrix} = \langle -2, -8, 14 \rangle$$

$$D = \frac{\left\| \overrightarrow{PQ} \times \mathbf{u} \right\|}{\|\mathbf{u}\|} = \frac{\sqrt{264}}{\sqrt{6}} = 2\sqrt{11}$$

55. $x + 2y + 3z = 6$

Plane

Intercepts: $(6, 0, 0), (0, 3, 0), (0, 0, 2),$

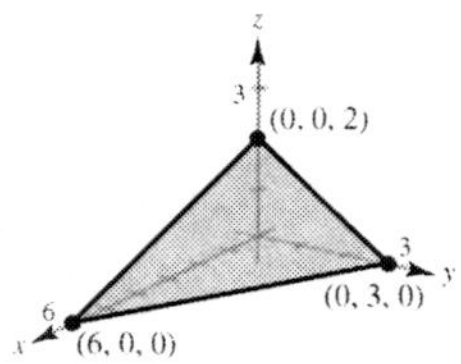

56. $y = z^2$

Because the x-coordinate is missing, you have a cylindrical surface with rulings parallel to the x-axis. The generating curve is a parabola in the yz-coordinate plane.

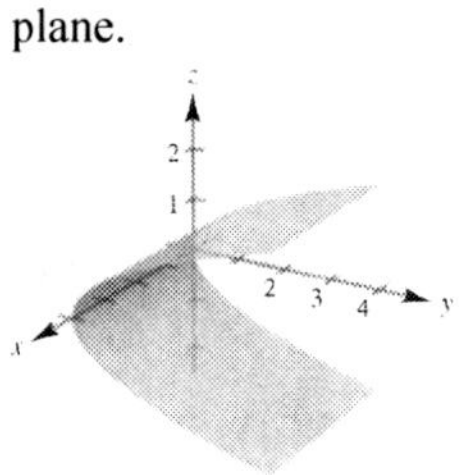

57. $y = \dfrac{1}{2}z$

Plane with rulings parallel to the x-axis.

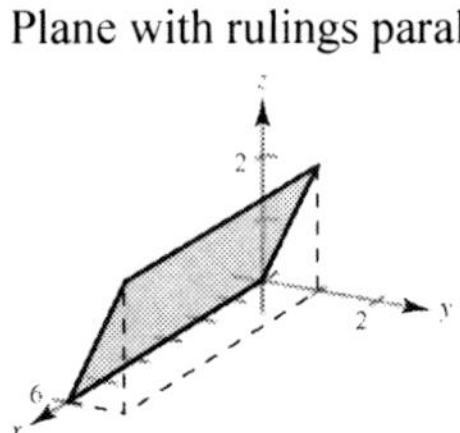

58. $y = \cos z$

Because the x-coordinate is missing, you have a cylindrical surface with rulings parallel to the x-axis. The generating curve is $y = \cos z$.

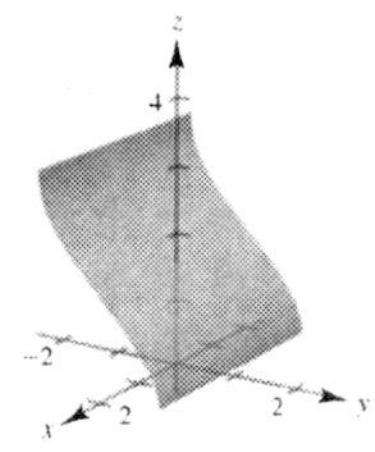

59. $\dfrac{x^2}{16} + \dfrac{y^2}{9} + z^2 = 1$

Ellipsoid

xy-trace: $\dfrac{x^2}{16} + \dfrac{y^2}{9} = 1$

xz-trace: $\dfrac{x^2}{16} + z^2 = 1$

yz-trace: $\dfrac{y^2}{9} + z^2 = 1$

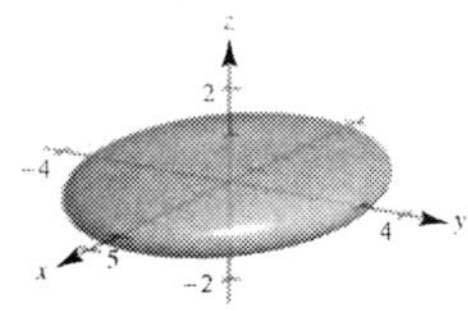

60. $16x^2 + 16y^2 - 9z^2 = 0$

Cone

xy-trace: point $(0, 0, 0)$

xz-trace: $z = \pm\dfrac{4x}{3}$

yz-trace: $z = \pm\dfrac{4y}{3}$

$z = 4, x^2 + y^2 = 9$

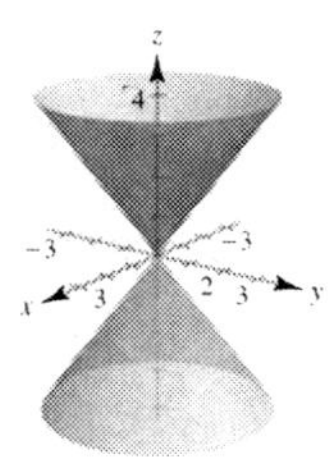

61. $\dfrac{x^2}{16} - \dfrac{y^2}{9} + z^2 = -1$

$\dfrac{y^2}{9} - \dfrac{x^2}{16} - z^2 = 1$

Hyperboloid of two sheets

xy-trace: $\dfrac{y^2}{9} - \dfrac{x^2}{16} = 1$

xz-trace: None

yz-trace: $\dfrac{y^2}{9} - z^2 = 1$

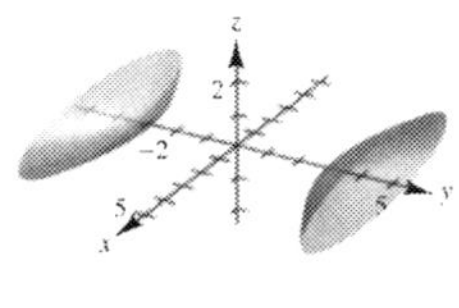

62. $\dfrac{x^2}{25} + \dfrac{y^2}{4} - \dfrac{z^2}{100} = 1$

Hyperboloid of one sheet

xy-trace: $\dfrac{x^2}{25} + \dfrac{y^2}{4} = 1$

xz-trace: $\dfrac{x^2}{25} - \dfrac{z^2}{100} = 1$

yz-trace: $\dfrac{y^2}{4} - \dfrac{z^2}{100} = 1$

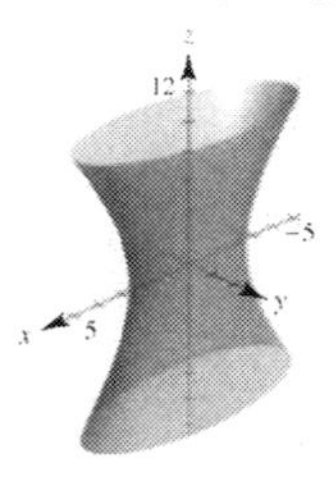

63. $x^2 + z^2 = 4.$

Cylinder of radius 2 about y-axis

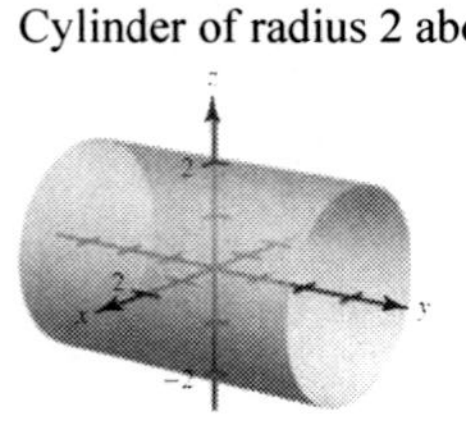

64. $y^2 + z^2 = 16.$

Cylinder of radius 4 about x-axis

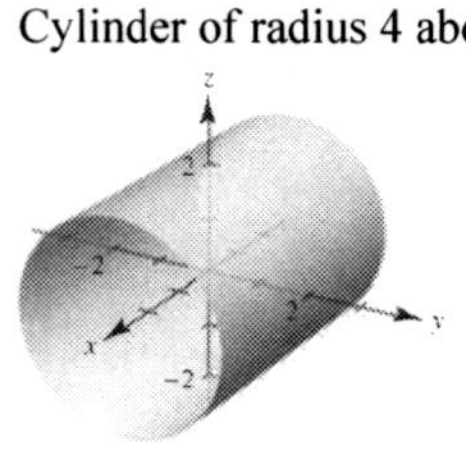

65. $z^2 = 2y$ revolved about y-axis

$z = \pm\sqrt{2y}$

$x^2 + z^2 = [r(y)]^2 = 2y$

$x^2 + z^2 = 2y$

66. $2x + 3z = 1$ revolved about the x-axis

$z = \dfrac{1-2x}{3}$

$y^2 + z^2 = [r(x)]^2 = \left(\dfrac{1-2x}{3}\right)^2$, Cone

67. $\left(-2\sqrt{2}, 2\sqrt{2}, 2\right)$, rectangular

(a) $r = \sqrt{\left(-2\sqrt{2}\right)^2 + \left(2\sqrt{2}\right)^2} = 4,$

$\theta = \arctan(-1) = \dfrac{3\pi}{4}, z = 2,$

$\left(4, \dfrac{3\pi}{4}, 2\right)$, cylindrical

(b) $\rho = \sqrt{\left(-2\sqrt{2}\right)^2 + \left(2\sqrt{2}\right)^2 + (2)^2} = 2\sqrt{5},$

$\theta = \dfrac{3\pi}{4}, \phi = \arccos\dfrac{2}{2\sqrt{5}} = \arccos\dfrac{1}{\sqrt{5}},$

$\left(2\sqrt{5}, \dfrac{3\pi}{4}, \arccos\dfrac{\sqrt{5}}{5}\right)$, spherical

68. $\left(\dfrac{\sqrt{3}}{4}, \dfrac{3}{4}, \dfrac{3\sqrt{3}}{2}\right)$, rectangular

(a) $r = \sqrt{\left(\dfrac{\sqrt{3}}{4}\right)^2 + \left(\dfrac{3}{4}\right)^2} = \dfrac{\sqrt{3}}{2},$

$\theta = \arctan\sqrt{3} = \dfrac{\pi}{3},$

$z = \dfrac{3\sqrt{3}}{2}, \left(\dfrac{\sqrt{3}}{2}, \dfrac{\pi}{2}, \dfrac{3\sqrt{3}}{2}\right)$, cylindrical

(b) $\rho = \sqrt{\left(\dfrac{\sqrt{3}}{4}\right)^2 + \left(\dfrac{3}{4}\right)^2 + \left(\dfrac{3\sqrt{3}}{2}\right)^2} = \dfrac{\sqrt{30}}{2}, \theta = \dfrac{\pi}{3},$

$\phi = \arccos\dfrac{3}{\sqrt{10}}, \left(\dfrac{\sqrt{30}}{2}, \dfrac{\pi}{3}, \arccos\dfrac{3}{\sqrt{10}}\right)$, spherical

69. $\left(100, -\frac{\pi}{6}, 50\right)$, cylindrical

$\rho = \sqrt{100^2 + 50^2} = 50\sqrt{5}$

$\theta = -\frac{\pi}{6}$

$\phi = \arccos\left(\frac{50}{50\sqrt{5}}\right) = \arccos\left(\frac{1}{\sqrt{5}}\right) \approx 63.4° \text{ or } 1.107$

$\left(50\sqrt{5}, -\frac{\pi}{6}, 63.4°\right)$, sperical or $\left(50\sqrt{5}, -\frac{\pi}{6}, 1.1071\right)$

70. $\left(81, -\frac{5\pi}{6}, 27\sqrt{3}\right)$, cylindrical

$\rho = \sqrt{6561 + 2187} = 54\sqrt{3}$

$\theta = -\frac{5\pi}{6}$

$\phi = \arccos\left(\frac{27\sqrt{3}}{54\sqrt{3}}\right) = \arccos\frac{1}{2} = \frac{\pi}{3}$

$\left(54\sqrt{3}, -\frac{5\pi}{6}, \frac{\pi}{3}\right)$, spherical

71. $\left(25, -\frac{\pi}{4}, \frac{3\pi}{4}\right)$, spherical

$r^2 = \left(25\sin\left(\frac{3\pi}{4}\right)\right)^2 \Rightarrow r = \frac{25\sqrt{2}}{2}$

$\theta = -\frac{\pi}{4}$

$z = \rho\cos\phi = 25\cos\frac{3\pi}{4} = -\frac{25\sqrt{2}}{2}$

$\left(\frac{25\sqrt{2}}{2}, -\frac{\pi}{4}, -\frac{25\sqrt{2}}{2}\right)$, cylindrical

72. $\left(12, -\frac{\pi}{2}, \frac{2\pi}{3}\right)$, spherical

$r^2 = \left(12\sin\left(\frac{2\pi}{3}\right)\right)^2 \Rightarrow r = 6\sqrt{3}$

$\theta = -\frac{\pi}{2}$

$z = \rho\cos\phi = 12\cos\left(\frac{2\pi}{3}\right) = -6$

$\left(6\sqrt{3}, -\frac{\pi}{2}, -6\right)$, cylindrical

73. $x^2 - y^2 = 2z$

(a) Cylindrical:

$r^2\cos^2\theta - r^2\sin^2\theta = 2z \Rightarrow r^2\cos 2\theta = 2z$

(b) Spherical:

$\rho^2\sin^2\phi\cos^2\theta - \rho^2\sin^2\phi\sin^2\theta = 2\rho\cos\phi$

$\rho\sin^2\phi\cos 2\theta - 2\cos\phi = 0$

$\rho = 2\sec 2\theta\cos\phi\csc^2\phi$

74. $x^2 + y^2 + z^2 = 16$

(a) Cylindrical: $r^2 + z^2 = 16$

(b) Spherical: $\rho = 4$

75. $r = 5\cos\theta$, cylindrical equation

$r^2 = 5r\cos\theta$

$x^2 + y^2 = 5x$

$x^2 - 5x + \frac{25}{4} + y^2 = \frac{25}{4}$

$\left(x - \frac{5}{2}\right)^2 + y^2 = \left(\frac{5}{2}\right)^2$, rectangular equation

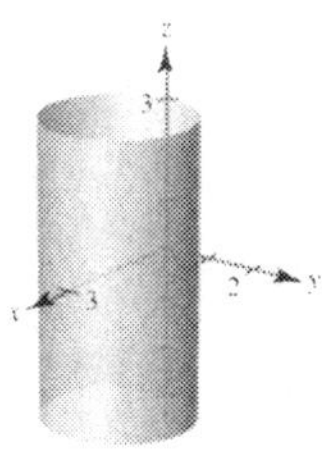

76. $z = 4$, cylindrical equation

$z = 4$, rectangular equation

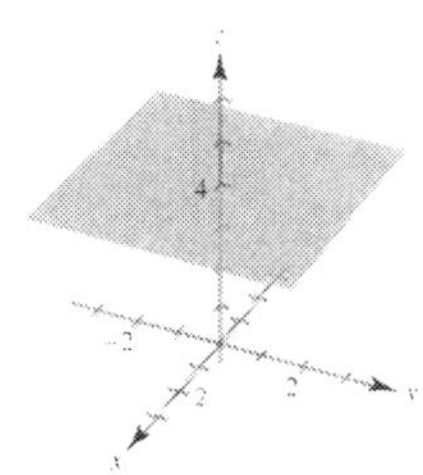

77. $\theta = \dfrac{\pi}{4}$, spherical coordinates

$$\tan\theta = \tan\frac{\pi}{4} = 1$$

$$\frac{y}{x} = 1$$

$y = x, x \ge 0$, rectangular coordinates, half-plane

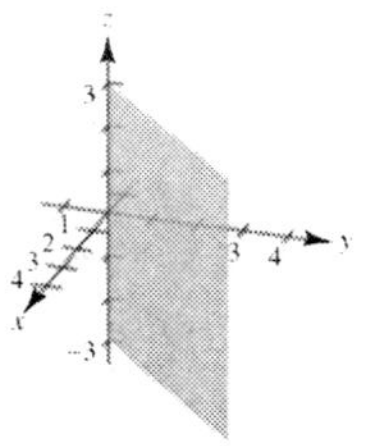

78. $\rho = 3\cos\theta$, spherical coordinates

$$\sqrt{x^2 + y^2 + z^2} = \frac{3z}{\sqrt{x^2 + y^2 + z^2}}$$

$$x^2 + y^2 + z^2 - 3z = 0$$

$x^2 + y^2 + \left(z - \dfrac{3}{2}\right)^2 = \left(\dfrac{3}{2}\right)^2$, rectangular coordinates, sphere

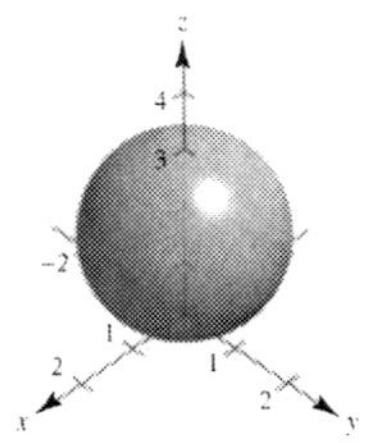

Problem Solving for Chapter 11

1.

$$\mathbf{a} + \mathbf{b} + \mathbf{c} = \mathbf{0}$$
$$\mathbf{b} \times (\mathbf{a} + \mathbf{b} + \mathbf{c}) = \mathbf{0}$$
$$(\mathbf{b} \times \mathbf{a}) + (\mathbf{b} \times \mathbf{c}) = \mathbf{0}$$
$$\|\mathbf{a} \times \mathbf{b}\| = \|\mathbf{b} \times \mathbf{c}\|$$
$$\|\mathbf{b} \times \mathbf{c}\| = \|\mathbf{b}\|\|\mathbf{c}\| \sin A$$
$$\|\mathbf{a} \times \mathbf{b}\| = \|\mathbf{a}\|\|\mathbf{b}\| \sin C$$

Then,

$$\frac{\sin A}{\|\mathbf{a}\|} = \frac{\|\mathbf{b} \times \mathbf{c}\|}{\|\mathbf{a}\|\|\mathbf{b}\|\|\mathbf{c}\|} = \frac{\|\mathbf{a} \times \mathbf{b}\|}{\|\mathbf{a}\|\|\mathbf{b}\|\|\mathbf{c}\|} = \frac{\sin C}{\|\mathbf{c}\|}.$$

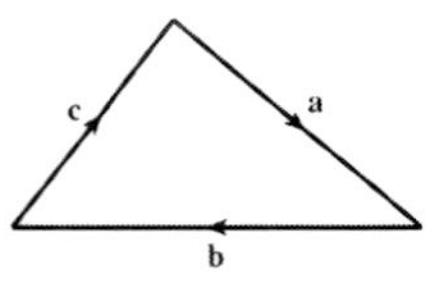

The other case, $\dfrac{\sin A}{\|\mathbf{a}\|} = \dfrac{\sin B}{\|\mathbf{b}\|}$ is similar.

2. $f(x) = \int_0^x \sqrt{t^4 + 1}\, dt$

(a)

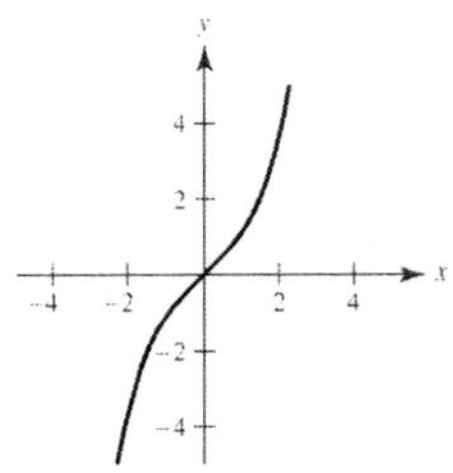

(b) $f'(x) = \sqrt{x^4 + 1}$

$f'(0) = 1 = \tan\theta$

$\theta = \dfrac{\pi}{4}$

$\mathbf{u} = \dfrac{1}{\sqrt{2}}(\mathbf{i} + \mathbf{j}) = \left\langle \dfrac{\sqrt{2}}{2}, \dfrac{\sqrt{2}}{2} \right\rangle$

(c) $\pm\left\langle \dfrac{\sqrt{2}}{2}, -\dfrac{\sqrt{2}}{2} \right\rangle$

(d) The line is $y = x$: $x = t, y = t$.

3. Label the figure as indicated.

From the figure, you see that

$$\overrightarrow{SP} = \frac{1}{2}\mathbf{a} - \frac{1}{2}\mathbf{b} = \overrightarrow{RQ} \text{ and } \overrightarrow{SR} = \frac{1}{2}\mathbf{a} + \frac{1}{2}\mathbf{b} = \overrightarrow{PQ}.$$

Because $\overrightarrow{SP} = \overrightarrow{RQ}$ and $\overrightarrow{SR} = \overrightarrow{PQ}$,

PSRQ is a parallelogram.

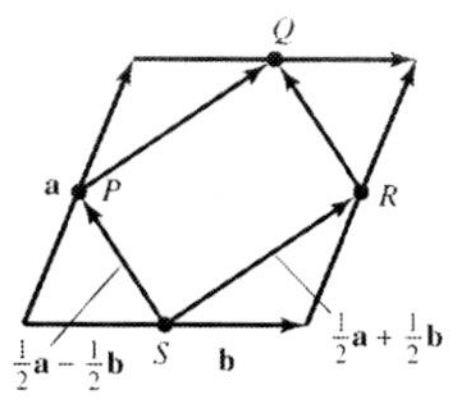

4. Label the figure as indicated.

$\overrightarrow{PR} = \mathbf{a} + \mathbf{b}$

$\overrightarrow{SQ} = \mathbf{b} - \mathbf{a}$

$(\mathbf{a} + \mathbf{b}) \cdot (\mathbf{b} - \mathbf{a}) = \|\mathbf{b}\|^2 - \|\mathbf{a}\|^2 = 0$, because

$\|\mathbf{a}\| = \|\mathbf{b}\|$ in a rhombus.

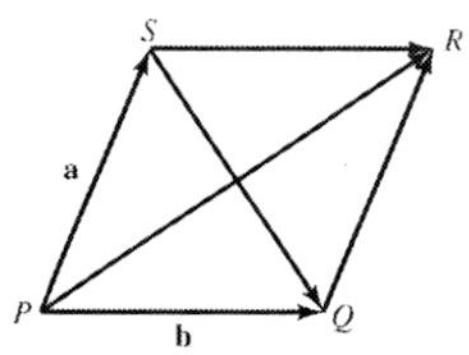

5. (a) $\mathbf{u} = \langle 0, 1, 1\rangle$ is the direction vector of the line determined by P_1 and P_2.

$$D = \frac{\|\overrightarrow{P_1Q} \times \mathbf{u}\|}{\|\mathbf{u}\|}$$

$$= \frac{\|\langle 2, 0, -1\rangle \times \langle 0, 1, 1\rangle\|}{\sqrt{2}}$$

$$= \frac{\|\langle 1, -2, 2\rangle\|}{\sqrt{2}} = \frac{3}{\sqrt{2}} = \frac{3\sqrt{2}}{2}$$

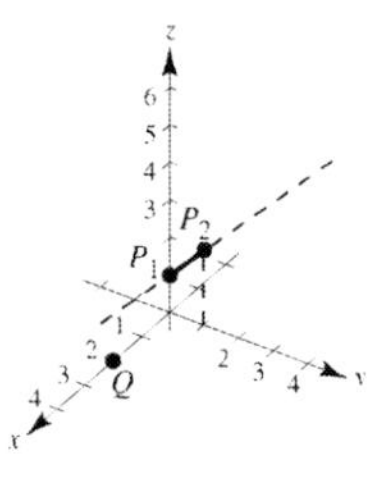

(b) The shortest distance to the line **segment** is $\|P_1Q\| = \|\langle 2, 0, -1\rangle\| = \sqrt{5}$.

6. $\left(\mathbf{n} + \overrightarrow{PP_0}\right) \perp \left(\mathbf{n} - \overrightarrow{PP_0}\right)$

Figure is a square.

So, $\|\overrightarrow{PP_0}\| = \|\mathbf{n}\|$ and the points P form a circle of radius $\|\mathbf{n}\|$ in the plane with center at P_0.

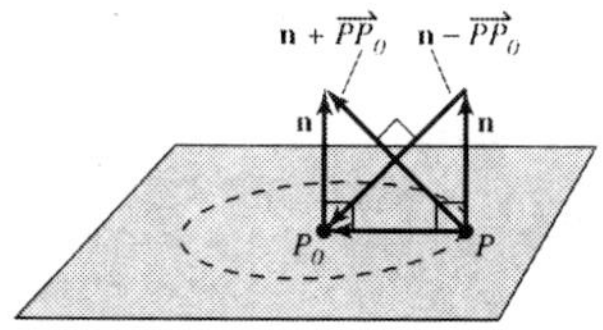

7. (a) $V = \pi\int_0^1 \left(\sqrt{z}\right)^2 dz = \left[\pi\frac{z^2}{2}\right]_0^1 = \frac{1}{2}\pi$

Note: $\frac{1}{2}$(base)(altitude) $= \frac{1}{2}\pi(1) = \frac{1}{2}\pi$

(b) $\frac{x^2}{a^2} + \frac{y^2}{b^2} = z$: (slice at $z = c$)

$$\frac{x^2}{\left(\sqrt{c}a\right)^2} + \frac{y^2}{\left(\sqrt{c}b\right)^2} = 1$$

At $z = c$, figure is ellipse of area

$\pi\left(\sqrt{c}a\right)\left(\sqrt{c}b\right) = \pi abc.$

$$V = \int_0^k \pi abc \cdot dc = \left[\frac{\pi abc^2}{2}\right]_0^k = \frac{\pi abk^2}{2}$$

(c) $V = \frac{1}{2}(\pi abk)k = \frac{1}{2}$(area of base)(height)

8. (a) $V = 2\int_0^r \pi\left(r^2 - x^2\right)dx = 2\pi\left[r^2x - \frac{x^3}{3}\right]_0^r = \frac{4}{3}\pi r^3$

(b) At height $z = d > 0$,

$$\frac{x^2}{a^2} + \frac{y^2}{b^2} + \frac{d^2}{c^2} = 1$$

$$\frac{x^2}{a^2} + \frac{y^2}{b^2} = 1 - \frac{d^2}{c^2} = \frac{c^2 - d^2}{c^2}$$

$$\frac{x^2}{\frac{a^2\left(c^2 - d^2\right)}{c^2}} + \frac{y^2}{\frac{b^2\left(c^2 - d^2\right)}{c^2}} = 1.$$

$$\text{Area} = \pi\sqrt{\left(\frac{a^2\left(c^2 - d^2\right)}{c^2}\right)\left(\frac{b^2\left(c^2 - d^2\right)}{c^2}\right)} = \frac{\pi ab}{c^2}\left(c^2 - d^2\right)$$

$$V = 2\int_0^c \frac{\pi ab}{c^2}\left(c^2 - d^2\right)dd$$

$$= \frac{2\pi ab}{c^2}\left[c^2d - \frac{d^3}{3}\right]_0^c = \frac{4}{3}\pi abc$$

9. From Exercise 58, Section 11.4,

$$(\mathbf{u} \times \mathbf{v}) \times (\mathbf{w} \times \mathbf{z}) = [(\mathbf{u} \times \mathbf{v}) \cdot \mathbf{z}]\mathbf{w} - [(\mathbf{u} \times \mathbf{v}) \cdot \mathbf{w}]\mathbf{z}.$$

10. $x = -t + 3, y = \frac{1}{2}t + 1, z = 2t - 1; Q = (4, 3, s)$

(a) $\mathbf{u} = \langle -2, 1, 4 \rangle$ direction vector for line

$P = (3, 1, -1)$ point on line

$\overrightarrow{PQ} = \langle 1, 2, s + 1 \rangle$

$$\overrightarrow{PQ} \times \mathbf{u} = \begin{vmatrix} \mathbf{i} & \mathbf{j} & \mathbf{k} \\ 1 & 2 & s+1 \\ -2 & 1 & 4 \end{vmatrix}$$

$$= (7 - s)\mathbf{i} + (-6 - 2s)\mathbf{j} + 5\mathbf{k}$$

$$D = \frac{\|\overrightarrow{PQ} \times \mathbf{u}\|}{\|\mathbf{u}\|} = \frac{\sqrt{(7 - s)^2 + (-6 - 2s)^2 + 25}}{\sqrt{21}}$$

(b)

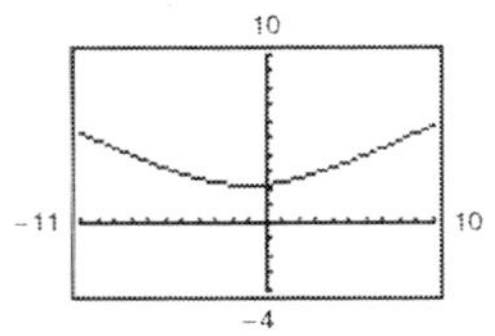

The minimum is $D \approx 2.2361$ at $s = -1$.

(c) Yes, there are slant asymptotes. Using $s = x$, you have

$$D(s) = \frac{1}{\sqrt{21}}\sqrt{5x^2 + 10x + 110} = \frac{\sqrt{5}}{\sqrt{21}}\sqrt{x^2 + 2x + 22} = \frac{\sqrt{5}}{\sqrt{21}}\sqrt{(x + 1)^2 + 21} \to \pm\sqrt{\frac{5}{21}}(x + 1)$$

$y = \pm\frac{\sqrt{105}}{21}(s + 1)$ slant asymptotes.

11. (a) $\rho = 2 \sin \phi$

Torus

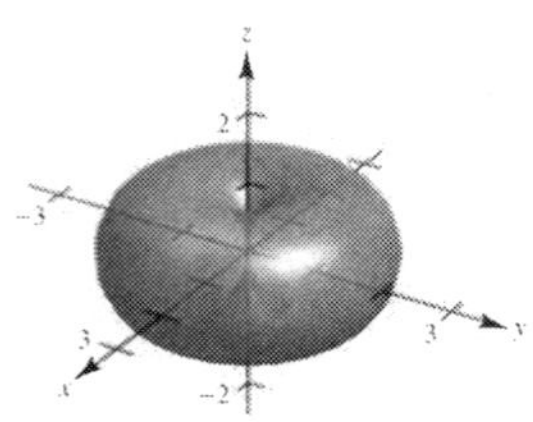

(b) $\rho = 2 \cos \phi$

Sphere

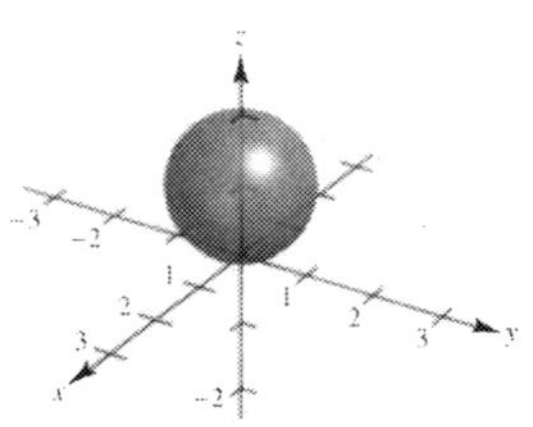

12. (a) $r = 2\cos\theta$

Cylinder

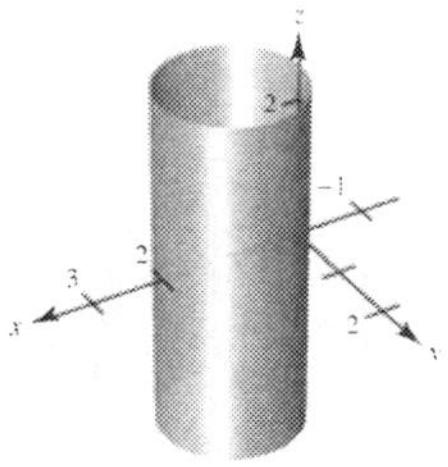

(b) $z = r^2 \cos 2\theta$

$= r^2\left(\cos^2\theta - \sin^2\theta\right)$

$= (r\cos\theta)^2 - (r\sin\theta)^2$

$= x^2 - y^2$

Hyperbolic paraboloid

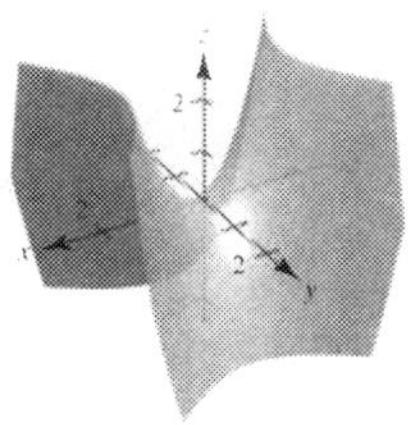

13. (a) $\mathbf{u} = \|\mathbf{u}\|(\cos 0\,\mathbf{i} + \sin 0\,\mathbf{j}) = \|\mathbf{u}\|\mathbf{i}$

Downward force $\mathbf{w} = -\mathbf{j}$

$\mathbf{T} = \|\mathbf{T}\|(\cos(90° + \theta)\mathbf{i} + \sin(90° + \theta)\mathbf{j})$

$= \|\mathbf{T}\|(-\sin\theta\,\mathbf{i} + \cos\theta\,\mathbf{j})$

$\mathbf{0} = \mathbf{u} + \mathbf{w} + \mathbf{T} = \|\mathbf{u}\|\mathbf{i} - \mathbf{j} + \|\mathbf{T}\|(-\sin\theta\,\mathbf{i} + \cos\theta\,\mathbf{j})$

$\|\mathbf{u}\| = \sin\theta\|\mathbf{T}\|$

$1 = \cos\theta\|\mathbf{T}\|$

If $\theta = 30°$, $\|\mathbf{u}\| = (1/2)\|\mathbf{T}\|$ and $1 = \left(\sqrt{3}/2\right)\|\mathbf{T}\| \Rightarrow \|\mathbf{T}\| = \dfrac{2}{\sqrt{3}} \approx 1.1547$ lb and $\|\mathbf{u}\| = \dfrac{1}{2}\left(\dfrac{2}{\sqrt{3}}\right) \approx 0.5774$ lb

(b) From part (a), $\|\mathbf{u}\| = \tan\theta$ and $\|\mathbf{T}\| = \sec\theta$.

Domain: $0 \le \theta \le 90°$

(c)

θ	0°	10°	20°	30°	40°	50°	60°
T	1	1.0154	1.0642	1.1547	1.3054	1.5557	2
$\|\mathbf{u}\|$	0	0.1763	0.3640	0.5774	0.8391	1.1918	1.7321

(d)

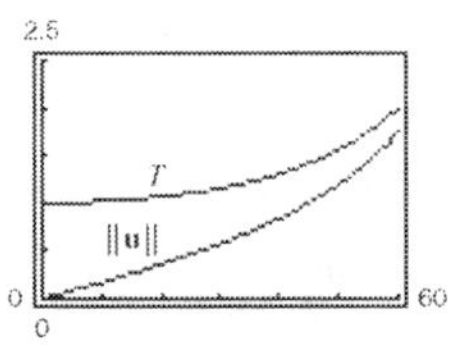

(e) Both are increasing functions.

(f) $\lim\limits_{\theta\to\pi/2^-} T = \infty$ and $\lim\limits_{\theta\to\pi/2^-} \|\mathbf{u}\| = \infty$.

Yes. As θ increases, both T and $\|\mathbf{u}\|$ increase.

14. (a) The tension T is the same in each tow line.

$$6000\mathbf{i} = T(\cos 20° + \cos(-20))\mathbf{i} + T(\sin 20° + \sin(-20°))\mathbf{j}$$
$$= 2T\cos 20°\mathbf{i}$$
$$\Rightarrow T = \frac{6000}{2\cos 20°} \approx 3192.5 \text{ lb}$$

(b) As in part (a), $6000\mathbf{i} = 2T\cos\theta$

$$\Rightarrow T = \frac{3000}{\cos\theta}$$

Domain: $0 < \theta < 90°$

(c)

θ	10°	20°	30°	40°	50°	60°
T	3046.3	3192.5	3464.1	3916.2	4667.2	6000.0

(d) 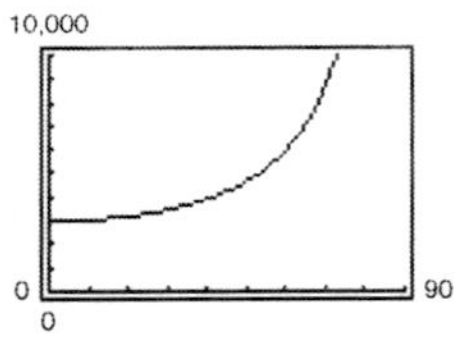

(e) As θ increases, there is less force applied in the direction of motion.

15. Let $\theta = \alpha - \beta$, the angle between **u** and **v**. Then

$$\sin(\alpha - \beta) = \frac{\|\mathbf{u}\times\mathbf{v}\|}{\|\mathbf{u}\|\|\mathbf{v}\|} = \frac{\|\mathbf{v}\times\mathbf{u}\|}{\|\mathbf{u}\|\|\mathbf{v}\|}.$$

For $\mathbf{u} = \langle\cos\alpha, \sin\alpha, 0\rangle$ and $\mathbf{v} = \langle\cos\beta, \sin\beta, 0\rangle$, $\|\mathbf{u}\| = \|\mathbf{v}\| = 1$ and

$$\mathbf{v}\times\mathbf{u} = \begin{vmatrix} \mathbf{i} & \mathbf{j} & \mathbf{k} \\ \cos\beta & \sin\beta & 0 \\ \cos\alpha & \sin\alpha & 0 \end{vmatrix} = (\sin\alpha\cos\beta - \cos\alpha\sin\beta)\mathbf{k}.$$

So, $\sin(\alpha - \beta) = \|\mathbf{v}\times\mathbf{u}\| = \sin\alpha\cos\beta - \cos\alpha\sin\beta$.

16. (a) Los Angeles: $(4000, -118.24°, 55.95°)$

Rio de Janeiro: $(4000, -43.23°, 112.90°)$

(b) Los Angeles:
$$x = 4000\sin(55.95°)\cos(-118.24°)$$
$$y = 4000\sin(55.95°)\sin(-118.24°)$$
$$z = 4000\cos(55.95°)$$
$$(x, y, z) \approx (-1568.2, -2919.7, 2239.7)$$

Rio de Janeiro:
$$x = 4000\sin(112.90°)\cos(-43.23°)$$
$$y = 4000\sin(112.90°)\sin(-43.23°)$$
$$z = 4000\cos(112.90°)$$
$$(x, y, z) \approx (2684.7, -2523.8, -1556.5)$$

(c) $$\cos\theta = \frac{\mathbf{u}\cdot\mathbf{v}}{\|\mathbf{u}\|\|\mathbf{v}\|} = \frac{(-1568.2)(2684.7) + (-2919.7)(-2523.8) + (2239.7)(-1556.5)}{(4000)(4000)} \approx -0.02047$$

$\theta \approx 91.17°$ or 1.59 radians

(d) $s = r\theta = 4000(1.59) \approx 6360$ miles

(e) For Boston and Honolulu:

a. Boston: $(4000, -71.06°, 47.64°)$ Honolulu: $(4000, -157.86°, 68.69°)$

b. Boston: $x = 4000 \sin 47.64° \cos(-71.06°)$ Honolulu: $x = 4000 \sin 68.69° \cos(-157.86°)$

$y = 4000 \sin 47.64° \sin(-71.06°)$ $y = 4000 \sin 68.69° \sin(-157.86°)$

$z = 4000 \cos 47.64°$ $z = 4000 \cos 68.69°$

$(959.4, -2795.7, 2695.1)$ $(-3451.7, -1404.4, 1453.7)$

c. $\cos\theta = \dfrac{\mathbf{u}\cdot\mathbf{v}}{\|\mathbf{u}\|\|\mathbf{v}\|} = \dfrac{(959.4)(-3451.7) + (-2795.7)(-1404.4) + (2695.1)(1453.7)}{(4000)(4000)} \approx 0.28329$

$\theta \approx 73.54°$ or 1.28 radians

d. $s = r\theta = 4000(1.28) \approx 5120$ miles

17. From Theorem 11.13 and Theorem 11.7 (6) you have

$$D = \frac{|\overrightarrow{PQ}\cdot\mathbf{n}|}{\|\mathbf{n}\|}$$

$$= \frac{|\mathbf{w}\cdot(\mathbf{u}\times\mathbf{v})|}{\|\mathbf{u}\times\mathbf{v}\|} = \frac{|(\mathbf{u}\times\mathbf{v})\cdot\mathbf{w}|}{\|\mathbf{u}\times\mathbf{v}\|} = \frac{|\mathbf{u}\cdot(\mathbf{v}\times\mathbf{w})|}{\|\mathbf{u}\times\mathbf{v}\|}.$$

18. Assume one of a, b, c, is not zero, say a. Choose a point in the first plane such as $(-d_1/a, 0, 0)$. The distance between this point and the second plane is

$$D = \frac{|a(-d_1/a) + b(0) + c(0) + d_2|}{\sqrt{a^2+b^2+c^2}}$$

$$= \frac{|-d_1 + d_2|}{\sqrt{a^2+b^2+c^2}} = \frac{|d_1 - d_2|}{\sqrt{a^2+b^2+c^2}}.$$

19. $x^2 + y^2 = 1$ cylinder

$z = 2y$ plane

Introduce a coordinate system in the plane $z = 2y$.

The new u-axis is the original x-axis.

The new v-axis is the line $z = 2y$, $x = 0$.

Then the intersection of the cylinder and plane satisfies the equation of an ellipse:

$$x^2 + y^2 = 1$$

$$x^2 + \left(\frac{z}{2}\right)^2 = 1$$

$$x^2 + \frac{z^2}{4} = 1 \quad \text{ellipse}$$

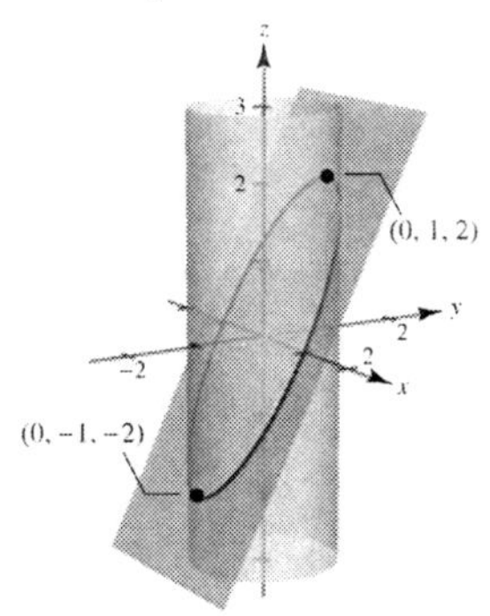

20. Essay

CHAPTER 12
Vector-Valued Functions

CHAPTER 12
Vector-Valued Functions

Section 12.1 Vector-Valued Functions

1. $\mathbf{r}(t) = \dfrac{1}{t+1}\mathbf{i} + \dfrac{t}{2}\mathbf{j} - 3t\mathbf{k}$

Component functions: $f(t) = \dfrac{1}{t+1}$

$g(t) = \dfrac{t}{2}$

$h(t) = -3t$

Domain: $(-\infty, -1) \cup (-1, \infty)$

2. $\mathbf{r}(t) = \sqrt{4-t^2}\mathbf{i} + t^2\mathbf{j} - 6t\mathbf{k}$

Component functions: $f(t) = \sqrt{4-t^2}$

$g(t) = t^2$

$h(t) = -6t$

Domain: $[-2, 2]$

3. $\mathbf{r}(t) = \ln t\mathbf{i} - e^t\mathbf{j} - t\mathbf{k}$

Component functions: $f(t) = \ln t$

$g(t) = -e^t$

$h(t) = -t$

Domain: $(0, \infty)$

4. $\mathbf{r}(t) = \sin t\mathbf{i} + 4\cos t\mathbf{j} + t\mathbf{k}$

Component functions: $f(t) = \sin t$

$g(t) = 4\cos t$

$h(t) = t$

Domain: $(-\infty, \infty)$

5. $\mathbf{r}(t) = \mathbf{F}(t) + \mathbf{G}(t) = \left(\cos t\mathbf{i} - \sin t\mathbf{j} + \sqrt{t}\mathbf{k}\right) + (\cos t\mathbf{i} + \sin t\mathbf{j}) = 2\cos t\mathbf{i} + \sqrt{t}\mathbf{k}$

Domain: $[0, \infty)$

6. $\mathbf{r}(t) = \mathbf{F}(t) - G(t) = \left(\ln t\mathbf{i} + 5t\mathbf{j} - 3t^2\mathbf{k}\right) - \left(\mathbf{i} + 4t\mathbf{j} - 3t^2\mathbf{k}\right)$

$= (\ln t - 1)\mathbf{i} + (5t - 4t)\mathbf{j} + \left(-3t^2 + 3t^2\right)\mathbf{k} = (\ln t - 1)\mathbf{i} + t\mathbf{j}$

Domain: $(0, \infty)$

7. $\mathbf{r}(t) = \mathbf{F}(t) \times \mathbf{G}(t) = \begin{vmatrix} \mathbf{i} & \mathbf{j} & \mathbf{k} \\ \sin t & \cos t & 0 \\ 0 & \sin t & \cos t \end{vmatrix} = \cos^2 t\mathbf{i} - \sin t\cos t\mathbf{j} + \sin^2 t\mathbf{k}$

Domain: $(-\infty, \infty)$

8. $\mathbf{r}(t) = \mathbf{F}(t) \times \mathbf{G}(t) = \begin{vmatrix} \mathbf{i} & \mathbf{j} & \mathbf{k} \\ t^3 & -t & t \\ \sqrt[3]{t} & \dfrac{1}{t+1} & t+2 \end{vmatrix} = \left(-t(t+2) - \dfrac{t}{t+1}\right)\mathbf{i} - \left(t^3(t+2) - t\sqrt[3]{t}\right)\mathbf{j} + \left(\dfrac{t^3}{t+1} + t\sqrt[3]{t}\right)\mathbf{k}$

Domain: $(-\infty, -1), (-1, \infty)$

9. $\mathbf{r}(t) = \frac{1}{2}t^2\mathbf{i} - (t-1)\mathbf{j}$

(a) $\mathbf{r}(1) = \frac{1}{2}\mathbf{i}$

(b) $\mathbf{r}(0) = \mathbf{j}$

(c) $\mathbf{r}(s+1) = \frac{1}{2}(s+1)^2\mathbf{i} - (s+1-1)\mathbf{j} = \frac{1}{2}(s+1)^2\mathbf{i} - s\mathbf{j}$

(d) $\mathbf{r}(2+\Delta t) - \mathbf{r}(2) = \frac{1}{2}(2+\Delta t)^2\mathbf{i} - (2+\Delta t-1)\mathbf{j} - (2\mathbf{i} - \mathbf{j}) = \left(2 + 2\Delta t + \frac{1}{2}(\Delta t)^2\right)\mathbf{i} - (1+\Delta t)\mathbf{j} - 2\mathbf{i} + \mathbf{j}$

$= \left(2\Delta t + \frac{1}{2}(\Delta t)^2\right)\mathbf{i} - (\Delta t)\mathbf{j} = \frac{1}{2}\Delta t(\Delta t + 4)\mathbf{i} - \Delta t\mathbf{j}$

10. $\mathbf{r}(t) = \cos t\mathbf{i} + 2\sin t\mathbf{j}$

(a) $\mathbf{r}(0) = \mathbf{i}$

(b) $\mathbf{r}\left(\frac{\pi}{4}\right) = \frac{\sqrt{2}}{2}\mathbf{i} + \sqrt{2}\mathbf{j}$

(c) $\mathbf{r}(\theta - \pi) = \cos(\theta - \pi)\mathbf{i} + 2\sin(\theta - \pi)\mathbf{j} = -\cos\theta\mathbf{i} - 2\sin\theta\mathbf{j}$

(d) $\mathbf{r}\left(\frac{\pi}{6} + \Delta t\right) - \mathbf{r}\left(\frac{\pi}{6}\right) = \cos\left(\frac{\pi}{6} + \Delta t\right)\mathbf{i} + 2\sin\left(\frac{\pi}{6} + \Delta t\right)\mathbf{j} - \left(\cos\left(\frac{\pi}{6}\right)\mathbf{i} + 2\sin\frac{\pi}{6}\mathbf{j}\right)$

11. $\mathbf{r}(t) = \ln t\mathbf{i} + \frac{1}{t}\mathbf{j} + 3t\mathbf{k}$

(a) $\mathbf{r}(2) = \ln 2\mathbf{i} + \frac{1}{2}\mathbf{j} + 6\mathbf{k}$

(b) $\mathbf{r}(-3)$ is not defined. $(\ln(-3)$ does not exist.$)$

(c) $\mathbf{r}(t - 4) = \ln(t - 4)\mathbf{i} + \frac{1}{t - 4}\mathbf{j} + 3(t - 4)\mathbf{k}$

(d) $\mathbf{r}(1 + \Delta t) - \mathbf{r}(1) = \ln(1 + \Delta t)\mathbf{i} + \frac{1}{1 + \Delta t}\mathbf{j} + 3(1 + \Delta t)\mathbf{k} - (0\mathbf{i} + \mathbf{j} + 3\mathbf{k}) = \ln(1 + \Delta t)\mathbf{i} + \left(\frac{-\Delta t}{1 + \Delta t}\right)\mathbf{j} + (3\Delta t)\mathbf{k}$

12. $\mathbf{r}(t) = \sqrt{t}\mathbf{i} + t^{3/2}\mathbf{j} + e^{-t/4}\mathbf{k}$

(a) $\mathbf{r}(0) = \mathbf{k}$

(b) $\mathbf{r}(4) = 2\mathbf{i} + 8\mathbf{j} + e^{-1}\mathbf{k}$

(c) $\mathbf{r}(c + 2) = \sqrt{c + 2}\mathbf{i} + (c + 2)^{3/2}\mathbf{j} + e^{-[(c+2)/4]}\mathbf{k}$

(d) $\mathbf{r}(9 + \Delta t) - \mathbf{r}(9) = \left(\sqrt{9 + \Delta t}\right)\mathbf{i} + (9 + \Delta t)^{3/2}\mathbf{j} + e^{-[9 + \Delta t/4]}\mathbf{k} - \left(3\mathbf{i} + 27\mathbf{j} + e^{-9/4}\mathbf{k}\right)$

$= \left(\sqrt{9 + \Delta t} - 3\right)\mathbf{i} + \left((9 + \Delta t)^{3/2} - 27\right)\mathbf{j} + \left(e^{-[(9 + \Delta t)/4]} - e^{-9/4}\mathbf{k}\right)$

13. $P(0, 0, 0), Q(3, 1, 2)$

$\mathbf{v} = \overrightarrow{PQ} = \langle 3, 1, 2\rangle$

$\mathbf{r}(t) = 3t\mathbf{i} + t\mathbf{j} + 2t\mathbf{k},\ 0 \le t \le 1$

$x = 3t, y = t, z = 2t, 0 \le t \le 1$, Parametric equation

(Answers may vary)

14. $P(0, 2, -1), Q(4, 7, 2)$

$\mathbf{v} = \overrightarrow{PQ} = \langle 4, 5, 3\rangle$

$\mathbf{r}(t) = 4t\mathbf{i} + (2 + 5t)\mathbf{j} + (-1 + 3t)\mathbf{k},\ 0 \le t \le 1$

$x = 4t, y = 2 + 5t, z = -1 + 3t,$

$0 \le t \le 1$, Parametric equation

(Answers may vary)

15. $P(-2, 5, -3), Q(-1, 4, 9)$

$\mathbf{v} = \overrightarrow{PQ} = \langle 1, -1, 12\rangle$

$\mathbf{r}(t) = (-2 + t)\mathbf{i} + (5 - t)\mathbf{j} + (-3 + 12t)\mathbf{k},\ 0 \le t \le 1$

$x = -2 + t, y = 5 - t, z = -3 + 12t,$

$0 \le t \le 1$, Parametric equation

(Answers may vary)

16. $P(1, -6, 8), Q(-3, -2, 5)$

$\mathbf{v} = \overrightarrow{PQ} = \langle -4, 4, -3\rangle$

$\mathbf{r}(t) = (1 - 4t)\mathbf{i} + (-6 + 4t)\mathbf{j} + (8 - 3t)\mathbf{k},\ 0 \le t \le 1$

$x = 1 - 4t,\ y = -6 + 4t,\ z = 8 - 3t,$

$0 \le t \le 1$, Parametric equation

(Answers may vary)

17. $\mathbf{r}(t) \cdot \mathbf{u}(t) = (3t - 1)(t^2) + \left(\frac{1}{4}t^3\right)(-8) + 4(t^3)$

$= 3t^3 - t^2 - 2t^3 + 4t^3 = 5t^3 - t^2$, a scalar.

No, the dot product is a scalar-valued function.

18. $\mathbf{r}(t) \cdot \mathbf{u}(t) = (3\cos t)(4\sin t) + (2\sin t)(-6\cos t) + (t-2)(t^2) = t^3 - 2t^2$, a scalar.

No, the dot product is a scalar-valued function.

19. $\mathbf{r}(t) = t\mathbf{i} + 2t\mathbf{j} + t^2\mathbf{k},\ -2 \le t \le 2$

$x = t, y = 2t, z = t^2$

So, $z = x^2$. Matches (b)

20. $\mathbf{r}(t) = \cos(\pi t)\mathbf{i} + \sin(\pi t)\mathbf{j} + t^2\mathbf{k},\ -1 \le t \le 1$

$x = \cos(\pi t),\ y = \sin(\pi t),\ z = t^2$

So, $x^2 + y^2 = 1$. Matches (c)

21. $\mathbf{r}(t) = t\mathbf{i} + t^2\mathbf{j} + e^{0.75t}\mathbf{k}, -2 \le t \le 2$

$x = t, y = t^2, z = e^{0.75t}$

So, $y = x^2$. Matches (d)

22. $\mathbf{r}(t) = t\mathbf{i} + \ln t\mathbf{j} + \dfrac{2t}{3}\mathbf{k}, 0.1 \le t \le 5$

$x = t, y = \ln t, z = \dfrac{2t}{3}$

So, $z = \frac{2}{3}x$ and $y = \ln x$. Matches (a)

23. $x = \dfrac{t}{4} \Rightarrow t = 4x$

$y = t - 1$

$y = 4x - 1$

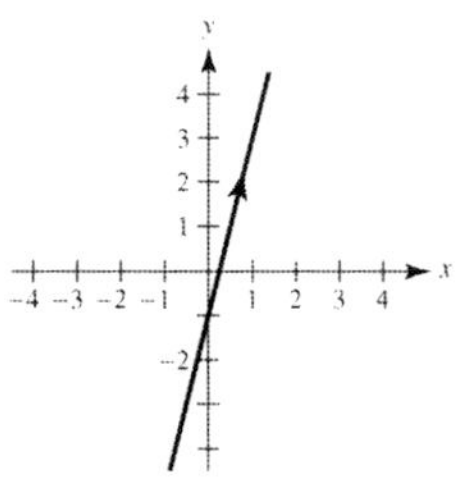

24. $x = 5 - t \Rightarrow t = 5 - x$

$y = \sqrt{t}$

$y = \sqrt{5 - x}$

Domain: $t \ge 0$

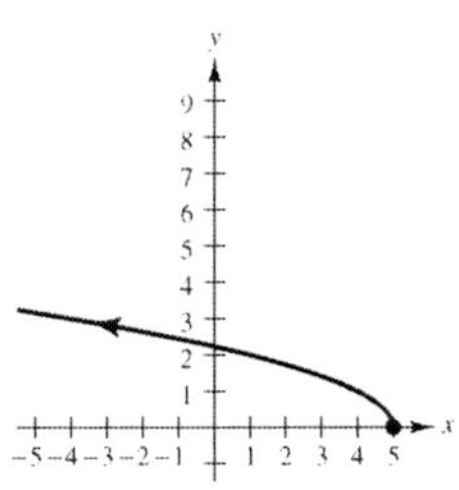

25. $x = t^3, y = t^2$

$y = x^{2/3}$

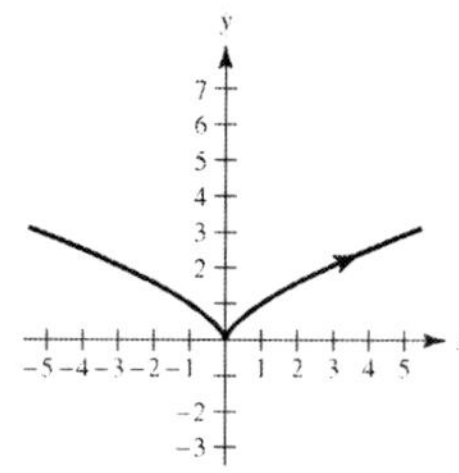

26. $x = t^2 + t, y = t^2 - t$

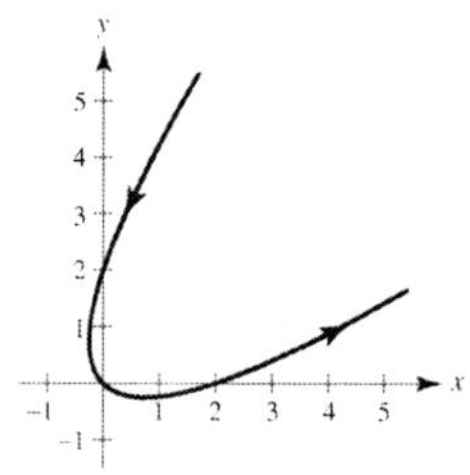

27. $x = \cos\theta, y = 3\sin\theta$

$x^2 + \dfrac{y^2}{9} = 1$, Ellipse

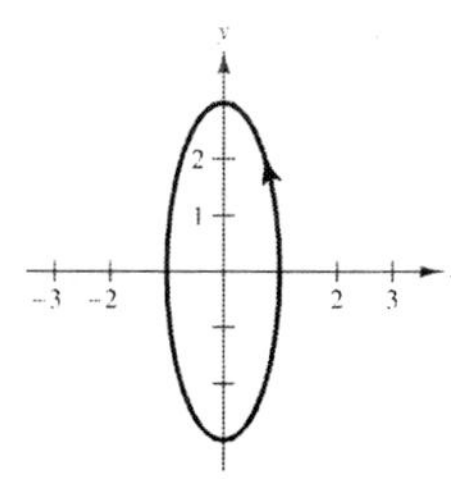

28. $x = 2\cos t$

$y = 2\sin t$

$x^2 + y^2 = 4$, circle

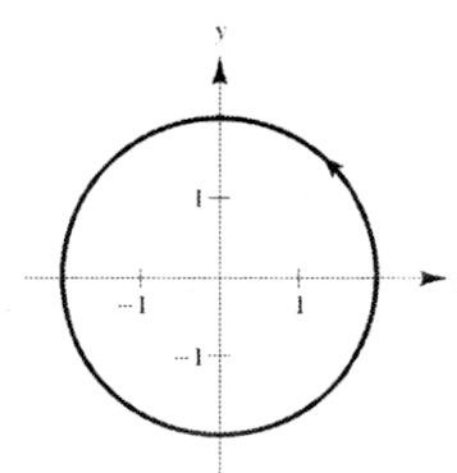

29. $x = 3\sec\theta, y = 2\tan\theta$

$\dfrac{x^2}{9} = \dfrac{y^2}{4} + 1$, Hyperbola

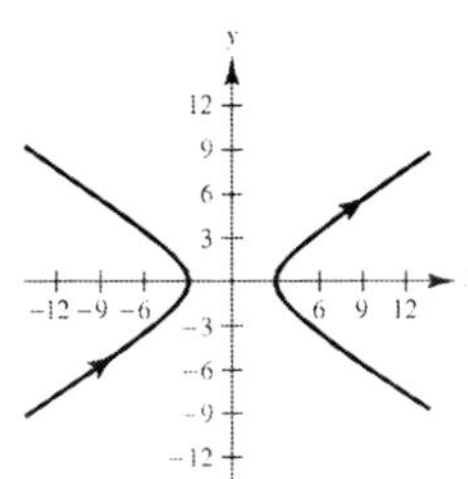

30. $x = 2\cos^3 t,\ y = 2\sin^3 t$

$$\left(\frac{x}{2}\right)^{2/3} + \left(\frac{y}{2}\right)^{2/3} = \cos^2 t + \sin^2 t$$
$$= 1$$
$$x^{2/3} + y^{2/3} = 2^{2/3}$$

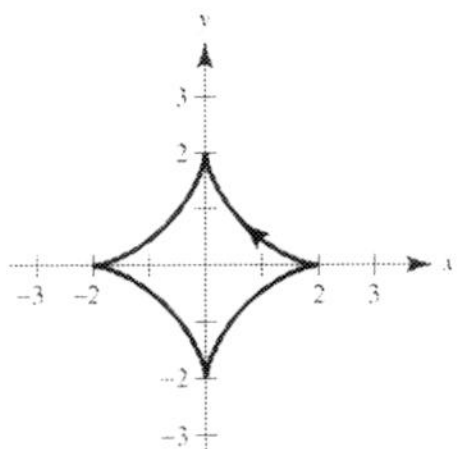

31. $x = -t + 1$

$y = 4t + 2$

$z = 2t + 3$

Line passing through the points: $(0, 6, 5), (1, 2, 3)$

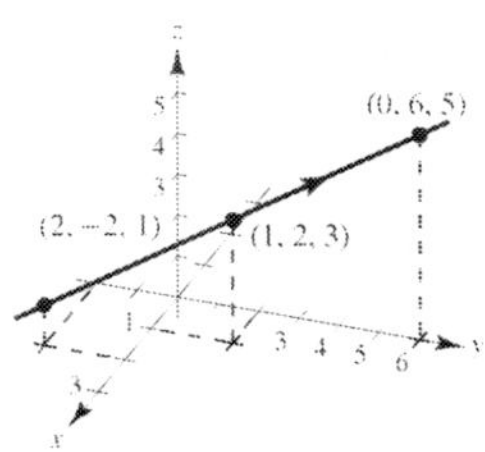

32. $x = t$

$y = 2t - 5$

$y = 3t$

Line passing through the points: $(0, -5, 0), \left(\frac{5}{2}, 0, \frac{15}{2}\right)$

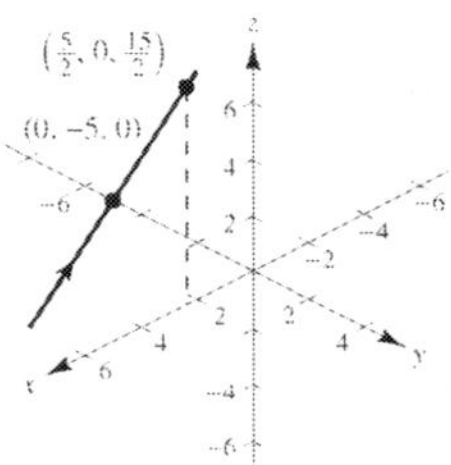

33. $x = 2\cos t,\ y = 2\sin t,\ z = t$

$$\frac{x^2}{4} + \frac{y^2}{4} = 1$$

$z = t$

Circular helix

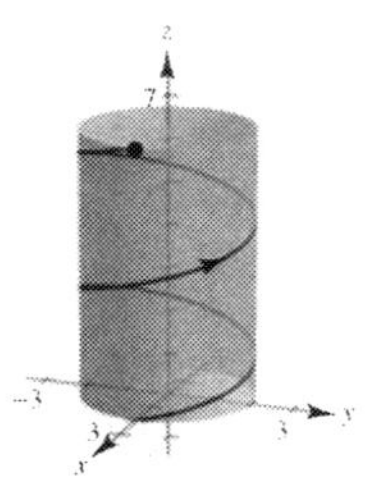

34. $x = t,\ y = 3\cos t,\ z = 3\sin t$

$$y^2 + z^2 = (3\cos t)^3 + (3\sin t)^3 = 9$$

Circular helix along cylinder $y^2 + z^2 = 9$

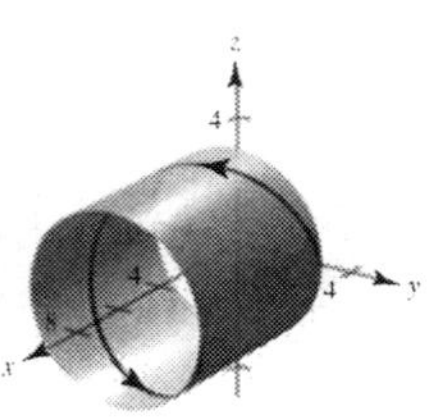

35. $x = 2\sin t,\ y = 2\cos t,\ z = e^{-t}$

$x^2 + y^2 = 4$

$z = e^{-t}$

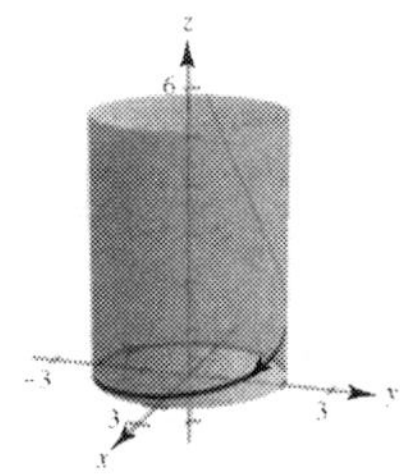

36. $x = t^2,\ y = 2t,\ z = \frac{3}{2}t$

$$x = \frac{y^2}{4},\ z = \frac{3}{4}y$$

t	-2	-1	0	1	2
x	4	1	0	1	4
y	-4	-2	0	2	4
z	-3	$-\frac{3}{2}$	0	$\frac{3}{2}$	3

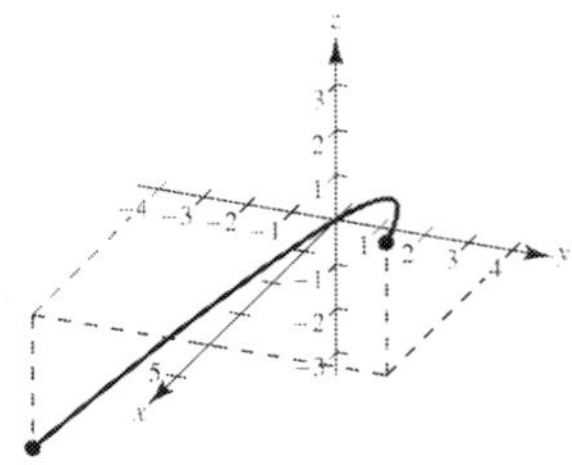

37. $x = t,\ y = t^2,\ z = \frac{2}{3}t^3$

$y = x^2,\ z = \frac{2}{3}x^3$

t	-2	-1	0	1	2
x	-2	-1	0	1	2
y	4	1	0	1	4
z	$-\frac{16}{3}$	$-\frac{2}{3}$	0	$\frac{2}{3}$	$\frac{16}{3}$

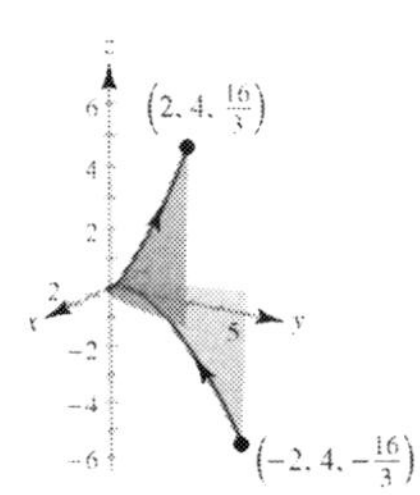

38. $x = \cos t + t \sin t$

$y = \sin t - t \cos t$

$z = t$

$x^2 + y^2 = 1 + t^2 = 1 + z^2$

or $x^2 + y^2 - z^2 = 1$

$z = t$

Helix along a hyperboloid of one sheet

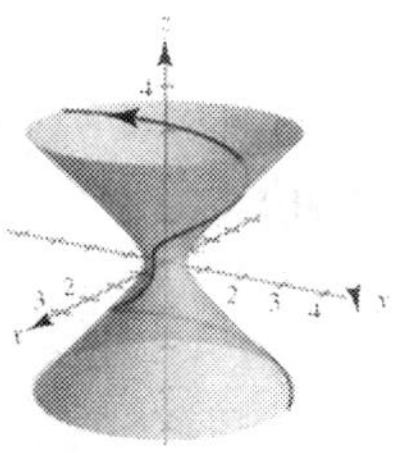

39. $\mathbf{r}(t) = -\frac{1}{2}t^2\mathbf{i} + t\mathbf{j} - \frac{\sqrt{3}}{2}t^2\mathbf{k}$

Parabola

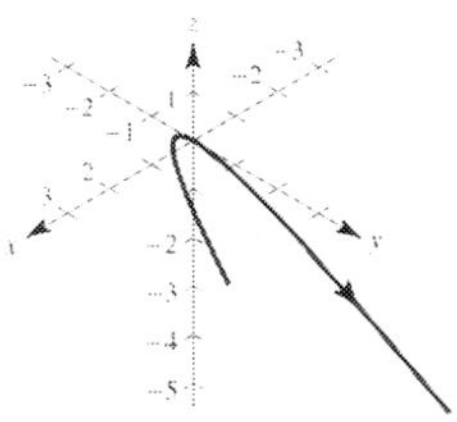

40. $\mathbf{r}(t) = t\mathbf{i} - \frac{\sqrt{3}}{2}t^2\mathbf{j} + \frac{1}{2}t^2\mathbf{k}$

Parabola

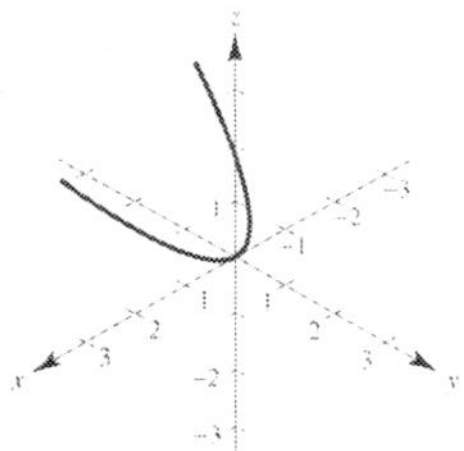

41. $\mathbf{r}(t) = \sin t\mathbf{i} + \left(\frac{\sqrt{3}}{2}\cos t - \frac{1}{2}t\right)\mathbf{j} + \left(\frac{1}{2}\cos t + \frac{\sqrt{3}}{2}\right)\mathbf{k}$

Helix

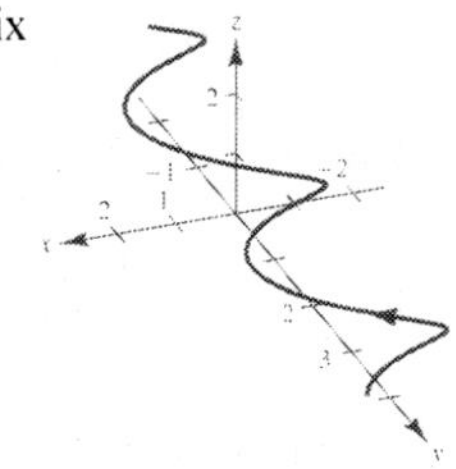

42. $\mathbf{r}(t) = -\sqrt{2}\sin t\mathbf{i} + 2\cos t\mathbf{j} + \sqrt{2}\sin t\mathbf{k}$

Ellipse

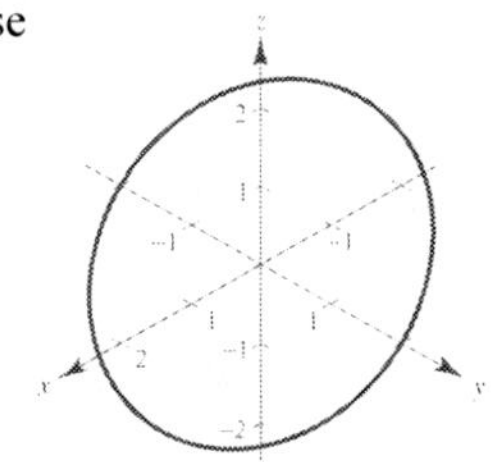

43.

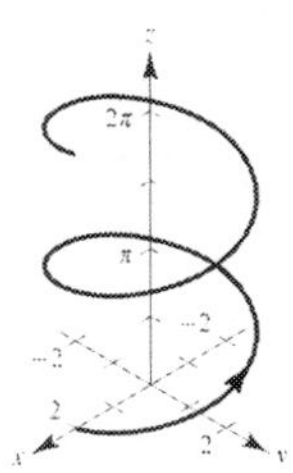

(a)

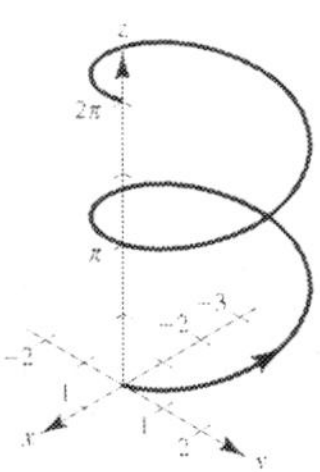

The helix is translated 2 units back on the x-axis.

(b)

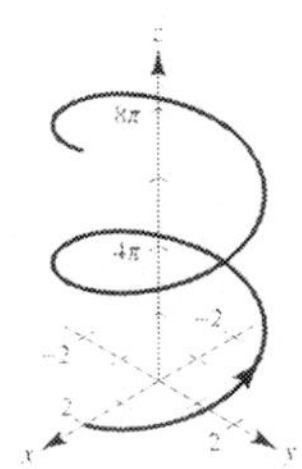

The height of the helix increases at a faster rate.

(c)

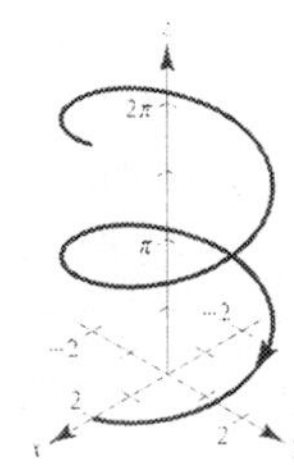

The orientation of the helix is reversed.

(d)

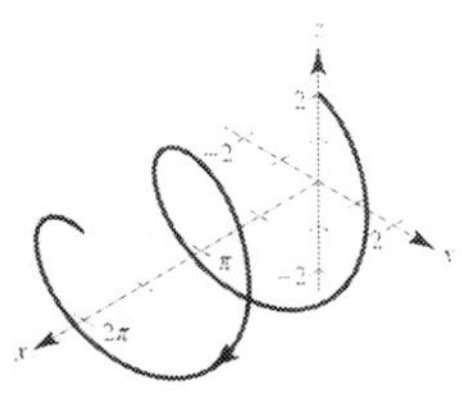

The axis of the helix is the x-axis.

(e)

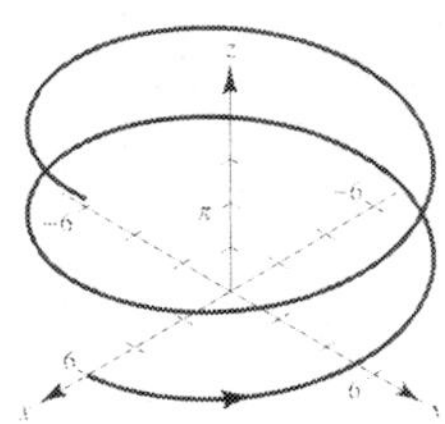

The radius of the helix is increased from 2 to 6.

44. $\mathbf{r}(t) = t\mathbf{i} + t^2\mathbf{j} + \frac{1}{2}t^3\mathbf{h}$

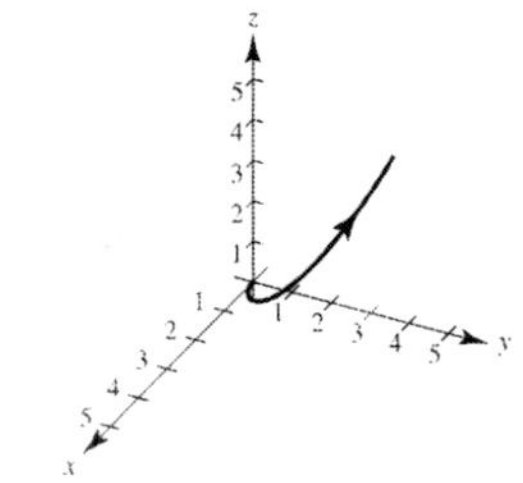

(a) $\mathbf{u}(t) = \mathbf{r}(t) - 2\mathbf{j}$ is a translation 2 units to the left along the y-axis.

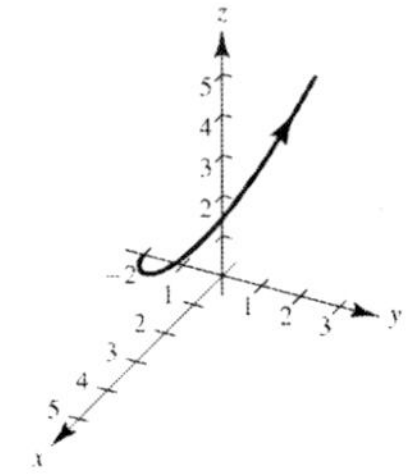

(b) $\mathbf{u}(t) = t^2\mathbf{i} + t\mathbf{j} + \frac{1}{2}t^3\mathbf{k}$ has the roles of x and y interchanged. The graph is a reflection in the plane $x = y$.

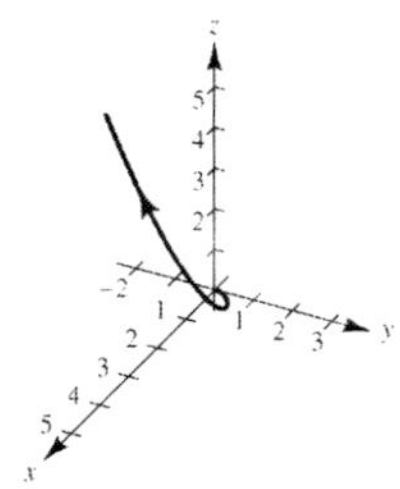

(c) $\mathbf{u}(t) = \mathbf{r}(t) + 4\mathbf{k}$ is an upward shift 4 units.

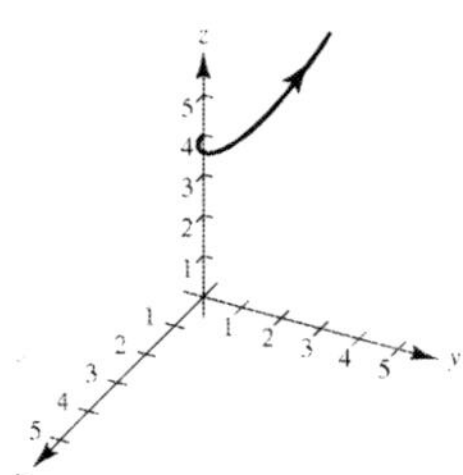

(d) $\mathbf{u}(t) = t\mathbf{i} + t^2\mathbf{j} + \frac{1}{8}t^3\mathbf{k}$ shrinks the z-value by a factor of 4. The curve rises more slowly.

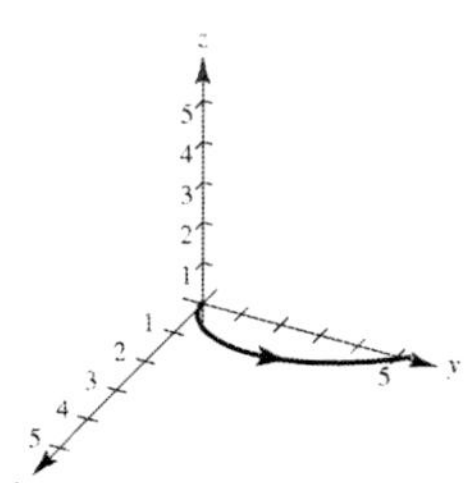

(e) $\mathbf{u}(t) = \mathbf{r}(-t)$ reverses the orientation.

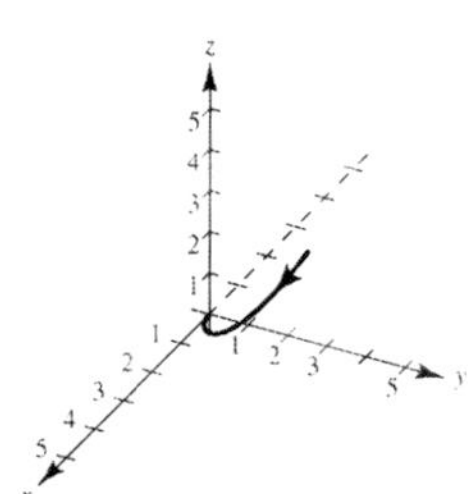

45. $y = x + 5$

Let $x = t$, then $y = t + 5$

$\mathbf{r}(t) = t\mathbf{i} + (t + 5)\mathbf{j}$

46. $2x - 3y + 5 = 0$

Let $x = t$, then $y = \frac{1}{3}(2t + 5)$.

$\mathbf{r}(t) = t\mathbf{i} + \frac{1}{3}(2\mathbf{t} + 5)\mathbf{j}$

47. $y = (x - 2)^2$

Let $x = t$, then $y = (t - 2)^2$.

$\mathbf{r}(t) = t\mathbf{i} + (t - 2)^2\mathbf{j}$

48. $y = 4 - x^2$

Let $x = t$, then $y = 4 - t^2$.

$\mathbf{r}(t) = t\mathbf{i} + (4 - t^2)\mathbf{j}$

49. $x^2 + y^2 = 25$

Let $x = 5\cos t$, then $y = 5\sin t$.

$\mathbf{r}(t) = 5\cos t\mathbf{i} + 5\sin t\mathbf{j}$

50. $(x - 2)^2 + y^2 = 4$

Let $x - 2 = 2\cos t$, $y = 2\sin t$.

$\mathbf{r}(t) = (2 + 2\cos t)\mathbf{i} + 2\sin t\mathbf{j}$

51. $\dfrac{x^2}{16} - \dfrac{y^2}{4} = 1$

Let $x = 4\sec t$, $y = 2\tan t$.

$\mathbf{r}(t) = 4\sec t\mathbf{i} + 2\tan t\mathbf{j}$

52. $\dfrac{x^2}{9} + \dfrac{y^2}{16} = 1$

Let $x = 3\cos t$ and $y = 4\sin t$

Then $\dfrac{x^2}{9} + \dfrac{y^2}{16} = \cos^2 t + \sin^2 t = 1$

$\mathbf{r}(t) = 3\cos t\mathbf{i} + 4\sin t\mathbf{j}$

53. $z = x^2 + y^2, x + y = 0$

Let $x = t$, then $y = -x = -t$ and $z = x^2 + y^2 = 2t^2$.

So, $x = t, y = -t, z = 2t^2$.

$\mathbf{r}(t) = t\mathbf{i} - t\mathbf{j} + 2t^2\mathbf{k}$

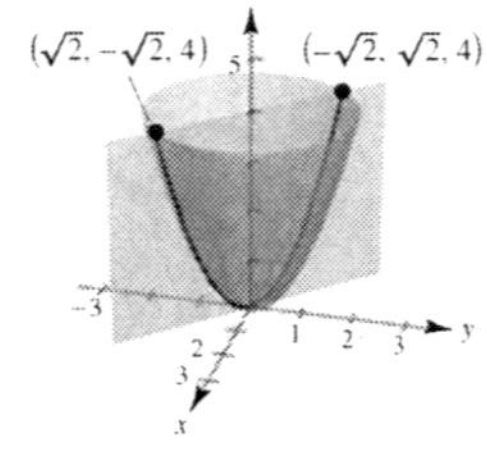

54. $z = x^2 + y^2, z = 4$

So, $x^2 + y^2 = 4$ or

$x = 2\cos t,\ y = 2\sin t,\ z = 4.$

$\mathbf{r}(t) = 2\cos t\mathbf{i} + 2\sin t\mathbf{j} + 4\mathbf{k}$

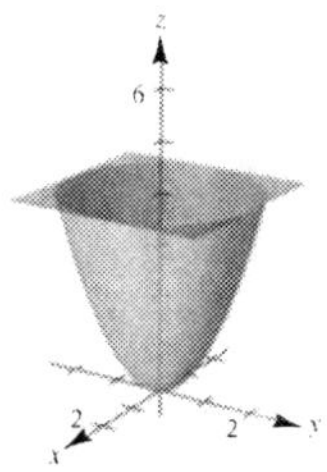

55. $x^2 + y^2 = 4,\ z = x^2$

$x = 2\sin t,\ y = 2\cos t$

$z = x^2 = 4\sin^2 t$

t	0	$\frac{\pi}{6}$	$\frac{\pi}{4}$	$\frac{\pi}{2}$	$\frac{3\pi}{4}$	π
x	0	1	$\sqrt{2}$	2	$\sqrt{2}$	0
y	2	$\sqrt{3}$	$\sqrt{2}$	0	$-\sqrt{2}$	-2
z	0	1	2	4	2	0

$\mathbf{r}(t) = 2\sin t\mathbf{i} + 2\cos t\mathbf{j} + 4\sin^2 t\mathbf{k}$

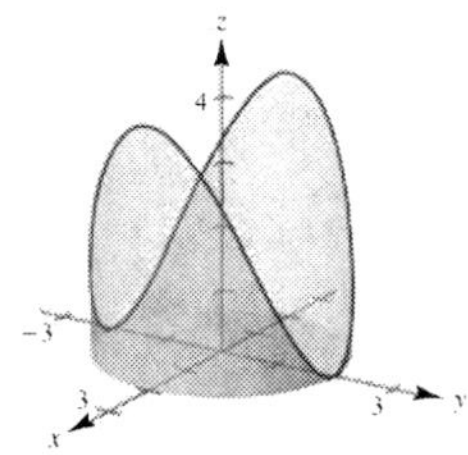

56. $4x^2 + 4y^2 + z^2 = 16,\ x = z^2$

If $z = t$, then $x = t^2$ and $y = \frac{1}{2}\sqrt{16 - 4t^4 - t^2}$.

t	-1.3	-1.2	-1	0	1	1.2
x	1.69	1.44	1	0	1	1.44
y	0.85	1.25	1.66	2	1.66	1.25
z	-1.3	-1.2	-1	0	1	1.2

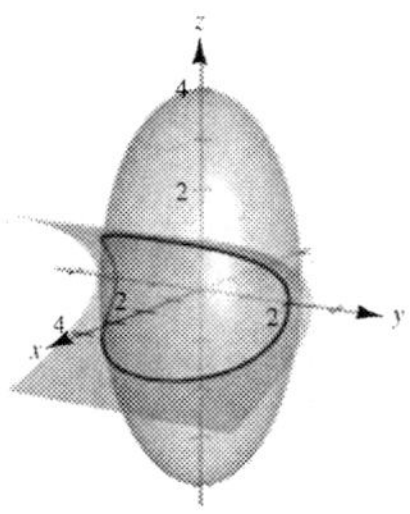

$\mathbf{r}(t) = t^2\mathbf{i} + \frac{1}{2}\sqrt{16 - 4t^4 - t^2}\mathbf{j} + t\mathbf{k}$

57. $x^2 + y^2 + z^2 = 4,\ x + z = 2$

Let $x = 1 + \sin t$, then $z = 2 - x = 1 - \sin t$ and $x^2 + y^2 + z^2 = 4$.

$(1 + \sin t)^2 + y^2 + (1 - \sin t)^2 = 2 + 2\sin^2 t + y^2 = 4$

$y^2 = 2\cos^2 t,\ y = \pm\sqrt{2}\cos t$

$x = 1 + \sin t,\ y = \pm\sqrt{2}\cos t$

$z = 1 - \sin t$

$\mathbf{r}(t) = (1 + \sin t)\mathbf{i} + \sqrt{2}\cos t\mathbf{j} - (1 - \sin t)\mathbf{k}$ and

$\mathbf{r}(t) = (1 + \sin t)\mathbf{i} - \sqrt{2}\cos t\mathbf{j} + (1 - \sin t)\mathbf{k}$

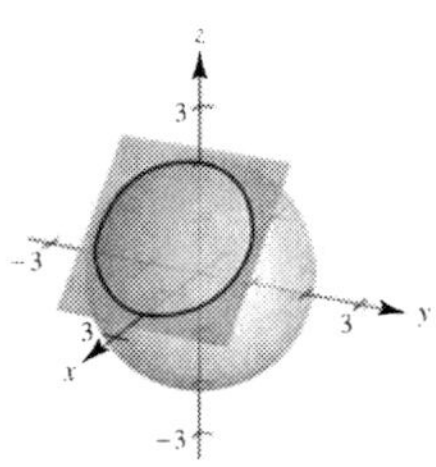

t	$-\frac{\pi}{2}$	$-\frac{\pi}{6}$	0	$\frac{\pi}{6}$	$\frac{\pi}{2}$
x	0	$\frac{1}{2}$	1	$\frac{3}{2}$	2
y	0	$\pm\frac{\sqrt{6}}{2}$	$\pm\sqrt{2}$	$\pm\frac{\sqrt{6}}{2}$	0
z	2	$\frac{3}{2}$	1	$\frac{1}{2}$	0

58. $x^2 + y^2 + z^2 = 10, x + y = 4$

Let $x = 2 + \sin t$, then $y = 2 - \sin t$ and $z = \sqrt{2(1 - \sin^2 t)} = \sqrt{2}\cos t$.

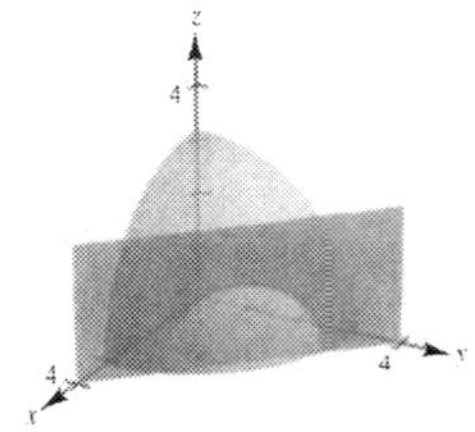

t	$-\frac{\pi}{2}$	$-\frac{\pi}{6}$	0	$\frac{\pi}{6}$	$\frac{\pi}{2}$	π
x	1	$\frac{3}{2}$	2	$\frac{5}{2}$	3	2
y	3	$\frac{5}{2}$	2	$\frac{3}{2}$	1	2
z	0	$\frac{\sqrt{6}}{2}$	$\sqrt{2}$	$\frac{\sqrt{6}}{2}$	0	$-\sqrt{2}$

$\mathbf{r}(t) = (2 + \sin t)\mathbf{i} + (2 - \sin t)\mathbf{j} + \sqrt{2}\cos t\mathbf{k}$

59. $x^2 + z^2 = 4, y^2 + z^2 = 4$

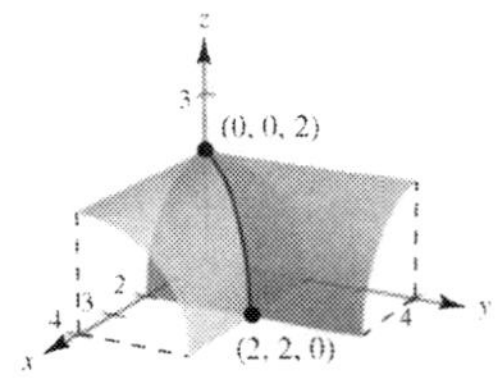

Subtracting, you have $x^2 - y^2 = 0$ or $y = \pm x$.

So, in the first octant, if you let $x = t$, then $x = t,\ y = t,\ z = \sqrt{4 - t^2}$.

$\mathbf{r}(t) = t\mathbf{i} + t\mathbf{j} + \sqrt{4 - t^2}\mathbf{k}$

60. $x^2 + y^2 + z^2 = 16, xy = 4$ (first octant)

Let $x = t$, then $y = \frac{4}{t}$ and $x^2 + y^2 + z^2 = t^2 + \frac{16}{t^2} + z^2 = 16$.

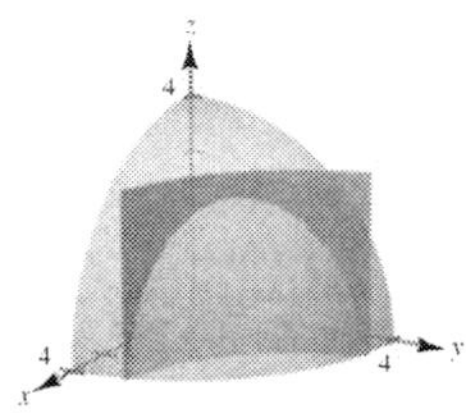

$z = \frac{1}{t}\sqrt{-t^4 + 16t^2 - 16}$

$\left(\sqrt{8 - 4\sqrt{3}} \le t \le \sqrt{8 + 4\sqrt{3}}\right)$

t	$\sqrt{8 + 4\sqrt{3}}$	1.5	2	2.5	3.0	3.5	$\sqrt{8 + 4\sqrt{3}}$
x	1.0	1.5	2	2.5	3.0	3.5	3.9
y	3.9	2.7	2	1.6	1.3	1.1	1.0
z	0	2.6	2.8	2.7	2.3	1.6	0

$\mathbf{r}(t) = t\mathbf{i} + \frac{4}{t}\mathbf{j} + \frac{1}{t}\sqrt{-t^4 + 16t^2 - 16}\mathbf{k}$

61. $y^2 + z^2 = (2t\cos t)^2 + (2t\sin t)^2 = 4t^2 = 4x^2$

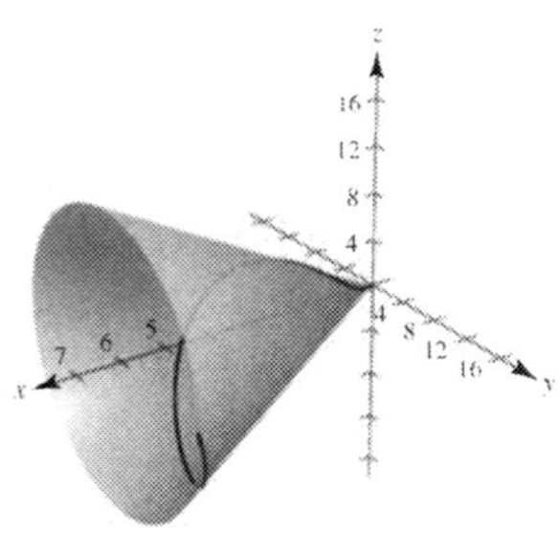

62. $x^2 + y^2 = (e^{-t}\cos t)^2 + (e^{-t}\sin t)^2 = e^{-2t} = z^2$

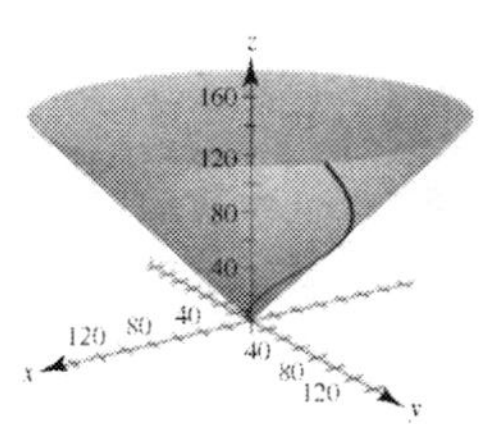

63. $\lim_{t \to \pi}(t\mathbf{i} + \cos t\mathbf{j} + \sin t\mathbf{k}) = \pi\mathbf{i} - \mathbf{j}$

64. $\lim_{t \to 2}\left(3t\mathbf{i} + \frac{2}{t^2 - 1}\mathbf{j} + \frac{1}{t}\mathbf{k}\right) = 6\mathbf{i} + \frac{2}{3}\mathbf{j} + \frac{1}{2}\mathbf{k}$

65. $\lim_{t\to 0}\left[t^2\mathbf{i} + 3t\mathbf{j} + \frac{1-\cos t}{t}\mathbf{k}\right] = 0$

because

$\lim_{t\to 0}\frac{1-\cos t}{t} = \lim_{t\to 0}\frac{\sin t}{1} = 0.$ (L'Hôpital's Rule)

66. $\lim_{t\to 1}\left(\sqrt{t}\,\mathbf{i} + \frac{\ln t}{t^2-1}\mathbf{j} + \frac{1}{t-1}\mathbf{k}\right)$

does not exist because $\lim_{t\to 1}\frac{1}{t-1}$ does not exist.

67. $\lim_{t\to 0}\left[e^t\mathbf{i} + \frac{\sin t}{t}\mathbf{j} + e^{-t}\mathbf{k}\right] = \mathbf{i} + \mathbf{j} + \mathbf{k}$

because

$\lim_{t\to 0}\frac{\sin t}{t} = \lim_{t\to 0}\frac{\cos t}{1} = 1$ (L'Hôpital's Rule)

68. $\lim_{t\to\infty}\left[e^{-t}\mathbf{i} + \frac{1}{t}\mathbf{j} + \frac{t}{t^2+1}\mathbf{k}\right] = 0$

because

$\lim_{t\to\infty} e^{-t} = 0,\ \lim_{t\to\infty}\frac{1}{t} = 0,$ and $\lim_{t\to\infty}\frac{t}{t^2+1} = 0.$

69. $\mathbf{r}(t) = t\mathbf{i} + \frac{1}{t}\mathbf{j}$

Continuous on $(-\infty, 0), (0, \infty)$

70. $\mathbf{r}(t) = \sqrt{t}\mathbf{i} + \sqrt{t-1}\mathbf{j}$

Continuous on $[1, \infty)$

71. $\mathbf{r}(t) = t\mathbf{i} + \arcsin t\mathbf{j} + (t-1)\mathbf{k}$

Continuous on $[-1, 1]$

72. $\mathbf{r}(t) = \langle 2e^{-t}, e^{-t}, \ln(t-1)\rangle$

Continuous on $t - 1 > 0$ or $t > 1$: $(1, \infty)$.

73. $\mathbf{r}(t) = \langle e^{-t}, t^2, \tan t\rangle$

Discontinuous at $t = \frac{\pi}{2} + n\pi$

Continuous on $\left(-\frac{\pi}{2} + n\pi, \frac{\pi}{2} + n\pi\right)$

74. $\mathbf{r}(t) = \langle 8, \sqrt{t}, \sqrt[3]{t}\rangle$

Continuous on $[0, \infty)$

75. $\mathbf{s}(t) = \mathbf{r}(t) + 3\mathbf{k} = t^2\mathbf{i} + (t-3)\mathbf{j} + (t+3)\mathbf{k}$

76. $\mathbf{s}(t) = \mathbf{r}(t) - 4\mathbf{k} = t^2\mathbf{i} + (t-3)\mathbf{j} + (t-4)\mathbf{k}$

77. $\mathbf{s}(t) = \mathbf{r}(t) - 2\mathbf{i} = (t^2-2)\mathbf{i} + (t-3)\mathbf{j} + t\mathbf{k}$

78. $\mathbf{s}(t) = \mathbf{r}(t) + 5\mathbf{j} = t^2\mathbf{i} + (t+2)\mathbf{j} + t\mathbf{k}$

79. A vector-valued function $\mathbf{r}$ is continuous at $t = a$ if the limit of $\mathbf{r}(t)$ exists as $t \to a$ and $\lim_{t\to a}\mathbf{r}(t) = \mathbf{r}(a)$.

The function $\mathbf{r}(t) = \begin{cases}\mathbf{i} + \mathbf{j} & t \ge 2\\ -\mathbf{i} + \mathbf{j} & t < 2\end{cases}$ is not continuous at $t = 0$.

80. (a) $x = -3\cos t + 1,\ y = 5\sin t + 2,\ z = 4$

$$\frac{(x-1)^2}{9} + \frac{(y-2)^2}{25} = 1, z = 4$$

(b) $x = 4,\ y = -3\cos t + 1,\ z = 5\sin t + 2$

$$\frac{(y-1)^2}{9} + \frac{(z-2)^2}{25} = 1, x = 4$$

(c) $x = 3\cos t - 1,\ y = -5\sin t - 2,\ z = 4$

$$\frac{(x+1)^2}{9} + \frac{(y+2)^2}{25} = 1, z = 4$$

(d) $x = -3\cos 2t + 1,\ y = 5\sin 2t + 2,\ z = 4$

$$\frac{(x-1)^2}{9} + \frac{(y-2)^2}{25} = 1, z = 4$$

(a) and (d) represent the same graph

81. One possible answer is

$\mathbf{r}(t) = 1.5\cos t\mathbf{i} + 1.5\sin t\mathbf{j} + \frac{1}{\pi}t\mathbf{k}, 0 \le t \le 2\pi$

Note that $\mathbf{r}(2\pi) = 1.5\mathbf{i} + 2\mathbf{k}$.

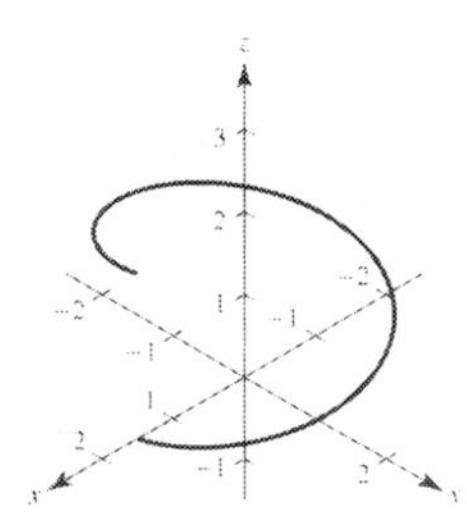

82. (a) View from the negative x-axis: $(-20, 0, 0)$

(b) View from above the first octant: $(10, 20, 10)$

(c) View from the z-axis: $(0, 0, 20)$

(d) View from the positive x-axis: $(20, 0, 0)$

83. Let $\mathbf{r}(t) = x_1(t)\mathbf{i} + y_1(t)\mathbf{j} + z_1(t)\mathbf{k}$ and $\mathbf{u}(t) = x_2(t)\mathbf{i} + y_2(t)\mathbf{j} + z_2(t)\mathbf{k}$. Then:

$$\begin{aligned}
\lim_{t\to c}\left[\mathbf{r}(t)\times\mathbf{u}(t)\right] &= \lim_{t\to c}\left\{\left[y_1(t)z_2(t) - y_2(t)z_1(t)\right]\mathbf{i} - \left[x_1(t)z_2(t) - x_2(t)z_1(t)\right]\mathbf{j} + \left[x_1(t)y_2(t) - x_2(t)y_1(t)\right]\mathbf{k}\right\} \\
&= \left[\lim_{t\to c} y_1(t)\lim_{t\to c} z_2(t) - \lim_{t\to c} y_2(t)\lim_{t\to c} z_1(t)\right]\mathbf{i} - \left[\lim_{t\to c} x_1(t)\lim_{t\to c} z_2(t) - \lim_{t\to c} x_2(t)\lim_{t\to c} z_1(t)\right]\mathbf{j} \\
&\quad + \left[\lim_{t\to c} x_1(t)\lim_{t\to c} y_2(t) - \lim_{t\to c} x_2(t)\lim_{t\to c} y_1(t)\right]\mathbf{k} \\
&= \left[\lim_{t\to c} x_1(t)\mathbf{i} + \lim_{t\to c} y_1(t)\mathbf{j} + \lim_{t\to c} z_1(t)\mathbf{k}\right]\times\left[\lim_{t\to c} x_2(t)\mathbf{i} + \lim_{t\to c} y_2(t)\mathbf{j} + \lim_{t\to c} z_2(t)\mathbf{k}\right] \\
&= \lim_{t\to c}\mathbf{r}(t)\times\lim_{t\to c}\mathbf{u}(t)
\end{aligned}$$

84. Let $\mathbf{r}(t) = x_1(t)\mathbf{i} + y_1(t)\mathbf{j} + z_1(t)\mathbf{k}$ and $\mathbf{u}(t) = x_2(t)\mathbf{i} + y_2(t)\mathbf{j} + z_2(t)\mathbf{k}$. Then:

$$\begin{aligned}
\lim_{t\to c}\left[\mathbf{r}(t)\cdot\mathbf{u}(t)\right] &= \lim_{t\to c}\left[x_1(t)x_2(t) + y_1(t)y_2(t) + z_1(t)z_2(t)\right] \\
&= \lim_{t\to c} x_1(t)\lim_{t\to c} x_2(t) + \lim_{t\to c} y_1(t)\lim_{t\to c} y_2(t) + \lim_{t\to c} z_1(t)\lim_{t\to c} z_2(t) \\
&= \left[\lim_{t\to c} x_1(t)\mathbf{i} + \lim_{t\to c} y_1(t)\mathbf{j} + \lim_{t\to c} z_1(t)\mathbf{k}\right]\cdot\left[\lim_{t\to c} x_2(t)\mathbf{i} + \lim_{t\to c} y_2(t)\mathbf{j} + \lim_{t\to c} z_2(t)\mathbf{k}\right] \\
&= \lim_{t\to c}\mathbf{r}(t)\cdot\lim_{t\to c}\mathbf{u}(t)
\end{aligned}$$

85. Let $\mathbf{r}(t) = x(t)\mathbf{i} + y(t)\mathbf{j} + z(t)\mathbf{k}$. Because $\mathbf{r}$ is continuous at $t = c$, then $\lim_{t\to c}\mathbf{r}(t) = \mathbf{r}(c)$.

$\mathbf{r}(c) = x(c)\mathbf{i} + y(c)\mathbf{j} + z(c)\mathbf{k} \Rightarrow x(c), y(c), z(c)$

are defined at c.

$$\|\mathbf{r}\| = \sqrt{(x(t))^2 + (y(t))^2 + (z(t))^2}$$

$$\lim_{t\to c}\|\mathbf{r}\| = \sqrt{(x(c))^2 + (y(c))^2 + (z(c))^2} = \|\mathbf{r}(c)\|$$

So, $\|\mathbf{r}\|$ is continuous at c.

86. Let

$$f(t) = \begin{cases} 1, & \text{if } t \ge 0 \\ -1, & \text{if } t < 0 \end{cases}$$

and $\mathbf{r}(t) = f(t)\mathbf{i}$. Then $\mathbf{r}$ is not continuous at $c = 0$, whereas, $\|\mathbf{r}\| = 1$ is continuous for all t.

87. $\mathbf{r}(t) = t^2\mathbf{i} + (9t - 20)\mathbf{j} + t^2\mathbf{k}$

$\mathbf{u}(s) = (3s + 4)\mathbf{i} + s^2\mathbf{j} + (5s - 4)\mathbf{k}$.

Equating components:

$$\begin{aligned}
t^2 &= 3s + 4 \\
9t - 20 &= s^2 \\
t^2 &= 5s - 4
\end{aligned}$$

So, $3s + 4 = 5s - 4 \Rightarrow s = 4$

$9t - 20 = s^2 = 16 \Rightarrow t = 4$.

The paths intersect at the same time $t = 4$ at the point $(16, 16, 16)$. The particles collide.

88. $\mathbf{r}(t) = t\mathbf{i} + t^2\mathbf{j} + t^3\mathbf{k}$

$\mathbf{u}(s) = (-2s + 3)\mathbf{i} + 8s\mathbf{j} + (12s + 2)\mathbf{k}$

Equating components

$t = -2s + 3$

$t^2 = 8s$

$t^3 = 12s + 2$

$(-2s + 3)^2 = 8s$

$4s^2 - 12s + 9 = 8s$

$4s^2 - 20s + 9 = 0$

$(2s - 9)(2s - 1) = 0$

For $s = \frac{1}{2}$, $t = -2\left(\frac{1}{2}\right) + 3 = 2$.

For $s = \frac{9}{2}$, $t = -2\left(\frac{9}{2}\right) + 3 = -6$ and

$t^2 = 8\left(\frac{9}{2}\right) = 36$ and $t^3 = 12\left(\frac{9}{2}\right) = 54$. Impossible.

The paths intersect at $(2, 4, 8)$, but at different times $\left(t = 2 \text{ and } s = \frac{1}{2}\right)$. No collision.

89. No, not necessarily. See Exercise 88.

90. Yes. See Exercise 87.

91. True

92. False. The graph of $x = y = z = t^3$ represents a line.

93. True. See Exercises 87 and 88.

94. True. $y^2 + z^2 = t^2 \sin^2 t + t^2 \cos^2 t = t^2 = x$

Section 12.2 Differentiation and Integration of Vector-Valued Functions

1. $\mathbf{r}(t) = t^2\mathbf{i} + t\mathbf{j}, t_0 = 2$

$x(t) = t^2, y(t) = t$

$x = y^2$

$\mathbf{r}(2) = 4\mathbf{i} + 2\mathbf{j}$

$\mathbf{r}'(t) = 2t\mathbf{i} + \mathbf{j}$

$\mathbf{r}'(2) = 4\mathbf{i} + \mathbf{j}$

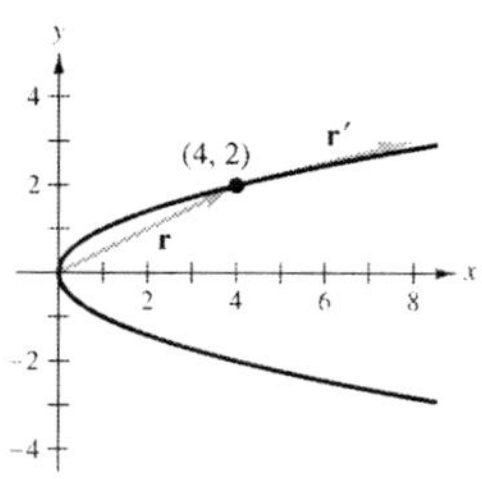

$\mathbf{r}'(t_0)$ is tangent to the curve at t_0.

2. (a) $\mathbf{r}(t) = (1 + t)\mathbf{i} + t^3\mathbf{j}, t_0 = 1$

$x = 1 + t$

$y = t^3 = (x - 1)^3$

(b) $\mathbf{r}(1) = 2\mathbf{i} + \mathbf{j}$

$\mathbf{r}'(t) = \mathbf{i} + 3t^2\mathbf{j}$

$\mathbf{r}'(1) = \mathbf{i} + 3\mathbf{j}$

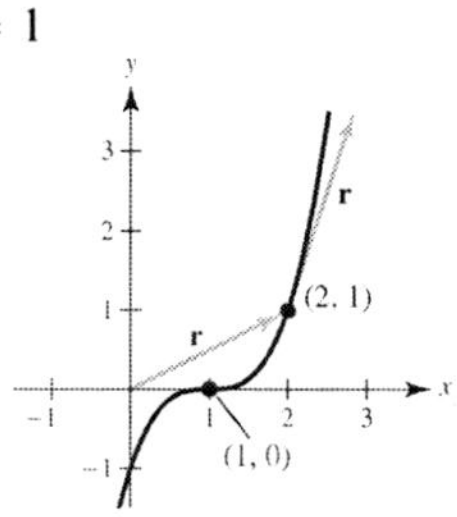

$\mathbf{r}'(t_0)$ is tangent to the curve at t_0.

3. $\mathbf{r}(t) = \cos t\mathbf{i} + \sin t\mathbf{j}, t_0 = \dfrac{\pi}{2}$

$x(t) = \cos t, y(t) = \sin t$

$x^2 + y^2 = 1$

$\mathbf{r}\left(\dfrac{\pi}{2}\right) = \mathbf{j}$

$\mathbf{r}'(t) = -\sin t\mathbf{i} + \cos t\mathbf{j}$

$\mathbf{r}'\left(\dfrac{\pi}{2}\right) = -\mathbf{i}$

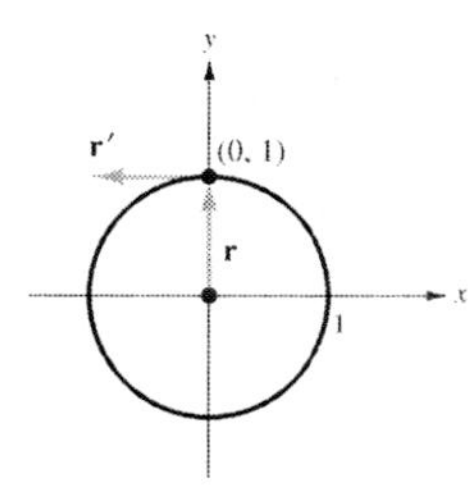

$\mathbf{r}'(t_0)$ is tangent to the curve at t_0.

4. $\mathbf{r}(t) = 3\sin t\mathbf{i} + 4\cos t\mathbf{j}, t_0 = \dfrac{\pi}{2}$

$x(t) = 3\sin t, y(t) = 4\cos t$

$\left(\dfrac{x}{3}\right)^2 + \left(\dfrac{y}{4}\right)^2 = 1$, ellipse

$\mathbf{r}\left(\dfrac{\pi}{2}\right) = 3i$

$\mathbf{r}'(t) = 3\cos t\mathbf{i} - 4\sin t\mathbf{j}$

$\mathbf{r}'\left(\dfrac{\pi}{2}\right) = -4\mathbf{j}$

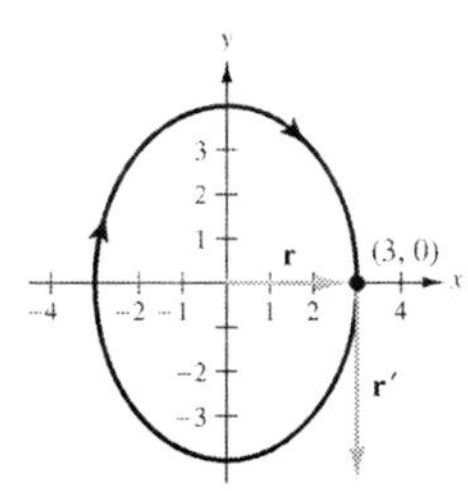

$\mathbf{r}'(t_0)$ is tangent to the curve at t_0.

5. $\mathbf{r}(t) = \langle e^t, e^{2t}\rangle,\ t_0 = 0$

$x(t) = e^t,\ y(t) = e^{2t} = (e^t)^2$

$y = x^2,\ x > 0$

$\mathbf{r}(0) = \langle 1, 1\rangle$

$\mathbf{r}'(t) = \langle e^t, 2e^{2t}\rangle$

$\mathbf{r}'(0) = \langle 1, 2\rangle$

$\mathbf{r}'(t_0)$ is tangent to the curve at t_0.

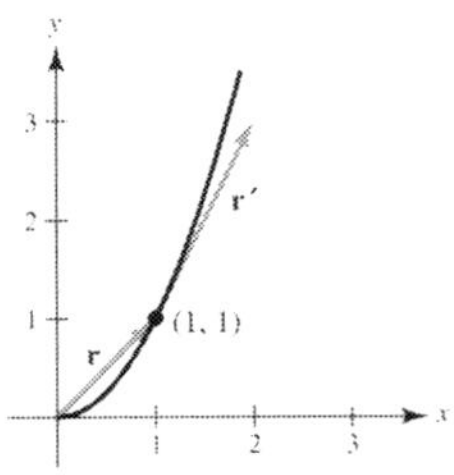

6. $\mathbf{r}(t) = \langle e^{-t}, e^t\rangle,\ t_0 = 0$

$x(t) = e^{-t} = \dfrac{1}{e^t},\ y(t) = e^t$

$y = \dfrac{1}{x},\ x > 0$

$\mathbf{r}(0) = \langle 1, 1\rangle$

$\mathbf{r}'(t) = \langle -e^{-t}, e^t\rangle$

$\mathbf{r}'(0) = \langle -1, 1\rangle$

$\mathbf{r}'(t_0)$ is tangent to the curve at t_0.

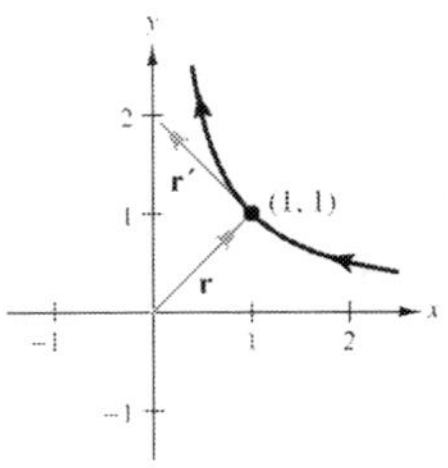

7. $\mathbf{r}(t) = 2\cos t\mathbf{i} + 2\sin t\mathbf{j} + t\mathbf{k},\ t_0 = \dfrac{3\pi}{2}$

$x^2 + y^2 = 4,\ z = t$

$\mathbf{r}'(t) = -2\sin t\mathbf{i} + 2\cos t\mathbf{j} + \mathbf{k}$

$\mathbf{r}\left(\dfrac{3\pi}{2}\right) = -2\mathbf{j} + \dfrac{3\pi}{2}\mathbf{k}$

$\mathbf{r}'\left(\dfrac{3\pi}{2}\right) = 2\mathbf{i} + \mathbf{k}$

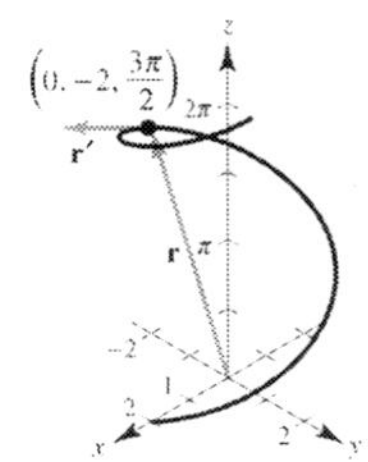

8. $\mathbf{r}(t) = t\mathbf{i} + t^2\mathbf{j} + \frac{3}{2}\mathbf{k},\ t_0 = 2$

$y = x^2,\ z = \frac{3}{2}$

$\mathbf{r}'(t) = \mathbf{i} + 2t\mathbf{j}$

$\mathbf{r}(2) = 2\mathbf{i} + 4\mathbf{j} + \frac{3}{2}\mathbf{k}$

$\mathbf{r}'(2) = \mathbf{i} + 4\mathbf{j}$

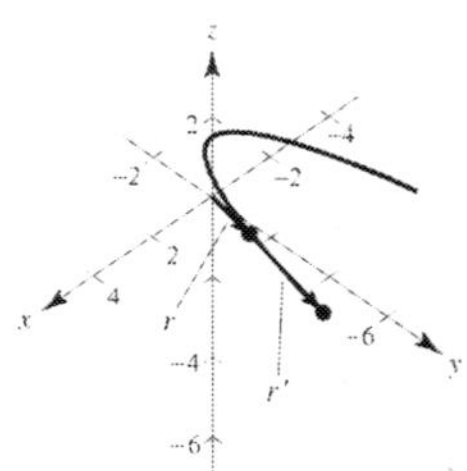

9. $\mathbf{r}(t) = t^3\mathbf{i} - 3t\mathbf{j}$

$\mathbf{r}'(t) = 3t^2\mathbf{i} - 3\mathbf{j}$

10. $\mathbf{r}(t) = \sqrt{t}\mathbf{i} + (1 - t^3)\mathbf{j}$

$\mathbf{r}'(t) = \dfrac{1}{2\sqrt{t}}\mathbf{i} - 3t^2\mathbf{j}$

11. $\mathbf{r}(t) = \langle 2\cos t, 5\sin t\rangle$

$\mathbf{r}'(t) = \langle -2\sin t, 5\cos t\rangle$

12. $\mathbf{r}(t) = \langle t\cos t, -2\sin t\rangle$

$\mathbf{r}'(t) = \langle -t\sin t + \cos t, -2\cos t\rangle$

13. $\mathbf{r}(t) = 6t\mathbf{i} - 7t^2\mathbf{j} + t^3\mathbf{k}$

$\mathbf{r}'(t) = 6\mathbf{i} - 14t\mathbf{j} + 3t^2\mathbf{k}$

14. $\mathbf{r}(t) = \dfrac{1}{t}\mathbf{i} + 16t\mathbf{j} + \dfrac{t^2}{2}\mathbf{k}$

$\mathbf{r}'(t) = -\dfrac{1}{t^2}\mathbf{i} + 16\mathbf{j} + t\mathbf{k}$

15. $\mathbf{r}(t) = a\cos^3 t\mathbf{i} + a\sin^3 t\mathbf{j} + \mathbf{k}$

$\mathbf{r}'(t) = -3a\cos^2 t\sin t\mathbf{i} + 3a\sin^2 t\cos t\mathbf{j}$

16. $\mathbf{r}(t) = 4\sqrt{t}\mathbf{i} + t^2\sqrt{t}\mathbf{j} + \ln t^2\mathbf{k}$

$$\mathbf{r}'(t) = \frac{2}{\sqrt{t}}\mathbf{i} + \left(2t\sqrt{t} + \frac{t^2}{2\sqrt{t}}\right)\mathbf{j} + \frac{2}{t}\mathbf{k}$$

$$= \frac{2}{\sqrt{t}}\mathbf{i} + \frac{5t^{3/2}}{2}\mathbf{j} + \frac{2}{t}\mathbf{k}$$

17. $\mathbf{r}(t) = e^{-t}\mathbf{i} + 4\mathbf{j} + 5te^t\mathbf{k}$

$\mathbf{r}'(t) = -e^{-t}\mathbf{i} + (5e^t + 5te^t)\mathbf{k}$

18. $\mathbf{r}(t) = \langle t^3, \cos 3t, \sin 3t\rangle$

$\mathbf{r}'(t) = \langle 3t^2, -3\sin 3t, 3\cos 3t\rangle$

19. $\mathbf{r}(t) = \langle t\sin t, t\cos t, t\rangle$

$\mathbf{r}'(t) = \langle \sin t + t\cos t, \cos t - t\sin t, 1\rangle$

20. $\mathbf{r}(t) = \langle \arcsin t, \arccos t, 0\rangle$

$$\mathbf{r}'(t) = \left\langle \frac{1}{\sqrt{1 - t^2}}, -\frac{1}{\sqrt{1 - t^2}}, 0\right\rangle$$

21. $\mathbf{r}(t) = t^3\mathbf{i} + \dfrac{1}{2}t^2\mathbf{j}$

(a) $\mathbf{r}'(t) = 3t^2\mathbf{i} + t\mathbf{j}$

(b) $\mathbf{r}''(t) = 6t\mathbf{i} + \mathbf{j}$

(c) $\mathbf{r}'(t)\cdot\mathbf{r}''(t) = 3t^2(6t) + t = 18t^3 + t$

22. $\mathbf{r}(t) = (t^2 + t)\mathbf{i} + (t^2 - t)\mathbf{j}$

(a) $\mathbf{r}'(\mathbf{t}) = (2t + 1)\mathbf{i} + (2t - 1)\mathbf{j}$

(b) $\mathbf{r}''(t) = 2\mathbf{i} + 2\mathbf{j}$

(c) $\mathbf{r}'(t)\cdot\mathbf{r}''(t) = (2t + 1)(2) + (2t - 1)(2) = 8t$

23. $\mathbf{r}(t) = 4\cos t\mathbf{i} + 4\sin t\mathbf{j}$

(a) $\mathbf{r}'(t) = -4\sin t\mathbf{i} + 4\cos t\mathbf{j}$

(b) $\mathbf{r}''(t) = -4\cos t\mathbf{i} - 4\sin t\mathbf{j}$

(c) $\mathbf{r}'(t) \cdot \mathbf{r}''(t) = (-4\sin t)(-4\cos t) + 4\cos t(-4\sin t) = 0$

24. $\mathbf{r}(t) = 8\cos t\mathbf{i} + 3\sin t\mathbf{j}$

(a) $\mathbf{r}'(t) = -8\sin t\mathbf{i} + 3\cos t\mathbf{j}$

(b) $\mathbf{r}''(t) = -8\cos t\mathbf{i} - 3\sin t\mathbf{j}$

(c) $\mathbf{r}'(t) \cdot \mathbf{r}''(t) = (-8\sin t)(-8\cos t) + 3\cos t(-3\sin t) = 55\sin t\cos t$

25. $\mathbf{r}(t) = \frac{1}{2}t^2\mathbf{i} - t\mathbf{j} + \frac{1}{6}t^3\mathbf{k}$

(a) $\mathbf{r}'(t) = t\mathbf{i} - \mathbf{j} + \frac{1}{2}t^2\mathbf{k}$

(b) $\mathbf{r}''(t) = \mathbf{i} + t\mathbf{k}$

(c) $\mathbf{r}'(t) \cdot \mathbf{r}''(t) = t(1) + (-1)(0) + \frac{1}{2}t^2(t) = t + \frac{1}{2}t^3$

(d) $\mathbf{r}'(t) \times \mathbf{r}''(t) = \begin{bmatrix} \mathbf{i} & \mathbf{j} & \mathbf{k} \\ t & -1 & \frac{1}{2}t^2 \\ 1 & 0 & t \end{bmatrix}$

$= (-t)\mathbf{i} - \left(t^2 - \frac{1}{2}t^2\right)\mathbf{j} + \mathbf{k}$

$= -t\mathbf{i} - \frac{1}{2}t^2\mathbf{j} + \mathbf{k}$

26. $\mathbf{r}(t) = t^3\mathbf{i} + (2t^2 + 3)\mathbf{j} + (3t - 5)\mathbf{k}$

(a) $\mathbf{r}'(t) = 3t^2\mathbf{i} + 4t\mathbf{j} + 3\mathbf{k}$

(b) $\mathbf{r}''(t) = 6t\mathbf{i} + 4\mathbf{j}$

(c) $\mathbf{r}'(t) \cdot \mathbf{r}''(t) = 3t^2(6t) + 4t(4) + 3(0)$

$= 18t^3 + 16t$

(d) $\mathbf{r}'(t) \times \mathbf{r}''(t) = \begin{bmatrix} \mathbf{i} & \mathbf{j} & \mathbf{k} \\ 3t^2 & 4t & 3 \\ 6t & 4 & 0 \end{bmatrix}$

$= (0 - 12)\mathbf{i} - (0 - 18t)\mathbf{j} + (12t^2 - 24t^2)\mathbf{k}$

$= -12\mathbf{i} + 18t\mathbf{j} - 12t^2\mathbf{k}$

27. $\mathbf{r}(t) = \langle \cos t + t\sin t, \sin t - t\cos t, t\rangle$

(a) $\mathbf{r}'(t) = \langle -\sin t + \sin t + t\cos t, \cos t - \cos t + t\sin t, 1\rangle$
$= \langle t\cos t, t\sin t, 1\rangle$

(b) $\mathbf{r}''(t) = \langle \cos t - t\sin t, \sin t + t\cos t, 0\rangle$

(c) $\mathbf{r}'(t)\cdot\mathbf{r}''(t) = (t\cos t)(\cos t - t\sin t) + t\sin t(\sin t + t\cos t) + 1(0)$
$= t\cos^2 t - t^2\cos t\sin t + t\sin^2 t + t^2\sin t\cos t$
$= t(\cos^2 t + \sin^2 t) = t$

(d) $$\mathbf{r}'(t)\times\mathbf{r}''(t) = \begin{bmatrix} \mathbf{i} & \mathbf{j} & \mathbf{k} \\ t\cos t & t\sin t & 1 \\ \cos t - t\sin t & \sin t + t\cos t & 0 \end{bmatrix}$$
$= (-\sin t - t\cos t)\mathbf{i} + (\cos t - t\sin t)\mathbf{j} + (t\cos t\sin t + t^2\cos^2 t - t\sin t\cos t + t^2\sin^2 t)\mathbf{k}$
$= (-\sin t - t\cos t)\mathbf{i} + (\cos t - t\sin t)\mathbf{j} + t^2\mathbf{k}$
$= \langle -\sin t - t\cos t, \cos t - t\sin t, t^2\rangle$

28. $\mathbf{r}(t) = \langle e^{-t}, t^2, \tan t\rangle$

(a) $\mathbf{r}'(t) = \langle -e^{-t}, 2t, \sec^2 t\rangle$

(b) $\mathbf{r}''(t) = \langle e^{-t}, 2, 2\sec^2 t\tan t\rangle$

(c) $\mathbf{r}'(t)\cdot\mathbf{r}''(t) = -e^{-2t} + 4t + 2\sec^4 t\tan t$

(d) $$\mathbf{r}'(t)\times\mathbf{r}''(t) = \begin{bmatrix} \mathbf{i} & \mathbf{j} & \mathbf{k} \\ -e^{-t} & 2t & \sec^2 t \\ e^{-t} & 2 & 2\sec^2 t\tan t \end{bmatrix}$$
$= \left[2t(2\sec^2 t\tan t) - 2\sec^2 t\right]\mathbf{i} - \left[-2e^{-t}\sec^2 t\tan t - e^{-t}\sec^2 t\right]\mathbf{j} + \left[-2e^{-t} - 2te^{-t}\right]\mathbf{k}$
$= \left[4t\sec^2 t\tan t - 2\sec^2 t\right]\mathbf{i} + \left[2e^{-t}\sec^2 t\tan t + e^{-t}\sec^2 t\right]\mathbf{j} - \left[2e^{-t} + 2te^{-t}\right]\mathbf{k}$

29. $\mathbf{r}(t) = t^2\mathbf{i} + t^3\mathbf{j}$

$\mathbf{r}'(t) = 2t\mathbf{i} + 3t^2\mathbf{j}$

$\mathbf{r}'(0) = \mathbf{0}$

Smooth on $(-\infty, 0), (0, \infty)$

30. $\mathbf{r}(t) = \dfrac{1}{t-1}\mathbf{i} + 3t\mathbf{j}$

$\mathbf{r}'(t) = -\dfrac{1}{(t-1)^2}\mathbf{i} + 3\mathbf{j}$

Not continuous when $t = 1$

Smooth on $(-\infty, 1), (1, \infty)$

31. $\mathbf{r}(\theta) = 2\cos^3\theta\mathbf{i} + 3\sin^3\theta\mathbf{j}$

$\mathbf{r}'(\theta) = -6\cos^2\theta\sin\theta\mathbf{i} + 9\sin^2\theta\cos\theta\mathbf{j}$

$\mathbf{r}'\left(\dfrac{n\pi}{2}\right) = \mathbf{0}$

Smooth on $\left(\dfrac{n\pi}{2}, \dfrac{(n+1)\pi}{2}\right)$, n any integer.

32. $\mathbf{r}(\theta) = (\theta + \sin\theta)\mathbf{i} + (1 - \cos\theta)\mathbf{j}$

$\mathbf{r}'(\theta) = (1 + \cos\theta)\mathbf{i} + \sin\theta\mathbf{j}$

$\mathbf{r}'((2n-1)\pi) = \mathbf{0}$, n any integer

Smooth on $((2n-1)\pi, (2n+1)\pi)$

33. $\mathbf{r}(\theta) = (\theta - 2\sin\theta)\mathbf{i} + (1 - 2\cos\theta)\mathbf{j}$

$\mathbf{r}'(\theta) = (1 - 2\cos\theta)\mathbf{i} + (2\sin\theta)\mathbf{j}$

$\mathbf{r}'(\theta) \neq \mathbf{0}$ for any value of θ

Smooth on $(-\infty, \infty)$

34. $\mathbf{r}(t) = \dfrac{2t}{8+t^3}\mathbf{i} + \dfrac{2t^2}{8+t^3}\mathbf{j}$

$\mathbf{r}'(t) = \dfrac{16-4t^3}{(t^3+8)^2}\mathbf{i} + \dfrac{32t-2t^4}{(t^3+8)^2}\mathbf{j}$

$\mathbf{r}'(t) \neq \mathbf{0}$ for any value of t.

$\mathbf{r}$ is not continuous when $t = -2$.

Smooth on $(-\infty, -2), (-2, \infty)$

35. $\mathbf{r}(t) = (t-1)\mathbf{i} + \dfrac{1}{t}\mathbf{j} - t^2\mathbf{k}$

$\mathbf{r}'(t) = \mathbf{i} - \dfrac{1}{t^2}\mathbf{j} - 2t\mathbf{k} \neq \mathbf{0}$

$\mathbf{r}$ is smooth or all $t \neq 0$: $(-\infty, 0), (0, \infty)$

36. $\mathbf{r}(t) = e^t\mathbf{i} - e^{-t}\mathbf{j} + 3t\mathbf{k}$

$\mathbf{r}'(t) = e^t\mathbf{i} + e^{-t}\mathbf{j} + 3\mathbf{k} \neq \mathbf{0}$

$\mathbf{r}$ is smooth for all t: $(-\infty, \infty)$

37. $\mathbf{r}(t) = t\mathbf{i} - 3t\mathbf{j} + \tan t\mathbf{k}$

$\mathbf{r}'(t) = \mathbf{i} - 3\mathbf{j} + \sec^2 t\mathbf{k} \neq \mathbf{0}$

$\mathbf{r}$ is smooth for all $t \neq \dfrac{\pi}{2} + n\pi = \dfrac{2n+1}{2}\pi$.

Smooth on intervals of form $\left(-\dfrac{\pi}{2} + n\pi, \dfrac{\pi}{2} + n\pi\right)$, n is an integer.

38. $\mathbf{r}(t) = \sqrt{t}\mathbf{i} + (t^2-1)\mathbf{j} + \dfrac{1}{4}t\mathbf{k}$

$\mathbf{r}'(t) = \dfrac{1}{2\sqrt{t}}\mathbf{i} + 2t\mathbf{j} + \dfrac{1}{4}\mathbf{k} \neq \mathbf{0}$

$\mathbf{r}$ is smooth for all $t > 0$: $(0, \infty)$

39. $\mathbf{r}(t) = t\mathbf{i} + 3t\mathbf{j} + t^2\mathbf{k}$, $\mathbf{u}(t) = 4t\mathbf{i} + t^2\mathbf{j} + t^3\mathbf{k}$

$\mathbf{r}'(t) = \mathbf{i} + 3\mathbf{j} + 2t\mathbf{k}$, $\mathbf{u}'(t) = 4\mathbf{i} + 2t\mathbf{j} + 3t^2\mathbf{k}$

(a) $\mathbf{r}'(t) = \mathbf{i} + 3\mathbf{j} + 2t\mathbf{k}$

(b)
$$\begin{aligned}\frac{d}{dt}\left[3\mathbf{r}(t) - \mathbf{u}(t)\right] &= 3\mathbf{r}'(t) - \mathbf{u}'(t)\\ &= 3(\mathbf{i} + 3\mathbf{j} + 2t\mathbf{k}) - (4\mathbf{i} + 2t\mathbf{j} + 3t^2\mathbf{k})\\ &= (3-4)\mathbf{i} + (9-2t)\mathbf{j} + (6t-3t^2)\mathbf{k}\\ &= -\mathbf{i} + (9-2t)\mathbf{j} + (6t-3t^2)\mathbf{k}\end{aligned}$$

(c)
$$\begin{aligned}\frac{d}{dt}(5t)\mathbf{u}(t) &= (5t)\mathbf{u}'(t) + 5\mathbf{u}(t)\\ &= 5t(4\mathbf{i} + 2t\mathbf{j} + 3t^2\mathbf{k}) + 5(4t\mathbf{i} + t^2\mathbf{j} + t^3\mathbf{k})\\ &= (20t + 20t)\mathbf{i} + (10t^2 + 5t^2)\mathbf{j} + (15t^3 + 5t^3)\mathbf{k}\\ &= 40t\mathbf{i} + 15t^2\mathbf{j} + 20t^3\mathbf{k}\end{aligned}$$

(d)
$$\begin{aligned}\frac{d}{dt}\left[\mathbf{r}(t)\cdot\mathbf{u}(t)\right] &= \mathbf{r}(t)\cdot\mathbf{u}'(t) + \mathbf{r}'(t)\cdot\mathbf{u}(t)\\ &= \left[(t)(4) + (3t)(2t) + (t^2)(3t^2)\right] + \left[(1)(4t) + (3)(t^2) + (2t)(t^3)\right]\\ &= (4t + 6t^2 + 3t^4) + (4t + 3t^2 + 2t^4)\\ &= 8t + 9t^2 + 5t^4\end{aligned}$$

(e)
$$\begin{aligned}\frac{d}{dt}\left[\mathbf{r}(t)\times\mathbf{u}(t)\right] &= \mathbf{r}(t)\times\mathbf{u}'(t) + \mathbf{r}'(t)\times\mathbf{u}(t)\\ &= \left[7t^3\mathbf{i} + (4t^2 - 3t^3)\mathbf{j} + (2t^2 - 12t)\mathbf{k}\right] + \left[t^3\mathbf{i} + (8t^2 - t^3)\mathbf{j} + (t^2 - 12t)\mathbf{k}\right]\\ &= 8t^3\mathbf{i} + (12t^2 - 4t^3)\mathbf{j} + (3t^2 - 24t)\mathbf{k}\end{aligned}$$

(f)
$$\begin{aligned}\frac{d}{dt}\mathbf{r}(2t) &= 2\mathbf{r}'(2t)\\ &= 2\left[\mathbf{i} + 3\mathbf{j} + 2(2t)\mathbf{k}\right]\\ &= 2\mathbf{i} + 6\mathbf{j} + 8t\mathbf{k}\end{aligned}$$

40. $\mathbf{r}(t) = t\mathbf{i} + 2\sin t\mathbf{j} + 2\cos t\mathbf{k},\ \mathbf{u}(t) = \frac{1}{t}\mathbf{i} + 2\sin t\mathbf{j} + 2\cos t\mathbf{k}$

$\mathbf{r}'(t) = \mathbf{i} + 2\cos t\mathbf{j} - 2\sin t\mathbf{k},\ \mathbf{u}'(t) = -\frac{1}{t^2}\mathbf{i} + 2\cos t\mathbf{j} - 2\sin t\mathbf{k}$

(a) $\mathbf{r}'(t) = \mathbf{i} + 2\cos t\mathbf{j} - 2\sin t\mathbf{k}$

(b) $$\begin{aligned}\frac{d}{dt}\left[3\mathbf{r}(t) - \mathbf{u}(t)\right] &= 3\mathbf{r}'(t) - \mathbf{u}'(t)\\ &= 3(\mathbf{i} + 2\cos t\mathbf{j} - 2\sin t\mathbf{k}) - \left(-\frac{1}{t^2}\mathbf{i} + 2\cos t\mathbf{j} - 2\sin t\mathbf{k}\right)\\ &= \left(3 + \frac{1}{t^2}\right)\mathbf{i} + (6\cos t - 2\cos t)\mathbf{j} + (-6\sin t + 2\sin t)\mathbf{k}\\ &= \left(3 + \frac{1}{t^2}\right)\mathbf{i} + 4\cos t\mathbf{j} - 4\sin t\mathbf{k}\end{aligned}$$

(c) $$\begin{aligned}\frac{d}{dt}\left[(5t)\mathbf{u}(t)\right] &= (5t)\mathbf{u}'(t) + 5\mathbf{u}(t)\\ &= 5t\left(-\frac{1}{t^2}\mathbf{i} + 2\cos t\mathbf{j} - 2\sin t\mathbf{k}\right) + 5\left(\frac{1}{t}\mathbf{i} + 2\sin t\mathbf{j} + 2\cos t\mathbf{k}\right)\\ &= \left(-\frac{5}{t}\mathbf{i} + 10t\cos t\mathbf{j} - 10t\sin t\mathbf{k}\right) + \left(\frac{5}{t}\mathbf{i} + 10\sin t\mathbf{j} + 10\cos t\mathbf{k}\right)\\ &= 10(t\cos t + \sin t)\mathbf{j} + 10(\cos t - t\sin t)\mathbf{k}\end{aligned}$$

(d) $$\begin{aligned}\frac{d}{dt}\left[\mathbf{r}(t)\cdot\mathbf{u}(t)\right] &= \mathbf{r}(t)\cdot\mathbf{u}'(t) + \mathbf{r}'(t)\cdot\mathbf{u}(t)\\ &= \left[(t)\left(-\frac{1}{t^2}\right) + (2\sin t)(2\cos t) + (2\cos t)(-2\sin t)\right] + \left[(1)\left(\frac{1}{t}\right) + (2\cos t)(2\sin t) + (-2\sin t)(2\cos t)\right]\\ &= \left(-\frac{1}{t} + 4\sin t\cos t - 4\sin t\cos t\right) + \left(\frac{1}{t} + 4\sin t\cos t - 4\sin t\cos t\right)\\ &= 0(t \neq 0)\end{aligned}$$

(e) $$\begin{aligned}\frac{d}{dt}\left[\mathbf{r}(t)\times\mathbf{u}(t)\right] &= \mathbf{r}(t)\times\mathbf{u}'(t) + \mathbf{r}'(t)\times\mathbf{u}(t)\\ &= \left[-4\mathbf{i} + \left(-\frac{2}{t^2}\cos t + 2t\sin t\right)\mathbf{j} + \left(2t\cos t + \frac{2}{t^2}\sin t\right)\mathbf{k}\right]\\ &\quad + \left[4\mathbf{i} + \left(-\frac{2}{t}\sin t - 2\cos t\right)\mathbf{j} + \left(2\sin t - \frac{2}{t}\cos t\right)\mathbf{k}\right]\\ &= 2\left[\left(t - \frac{1}{t}\right)\sin t - \left(\frac{1}{t^2} + 1\right)\cos t\right]\mathbf{j} + 2\left[\left(1 + \frac{1}{t^2}\right)\sin t + \left(t - \frac{1}{t}\right)\cos t\right]\mathbf{k}\end{aligned}$$

(f) $$\begin{aligned}\frac{d}{dt}\mathbf{r}(2t) &= 2\mathbf{r}'(2t)\\ &= 2(\mathbf{i} + 2\cos(2t)\mathbf{j} - 2\sin(2t)\mathbf{k})\\ &= 2\mathbf{i} + 4\cos(2t)\mathbf{j} - 4\sin(2t)\mathbf{k}\end{aligned}$$

41. $\mathbf{r}(t) = t\mathbf{i} + 2t^2\mathbf{j} + t^3\mathbf{k}, \mathbf{u}(t) = t^4\mathbf{k}$

(a) $\mathbf{r}(t) \cdot \mathbf{u}(t) = t^7$

(i) $D_t[\mathbf{r}(t) \cdot \mathbf{u}(t)] = 7t^6$

(ii) Alternate Solution:

$$D_t[\mathbf{r}(t) \cdot \mathbf{u}(t)] = \mathbf{r}(t) \cdot \mathbf{u}'(t) + \mathbf{r}'(t) \cdot \mathbf{u}(t) = \left(t\mathbf{i} + 2t^2\mathbf{j} + t^3\mathbf{k}\right) \cdot \left(4t^3\mathbf{k}\right) + \left(\mathbf{i} + 4t\mathbf{j} + 3t^2\mathbf{k}\right) \cdot \left(t^4\mathbf{k}\right) = 4t^6 + 3t^6 = 7t^6$$

(b) $$\mathbf{r}(t) \times \mathbf{u}(t) = \begin{vmatrix} \mathbf{i} & \mathbf{j} & \mathbf{k} \\ t & 2t^2 & t^3 \\ 0 & 0 & t^4 \end{vmatrix} = 2t^6\mathbf{i} - t^5\mathbf{j}$$

(i) $D_t[\mathbf{r}(t) \times \mathbf{u}(t)] = 12t^5\mathbf{i} - 5t^4\mathbf{j}$

(ii) Alternate Solution: $$D_t[\mathbf{r}(t) \times \mathbf{u}(t)] = \mathbf{r}(t) \times \mathbf{u}'(t) \times \mathbf{r}'(t) \times \mathbf{u}(t) = \begin{vmatrix} \mathbf{i} & \mathbf{j} & \mathbf{k} \\ t & 2t^2 & t^3 \\ 0 & 0 & 4t^3 \end{vmatrix} + \begin{vmatrix} \mathbf{i} & \mathbf{j} & \mathbf{k} \\ 1 & 4t & 3t^2 \\ 0 & 0 & t^4 \end{vmatrix} = 12t^5\mathbf{i} - 5t^4\mathbf{j}$$

42. $\mathbf{r}(t) = \cos t\mathbf{i} + \sin t\mathbf{j} + t\mathbf{k}, \mathbf{u}(t) = \mathbf{j} + t\mathbf{k}$

(a) $\mathbf{r}(t) \cdot \mathbf{u}(t) = \sin t + t^2$

(i) $D_t[\mathbf{r}(t) \cdot \mathbf{u}(t)] = \cos t + 2t$

(ii) Alternate Solution:

$$\begin{aligned} D_t[\mathbf{r}(t) \cdot \mathbf{u}(t)] &= \mathbf{r}(t) \cdot \mathbf{u}'(t) + \mathbf{r}'(t) \cdot \mathbf{u}(t) \\ &= (\cos t\mathbf{i} + \sin t\mathbf{j} + t\mathbf{k}) \cdot \mathbf{k} + (-\sin t\mathbf{i} + \cos t\mathbf{j} + \mathbf{k}) \cdot (\mathbf{j} + t\mathbf{k}) = t + \cos t + t = 2t + \cos t \end{aligned}$$

(b) $$\mathbf{r}(t) \times \mathbf{u}(t) = \begin{vmatrix} \mathbf{i} & \mathbf{j} & \mathbf{k} \\ \cos t & \sin t & t \\ 0 & 1 & t \end{vmatrix} = (t \sin t - t)\mathbf{i} - (t \cos t)\mathbf{j} + \cos t\mathbf{k}$$

(i) $D_t[\mathbf{r}(t) \times \mathbf{u}(t)] = (t \cos t + \sin t - 1)\mathbf{i} - (\cos t - t \sin t)\mathbf{j} - \sin t\mathbf{k}$

(ii) Alternate Solution:

$$\begin{aligned} D_t[\mathbf{r}(t) \times \mathbf{u}(t)] &= \mathbf{r}(t) \times \mathbf{u}'(t) + \mathbf{r}'(t) \times \mathbf{u}(t) \\ &= \begin{vmatrix} \mathbf{i} & \mathbf{j} & \mathbf{k} \\ \cos t & \sin t & t \\ 0 & 0 & 1 \end{vmatrix} + \begin{vmatrix} \mathbf{i} & \mathbf{j} & \mathbf{k} \\ -\sin t & \cos t & 1 \\ 0 & 1 & t \end{vmatrix} = (\sin t + t \cos t - 1)\mathbf{i} + (t \sin t - \cos t)\mathbf{j} - \sin t\mathbf{k} \end{aligned}$$

43. $\int (2t\mathbf{i} + \mathbf{j} + \mathbf{k})dt = t^2\mathbf{i} + t\mathbf{j} + t\mathbf{k} + \mathbf{C}$

44. $\int \left(4t^3\mathbf{i} + 6t\mathbf{j} - 4\sqrt{t}\mathbf{k}\right) dt = t^4\mathbf{i} + 3t^2\mathbf{j} - \frac{8}{3}t^{3/2}\mathbf{k} + \mathbf{C}$

45. $\int \left(\frac{1}{t}\mathbf{i} + \mathbf{j} - t^{3/2}\mathbf{k}\right) dt = \ln|t|\mathbf{i} + t\mathbf{j} - \frac{2}{5}t^{5/2}\mathbf{k} + \mathbf{C}$

46. $\int \left[\ln t\mathbf{i} + \frac{1}{t}\mathbf{j} + \mathbf{k}\right] dt = \left(t \ln|t| - t\right)\mathbf{i} + \ln t\mathbf{j} + t\mathbf{k} + \mathbf{C}$

(Integration by parts)

47. $\int \left[(2t - 1)\mathbf{i} + 4t^3\mathbf{j} + 3\sqrt{t}\mathbf{k}\right]dt = \left(t^2 - t\right)\mathbf{i} + t^4\mathbf{j} + 2t^{3/2}\mathbf{k} + \mathbf{C}$

48. $\int \left[e^t\mathbf{i} + \sin t\mathbf{j} + \cos t\mathbf{k}\right]dt = e^t\mathbf{i} - \cos t\mathbf{j} + \sin t\mathbf{k} + \mathbf{C}$

49. $\int \left[\sec^2 t\mathbf{i} + \frac{1}{1 + t^2}\mathbf{j}\right]dt = \tan t\mathbf{i} + \arctan t\mathbf{j} + \mathbf{C}$

50. $\int\left[e^{-t}\sin t\mathbf{i} + e^{-t}\cos t\mathbf{j}\right]dt = \frac{e^{-t}}{2}(-\sin t - \cos t)\mathbf{i} + \frac{e^{-t}}{2}(-\cos t + \sin t)\mathbf{j} + \mathbf{C}$

51. $\int_0^1 (8t\mathbf{i} + t\mathbf{j} - \mathbf{k})\,dt = \left[4t^2\mathbf{i}\right]_0^1 + \left[\frac{t^2}{2}\mathbf{j}\right]_0^1 - \left[t\mathbf{k}\right]_0^1 = 4\mathbf{i} + \frac{1}{2}\mathbf{j} - \mathbf{k}$

52. $\int_{-1}^1 \left(t\mathbf{i} + t^3\mathbf{j} + \sqrt[3]{t}\mathbf{k}\right)dt = \left[\frac{t^2}{2}\mathbf{i}\right]_{-1}^1 + \left[\frac{t^4}{4}\mathbf{j}\right]_{-1}^1 + \left[\frac{3}{4}t^{4/3}\mathbf{k}\right]_{-1}^1 = \mathbf{0}$

53. $\int_0^{\pi/2}\left[(a\cos t)\mathbf{i} + (a\sin t)\mathbf{j} + \mathbf{k}\right]dt = \left[a\sin t\mathbf{i}\right]_0^{\pi/2} - \left[a\cos t\mathbf{j}\right]_0^{\pi/2} + \left[t\mathbf{k}\right]_0^{\pi/2} = a\mathbf{i} + a\mathbf{j} + \frac{\pi}{2}\mathbf{k}$

54. $\int_0^{\pi/4}\left[(\sec t\tan t)\mathbf{i} + (\tan t)\mathbf{j} + (2\sin t\cos t)\mathbf{k}\right]dt = \left[\sec t\mathbf{i} + \ln\left|\sec t\right|\mathbf{j} + \sin^2 t\mathbf{k}\right]_0^{\pi/4} = \left(\sqrt{2} - 1\right)\mathbf{i} + \ln\sqrt{2}\mathbf{j} + \frac{1}{2}\mathbf{k}$

55. $\int_0^2 \left(t\mathbf{i} + e^t\mathbf{j} - te^t\mathbf{k}\right)dt = \left[\frac{t^2}{2}\mathbf{i}\right]_0^2 + \left[e^t\mathbf{j}\right]_0^2 - \left[(t-1)e^t\mathbf{k}\right]_0^2 = 2\mathbf{i} + \left(e^2 - 1\right)\mathbf{j} - \left(e^2 + 1\right)\mathbf{k}$

56. $\left\|t\mathbf{i} + t^2\mathbf{j}\right\| = \sqrt{t^2 + t^4} = t\sqrt{1 + t^2}$ for $t \geq 0$

$\int_0^3\left\|t\mathbf{i} + t^2\mathbf{j}\right\|dt = \int_0^3 t\sqrt{1 + t^2}\,dt = \left[\frac{1}{3}\left(1 + t^2\right)^{3/2}\right]_0^3 = \frac{1}{3}\left(10^{3/2} - 1\right)$

57. $\mathbf{r}(t) = \int\left(4e^{2t}\mathbf{i} + 3e^t\mathbf{j}\right)dt = 2e^{2t}\mathbf{i} + 3e^t\mathbf{j} + \mathbf{C}$

$\mathbf{r}(0) = 2\mathbf{i} + 3\mathbf{j} + \mathbf{C} = 2\mathbf{i} \Rightarrow \mathbf{C} = -3\mathbf{j}$

$\mathbf{r}(t) = 2e^{2t}\mathbf{i} + 3\left(e^t - 1\right)\mathbf{j}$

58. $\mathbf{r}(t) = \int\left(3t^2\mathbf{j} + 6\sqrt{t}\mathbf{k}\right)dt = t^3\mathbf{j} + 4t^{3/2}\mathbf{k} + \mathbf{C}$

$\mathbf{r}(0) = C = \mathbf{i} + 2\mathbf{j}$

$\mathbf{r}(t) = \mathbf{i} + \left(2 + t^3\right)\mathbf{j} + 4t^{3/2}\mathbf{k}$

59. $\mathbf{r}'(t) = \int -32\mathbf{j}\,dt = -32t\mathbf{j} + \mathbf{C}_1$

$\mathbf{r}'(0) = \mathbf{C}_1 = 600\sqrt{3}\mathbf{i} + 600\mathbf{j}$

$\mathbf{r}'(t) = 600\sqrt{3}\mathbf{i} + (600 - 32t)\mathbf{j}$

$\mathbf{r}(t) = \int\left[600\sqrt{3}\mathbf{i} + (600 - 32t)\mathbf{j}\right]dt$

$= 600\sqrt{3}t\mathbf{i} + \left(600t - 16t^2\right)\mathbf{j} + \mathbf{C}$

$\mathbf{r}(0) = \mathbf{C} = \mathbf{0}$

$\mathbf{r}(t) = 600\sqrt{3}t\mathbf{i} + \left(600t - 16t^2\right)\mathbf{j}$

60. $\mathbf{r}''(t) = -4\cos t\mathbf{j} - 3\sin t\mathbf{k}$

$\mathbf{r}'(t) = -4\sin t\mathbf{j} + 3\cos t\mathbf{k} + \mathbf{C}_1$

$\mathbf{r}'(0) = 3\mathbf{k} = 3\mathbf{k} + \mathbf{C}_1 \Rightarrow \mathbf{C}_1 = \mathbf{0}$

$\mathbf{r}(t) = 4\cos t\mathbf{j} + 3\sin t\mathbf{k} + \mathbf{C}_2$

$\mathbf{r}(0) = 4j + \mathbf{C}_2 = 4\mathbf{j} \Rightarrow \mathbf{C}_2 = 0$

$\mathbf{r}(t) = 4\cos t\mathbf{j} + 3\sin t\mathbf{k}$

61. $\mathbf{r}(t) = \int\left(te^{-t^2}\mathbf{i} - e^{-t}\mathbf{j} + \mathbf{k}\right)dt = -\frac{1}{2}e^{-t^2}\mathbf{i} + e^{-t}\mathbf{j} + t\mathbf{k} + \mathbf{C}$

$\mathbf{r}(0) = -\frac{1}{2}\mathbf{i} + \mathbf{j} + \mathbf{C} = \frac{1}{2}\mathbf{i} - \mathbf{j} + \mathbf{k} \Rightarrow \mathbf{C} = \mathbf{i} - 2\mathbf{j} + \mathbf{k}$

$\mathbf{r}(t) = \left(1 - \frac{1}{2}e^{-t^2}\right)\mathbf{i} + \left(e^{-t} - 2\right)\mathbf{j} + (t + 1)\mathbf{k}$

$= \left(\frac{2 - e^{-t^2}}{2}\right)\mathbf{i} + \left(e^{-t} - 2\right)\mathbf{j} + (t + 1)\mathbf{k}$

62. $\mathbf{r}(t) = \int\left[\frac{1}{1 + t^2}\mathbf{i} + \frac{1}{t^2}\mathbf{j} + \frac{1}{t}\mathbf{k}\right]dt$

$= \arctan t\mathbf{i} - \frac{1}{t}\mathbf{j} + \ln t\mathbf{k} + \mathbf{C}$

$\mathbf{r}(1) = \frac{\pi}{4}\mathbf{i} - \mathbf{j} + \mathbf{C} = 2\mathbf{i} \Rightarrow \mathbf{C} = \left(2 - \frac{\pi}{4}\right)\mathbf{i} + \mathbf{j}$

$\mathbf{r}(t) = \left[2 - \frac{\pi}{4} + \arctan t\right]\mathbf{i} + \left(1 - \frac{1}{t}\right)\mathbf{j} + \ln t\mathbf{k}$

63. See "Definition of the Derivative of a Vector-Valued Function" and Figure 12.8 on page 824.

64. To find the integral of a vector-valued function, you integrate each component function separately. The constant of integration **C** is a constant vector.

65. At $t = t_0$, the graph of $\mathbf{u}(t)$ is increasing in the x, y, and z directions simultaneously.

66. The graph of $\mathbf{u}(t)$ does not change position relative to the xy-plane.

67. Let $\mathbf{r}(t) = x(t)\mathbf{i} + y(t)\mathbf{j} + z(t)\mathbf{k}$. Then $c\mathbf{r}(t) = cx(t)\mathbf{i} + cy(t)\mathbf{j} + cz(t)\mathbf{k}$ and

$$\frac{d}{dt}\left[c\mathbf{r}(t)\right] = cx'(t)\mathbf{i} + cy'(t)\mathbf{j} + cz'(t)\mathbf{k}$$
$$= c\left[x'(t)\mathbf{i} + y'(t)\mathbf{j} + z'(t)\mathbf{k}\right] = c\mathbf{r}'(t).$$

68. Let $\mathbf{r}(t) = x_1(t)\mathbf{i} + y_1(t)\mathbf{j} + z_1(t)\mathbf{k}$ and $\mathbf{u}(t) = x_2(t)\mathbf{i} + y_2(t)\mathbf{j} + z_2(t)\mathbf{k}$.

$$\mathbf{r}(t) \pm \mathbf{u}(t) = \left[x_1(t) \pm x_2(t)\right]\mathbf{i} + \left[y_1(t) \pm y_2(t)\right]\mathbf{j} + \left[z_1(t) \pm z_2(t)\right]\mathbf{k}$$
$$\frac{d}{dt}\left[\mathbf{r}(t) \pm \mathbf{u}(t)\right] = \left[x_1'(t) \pm x_2'(t)\right]\mathbf{i} + \left[y_1'(t) \pm y_2'(t)\right]\mathbf{j} + \left[z_1'(t) \pm z_2'(t)\right]\mathbf{k}$$
$$= \left[x_1'(t)\mathbf{i} + y_1'(t)\mathbf{j} + z_1'(t)\mathbf{k}\right] \pm \left[x_2'(t)\mathbf{i} + y_2'(t)\mathbf{j} + z_2'(t)\mathbf{k}\right] = \mathbf{r}'(t) \pm \mathbf{u}'(t)$$

69. Let $\mathbf{r}(t) = x(t)\mathbf{i} + y(t)\mathbf{j} + z(t)\mathbf{k}$, then $w(t)\mathbf{r}(t) = w(t)x(t)\mathbf{i} + w(t)y(t)\mathbf{j} + w(t)z(t)\mathbf{k}$.

$$\frac{d}{dt}\left[w(t)\mathbf{r}(t)\right] = \left[w(t)x'(t) + w'(t)x(t)\right]\mathbf{i} + \left[w(t)y'(t) + w'(t)y(t)\right]\mathbf{j} + \left[w(t)z'(t) + w'(t)z(t)\right]\mathbf{k}$$
$$= w(t)\left[x'(t)\mathbf{i} + y'(t)\mathbf{j} + z'(t)\mathbf{k}\right] + w'(t)\left[x(t)\mathbf{i} + y(t)\mathbf{j} + z(t)\mathbf{k}\right] = w(t)\mathbf{r}'(t) + w'(t)\mathbf{r}(t)$$

70. Let $\mathbf{r}(t) = x_1(t)\mathbf{i} + y_1(t)\mathbf{j} + z_1(t)\mathbf{k}$ and $\mathbf{u}(t) = x_2(t)\mathbf{i} + y_2(t)\mathbf{j} + z_2(t)\mathbf{k}$.

$$\mathbf{r}(t) \times \mathbf{u}(t) = \left[y_1(t)z_2(t) - z_1(t)y_2(t)\right]\mathbf{i} - \left[x_1(t)z_2(t) - z_1(t)x_2(t)\right]\mathbf{j} + \left[x_1(t)y_2(t) - y_1(t)x_2(t)\right]\mathbf{k}$$
$$\frac{d}{dt}\left[\mathbf{r}(t) \times \mathbf{u}(t)\right] = \left[y_1(t)z_2'(t) + y_1'(t)z_2(t) - z_1(t)y_2'(t) - z_1'(t)y_2(t)\right]\mathbf{i} - \left[x_1(t)z_2'(t) + x_1'(t)z_2(t) - z_1(t)x_2'(t) - z_1'(t)x_2(t)\right]\mathbf{j}$$
$$+ \left[x_1(t)y_2'(t) + x_1'(t)y_2(t) - y_1(t)x_2'(t) - y_1'(t)x_2(t)\right]\mathbf{k}$$
$$= \left\{\left[y_1(t)z_2'(t) - z_1(t)y_2'(t)\right]\mathbf{i} - \left[x_1(t)z_2'(t) - z_1(t)x_2'(t)\right]\mathbf{j} + \left[x_1(t)y_2'(t) - y_1(t)x_2'(t)\right]\mathbf{k}\right\}$$
$$+ \left\{\left[y_1'(t)z_2(t) - z_1'(t)y_2(t)\right]\mathbf{i} - \left[x_1'(t)z_2(t) - z_1'(t)x_2(t)\right]\mathbf{j} + \left[x_1'(t)y_2(t) - y_1'(t)x_2(t)\right]\mathbf{k}\right\}$$
$$= \mathbf{r}(t) \times \mathbf{u}'(t) + \mathbf{r}'(t) \times \mathbf{u}(t)$$

71. Let $\mathbf{r}(t) = x(t)\mathbf{i} + y(t)\mathbf{j} + z(t)\mathbf{k}$. Then $\mathbf{r}(w(t)) = x(w(t))\mathbf{i} + y(w(t))\mathbf{j} + z(w(t))\mathbf{k}$ and

$$\frac{d}{dt}\left[\mathbf{r}(w(t))\right] = x'(w(t))w'(t)\mathbf{i} + y'(w(t))w'(t)\mathbf{j} + z'(w(t))w'(t)\mathbf{k} \quad \text{(Chain Rule)}$$
$$= w'(t)\left[x'(w(t))\mathbf{i} + y'(w(t))\mathbf{j} + z'(w(t))\mathbf{k}\right] = w'(t)\mathbf{r}'(w(t)).$$

72. Let $\mathbf{r}(t) = x(t)\mathbf{i} + y(t)\mathbf{j} + z(t)\mathbf{k}$. Then $\mathbf{r}'(t) = x'(t)\mathbf{i} + y'(t)\mathbf{j} + z'(t)\mathbf{k}$.

$$\mathbf{r}(t) \times \mathbf{r}'(t) = \left[y(t)z'(t) - z(t)y'(t)\right]\mathbf{i} - \left[x(t)z'(t) - z(t)x'(t)\right]\mathbf{j} + \left[x(t)y'(t) - y(t)x'(t)\right]\mathbf{k}$$
$$\frac{d}{dt}\left[\mathbf{r}(t) \times \mathbf{r}'(t)\right] = \left[y(t)z''(t) + y'(t)z'(t) - z(t)y''(t) - z'(t)y'(t)\right]\mathbf{i} - \left[x(t)z''(t) + x'(t)z'(t) - z(t)x''(t) - z'(t)x'(t)\right]\mathbf{j}$$
$$+ \left[x(t)y''(t) + x'(t)y'(t) - y(t)x''(t) - y'(t)x'(t)\right]\mathbf{k}$$
$$= \left[y(t)z''(t) - z(t)y''(t)\right]\mathbf{i} - \left[x(t)z''(t) - z(t)x''(t)\right]\mathbf{j} + \left[x(t)y''(t) - y(t)x''(t)\right]\mathbf{k} = \mathbf{r}(t) \times \mathbf{r}''(t)$$

73. Let $\mathbf{r}(t) = x_1(t)\mathbf{i} + y_1(t)\mathbf{j} + z_1(t)\mathbf{k}$, $\mathbf{u}(t) = x_2(t)\mathbf{i} + y_2(t)\mathbf{j} + z_2(t)\mathbf{k}$, and $\mathbf{v}(t) = x_3(t)\mathbf{i} + y_3(t)\mathbf{j} + z_3(t)\mathbf{k}$. Then:

$$\mathbf{r}(t)\cdot\left[\mathbf{u}(t)\times\mathbf{v}(t)\right] = x_1(t)\left[y_2(t)z_3(t) - z_2(t)y_3(t)\right] - y_1(t)\left[x_2(t)z_3(t) - z_2(t)x_3(t)\right] + z_1(t)\left[x_2(t)y_3(t) - y_2(t)x_3(t)\right]$$

$$\begin{aligned}\frac{d}{dt}\left[\mathbf{r}(t)\cdot(\mathbf{u}(t)\times\mathbf{v}(t))\right] &= x_1(t)y_2(t)z_3'(t) + x_1(t)y_2'(t)z_3(t) + x_1'(t)y_2(t)z_3(t) - x_1(t)y_3(t)z_2'(t)\\ &\quad - x_1(t)y_3'(t)z_2(t) - x_1'(t)y_3(t)z_2(t) - y_1(t)x_2(t)z_3'(t) - y_1(t)x_2'(t)z_3(t) - y_1'(t)x_2(t)z_3(t)\\ &\quad + y_1(t)z_2(t)x_3'(t) + y_1(t)z_2'(t)x_3(t) + y_1'(t)z_2(t)x_3(t) + z_1(t)x_2(t)y_3'(t) + z_1(t)x_2'(t)y_3(t)\\ &\quad + z_1'(t)x_2(t)y_3(t) - z_1(t)y_2(t)x_3'(t) - z_1(t)y_2'(t)x_3(t) - z_1'(t)y_2(t)x_3(t)\\ &= \left\{x_1'(t)\left[y_2(t)z_3(t) - y_3(t)z_2(t)\right] + y_1'(t)\left[-x_2(t)z_3(t) + z_2(t)x_3(t)\right] + z_1'(t)\left[x_2(t)y_3(t) - y_2(t)x_3(t)\right]\right\}\\ &\quad + \left\{x_1(t)\left[y_2'(t)z_3(t) - y_3(t)z_2'(t)\right] + y_1(t)\left[-x_2'(t)z_3(t) + z_2'(t)x_3(t)\right] + z_1(t)\left[x_2'(t)y_3(t) - y_2'(t)x_3(t)\right]\right\}\\ &\quad + \left\{x_1(t)\left[y_2(t)z_3'(t) - y_3'(t)z_2(t)\right] + y_1(t)\left[-x_2(t)z_3'(t) + z_2(t)x_3'(t)\right] + z_1(t)\left[x_2(t)y_3'(t) - y_2(t)x_3'(t)\right]\right\}\\ &= \mathbf{r}'(t)\cdot\left[\mathbf{u}(t)\times\mathbf{v}(t)\right] + \mathbf{r}(t)\cdot\left[\mathbf{u}'(t)\times\mathbf{v}(t)\right] + \mathbf{r}(t)\cdot\left[\mathbf{u}(t)\times\mathbf{v}'(t)\right]\end{aligned}$$

74. Let $\mathbf{r}(t) = x(t)\mathbf{i} + y(t)\mathbf{j} + z(t)\mathbf{k}$. If $\mathbf{r}(t)\cdot\mathbf{r}(t)$ is constant, then:

$$\begin{aligned} x^2(t) + y^2(t) + z^2(t) &= C\\ \frac{d}{dt}\left[x^2(t) + y^2(t) + z^2(t)\right] &= D_t[C]\\ 2x(t)x'(t) + 2y(t)y'(t) + 2z(t)z'(t) &= 0\\ 2\left[x(t)x'(t) + y(t)y'(t) + z(t)z'(t)\right] &= 0\\ 2\left[\mathbf{r}(t)\cdot\mathbf{r}'(t)\right] &= 0.\end{aligned}$$

So, $\mathbf{r}(t)\cdot\mathbf{r}'(t) = 0$.

75. $\mathbf{r}(t) = (t - \sin t)\mathbf{i} + (1 - \cos t)\mathbf{j}$

(a)

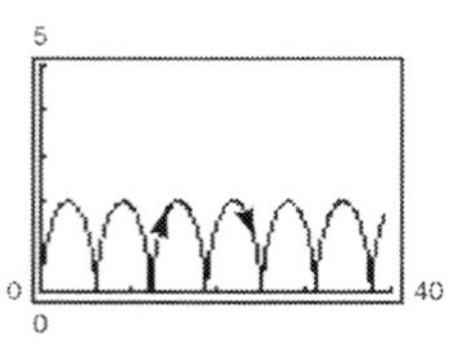

The curve is a cycloid.

(b) $\mathbf{r}'(t) = (1 - \cos t)\mathbf{i} + \sin t\mathbf{j}$

$\mathbf{r}''(t) = \sin t\mathbf{i} + \cos t\mathbf{j}$

$$\begin{aligned}\|\mathbf{r}'(t)\| &= \sqrt{1 - 2\cos t + \cos^2 t + \sin^2 t}\\ &= \sqrt{2 - 2\cos t}\end{aligned}$$

Minimum of $\|\mathbf{r}'(t)\|$ is 0, $(t = 0)$.

Maximum of $\|\mathbf{r}'(t)\|$ is 2, $(t = \pi)$.

$\|\mathbf{r}''(t)\| = \sqrt{\sin^2 t + \cos^2 t} = 1$

Minimum and maximum of $\|\mathbf{r}'(t)\|$ is 1.

76. $r(t) = 2\cos t\mathbf{i} + 3\sin t\mathbf{j}$

(a) Ellipse

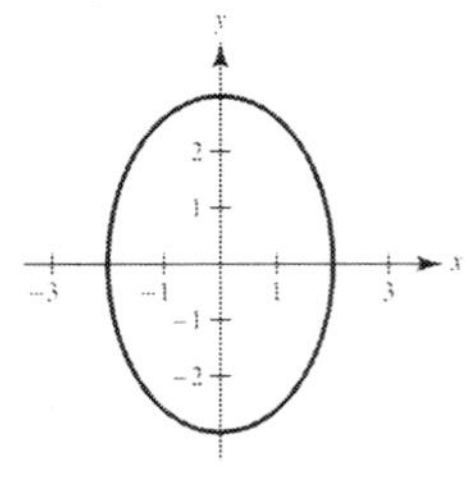

(b) $\mathbf{r}'(t) = -2\sin t\mathbf{i} + 3\cos t\mathbf{j}$

$\mathbf{r}''(t) = -2\cos t\mathbf{i} - 3\sin t\mathbf{j}$

$\|\mathbf{r}'(t)\| = \sqrt{4\sin^2 t + 9\cos^2 t}$

Minimum of $\|\mathbf{r}'(t)\|$ is 2, $(t = \pi/2)$.

Maximum of $\|\mathbf{r}'(t)\|$ is 3, $(t = 0)$.

77. $\mathbf{r}(t) = e^t \sin t\mathbf{i} + e^t \cos t\mathbf{j}$

$\mathbf{r}'(t) = (e^t \cos t + e^t \sin t)\mathbf{i} + (e^t \cos t - e^t \sin t)\mathbf{j}$

$\mathbf{r}''(t) = (-e^t \sin t + e^t \cos t + e^t \sin t + e^t \cos t)\mathbf{i} + (e^t \cos t - e^t \sin t - e^t \sin t - e^t \cos t)\mathbf{j} = 2e^t \cos t\mathbf{i} - 2e^t \sin t\mathbf{j}$

$\mathbf{r}(t) \cdot \mathbf{r}''(t) = 2e^{2t} \sin t \cos t - 2e^{2t} \sin t \cos t = 0$

So, $\mathbf{r}(t)$ is always perpendicular to $\mathbf{r}''(t)$.

78. (a) $t = \frac{\pi}{4}$: both components positive

$t = \frac{5\pi}{6}$: x-component negative, y-component positive

$t = \frac{5\pi}{4}$: x-component positive, y-component negative

(b) No. There is a cusp when $t = 0$, at $(1, 0)$.

79. True

80. False. The definite integral is a vector, not a real number.

81. False. Let $\mathbf{r}(t) = \cos t\mathbf{i} + \sin t\mathbf{j} + \mathbf{k}$.

$\|\mathbf{r}(t)\| = \sqrt{2}$

$\frac{d}{dt}[\|\mathbf{r}(t)\|] = 0$

$\mathbf{r}'(t) = -\sin t\mathbf{i} + \cos t\mathbf{j}$

$\|\mathbf{r}'(t)\| = 1$

82. False.

$\frac{d}{dt}[\mathbf{r}(t) \cdot \mathbf{u}(t)] = \mathbf{r}(t) \cdot \mathbf{u}'(t) + \mathbf{r}'(t) \cdot \mathbf{u}(t)$

(See Theorem 2.2, part 4)

Section 12.3 Velocity and Acceleration

1. $\mathbf{r}(t) = 3t\mathbf{i} + (t - 1)\mathbf{j}$, $(3, 0)$

(a) $\mathbf{v}(t) = \mathbf{r}'(t) = 3\mathbf{i} + \mathbf{j}$

Speed $= \|\mathbf{v}(t)\| = \sqrt{3^2 + 1^2} = \sqrt{10}$

$\mathbf{a}(t) = \mathbf{r}''(t) = \mathbf{0}$

(b) At $(3, 0)$, $t = 1$.

$\mathbf{v}(1) = 3\mathbf{i} + \mathbf{j}$, $\mathbf{a}(1) = \mathbf{0}$

(c) $x = 3t$, $y = t - 1$

$y = \frac{x}{3} - 1$, line

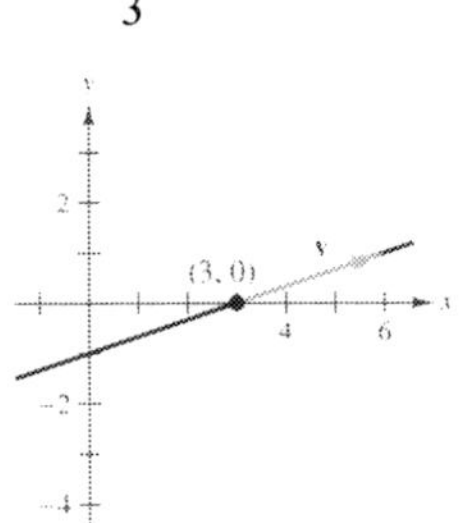

2. $\mathbf{r}(t) = t\mathbf{i} + (4 - t^2)\mathbf{j}$, $(1, 3)$

(a) $\mathbf{v}(t) = \mathbf{r}'(t) = \mathbf{i} - 2t\mathbf{j}$

Speed $= \|\mathbf{v}(t)\| = \sqrt{1 + (-2t)^2} = \sqrt{1 + 4t^2}$

$\mathbf{a}(t) = \mathbf{r}''(t) = -2\mathbf{j}$

(b) At $(1, 3)$, $t = 1$.

$\mathbf{v}(1) = \mathbf{i} - 2\mathbf{j}$, $\mathbf{a}(1) = -2\mathbf{j}$

(c) $x = t$, $y = 4 - t^2$

$y = 4 - x^2$, parabola

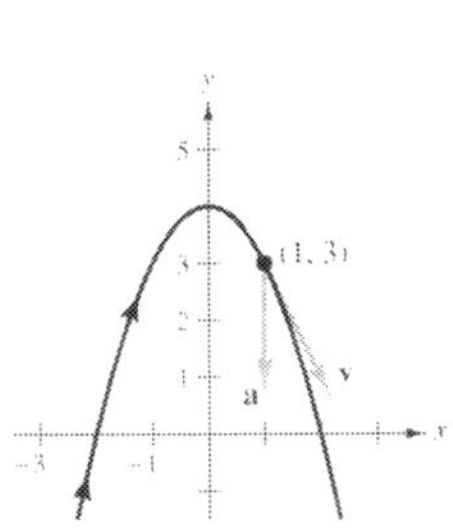

3. $\mathbf{r}(t) = t^2\mathbf{i} + t\mathbf{j}, (4, 2)$

(a) $\mathbf{v}(t) = \mathbf{r}'(t) = 2t\mathbf{i} + \mathbf{j}$

Speed $= \|\mathbf{v}(t)\| = \sqrt{4t^2 + 1}$

$\mathbf{a}(t) = \mathbf{r}''(t) = 2\mathbf{i}$

(b) At $(4, 2)$, $t = 2$.

$\mathbf{v}(2) = 4\mathbf{i} + \mathbf{j}, \mathbf{a}(2) = 2\mathbf{j}$

(c) $x = t^2, y = t$

$x = y^2$, parabola

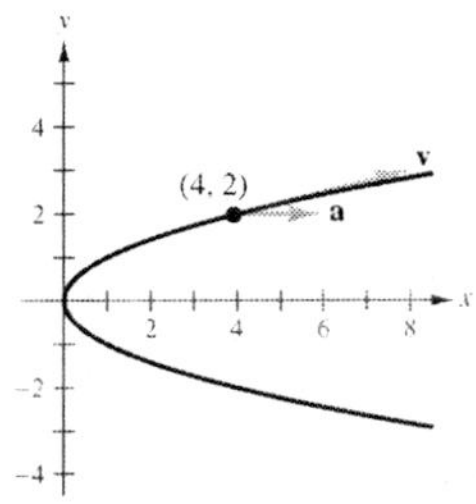

4. $\mathbf{r}(t) = \left(\frac{1}{4}t^3 + 1\right)\mathbf{i} + t\mathbf{j}, (3, 2)$

(a) $\mathbf{v}(t) = \mathbf{r}'(t) = \frac{3}{4}t^2\mathbf{i} + \mathbf{j}$

Speed $= \|\mathbf{v}(t)\| = \sqrt{\frac{9}{16}t^4 + 1}$

$\mathbf{a}(t) = \frac{3}{2}t\mathbf{i}$

(b) At $(3, 2)$, $t = 2$.

$\mathbf{v}(2) = 3\mathbf{i} + \mathbf{j}, \mathbf{a}(2) = 3\mathbf{i}$

(c) $x = \frac{1}{4}t^3 + 1, y = t$

$x = \frac{1}{4}y^3 + 1 \Rightarrow y = \sqrt[3]{4(x - 1)}$

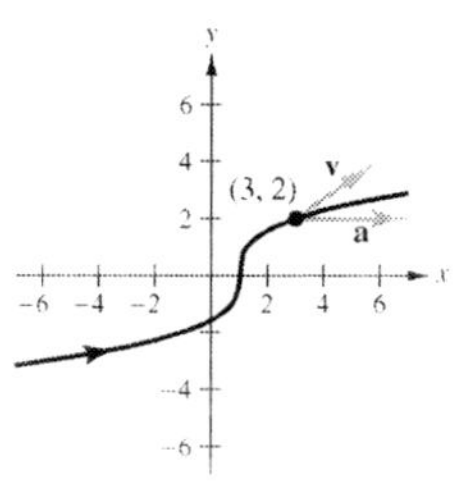

5. $\mathbf{r}(t) = 2\cos t\mathbf{i} + 2\sin t\mathbf{j}, \left(\sqrt{2}, \sqrt{2}\right)$

(a) $\mathbf{v}(t) = \mathbf{r}'(t) = -2\sin t\mathbf{i} + 2\cos t\mathbf{j}$

Speed $= \|\mathbf{v}(t)\| = \sqrt{4\sin^2 t + 4\cos^2 t} = 2$

$\mathbf{a}(t) = -2\cos t\mathbf{i} - 2\sin t\,\mathbf{j}$

(b) At $\left(\sqrt{2}, \sqrt{2}\right)$, $t = \frac{\pi}{4}$.

$\mathbf{v}\left(\frac{\pi}{4}\right) = -\sqrt{2}\mathbf{i} + \sqrt{2}\mathbf{j}$

$\mathbf{a}\left(\frac{\pi}{4}\right) = -\sqrt{2}\mathbf{i} - \sqrt{2}\mathbf{j}$

(c) $x = 2\cos t, y = 2\sin t$

$x^2 + y^2 = 4$, circle

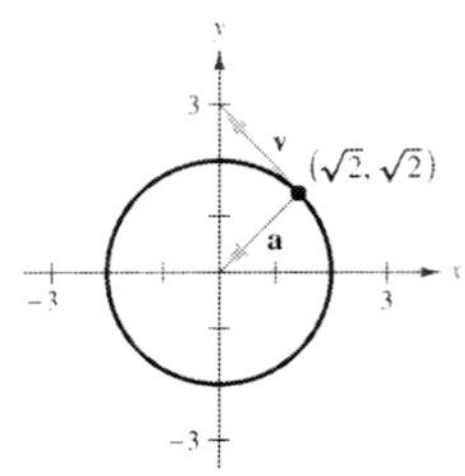

6. $\mathbf{r}(t) = 3\cos t\mathbf{i} + 2\sin t\mathbf{j}, (3, 0)$

(a) $\mathbf{v}(t) = \mathbf{r}'(t) = -3\sin t\mathbf{i} + 2\cos t\mathbf{j}$

Speed $= \|\mathbf{v}(t)\| = \sqrt{9\sin^2 t + 4\cos^2 t}$

$\mathbf{a}(t) = -3\cos t\mathbf{i} - 2\sin t\mathbf{j}$

(b) At $(3, 0)$, $t = 0$.

$\mathbf{v}(0) = 2\mathbf{j}, \mathbf{a}(0) = -3\mathbf{i}$

(c) $x = 3\cos t, y = 2\sin t$

$\left(\frac{x}{3}\right)^2 + \left(\frac{y}{2}\right)^2 = \cos^2 t + \sin^2 t = 1$, ellipse

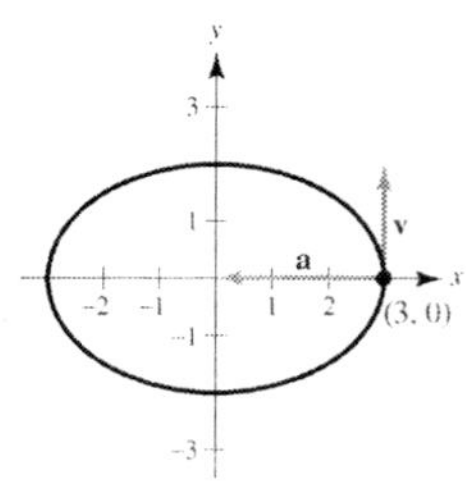

7. $\mathbf{r}(t) = \langle t - \sin t, 1 - \cos t\rangle, (\pi, 2)$

(a) $\mathbf{v}(t) = \mathbf{v}'(t) = \langle 1 - \cos t, \sin t\rangle$

Speed $= \|\mathbf{v}(t)\| = \sqrt{(1 - \cos t)^2 + \sin^2 t} = \sqrt{2 - 2\cos t}$

$\mathbf{a}(t) = \langle \sin t, \cos t\rangle$

(b) At $(\pi, 2), t = \pi$.

$\mathbf{v}(\pi) = \langle 2, 0\rangle, \mathbf{a}(\pi) = \langle 0, -1\rangle$

(c) $x = t - \sin t, y = 1 - \cos t$

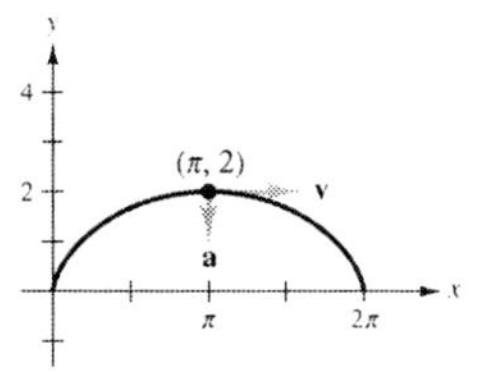

8. $\mathbf{r}(t) = \langle e^{-t}, e^{t}\rangle, (1, 1)$

(a) $\mathbf{v}(t) = \mathbf{r}'(t) = \langle -e^{-t}, e^{t}\rangle$

Speed $= \|\mathbf{v}(t)\| = \sqrt{(-e^{-t})^2 + (e^{t})^2} = \sqrt{e^{-2t} + e^{2t}}$

$\mathbf{a}(t) = \mathbf{r}''(t) = \langle e^{-t}, e^{t}\rangle$

(b) At $(1, 1), t = 0$.

$\mathbf{v}(0) = \langle -1, 1\rangle, \mathbf{a}(0) = \langle 1, 1\rangle$

(c) $x = e^{-t}, y = e^{t}, y = \dfrac{1}{x}$

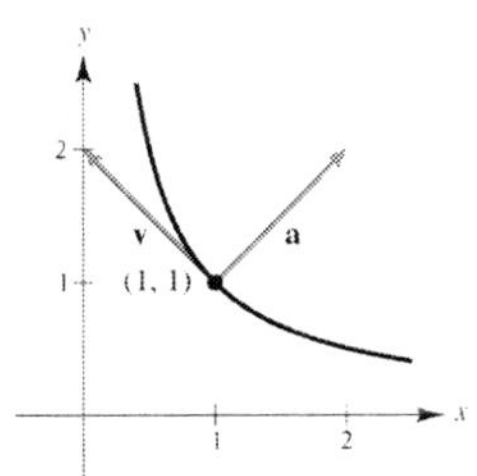

9. $\mathbf{r}(t) = t\mathbf{i} + 5t\mathbf{j} + 3t\mathbf{k}, t = 1$

(a) $\mathbf{v}(t) = \mathbf{r}'(t) = \mathbf{i} + 5\mathbf{j} + 3\mathbf{k}$

Speed $= \|\mathbf{v}(t)\| = \sqrt{1^2 + 5^2 + 3^2} = \sqrt{35}$

$\mathbf{a}(t) = \mathbf{r}''(t) = \mathbf{0}$

(b) $\mathbf{v}(1) = \mathbf{i} + 5\mathbf{j} + 3\mathbf{k}$

$\mathbf{a}(1) = \mathbf{0}$

10. $\mathbf{r}(t) = 4t\mathbf{i} + 4t\mathbf{j} + 2t\mathbf{k}, t = 3$

(a) $\mathbf{v}(t) = \mathbf{r}'(t) = 4\mathbf{i} + 4\mathbf{j} + 2\mathbf{k}$

Speed $= \|\mathbf{v}(t)\| = \sqrt{4^2 + 4^2 + 2^2} = \sqrt{36} = 6$

$\mathbf{a}(t) = \mathbf{r}''(t) = \mathbf{0}$

(b) $\mathbf{v}(3) = 4\mathbf{i} + 4\mathbf{j} + 2\mathbf{k}$

$\mathbf{a}(3) = \mathbf{0}$

11. $\mathbf{r}(t) = t\mathbf{i} + t^2\mathbf{j} + \frac{1}{2}t^2\mathbf{k}, t = 4$

(a) $\mathbf{v}(t) = \mathbf{r}'(t) = \mathbf{i} + 2t\mathbf{j} + t\mathbf{k}$

Speed $= \|\mathbf{v}(t)\| = \sqrt{1 + 4t^2 + t^2} = \sqrt{1 + 5t^2}$

$\mathbf{a}(t) = \mathbf{r}''(t) = 2\mathbf{j} + \mathbf{k}$

(b) $\mathbf{v}(4) = \mathbf{i} + 8\mathbf{j} + 4\mathbf{k}$

$\mathbf{a}(4) = 2\mathbf{j} + \mathbf{k}$

12. $\mathbf{r}(t) = 3t\mathbf{i} + t\mathbf{j} + \frac{1}{4}t^2\mathbf{k}, t = 2$

(a) $\mathbf{v}(t) = \mathbf{r}'(t) = 3\mathbf{i} + \mathbf{j} + \frac{1}{2}t\mathbf{k}$

Speed $= \|\mathbf{v}(t)\| = \sqrt{9 + 1 + \frac{1}{4}t^2} = \sqrt{10 + \frac{1}{4}t^2}$

$\mathbf{a}(t) = \frac{1}{2}\mathbf{k}$

(b) $\mathbf{v}(2) = 3\mathbf{i} + \mathbf{j} + \mathbf{k}$

$\mathbf{a}(2) = \frac{1}{2}\mathbf{k}$

13. $\mathbf{r}(t) = t\mathbf{i} + t\mathbf{j} + \sqrt{9 - t^2}\mathbf{k}, t = 0$

(a) $\mathbf{v}(t) = \mathbf{r}'(t) = \mathbf{i} + \mathbf{j} - \dfrac{t}{\sqrt{9 - t^2}}\mathbf{k}$

$\text{Speed} = \|\mathbf{v}(t)\| = \sqrt{1 + 1 + \dfrac{t^2}{9 - t^2}} = \sqrt{\dfrac{18 - t^2}{9 - t^2}}$

$\mathbf{a}(t) = \mathbf{r}''(t) = -\dfrac{9}{(9 - t^2)^{3/2}}\mathbf{k}$

(b) $\mathbf{v}(0) = \mathbf{i} + \mathbf{j}$

$\mathbf{a}(0) = -\dfrac{9}{9^{3/2}}\mathbf{k} = -\dfrac{1}{3}\mathbf{k}$

14. $\mathbf{r}(t) = t^2\mathbf{i} + t\mathbf{j} + 2t^{3/2}\mathbf{k}, t = 4$

(a) $\mathbf{v}(t) = \mathbf{r}'(t) = 2t\mathbf{i} + \mathbf{j} + 3\sqrt{t}\,\mathbf{k}$

$\text{Speed} = \|\mathbf{v}(t)\| = \sqrt{4t^2 + 1 + 9t}$

$\mathbf{a}(t) = 2\mathbf{i} + \dfrac{3}{2\sqrt{t}}\mathbf{k}$

(b) $\mathbf{v}(4) = 8\mathbf{i} + \mathbf{j} + 6\mathbf{k}$

$\mathbf{a}(4) = 2\mathbf{i} + \dfrac{3}{4}\mathbf{k}$

15. $\mathbf{r}(t) = \langle 4t, 3\cos t, 3\sin t\rangle, t = \pi$

(a) $\mathbf{v}(t) = \mathbf{r}'(t) = \langle 4, -3\sin t, 3\cos t\rangle$

$\text{Speed} = \|\mathbf{v}(t)\| = \sqrt{4^2 + (-3\sin t)^2 + (3\cos t)^2} = \sqrt{16 + 9} = 5$

$\mathbf{a}(t) = \mathbf{r}''(t) = \langle 0, -3\cos t, -3\sin t\rangle$

(b) $\mathbf{v}(\pi) = \langle 4, 0, -3\rangle$

$\mathbf{a}(\pi) = \langle 0, 3, 0\rangle$

16. $\mathbf{r}(t) = \langle 2\cos t, 2\sin t, t^2\rangle, t = \dfrac{\pi}{4}$

(a) $\mathbf{v}(t) = \mathbf{r}'(t) = \langle -2\sin t, 2\cos t, 2t\rangle$

$\text{Speed} = \|\mathbf{v}(t)\| = \sqrt{(-2\sin t)^2 + (2\cos t)^2 + 4t^2} = \sqrt{4 + 4t^2} = 2\sqrt{1 + t^2}$

$\mathbf{a}(t) = \langle -2\cos t, -2\sin t, 2\rangle$

(b) $\mathbf{v}\left(\dfrac{\pi}{4}\right) = \left\langle -2\left(\dfrac{\sqrt{2}}{2}\right), 2\left(\dfrac{\sqrt{2}}{2}\right), 2\left(\dfrac{\pi}{4}\right)\right\rangle$

$= \left\langle -\sqrt{2}, \sqrt{2}, \dfrac{\pi}{2}\right\rangle$

$\mathbf{a}\left(\dfrac{\pi}{4}\right) = \left\langle -2\left(\dfrac{\sqrt{2}}{2}\right), -2\left(\dfrac{\sqrt{2}}{2}\right), 2\right\rangle$

$= \langle -\sqrt{2}, -\sqrt{2}, 2\rangle$

17. $\mathbf{r}(t) = \langle e^t\cos t, e^t\sin t, e^t\rangle, t = 0$

(a) $\mathbf{v}(t) = \mathbf{r}'(t) = \langle e^t\cos t - e^t\sin t, e^t\sin t + e^t\cos t, e^t\rangle$

$\text{Speed} = \|\mathbf{v}(t)\| = \sqrt{e^{2t}(\cos t - \sin t)^2 + e^{2t}(\cos t + \sin t)^2 + e^{2t}}$

$= e^t\sqrt{3}$

$\mathbf{a}(t) = \mathbf{r}''(t)$

$= \langle e^t\cos t - e^t\sin t - e^t\sin t - e^t\cos t, e^t\sin t + e^t\cos t + e^t\cos t - e^t\sin t, e^t\rangle$

$= \langle -2e^t\sin t, 2e^t\cos t, e^t\rangle$

(b) $\mathbf{v}(0) = \langle 1, 1, 1\rangle$

$\mathbf{a}(0) = \langle 0, 2, 1\rangle$

18. $\mathbf{r}(t) = \left\langle \ln t, \dfrac{1}{t}, t^4 \right\rangle, t = 2$

(a) $\mathbf{v}(t) = \mathbf{r}'(t) = \left\langle \dfrac{1}{t}, -\dfrac{1}{t^2}, 4t^3 \right\rangle$

$\text{speed} = \|\mathbf{v}(t)\| = \sqrt{\dfrac{1}{t^2} + \dfrac{1}{t^4} + 16t^6}$

$\mathbf{a}(t) = \mathbf{r}''(t) = \left\langle -\dfrac{1}{t^2}, \dfrac{2}{t^3}, 12t^2 \right\rangle$

(b) $\mathbf{v}(2) = \left\langle \frac{1}{2}, -\frac{1}{4}, 32 \right\rangle$

$\mathbf{a}(2) = \left\langle -\frac{1}{4}, \frac{1}{4}, 48 \right\rangle$

19. $\mathbf{a}(t) = \mathbf{i} + \mathbf{j} + \mathbf{k}, \mathbf{v}(0) = \mathbf{0}, \mathbf{r}(0) = \mathbf{0}$

$\mathbf{v}(t) = \int(\mathbf{i} + \mathbf{j} + \mathbf{k})\,dt = t\mathbf{i} + t\mathbf{j} + t\mathbf{k} + \mathbf{C}$

$\mathbf{v}(0) = \mathbf{C} = \mathbf{0}, \mathbf{v}(t) = t\mathbf{i} + t\mathbf{j} + t\mathbf{k}, \mathbf{v}(t) = t(\mathbf{i} + \mathbf{j} + \mathbf{k})$

$\mathbf{r}(t) = \int(t\mathbf{i} + t\mathbf{j} + t\mathbf{k})\,dt = \dfrac{t^2}{2}(\mathbf{i} + \mathbf{j} + \mathbf{k}) + \mathbf{C}$

$\mathbf{r}(0) = \mathbf{C} = \mathbf{0}, \mathbf{r}(t) = \dfrac{t^2}{2}(\mathbf{i} + \mathbf{j} + \mathbf{k}),$

$\mathbf{r}(2) = 2(\mathbf{i} + \mathbf{j} + \mathbf{k}) = 2\mathbf{i} + 2\mathbf{j} + 2\mathbf{k}$

20. $\mathbf{a}(t) = 2\mathbf{i} + 3\mathbf{k}, \mathbf{v}(0) = 4\mathbf{j}, \mathbf{r}(0) = \mathbf{0}$

$\mathbf{v}(t) = \int(2\mathbf{i} + 3\mathbf{k})\,dt = 2t\mathbf{i} + 3t\mathbf{k} + \mathbf{C}$

$\mathbf{v}(0) = \mathbf{C} = 4\mathbf{j} \Rightarrow \mathbf{v}(t) = 2t\mathbf{i} + 4\mathbf{j} + 3t\mathbf{k}$

$\mathbf{r}(t) = \int(2t\mathbf{i} + 4\mathbf{j} + 3t\mathbf{k})\,dt = t^2\mathbf{i} + 4t\mathbf{j} + \frac{3}{2}t^2\mathbf{k} + \mathbf{C}$

$\mathbf{r}(0) = \mathbf{C} = 0 \Rightarrow \mathbf{r}(t) = t^2\mathbf{i} + 4t\mathbf{j} + \frac{3}{2}t^2\mathbf{k}$

$\mathbf{r}(2) = 4\mathbf{i} + 8\mathbf{j} + 6\mathbf{k}$

21. $\mathbf{a}(t) = t\mathbf{j} + t\mathbf{k}, \mathbf{v}(1) = 5\mathbf{j}, \mathbf{r}(1) = \mathbf{0}$

$\mathbf{v}(t) = \int(t\mathbf{j} + t\mathbf{k})\,dt = \dfrac{t^2}{2}\mathbf{j} + \dfrac{t^2}{2}\mathbf{k} + \mathbf{C}$

$\mathbf{v}(1) = \dfrac{1}{2}\mathbf{j} + \dfrac{1}{2}\mathbf{k} + \mathbf{C} = 5\mathbf{j} \Rightarrow \mathbf{C} = \dfrac{9}{2}\mathbf{j} - \dfrac{1}{2}\mathbf{k}$

$\mathbf{v}(t) = \left(\dfrac{t^2}{2} + \dfrac{9}{2}\right)\mathbf{j} + \left(\dfrac{t^2}{2} - \dfrac{1}{2}\right)\mathbf{k}$

$\mathbf{r}(t) = \int\left[\left(\dfrac{t^2}{2} + \dfrac{9}{2}\right)\mathbf{j} + \left(\dfrac{t^2}{2} - \dfrac{1}{2}\right)\mathbf{k}\right]dt = \left(\dfrac{t^3}{6} + \dfrac{9}{2}t\right)\mathbf{j} + \left(\dfrac{t^3}{6} - \dfrac{1}{2}t\right)\mathbf{k} + \mathbf{C}$

$\mathbf{r}(1) = \dfrac{14}{3}\mathbf{j} - \dfrac{1}{3}\mathbf{k} + \mathbf{C} = \mathbf{0} \Rightarrow \mathbf{C} = -\dfrac{14}{3}\mathbf{j} + \dfrac{1}{3}\mathbf{k}$

$\mathbf{r}(t) = \left(\dfrac{t^3}{6} + \dfrac{9}{2}t - \dfrac{14}{3}\right)\mathbf{j} + \left(\dfrac{t^3}{6} - \dfrac{1}{2}t + \dfrac{1}{3}\right)\mathbf{k}$

$\mathbf{r}(2) = \dfrac{17}{3}\mathbf{j} + \dfrac{2}{3}\mathbf{k}$

22. $\mathbf{a}(t) = -32\mathbf{k}, \mathbf{v}(0) = 3\mathbf{i} - 2\mathbf{j} + \mathbf{k}, \mathbf{r}(0) = 5\mathbf{j} + 2\mathbf{k}$

$\mathbf{v}(t) = \int -32\mathbf{k}\,dt = -32t\mathbf{k} + \mathbf{C}$

$\mathbf{v}(0) = \mathbf{C} = 3\mathbf{i} - 2\mathbf{j} + \mathbf{k} \Rightarrow \mathbf{v}(t) = 3\mathbf{i} - 2\mathbf{j} + (1 - 32t)\mathbf{k}$

$\mathbf{r}(t) = \int[3\mathbf{i} - 2\mathbf{j} + (1 - 32t)\mathbf{k}]\,dt = 3t\mathbf{i} - 2t\mathbf{j} + (t - 16t^2)\mathbf{k} + \mathbf{C}$

$\mathbf{r}(0) = \mathbf{C} = 5\mathbf{j} + 2\mathbf{k} \Rightarrow \mathbf{r}(t) = 3t\mathbf{i} + (5 - 2t)\mathbf{j} + (2 + t - 16t^2)\mathbf{k}$

$\mathbf{r}(2) = 6\mathbf{i} + \mathbf{j} - 60\mathbf{k}$

23. $\mathbf{a}(t) = -\cos t\mathbf{i} - \sin t\mathbf{j}, \mathbf{v}(0) = \mathbf{j} + \mathbf{k}, \mathbf{r}(0) = \mathbf{i}$

$\mathbf{v}(t) = \int(-\cos t\mathbf{i} - \sin t\mathbf{j})\,dt = -\sin t\mathbf{i} + \cos t\mathbf{j} + \mathbf{C}$

$\mathbf{v}(0) = \mathbf{j} + \mathbf{C} = \mathbf{j} + \mathbf{k} \Rightarrow \mathbf{C} = \mathbf{k}$

$\mathbf{v}(t) = -\sin t\mathbf{i} + \cos t\mathbf{j} + \mathbf{k}$

$\mathbf{r}(t) = \int(-\sin t\mathbf{i} + \cos t\mathbf{j} + \mathbf{k})\,dt = \cos t\mathbf{i} + \sin t\mathbf{j} + t\mathbf{k} + \mathbf{C}$

$\mathbf{r}(0) = \mathbf{i} + \mathbf{C} = \mathbf{i} \Rightarrow \mathbf{C} = \mathbf{0}$

$\mathbf{r}(t) = \cos t\mathbf{i} + \sin t\mathbf{j} + t\mathbf{k}$

$\mathbf{r}(2) = (\cos 2)\mathbf{i} + (\sin 2)\mathbf{j} + 2\mathbf{k}$

24. $\mathbf{a}(t) = e^t\mathbf{i} - 8\mathbf{k}, \mathbf{v}(0) = 2\mathbf{i} + 3\mathbf{j} + \mathbf{k}, \mathbf{r}(0) = \mathbf{0}$

$\mathbf{v}(t) = \int(e^t\mathbf{i} - 8\mathbf{k})\,dt = e^t\mathbf{i} - 8t\mathbf{k} + \mathbf{C}$

$\mathbf{v}(0) = \mathbf{i} + \mathbf{C} = 2\mathbf{i} + 3\mathbf{j} + \mathbf{k} \Rightarrow \mathbf{C} = \mathbf{i} + 3\mathbf{j} + \mathbf{k}$

$\mathbf{v}(t) = (e^t + 1)\mathbf{i} + 3\mathbf{j} + (1 - 8t)\mathbf{k}$

$\mathbf{r}(t) = \int\left[(e^t + 1)\mathbf{i} + 3\mathbf{j} + (1 - 8t)\mathbf{k}\right]dt$

$= (e^t + t)\mathbf{i} + 3t\mathbf{j} + (t - 4t^2)\mathbf{k} + \mathbf{C}$

$\mathbf{r}(0) = \mathbf{i} + \mathbf{C} = \mathbf{0} \Rightarrow \mathbf{C} = -\mathbf{i}$

$\mathbf{r}(t) = (e^t + t - 1)\mathbf{i} + 3t\mathbf{j} + (t - 4t^2)\mathbf{k}$

25. $\mathbf{r}(t) = 140(\cos 22^\circ)t\mathbf{i} + (2.5 + 140(\sin 22^\circ)t - 16t^2)\mathbf{j}$

$\mathbf{v}(t) = \mathbf{r}'(t) = 140(\cos 22^\circ)\mathbf{i} + (140(\sin 22^\circ) - 32t)\mathbf{j}$

The maximum height occurs when

$$y'(t) = 140(\sin 22^\circ) - 32t = 0 \Rightarrow t = \frac{140 \sin 22^\circ}{32} = \frac{35}{8}\sin 22^\circ \approx 1.639.$$

The maximum height is

$$y = 2.5 + 140(\sin 22^\circ)\left(\frac{35}{8}\sin 22^\circ\right) - 16\left(\frac{35}{8}\sin 22^\circ\right)^2 \approx 45.5 \text{ feet.}$$

When $x = 375$, $t = \dfrac{375}{140 \cos 22^\circ} \approx 2.889$.

For this value of t, $y \approx 20.47$ feet.

So the ball clears the 10-foot fence.

26. $\mathbf{r}(t) = (900 \cos 45^\circ)t\mathbf{i} + \left[3 + (900 \sin 45^\circ)t - 16t^2\right]\mathbf{j} = 450\sqrt{2}t\mathbf{i} + (3 + 450\sqrt{2}t - 16t^2)\mathbf{j}$

The maximum height occurs when $y'(t) = 450\sqrt{2} - 32t = 0$, which implies that $t = (225\sqrt{2})/16$.

The maximum height reached by the projectile is

$$y = 3 + 450\sqrt{2}\left(\frac{225\sqrt{2}}{16}\right) - 16\left(\frac{225\sqrt{2}}{16}\right)^2 = \frac{50{,}649}{8} = 6331.125 \text{ feet.}$$

The range is determined by setting $y(t) = 3 + 450\sqrt{2}t - 16t^2 = 0$ which implies that

$$t = \frac{-450\sqrt{2} - \sqrt{405{,}192}}{-32} \approx 39.779 \text{ seconds}$$

Range: $x = 450\sqrt{2}\left(\dfrac{-450\sqrt{2} - \sqrt{405{,}192}}{-32}\right) \approx 25{,}315.500$ feet

27. $\mathbf{r}(t) = (v_0 \cos\theta)t\mathbf{i} + \left[h + (v_0 \sin\theta)t - \frac{1}{2}gt^2\right]\mathbf{j} = \frac{v_0}{\sqrt{2}}t\mathbf{i} + \left(3 + \frac{v_0}{\sqrt{2}}t - 16t^2\right)\mathbf{j}$

$\frac{v_0}{\sqrt{2}}t = 300$ when $3 + \frac{v_0}{\sqrt{2}}t - 16t^2 = 3.$

$$t = \frac{300\sqrt{2}}{v_0}, \ \frac{v_0}{\sqrt{2}}\left(\frac{300\sqrt{2}}{v_0}\right) - 16\left(\frac{300\sqrt{2}}{v_0}\right)^2 = 0, \ 300 - \frac{300^2(32)}{v_0^2} = 0$$

$v_0^2 = 300(32)$, $v_0 = \sqrt{9600} = 40\sqrt{6}$, $v_0 = 40\sqrt{6} \approx 97.98$ ft/sec

The maximum height is reached when the derivative of the vertical component is zero.

$$y(t) = 3 + \frac{tv_0}{\sqrt{2}} - 16t^2 = 3 + \frac{40\sqrt{6}}{\sqrt{2}}t - 16t^2 = 3 + 40\sqrt{3}t - 16t^2$$

$$y'(t) = 40\sqrt{3} - 32t = 0$$

$$t = \frac{40\sqrt{3}}{32} = \frac{5\sqrt{3}}{4}$$

Maximum height: $y\left(\frac{5\sqrt{3}}{4}\right) = 3 + 40\sqrt{3}\left(\frac{5\sqrt{3}}{4}\right) - 16\left(\frac{5\sqrt{3}}{4}\right)^2 = 78$ feet

28. 50 mi/h $= \frac{220}{3}$ ft/sec

$$\mathbf{r}(t) = \left(\frac{220}{3}\cos 15^\circ\right)t\mathbf{i} + \left[5 + \left(\frac{220}{3}\sin 15^\circ\right)t - 16t^2\right]\mathbf{j}$$

The ball is 90 feet from where it is thrown when $x = \frac{220}{3}\cos 15^\circ t = 90 \Rightarrow t = \frac{27}{22\cos 15^\circ} \approx 1.2706$ seconds.

The height of the ball at this time is $y = 5 + \left(\frac{220}{3}\sin 15^\circ\right)\left(\frac{27}{22\cos 15^\circ}\right) - 16\left(\frac{27}{22\cos 15^\circ}\right)^2 \approx 3.286$ feet.

29. $x(t) = t(v_0 \cos\theta)$ or $t = \frac{x}{v_0 \cos\theta}$

$$y(t) = t(v_0 \sin\theta) - 16t^2 + h$$

$$y = \frac{x}{v_0\cos\theta}(v_0\sin\theta) - 16\left(\frac{x^2}{v_0^2\cos^2\theta}\right) + h = (\tan\theta)x - \left(\frac{16}{v_0^2}\sec^2\theta\right)x^2 + h$$

30. $y = x - 0.005x^2$

From Exercise 29 we know that tan θ is the coefficient of x. So, $\tan\theta = 1$, $\theta = (\pi/4)$ rad $= 45°$. Also

$$\frac{16}{v_0^2}\sec^2\theta = \text{negative of coefficient of } x^2$$

$$\frac{16}{v_0^2}(2) = 0.005 \text{ or } v_0 = 80 \text{ ft/sec}$$

$\mathbf{r}(t) = \left(40\sqrt{2}t\right)\mathbf{i} + \left(40\sqrt{2}t - 16t^2\right)\mathbf{j}$. Position function

When $40\sqrt{2}t = 60$,

$$t = \frac{60}{40\sqrt{2}} = \frac{3\sqrt{2}}{4}$$

$$\mathbf{v}(t) = 40\sqrt{2}\mathbf{i} + \left(40\sqrt{2} - 32t\right)\mathbf{j}$$

$$\mathbf{v}\left(\frac{3\sqrt{2}}{4}\right) = 40\sqrt{2}\mathbf{i} + \left(40\sqrt{2} - 24\sqrt{2}\right)\mathbf{j}$$

$= 8\sqrt{2}(5\mathbf{i} + 2\mathbf{j})$. Direction

$$\text{Speed} = \left\|\mathbf{v}\left(\frac{3\sqrt{2}}{4}\right)\right\| = 8\sqrt{2}\sqrt{25 + 4} = 8\sqrt{58} \text{ ft/sec}$$

31. $100 \text{ mi/h} = \left(100\frac{\text{miles}}{\text{hr}}\right)\left(5280\frac{\text{feet}}{\text{mile}}\right)\Big/\left(3600\frac{\text{sec}}{\text{hour}}\right) = \frac{440}{3}\text{ft/sec}$

(a) $\mathbf{r}(t) = \left(\frac{440}{3}\cos\theta_0\right)t\mathbf{i} + \left[3 + \left(\frac{440}{3}\sin\theta_0\right)t - 16t^2\right]\mathbf{j}$

Graphing these curves together with $y = 10$ shows that $\theta_0 = 20°$.

(b)

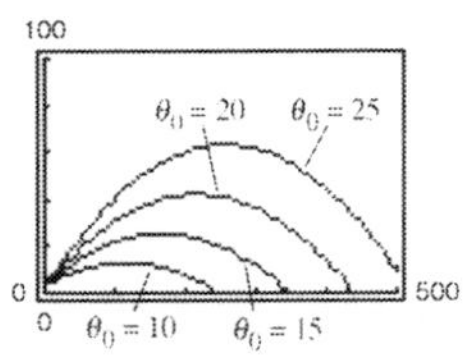

(c) You want

$$x(t) = \left(\frac{440}{3}\cos\theta\right)t \geq 400 \text{ and } y(t) = 3 + \left(\frac{440}{3}\sin\theta\right)t - 16t^2 \geq 10.$$

From $x(t)$, the minimum angle occurs when $t = 30/(11\cos\theta)$. Substituting this for t in $y(t)$ yields:

$$3 + \left(\frac{440}{3}\sin\theta\right)\left(\frac{30}{11\cos\theta}\right) - 16\left(\frac{30}{11\cos\theta}\right)^2 = 10$$

$$400\tan\theta - \frac{14{,}400}{121}\sec^2\theta = 7$$

$$\frac{14{,}400}{121}\left(1 + \tan^2\theta\right) - 400\tan\theta + 7 = 0$$

$$14{,}400\tan^2\theta - 48{,}400\tan\theta + 15{,}247 = 0$$

$$\tan\theta = \frac{48{,}400 \pm \sqrt{48{,}400^2 - 4(14{,}400)(15{,}247)}}{2(14{,}400)}$$

$$\theta = \tan^{-1}\left(\frac{48{,}400 - \sqrt{1{,}464{,}332{,}800}}{28{,}800}\right) \approx 19.38°$$

32. $h = 7$ feet, $\theta = 35°$, 30 yards $= 90$ feet

$\mathbf{r}(t) = (v_0 \cos 35°)t\mathbf{i} + \left[7 + (v_0 \sin 35°)t - 16t^2\right]\mathbf{j}$

(a) $v_0 \cos 35° t = 90$ when $7 + (v_0 \sin 35°)t - 16t^2 = 4$

$$t = \frac{90}{v_0 \cos 35°}$$

$$7 + (v_0 \sin 35°)\left(\frac{90}{v_0 \cos 35°}\right) - 16\left(\frac{90}{v_0 \cos 35°}\right)^2 = 4$$

$$90 \tan 35° + 3 = \frac{129{,}600}{v_0^2 \cos^2 35°}$$

$$v_0^2 = \frac{129{,}600}{\cos^2 35°(90 \tan 35° + 3)}$$

$$v_0 \approx 54.088 \text{ ft/sec}$$

(b) The maximum height occurs when $y'(t) = v_0 \sin 35° - 32t = 0.$

$$t = \frac{v_0 \sin 35}{32} \approx 0.969 \text{ sec}$$

At this time, the height is $y(0.969) \approx 22.0$ ft.

(c) $x(t) = 90 \Rightarrow (v_0 \cos 35°)t = 90$

$$t = \frac{90}{54.088 \cos 35°} \approx 2.0 \text{ sec}$$

33. $\mathbf{r}(t) = (v \cos \theta)t\mathbf{i} + \left[(v \sin \theta)t - 16t^2\right]\mathbf{j}$

(a) You want to find the minimum initial speed v as a function of the angle θ. Because the bale must be thrown to the position $(16, 8)$, you have

$$16 = (v \cos \theta)t$$
$$8 = (v \sin \theta)t - 16t^2.$$

$t = 16/(v \cos \theta)$ from the first equation. Substituting into the second equation and solving for v, you obtain:

$$8 = (v \sin \theta)\left(\frac{16}{v \cos \theta}\right) - 16\left(\frac{16}{v \cos \theta}\right)^2$$

$$1 = 2\left(\frac{\sin \theta}{\cos \theta}\right) - 512\left(\frac{1}{v^2 \cos^2 \theta}\right)$$

$$512\left(\frac{1}{v^2 \cos^2 \theta}\right) = 2\left(\frac{\sin \theta}{\cos \theta}\right) - 1$$

$$\frac{1}{v^2} = \left(2\frac{\sin \theta}{\cos \theta} - 1\right)\frac{\cos^2 \theta}{512} = \frac{2 \sin \theta \cos \theta - \cos^2 \theta}{512}$$

$$v^2 = \frac{512}{2 \sin \theta \cos \theta - \cos^2 \theta}$$

You minimize $f(\theta) = \dfrac{512}{2 \sin \theta \cos \theta - \cos^2 \theta}$.

$$f'(\theta) = -512\left(\frac{2\cos^2 \theta - 2 \sin^2 \theta + 2 \sin \theta \cos \theta}{\left(2 \sin \theta \cos \theta - \cos^2 \theta\right)^2}\right)$$

$$f'(\theta) = 0 \Rightarrow 2 \cos(2\theta) + \sin(2\theta) = 0$$
$$\tan(2\theta) = -2$$
$$\theta \approx 1.01722 \approx 58.28°$$

Substituting into the equation for v, $v \approx 28.78$ ft/sec.

(b) If $\theta = 45°$,

$$16 = (v\cos\theta)t = v\frac{\sqrt{2}}{2}t$$

$$8 = (v\sin\theta)t - 16t^2 = v\frac{\sqrt{2}}{2}t - 16t^2$$

From part (a), $v^2 = \dfrac{512}{2(\sqrt{2}/2)(\sqrt{2}/2) - (\sqrt{2}/2)^2} = \dfrac{512}{1/2} = 1024 \Rightarrow v = 32$ ft/sec.

34. Place the origin directly below the plane. Then $\theta = 0$, $v_0 = 792$ and

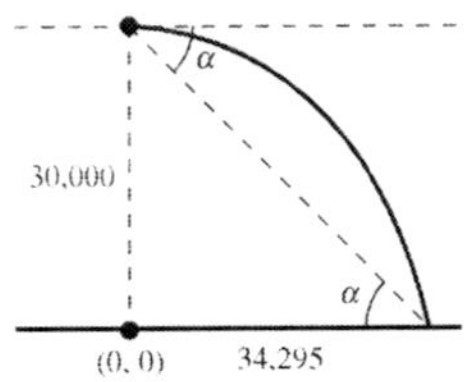

$\mathbf{r}(t) = (v_0\cos\theta)t\mathbf{i} + (30{,}000 + (v_0\sin\theta)t - 16t^2)\mathbf{j} = 792t\mathbf{i} + (30{,}000 - 16t^2)\mathbf{j}$

$\mathbf{v}(t) = 792\mathbf{i} - 32t\mathbf{j}$.

At time of impact, $30{,}000 - 16t^2 = 0 \Rightarrow t^2 = 1875 \Rightarrow t \approx 43.3$ seconds.

$\mathbf{r}(43.3) = 34{,}294.6\mathbf{i}$

$\mathbf{v}(43.3) = 792\mathbf{i} - 1385.6\mathbf{j}$

$\|\mathbf{v}(43.3)\| = 1596$ ft/sec $= 1088$ mi/h

$\tan\alpha = \dfrac{30{,}000}{34{,}294.6} \approx 0.8748 \Rightarrow \alpha \approx 0.7187(41.18°)$

35. $\mathbf{r}(t) = (v_0\cos\theta)t\mathbf{i} + [(v_0\sin\theta)t - 16t^2]\mathbf{j}$

$(v_0\sin\theta)t - 16t^2 = 0$ when $t = 0$ and $t = \dfrac{v_0\sin\theta}{16}$.

The range is

$$x = (v_0\cos\theta)t = (v_0\cos\theta)\frac{v_0\sin\theta}{16} = \frac{v_0^2}{32}\sin 2\theta.$$

So,

$$x = \frac{1200^2}{32}\sin(2\theta) = 3000 \Rightarrow \sin 2\theta = \frac{1}{15} \Rightarrow \theta \approx 1.91°.$$

36. From Exercise 35, the range is

$$x = \frac{v_0^2}{32}\sin 2\theta$$

So, $x = 200 = \dfrac{v_0^2}{32}\sin(24°)$

$\Rightarrow v_0^2 = 6400/\sin(24°)$

$\Rightarrow v_0 \approx 125.4$ ft/sec

37. (a) $\theta = 10°$, $v_0 = 66$ ft/sec

$\mathbf{r}(t) = (66\cos 10°)t\mathbf{i} + [0 + (66\sin 10°)t - 16t^2]\mathbf{j}$

$\mathbf{r}(t) \approx (65t)\mathbf{i} + (11.46t - 16t^2)\mathbf{j}$

Maximum height: 2.052 feet

Range: 46.557 feet

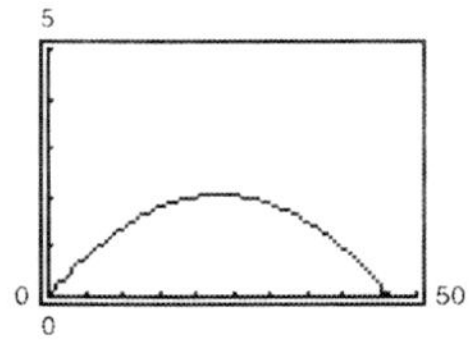

(b) $\theta = 10°$, $v_0 = 146$ ft/sec

$\mathbf{r}(t) = (146\cos 10°)t\mathbf{i} + [0 + (146\sin 10°)t - 16t^2]\mathbf{j}$

$\mathbf{r}(t) \approx (143.78t)\mathbf{i} + (25.35t - 16t^2)\mathbf{j}$

Maximum height: 10.043 feet

Range: 227.828 feet

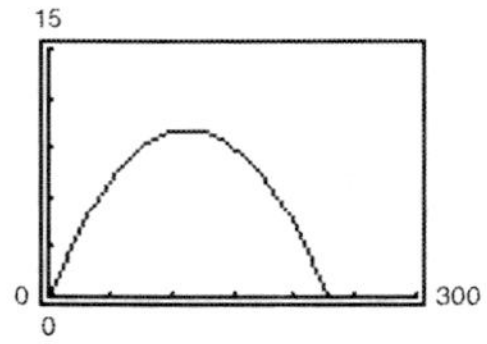

(c) $\theta = 45°$, $v_0 = 66$ ft/sec

$\mathbf{r}(t) = (66\cos 45°)t\mathbf{i} + [0 + (66\sin 45°)t - 16t^2]\mathbf{j}$

$\mathbf{r}(t) \approx (46.67t)\mathbf{i} + (46.67t - 16t^2)\mathbf{j}$

Maximum height: 34.031 feet

Range: 136.125 feet

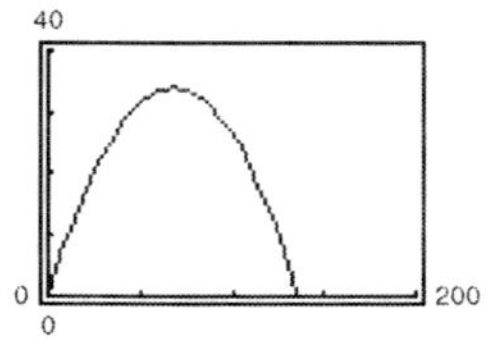

(d) $\theta = 45°$, $v_0 = 146$ ft/sec

$\mathbf{r}(t) = (146 \cos 45°)t\mathbf{i} + \left[0 + (146 \sin 45°)t - 16t^2\right]\mathbf{j}$

$\mathbf{r}(t) \approx (103.24t)\mathbf{i} + (103.24t - 16t^2)\mathbf{j}$

Maximum height: 166.531 feet

Range: 666.125 feet

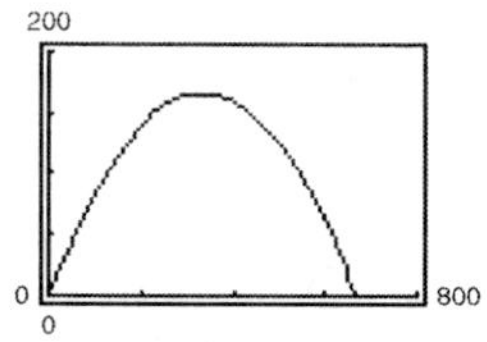

(e) $\theta = 60°$, $v_0 = 66$ ft/sec

$\mathbf{r}(t) = (66 \cos 60°)t\mathbf{i} + \left[0 + (66 \sin 60°)t - 16t^2\right]\mathbf{j}$

$\mathbf{r}(t) \approx (33t)\mathbf{i} + (57.16t - 16t^2)\mathbf{j}$

Maximum height: 51.047 feet

Range: 117.888 feet

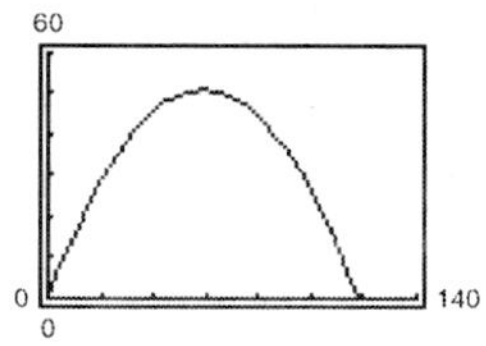

(f) $\theta = 60°$, $v_0 = 146$ ft/sec

$\mathbf{r}(t) = (146 \cos 60°)t\mathbf{i} + \left[0 + (146 \sin 60°)t - 16t^2\right]\mathbf{j}$

$\mathbf{r}(t) \approx (73t)\mathbf{i} + (126.44t - 16t^2)\mathbf{j}$

Maximum height: 249.797 feet

Range: 576.881 feet

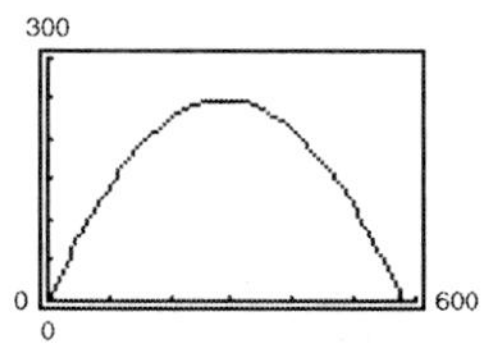

38. (a) $\mathbf{r}(t) = t(v_0 \cos \theta)\mathbf{i} + (tv_0 \sin \theta - 16t^2)\mathbf{j}$

$t(v_0 \sin \theta - 16t) = 0$ when $t = \dfrac{v_0 \sin \theta}{16}$.

Range: $x = v_0 \cos \theta\left(\dfrac{v_0 \sin \theta}{32}\right) = \left(\dfrac{v_0^2}{32}\right)\sin 2\theta$

The range will be maximum when

$$\frac{dx}{dt} = \left(\frac{v_0^2}{32}\right)2 \cos 2\theta = 0$$

or

$$2\theta = \frac{\pi}{2}, \theta = \frac{\pi}{4} \text{ rad.}$$

(b) $y(t) = tv_0 \sin \theta - 16t^2$

$$\frac{dy}{dt} = v_0 \sin \theta - 32t = 0 \text{ when } t = \frac{v_0 \sin \theta}{32}.$$

Maximum height:

$$y\left(\frac{v_0 \sin \theta}{32}\right) = \frac{v_0^2 \sin^2 \theta}{32} - 16\frac{v_0^2 \sin^2 \theta}{32^2} = \frac{v_0^2 \sin^2 \theta}{64}$$

Maximum height when $\sin \theta = 1$, or $\theta = \dfrac{\pi}{2}$.

39. $\mathbf{r}(t) = (v_0 \cos \theta)t\mathbf{i} + \left[h + (v_0 \sin \theta)t - 4.9t^2\right]\mathbf{j}$

$= (100 \cos 30°)t\mathbf{i} + \left[1.5 + (100 \sin 30°)t - 4.9t^2\right]\mathbf{j}$

The projectile hits the ground when

$-4.9t^2 + 100\left(\frac{1}{2}\right)t + 1.5 = 0 \Rightarrow t \approx 10.234$ seconds.

So the range is $(100 \cos 30°)(10.234) \approx 886.3$ meters.

The maximum height occurs when $dy/dt = 0$.

$100 \sin 30 = 9.8t \Rightarrow t \approx 5.102$ sec

The maximum height is

$y = 1.5 + (100 \sin 30°)(5.102) - 4.9(5.102)^2$

≈ 129.1 meters.

40. $\mathbf{r}(t) = (v_0 \cos \theta)t\mathbf{i} + \left[h + (v_0 \sin \theta)t - 4.9t^2\right]\mathbf{j}$

$= (v_0 \cos 8°)t\mathbf{i} + \left[(v_0 \sin 8°)t - 4.9t^2\right]\mathbf{j}$

$x = 50$ when $(v_0 \cos 8°)t = 50 \Rightarrow t = \dfrac{50}{v_0 \cos 8°}$. For this value of t, $y = 0$:

$$(v_0 \sin 8°)\left(\frac{50}{v_0 \cos 8°}\right) - 4.9\left(\frac{50}{v_0 \cos 8°}\right)^2 = 0$$

$$50 \tan 8° = \frac{(4.9)(2500)}{v_0^2 \cos^2 8°} \Rightarrow v_0^2 = \frac{(4.9)50}{\tan 8° \cos^2 8°} \approx 1777.698 \Rightarrow v_0 \approx 42.2 \text{ m/sec}$$

41. To find the range, set

$y(t) = h + (v_0 \sin\theta)t - \frac{1}{2}gt^2 = 0$ then

$0 = \left(\frac{1}{2}g\right)t^2 - (v_0 \sin\theta)t - h$. By the Quadratic Formula, (discount the negative value)

$$t = \frac{v_0 \sin\theta + \sqrt{(-v_0 \sin\theta)^2 - 4[(1/2)g](-h)}}{2[(1/2)g]}$$

$$= \frac{v_0 \sin\theta + \sqrt{v_0^2 \sin^2\theta + 2gh}}{g} \text{ second.}$$

At this time,

$$x(t) = v_0 \cos\theta\left(\frac{v_0 \sin\theta + \sqrt{v_0^2 \sin^2\theta + 2gh}}{g}\right)$$

$$= \frac{v_0 \cos\theta}{g}\left(v_0 \sin\theta + \sqrt{v_0^2\left(\sin^2\theta + \frac{2gh}{v_0^2}\right)}\right)$$

$$= \frac{v_0^2 \cos\theta}{g}\left(\sin\theta + \sqrt{\sin^2\theta + \frac{2gh}{v_0^2}}\right) \text{ feet.}$$

42. $h = 6$ feet, $v_0 = 45$ feet per second, $\theta = 42.5°$. From Exercise 41,

$$t = \frac{45 \sin 42.5° + \sqrt{(45)^2 \sin^2 42.5° + 2(32)(6)}}{32}$$

≈ 2.08 seconds.

At this time, $x(t) \approx 69.02$ feet.

43. $\mathbf{r}(t) = b(\omega t - \sin\omega t)\mathbf{i} + b(1 - \cos\omega t)\mathbf{j}$

$\mathbf{v}(t) = b(\omega - \omega\cos\omega t)\mathbf{i} + b\omega\sin\omega t\mathbf{j} = b\omega(1 - \cos\omega t)\mathbf{i} + b\omega\sin\omega t\mathbf{j}$

$\mathbf{a}(t) = (b\omega^2 \sin\omega t)\mathbf{i} + (b\omega^2 \cos\omega t)\mathbf{j} = b\omega^2[\sin(\omega t)\mathbf{i} + \cos(\omega t)\mathbf{j}]$

$\|\mathbf{v}(t)\| = \sqrt{2}b\omega\sqrt{1 - \cos(\omega t)}$

$\|\mathbf{a}(t)\| = b\omega^2$

(a) $\|\mathbf{v}(t)\| = 0$ when $\omega t = 0, 2\pi, 4\pi, \ldots$.

(b) $\|\mathbf{v}(t)\|$ is maximum when $\omega t = \pi, 3\pi, \ldots$, then $\|\mathbf{v}(t)\| = 2b\omega$.

44. $\mathbf{r}(t) = b(\omega t - \sin\omega t)\mathbf{i} + b(1 - \cos\omega t)\mathbf{j}$

$\mathbf{v}(t) = b\omega[(1 - \cos\omega t)\mathbf{i} + (\sin\omega t)\mathbf{j}]$

Speed $= \|\mathbf{v}(t)\| = b\omega\sqrt{1 - 2\cos\omega t + \cos^2\omega t + \sin^2\omega t} = \sqrt{2}\,b\omega\sqrt{1 - \cos\omega t}$.

The speed has a maximum value of $2b\omega$ when $\omega t = \pi, 3\pi, \ldots$

60 mi/h = 88 ft/sec = 88 rad/sec (since $b = 1$).

So, the maximum speed of a point on the tire is twice the speed of the car:
2(88) ft/sec = 120 mi/h

45. $\mathbf{v}(t) = -b\omega\sin(\omega t)\mathbf{i} + b\omega\cos(\omega t)\mathbf{j}$

$\mathbf{r}(t)\cdot\mathbf{v}(t) = -b^2\omega\sin(\omega t)\cos(\omega t) + b^2\omega\sin(vt)\cos(\omega t) = 0$

So, $\mathbf{r}(t)$ and $\mathbf{v}(t)$ are orthogonal.

46. (a) Speed $= \|\mathbf{v}\| = \sqrt{b^2\omega^2 \sin^2(\omega t) + b^2\omega^2 \cos^2(\omega t)} = \sqrt{b^2\omega^2[\sin^2(\omega t) + \cos^2(\omega t)]} = b\omega$

(b)

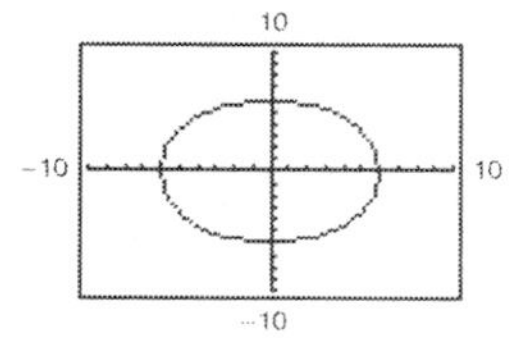

The graphing utility draws the circle faster for greater values of ω.

47. $\mathbf{a}(t) = -b\omega^2 \cos(\omega t)\mathbf{i} - b\omega^2 \sin(\omega t)\mathbf{j} = -b\omega^2[\cos(\omega t)\mathbf{i} + \sin(\omega t)\mathbf{j}] = -\omega^2\mathbf{r}(t)$

$\mathbf{a}(t)$ is a negative multiple of a unit vector from $(0, 0)$ to $(\cos \omega t, \sin \omega t)$ and so $\mathbf{a}(t)$ is directed toward the origin.

48. $\|\mathbf{a}(t)\| = b\omega^2\|\cos(\omega t)\mathbf{i} + \sin(vt)\mathbf{j}\| = b\omega^2$

49. $\|\mathbf{a}(t)\| = \omega^2 b,\ b = 2$

$1 = m(32)$

$\mathbf{F} = m(\omega^2 b) = \frac{1}{32}(2\omega^2) = 10$

$\omega = 4\sqrt{10}$ rad/sec

$\|\mathbf{v}(t)\| = b\omega = 8\sqrt{10}$ ft/sec

50. $\|\mathbf{v}(t)\| = 30$ mi/h $= 44$ ft/sec

$\omega = \dfrac{\|\mathbf{v}(t)\|}{b} = \dfrac{44}{300}$ rad/sec

$\|a(t)\| = b\omega^2$

$\|\mathbf{F}\| = m(b\omega^2) = \dfrac{3400}{32}(300)\left(\dfrac{44}{300}\right)^2 = \dfrac{2057}{3}$ lb

Let n be normal to the road.

$\|\mathbf{n}\|\cos\theta = 3400$

$\|\mathbf{n}\|\sin\theta = \dfrac{2057}{3}$

Dividing, $\tan\theta = \dfrac{121}{600}$

$\theta \approx 11.4°$

51. The velocity of an object involves both magnitude and direction of motion, whereas speed involves only magnitude.

52. (a) $\mathbf{r}_1(t) = x(t)\mathbf{i} + y(t)\mathbf{j} + z(t)\mathbf{k}$

$\mathbf{r}_2(t) = \mathbf{r}_1(2t)$

Velocity: $\mathbf{r}_2'(t) = 2\mathbf{r}_1'(2t)$

Acceleration: $\mathbf{r}_2''(t) = 4\mathbf{r}_1''(2t)$

(b) In general, if $\mathbf{r}_3(t) = \mathbf{r}_1(\omega t)$, then:

Velocity: $\mathbf{r}_3'(t) = \omega r_1'(\omega t)$

Acceleration: $\mathbf{r}_3''(t) = \omega^2\mathbf{r}_1''(\omega t)$

53. $\mathbf{r}(t) = x(t)\mathbf{i} + y(t)\mathbf{j} + z(t)\mathbf{k}$ Position vector

$\mathbf{v}(t) = x'(t)\mathbf{i} + y'(t)\mathbf{j} + z'(t)\mathbf{k}$ Velocity vector

$\mathbf{a}(t) = x''(t)\mathbf{i} + y''(t)\mathbf{j} + z''(t)\mathbf{k}$ Acceleration vector

Speed $= \|\mathbf{v}(t)\| = \sqrt{x'(t)^2 + y'(t)^2 + z'(t)^2}$

$= C$, C is a constant.

$$\frac{d}{dt}\left[x'(t)^2 + y'(t)^2 + z'(t)^2\right] = 0$$

$$2x'(t)x''(t) + 2y'(t)y''(t) + 2z'(t)z''(t) = 0$$

$$2[x'(t)x''(t) + y'(t)y''(t) + z'(t)z''(t)] = 0$$

$$\mathbf{v}(t) \cdot \mathbf{a}(t) = 0$$

Orthogonal

54. $\mathbf{r}(t) = x(t)\mathbf{i} + y(t)\mathbf{j}$

$y(t) = m(x(t)) + b$, m and b are constants.

$\mathbf{r}(t) = x(t)\mathbf{i} + [m(x(t)) + b]\mathbf{j}$

$\mathbf{v}(t) = x'(t)\mathbf{i} + mx'(t)\mathbf{j}$

$\mathbf{s}(t) = \sqrt{[x'(t)]^2 + [mx'(t)]^2} = C$, C is a constant.

So, $x'(t) = \dfrac{C}{\sqrt{1 + m^2}}$

$x''(t) = 0$

$\mathbf{a}(t) = x''(t)\mathbf{i} + mx''(t)\mathbf{j} = \mathbf{0}.$

55. $\mathbf{r}(t) = 6\cos t\mathbf{i} + 3\sin t\mathbf{j}$

(a) $\mathbf{v}(t) = \mathbf{r}'(t) = -6\sin t\mathbf{i}\ 3\cos t\mathbf{j}$

$\|\mathbf{v}(t)\| = \sqrt{36\sin^2 t + 9\cos^2 t}$

$= 3\sqrt{4\sin^2 t + \cos^2 t} = 3\sqrt{3\sin^2 t + 1}$

$\mathbf{a}(t) = \mathbf{v}'(t) = -6\cos t\mathbf{i} - 3\sin t\mathbf{j}$

(b)

t	0	$\frac{\pi}{4}$	$\frac{\pi}{2}$	$\frac{2\pi}{3}$	π
Speed	3	$\frac{3}{2}\sqrt{10}$	6	$\frac{3}{2}\sqrt{13}$	3

(c)

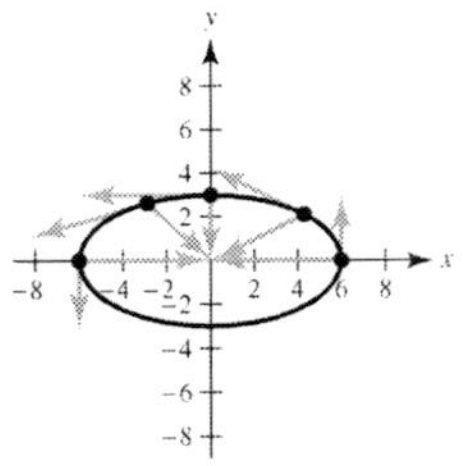

(d) The speed is increasing when the angle between $\mathbf{v}$ and $\mathbf{a}$ is in the interval

$\left[0, \frac{\pi}{2}\right).$

The speed is decreasing when the angle is in the interval

$\left(\frac{\pi}{2}, \pi\right].$

56. $\mathbf{r}(t) = a\cos\omega t\mathbf{i} + b\sin\omega t\mathbf{j}$

(a) $\mathbf{r}'(t) = \mathbf{v}(t) = -a\omega\sin\omega t\mathbf{i} + b\omega\cos\omega t\mathbf{j}$

Speed $= \|\mathbf{v}(t)\| = \sqrt{a^2\omega^2\sin^2\omega t + b^2\omega^2\cos^2\omega t}$

(b) $\mathbf{a}(t) = \mathbf{v}'(t) = -a\omega^2\cos\omega t\mathbf{i} - b\omega^2\sin\omega t\mathbf{j}$

$= \omega^2(-a\cos\omega t\mathbf{i} - b\sin\omega t\mathbf{j})$

$= -\omega^2\mathbf{r}(t)$

57. $\mathbf{a}(t) = \sin t\mathbf{i} - \cos t\mathbf{j}$

$\mathbf{v}(t) = \int \mathbf{a}(t)\,dt = -\cos t\mathbf{i} = \sin t\mathbf{j} + \mathbf{C}_1$

$\mathbf{v}(0) = -\mathbf{i} = -\mathbf{i} + \mathbf{C}_1 \Rightarrow \mathbf{C}_1 = \mathbf{0}$

$\mathbf{v}(t) = -\cos t\mathbf{i} - \sin t\mathbf{j}$

$\mathbf{r}(t) = \int \mathbf{v}(t)\,dt = -\sin t\mathbf{i} + \cos t\mathbf{j} + \mathbf{C}_2$

$\mathbf{r}(0) = \mathbf{j} = \mathbf{j} + \mathbf{C}_2 \Rightarrow \mathbf{C}_2 = \mathbf{0}$

$\mathbf{r}(t) = -\sin t\mathbf{i} + \cos t\mathbf{j}$

The path is a circle.

58. The angle at time t_1 is obtuse.

The angle at time t_2 is acute.

The speed is decreasing at time t_1 because the projectile is reaching its maximum height.

The speed is increasing at time t_2 because the object is accelerating due to gravity.

59. False. The acceleration is the derivative of the velocity.

60. True

61. True

62. False. For example, $6t\mathbf{r}(t) = t^3\mathbf{i}$. Then $\mathbf{v}(t) = 3t^2\mathbf{i}$ and $\mathbf{a}(t) = 6t\mathbf{i}$. $\mathbf{v}(t)$ is not orthogonal to $\mathbf{a}(t)$.

Section 12.4 Tangent Vectors and Normal Vectors

1. $\mathbf{r}(t) = t^2\mathbf{i} + 2t\mathbf{j},\ t = 1$

$\mathbf{r}'(t) = 2t\mathbf{i} + 2\mathbf{j},\ \|\mathbf{r}'(t)\| = \sqrt{4t^2 + 4} = 2\sqrt{t^2 + 1}$

$\mathbf{T}(1) = \frac{\mathbf{r}'(1)}{\|\mathbf{r}'(1)\|} = \frac{1}{\sqrt{2}}(\mathbf{i} + \mathbf{j}) = \frac{\sqrt{2}}{2}\mathbf{i} + \frac{\sqrt{2}}{2}\mathbf{j}$

2. $\mathbf{r}(t) = t^3\mathbf{i} + 2t^2\mathbf{j},\ t = 1$

$\mathbf{r}'(t) = 3t^2\mathbf{i} + 4t\mathbf{j}$

$\|\mathbf{r}'(t)\| = \sqrt{9t^4 + 16t^2}$

$\mathbf{T}(1) = \frac{\mathbf{r}'(1)}{\|\mathbf{r}'(1)\|} = \frac{1}{\sqrt{9 + 16}}(3\mathbf{i} + 4\mathbf{j}) = \frac{3}{5}\mathbf{i} + \frac{4}{5}\mathbf{j}$

3. $\mathbf{r}(t) = 4\cos t\mathbf{i} + 4\sin t\mathbf{j},\ t = \frac{\pi}{4}$

$\mathbf{r}'(t) = -4\sin t\mathbf{i} + 4\cos t\mathbf{j}$

$\|\mathbf{r}'(t)\| = \sqrt{16\sin^2 t + 16\cos^2 t} = 4$

$\mathbf{T}\left(\frac{\pi}{4}\right) = \frac{\mathbf{r}'\left(\frac{\pi}{4}\right)}{\left\|\mathbf{r}'\left(\frac{\pi}{4}\right)\right\|} = -\frac{\sqrt{2}}{2}\mathbf{i} + \frac{\sqrt{2}}{2}\mathbf{j}$

4. $\mathbf{r}(t) = 6\cos t\mathbf{i} + 2\sin t\mathbf{j},\ t = \frac{\pi}{3}$

$\mathbf{r}'(t) = -6\sin t\mathbf{i} + 2\cos t\mathbf{j}$

$\|\mathbf{r}'(t)\| = \sqrt{36\sin^2 t + 4\cos^2 t}$

$$\mathbf{T}\left(\frac{\pi}{3}\right) = \frac{\mathbf{r}'\left(\frac{\pi}{3}\right)}{\left\|\mathbf{r}'\left(\frac{\pi}{3}\right)\right\|} = \frac{-3\sqrt{3}\mathbf{i} + \mathbf{j}}{\sqrt{36(3/4) + (1/4)}} = \frac{1}{\sqrt{28}}\left(-3\sqrt{3}\mathbf{i} + \mathbf{j}\right)$$

5. $\mathbf{r}(t) = 3t\mathbf{i} - \ln t\mathbf{j},\ t = e$

$\mathbf{r}'(t) = 3\mathbf{i} - \frac{1}{t}\mathbf{j}$

$\mathbf{r}'(e) = 3\mathbf{i} - \frac{1}{e}\mathbf{j}$

$$\mathbf{T}(e) = \frac{\mathbf{r}'(e)}{\|\mathbf{r}'(e)\|}$$

$$= \frac{3\mathbf{i} - \frac{1}{e}\mathbf{j}}{\sqrt{9 + \frac{1}{e^2}}} = \frac{3e\mathbf{i} - \mathbf{j}}{\sqrt{9e^2 + 1}} \approx 0.9926\mathbf{i} - 0.1217\mathbf{j}$$

6. $\mathbf{r}(t) = e^t\cos t\mathbf{i} + e^t\mathbf{j},\ t = 0$

$\mathbf{r}'(t) = \left(e^t\cos t - e^t\sin t\right)\mathbf{i} + e^t\mathbf{j}$

$\mathbf{r}'(0) = \mathbf{i} + \mathbf{j}$

$$\mathbf{T}(0) = \frac{\mathbf{r}'(0)}{\|\mathbf{r}'(0)\|} = \frac{\mathbf{i} + \mathbf{j}}{\sqrt{2}} = \frac{\sqrt{2}}{2}\mathbf{i} + \frac{\sqrt{2}}{2}\mathbf{j}$$

7. $\mathbf{r}(t) = t\mathbf{i} + t^2\mathbf{j} + t\mathbf{k},\ P(0, 0, 0)$

$\mathbf{r}'(t) = \mathbf{i} + 2t\mathbf{j} + \mathbf{k}$

When $t = 0$, $\mathbf{r}'(0) = \mathbf{i} + \mathbf{k}$, $\left[t = 0 \text{ at } (0, 0, 0)\right]$.

$$\mathbf{T}(0) = \frac{\mathbf{r}'(0)}{\|\mathbf{r}'(0)\|} = \frac{\sqrt{2}}{2}(\mathbf{i} + \mathbf{k})$$

Direction numbers: $a = 1,\ b = 0,\ c = 1$

Parametric equations: $x = t,\ y = 0,\ z = t$

8. $\mathbf{r}(t) = t^2\mathbf{i} + t\mathbf{j} + \frac{4}{3}\mathbf{k},\ P\left(1, 1, \frac{4}{3}\right)$

$\mathbf{r}'(t) = 2t\mathbf{i} + \mathbf{j}$

When $t = 1$, $\mathbf{r}'(t) = \mathbf{r}'(1) = 2\mathbf{i} + \mathbf{j}\ \left[t = 1\ at\left(1, 1, \frac{4}{3}\right)\right]$.

$$\mathbf{T}(1) = \frac{\mathbf{r}'(1)}{\|\mathbf{r}'(1)\|} = \frac{2\mathbf{i} + \mathbf{j}}{\sqrt{5}} = \frac{\sqrt{5}}{5}(2\mathbf{i} + \mathbf{j})$$

Direction numbers: $a = 2,\ b = 1,\ c = 0$

Parametric equations: $x = 2t + 1,\ y = t + 1,\ z = \frac{4}{3}$

9. $\mathbf{r}(t) = 3\cos t\mathbf{i} + 3\sin t\mathbf{j} + t\mathbf{k},\ P(3, 0, 0)$

$\mathbf{r}'(t) = -3\sin t\mathbf{i} + 3\cos t\mathbf{j} + \mathbf{k}$

$t = 0$ at $P(3, 0, 0)$

$\mathbf{r}'(0) = 3\mathbf{j} + \mathbf{k}$

$$\mathbf{T}(0) = \frac{\mathbf{r}'(0)}{\|\mathbf{r}'(0)\|} = \frac{3\mathbf{j} + \mathbf{k}}{\sqrt{10}}$$

Direction numbers: $a = 0,\ b = 3,\ c = 1$

Parametric equations: $x = 3,\ y = 3t,\ z = t$

10. $\mathbf{r}(t) = \left\langle t, t, \sqrt{4 - t^2}\right\rangle,\ P\left(1, 1, \sqrt{3}\right)$

$$\mathbf{r}'(t) = \left\langle 1, 1, -\frac{t}{4 - t^2}\right\rangle$$

When $t = 1$, $\mathbf{r}'(1) = \left\langle 1, 1, -\frac{1}{\sqrt{3}}\right\rangle$, $\left[t = 1 \text{ at } \left(1, 1, \sqrt{3}\right)\right]$.

$$\mathbf{T}(1) = \frac{\mathbf{r}'(1)}{\|\mathbf{r}'(1)\|} = \frac{\sqrt{21}}{7}\left\langle 1, 1, -\frac{1}{\sqrt{3}}\right\rangle$$

Direction numbers: $a = 1, b = 1, c = -\frac{1}{\sqrt{3}}$

Parametric equations: $x = t + 1,\ y = t + 1,$

$z = -\frac{1}{\sqrt{3}}t + \sqrt{3}$

11. $\mathbf{r}(t) = \langle 2\cos t, 2\sin t, 4\rangle,\ P\left(\sqrt{2}, \sqrt{2}, 4\right)$

$\mathbf{r}'(t) = \langle -2\sin t, 2\cos t, 0\rangle$

When $t = \frac{\pi}{4}$, $\mathbf{r}'\left(\frac{\pi}{4}\right) = \left\langle -\sqrt{2}, \sqrt{2}, 0\right\rangle$,

$\left[t = \frac{\pi}{4} \text{ at } \left(\sqrt{2}, \sqrt{2}, 4\right)\right]$.

$$\mathbf{T}\left(\frac{\pi}{4}\right) = \frac{\mathbf{r}'(\pi/4)}{\|\mathbf{r}'(\pi/4)\|} = \frac{1}{2}\left\langle -\sqrt{2}, \sqrt{2}, 0\right\rangle$$

Direction numbers: $a = -\sqrt{2}, b = \sqrt{2}, c = 0$

Parametric equations: $x = -\sqrt{2}t + \sqrt{2},\ y = \sqrt{2}t + \sqrt{2},$

$z = 4$

12. $\mathbf{r}(t) = \left\langle 2\sin t, 2\cos t, 4\sin^2 t\right\rangle,\ P\left(1, \sqrt{3}, 1\right)$

$\mathbf{r}'(t) = \langle 2\cos t, -2\sin t, 8\sin t\cos t\rangle$

When $t = \frac{\pi}{6}$, $\mathbf{r}'\left(\frac{\pi}{6}\right) = \left\langle \sqrt{3}, -1, 2\sqrt{3}\right\rangle$,

$\left[t = \frac{\pi}{6} \text{ at } \left(1, \sqrt{3}, 1\right)\right]$.

$$\mathbf{T}\left(\frac{\pi}{6}\right) = \frac{\mathbf{r}'(\pi/6)}{\|\mathbf{r}'(\pi/6)\|} = \frac{1}{4}\left\langle \sqrt{3}, -1, 2\sqrt{3}\right\rangle$$

Direction numbers: $a = \sqrt{3},\ b = -1,\ c = 2\sqrt{3}$

Parametric equations: $x = \sqrt{3}t + 1,\ y = -t + \sqrt{3},$

$z = 2\sqrt{3}t + 1$

13. $\mathbf{r}(t) = t\mathbf{i} + \frac{1}{2}t^2\mathbf{j},\ t = 2$

$$\mathbf{r}'(t) = \mathbf{i} + t\mathbf{j}$$

$$\mathbf{T}(t) = \frac{\mathbf{r}'(t)}{\|\mathbf{r}'(t)\|} = \frac{\mathbf{i} + t\mathbf{j}}{\sqrt{1 + t^2}}$$

$$\mathbf{T}'(t) = \frac{-t}{\left(t^2 + 1\right)^{3/2}}\mathbf{i} + \frac{1}{\left(t^2 + 1\right)^{3/2}}\mathbf{j}$$

$$\mathbf{T}'(2) = \frac{-2}{5^{3/2}}\mathbf{i} + \frac{1}{5^{3/2}}\mathbf{j}$$

$$\mathbf{N}(2) = \frac{\mathbf{T}'(2)}{\|\mathbf{T}'(2)\|} = \frac{1}{\sqrt{5}}(-2\mathbf{i} + \mathbf{j}) = \frac{-2\sqrt{5}}{5}\mathbf{i} + \frac{\sqrt{5}}{5}\mathbf{j}$$

14. $\mathbf{r}(t) = t\mathbf{i} + \frac{6}{t}\mathbf{j},\ t = 3$

$$\mathbf{r}'(t) = \mathbf{i} - \frac{6}{t^2}\mathbf{j}$$

$$\mathbf{T}(t) = \frac{\mathbf{r}'(t)}{\|\mathbf{r}'(t)\|} = \frac{1}{\sqrt{1 + \left(36/t^4\right)}}\left(i - \frac{6}{t^2}\mathbf{j}\right)$$

$$= \frac{t^2}{\sqrt{t^4 + 36}}\left(\mathbf{i} - \frac{6}{t^2}\mathbf{j}\right)$$

$$\mathbf{T}'(t) = \frac{72t}{\left(t^4 + 36\right)^{3/2}}\mathbf{i} + \frac{12t^3}{\left(t^4 + 36\right)^{3/2}}\mathbf{j}$$

$$\mathbf{T}'(2) = \frac{144}{52^{3/2}}\mathbf{i} + \frac{96}{52^{3/2}}\mathbf{j}$$

$$\mathbf{N}(2) = \frac{\mathbf{T}'(2)}{\|\mathbf{T}'(2)\|} + \frac{1}{\sqrt{13}}(3\mathbf{i} + 2\mathbf{j})$$

15. $\mathbf{r}(t) = \ln t\mathbf{i} + (t + 1)\mathbf{j},\ t = 2$

$$\mathbf{r}'(t) = \frac{1}{t}\mathbf{i} + \mathbf{j}$$

$$\mathbf{T}(t) = \frac{\mathbf{r}'(t)}{\|\mathbf{r}'(t)\|} = \frac{\frac{1}{t}\mathbf{i} + \mathbf{j}}{\sqrt{\frac{1}{t^2} + 1}} = \frac{\mathbf{i} + t\mathbf{j}}{\sqrt{1 + t^2}}$$

$$\mathbf{T}'(t) = \frac{-t}{\left(1 + t^2\right)^{3/2}}\mathbf{i} + \frac{1}{\left(1 + t^2\right)^{3/2}}\mathbf{j}$$

$$\mathbf{T}'(2) = \frac{-2}{5^{3/2}}\mathbf{i} + \frac{1}{5^{3/2}}\mathbf{j}$$

$$\mathbf{N}(2) = \frac{\mathbf{T}'(2)}{\|\mathbf{T}'(2)\|} = \frac{-2\sqrt{5}}{5}\mathbf{i} + \frac{\sqrt{5}}{5}\mathbf{j}$$

16. $\mathbf{r}(t) = \pi\cos t\mathbf{i} + \pi\sin t\mathbf{j},\ t = \frac{\pi}{6}$

$$\mathbf{r}'(t) = -\pi\sin t\mathbf{i} + \pi\cos t\mathbf{j}$$

$$\|\mathbf{r}'(t)\| = \pi$$

$$\mathbf{T}(t) = \frac{\mathbf{r}'(t)}{\|\mathbf{r}'(t)\|} = -\sin t\mathbf{i} + \cos t\mathbf{j}$$

$$\mathbf{T}'(t) = -\cos t\mathbf{i} - \sin t\mathbf{j},\ \|\mathbf{T}'(t)\| = 1$$

$$\mathbf{T}'\left(\frac{\pi}{6}\right) = -\frac{\sqrt{3}}{2}\mathbf{i} - \frac{1}{2}\mathbf{j}$$

$$\mathbf{N}\left(\frac{\pi}{6}\right) = \frac{\mathbf{T}'\left(\frac{\pi}{6}\right)}{\left\|\mathbf{T}'\left(\frac{\pi}{6}\right)\right\|} = -\frac{\sqrt{3}}{2}\mathbf{i} - \frac{1}{2}\mathbf{j}$$

17. $\mathbf{r}(t) = t\mathbf{i} + t^2\mathbf{j} + \ln t\mathbf{k},\ t = 1$

$$\mathbf{r}'(t) = \mathbf{i} + 2t\mathbf{j} + \frac{1}{t}\mathbf{k}$$

$$\mathbf{T}(t) = \frac{\mathbf{r}'(t)}{\|\mathbf{r}'(t)\|} = \frac{\mathbf{i} + 2t\mathbf{j} + \frac{1}{t}\mathbf{k}}{\sqrt{1 + 4t^2 + \frac{1}{t^2}}} = \frac{t\mathbf{i} + 2t^2\mathbf{j} + \mathbf{k}}{\sqrt{4t^4 + t^2 + 1}}$$

$$\mathbf{T}'(t) = \frac{1 - 4t^4}{\left(4t^4 + t^2 + 1\right)^{3/2}}\mathbf{i} + \frac{2t^3 + 4t}{\left(4t^4 + t^2 + 1\right)^{3/2}}\mathbf{j} + \frac{-8t^3 - t}{\left(4t^4 + t^2 + 1\right)^{3/2}}\mathbf{k}$$

$$\mathbf{T}'(1) = \frac{-3}{6^{3/2}}\mathbf{i} + \frac{6}{6^{3/2}}\mathbf{j} + \frac{-9}{6^{3/2}}\mathbf{k} = \frac{3}{6^{3/2}}\left[-\mathbf{i} + 2\mathbf{j} - 3\mathbf{k}\right]$$

$$\mathbf{N}(1) = \frac{-\mathbf{i} + 2\mathbf{j} - 3\mathbf{k}}{\sqrt{14}} = \frac{-\sqrt{14}}{14}\mathbf{i} + \frac{2\sqrt{14}}{14}\mathbf{j} - \frac{3\sqrt{14}}{14}\mathbf{k}$$

18. $\mathbf{r}(t) = \sqrt{2}t\mathbf{i} + e^t\mathbf{j} + e^{-t}\mathbf{k},\ t = 0$

$\mathbf{r}'(t) = \sqrt{2}\mathbf{i} + e^t\mathbf{j} - e^{-t}\mathbf{k}$

$\|\mathbf{r}'(t)\| = \sqrt{2 + e^{2t} + e^{-2t}} = \sqrt{\left(e^t + e^{-t}\right)^2} = e^t + e^{-t}$

$\mathbf{T}(t) = \dfrac{\mathbf{r}'(t)}{\|\mathbf{r}'(t)\|} = \dfrac{\sqrt{2}\mathbf{i} + e^t\mathbf{j} - e^{-t}\mathbf{k}}{e^t + e^{-t}}$

$\mathbf{T}'(t) = \dfrac{\sqrt{2}\left(e^{-t} - e^t\right)}{\left(e^t + e^{-t}\right)^2}\mathbf{i} + \dfrac{2}{\left(e^t + e^{-t}\right)^2}\mathbf{j} + \dfrac{2}{\left(e^t + e^{-t}\right)^2}\mathbf{k}$

$\mathbf{T}'(0) = \dfrac{1}{2}\mathbf{j} + \dfrac{1}{2}\mathbf{k}$

$\mathbf{N}(0) = \dfrac{\sqrt{2}}{2}\mathbf{j} + \dfrac{\sqrt{2}}{2}\mathbf{k}$

19. $\mathbf{r}(t) = 6\cos t\mathbf{i} + 6\sin t\mathbf{j} + \mathbf{k},\ t = \dfrac{3\pi}{4}$

$\mathbf{r}'(t) = -6\sin t\mathbf{i} + 6\cos t\mathbf{j}$

$\mathbf{T}(t) = \dfrac{\mathbf{r}'(t)}{\|\mathbf{r}'(t)\|} = -\sin t\mathbf{i} + \cos t\mathbf{j}$

$\mathbf{T}'(t) = -\cos t\mathbf{i} - \sin t\mathbf{j},\ \|\mathbf{T}'(t)\| = 1$

$\mathbf{N}\left(\dfrac{3\pi}{4}\right) = \dfrac{\mathbf{T}'(3\pi/4)}{\|\mathbf{T}'(3\pi/4)\|} = \dfrac{\sqrt{2}}{2}\mathbf{i} - \dfrac{\sqrt{2}}{2}\mathbf{j}$

20. $\mathbf{r}(t) = \cos 3t\mathbf{i} + 2\sin 3t\mathbf{j} + \mathbf{k},\ t = \pi$

$\mathbf{r}'(t) = -3\sin 3t\mathbf{i} + 6\cos 3t\mathbf{j}$

$\mathbf{T}(t) = \dfrac{\mathbf{r}'(t)}{\|\mathbf{r}'(t)\|} = \dfrac{-3\sin 3t\mathbf{i} + 6\cos 3t\mathbf{j}}{\sqrt{9\sin^2 3t + 36\cos^2 3t}}$

The normal vector is perpendicular to $\mathbf{T}(t)$ and points toward the z-axis:

$\mathbf{N}(t) = \dfrac{-6\cos 3t\mathbf{i} - 3\sin 3t\mathbf{j}}{\sqrt{9\sin^2 3t + 36\cos^2 3t}}$

$\mathbf{N}(\pi) = \dfrac{6\mathbf{i}}{\sqrt{36}} = \mathbf{i}$

21. $\mathbf{r}(t) = t\mathbf{i} + \dfrac{1}{t}\mathbf{j},\ t = 1$

$\mathbf{v}(t) = \mathbf{i} - \dfrac{1}{t^2}\mathbf{j},\ \mathbf{v}(1) = \mathbf{i} - \mathbf{j},$

$\mathbf{a}(t) = \dfrac{2}{t^3}\mathbf{j},\ \mathbf{a}(1) = 2\mathbf{j}$

$\mathbf{T}(t) = \dfrac{\mathbf{v}(t)}{\|\mathbf{v}(t)\|} = \dfrac{t^2}{\sqrt{t^4 + 1}}\left(\mathbf{i} - \dfrac{1}{t^2}\mathbf{j}\right) = \dfrac{1}{\sqrt{t^4 + 1}}\left(t^2\mathbf{i} - \mathbf{j}\right)$

$\mathbf{T}(1) = \dfrac{1}{\sqrt{2}}(\mathbf{i} - \mathbf{j}) = \dfrac{\sqrt{2}}{2}(\mathbf{i} - \mathbf{j})$

$\mathbf{N}(t) = \dfrac{\mathbf{T}'(t)}{\|\mathbf{T}'(t)\|} = \dfrac{\dfrac{2t}{\left(t^4 + 1\right)^{3/2}}\mathbf{i} + \dfrac{2t^3}{\left(t^4 + 1\right)^{3/2}}\mathbf{j}}{\dfrac{2t}{\left(t^4 + 1\right)}}$

$= \dfrac{1}{\sqrt{t^4 + 1}}\left(\mathbf{i} + t^2\mathbf{j}\right)$

$\mathbf{N}(1) = \dfrac{1}{\sqrt{2}}(\mathbf{i} + \mathbf{j}) = \dfrac{\sqrt{2}}{2}(\mathbf{i} + \mathbf{j})$

$a_{\mathbf{T}} = \mathbf{a} \cdot \mathbf{T} = -\sqrt{2}$

$a_{\mathbf{N}} = \mathbf{a} \cdot \mathbf{N} = \sqrt{2}$

22. $\mathbf{r}(t) = t^2\mathbf{i} + 2t\mathbf{j},\ t = 1$

$\mathbf{v}(t) = 2t\mathbf{i} + 2\mathbf{j},\ \mathbf{v}(1) = 2\mathbf{i} + 2\mathbf{j}$

$\mathbf{a}(t) = 2\mathbf{i},\ \mathbf{a}(1) = 2\mathbf{i}$

$\mathbf{T}(t) = \dfrac{\mathbf{v}(t)}{\|\mathbf{v}(t)\|} = \dfrac{1}{\sqrt{4t^2 + 4}}(2t\mathbf{i} + 2\mathbf{j}) = \dfrac{1}{\sqrt{t^2 + 1}}(t\mathbf{i} + \mathbf{j})$

$\mathbf{T}(1) = \dfrac{1}{\sqrt{2}}(\mathbf{i} + \mathbf{j}) = \dfrac{\sqrt{2}}{2}\mathbf{i} + \dfrac{\sqrt{2}}{2}\mathbf{j}$

$\mathbf{N}(t) = \dfrac{\mathbf{T}'(t)}{\|\mathbf{T}'(t)\|} = \dfrac{\dfrac{1}{\left(t^2 + 1\right)^{3/2}}\mathbf{i} + \dfrac{-t}{\left(t^2 + 1\right)^{3/2}}\mathbf{j}}{\dfrac{1}{t^2 + 1}}$

$= \dfrac{1}{\sqrt{t^2 + 1}}(\mathbf{i} + t\mathbf{j})$

$\mathbf{N}(1) = \dfrac{\sqrt{2}}{2}\mathbf{i} - \dfrac{\sqrt{2}}{2}\mathbf{j}$

$a_{\mathbf{T}} = \mathbf{a} \cdot \mathbf{T} = \sqrt{2}$

$a_{\mathbf{N}} = \mathbf{a} \cdot \mathbf{N} = \sqrt{2}$

23. $\mathbf{r}(t) = (t - t^3)\mathbf{i} + 2t^2\mathbf{j},\ t = 1$

$\mathbf{v}(t) = (1 - 3t^2)\mathbf{i} + 4t\mathbf{j},\ \mathbf{v}(1) = -2\mathbf{i} + 4\mathbf{j}$

$\mathbf{a}(t) = -6t\mathbf{i} + 4\mathbf{j},\ \mathbf{a}(1) = -6\mathbf{i} + 4\mathbf{j}$

$$\mathbf{T}(t) = \frac{\mathbf{v}(t)}{\|\mathbf{v}(t)\|} = \frac{(1 - 3t^2)\mathbf{i} + 4t\mathbf{j}}{\sqrt{9t^4 + 10t^2 + 1}}$$

$$\mathbf{T}(1) = \frac{-2\mathbf{i} + 4\mathbf{j}}{\sqrt{20}} = \frac{-\mathbf{i} + 2\mathbf{j}}{\sqrt{5}} = \frac{-\sqrt{5}}{5}(\mathbf{i} - 2\mathbf{j})$$

$$\mathbf{T}'(t) = \frac{-16t(3t^2 + 1)}{(9t^4 + 10t^2 + 1)^{3/2}}\mathbf{i} + \frac{4 - 36t^4}{(9t^4 + 10t^2 + 1)^{3/2}}\mathbf{j}$$

$$\mathbf{T}'(1) = \frac{-64}{20^{3/2}}\mathbf{i} + \frac{-32}{20^{3/2}}\mathbf{j}$$

$$\mathbf{N}(1) = \frac{-2\mathbf{i} - \mathbf{j}}{\sqrt{5}}$$

$$a_{\mathbf{T}} = \mathbf{a}\cdot\mathbf{T} = \frac{1}{\sqrt{5}}(6 + 8) = \frac{14\sqrt{5}}{5}$$

$$a_{\mathbf{N}} = \mathbf{a}\cdot\mathbf{N} = \frac{1}{\sqrt{5}}(12 - 4) = \frac{8\sqrt{5}}{5}$$

24. $\mathbf{r}(t) = (t^3 - 4t)\mathbf{i} + (t^2 - 1)\mathbf{j},\ t = 0$

$\mathbf{v}(t) = (3t^2 - 4)\mathbf{i} + 2t\mathbf{j},\ \mathbf{v}(0) = -4\mathbf{i}$

$\mathbf{a}(t) = 6t\mathbf{i} + 2\mathbf{j},\ \mathbf{a}(0) = 2\mathbf{j}$

$$\mathbf{T}(t) = \frac{\mathbf{v}(t)}{\|\mathbf{v}(t)\|} = \frac{(3t^2 - 4)\mathbf{i} + 2t\mathbf{j}}{\sqrt{9t^4 - 20t^2 + 16}}$$

$$\mathbf{T}(0) = \frac{-4\mathbf{i}}{\sqrt{16}} = -\mathbf{i}$$

$$\mathbf{T}'(t) = \frac{4t(3t^2 + 4)}{(9t^4 - 20t^2 + 16)^{3/2}}\mathbf{i} + \frac{32 - 18t^4}{(9t^4 - 20t^2 + 16)^{3/2}}\mathbf{j}$$

$$\mathbf{T}'(1) = \frac{32}{16^{3/2}}\mathbf{j} = \frac{1}{2}\mathbf{j}$$

$\mathbf{N}(1) = \mathbf{j}$

$a_{\mathbf{T}} = \mathbf{a}\cdot\mathbf{T} = 0$

$a_{\mathbf{N}} = \mathbf{a}\cdot\mathbf{N} = \sqrt{2}$

25. $\mathbf{r}(t) = e^t\mathbf{i} + e^{-2t}\mathbf{j},\ t = 0$

$\mathbf{v}(t) = e^t\mathbf{i} - 2e^{-2t}\mathbf{j},\ \mathbf{v}(0) = \mathbf{i} - 2\mathbf{j}$

$\mathbf{a}(t) = e^t\mathbf{i} + 4e^{-2t}\mathbf{j},\ \mathbf{a}(0) = \mathbf{i} + 4\mathbf{j}$

$$\mathbf{T}(t) = \frac{\mathbf{v}(t)}{\|\mathbf{v}(t)\|} = \frac{e^t\mathbf{i} - 2e^{-2t}\mathbf{j}}{\sqrt{4e^{-4t} + e^{2t}}}$$

$$\mathbf{T}(0) = \frac{\mathbf{i} - 2\mathbf{j}}{\sqrt{5}}$$

$$\mathbf{N}(0) = \frac{2\mathbf{i} + \mathbf{j}}{\sqrt{5}}$$

$$a_{\mathbf{T}} = \mathbf{a}\cdot\mathbf{T} = \frac{1}{\sqrt{5}}(1 - 8) = \frac{-7\sqrt{5}}{5}$$

$$a_{\mathbf{N}} = \mathbf{a}\cdot\mathbf{N} = \frac{1}{\sqrt{5}}(2 + 4) = \frac{6\sqrt{5}}{5}$$

26. $\mathbf{r}(t) = e^t\mathbf{i} + e^{-t}\mathbf{j} + t\mathbf{k},\ t = 0$

$\mathbf{v}(t) = e^t\mathbf{i} - e^{-t}\mathbf{j} + \mathbf{k},\ \mathbf{v}(0) = \mathbf{i} - \mathbf{j} + \mathbf{k}$

$\mathbf{a}(t) = e^t\mathbf{i} + e^{-t}\mathbf{j},\ \mathbf{a}(0) = \mathbf{i} + \mathbf{j}$

$$\mathbf{T}(t) = \frac{\mathbf{v}(t)}{\|\mathbf{v}(t)\|} = \frac{e^t\mathbf{i} - e^{-t}\mathbf{j} + \mathbf{k}}{\sqrt{e^{2t} + e^{-2t} + 1}}$$

$$\mathbf{T}(0) = \frac{\mathbf{i} - \mathbf{j} + \mathbf{k}}{\sqrt{3}}$$

$$\mathbf{T}'(t) = \frac{e^{2t}(e^{2t} + 2)}{(e^{4t} + e^{2t} + 1)^{3/2}}\mathbf{i} + \frac{e^{2t}(2e^{2t} + 1)}{(e^{4t} + e^{2t} + 1)^{3/2}}\mathbf{j} + \frac{e^t(1 - e^{4t})}{(e^{4t} + e^{2t} + 1)^{3/2}}\mathbf{k}$$

$$\mathbf{T}'(0) = \frac{3}{3^{3/2}}\mathbf{i} + \frac{3}{3^{3/2}}\mathbf{j}$$

$$\mathbf{N}(0) = \frac{\sqrt{2}}{2}\mathbf{i} + \frac{\sqrt{2}}{2}\mathbf{j}$$

$a_{\mathbf{T}} = \mathbf{a}\cdot\mathbf{T} = 0$

$a_{\mathbf{N}} = \mathbf{a}\cdot\mathbf{N} = \sqrt{2}$

27. $\mathbf{r}(t) = (e^t \cos t)\mathbf{i} + (e^t \sin t)\mathbf{j},\ t = \frac{\pi}{2}$

$\mathbf{v}(t) = e^t(\cos t - \sin t)\mathbf{i} + e^t(\cos t + \sin t)\mathbf{j}$

$\mathbf{a}(t) = e^t(-2\sin t)\mathbf{i} + e^t(2\cos t)\mathbf{j}$

At $t = \frac{\pi}{2}$, $\mathbf{T} = \frac{\mathbf{v}}{\|\mathbf{v}\|} = \frac{1}{\sqrt{2}}(-\mathbf{i} + \mathbf{j}) = \frac{\sqrt{2}}{2}(-\mathbf{i} + \mathbf{j})$.

Motion along $\mathbf{r}$ is counterclockwise. So,

$\mathbf{N} = \frac{1}{\sqrt{2}}(-\mathbf{i} - \mathbf{j}) = -\frac{\sqrt{2}}{2}(\mathbf{i} + \mathbf{j})$.

$a_{\mathbf{T}} = \mathbf{a} \cdot \mathbf{T} = \sqrt{2}e^{\pi/2}$

$a_{\mathbf{N}} = \mathbf{a} \cdot \mathbf{N} = \sqrt{2}e^{\pi/2}$

28. $\mathbf{r}(t) = 4\cos 3t\mathbf{i} + 4\sin 3t\mathbf{j},\ t = \pi$

$\mathbf{v}(t) = -12\sin 3t\mathbf{i} + 12\cos 3t\mathbf{j}$

$\mathbf{a}(t) = -36\cos 3t\mathbf{i} - 36\sin 3t\mathbf{j}$

$\mathbf{a}(\pi) = 36\mathbf{i}$

At $t = \pi$, $\mathbf{v}\pi = -12\mathbf{j}$ and $\mathbf{T}(\pi) = -\mathbf{j}$

Movement is counterclockwise around a circle. So, $\mathbf{N}(\pi) = \mathbf{i}$.

$a_{\mathbf{T}} = \mathbf{a} \cdot \mathbf{T} = 0$

$a_{\mathbf{N}} = \mathbf{a} \cdot \mathbf{N} = 36$

29. $\mathbf{r}(t) = a\cos\omega t\mathbf{i} + a\sin\omega t\mathbf{j}$

$\mathbf{v}(t) = -a\omega\sin\omega t\mathbf{i} + a\omega\cos\omega t\mathbf{j}$

$\mathbf{a}(t) = -a\omega^2\cos\omega t\mathbf{i} - a\omega^2\sin\omega t\mathbf{j}$

$\mathbf{T}(t) = \frac{\mathbf{v}(t)}{\|\mathbf{v}(t)\|} = -\sin\omega t\mathbf{i} + \cos\omega t\mathbf{j}$

$\mathbf{N}(t) = \frac{\mathbf{T}'(t)}{\|\mathbf{T}'(t)\|} = -\cos\omega t\mathbf{i} - \sin\omega t\mathbf{j}$

$a_{\mathbf{T}} = \mathbf{a} \cdot \mathbf{T} = 0$

$a_{\mathbf{N}} = \mathbf{a} \cdot \mathbf{N} = a\omega^2$

30. $\mathbf{T}(t)$ points in the direction that $\mathbf{r}$ is moving. $\mathbf{N}(t)$ points in the direction that $\mathbf{r}$ is turning, toward the concave side of the curve.

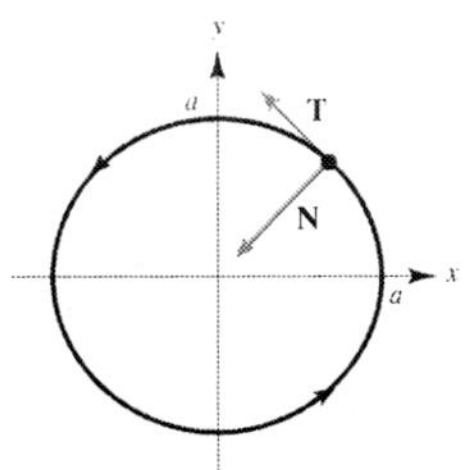

31. Speed: $\|\mathbf{v}(t)\| = a\omega$

The speed is constant because $a_{\mathbf{T}} = 0$.

32. If the angular velocity ω is halved,

$a_{\mathbf{N}} = a\left(\frac{\omega}{2}\right)^2 = \frac{a\omega^2}{4}$.

$a_{\mathbf{N}}$ is changed by a factor of $\frac{1}{4}$.

33. $\mathbf{r}(t) = t\mathbf{i} + \frac{1}{t}\mathbf{j},\ t_0 = 2$

$x = t,\ y = \frac{1}{t} \Rightarrow xy = 1$

$\mathbf{r}'(t) = \mathbf{i} - \frac{1}{t^2}\mathbf{j}$

$\mathbf{T}(t) = \frac{t^2\mathbf{i} - \mathbf{j}}{\sqrt{t^4 + 1}}$

$\mathbf{N}(t) = \frac{\mathbf{i} + t^2\mathbf{j}}{\sqrt{t^4 + 1}}$

$\mathbf{r}(2) = 2\mathbf{i} + \frac{1}{2}\mathbf{j}$

$\mathbf{T}(2) = \frac{\sqrt{17}}{17}(4\mathbf{i} - \mathbf{j})$

$\mathbf{N}(2) = \frac{\sqrt{17}}{17}(\mathbf{i} + 4\mathbf{j})$

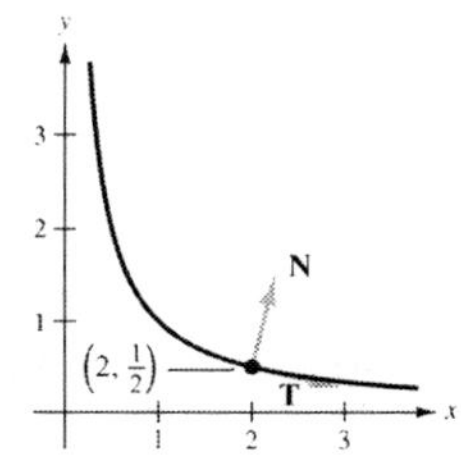

34. $\mathbf{r}(t) = t^3\mathbf{i} + t\mathbf{j},\ t_0 = 1$

$x = t^3,\ y = t \Rightarrow x = y^3$ or $y = x^{1/3}$

$\mathbf{r}'(t) = 3t^2\mathbf{i} + \mathbf{j}$

$\mathbf{T}(t) = \frac{\mathbf{r}'(t)}{\|\mathbf{r}'(t)\|} = \frac{3t^2\mathbf{i} + \mathbf{j}}{\sqrt{9t^4 + 1}}$

$\mathbf{T}(1) = \frac{3\mathbf{i} + \mathbf{j}}{\sqrt{10}}$

$= \frac{3\sqrt{10}}{10}\mathbf{i} + \frac{\sqrt{10}}{10}\mathbf{j}$

$\mathbf{N}(1) = \frac{\sqrt{10}}{10}\mathbf{i} - \frac{3\sqrt{10}}{10}\mathbf{j}$

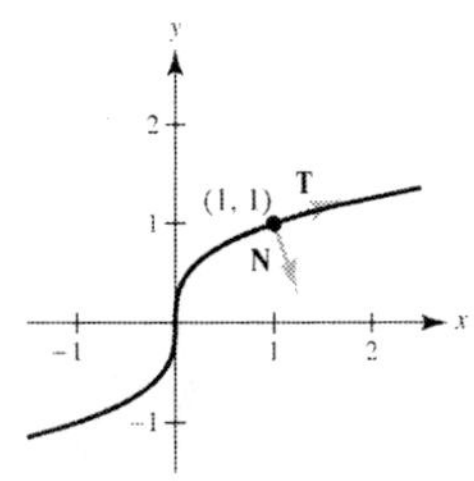

35. $\mathbf{r}(t) = (2t + 1)\mathbf{i} - t^2\mathbf{j}, t_0 = 2$

$x = 2t + 1,$

$y = -t^2 = -\left(\frac{x-1}{2}\right)^2$

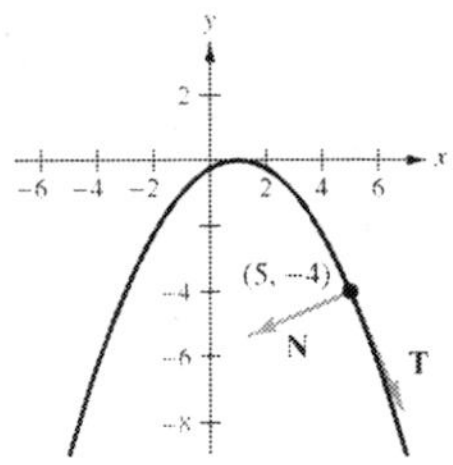

$\mathbf{r}(2) = 5\mathbf{i} - 4\mathbf{j}$

$\mathbf{r}'(t) = 2\mathbf{i} - 2t\mathbf{j}$

$\mathbf{T}(t) = \frac{2\mathbf{i} - 2t\mathbf{j}}{\sqrt{4 + 4t^2}} = \frac{\mathbf{i} - t\mathbf{j}}{\sqrt{1 + t^2}}$

$\mathbf{T}(2) = \frac{\mathbf{i} - 2\mathbf{j}}{\sqrt{5}}$

$\mathbf{N}(2) = \frac{-2\mathbf{i} - \mathbf{j}}{\sqrt{5}}$, perpendicular to $\mathbf{T}(2)$

36. $\mathbf{r}(t) = 2\cos t\mathbf{i} + 2\sin t\mathbf{j}, t_0 = \frac{\pi}{4}$

$x = 2\cos t,\ y = 2\sin t \Rightarrow x^2 + y^2 = 4$

$\mathbf{r}'(t) = -2\sin t\mathbf{i} + 2\cos t\mathbf{j}$

$\mathbf{T}(t) = \frac{1}{2}(-2\sin t\mathbf{i} + 2\cos t\mathbf{j}) = -\sin t\mathbf{i} + \cos t\mathbf{j}$

$\mathbf{N}(t) = -\cos t\mathbf{i} - \sin t\mathbf{j}$

$\mathbf{r}\left(\frac{\pi}{4}\right) = \sqrt{2}\mathbf{i} + \sqrt{2}\mathbf{j}$

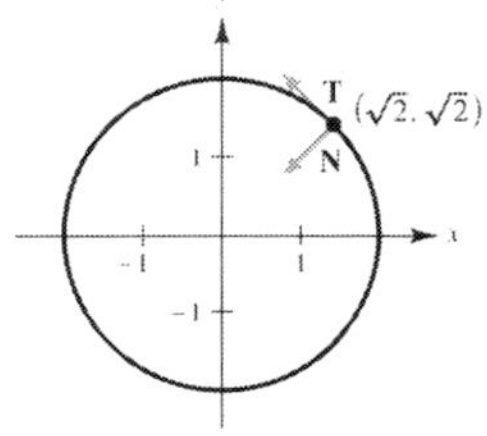

$\mathbf{T}\left(\frac{\pi}{4}\right) = \frac{\sqrt{2}}{2}(-\mathbf{i} + \mathbf{j})$

$\mathbf{N}\left(\frac{\pi}{4}\right) = \frac{\sqrt{2}}{2}(-\mathbf{i} - \mathbf{j})$

37. $\mathbf{r}(t) = t\mathbf{i} + 2t\mathbf{j} - 3t\mathbf{k}, t = 1$

$\mathbf{v}(t) = \mathbf{i} + 2\mathbf{j} - 3\mathbf{k}$

$\mathbf{a}(t) = \mathbf{0}$

$\mathbf{T}(t) = \frac{\mathbf{v}}{\|\mathbf{v}\|}$

$= \frac{1}{\sqrt{14}}(\mathbf{i} + 2\mathbf{j} - 3\mathbf{k})$

$= \frac{\sqrt{14}}{14}(\mathbf{i} + 2\mathbf{j} - 3\mathbf{k}) = \mathbf{T}(1)$

$\mathbf{N}(t) = \frac{\mathbf{T}'}{\|\mathbf{T}'\|}$ is undefined.

$a_{\mathbf{T}}, a_{\mathbf{N}}$ are not defined.

38. $\mathbf{r}(t) = \cos t\mathbf{i} + \sin t\mathbf{j} + 2t\mathbf{k}, t = \frac{\pi}{3}$

$\mathbf{v}(t) = -\sin t\mathbf{i} + \cos t\mathbf{j} + 2\mathbf{k}$

$\mathbf{a}(t) = -\cos t\mathbf{i} - \sin t\mathbf{j}, \|\mathbf{a}\| = 1$

$\|\mathbf{v}(t)\| = \sqrt{\sin^2 t + \cos^2 t + 4} = \sqrt{5}$

$\mathbf{T}(t) = \frac{\mathbf{v}(t)}{\|\mathbf{v}(t)\|} = \frac{1}{\sqrt{5}}(-\sin t\mathbf{i} + \cos t\mathbf{j} + 2\mathbf{k})$

$a_{\mathbf{T}} = \mathbf{a}(t) \cdot \mathbf{T}(t) = 0$

$a_{\mathbf{N}} = \sqrt{\|\mathbf{a}\|^2 - a_{\mathbf{T}}^2} = \sqrt{1 - 0} = 1$

$\mathbf{T}\left(\frac{\pi}{3}\right) = \frac{1}{\sqrt{5}}\left(-\frac{\sqrt{3}}{2}\mathbf{i} + \frac{1}{2}\mathbf{j} + 2\mathbf{k}\right)$

$\mathbf{a}\left(\frac{\pi}{3}\right) = -\frac{1}{2}\mathbf{i} - \frac{\sqrt{3}}{2}\mathbf{j} = a_{\mathbf{T}}\mathbf{T} + a_{\mathbf{n}}\mathbf{N} = \mathbf{N}$

39. $\mathbf{r}(t) = t\mathbf{i} + t^2\mathbf{j} + \frac{t^2}{2}\mathbf{k}, t = 1$

$\mathbf{v}(t) = \mathbf{i} + 2t\mathbf{j} + t\mathbf{k}$

$\mathbf{v}(1) = \mathbf{i} + 2\mathbf{j} + \mathbf{k}$

$\mathbf{a}(t) = 2\mathbf{j} + \mathbf{k}$

$\mathbf{T}(t) = \frac{\mathbf{v}}{\|\mathbf{v}\|} = \frac{1}{\sqrt{1 + 5t^2}}(\mathbf{i} + 2t\mathbf{j} + t\mathbf{k})$

$\mathbf{T}(1) = \frac{\sqrt{6}}{6}(\mathbf{i} + 2\mathbf{j} + \mathbf{k})$

$\mathbf{N}(t) = \frac{\mathbf{T}'}{\|\mathbf{T}'\|} = \frac{\dfrac{-5t\mathbf{i} + 2\mathbf{j} + \mathbf{k}}{\left(1 + 5t^2\right)^{3/2}}}{\dfrac{\sqrt{5}}{1 + 5t^2}} = \frac{-5t\mathbf{i} + 2\mathbf{j} + \mathbf{k}}{\sqrt{5}\sqrt{1 + 5t^2}}$

$\mathbf{N}(1) = \frac{\sqrt{30}}{30}(-5\mathbf{i} + 2\mathbf{j} + \mathbf{k})$

$a_{\mathbf{T}} = \mathbf{a} \cdot \mathbf{T} = \frac{5\sqrt{6}}{6}$

$a_{\mathbf{N}} = \mathbf{a} \cdot \mathbf{N} = \frac{\sqrt{30}}{6}$

40. $\mathbf{r}(t) = (2t - 1)\mathbf{i} + t^2\mathbf{j} - 4t\mathbf{k},\ t = 2$

$\mathbf{v}(t) = 2\mathbf{i} + 2t\mathbf{j} - 4\mathbf{k},$

$\|\mathbf{v}(t)\| = \sqrt{20 + 4t^2} = 2\sqrt{5 + t^2}$

$\mathbf{a}(t) = 2\mathbf{j},\ \|\mathbf{a}\| = 2$

$$\mathbf{T}(t) = \frac{2\mathbf{i} + 2t\mathbf{j} - 4\mathbf{k}}{2\sqrt{5 + t^2}} = \frac{1}{\sqrt{5 + t^2}}(\mathbf{i} + t\mathbf{j} - 2\mathbf{k})$$

$$\mathbf{a}(2) = 2\mathbf{j},\ \mathbf{T}(2) = \frac{1}{3}(\mathbf{i} + 2\mathbf{j} - 2\mathbf{k})$$

$$a_{\mathbf{T}} = \mathbf{a}(2)\cdot\mathbf{T}(2) = \frac{4}{3}$$

$$a_{\mathbf{N}} = \sqrt{\|\mathbf{a}\|^2 - a^2_{\mathbf{T}}} = \sqrt{4 - \frac{16}{9}} = \frac{2}{3}\sqrt{5}$$

$$\mathbf{a} = 2\mathbf{j} = a_{\mathbf{T}}\mathbf{T} + a_{\mathbf{N}}\mathbf{N} = \frac{4}{9}(\mathbf{i} + 2\mathbf{j} - 2\mathbf{k}) + \frac{2}{3}\sqrt{5}\mathbf{N}$$

$$\mathbf{N} = \frac{3}{2\sqrt{5}}\left[2\mathbf{j} - \frac{4}{9}(\mathbf{i} + 2\mathbf{j} - 2\mathbf{k})\right]$$

$$= \frac{-2\sqrt{5}}{15}\mathbf{i} + \frac{\sqrt{5}}{3}\mathbf{j} + \frac{4\sqrt{5}}{15}\mathbf{k}$$

41. $\mathbf{r}(t) = e^t \sin t\mathbf{i} + e^t \cos t\mathbf{j} + e^t\mathbf{k},\ t = 0$

$\mathbf{v}(t) = (e^t \cos t + e^t \sin t)\mathbf{i} + (-e^t \sin t + e^t \cos t)\mathbf{j} + e^t\mathbf{k}$

$\mathbf{v}(0) = \mathbf{i} + \mathbf{j} + \mathbf{k}$

$\mathbf{a}(t) = 2e^t \cos t\mathbf{i} - 2e^t \sin t\mathbf{j} + e^t\mathbf{k}$

$\mathbf{a}(0) = 2\mathbf{i} + \mathbf{k}$

$$\mathbf{T}(t) = \frac{\mathbf{v}}{\|\mathbf{v}\|}$$

$$= \frac{1}{\sqrt{3}}\left[(\cos t + \sin t)\mathbf{i} + (-\sin t + \cos t)\mathbf{j} + \mathbf{k}\right]$$

$$\mathbf{T}(0) = \frac{1}{\sqrt{3}}[\mathbf{i} + \mathbf{j} + \mathbf{k}]$$

$$\mathbf{N}(t) = \frac{1}{\sqrt{2}}\left[(-\sin t + \cos t)\mathbf{i} + (-\cos t - \sin t)\mathbf{j}\right]$$

$$\mathbf{N}(0) = \frac{\sqrt{2}}{2}\mathbf{i} - \frac{\sqrt{2}}{2}\mathbf{j}$$

$a_{\mathbf{T}} = \mathbf{a}\cdot\mathbf{T} = \sqrt{3}$

$a_{\mathbf{N}} = \mathbf{a}\cdot\mathbf{N} = \sqrt{2}$

42. $\mathbf{r}(t) = e^t\mathbf{i} + 2t\mathbf{j} + e^{-t}\mathbf{k},\ t = 0$

$\mathbf{v}(t) = e^t\mathbf{i} + 2\mathbf{j} - e^{-t}\mathbf{k},\ \|\mathbf{v}(t)\| = \sqrt{e^{2t} + 4 + e^{-2t}}$

$\mathbf{a}(t) = e^t\mathbf{i} + e^{-t}\mathbf{k},\ \|\mathbf{a}(t)\| = \sqrt{e^{2t} + e^{-2t}}$

$\mathbf{v}(0) = \mathbf{i} + 2\mathbf{j} - \mathbf{k},\ \mathbf{a}(0) = \mathbf{i} + \mathbf{k}$

$$\mathbf{T}(0) = \frac{\mathbf{i} + 2\mathbf{j} - \mathbf{k}}{\sqrt{6}}$$

$a_{\mathbf{T}} = \mathbf{a}\cdot\mathbf{T} = 0$

$a_{\mathbf{N}} = \sqrt{\|\mathbf{a}\|^2 - a_{\mathbf{T}}^2} = \sqrt{2}$

$\mathbf{a} = \mathbf{i} + \mathbf{k} = a_{\mathbf{T}}\mathbf{T} + a_{\mathbf{N}}\mathbf{N} = \sqrt{2}\mathbf{N}$

$$\Rightarrow \mathbf{N} = \frac{1}{\sqrt{2}}(\mathbf{i} + \mathbf{k})$$

43. Let C be a smooth curve represented by $\mathbf{r}$ on an open interval I. The unit tangent vector $\mathbf{T}(t)$ at t is defined as

$$\mathbf{T}(t) = \frac{\mathbf{r}'(t)}{\|\mathbf{r}'(t)\|},\ \mathbf{r}'(t) \neq 0.$$

The principal unit normal vector $\mathbf{N}(t)$ at t is defined as

$$\mathbf{N}(t) = \frac{\mathbf{T}'(t)}{\|\mathbf{T}'(t)\|},\ \mathbf{T}'(t) \neq 0.$$

The tangential and normal components of acceleration are defined as $\mathbf{a}(t) = a_{\mathbf{T}}\mathbf{T}(t) + a_{\mathbf{N}}\mathbf{N}(t)$.

44. The unit tangent vector points in the direction of motion.

45. If $a_{\mathbf{N}} = 0$, then the motion is in a straight line.

46. If $a_{\mathbf{T}} = 0$, then the speed is constant.

47. $\mathbf{r}(t) = 3t\mathbf{i} + 4t\mathbf{j}$

$\mathbf{v}(t) = \mathbf{r}'(t) = 3\mathbf{i} + 4\mathbf{j},\ \|\mathbf{v}(t)\| = \sqrt{9 + 16} = 5$

$\mathbf{a}(t) = \mathbf{v}'(t) = \mathbf{0}$

$$\mathbf{T}(t) = \frac{\mathbf{v}(t)}{\|\mathbf{v}(t)\|} = \frac{3}{5}\mathbf{i} + \frac{4}{5}\mathbf{j}$$

$\mathbf{T}'(t) = 0 \Rightarrow \mathbf{N}(t)$ does not exist.

The path is a line. The speed is constant (5).

48. (a)(i) The vector **s** represents the unit tangent vector because it points in the direction of motion.

(ii) The vector **t** represents the unit tangent vector because it points in the direction of motion.

(b)(i) The vector **z** represents the unit normal vector because it points in the direction that the curve is bending.

(ii) The vector **z** represents the unit normal vector because it points in the direction that the curve is bending.

49. $\mathbf{r}(t) = \langle \pi t - \sin \pi t, 1 - \cos \pi t \rangle$

The graph is a cycloid.

(a) $\mathbf{r}(t) = \langle \pi t - \sin \pi t, 1 - \cos \pi t \rangle$

$\mathbf{v}(t) = \langle \pi - \pi \cos \pi t, \pi \sin \pi t \rangle$

$\mathbf{a}(t) = \langle \pi^2 \sin \pi t, \pi^2 \cos \pi t \rangle$

$$\mathbf{T}(t) = \frac{\mathbf{v}(t)}{\|\mathbf{v}(t)\|} = \frac{1}{\sqrt{2(1 - \cos \pi t)}}\langle 1 - \cos \pi t, \sin \pi t \rangle$$

$$\mathbf{N}(t) = \frac{\mathbf{T}'(t)}{\|\mathbf{T}'(t)\|} = \frac{1}{\sqrt{2(1 - \cos \pi t)}}\langle \sin \pi t, -1 + \cos \pi t \rangle$$

$$a_{\mathbf{T}} = \mathbf{a} \cdot \mathbf{T} = \frac{1}{\sqrt{2(1 - \cos \pi t)}}\left[\pi^2 \sin \pi t(1 - \cos \pi t) + \pi^2 \cos \pi t \sin \pi t\right] = \frac{\pi^2 \sin \pi t}{\sqrt{2(1 - \cos \pi t)}}$$

$$a_{\mathbf{N}} = \mathbf{a} \cdot \mathbf{N} = \frac{1}{\sqrt{2(1 - \cos \pi t)}}\left[\pi^2 \sin^2 \pi t + \pi^2 \cos \pi t(-1 + \cos \pi t)\right] = \frac{\pi^2(1 - \cos \pi t)}{\sqrt{2(1 - \cos \pi t)}} = \frac{\pi^2\sqrt{2(1 - \cos \pi t)}}{2}$$

When $t = \frac{1}{2}$: $a_{\mathbf{T}} = \frac{\pi^2}{\sqrt{2}} = \frac{\sqrt{2}\pi^2}{2}$, $a_{\mathbf{N}} = \frac{\sqrt{2}\pi^2}{2}$

When $t = 1$: $a_{\mathbf{T}} = 0$, $a_{\mathbf{N}} = \pi^2$

When $t = \frac{3}{2}$: $a_{\mathbf{T}} = -\frac{\sqrt{2}\pi^2}{2}$, $a_{\mathbf{N}} = \frac{\sqrt{2}\pi^2}{2}$

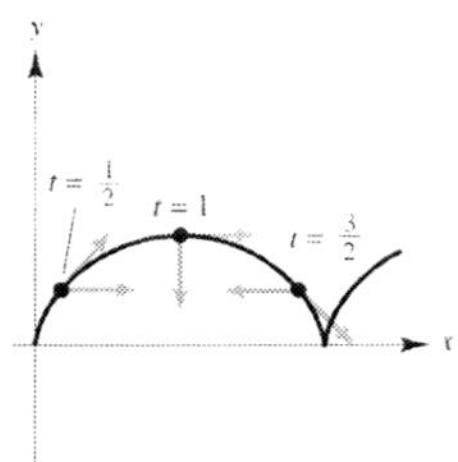

(b) Speed: $s = \|\mathbf{v}(t)\| = \pi\sqrt{2(1 - \cos \pi t)}$

$$\frac{ds}{dt} = \frac{\pi^2 \sin \pi t}{\sqrt{2(1 - \cos \pi t)}} = a_{\mathbf{T}}$$

When $t = \frac{1}{2}$: $a_{\mathbf{T}} = \frac{\sqrt{2}\pi^2}{2} > 0 \Rightarrow$ the speed in increasing.

When $t = 1$: $a_{\mathbf{T}} = 0 \Rightarrow$ the height is maximum.

When $t = \frac{3}{2}$: $a_{\mathbf{T}} = -\frac{\sqrt{2}\pi^2}{2} < 0 \Rightarrow$ the speed is decreasing.

50. (a) $\mathbf{r}(t) = \langle \cos \pi t + \pi t \sin \pi t, \sin \pi t - \pi t \cos \pi t \rangle$

$\mathbf{v}(t) = \langle -\pi \sin \pi t + \pi \sin \pi t + \pi^2 t \cos \pi t, \pi \cos \pi t - \pi \cos \pi t + \pi^2 t \sin \pi t \rangle = \langle \pi^2 t \cos \pi t, \pi^2 t \sin \pi t \rangle$

$\mathbf{a}(t) = \langle \pi^2 \cos \pi t - \pi^3 t \sin \pi t, \pi^2 \sin \pi t + \pi^3 t \cos \pi t \rangle$

$$\mathbf{T}(t) = \frac{\mathbf{v}(t)}{\|\mathbf{v}(t)\|} = \langle \cos \pi t, \sin \pi t \rangle$$

$a_{\mathbf{T}} = \mathbf{a} \cdot \mathbf{T} = \cos \pi t(\pi^2 \cos \pi t - \pi^3 t \sin \pi t) + \sin \pi t(\pi^2 \sin \pi t + \pi^3 t \cos \pi t) = \pi^2$

$a_{\mathbf{N}} = \sqrt{\|\mathbf{a}\|^2 - a_{\mathbf{T}}^2} = \sqrt{\pi^4(1 + \pi^2 t^2) - \pi^4} = \pi^3 t$

When $t = 1$, $a_{\mathbf{T}} = \pi^2$, $a_{\mathbf{N}} = \pi^3$. When $t = 2$, $a_{\mathbf{T}} = \pi^2$, $a_{\mathbf{N}} = 2\pi^3$.

(b) Because $a_{\mathbf{T}} = \pi^2 > 0$ for all values of t, the speed is increasing when $t = 1$ and $t = 2$.

51. $\mathbf{r}(t) = 2\cos t\mathbf{i} + 2\sin t\mathbf{j} + \dfrac{t}{2}\mathbf{k},\ t_0 = \dfrac{\pi}{2}$

$$\mathbf{r}'(t) = -2\sin t\mathbf{i} + 2\cos t\mathbf{j} + \frac{1}{2}\mathbf{k}$$

$$\mathbf{T}(t) = \frac{2\sqrt{17}}{17}\left(-2\sin t\mathbf{i} + 2\cos t\mathbf{j} + \frac{1}{2}\mathbf{k}\right)$$

$$\mathbf{N}(t) = -\cos t\mathbf{i} - \sin t\mathbf{j}$$

$$\mathbf{r}\left(\frac{\pi}{2}\right) = 2\mathbf{j} + \frac{\pi}{4}\mathbf{k}$$

$$\mathbf{T}\left(\frac{\pi}{2}\right) = \frac{2\sqrt{17}}{17}\left(-2\mathbf{i} + \frac{1}{2}\mathbf{k}\right) = \frac{\sqrt{17}}{17}(-4\mathbf{i} + \mathbf{k})$$

$$\mathbf{N}\left(\frac{\pi}{2}\right) = -\mathbf{j}$$

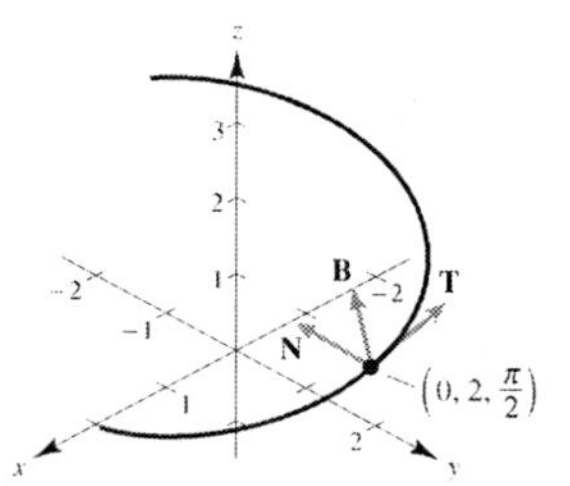

$$\mathbf{B}\left(\frac{\pi}{2}\right) = \mathbf{T}\left(\frac{\pi}{2}\right) \times \mathbf{N}\left(\frac{\pi}{2}\right) = \begin{vmatrix} \mathbf{i} & \mathbf{j} & \mathbf{k} \\ -\frac{4\sqrt{17}}{17} & 0 & \frac{\sqrt{17}}{17} \\ 0 & -1 & 0 \end{vmatrix} = \frac{\sqrt{17}}{17}\mathbf{i} + \frac{4\sqrt{17}}{17}\mathbf{k} = \frac{\sqrt{17}}{17}(\mathbf{i} + 4\mathbf{k})$$

52. $\mathbf{r}(t) = t\mathbf{i} + t^2\mathbf{j} + \dfrac{t^3}{3}\mathbf{k},\ t_0 = 1$

$$\mathbf{r}'(t) = \mathbf{i} + 2t\mathbf{j} + t^2\mathbf{k}$$

$$\mathbf{T}(t) = \frac{1}{\sqrt{1 + 4t^2 + t^4}}\left(\mathbf{i} + 2t\mathbf{j} + t^2\mathbf{k}\right)$$

$$\mathbf{N}(t) = \frac{1}{\sqrt{1 + 4t^2 + t^4}\sqrt{1 + t^2 + t^4}}\left[\left(-2t - t^3\right)\mathbf{i} + \left(1 - t^4\right)\mathbf{j} + \left(t + 2t^3\right)\mathbf{k}\right]$$

$$\mathbf{r}(1) = \mathbf{i} + \mathbf{j} + \frac{1}{3}\mathbf{k}$$

$$\mathbf{T}(1) = \frac{1}{\sqrt{6}}(\mathbf{i} + 2\mathbf{j} + \mathbf{k})$$

$$\mathbf{N}(1) = \frac{1}{\sqrt{6}\sqrt{3}}(-3\mathbf{i} + 3\mathbf{k}) = \frac{\sqrt{2}}{2}(-\mathbf{i} + \mathbf{k})$$

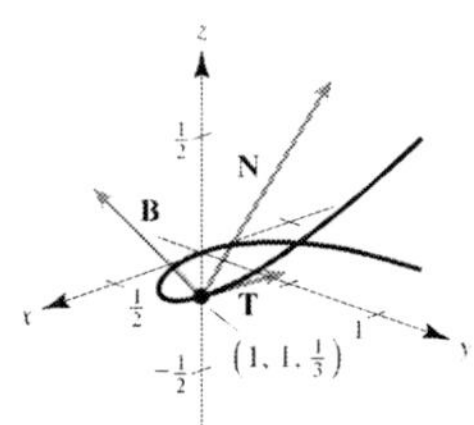

$$\mathbf{B}(1) = \mathbf{T}(1) \times \mathbf{N}(1) = \begin{vmatrix} \mathbf{i} & \mathbf{j} & \mathbf{k} \\ \frac{\sqrt{6}}{6} & \frac{\sqrt{6}}{3} & \frac{\sqrt{6}}{6} \\ -\frac{\sqrt{2}}{2} & 0 & \frac{\sqrt{2}}{2} \end{vmatrix} = \frac{\sqrt{3}}{3}\mathbf{i} - \frac{\sqrt{3}}{3}\mathbf{j} + \frac{\sqrt{3}}{3}\mathbf{k} = \frac{\sqrt{3}}{3}(\mathbf{i} - \mathbf{j} + \mathbf{k})$$

53. $\mathbf{r}(t) = \mathbf{i} + \sin t\mathbf{j} + \cos t\mathbf{k},\ t_0 = \dfrac{\pi}{4}$

$\mathbf{r}'(t) = \cos t\mathbf{j} - \sin t\mathbf{k},$

$\|\mathbf{r}'(t)\| = 1$

$\mathbf{r}'\left(\dfrac{\pi}{4}\right) = \mathbf{T}\left(\dfrac{\pi}{4}\right) = \dfrac{\sqrt{2}}{2}\mathbf{j} - \dfrac{\sqrt{2}}{2}\mathbf{k}$

$\mathbf{T}'(t) = -\sin t\mathbf{j} - \cos t\mathbf{k},$

$\mathbf{N}\left(\dfrac{\pi}{4}\right) = -\dfrac{\sqrt{2}}{2}\mathbf{j} - \dfrac{\sqrt{2}}{2}\mathbf{k}$

$$\mathbf{B}\left(\frac{\pi}{4}\right) = \mathbf{T}\left(\frac{\pi}{4}\right) \times \mathbf{N}\left(\frac{\pi}{4}\right) = \begin{vmatrix} \mathbf{i} & \mathbf{j} & \mathbf{k} \\ 0 & \frac{\sqrt{2}}{2} & -\frac{\sqrt{2}}{2} \\ 0 & -\frac{\sqrt{2}}{2} & -\frac{\sqrt{2}}{2} \end{vmatrix} = -\mathbf{i}$$

54. $\mathbf{r}(t) = 2e^t\mathbf{i} + e^t\cos t\mathbf{j} + e^t\sin t\mathbf{k},\ t_0 = 0$

$\mathbf{r}'(t) = 2e^t\mathbf{i} + \left(e^t\cos t - e^t\sin t\right)\mathbf{j} + \left(e^t\sin t + e^t\cos t\right)\mathbf{k}$

$\mathbf{r}'(0) = 2\mathbf{i} + \mathbf{j} + \mathbf{k} \Rightarrow \mathbf{T}(0) = \dfrac{1}{\sqrt{6}}(2\mathbf{i} + \mathbf{j} + \mathbf{k})$

$$\begin{aligned}\|\mathbf{r}'(t)\|^2 &= 4e^{2t} + e^{2t}\cos^2 t + e^{2t}\sin^2 t - 2e^{2t}\cos t\sin t + e^{2t}\sin^2 t + e^{2t}\cos^2 t + 2e^{2t}\sin t\cos t \\ &= 4e^{2t} + 2e^{2t}\left(\cos^2 t + \sin^2 t\right) = 6e^{2t}\end{aligned}$$

$\|\mathbf{r}'(t)\| = \sqrt{6}e^t$

$\mathbf{T}(t) = \dfrac{\mathbf{r}'(t)}{\|\mathbf{r}'(t)\|} = \dfrac{1}{\sqrt{6}}\left[2\mathbf{i} + (\cos t - \sin t)\mathbf{j} + (\sin t + \cos t)\mathbf{k}\right]$

$\mathbf{T}'(t) = \dfrac{1}{\sqrt{6}}\left[(-\sin t - \cos t)\mathbf{j} + (\cos t - \sin t)\mathbf{k}\right]$

$\mathbf{T}'(0) = \dfrac{1}{\sqrt{6}}[-\mathbf{j} + \mathbf{k}] \Rightarrow \mathbf{N}(0) = \dfrac{-\sqrt{2}}{2}\mathbf{j} + \dfrac{\sqrt{2}}{2}\mathbf{k}$

$$\mathbf{B}(0) = \mathbf{T}(0) \times \mathbf{N}(0) = \begin{vmatrix} \mathbf{i} & \mathbf{j} & \mathbf{k} \\ \frac{2}{\sqrt{6}} & \frac{1}{\sqrt{6}} & \frac{1}{\sqrt{6}} \\ 0 & -\frac{\sqrt{2}}{2} & \frac{\sqrt{2}}{2} \end{vmatrix} = \frac{\sqrt{3}}{3}\mathbf{i} - \frac{\sqrt{3}}{3}\mathbf{j} - \frac{\sqrt{3}}{3}\mathbf{k}$$

55. $\mathbf{r}(t) = 4\sin t\mathbf{i} + 4\cos t\mathbf{j} + 2t\mathbf{k},\ t_0 = \dfrac{\pi}{3}$

$$\mathbf{r}'(t) = 4\cos t\mathbf{i} - 4\sin t\mathbf{j} + 2\mathbf{k},$$

$$\|\mathbf{r}'(t)\| = \sqrt{16\cos^2 t + 16\sin^2 t + 4} = \sqrt{20} = 2\sqrt{5}$$

$$\mathbf{r}'\left(\frac{\pi}{3}\right) = 2\mathbf{i} - 2\sqrt{3}\mathbf{j} + 2\mathbf{k}$$

$$\mathbf{T}\left(\frac{\pi}{3}\right) = \frac{1}{2\sqrt{5}}\left(2\mathbf{i} - 2\sqrt{3}\mathbf{j} + 2\mathbf{k}\right) = \frac{\sqrt{5}}{5}\mathbf{i} - \frac{\sqrt{15}}{5}\mathbf{j} + \frac{\sqrt{5}}{5}\mathbf{k} = \frac{\sqrt{5}}{5}\left(\mathbf{i} - \sqrt{3}\mathbf{j} + \mathbf{k}\right)$$

$$\mathbf{T}'(t) = \frac{1}{2\sqrt{5}}(-4\sin t\mathbf{i} - 4\cos t\mathbf{j})$$

$$\mathbf{N}\left(\frac{\pi}{3}\right) = -\frac{\sqrt{3}}{2}\mathbf{i} - \frac{1}{2}\mathbf{j}$$

$$\mathbf{B}\left(\frac{\pi}{3}\right) = \mathbf{T}\left(\frac{\pi}{3}\right) \times \mathbf{N}\left(\frac{\pi}{3}\right) = \begin{vmatrix} \mathbf{i} & \mathbf{j} & \mathbf{k} \\ \frac{\sqrt{5}}{5} & -\frac{\sqrt{15}}{5} & \frac{\sqrt{5}}{5} \\ -\frac{\sqrt{3}}{2} & -\frac{1}{2} & 0 \end{vmatrix} = \frac{\sqrt{5}}{10}\mathbf{i} - \frac{\sqrt{15}}{10}\mathbf{j} - \frac{4\sqrt{5}}{10}\mathbf{k} = \frac{\sqrt{5}}{10}\left(\mathbf{i} - \sqrt{3}\mathbf{j} - 4\mathbf{k}\right)$$

56. $\mathbf{r}(t) = 3\cos 2t\mathbf{i} + 3\sin 2t\mathbf{j} + t\mathbf{k},\ t = \dfrac{\pi}{4}$

$$\mathbf{r}'(t) = -6\sin 2t\mathbf{i} + 6\cos 2t\mathbf{j} + \mathbf{k}$$

$$\|\mathbf{r}'(t)\| = \sqrt{37}$$

$$\mathbf{r}'\left(\frac{\pi}{4}\right) = -6\mathbf{i} + \mathbf{k}$$

$$\mathbf{T}(t) = \frac{r'(t)}{\|r'(t)\|} = \frac{1}{\sqrt{37}}(-6\sin 2t\mathbf{i} + 6\cos 2t\mathbf{j} + \mathbf{k})$$

$$\mathbf{T}'(t) = \frac{1}{\sqrt{37}}(-12\cos 2t\mathbf{i} - 12\sin 2t\mathbf{j})$$

$$\mathbf{T}\left(\frac{\pi}{4}\right) = \frac{1}{\sqrt{37}}(-6\mathbf{i} + \mathbf{k})$$

$$\mathbf{N}(t) = \frac{\mathbf{T}'(t)}{\|\mathbf{T}'(t)\|} = -\cos 2t\mathbf{i} - \sin 2t\mathbf{j}$$

$$\mathbf{N}\left(\frac{\pi}{4}\right) = -j$$

$$\mathbf{B}\left(\frac{\pi}{4}\right) = \mathbf{T}\left(\frac{\pi}{4}\right) \times \mathbf{N}\left(\frac{\pi}{4}\right) = \begin{vmatrix} \mathbf{i} & \mathbf{j} & \mathbf{k} \\ \frac{-6}{\sqrt{37}} & 0 & \frac{1}{\sqrt{37}} \\ 0 & -1 & 0 \end{vmatrix} = \frac{1}{\sqrt{37}}\mathbf{i} + \frac{6}{\sqrt{37}}\mathbf{k}$$

57. $\mathbf{r}(t) = 3t\mathbf{i} + 2t^2\mathbf{j}$

$\mathbf{v}(t) = 3\mathbf{i} + 4t\mathbf{j}$

$\mathbf{a}(t) = 4\mathbf{j}$

$\mathbf{v}\cdot\mathbf{v} = 9 + 16t^2$

$\mathbf{v}\cdot\mathbf{a} = 16t$

$$(\mathbf{v}\cdot\mathbf{v})\mathbf{a} - (\mathbf{v}\cdot\mathbf{a})\mathbf{v} = (9 + 16t^2)4\mathbf{j} - (16t)(3\mathbf{i} + 4t\mathbf{j}) = -48t\mathbf{i} + 36\mathbf{j}$$

$$\mathbf{N} = \frac{(\mathbf{v}\cdot\mathbf{v})\mathbf{a} - (\mathbf{v}\cdot\mathbf{a})\mathbf{v}}{\left\|(\mathbf{v}\cdot\mathbf{v})\mathbf{a} - (\mathbf{v}\cdot\mathbf{a})\mathbf{v}\right\|} = \frac{1}{\sqrt{9 + 16t^2}}(-4t\mathbf{i} + 3\mathbf{j})$$

58. $\mathbf{r}(t) = 3\cos 2t\mathbf{i} + 3\sin 2t\mathbf{j}$

$\mathbf{v}(t) = -6\sin 2t\mathbf{i} + 6\cos 2t\mathbf{j}$

$\mathbf{a}(t) = -12\cos 2t\mathbf{i} - 12\sin 2t\mathbf{j}$

$\mathbf{v}\cdot\mathbf{v} = 36\sin^2 2t + 36\cos^2 2t = 36$

$\mathbf{v}\cdot\mathbf{a} = 0$

$$\mathbf{N} = \frac{(\mathbf{v}\cdot\mathbf{v})\mathbf{a} - (\mathbf{v}\cdot\mathbf{a})\mathbf{v}}{\left\|(\mathbf{v}\cdot\mathbf{v})\mathbf{a} - (\mathbf{v}\cdot\mathbf{a})\mathbf{v}\right\|} = \frac{36(-12\cos 2t\mathbf{i} - 12\sin 2t\mathbf{j})}{\left\|36(-12\cos 2t\mathbf{i} - 12\sin 2t\mathbf{j})\right\|} = -\cos 2t\mathbf{i} - \sin 2t\mathbf{j}$$

59. $\mathbf{r}(t) = 2t\mathbf{i} + 4t\mathbf{j} + t^2\mathbf{k}$

$\mathbf{v}(t) = 2\mathbf{i} + 4\mathbf{j} + 2t\mathbf{k}$

$\mathbf{a}(t) = 2\mathbf{k}$

$\mathbf{v}\cdot\mathbf{v} = 4 + 16 + 4t^2 = 20 + 4t^2$

$\mathbf{v}\cdot\mathbf{a} = 4t$

$$(\mathbf{v}\cdot\mathbf{v})\mathbf{a} - (\mathbf{v}\cdot\mathbf{a})\mathbf{v} = (20 + 4t^2)2\mathbf{k} - 4t(2\mathbf{i} + 4\mathbf{j} + 2t\mathbf{k}) = -8t\mathbf{i} - 16t\mathbf{j} + 40\mathbf{k}$$

$$\mathbf{N} = \frac{(\mathbf{v}\cdot\mathbf{v})\mathbf{a} - (\mathbf{v}\cdot\mathbf{a})\mathbf{v}}{\left\|(\mathbf{v}\cdot\mathbf{v})\mathbf{a} - (\mathbf{v}\cdot\mathbf{a})\mathbf{v}\right\|} = \frac{1}{\sqrt{5t^2 + 25}}(-t\mathbf{i} - 2t\mathbf{j} + 5\mathbf{k})$$

60. $\mathbf{r}(t) = 5\cos t\mathbf{i} + 5\sin t\mathbf{j} + 3t\mathbf{k}$

$\mathbf{v}(t) = -5\sin t\mathbf{i} + 5\cos t\mathbf{j} + 3\mathbf{k}$

$\mathbf{a}(t) = -5\cos t\mathbf{i} - 5\sin t\mathbf{j}$

$\mathbf{v}\cdot\mathbf{v} = 25 + 9 = 34$

$\mathbf{v}\cdot\mathbf{a} = 0$

$(\mathbf{v}\cdot\mathbf{v})\mathbf{a} - (\mathbf{v}\cdot\mathbf{a})\mathbf{v} = 34(-5\cos t\mathbf{i} - 5\sin t\mathbf{j})$

$$\mathbf{N} = \frac{(\mathbf{v}\cdot\mathbf{v})\mathbf{a} - (\mathbf{v}\cdot\mathbf{a})\mathbf{v}}{\left\|(\mathbf{v}\cdot\mathbf{v})\mathbf{a} - (\mathbf{v}\cdot\mathbf{a})\mathbf{v}\right\|} = \frac{34(-5\cos t\mathbf{i} - 5\sin t\mathbf{j})}{\left\|34(-5\cos t\mathbf{i} - 5\sin t\mathbf{j})\right\|} = -\cos t\mathbf{i} - \sin t\mathbf{j}$$

61. From Theorem 12.3 you have:

$\mathbf{r}(t) = (v_0 t\cos\theta)\mathbf{i} + (h + v_0 t\sin\theta - 16t^2)\mathbf{j}$

$\mathbf{v}(t) = v_0\cos\theta\mathbf{i} + (v_0\sin\theta - 32t)\mathbf{j}$

$\mathbf{a}(t) = -32\mathbf{j}$

$$\mathbf{T}(t) = \frac{(v_0\cos\theta)\mathbf{i} + (v_0\sin\theta - 32t)\mathbf{j}}{\sqrt{v_0^2\cos^2\theta + (v_0\sin\theta - 32t)^2}}$$

$$\mathbf{N}(t) = \frac{(v_0\sin\theta - 32t)\mathbf{i} - v_0\cos\theta\mathbf{j}}{\sqrt{v_0^2\cos^2\theta + (v_0\sin\theta - 32t)^2}} \quad \text{(Motion is clockwise.)}$$

$$a_{\mathbf{T}} = \mathbf{a}\cdot\mathbf{T} = \frac{-32(v_0\sin\theta - 32t)}{\sqrt{v_0^2\cos^2\theta + (v_0\sin\theta - 32t)^2}}$$

$$a_{\mathbf{N}} = \mathbf{a}\cdot\mathbf{N} = \frac{32v_0\cos\theta}{\sqrt{v_0^2\cos^2\theta + (v_0\sin\theta - 32t)^2}}$$

Maximum height when $v_0\sin\theta - 32t = 0$; (vertical component of velocity)

At maximum height, $a_{\mathbf{T}} = 0$ and $a_{\mathbf{N}} = 32$.

62. $\theta = 45°, v_0 = 150$

$$v_0 \cos\theta = 150 \cdot \frac{\sqrt{2}}{2} = 75\sqrt{2}$$

$$v_0 \sin\theta - 32t = 150 \cdot \frac{\sqrt{2}}{2} - 32t = 75\sqrt{2} - 32t$$

$$a_{\mathbf{T}} = \frac{-32\left(75\sqrt{2} - 32t\right)}{\sqrt{11250 + \left(75\sqrt{2} - 32t\right)^2}} = \frac{16\left(32t - 75\sqrt{2}\right)}{\sqrt{256t^2 - 1200\sqrt{2}t + 5625}}$$

$$a_{\mathbf{N}} = \frac{32\left(75\sqrt{2}\right)}{\sqrt{11250 + \left(75\sqrt{2} - 32t\right)^2}} = \frac{1200\sqrt{2}}{\sqrt{256t^2 - 1200\sqrt{2}t + 5625}}$$

At the maximum height, $a_{\mathbf{T}} = 0$ and $a_{\mathbf{N}} = 32$.

63. (a) $\mathbf{r}(t) = (v_0 \cos\theta)t\mathbf{i} + \left[h + (v_0 \sin\theta)t - \frac{1}{2}gt^2\right]\mathbf{j}$

$= (120 \cos 30°)t\mathbf{i} + \left[5 + (120 \sin 30°)t - 16t^2\right]\mathbf{j} = 60\sqrt{3}t\mathbf{i} + \left[5 + 60t - 16t^2\right]\mathbf{j}$

(b)

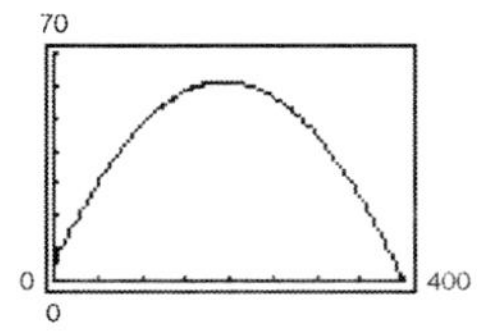

Maximum height ≈ 61.25 feet

range ≈ 398.2 feet

(c) $\mathbf{v}(t) = 60\sqrt{3}\mathbf{i} + (60 - 32t)\mathbf{j}$

Speed $= \|\mathbf{v}(t)\| = \sqrt{3600(3) + (60 - 32t)^2} = 8\sqrt{16t^2 - 60t + 225}$

$\mathbf{a}(t) = -32\mathbf{j}$

(d)

t	0.5	1.0	1.5	2.0	2.5	3.0
Speed	112.85	107.63	104.61	104.0	105.83	109.98

(e) From Exercise 61, using $\mathbf{v}_0 = 120$ and $\theta = 30°$,

$$a_{\mathbf{T}} = \frac{-32(60 - 32t)}{\sqrt{\left(60\sqrt{3}\right)^2 + (60 - 32t)^2}}$$

$$a_{\mathbf{N}} = \frac{32\left(60\sqrt{3}\right)}{\sqrt{\left(60\sqrt{3}\right)^2 + (60 - 32t)^2}}$$

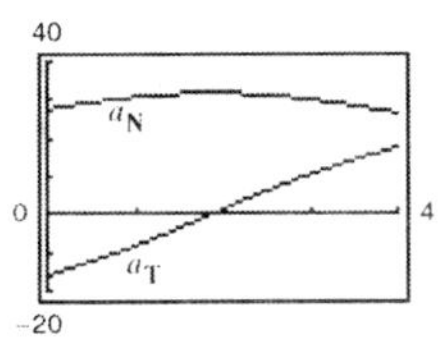

At $t = 1.875$, $a_{\mathbf{T}} = 0$ and the projectile is at its maximum height. When $a_{\mathbf{T}}$ and $a_{\mathbf{N}}$ have opposite signs, the speed is decreasing.

64. (a) $\mathbf{r}(t) = (v_0 \cos\theta)t\mathbf{i} + \left[h + (v_0 \sin\theta)t - \frac{1}{2}gt^2\right]\mathbf{j}$

$= (220 \cos 45°)t\mathbf{i} + \left[4 + (220 \sin 45°)t - 16t^2\right]\mathbf{j}$

$= 110\sqrt{2}t\mathbf{i} + \left[4 + 110\sqrt{2}t - 16t^2\right]\mathbf{j}$

(b)

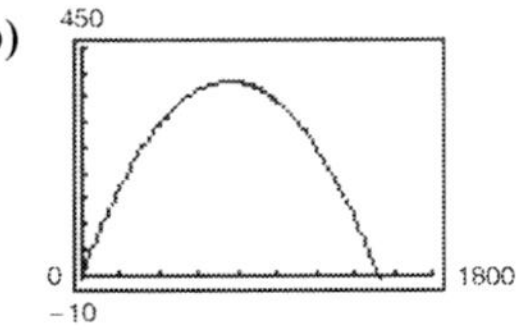

Maximum height $\approx$ 382.125 at $t \approx 4.86$

Range $\approx$ 1516.4

(c) $\mathbf{v}(t) = 110\sqrt{2}\mathbf{i} + \left[110\sqrt{2} - 32t\right]\mathbf{j}$

$\|\mathbf{v}(t)\| = \sqrt{\left(110\sqrt{2}\right)^2 + \left(110\sqrt{2} - 32t\right)^2}$

$\mathbf{a}(t) = -32\mathbf{j}$

(d)

t	0.5	1.0	1.5	2.0	2.5	3.0
Speed	208.99	198.67	189.13	180.51	172.94	166.58

65. $\mathbf{r}(t) = \langle 10 \cos 10\pi t, 10 \sin 10\pi t, 4 + 4t\rangle, 0 \le t \le \frac{1}{20}$

(a) $\mathbf{r}'(t) = \langle -100\pi \sin(10\pi t), 100\pi \cos(10\pi t), 4\rangle$

$\|\mathbf{r}'(t)\| = \sqrt{(100\pi)^2 \sin^2(10\pi t) + (100\pi)^2 \cos^2(10\pi t) + 16}$

$= \sqrt{(100\pi)^2 + 16} = 4\sqrt{625\pi^2 + 1} \approx 314$ mi/h

(b) $a_{\mathbf{T}} = 0$ and $a_{\mathbf{N}} = 1000\pi^2$

$a_{\mathbf{T}} = 0$ because the speed is constant.

66. 600 mi/h = 880 ft/sec

$\mathbf{r}(t) = 880t\mathbf{i} + \left(-16t^2 + 36{,}000\right)\mathbf{j}$

$\mathbf{v}(t) = 880\mathbf{i} - 32t\mathbf{j}$

$\mathbf{a}(t) = -32\mathbf{j}$

$\mathbf{T}(t) = \dfrac{880\mathbf{i} - 32t\mathbf{j}}{16\sqrt{4t^2 + 3025}} = \dfrac{55\mathbf{i} - 2t\mathbf{j}}{\sqrt{4t^2 + 3025}}$

Motion along $\mathbf{r}$ is clockwise, therefore

$\mathbf{N}(t) = \dfrac{-2t\mathbf{i} - 55\mathbf{j}}{\sqrt{4t^2 + 3025}}$

$a_{\mathbf{T}} = \mathbf{a} \cdot \mathbf{T} = \dfrac{64t}{\sqrt{4t^2 + 3025}}$

$a_{\mathbf{N}} = \mathbf{a} \cdot \mathbf{N} = \dfrac{1760}{\sqrt{4t^2 + 3025}}$

67. $\mathbf{r}(t) = (a \cos \omega t)\mathbf{i} + (a \sin \omega t)\mathbf{j}$

From Exercise 29, we know $\mathbf{a} \cdot \mathbf{T} = 0$ and $\mathbf{a} \cdot \mathbf{N} = a\omega^2$.

(a) Let $\omega_0 = 2\omega$. Then

$\mathbf{a} \cdot \mathbf{N} = a\omega_0^2 = a(2\omega)^2 = 4a\omega^2$

or the centripetal acceleration is increased by a factor of 4 when the velocity is doubled.

(b) Let $a_0 = a/2$. Then

$\mathbf{a} \cdot \mathbf{N} = a_0\omega^2 = \left(\dfrac{a}{2}\right)\omega^2 = \left(\dfrac{1}{2}\right)a\omega^2$

or the centripetal acceleration is halved when the radius is halved.

68. $\mathbf{r}(t) = (r\cos\omega t)\mathbf{i} + (r\sin\omega t)\mathbf{j}$

$\mathbf{v}(t) = (-r\omega\sin\omega t)\mathbf{i} + (r\omega\cos\omega t)\mathbf{j}$

$\|\mathbf{v}(t)\| = r\omega\sqrt{1} = r\omega = v$

$\mathbf{a}(t) = (-r\omega^2\cos\omega t)\mathbf{i} + (r\omega\cos\omega t)\mathbf{j}$

$\|\mathbf{a}(t)\| = r\omega^2$

(a) $F = m\|\mathbf{a}(t)\| = m(r\omega^2) = \dfrac{m}{r}(r^2\omega^2) = \dfrac{mv^2}{r}$

(b) By Newton's Law:

$$\frac{mv^2}{r} = \frac{GMm}{r^2},\ v^2 = \frac{GM}{r},\ v = \sqrt{\frac{GM}{r}}$$

69. $\mathbf{v} = \sqrt{\dfrac{GM}{r}} = \sqrt{\dfrac{9.56\times10^4}{4000+255}} \approx 4.74$ mi/sec

70. $\mathbf{v} = \sqrt{\dfrac{GM}{r}} = \sqrt{\dfrac{9.56\times10^4}{4000+360}} \approx 4.68$ mi/sec

71. $v = \sqrt{\dfrac{9.56\times10^4}{4000+385}} \approx 4.67$ mi/sec

72. Let x = distance from the satellite to the center of the earth $(x = r + 4000)$. Then:

$$v = \frac{2\pi x}{t} = \frac{2\pi x}{24(3600)} = \sqrt{\frac{9.56\times10^4}{x}}$$

$$\frac{4\pi^2x^2}{(24)^2(3600)^2} = \frac{9.56\times10^4}{x}$$

$$x^3 = \frac{(9.56\times10^4)(24)^2(3600)^2}{4\pi^2} \Rightarrow x \approx 26{,}245 \text{ mi}$$

$$v \approx \frac{2\pi(26{,}245)}{24(3600)} \approx 1.92 \text{ mi/sec} \approx 6871 \text{ mi/h}$$

73. False. You could be turning.

74. True. All the motion is in the tangential direction.

75. (a) $\mathbf{r}(t) = \cosh(bt)\mathbf{i} + \sinh(bt)\mathbf{j},\ b > 0$

$x = \cosh(bt),\ y = \sinh(bt)$

$x^2 - y^2 = \cosh^2(bt) - \sinh^2(bt) = 1$, hyperbola

(b) $\mathbf{v}(t) = b\sinh(bt)\mathbf{i} + b\cosh(bt)\mathbf{j}$

$\mathbf{a}(t) = b^2\cosh(bt)\mathbf{i} + b^2\sinh(bt)\mathbf{j} = b^2\mathbf{r}(t)$

76. Let $\mathbf{T}(t) = \cos\phi\mathbf{i} + \sin\phi\mathbf{j}$ be the unit tangent vector. Then

$$\mathbf{T}'(t) = \frac{d\mathbf{T}}{dt}$$
$$= \frac{d\mathbf{T}}{d\phi}\frac{d\phi}{dt} = -(\sin\phi\mathbf{i} - \cos\phi\mathbf{j})\frac{d\phi}{dt} = \mathbf{M}\frac{d\phi}{dt}.$$

$$\mathbf{M} = -\sin\phi\mathbf{i} + \cos\phi\mathbf{j}$$
$$= \cos[\phi + (\pi/2)]\mathbf{i} + \sin[\phi + (\pi/2)]\mathbf{j}$$

and is rotated counterclockwise through an angle of $\pi/2$ from $\mathbf{T}$.

If $d\phi/dt > 0$, then the curve bends to the left and $\mathbf{M}$ has the same direction as $\mathbf{T}'$. So, $\mathbf{M}$ has the same direction as

$$\mathbf{N} = \frac{\mathbf{T}'}{\|\mathbf{T}'\|},$$

which is toward the concave side of the curve.

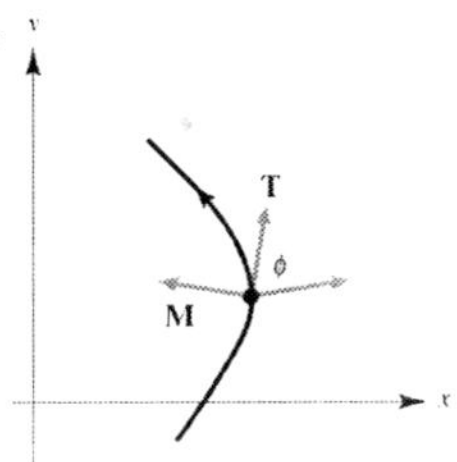

If $d\phi/dt < 0$, then the curve bends to the right and $\mathbf{M}$ has the opposite direction as $\mathbf{T}'$. Thus,

$$\mathbf{N} = \frac{\mathbf{T}'}{\|\mathbf{T}'\|}$$

again points to the concave side of the curve.

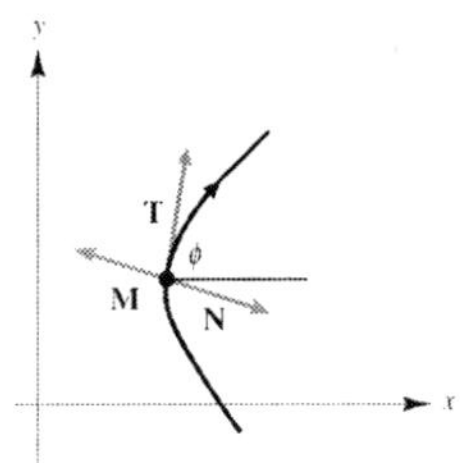

77. $\mathbf{r}(t) = x(t)\mathbf{i} + y(t)\mathbf{j}$

$y(t) = m(x(t)) + b$, m and b are constants.

$$\mathbf{r}(t) = x(t)\mathbf{i} + [m(x(t)) + b]\mathbf{j}$$

$$\mathbf{v}(t) = x'(t)\mathbf{i} + mx'(t)\mathbf{j}$$

$$\|\mathbf{v}(t)\| = \sqrt{[x'(t)]^2 + [mx'(t)]^2} = |x'(t)|\sqrt{1 + m^2}$$

$$\mathbf{T}(t) = \frac{\mathbf{v}(t)}{\|\mathbf{v}(t)\|} = \frac{\pm(\mathbf{i} + m\mathbf{j})}{\sqrt{1 + m^2}}, \text{ constant}$$

So, $\mathbf{T}'(t) = \mathbf{0}$.

78. Using $\mathbf{a} = a_{\mathbf{T}}\mathbf{T} + a_{\mathbf{N}}\mathbf{N}$, $\mathbf{T} \times \mathbf{T} = \mathbf{0}$, and $\|\mathbf{T} \times \mathbf{N}\| = 1$, you have:

$$\mathbf{v} \times \mathbf{a} = \|\mathbf{v}\|\mathbf{T} \times (a_{\mathbf{T}}\mathbf{T} + a_{\mathbf{N}}\mathbf{N})$$
$$= \|\mathbf{v}\|a_{\mathbf{T}}(\mathbf{T} \times \mathbf{T}) + \|\mathbf{v}\|a_{\mathbf{N}}(\mathbf{T} \times \mathbf{N})$$
$$= \|\mathbf{v}\|a_{\mathbf{N}}(\mathbf{T} \times \mathbf{N})$$

$$\|\mathbf{v} \times \mathbf{a}\| = \|\mathbf{v}\|a_{\mathbf{N}}\|\mathbf{T} \times \mathbf{N}\| = \|\mathbf{v}\|a_{\mathbf{N}}$$

So, $a_{\mathbf{N}} = \dfrac{\|\mathbf{v} \times \mathbf{a}\|}{\|\mathbf{v}\|}$.

79. $\|\mathbf{a}\|^2 = \mathbf{a} \cdot \mathbf{a}$

$$= (a_{\mathbf{T}}\mathbf{T} + a_{\mathbf{N}}\mathbf{N}) \cdot (a_{\mathbf{T}}\mathbf{T} + a_{\mathbf{N}}\mathbf{N})$$
$$= a_{\mathbf{T}}^2\|\mathbf{T}\|^2 + 2a_{\mathbf{T}}a_{\mathbf{N}}\mathbf{T} \cdot \mathbf{N} + a_{\mathbf{N}}^2\|\mathbf{N}\|^2$$
$$= a_{\mathbf{T}}^2 + a_{\mathbf{N}}^2$$

$$a_{\mathbf{N}}^2 = \|\mathbf{a}\|^2 - a_{\mathbf{T}}^2$$

Because $a_{\mathbf{N}} > 0$, we have $a_{\mathbf{N}} = \sqrt{\|\mathbf{a}\|^2 - a_{\mathbf{T}}^2}$.

80. $F = ma = (1)\dfrac{dv}{dt} = \dfrac{dv}{dt}$ Force

$$x = at + bt^2 + ct^3$$

$$v = \frac{dx}{dt} = a + 2bt + 3ct^2$$

$$\frac{dv}{dt} = 2b + 6ct$$

$$F^2 = 4b^2 + 24bct + 36c^2t^2$$
$$= 4b^2 + 12c + (2bt + 3ct^2)$$
$$= 4b^2 + 12c + (v - a)$$

$$F = f(v) = \pm\sqrt{4b^2 - 12ac + 12cv}$$

The sign of the radical is the sign of $2b + 6ct$, which cannot change.

Section 12.5 Arc Length and Curvature

1. $\mathbf{r}(t) = 3t\mathbf{i} - t\mathbf{j}, [0, 3]$

$$\frac{dx}{dt} = 3, \frac{dy}{dt} = -1, \frac{dz}{dt} = 0$$

$$s = \int_0^3 \sqrt{3^2 + (-1)^2}\, dt = \left[\sqrt{10}t\right]_0^3 = 3\sqrt{10}$$

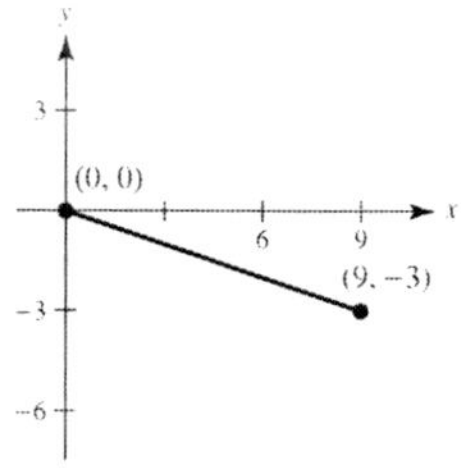

2. $\mathbf{r}(t) = t\mathbf{i} + t^2\mathbf{j}, [0, 4]$

$$\frac{dx}{dt} = 1, \frac{dy}{dt} = 2t, \frac{dz}{dt} = 0$$

$$s = \int_0^4 \sqrt{1 + 4t^2}\, dt$$

$$= \frac{1}{4}\left[2t\sqrt{1 + 4t^2} + \ln\left|2t + \sqrt{1 + 4t}\right|\right]_0^4 = \frac{1}{4}\left[8\sqrt{65} + \ln\left(8 + \sqrt{65}\right)\right] \approx 16.819$$

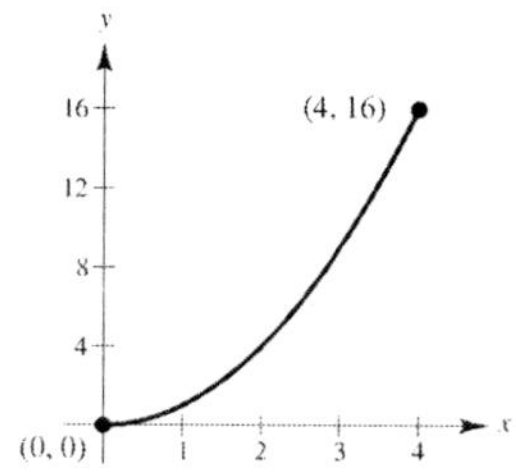

3. $\mathbf{r}(t) = t^3\mathbf{i} + t^2\mathbf{j}, [0, 1]$

$$\frac{dx}{dt} = 3t^2, \frac{dy}{dt} = 2t, \frac{dz}{dt} = 0$$

$$s = \int_0^1 \sqrt{9t^4 + 4t^2}\, dt = \int_0^1 \sqrt{9t^2 + 4}\, t\, dt$$

$$= \frac{1}{18}\int_0^1 \left(9t^2 + 4\right)^{1/2}(18t)\, dt = \frac{1}{27}\left[\left(9t^2 + 4\right)^{3/2}\right]_0^1 = \frac{1}{27}\left(13^{3/2} - 8\right) \approx 1.4397$$

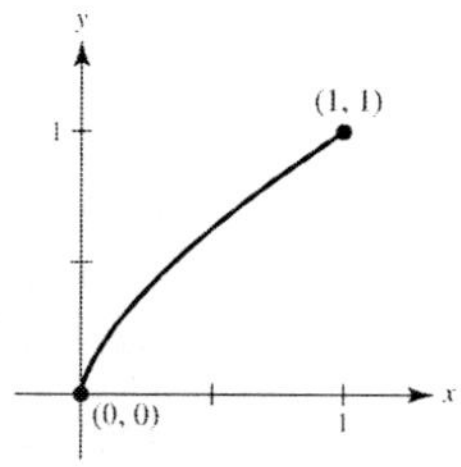

4. $\mathbf{r}(t) = (t + 1)\mathbf{i} + t^2\mathbf{j},\ 0 \le t \le 6$

$$\frac{dx}{dt} = 1, \frac{dy}{dt} = 2t$$

$$s = \int_0^6 \sqrt{1 + 4t^2}\, dt$$

$$= \left[\frac{1}{4}\ln\left(\sqrt{4t^2 + 1} + 2t\right) + \frac{1}{2}t\sqrt{4t^2 + 1}\right]_0^6 = \frac{1}{4}\ln\left(\sqrt{145} + 12\right) + 3\sqrt{145}$$

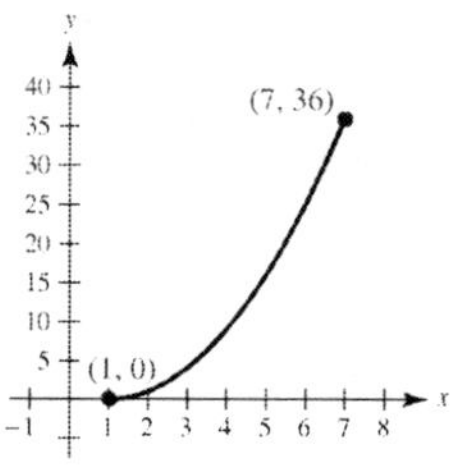

5. $\mathbf{r}(t) = a\cos^3 t\mathbf{i} + a\sin^3 t\mathbf{j}, [0, 2\pi]$

$$\frac{dx}{dt} = -3a\cos^2 t\sin t, \frac{dy}{dt} = 3a\sin^2 t\cos t$$

$$s = 4\int_0^{\pi/2} \sqrt{\left[-3a\cos^2 t\sin t\right]^2 + \left[3a\sin^2 t\cos t\right]^2}\, dt$$

$$= 12a\int_0^{\pi/2} \sin t\cos t\, dt = 3a\int_0^{\pi/2} 2\sin 2t\, dt = \left[-3a\cos 2t\right]_0^{\pi/2} = 6a$$

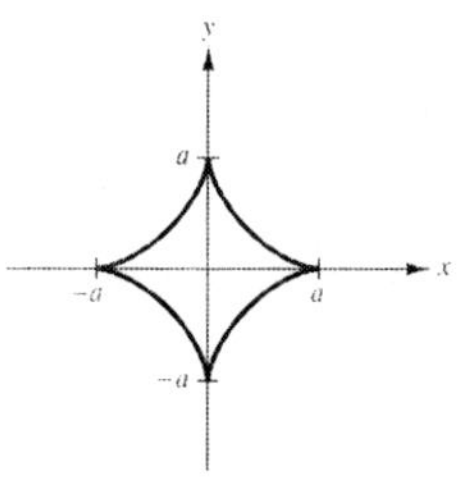

6. $\mathbf{r}(t) = a\cos t\mathbf{i} + a\sin t\mathbf{j}, [0, 2\pi]$

$$\frac{dx}{dt} = -a\sin t, \frac{dy}{dt} = a\cos t$$

$$s = \int_0^{2\pi} \sqrt{a^2\sin^2 t + a^2\cos^2 t}\, dt = \int_0^{2\pi} a\, dt = \left[at\right]_0^{2\pi} = 2\pi a$$

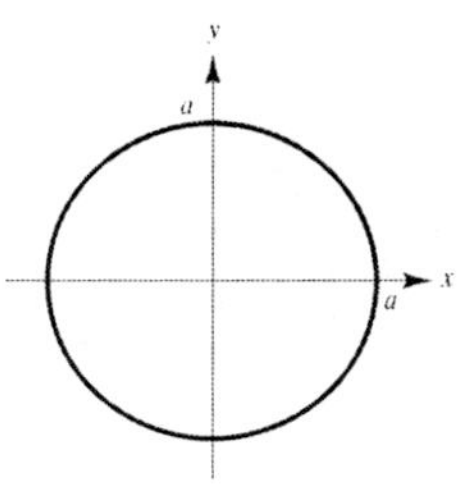

7. (a) $\mathbf{r}(t) = (v_0 \cos\theta)t\mathbf{i} + \left[h + (v_0 \sin\theta)t - \frac{1}{2}gt^2\right]\mathbf{j}$

$= (100\cos 45°)t\mathbf{i} + \left[3 + (100 \sin 45°)t - \frac{1}{2}(32)t^2\right]\mathbf{j} = 50\sqrt{2}t\mathbf{i} + \left[3 + 50\sqrt{2}t - 16t^2\right]\mathbf{j}$

(b) $\mathbf{v}(t) = 50\sqrt{2}\mathbf{i} + \left(50\sqrt{2} - 32t\right)\mathbf{j}$

$50\sqrt{2} - 32t = 0 \Rightarrow t = \frac{25\sqrt{2}}{16}$

Maximum height: $3 + 50\sqrt{2}\left(\frac{25\sqrt{2}}{16}\right) - 16\left(\frac{15\sqrt{2}}{16}\right)^2 = 81.125$ feet

(c) $3 + 50\sqrt{2}t - 16t^2 = 0 \Rightarrow t \approx 4.4614$

Range: $50\sqrt{2}(4.4614) \approx 315.5$ feet

(d) $s = \int_0^{4.4614} \sqrt{\left(50\sqrt{2}\right)^2 + \left(50\sqrt{2} - 32t\right)^2}\, dt \approx 362.9$ feet

8. (a) $\mathbf{r}(t) = (v_0 \cos\theta)t\mathbf{i} + \left[h + (v_0 \sin\theta)t - \frac{1}{2}gt^2\right]\mathbf{j}$

$= (80\cos 30°)t\mathbf{i} + \left[4 + (80 \sin 30°)t - \frac{1}{2}(32)t^2\right]\mathbf{j}$

$= 40\sqrt{3}t\mathbf{i} + \left[4 + 40t - 16t^2\right]\mathbf{j}$

(b) $\mathbf{v}(t) = 40\sqrt{3}\mathbf{i} + (40 - 32t)\mathbf{j}$

$40 - 32t = 0 \Rightarrow t = \frac{5}{4}$

Maximum height: $4 + 40\left(\frac{5}{4}\right) - 16\left(\frac{5}{4}\right)^2 = 29$ feet

(c) $4 + 40t - 16t^2 = 0 \Rightarrow t \approx 2.596$

Range: $40\sqrt{3}(2.596) \approx 179.9$ feet

(d) $s = \int_0^{2.596} \sqrt{\left(40\sqrt{3}\right)^2 + (40 - 32t)^2}\, dt \approx 190.15$ feet

9. $\mathbf{r}(t) = -t\mathbf{i} + 4t\mathbf{j} + 3t\mathbf{k}, [0, 1]$

$\frac{dx}{dt} = -1, \frac{dy}{dt} = 4, \frac{dz}{dt} = 3$

$s = \int_0^1 \sqrt{1 + 16 + 9}\, dt = \left[\sqrt{26}t\right]_0^1 = \sqrt{26}$

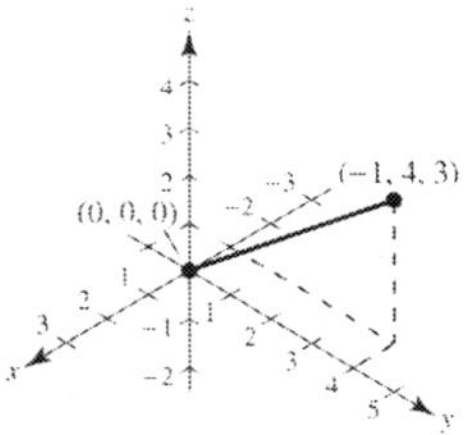

10. $\mathbf{r}(t) = \mathbf{i} + t^2\mathbf{j} + t^3\mathbf{k}, [0, 2]$

$\frac{dx}{dt} = 0, \frac{dy}{dt} = 2t, \frac{dz}{dt} = 3t^2$

$s = \int_0^2 \sqrt{4t^2 + 9t^4}\, dt$

$= \int_0^2 \sqrt{4 + 9t^2}\, t\, dt = \frac{1}{27}\left(4 + 9t^2\right)^{3/2}\Big]_0^2 = \frac{1}{27}\left(40^{3/2} - 4^{3/2}\right) = \frac{1}{27}\left[80\sqrt{10} - 8\right]$

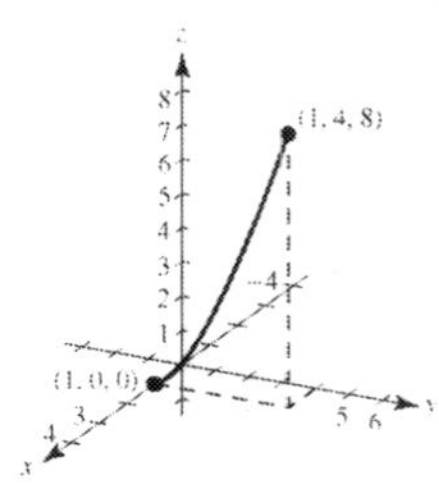

11. $\mathbf{r}(t) = \langle 4t, -\cos t, \sin t\rangle, \left[0, \dfrac{3\pi}{2}\right]$

$$\frac{dx}{dt} = 4, \frac{dy}{dt} = \sin t, \frac{dz}{dt} = \cos t$$

$$s = \int_0^{3\pi/2} \sqrt{16 + \sin^2 t + \cos^2 t}\, dt = \int_0^{3\pi/2} \sqrt{17}\, dt = \left[\sqrt{17}t\right]_0^{3\pi/2} = \frac{3\pi}{2}\sqrt{17}$$

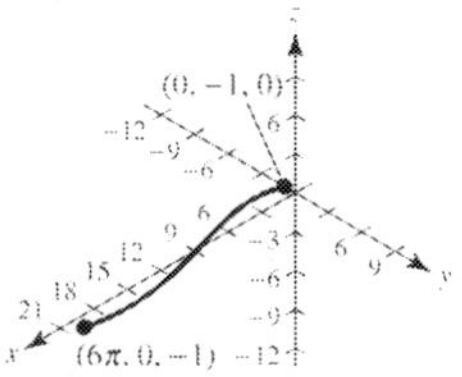

12. $\mathbf{r}(t) = \langle 2\sin t, 5t, 2\cos t\rangle, [0, \pi]$

$$\frac{dx}{dt} = 2\cos t, \frac{dy}{dt} = 5, \frac{dz}{dt} = -2\sin t$$

$$s = \int_0^{\pi} \sqrt{4\cos^2 t + 25 + 4\sin^2 t}\, dt = \int_0^{\pi} \sqrt{29}\, dt = \sqrt{29}\pi$$

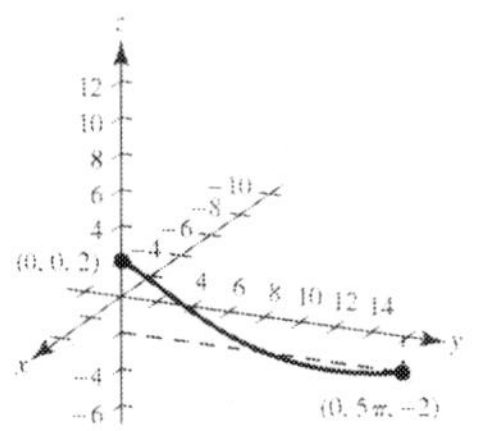

13. $\mathbf{r}(t) = a\cos t\mathbf{i} + a\sin t\mathbf{j} + bt\mathbf{k}, [0, 2\pi]$

$$\frac{dx}{dt} = -a\sin t, \frac{dy}{dt} = a\cos t, \frac{dz}{dt} = b$$

$$s = \int_0^{2\pi} \sqrt{a^2\sin^2 t + a^2\cos^2 t + b^2}\, dt$$

$$= \int_0^{2\pi} \sqrt{a^2 + b^2}\, dt = \left[\sqrt{a^2 + b^2}t\right]_0^{2\pi} = 2\pi\sqrt{a^2 + b^2}$$

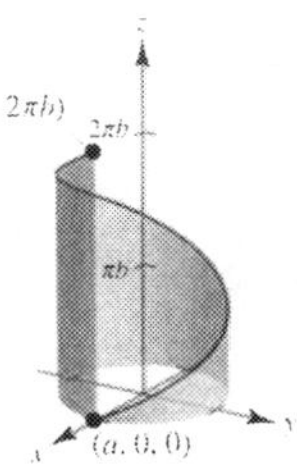

14. $\mathbf{r}(t) = \left\langle \cos t + t\sin t, \sin t - t\cos t, t^2\right\rangle, \left[0, \dfrac{\pi}{2}\right]$

$$\frac{dx}{dt} = t\cos t, \frac{dy}{dt} = t\sin t, \frac{dz}{dt} = 2t$$

$$s = \int_0^{\pi/2} \sqrt{(t\cos t)^2 + (t\sin t)^2 + (2t)^2}\, dt$$

$$= \int_0^{\pi/2} \sqrt{5t^2}\, dt = \left[\sqrt{5}\frac{t^2}{2}\right]_0^{\pi/2} = \frac{\sqrt{5}\pi^2}{8}$$

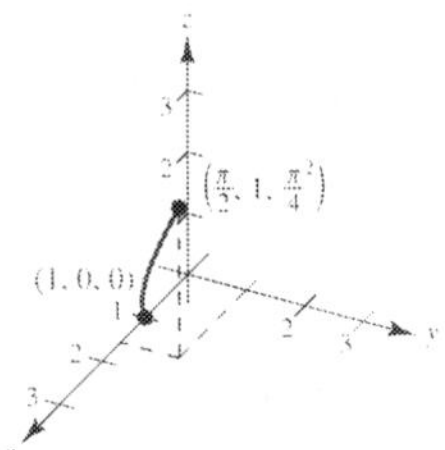

15. $\mathbf{r}(t) = t\mathbf{i} + (4 - t^2)\mathbf{j} + t^3\mathbf{k}, [0, 2]$

(a) $\mathbf{r}(0) = \langle 0, 4, 0\rangle, \mathbf{r}(2) = \langle 2, 0, 8\rangle$

$$\text{distance} = \sqrt{2^2 + 4^2 + 8^2} = \sqrt{84} = 2\sqrt{21} \approx 9.165$$

(b) $\mathbf{r}(0) = \langle 0, 4, 0\rangle$

$\mathbf{r}(0.5) = \langle 0.5, 3.75, 0.125\rangle$

$\mathbf{r}(1) = \langle 1, 3, 1\rangle$

$\mathbf{r}(1.5) = \langle 1.5, 1.75, 3.375\rangle$

$\mathbf{r}(2) = \langle 2, 0, 8\rangle$

$$\begin{aligned}\text{distance} &\approx \sqrt{(0.5)^2 + (0.25)^2 + (0.125)^2} \\ &+ \sqrt{(0.5)^2 + (0.75)^2 + (0.875)^2} \\ &+ \sqrt{(0.5)^2 + (1.25)^2 + (2.375)^2} \\ &+ \sqrt{(0.5)^2 + (1.75)^2 + (4.625)^2} \\ &\approx 0.5728 + 1.2562 + 2.7300 + 4.9702 \\ &\approx 9.529\end{aligned}$$

(c) Increase the number of line segments.

(d) Using a graphing utility, you obtain 9.57057.

16. $\mathbf{r}(t) = 6\cos\left(\frac{\pi t}{4}\right)\mathbf{i} + 2\sin\left(\frac{\pi t}{4}\right)\mathbf{j} + t\mathbf{k}, [0, 2]$

(a) $\mathbf{r}(0) = 6\mathbf{i} = \langle 6, 0, 0\rangle$

$\mathbf{r}(2) = 2\mathbf{j} + 2\mathbf{k} = \langle 0, 2, 2\rangle$

$$\text{distance} = \sqrt{6^2 + 2^2 + 2^2} = \sqrt{44} = 2\sqrt{11} \approx 6.633$$

(b) $\mathbf{r}(0) = \langle 6, 0, 0\rangle$

$\mathbf{r}(0.5) = \langle 5.543, 0.765, 0.5\rangle$

$\mathbf{r}(1.0) = \langle 4.243, 1.414, 1.0\rangle$

$\mathbf{r}(1.5) = \langle 2.296, 1.848, 1.5\rangle$

$\mathbf{r}(2.0) = \langle 0, 2, 2\rangle$

distance ≈ 6.9698

(c) Increase the number of line segments.

(d) Using a graphing utility, you obtain

$$s = \int_0^2 \|\mathbf{r}'(t)\|\,dt \approx 7.0105.$$

17. $\mathbf{r}(t) = \langle 2\cos t, 2\sin t, t\rangle$

(a) $s = \int_0^t \sqrt{[x'(u)]^2 + [y'(u)]^2 + [z'(u)]^2}\,du = \int_0^t \sqrt{(-2\sin u)^2 + (2\cos u)^2 + (1)^2}\,du = \int_0^t \sqrt{5}\,du = \left[\sqrt{5}u\right]_0^t = \sqrt{5}t$

(b) $\frac{s}{\sqrt{5}} = t$

$$x = 2\cos\left(\frac{s}{\sqrt{5}}\right), y = 2\sin\left(\frac{s}{\sqrt{5}}\right), z = \frac{s}{\sqrt{5}}$$

$$\mathbf{r}(s) = 2\cos\left(\frac{s}{\sqrt{5}}\right)\mathbf{i} + 2\sin\left(\frac{s}{\sqrt{5}}\right)\mathbf{j} + \frac{s}{\sqrt{5}}\mathbf{k}$$

(c) When $s = \sqrt{5}$: $x = 2\cos 1 \approx 1.081$

$y = 2\sin 1 \approx 1.683$

$z = 1$

$(1.081, 1.683, 1.000)$

When $s = 4$: $x = 2\cos\frac{4}{\sqrt{5}} \approx -0.433$

$y = 2\sin\frac{4}{\sqrt{5}} \approx 1.953$

$z = \frac{4}{\sqrt{5}} \approx 1.789$

$(-0.433, 1.953, 1.789)$

(d) $$\|\mathbf{r}'(s)\| = \sqrt{\left(-\frac{2}{\sqrt{5}}\sin\left(\frac{s}{\sqrt{5}}\right)\right)^2 + \left(\frac{2}{\sqrt{5}}\cos\left(\frac{\sqrt{s}}{\sqrt{5}}\right)\right)^2 + \left(\frac{1}{\sqrt{5}}\right)^2} = \sqrt{\frac{4}{5} + \frac{1}{5}} = 1$$

18. $\mathbf{r}(t) = \left\langle 4(\sin t - t\cos t), 4(\cos t + t\sin t), \frac{3}{2}t^2 \right\rangle$

(a) $s = \int_0^t \sqrt{[x'(u)]^2 + [y'(u)]^2 + [z'(u)]^2}\, du$

$= \int_0^t \sqrt{(4u\sin u)^2 + (4u\cos u)^2 + (3u)^2}\, du = \int_0^t \sqrt{16u + 9u^2}\, du = \int_0^t 5u\, du = \frac{5}{2}t^2$

(b) $t = \sqrt{\frac{2s}{5}}$

$x = 4\left(\sin\sqrt{\frac{2s}{5}} - \sqrt{\frac{2s}{5}}\cos\sqrt{\frac{2s}{5}}\right)$

$y = 4\left(\cos\sqrt{\frac{2s}{5}} + \sqrt{\frac{2s}{5}}\sin\sqrt{\frac{2s}{5}}\right)$

$z = \frac{3}{2}\left(\sqrt{\frac{2s}{5}}\right)^2 = \frac{3s}{5}$

$\mathbf{r}(s) = 4\left(\sin\sqrt{\frac{2s}{5}} - \sqrt{\frac{2s}{5}}\cos\sqrt{\frac{2s}{5}}\right)\mathbf{i} + 4\left(\cos\sqrt{\frac{2s}{5}} + \sqrt{\frac{2s}{5}}\sin\sqrt{\frac{2s}{5}}\right)\mathbf{j} + \frac{3s}{5}\mathbf{k}$

(c) When $s = \sqrt{5}$:

$x = 4\left(\sin\sqrt{\frac{2\sqrt{5}}{5}} - \sqrt{\frac{2\sqrt{5}}{5}}\cos\sqrt{\frac{2\sqrt{5}}{5}}\right) \approx -1.030$

$y = 4\left(\cos\sqrt{\frac{2\sqrt{5}}{5}} + \sqrt{\frac{2\sqrt{5}}{5}}\sin\sqrt{\frac{2\sqrt{5}}{5}}\right) \approx 5.408$

$z = \frac{3\sqrt{5}}{5} \approx 1.342$

$(-1.030, 5.408, 1.342)$

When $s = 4$:

$x = 4\left(\sin\sqrt{\frac{8}{5}} - \sqrt{\frac{8}{5}}\cos\sqrt{\frac{8}{5}}\right) \approx 2.291$

$y = 4\left(\cos\sqrt{\frac{8}{5}} + \sqrt{\frac{8}{5}}\sin\sqrt{\frac{8}{5}}\right) \approx 6.029$

$z = \frac{12}{5} = 2.4$

$(2.291, 6.029, 2.400)$

(d) $\|\mathbf{r}'(s)\| = \sqrt{\left(\frac{4}{5}\sin\sqrt{\frac{2s}{5}}\right)^2 + \left(\frac{4}{5}\cos\sqrt{\frac{2s}{5}}\right)^2 + \left(\frac{3}{5}\right)^2} = \sqrt{\frac{16}{25} + \frac{9}{25}} = 1$

19. $\mathbf{r}(s) = \left(1 + \frac{\sqrt{2}}{2}s\right)\mathbf{i} + \left(1 - \frac{\sqrt{2}}{2}s\right)\mathbf{j}$

$\mathbf{r}'(s) = \frac{\sqrt{2}}{2}\mathbf{i} - \frac{\sqrt{2}}{2}\mathbf{j}$ and $\|\mathbf{r}'(s)\| = \sqrt{\frac{1}{2} + \frac{1}{2}} = 1$

$\mathbf{T}(s) = \frac{\mathbf{r}'(s)}{\|\mathbf{r}'(s)\|} = \mathbf{r}'(s)$

$\mathbf{T}'(s) = \mathbf{0} \Rightarrow K = \|\mathbf{T}'(s)\| = 0$ (The curve is a line.)

20. $\mathbf{r}(s) = (3 + s)\mathbf{i} + \mathbf{j}$

$\mathbf{r}'(s) = \mathbf{i}$ and $\|\mathbf{r}'(s)\| = 1$

$\mathbf{T}(s) = \mathbf{r}'(s)$

$\mathbf{T}'(s) = \mathbf{0} \Rightarrow K = \|\mathbf{T}'(s)\| = 0$ (The curve is a line.)

21. $\mathbf{r}(s) = 2\cos\left(\frac{s}{\sqrt{5}}\right)\mathbf{i} + 2\sin\left(\frac{s}{\sqrt{5}}\right)\mathbf{j} + \frac{s}{\sqrt{5}}\mathbf{k}$

$\mathbf{T}(s) = \mathbf{r}'(s)$

$= -\frac{2}{\sqrt{5}}\sin\left(\frac{s}{\sqrt{5}}\right)\mathbf{i} + \frac{2}{\sqrt{5}}\cos\left(\frac{s}{\sqrt{5}}\right)\mathbf{j} + \frac{1}{\sqrt{5}}\mathbf{k}$

$\mathbf{T}'(s) = -\frac{2}{5}\cos\left(\frac{s}{\sqrt{5}}\right)\mathbf{i} - \frac{2}{5}\sin\left(\frac{s}{\sqrt{5}}\right)\mathbf{j}$

$K = \left\|\mathbf{T}'(s)\right\| = \frac{2}{5}$

22. $\mathbf{r}(s) = 4\left(\sin\sqrt{\frac{2s}{5}} - \sqrt{\frac{2s}{5}}\cos\sqrt{\frac{2s}{5}}\right)\mathbf{i} + 4\left(\cos\sqrt{\frac{2s}{5}} + \sqrt{\frac{2s}{5}}\sin\sqrt{\frac{2s}{5}}\right)\mathbf{j} + \frac{3s}{5}\mathbf{k}$

$\mathbf{T}(s) = \mathbf{r}'(s) = \frac{4}{5}\sin\sqrt{\frac{2s}{5}}\mathbf{i} + \frac{4}{5}\cos\sqrt{\frac{2s}{5}}\mathbf{j} + \frac{3}{5}\mathbf{k}$

$\mathbf{T}'(s) = \frac{4}{25}\sqrt{\frac{5}{2s}}\cos\sqrt{\frac{2s}{5}}\mathbf{i} - \frac{4}{25}\sqrt{\frac{5}{2s}}\sin\sqrt{\frac{2s}{5}}\mathbf{j}$

$K = \left\|\mathbf{T}'(s)\right\| = \frac{4}{25}\sqrt{\frac{5}{2s}} = \frac{2\sqrt{10s}}{25s}$

23. $\mathbf{r}(t) = 4t\mathbf{i} - 2t\mathbf{j},\ t = 1$

$\mathbf{v}(t) = 4\mathbf{i} - 2\mathbf{j}$

$\mathbf{T}(t) = \frac{1}{\sqrt{5}}(2\mathbf{i} - \mathbf{j})$

$\mathbf{T}'(t) = \mathbf{0}$

$K = \frac{\left\|\mathbf{T}'(t)\right\|}{\left\|\mathbf{r}'(t)\right\|} = 0$

(The curve is a line.)

24. $\mathbf{r}(t) = t^2\mathbf{i} + \mathbf{j},\ t = 2$

$\mathbf{v}(t) = 2t\mathbf{i}$

$\mathbf{T}(t) = \mathbf{i}$

$\mathbf{T}'(t) = \mathbf{0}$

$K = \frac{\left\|\mathbf{T}'(t)\right\|}{\left\|\mathbf{r}'(t)\right\|} = 0$

25. $\mathbf{r}(t) = t\mathbf{i} + \frac{1}{t}\mathbf{j},\ t = 1$

$\mathbf{v}(t) = \mathbf{i} - \frac{1}{t^2}\mathbf{j}$

$\mathbf{v}(1) = \mathbf{i} - \mathbf{j}$

$\mathbf{a}(t) = \frac{2}{t^3}\mathbf{j}$

$\mathbf{a}(1) = 2\mathbf{j}$

$\mathbf{T}(t) = \frac{t^2\mathbf{i} - \mathbf{j}}{\sqrt{t^4 + 1}}$

$\mathbf{N}(t) = \frac{1}{\left(t^4 + 1\right)^{1/2}}\left(\mathbf{i} + t^2\mathbf{j}\right)$

$\mathbf{N}(1) = \frac{1}{\sqrt{2}}(\mathbf{i} + \mathbf{j})$

$K = \frac{\mathbf{a}\cdot\mathbf{N}}{\left\|\mathbf{v}\right\|^2} = \frac{\sqrt{2}}{2}$

26. $\mathbf{r}(t) = t\mathbf{i} + \frac{1}{9}t^3\mathbf{j},\ t = 2$

$\mathbf{v}(t) = \mathbf{i} + \frac{1}{3}t^2\mathbf{j}$

$\mathbf{v}(2) = \mathbf{i} + \frac{4}{3}\mathbf{j},\ \|\mathbf{v}(2)\| = \sqrt{1 + \frac{16}{9}} = \frac{5}{3}$

$\mathbf{a}(t) = \frac{2}{3}t\mathbf{j}$

$\mathbf{a}(2) = \frac{4}{3}\mathbf{j}$

$$\mathbf{T}(t) = \frac{\mathbf{i} + \frac{1}{3}t^2\mathbf{j}}{\sqrt{1 + \frac{t^4}{9}}} = \frac{3\mathbf{i} + t^2\mathbf{j}}{\sqrt{9 + t^4}}$$

$$\mathbf{T}(2) = \frac{3\mathbf{i} + 4\mathbf{j}}{5}$$

$$\mathbf{N}(2) = \frac{-4\mathbf{i} + 3\mathbf{j}}{5}$$

$$K = \frac{\mathbf{a}\cdot\mathbf{N}}{\|\mathbf{v}\|^2} = \frac{4/5}{25/9} = \frac{36}{125}$$

27. $\mathbf{r}(t) = \langle t, \sin t\rangle,\ t = \frac{\pi}{2}$

$\mathbf{r}'(t) = \langle 1, \cos t\rangle,\ \|\mathbf{r}'(t)\| = \sqrt{1 + \cos^2 t}$

$\mathbf{r}'\left(\frac{\pi}{2}\right) = \langle 1, 0\rangle,\ \left\|\mathbf{r}'\left(\frac{\pi}{2}\right)\right\| = 1$

$\mathbf{a}(t) = \langle 0, -\sin t\rangle,\ \mathbf{a}\left(\frac{\pi}{2}\right) = \langle 0, -1\rangle$

$$\mathbf{T}(t) = \frac{1}{\sqrt{1 + \cos^2 t}}\langle 1, \cos t\rangle$$

$\mathbf{T}\left(\frac{\pi}{2}\right) = \langle 1, 0\rangle$

$\mathbf{N}\left(\frac{\pi}{2}\right) = \langle 0, -1\rangle$

$$K = \frac{\mathbf{a}\cdot\mathbf{N}}{\|\mathbf{v}\|^2} = \frac{1}{1} = 1$$

28. $\mathbf{r}(t) = \langle 5\cos t, 4\sin t\rangle,\ t = \frac{\pi}{3}$

$x(t) = 5\cos t,\ y(t) = 4\sin t$

$$\begin{aligned} K &= \frac{|x'y'' - y'x''|}{\left[(x')^2 + (y')^2\right]^{3/2}} \\ &= \frac{|(-5\sin t)(-4\sin t) - (4\cos t)(-5\cos t)|}{\left[25\sin^2 t + 16\cos^2 t\right]^{3/2}} \\ &= \frac{20}{\left[25\sin^2 t + 16\cos^2 t\right]^{3/2}} \end{aligned}$$

$$K\left(\frac{\pi}{3}\right) = \frac{20}{\left[25(3/4) + 16(1/4)\right]^{3/2}} = \frac{160\sqrt{91}}{8281}$$

29. $\mathbf{r}(t) = 4\cos 2\pi t\mathbf{i} + 4\sin 2\pi t\mathbf{j}$

$\mathbf{r}'(t) = -8\pi\sin 2\pi t\mathbf{i} + 8\pi\cos 2\pi t\mathbf{j}$

$\mathbf{T}(t) = -\sin 2\pi t\mathbf{i} + \cos 2t\mathbf{j}$

$\mathbf{T}'(t) = -2\pi\cos 2\pi t\mathbf{i} - 2\pi\sin 2\pi t\mathbf{j}$

$$K = \frac{\|\mathbf{T}'(t)\|}{\|\mathbf{r}'(t)\|} = \frac{2\pi}{8\pi} = \frac{1}{4}$$

30. $\mathbf{r}(t) = 2\cos\pi t\mathbf{i} + \sin\pi t\mathbf{j}$

$\mathbf{r}'(t) = -2\pi\sin\pi t\mathbf{i} + \pi\cos\pi t\mathbf{j}$

$\|\mathbf{r}'(t)\| = \pi\sqrt{4\sin^2\pi t + \cos^2\pi t}$

$$\mathbf{T}(t) = \frac{-2\sin\pi t\mathbf{i} + \cos\pi t\mathbf{j}}{\sqrt{4\sin^2\pi t + \cos^2\pi t}}$$

$$\mathbf{T}'(t) = \frac{-2\pi\cos\pi t\mathbf{i} - 4\pi\sin\pi t\mathbf{j}}{\left(4\sin^2\pi t + \cos^2\pi t\right)^{3/2}}$$

$$\begin{aligned} K &= \frac{\|\mathbf{T}'(t)\|}{\|\mathbf{r}'(t)\|} = \frac{\dfrac{2\pi}{4\sin^2\pi t + \cos^2\pi t}}{\pi\sqrt{4\sin^2\pi t + \cos^2\pi t}} \\ &= \frac{2}{\left(4\sin^2\pi t + \cos^2\pi t\right)^{3/2}} \end{aligned}$$

31. $\mathbf{r}(t) = a\cos\omega t\mathbf{i} + a\sin\omega t\mathbf{j}$

$\mathbf{r}'(t) = -a\omega\sin\omega t\mathbf{i} + a\omega\cos\omega t\mathbf{j}$

$\mathbf{T}(t) = -\sin\omega t\mathbf{i} + \cos\omega t\mathbf{j}$

$\mathbf{T}'(t) = -\omega\cos\omega t\mathbf{i} - \omega\sin\omega t\mathbf{j}$

$$K = \frac{\|\mathbf{T}'(t)\|}{\|\mathbf{r}'(t)\|} = \frac{\omega}{a\omega} = \frac{1}{a}$$

32. $\mathbf{r}(t) = a\cos(\omega t)\mathbf{i} + b\sin(\omega t)\mathbf{j}$

$$\mathbf{r}'(t) = -a\omega\sin(\omega t)\mathbf{i} + b\omega\cos(\omega t)\mathbf{j}$$

$$\mathbf{T}(t) = \frac{-a\sin(\omega t)\mathbf{i} + b\cos(\omega t)\mathbf{j}}{\sqrt{a^2\sin^2(\omega t) + b^2\cos^2(\omega t)}}$$

$$\mathbf{T}'(t) = \frac{-ab^2\omega\cos(\omega t)\mathbf{i} - a^2b\omega\sin(\omega t)\mathbf{j}}{\left[a^2\sin^2(\omega t) + b^2\cos^2(\omega t)\right]^{3/2}}$$

$$K = \frac{\|\mathbf{T}'(t)\|}{\|\mathbf{r}'(t)\|} = \frac{\dfrac{ab\omega}{a^2\sin^2(\omega t) + b^2\cos^2(\omega t)}}{\omega\sqrt{a^2\sin^2(\omega t) + b^2\cos^2(\omega t)}}$$

$$= \frac{ab}{\left[a^2\sin^2(\omega t) + b^2\cos^2(\omega t)\right]^{3/2}}$$

33. $\mathbf{r}(t) = t\mathbf{i} + t^2\mathbf{j} + \dfrac{t^2}{2}\mathbf{k}$

$$\mathbf{r}'(t) = \mathbf{i} + 2t\mathbf{j} + t\mathbf{k}$$

$$\mathbf{T}(t) = \frac{\mathbf{i} + 2t\mathbf{j} + t\mathbf{k}}{\sqrt{1 + 5t^2}}$$

$$\mathbf{T}'(t) = \frac{-5t\mathbf{i} + 2\mathbf{j} + \mathbf{k}}{\left(1 + 5t^2\right)^{3/2}}$$

$$K = \frac{\|\mathbf{T}'(t)\|}{\|\mathbf{r}'(t)\|} = \frac{\dfrac{\sqrt{5}}{\left(1 + 5t^2\right)}}{\sqrt{1 + 5t^2}} = \frac{\sqrt{5}}{\left(1 + 5t^2\right)^{3/2}}$$

34. $\mathbf{r}(t) = 2t^2\mathbf{i} + t\mathbf{j} + \dfrac{1}{2}t^2\mathbf{k}$

$$\mathbf{r}'(t) = 4t\mathbf{i} + \mathbf{j} + t\mathbf{k}$$

$$\mathbf{T}(t) = \frac{4t\mathbf{i} + \mathbf{j} + t\mathbf{k}}{\sqrt{1 + 17t^2}}$$

$$\mathbf{T}'(t) = \frac{4\mathbf{i} - 17t\mathbf{j} + \mathbf{k}}{\left(1 + 17t^2\right)^{3/2}}$$

$$K = \frac{\|\mathbf{T}'(t)\|}{\mathbf{r}'(t)} = \frac{\sqrt{289t^2 + 17}}{\left(1 + 17t^2\right)^{3/2}}\Bigg/\left(1 + 17t^2\right)^{1/2}$$

$$= \frac{\sqrt{17}}{\left(1 + 17t^2\right)^{3/2}}$$

35. $\mathbf{r}(t) = 4t\mathbf{i} + 3\cos t\mathbf{j} + 3\sin t\mathbf{k}$

$$\mathbf{r}'(t) = 4\mathbf{i} - 3\sin t\mathbf{j} + 3\cos t\mathbf{k}$$

$$\mathbf{T}(t) = \frac{1}{5}\left[4\mathbf{i} - 3\sin t\mathbf{j} + 3\cos t\mathbf{k}\right]$$

$$\mathbf{T}'(t) = \frac{1}{5}\left[-3\cos t\mathbf{j} - 3\sin t\mathbf{k}\right]$$

$$K = \frac{\|\mathbf{T}'(t)\|}{\mathbf{r}'(t)} = \frac{3/5}{5} = \frac{3}{25}$$

36. $\mathbf{r}(t) = e^{2t}\mathbf{i} + e^{2t}\cos t\mathbf{j} + e^{2t}\sin t\mathbf{k}$

$$\mathbf{r}'(t) = 2e^{2t}\mathbf{i} + \left(2e^{2t}\cos t - e^{2t}\sin t\right)\mathbf{j} + \left(2e^{2t}\sin t + e^{2t}\cos t\right)\mathbf{k} = e^{2t}\left[2\mathbf{i} + (2\cos t - \sin t)\mathbf{j} + (2\sin t + \cos t)\mathbf{k}\right]$$

$$\|\mathbf{r}'(t)\| = e^{2t}\left[4 + \left(4\cos^2 t - 4\cos t\sin t + \sin^2 t\right) + \left(4\sin^2 t + 4\sin t\cos t + \cos^2 t\right)\right]^{1/2} = e^{2t}[9]^{1/2} = 3e^{2t}$$

$$\mathbf{T}(t) = \frac{\mathbf{r}'(t)}{\|\mathbf{r}'(t)\|} = \frac{2}{3}\mathbf{i} + \left(\frac{2}{3}\cos t - \frac{1}{3}\sin t\right)\mathbf{j} + \left(\frac{2}{3}\sin t + \frac{1}{3}\cos t\right)\mathbf{k}$$

$$\mathbf{T}'(t) = \left(-\frac{2}{3}\sin t - \frac{1}{3}\cos t\right)\mathbf{j} + \left(\frac{2}{3}\cos t - \frac{1}{3}\sin t\right)\mathbf{k}$$

$$\|\mathbf{T}'(t)\| = \left[\left(\frac{4}{9}\sin^2 t + \frac{1}{9}\cos^2 t + \frac{4}{9}\sin t\cos t\right) + \left(\frac{4}{9}\cos^2 t + \frac{1}{9}\sin^2 t - \frac{4}{9}\cos t\sin t\right)\right]^{1/2} = \frac{\sqrt{5}}{3}$$

$$K = \frac{\|\mathbf{T}'(t)\|}{\|\mathbf{r}'(t)\|} = \frac{\dfrac{\sqrt{5}}{3}}{3e^{2t}} = \frac{\sqrt{5}}{9e^{2t}}$$

37. $\mathbf{r}(t) = 3t\mathbf{i} + 2t^2\mathbf{j},\ P(-3, 2) \Rightarrow t = -1$

$x = 3t,\ x' = 3,\ x'' = 0$

$y = 2t^2,\ y' = 4t,\ y'' = 4$

$$K = \frac{|x'y'' - y'x''|}{\left[(x')^2 + (y')^2\right]^{3/2}} = \frac{|3(4) - 0|}{\left[9 + (4t)^2\right]^{3/2}}$$

$$\text{At } t = -1,\ K = \frac{12}{(9 + 16)^{3/2}} = \frac{12}{125}$$

38. $\mathbf{r}(t) = e^t\mathbf{i} + 4t\mathbf{j},\ P(1, 0) \Rightarrow t = 0$

$x = e^t,\ x' = e^t,\ x'' = e^t$

$y = 4t,\ y' = 4,\ y'' = 0$

$$K = \frac{|x'y'' - y'x''|}{\left[(x')^2 + (y')^2\right]^{3/2}} = \frac{|0 - 4|}{(1 + 16)^{3/2}} = \frac{4}{17^{3/2}}$$

39. $\mathbf{r}(t) = t\mathbf{i} + t^2\mathbf{j} + \frac{t^3}{4}\mathbf{k},\ P(2, 4, 2) \Rightarrow t = 2$

$\mathbf{r}'(t) = \mathbf{i} + 2t\mathbf{j} + \frac{3}{4}t^2\mathbf{k}$

$\mathbf{r}'(2) = \mathbf{i} + 4\mathbf{j} + 3\mathbf{k},\ \|\mathbf{r}'(2)\| = \sqrt{26}$

$\mathbf{r}''(t) = 2\mathbf{j} + \frac{3}{2}t\mathbf{k}$

$\mathbf{r}''(2) = 2\mathbf{j} + 3\mathbf{k}$

$$\mathbf{r}'(2) \times \mathbf{r}'(2) = \begin{vmatrix} \mathbf{i} & \mathbf{j} & \mathbf{k} \\ 1 & 4 & 3 \\ 0 & 2 & 3 \end{vmatrix} = 6\mathbf{i} - 3\mathbf{j} + 2\mathbf{k}$$

$\|\mathbf{r}'(2) \times \mathbf{r}''(2)\| = \sqrt{49} = 7$

$$K = \frac{\|\mathbf{r}' \times \mathbf{r}''\|}{\|\mathbf{r}'\|^3} = \frac{7}{26^{3/2}} = \frac{7\sqrt{26}}{676}$$

40. $\mathbf{r}(t) = e^t \cos t\mathbf{i} + e^t \sin t\mathbf{j} + e^t\mathbf{k},\ P(1, 0, 1) \Rightarrow t = 0$

$\mathbf{r}'(t) = \left(e^t \cos t - e^t \sin t\right)\mathbf{i} + \left(e^t \sin t + e^t \cos t\right)\mathbf{j} + e^t\mathbf{k}$

$\|\mathbf{r}'(t)\| = e^t\sqrt{\left(\cos^2 t - 2\cos t \sin t + \sin^2 t\right) + \left(\sin^2 t + 2\sin t\cos t + \cos^2 t\right) + 1} = \sqrt{3}e^t$

$\mathbf{T}(t) = \frac{1}{\sqrt{3}}\left[(\cos t - \sin t)\mathbf{i} + (\sin t + \cos t)\mathbf{j} + \mathbf{k}\right]$

$\mathbf{T}'(t) = \frac{1}{\sqrt{3}}\left[(-\sin t - \cos t)\mathbf{i} + (\cos t - \sin t)\mathbf{j}\right]$

$\mathbf{r}'(0) = \mathbf{i} + \mathbf{j} + \mathbf{k} \Rightarrow \|\mathbf{r}'(0)\| = \sqrt{3}$

$\mathbf{T}'(0) = \frac{1}{\sqrt{3}}(-\mathbf{i} + \mathbf{j}) \Rightarrow \|\mathbf{T}'(0)\| = \frac{\sqrt{2}}{\sqrt{3}}$

$$K = \frac{\|\mathbf{T}'(0)\|}{\|\mathbf{r}'(0)\|} = \frac{\sqrt{2}/\sqrt{3}}{\sqrt{3}} = \frac{\sqrt{2}}{3}$$

41. $y = 3x - 2,\ x = a$

Because $y'' = 0$, $K = 0$, and the radius of curvature is undefined.

42. $y = 2x + \frac{4}{x},\ x = 1$

$y' = 2 - \frac{4}{x^2},\ y'(1) = -2$

$y'' = \frac{8}{x^3},\ y''(1) = 8$

$$K = \frac{|y''|}{\left[1 + (y')^2\right]^{3/2}} = \frac{8}{(1 + 4)^{3/2}} = \frac{8}{5^{3/2}}$$

$\frac{1}{K} = \frac{5^{3/2}}{8}$ (radius of curvature)

43. $y = 2x^2 + 3,\ x = -1$

$y' = 4x$

$y'' = 4$

$$K = \frac{4}{\left[1 + (-4)^2\right]^{3/2}} = \frac{4}{17^{3/2}} \approx 0.057$$

$\frac{1}{K} = \frac{17^{3/2}}{4} \approx 17.523$ (radius of curvature)

44. $y = \frac{3}{4}\sqrt{16 - x^2}, x = 0$

$y' = \frac{-9x}{16y}$

$y'' = \frac{-\left[9 + (16y')^2\right]}{16y}$

At $x = 0$: $y' = 0$

$y'' = -\frac{3}{16}$

$K = \left|\frac{-3/16}{\left(1 + 0^2\right)^{3/2}}\right| = \frac{3}{16}$

$\frac{1}{K} = \frac{16}{3}$ (radius of curvature)

45. $y = \cos 2x, x = 2\pi$

$y' = -2\sin 2x$

$y'' = -4\cos 2x$

At $x = 2\pi$, $y = 1$, $y' = 0$, $y'' = -4$

$K = \frac{|-4|}{\left[1 + 0^2\right]^{3/2}} = 4$

$\frac{1}{K} = \frac{1}{4}$

46. $y = e^{3x}, x = 0$

$y' = 3e^{3x}, y'' = 9e^{3x}$

At $x = 0$, $y = 1$, $y' = 3$, $y'' = 9$

$K = \frac{9}{\left[1 + 3^2\right]^{3/2}} = \frac{9}{10^{3/2}}$

$\frac{1}{K} = \frac{10\sqrt{10}}{9}$

47. $y = x^3, x = 2$

$y' = 3x^2, y'' = 6x$

At $x = 2$, $y = 8$, $y' = 12$, $y'' = 12$

$K = \frac{12}{\left[1 + (12)^2\right]^{3/2}} = \frac{12}{(145)^{3/2}}$

$\frac{1}{K} = \frac{145\sqrt{145}}{12}$

48. $y = x^n$, $x = 1$, $n \ge 2$

$y' = nx^{n-1}$

$y'' = n(n - 1)x^{n-2}$

At $x = 1$, $y = 1$, $y' = n$, $y'' = n(n - 1)$

$K = \frac{n(n - 1)}{\left[1 + n^2\right]^{3/2}}$

49. $y = (x - 1)^2 + 3$, $y' = 2(x - 1)$, $y'' = 2$

$K = \frac{2}{\left(1 + \left[2(x - 1)\right]^2\right)^{3/2}} = \frac{2}{\left[1 + 4(x - 1)^2\right]^{3/2}}$

(a) K is maximum when $x = 1$ or at the vertex $(1, 3)$.

(b) $\lim_{x\to\infty} K = 0$

50. $y = x^3$, $y' = 3x^2$, $y'' = 6x$

$K = \left|\frac{6x}{\left(1 + 9x^4\right)^{3/2}}\right|$

(a) K is maximum at $\left(\frac{1}{\sqrt[4]{45}}, \frac{1}{\sqrt[4]{45^3}}\right), \left(\frac{-1}{\sqrt[4]{45}}, \frac{-1}{\sqrt[4]{45^3}}\right)$.

(b) $\lim_{x\to\infty} K = 0$

51. $y = x^{2/3}$, $y' = \frac{2}{3}x^{-1/3}$, $y'' = -\frac{2}{9}x^{-4/3}$

$K = \left|\frac{(-2/9)x^{-4/3}}{\left[1 + (4/9)x^{-2/3}\right]^{3/2}}\right| = \left|\frac{6}{x^{1/3}\left(9x^{2/3} + 4\right)^{3/2}}\right|$

(a) $K \to \infty$ as $x \to 0$. No maximum

(b) $\lim_{x\to\infty} K = 0$

52. $y = \frac{1}{x}$, $y' = -\frac{1}{x^2}$, $y'' = \frac{2}{x^3}$. Assume $x > 0$.

$K = \frac{|y''|}{\left[1 + (y')^2\right]^{3/2}} = \frac{\left|2/x^3\right|}{\left(1 + 1/x^4\right)^{3/2}} = \frac{2x^3}{\left(x^4 + 1\right)^{3/2}}$

$\frac{dK}{dx} = \frac{6x^2\left(1 - x^4\right)}{\left(x^4 + 1\right)^{5/2}}$

(a) K has a maximum at $x = 1$ (and $x = -1$ by symmetry).

(b) $\lim_{x\to\infty} K = 0$

53. $y = \ln x,\ y' = \dfrac{1}{x},\ y'' = -\dfrac{1}{x^2}$

$$K = \left|\frac{-1/x^2}{\left[1 + (1/x)^2\right]^{3/2}}\right| = \frac{x}{\left(x^2 + 1\right)^{3/2}}$$

$$\frac{dK}{dx} = \frac{-2x^2 + 1}{\left(x^2 + 1\right)^{5/2}}$$

(a) K has a maximum when $x = \dfrac{1}{\sqrt{2}}$.

(b) $\lim\limits_{x\to\infty} K = 0$

54. $y = e^x,\ y' = y'' = e^x$

$$K = \frac{|y''|}{\left[1 + (y')^2\right]^{3/2}} = \frac{e^x}{\left(1 + e^{2x}\right)^{3/2}}$$

$$\frac{dK}{dx} = \frac{e^x\left(1 - 2e^{2x}\right)}{\left(1 + e^{2x}\right)^{5/2}}$$

(a) $1 - 2e^{2x} = 0 \Rightarrow e^{2x} = \dfrac{1}{2} \Rightarrow x = \dfrac{1}{2}\ln\left(\dfrac{1}{2}\right) = -\dfrac{1}{2}\ln 2$

K has maximum curvature at $x = -\dfrac{1}{2}\ln 2$.

(b) $\lim\limits_{x\to\infty} K = 0$

55. $y = 1 - x^3,\ y' = -3x^2,\ y'' = -6x$

$$K = \frac{|-6x|}{\left[1 + 9x^4\right]^{3/2}}$$

Curvature is 0 at $x = 0$: $(0, 1)$.

56. $y = (x - 1)^3 + 3,\ y' = 3(x - 1)^2,\ y'' = 6(x - 1)$

$$K = \frac{|y''|}{\left[1 + (y')^2\right]^{3/2}} = \frac{|6(x - 1)|}{\left[1 + 9(x - 1)^4\right]^{3/2}} = 0 \text{ at } x = 1.$$

Curvature is 0 at $(1, 3)$.

57. $y = \cos x,\ y' = -\sin x,\ y'' = -\cos x$

$$K = \frac{|y''|}{\left[1 + (y')^2\right]^{3/2}} = \frac{|-\cos x|}{\left(1 + \sin^2 x\right)^{3/2}} = 0 \text{ for}$$

$x = \dfrac{\pi}{2} + K\pi.$

Curvature is 0 at $\left(\dfrac{\pi}{2} + K\pi, 0\right)$.

58. $y = \sin x,\ y' = \cos x,\ y'' = -\sin x$

$$K = \frac{|\sin x|}{\left(1 + \cos^2 x\right)^{3/2}} = 0 \text{ for } x = n\pi.$$

Curvature is 0 for $x = n\pi$: $(n\pi, 0)$

59. $s = \displaystyle\int_a^b \sqrt{x'(t)^2 + y'(t)^2 + z'(t)^2}\,dt = \int_a^b \left\|r'(t)\right\|\,dt$

60. Plane: $K = \left\|\dfrac{d\mathbf{T}}{dS}\right\| = \left\|\mathbf{T}'(s)\right\|$

Space: $K = \dfrac{\left\|\mathbf{T}'(t)\right\|}{\left\|\mathbf{r}'(t)\right\|} = \dfrac{\left\|\mathbf{r}'(t) \times \mathbf{r}''(t)\right\|}{\left\|\mathbf{r}'(t)\right\|^3}$

Answers will vary.

61. The curve is a line.

62. $K = \dfrac{|y''|}{\left[1 + (y')^2\right]^{3/2}}$

At the smooth relative extremum $y' = 0$, so $K = |y''|$.

Yes, for example, $y = x^4$ has a curvature of 0 at its relative minimum $(0, 0)$. The curvature is positive at any other point on the curve.

63. $f(x) = x^4 - x^2$

(a) $K = \dfrac{2|6x^2 - 1|}{|16x^6 - 16x^4 + 4x^2 + 1|^{3/2}}$

(b) For $x = 0$, $K = 2$. $f(0) = 0$. At $(0, 0)$, the circle of curvature has radius $\dfrac{1}{2}$. Using the symmetry of the graph of f, you obtain $x^2 + \left(y + \dfrac{1}{2}\right)^2 = \dfrac{1}{4}$.

For $x = 1$, $K = (2\sqrt{5})/5$. $f(1) = 0$. At $(1, 0)$, the circle of curvature has radius $\dfrac{\sqrt{5}}{2} = \dfrac{1}{K}$.

Using the graph of f, you see that the center of curvature is $\left(0, \dfrac{1}{2}\right)$. So,

$$x^2 + \left(y - \frac{1}{2}\right)^2 = \frac{5}{4}.$$

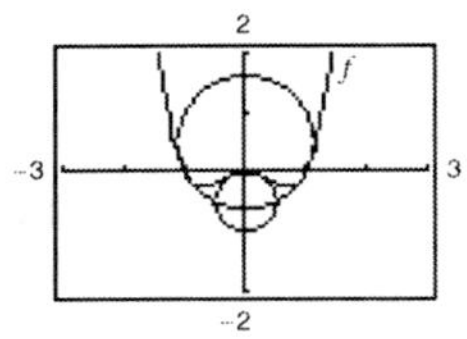

To graph these circles, use

$$y = -\frac{1}{2} \pm \sqrt{\frac{1}{4} - x^2} \text{ and } y = \frac{1}{2} \pm \sqrt{\frac{5}{4} - x^2}.$$

(c) The curvature tends to be greatest near the extrema of f, and K decreases as $x \to \pm\infty$. f and K, however, do not have the same critical numbers.

Critical numbers of f:

$x = 0, \pm\dfrac{\sqrt{2}}{2} \approx \pm 0.7071$

Critical numbers of K:

$x = 0, \pm 0.7647, \pm 0.4082$

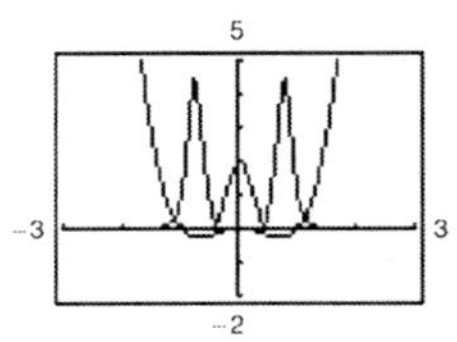

64. $\mathbf{r}(t) = t\mathbf{i} + t^2\mathbf{j}$

(a) $\dfrac{dx}{dt} = 1, \dfrac{dy}{dt} = 2t$

$$s = \int_0^2 \sqrt{1 + 4t^2}\, dt = \frac{1}{2}\int_0^2 \sqrt{1 + 4t^2}(2)\, dt (u = 2t) = \frac{1}{2} \cdot \frac{1}{2}\left[2t\sqrt{1 + 4t^2} + \ln\left|2t + \sqrt{1 + 4t^2}\right|\right]_0^2 \text{ (Theorem 8.2)}$$

$$= \frac{1}{4}\left[4\sqrt{17} + \ln\left|4 + \sqrt{17}\right|\right] \approx 4.647$$

(b) Let $y = x^2$, $y' = 2x$, $y'' = 2$

At $t = 0$, $x = 0$, $y = 0$, $y' = 0$, $y'' = 2$

$$K = \frac{2}{[1 + 0]^{3/2}} = 2$$

At $t = 1$, $x = 1$, $y = 1$, $y' = 2$, $y'' = 2$

$$K = \frac{2}{\left[1 + (2)^2\right]^{3/2}} = \frac{2}{5^{3/2}} \approx 0.179$$

At $t = 2$, $x = 2$, $y = 4$, $y' = 4$ $y'' = 2$

$$K = \frac{2}{[1 + 16]^{3/2}} = \frac{2}{17^{3/2}} \approx 0.0285$$

(c) As t changes from 0 to 2, the curvature decreases.

65. $y_1 = ax(b - x),\ y_2 = \dfrac{x}{x + 2}$

You observe that $(0, 0)$ is a solution point to both equations. So, the point P is origin.

$$y_1 = ax(b - x),\ y_1' = a(b - 2x),\ y_1'' = -2a$$

$$y_2 = \frac{x}{x + 2},\ y_2' = \frac{2}{(x + 2)^2},\ y_2'' = \frac{-4}{(x + 2)^3}$$

At P, $y_1'(0) = ab$ and $y_2'(0) = \dfrac{2}{(0 + 2)^2} = \dfrac{1}{2}$.

Because the curves have a common tangent at P, $y_1'(0) = y_2'(0)$ or $ab = \dfrac{1}{2}$. So, $y_1'(0) = \dfrac{1}{2}$.

Because the curves have the same curvature at P, $K_1(0) = K_2(0)$.

$$K_1(0) = \left|\frac{y_1''(0)}{\left[1 + (y_1'(0))^2\right]^{3/2}}\right| = \left|\frac{-2a}{\left[1 + (1/2)^2\right]^{3/2}}\right|$$

$$K_2(0) = \left|\frac{y_2''(0)}{\left[1 + (y_2'(0))^2\right]^{3/2}}\right| = \left|\frac{-1/2}{\left[1 + (1/2)^2\right]^{3/2}}\right|$$

So, $2a = \pm\dfrac{1}{2}$ or $a = \pm\dfrac{1}{4}$. In order that the curves intersect at only one point, the parabola must be concave downward. So,

$$a = \frac{1}{4} \text{ and } b = \frac{1}{2a} = 2.$$

$$y_1 = \frac{1}{4}x(2 - x) \text{ and } y_2 = \frac{x}{x + 2}$$

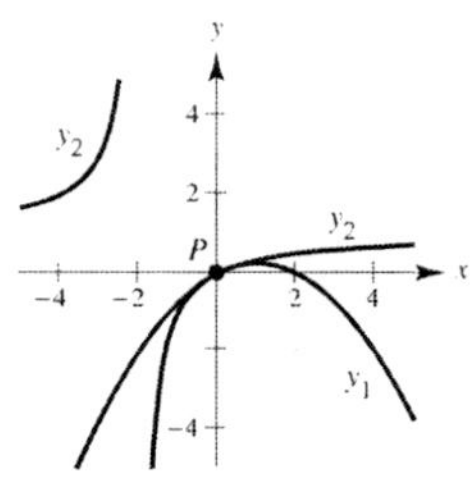

66. From the shape of the ellipse, you see that the curvature is greatest at the endpoints of the major axis, $(\pm 2, 0)$, and least at the endpoints of the minor axis, $(0, \pm 1)$.

67. (a) Imagine dropping the circle $x^2 + (y - k)^2 = 16$ into the parabola $y = x^2$. The circle will drop to the point where the tangents to the circle and parabola are equal.

$$y = x^2 \text{ and } x^2 + (y - k)^2 = 16 \Rightarrow x^2 + (x^2 - k)^2 = 16$$

Taking derivatives, $2x + 2(y - k)y' = 0$ and $y' = 2x$. So,

$$(y - k)y' = -x \Rightarrow y' = \frac{-x}{y - k}.$$

$$\text{So, } \frac{-x}{y - k} = 2x \Rightarrow -x = 2x(y - k) \Rightarrow -1 = 2(x^2 - k) \Rightarrow x^2 - k = -\frac{1}{2}.$$

$$\text{So, } x^2 + (x^2 - k)^2 = x^2 + \left(-\frac{1}{2}\right)^2 = 16 \Rightarrow x^2 = 15.75.$$

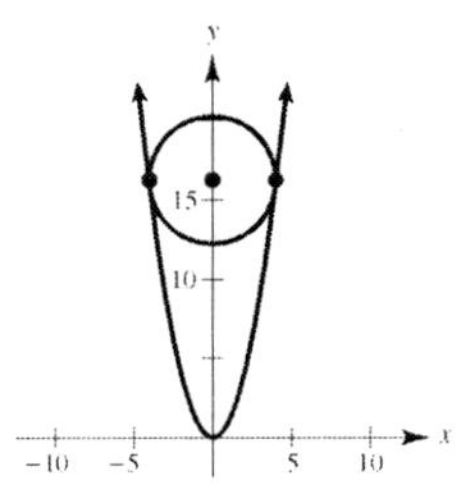

Finally, $k = x^2 + \dfrac{1}{2} = 16.25$, and the center of the circle is 16.25 units from the vertex of the parabola. Because the radius of the circle is 4, the circle is 12.25 units from the vertex.

(b) In 2-space, the parabola $z = y^2$ (or $z = x^2$) has a curvature of $K = 2$ at $(0, 0)$. The radius of the largest sphere that will touch the vertex has radius $= 1/K = \dfrac{1}{2}$.

68. $s = \dfrac{c}{\sqrt{K}}$

$$y = \frac{1}{3}x^3$$

$$y' = x^2$$

$$y'' = 2x$$

$$K = \left|\frac{2x}{\left(1 + x^4\right)^{3/2}}\right|$$

When $x = 1$: $K = \dfrac{1}{\sqrt{2}}$

$$s = \frac{c}{\sqrt{1/\sqrt{2}}} = \sqrt[4]{2}c$$

$$30 = \sqrt[4]{2}c \Rightarrow c = \frac{30}{\sqrt[4]{2}}$$

At $x = \dfrac{3}{2}$, $K = \dfrac{3}{\left[1 + (81/16)\right]^{3/2}} \approx 0.201$

$$s = \left(\frac{3}{2}\right) = \frac{c}{\sqrt{K}} = \frac{30/\sqrt[4]{2}}{\sqrt{K}} \approx 56.27 \text{ mi/h}$$

69. $P(x_0, y_0)$ point on curve $y = f(x)$. Let (α, β) be the center of curvature. The radius of curvature is $\dfrac{1}{K}$.

$y' = f'(x)$. Slope of normal line at (x_0, y_0) is $\dfrac{-1}{f'(x_0)}$.

Equation of normal line: $y - y_0 = \dfrac{-1}{f'(x_0)}(x - x_0)$

(α, β) is on the normal line: $-f'(x_0)(\beta - y_0) = \alpha - x_0$ Equation 1

(x_0, y_0) lies on the circle: $(x_0 - \alpha)^2 + (y_0 - \beta)^2 = \left(\dfrac{1}{K}\right)^2 = \left[\dfrac{\left(1 + f'(x_0)^2\right)^{3/2}}{\left|f''(x_0)\right|}\right]^2$ Equation 2

Substituting Equation 1 into Equation 2:

$$\left[f'(x_0)(\beta - y_0)\right]^2 + (y_0 - \beta)^2 = \left(\frac{1}{K}\right)^2$$

$$(\beta - y_0)^2 + \left[1 + f'(x_0)^2\right] = \frac{\left(1 + f'(x_0)^2\right)^3}{\left(f''(x_0)\right)^2}$$

$$(\beta - y_0)^2 = \frac{\left[1 + f'(x_0)^2\right]^2}{f''(x_0)^2}$$

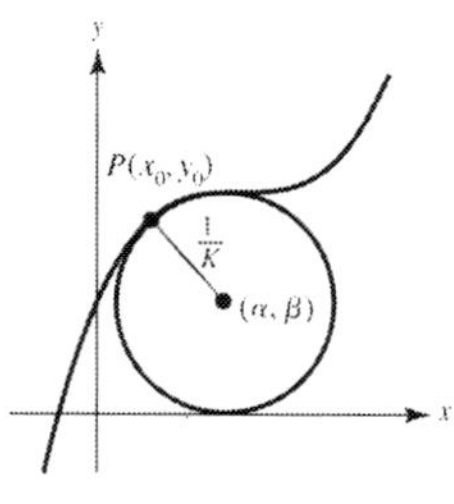

When $f''(x_0) > 0$, $\beta - y_0 > 0$, and if $f''(x_0) < 0$, then $\beta - y_0 < 0$.

So $\beta - y_0 = \dfrac{1 + f'(x_0)^2}{f''(x_0)}$

$$\beta = y_0 + \frac{1 + f'(x_0)^2}{f''(x_0)} = y_0 + z$$

Similarly, $\alpha = x_0 - f'(x_0)z$.

70. (a) $y = f(x) = e^x,\ f'(x) = f''(x) = e^x, (0, 1)$

$$z = \frac{1 + f'(0)^2}{f''(0)} = 2$$

$$(\alpha, \beta) = (0 - 2, 1 + 2) = (-2, 3)$$

(b) $y = \dfrac{x^2}{2}, y' = x, y'' = 1, \left(1, \dfrac{1}{2}\right)$

$$z = \frac{1 + f'(1)^2}{f''(1)} = 2$$

$$(\alpha, \beta) = \left(1 - 2, \frac{1}{2} + 2\right) = \left(-1, \frac{5}{2}\right)$$

(c) $y = x^2, y' = 2x, y'' = 2, (0, 0)$

$$z = \frac{1 + f'(0)^2}{f''(0)} = \frac{1}{2}$$

$$(\alpha, \beta) = \left(0, 0 + \frac{1}{2}\right) = \left(0, \frac{1}{2}\right)$$

71. $r(\theta) = r\cos\theta\mathbf{i} + r\sin\theta\mathbf{j} = f(\theta)\cos\theta\mathbf{i} + f(\theta)\sin\theta\mathbf{j}$

$$x(\theta) = f(\theta)\cos\theta$$

$$y(\theta) = f(\theta)\sin\theta$$

$$x'(\theta) = -f(\theta)\sin\theta + f'(\theta)\cos\theta$$

$$y'(\theta) = f(\theta)\cos\theta + f'(\theta)\sin\theta$$

$$x''(\theta) = -f(\theta)\cos\theta - f'(\theta)\sin\theta - f'(\theta)\sin\theta + f''(\theta)\cos\theta = -f(\theta)\cos\theta - 2f'(\theta)\sin\theta + f''(\theta)\cos\theta$$

$$y''(\theta) = -f(\theta)\sin\theta + f'(\theta)\cos\theta + f'(\theta)\cos\theta + f''(\theta)\sin\theta = -f(\theta)\sin\theta + 2f'(\theta)\cos\theta + f''(\theta)\sin\theta$$

$$K = \frac{|x'y'' - y'x''|}{\left[(x')^2 + (y')^2\right]^{3/2}} = \frac{\left|f^2(\theta) - f(\theta)f''(\theta) + 2\left(f'(\theta)\right)^2\right|}{\left[f^2(\theta) + \left(f'(\theta)\right)^2\right]^{3/2}} = \frac{\left|r^2 - rr'' + 2(r')^2\right|}{\left[r^2 + (r')^2\right]^{3/2}}$$

72. (a) $r = 1 + \sin\theta$

$$r' = \cos\theta$$

$$r'' = -\sin\theta$$

$$K = \frac{\left|2(r')^2 - rr'' + r^2\right|}{\left[(r')^2 + r^2\right]^{3/2}} = \frac{\left|2\cos^2\theta - (1 + \sin\theta)(-\sin\theta) + (1 + \sin\theta)^2\right|}{\sqrt{\left[\cos^2\theta + (1 + \sin\theta)^2\right]^3}} = \frac{3(1 + \sin\theta)}{\sqrt{8(1 + \sin\theta)^3}} = \frac{3}{2\sqrt{2(1 + \sin\theta)}}$$

(b) $r = \theta$

$$r' = 1$$

$$r'' = 0$$

$$K = \frac{\left|2(r')^2 - rr'' + r^2\right|}{\left[(r')^2 + r^2\right]^{3/2}} = \frac{2 + \theta^2}{\left(1 + \theta^2\right)^{3/2}}$$

(c) $r = a\sin\theta$

$r' = a\cos\theta$

$r = -a\sin\theta$

$$K = \frac{\left|2(r')^2 - rr'' + r^2\right|}{\left[(r')^2 + r^2\right]^{3/2}} = \frac{\left|2a^2\cos^2\theta + a^2\sin^2\theta + a^2\sin^2\theta\right|}{\sqrt{\left[a^2\cos^2\theta + a^2\sin^2\theta\right]^3}} = \frac{2a^2}{a^3} = \frac{2}{a}, a > 0$$

(d) $r = e^\theta$

$r' = e^\theta$

$r'' = e^\theta$

$$K = \frac{\left|2(r')^2 - rr'' + r^2\right|}{\left[(r')^2 + r^2\right]^{3/2}} = \frac{2e^{2\theta}}{\left(2e^{2\theta}\right)^{3/2}} = \frac{1}{\sqrt{2}e^\theta}$$

73. $r = e^{a\theta}, a > 0$

$r' = ae^{a\theta}$

$r'' = a^2e^{a\theta}$

$$K = \frac{\left|2(r')^2 - rr'' + r^2\right|}{\left[(r')^2 + r^2\right]^{3/2}} = \frac{\left|2a^2e^{2a\theta} - a^2e^{2a\theta} + e^{2a\theta}\right|}{\left[a^2e^{2a\theta} + e^{2a\theta}\right]^{3/2}} = \frac{1}{e^{a\theta}\sqrt{a^2+1}}$$

(a) As $\theta \to \infty$, $K \to 0$.

(b) As $a \to \infty$, $K \to 0$.

74. At the pole, $r = 0$.

$$K = \frac{\left|2(r')^2 - rr'' + r^2\right|}{\left[(r')^2 + r^2\right]^{3/2}} = \frac{\left|2(r')^2\right|}{|r'|^3} = \frac{2}{|r'|}$$

75. $r = 4\sin 2\theta$

$r' = 8\cos 2\theta$

At the pole: $K = \dfrac{2}{|r'(0)|} = \dfrac{2}{8} = \dfrac{1}{4}$

76. $r = 6\cos 3\theta$

$r' = -18\sin 3\theta$

At the pole,

$\theta = \dfrac{\pi}{6}, r'\left(\dfrac{\pi}{6}\right) = -18,$

and

$$K = \frac{2}{|r'(\pi/6)|} = \frac{2}{|-18|} = \frac{1}{9}.$$

77. $x = f(t), y = g(t)$

$$y' = \frac{dy}{dx} = \frac{\dfrac{dy}{dt}}{\dfrac{dx}{dt}} = \frac{g'(t)}{f'(t)}$$

$$y'' = \frac{\dfrac{d}{dt}\left[\dfrac{g'(t)}{f'(t)}\right]}{\dfrac{dx}{dt}} = \frac{\dfrac{f'(t)g''(t) - g'(t)f''(t)}{\left[f'(t)\right]^2}}{f'(t)} = \frac{f'(t)g''(t) - g'(t)f''(t)}{\left[f'(t)\right]^3}$$

$$K = \frac{|y''|}{\left[1 + (y')^2\right]^{3/2}} = \frac{\left|\dfrac{f'(t)g''(t) - g'(t)f''(t)}{\left[f'(t)\right]^3}\right|}{\left[1 + \left(\dfrac{g'(t)}{f'(t)}\right)^2\right]^{3/2}}$$

$$= \frac{\left|\dfrac{f'(t)g''(t) - g'(t)f''(t)}{\left[f'(t)\right]^3}\right|}{\sqrt{\left\{\dfrac{\left[f'(t)\right]^2 + \left[g'(t)\right]^2}{\left[f'(t)\right]^2}\right\}^3}} = \frac{\left|f'(t)g''(t) - g'(t)f''(t)\right|}{\left(\left[f'(t)\right]^2 + \left[g'(t)\right]^2\right)^{3/2}}$$

78. $x(t) = t^3,\ x'(t) = 3t^2,\ x''(t) = 6t$

$y(t) = \frac{1}{2}t^2,\ y'(t) = t,\ y''(t) = 1$

$$K = \frac{\left|(3t^2)(1) - (t)(6t)\right|}{\left[(3t^2)^2 + (t)^2\right]^{3/2}} = \frac{3t^2}{|t^3|(9t^2+1)^{3/2}} = \frac{3}{|t|(9t^2+1)^{3/2}}$$

$K \to 0$ as $t \to \pm\infty$

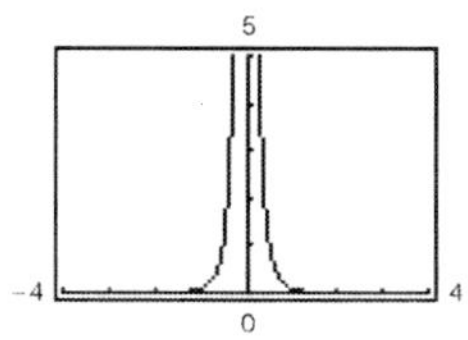

79. $x(\theta) = a(\theta - \sin\theta)$ $\quad y(\theta) = a(1 - \cos\theta)$

$x'(\theta) = a(1 - \cos\theta)$ $\quad y'(\theta) = a\sin\theta$

$x''(\theta) = a\sin\theta$ $\quad y''(\theta) = a\cos\theta$

$$K = \frac{\left|x'(\theta)y''(\theta) - y'(\theta)x''(\theta)\right|}{\left[x'(\theta)^2 + y'(\theta)^2\right]^{3/2}}$$

$$= \frac{\left|a^2(1 - \cos\theta)\cos\theta - a^2\sin^2\theta\right|}{\left[a^2(1 - \cos\theta)^2 + a^2\sin^2\theta\right]^{3/2}}$$

$$= \frac{1}{a}\frac{|\cos\theta - 1|}{[2 - 2\cos\theta]^{3/2}}$$

$$= \frac{1}{a}\frac{1 - \cos\theta}{2\sqrt{2}[1 - \cos\theta]^{3/2}} \quad (1 - \cos \ge 0)$$

$$= \frac{1}{2a\sqrt{2 - 2\cos\theta}} = \frac{1}{4a}\csc\left(\frac{\theta}{2}\right)$$

Minimum: $\frac{1}{4a}$ $\quad (\theta = \pi)$

Maximum: none $\left(K \to \infty \text{ as } \theta \to 0\right)$

80. (a) $\mathbf{r}(t) = 3t^2\mathbf{i} + (3t - t^3)\mathbf{j}$

$\mathbf{v}(t) = 6t\mathbf{i} + (3 - 3t^2)\mathbf{j}$

$\frac{ds}{dt} = \|\mathbf{v}(t)\| = 3(1 + t^2),\ \frac{d^2s}{dt^2} = 6t$

$$K = \frac{2}{3(1 + t^2)^2}$$

$$a_{\mathbf{T}} = \frac{d^2s}{dt^2} = 6t$$

$$a_{\mathbf{N}} = K\left(\frac{ds}{dt}\right)^2 = \frac{2}{3(1+t^2)^2}\cdot 9(1+t^2)^2 = 6$$

(b) $\mathbf{r}(t) = t\mathbf{i} + t^2\mathbf{j} + \frac{1}{2}t^2\mathbf{k}$

$\mathbf{v}(t) = \mathbf{i} + 2t\mathbf{j} + t\mathbf{k}$

$$\frac{ds}{dt} = \|\mathbf{v}(t)\| = \sqrt{5t^2 + 1}$$

$$\frac{d^2s}{dt^2} = \frac{5t}{\sqrt{5t^2 + 1}}$$

$\mathbf{a}(t) = 2\mathbf{j} + \mathbf{k}$

$$\mathbf{r}'(t) \times \mathbf{r}''(t) = \mathbf{v}(t) \times \mathbf{a}(t) = \begin{vmatrix} \mathbf{i} & \mathbf{j} & \mathbf{k} \\ 1 & 2t & t \\ 0 & 2 & 1 \end{vmatrix} = -\mathbf{j} + 2\mathbf{k}$$

$$K = \frac{\|\mathbf{r}'(t) \times \mathbf{r}''(t)\|}{\|\mathbf{r}'(t)\|^3} = \frac{\sqrt{5}}{(5t^2 + 1)^{3/2}}$$

$$a_{\mathbf{T}} = \frac{d^2s}{dt^2} = \frac{5t}{\sqrt{5t^2 + 1}}$$

$$a_{\mathbf{N}} = K\left(\frac{ds}{dt}\right)^2 = \frac{\sqrt{5}}{(5t^2+1)^{3/2}}(5t^2 + 1) = \frac{\sqrt{5}}{\sqrt{5t^2 + 1}}$$

81. $F = ma_{\mathbf{N}} = mK\left(\frac{ds}{dt}\right)^2$

$$= \left(\frac{5500 \text{ lb}}{32 \text{ ft/sec}^2}\right)\left(\frac{1}{100 \text{ ft}}\right)\left(\frac{30(5280) \text{ ft}}{3600 \text{ sec}}\right)^2 = 3327.5 \text{ lb}$$

82. $F = ma_{\mathbf{N}} = mK\left(\dfrac{ds}{dt}\right)^2$

$$= \left(\frac{6400 \text{ lb}}{32 \text{ ft/sec}^2}\right)\left(\frac{1}{250 \text{ ft}}\right)\left(\frac{35(5280) \text{ ft}}{3600 \text{ sec}}\right)^2$$

$$= \frac{94864}{45} \approx 2108.1 \text{ lb}$$

83. $y = \cosh x = \dfrac{e^x + e^{-x}}{2}$

$$y' = \frac{e^x - e^{-x}}{2} = \sinh x$$

$$y'' = \frac{e^x + e^{-x}}{2} = \cosh x$$

$$K = \frac{|\cosh x|}{\left[1 + (\sinh x)^2\right]^{3/2}} = \frac{\cosh x}{\left(\cosh^2 x\right)^{3/2}} = \frac{1}{\cosh^2 x} = \frac{1}{y^2}$$

84. (a) $K = \|\mathbf{T}'(s)\| = \left\|\dfrac{d\mathbf{T}}{ds}\right\| = \left\|\dfrac{d\mathbf{T}}{dt} \cdot \dfrac{dt}{ds}\right\|$, by the Chain Rule

$$= \left\|\frac{d\mathbf{T}/dt}{ds/dt}\right\| = \frac{\|\mathbf{T}'(t)\|}{\|\mathbf{v}(t)\|} = \frac{\|\mathbf{T}'(t)\|}{\|\mathbf{r}'(t)\|}$$

(b) $\mathbf{T}(t) = \dfrac{\mathbf{r}'(t)}{\|\mathbf{r}'(t)\|} = \dfrac{\mathbf{r}'(t)}{ds/dt}$

$$\mathbf{r}'(t) = \frac{ds}{dt}\mathbf{T}(t)$$

$$\mathbf{r}''(t) = \left(\frac{d^2s}{dt^2}\right)\mathbf{T}(t) + \frac{ds}{dt}\mathbf{T}'(t)$$

$$\mathbf{r}'(t) \times \mathbf{r}''(t) = \left(\frac{ds}{dt}\right)\left(\frac{d^2s}{dt^2}\right)\left[\mathbf{T}(t) \times \mathbf{T}(t)\right] + \left(\frac{ds}{dt}\right)^2\left[\mathbf{T}(t) \times \mathbf{T}'(t)\right]$$

Because $\mathbf{T}(t) \times \mathbf{T}(t) = \mathbf{0}$ and $\dfrac{ds}{dt} = \|\mathbf{r}'(t)\|$, you have:

$$\mathbf{r}'(t) \times \mathbf{r}''(t) = \|\mathbf{r}'(t)\|^2\left[\mathbf{T}(t) \times \mathbf{T}'(t)\right]$$

$$\|\mathbf{r}'(t) \times \mathbf{r}''(t)\| = \|\mathbf{r}'(t)\|^2\|\mathbf{T}(t) \times \mathbf{T}'(t)\| = \|\mathbf{r}'(t)\|^2\|\mathbf{T}(t)\|\|\mathbf{T}'(t)\| = \|\mathbf{r}'(t)\|^2(1)K\|\mathbf{r}'(t)\| \text{ from } (a)$$

So, $\dfrac{\|\mathbf{r}'(t) \times \mathbf{r}''(t)\|}{\|\mathbf{r}'(t)\|^3} = K.$

(c) $K = \dfrac{\|\mathbf{r}'(t) \times \mathbf{r}''(t)\|}{\left\|\mathbf{r}'(t)^3\right\|} = \dfrac{\dfrac{\|\mathbf{r}'(t) \times \mathbf{r}''(t)\|}{\|\mathbf{r}'(t)\|}}{\|\mathbf{r}'(t)\|^2} = \dfrac{\dfrac{\|\mathbf{v}(t) \times \mathbf{a}(t)\|}{\|\mathbf{v}(t)\|}}{\|\mathbf{r}'(t)\|^2} = \dfrac{\mathbf{a}(t) \cdot \mathbf{N}(t)}{\|\mathbf{r}'(t)\|^2}$

85. False

86. False

Curvature $= \dfrac{1}{\text{radius}}$

87. True

88. True

$$a_{\mathbf{N}} = K\left(\frac{ds}{dt}\right)^2$$

89. Let $\mathbf{r} = x(t)\mathbf{i} + y(t)\mathbf{j} + z(t)\mathbf{k}$. Then $r = \|\mathbf{r}\| = \sqrt{\left[x(t)\right]^2 + \left[y(t)\right]^2 + \left[z(t)\right]^2}$ and $r' = x'(t)\mathbf{i} + y'(t)\mathbf{j} + z'(t)\mathbf{k}$. Then,

$$r\left(\frac{dr}{dt}\right) = \sqrt{\left[x(t)\right]^2 + \left[y(t)\right]^2 + \left[z(t)\right]^2}\left[\frac{1}{2}\left\{\left[x(t)\right]^2 + \left[y(t)\right]^2 + \left[z(t)\right]^2\right\}^{-1/2} \cdot \left(2x(t)x'(t) + 2y(t)y'(t) + 2z(t)z'(t)\right)\right]$$

$$= x(t)x'(t) + y(t)y'(t) + z(t)z'(t) = \mathbf{r} \cdot \mathbf{r}'.$$

90. $\mathbf{F} = m\mathbf{a} \Rightarrow m\mathbf{a} = \dfrac{-GmM}{r^3}\mathbf{r}$

$$\mathbf{a} = -\frac{GM}{r^3}\mathbf{r}$$

Because $\mathbf{r}$ is a constant multiple of $\mathbf{a}$, they are parallel. Because $\mathbf{a} = \mathbf{r}''$ is parallel to $\mathbf{r}$, $\mathbf{r} \times \mathbf{r}'' = \mathbf{0}$. Also,

$\left(\dfrac{d}{dt}\right)(\mathbf{r} \times \mathbf{r}') = \mathbf{r}' \times \mathbf{r}' + \mathbf{r} \times \mathbf{r}'' = \mathbf{0} + \mathbf{0} = \mathbf{0}$. So, $\mathbf{r} \times \mathbf{r}'$ is a constant vector which we will denote by $\mathbf{L}$.

91. Let $\mathbf{r} = x\mathbf{i} + y\mathbf{j} + z\mathbf{k}$ where x, y, and z are function of t, and $r = \|\mathbf{r}\|$.

$$\begin{aligned}
\frac{d}{dt}\left[\frac{\mathbf{r}}{r}\right] &= \frac{r\mathbf{r}' - \mathbf{r}(dr/dt)}{r^2} = \frac{r\mathbf{r}' - \mathbf{r}\left[(\mathbf{r} \cdot \mathbf{r}')/r\right]}{r^2} \\
&= \frac{r^2\mathbf{r}' - (\mathbf{r} \cdot \mathbf{r}')\mathbf{r}}{r^3} \text{ (using Exercise 105)} \\
&= \frac{(x^2 + y^2 + z^2)(x'\mathbf{i} + y'\mathbf{j} + z'\mathbf{k}) - (xx' + yy' + zz')(x\mathbf{i} + y\mathbf{j} + z\mathbf{k})}{r^3} \\
&= \frac{1}{r^3}\left[(x'y^2 + x'z^2 - xyy' - xzz')\mathbf{i} + (x^2y' + z^2y' - xx'y - zz'y)\mathbf{j} + (x^2z' + y^2z' - xx'z - yy'z)\mathbf{k}\right] \\
&= \frac{1}{r^3}\begin{vmatrix} \mathbf{i} & \mathbf{j} & \mathbf{k} \\ yz' - y'z & -(xz' - x'z) & xy' - x'y \\ x & y & z \end{vmatrix} = \frac{1}{r^3}\{[\mathbf{r} \times \mathbf{r}'] \times \mathbf{r}\}
\end{aligned}$$

92.
$$\begin{aligned}
\frac{d}{dt}\left[\frac{\mathbf{r}'}{GM} \times \mathbf{L} - \frac{\mathbf{r}}{r}\right] &= \frac{1}{GM}[\mathbf{r}' \times \mathbf{0} + \mathbf{r}'' \times \mathbf{L}] - \frac{1}{r^3}\{[\mathbf{r} \times \mathbf{r}'] \times \mathbf{r}\} \\
&= \frac{1}{GM}\left[\mathbf{0} + \left(\frac{-GM\mathbf{r}}{r^3}\right) \times [\mathbf{r} \times \mathbf{r}']\right] - \frac{1}{r^3}\{[\mathbf{r} \times \mathbf{r}'] \times \mathbf{r}\} \\
&= -\frac{\mathbf{r}}{r^3} \times [\mathbf{r} \times \mathbf{r}'] - \frac{1}{r^3}\{[\mathbf{r} \times \mathbf{r}'] \times \mathbf{r}\} \\
&= \frac{1}{r^3}\{[\mathbf{r} \times \mathbf{r}'] \times \mathbf{r} - [\mathbf{r} \times \mathbf{r}'] \times \mathbf{r}\} = \mathbf{0}
\end{aligned}$$

So, $\left(\dfrac{\mathbf{r}'}{GM}\right) \times \mathbf{L} - \left(\dfrac{\mathbf{r}}{r}\right)$ is a constant vector which we will denote by $\mathbf{e}$.

93. From Exercise 90, you have concluded that planetary motion is planar. Assume that the planet moves in the xy-plane with the sum at the origin. From Exercise 92, you have

$$\mathbf{r}' \times \mathbf{L} = GM\left(\frac{\mathbf{r}}{r} + \mathbf{e}\right).$$

Because $\mathbf{r}' \times \mathbf{L}$ and $\mathbf{r}$ are both perpendicular to $\mathbf{L}$, so is $\mathbf{e}$.

So, $\mathbf{e}$ lies in the xy-plane. Situate the coordinate system so that $\mathbf{e}$ lies along the positive x-axis and θ is the angle between $\mathbf{e}$ and $\mathbf{r}$. Let $e = \|\mathbf{e}\|$. Then $\mathbf{r} \cdot \mathbf{e} = \|\mathbf{r}\|\|\mathbf{e}\|\cos\theta = re\cos\theta$. Also,

$$\begin{aligned}
\|\mathbf{L}\|^2 &= \mathbf{L} \cdot \mathbf{L} \\
&= (\mathbf{r} \times \mathbf{r}') \cdot \mathbf{L} \\
&= \mathbf{r} \cdot (\mathbf{r}' \times \mathbf{L}) \\
&= \mathbf{r} \cdot \left[GM\left(\mathbf{e} + \frac{\mathbf{r}}{r}\right)\right] \\
&= GM\left[\mathbf{r} \cdot \mathbf{e} + \frac{\mathbf{r} \cdot \mathbf{r}}{r}\right] \\
&= GM[re\cos\theta + r].
\end{aligned}$$

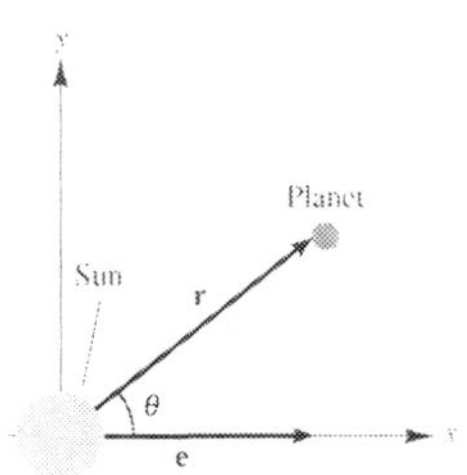

So, $\dfrac{\|\mathbf{L}\|^2/GM}{1 + e\cos\theta} = r$

and the planetary motion is a conic section. Because the planet returns to its initial position periodically, the conic is an ellipse.

94. $\|\mathbf{L}\| = \|\mathbf{r} \times \mathbf{r}'\|$

Let: $\mathbf{r} = r(\cos\theta\mathbf{i} + \sin\theta\mathbf{j})$

$$\mathbf{r}' = r(-\sin\theta\mathbf{i} + \cos\theta\mathbf{j})\frac{d\theta}{dt}\left(\frac{d\mathbf{r}}{dt} = \frac{d\mathbf{r}}{d\theta}\cdot\frac{d\theta}{dt}\right)$$

Then: $\mathbf{r} \times \mathbf{r}' = \begin{vmatrix} \mathbf{i} & \mathbf{j} & \mathbf{k} \\ r\cos\theta & r\sin\theta & 0 \\ -r\sin\theta\frac{d\theta}{dt} & r\cos\theta\frac{d\theta}{dt} & 0 \end{vmatrix} = r^2\frac{d\theta}{dt}\mathbf{k}$ and $\|\mathbf{L}\| = \|\mathbf{r} \times \mathbf{r}'\| = r^2\frac{d\theta}{dt}$.

95. $A = \frac{1}{2}\int_{\alpha}^{\beta} r^2\, d\theta$

So,

$$\frac{dA}{dt} = \frac{dA}{d\theta}\frac{d\theta}{dt} = \frac{1}{2}r^2\frac{d\theta}{dt} = \frac{1}{2}\|\mathbf{L}\|$$

and $\mathbf{r}$ sweeps out area at a constant rate.

96. Let P denote the period. Then

$$A = \int_0^P \frac{dA}{dt}\, dt = \frac{1}{2}\|\mathbf{L}\|P.$$

Also, the area of an ellipse is πab where $2a$ and $2b$ are the lengths of the major and minor axes.

$$\pi ab = \frac{1}{2}\|\mathbf{L}\|P$$

$$P = \frac{2\pi ab}{\|\mathbf{L}\|}$$

$$P^2 = \frac{4\pi^2a^2}{\|\mathbf{L}\|^2}(a^2 - c^2) = \frac{4\pi^2a^2}{\|\mathbf{L}\|^2}a^2(1 - e^2) = \frac{4\pi^2a^4}{\|\mathbf{L}\|^2}\left(\frac{ed}{a}\right) = \frac{4\pi^2 ed}{\|\mathbf{L}\|^2}a^3 = \frac{4\pi^2(\|\mathbf{L}\|^2/GM)}{\|\mathbf{L}\|^2}a^3 = \frac{4\pi^2}{GM}a^3 = Ka^3$$

Review Exercises for Chapter 12

1. $\mathbf{r}(t) = \tan t\mathbf{i} + \mathbf{j} + t\mathbf{k}$

(a) Domain: $t \neq \frac{\pi}{2} + n\pi$, n an integer

(b) Continuous for all $t \neq \frac{\pi}{2} + n\pi$, n an integer

2. $\mathbf{r}(t) = \sqrt{t}\mathbf{i} + \frac{1}{t-4}\mathbf{j} + \mathbf{k}$

(a) Domain: $[0, 4)$ and $(4, \infty)$

(b) Continuous except at $t = 4$

3. $\mathbf{r}(t) = \ln t\mathbf{i} + t\mathbf{j} + t\mathbf{k}$

(a) Domain: $(0, \infty)$

(b) Continuous for all $t > 0$

4. $\mathbf{r}(t) = (2t + 1)\mathbf{i} + t^2\mathbf{j} + t\mathbf{k}$

(a) Domain: $(-\infty, \infty)$

(b) Continuous for all t

5. $\mathbf{r}(t) = (2t + 1)\mathbf{i} + t^2\mathbf{j} - \sqrt{t+2}\mathbf{k}$

(a) $\mathbf{r}(0) = \mathbf{i} - \sqrt{2}\mathbf{k}$

(b) $\mathbf{r}(-2) = -3\mathbf{i} + 4\mathbf{j}$

(c) $\mathbf{r}(c - 1) = (2c - 1)\mathbf{i} + (c - 1)^2\mathbf{j} - \sqrt{c+1}\mathbf{k}$

(d) $\mathbf{r}(1 + \Delta t) - \mathbf{r}(1) = \left[2(1 + \Delta t) + 1\right]\mathbf{i} + (1 + \Delta t)^2\mathbf{j} - \sqrt{3 + \Delta t}\mathbf{k} - \left(3\mathbf{i} + \mathbf{j} - \sqrt{3}\mathbf{k}\right)$

$= 2\Delta t\,\mathbf{i} + \left(\Delta t^2 + 2\Delta t\right)\mathbf{j} - \left(\sqrt{3 + \Delta t} - \sqrt{3}\right)\mathbf{k}$

6. (a) $\mathbf{r}(0) = 3\mathbf{i} + \mathbf{j}$

(b) $\mathbf{r}\left(\frac{\pi}{2}\right) = -\frac{\pi}{2}\mathbf{k}$

(c) $\mathbf{r}(s - \pi) = 3\cos(s - \pi)\mathbf{i} + (1 - \sin(s - \pi))\mathbf{j} - (s - \pi)\mathbf{k}$

(d) $\mathbf{r}(\pi + \Delta t) - \mathbf{r}(\pi) = (3\cos(\pi + \Delta t)\mathbf{i} + (1 - \sin(\pi + \Delta t))\mathbf{j} - (\pi + \Delta t)\mathbf{k}) - (-3\mathbf{i} + \mathbf{j} - \pi\mathbf{k})$

$= (-3\cos\Delta t + 3)\mathbf{i} + \sin\Delta t - \Delta t\mathbf{k}$

7. $P(3, 0, 5), Q(2, -2, 3)$

$\mathbf{v} = \overrightarrow{PQ} = \langle -1, -2, -2 \rangle$

$\mathbf{r}(t) = (3 - t)\mathbf{i} - 2t\mathbf{j} + (5 - 2t)\mathbf{k},\ 0 \le t \le 1$

$x = 3 - t,\ y = -2t,\ z = 5 - 2t,\ 0 \le t \le 1$

(Answers may vary)

8. $P(-2, -3, 8), Q(5, 1, -2)$

$\mathbf{v} = \overrightarrow{PQ} = \langle 7, 4, -10 \rangle$

$\mathbf{r}(t) = (-2 + 7t)\mathbf{i} + (-3 + 4t)\mathbf{j} + (8 - 10t)\mathbf{k},\ 0 \le t \le 1$

$x = -2 + 7t,\ y = -3 + 4t,\ z = 8 - 10t,\ 0 \le t \le 1$

(Answers may vary)

9. $\mathbf{r}(t) = \langle \pi\cos t, \pi\sin t \rangle$

$x = \pi\cos t,\ y = \pi\sin t$

$x^2 + y^2 = \pi^2$, circle

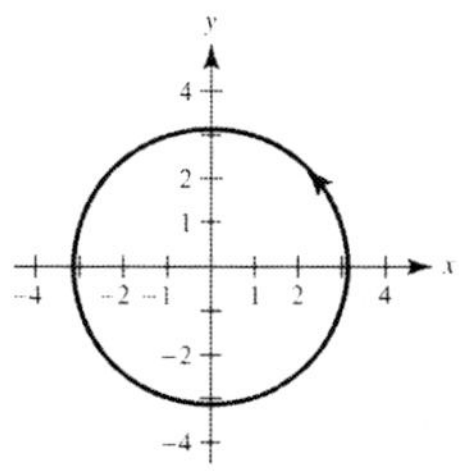

10. $\mathbf{r}(t) = \langle t + 2, t^2 - 1 \rangle$

$x = t + 2 \Rightarrow t = x - 2$

$y = t^2 - 1 = (x - 2)^2 - 1$, parabola

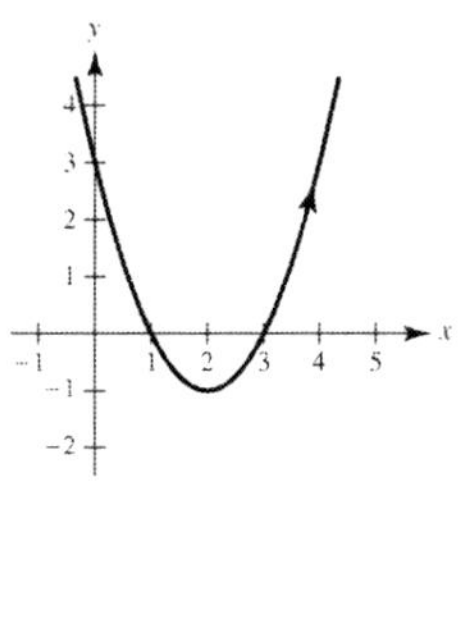

11. $\mathbf{r}(t) = (t + 1)\mathbf{i} + (3t - 1)\mathbf{j} + 2t\mathbf{k}$

$x = t + 1,\ y = 3t - 1,\ z = 2t$

This is a line passing through the points $(1, -1, 0)$ and $(2, 2, 2)$.

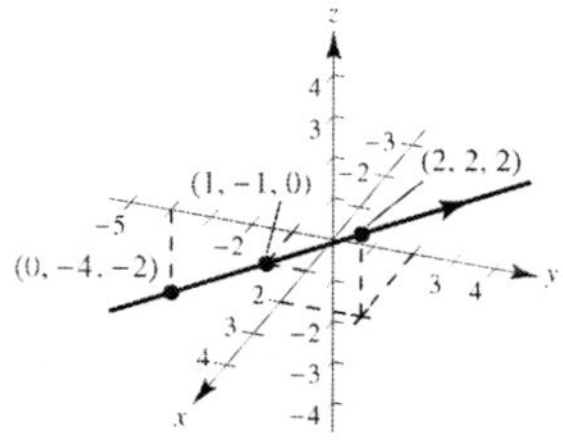

12. $\mathbf{r}(t) = 2\cos t\mathbf{i} + t\mathbf{j} + 2\sin t\mathbf{k}$

$x = 2\cos t,\ y = t,\ z = 2\sin t$

$x^2 + z^2 = 4$, Circular helix

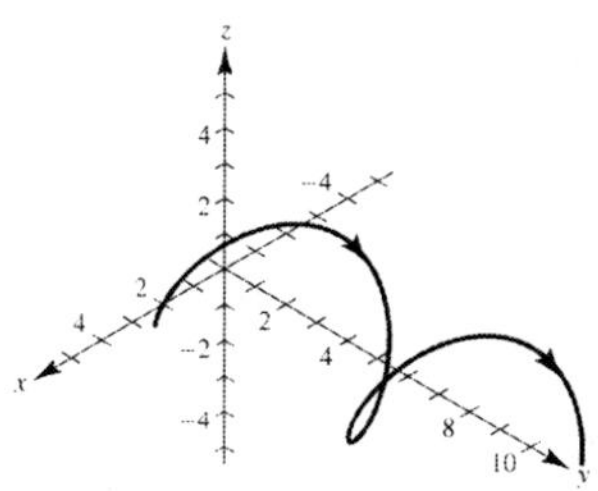

13. $3x + 4y - 12 = 0$

Let $x = t$, then $y = \dfrac{12 - 3t}{4}$.

$r(t) = t\mathbf{i} + \dfrac{12 - 3t}{4}\mathbf{j}$

Alternate solution: $x = 4t,\ y = 3 - 3t$

$r(t) = 4t\mathbf{i} + (3 - 3t)\mathbf{j}$

14. $y = 9 - x^2$

Let $x = t$, then $y = 9 - t^2$.

$r(t) = t\mathbf{i} + (9 - t^2)\mathbf{j}$

15. $z = x^2 + y^2, x + y = 0, t = x$

$x = t, y = -t, z = 2t^2$

$\mathbf{r}(t) = t\mathbf{i} - t\mathbf{j} + 2t^2\mathbf{k}$

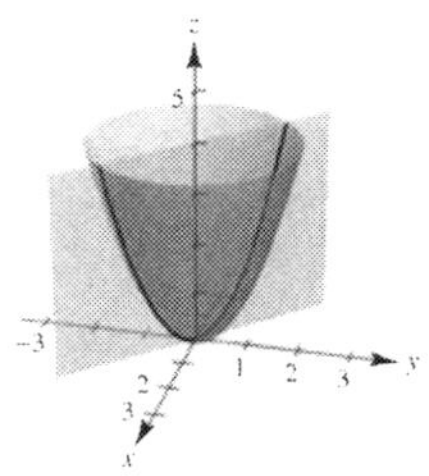

16. $x^2 + z^2 = 4, x - y = 0, t = x$

$x = t, y = t, z = \pm\sqrt{4 - t^2}$

$\mathbf{r}(t) = t\mathbf{i} + t\mathbf{j} + \sqrt{4 - t^2}\,\mathbf{k}$

$\mathbf{r}(t) = t\mathbf{i} + t\mathbf{j} - \sqrt{4 - t^2}\,\mathbf{k}$

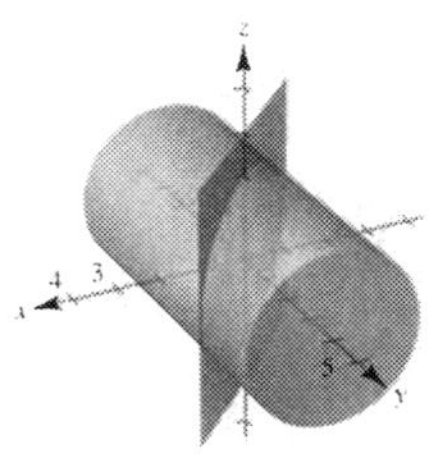

17. $\lim_{t \to 4^-}\left(t\mathbf{i} + \sqrt{4 - t}\,\mathbf{j} + \mathbf{k}\right) = 4\mathbf{i} + \mathbf{k}$

18. $\lim_{t \to 0}\left(\dfrac{\sin 2t}{t}\mathbf{i} + e^{-t}\mathbf{j} + e^t\mathbf{k}\right) = \left(\lim_{t \to 0}\dfrac{2\cos 2t}{1}\right)\mathbf{i} + \mathbf{j} + \mathbf{k}$

$= 2\mathbf{i} + \mathbf{j} + \mathbf{k}$

19. $\mathbf{r}(t) = \left(t^2 + 4t\right)\mathbf{i} - 3t^2\mathbf{j}$

(a) $\mathbf{r}'(t) = (2t + 4)\mathbf{i} - 6t\mathbf{j}$

(b) $\mathbf{r}''(t) = 2\mathbf{i} - 6\mathbf{j}$

(c) $\mathbf{r}'(t) \cdot \mathbf{r}''(t) = (2t + 4)(2) + (-6t)(-6)$

$= 40t + 8$

20. $\mathbf{r}(t) = 5\cos t\mathbf{i} + 2\sin t\mathbf{j}$

(a) $\mathbf{r}'(t) = -5\sin t\mathbf{i} + 2\cos t\mathbf{j}$

(b) $\mathbf{r}''(t) = -5\cos t\mathbf{i} - 2\sin t\mathbf{j}$

(c) $\mathbf{r}'(t) \cdot \mathbf{r}''(t) = (-5\sin t)(-5\cos t) + (2\cos t)(-2\sin t)$

$= 21\sin t\cos t$

21. $\mathbf{r}(t) = 2t^3\mathbf{i} + 4t\mathbf{j} - t^2\mathbf{k}$

(a) $\mathbf{r}'(t) = 6t^2\mathbf{i} + 4\mathbf{j} - 2t\mathbf{k}$

(b) $\mathbf{r}''(t) = 12t\mathbf{i} - 2\mathbf{k}$

(c) $\mathbf{r}'(t) \cdot \mathbf{r}''(t) = 6t^2(12t) + (-2t)(-2)$

$= 72t^3 + 4t$

(d) $\mathbf{r}'(t) \times \mathbf{r}''(t) = \begin{bmatrix} \mathbf{i} & \mathbf{j} & \mathbf{k} \\ 6t^2 & 4 & -2t \\ 12t & 0 & -2 \end{bmatrix}$

$= -8\mathbf{i} - \left(-12t^2 + 24t^2\right)\mathbf{j} + (-48t)\mathbf{k}$

$= -8\mathbf{i} - 12t^2\mathbf{j} - 48t\mathbf{k}$

22. $\mathbf{r}(t) = (4t + 3)\mathbf{i} + t^2\mathbf{j} + \left(2t^2 + 4\right)\mathbf{k}$

(a) $\mathbf{r}'(t) = 4\mathbf{i} + 2t\mathbf{j} + 4t\mathbf{k}$

(b) $\mathbf{r}''(t) = 2\mathbf{j} + 4\mathbf{k}$

(c) $\mathbf{r}'(t) \cdot \mathbf{r}''(t) = (2t)(2) + (4t)(4)$

$= 4t + 16t = 20t$

(d) $\mathbf{r}'(t) \times \mathbf{r}''(t) = \begin{bmatrix} \mathbf{i} & \mathbf{j} & \mathbf{k} \\ 4 & 2t & 4t \\ 0 & 2 & 4 \end{bmatrix}$

$= (8t - 8t)\mathbf{i} - (16 - 0)\mathbf{j} + (8 - 0)\mathbf{k}$

$= -16\mathbf{j} + 8\mathbf{k}$

23. $\mathbf{r}(t) = 3t\mathbf{i} + (t - 1)\mathbf{j}, \mathbf{u}(t) = t\mathbf{i} + t^2\mathbf{j} + \frac{2}{3}t^3\mathbf{k}$

$\mathbf{r}'(t) = 3\mathbf{i} + \mathbf{j}, \mathbf{u}'(t) = \mathbf{i} + 2t\mathbf{j} + 2t^2\mathbf{k}$

(a) $\mathbf{r}'(t) = 3\mathbf{i} + \mathbf{j}$

(b) $\dfrac{d}{dt}\left[\mathbf{u}(t) - 2\mathbf{r}(t)\right] = \mathbf{u}'(t) - 2\mathbf{r}'(t)$

$= \left(\mathbf{i} + 2t\mathbf{j} + 2t^2\mathbf{k}\right) - 2(3\mathbf{i} + \mathbf{j})$

$= (1 - 6)\mathbf{i} + (2t - 2)\mathbf{j} + 2t^2\mathbf{k}$

$= -5\mathbf{i} + (2t - 2)\mathbf{j} + 2t^2\mathbf{k}$

(c) $\dfrac{d}{dt}(3t)\mathbf{r}(t) = (3t)\mathbf{r}'(t) + 3\mathbf{r}(t)$

$= (3t)(3\mathbf{i} + \mathbf{j}) + 3\left[3t\mathbf{i} + (t - 1)\mathbf{j}\right]$

$= 9t\mathbf{i} + 3t\mathbf{j} + 9t\mathbf{i} + (3t - 3)\mathbf{j}$

$= 18t\mathbf{i} + (6t - 3)\mathbf{j}$

(d) $\dfrac{d}{dt}\left[\mathbf{r}(t)\cdot\mathbf{u}(t)\right] = \mathbf{r}(t)\cdot\mathbf{u}'(t) + \mathbf{r}'(t)\cdot\mathbf{u}(t)$

$$= \left[(3t)(1) + (t-1)(2t) + (0)\left(2t^2\right)\right] + \left[(3)(t) + (1)\left(t^2\right) + (0)\left(\tfrac{2}{3}t^3\right)\right]$$

$$= \left(3t + 2t^2 - 2t\right) + \left(3t + t^2\right)$$

$$= 4t + 3t^2$$

(e) $\dfrac{d}{dt}\left[\mathbf{r}(t)\times\mathbf{u}(t)\right] = \mathbf{r}(t)\times\mathbf{u}'(t) + \mathbf{r}'(t)\times\mathbf{u}(t)$

$$= \left[\left(2t^3 - 2t^2\right)\mathbf{i} - 6t^3\mathbf{j} + \left(6t^2 - t + 1\right)\mathbf{k}\right] + \left[\tfrac{2}{3}t^3\mathbf{i} - 2t^3\mathbf{j} + \left(3t^2 - t\right)\mathbf{k}\right]$$

$$= \left(\tfrac{8}{3}t^3 - 2t^2\right)\mathbf{i} - 8t^3\mathbf{j} + \left(9t^2 - 2t + 1\right)\mathbf{k}$$

(f) $\dfrac{d}{dt}\left[\mathbf{u}(2t)\right] = 2\mathbf{u}'(2t)$

$$= 2\left[\mathbf{i} + 2(2t)\mathbf{j} + 2(2t)^2\mathbf{k}\right]$$

$$= 2\mathbf{i} + 8t\mathbf{j} + 16t^2\mathbf{k}$$

24. $\mathbf{r}(t) = \sin t\,\mathbf{i} + \cos t\,\mathbf{j} + t\mathbf{k},\ \mathbf{u}(t) = \sin t\,\mathbf{i} + \cos t\,\mathbf{j} + \dfrac{1}{t}\mathbf{k}$

$\mathbf{r}'(t) = \cos t\,\mathbf{i} - \sin t\,\mathbf{j} + \mathbf{k},\ \mathbf{u}'(t) = \cos t\,\mathbf{i} - \sin t\,\mathbf{j} - \dfrac{1}{t^2}\mathbf{k}$

(a) $\mathbf{r}'(t) = \cos t\,\mathbf{i} - \sin t\,\mathbf{j} + \mathbf{k}$

(b) $\dfrac{d}{dt}\left[\mathbf{u}(t) - 2\mathbf{r}(t)\right] = \mathbf{u}'(t) - 2\mathbf{r}'(t)$

$$= \left(\cos t\,\mathbf{i} - \sin t\,\mathbf{j} - \frac{1}{t^2}\mathbf{k}\right) - 2(\cos t\,\mathbf{i} - \sin t\,\mathbf{j} + \mathbf{k})$$

$$= -\cos t\,\mathbf{i} + \sin t\,\mathbf{j} - \left(2 + \frac{1}{t^2}\right)\mathbf{j}$$

(c) $\dfrac{d}{dt}\left[(3t)\mathbf{r}(t)\right] = (3t)\mathbf{r}'(t) + 3\mathbf{r}(t)$

$$= (3t)(\cos t\,\mathbf{i} - \sin t\,\mathbf{j} + \mathbf{k}) + 3(\sin t\,\mathbf{i} + \cos t\,\mathbf{j} + t\mathbf{k})$$

$$= (3t\cos t + 3\sin t)\mathbf{i} + (3\cos t - 3t\sin t)\mathbf{j} + 6t\,\mathbf{k}$$

(d) $\dfrac{d}{dt}\left[\mathbf{r}(t)\cdot\mathbf{u}(t)\right] = \mathbf{r}(t)\cdot\mathbf{u}'(t) + \mathbf{r}'(t)\cdot\mathbf{u}(t)$

$$= \left[(\sin t)(\cos t) + (\cos t)(-\sin t) + (t)\left(-\frac{1}{t^2}\right)\right] + \left[(\cos t)(\sin t) + (-\sin t)(\cos t) + (1)\left(\frac{1}{t}\right)\right]$$

$$= \sin t\cos t - \sin t\cos t - \frac{1}{t} + \sin t\cos t - \sin t\cos t + \frac{1}{t}$$

$$= 0$$

(e) $\dfrac{d}{dt}\left[\mathbf{r}(t)\times\mathbf{u}(t)\right] = \mathbf{r}(t)\times\mathbf{u}'(t) + \mathbf{r}'(t)\times\mathbf{u}(t)$

$$= \left[\left(-\frac{1}{t^2}\cos t + t\sin t\right)\mathbf{i} - \left(-\frac{1}{t^2}\sin t + t\cos t\right)\mathbf{j} + \left(-\sin^2 t - \cos^2 t\right)\mathbf{k}\right]$$

$$+ \left[\left(-\frac{1}{t}\sin t - \cos t\right)\mathbf{i} - \left(\frac{1}{t}\cos t - \sin t\right)\mathbf{j} + \left(\cos^2 t + \sin^2 t\right)\mathbf{k}\right]$$

$$= \left[\left(t - \frac{1}{t}\right)\sin t - \left(1 + \frac{1}{t^2}\right)\cos t\right]\mathbf{i} - \left[\left(\frac{1}{t} - t\right)\cos t - \left(1 + \frac{1}{t^2}\right)\sin t\right]\mathbf{j}$$

(f) $\frac{d}{dt}\left[\mathbf{u}(2t)\right] = 2\mathbf{u}'(2t)$

$$= 2\left[\cos(2t)\mathbf{i} - \sin(2t)\mathbf{j} - \frac{1}{(2t)^2}\mathbf{k}\right]$$

$$= 2\cos(2t)\mathbf{i} - 2\sin(2t)\mathbf{j} - \frac{1}{2t^2}\mathbf{k}$$

25. $\int(\mathbf{i} + 3\mathbf{j} + 4t\mathbf{k})\,dt = t\,\mathbf{i} + 3t\mathbf{j} + 2t^2\mathbf{k} + \mathbf{C}$

26. $\int\left(t^2\mathbf{i} + 5t\mathbf{j} + 8t^3\mathbf{k}\right)dt = \frac{t^3}{3}\mathbf{i} + \frac{5}{2}t^2\mathbf{j} + 2t^4\mathbf{k} + \mathbf{C}$

27. $\int\left(3\sqrt{t}\,\mathbf{i} + \frac{2}{t}\mathbf{j} + \mathbf{k}\right)dt = 2t^{3/2}\,\mathbf{i} + 2\ln|t|\,\mathbf{j} + t\mathbf{k} + \mathbf{C}$

28. $\int\left(\sin t\,\mathbf{i} + \cos t\,\mathbf{j} + e^{2t}\mathbf{k}\right)dt = -\cos t\,\mathbf{i} + \sin t\,\mathbf{j} + \frac{1}{2}e^{2t}\mathbf{k} + \mathbf{C}$

29. $\int_{-2}^{2}\left(3t\,\mathbf{i} + 2t^2\,\mathbf{j} - t^3\,\mathbf{k}\right)dt = \left[\frac{3t^2}{2}\mathbf{i} + \frac{2t^3}{3}\mathbf{j} - \frac{t^4}{4}\mathbf{k}\right]_{-2}^{2} = \frac{32}{3}\mathbf{j}$

30. $\int_0^1\left(t\,\mathbf{i} + \sqrt{t}\,\mathbf{j} + 4t\mathbf{k}\right)dt = \left[\frac{t^2}{2}\mathbf{i} + \frac{2}{3}t^{3/2}\mathbf{j} + 2t^2\mathbf{k}\right]_0^1 = \frac{1}{2}\mathbf{i} + \frac{2}{3}\mathbf{j} + 2\mathbf{k}$

31. $\int_0^2\left(e^{t/2}\mathbf{i} - 3t^2\,\mathbf{j} - \mathbf{k}\right)dt = \left[2e^{t/2}\mathbf{i} - t^3\,\mathbf{j} - t\mathbf{k}\right]_0^2 = (2e - 2)\mathbf{i} - 8\mathbf{j} - 2\mathbf{k}$

32. $\int_0^{\pi/3}(2\cos t\,\mathbf{i} + \sin t\,\mathbf{j} + 3\mathbf{k})\,dt = \left[2\sin t\,\mathbf{i} - \cos t\,\mathbf{j} + 3t\,\mathbf{k}\right]_0^{\pi/3} = \left(\sqrt{3}\,\mathbf{i} - \frac{1}{2}\mathbf{j} + \pi\,\mathbf{k}\right) - (-\mathbf{j}) = \sqrt{3}\,\mathbf{i} + \frac{1}{2}\mathbf{j} + \pi\,\mathbf{k}$

33. $\mathbf{r}(t) = \int\left(2t\,\mathbf{i} + e^t\,\mathbf{j} + e^{-t}\,\mathbf{k}\right)dt = t^2\,\mathbf{i} + e^t\,\mathbf{j} - e^{-t}\,\mathbf{k} + \mathbf{C}$

$\mathbf{r}(0) = \mathbf{j} - \mathbf{k} + \mathbf{C} = \mathbf{i} + 3\mathbf{j} - 5\mathbf{k} \Rightarrow \mathbf{C} = \mathbf{i} + 2\mathbf{j} - 4\mathbf{k}$

$\mathbf{r}(t) = \left(t^2 + 1\right)\mathbf{i} + \left(e^t + 2\right)\mathbf{j} - \left(e^{-t} + 4\right)\mathbf{k}$

34. $\mathbf{r}(t) = \int\left(\sec t\,\mathbf{i} + \tan t\,\mathbf{j} + t^2\,\mathbf{k}\right)dt$

$= \ln|\sec t + \tan t|\mathbf{i} - \ln|\cos t|\mathbf{j} + \frac{t^3}{3}\mathbf{k} + \mathbf{C}$

$\mathbf{r}(0) = \mathbf{C} = 3\mathbf{k}$

$\mathbf{r}(t) = \ln|\sec t + \tan t|\mathbf{i} - \ln|\cos t|\mathbf{j} + \left(\frac{t^3}{3} + 3\right)\mathbf{k}$

35. $\mathbf{r}(t) = 4t\mathbf{i} + t^3\mathbf{j} - t\mathbf{k},\ t = 1$

(a) $\mathbf{v}(t) = \mathbf{r}'(t) = 4\mathbf{i} + 3t^2\mathbf{j} - \mathbf{k}$

Speed $= \|\mathbf{v}(t)\| = \sqrt{16 + 9t^4 + 1} = \sqrt{17 + 9t^4}$

$\mathbf{a}(t) = \mathbf{r}''(t) = 6t\,\mathbf{j}$

(b) $\mathbf{v}(1) = 4\mathbf{i} + 3\mathbf{j} - \mathbf{k}$

$\mathbf{a}(1) = 6\mathbf{j}$

36. $\mathbf{r}(t) = \sqrt{t}\,\mathbf{i} + 5t\mathbf{j} + 2t^2\mathbf{k},\ t = 4$

(a) $\mathbf{v}(t) = \mathbf{r}'(t) = \frac{1}{2\sqrt{t}}\mathbf{i} + 5\mathbf{j} + 4t\mathbf{k}$

Speed $= \|\mathbf{v}(t)\| = \sqrt{\frac{1}{4t} + 25 + 16t^2}$

$\mathbf{a}(t) = \mathbf{r}''(t) = -\frac{1}{4t^{3/2}}\mathbf{i} + 4\mathbf{k}$

(b) $\mathbf{v}(4) = \frac{1}{4}\mathbf{i} + 5\mathbf{j} + 16\mathbf{k}$

$\mathbf{a}(4) = -\frac{1}{32}\mathbf{i} + 4\mathbf{k}$

37. $\mathbf{r}(t) = \langle \cos^3 t, \sin^3 t, 3t \rangle, t = \pi$

(a) $\mathbf{v}(t) = \mathbf{r}'(t) = \langle -3\cos^2 t \sin t, 3\sin^2 t \cos t, 3 \rangle$

$$\text{Speed} = \|\mathbf{v}(t)\| = \sqrt{9\cos^4 t \sin^2 t + 9\sin^4 t \cos^2 t + 9}$$
$$= 3\sqrt{\cos^2 t \sin^2 t(\cos^2 t + \sin^2 t) + 1}$$
$$= 3\sqrt{\cos^2 t \sin^2 t + 1}$$

$$\mathbf{a}(t) = \langle -6\cos t(-\sin^2 t) + (-3\cos^2 t)\cos t, 6\sin t\cos^2 t + 3\sin^2 t(-\sin t), 0 \rangle$$
$$= \langle 3\cos t(2\sin^2 t - \cos^2 t), 3\sin t(2\cos^2 t - \sin^2 t), 0 \rangle$$

(b) $\mathbf{v}(\pi) = \langle 0, 0, 3 \rangle$

$\mathbf{a}(\pi) = \langle 3, 0, 0 \rangle$

38. $\mathbf{r}(t) = \langle t, -\tan t, e^t \rangle, t = 0$

(a) $\mathbf{v}(t) = \mathbf{r}'(t) = \langle 1, -\sec^2 t, e^t \rangle$

$\text{Speed} = \|\mathbf{v}(t)\| = \sqrt{1 + \sec^4 t + e^{2t}}$

$\mathbf{a}(t) = \langle 0, -2\sec^2 t \tan t, e^t \rangle$

(b) $\mathbf{v}(0) = \langle 1, -1, 1 \rangle$

$\mathbf{a}(0) = \langle 0, 0, 1 \rangle$

39. $\mathbf{r}(t) = \left\langle (v_0 \cos\theta)t, (v_0 \sin\theta)t - \frac{1}{2}gt^2 \right\rangle$

$$= \langle 42\sqrt{3}t, 42t - 16t^2 \rangle$$

$$42t = 16t^2 \Rightarrow t = 0, \frac{21}{8}$$

$$\text{Range} = 42\sqrt{3}\left(\frac{21}{8}\right) = \frac{441\sqrt{3}}{4} \approx 190.96 \text{ ft}$$

40. $\mathbf{r}(t) = (v_0 \cos\theta)t\mathbf{i} + \left[h + (v_0 \sin\theta)t - 16t^2\right]\mathbf{j}$

$$= (120\cos 30°)t\mathbf{i} + (3.5 + (120\sin 30°)t - 16t^2)\mathbf{j}$$
$$= 60\sqrt{3}t\mathbf{i} + (3.5 + 60t - 16t^2)\mathbf{j}$$

$$\mathbf{r}'(t) = \mathbf{v}(t) = 60\sqrt{3}\mathbf{i} + (60 - 32t)\mathbf{j}$$

To find the maximum height,

$$y'(t) = 60 - 32t = 0 \Rightarrow t = \frac{60}{32} = \frac{15}{8} = 1.875$$

$$y\left(\frac{15}{8}\right) = 3.5 + 60\left(\frac{15}{8}\right) - 16\left(\frac{15}{8}\right)^2 = 59.75 \text{ feet, Maximum height}$$

$$375 = 60\sqrt{3}t \Rightarrow t = \frac{375}{60\sqrt{3}} = \frac{25}{4\sqrt{3}} = \frac{25\sqrt{3}}{12} \approx 3.608$$

$$y\left(\frac{25\sqrt{3}}{12}\right) \approx 11.67 \text{ feet}$$

The baseball clears the 8-foot fence.

41. $\text{Range} = x = \frac{{v_0}^2}{9.8}\sin 2\theta = 95$

$${v_0}^2 = \frac{9.8(95)}{\sin(40°)}$$

$$v_0 \approx 38.06 \text{ m/sec}$$

42. $\mathbf{r}(t) = \left[(v_0 \cos\theta)t\right]\mathbf{i} + \left[(v_0 \sin\theta)t - \frac{1}{2}(9.8)t^2\right]\mathbf{j}$

(a) $\mathbf{r}(t) = \left[(20\cos 30°)t\right]\mathbf{i} + \left[(20\sin 30°)t - 4.9t^2\right]\mathbf{j}$

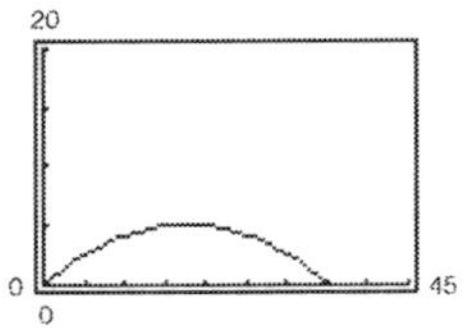

Maximum height ≈ 5.1 m; Range ≈ 35.3 m

(b) $\mathbf{r}(t) = \left[(20\cos 45°)t\right]\mathbf{i} + \left[(20\sin 45°)t - 4.9t^2\right]\mathbf{j}$

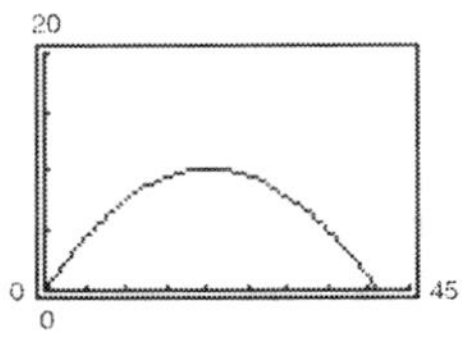

Maximum height ≈ 10.2 m; Range ≈ 40.8 m

(c) $\mathbf{r}(t) = \left[(20\cos 60°)t\right]\mathbf{i} + \left[(20\sin 60°)t - 4.9t^2\right]\mathbf{j}$

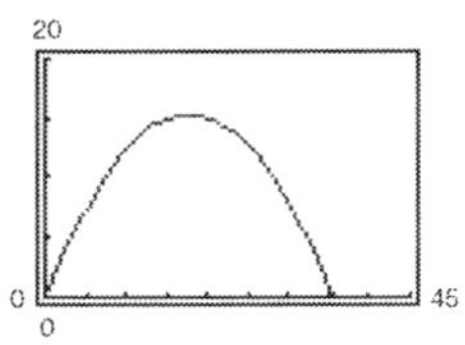

Maximum height ≈ 15.3 m; Range ≈ 35.3 m

(Note that 45° gives the longest range)

43. $\mathbf{r}(t) = 3t\mathbf{i} + 3t^3\mathbf{j},\ t = 1$

$\mathbf{r}'(t) = 3\mathbf{i} + 9t^2\mathbf{j}$

$\mathbf{r}'(1) = 3\mathbf{i} + 9\mathbf{j}, \left\|\mathbf{r}'(1)\right\| = \sqrt{9 + 81} = \sqrt{90} = 3\sqrt{10}$

$\mathbf{T}(1) = \frac{\mathbf{r}'(1)}{\left\|\mathbf{r}'(1)\right\|} = \frac{1}{3\sqrt{10}}(3\mathbf{i} + 9\mathbf{j}) = \frac{1}{\sqrt{10}}\mathbf{i} + \frac{3}{\sqrt{10}}\mathbf{j}$

44. $\mathbf{r}(t) = 2\sin t\mathbf{i} + 4\cos t\mathbf{j},\ t = \frac{\pi}{6}$

$\mathbf{r}'(t) = 2\cos t\mathbf{i} - 4\sin t\mathbf{j}$

$\mathbf{r}'\left(\frac{\pi}{6}\right) = \sqrt{3}\mathbf{i} - 2\mathbf{j}$

$\left\|\mathbf{r}'\left(\frac{\pi}{6}\right)\right\| = \sqrt{3 + 4} = \sqrt{7}$

$\mathbf{T}\left(\frac{\pi}{6}\right) = \frac{\mathbf{r}'\left(\frac{\pi}{6}\right)}{\left\|\mathbf{r}'\left(\frac{\pi}{6}\right)\right\|} = \frac{1}{\sqrt{7}}\left(\sqrt{3}\mathbf{i} - 2\mathbf{j}\right)$

$= \frac{\sqrt{21}}{7}\mathbf{i} - \frac{2\sqrt{7}}{7}\mathbf{j}$

45. $\mathbf{r}(t) = 2\cos t\mathbf{i} + 2\sin t\mathbf{j} + t\mathbf{k},\ P\left(1, \sqrt{3}, \frac{\pi}{3}\right)$

$\mathbf{r}'(t) = -2\sin t\mathbf{i} + 2\cos t\mathbf{j} + \mathbf{k}$

$t = \frac{\pi}{3}$ at $P\left(1, \sqrt{3}, \frac{\pi}{3}\right)$

$\mathbf{r}'\left(\frac{\pi}{3}\right) = -\sqrt{3}\mathbf{i} + \mathbf{j} + \mathbf{k}$

$\mathbf{T}\left(\frac{\pi}{3}\right) = \frac{\mathbf{r}'\left(\frac{\pi}{3}\right)}{\left\|\mathbf{r}'\left(\frac{\pi}{3}\right)\right\|} = \frac{-\sqrt{3}\mathbf{i} + \mathbf{j} + \mathbf{k}}{\sqrt{5}} = \frac{\sqrt{15}}{5}\mathbf{i} + \frac{\sqrt{5}}{5}\mathbf{j} + \frac{\sqrt{5}}{5}\mathbf{k}$

Direction numbers: $-\sqrt{3}, 1, 1$

$x = 1 - \sqrt{3}t,\ y = \sqrt{3} + t,\ z = \frac{\pi}{3} + t$

46. $\mathbf{r}(t) = t\mathbf{i} + t^2\mathbf{j} + \frac{2}{3}t^3\mathbf{k},\ P\left(2, 4, \frac{16}{3}\right)$

$\mathbf{r}'(t) = \mathbf{i} + 2t\mathbf{j} + 2t^2\mathbf{k}$

$t = 2$ at $P\left(2, 4, \frac{16}{3}\right)$

$\mathbf{r}'(2) = \mathbf{i} + 4\mathbf{j} + 8\mathbf{k}$

$$\mathbf{T}(2) = \frac{\mathbf{r}'(2)}{\|\mathbf{r}'(2)\|} = \frac{\mathbf{i} + 4\mathbf{j} + 8\mathbf{k}}{9} = \frac{1}{9}\mathbf{i} + \frac{4}{9}\mathbf{j} + \frac{8}{9}\mathbf{k}$$

Direction numbers when 1, 4, 8

$x = t + 2,\ y = 4t + 4,\ z = 8t + \frac{16}{3}$

47. $\mathbf{r}(t) = 2t\mathbf{i} + 3t^2\mathbf{j},\ t = 1$

$\mathbf{r}'(t) = 2\mathbf{i} + 6t\mathbf{j},\ \mathbf{r}'(1) = 2\mathbf{i} + 6\mathbf{j}$

$\|\mathbf{r}'(t)\| = \sqrt{4 + 36t^2},\ \|\mathbf{r}'(1)\| = \sqrt{40} = 2\sqrt{10}$

$$\mathbf{T}(1) = \frac{\mathbf{r}'(1)}{\|\mathbf{r}'(1)\|} = \frac{2\mathbf{i} + 6\mathbf{j}}{2\sqrt{10}} = \frac{1}{\sqrt{10}}(\mathbf{i} + 3\mathbf{j})$$

$\mathbf{N}(1)$ is orthogonal to $\mathbf{T}(1)$ and points towards the concave side. Hence,

$$\mathbf{N}(1) = \frac{1}{\sqrt{10}}(-3\mathbf{i} + \mathbf{j}).$$

48. $\mathbf{r}(t) = t\mathbf{i} + \ln t\,\mathbf{j},\ t = 2$

$\mathbf{r}'(t) = \mathbf{i} + \frac{1}{t}\mathbf{j},\ \mathbf{r}'(2) = \mathbf{i} + \frac{1}{2}\mathbf{j}$

$\|\mathbf{r}'(t)\| = \sqrt{1 + \frac{1}{t^2}} = \frac{1}{t}\sqrt{t^2 + 1},\ \|\mathbf{r}'(2)\| = \frac{1}{2}\sqrt{5}$

$$\mathbf{T}(2) = \frac{\mathbf{r}'(2)}{\|\mathbf{r}'(2)\|} = \frac{2}{\sqrt{5}}\left(\mathbf{i} + \frac{1}{2}\mathbf{j}\right) = \frac{2}{\sqrt{5}}\mathbf{i} + \frac{1}{\sqrt{5}}\mathbf{j}$$

$\mathbf{N}(2)$ is orthogonal to $\mathbf{T}(2)$ and points towards the concave side. Hence,

$$\mathbf{N}(t) = \frac{1}{\sqrt{5}}\mathbf{i} - \frac{2}{\sqrt{5}}\mathbf{j} = \frac{\sqrt{5}}{5}\mathbf{i} - \frac{2\sqrt{5}}{5}\mathbf{j}.$$

49. $\mathbf{r}(t) = 3\cos 2t\mathbf{i} + 3\sin 2t\,\mathbf{j} + 3\mathbf{k},\ t = \frac{\pi}{4}$

$\mathbf{r}'(t) = -6\sin 2t\mathbf{i} + 6\cos 2t\,\mathbf{j},\ \|\mathbf{r}'(t)\| = 6$

$\mathbf{T}(t) = -\sin 2t\mathbf{i} + \cos 2t\,\mathbf{j},\ \mathbf{T}(\pi/4) = -\mathbf{i}$

$\mathbf{T}'(t) = -2\cos 2t\mathbf{i} - 2\sin 2t\,\mathbf{j},\ \|\mathbf{T}'(t)\| = 2$

$\mathbf{N}(t) = -\cos 2t\mathbf{i} - \sin 2t\,\mathbf{j}$

$\mathbf{N}(\pi/4) = -\mathbf{j}$

50. $\mathbf{r}(t) = 4\cos t\mathbf{i} + 4\sin t\,\mathbf{j} + \mathbf{k},\ t = \frac{2\pi}{3}$

$\mathbf{r}'(t) = -4\sin t\mathbf{i} + 4\cos t\,\mathbf{j},\ \|\mathbf{r}'(t)\| = 4$

$\mathbf{T}(t) = -\sin t\mathbf{i} + \cos t\,\mathbf{j},\ \mathbf{T}\left(\frac{2\pi}{3}\right) = -\frac{\sqrt{3}}{2}\mathbf{i} - \frac{1}{2}\mathbf{j}$

$\mathbf{T}'(t) = -\cos t\mathbf{i} - \sin t\,\mathbf{j},\ \|\mathbf{T}'(t)\| = 1$

$\mathbf{N}(t) = -\cos t\mathbf{i} - \sin t\,\mathbf{j},\ \mathbf{N}\left(\frac{2\pi}{3}\right) = \frac{1}{2}\mathbf{i} - \frac{\sqrt{3}}{2}\mathbf{j}$

51. $\mathbf{r}(t) = \frac{3}{t}\mathbf{i} - 6t\,\mathbf{j},\ t = 3$

$\mathbf{v}(t) = -\frac{3}{t^2}\mathbf{i} - 6\mathbf{j},\ \mathbf{v}(3) = -\frac{1}{3}\mathbf{i} - 6\mathbf{j}$

$\mathbf{a}(t) = \frac{6}{t^3}\mathbf{i},\ \mathbf{a}(3) = \frac{2}{9}\mathbf{i}$

$$\mathbf{T}(t) = \frac{\mathbf{v}(t)}{\|\mathbf{v}(t)\|} = \frac{\left(-\frac{3}{t^2}\right)\mathbf{i} - 6\mathbf{j}}{\sqrt{\frac{9}{t^4} + 36}} = \frac{-3\mathbf{i} - 6t^2\mathbf{j}}{3\sqrt{1 + 4t^4}}$$

$$\mathbf{T}(3) = \frac{-3\mathbf{i} - 54\mathbf{j}}{3\sqrt{1 + 324}} = \frac{-\mathbf{i} - 18\mathbf{j}}{\sqrt{325}} = -\frac{\sqrt{13}}{65}\mathbf{i} - \frac{18\sqrt{13}}{65}\mathbf{j}$$

$\mathbf{N}(3)$ is orthogonal to $\mathbf{T}(3)$, and points in the direction the curve is bending. Hence,

$$\mathbf{N}(3) = \frac{18\mathbf{i} - \mathbf{j}}{\sqrt{325}} = \frac{18\sqrt{13}}{65}\mathbf{i} - \frac{\sqrt{13}}{65}\mathbf{j}.$$

$$a_{\mathbf{T}} = \mathbf{a}\cdot\mathbf{T} = -\frac{2}{9\sqrt{325}} = -\frac{2\sqrt{13}}{585}$$

$$a_{\mathbf{N}} = \mathbf{a}\cdot\mathbf{N} = \frac{4}{\sqrt{325}} = \frac{4\sqrt{13}}{65}$$

52. $\mathbf{r}(t) = 3\cos 2t\mathbf{i} + 3\sin 2t\,\mathbf{j},\ t = \dfrac{\pi}{6}$

$\mathbf{v}(t) = -6\sin 2t\mathbf{i} + 6\cos 2\,t\,\mathbf{j},\ \mathbf{v}\left(\dfrac{\pi}{6}\right) = -3\sqrt{3}\mathbf{i} + 3\mathbf{j}$

$\mathbf{a}(t) = -12\cos 2t\mathbf{i} - 12\sin 2t\,\mathbf{j},\ \mathbf{a}\left(\dfrac{\pi}{6}\right) = -6\mathbf{i} - 6\sqrt{3}\,\mathbf{j}$

$\left\|\mathbf{v}(t)\right\| = \sqrt{36\sin^2 2t + 36\cos^2 2t} = 6$

$\mathbf{T}(t) = \dfrac{\mathbf{v}(t)}{\left\|\mathbf{v}(t)\right\|} = \dfrac{1}{6}(-6\sin 2t\mathbf{i} + 6\cos 2t\mathbf{j})$

$= -\sin 2t\mathbf{i} + \cos 2\,t\,\mathbf{j}$

$\mathbf{T}\left(\dfrac{\pi}{6}\right) = -\dfrac{\sqrt{3}}{2}\mathbf{i} + \dfrac{1}{2}\mathbf{j}$

$\mathbf{T}'(t) = -2\cos 2t\mathbf{i} - 2\sin 2t\,\mathbf{j},\ \left\|\mathbf{T}'(t)\right\| = 2$

$\mathbf{N}(t) = \dfrac{\mathbf{T}'(t)}{\left\|\mathbf{T}'(t)\right\|} = -\cos 2t\mathbf{i} - \sin 2t\,\mathbf{j}$

$\mathbf{N}\left(\dfrac{\pi}{6}\right) = -\dfrac{1}{2}\mathbf{i} - \dfrac{\sqrt{3}}{2}\mathbf{j}$

$a_{\mathbf{T}} = \mathbf{a}\cdot\mathbf{T} = \left(-6\mathbf{i} - 6\sqrt{3}\mathbf{j}\right)\cdot\left(-\dfrac{\sqrt{3}}{2}\mathbf{i} + \dfrac{1}{2}\mathbf{j}\right) = 3\sqrt{3} - 3\sqrt{3} = 0$

$a_{\mathbf{N}} = \mathbf{a}\cdot\mathbf{N} = \left(-6\mathbf{i} - 6\sqrt{3}\mathbf{j}\right)\cdot\left(-\dfrac{1}{2}\mathbf{i} - \dfrac{\sqrt{3}}{2}\mathbf{j}\right) = 3 + 9 = 12$

53. $\mathbf{r}(t) = 2t\mathbf{i} - 3t\mathbf{j},\ [0, 5]$

$\mathbf{r}'(t) = 2\mathbf{i} - 3\mathbf{j}$

$s = \int_a^b \left\|\mathbf{r}'(t)\right\|\,dt = \int_0^5 \sqrt{4 + 9}\,dt = \left[\sqrt{13}\,t\right]_0^5 = 5\sqrt{13}$

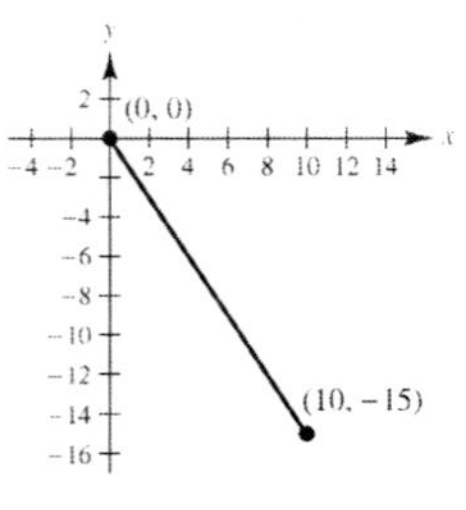

54. $\mathbf{r}(t) = t^2\mathbf{i} + 2t\mathbf{k},\ [0, 3]$

$\mathbf{r}'(t) = 2t\mathbf{i} + 2\mathbf{k}$

$s = \int_a^b \left\|\mathbf{r}'(t)\right\|\,dt = \int_0^3 \sqrt{4t^2 + 4}\,dt$

$= \left[\ln\left|\sqrt{t^2 + 1} + t\right| + t\sqrt{t^2 + 1}\right]_0^3$

$= \ln\left(\sqrt{10} + 3\right) + 3\sqrt{10} \approx 11.3053$

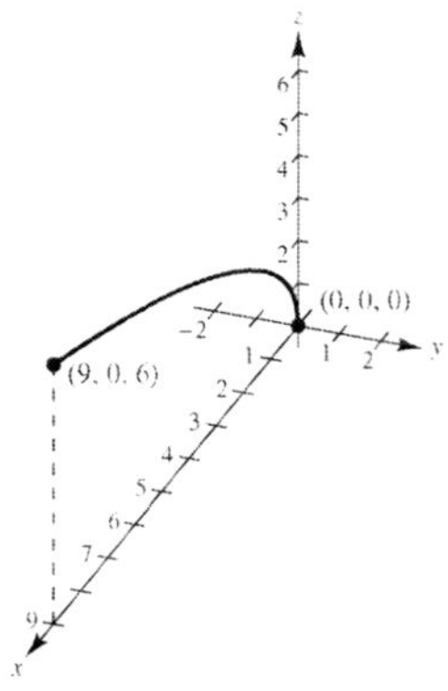

55. $\mathbf{r}(t) = 10\cos^3 t\,\mathbf{i} + 10\sin^3 t\,\mathbf{j}, [0, 2\pi]$

$\mathbf{r}'(t) = -30\cos^2 t \sin t\,\mathbf{i} + 30\sin^2 t \cos t\,\mathbf{j}$

$\|\mathbf{r}'(t)\| = 30\sqrt{\cos^4 t \sin^2 t + \sin^4 t \cos^2 t}$

$= 30|\cos t \sin t|$

$s = 4\int_0^{\pi/2} 30\cos t \cdot \sin t\, dt = \left[120\dfrac{\sin^2 t}{2}\right]_0^{\pi/2} = 60$

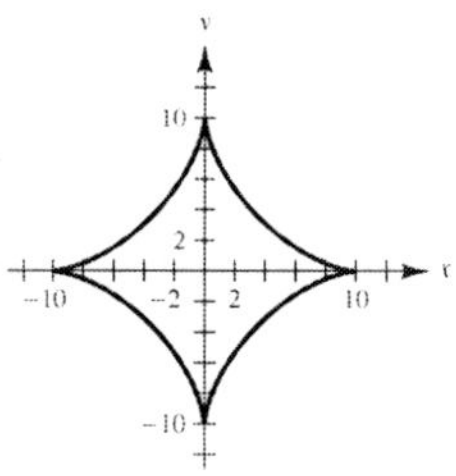

56. $\mathbf{r}(t) = 10\cos t\,\mathbf{i} + 10\sin t\,\mathbf{j}, [0, 2\pi]$

$\mathbf{r}'(t) = -10\sin t\,\mathbf{i} + 10\cos t\,\mathbf{j}$

$\|\mathbf{r}'(t)\| = 10$

$s = \int_0^{2\pi} 10\, dt = 20\pi$

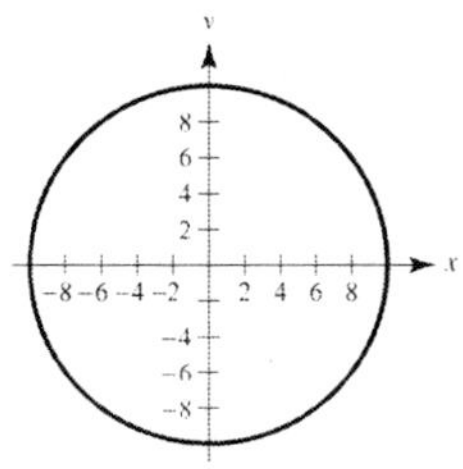

57. $\mathbf{r}(t) = -3t\,\mathbf{i} + 2t\,\mathbf{j} + 4t\,\mathbf{k}, [0, 3]$

$\mathbf{r}'(t) = -3\mathbf{i} + 2\mathbf{j} + 4\mathbf{k}$

$s = \int_a^b \|\mathbf{r}'(t)\|\, dt = \int_0^3 \sqrt{9 + 4 + 16}\, dt$

$= \int_0^3 \sqrt{29}\, dt = 3\sqrt{29}$

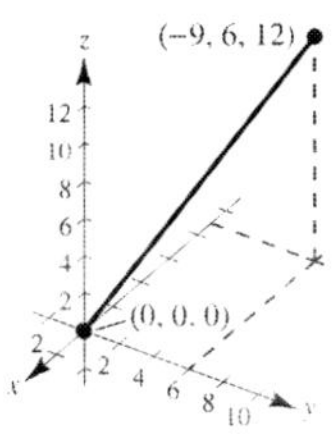

58. $\mathbf{r}(t) = t\,\mathbf{i} + t^2\,\mathbf{j} + 2t\,\mathbf{k}, [0, 2]$

$\mathbf{r}'(t) = \mathbf{i} + 2t\,\mathbf{j} + 2\mathbf{k}, \|\mathbf{r}'(t)\| = \sqrt{5 + 4t^2}$

$s = \int_a^b \|\mathbf{r}'(t)\|\, dt = \int_0^2 \sqrt{5 + 4t^2}\, dt$

$= \sqrt{21} + \frac{5}{4}\ln 5 - \frac{5}{4}\ln\left(\sqrt{105} - 4\sqrt{5}\right) \approx 6.2638$

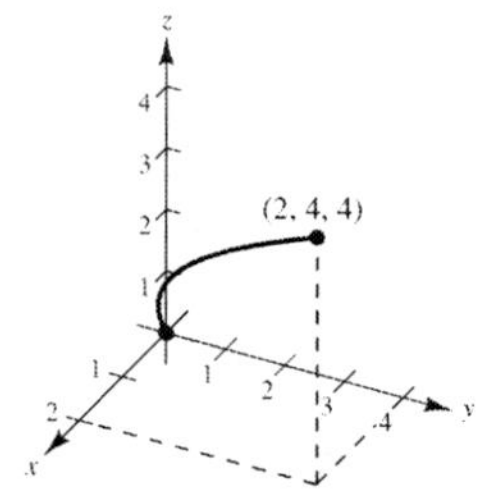

59. $\mathbf{r}(t) = \langle 8\cos t, 8\sin t, t\rangle, \left[0, \dfrac{\pi}{2}\right]$

$\mathbf{r}'(t) = \langle -8\sin t, 8\cos t, 1\rangle, \|\mathbf{r}'(t)\| = \sqrt{65}$

$s = \int_a^b \|\mathbf{r}'(t)\|\, dt = \int_0^{\pi/2} \sqrt{65}\, dt = \dfrac{\pi\sqrt{65}}{2}$

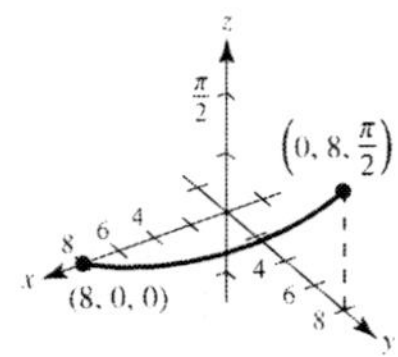

60. $\mathbf{r}(t) = \langle 2(\sin t - t\cos t), 2(\cos t + t\sin t), t\rangle, \left(0, \dfrac{\pi}{2}\right)$

$\mathbf{r}'(t) = \langle 2t\sin t, 2t\cos t, 1\rangle, \|\mathbf{r}'(t)\| = \sqrt{4t^2 + 1}$

$s = \int_a^b \|\mathbf{r}'(t)\|\, dt = \int_0^{\pi/2} \sqrt{4t^2 + 1}\, dt$

$= \dfrac{1}{4}\ln\left(\sqrt{\pi^2 + 1} + \pi\right) + \dfrac{\pi}{4}\sqrt{\pi^2 + 1} \approx 3.055$

61. $\mathbf{r}(t) = 3t\,\mathbf{i} + 2t\,\mathbf{j}$

Line

$K = 0$

62. $\mathbf{r}(t) = 2\sqrt{t}\,\mathbf{i} + 3t\mathbf{j}$

$$\mathbf{r}'(t) = \frac{1}{\sqrt{t}}\mathbf{i} + 3\mathbf{j}, \|\mathbf{r}'(t)\| = \sqrt{\frac{1}{t} + 9} = \sqrt{\frac{1 + 9t}{t}}$$

$$\mathbf{r}''(t) = -\frac{1}{2}t^{-3/2}\mathbf{i}$$

$$\mathbf{r}' \times \mathbf{r}'' = \begin{vmatrix} \mathbf{i} & \mathbf{j} & \mathbf{k} \\ \frac{1}{\sqrt{t}} & 3 & 0 \\ -\frac{1}{2}t^{-3/2} & 0 & 0 \end{vmatrix} = \frac{3}{2}t^{-3/2}\mathbf{k}; \|\mathbf{r}' \times \mathbf{r}''\| = \frac{3}{2t^{3/2}}$$

$$K = \frac{\|\mathbf{r}'(t) \times \mathbf{r}''(t)\|}{\|\mathbf{r}'(t)\|^3} = \frac{3/2t^{3/2}}{(1 + 9t)^{3/2}/t^{3/2}} = \frac{3}{2(1 + 9t)^{3/2}}$$

63. $\mathbf{r}(t) = 2t\mathbf{i} + \frac{1}{2}t^2\mathbf{j} + t^2\mathbf{k}$

$$\mathbf{r}'(t) = 2\mathbf{i} + t\mathbf{j} + 2t\mathbf{k}, \|\mathbf{r}'\| = \sqrt{5t^2 + 4}$$

$$\mathbf{r}''(t) = \mathbf{j} + 2\mathbf{k}$$

$$\mathbf{r}' \times \mathbf{r}'' = \begin{vmatrix} \mathbf{i} & \mathbf{j} & \mathbf{k} \\ 2 & t & 2t \\ 0 & 1 & 2 \end{vmatrix} = -4\mathbf{j} + 2\mathbf{k}, \|\mathbf{r}' \times \mathbf{r}''\| = \sqrt{20}$$

$$K = \frac{\|\mathbf{r}' \times \mathbf{r}''\|}{\|\mathbf{r}'\|^3} = \frac{\sqrt{20}}{(5t^2 + 4)^{3/2}} = \frac{2\sqrt{5}}{(4 + 5t^2)^{3/2}}$$

64. $\mathbf{r}(t) = 2t\mathbf{i} + 5\cos t\mathbf{j} + 5\sin t\mathbf{k}$

$$\mathbf{r}'(t) = 2\mathbf{i} - 5\sin t\mathbf{j} + 5\cos t\mathbf{k}, \|\mathbf{r}'(t)\| = \sqrt{29}$$

$$\mathbf{r}''(t) = 5\cos t\mathbf{j} - 5\sin t\mathbf{k}$$

$$\mathbf{r}' \times \mathbf{r}'' = \begin{vmatrix} \mathbf{i} & \mathbf{j} & \mathbf{k} \\ 2 & -5\sin t & 5\cos t \\ 0 & -5\cos t & -5\sin t \end{vmatrix}$$

$$= 25\mathbf{i} + 10\sin t\mathbf{j} - 10\cos t\mathbf{k}$$

$$\|\mathbf{r}' \times \mathbf{r}''\| = \sqrt{725}$$

$$K = \frac{\|\mathbf{r}' \times \mathbf{r}''\|}{\|\mathbf{r}'\|^3} = \frac{\sqrt{725}}{(29)^{3/2}} = \frac{\sqrt{25 \cdot 29}}{29\sqrt{29}} = \frac{5}{29}$$

65. $\mathbf{r}(t) = \frac{1}{2}t^2\mathbf{i} + t\mathbf{j} + \frac{1}{3}t^3\mathbf{k}, P\left(\frac{1}{2}, 1, \frac{1}{3}\right) \Rightarrow t = 1$

$$\mathbf{r}'(t) = t\mathbf{i} + \mathbf{j} + t^2\mathbf{k}, \mathbf{r}'(1) = \mathbf{i} + \mathbf{j} + \mathbf{k}$$

$$\mathbf{r}''(t) = \mathbf{i} + 2t\mathbf{k}, \mathbf{r}''(1) = \mathbf{i} + 2\mathbf{k}$$

$$\mathbf{r}' \times \mathbf{r}'' = \begin{vmatrix} \mathbf{i} & \mathbf{j} & \mathbf{k} \\ 1 & 1 & 1 \\ 1 & 0 & 2 \end{vmatrix} = 2\mathbf{i} - \mathbf{j} - \mathbf{k}$$

$$K = \frac{\|\mathbf{r}' \times \mathbf{r}''\|}{\|\mathbf{r}'\|^3} = \frac{\sqrt{4 + 1 + 1}}{(\sqrt{3})^3} = \frac{\sqrt{6}}{3\sqrt{3}} = \frac{\sqrt{2}}{3}$$

66. $\mathbf{r}(t) = 4\cos t\mathbf{i} + 3\sin t\mathbf{j} + t\mathbf{k}, P(-4, 0, \pi) \Rightarrow t = \pi$

$$\mathbf{r}'(t) = -4\sin t\mathbf{i} + 3\cos t\mathbf{j} + \mathbf{k}, \mathbf{r}'(\pi) = -3\mathbf{j} + \mathbf{k}$$

$$\mathbf{r}''(t) = -4\cos t\mathbf{i} - 3\sin t\mathbf{j}, \mathbf{r}''(\pi) = 4\mathbf{i}$$

$$\mathbf{r}' \times \mathbf{r}'' = \begin{vmatrix} \mathbf{i} & \mathbf{j} & \mathbf{k} \\ 0 & -3 & 1 \\ 4 & 0 & 0 \end{vmatrix} = 4\mathbf{j} + 12\mathbf{k}$$

$$K = \frac{\|\mathbf{r}' \times \mathbf{r}''\|}{\|\mathbf{r}'\|^3} = \frac{\sqrt{16 + 144}}{(9 + 1)^{3/2}} = \frac{2}{5}$$

67. $y = \frac{1}{2}x^2 + 2, x = 4$

$$y' = x$$

$$y'' = 1$$

$$K = \frac{|y''|}{\left[1 + (y')^2\right]^{3/2}} = \frac{1}{(1 + x^2)^{3/2}}$$

At $x = 4$, $K = \frac{1}{17^{3/2}}$ and $r = 17^{3/2} = 17\sqrt{17}$.

68. $y = e^{-x/2}, x = 0$

$$y' = -\frac{1}{2}e^{-x/2}, y'' = \frac{1}{4}e^{-x/2}$$

$$K = \frac{|y''|}{\left[1 + (y')^2\right]^{3/2}} = \frac{\frac{1}{4}e^{-x/2}}{\left[1 + \frac{1}{4}e^{-x}\right]^{3/2}}$$

At $x = 0$, $K = \frac{1/4}{(5/4)^{3/2}} = \frac{2}{5^{3/2}} = \frac{2}{5\sqrt{5}} = \frac{2\sqrt{5}}{25}$,

$$r = \frac{5\sqrt{5}}{2}.$$

69. $y = \ln x, x = 1$

$$y' = \frac{1}{x}$$

$$y'' = -\frac{1}{x^2}$$

$$K = \frac{|y''|}{\left[1 + (y')^2\right]^{3/2}} = \frac{1/x^2}{\left[1 + (1/x)^2\right]^{3/2}}$$

At $x = 1$, $K = \frac{1}{2^{3/2}} = \frac{1}{2\sqrt{2}} = \frac{\sqrt{2}}{4}$ and $r = 2\sqrt{2}$.

70. $y = \tan x,\ x = \dfrac{\pi}{4}$

$y' = \sec^2 x$

$y'' = 2\sec^2 x \tan x$

$$K = \frac{|y''|}{\left[1 + (y')^2\right]^{3/2}} = \frac{\left|2\sec^2 x \tan x\right|}{\left[1 + \sec^4 x\right]^{3/2}}$$

At $x = \dfrac{\pi}{4}$, $K = \dfrac{4}{5^{3/2}} = \dfrac{4}{5\sqrt{5}} = \dfrac{4\sqrt{5}}{25}$ and $r = \dfrac{5\sqrt{5}}{4}$.

71. $\mathbf{F} = ma_{\mathbf{N}} = mk\left(\dfrac{ds}{dt}\right)^2$

$$= \left(\frac{7200 \text{ lb}}{32 \text{ ft/sec}^2}\right)\left(\frac{1}{150 \text{ ft}}\right)\left(\frac{25(5280)\text{ft}}{3600 \text{ sec}}\right)^2$$

≈ 2016.67 pounds

Problem Solving for Chapter 12

1. $x(t) = \int_0^t \cos\left(\dfrac{\pi u^2}{2}\right)du,\ y(t) = \int_0^t \sin\left(\dfrac{\pi u^2}{2}\right)du$

$x'(t) = \cos\left(\dfrac{\pi t^2}{2}\right),\ y'(t) = \sin\left(\dfrac{\pi t^2}{2}\right)$

(a) $s = \int_0^a \sqrt{x'(t)^2 + y'(t)^2}\,dt = \int_0^a dt = a$

(b) $x''(t) = -\pi t \sin\left(\dfrac{\pi t^2}{2}\right),\ y''(t) = \pi t \cos\left(\dfrac{\pi t^2}{2}\right)$

$$K = \frac{\left|\pi t \cos^2\left(\frac{\pi t^2}{2}\right) + \pi t \sin^2\left(\frac{\pi t^2}{2}\right)\right|}{1} = \pi t$$

At $t = a$, $K = \pi a$.

(c) $K = \pi a = \pi$ (length)

2. $x^{2/3} + y^{2/3} = a^{2/3}$

$\dfrac{2}{3}x^{-1/3} + \dfrac{2}{3}y^{-1/3}y' = 0$

$y' = \dfrac{-y^{1/3}}{x^{1/3}}$ Slope at $P(x, y)$.

$\mathbf{r}(t) = \cos^3 t\,\mathbf{i} + \sin^3 t\,\mathbf{j}$

$\mathbf{r}'(t) = -3\cos^2 t \sin t\,\mathbf{i} + 3\sin^2 t \cos t\,\mathbf{j}$

$\|\mathbf{r}'(t)\|\mathbf{i} = |3\cos t \sin t|$

$\mathbf{T}(t) = \dfrac{\mathbf{r}'(t)}{\|\mathbf{r}'(t)\|} = -\cos t\,\mathbf{i} + \sin t\,\mathbf{j}$

$\mathbf{T}'(t) = \sin t\,\mathbf{i} + \cos t\,\mathbf{j}$

$Q(0, 0, 0)$ origin

$P = (\cos^3 t, \sin^3 t, 0)$ on curve.

$$\overline{PQ} \times \mathbf{T} = \begin{vmatrix} \mathbf{i} & \mathbf{j} & \mathbf{k} \\ \cos^3 t & \sin^3 t & 0 \\ -\cos t & \sin t & 0 \end{vmatrix} = (\cos^3 t \sin t - \sin^3 t \cos t)\mathbf{k}$$

$$D = \frac{\|\overline{PQ} \times \mathbf{T}\|}{\|\mathbf{T}\|} = |\cos t \sin t|$$

$$K = \frac{\|\mathbf{T}'(t)\|}{\|\mathbf{r}'(t)\|} = \frac{1}{|3\cos t \sin t|}$$

So, the radius of curvature, $\dfrac{1}{K}$, is three times the distance from the origin to the tangent line.

3. Bomb: $\mathbf{r}_1(t) = \langle 5000 - 400t, 3200 - 16t^2 \rangle$

Projectile: $\mathbf{r}_2(t) = \langle (v_0 \cos\theta)t, (v_0 \sin\theta)t - 16t^2 \rangle$

At 1600 feet: Bomb:

$3200 - 16t^2 = 1600 \Rightarrow t = 10$ seconds.

Projectile will travel 5 seconds:

$$5(v_0 \sin\theta) - 16(25) = 1600$$
$$v_0 \sin\theta = 400.$$

Horizontal position:

At $t = 10$, bomb is at $5000 - 400(10) = 1000$.

At $t = 5$, projectile is at $5v_0 \cos\theta$.

So, $v_0 \cos\theta = 200$.

Combining,

$$\frac{v_0 \sin\theta}{v_0 \cos\theta} = \frac{400}{200} \Rightarrow \tan\theta = 2 \Rightarrow \theta \approx 63.43°.$$

$$v_0 = \frac{200}{\cos\theta} \approx 447.2 \text{ ft/sec}$$

4. Bomb: $\mathbf{r}_1(t) = \langle 5000 + 400t, 3200 - 16t^2 \rangle$

Projectile: $\mathbf{r}_2(t) = \langle (v_0 \cos\theta)t, (v_0 \sin\theta)t - 16t^2 \rangle$

At 1600 feet: Bomb:

$3200 - 16t^2 = 1600 \Rightarrow t = 10$

Projectile will travel 5 seconds:

$$5(v_0 \sin\theta) - 16(25) = 1600$$
$$v_0 \sin\theta = 400.$$

Horizontal position:

At $t = 10$, bomb is at $5000 + 400(10) = 9000$.

At $t = 5$, projectile is at $(v_0 \cos\theta)5$.

So,

$$5v_0 \cos\theta = 9000$$
$$v_0 \cos\theta = 1800.$$

Combining,

$$\frac{v_0 \sin\theta}{v_0 \cos\theta} = \frac{400}{1800} \Rightarrow \tan\theta = \frac{2}{9} \Rightarrow \theta \approx 12.5°.$$

$$v_0 = \frac{1800}{\cos\theta} \approx 1843.9 \text{ ft/sec}$$

5. $x'(\theta) = 1 - \cos\theta,\ y'(\theta) = \sin\theta,\ 0 \le \theta \le 2\pi$

$$\sqrt{x'(\theta)^2 y'(\theta)^2} = \sqrt{(1 - \cos\theta)^2 + \sin^2\theta}$$
$$= \sqrt{2 - 2\cos\theta} = \sqrt{4\sin^2\frac{\theta}{2}}$$

$$s(t) = \int_\pi^t 2\sin\frac{\theta}{2}\,d\theta = \left[-4\cos\frac{\theta}{2}\right]_\pi^t = -4\cos\frac{t}{2}$$

$$x''(\theta) = \sin\theta,\ y''(\theta) = \cos\theta$$

$$K = \frac{\left|(1 - \cos\theta)\cos\theta - \sin\theta\sin\theta\right|}{\left(2\sin\frac{\theta}{2}\right)^3}$$
$$= \frac{\left|\cos\theta - 1\right|}{8\sin^3\frac{\theta}{2}}$$
$$= \frac{1}{4\sin\frac{\theta}{2}}$$

So, $\rho = \dfrac{1}{K} = 4\sin\dfrac{t}{2}$ and

$$s^2 + \rho^2 = 16\cos^2\left(\frac{t}{2}\right) + 16\sin^2\left(\frac{t}{2}\right) = 16.$$

6. $r = 1 - \cos\theta$

$r' = \sin\theta$

$$s(t) = \int_\pi^t \sqrt{(1 - \cos\theta)^2 + \sin^2\theta}\,d\theta = \int_\pi^t \sqrt{2 - 2\cos\theta}\,d\theta$$
$$= \int_\pi^t 2\sin\frac{\theta}{2}\,d\theta = \left[-4\cos\frac{\theta}{2}\right]_\pi^t = -4\cos\frac{t}{2}$$

$$K = \frac{\left|2(r')^2 - rr'' + r^2\right|}{\left[(r')^2 + r^2\right]^{3/2}}$$
$$= \frac{\left|2\sin^2\theta - (1 - \cos\theta)(\cos\theta) + (1 - \cos\theta)^2\right|}{8\sin^3\frac{\theta}{2}}$$
$$= \frac{\left|3 - 3\cos\theta\right|}{8\sin^3\frac{\theta}{2}} = \frac{3}{4}\frac{\sin^2\frac{\theta}{2}}{\sin^3\frac{\theta}{2}} = \frac{3}{4\sin\frac{\theta}{2}}$$

$$\rho = \frac{1}{K} = \frac{4\sin\frac{\theta}{2}}{3}$$

$$s^2 + 9\rho^2 = 16\cos^2\frac{\theta}{2} + 16\sin^2\frac{\theta}{2} = 16$$

7. $\|\mathbf{r}(t)\|^2 = \mathbf{r}(t)\cdot\mathbf{r}(t)$

$$\frac{d}{dt}\left(\|\mathbf{r}(t)\|\right)^2 = 2\|\mathbf{r}(t)\|\frac{d}{dt}\|\mathbf{r}(t)\| = \mathbf{r}(t)\cdot\mathbf{r}'(t) + \mathbf{r}'(t)\cdot\mathbf{r}(t) \Rightarrow \frac{d}{dt}\|\mathbf{r}(t)\| = \frac{\mathbf{r}(t)\cdot\mathbf{r}'(t)}{\|\mathbf{r}(t)\|}$$

8. (a) $\mathbf{r} = x\mathbf{i} + y\mathbf{j}$ position vector

$$\mathbf{r} = r\cos\theta\,\mathbf{i} + r\sin\theta\,\mathbf{j}$$

$$\frac{d\mathbf{r}}{dt} = \left[\frac{dr}{dt}\cos\theta - r\sin\theta\frac{d\theta}{dt}\right]\mathbf{i} + \left[\frac{dr}{dt}\sin\theta + r\cos\theta\frac{d\theta}{dt}\right]\mathbf{j}$$

$$\mathbf{a} = \frac{d^2\mathbf{r}}{dt^2} = \left[\frac{d^2r}{dt^2}\cos\theta - \frac{dr}{dt}\sin\theta\frac{d\theta}{dt} - \frac{dr}{dt}\sin\theta\frac{d\theta}{dt} - r\cos\theta\left(\frac{d\theta}{dt}\right)^2 - r\sin\theta\frac{d^2\theta}{dt^2}\right]\mathbf{i}$$

$$+ \left[\frac{d^2r}{dt^2}\sin\theta + \frac{dr}{dt}\cos\theta\frac{d\theta}{dt} + \frac{dr}{dt}\cos\theta\frac{d\theta}{dt} - r\sin\theta\left(\frac{d\theta}{dt}\right)^2 + r\cos\theta\frac{d^2\theta}{dt^2}\right]$$

$$a_r = \mathbf{a}\cdot\mathbf{u}_r = \mathbf{a}\cdot(\cos\theta\,\mathbf{i} + \sin\theta\,\mathbf{j})$$

$$= \left[\frac{d^2r}{dt^2}\cos^2\theta - 2\frac{dr}{dt}\sin\theta\cos\theta\frac{d\theta}{dt} - r\cos^2\theta\left(\frac{d\theta}{dt}\right)^2 - r\cos\theta\sin\theta\frac{d^2\theta}{dt^2}\right]$$

$$+ \left[\frac{d^2r}{dt^2}\sin^2\theta + 2\frac{dr}{dt}\sin\theta\cos\theta\frac{d\theta}{dt} - r\sin^2\theta\left(\frac{d\theta}{dt}\right)^2 + r\cos\theta\sin\theta\frac{d^2\theta}{dt^2}\right] = \frac{d^2r}{dt^2} - r\left(\frac{d\theta}{dt}\right)^2$$

$$a_\theta = \mathbf{a}\cdot u_\theta = \mathbf{a}\cdot(-\sin\theta\,\mathbf{i} + \cos\theta\,\mathbf{j}) = 2\frac{dr}{dr}\frac{d\theta}{dt} + r\frac{d^2\theta}{dt^2}$$

$$\mathbf{a} = (\mathbf{a}\cdot\mathbf{u}_r)\mathbf{u}_r + (\mathbf{a}\cdot\mathbf{u}_\theta)u_\theta = \left[\frac{d^2r}{dt^2} - r\left(\frac{d\theta}{dt}\right)^2\right]\mathbf{u}_r + \left[2\frac{dr}{dt}\frac{d\theta}{dt} + r\frac{d^2\theta}{dt^2}\right]\mathbf{u}_\theta$$

(b) $\mathbf{r} = 42{,}000\cos\left(\frac{\pi t}{12}\right)\mathbf{i} + 42{,}000\sin\left(\frac{\pi t}{12}\right)\mathbf{j}$

$$\mathbf{r} = 42{,}000,\ \frac{dr}{dt} = 0,\ \frac{d^2r}{dt^2} = 0$$

$$\frac{d\theta}{dt} = \frac{\pi}{12},\ \frac{d^2\theta}{dt^2} = 0$$

So, $\mathbf{a} = -42{,}000\left(\frac{\pi}{12}\right)^2\mathbf{u}_r = -\frac{875}{3}\pi^2\mathbf{u}_r.$

Radial component: $-\frac{875}{3}\pi^2$

Angular component: 0

9. $\mathbf{r}(t) = 4\cos t\,\mathbf{i} + 4\sin t\,\mathbf{j} + 3t\mathbf{k},\ t = \frac{\pi}{2}$

$$\mathbf{r}'(t) = -4\sin t\,\mathbf{i} + 4\cos t\,\mathbf{j} + 3\mathbf{k},\ \|\mathbf{r}'(t)\| = 5$$

$$\mathbf{r}''(t) = -4\cos t\,\mathbf{i} - 4\sin t\,\mathbf{j}$$

$$\mathbf{T} = -\frac{4}{5}\sin t\,\mathbf{i} + \frac{4}{5}\cos t\,\mathbf{j} + \frac{3}{5}\mathbf{k}$$

$$\mathbf{T}' = -\frac{4}{5}\cos t\,\mathbf{i} - \frac{4}{5}\sin t\,\mathbf{j}$$

$$\mathbf{N} = -\cos t\mathbf{i} - \sin t\mathbf{j}$$

$$\mathbf{B} = \mathbf{T}\times\mathbf{N} = \frac{3}{5}\sin t\,\mathbf{i} - \frac{3}{5}\cos t\,\mathbf{j} + \frac{4}{5}\mathbf{k}$$

At $t = \frac{\pi}{2}$, $\mathbf{T}\left(\frac{\pi}{2}\right) = -\frac{4}{5}\mathbf{i} + \frac{3}{5}\mathbf{k}$

$$\mathbf{N}\left(\frac{\pi}{2}\right) = -\mathbf{j}$$

$$\mathbf{B}\left(\frac{\pi}{2}\right) = \frac{3}{5}\mathbf{i} + \frac{4}{5}\mathbf{k}$$

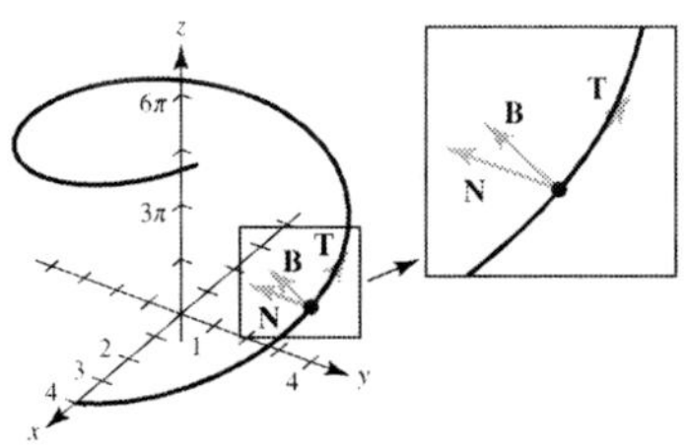

10. $\mathbf{r}(t) = \cos t\mathbf{i} + \sin t\mathbf{j} - \mathbf{k},\ t = \dfrac{\pi}{4}$

$\mathbf{r}'(t) = -\sin t\mathbf{i} + \cos t\mathbf{j},\ \|\mathbf{r}'(t)\| = 1$

$\mathbf{T} = -\sin t\mathbf{i} + \cos t\mathbf{j}$

$\mathbf{T}' = -\cos t\mathbf{i} - \sin t\mathbf{j}$

$\mathbf{N} = -\cos t\mathbf{i} - \sin t\mathbf{j}$

$\mathbf{B} = \mathbf{T} \times \mathbf{N} = \mathbf{k}$

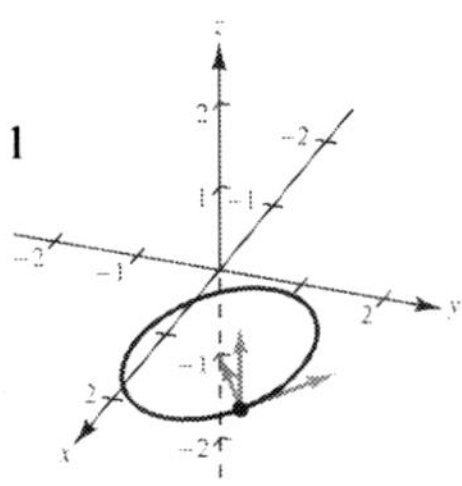

At $t = \dfrac{\pi}{4},\ \mathbf{T}\left(\dfrac{\pi}{4}\right) = -\dfrac{\sqrt{2}}{2}\mathbf{i} + \dfrac{\sqrt{2}}{2}\mathbf{j}$

$\mathbf{N}\left(\dfrac{\pi}{4}\right) = -\dfrac{\sqrt{2}}{2}\mathbf{i} - \dfrac{\sqrt{2}}{2}\mathbf{j}$

$\mathbf{B}\left(\dfrac{\pi}{4}\right) = \mathbf{k}$

11. (a) $\|\mathbf{B}\| = \|\mathbf{T} \times \mathbf{N}\| = 1$ constant length $\Rightarrow \dfrac{d\mathbf{B}}{ds} \perp \mathbf{B}$

$$\frac{d\mathbf{B}}{ds} = \frac{d}{ds}(\mathbf{T} \times \mathbf{N}) = (\mathbf{T} \times \mathbf{N}') + (\mathbf{T}' \times \mathbf{N})$$

$$\mathbf{T} \cdot \frac{d\mathbf{B}}{ds} = \mathbf{T} \cdot (\mathbf{T} \times \mathbf{N}') + \mathbf{T} \cdot (\mathbf{T}' \times \mathbf{N})$$

$$= (\mathbf{T} \times \mathbf{T}) \cdot \mathbf{N}' + \mathbf{T} \cdot \left(\mathbf{T}' \times \frac{\mathbf{T}'}{\|\mathbf{T}'\|}\right) = 0$$

So, $\dfrac{d\mathbf{B}}{ds} \perp \mathbf{B}$ and $\dfrac{d\mathbf{B}}{ds} \perp \mathbf{T} \Rightarrow \dfrac{d\mathbf{B}}{ds} = \tau\mathbf{N}$

for some scalar τ.

(b) $\mathbf{B} = \mathbf{T} \times \mathbf{N}$. Using Section 11.4, exercise 58,

$$\mathbf{B} \times \mathbf{N} = (\mathbf{T} \times \mathbf{N}) \times \mathbf{N} = -\mathbf{N} \times (\mathbf{T} \times \mathbf{N})$$
$$= -[(\mathbf{N} \cdot \mathbf{N})\mathbf{T} - (\mathbf{N} \cdot \mathbf{T})\mathbf{N}]$$
$$= -\mathbf{T}$$

$$\mathbf{B} \times \mathbf{T} = (\mathbf{T} \times \mathbf{N}) \times \mathbf{T} = -\mathbf{T} \times (\mathbf{T} \times \mathbf{N})$$
$$= -[(\mathbf{T} \cdot \mathbf{N})\mathbf{T} - (\mathbf{T} \cdot \mathbf{T})\mathbf{N}]$$
$$= \mathbf{N}.$$

Now, $K\mathbf{N} = \left\|\dfrac{d\mathbf{T}}{ds}\right\| \dfrac{\mathbf{T}'(s)}{\|\mathbf{T}'(s)\|} = \mathbf{T}'(s) = \dfrac{d\mathbf{T}}{ds}$

Finally,

$$\mathbf{N}'(s) = \frac{d}{ds}(\mathbf{B} \times \mathbf{T}) = (\mathbf{B} \times \mathbf{T}') + (\mathbf{B}' \times \mathbf{T})$$
$$= (\mathbf{B} \times K\mathbf{N}) + (-\tau\mathbf{N} \times \mathbf{T})$$
$$= -K\mathbf{T} + \tau\mathbf{B}.$$

12. $y = \dfrac{1}{32}x^{5/2}$

$y' = \dfrac{5}{64}x^{3/2}$

$y'' = \dfrac{15}{128}x^{1/2}$

$$K = \left|\frac{\frac{15}{128}x^{1/2}}{\left(1 + \frac{25}{4096}x^3\right)^{3/2}}\right|$$

At the point $(4, 1)$, $K = \dfrac{120}{(89)^{3/2}} \Rightarrow r = \dfrac{1}{K} = \dfrac{(89)^{3/2}}{120} \approx 7.$

13. $\mathbf{r}(t) = \langle t\cos \pi t, t\sin \pi t\rangle,\ 0 \le t \le 2$

(a)

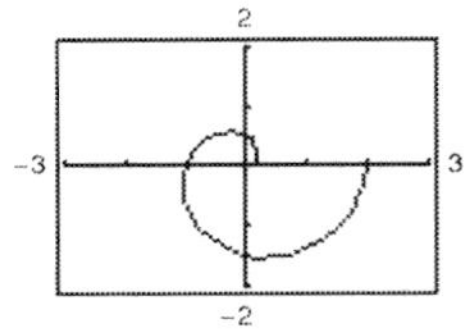

(b) Length $= \displaystyle\int_0^2 \|\mathbf{r}'(t)\|\,dt$

$= \displaystyle\int_0^2 \sqrt{\pi^2 t^2 + 1}\,dt$

≈ 6.766 (graphing utility)

(c) $K = \dfrac{\pi(\pi^2 t^2 + 2)}{[\pi^2 t^2 + 1]^{3/2}}$

$K(0) = 2\pi$

$K(1) = \dfrac{\pi(\pi^2 + 2)}{(\pi^2 + 1)^{3/2}} \approx 1.04$

$K(2) \approx 0.51$

(d)

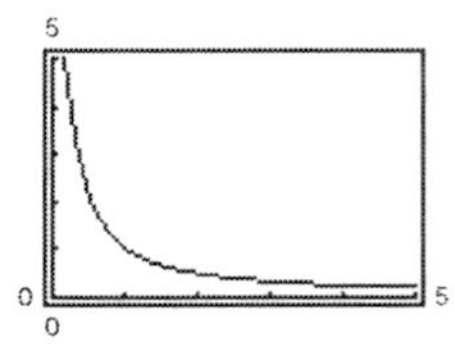

(e) $\displaystyle\lim_{t \to \infty} K = 0$

(f) As $t \to \infty$, the graph spirals outward and the curvature decreases.

14. (a) Eliminate the parameter to see that the Ferris wheel has a radius of 15 meters and is centered at $16\mathbf{j}$. At $t = 0$, the friend is located at $\mathbf{r}_1(0) = \mathbf{j}$, which is the low point on the Ferris wheel.

(b) If a revolution takes Δt seconds, then

$$\frac{\pi(t + \Delta t)}{10} = \frac{\pi t}{10} + 2\pi$$

and so $\Delta t = 20$ seconds. The Ferris wheel makes three revolutions per minute.

(c) The initial velocity is $r_2'(t_0) = -8.03\mathbf{i} + 11.47\mathbf{j}$. The speed is $\sqrt{8.03^2 + 11.47^2} \approx 14$ m/sec. The angle of inclination is $\arctan\left(\frac{11.47}{8.03}\right) \approx 0.96$ radians or 55°.

(d) Although you may start with other values, $t_0 = 0$ is a fine choice. The graph at the right shows two points of intersection. At $t = 3.15$ sec the friend is near the vertex of the parabola, which the object reaches when

$$t - t_0 = -\frac{11.47}{2(-4.9)} \approx 1.17 \text{ sec.}$$

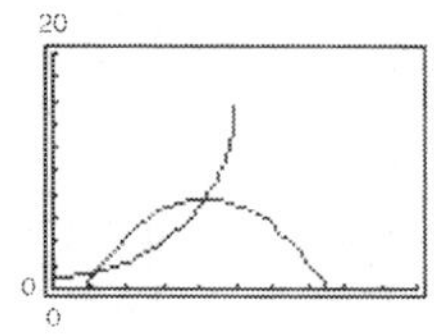

So, after the friend reaches the low point on the Ferris wheel, wait $t_0 = 2$ sec before throwing the object in order to allow it to be within reach.

(e) The approximate time is 3.15 seconds after starting to rise from the low point on the Ferris wheel. The friend has a constant speed of $\|\mathbf{r}_1'(t)\| = 15$ m/sec. The speed of the object at that time is

$$\|\mathbf{r}_2'(3.15)\| = \sqrt{8.03^2 + [11.47 - 9.8(3.15 - 2)]^2} \approx 8.03 \text{ m/sec.}$$

CHAPTER 13
Functions of Several Variables

CHAPTER 13
Functions of Several Variables

Section 13.1 Introduction to Functions of Several Variables

1. No, it is not the graph of a function. For some values of x and y (for example, $(x, y) = (0, 0)$), there are 2 z-values.

2. Yes, it is the graph of a function.

3. $x^2z + 3y^2 - xy = 10$

$$x^2z = 10 + xy - 3y^2$$

$$z = \frac{10 + xy - 3y^2}{x^2}$$

Yes, z is a function of x and y.

4. $xz^2 + 2xy - y^2 = 4$

No, z is not a function of x and y. For example, $(x, y) = (1, 0)$ corresponds to both $z = \pm 2$.

5. $\frac{x^2}{4} + \frac{y^2}{9} + z^2 = 1$

No, z is not a function of x and y. For example, $(x, y) = (0, 0)$ corresponds to both $z = \pm 1$.

6. $z + x \ln y - 8yz = 0$

$$z(1 - 8y) = -x \ln y$$

$$z = \frac{x \ln y}{8y - 1}$$

Yes, z is a function of x and y.

7. $f(x, y) = xy$

(a) $f(3, 2) = 3(2) = 6$

(b) $f(-1, 4) = -1(4) = -4$

(c) $f(30, 5) = 30(5) = 150$

(d) $f(5, y) = 5y$

(e) $f(x, 2) = 2x$

(f) $f(5, t) = 5t$

8. $f(x, y) = 4 - x^2 - 4y^2$

(a) $f(0, 0) = 4$

(b) $f(0, 1) = 4 - 0 - 4 = 0$

(c) $f(2, 3) = 4 - 4 - 36 = -36$

(d) $f(1, y) = 4 - 1 - 4y^2 = 3 - 4y^2$

(e) $f(x, 0) = 4 - x^2 - 0 = 4 - x^2$

(f) $f(t, 1) = 4 - t^2 - 4 = -t^2$

9. $f(x, y) = xe^y$

(a) $f(5, 0) = 5e^0 = 5$

(b) $f(3, 2) = 3e^2$

(c) $f(2, -1) = 2e^{-1} = \frac{2}{e}$

(d) $f(5, y) = 5e^y$

(e) $f(x, 2) = xe^2$

(f) $f(t, t) = te^t$

10. $g(x, y) = \ln|x + y|$

(a) $g(1, 0) = \ln|1 + 0| = 0$

(b) $g(0, -1) = \ln|0 - 1| = \ln 1 = 0$

(c) $g(0, e) = \ln|0 + e| = 1$

(d) $g(1, 1) = \ln|1 + 1| = \ln 2$

(e) $g\left(e, \frac{e}{2}\right) = \ln\left|e + \frac{e}{2}\right| = \ln\left(\frac{3e}{2}\right) = \ln 3 + \ln e - \ln 2$

$$= 1 + \ln 3 - \ln 2$$

(f) $g(2, 5) = \ln|2 + 5| = \ln 7$

11. $h(x, y, z) = \dfrac{xy}{z}$

(a) $h(2, 3, 9) = \dfrac{2(3)}{9} = \dfrac{2}{3}$

(b) $h(1, 0, 1) = \dfrac{1(0)}{1} = 0$

(c) $h(-2, 3, 4) = \dfrac{(-2)(3)}{4} = -\dfrac{3}{2}$

(d) $h(5, 4, -6) = \dfrac{5(4)}{-6} = -\dfrac{10}{3}$

12. $f(x, y, z) = \sqrt{x + y + z}$

(a) $f(0, 5, 4) = \sqrt{0 + 5 + 4} = 3$

(b) $f(6, 8, -3) = \sqrt{6 + 8 - 3} = \sqrt{11}$

(c) $f(4, 6, 2) = \sqrt{4 + 6 + 2} = \sqrt{12} = 2\sqrt{3}$

(d) $f(10, -4, -3) = \sqrt{10 - 4 - 3} = \sqrt{3}$

13. $f(x, y) = x \sin y$

(a) $f\left(2, \dfrac{\pi}{4}\right) = 2 \sin \dfrac{\pi}{4} = \sqrt{2}$

(b) $f(3, 1) = 3 \sin(1)$

(c) $f\left(-3, \dfrac{\pi}{3}\right) = -3 \sin \dfrac{\pi}{3} = -3\left(\dfrac{\sqrt{3}}{2}\right) = \dfrac{-3\sqrt{3}}{2}$

(d) $f\left(4, \dfrac{\pi}{2}\right) = 4 \sin \dfrac{\pi}{2} = 4$

14. $V(r, h) = \pi r^2 h$

(a) $V(3, 10) = \pi(3^2)10 = 90\pi$

(b) $V(5, 2) = \pi(5^2)2 = 50\pi$

(c) $V(4, 8) = \pi(4^2)8 = 128\pi$

(d) $V(6, 4) = \pi(6^2)4 = 144\pi$

15. $g(x, y) = \int_x^y (2t - 3)\, dt$

$= \left[t^2 - 3t\right]_x^y = y^2 - 3y - x^2 + 3x$

(a) $g(4, 0) = 0 - 16 + 12 = -4$

(b) $g(4, 1) = (1 - 3) - 16 + 12 = -6$

(c) $g\left(4, \frac{3}{2}\right) = \left(\frac{9}{4} - \frac{9}{2}\right) - 16 + 12 = -\frac{25}{4}$

(d) $g\left(\frac{3}{2}, 0\right) = 0 - \frac{9}{4} + \frac{9}{2} = \frac{9}{4}$

16. $g(x, y) = \int_x^y \dfrac{1}{t}\, dt = \ln|t|\Big]_x^y = \ln|y| - \ln|x| = \ln\left|\dfrac{y}{x}\right|$

(a) $g(4, 1) = \ln \dfrac{1}{4} = -\ln 4$

(b) $g(6, 3) = \ln \dfrac{3}{6} = -\ln 2$

(c) $g(2, 5) = \ln \dfrac{5}{2}$

(d) $g\left(\dfrac{1}{2}, 7\right) = \ln \dfrac{7}{\left(\frac{1}{2}\right)} = \ln 14$

17. $f(x, y) = 2x + y^2$

(a) $\dfrac{f(x + \Delta x, y) - f(x, y)}{\Delta x} = \dfrac{2(x + \Delta x) + y^2 - (2x + y^2)}{\Delta x} = \dfrac{2\Delta x}{\Delta x} = 2,\ \Delta x \neq 0$

(b) $\dfrac{f(x, y + \Delta y) - f(x, y)}{\Delta y} = \dfrac{2x + (y + \Delta y)^2 - 2x - y^2}{\Delta y} = \dfrac{2y\Delta y + (\Delta y^2)}{\Delta y} = 2y + \Delta y,\ \Delta y \neq 0$

18. $f(x, y) = 3x^2 - 2y$

(a) $\dfrac{f(x + \Delta x, y) - f(x, y)}{\Delta x} = \dfrac{3(x + \Delta x)^2 - 2y - (3x^2 - 2y)}{\Delta x} = \dfrac{6x\Delta x + 3(\Delta x)^2}{\Delta x} = 6x + 3\Delta x,\ \Delta x \neq 0$

(b) $\dfrac{f(x, y + \Delta y) - f(x, y)}{\Delta y} = \dfrac{3x^2 - 2(y + \Delta y) - (3x^2 - 2y)}{\Delta y} = \dfrac{-2\Delta y}{\Delta y} = -2,\ \Delta y \neq 0$

19. $f(x, y) = x^2 + y^2$

Domain:

$\{(x, y)$: x is any real number, y is any real number$\}$

Range: $z \geq 0$

20. $f(x, y) = e^{xy}$

Domain: Entire xy-plane

Range: $z > 0$

21. $g(x, y) = x\sqrt{y}$

Domain: $\{(x, y): y \geq 0\}$

Range: all real numbers

22. $f(x, y) = \dfrac{y}{\sqrt{x}}$

Domain: $\{(x, y): x > 0\}$

Range: all real numbers

23. $z = \dfrac{x + y}{xy}$

Domain: $\{(x, y): x \neq 0 \text{ and } y \neq 0\}$

Range: all real numbers

24. $z = \dfrac{xy}{x - y}$

Domain: $\{(x, y): x \neq y\}$

Range: all real numbers

25. $f(x, y) = \sqrt{4 - x^2 - y^2}$

Domain: $4 - x^2 - y^2 \geq 0$

$x^2 + y^2 \leq 4$

$\{(x, y): x^2 + y^2 \leq 4\}$

Range: $0 \leq z \leq 2$

26. $f(x, y) = \sqrt{4 - x^2 - 4y^2}$

Domain: $4 - x^2 - 4y^2 \geq 0$

$x^2 + 4y^2 \leq 4$

$\dfrac{x^2}{4} + \dfrac{y^2}{1} \leq 1$

$\left\{(x, y): \dfrac{x^2}{4} + \dfrac{y^2}{1} \leq 1\right\}$

Range: $0 \leq z \leq 2$

27. $f(x, y) = \arccos(x + y)$

Domain: $\{(x, y): -1 \leq x + y \leq 1\}$

Range: $0 \leq z \leq \pi$

28. $f(x, y) = \arcsin\left(\dfrac{y}{x}\right)$

Domain: $\left\{(x, y): -1 \leq \dfrac{y}{x} \leq 1\right\}$

Range: $-\dfrac{\pi}{2} \leq z \leq \dfrac{\pi}{2}$

29. $f(x, y) = \ln(4 - x - y)$

Domain: $4 - x - y > 0$

$x + y < 4$

$\{(x, y): y < -x + 4\}$

Range: all real numbers

30. $f(x, y) = \ln(xy - 6)$

Domain: $xy - 6 > 0$

$xy > 6$

$\{(x, y): xy > 6\}$

Range: all real numbers

31. $f(x, y) = \dfrac{-4x}{x^2 + y^2 + 1}$

(a) View from the positive x-axis: $(20, 0, 0)$

(b) View where x is negative, y and z are positive: $(-15, 10, 20)$

(c) View from the first octant: $(20, 15, 25)$

(d) View from the line $y = x$ in the xy-plane: $(20, 20, 0)$

32. (a) Domain:

$\{(x, y): x \text{ is any real number}, y \text{ is any real number}\}$

Range: $-2 \leq z \leq 2$

(b) $z = 0$ when $x = 0$ which represents points on the y-axis.

(c) No. When x is positive, z is negative. When x is negative, z is positive. The surface does not pass through the first octant, the octant where y is negative and x and z are positive, the octant where y is positive and x and z are negative, and the octant where x, y and z are all negative.

33. $f(x, y) = 4$

Plane: $z = 4$

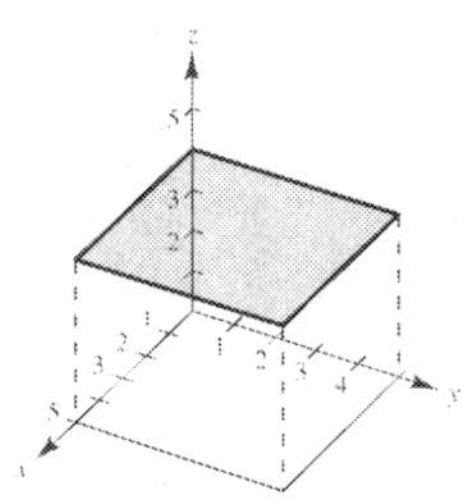

34. $f(x, y) = 6 - 2x - 3y$

Plane

Domain: entire xy-plane

Range: $-\infty < z < \infty$

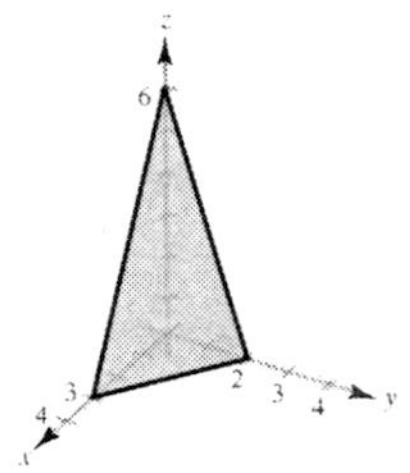

35. $f(x, y) = y^2$

Because the variable x is missing, the surface is a cylinder with rulings parallel to the x-axis. The generating curve is $z = y^2$. The domain is the entire xy-plane and the range is $z \geq 0$.

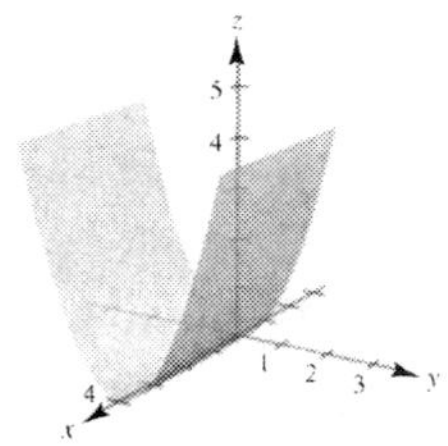

36. $g(x, y) = \frac{1}{2}y$

Plane: $z = \frac{1}{2}y$

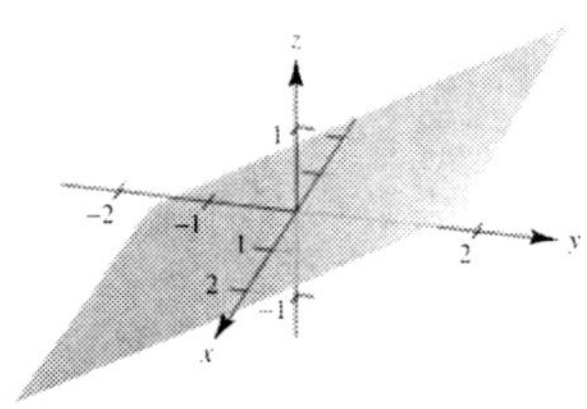

37. $z = -x^2 - y^2$

Paraboloid

Domain: entire xy-plane

Range: $z \leq 0$

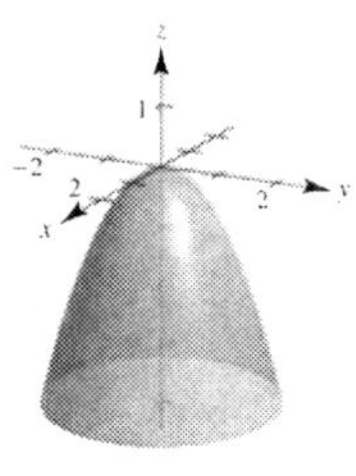

38. $z = \frac{1}{2}\sqrt{x^2 + y^2}$

Cone

Domain of f: entire xy-plane

Range: $z \geq 0$

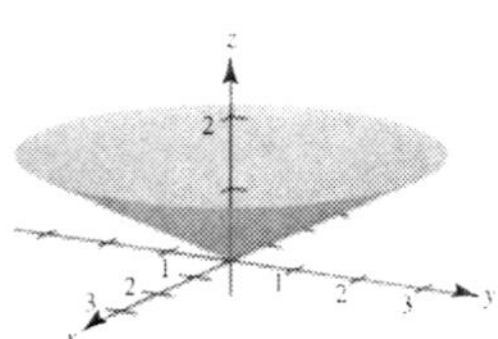

39. $f(x, y) = e^{-x}$

Because the variable y is missing, the surface is a cylinder with rulings parallel to the y-axis. The generating curve is $z = e^{-x}$.

The domain is the entire xy-plane and the range is $z > 0$.

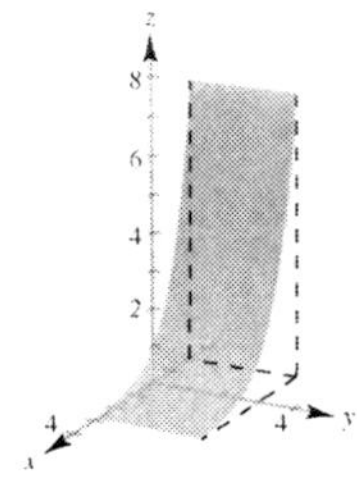

40. $f(x, y) = \begin{cases} xy, & x \geq 0, y \geq 0 \\ 0, & \text{elsewhere} \end{cases}$

Domain of f: entire xy-plane

Range: $z \geq 0$

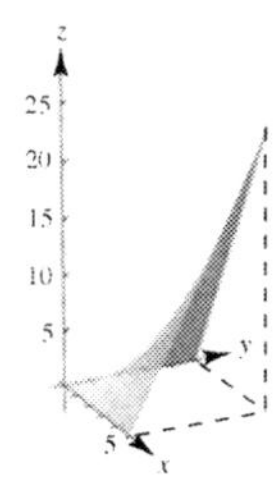

41. $z = y^2 - x^2 + 1$

Hyperbolic paraboloid

Domain: entire xy-plane

Range: $-\infty < z < \infty$

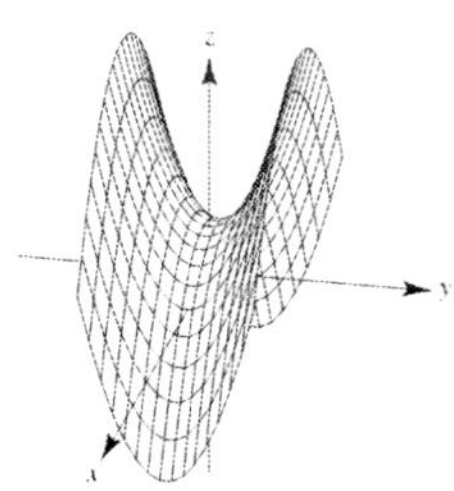

42. $f(x, y) = \frac{1}{12}\sqrt{144 - 16x^2 - 9y^2}$

Semi-ellipsoid

Domain: set of all points lying on or inside the ellipse

$$\left(\frac{x^2}{9}\right) + \left(\frac{y^2}{16}\right) = 1$$

Range: $0 \leq z \leq 1$

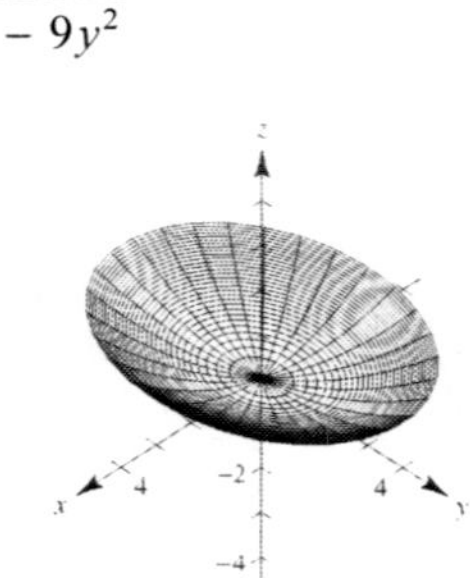

43. $f(x, y) = x^2e^{(-xy/2)}$

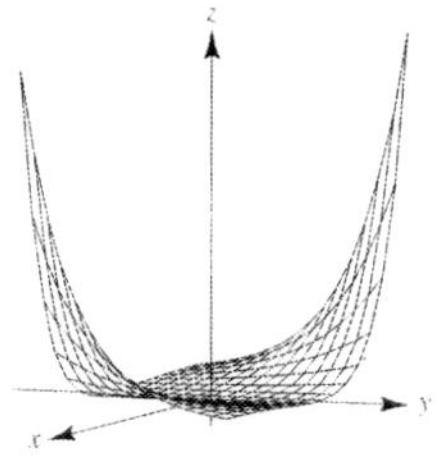

44. $f(x, y) = x\sin y$

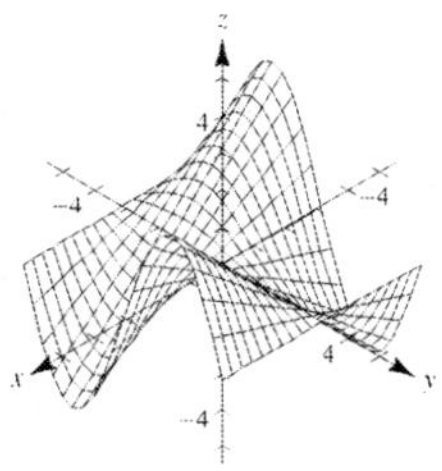

45. $z = e^{1-x^2-y^2}$

Level curves:

$$c = e^{1-x^2-y^2}$$
$$\ln c = 1 - x^2 - y^2$$
$$x^2 + y^2 = 1 - \ln c$$

Circles centered at $(0, 0)$

Matches (c)

46. $z = e^{1-x^2+y^2}$

Level curves:

$$c = e^{1-x^2+y^2}$$
$$\ln c = 1 - x^2 + y^2$$
$$x^2 - y^2 = 1 - \ln c$$

Hyperbolas centered at $(0, 0)$

Matches (d)

47. $z = \ln|y - x^2|$

Level curves:

$$c = \ln|y - x^2|$$
$$\pm e^c = y - x^2$$
$$y = x^2 \pm e^c$$

Parabolas

Matches (b)

48. $z = \cos\left(\dfrac{x + 2y^2}{4}\right)$

Level curves:

$$c = \cos\left(\frac{x^2 + 2y^2}{4}\right)$$
$$\cos^{-1} c = \frac{x^2 + 2y^2}{4}$$
$$x^2 + 2y^2 = 4\cos^{-1} c$$

Ellipses

Matches (a)

49. $z = x + y$

Level curves are parallel lines of the form $x + y = c$.

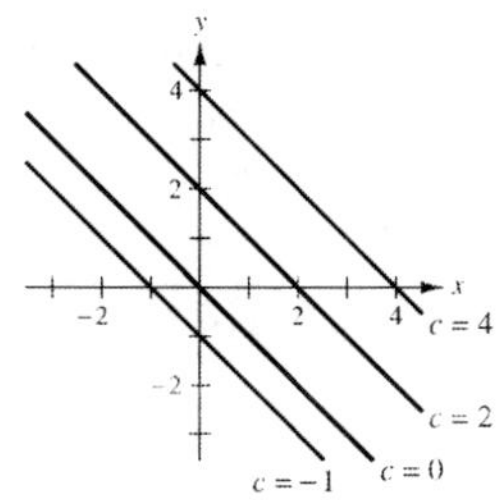

50. $f(x, y) = 6 - 2x - 3y$

The level curves are of the form $6 - 2x - 3y = c$ or $2x + 3y = 6 - c$. So, the level curves are straight lines with a slope of $-\frac{2}{3}$.

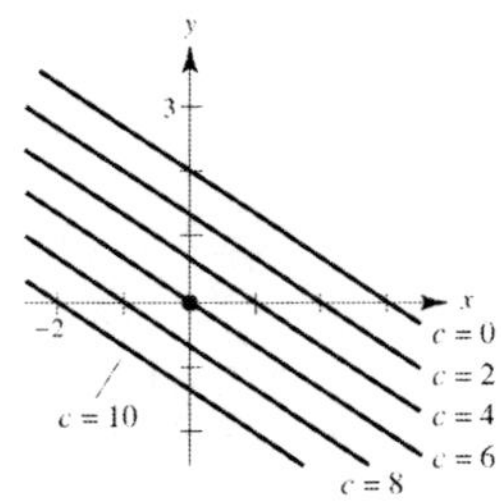

51. $z = x^2 + 4y^2$

The level curves are ellipses of the form

$x^2 + 4y^2 = c$

(except $x^2 + 4y^2 = 0$ is the point $(0, 0)$).

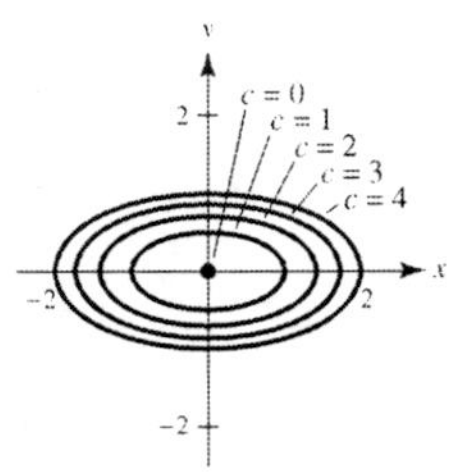

52. $f(x, y) = \sqrt{9 - x^2 - y^2}$

The level curves are of the form

$c = \sqrt{9 - x^2 - y^2}$

$x^2 + y^2 = 9 - c^2$, circles.

$(x^2 + y^2 = 0$ is the point $(0, 0).)$

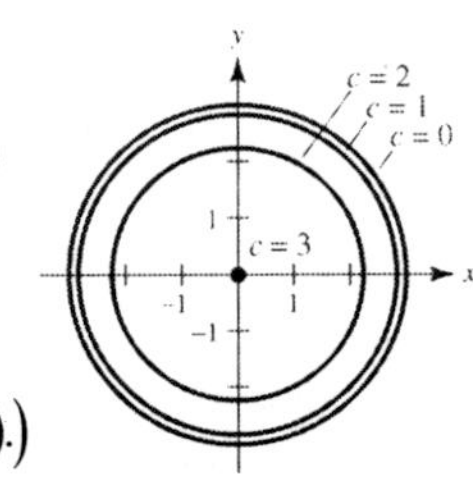

53. $f(x, y) = xy$

The level curves are hyperbolas of the form $xy = c$.

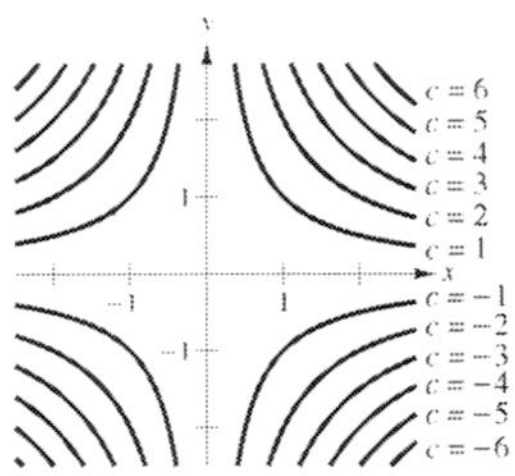

54. $f(x, y) = e^{xy/2}$

The level curves are of the form

$e^{xy/2} = c$, or $\ln c = \dfrac{xy}{2}$.

So, the level curves are hyperbolas.

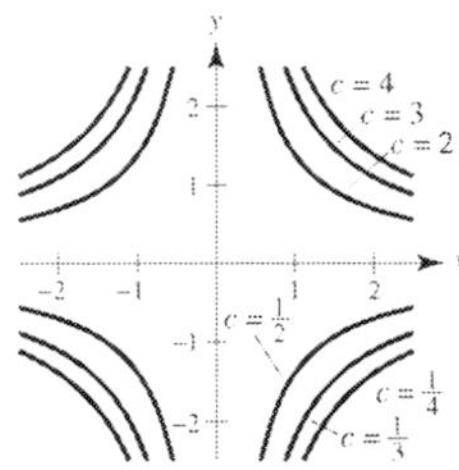

55. $f(x, y) = \dfrac{x}{x^2 + y^2}$

The level curves are of the form

$$c = \frac{x}{x^2 + y^2}$$

$$x^2 - \frac{x}{c} + y^2 = 0$$

$$\left(x - \frac{1}{2c}\right)^2 + y^2 = \left(\frac{1}{2c}\right)^2.$$

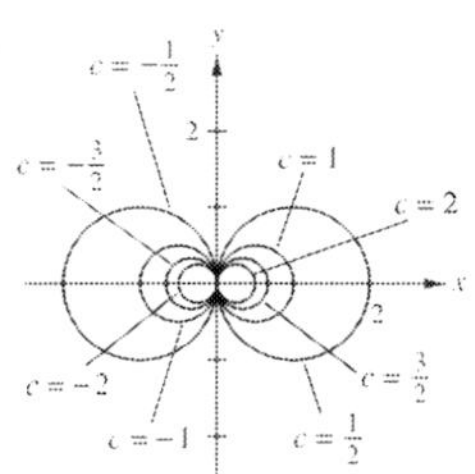

So, the level curves are circles passing through the origin and centered at $(\pm 1/2c, 0)$.

56. $f(x, y) = \ln(x - y)$

The level curves are of the form

$c = \ln(x - y)$

$e^c = x - y$

$y = x - e^c$.

So, the level curves are parallel lines of slope 1 passing through the fourth quadrant.

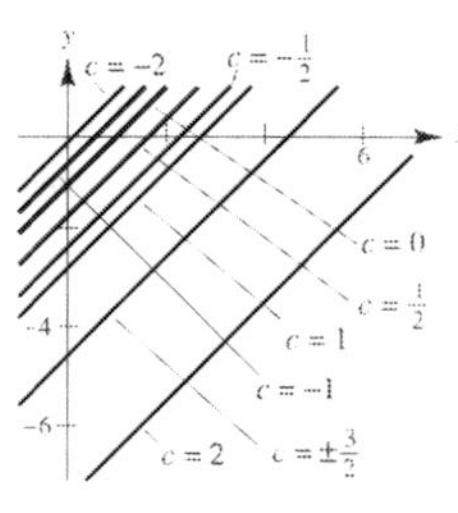

57. $f(x, y) = x^2 - y^2 + 2$

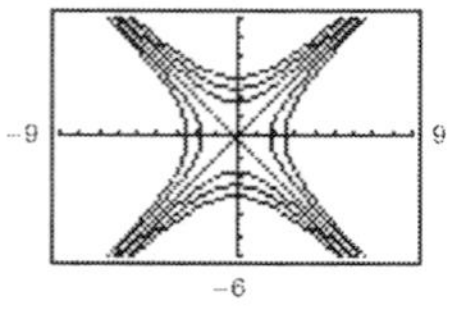

58. $f(x, y) = |xy|$

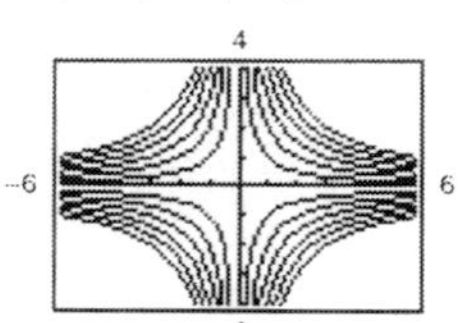

59. $g(x, y) = \dfrac{8}{1 + x^2 + y^2}$

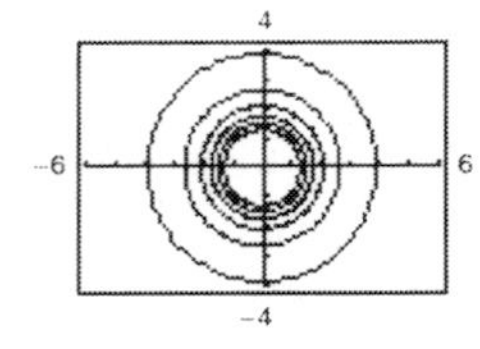

60. $h(x, y) = 3\sin(|x| + |y|)$

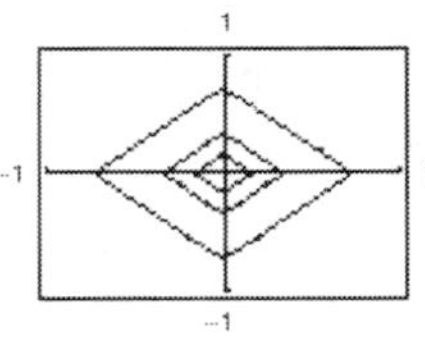

61. The graph of a function of two variables is the set of all points (x, y, z) for which $z = f(x, y)$ and (x, y) is in the domain of f. The graph can be interpreted as a surface in space. Level curves are the scalar fields $f(x, y) = c$, where c is a constant.

62. No, the following graphs are not hemispheres.

$z = e^{-(x^2 + y^2)}$

$z = x^2 + y^2$

63. $f(x, y) = \dfrac{x}{y}$

The level curves are the lines $c = \dfrac{x}{y}$ or $y = \dfrac{1}{c}x$.

These lines all pass through the origin.

64. $f(x, y) = xy$, $x \ge 0$, $y \ge 0$

(a)

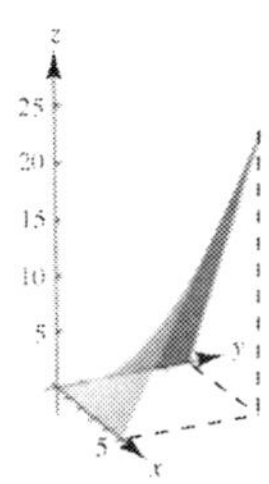

(b) g is a vertical translation of f three units downward.

(c) g is a reflection of f in the xy-plane.

(d) The graph of g is lower than the graph of f. If $z = f(x, y)$ is on the graph of f, then $\frac{1}{2}z$ is on the graph of g.

(e)

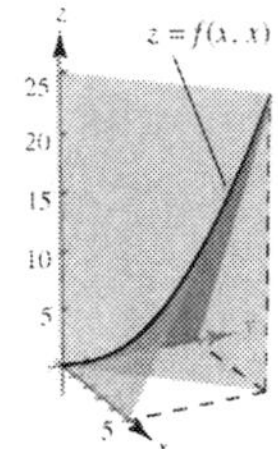

65. The surface is sloped like a saddle. The graph is not unique. Any vertical translation would have the same level curves.

One possible function is

$f(x, y) = |xy|.$

66. The surface could be an ellipsoid centered at $(0, 1, 0)$.

One possible function is

$$f(x, y) = x^2 + \frac{(y - 1)^2}{4} - 1.$$

67. $V(I, R) = 1000\left[\dfrac{1 + 0.06(1 - R)}{1 + I}\right]^{10}$

	Inflation Rate		
Tax Rate	0	0.03	0.05
0	1790.85	1332.56	1099.43
0.28	1526.43	1135.80	937.09
0.35	1466.07	1090.90	900.04

68. $A(r, t) = 5000e^{rt}$

	Number of Year			
Rate	5	10	15	20
0.02	5525.85	6107.01	6749.29	7459.12
0.03	5809.17	6749.29	7841.56	9110.59
0.04	6107.01	7459.12	9110.59	11,127.70
0.05	6420.13	8243.61	10,585.00	13,591.41

69. $f(x, y, z) = x - y + z, c = 1$

$1 = x - y + z$, Plane

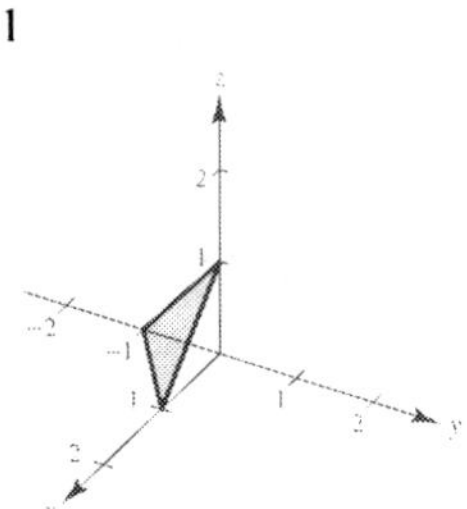

70. $f(x, y, z) = 4x + y + 2z$

$c = 4$

$4 = 4x + y + 2z$

Plane

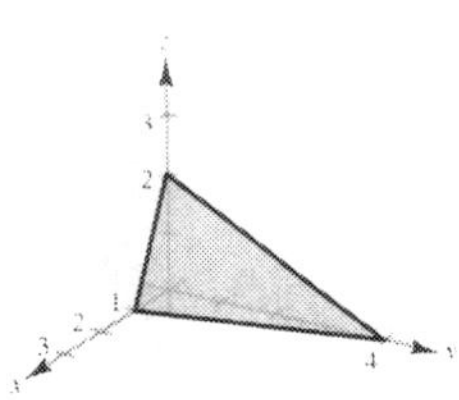

71. $f(x, y, z) = x^2 + y^2 + z^2$

$c = 9$

$9 = x^2 + y^2 + z^2$

Sphere

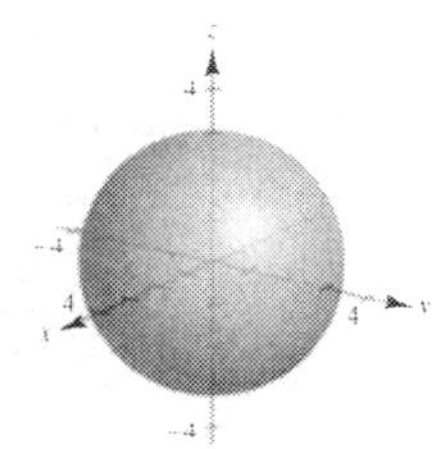

72. $f(x, y, z) = x^2 + \frac{1}{4}y^2 - z$

$c = 1$

$1 = x^2 + \frac{1}{4}y^2 - z$

Elliptic paraboloid

Vertex: $(0, 0, -1)$

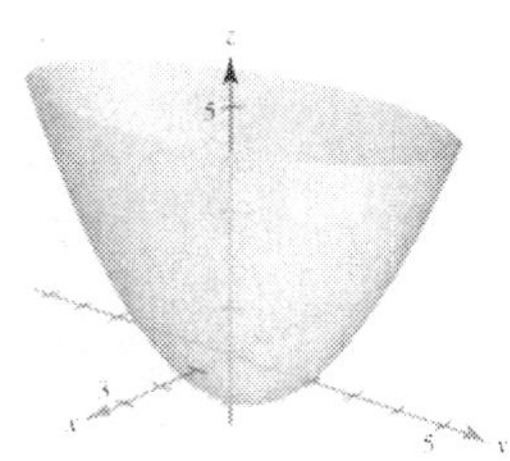

73. $f(x, y, z) = 4x^2 + 4y^2 - z^2$

$c = 0$

$0 = 4x^2 + 4y^2 - z^2$

Elliptic cone

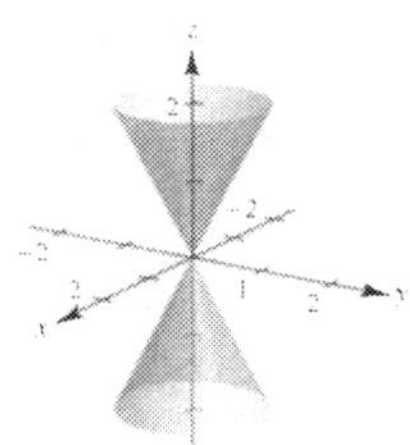

74. $f(x, y, z) = \sin x - z$

$c = 0$

$0 = \sin x - z$ or $z = \sin x$

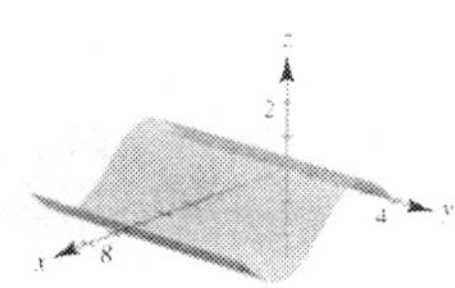

75. $N(d, L) = \left(\dfrac{d - 4}{4}\right)^2 L$

(a) $N(22, 12) = \left(\dfrac{22 - 4}{4}\right)^2(12) = 243$ board-feet

(b) $N(30, 12) = \left(\dfrac{30 - 4}{4}\right)^2(12) = 507$ board-feet

76. $w = \dfrac{1}{x - y}, \; y < x$

(a) $w(15, 9) = \dfrac{1}{15 - 9} = \dfrac{1}{6}\text{h} = 10$ min

(b) $w(15, 13) = \dfrac{1}{15 - 13} = \dfrac{1}{2}\text{h} = 30$ min

(c) $w(12, 7) = \dfrac{1}{12 - 7} = \dfrac{1}{5}\text{h} = 12$ min

(d) $w(5, 2) = \dfrac{1}{5 - 2} = \dfrac{1}{3}\text{h} = 20$ min

77. $T = 600 - 0.75x^2 - 0.75y^2$

The level curves are of the form

$$c = 600 - 0.75x^2 - 0.75y^2$$

$$x^2 + y^2 = \frac{600 - c}{0.75}.$$

The level curves are circles centered at the origin.

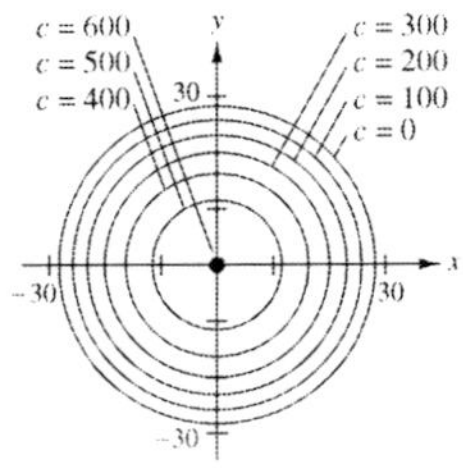

78. $V(x, y) = \dfrac{5}{\sqrt{25 + x^2 + y^2}}$

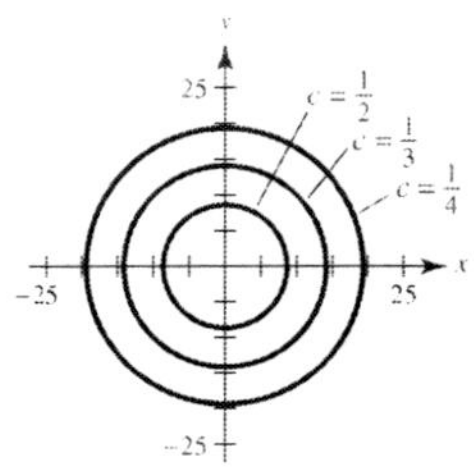

79.

$$\begin{aligned} f(x, y) &= 100x^{0.6}y^{0.4} \\ f(2x, 2y) &= 100(2x)^{0.6}(2y)^{0.4} \\ &= 100(2)^{0.6}x^{0.6}(2)^{0.4}y^{0.4} \\ &= 100(2)^{0.6}(2)^{0.4}x^{0.6}y^{0.4} \\ &= 2\left[100x^{0.6}y^{0.4}\right] = 2f(x, y) \end{aligned}$$

80. $z = Cx^a y^{1-a}$

$$\ln z = \ln C + a \ln x + (1 - a) \ln y$$

$$\ln z - \ln y = \ln C + a \ln x - a \ln y$$

$$\ln \frac{z}{y} = \ln C + a \ln \frac{x}{y}$$

81. $PV = kT$

(a) $26(2000) = k(300) \Rightarrow k = \dfrac{520}{3}$

(b) $P = \dfrac{kT}{V} = \dfrac{520}{3}\left(\dfrac{T}{V}\right)$

The level curves are of the form

$$c = \frac{520}{3}\left(\frac{T}{V}\right), \text{ or } V = \frac{520}{3c}T.$$

These are lines through the origin with slope $\dfrac{520}{3c}$.

82. $z = f(x, y) = 0.035x + 0.640y - 1.77$

(a)

Year	2006	2007	2008	2009	2010	2011
z	10.0	14.5	22.3	31.6	47.8	76.6
Model	9.9	15.0	22.7	30.1	48.6	76.5

(b) y has the greater influence because its coefficient (0.640) is greater than the coefficient of x (0.035).

(c) $f(x, 150) = 0.035x + 0.640(150) - 1.77$
$= 0.035x + 94.23$

This gives the shareholder's equity z in terms of net sales x, assuming total assets of $150 billion.

83. (a) Highest pressure at C

(b) Lowest pressure at A

(c) Highest wind velocity at B

84. Southwest

85. $C = \underbrace{1.20xy}_{\text{base}} + \underbrace{2(0.75)xz}_{\text{front and back}} + \underbrace{2(0.75)yz}_{\text{2 ends}}$

$= 1.20xy + 1.50(xz + yz)$

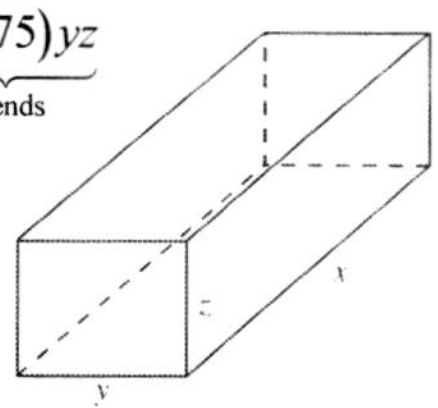

86. (a) No; the level curves are uneven and sporadically spaced.

(b) Use more colors.

87. False. Let

$f(x, y) = 2xy$

$f(1, 2) = f(2, 1)$, but $1 \neq 2$.

88. False. Let

$f(x, y) = 5.$

Then, $f(2x, 2y) = 5 \neq 2^2 f(x, y)$.

89. True

90. False. If there were a point (x, y) on the level curves $f(x, y) = C_1$ and $f(x, y) = C_2$, then $C_1 = C_2$.

91. We claim that $g(x) = f(x, 0)$. First note that $x = y = z = 0$ implies $3f(0, 0) = 0 \Rightarrow f(0, 0) = 0$.

Letting $y = z = 0$ implies $f(x, 0) + f(0, 0) + f(0, x) = 0 \Rightarrow -f(0, x) = f(x, 0)$.

Letting $z = 0$ implies $f(x, y) + f(y, 0) + f(0, x) = 0 \Rightarrow f(x, y) = -f(y, 0) - f(0, x) = f(x, 0) - f(y, 0)$.

Hence, $f(x, y) = g(x) - g(y)$, as desired.

Section 13.2 Limits and Continuity

1. $\lim\limits_{(x, y)\to(1, 0)} x = 1$

$f(x, y) = x, L = 1$

We need to show that for all $\varepsilon > 0$, there exists a δ-neighborhood about $(1, 0)$ such that

$|f(x, y) - L| = |x - 1| < \varepsilon$

Whenever $(x, y) \neq (1, 0)$ lies in the neighborhood.

From $0 < \sqrt{(x-1)^2 + (y-0)^2} < \delta$, it follows that

$|x - 1| = \sqrt{(x-1)^2} \le \sqrt{(x-1)^2 + (y-0)^2} < \delta.$

So, choose $\delta = \varepsilon$ and the limit is verified.

2. $\lim\limits_{(x, y)\to(4, -1)} x = 4$

Let $\varepsilon > 0$ be given. We need to find $\delta > 0$ such that

$|f(x, y) - L| = |x - 4| < \varepsilon$

whenever

$0 < \sqrt{(x-a)^2 + (y-b)^2} = \sqrt{(x-4)^2 + (y+1)^2} < \delta.$

Take $\delta = \varepsilon$.

Then if $0 < \sqrt{(x-4)^2 + (y+1)^2} < \delta = \varepsilon$, we have

$\sqrt{(x-4)^2} < \varepsilon$

$|x - 4| < \varepsilon.$

3. $\lim\limits_{(x, y)\to(1, -3)} y = -3$. $f(x, y) = y$, $L = -3$

We need to show that for all $\varepsilon > 0$, there exists a δ-neighborhood about $(1, -3)$ such that

$|f(x, y) - L| = |y + 3| < \varepsilon$

whenever $(x, y) \neq (1, -3)$ lies in the neighborhood.

From $0 < \sqrt{(x-1)^2 + (y+3)^2} < \delta$ it follows that

$|y + 3| = \sqrt{(y+3)^2} \le \sqrt{(x-1)^2 + (y+3)^2} < \delta.$

So, choose $\delta = \varepsilon$ and the limit is verified.

4. $\lim\limits_{(x, y)\to(a, b)} y = b$

Let $\varepsilon > 0$ be given. We need to find $\delta > 0$ such that

$|f(x, y) - L| = |y - b| < \varepsilon$

whenever $0 < \sqrt{(x-a)^2 + (y-b)^2} < \delta$. Take $\delta = \varepsilon$.

Then if $0 < \sqrt{(x-a)^2 + (y-b)^2} < \delta = \varepsilon$, we have

$\sqrt{(y-b)^2} < \varepsilon$

$|y - b| < \varepsilon.$

5. $\lim_{(x, y)\to(a, b)}\left[f(x, y) - g(x, y)\right] = \lim_{(x, y)\to(a, b)} f(x, y) - \lim_{(x, y)\to(a, b)} g(x, y) = 4 - 3 = 1$

6. $\lim_{(x, y)\to(a, b)}\left[\dfrac{5f(x, y)}{g(x, y)}\right] = \dfrac{5\left[\lim_{(x, y)\to(a, b)} f(x, y)\right]}{\lim_{(x, y)\to(a, b)} g(x, y)} = \dfrac{5(4)}{3} = \dfrac{20}{3}$

7. $\lim_{(x, y)\to(a, b)}\left[f(x, y)g(x, y)\right] = \left[\lim_{(x, y)\to(a, b)} f(x, y)\right]\left[\lim_{(x, y)\to(a, b)} g(x, y)\right] = 4(3) = 12$

8. $\lim_{(x, y)\to(a, b)}\left[\dfrac{f(x, y) + g(x, y)}{f(x, y)}\right] = \dfrac{\lim_{(x, y)\to(a, b)} f(x, y) + \lim_{(x, y)\to(a, b)} g(x, y)}{\lim_{(x, y)\to(a, b)} f(x, y)} = \dfrac{4 + 3}{4} = \dfrac{7}{4}$

9. $\lim_{(x, y)\to(2, 1)}\left(2x^2 + y\right) = 8 + 1 = 9$

Continuous everywhere

10. $\lim_{(x, y)\to(0, 0)}\left(x + 4y + 1\right) = 0 + 4(0) + 1 = 1$

Continuous everywhere

11. $\lim_{(x, y)\to(1, 2)} e^{xy} = e^{1(2)} = e^2$

Continuous everywhere

12. $\lim_{(x, y)\to(2, 4)} \dfrac{x + y}{x^2 + 1} = \dfrac{2 + 4}{2^2 + 1} = \dfrac{6}{5}$

Continuous everywhere

13. $\lim_{(x, y)\to(0, 2)} \dfrac{x}{y} = \dfrac{0}{2} = 0$

Continuous for all $y \neq 0$

14. $\lim_{(x, y)\to(-1, 2)} \dfrac{x + y}{x - y} = \dfrac{-1 + 2}{-1 - 2} = -\dfrac{1}{3}$

Continuous for all $x \neq y$.

15. $\lim_{(x, y)\to(1, 1)} \dfrac{xy}{x^2 + y^2} = \dfrac{1}{2}$

Continuous except at $(0, 0)$

16. $\lim_{(x, y)\to(1, 1)} \dfrac{x}{\sqrt{x + y}} = \dfrac{1}{\sqrt{1 + 1}} = \dfrac{\sqrt{2}}{2}$

Continuous for $x + y > 0$

17. $\lim_{(x, y)\to(\pi/4, 2)} y\cos(xy) = 2\cos\dfrac{\pi}{2} = 0$

Continuous everywhere

18. $\lim_{(x, y)\to(2\pi, 4)} \sin\dfrac{x}{y} = \sin\dfrac{2\pi}{4} = 1$

Continuous for all $y \neq 0$

19. $\lim_{(x, y)\to(0, 1)} \dfrac{\arcsin xy}{1 - xy} = \dfrac{\arcsin 0}{1} = 0$

Continuous for $xy \neq 1$, $|xy| \leq 1$

20. $\lim_{(x, y)\to(0, 1)} \dfrac{\arccos\left(\dfrac{x}{y}\right)}{1 + xy} = \dfrac{\arccos 0}{1} = \dfrac{\pi}{2}$

Continuous for $xy \neq -1$, $y \neq 0$, $0 \leq \dfrac{x}{y} \leq \pi$

21. $\lim_{(x, y, z)\to(1, 3, 4)} \sqrt{x + y + z} = \sqrt{1 + 3 + 4} = 2\sqrt{2}$

Continuous for $x + y + z \geq 0$

22. $\lim_{(x, y, z)\to(-2, 1, 0)} xe^{yz} = (-2)e^{1(0)} = -2$

Continuous everywhere

23. $\lim_{(x, y)\to(1, 1)} \dfrac{xy - 1}{1 + xy} = \dfrac{1 - 1}{1 + 1} = 0$

24. $\lim_{(x, y)\to(1, -1)} \dfrac{x^2y}{1 + xy^2} = \dfrac{-1}{1 + 1} = -\dfrac{1}{2}$

25. $\lim_{(x, y)\to(0, 0)} \dfrac{1}{x + y}$ does not exist

Because the denominator $x + y$ approaches 0 as $(x, y) \to (0, 0)$.

26. $\lim_{(x, y)\to(0, 0)} \dfrac{1}{x^2y^2}$ does not exist because the denominator xy approaches 0 as $(x, y) \to (0, 0)$.

27. $\lim_{(x, y)\to(0, 0)} \dfrac{x - y}{\sqrt{x} - \sqrt{y}}$

does not exist because you can't approach $(0, 0)$ from negative values of x and y.

28. $\lim_{(x,y)\to(2,1)} \dfrac{x-y-1}{\sqrt{x-y}-1}\cdot\dfrac{\sqrt{x-y}+1}{\sqrt{x-y}+1}$

$= \lim_{(x,y)\to(2,1)} \dfrac{(x-y-1)(\sqrt{x-y}+1)}{(x-y)-1}$

$= \lim_{(x,y)\to(2,1)} (\sqrt{x-y}+1) = 2$

29. The limit does not exist because along the line $y = 0$ you have

$$\lim_{(x,y)\to(0,0)} \frac{x+y}{x^2+y} = \lim_{(x,0)\to(0,0)} \frac{x}{x^2} = \lim_{(x,0)\to(0,0)} \frac{1}{x}$$

which does not exist.

30. The limit does not exist because along the line $x = y$ you have

$$\lim_{(x,y)\to(0,0)} \frac{x}{x^2-y^2} = \lim_{(x,x)\to(0,0)} \frac{x}{x^2-x^2} = \lim_{(x,x)\to(0,0)} \frac{x}{0}.$$

Because the denominator is 0, the limit does not exist.

31. $\lim_{(x,y)\to(0,0)} \dfrac{x^2}{(x^2+1)(y^2+1)} = \dfrac{0}{(1)(1)} = 0$

32. $\lim_{(x,y)\to(0,0)} \ln(x^2+y^2)$ does not exist

because $\ln(x^2+y^2) \to -\infty$ as $(x, y) \to (0, 0)$.

33. The limit does not exist because along the path $x = 0,\ y = 0$, you have

$$\lim_{(x,y,z)\to(0,0,0)} \frac{xy+yz+xz}{x^2+y^2+z^2} = \lim_{(0,0,z)\to(0,0,0)} \frac{0}{z^2} = 0$$

whereas along the path $x = y = z$, you have

$$\lim_{(x,y,z)\to(0,0,0)} \frac{xy+yz+xz}{x^2+y^2+z^2} = \lim_{(x,x,x)\to(0,0,0)} \frac{x^2+x^2+x^2}{x^2+x^2+x^2} = 1$$

34. The limit does not exist because along the path $y = z = 0$, you have

$$\lim_{(x,y,z)\to(0,0,0)} \frac{xy+yz^2+xz^2}{x^2+y^2+z^2} = \lim_{(x,0,0)\to(0,0,0)} \frac{0}{x^2} = 0$$

However, along the path $z = 0,\ x = y$, you have

$$\lim_{(x,y,z)\to(0,0,0)} \frac{xy+yz^2+xz^2}{x^2+y^2+z^2} = \lim_{(x,x,0)\to(0,0,0)} \frac{x^2}{x^2+x^2} = \frac{1}{2}$$

35. $\lim_{(x,y)\to(0,0)} e^{xy} = 1$

Continuous everywhere

36. $\lim_{(x,y)\to(0,0)} \left[1 - \dfrac{\cos(x^2+y^2)}{x^2+y^2}\right] = -\infty$

The limit does not exist.

Continuous except at $(0, 0)$

37. $f(x, y) = \dfrac{xy}{x^2+y^2}$

Continuous except at $(0, 0)$

Path: $y = 0$

(x, y)	$(1, 0)$	$(0.5, 0)$	$(0.1, 0)$	$(0.01, 0)$	$(0.001, 0)$
$f(x, y)$	0	0	0	0	0

Path: $y = x$

(x, y)	$(1, 1)$	$(0.5, 0.5)$	$(0.1, 0.1)$	$(0.01, 0.01)$	$(0.001, 0.001)$
$f(x, y)$	$\frac{1}{2}$	$\frac{1}{2}$	$\frac{1}{2}$	$\frac{1}{2}$	$\frac{1}{2}$

The limit does not exist because along the path $y = 0$ the function equals 0, whereas along the path $y = x$ the function equals $\frac{1}{2}$.

38. $f(x, y) = -\dfrac{xy^2}{x^2 + y^4}$

Continuous except at $(0, 0)$

Path: $x = y^2$

(x, y)	$(1, 1)$	$(0.25, 0.5)$	$(0.01, 0.1)$	$(0.0001, 0.01)$	$(0.000001, 0.001)$
$f(x, y)$	$-\frac{1}{2}$	$-\frac{1}{2}$	$-\frac{1}{2}$	$-\frac{1}{2}$	$-\frac{1}{2}$

Path: $x = -y^2$

(x, y)	$(-1, 1)$	$(-0.25, 0.5)$	$(-0.01, 0.1)$	$(-0.0001, 0.01)$	$(-0.000001, 0.001)$
$f(x, y)$	$\frac{1}{2}$	$\frac{1}{2}$	$\frac{1}{2}$	$\frac{1}{2}$	$\frac{1}{2}$

The limit does not exist because along the path $x = y^2$ the function equals $-\frac{1}{2}$, whereas along the path $x = -y^2$ the function equals $\frac{1}{2}$.

39. $f(x, y) = \dfrac{y}{x^2 + y^2}$

Continuous except at $(0, 0)$

Path: $y = 0$

(x, y)	$(1, 0)$	$(0.5, 0)$	$(0.1, 0)$	$(0.01, 0)$	$(0.001, 0)$
$f(x, y)$	0	0	0	0	0

Path: $y = x$

(x, y)	$(1, 1)$	$(0.5, 0.5)$	$(0.1, 0.1)$	$(0.01, 0.01)$	$(0.001, 0.001)$
$f(x, y)$	$\frac{1}{2}$	1	5	50	500

The limit does not exist because along the path $y = 0$ the function equals 0, whereas along the path $y = x$ the function tends to infinity.

40. $f(x, y) = \dfrac{2x - y^2}{2x^2 + y}$

Continuous except at $(0, 0)$

Path: $y = 0$

(x, y)	$(1, 0)$	$(0.25, 0)$	$(0.01, 0)$	$(0.001, 0)$	$(0.000001, 0)$
$f(x, y)$	1	4	100	1000	1,000,000

Path: $y = x$

(x, y)	$(1, 1)$	$(0.25, 0.25)$	$(0.01, 0.01)$	$(0.001, 0.001)$	$(0.0001, 0.0001)$
$f(x, y)$	$\frac{1}{3}$	1.17	1.95	1.995	2.0

The limit does not exist because along the line $y = 0$ the function tends to infinity, whereas along the line $y = x$ the function tends to 2.

41. $\lim_{(x,y)\to(0,0)} \dfrac{x^4 - y^4}{x^2 + y^2} = \lim_{(x,y)\to(0,0)} \dfrac{(x^2 + y^2)(x^2 - y^2)}{x^2 + y^2} = \lim_{(x,y)\to(0,0)} (x^2 - y^2) = 0$

So, f is continuous everywhere, whereas g is continuous everywhere except at $(0, 0)$. g has a removable discontinuity at $(0, 0)$.

42. $\lim_{(x,y)\to(0,0)} f(x, y) = \lim_{(x,y)\to(0,0)} \left(\dfrac{x^2 + 2xy^2 + y^2}{x^2 + y^2}\right)$

$= \lim_{(x,y)\to(0,0)} \left(1 + \dfrac{2xy^2}{x^2 + y^2}\right) = 1$

(same limit for g)

So, f is not continuous at $(0, 0)$, whereas g is continuous at $(0, 0)$.

43. $\lim_{(x,y)\to(0,0)} \dfrac{xy^2}{x^2 + y^2} = \lim_{r\to 0} \dfrac{(r\cos\theta)(r^2\sin^2\theta)}{r^2}$

$= \lim_{r\to 0}(r\cos\theta\sin^2\theta) = 0$

44. $\lim_{(x,y)\to(0,0)} \dfrac{x^3 + y^3}{x^2 + y^2} = \lim_{r\to 0} \dfrac{r^3(\cos^3\theta + \sin^3\theta)}{r^2}$

$= \lim_{r\to 0} r(\cos^3\theta + \sin^3\theta) = 0$

45. $\lim_{(x,y)\to(0,0)} \dfrac{x^2y^2}{x^2 + y^2} = \lim_{r\to 0} \dfrac{r^4\cos^2\theta\sin^2\theta}{r^2}$

$= \lim_{r\to 0} r^2\cos^2\theta\sin^2\theta = 0$

46. $x = r\cos\theta,\ y = r\sin\theta,\ \sqrt{x^2 + y^2} = r,\ x^2 - y^2 = r^2(\cos^2\theta - \sin^2\theta)$

$\lim_{(x,y)\to(0,0)} \dfrac{x^2 - y^2}{\sqrt{x^2 + y^2}} = \lim_{r\to 0} \dfrac{r^2(\cos^2\theta - \sin^2\theta)}{r} = \lim_{r\to 0} r(\cos^2\theta - \sin^2\theta) = 0$

47. $\lim_{(x,y)\to(0,0)} \cos(x^2 + y^2) = \lim_{r\to 0}\cos(r^2) = \cos(0) = 1$

48. $\lim_{(x,y)\to(0,0)} \sin\sqrt{x^2 + y^2} = \lim_{r\to 0}\sin(r) = \sin(0) = 0$

49. $\sqrt{x^2 + y^2} = r$

$\lim_{(x,y)\to(0,0)} \dfrac{\sin\sqrt{x^2 + y^2}}{\sqrt{x^2 + y^2}} = \lim_{r\to 0^+} \dfrac{\sin(r)}{r} = 1$

50. $\lim_{(x,y)\to(0,0)} \dfrac{\sin(x^2 + y^2)}{x^2 + y^2} = \lim_{r\to 0} \dfrac{\sin r^2}{r^2} = \lim_{r\to 0} \dfrac{2r\cos r^2}{2r} = \lim_{r\to 0}\cos r^2 = 1$

51. $x^2 + y^2 = r^2$

$\lim_{(x,y)\to(0,0)} \dfrac{1 - \cos(x^2 + y^2)}{x^2 + y^2} = \lim_{x\to 0} \dfrac{1 - \cos(r^2)}{r^2} = 0$

52. $x^2 + y^2 = r^2$

$$\lim_{(x, y)\to(0,0)} \left(x^2 + y^2\right)\ln\left(x^2 + y^2\right) = \lim_{r\to 0} r^2 \ln\left(r^2\right) = \lim_{r\to 0^+} 2r^2 \ln(r)$$

By L'Hôpital's Rule, $\lim_{r\to 0^+} 2r^2 \ln(r) = \lim_{r\to 0^+} \dfrac{2\ln(r)}{1/r^2} = \lim_{r\to 0^+} \dfrac{2/r}{-2/r^3} = \lim_{r\to 0^+} \left(-r^2\right) = 0$

53. $f(x, y, z) = \dfrac{1}{\sqrt{x^2 + y^2 + z^2}}$

Continuous except at $(0, 0, 0)$

54. $f(x, y, z) = \dfrac{z}{x^2 + y^2 - 4}$

Continuous for $x^2 + y^2 \neq 4$.

55. $f(x, y, z) = \dfrac{\sin z}{e^x + e^y}$

Continuous everywhere

56. $f(x, y, z) = xy \sin z$

Continuous everywhere

57. For $xy \neq 0$, the function is clearly continuous.

For $xy \neq 0$, let $z = xy$. Then

$$\lim_{z\to 0} \frac{\sin z}{z} = 1$$

implies that f is continuous for all x, y.

58. For $x^2 \neq y^2$, the function is clearly continuous.

For $x^2 \neq y^2$, let $z = x^2 - y^2$. Then

$$\lim_{z\to 0} \frac{\sin(z)}{z} = 1$$

implies that f is continuous for all x, y.

59. $f(t) = t^2,\ g(x, y) = 2x - 3y$

$f(g(x, y)) = f(2x - 3y) = (2x - 3y)^2$

Continuous everywhere

60. $f(t) = \dfrac{1}{t}$

$g(x, y) = x^2 + y^2$

$f(g(x, y)) = f\left(x^2 + y^2\right) = \dfrac{1}{x^2 + y^2}$

Continuous except at $(0, 0)$

61. $f(t) = \dfrac{1}{t},\ g(x, y) = 2x - 3y$

$f(g(x, y)) = f(2x - 3y) = \dfrac{1}{2x - 3y}$

Continuous for all $y \neq \dfrac{2}{3}x$

62. $f(t) = \dfrac{1}{1 - t},\ g(x, y) = x^2 + y^2$

$f(g(x, y)) = f\left(x^2 + y^2\right) = \dfrac{1}{1 - x^2 - y^2}$

Continuous for $x^2 + y^2 \neq 1$

63. $f(x, y) = x^2 - 4y$

(a) $\displaystyle \lim_{\Delta x\to 0} \frac{f(x + \Delta x, y) - f(x, y)}{\Delta x} = \lim_{\Delta x\to 0} \frac{\left[(x + \Delta x)^2 - 4y\right] - \left(x^2 - 4y\right)}{\Delta x} = \lim_{\Delta x\to 0} \frac{2x\Delta x + (\Delta x)^2}{\Delta x} = \lim_{\Delta x\to 0} (2x + \Delta x) = 2x$

(b) $\displaystyle \lim_{\Delta y\to 0} \frac{f(x, y + \Delta y) - f(x, y)}{\Delta y} = \lim_{\Delta y\to 0} \frac{\left[x^2 - 4(y + \Delta y)\right] - \left(x^2 - 4y\right)}{\Delta y} = \lim_{\Delta y\to 0} \frac{-4\Delta y}{\Delta y} = \lim_{\Delta y\to 0} (-4) = -4$

64. $f(x, y) = x^2 + y^2$

(a) $\displaystyle \lim_{\Delta x\to 0} \frac{f(x + \Delta x, y) - f(x, y)}{\Delta x} = \lim_{\Delta x\to 0} \frac{\left[(x + \Delta x)^2 + y^2\right] - \left(x^2 + y^2\right)}{\Delta x} = \lim_{\Delta x\to 0} \frac{2x\Delta x + (\Delta x)^2}{\Delta x} = \lim_{\Delta x\to 0} (2x + \Delta x) = 2x$

(b) $\displaystyle \lim_{\Delta y\to 0} \frac{f(x, y + \Delta y) - f(x, y)}{\Delta y} = \lim_{\Delta y\to 0} \frac{\left[x^2 + (y + \Delta y)^2\right] - \left(x^2 + y^2\right)}{\Delta y} = \lim_{\Delta y\to 0} \frac{2y\Delta y + (\Delta y)^2}{\Delta y} = \lim_{\Delta y\to 0} (2y + \Delta y) = 2y$

65. $f(x, y) = \dfrac{x}{y}$

(a) $$\lim_{\Delta x \to 0} \frac{f(x + \Delta x, y) - f(x, y)}{\Delta x} = \lim_{\Delta x \to 0} \frac{\dfrac{x + \Delta x}{y} - \dfrac{x}{y}}{\Delta x} = \lim_{\Delta x \to 0} \frac{\dfrac{\Delta x}{y}}{\Delta x} = \lim_{\Delta x \to 0} \frac{1}{y} = \frac{1}{y}$$

(b) $$\lim_{\Delta y \to 0} \frac{f(x, y + \Delta y) - f(x, y)}{\Delta y} = \lim_{\Delta y \to 0} \frac{\dfrac{x}{y + \Delta y} - \dfrac{x}{y}}{\Delta y} = \lim_{\Delta y \to 0} \frac{xy - (xy + x\Delta y)}{(y + \Delta y)y\Delta y} = \lim_{\Delta y \to 0} \frac{-x\Delta y}{(y + \Delta y)y\Delta y} = \lim_{\Delta y \to 0} \frac{-x}{(y + \Delta y)y} = \frac{-x}{y^2}$$

66. $f(x, y) = \dfrac{1}{x + y}$

(a) $$\lim_{\Delta x \to 0} \frac{f(x + \Delta x, y) - f(x, y)}{\Delta x} = \lim_{\Delta x \to 0} \frac{\dfrac{1}{x + \Delta x + y} - \dfrac{1}{x + y}}{\Delta x} = \lim_{\Delta x \to 0} \frac{(x + y) - (x + \Delta x + y)}{(x + \Delta x + y)(x + y)\Delta x}$$
$$= \lim_{\Delta x \to 0} \frac{-\Delta x}{(x + \Delta x + y)(x + y)\Delta x} = \lim_{\Delta x \to 0} \frac{-1}{(x + \Delta x + y)(x + y)} = \frac{-1}{(x + y)^2}$$

(b) By symmetry, $\displaystyle\lim_{\Delta y \to 0} \frac{f(x, y + \Delta y) - f(x, y)}{\Delta y} = \frac{-1}{(x + y)^2}$.

67. $f(x, y) = 3x + xy - 2y$

(a) $$\lim_{\Delta x \to 0} \frac{f(x + \Delta x, y) - f(x, y)}{\Delta x} = \lim_{\Delta x \to 0} \frac{3(x + \Delta x) + (x + \Delta x)y - 2y - (3x + xy - 2y)}{\Delta x}$$
$$= \lim_{\Delta x \to 0} \frac{3\Delta x + y\Delta x}{\Delta x} = \lim_{\Delta x \to 0} (3 + y) = 3 + y$$

(b) $$\lim_{\Delta y \to 0} \frac{f(x, y + \Delta y) - f(x, y)}{\Delta y} = \lim_{\Delta y \to 0} \frac{3x + x(y + \Delta y) - 2(y + \Delta y) - (3x + xy - 2y)}{\Delta y}$$
$$= \lim_{\Delta y \to 0} \frac{x\Delta y - 2\Delta y}{\Delta y} = \lim_{\Delta y \to 0} (x - 2) = x - 2$$

68. $f(x, y) = \sqrt{y}(y + 1)$

(a) $$\lim_{\Delta x \to 0} \frac{f(x + \Delta x, y) - f(x, y)}{\Delta x} = \lim_{\Delta x \to 0} \frac{\sqrt{y}(y + 1) - \sqrt{y}(y + 1)}{\Delta x} = 0$$

(b) $$\lim_{\Delta y \to 0} \frac{f(x, y + \Delta y) - f(x, y)}{\Delta y} = \lim_{\Delta y \to 0} \frac{(y + \Delta y)^{3/2} + (y + \Delta y)^{1/2} - (y^{3/2} + y^{1/2})}{\Delta y}$$
$$= \lim_{\Delta y \to 0} \frac{(y + \Delta y)^{3/2} - y^{3/2}}{\Delta y} + \lim_{\Delta y \to 0} \frac{(y + \Delta y)^{1/2} - y^{1/2}}{\Delta y}$$
$$= \frac{3}{2}y^{1/2} + \frac{1}{2}y^{-1/2} \quad \text{(L'Hôpital's Rule)}$$
$$= \frac{3y + 1}{2\sqrt{y}}$$

69. True. Assuming $f(x, 0)$ exists for $x \neq 0$.

70. False. Let $f(x, y) = \dfrac{xy}{x^2 + y^2}$.

See Exercise 37.

71. False. Let $f(x, y) = \begin{cases} \ln(x^2 + y^2), & (x, y) \neq (0, 0) \\ 0, & x = 0, y = 0 \end{cases}$.

72. True

73. $\lim\limits_{(x, y)\to(0, 0)} \dfrac{x^2 + y^2}{xy}$

(a) Along $y = ax$:

$$\lim_{(x, ax)\to(0, 0)} \frac{x^2 + (ax)^2}{x(ax)} = \lim_{x\to 0} \frac{x^2(1 + a^2)}{ax^2} = \frac{1 + a^2}{a}, a \neq 0$$

If $a = 0$, then $y = 0$ and the limit does not exist.

(b) Along

$$y = x^2: \lim_{(x, x^2)\to(0, 0)} \frac{x^2 + (x^2)^2}{x(x^2)} = \lim_{x\to 0} \frac{1 + x^2}{x}$$

Limit does not exist.

(c) No, the limit does not exist. Different paths result in different limits.

74. $f(x, y) = \dfrac{x^2 y}{x^4 + y^2}$

(a) $y = ax$: $f(x, ax) = \dfrac{x^2(ax)}{x^4 + (ax)^2} = \dfrac{ax}{x^2 + a^2}$

If $a \neq 0$, $\lim\limits_{(x, ax)\to(0, 0)} \dfrac{ax}{x^2 + a^2} = 0$.

(b) $y = x^2$: $f(x, x^2) = \dfrac{x^2(x^2)}{x^4 + (x^2)^2} = \dfrac{x^4}{2x^4}$

$$\lim_{(x, x^2)} \frac{x^4}{2x^4} = \frac{1}{2}$$

(c) No, the limit does not exist. f approaches different numbers along different paths.

75. $$\lim_{(x, y, z)\to(0, 0, 0)} \frac{xyz}{x^2 + y^2 + z^2} = \lim_{\rho\to 0^+} \frac{(\rho \sin\phi \cos\theta)(\rho \sin\phi \sin\theta)(\rho \cos\phi)}{\rho^2} = \lim_{\rho\to 0^+} \rho\left[\sin^2\phi \cos\theta \sin\theta \cos\phi\right] = 0$$

76. $$\lim_{(x, y, z)\to(0, 0, 0)} \tan^{-1}\left[\frac{1}{x^2 + y^2 + z^2}\right] = \lim_{\rho\to 0^+} \tan^{-1}\left[\frac{1}{\rho^2}\right] = \frac{\pi}{2}$$

77. As $(x, y) \to (0, 1)$, $x^2 + 1 \to 1$ and $x^2 + (y - 1)^2 \to 0$.

So, $\lim\limits_{(x, y)\to(0, 1)} \tan^{-1}\left[\dfrac{x^2 + 1}{x^2 + (y - 1)^2}\right] = \dfrac{\pi}{2}$.

78. $$\lim_{(x, y)\to(0, 0)} f(x, y) = \lim_{r\to 0} (r\cos\theta)(r\sin\theta)\frac{r^2\cos^2\theta - r^2\sin^2\theta}{r^2} = \lim_{r\to 0} r^2\left[\cos\theta \sin\theta(\cos^2\theta - \sin^2\theta)\right] = 0$$

So, define $f(0, 0) = 0$.

79. Because $\lim\limits_{(x, y)\to(a, b)} f(x, y) = L_1$, then for $\varepsilon/2 > 0$, there corresponds $\delta_1 > 0$ such that $|f(x, y) - L_1| < \varepsilon/2$ whenever $0 < \sqrt{(x - a)^2 + (y - b)^2} < \delta_1$.

Because $\lim\limits_{(x, y)\to(a, b)} g(x, y) = L_2$, then for $\varepsilon/2 > 0$, there corresponds $\delta_2 > 0$ such that $|g(x, y) - L_2| < \varepsilon/2$ whenever $0 < \sqrt{(x - a)^2 + (y - b)^2} < \delta_2$.

Let δ be the smaller of δ_1 and δ_2. By the triangle inequality, whenever $\sqrt{(x - a)^2 + (y - b)^2} < \delta$, we have

$$|f(x, y) + g(x, y) - (L_1 + L_2)| = |(f(x, y) - L_1) + (g(x, y) - L_2)| \le |f(x, y) - L_1| + |g(x, y) - L_2| < \frac{\varepsilon}{2} + \frac{\varepsilon}{2} = \varepsilon.$$

So, $\lim\limits_{(x, y)\to(a, b)} \left[f(x, y) + g(x, y)\right] = L_1 + L_2$.

80. Given that $f(x, y)$ is continuous, then $\lim_{(x, y)\to(a, b)} f(x, y) = f(a, b) < 0$, which means that for each $\varepsilon > 0$, there corresponds a $\delta > 0$ such that $|f(x, y) - f(a, b)| < \varepsilon$ whenever

$0 < \sqrt{(x-a)^2 + (y-b)^2} < \delta.$

Let $\varepsilon = |f(a, b)|/2$, then $f(x, y) < 0$ for every point in the corresponding δ neighborhood because

$$|f(x, y) - f(a, b)| < \frac{|f(a, b)|}{2} \Rightarrow -\frac{|f(a, b)|}{2} < f(x, y) - f(a, b) < \frac{|f(a, b)|}{2}$$

$$\Rightarrow \frac{3}{2}f(a, b) < f(x, y) < \frac{1}{2}f(a, b) < 0.$$

81. See the definition on page 881. Show that the value of $\lim_{(x, y)\to(x_0, y_0)} f(x, y)$ is not the same for two different paths to (x_0, y_0).

82. See the definition on page 884.

83. (a) No. The existence of $f(2, 3)$ has no bearing on the existence of the limit as $(x, y) \to (2, 3)$.

(b) No, $f(2, 3)$ can equal any number, or not even be defined.

84. The limit appears to exist at all the points except (c) $(0, 0)$. Near this point, the graph tends to $-\infty$.

Section 13.3 Partial Derivatives

1. No, x only occurs in the numerator.

2. Yes, y occurs in both the numerator and denominator.

3. No, y only occurs in the numerator.

4. Yes, x occurs in both the numerator and denominator.

5. Yes, x occurs in both the numerator and denominator.

6. No, y only occurs in the numerator.

7. $f(x, y) = 2x - 5y + 3$

$f_x(x, y) = 2$

$f_y(x, y) = -5$

8. $f(x, y) = x^2 - 2y^2 + 4$

$f_x(x, y) = 2x$

$f_y(x, y) = -4y$

9. $f(x, y) = x^2y^3$

$f_x(x, y) = 2xy^3$

$f_y(x, y) = 3x^2y^2$

10. $f(x, y) = 4x^3y^{-2}$

$f_x(x, y) = 12x^2y^{-2}$

$f_y(x, y) = -8x^3y^{-3}$

11. $z = x\sqrt{y}$

$\frac{\partial z}{\partial x} = \sqrt{y}$

$\frac{\partial z}{\partial y} = \frac{x}{2\sqrt{y}}$

12. $z = 2y^2\sqrt{x}$

$\frac{\partial z}{\partial x} = \frac{y^2}{\sqrt{x}}$

$\frac{\partial z}{\partial y} = 4y\sqrt{x}$

13. $z = x^2 - 4xy + 3y^2$

$\frac{\partial z}{\partial x} = 2x - 4y$

$\frac{\partial z}{\partial y} = -4x + 6y$

14. $z = y^3 - 2xy^2 - 1$

$\frac{\partial z}{\partial x} = -2y^2$

$\frac{\partial z}{\partial y} = 3y^2 - 4xy$

15. $z = e^{xy}$

$$\frac{\partial z}{\partial x} = ye^{xy}$$

$$\frac{\partial z}{\partial y} = xe^{xy}$$

16. $z = e^{x/y} = e^{xy^{-1}}$

$$\frac{\partial z}{\partial x} = \frac{1}{y}e^{x/y}$$

$$\frac{\partial z}{\partial y} = \frac{-x}{y^2}e^{x/y}$$

17. $z = x^2e^{2y}$

$$\frac{\partial z}{\partial x} = 2xe^{2y}$$

$$\frac{\partial z}{\partial y} = 2x^2e^{2y}$$

18. $z = ye^{y/x} = ye^{yx^{-1}}$

$$\frac{\partial z}{\partial x} = ye^{yx^{-1}}\left[-yx^{-2}\right] = \frac{-y^2}{x^2}e^{y/x}$$

$$\frac{\partial z}{\partial y} = e^{y/x} + \frac{1}{x}ye^{y/x} = e^{y/x}\left(1 + \frac{y}{x}\right)$$

19. $z = \ln\frac{x}{y} = \ln x - \ln y$

$$\frac{\partial z}{\partial x} = \frac{1}{x}$$

$$\frac{\partial z}{\partial y} = -\frac{1}{y}$$

20. $z = \ln\sqrt{xy} = \frac{1}{2}\ln(xy)$

$$\frac{\partial z}{\partial x} = \frac{1}{2}\frac{y}{xy} = \frac{1}{2x}$$

$$\frac{\partial z}{\partial y} = \frac{1}{2}\frac{x}{xy} = \frac{1}{2y}$$

21. $z = \ln(x^2 + y^2)$

$$\frac{\partial z}{\partial x} = \frac{2x}{x^2 + y^2}$$

$$\frac{\partial z}{\partial y} = \frac{2y}{x^2 + y^2}$$

22. $z = \ln\frac{x + y}{x - y} = \ln(x + y) - \ln(x - y)$

$$\frac{\partial z}{\partial x} = \frac{1}{x + y} - \frac{1}{x - y} = \frac{-2y}{(x + y)(x - y)}$$

$$\frac{\partial z}{\partial y} = \frac{1}{x + y} + \frac{1}{x - y} = \frac{2x}{(x + y)(x - y)}$$

23. $z = \frac{x^2}{2y} + \frac{3y^2}{x}$

$$\frac{\partial z}{\partial x} = \frac{2x}{2y} - \frac{3y^2}{x^2} = \frac{x^3 - 3y^3}{x^2y}$$

$$\frac{\partial z}{\partial y} = \frac{-x^2}{2y^2} + \frac{6y}{x} = \frac{12y^3 - x^3}{2xy^2}$$

24. $f(x, y) = \frac{xy}{x^2 + y^2}$

$$f_x(x, y) = \frac{(x^2 + y^2)(y) - (xy)(2x)}{(x^2 + y^2)^2} = \frac{y^3 - x^2y}{(x^2 + y^2)^2}$$

$$f_y(x, y) = \frac{(x^2 + y^2)(x) - (xy)(2y)}{(x^2 + y^2)^2} = \frac{x^3 - xy^2}{(x^2 + y^2)^2}$$

25. $h(x, y) = e^{-(x^2+y^2)}$

$$h_x(x, y) = -2xe^{-(x^2+y^2)}$$

$$h_y(x, y) = -2ye^{-(x^2+y^2)}$$

26. $g(x, y) = \ln\sqrt{x^2 + y^2} = \frac{1}{2}\ln(x^2 + y^2)$

$$g_x(x, y) = \frac{1}{2}\frac{2x}{x^2 + y^2} = \frac{x}{x^2 + y^2}$$

$$g_y(x, y) = \frac{1}{2}\frac{2y}{x^2 + y^2} = \frac{y}{x^2 + y^2}$$

27. $f(x, y) = \sqrt{x^2 + y^2}$

$$f_x(x, y) = \frac{1}{2}(x^2 + y^2)^{-1/2}(2x) = \frac{x}{\sqrt{x^2 + y^2}}$$

$$f_y(x, y) = \frac{1}{2}(x^2 + y^2)^{-1/2}(2y) = \frac{y}{\sqrt{x^2 + y^2}}$$

28. $f(x, y) = \sqrt{2x + y^3}$

$$\frac{\partial f}{\partial x} = \frac{1}{2}(2x + y^3)^{-1/2}(2) = \frac{1}{\sqrt{2x + y^3}}$$

$$\frac{\partial f}{\partial y} = \frac{1}{2}(2x + y^3)^{-1/2}(3y^2) = \frac{3y^2}{2\sqrt{2x + y^3}}$$

29. $z = \cos xy$

$$\frac{\partial z}{\partial x} = -y \sin xy$$

$$\frac{\partial z}{\partial y} = -x \sin xy$$

30. $z = \sin(x + 2y)$

$$\frac{\partial z}{\partial x} = \cos(x + 2y)$$

$$\frac{\partial z}{\partial y} = 2 \cos(x + 2y)$$

31. $z = \tan(2x - y)$

$$\frac{\partial z}{\partial x} = 2 \sec^2(2x - y)$$

$$\frac{\partial z}{\partial y} = -\sec^2(2x - y)$$

32. $z = \sin 5x \cos 5y$

$$\frac{\partial z}{\partial x} = 5 \cos 5x \cos 5y$$

$$\frac{\partial z}{\partial y} = -5 \sin 5x \sin 5y$$

33. $z = e^y \sin xy$

$$\frac{\partial z}{\partial x} = ye^y \cos xy$$

$$\frac{\partial z}{\partial y} = e^y \sin xy + xe^y \cos x$$

$$= e^y(x \cos xy + \sin xy)$$

34. $z = \cos(x^2 + y^2)$

$$\frac{\partial z}{\partial x} = -2x \sin(x^2 + y^2)$$

$$\frac{\partial z}{\partial y} = -2y \sin(x^2 + y^2)$$

35. $z = \sinh(2x + 3y)$

$$\frac{\partial z}{\partial x} = 2 \cosh(2x + 3y)$$

$$\frac{\partial z}{\partial y} = 3 \cosh(2x + 3y)$$

36. $z = \cosh xy^2$

$$\frac{\partial z}{\partial x} = y^2 \sinh xy^2$$

$$\frac{\partial z}{\partial y} = 2xy \sinh xy^2$$

37. $f(x, y) = \int_x^y (t^2 - 1)\, dt$

$$= \left[\frac{t^3}{3} - t\right]_x^y = \left(\frac{y^3}{3} - y\right) - \left(\frac{x^3}{3} - x\right)$$

$$f_x(x, y) = -x^2 + 1 = 1 - x^2$$

$$f_y(x, y) = y^2 - 1$$

[You could also use the Second Fundamental Theorem of Calculus.]

38. $f(x, y) = \int_x^y (2t + 1)\, dt + \int_y^x (2t - 1)\, dt$

$$= \int_x^y (2t + 1)\, dt - \int_x^y (2t - 1)\, dt$$

$$= \int_x^y 2\, dt = \left[2t\right]_x^y = 2y - 2x$$

$$f_x(x, y) = -2$$

$$f_x(x, y) = 2$$

39. $f(x, y) = 3x + 2y$

$$\frac{\partial f}{\partial x} = \lim_{\Delta x \to 0} \frac{f(x + \Delta x, y) - f(x, y)}{\Delta x}$$

$$= \lim_{\Delta x \to 0} \frac{3(x + \Delta x) + 2y - (3x + 2y)}{\Delta x}$$

$$= \lim_{\Delta x \to 0} \frac{3\Delta x}{\Delta x} = 3$$

$$\frac{\partial f}{\partial y} = \lim_{\Delta y \to 0} \frac{f(x, y + \Delta y) - f(x, y)}{\Delta y}$$

$$= \lim_{\Delta y \to 0} \frac{3x + 2(y + \Delta y) - (3x + 2y)}{\Delta y}$$

$$= \lim_{\Delta y \to 0} \frac{2\Delta y}{\Delta y} = 2$$

40. $f(x, y) = x^2 - 2xy + y^2 = (x - y)^2$

$$\frac{\partial f}{\partial x} = \lim_{\Delta x \to 0} \frac{f(x + \Delta x, y) - f(x, y)}{\Delta x}$$

$$= \lim_{\Delta x \to 0} \frac{(x + \Delta x)^2 - 2(x + \Delta x)y + y^2 - x^2 + 2xy - y^2}{\Delta x} = \lim_{\Delta x \to 0} (2x + \Delta x - 2y) = 2(x - y)$$

$$\frac{\partial f}{\partial y} = \lim_{\Delta y \to 0} \frac{f(x, y + \Delta y) - f(x, y)}{\Delta y}$$

$$= \lim_{\Delta y \to 0} \frac{x^2 - 2x(y + \Delta y) + (y + \Delta y)^2 - x^2 + 2xy - y^2}{\Delta y} = \lim_{\Delta y \to 0} (-2x + 2y + \Delta y) = 2(y - x)$$

41. $f(x, y) = \sqrt{x + y}$

$$\frac{\partial f}{\partial x} = \lim_{\Delta x \to 0} \frac{f(x + \Delta x, y) - f(x, y)}{\Delta x}$$

$$= \lim_{\Delta x \to 0} \frac{\sqrt{x + \Delta x + y} - \sqrt{x + y}}{\Delta x}$$

$$= \lim_{\Delta x \to 0} \frac{\left(\sqrt{x + \Delta x + y} - \sqrt{x + y}\right)\left(\sqrt{x + \Delta x + y} + \sqrt{x + y}\right)}{\Delta x\left(\sqrt{x + \Delta x + y} + \sqrt{x + y}\right)} = \lim_{\Delta x \to 0} \frac{1}{\sqrt{x + \Delta x + y} + \sqrt{x + y}} = \frac{1}{2\sqrt{x + y}}$$

$$\frac{\partial f}{\partial y} = \lim_{\Delta y \to 0} \frac{f(x, y + \Delta y) - f(x, y)}{\Delta y} = \lim_{\Delta y \to 0} \frac{\sqrt{x + y + \Delta y} - \sqrt{x + y}}{\Delta y}$$

$$= \lim_{\Delta y \to 0} \frac{\left(\sqrt{x + y + \Delta y} - \sqrt{x + y}\right)\left(\sqrt{x + y + \Delta y} + \sqrt{x + y}\right)}{\Delta y\left(\sqrt{x + y + \Delta y} + \sqrt{x + y}\right)}$$

$$= \lim_{\Delta y \to 0} \frac{1}{\sqrt{x + y + \Delta y} + \sqrt{x + y}} = \frac{1}{2\sqrt{x + y}}$$

42. $f(x, y) = \dfrac{1}{x + y}$

$$\frac{\partial f}{\partial x} = \lim_{\Delta x \to 0} \frac{f(x + \Delta x, y) - f(x, y)}{\Delta x} = \lim_{\Delta x \to 0} \frac{\dfrac{1}{x + \Delta x + y} - \dfrac{1}{x + y}}{\Delta x} = \lim_{\Delta x \to 0} \frac{-1}{(x + \Delta x + y)(x + y)} = \frac{-1}{(x + y)^2}$$

$$\frac{\partial f}{\partial y} = \lim_{\Delta y \to 0} \frac{f(x, y + \Delta y) - f(x, y)}{\Delta y} = \lim_{\Delta y \to 0} \frac{\dfrac{1}{x + y + \Delta} - \dfrac{1}{x + y}}{\Delta y} = \lim_{\Delta y \to 0} \frac{-1}{(x + y + \Delta y)(x + y)} = \frac{-1}{(x + y)^2}$$

43. $f(x, y) = e^y \sin x$

$f_x(x, y) = e^y \cos x$

At $(\pi, 0)$, $f_x(\pi, 0) = -1$.

$f_y(x, y) = e^y \sin x$

At $(\pi, 0)$, $f_y(\pi, 0) = 0$.

44. $f(x, y) = e^{-x} \cos y$

$f_x(x, y) = -e^{-x} \cos y$

At $(0, 0)$, $f_x(0, 0) = -1$.

$f_y(x, y) = -e^{-x} \sin y$

At $(0, 0)$, $f_y(0, 0) = 0$.

45. $f(x, y) = \cos(2x - y)$

$f_x(x, y) = -2\sin(2x - y)$

At $\left(\frac{\pi}{4}, \frac{\pi}{3}\right)$, $f_x\left(\frac{\pi}{4}, \frac{\pi}{3}\right) = -2\sin\left(\frac{\pi}{2} - \frac{\pi}{3}\right) = -1.$

$f_y(x, y) = \sin(2x - y)$

At $\left(\frac{\pi}{4}, \frac{\pi}{3}\right)$, $f_y\left(\frac{\pi}{4}, \frac{\pi}{3}\right) = \sin\left(\frac{\pi}{2} - \frac{\pi}{3}\right) = \frac{1}{2}.$

46. $f(x, y) = \sin xy$

$f_x(x, y) = y\cos xy$

At $\left(2, \frac{\pi}{4}\right)$, $f_x\left(2, \frac{\pi}{4}\right) = \frac{\pi}{4}\cos\frac{\pi}{2} = 0.$

$f_y(x, y) = x\cos xy$

At $\left(2, \frac{\pi}{4}\right)$, $f_y\left(2, \frac{\pi}{4}\right) = 2\cos\frac{\pi}{2} = 0.$

47. $f(x, y) = \arctan\frac{y}{x}$

$f_x(x, y) = \frac{1}{1 + (y^2/x^2)}\left(-\frac{y}{x^2}\right) = \frac{-y}{x^2 + y^2}$

At $(2, -2)$: $f_x(2, -2) = \frac{1}{4}$

$f_y(x, y) = \frac{1}{1 + (y^2/x^2)}\left(\frac{1}{x}\right) = \frac{x}{x^2 + y^2}$

At $(2, -2)$: $f_y(2, -2) = \frac{1}{4}$

48. $f(x, y) = \arccos(xy)$

$f_x(x, y) = \frac{-y}{\sqrt{1 - x^2y^2}}$

At $(1, 1)$, f_x is undefined.

$f_y(x, y) = \frac{-x}{\sqrt{1 - x^2y^2}}$

At $(1, 1)$, f_y is undefined.

49. $f(x, y) = \frac{xy}{x - y}$

$f_x(x, y) = \frac{y(x - y) - xy}{(x - y)^2} = \frac{-y^2}{(x - y)^2}$

At $(2, -2)$: $f_x(2, -2) = -\frac{1}{4}$

$f_y(x, y) = \frac{x(x - y) + xy}{(x - y)^2} = \frac{x^2}{(x - y)^2}$

At $(2, -2)$: $f_y(2, -2) = \frac{1}{4}$

50. $f(x, y) = \frac{2xy}{\sqrt{4x^2 + 5y^2}}$

$f_x(x, y) = \frac{10y^3}{(4x^2 + 5y^2)^{3/2}}$

At $(1, 1)$, $f_x(1, 1) = \frac{10}{9^{3/2}} = \frac{10}{27}.$

$f_y(x, y) = \frac{8x^3}{(4x^2 + 5y^2)^{3/2}}$

At $(1, 1)$, $f_y(1, 1) = \frac{8}{9^{3/2}} = \frac{8}{27}.$

51. $g(x, y) = 4 - x^2 - y^2$

$g_x(x, y) = -2x$

At $(1, 1)$: $g_x(1, 1) = -2$

$g_y(x, y) = -2y$

At $(1, 1)$: $g_y(1, 1) = -2$

52. $h(x, y) = x^2 - y^2$

$h_x(x, y) = 2x$

At $(-2, 1)$: $h_x(-2, 1) = -4$

$h_y(x, y) = -2y$

At $(-2, 1)$: $h_y(-2, 1) = -2$

53. $H(x, y, z) = \sin(x + 2y + 3z)$

$H_x(x, y, z) = \cos(x + 2y + 3z)$

$H_y(x, y, z) = 2\cos(x + 2y + 3z)$

$H_z(x, y, z) = 3\cos(x + 2y + 3z)$

54. $f(x, y, z) = 3x^2y - 5xyz + 10yz^2$

$f_x(x, y, z) = 6xy - 5yz$

$f_y(x, y, z) = 3x^2 - 5xz + 10z^2$

$f_z(x, y, z) = -5xy + 20yz$

55. $w = \sqrt{x^2 + y^2 + z^2}$

$\frac{\partial w}{\partial x} = \frac{x}{\sqrt{x^2 + y^2 + z^2}}$

$\frac{\partial w}{\partial y} = \frac{y}{\sqrt{x^2 + y^2 + z^2}}$

$\frac{\partial w}{\partial z} = \frac{z}{\sqrt{x^2 + y^2 + z^2}}$

56. $w = \dfrac{7xz}{x+y} = 7xz(x+y)^{-1}$

$$\frac{\partial w}{\partial x} = \frac{(x+y)(7z) - 7xz}{(x+y)^2} = \frac{7yz}{(x+y)^2}$$

$$\frac{\partial w}{\partial y} = \frac{-7xz}{(x+y)^2}$$

$$\frac{\partial w}{\partial z} = \frac{7x}{x+y}$$

57. $F(x, y, z) = \ln\sqrt{x^2 + y^2 + z^2} = \dfrac{1}{2}\ln(x^2 + y^2 + z^2)$

$$F_x(x, y, z) = \frac{x}{x^2 + y^2 + z^2}$$

$$F_y(x, y, z) = \frac{y}{x^2 + y^2 + z^2}$$

$$F_z(x, y, z) = \frac{z}{x^2 + y^2 + z^2}$$

58. $G(x, y, z) = \dfrac{1}{\sqrt{1 - x^2 - y^2 - z^2}}$

$$G_x(x, y, z) = \frac{x}{(1 - x^2 - y^2 - z^2)^{3/2}}$$

$$G_y(x, y, z) = \frac{y}{(1 - x^2 - y^2 - z^2)^{3/2}}$$

$$G_z(x, y, z) = \frac{z}{(1 - x^2 - y^2 - z^2)^{3/2}}$$

59. $f(x, y, z) = x^3yz^2$

$$f_x(x, y, z) = 3x^2yz^2$$

$$f_x(1, 1, 1) = 3$$

$$f_y(x, y, z) = x^3z^2$$

$$f_y(1, 1, 1) = 1$$

$$f_z(x, y, z) = 2x^3yz$$

$$f_z(1, 1, 1) = 2$$

60. $f(x, y, z) = x^2y^3 + 2xyz - 3yz$

$$f_x(x, y, z) = 2xy^3 + 2yz$$

$$f_x(-2, 1, 2) = -4 + 4 = 0$$

$$f_y(x, y, z) = 3x^2y^2 + 2xz - 3z$$

$$f_y(-2, 1, 2) = 12 - 8 - 6 = -2$$

$$f_z(x, y, z) = 2xy - 3y$$

$$f_z(-2, 1, 2) = -4 - 3 = -7$$

61. $f(x, y, z) = \dfrac{x}{yz}$

$$f_x(x, y, z) = \frac{1}{yz}$$

$$f_x(1, -1, -1) = 1$$

$$f_y(x, y, z) = \frac{-x}{y^2z}$$

$$f_y(1, -1, -1) = 1$$

$$f_z(x, y, z) = \frac{-x}{yz^2}$$

$$f_z(1, -1, -1) = 1$$

62. $f(x, y, z) = \dfrac{xy}{x + y + z}$

$$f_x(x, y, z) = \frac{(x + y + z)y - xy}{(x + y + z)^2} = \frac{y^2 + yz}{(x + y + z)^2}$$

$$f_x(3, 1, -1) = \frac{1 - 1}{3^2} = 0$$

$$f_y(x, y, z) = \frac{(x + y + z)x - xy}{(x + y + z)^2} = \frac{x^2 + xz}{(x + y + z)^2}$$

$$f_y(3, 1, -1) = \frac{9 - 3}{3^2} = \frac{2}{3}$$

$$f_z(x, y, z) = \frac{(x + y + z)(0) - xy}{(x + y + z)^2} = \frac{-xy}{(x + y + z)^2}$$

$$f_z(3, 1, -1) = \frac{-3}{9} = \frac{-1}{3}$$

63. $f(x, y, z) = z\sin(x + y)$

$$f_x(x, y, z) = z\cos(x + y)$$

$$f_x\left(0, \frac{\pi}{2}, -4\right) = -4\cos\frac{\pi}{2} = 0$$

$$f_y(x, y, z) = z\cos(x + y)$$

$$f_y\left(0, \frac{\pi}{2}, -4\right) = -4\cos\frac{\pi}{2} = 0$$

$$f_z(x, y, z) = \sin(x + y)$$

$$f_z\left(0, \frac{\pi}{2}, -4\right) = \sin\frac{\pi}{2} = 1$$

64. $\sqrt{3x^2 + y^2 - 2z^2}$

$$f_x(x, y, z) = \frac{6x}{2\sqrt{3x^2 + y^2 - 2z^2}} = \frac{3x}{\sqrt{3x^2 + y^2 - 2z^2}}$$

$$f_x(1, -2, 1) = \frac{6}{2\sqrt{3 + 4 - 2}} = \frac{3}{\sqrt{5}} = \frac{3\sqrt{5}}{5}$$

$$f_y(x, y, z) = \frac{2y}{2\sqrt{3x^2 + y^2 - 2z^2}} = \frac{y}{\sqrt{3x^2 + y^2 - 2z^2}}$$

$$f_y(1, -2, 1) = \frac{-2}{\sqrt{5}} = \frac{-2\sqrt{5}}{5}$$

$$f_z(x, y, z) = \frac{-4z}{2\sqrt{3x^2 + y^2 - 2z^2}} = \frac{-2z}{\sqrt{3x^2 + y^2 - 2z^2}}$$

$$f_z(1, -2, 1) = \frac{-2}{\sqrt{5}} = \frac{-2\sqrt{5}}{5}$$

65. $f_x(x, y) = 2x + y - 2 = 0$

$f_y(x, y) = x + 2y + 2 = 0$

$2x + y - 2 = 0 \Rightarrow y = 2 - 2x$

$x + 2(2 - 2x) + 2 = 0 \Rightarrow -3x + 6 = 0 \Rightarrow x = 2,$

$y = -2$

Point: $(2, -2)$

66. $f_x(x, y) = 2x - y - 5 = 0$

$f_y(x, y) = -x + 2y + 1 = 0$

$2x - y - 5 = 0 \Rightarrow y = 2x - 5$

$-x + 2(2x - 5) + 1 = 0 \Rightarrow 3x - 9 = 0 \Rightarrow x = 3,$

$y = 1$

Point: $(3, 1)$

67. $f_x(x, y) = 2x + 4y - 4,\ f_y(x, y) = 4x + 2y + 16$

$f_x = f_y = 0$: $2x + 4y = 4$

$4x + 2y = -16$

Solving for x and y,

$x = -6$ and $y = 4$.

68. $f_x(x, y) = 2x - y = 0$

$f_y(x, y) = -x + 2y = 0$

$2x - y = 0 \Rightarrow y = 2x$

$-x + 2(2x) = 0 \Rightarrow x = 0,\ y = 0$

Point: $(0, 0)$

69. $f_x(x, y) = -\dfrac{1}{x^2} + y,\ f_y(x, y) = -\dfrac{1}{y^2} + x$

$f_x = f_y = 0$: $-\dfrac{1}{x^2} + y = 0$ and $-\dfrac{1}{y^2} + x = 0$

$y = \dfrac{1}{x^2}$ and $x = \dfrac{1}{y^2}$

$y = y^4 \Rightarrow y = 1 = x$

Points: $(1, 1)$

70. $f_x(x, y) = 9x^2 - 12y,\ f_y(x, y) = -12x + 3y^2$

$f_x = f_y = 0$: $9x^2 - 12y = 0 \Rightarrow 3x^2 = 4y$

$3y^2 - 12x = 0 \Rightarrow y^2 = 4x$

Solving for x in the second equation, $x = y^2/4$, you obtain $3\left(y^2/4\right)^2 = 4y$.

$3y^4 = 64y \Rightarrow y = 0$ or $y = \dfrac{4}{3^{1/3}}$

$\Rightarrow x = 0$ or $x = \dfrac{1}{4}\left(\dfrac{16}{3^{2/3}}\right)$

Points: $(0, 0), \left(\dfrac{4}{3^{2/3}}, \dfrac{4}{3^{1/3}}\right)$

71. $f_x(x, y) = (2x + y)e^{x^2 + xy + y^2} = 0$

$f_y(x, y) = (x + 2y)e^{x^2 + xy + y^2} = 0$

$2x + y = 0 \Rightarrow y = -2x$

$x + 2(-2x) = 0 \Rightarrow x = 0 \Rightarrow y = 0$

Point: $(0, 0)$

72. $f_x(x, y) = \dfrac{2x}{x^2 + y^2 + 1} = 0 \Rightarrow x = 0$

$f_y(x, y) = \dfrac{2y}{x^2 + y^2 + 1} = 0 \Rightarrow y = 0$

Points: $(0, 0)$

73. $z = 3xy^2$

$$\frac{\partial z}{\partial x} = 3y^2,\ \frac{\partial^2 z}{\partial x^2} = 0,\ \frac{\partial^2 z}{\partial y \partial x} = 6y$$

$$\frac{\partial z}{\partial y} = 6xy,\ \frac{\partial^2 z}{\partial y^2} = 6x,\ \frac{\partial^2 z}{\partial x \partial y} = 6y$$

74. $z = x^2 + 3y^2$

$$\frac{\partial z}{\partial x} = 2x, \frac{\partial^2 z}{\partial x^2} = 2, \frac{\partial^2 z}{\partial y \partial x} = 0$$

$$\frac{\partial z}{\partial y} = 6y, \frac{\partial^2 z}{\partial y^2} = 6, \frac{\partial^2 z}{\partial x \partial y} = 0$$

75. $z = x^2 - 2xy + 3y^2$

$$\frac{\partial z}{\partial x} = 2x - 2y$$

$$\frac{\partial^2 z}{\partial x^2} = 2$$

$$\frac{\partial^2 z}{\partial y \partial x} = -2$$

$$\frac{\partial z}{\partial y} = -2x + 6y$$

$$\frac{\partial^2 z}{\partial y^2} = 6$$

$$\frac{\partial^2 z}{\partial x \partial y} = -2$$

76. $z = x^4 - 3x^2y^2 + y^4$

$$\frac{\partial z}{\partial x} = 4x^3 - 6xy^2$$

$$\frac{\partial^2 z}{\partial x^2} = 12x^2 - 6y^2$$

$$\frac{\partial^2 z}{\partial y \partial x} = -12xy$$

$$\frac{\partial z}{\partial y} = -6x^2y + 4y^3$$

$$\frac{\partial^2 z}{\partial y^2} = -6x^2 + 12y^2$$

$$\frac{\partial^2 z}{\partial x \partial y} = -12xy$$

77. $z = \sqrt{x^2 + y^2}$

$$\frac{\partial z}{\partial x} = \frac{x}{\sqrt{x^2 + y^2}}$$

$$\frac{\partial^2 z}{\partial x^2} = \frac{y^2}{\left(x^2 + y^2\right)^{3/2}}$$

$$\frac{\partial^2 z}{\partial y \partial x} = \frac{-xy}{\left(x^2 + y^2\right)^{3/2}}$$

$$\frac{\partial z}{\partial y} = \frac{y}{\sqrt{x^2 + y^2}}$$

$$\frac{\partial^2 z}{\partial y^2} = \frac{x^2}{\left(x^2 + y^2\right)^{3/2}}$$

$$\frac{\partial^2 z}{\partial x \partial y} = \frac{-xy}{\left(x^2 + y^2\right)^{3/2}}$$

78. $z = \ln(x - y)$

$$\frac{\partial z}{\partial x} = \frac{1}{x - y}$$

$$\frac{\partial^2 z}{\partial x^2} = -\frac{1}{(x - y)^2}$$

$$\frac{\partial^2 z}{\partial y \partial x} = \frac{1}{(x - y)^2}$$

$$\frac{\partial z}{\partial y} = \frac{-1}{x - y} = \frac{1}{y - x}$$

$$\frac{\partial^2 z}{\partial y^2} = -\frac{1}{(x - y)^2}$$

$$\frac{\partial^2 z}{\partial x \partial y} = \frac{1}{(x - y)^2}$$

So, $\dfrac{\partial^2 z}{\partial y \partial x} = \dfrac{\partial^2 z}{\partial x \partial y}$.

79. $z = e^x \tan y$

$$\frac{\partial z}{\partial x} = e^x \tan y$$

$$\frac{\partial^2 z}{\partial x^2} = e^x \tan y$$

$$\frac{\partial^2 z}{\partial y \partial x} = e^x \sec^2 y$$

$$\frac{\partial z}{\partial y} = e^x \sec^2 y$$

$$\frac{\partial^2 z}{\partial y^2} = 2e^x \sec^2 y \tan y$$

$$\frac{\partial^2 z}{\partial x \partial y} = e^x \sec^2 y$$

80. $z = 2xe^y - 3ye^{-x}$

$$\frac{\partial z}{\partial x} = 2e^y + 3ye^{-x}$$

$$\frac{\partial^2 z}{\partial x^2} = -3ye^{-x}$$

$$\frac{\partial^2 z}{\partial y \partial x} = 2e^y + 3ye^{-x}$$

$$\frac{\partial z}{\partial y} = 2xe^y - 3e^{-x}$$

$$\frac{\partial^2 z}{\partial y^2} = 2xe^y$$

$$\frac{\partial^2 z}{\partial x \partial y} = 2e^y + 3e^{-x}$$

81. $z = \cos xy$

$\dfrac{\partial z}{\partial x} = -y \sin xy, \dfrac{\partial^2 z}{\partial x^2} = -y^2 \cos xy$

$\dfrac{\partial^2 z}{\partial y \partial x} = -yx \cos xy - \sin xy$

$\dfrac{\partial z}{\partial y} = -x \sin xy, \dfrac{\partial^2 z}{\partial y^2} = -x^2 \cos xy$

$\dfrac{\partial^2 z}{\partial x \partial y} = -xy \cos xy - \sin xy$

82. $z = \arctan \dfrac{y}{x}$

$\dfrac{\partial z}{\partial x} = \dfrac{1}{1 + (y^2/x^2)}\left(-\dfrac{y}{x^2}\right) = \dfrac{-y}{x^2 + y^2}$

$\dfrac{\partial^2 z}{\partial x^2} = \dfrac{2xy}{(x^2 + y^2)^2}$

$\dfrac{\partial^2 z}{\partial y \partial x} = \dfrac{-(x^2 + y^2) + y(2y)}{(x^2 + y^2)^2} = \dfrac{y^2 - x^2}{(x^2 + y^2)^2}$

$\dfrac{\partial z}{\partial y} = \dfrac{1}{1 + (y^2/x^2)}\left(\dfrac{1}{x}\right) = \dfrac{x}{x^2 + y^2}$

$\dfrac{\partial^2 z}{\partial y^2} = \dfrac{-2xy}{(x^2 + y^2)^2}$

$\dfrac{\partial^2 z}{\partial x \partial y} = \dfrac{(x^2 + y^2) - x(2x)}{(x^2 + y^2)^2} = \dfrac{y^2 - x^2}{(x^2 + y^2)^2}$

83. $z = x \sec y$

$\dfrac{\partial z}{\partial x} = \sec y$

$\dfrac{\partial^2 z}{\partial x^2} = 0$

$\dfrac{\partial^2 z}{\partial y \partial x} = \sec y \tan y$

$\dfrac{\partial z}{\partial y} = x \sec y \tan y$

$\dfrac{\partial^2 z}{\partial y^2} = x \sec y(\sec^2 y + \tan^2 y)$

$\dfrac{\partial^2 z}{\partial x \partial y} = \sec y \tan y$

So, $\dfrac{\partial^2 z}{\partial y \partial x} = \dfrac{\partial^2 z}{\partial x \partial y}$

There are no points for which $z_x = 0 = z_y$, because

$\dfrac{\partial z}{\partial x} = \sec y \neq 0.$

84. $z = \sqrt{25 - x^2 - y^2}$

$\dfrac{\partial z}{\partial x} = \dfrac{-x}{\sqrt{25 - x^2 - y^2}}$

$\dfrac{\partial^2 z}{\partial x^2} = \dfrac{y^2 - 25}{(25 - x^2 - y^2)^{3/2}}$

$\dfrac{\partial^2 z}{\partial y \partial x} = \dfrac{-xy}{(25 - x^2 - y^2)^{3/2}}$

$\dfrac{\partial z}{\partial y} = \dfrac{-y}{\sqrt{25 - x^2 - y^2}}$

$\dfrac{\partial^2 z}{\partial y^2} = \dfrac{x^2 - 25}{(25 - x^2 - y^2)^{3/2}}$

$\dfrac{\partial^2 z}{\partial x \partial y} = \dfrac{-xy}{(25 - x^2 - y^2)^{3/2}}$

$\dfrac{\partial z}{\partial x} = \dfrac{\partial z}{\partial y} = 0 \text{ if } x = y = 0$

85. $z = \ln\left(\dfrac{x}{x^2 + y^2}\right) = \ln x - \ln(x^2 + y^2)$

$\dfrac{\partial z}{\partial x} = \dfrac{1}{x} - \dfrac{2x}{x^2 + y^2} = \dfrac{y^2 - x^2}{x(x^2 + y^2)}$

$\dfrac{\partial^2 z}{\partial x^2} = \dfrac{x^4 - 4x^2y^2 - y^4}{x^2(x^2 + y^2)^2}$

$\dfrac{\partial^2 z}{\partial y \partial x} = \dfrac{4xy}{(x^2 + y^2)^2}$

$\dfrac{\partial z}{\partial y} = -\dfrac{2y}{x^2 + y^2}$

$\dfrac{\partial^2 z}{\partial y^2} = \dfrac{2(y^2 - x^2)}{(x^2 + y^2)^2}$

$\dfrac{\partial^2 z}{\partial x \partial y} = \dfrac{4xy}{(x^2 + y^2)^2}$

There are no points for which $z_x = z_y = 0$.

86. $z = \dfrac{xy}{x - y}$

$\dfrac{\partial z}{\partial x} = \dfrac{y(x - y) - xy}{(x - y)^2} = \dfrac{-y^2}{(x - y)^2}$

$\dfrac{\partial^2 z}{\partial x^2} = \dfrac{2y^2}{(x - y)^3}$

$\dfrac{\partial^2 z}{\partial y \partial x} = \dfrac{(x - y)^2(-2y) + y^2(2)(x - y)(-1)}{(x - y)^4} = \dfrac{-2xy}{(x - y)^3}$

$\dfrac{\partial z}{\partial y} = -\dfrac{x(x - y) + xy}{(x - y)^2} = \dfrac{x^2}{(x - y)^2}$

$\dfrac{\partial^2 z}{\partial y^2} = \dfrac{2x^2}{(x - y)^3}$

$\dfrac{\partial^2 z}{\partial x \partial y} = \dfrac{(x - y)^2(2x) - x^2(2)(x - y)}{(x - y)^4} = \dfrac{-2xy}{(x - y)^3}$

There are no points for which $z_x = z_y = 0$.

87. $f(x, y, z) = xyz$

$f_x(x, y, z) = yz$

$f_y(x, y, z) = xz$

$f_{yy}(x, y, z) = 0$

$f_{xy}(x, y, z) = z$

$f_{yx}(x, y, z) = z$

$f_{yyx}(x, y, z) = 0$

$f_{xyy}(x, y, z) = 0$

$f_{yxy}(x, y, z) = 0$

So, $f_{xyy} = f_{yxy} = f_{yyx} = 0$.

88. $f(x, y, z) = x^2 - 3xy + 4yz + z^3$

$f_x(x, y, z) = 2x - 3y$

$f_y(x, y, z) = -3x + 4z$

$f_{yy}(x, y, z) = 0$

$f_{xy}(x, y, z) = -3$

$f_{yx}(x, y, z) = -3$

$f_{yyx}(x, y, z) = 0$

$f_{xyy}(x, y, z) = 0$

$f_{yxy}(x, y, z) = 0$

So, $f_{xyy} = f_{yxy} = f_{yyx} = 0$.

89. $f(x, y, z) = e^{-x} \sin yz$

$f_x(x, y, z) = -e^{-x} \sin yz$

$f_y(x, y, z) = ze^{-x} \cos yz$

$f_{yy}(x, y, z) = -z^2e^{-x} \sin yz$

$f_{xy}(x, y, z) = -ze^{-x} \cos yz$

$f_{yx}(x, y, z) = -ze^{-x} \cos yz$

$f_{yyx}(x, y, z) = z^2e^{-x} \sin yz$

$f_{xyy}(x, y, z) = z^2e^{-x} \sin yz$

$f_{yxy}(x, y, z) = z^2e^{-x} \sin yz$

So, $f_{xyy} = f_{yxy} = f_{yyz}$.

90. $f(x, y, z) = \dfrac{2z}{x + y}$

$f_x(x, y, z) = \dfrac{-2z}{(x + y)^2}$

$f_y(x, y, z) = \dfrac{-2z}{(x + y)^2}$

$f_{yy}(x, y, z) = \dfrac{4z}{(x + y)^3}$

$f_{xy}(x, y, z) = \dfrac{4z}{(x + y)^3}$

$f_{yx}(x, y, z) = \dfrac{4z}{(x + y)^3}$

$f_{yyx}(x, y, z) = \dfrac{-12z}{(x + y)^4}$

$f_{xyy}(x, y, z) = \dfrac{-12z}{(x + y)^4}$

$f_{yxy}(x, y, z) = \dfrac{-12z}{(x + y)^4}$

91. $z = 5xy$

$\dfrac{\partial z}{\partial x} = 5y$

$\dfrac{\partial^2 z}{\partial x^2} = 0$

$\dfrac{\partial z}{\partial y} = 5x$

$\dfrac{\partial^2 z}{\partial y^2} = 0$

So, $\dfrac{\partial^2 z}{\partial x^2} + \dfrac{\partial^2 z}{\partial y^2} = 0 + 0 = 0$.

92. $z = \sin x\left(\dfrac{e^y - e^{-y}}{2}\right)$

$$\frac{\partial z}{\partial x} = \cos x\left(\frac{e^y - e^{-y}}{2}\right)$$

$$\frac{\partial^2 z}{\partial x^2} = -\sin x\left(\frac{e^y - e^{-y}}{2}\right)$$

$$\frac{\partial z}{\partial y} = \sin x\left(\frac{e^y + e^{-y}}{2}\right)$$

$$\frac{\partial^2 z}{\partial y^2} = \sin x\left(\frac{e^y - e^{-y}}{2}\right)$$

So,

$$\frac{\partial^2 z}{\partial x^2} + \frac{\partial^2 z}{\partial y^2} = -\sin x\left(\frac{e^y - e^{-y}}{2}\right) + \sin x\left(\frac{e^y - e^{-y}}{2}\right) = 0.$$

93. $z = e^x \sin y$

$$\frac{\partial z}{\partial x} = e^x \sin y$$

$$\frac{\partial^2 z}{\partial x^2} = e^x \sin y$$

$$\frac{\partial z}{\partial y} = e^x \cos y$$

$$\frac{\partial^2 z}{\partial y^2} = -e^x \sin y$$

So, $\dfrac{\partial^2 z}{\partial x^2} + \dfrac{\partial^2 z}{\partial y^2} = e^x \sin y - e^x \sin y = 0.$

94. $z = \arctan \dfrac{y}{x}$

From Exercise 82, we have

$$\frac{\partial^2 z}{\partial x^2} + \frac{\partial^2 z}{\partial y^2} = \frac{2xy}{\left(x^2 + y^2\right)^2} + \frac{-2xy}{\left(x^2 + y^2\right)^2} = 0.$$

95. $z = \sin(x - ct)$

$$\frac{\partial z}{\partial t} = -c\cos(x - ct)$$

$$\frac{\partial^2 z}{\partial t^2} = -c^2 \sin(x - ct)$$

$$\frac{\partial z}{\partial x} = \cos(x - ct)$$

$$\frac{\partial^2 z}{\partial x^2} = -\sin(x - ct)$$

So, $\dfrac{\partial^2 z}{\partial t^2} = c^2\left(\dfrac{\partial^2 z}{\partial x^2}\right).$

96. $z = \cos(4x + 4ct)$

$$\frac{\partial z}{\partial t} = -4c\sin(4x + 4ct)$$

$$\frac{\partial^2 z}{\partial t^2} = -16c^2 \cos(4x + 4ct)$$

$$\frac{\partial z}{\partial x} = -4\sin(4x + 4ct)$$

$$\frac{\partial^2 z}{\partial x^2} = -16\cos(4x + 4ct)$$

$$\frac{\partial^2 z}{\partial t^2} = c^2\left(-16\cos(4x + 4ct)\right) = c^2\left(\frac{\partial^2 z}{\partial x^2}\right)$$

97. $z = \ln(x + ct)$

$$\frac{\partial z}{\partial t} = \frac{c}{x + ct}$$

$$\frac{\partial^2 z}{\partial t^2} = \frac{-c^2}{(x + ct)^2}$$

$$\frac{\partial z}{\partial x} = \frac{1}{x + ct}$$

$$\frac{\partial^2 z}{\partial x^2} = \frac{-1}{(x + ct)^2}$$

$$\frac{\partial^2 z}{\partial t^2} = \frac{-c^2}{(x + ct)^2} = c^2\left(\frac{\partial^2 z}{\partial x^2}\right)$$

98. $z = \sin(\omega ct)\sin(\omega x)$

$$\frac{\partial z}{\partial t} = \omega c\cos(\omega ct)\sin(\omega x)$$

$$\frac{\partial^2 z}{\partial t^2} = -\omega^2 c^2 \sin(\omega ct)\sin(\omega x)$$

$$\frac{\partial z}{\partial x} = \omega\sin(\omega ct)\cos(\omega x)$$

$$\frac{\partial^2 z}{\partial x^2} = -\omega^2 \sin(\omega ct)\sin(\omega x)$$

So, $\dfrac{\partial^2 z}{\partial t^2} = c^2\left(\dfrac{\partial^2 z}{\partial x^2}\right).$

99. $z = e^{-t}\cos\dfrac{x}{c}$

$$\frac{\partial z}{\partial t} = -e^{-t}\cos\frac{x}{c}$$

$$\frac{\partial z}{\partial x} = -\frac{1}{c}e^{-t}\sin\frac{x}{c}$$

$$\frac{\partial^2 z}{\partial x^2} = -\frac{1}{c^2}e^{-t}\cos\frac{x}{c}$$

So, $\dfrac{\partial z}{\partial t} = c^2\left(\dfrac{\partial^2 z}{\partial x^2}\right).$

100. $z = e^{-t} \sin \frac{x}{c}$

$$\frac{\partial z}{\partial t} = -e^{-t} \sin \frac{x}{c}$$

$$\frac{\partial z}{\partial x} = \frac{1}{c}e^{-t} \cos \frac{x}{c}$$

$$\frac{\partial^2 z}{\partial x^2} = -\frac{1}{c^2}e^{-t} \sin \frac{x}{c}$$

So, $\frac{\partial z}{\partial t} = c^2\left(\frac{\partial^2 z}{\partial x^2}\right)$.

101. Yes. The function $f(x, y) = \cos(3x - 2y)$ satisfies both equations.

102. A function $f(x, y)$ with the given partial derivatives does not exist.

103. If $z = f(x, y)$, then to find f_x you consider y constant and differentiate with respect to x. Similarly, to find f_y, you consider x constant and differentiate with respect to y.

104.

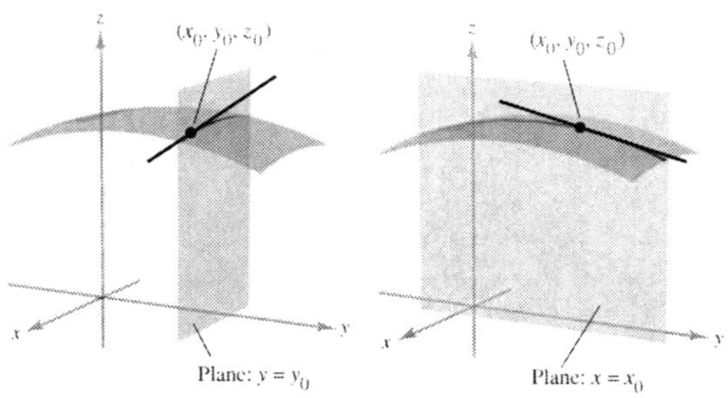

$\frac{\partial f}{\partial x}$ denotes the slope of surface in the x-direction.

$\frac{\partial f}{\partial y}$ denotes the slope of the surface in the y-direction.

105. The plane $z = -x + y = f(x, y)$ satisfies

$$\frac{\partial f}{\partial x} < 0 \text{ and } \frac{\partial f}{\partial y} > 0.$$

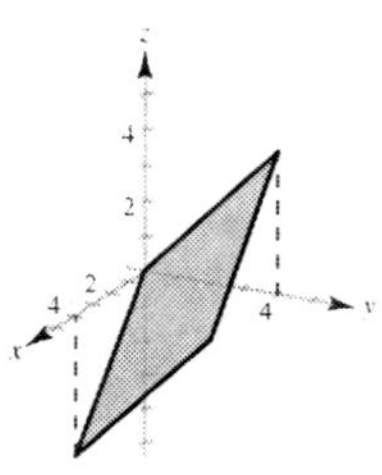

106. The plane $z = x + y = f(x, y)$ satisfies

$$\frac{\partial f}{\partial x} > 0 \text{ and } \frac{\partial f}{\partial y} > 0.$$

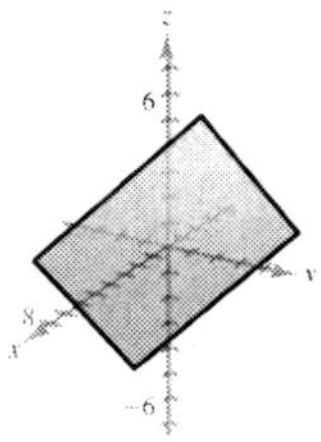

107. In this case, the mixed partials are equal, $f_{xy} = f_{yx}$. See Theorem 13.3.

108. (a) $f_x(4, 1) < 0$

(b) $f_y(4, 1) > 0$

(c) $f_x(-1, -2) < 0$

(d) $f_y(-1, -2) > 0$

109. $R = 200x_1 + 200x_2 - 4x_1^2 - 8x_1x_2 - 4x_2^2$

(a) $\frac{\partial r}{\partial x_1} = 200 - 8x_1 - 8x_2$

At $(x_1, x_2) = (4, 12)$, $\frac{\partial R}{\partial x_1} = 200 - 32 - 96 = 72.$

(b) $\frac{\partial R}{\partial x_2} = 200 - 8x_1 - 8x_2$

At $(x_1, x_2) = (4, 12)$, $\frac{\partial R}{\partial x^2} = 72.$

110. (a) $C = 32\sqrt{xy} + 175x + 205y + 1050$

$$\frac{\partial C}{\partial x} = 16\sqrt{\frac{y}{x}} + 175$$

$$\left.\frac{\partial C}{\partial x}\right]_{(80, 20)} = 16\sqrt{\frac{1}{4}} + 175 = 183$$

$$\frac{\partial C}{\partial y} = 16\sqrt{\frac{x}{y}} + 205$$

$$\left.\frac{\partial C}{\partial y}\right]_{(80, 20)} = 16\sqrt{4} + 205 = 237$$

(b) The fireplace-insert stove results in the cost increasing at a faster rate because $\frac{\partial C}{\partial y} > \frac{\partial C}{\partial x}$.

111. $IQ(M, C) = 100\dfrac{M}{C}$

$$IQ_M = \frac{100}{C},\ IQ_M(12, 10) = 10$$

$$IQ_c = \frac{-100M}{C^2},\ IQ_c(12, 10) = -12$$

When the chronological age is constant, IQ increases at a rate of 10 points per mental age year.

When the mental age is constant, IQ decreases at a rate of 12 points per chronological age year.

112. $f(x, y) = 200x^{0.7}y^{0.3}$

(a) $$\frac{\partial f}{\partial x} = 140x^{-0.3}y^{0.3} = 140\left(\frac{y}{x}\right)^{0.3}$$

At $(x, y) = (1000, 500)$,

$$\frac{\partial f}{\partial x} = 140\left(\frac{500}{1000}\right)^{0.3} = 140\left(\frac{1}{2}\right)^{0.3} \approx 113.72.$$

(b) $$\frac{\partial f}{\partial y} = 60x^{0.7}y^{-0.7} = 60\left(\frac{x}{y}\right)^{0.7}$$

At $(x, y) = (1000, 500)$,

$$\frac{\partial f}{\partial y} = 60\left(\frac{1000}{500}\right)^{0.7} = 60(2)^{0.7} \approx 97.47.$$

113. An increase in either price will cause a decrease in demand.

114. $$V(I, R) = 1000\left[\frac{1 + 0.06(1 - R)}{1 + I}\right]^{10}$$

$$V_I(I, R) = 10{,}000\left[\frac{1 + 0.06(1 - R)}{1 + I}\right]^{9}\left[-\frac{1 + 0.06(1 - R)}{(1 + I)^2}\right] = -10{,}000\left[\frac{(1 + 0.06(1 - R))^{10}}{(1 + I)^{11}}\right]$$

$$V_I(0.03, 0.28) = -11{,}027.20$$

$$V_R(I, R) = 10{,}000\left[\frac{1 + 0.06(1 - R)}{1 + I}\right]^{9}\left[-\frac{0.06}{1 + I}\right] = -600\left[\frac{(1 + 0.06(1 - R))^{9}}{(1 + I)^{10}}\right]$$

$$V_R(0.03, 0.28) = -653.26$$

The rate of inflation has the greater negative influence.

115. $T = 500 - 0.6x^2 - 1.5y^2$

$$\frac{\partial T}{\partial x} = -1.2x,\ \frac{\partial T}{\partial x}(2, 3) = -2.4°/\text{m}$$

$$\frac{\partial T}{\partial y} = -3y = \frac{\partial T}{\partial y}(2, 3) = -9°/\text{m}$$

116. $A = 0.885t - 22.4h + 1.20th - 0.544$

(a) $$\frac{\partial A}{\partial t} = 0.885 + 1.20h$$

$$\frac{\partial A}{\partial t}(30°, 0.80) = 0.885 + 1.20(0.80) = 1.845$$

$$\frac{\partial A}{\partial h} = -22.4 + 1.20t$$

$$\frac{\partial A}{\partial h}(30°, 0.80) = -22.4 + 1.20(30°) = 13.6$$

(b) The humidity has a greater effect on A because its coefficient -22.4 is larger than that of t.

117. $$PV = \frac{n}{xB}RT$$

$$T = \frac{PV}{\frac{n}{xB}R} \Rightarrow \frac{\partial T}{\partial P} = \frac{V}{\frac{n}{xB}R}$$

$$P = \frac{\frac{n}{xB}RT}{V} \Rightarrow \frac{\partial P}{\partial V} = -\frac{\frac{n}{xB}RT}{V^2}$$

$$V = \frac{\frac{n}{xB}RT}{P} \Rightarrow \frac{\partial V}{\partial T} = \frac{\frac{n}{xB}R}{P}$$

$$\frac{\partial T}{\partial P}\cdot\frac{\partial P}{\partial V}\cdot\frac{\partial V}{\partial T} = \left(\frac{V}{\frac{n}{xB}R}\right)\left(-\frac{\frac{n}{xB}RT}{V^2}\right)\left(\frac{\frac{n}{xB}R}{P}\right)$$

$$= -\frac{\frac{n}{xB}RT}{VP} = -\frac{\frac{n}{xB}RT}{\frac{n}{xB}RT} = -1$$

118. $U = -5x^2 + xy - 3y^2$

(a) $U_x = -10x + y$

(b) $U_y = x - 6y$

(c) $U_x(2, 3) = -17$ and $U_y(2, 3) = -16$. The person should consume one more unit of y because the rate of decrease of satisfaction is less for y.

(d)

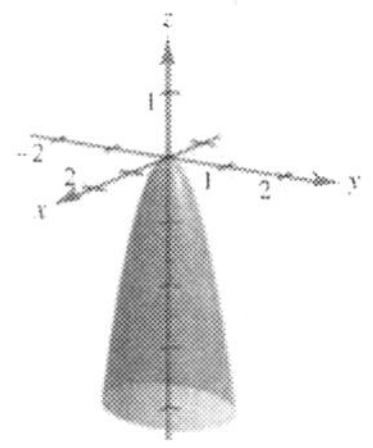

119. $z = 0.461x + 0.301y - 494$

(a) $\frac{\partial z}{\partial x} = 0.461 \qquad \frac{\partial z}{\partial y} = 0.301$

(b) As the expenditures on amusement parks and campgrounds (x) increase, the expenditures on spectator sports (z) increase. As the expenditures on live entertainment (y) increase, the expenditures on spectator sports (z) increase.

120. $z = 11.734x^2 - 0.028y^2 - 888.24x + 23.09y + 12{,}573.9$

(a) $\frac{\partial z}{\partial x} = 23.468x - 888.24$

$\frac{\partial^2 z}{\partial x^2} = 23.468$

$\frac{\partial z}{\partial y} = -0.056y + 23.09$

$\frac{\partial^2 z}{\partial y^2} = -0.056$

(b) Traces parallel to the xz-plane are concave upward $\left(\frac{\partial^2 z}{\partial x^2} > 0\right)$. The rate of change of Medicare expenses is increasing with respect to worker's compensation (x).

(c) Traces parallel to the yz-plane are concave downward $\left(\frac{\partial^2 z}{\partial y^2} < 0\right)$. The rate of change of Medicare expenses is decreasing with respect to Medicaid (y).

121. False

Let $z = x + y + 1$.

122. True

123. True

124. True

125. $f(x, y) = \begin{cases} \dfrac{xy(x^2 - y^2)}{x^2 + y^2}, & (x, y) \neq (0, 0) \\ 0, & (x, y) = (0, 0) \end{cases}$

(a) $f_x(x, y) = \dfrac{(x^2 + y^2)(3x^2y - y^3) - (x^3y - xy^3)(2x)}{(x^2 + y^2)^2} = \dfrac{y(x^4 + 4x^2y^2 - y^4)}{(x^2 + y^2)^2}$

$f_y(x, y) = \dfrac{(x^2 + y^2)(x^3 - 3xy^2) - (x^3y - xy^3)(2y)}{(x^2 + y^2)^2} = \dfrac{x(x^4 - 4x^2y^2 - y^4)}{(x^2 + y^2)^2}$

(b) $f_x(0, 0) = \lim\limits_{\Delta x \to 0} \dfrac{f(\Delta x, 0) - f(0, 0)}{\Delta x} = \lim\limits_{\Delta x \to 0} \dfrac{0/[(\Delta x)^2] - 0}{\Delta x} = 0$

$f_y(0, 0) = \lim\limits_{\Delta y \to 0} \dfrac{f(0, \Delta y) - f(0, 0)}{\Delta y} = \lim\limits_{\Delta y \to 0} \dfrac{0/[(\Delta y)^2] - 0}{\Delta y} = 0$

(c) $f_{xy}(0, 0) = \dfrac{\partial}{\partial y}\left(\dfrac{\partial f}{\partial x}\right)\Bigg|_{(0, 0)} = \lim\limits_{\Delta y \to 0} \dfrac{f_x(0, \Delta y) - f_x(0, 0)}{\Delta y} = \lim\limits_{\Delta y \to 0} \dfrac{\Delta y(-(\Delta y)^4)}{((\Delta y)^2)^2(\Delta y)} = \lim\limits_{\Delta y \to 0} (-1) = -1$

$f_{yx}(0, 0) = \dfrac{\partial}{\partial x}\left(\dfrac{\partial f}{\partial y}\right)\Bigg|_{(0, 0)} = \lim\limits_{\Delta x \to 0} \dfrac{f_y(\Delta x, 0) - f_y(0, 0)}{\Delta x} = \lim\limits_{\Delta x \to 0} \dfrac{\Delta x((\Delta x)^4)}{((\Delta x)^2)^2(\Delta x)} = \lim\limits_{\Delta x \to 0} 1 = 1$

(d) f_{yx} or f_{xy} or both are not continuous at $(0, 0)$.

126. $f(x, y) = (x^3 + y^3)^{1/3}$

(a) $f_x(0, 0) = \lim\limits_{\Delta x \to 0} \dfrac{f(0 + \Delta x, 0) - f(0, 0)}{\Delta x}$

$= \lim\limits_{\Delta x \to 0} \dfrac{\Delta x}{\Delta x} = 1$

$f_y(0, 0) = \lim\limits_{\Delta y \to 0} \dfrac{f(0, 0 + \Delta y) - f(0, 0)}{\Delta y}$

$= \lim\limits_{\Delta y \to 0} \dfrac{\Delta y}{\Delta y} = 1$

(b) $f_x(x, y)$ and $f_y(x, y)$ fail to exist for $y = -x, x \neq 0$.

127. $f(x, y) = (x^2 + y^2)^{2/3}$

For $(x, y) \neq (0, 0)$, $f_x(x, y) = \dfrac{2}{3}(x^2 + y^2)^{-1/3}(2x) = \dfrac{4x}{3(x^2 + y^2)^{1/3}}$.

For $(x, y) = (0, 0)$, use the definition of partial derivative.

$f_x(0, 0) = \lim\limits_{\Delta x \to 0} \dfrac{f(0 + \Delta x) - f(0, 0)}{\Delta x} = \lim\limits_{\Delta x \to 0} \dfrac{(\Delta x)^{4/3}}{\Delta x} = \lim\limits_{\Delta x \to 0} (\Delta x)^{1/3} = 0$

Section 13.4 Differentials

1. $z = 2x^2y^3$

$dz = 4xy^3\,dx + 6x^2y^2\,dy$

2. $z = 2x^4y - 8x^2y^3$

$dz = \left(8x^3y - 16xy^3\right)dx + \left(2x^4 - 24x^2y^2\right)dy$

3. $z = \dfrac{-1}{x^2 + y^2}$

$$dz = \frac{2x}{\left(x^2 + y^2\right)^2}dx + \frac{2y}{\left(x^2 + y^2\right)^2}dy$$

$$= \frac{2}{\left(x^2 + y^2\right)^2}(x\,dx + y\,dy)$$

4. $w = \dfrac{x + y}{z - 3y}$

$$dw = \frac{1}{z - 3y}dx + \frac{3x + z}{(z - 3y)^2}dy - \frac{x + y}{(z - 3y)^2}dz$$

5. $z = x\cos y - y\cos x$

$dz = (\cos y + y\sin x)\,dx + (-x\sin y - \cos x)\,dy$

$= (\cos y + y\sin x)\,dx - (x\sin y + \cos x)\,dy$

6. $z = \left(\dfrac{1}{2}\right)\left(e^{x^2+y^2} - e^{-x^2-y^2}\right)$

$$dz = 2x\left(\frac{e^{x^2+y^2} + e^{-x^2-y^2}}{2}\right)dx + 2y\left(\frac{e^{x^2+y^2} + e^{-x^2-y^2}}{2}\right)dy$$

$$= \left(e^{x^2+y^2} + e^{-x^2-y^2}\right)(x\,dx + y\,dy)$$

7. $z = e^x\sin y$

$dz = \left(e^x\sin y\right)dx + \left(e^x\cos y\right)dy$

8. $w = e^y\cos x + z^2$

$dw = -e^y\sin x\,dx + e^y\cos x\,dy + 2z\,dz$

9. $w = 2z^3y\sin x$

$dw = 2z^3y\cos x\,dx + 2z^3\sin x\,dy + 6z^2y\sin x\,dz$

10. $w = x^2yz^2 + \sin yz$

$dw = 2xyz^2\,dx + \left(x^2z^2 + z\cos yz\right)dy + \left(2x^2yz + y\cos yz\right)dz$

11. $f(x, y) = 2x - 3y$

(a) $f(2, 1) = 1$

$f(2.1, 1.05) = 1.05$

$\Delta z = f(2.1, 1.05) - f(2, 1) = 0.05$

(b) $dz = 2\,dx - 3\,dy = 2(0.1) - 3(0.05) = 0.05$

12. $f(x, y) = x^2 + y^2$

(a) $f(2, 1) = 5$

$f(2.1, 1.05) = 5.5125$

$\Delta z = f(2.1, 1.05) - f(2, 1) = 0.5125$

(b) $dz = 2x\,dx + 2y\,dy$

$= 2(2)(0.1) + 2(1)(0.05) = 0.5$

13. $f(x, y) = 16 - x^2 - y^2$

(a) $f(2, 1) = 11$

$f(2.1, 1.05) = 10.4875$

$\Delta z = f(2.1, 1.05) - f(2.1) = -0.5125$

(b) $dz = -2x\,dx - 2y\,dy$

$= -2(2)(0.1) - 2(1)(0.05) = -0.5$

14. $f(x, y) = \dfrac{y}{x}$

(a) $f(2, 1) = 0.5$

$f(2.1, 1.05) = 0.5$

$\Delta z = f(2.1, 1.05) - f(2, 1) = 0$

(b) $dz = \dfrac{-y}{x^2}dx + \dfrac{1}{x}dy = \dfrac{-1}{4}(0.1) + \dfrac{1}{2}(0.05) = 0$

15. $f(x, y) = ye^x$

(a) $f(2, 1) = e^2 \approx 7.3891$

$f(2.1, 1.05) = 1.05e^{2.1} \approx 8.5745$

$\Delta z = f(2.1, 1.05) - f(2, 1) = 1.1854$

(b) $dz = ye^x\,dx + e^x\,dy$

$= e^2(0.1) + e^2(0.05) \approx 1.1084$

16. $f(x, y) = x \cos y$

(a) $f(2, 1) = 2 \cos 1 \approx 1.0806$

$f(2.1, 1.05) = 2.1 \cos 1.05 \approx 1.0449$

$\Delta z = f(2.1, 1.05) - f(2, 1) = -0.0357$

(b) $dz = \cos y\, dx - x \sin y\, dy$

$= \cos 1(0.1) - 2 \sin 1(0.05) \approx -0.0301$

17. Let $z = x^2 y$, $x = 2$, $y = 9$, $dx = 0.01$, $dy = 0.02$.

Then: $dz = 2xy\, dx + x^2\, dy$

$(2.01)^2(9.02) - 2^2 \cdot 9 \approx 2(2)(9)(0.01) + 2^2(0.02) = 0.44$

18. Let $z = (1 - x^2)/y^2$, $x = 3$, $y = 6$, $dx = 0.05$, $dy = -0.05$. Then:

$$dz = -\frac{2x}{y^2}\, dx + \frac{-2(1 - x^2)}{y^3}\, dy$$

$$\frac{1 - (3.05)^2}{(5.95)^2} - \frac{1 - 3^2}{6^2} \approx -\frac{2(3)}{6^2}(0.05) - \frac{2(1 - 3^2)}{6^3}(-0.05) \approx -0.012$$

19. Let $z = \sqrt{x^2 + y^2}$, $x = 5$, $y = 3$, $dx = 0.05$, $dy = 0.1$.

Then:

$$dz = \frac{x}{\sqrt{x^2 + y^2}}\, dx + \frac{y}{\sqrt{x^2 + y^2}}\, dy$$

$$\sqrt{(5.05)^2 + (3.1)^2} - \sqrt{5^2 + 3^2} \approx \frac{5}{\sqrt{5^2 + 3^2}}(0.05) + \frac{3}{\sqrt{5^2 + 3^2}}(0.1) = \frac{0.55}{\sqrt{34}} \approx 0.094$$

20. Let $z = \sin(x^2 + y^2)$, $x = y = 1$, $dx = 0.05$, $dy = -0.05$. Then: $dz = 2x \cos(x^2 + y^2)\, dx + 2y \cos(x^2 + y^2)\, dy$

$$\sin\left[(1.05)^2 + (0.95)^2\right] - \sin 2 \approx 2(1) \cos(1^2 + 1^2)(0.05) + 2(1) \cos(1^2 + 1^2)(-0.05) = 0$$

21. In general, the accuracy worsens as Δx and Δy increase.

22. The tangent plane to the surface $z = f(x, y)$ at the point P is a linear approximation of z.

23. If $z = f(x, y)$, then $\Delta z \approx dz$ is the propagated error, and $\frac{\Delta z}{z} \approx \frac{dz}{z}$ is the relative error.

24. The differential is greater at $\left(\frac{1}{2}, \frac{1}{2}\right)$ than at $(2, 2)$ because the surface is increasing faster there.

25. $A = lh$

$dA = l\, dh + h\, dl$

$\Delta A = (l + dl)(h + dh) - lh$

$= h\, dl + l\, dh + dl\, dh$

$\Delta A - dA = dl\, dh$

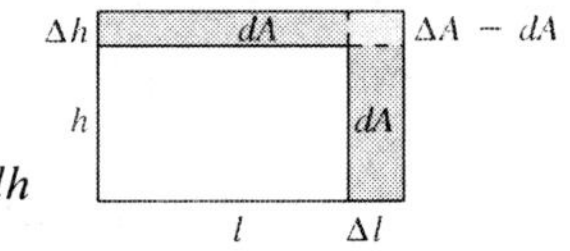

26. $V = \pi r^2 h$

$dV = 2\pi rh\, dr + \pi r^2\, dh$

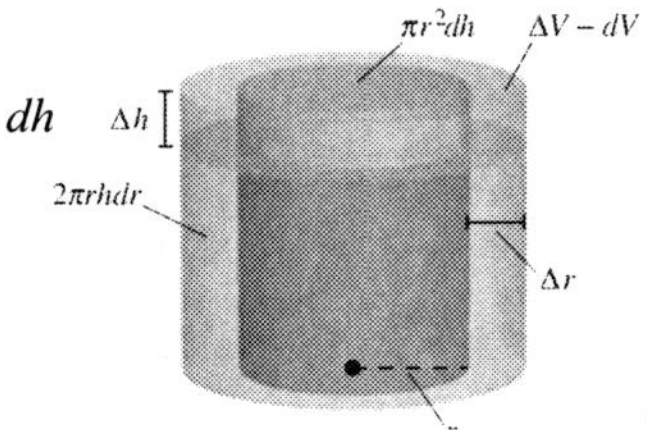

27. $V = \dfrac{\pi r^2 h}{3}, r = 4, h = 8$

$$dV = \frac{2\pi rh}{3}dr + \frac{\pi r^2}{3}dh = \frac{\pi r}{3}(2h\,dr + r\,dh) = \frac{4\pi}{3}(16\,dr + 4\,dh)$$

$$\Delta V = \frac{\pi}{3}\left[(r + \Delta r)^2(h + \Delta h) - r^2 h\right] = \frac{\pi}{3}\left[(4 + \Delta r)^2(8 + \Delta h) - 128\right]$$

Δr	Δh	dV	ΔV	$\Delta V - dV$
0.1	0.1	8.3776	8.5462	0.1686
0.1	−0.1	5.0265	5.0255	−0.0010
0.001	0.002	0.1005	0.1006	0.0001
−0.0001	0.0002	−0.0034	−0.0034	0.0000

28. $S = \pi r\sqrt{r^2 + h^2}, r = 6, h = 16$

$$\frac{dS}{dr} = \pi\left(r^2 + h^2\right)^{1/2} + \pi r^2\left(r^2 + h^2\right)^{-1/2} = \pi\frac{2r^2 + h^2}{\sqrt{r^2 + h^2}}$$

$$\frac{dS}{dh} = \pi\frac{rh}{\sqrt{r^2 + h^2}}$$

$$dS = \frac{\pi}{\sqrt{r^2 + h^2}}\left[\left(2r^2 + h^2\right)dr + (rh)dh\right] = \frac{\pi}{\sqrt{292}}[328\,dr + 96\,dh]$$

$S(6, 16) = 322.101353$

$$\Delta S = \pi(r + \Delta r)\sqrt{(r + \Delta r)^2 + (h + \Delta h)^2} = \pi(6 + \Delta r)\sqrt{(6 + \Delta r)^2 + (16 + \Delta h)^2} - 322.101353$$

Δr	Δh	dS	ΔS	$\Delta S - dS$
0.1	0.1	7.7951	7.8375	0.0424
0.1	−0.1	4.2653	4.2562	−0.0091
0.001	0.002	0.0956	0.0956	0.0000
−0.0001	0.0002	−0.0025	−0.0025	−0.0000

29. $V = xyz, dV = yz\,dx + xz\,dy + xy\,dz$

$$\text{Propagated error} = dV = 5(12)(\pm 0.02) + 8(12)(\pm 0.02) + 8(5)(\pm 0.02)$$
$$= (60 + 96 + 40)(\pm 0.02) = 196(\pm 0.02) = \pm 3.92 \text{ in.}^3$$

The measured volume is $V = 8(5)(12) = 480 \text{ in.}^3$

$$\text{Relative error} = \frac{\Delta V}{V} \approx \frac{dV}{V} = \frac{3.92}{480} \approx 0.008167 \approx 0.82\%$$

30. $V = \pi r^2 h, dV = 2\pi rh\,dr + \pi r^2\,dh$

$$\text{Propagated error} = dV = 2\pi(3)(10)(\pm 0.05) + \pi(3)^2(\pm 0.05)$$
$$= (60\pi + 9\pi)(\pm 0.05) = \pm 3.45\pi \text{ cm}^3$$

The measured volume is $V = \pi(3^2)(10) = 90\pi \text{ cm}^3$.

$$\text{Relative error} = \frac{\Delta V}{V} \approx \frac{dV}{V} = \frac{3.45\pi}{90\pi} \approx 0.0383 = 3.83\%$$

31. $C = 35.74 + 0.6215T - 35.75v^{0.16} + 0.4275Tv^{0.16}$

$$\frac{\partial C}{\partial T} = 0.6215 + 0.4275v^{0.16}$$

$$\frac{\partial C}{\partial v} = -5.72v^{-0.84} + 0.0684Tv^{-0.84}$$

$$dC = \frac{\partial C}{\partial T}dT + \frac{\partial C}{\partial v}dv = \left(0.6215 + 0.4275(23)^{0.16}\right)(\pm 1) + \left(-5.72(23)^{-0.84} + 0.0684(8)(23)^{-0.84}\right)(\pm 3)$$

$$= \pm 1.3275 \pm 1.1143 = \pm 2.4418 \text{ Maximum propagated error}$$

$$\frac{dC}{C} = \frac{2.4418}{-12.6807} \approx 0.19 = 19\% \text{ Maximum relative error}$$

32. $\dfrac{1}{R} = \dfrac{1}{R_1} + \dfrac{1}{R_2}$

$$R = \frac{R_1 R_2}{R_1 + R_2}$$

$$dR_1 = \Delta R_1 = 0.5$$

$$dR_2 = \Delta R_2 = -2$$

$$\Delta R \approx dR = \frac{\partial R}{\partial R_1}dR + \frac{\partial R}{\partial R_2}dR_2 = \frac{R_2^2}{(R_1 + R_2)^2}\Delta R_1 + \frac{R_1^2}{(R_1 + R_2)^2}\Delta R_2$$

When $R_1 = 10$ and $R_2 = 15$, we have $\Delta R \approx \dfrac{15^2}{(10 + 15)^2}(0.5) + \dfrac{10^2}{(10 + 15)^2}(-2) = -0.14$ ohm.

33. $P = \dfrac{E^2}{R}, \left|\dfrac{dE}{E}\right| = 3\% = 0.03, \left|\dfrac{dR}{R}\right| = 4\% = 0.04$

$$dP = \frac{2E}{R}dE - \frac{E^2}{R^2}dR$$

$$\frac{dP}{P} = \left[\frac{2E}{R}dE - \frac{E^2}{R^2}dR\right]\Big/P = \left[\frac{2E}{R}dE - \frac{E^2}{R^2}dR\right]\Big/\left(E^2/R\right) = \frac{2}{E}dE - \frac{1}{R}dR$$

Using the worst case scenario, $\dfrac{dE}{E} = 0.03$ and $\dfrac{dR}{R} = -0.04$: $\dfrac{dP}{P} \le 2(0.03) - (-0.04) = 0.10 = 10\%$.

34. $a = \dfrac{v^2}{r}$

$$da = \frac{2v}{r}dv - \frac{v^2}{r^2}dr$$

$$\frac{da}{a} = 2\frac{dv}{v} - \frac{dr}{r} = 2(0.03) - (-0.02) = 0.08 = 8\%$$

Note: The maximum error will occur when dv and dr differ in signs.

35. (a) $V = \dfrac{1}{2}bhl = \left(18\sin\dfrac{\theta}{2}\right)\left(18\cos\dfrac{\theta}{2}\right)(16)(12) = 31{,}104\sin\theta \text{ in.}^3 = 18\sin\theta \text{ ft}^3$

V is maximum when $\sin\theta = 1$ or $\theta = \pi/2$.

(b) $V = \dfrac{s^2}{2}(\sin\theta)l$

$$dV = s(\sin\theta)l\,ds + \frac{s^2}{2}l(\cos\theta)\,d\theta + \frac{s^2}{2}(\sin\theta)\,dl$$

$$= 18\left(\sin\frac{\pi}{2}\right)(16)(12)\left(\frac{1}{2}\right) + \frac{18^2}{2}(16)(12)\left(\cos\frac{\pi}{2}\right)\left(\frac{\pi}{90}\right) + \frac{18^2}{2}\left(\sin\frac{\pi}{2}\right)\left(\frac{1}{2}\right) = 1809 \text{ in.}^3 \approx 1.047 \text{ ft}^3$$

36. (a) Using the Law of Cosines:

$$a^2 = b^2 + c^2 - 2bc\cos A = 330^2 + 420^2 - 2(330)(420)\cos 9°$$

$$a \approx 107.3 \text{ ft.}$$

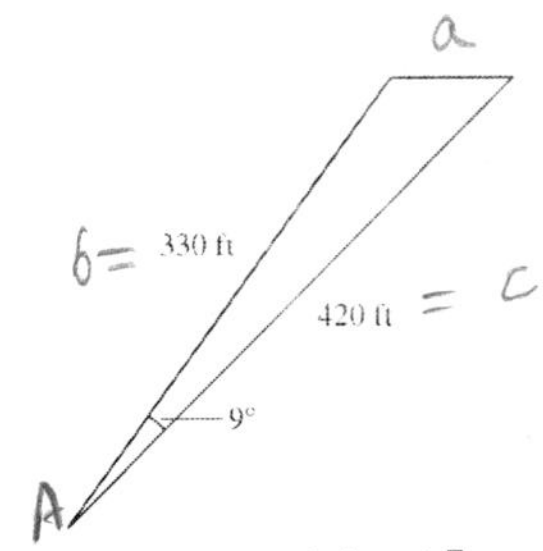

(b) $a = \sqrt{b^2 + 420^2 - 2b(420)\cos\theta}$

$$da = \frac{1}{2}\left[b^2 + 420^2 - 840b\cos\theta\right]^{-1/2}\left[(2b - 840\cos\theta)\,db + 840b\sin\theta\,d\theta\right]$$

$$= \frac{1}{2}\left[330^2 + 420^2 - 840(330)\left(\cos\frac{\pi}{20}\right)\right]^{-1/2}\left[\left(2(330) - 840\cos\frac{\pi}{20}\right)(6) + 840(330)\left(\sin\frac{\pi}{20}\right)\left(\frac{\pi}{180}\right)\right]$$

$$\approx \frac{1}{2}[11512.79]^{-1/2}[\pm 1774.79] \approx \pm 8.27 \text{ ft}$$

37. $L = 0.00021\left(\ln\frac{2h}{r} - 0.75\right)$

$$dL = 0.00021\left[\frac{dh}{h} - \frac{dr}{r}\right] = 0.00021\left[\frac{(\pm 1/100)}{100} - \frac{(\pm 1/16)}{2}\right] \approx (\pm 6.6) \times 10^{-6}$$

$$L = 0.00021(\ln 100 - 0.75) \pm dL \approx 8.096 \times 10^{-4} \pm 6.6 \times 10^{-6} \text{ micro henrys}$$

38. $T = 2\pi\sqrt{\frac{L}{g}}$

$$dg = 32.23 - 32.09 = 0.14$$

$$dL = 2.48 - 2.50 = -0.02$$

$$\Delta T \approx dT = \frac{\partial T}{\partial g}dg + \frac{\partial T}{\partial L}dL = \frac{-\pi}{g}\sqrt{\frac{L}{g}}\,dg + \frac{\pi}{\sqrt{Lg}}\,dL$$

When $g = 32.09$ and $L = 2.50$, $\Delta T \approx \frac{-\pi}{32.09}\sqrt{\frac{2.5}{32.09}}(0.14) + \frac{\pi}{\sqrt{(2.5)(32.09)}}(-0.02) \approx -0.0108$ seconds.

39. $z = f(x, y) = x^2 - 2x + y$

$$\Delta z = f(x + \Delta x, y + \Delta y) - f(x, y) = \left(x^2 + 2x(\Delta x) + (\Delta x)^2 - 2x - 2(\Delta x) + y + (\Delta y)\right) - \left(x^2 - 2x + y\right)$$

$$= 2x(\Delta x) + (\Delta x)^2 - 2(\Delta x) + (\Delta y) = (2x - 2)\,\Delta x + \Delta y + \Delta x(\Delta x) + 0(\Delta y)$$

$$= f_x(x, y)\,\Delta x + f_y(x, y)\,\Delta y + \varepsilon_1\Delta x + \varepsilon_2\Delta y \text{ where } \varepsilon_1 = \Delta x \text{ and } \varepsilon_2 = 0.$$

As $(\Delta x, \Delta y) \to (0, 0)$, $\varepsilon_1 \to 0$ and $\varepsilon_2 \to 0$.

40. $z = f(x, y) = x^2 + y^2$

$$\Delta z = f(x + \Delta x, y + \Delta y) - f(x, y) = x^2 + 2x(\Delta x) + (\Delta x)^2 + y^2 + 2y(\Delta y) + (\Delta y)^2 - \left(x^2 + y^2\right)$$

$$= 2x(\Delta x) + 2y(\Delta y) + \Delta x(\Delta x) + \Delta y(\Delta y) = f_x(x, y)\,\Delta x + f_y(x, y)\,\Delta y + \varepsilon_1\Delta x + \varepsilon_2\Delta y \text{ where } \varepsilon_1 = \Delta x \text{ and } \varepsilon_2 = \Delta y.$$

As $(\Delta x, \Delta y) \to (0, 0)$, $\varepsilon_1 \to 0$ and $\varepsilon_2 \to 0$.

41. $z = f(x, y) = x^2y$

$$\Delta z = f(x + \Delta x, y + \Delta y) - f(x, y) = \left(x^2 + 2x(\Delta x) + (\Delta x)^2\right)(y + \Delta y) - x^2y$$

$$= 2xy(\Delta x) + y(\Delta x)^2 + x^2\Delta y + 2x(\Delta x)(\Delta y) + (\Delta x)^2\Delta y = 2xy(\Delta x) + x^2\Delta y + (y\Delta x)\Delta x + \left[2x\Delta x + (\Delta x)^2\right]\Delta y$$

$$= f_x(x, y)\,\Delta x + f_y(x, y)\,\Delta y + \varepsilon_1\Delta x + \varepsilon_2\Delta y \text{ where } \varepsilon_1 = y(\Delta x) \text{ and } \varepsilon_2 = 2x\Delta x + (\Delta x)^2.$$

As $(\Delta x, \Delta y) \to (0, 0)$, $\varepsilon_1 \to 0$ and $\varepsilon_2 \to 0$.

42. $z = f(x, y) = 5x - 10y + y^3$

$$\Delta z = f(x + \Delta x, y + \Delta y) - f(x, y)$$

$$= 5x + 5\Delta x - 10y - 10\Delta y + y^3 + 3y^2(\Delta y) + 3y(\Delta y)^2 + (\Delta y)^3 - (5x - 10y + y^3)$$

$$= 5(\Delta x) + (3y^2 - 10)(\Delta y) + 0(\Delta x) + (3y(\Delta y) + (\Delta y)^2)\Delta y$$

$$= f_x(x, y)\,\Delta x + f_y(x, y)\,\Delta y + \varepsilon_1 \Delta x + \varepsilon_2 \Delta y \text{ where } \varepsilon_1 = 0 \text{ and } \varepsilon_2 = 3y(\Delta y) + (\Delta y)^2.$$

As $(\Delta x, \Delta y) \to (0, 0)$, $\varepsilon_1 \to 0$ and $\varepsilon_2 \to 0$.

43. $f(x, y) = \begin{cases} \dfrac{3x^2y}{x^4 + y^2}, & (x, y) \neq (0, 0) \\ 0, & (x, y) = (0, 0) \end{cases}$

$$f_x(0, 0) = \lim_{\Delta x \to 0} \frac{f(\Delta x, 0) - f(0, 0)}{\Delta x} = \lim_{\Delta x \to 0} \frac{\dfrac{0}{(\Delta x)^4} - 0}{\Delta x} = 0$$

$$f_y(0, 0) = \lim_{\Delta y \to 0} \frac{f(0, \Delta y) - f(0, 0)}{\Delta y} = \lim_{\Delta y \to 0} \frac{\dfrac{0}{(\Delta y)^2} - 0}{\Delta y} = 0$$

So, the partial derivatives exist at $(0, 0)$.

Along the line $y = x$: $\displaystyle\lim_{(x, y) \to (0, 0)} f(x, y) = \lim_{x \to 0} \frac{3x^3}{x^4 + x^2} = \lim_{x \to 0} \frac{3x}{x^2 + 1} = 0$

Along the curve $y = x^2$: $\displaystyle\lim_{(x, y) \to (0, 0)} f(x, y) = \frac{3x^4}{2x^4} = \frac{3}{2}$

f is not continuous at $(0, 0)$. So, f is not differentiable at $(0, 0)$. (See Theorem 12.5)

44. $f(x, y) = \begin{cases} \dfrac{5x^2y}{x^3 + y^3}, & (x, y) \neq (0, 0) \\ 0, & (x, y) = (0, 0) \end{cases}$

$$f_x(0, 0) = \lim_{\Delta x \to 0} \frac{f(\Delta x, 0) - f(0, 0)}{\Delta x} = \lim_{\Delta x \to 0} \frac{0 - 0}{\Delta x} = 0$$

$$f_y(0, 0) = \lim_{\Delta y \to 0} \frac{f(0, \Delta y) - f(0, 0)}{\Delta y} = \lim_{\Delta y \to 0} \frac{0 - 0}{\Delta y} = 0$$

So, the partial derivatives exist at $(0, 0)$.

Along the line $y = x$: $\displaystyle\lim_{(x, y) \to (0, 0)} f(x, y) = \lim_{x \to 0} \frac{5x^3}{2x^3} = \frac{5}{2}.$

Along the line $x = 0$, $\displaystyle\lim_{(x, y) \to (0, 0)} f(x, y) = 0.$

So, f is not continuous at $(0, 0)$. Therefore f is not differentiable at $(0, 0)$.

Section 13.5 Chain Rules for Functions of Several Variables

1. $w = x^2 + y^2$

$x = 2t,\ y = 3t$

$$\frac{dw}{dt} = \frac{\partial w}{\partial x}\frac{dx}{dt} + \frac{\partial w}{\partial y}\frac{dy}{dt} = (2x)(2) + (2y)(3)$$
$$= 4x + 6y = 8t + 18t = 26t$$

When $t = 2$, $\dfrac{dw}{dt} = 26(2) = 52$.

2. $w = \sqrt{x^2 + y^2}$

$x = \cos t,\ y = e^t$

$$\frac{dw}{dt} = \frac{\partial w}{\partial x}\frac{dx}{dt} + \frac{\partial w}{\partial y}\frac{dy}{dt}$$
$$= \frac{x}{\sqrt{x^2 + y^2}}(-\sin t) + \frac{y}{\sqrt{x^2 + y^2}}e^t$$
$$= \frac{-x\sin t + ye^t}{\sqrt{x^2 + y^2}} = \frac{-\cos t\sin t + e^{2t}}{\sqrt{\cos^2 t + e^{2t}}}$$

When $t = 0$, $\dfrac{dw}{dt} = \dfrac{-(1)(0) + 1}{\sqrt{1^2 + 1}} = \dfrac{1}{\sqrt{2}} = \dfrac{\sqrt{2}}{2}$.

3. $w = x\sin y$

$x = e^t,\ y = \pi - t$

$$\frac{dw}{dt} = \frac{\partial w}{\partial x}\frac{dx}{\partial t} + \frac{\partial w}{\partial y}\frac{dy}{dt} = \sin y(e^t) + x\cos y(-1)$$
$$= \sin(\pi - t)e^t - e^t\cos(\pi - t) = e^t\sin t + e^t\cos t$$

When $t = 0$, $\dfrac{dw}{dt} = (1)(0) + (1)(1) = 0 + 1 = 1$.

4. $w = \ln\dfrac{y}{x}$

$x = \cos t$

$y = \sin t$

$$\frac{dw}{dt} = \left(\frac{-1}{x}\right)(-\sin t) + \left(\frac{1}{y}\right)(\cos t)$$
$$= \tan t + \cot t = \frac{1}{\sin t\cos t}$$

When $t = \dfrac{\pi}{4}$, $\dfrac{dw}{dt} = \dfrac{1}{\left(\frac{1}{\sqrt{2}}\right)\left(\frac{1}{\sqrt{2}}\right)} = \dfrac{1}{2}$.

5. $w = xy,\ x = e^t,\ y = e^{-2t}$

(a) $$\frac{dw}{dt} = \frac{\partial w}{\partial x}\frac{dx}{dt} + \frac{\partial w}{\partial y}\frac{dy}{dt}$$
$$= y(e^t) + x(-2e^{-2t}) = e^{-2t}e^t - e^t2e^{-2t} = -e^{-t}$$

(b) $w = e^te^{-2t} = e^{-t}$

$$\frac{dw}{dt} = -e^{-t}$$

6. $w = \cos(x - y),\ x = t^2,\ y = 1$

(a) $$\frac{dw}{dt} = -\sin(x - y)(2t) + \sin(x - y)(0)$$
$$= -2t\sin(x - y) = -2t\sin(t^2 - 1)$$

(b) $w = \cos(t^2 - 1)$, $\dfrac{dw}{dt} = -2t\sin(t^2 - 1)$

7. $w = x^2 + y^2 + z^2,\ x = \cos t,\ y = \sin t,\ z = e^t$

(a) $$\frac{dw}{dt} = \frac{\partial w}{\partial x}\frac{dx}{dt} + \frac{\partial w}{\partial y}\frac{dy}{dt} + \frac{\partial w}{\partial z}\frac{dz}{dt}$$
$$= 2x(-\sin t) + 2y(\cos t) + 2z(e^t)$$
$$= -2\cos t\sin t + 2\sin t\cos t + 2e^{2t} = 2e^{2t}$$

(b) $w = \cos^2 t + \sin^2 t + e^{2t} = 1 + e^{2t}$

$$\frac{dw}{dt} = 2e^{2t}$$

8. $w = xy\cos z$

$x = t$

$y = t^2$

$z = \arccos t$

(a) $\dfrac{dw}{dt} = (y\cos z)(1) + (x\cos z)(2t) + (-xy\sin z)\left(-\dfrac{1}{\sqrt{1-t^2}}\right) = t^2(t) + t(t)(2t) - t(t^2)\sqrt{1-t^2}\left(\dfrac{-1}{\sqrt{1-t^2}}\right)$

$= t^3 + 2t^3 + t^3 = 4t^3$

(b) $w = t^4, \dfrac{dw}{dt} = 4t^3$

9. $w = xy + xz + yz,\ x = t - 1,\ y = t^2 - 1,\ z = t$

(a) $\dfrac{dw}{dt} = \dfrac{\partial w}{\partial x}\dfrac{dx}{dt} + \dfrac{\partial w}{\partial y}\dfrac{dy}{dt} + \dfrac{\partial w}{\partial z}\dfrac{dz}{dt} = (y+z) + (x+z)(2t) + (x+y)$

$= (t^2 - 1 + t) + (t - 1 + t)(2t) + (t - 1 + t^2 - 1) = 3(2t^2 - 1)$

(b) $w = (t-1)(t^2-1) + (t-1)t + (t^2-1)t$

$\dfrac{dw}{dt} = 2t(t-1) + (t^2-1) + 2t - 1 + 3t^2 - 1 = 3(2t^2 - 1)$

10. $w = xy^2 + x^2z + yz^2,\ x = t^2, y = 2t, z = 2$

(a) $\dfrac{dw}{dt} = \dfrac{\partial w}{\partial x}\dfrac{dx}{dt} + \dfrac{\partial w}{\partial y}\dfrac{dy}{dt} + \dfrac{\partial w}{\partial z}\dfrac{dz}{dt}$

$= (y^2 + 2xz)(2t) + (2xy + z^2)(2) + (x^2 + 2yz)(0) = (4t^2 + 4t^2)(2t) + (4t^3 + 4)(2) = 24t^3 + 8$

(b) $w = t^2(4t^2) + t^4(2) + 2t(4) = 6t^4 + 8t$

$\dfrac{dw}{dt} = 24t^3 + 8$

11. Distance $= f(t) = \sqrt{(x_1 - x_2)^2 + (y_1 - y_2)^2} = \sqrt{(10\cos 2t - 7\cos t)^2 + (6\sin 2t - 4\sin t)^2}$

$f'(t) = \dfrac{1}{2}\left[(10\cos 2t - 7\cos t)^2 + (6\sin 2t - 4\sin t)^2\right]^{-1/2}$

$\left[\left[2(10\cos 2t - 7\cos t)(-20\sin 2t + 7\sin t)\right] + \left[2(6\sin 2t - 4\sin t)(12\cos 2t - 4\cos t)\right]\right]$

$f'\left(\dfrac{\pi}{2}\right) = \dfrac{1}{2}\left[(-10)^2 + 4^2\right]^{-1/2}\left[\left[2(-10)(7)\right] + \left(2(-4)(-12)\right)\right] = \dfrac{1}{2}(116)^{-1/2}(-44) = \dfrac{-22}{2\sqrt{29}} = \dfrac{-11\sqrt{29}}{29} \approx -2.04$

12. Distance $= f(t) = \sqrt{(x_2 - x_1)^2 + (y_2 - y_1)^2} = \sqrt{\left[48t\left(\sqrt{3} - \sqrt{2}\right)\right]^2 + \left[48t\left(1 - \sqrt{2}\right)\right]^2} = 48t\sqrt{8 - 2\sqrt{2} - 2\sqrt{6}}$

$f'(t) = 48\sqrt{8 - 2\sqrt{2} - 2\sqrt{6}} = f'(1)$

13. $w = x^2 + y^2$

$x = s + t, y = s - t$

$\dfrac{\partial w}{\partial s} = 2x(1) + 2y(1) = 2(s+t) + 2(s-t) = 4s$

$\dfrac{\partial w}{\partial t} = 2x(1) + 2y(-1) = 2(s+t) - 2(s-t) = 4t$

When $s = 1$ and $t = 0$, $\dfrac{\partial w}{\partial s} = 4$ and $\dfrac{\partial w}{\partial t} = 0$.

14. $w = y^3 - 3x^2y$

$x = e^s, y = e^t$

$\dfrac{\partial w}{\partial s} = -6xy(e^s) + (3y^2 - 3x^2)(0) = -6e^se^te^s = -6e^{2s+t}$

$\dfrac{\partial w}{\partial t} = (-6xy)(0) + (3y^2 - 3x^2)e^t = (3e^{2t} - 3e^{2s})e^t$

$= 3e^{3t} - 3e^{2s+t}$

When $s = -1$ and $t = 2$, $\dfrac{\partial w}{\partial s} = -6$ and $\dfrac{\partial w}{\partial t} = 3e^6 - 3$.

15. $w = \sin(2x + 3y)$

$x = s + t$

$y = s - t$

$$\frac{\partial w}{\partial s} = 2\cos(2x + 3y) + 3\cos(2x + 3y)$$
$$= 5\cos(2x + 3y) = 5\cos(5s - t)$$

$$\frac{\partial w}{\partial t} = 2\cos(2x + 3y) - 3\cos(2x + 3y)$$
$$= -\cos(2x + 3y) = -\cos(5s - t)$$

When $s = 0$ and $t = \frac{\pi}{2}$, $\frac{\partial w}{\partial s} = 0$ and $\frac{\partial w}{\partial t} = 0$.

16. $w = x^2 - y^2$

$x = s\cos t$

$y = s\sin t$

$$\frac{\partial w}{\partial s} = 2x\cos t - 2y\sin t$$
$$= 2s\cos^2 t - 2s\sin^2 t = 2s\cos 2t$$

$$\frac{\partial w}{\partial t} = 2x(-s\sin t) - 2y(s\cos t) = -2s^2\sin 2t$$

When $s = 3$ and $t = \frac{\pi}{4}$, $\frac{\partial w}{\partial s} = 0$ and $\frac{\partial w}{\partial t} = -18$.

17. (a) $w = xyz,\ x = s + t,\ y = s - t,\ z = st^2$

$$\frac{\partial w}{\partial s} = yz(1) + xz(1) + xy(t^2)$$
$$= (s - t)st^2 + (s + t)st^2 + (s + t)(s - t)t^2 = 2s^2t^2 + s^2t^2 - t^4 = 3s^2t^2 - t^4 = t^2(3s^2 - t^2)$$

$$\frac{\partial w}{\partial t} = yz(1) + xz(-1) + xy(2st) = (s - t)st^2 - (s + t)st^2 + (s + t)(s - t)(2st) = -2st^3 + 2s^3t - 2st^3 = 2s^3t - 4st^3$$
$$= 2st(s^2 - 2t^2)$$

(b) $w = xyz = (s + t)(s - t)st^2 = (s^2 - t^2)st^2 = s^3t^2 - st^4$

$$\frac{\partial w}{\partial s} = 3s^2t^2 - t^4 = t^2(3s^2 - t^2)$$

$$\frac{\partial w}{\partial t} = 2s^3t - 4st^3 = 2st(s^2 - 2t^2)$$

18. (a) $w = x^2 + y^2 + z^2,\ x = t\sin s,\ y = t\cos s,\ z = st^2$

$$\frac{\partial w}{\partial s} = 2x + \cos s + 2y(-t\sin s) + 2z(t^2)$$
$$= 2t^2\sin s\cos s - 2t^2\sin s\cos s + 2st^4 = 2st^4$$

$$\frac{\partial w}{\partial t} = 2x\sin s + 2y\cos s + 2z(2st)$$
$$= 2t\sin^2 s + 2t\cos^2 s + 4s^2t^3 = 2t + 4s^2t^3$$

(b) $w = x^2 + y^2 + z^2 = (t\sin s)^2 + (t\cos s)^2 + (st^2)^2$

$$= t^2(\sin^2 s + \cos^2 s) + s^2t^4$$
$$= t^2 + s^2t^4$$

$$\frac{\partial w}{\partial s} = 2st^4$$

$$\frac{\partial w}{\partial t} = 2t + 4s^2t^3$$

19. (a) $w = ze^{xy}, x = s - t, y = s + t, z = st$

$$\frac{\partial w}{\partial s} = yze^{xy}(1) + xze^{xy}(1) + e^{xy}(t)$$
$$= e^{(s-t)(s+t)}\left[(s + t)st + (s - t)st + t\right]$$
$$= e^{(s-t)(s+t)}\left[2s^2t + t\right] = te^{s^2-t^2}\left(2s^2 + 1\right)$$

$$\frac{\partial w}{\partial t} = yze^{xy}(-1) + xze^{xy}(1) + e^{xy}(s)$$
$$= e^{(s-t)(s+t)}\left[-(s + t)(st) + (s - t)st + s\right]$$
$$= e^{(s-t)(s+t)}\left[-2st^2 + s\right] = se^{s^2-t^2}\left(1 - 2t^2\right)$$

(b) $w = ze^{xy} = ste^{(s-t)(s+t)} = ste^{s^2-t^2}$

$$\frac{\partial w}{\partial s} = te^{s^2-t^2} + st(2s)e^{s^2-t^2} = te^{s^2-t^2}\left(1 + 2s^2\right)$$
$$\frac{\partial w}{\partial t} = se^{s^2-t^2} + st(-2t)e^{s^2-t^2} = se^{s^2-t^2}\left(1 - 2t^2\right)$$

20. (a) $w = x\cos yz, x = s^2, y = t^2, z = s - 2t$

$$\frac{\partial w}{\partial s} = \cos(yz)(2s) - xz\sin(yz)(0) - xy\sin(yz)(1)$$
$$= \cos\left(st^2 - 2t^3\right)2s - s^2t^2\sin\left(st^2 - 2t^3\right)$$

$$\frac{\partial w}{\partial t} = \cos(yz)(0) - xz\sin(yz)(2t) - xy\sin(yz)(-2)$$
$$= -2s^2t(s - 2t)\sin\left(st^2 - 2t^3\right) + 2s^2t^2\sin\left(st^2 - 2t^3\right)$$
$$= \left(6s^2t^2 - 2s^3t\right)\sin\left(st^2 - 2t^3\right)$$

(b) $w = x\cos yz = s^2\cos\left(t^2(s - 2t)\right) = s^2\cos\left(st^2 - 2t^3\right)$

$$\frac{\partial w}{\partial s} = s^2\left(-\sin\left(st^2 - 2t^3\right)\right)\left(t^2\right) + 2s\cos\left(st^2 - 2t^3\right)$$
$$= 2s\cos\left(st^2 - 2t^3\right) - s^2t^2\sin\left(st^2 - 2t^3\right)$$
$$\frac{\partial w}{\partial t} = -s^2\sin\left(st^2 - 2t^3\right)\left(2st - 6t^2\right)$$
$$= \left(6t^2s^2 - 2s^3t\right)\sin\left(st^2 - 2t^3\right)$$

21. $x^2 - xy + y^2 - x + y = 0$

$$\frac{dy}{dx} = -\frac{F_x(x, y)}{F_y(x, y)} = -\frac{2x - y - 1}{-x + 2y + 1} = \frac{y - 2x + 1}{2y - x + 1}$$

22. $\sec xy + \tan xy + 5 = 0$

$$\frac{dy}{dx} = -\frac{F_x(x, y)}{F_y(x, y)} = -\frac{y\sec xy\tan xy + y\sec^2 xy}{x\sec xy\tan xy + x\sec^2 xy}$$
$$= \frac{-y\left(\sec xy\tan xy + \sec^2 xy\right)}{x\left(\sec xy\tan xy + \sec^2 xy\right)} = -\frac{y}{x}$$

23. $\ln\sqrt{x^2 + y^2} + x + y = 4$

$$\frac{1}{2}\ln\left(x^2 + y^2\right) + x + y - 4 = 0$$

$$\frac{dy}{dx} = -\frac{F_x(x, y)}{F_y(x, y)} = -\frac{\dfrac{x}{x^2 + y^2} + 1}{\dfrac{y}{x^2 + y^2} + 1} = -\frac{x + x^2 + y^2}{y + x^2 + y^2}$$

24. $\dfrac{x}{x^2 + y^2} - y^2 - 6 = 0$

$$\frac{dy}{dx} = -\frac{F_x(x, y)}{F_y(x, y)}$$
$$= -\frac{(y^2 - x^2)/(x^2 + y^2)^2}{(-2xy)/(x^2 + y^2)^2 - 2y}$$
$$= \frac{y^2 - x^2}{2xy + 2y(x^2 + y^2)^2}$$
$$= \frac{y^2 - x^2}{2xy + 2yx^4 + 4x^2y^3 + 2y^5}$$

25. $F(x, y, z) = x^2 + y^2 + z^2 - 1$

$F_x = 2x, F_y = 2y, F_z = 2z$

$$\frac{\partial_z}{\partial_x} = -\frac{F_x}{F_z} = -\frac{x}{z}$$
$$\frac{\partial z}{\partial y} = -\frac{F_y}{F_z} = -\frac{y}{z}$$

26. $F(x, y, z) = xz + yz + xy$

$F_x = z + y$

$F_y = z + x$

$F_z = x + y$

$$\frac{\partial z}{\partial x} = -\frac{F_x}{F_z} = -\frac{y + z}{x + y}$$
$$\frac{\partial z}{\partial y} = -\frac{F_y}{F_z} = -\frac{x + z}{x + y}$$

27. $F(x, y, z) = x^2 + 2yz + z^2 - 1 = 0$

$$\frac{\partial z}{\partial x} = -\frac{F_x(x, y, z)}{F_z(x, y, z)} = \frac{-2x}{2y + 2z} = \frac{-x}{y + z}$$
$$\frac{\partial z}{\partial y} = -\frac{F_y(x, y, z)}{F_z(x, y, z)} = \frac{-2z}{2y + 2z} = \frac{-z}{y + z}$$

28. $x + \sin(y + z) = 0$

(i) $1 + \dfrac{\partial z}{\partial x}\cos(y + z) = 0$ implies

$$\frac{\partial z}{\partial x} = -\frac{1}{\cos(y + z)} = -\sec(y + z).$$

(ii) $\left(1 + \dfrac{\partial z}{\partial y}\right)\cos(y + z) = 0$ implies $\dfrac{\partial z}{\partial y} = -1.$

29. $F(x, y, z) = \tan(x + y) + \tan(y + z) - 1$

$F_x = \sec^2(x + y)$

$F_y = \sec^2(x + y) + \sec^2(y + z)$

$F_z = \sec^2(y + z)$

$$\frac{\partial z}{\partial x} = -\frac{F_x}{F_z} = -\frac{\sec^2(x + y)}{\sec^2(y + z)}$$
$$\frac{\partial z}{\partial y} = -\frac{F_y}{F_z} = -\frac{\sec^2(x + y) + \sec^2(y + z)}{\sec^2(y + z)}$$
$$= -\left(\frac{\sec^2(x + y)}{\sec^2(y + z)} + 1\right)$$

30. $F(x, y, z) = e^x \sin(y + z) - z$

$F_x = e^x \sin(y + z)$

$F_y = e^x \cos(y + z)$

$F_z = e^x \cos(y + z) - 1$

$$\frac{\partial z}{\partial x} = -\frac{F_x}{F_z} = \frac{e^x \sin(y + z)}{1 - e^x \cos(y + x)}$$
$$\frac{\partial z}{\partial y} = -\frac{F_y}{F_z} = \frac{e^x \cos(y + z)}{1 - e^x \cos(y + z)}$$

31. $F(x, y, z) = e^{xz} + xy = 0$

$$\frac{\partial z}{\partial x} = -\frac{F_x(x, y, z)}{F_z(x, y, z)} = -\frac{ze^{xz} + y}{xe^{xz}}$$
$$\frac{\partial z}{\partial y} = -\frac{F_y(x, y, z)}{F_z(x, y, z)} = \frac{-x}{xe^{xz}} = \frac{-1}{e^{xz}} = -e^{-xz}$$

32. $x \ln y + y^2z + z^2 - 8 = 0$

(i) $$\frac{\partial z}{\partial x} = \frac{-F_x(x, y, z)}{F_z(x, y, z)} = \frac{-\ln y}{y^2 + 2z}$$

(ii) $$\frac{\partial z}{\partial y} = \frac{-F_y(x, y, z)}{F_z(x, y, z)} = -\frac{\frac{x}{y} + 2yz}{y^2 + 2z} = -\frac{x + 2y^2z}{y^3 + 2yz}$$

33. $F(x, y, z, w) = xy + yz - wz + wx - s$

$F_x = y + w$

$F_y = x + z$

$F_z = y - w$

$F_w = -z + x$

$$\frac{\partial w}{\partial x} = -\frac{F_x}{F_w} = -\frac{y + w}{-z + x} = \frac{y + w}{z - x}$$
$$\frac{\partial w}{\partial y} = -\frac{F_y}{F_w} = -\frac{x + z}{-z + x} = \frac{x + z}{z - x}$$
$$\frac{\partial w}{\partial z} = -\frac{F_z}{F_w} = -\frac{y - w}{-z + x} = \frac{y - w}{z - x}$$

34. $x^2 + y^2 - z^2 - 5yw + 10w^2 - 2 = F(x, y, z, w)$

$F_x = 2x, F_y = 2y - 5w, F_z = 2z, F_w = -5y + 20w$

$$\frac{\partial w}{\partial x} = -\frac{F_x}{F_w} = \frac{-2x}{-5y + 20w} = \frac{2x}{5y - 20w}$$

$$\frac{\partial w}{\partial y} = -\frac{F_y}{F_w} = \frac{5w - 2y}{20w - 5y}$$

$$\frac{\partial w}{\partial z} = -\frac{F_z}{F_w} = \frac{2z}{5y - 20w}$$

35. $F(x, y, z, w) = \cos xy + \sin yz + wz - 20$

$$\frac{\partial w}{\partial x} = \frac{-F_x}{F_w} = \frac{y \sin xy}{z}$$

$$\frac{\partial w}{\partial y} = \frac{-F_y}{F_w} = \frac{x \sin xy - z \cos yz}{z}$$

$$\frac{\partial w}{\partial z} = \frac{-F_z}{F_w} = -\frac{y \cos zy + w}{z}$$

36. $F(x, y, z, w) = w - \sqrt{x - y} - \sqrt{y - z} = 0$

$$\frac{\partial w}{\partial x} = \frac{-F_x}{F_w} = \frac{\frac{1}{2}(x - y)^{-1/2}}{1} = \frac{1}{2\sqrt{x - y}}$$

$$\frac{\partial w}{\partial y} = \frac{-F_y}{F_w} = \frac{-1}{2}(x - y)^{-1/2} + \frac{1}{2}(y - z)^{-1/2} = \frac{-1}{2\sqrt{x - y}} + \frac{1}{2\sqrt{y - z}}$$

$$\frac{\partial w}{\partial z} = \frac{-F_z}{F_w} = \frac{-1}{2\sqrt{y - z}}$$

37. (a) $f(x, y) = \dfrac{xy}{\sqrt{x^2 + y^2}}$

$$f(tx, ty) = \frac{(tx)(ty)}{\sqrt{(tx)^2 + (ty)^2}} = t\left(\frac{xy}{\sqrt{x^2 + y^2}}\right) = tf(x, y)$$

Degree: 1

(b) $xf_x(x, y) + yf_y(x, y) = x\left(\dfrac{y^3}{\left(x^2 + y^2\right)^{3/2}}\right) + y\left(\dfrac{x^3}{\left(x^2 + y^2\right)^{3/2}}\right) = \dfrac{xy}{\sqrt{x^2 + y^2}} = 1f(x, y)$

38. (a) $f(x, y) = x^3 - 3xy^2 + y^3$

$$f(tx, ty) = (tx)^3 - 3(tx)(ty)^2 + (ty)^3 = t^3\left(x^3 - 3xy^2 + y^3\right) = t^3 f(x, y)$$

Degree: 3

(b) $xf_x(x, y) + yf_y(x, y) = x\left(3x^2 - 3y^2\right) + y\left(-6xy + 3y^2\right) = 3x^3 - 9xy^2 + 3y^3 = 3f(x, y)$

39. (a) $f(x, y) = e^{x/y}$

$$f(tx, ty) = e^{tx/ty} = e^{x/y} = f(x, y)$$

Degree: 0

(b) $xf_x(x, y) + yf_y(x, y) = x\left(\dfrac{1}{y}e^{x/y}\right) + y\left(-\dfrac{x}{y^2}e^{x/y}\right) = 0$

40. (a) $f(x, y) = \dfrac{x^2}{\sqrt{x^2 + y^2}}$

$$f(tx, ty) = \frac{(tx)^2}{\sqrt{(tx)^2 + (ty)^2}} = t\left(\frac{x^2}{\sqrt{x^2 + y^2}}\right) = tf(x, y)$$

Degree: 1

(b) $$xf_x(x, y) + yf_y(x, y) = x\left[\frac{x^3 + 2xy^2}{(x^2 + y^2)^{3/2}}\right] + y\left[\frac{-x^2y}{(x^2 + y^2)^{3/2}}\right] = \frac{x^4 + x^2y^2}{(x^2 + y^2)^{3/2}} = \frac{x^2(x^2 + y^2)}{(x^2 + y^2)^{3/2}} = \frac{x^2}{\sqrt{x^2 + y^2}} = f(x, y)$$

41. $$\frac{dw}{dt} = \frac{\partial w}{\partial x}\frac{dx}{dt} + \frac{\partial w}{\partial y}\frac{dy}{dt} = \frac{\partial f}{\partial x}\frac{dg}{dt} + \frac{\partial f}{\partial y}\frac{dh}{dt}$$

At $t = 2$, $x = 4$, $y = 3$, $f_x(4, 3) = -5$ and $f_y(4, 3) = 7$.

So, $\dfrac{dw}{dt} = (-5)(-1) + (7)(6) = 47$

42. $$\frac{\partial w}{\partial s} = \frac{\partial w}{\partial x}\frac{\partial x}{\partial s} + \frac{\partial w}{\partial y}\frac{\partial y}{\partial s} = \frac{\partial f}{\partial x}\frac{\partial g}{\partial s} + \frac{\partial f}{\partial y}\frac{\partial h}{\partial s} = (-5)(-3) + (7)(5) = 50$$

$$\frac{\partial w}{\partial t} = \frac{\partial w}{\partial x}\frac{\partial x}{\partial t} + \frac{\partial w}{\partial y}\frac{\partial y}{\partial t} = \frac{\partial f}{\partial x}\frac{\partial g}{\partial t} + \frac{\partial f}{\partial y}\frac{\partial h}{\partial t} = (-5)(-2) + (7)(8) = 66$$

43. $\dfrac{dw}{dt} = \dfrac{\partial w}{\partial x}\dfrac{dx}{dt} + \dfrac{\partial w}{\partial y}\dfrac{dy}{dt}$ (Page 907)

44. $$\frac{\partial w}{\partial s} = \frac{\partial w}{\partial x}\frac{\partial x}{\partial s} + \frac{\partial w}{\partial y}\frac{\partial y}{\partial s}$$

$\dfrac{\partial w}{\partial t} = \dfrac{\partial w}{\partial x}\dfrac{\partial x}{\partial t} + \dfrac{\partial w}{\partial y}\dfrac{\partial y}{\partial t}$ (Page 909)

45. $$\frac{dy}{dx} = -\frac{f_x(x, y)}{f_y(x, y)}$$

$$\frac{\partial z}{\partial x} = -\frac{f_x(x, y, z)}{f_z(x, y, z)}$$

$\dfrac{\partial z}{\partial y} = -\dfrac{f_y(x, y, z)}{f_z(x, y, z)}$ (page 912)

46. (a) $\dfrac{dw}{dr} = \dfrac{\partial w}{\partial x}\dfrac{dx}{dr} + \dfrac{\partial w}{\partial y}\dfrac{dy}{dr}$

(b) $\dfrac{\partial w}{\partial r} = \dfrac{\partial w}{\partial x}\dfrac{\partial x}{\partial r} + \dfrac{\partial w}{\partial y}\dfrac{\partial y}{\partial r}$

$$\frac{\partial w}{\partial \theta} = \frac{\partial w}{\partial x}\frac{\partial x}{\partial \theta} + \frac{\partial w}{\partial y}\frac{\partial y}{\partial \theta}$$

47. $V = \pi r^2 h$

$$\frac{dV}{dt} = \pi\left(2rh\frac{dr}{dt} + r^2\frac{dh}{dt}\right) = \pi r\left(2h\frac{dr}{dt} + r\frac{dh}{dt}\right) = \pi(12)\left[2(36)(6) + 12(-4)\right] = 4608\pi \text{ in.}^3/\text{min}$$

$$S = 2\pi r(r + h)$$

$$\frac{dS}{dt} = 2\pi\left[(2r + h)\frac{dr}{dt} + r\frac{dh}{dt}\right] = 2\pi\left[(24 + 36)(6) + 12(-4)\right] = 624\pi \text{ in.}^2/\text{min}$$

48. $pV = mRT$

$$T = \frac{1}{mR}(pV)$$

$$\frac{dT}{dt} = \frac{1}{mR}\left[V\frac{dp}{dt} + p\frac{dV}{dt}\right]$$

49. $I = \dfrac{1}{2}m\left(r_1^2 + r_2^2\right)$

$$\frac{dI}{dt} = \frac{1}{2}m\left[2r_1\frac{dr_1}{dt} + 2r_2\frac{dr_2}{dt}\right] = m\left[(6)(2) + (8)(2)\right] = 28m \text{ cm}^2/\text{sec}$$

50. $V = \frac{\pi}{3}\left(r^2 + rR + R^2\right)h$

$$\frac{dV}{dt} = \frac{\pi}{3}\left[(2r + R)h\frac{dr}{dt} + (r + 2R)h\frac{dR}{dt} + \left(r^2 + rR + R^2\right)\frac{dh}{dt}\right]$$

$$= \frac{\pi}{3}\left[\left[2(15) + 25\right](10)(4) + \left[15 + 2(25)\right](10)(4) + \left[(15)^2 + (15)(25) + (25)^2\right](12)\right]$$

$$= \frac{\pi}{3}(19{,}500)$$

$$= 6{,}500\pi \text{ cm}^3/\text{min}$$

$$S = \pi(R + r)\sqrt{(R - r)^2 + h^2}$$

$$\frac{dS}{dt} = \pi\left\{\left[\sqrt{(R - r)^2 + h^2} - (R + r)\frac{(R - r)}{\sqrt{(R - r)^2 + h^2}}\right]\frac{dr}{dt} + \left[\sqrt{(R - r)^2 + h^2} + (R + r)\frac{(R - r)}{\sqrt{(R - r)^2 + h^2}}\right]\frac{dR}{dt} + (R + r)\frac{h}{\sqrt{(R - r)^2 + h^2}}\frac{dh}{dt}\right\}$$

$$= \pi\left\{\left[\sqrt{(25 - 15)^2 + 10^2} - (25 + 15)\frac{25 - 15}{\sqrt{(25 - 15)^2 + 10^2}}\right](4) + \left[\sqrt{(25 - 15)^2 + 10^2} + (25 + 15)\frac{25 - 15}{\sqrt{(25 - 15)^2 + 10^2}}\right](4) + (25 + 15)\left[\frac{10}{\sqrt{(25 - 15)^2 + 10^2}}(12)\right]\right\}$$

$$= 320\sqrt{2}\pi \text{ cm}^2/\text{min}$$

51.

$$w = f(x, y)$$
$$x = u - v$$
$$y = v - u$$
$$\frac{\partial w}{\partial u} = \frac{\partial w}{\partial x}\frac{dx}{du} + \frac{\partial w}{\partial y}\frac{dy}{du} = \frac{\partial w}{\partial x} - \frac{\partial w}{\partial y}$$
$$\frac{\partial w}{\partial v} = \frac{\partial w}{\partial x}\frac{dx}{dv} + \frac{\partial w}{\partial y}\frac{dy}{dv} = -\frac{\partial w}{\partial x} + \frac{\partial w}{\partial y}$$
$$\frac{\partial w}{\partial u} + \frac{\partial w}{\partial v} = 0$$

52.

$$w = (x - y)\sin(y - x)$$
$$\frac{\partial w}{\partial x} = -(x - y)\cos(y - x) + \sin(y - x)$$
$$\frac{\partial w}{\partial y} = (x - y)\cos(y - x) - \sin(y - x)$$
$$\frac{\partial w}{\partial x} + \frac{\partial w}{\partial y} = 0$$

53. Given $\frac{\partial u}{\partial x} = \frac{\partial v}{\partial y}$ and $\frac{\partial u}{\partial y} = -\frac{\partial v}{\partial x}$, $x = r\cos\theta$ and $y = r\sin\theta$.

$$\frac{\partial u}{\partial r} = \frac{\partial u}{\partial x}\cos\theta + \frac{\partial u}{\partial y}\sin\theta = \frac{\partial v}{\partial y}\cos\theta - \frac{\partial v}{\partial x}\sin\theta$$

$$\frac{\partial v}{\partial \theta} = \frac{\partial v}{\partial x}(-r\sin\theta) + \frac{\partial v}{\partial y}(r\cos\theta) = r\left[\frac{\partial v}{\partial y}\cos\theta - \frac{\partial v}{\partial x}\sin\theta\right]$$

So, $\frac{\partial u}{\partial r} = \frac{1}{r}\frac{\partial v}{\partial \theta}$.

$$\frac{\partial v}{\partial r} = \frac{\partial v}{\partial x}\cos\theta + \frac{\partial v}{\partial y}\sin\theta = -\frac{\partial u}{\partial y}\cos\theta + \frac{\partial u}{\partial x}\sin\theta$$

$$\frac{\partial u}{\partial \theta} = \frac{\partial u}{\partial x}(-r\sin\theta) + \frac{\partial u}{\partial y}(r\cos\theta) = -r\left[-\frac{\partial u}{\partial y}\cos\theta + \frac{\partial u}{\partial x}\sin\theta\right]$$

So, $\frac{\partial v}{\partial r} = -\frac{1}{r}\frac{\partial u}{\partial \theta}$.

54. Note first that

$$\frac{\partial u}{\partial x} = \frac{\partial v}{\partial y} = \frac{x}{x^2 + y^2}$$

$$\frac{\partial u}{\partial y} = -\frac{\partial v}{\partial x} = \frac{y}{x^2 + y^2}.$$

$$\frac{\partial u}{\partial r} = \frac{x}{x^2 + y^2}\cos\theta + \frac{y}{x^2 + y^2}\sin\theta = \frac{r\cos^2\theta + r\sin^2\theta}{r^2} = \frac{1}{r}$$

$$\frac{\partial v}{\partial \theta} = \frac{-y}{x^2 + y^2}(-r\sin\theta) + \frac{x}{x^2 + y^2}(r\cos\theta) = \frac{r^2\sin^2\theta + r^2\cos^2\theta}{r^2} = 1$$

So, $\dfrac{\partial u}{\partial r} = \dfrac{1}{r}\dfrac{\partial v}{\partial \theta}$.

$$\frac{\partial v}{\partial r} = \frac{-y}{x^2 + y^2}\cos\theta + \frac{x}{x^2 + y^2}\sin\theta = \frac{-r\sin\theta\cos\theta + r\sin\theta\cos\theta}{r^2} = 0$$

$$\frac{\partial u}{\partial \theta} = \frac{x}{x^2 + y^2}(-r\sin\theta) + \frac{y}{x^2 + y^2}(r\cos\theta) = \frac{-r^2\sin\theta\cos\theta + r^2\sin\theta\cos\theta}{r^2} = 0$$

So, $\dfrac{\partial v}{\partial r} = -\dfrac{1}{r}\dfrac{\partial u}{\partial \theta}$.

55. $g(t) = f(xt, yt) = t^n f(x, y)$

Let $u = xt$, $v = yt$, then

$$g'(t) = \frac{\partial f}{\partial u}\cdot\frac{du}{dt} + \frac{\partial f}{\partial v}\cdot\frac{dv}{dt} = \frac{\partial f}{\partial u}x + \frac{\partial f}{\partial v}y$$

and $g'(t) = nt^{n-1}f(x, y)$.

Now, let $t = 1$ and we have $u = x$, $v = y$. Thus,

$$\frac{\partial f}{\partial x}x + \frac{\partial f}{\partial y}y = nf(x, y).$$

Section 13.6 Directional Derivatives and Gradients

1. $f(x, y) = x^2 + y^2$, $P(1, -2)$, $\theta = \pi/4$

$$\begin{aligned} D_{\mathbf{u}} f(x, y) &= f_x(x, y)\cos\theta + f_y(x, y)\sin\theta \\ &= 2x\cos\theta + 2y\sin\theta \end{aligned}$$

At $\theta = \pi/4$, $x = 1$, and $y = -2$,

$$\begin{aligned} D_{\mathbf{u}} f(1, -2) &= 2(1)\cos\pi/4 + 2(-2)\sin\pi/4 \\ &= \sqrt{2} - 2\sqrt{2} = -\sqrt{2}. \end{aligned}$$

2. $f(x, y) = \dfrac{y}{x + y}$, $P(3, 0)$, $\theta = -\pi/6$

$$\begin{aligned} D_{\mathbf{u}} f(x, y) &= f_x(x, y)\cos\theta + f_y(x, y)\sin\theta \\ &= \frac{-y}{(x + y)^2}\cos\theta + \frac{x}{(x + y)^2}\sin\theta \end{aligned}$$

At $\theta = -\pi/6$, $x = 3$, and $y = 0$,

$$D_{\mathbf{u}} f(3, 0) = \frac{3}{3^2}\sin\left(\frac{-\pi}{6}\right) = -\frac{1}{6}.$$

3. $f(x, y) = \sin(2x + y)$, $P(0, 0)$, $\theta = \pi/3$

$$\begin{aligned} D_{\mathbf{u}} f(x, y) &= f_x(x, y)\cos\theta + f_y(x, y)\sin\theta \\ &= 2\cos(2x + y)\cos\theta + \cos(2x + y)\sin\theta \end{aligned}$$

At $\theta = \pi/3$ and $x = y = 0$,

$$D_{\mathbf{u}} f(0, 0) = 2\cos\pi/3 + \sin\pi/3 = 1 + \sqrt{3}/2.$$

4. $g(x, y) = xe^y,\ P(0, 2),\ \theta = \dfrac{2\pi}{3}$

$$D_{\mathbf{u}}g(x, y) = g_x(x, y)\cos\theta + g_y(x, y)\sin\theta$$
$$= e^y\cos\theta + xe^y\sin\theta$$

At $\theta = \dfrac{2\pi}{3}$, $x = 0$, and $y = 2$,

$$D_{\mathbf{u}}g(0, 2) = e^2\cos\frac{2\pi}{3} = -\frac{1}{2}e^2.$$

5. $f(x, y) = 3x - 4xy + 9y,\ P(1, 2),\ \mathbf{v} = \dfrac{3}{5}\mathbf{i} + \dfrac{4}{5}\mathbf{j}$

$$\mathbf{u} = \frac{\mathbf{v}}{\|\mathbf{v}\|} = \frac{3}{5}\mathbf{i} + \frac{4}{5}\mathbf{j} = \cos\theta\,\mathbf{i} + \sin\theta\,\mathbf{j}$$

$$D_{\mathbf{u}}f(x, y) = (3 - 4y)\cos\theta + (-4x + 9)\sin\theta$$

$$D_{\mathbf{u}}(1, 2) = (3 - 4(2))\frac{3}{5} + (-4(1) + 9)\frac{4}{5}$$
$$= -3 + 4 = 1$$

6. $f(x, y) = x^3 - y^3,\ P(4, 3),\ \mathbf{v} = \dfrac{\sqrt{2}}{2}(\mathbf{i} + \mathbf{j})$

$$\mathbf{u} = \frac{\mathbf{v}}{\|\mathbf{v}\|} = \frac{\sqrt{2}}{2}\mathbf{i} + \frac{\sqrt{2}}{2}\mathbf{j} = \cos\theta\,\mathbf{i} + \sin\theta\,\mathbf{j}$$

$$D_{\mathbf{u}}f(x, y) = (3x^2)\left(\frac{\sqrt{2}}{2}\right) + (-3y^2)\left(\frac{\sqrt{2}}{2}\right)$$

$$D_{\mathbf{u}}f(4, 3) = 3(16)\frac{\sqrt{2}}{2} - 3(9)\frac{\sqrt{2}}{2}$$
$$= \frac{21\sqrt{2}}{2}$$

7. $g(x, y) = \sqrt{x^2 + y^2},\ P(3, 4),\ \mathbf{v} = 3\mathbf{i} - 4\mathbf{j}$

$$\mathbf{u} = \frac{\mathbf{v}}{\|\mathbf{v}\|} = \frac{3}{5}\mathbf{i} - \frac{4}{5}\mathbf{j}$$

$$D_{\mathbf{u}}g(x, y) = \frac{x}{\sqrt{x^2 + y^2}}\left(\frac{3}{5}\right) + \frac{y}{\sqrt{x^2 + y^2}}\left(-\frac{4}{5}\right)$$

$$D_{\mathbf{u}}g(3, 4) = \frac{3}{5}\left(\frac{3}{5}\right) + \frac{4}{5}\left(-\frac{4}{5}\right) = -\frac{7}{25}$$

8. $h(x, y) = e^{-(x^2 + y^2)},\ P(0, 0),\ \mathbf{v} = \mathbf{i} + \mathbf{j}$

$$\mathbf{u} = \frac{\mathbf{v}}{\|\mathbf{v}\|} = \frac{\sqrt{2}}{2}\mathbf{i} - \frac{\sqrt{2}}{2}\mathbf{j}$$

$$D_{\mathbf{u}}h(x, y) = -2xe^{-(x^2 + y^2)}\left(\frac{\sqrt{2}}{2}\right) + \left(-2ye^{-(x^2 + y^2)}\right)\left(\frac{\sqrt{2}}{2}\right)$$

$$D_{\mathbf{u}}h(0, 0) = 0$$

9. $f(x, y) = x^2 + 3y^2,\ P(1, 1),\ Q(4, 5)$

$$\mathbf{v} = (4 - 1)\mathbf{i} + (5 - 1)\mathbf{j} = 3\mathbf{i} + 4\mathbf{j}$$

$$\mathbf{u} = \frac{\mathbf{v}}{\|\mathbf{v}\|} = \frac{3}{5}\mathbf{i} + \frac{4}{5}\mathbf{j}$$

$$D_{\mathbf{u}}f(x, y) = 2x\left(\frac{3}{5}\right) + 6y\left(\frac{4}{5}\right)$$

$$D_{\mathbf{u}}f(1, 1) = 2\left(\frac{3}{5}\right) + 6\left(\frac{4}{5}\right) = 6$$

10. $f(x, y) = \cos(x + y),\ P(0, \pi),\ Q\left(\dfrac{\pi}{2}, 0\right)$

$$\mathbf{v} = \left(\frac{\pi}{2} - 0\right)\mathbf{i} + (0 - \pi)\mathbf{j}$$

$$\mathbf{v} = \frac{\pi}{2}\mathbf{i} - \pi\mathbf{j}$$

$$\mathbf{u} = \frac{\mathbf{v}}{\|\mathbf{v}\|} = \frac{1}{\sqrt{5}}\mathbf{i} - \frac{2}{\sqrt{5}}\mathbf{j}$$

$$D_{\mathbf{u}}f(x, y) = -\sin(x + y)\left(\frac{1}{\sqrt{5}}\right) - \sin(x + y)\left(\frac{-2}{\sqrt{5}}\right)$$

$$D_{\mathbf{u}}f(0, \pi) = 0$$

11. $f(x, y) = e^y \sin x,\ P(0, 0),\ Q(2, 1)$

$$\mathbf{v} = (2 - 0)\mathbf{i} + (1 - 0)\mathbf{j}$$

$$\mathbf{v} = 2\mathbf{i} + \mathbf{j},\ \mathbf{u} = \frac{\mathbf{v}}{\|\mathbf{v}\|} = \frac{2}{\sqrt{5}}\mathbf{i} + \frac{1}{\sqrt{5}}\mathbf{j}$$

$$D_{\mathbf{u}} f(x, y) = e^y \cos x\left(\frac{2}{\sqrt{5}}\right) + e^y \sin x\left(\frac{1}{\sqrt{5}}\right)$$

$$D_{\mathbf{u}} f(0, 0) = \frac{2}{\sqrt{5}} = \frac{2\sqrt{5}}{5}$$

12. $f(x, y) = \sin 2x \cos y,\ P(\pi, 0),\ Q\left(\frac{\pi}{2}, \pi\right)$

$$\mathbf{v} = \left(\frac{\pi}{2} - \pi\right)\mathbf{i} + (\pi - 0)\mathbf{j}$$

$$\mathbf{v} = -\frac{\pi}{2}\mathbf{i} + \pi\mathbf{j}$$

$$\mathbf{u} = \frac{\mathbf{v}}{\|\mathbf{v}\|} = -\frac{1}{\sqrt{5}}\mathbf{i} + \frac{2}{\sqrt{5}}\mathbf{j}$$

$$D_{\mathbf{u}} f(x, y) = 2\cos 2x \cos y\left(-\frac{1}{\sqrt{5}}\right) + (-\sin 2x \sin y)\left(\frac{2}{\sqrt{5}}\right)$$

$$D_{\mathbf{u}} f(\pi, 0) = -\frac{2}{\sqrt{5}} = -\frac{2\sqrt{5}}{5}$$

13. $f(x, y) = 3x + 5y^2 + 1$

$$\nabla f(x, y) = 3\mathbf{i} + 10y\mathbf{j}$$

$$\nabla f(2, 1) = 3\mathbf{i} + 10\mathbf{j}$$

14. $g(x, y) = 2xe^{y/x}$

$$\nabla g(x, y) = \left(-\frac{2y}{x}e^{y/x} + 2e^{y/x}\right)\mathbf{i} + 2e^{y/x}\mathbf{j}$$

$$\nabla g(2, 0) = 2\mathbf{i} + 2\mathbf{j}$$

15. $z = \ln(x^2 - y)$

$$\nabla z(x, y) = \frac{2x}{x^2 - y}\mathbf{i} - \frac{1}{x^2 - y}\mathbf{j}$$

$$\nabla z(2, 3) = 4\mathbf{i} - \mathbf{j}$$

16. $z = \cos(x^2 + y^2)$

$$\nabla z(x, y) = -2x \sin(x^2 + y^2)\mathbf{i} - 2y \sin(x^2 + y^2)\mathbf{j}$$

$$\nabla z(3, -4) = -6 \sin 25\mathbf{i} + 8 \sin 25\mathbf{j} \approx 0.7941\mathbf{i} - 1.0588\mathbf{j}$$

17. $w = 3x^2 - 5y^2 + 2z^2$

$$\nabla w(x, y, z) = 6x\mathbf{i} - 10y\mathbf{j} + 4z\mathbf{k}$$

$$\nabla w(1, 1, -2) = 6\mathbf{i} - 10\mathbf{j} - 8\mathbf{k}$$

18. $w = x \tan(y + z)$

$$\nabla w(x, y, z) = \tan(y + z)\mathbf{i} + x \sec^2(y + z)\mathbf{j} + x \sec^2(y + z)\mathbf{k}$$

$$\nabla w(4, 3, -1) = \tan 2\mathbf{i} + 4 \sec^2 2\mathbf{j} + 4 \sec^2 2\mathbf{k}$$

19. $f(x, y) = xy$

$$\mathbf{v} = \frac{1}{2}\left(\mathbf{i} + \sqrt{3}\mathbf{j}\right)$$

$$\nabla f(x, y) = y\mathbf{i} + x\mathbf{j}$$

$$\nabla f(0, -2) = -2\mathbf{i}$$

$$\mathbf{u} = \frac{\mathbf{v}}{\|\mathbf{v}\|} = \frac{1}{2}\mathbf{i} + \frac{\sqrt{3}}{2}\mathbf{j}$$

$$D_{\mathbf{u}} f(0, -2) = \nabla f(0, -2) \cdot \mathbf{u} = -1$$

20. $h(x, y) = e^x \sin y$

$\mathbf{v} = -\mathbf{i}$

$\nabla h = e^x \sin y\mathbf{i} + e^x \cos y\mathbf{j}$

$\nabla h\left(1, \frac{\pi}{2}\right) = e\mathbf{i}$

$\mathbf{u} = \frac{\mathbf{v}}{\|\mathbf{v}\|} = -\mathbf{i}$

$D_{\mathbf{u}}h\left(1, \frac{\pi}{2}\right) = \nabla h\left(1, \frac{\pi}{2}\right) \cdot \mathbf{u} = -e$

21. $f(x, y, z) = x^2 + y^2 + z^2$

$\mathbf{v} = \frac{\sqrt{3}}{3}(\mathbf{i} - \mathbf{j} + \mathbf{k})$

$\nabla f(x, y, z) = 2x\mathbf{i} + 2y\mathbf{j} + 2z\mathbf{k}$

$\nabla f(1, 1, 1) = 2\mathbf{i} + 2\mathbf{j} + 2\mathbf{k}$

$\mathbf{u} = \frac{\mathbf{v}}{\|\mathbf{v}\|} = \frac{\sqrt{3}}{3}\mathbf{i} - \frac{\sqrt{3}}{3}\mathbf{j} + \frac{\sqrt{3}}{3}\mathbf{k}$

$D_{\mathbf{u}} f(1, 1, 1) = \nabla f(1, 1, 1) \cdot \mathbf{u} = \frac{2}{3}\sqrt{3}$

22. $f(x, y, z) = xy + yz + xz$

$\mathbf{v} = 2\mathbf{i} + \mathbf{j} - \mathbf{k}$

$\nabla f(x, y, z) = (y + z)\mathbf{i} + (x + z)\mathbf{j} + (y + x)\mathbf{k}$

$\nabla f(1, 2, -1) = \mathbf{i} + 3\mathbf{k}$

$\mathbf{u} = \frac{\mathbf{v}}{\|\mathbf{v}\|} = \frac{1}{\sqrt{6}}(2\mathbf{i} + \mathbf{j} - \mathbf{k})$

$D_{\mathbf{u}} f(1, 2, -1) = \nabla f(1, 2, -1) \cdot \mathbf{u}$

$= \frac{2}{\sqrt{6}} - \frac{3}{\sqrt{6}} = \frac{-\sqrt{6}}{6}$

23. $\overrightarrow{PQ} = \mathbf{i} + \mathbf{j}$, $\mathbf{u} = \frac{\sqrt{2}}{2}\mathbf{i} + \frac{\sqrt{2}}{2}\mathbf{j}$

$\nabla g(x, y) = 2x\mathbf{i} + 2y\mathbf{j}$, $\nabla g(1, 2) = 2\mathbf{i} + 4\mathbf{j}$

$D_{\mathbf{u}}g = \nabla g \cdot \mathbf{u} = \sqrt{2} + 2\sqrt{2} = 3\sqrt{2}$

24. $\overrightarrow{PQ} = 4\mathbf{i} + 2\mathbf{j}$, $\mathbf{u} = \frac{2}{\sqrt{5}}\mathbf{i} + \frac{1}{\sqrt{5}}\mathbf{j}$

$\nabla f = 6x\mathbf{i} - 2y\mathbf{j}$, $\nabla f(-1, 4) = -6\mathbf{i} - 8\mathbf{j}$

$D_{\mathbf{u}} f = \nabla f \cdot \mathbf{u} = -\frac{12}{\sqrt{5}} - \frac{8}{\sqrt{5}} = -4\sqrt{5}$

25. $g(x, y, z) = xye^z$

$\mathbf{v} = -2\mathbf{i} - 4\mathbf{j}$

$\nabla g = ye^z\mathbf{i} + xe^z\mathbf{j} + xye^z\mathbf{k}$

At $(2, 4, 0)$, $\nabla g = 4\mathbf{i} + 2\mathbf{j} + 8\mathbf{k}$.

$\mathbf{u} = \frac{\mathbf{v}}{\|\mathbf{v}\|} = -\frac{1}{\sqrt{5}}\mathbf{i} - \frac{2}{\sqrt{5}}\mathbf{j}$

$D_{\mathbf{u}}g = \nabla g \cdot \mathbf{u} = -\frac{4}{\sqrt{5}} - \frac{4}{\sqrt{5}} = -\frac{8}{\sqrt{5}}$

26. $h(x, y, z) = \ln(x + y + z)$

$\mathbf{v} = 3\mathbf{i} + 3\mathbf{j} + \mathbf{k}$

$\nabla h = \frac{1}{x + y + z}(\mathbf{i} + \mathbf{j} + \mathbf{k})$

At $(1, 0, 0)$, $\nabla h = \mathbf{i} + \mathbf{j} + \mathbf{k}$.

$\mathbf{u} = \frac{\mathbf{v}}{\|\mathbf{v}\|} = \frac{1}{\sqrt{19}}(3\mathbf{i} + 3\mathbf{j} + \mathbf{k})$

$D_{\mathbf{u}}h = \nabla h \cdot \mathbf{u} = \frac{7}{\sqrt{19}} = \frac{7\sqrt{19}}{19}$

27. $f(x, y) = x^2 + 2xy$

$\nabla f(x, y) = (2x + 2y)\mathbf{i} + 2x\mathbf{j}$

$\nabla f(1, 0) = 2\mathbf{i} + 2\mathbf{j}$

$\|\nabla f(1, 0)\| = 2\sqrt{2}$

28. $f(x, y) = \frac{x + y}{y + 1}$

$\nabla f(x, y) = \frac{1}{y + 1}\mathbf{i} + \frac{1 - x}{(y + 1)^2}\mathbf{j}$

$\nabla f(0, 1) = \frac{1}{2}\mathbf{i} + \frac{1}{4}\mathbf{j}$

$\|\nabla f(0, 1)\| = \sqrt{\frac{1}{4} + \frac{1}{16}} = \frac{1}{4}\sqrt{5}$

29. $h(x, y) = x \tan y$

$\nabla h(x, y) = \tan y\mathbf{i} + x \sec^2 y\mathbf{j}$

$\nabla h\left(2, \frac{\pi}{4}\right) = \mathbf{i} + 4\mathbf{j}$

$\left\|\nabla h\left(2, \frac{\pi}{4}\right)\right\| = \sqrt{17}$

30. $h(x, y) = y\cos(x - y)$

$\nabla h(x, y) = -y\sin(x - y)\mathbf{i} + [\cos(x - y) + y\sin(x - y)]\mathbf{j}$

$\nabla h\left(0, \frac{\pi}{3}\right) = \frac{\sqrt{3}\pi}{6}\mathbf{i} + \left(\frac{3 - \sqrt{3}\pi}{6}\right)\mathbf{j}$

$\left\|\nabla h\left(0, \frac{\pi}{3}\right)\right\| = \sqrt{\frac{3\pi^2}{36} + \frac{9 - 6\sqrt{3}\pi + 3\pi^2}{36}} = \frac{\sqrt{3\left(2\pi^2 - 2\sqrt{3}\pi + 3\right)}}{6}$

31. $g(x, y) = ye^{-x}$

$\nabla g(x, y) = -ye^{-x}\mathbf{i} + e^{-x}\mathbf{j}$

$\nabla g(0, 5) = -5\mathbf{i} + \mathbf{j}$

$\|\nabla g(0, 5)\| = \sqrt{26}$

32. $g(x, y) = \ln\sqrt[3]{x^2 + y^2} = \frac{1}{3}\ln(x^2 + y^2)$

$\nabla g(x, y) = \frac{1}{3}\left[\frac{2x}{x^2 + y^2}\mathbf{i} + \frac{2y}{x^2 + y^2}\mathbf{j}\right]$

$\nabla g(1, 2) = \frac{1}{3}\left(\frac{2}{5}\mathbf{i} + \frac{4}{5}\mathbf{j}\right) = \frac{2}{15}(\mathbf{i} + 2\mathbf{j})$

$\|\nabla g(1, 2)\| = \frac{2\sqrt{5}}{15}$

33. $f(x, y, z) = \sqrt{x^2 + y^2 + z^2}$

$\nabla f(x, y, z) = \frac{1}{\sqrt{x^2 + y^2 + z^2}}(x\mathbf{i} + y\mathbf{j} + z\mathbf{k})$

$\nabla f(1, 4, 2) = \frac{1}{\sqrt{21}}(\mathbf{i} + 4\mathbf{j} + 2\mathbf{k})$

$\|\nabla f(1, 4, 2)\| = 1$

34. $w = \frac{1}{\sqrt{1 - x^2 - y^2 - z^2}}$

$\nabla w = \frac{1}{\left(\sqrt{1 - x^2 - y^2 - z^2}\right)^3}(x\mathbf{i} + y\mathbf{j} + z\mathbf{k})$

$\nabla w(0, 0, 0) = \mathbf{0}$

$\|\nabla w(0, 0, 0)\| = 0$

35. $w = xy^2z^2$

$\nabla w = y^2z^2\mathbf{i} + 2xyz^2\mathbf{j} + 2xy^2z\mathbf{k}$

$\nabla w(2, 1, 1) = \mathbf{i} + 4\mathbf{j} + 4\mathbf{k}$

$\|\nabla w(2, 1, 1)\| = \sqrt{33}$

36. $f(x, y, z) = xe^{yz}$

$\nabla f(x, y, z) = e^{yz}\mathbf{i} + xze^{yz}\mathbf{j} + xye^{yz}\mathbf{k}$

$\nabla f(2, 0, -4) = \mathbf{i} - 8\mathbf{j}$

$\|\nabla f(2, 0, -4)\| = \sqrt{65}$

For exercises 37–42, $f(x, y) = 3 - \frac{x}{3} - \frac{y}{2}$ and $D_{\mathbf{u}}f(x, y) = -\left(\frac{1}{3}\right)\cos\theta - \left(\frac{1}{2}\right)\sin\theta$.

37. $f(x, y) = 3 - \frac{x}{3} - \frac{y}{2}$

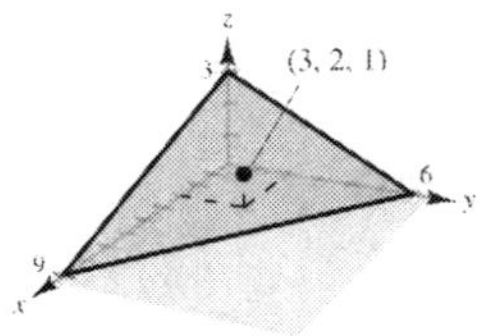

38. (a) $D_{\mathbf{u}}f(3, 2) = -\left(\frac{1}{3}\right)\frac{\sqrt{2}}{2} - \left(\frac{1}{2}\right)\frac{\sqrt{2}}{2} = -\frac{5\sqrt{2}}{12}$

(b) $D_{\mathbf{u}}f(3, 2) = -\left(\frac{1}{3}\right)\left(-\frac{1}{2}\right) - \left(\frac{1}{2}\right)\frac{\sqrt{3}}{2} = \frac{2 - 3\sqrt{3}}{12}$

(c) $D_{\mathbf{u}}f(3, 2) = -\left(\frac{1}{3}\right)\left(-\frac{1}{2}\right) - \left(\frac{1}{2}\right)\left(-\frac{\sqrt{3}}{2}\right) = \frac{2 + 3\sqrt{3}}{12}$

(d) $D_{\mathbf{u}}f(3, 2) = -\left(\frac{1}{3}\right)\left(\frac{\sqrt{3}}{2}\right) - \left(\frac{1}{2}\right)\left(-\frac{1}{2}\right) = \frac{3 - 2\sqrt{3}}{12}$

39. (a) $\mathbf{u} = \left(\frac{1}{\sqrt{2}}\right)(\mathbf{i} + \mathbf{j})$

$D_{\mathbf{u}}f = \nabla f \cdot \mathbf{u} = -\left(\frac{1}{3}\right)\frac{1}{\sqrt{2}} - \left(\frac{1}{2}\right)\frac{1}{\sqrt{2}} = -\frac{5\sqrt{2}}{12}$

(b) $\mathbf{v} = -3\mathbf{i} - 4\mathbf{j}$

$\|\mathbf{v}\| = \sqrt{9 + 16} = 5$

$\mathbf{u} = -\frac{3}{5}\mathbf{i} - \frac{4}{5}\mathbf{j}$

$D_{\mathbf{u}}f = \nabla f \cdot \mathbf{u} = \frac{1}{5} + \frac{2}{5} = \frac{3}{5}$

(c) $\mathbf{v} - 3\mathbf{i} + 4\mathbf{j}$

$\|\mathbf{v}\| = \sqrt{9 + 16} = 5$

$\mathbf{u} = -\frac{3}{5}\mathbf{i} + \frac{4}{5}\mathbf{j}$

$D_{\mathbf{u}}f = \nabla f \cdot \mathbf{u} = \frac{1}{5} - \frac{2}{5} = -\frac{1}{5}$

(d) $\mathbf{v} = \mathbf{i} + 3\mathbf{j}$

$\|\mathbf{v}\| = \sqrt{10}$

$\mathbf{u} = \frac{1}{\sqrt{10}}\mathbf{i} + \frac{3}{\sqrt{10}}\mathbf{j}$

$D_{\mathbf{u}}f = \nabla f \cdot \mathbf{u} = \frac{-11}{6\sqrt{10}} = -\frac{11\sqrt{10}}{60}$

40. $\nabla f = -\left(\frac{1}{3}\right)\mathbf{i} - \left(\frac{1}{2}\right)\mathbf{j}$

41. $\|\nabla f\| = \sqrt{\frac{1}{9} + \frac{1}{4}} = \frac{1}{6}\sqrt{13}$

42. $\nabla f = -\frac{1}{3}\mathbf{i} - \frac{1}{2}\mathbf{j}$

$\frac{\nabla f}{\|\nabla f\|} = \frac{1}{\sqrt{13}}(-2\mathbf{i} - 3\mathbf{j})$

So, $\mathbf{u} = \left(1/\sqrt{13}\right)(3\mathbf{i} - 2\mathbf{j})$ and

$D_{\mathbf{u}}f(3, 2) = \nabla f \cdot \mathbf{u} = 0$. ∇f is the direction of greatest rate of change of f. So, in a direction orthogonal to ∇f, the rate of change of f is 0.

43. (a) In the direction of the vector $-4\mathbf{i} + \mathbf{j}$

(b) $\nabla f = \frac{1}{10}(2x - 3y)\mathbf{i} + \frac{1}{10}(-3x + 2y)\mathbf{j}$

$\nabla f(1, 2) = \frac{1}{10}(-4)\mathbf{i} + \frac{1}{10}(1)\mathbf{j} = -\frac{2}{5}\mathbf{i} + \frac{1}{10}\mathbf{j}$

(Same direction as in part (a))

(c) $-\nabla f = \frac{2}{5}\mathbf{i} - \frac{1}{10}\mathbf{j}$, the direction opposite that of the gradient

44. (a) In the direction of the vector $\mathbf{i} + \mathbf{j}$

(b) $\nabla f = \frac{1}{2}y\frac{1}{2\sqrt{x}}\mathbf{i} + \frac{1}{2}\sqrt{x}\mathbf{j} = \frac{y}{4\sqrt{x}}\mathbf{i} + \frac{1}{2}\sqrt{x}\mathbf{j}$

$\nabla f(1, 2) = \frac{1}{2}\mathbf{i} + \frac{1}{2}\mathbf{j}$

(Same direction as in part (a))

(c) $-\nabla f = -\frac{1}{2}\mathbf{i} - \frac{1}{2}\mathbf{j}$, the direction opposite that of the gradient

45. $f(x, y) = x^2 - y^2$, $(4, -3, 7)$

(a)

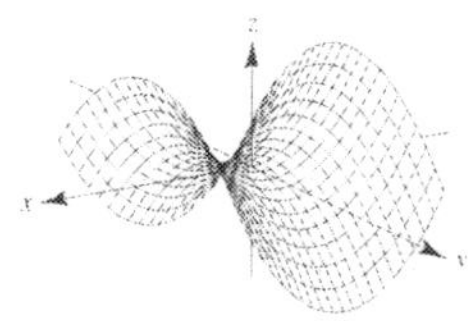

(b) $D_{\mathbf{u}}f(x, y) = \nabla f(x, y) \cdot \mathbf{u} = 2x\cos\theta - 2y\sin\theta$

$D_{\mathbf{u}}f(4, -3) = 8\cos\theta + 6\sin\theta$

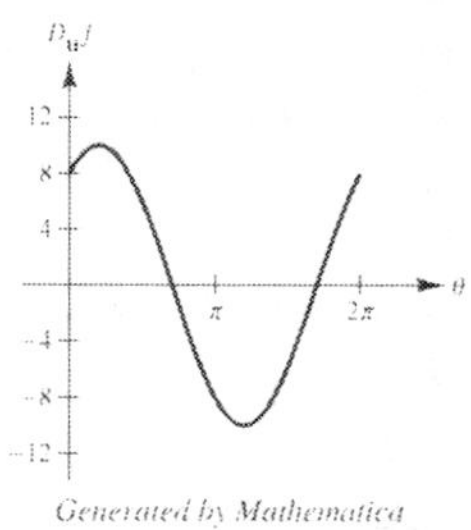

(c) Zeros: $\theta \approx 2.21, 5.36$

These are the angles θ for which $D_{\mathbf{u}}f(4, 3)$ equals zero.

(d) $g(\theta) = D_{\mathbf{u}}f(4, -3) = 8\cos\theta + 6\sin\theta$

$g'(\theta) = -8\sin\theta + 6\cos\theta$

Critical numbers: $\theta \approx 0.64, 3.79$

These are the angels for which $D_{\mathbf{u}}f(4, -3)$ is a maximum (0.64) and minimum (3.79).

(e) $\|\nabla f(4, -3)\| = \|2(4)\mathbf{i} - 2(-3)\mathbf{j}\| = \sqrt{64 + 36} = 10$, the maximum value of $D_{\mathbf{u}}f(4, -3)$, at $\theta \approx 0.64$.

(f) $f(x, y) = x^2 - y^2 = 7$

$\nabla f(4, -3) = 8\mathbf{i} + 6\mathbf{j}$ is perpendicular to the level curve at $(4, -3)$.

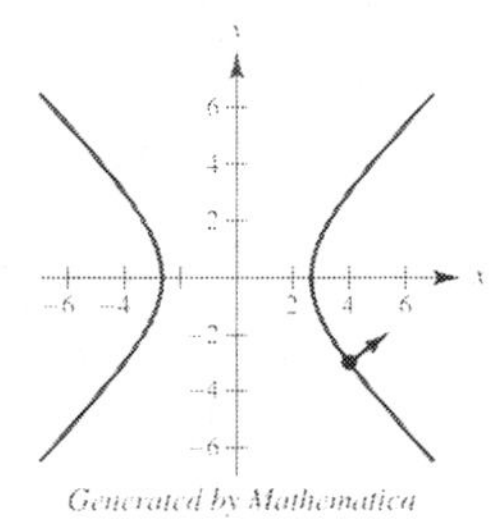

46. (a) $f(x, y) = \dfrac{8y}{1 + x^2 + y^2} = 2$

$\Rightarrow 4y = 1 + x^2 + y^2$

$4 = y^2 - 4y + 4 + x^2 + 1$

$(y - 2)^2 + x^2 = 3$

Circle: center: $(0, 2)$, radius: $\sqrt{3}$

(b) $\nabla f = \dfrac{-16xy}{(1 + x^2 + y^2)^2}\mathbf{i} + \dfrac{8 + 8x^2 - 8y^2}{(1 + x^2 + y^2)^2}\mathbf{j}$

$\nabla f(\sqrt{3}, 2) = \dfrac{-\sqrt{3}}{2}\mathbf{i}$

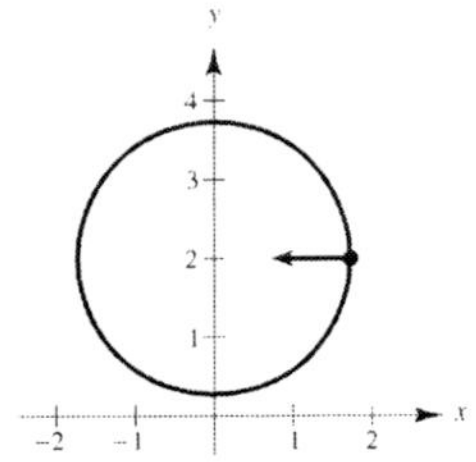

(c) The directional derivative of f is 0 in the direction $\pm\mathbf{j}$.

(d)

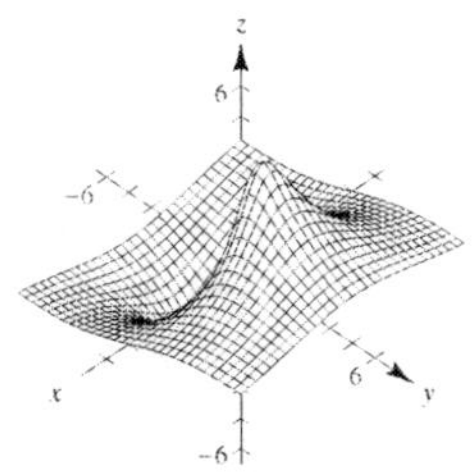

47. $f(x, y) = 6 - 2x - 3y$

$c = 6, P = (0, 0)$

$\nabla f(x, y) = -2\mathbf{i} - 3\mathbf{j}$

$6 - 2x - 3y = 6$

$0 = 2x + 3y$

$\nabla f(0, 0) = -2\mathbf{i} - 3\mathbf{j}$

48. $f(x, y) = x^2 + y^2$

$c = 25, P = (3, 4)$

$\nabla f(x, y) = 2x\mathbf{i} + 2y\mathbf{j}$

$x^2 + y^2 = 25$

$\nabla f(3, 4) = 6\mathbf{i} + 8\mathbf{j}$

49. $f(x, y) = xy$

$c = -3, P = (-1, 3)$

$\nabla f(x, y) = y\mathbf{i} + x\mathbf{j}$

$xy = -3$

$\nabla f(-1, 3) = 3\mathbf{i} - \mathbf{j}$

50. $f(x, y) = \dfrac{x}{x^2 + y^2}$

$c = \dfrac{1}{2}, P = (1, 1)$

$\nabla f(x, y) = \dfrac{y^2 - x^2}{(x^2 + y^2)^2}\mathbf{i} - \dfrac{2xy}{(x^2 + y^2)^2}\mathbf{j}$

$\dfrac{x}{x^2 + y^2} = \dfrac{1}{2}$

$x^2 + y^2 - 2x = 0$

$\nabla f(1, 1) = -\dfrac{1}{2}\mathbf{j}$

51. $f(x, y) = 4x^2 - y$

(a) $\nabla f(x, y) = 8x\mathbf{i} - \mathbf{j}$

$\nabla f(2, 10) = 16\mathbf{i} - \mathbf{j}$

(b) $\|16\mathbf{i} - \mathbf{j}\| = \sqrt{257}$

$\dfrac{1}{\sqrt{257}}(16\mathbf{i} - \mathbf{j})$ is a unit vector normal to the level curve $4x^2 - y = 6$ at $(2, 10)$.

(c) The vector $\mathbf{i} + 16\mathbf{j}$ is tangent to the level curve.

Slope $= \dfrac{16}{1} = 16$

$y - 10 = 16(x - 2)$

$y = 16x - 22$ Tangent line

(d)

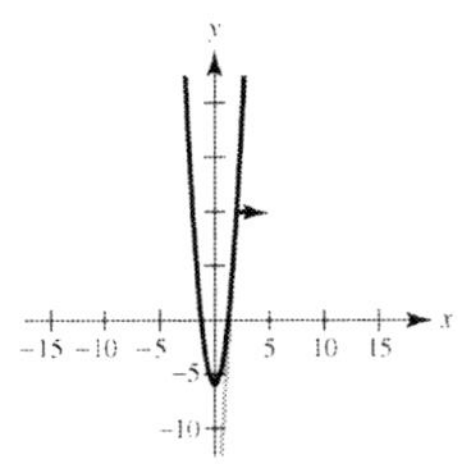

52. $f(x, y) = x - y^2$

(a) $\nabla f(x, y) = \mathbf{i} - 2y\mathbf{j}$

$\nabla f(4, -1) = \mathbf{i} + 2\mathbf{j}$

(b) $\|\nabla f(4, -1)\| = \sqrt{5}$

$\dfrac{1}{\sqrt{5}}(\mathbf{i} + 2\mathbf{j})$ is a unit vector normal to the level curve $x - y^2 = 3$ at $(4, -1)$.

(c) The vector $2\mathbf{i} - \mathbf{j}$ is tangent to the level curve.

Slope $= -\dfrac{1}{2}$.

$y + 1 = -\dfrac{1}{2}(x - 4)$

$y = -\dfrac{1}{2}x + 1$ Tangent line

(d)

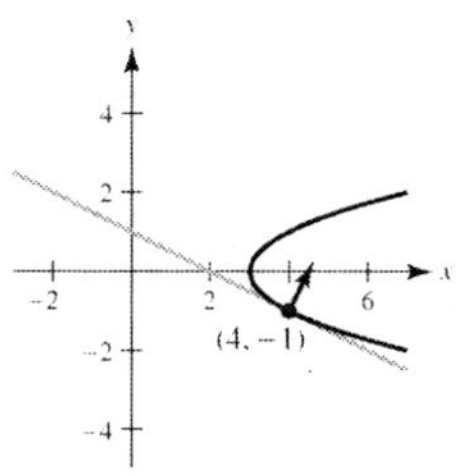

53. $f(x, y) = 3x^2 - 2y^2$

(a) $\nabla f = 6x\mathbf{i} - 4y\mathbf{j}$

$\nabla f(1, 1) = 6\mathbf{i} - 4\mathbf{j}$

(b) $\|\nabla f(1, 1)\| = \sqrt{36 + 16} = 2\sqrt{13}$

$\dfrac{1}{\sqrt{13}}(3\mathbf{i} - 2\mathbf{j})$ is a unit vector normal to the level curve $3x^2 - 2y^2 = 1$ at $(1, 1)$.

(c) The vector $2\mathbf{i} + 3\mathbf{j}$ is tangent to the level curve.

Slope $= \dfrac{3}{2}$.

$y - 1 = \frac{3}{2}(x - 1)$

$y = \frac{3}{2}x - \frac{1}{2}$ tangent line

(d)

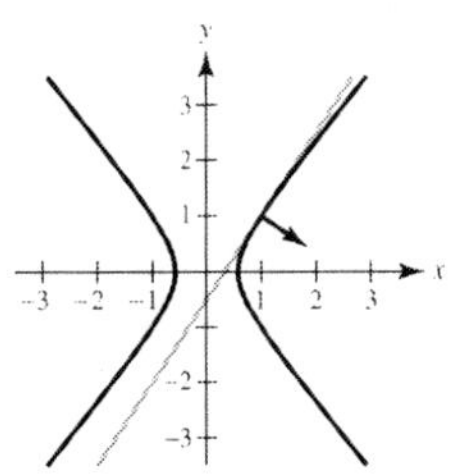

54. $f(x, y) = 9x^2 + 4y^2$

(a) $\nabla f = 18x\mathbf{i} + 8y\mathbf{j}$

$\nabla f(2, -1) = 36\mathbf{i} - 8\mathbf{j}$

(b) $\|\nabla f(2, -1)\| = \sqrt{1360} = 4\sqrt{85}$

$\dfrac{1}{\sqrt{85}}(9\mathbf{i} - 2\mathbf{j})$ is a unit vector normal to the level curve $9x^2 + 4y^2 = 40$ at $(2, -1)$.

(c) The vector $2\mathbf{i} + 9\mathbf{j}$ is tangent to the level curve.

Slope $= \dfrac{9}{2}$.

$y + 1 = \dfrac{9}{2}(x - 2)$

$y = \dfrac{9}{2}x - 10$ Tangent line

(d)

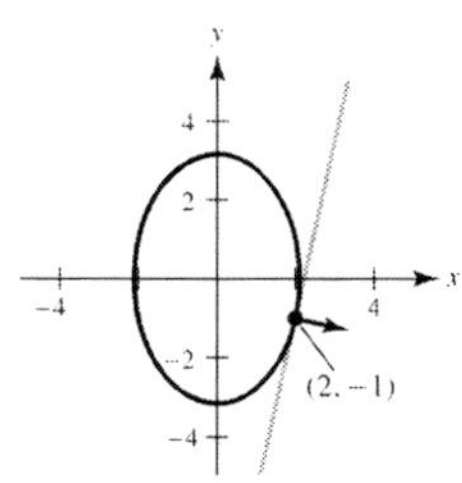

55. See the definition, page 916.

56. Let $f(x, y)$ be a function of two variables and $\mathbf{u} = \cos\theta\mathbf{i} + \sin\theta\mathbf{j}$ a unit vector.

(a) If $\theta = 0°$, then $D_{\mathbf{u}}f = \dfrac{\partial f}{\partial x}$.

(b) If $\theta = 90°$, then $D_{\mathbf{u}}f = \dfrac{\partial f}{\partial y}$.

57. See the definition, pages 918 and 919.

58.

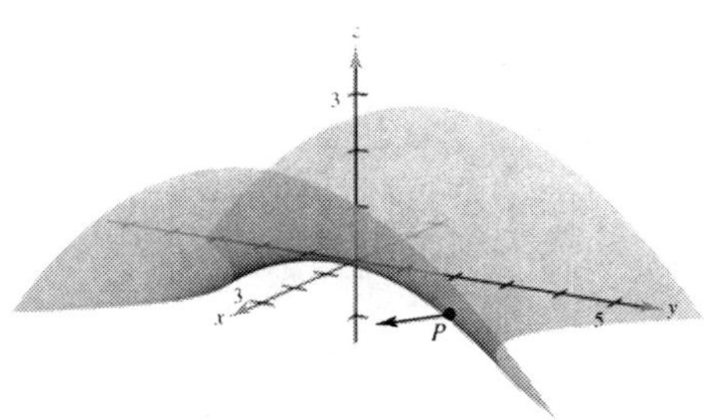

59. The gradient vector is normal to the level curves. See Theorem 13.12.

60. $f(x, y) = 9 - x^2 - y^2$ and

$$D_{\mathbf{u}} f(x, y) = -2x \cos\theta - 2y \sin\theta = -2(x \cos\theta + y \sin\theta)$$

(a) $f(x, y) = 9 - x^2 - y^2$

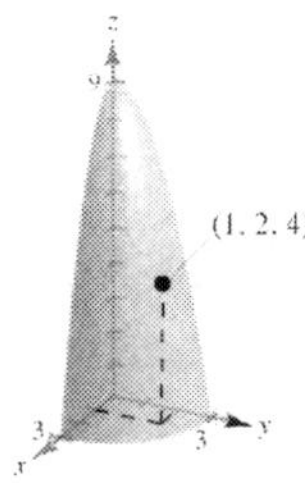

(b) $D_{\mathbf{u}} f(1, 2) = -2\left(\dfrac{\sqrt{2}}{2} - \sqrt{2}\right) = \sqrt{2}$

(c) $D_{\mathbf{u}} f(1, 2) = -2\left(\dfrac{1}{2} + \sqrt{3}\right) = -\left(1 + 2\sqrt{3}\right)$

(d) $\nabla f(1, 2) = -2\mathbf{i} - 4\mathbf{j}$

$\left\|\nabla f(1, 2)\right\| = \sqrt{4 + 16} = \sqrt{20} = 2\sqrt{5}$

(e) $\nabla f(1, 2) = -2\mathbf{i} - 4\mathbf{j}$

$$\frac{\nabla f(1, 2)}{\left\|\nabla f(1, 2)\right\|} = \frac{1}{\sqrt{5}}(-\mathbf{i} - 2\mathbf{j})$$

Therefore, $\mathbf{u} = \left(1/\sqrt{5}\right)(-2\mathbf{i} + \mathbf{j})$ and

$D_{\mathbf{u}} f(1, 2) = \nabla f(1, 2) \cdot \mathbf{u} = 0.$

61. $h(x, y) = 5000 - 0.001x^2 - 0.004y^2$

$\nabla h = -0.002x\mathbf{i} - 0.008y\mathbf{j}$

$\nabla h(500, 300) = -\mathbf{i} - 2.4\mathbf{j}$ or

$5\nabla h = -(5\mathbf{i} + 12\mathbf{j})$

62.

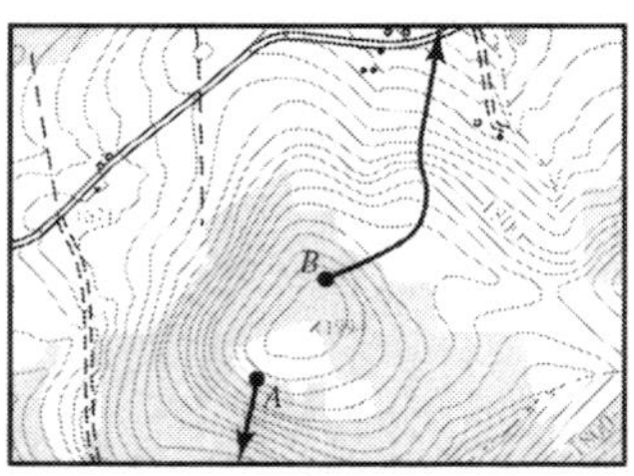

63. $T = \dfrac{x}{x^2 + y^2}$

$$\nabla T = \frac{y^2 - x^2}{\left(x^2 + y^2\right)^2}\mathbf{i} - \frac{2xy}{\left(x^2 + y^2\right)^2}\mathbf{j}$$

$$\nabla T(3, 4) = \frac{7}{625}\mathbf{i} - \frac{24}{625}\mathbf{j} = \frac{1}{625}(7\mathbf{i} - 24\mathbf{j})$$

64. (a)

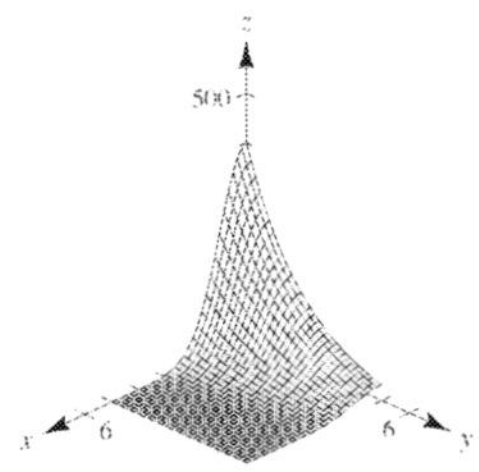

(b) $\nabla T(x, y) = 400e^{-\left(x^2 + y\right)/2}\left[(-x)\mathbf{i} - \frac{1}{2}\mathbf{j}\right]$

$\nabla T(3, 5) = 400e^{-7}\left[-3\mathbf{i} - \frac{1}{2}\mathbf{j}\right]$

There will be no change in directions perpendicular to the gradient: $\pm(\mathbf{i} - 6\mathbf{j})$

(c) The greatest increase is in the direction of the gradient: $-3\mathbf{i} - \frac{1}{2}\mathbf{j}$

65. $T(x, y) = 80 - 3x^2 - y^2$, $P(-1, 5)$

$\nabla T(x, y) = -6x\mathbf{i} - 2y\mathbf{j}$

Maximum increase in direction:

$\nabla T(-1, 5) = (-6)(-1)\mathbf{i} - 2(5)\mathbf{j} = 6\mathbf{i} - 10\mathbf{j}$

Maximum rate:

$\left\|\nabla T(-1, 5)\right\| = \sqrt{6^2 + (-10)^2} = 2\sqrt{34}$

$\approx 11.66°$ per centimeter

66. $T(x, y) = 50 - x^2 - 4y^2$, $P(2, -1)$

$\nabla T(x, y) = -2x\mathbf{i} - 8y\mathbf{j}$

Maximum increase in direction:

$\nabla T(2, -1) = -2(2)\mathbf{i} - 8(-1)\mathbf{j} = -4\mathbf{i} + 8\mathbf{j}$

Maximum rate:

$\left\|\nabla T(2, -1)\right\| = \sqrt{16 + 64} = 4\sqrt{5}$

$\approx 8.94°$ per centimeter

67. $T(x, y) = 400 - 2x^2 - y^2$, $P = (10, 10)$

$\dfrac{dx}{dt} = -4x \qquad \dfrac{dy}{dt} = -2y$

$x(t) = C_1e^{-4t} \qquad y(t) = C_2e^{-2t}$

$10 = x(0) = C_1 \qquad 10 = y(0) = C_2$

$x(t) = 10e^{-4t} \qquad y(t) = 10e^{-2t}$

$x = \dfrac{y^2}{10} \qquad y^2(t) = 100e^{-4t}$

$y^2 = 10x$

68. $T(x, y) = 100 - x^2 - 2y^2,\ P = (4, 3)$

$$\frac{dx}{dt} = -2x \qquad \frac{dy}{dt} = -4y$$

$$x(t) = C_1 e^{-2t} \qquad y(t) = C_2 e^{-4t}$$

$$4 = x(0) = C_1 \qquad 3 = y(0) = C_2$$

$$x(t) = 4e^{-2t} \qquad y(t) = 3e^{-4t}$$

$$\frac{3x^2}{16} = e^{-4t} = y \Rightarrow u = \frac{3}{16}x^2$$

69. True

70. False

$$D_{\mathbf{u}} f(x, y) = \sqrt{2} > 1 \text{ when } \mathbf{u} = \left(\cos\frac{\pi}{4}\right)\mathbf{i} + \left(\sin\frac{\pi}{4}\right)\mathbf{j}.$$

71. True

72. True

73. Let $f(x, y, z) = e^x \cos y + \frac{z^2}{2} + C$. Then

$$\nabla f(x, y, z) = e^x \cos y\mathbf{i} - e^x \sin y\mathbf{j} + z\mathbf{k}.$$

74. (a)

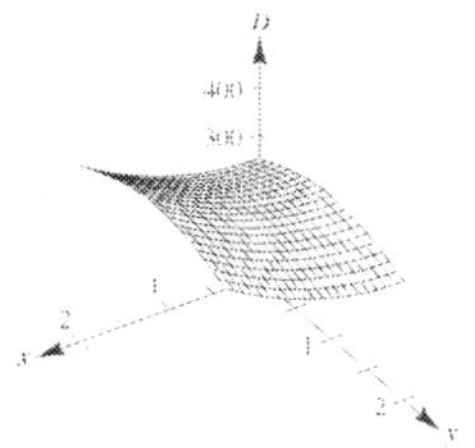

(b) The graph of $-D = -250 - 30x^2 - 50\sin(\pi y/2)$ would model the ocean floor.

(c) $D(1, 0.5) = 250 + 30(1) + 50\sin\frac{\pi}{4} \approx 315.4$ ft

(d) $\frac{\partial D}{\partial x} = 60x$ and $\frac{\partial D}{\partial x}(1, 0.5) = 60$

(e) $\frac{\partial D}{\partial y} = 25\pi\cos\frac{\pi y}{2}$ and

$$\frac{\partial D}{\partial y}(1, 0.5) = 25\pi\cos\frac{\pi}{4} \approx 55.5$$

(f) $\nabla D = 60x\mathbf{i} + 25\pi\cos\left(\frac{\pi y}{2}\right)\mathbf{j}$

$$\nabla D(1, 0.5) = 60\mathbf{i} + 55.5\mathbf{j}$$

75. (a) $f(x, y) = \sqrt[3]{xy}$ is the composition of two continuous functions, $h(x, y) = xy$ and $g(z) = z^{1/3}$, and therefore continuous by Theorem 13.2.

(b) $$f_x(0, 0) = \lim_{\Delta x \to 0} \frac{f(0 + \Delta x, 0) - f(0, 0)}{\Delta x} = \lim_{\Delta x \to 0} \frac{(0 \cdot \Delta x)^{1/3} - 0}{\Delta x} = 0$$

$$f_y(0, 0) = \lim_{\Delta y \to 0} \frac{f(0, 0 + \Delta y) - f(0, 0)}{\Delta y} = \lim_{\Delta y \to 0} \frac{(0 \cdot \Delta y)^{1/3} - 0}{\Delta y} = 0$$

Let $\mathbf{u} = \cos\theta i + \sin\theta\mathbf{j},\ \theta \neq 0, \frac{\pi}{2}, \pi, \frac{3\pi}{2}$. Then

$$D_{\mathbf{u}} f(0, 0) = \lim_{t \to 0} \frac{f(0 + t\cos\theta, 0 + t\sin\theta) - f(0, 0)}{t} = \lim_{t \to 0} \frac{\sqrt[3]{t^2 \cos\theta \sin\theta}}{t} = \lim_{t \to 0} \frac{\sqrt[3]{\cos\theta\sin\theta}}{t^{1/3}}, \text{ does not exist.}$$

(c)

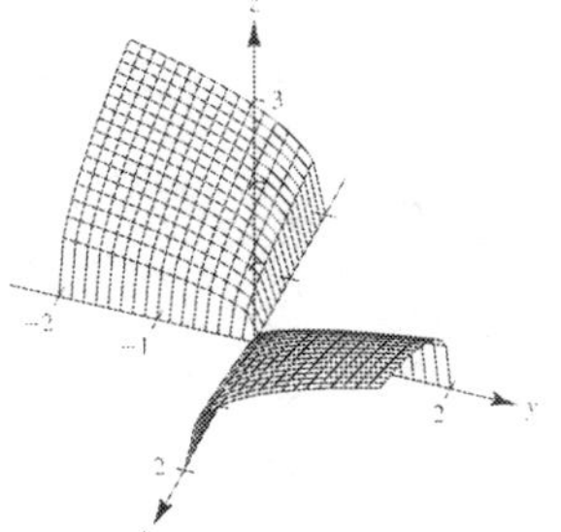

76. We cannot use Theorem 13.9 because f is not a differentiable function of x and y. So, we use the definition of directional derivatives.

$$D_{\mathbf{u}} f(x, y) = \lim_{t \to 0} \frac{f(x + t\cos\theta, y + t\sin\theta) - f(x, y)}{t}$$

$$D_{\mathbf{u}} f(0, 0) = \lim_{t \to 0} \frac{f\left[0 + \left(\frac{t}{\sqrt{2}}\right), 0 + \left(\frac{t}{\sqrt{2}}\right)\right] - f(0, 0)}{t} = \lim_{t \to 0} \frac{1}{t}\left[\frac{4\left(\frac{t}{\sqrt{2}}\right)\left(\frac{t}{\sqrt{2}}\right)}{\left(\frac{t^2}{2}\right) + \left(\frac{t^2}{2}\right)}\right] = \lim_{t \to 0} \frac{1}{t}\left[\frac{2t^2}{t^2}\right] = \lim_{t \to 0} \frac{2}{t} \text{ which does not exist.}$$

If $f(0, 0) = 2$, then $D_{\mathbf{u}} f(0, 0) = \lim_{t \to 0} \dfrac{f\left(0 + \frac{t}{\sqrt{2}}, 0 + \frac{t}{\sqrt{2}}\right) - 2}{t} = \lim_{t \to 0} \dfrac{1}{t}\left[\dfrac{2t^2}{t^2} - 2\right] = 0$

which implies that the directional derivative exists.

Section 13.7 Tangent Planes and Normal Lines

1. $F(x, y, z) = 3x - 5y + 3z - 15 = 0$

$3x - 5y + 3z = 15$ Plane

2. $F(x, y, z) = x^2 + y^2 + z^2 - 25 = 0$

$x^2 + y^2 + z^2 = 25$

Sphere, radius 5, centered at origin.

3. $F(x, y, z) = 4x^2 + 9y^2 - 4z^2 = 0$

$4x^2 + 9y^2 = 4z^2$ Elliptic cone

4. $F(x, y, z) = 16x^2 - 9y^2 + 36z = 0$

$16x^2 - 9y^2 + 36z = 0$ Hyperbolic paraboloid

5. $F(x, y, z) = 3x + 4y + 12z = 0$

$\nabla F = 3\mathbf{i} + 4\mathbf{j} + 12\mathbf{k}, \ \|\nabla F\| = \sqrt{9 + 16 + 144} = 13$

$\mathbf{n} = \dfrac{\nabla F}{\|\nabla F\|} = \dfrac{3}{13}\mathbf{i} + \dfrac{4}{13}\mathbf{j} + \dfrac{12}{13}\mathbf{k}$

6. $F(x, y, z) = x^2 + y^2 + z^2 - 6$

$\nabla F = 2x\mathbf{i} + 2y\mathbf{j} + 2z\mathbf{k}$

$\nabla F(1, 1, 2) = 2\mathbf{i} + 2\mathbf{j} + 4\mathbf{k}$

$\|\nabla F(1, 1, 2)\| = \sqrt{4 + 4 + 16} = 2\sqrt{6}$

$\mathbf{n} = \dfrac{\nabla F}{\|\nabla F\|} = \dfrac{1}{\sqrt{6}}\mathbf{i} + \dfrac{1}{\sqrt{6}}\mathbf{j} + \dfrac{2}{\sqrt{6}}\mathbf{k}$

7. $F(x, y, z) = x^2 + 3y + z^3 - 9$

$\nabla F(x, y, z) = 2x\mathbf{i} + 3\mathbf{j} + 3z^2\mathbf{k}$

$\nabla F(2, -1, 2) = 4\mathbf{i} + 3\mathbf{j} + 12\mathbf{k}$

$\mathbf{n} = \dfrac{\nabla F}{\|\nabla F\|} = \dfrac{1}{13}(4\mathbf{i} + 3\mathbf{j} + 12\mathbf{k})$

8. $F(x, y, z) = x^2y^3 - y^2z + 2xz^3 - 4$

$\nabla F = (2xy^3 + 2z^3)\mathbf{i} + (3x^2y^2 - 2yz)\mathbf{j} + (6xz^2 - y^2)\mathbf{k}$

$\nabla F(-1, 1, -1) = -4\mathbf{i} + 5\mathbf{j} - 7\mathbf{k}$

$\|\nabla F(-1, 1, -1)\| = 3\sqrt{10}$

$\mathbf{n} = \dfrac{\nabla F}{\|\nabla F\|} = \dfrac{1}{3\sqrt{10}}(-4\mathbf{i} + 5\mathbf{j} - 7\mathbf{k})$

9. $z = x^2 + y^2 + 3, (2, 1, 8)$

$F(x, y, z) = x^2 + y^2 + 3 - z$

$F_x(x, y, z) = 2x \quad F_y(x, y, z) = 2y \quad F_z(x, y, z) = -1$

$F_x(2, 1, 8) = 4 \quad F_y(2, 1, 8) = 2 \quad F_z(2, 1, 8) = -1$

$4(x - 2) + 2(y - 1) - 1(z - 8) = 0$

$4x + 2y - z = 2$

10. $f(x, y) = \dfrac{y}{x}, (1, 2, 2)$

$F(x, y, z) = \dfrac{y}{x} - z$

$F_x(x, y, z) = -\dfrac{y}{x^2} \quad F_y(x, y, z) = \dfrac{1}{x} \quad F_z(x, y, z) = -1$

$F_x(1, 2, 2) = -2 \quad F_y(1, 2, 2) = 1 \quad F_z(1, 2, 2) = -1$

$-2(x - 1) + (y - 2) - (z - 2) = 0$

$-2x + y - z + 2 = 0$

$2x - y + z = 2$

11. $z = \sqrt{x^2 + y^2}, (3, 4, 5)$

$$F(x, y, z) = \sqrt{x^2 + y^2} - z$$

$$F_x(x, y, z) = \frac{x}{\sqrt{x^2 + y^2}} \qquad F_y(x, y, z) = \frac{y}{\sqrt{x^2 + y^2}} \qquad F_z(x, y, z) = -1$$

$$F_x(3, 4, 5) = \frac{3}{5} \qquad F_y(3, 4, 5) = \frac{4}{5} \qquad F_z(3, 4, 5) = -1$$

$$\frac{3}{5}(x - 3) + \frac{4}{5}(y - 4) - (z - 5) = 0$$

$$3(x - 3) + 4(y - 4) - 5(z - 5) = 0$$

$$3x + 4y - 5z = 0$$

12. $g(x, y) = \arctan\frac{y}{x}, (1, 0, 0)$

$$G(x, y, z) = \arctan\frac{y}{x} - z$$

$$G_x(x, y, z) = \frac{-(y/x^2)}{1 + (y^2/x^2)} = \frac{-y}{x^2 + y^2} \qquad G_y(x, y, z) = \frac{1/x}{1 + (y^2/x^2)} = \frac{x}{x^2 + y^2} \qquad G_z(x, y, z) = -1$$

$$G_x(1, 0, 0) = 0 \qquad G_y(1, 0, 0) = 1 \qquad G_z(1, 0, 0) = -1$$

$$y - z = 0$$

13. $g(x, y) = x^2 + y^2, (1, -1, 2)$

$$G(x, y, z) = x^2 + y^2 - z$$

$$G_x(x, y, z) = 2x \qquad G_y(x, y, z) = 2y \qquad G_z(x, y, z) = -1$$

$$G_x(1, -1, 2) = 2 \qquad G_y(1, -1, 2) = -2 \qquad G_z(1, -1, 2) = -1$$

$$2(x - 1) - 2(y + 1) - 1(z - 2) = 0$$

$$2x - 2y - z = 2$$

14. $f(x, y) = x^2 - 2xy + y^2, (1, 2, 1)$

$$F(x, y, z) = x^2 - 2xy + y^2 - z$$

$$F_x(x, y, z) = 2x - 2y \qquad F_y(x, y, z) = -2x + 2y \qquad F_z(x, y, z) = -1$$

$$F_x(1, 2, 1) = -2 \qquad F_y(1, 2, 1) = 2 \qquad F_z(1, 2, 1) = -1$$

$$-2(x - 1) + 2(y - 2) - (z - 1) = 0$$

$$-2x + 2y - z - 1 = 0$$

$$2x - 2y + z = -1$$

15. $h(x, y) = \ln\sqrt{x^2 + y^2}, (3, 4, \ln 5)$

$$H(x, y, z) = \ln\sqrt{x^2 + y^2} - z = \frac{1}{2}\ln(x^2 + y^2) - z$$

$$H_x(x, y, z) = \frac{x}{x^2 + y^2} \qquad H_y(x, y, z) = \frac{y}{x^2 + y^2} \qquad H_z(x, y, z) = -1$$

$$H_x(3, 4, \ln 5) = \frac{3}{25} \qquad H_y(3, 4, \ln 5) = \frac{4}{25} \qquad H_z(3, 4, \ln 5) = -1$$

$$\frac{3}{25}(x - 3) + \frac{4}{25}(y - 4) - (z - \ln 5) = 0$$

$$3(x - 3) + 4(y - 4) - 25(z - \ln 5) = 0$$

$$3x + 4y - 25z = 25(1 - \ln 5)$$

16. $h(x, y) = \cos y, \left(5, \frac{\pi}{4}, \frac{\sqrt{2}}{2}\right)$

$$H(x, y, z) = \cos y - z$$

$$H_x(x, y, z) = 0 \qquad H_y(x, y, z) = -\sin y \qquad H_z(x, y, z) = -1$$

$$H_x\left(5, \frac{\pi}{4}, \frac{\sqrt{2}}{2}\right) = 0 \qquad H_y\left(5, \frac{\pi}{4}, \frac{\sqrt{2}}{2}\right) = -\frac{\sqrt{2}}{2} \qquad H_z\left(5, \frac{\pi}{4}, \frac{\sqrt{2}}{2}\right) = -1$$

$$-\frac{\sqrt{2}}{2}\left(y - \frac{\pi}{4}\right) - \left(z - \frac{\sqrt{2}}{2}\right) = 0$$

$$-\frac{\sqrt{2}}{2}y - z + \frac{\sqrt{2}\pi}{8} + \frac{\sqrt{2}}{2} = 0$$

$$4\sqrt{2}y + 8z = \sqrt{2}(\pi + 4)$$

17. $x^2 + 4y^2 + z^2 = 36, (2, -2, 4)$

$$F(x, y, z) = x^2 + 4y^2 + z^2 - 36$$

$$F_x(x, y, z) = 2x \qquad F_y(x, y, z) = 8y \qquad F_z(x, y, z) = 2z$$

$$F_x(2, -2, 4) = 4 \qquad F_y(2, -2, 4) = -16 \qquad F_z(2, -2, 4) = 8$$

$$4(x - 2) - 16(y + 2) + 8(z - 4) = 0$$

$$(x - 2) - 4(y + 2) + 2(z - 4) = 0$$

$$x - 4y + 2z = 18$$

18. $x^2 + 2z^2 = y^2, (1, 3, -2)$

$$F(x, y, z) = x^2 - y^2 + 2z^2$$

$$F_x(x, y, z) = 2x \qquad F_y(x, y, z) = -2y \qquad F_z(x, y, z) = 4z$$

$$F_x(1, 3, -2) = 2 \qquad F_y(1, 3, -2) = -6 \qquad F_z(1, 3, -2) = -8$$

$$2(x - 1) - 6(y - 3) - 8(z + 2) = 0$$

$$(x - 1) - 3(y - 3) - 4(z + 2) = 0$$

$$x - 3y - 4z = 0$$

19. $xy^2 + 3x - z^2 = 8, (1, -3, 2)$

$F(x, y, z) = xy^2 + 3x - z^2 - 8$

$F_x(x, y, z) = y^2 + 3 \quad F_y(x, y, z) = 2xy \quad F_z(x, y, z) = -2z$

$F_x(1, -3, 2) = 12 \quad F_y(1, -3, 2) = -6 \quad F_z(1, -3, 2) = -4$

$$12(x - 1) - 6(y + 3) - 4(z - 2) = 0$$
$$12x - 6y - 4z = 22$$
$$6x - 3y - 2z = 11$$

20. $z = e^x(\sin y + 1), \left(0, \frac{\pi}{2}, 2\right)$

$F(x, y, z) = e^x(\sin y + 1) - z$

$F_x(x, y, z) = e^x(\sin y + 1) \quad F_y(x, y, z) = e^x \cos y \quad F_z(x, y, z) = -1$

$F_x\left(0, \frac{\pi}{2}, 2\right) = 2 \quad F_y\left(0, \frac{\pi}{2}, 2\right) = 0 \quad F_z\left(0, \frac{\pi}{2}, 2\right) = -1$

$2x - z = -2$

21. $x + y + z = 9, (3, 3, 3)$

$F(x, y, z) = x + y + z - 9$

$F_x(x, y, z) = 1 \quad F_y(x, y, z) = 1 \quad F_z(x, y, z) = 1$

$F_x(3, 3, 3) = 1 \quad F_y(3, 3, 3) = 1 \quad F_z(3, 3, 3) = 1$

$$(x - 3) + (y - 3) + (z - 3) = 0$$
$$x + y + z = 9 \text{ (same plane!)}$$

Direction numbers: 1, 1, 1

Line: $x - 3 = y - 3 = z - 3$

22. $x^2 + y^2 + z^2 = 9, (1, 2, 2)$

$F(x, y, z) = x^2 + y^2 + z^2 - 9$

$F_x(x, y, z) = 2x \quad F_y(x, y, z) = 2y \quad F_z(x, y, z) = 2z$

$F_x(1, 2, 2) = 2 \quad F_y(1, 2, 2) = 4 \quad F_z(1, 2, 2) = 4$

Direction numbers: 1, 2, 2

Plane: $(x - 1) + 2(y - 2) + 2(z - 2) = 0, x + 2y + 2z = 9$

Line: $\dfrac{x - 1}{1} = \dfrac{y - 2}{2} = \dfrac{z - 2}{2}$

23. $x^2 + y^2 + z = 9, (1, 2, 4)$

$F(x, y, z) = x^2 + y^2 + z - 9$

$F_x(x, y, z) = 2x \quad F_y(x, y, z) = 2y \quad F_z(x, y, z) = 1$

$F_x(1, 2, 4) = 2 \quad F_y(1, 2, 4) = 4 \quad F_z(1, 2, 4) = 1$

Direction numbers: 2, 4, 1

Plane: $2(x - 1) + 4(y - 2) + (z - 4) = 0, \ 2x + 4y + z = 14$

Line: $\dfrac{x - 1}{2} = \dfrac{y - 2}{4} = \dfrac{z - 4}{1}$

24. $z = 16 - x^2 - y^2, (2, 2, 8)$

$F(x, y, z) = 16 - x^2 - y^2 - z$

$F_x(x, y, z) = -2x \quad F_y(x, y, z) = -2y \quad F_z(x, y, z) = -1$

$F_x(2, 2, 8) = -4 \quad F_y(2, 2, 8) = -4 \quad F_z(2, 2, 8) = -1$

$$-4(x - 2) - 4(y - 2) - (z - 8) = 0$$
$$-4x - 4y - z = -24$$
$$4x + 4y + z = 24$$

Direction numbers: 4, 4, 1

Line: $\dfrac{x - 2}{4} = \dfrac{y - 2}{4} = z - 8$

25. $z = x^2 - y^2, (3, 2, 5)$

$F(x, y, z) = x^2 - y^2 - z$

$F_x(x, y, z) = 2x \quad F_y(x, y, z) = -2y \quad F_z(x, y, z) = -1$

$F_x(3, 2, 5) = 6 \quad F_y(3, 2, 5) = -4 \quad F_z(3, 2, 5) = -1$

$$6(x - 3) - 4(y - 2) - (z - 5) = 0$$
$$6x - 4y - z = 5$$

Direction numbers: 6, −4, −1

Line: $\dfrac{x - 3}{6} = \dfrac{y - 2}{-4} = \dfrac{z - 5}{-1}$

26. $xy - z = 0, (-2, -3, 6)$

$F(x, y, z) = xy - z$

$F_x(x, y, z) = y \quad F_y(x, y, z) = x \quad F_z(x, y, z) = -1$

$F_x(-2, -3, 6) = -3 \quad F_y(-2, -3, 6) = -2 \quad F_z(-2, -3, 6) = -1$

Direction numbers: 3, 2, 1

Plane: $3(x + 2) + 2(y + 3) + (z - 6) = 0, 3x + 2y + z = -6$

Line: $\dfrac{x + 2}{3} = \dfrac{y + 3}{2} = \dfrac{z - 6}{1}$

27. $xyz = 10, (1, 2, 5)$

$F(x, y, z) = xyz - 10$

$F_x(x, y, z) = yz \quad F_y(x, y, z) = xz \quad F_z(x, y, z) = xy$

$F_x(1, 2, 5) = 10 \quad F_y(1, 2, 5) = 5 \quad F_z(1, 2, 5) = 2$

Direction numbers: 10, 5, 2

Plane: $10(x - 1) + 5(y - 2) + 2(z - 5) = 0, 10x + 5y + 2z = 30$

Line: $\dfrac{x - 1}{10} = \dfrac{y - 2}{5} = \dfrac{z - 5}{2}$

28. $z = ye^{2xy}, (0, 2, 2)$

$F(x, y, z) = ye^{2xy} - z$

$F_x(x, y, z) = 2y^2e^{2xy}$ $\quad F_y(x, y, z) = (1 + 2xy)e^{2xy}$ $\quad F_z(x, y, z) = -1$

$F_x(0, 2, 2) = 8$ $\quad F_y(0, 2, 2) = 1$ $\quad F_z(0, 2, 2) = -1$

$$8(x - 0) + (y - 2) - (z - 2) = 0$$
$$8x + y - z = 0$$

Direction number: 8, 1, –1

Line: $\dfrac{x}{8} = \dfrac{y-2}{1} = \dfrac{z-2}{-1}$

29. $z = \arctan \dfrac{y}{x}, \left(1, 1, \dfrac{\pi}{4}\right)$

$F(x, y, z) = \arctan \dfrac{y}{x} - z$

$F_x(x, y, z) = \dfrac{-y}{x^2 + y^2}$ $\quad F_y(x, y, z) = \dfrac{x}{x^2 + y^2}$ $\quad F_z(x, y, z) = -1$

$F_x\left(1, 1, \dfrac{\pi}{4}\right) = -\dfrac{1}{2}$ $\quad F_y\left(1, 1, \dfrac{\pi}{4}\right) = \dfrac{1}{2}$ $\quad F_z\left(1, 1, \dfrac{\pi}{4}\right) = -1$

Direction numbers: 1, –1, 2

Plane: $(x - 1) - (y - 1) + 2\left(z - \dfrac{\pi}{4}\right) = 0, x - y + 2z = \dfrac{\pi}{2}$

Line: $\dfrac{x-1}{1} = \dfrac{y-1}{-1} = \dfrac{z - (\pi/4)}{2}$

30. $y \ln(xz^2) = 2, (e, 2, 1)$

$F(x, y, z) = y[\ln x + 2 \ln z] - 2$

$F_x(x, y, z) = \dfrac{y}{x}$ $\quad F_y(x, y, z) = \ln x + 2 \ln z$ $\quad F_z(x, y, z) = \dfrac{2y}{z}$

$F_x(e, 2, 1) = \dfrac{2}{e}$ $\quad F_y(e, 2, 1) = 1$ $\quad F_z(e, 2, 1) = 4$

$$\frac{2}{e}(x - e) + (y - 2) + 4(z - 1) = 0$$
$$\frac{2}{e}x + y + 4z = 8$$

Direction numbers: $\dfrac{2}{e}$, 1, 4

$$\frac{x - e}{(2/e)} = \frac{y-2}{1} = \frac{z-1}{4}$$

31. $F(x, y, z) = x^2 + y^2 - 2 \quad G(x, y, z) = x - z$

$\nabla F(x, y, z) = 2x\mathbf{i} + 2y\mathbf{j} \quad \nabla G(x, y, z) = \mathbf{i} - \mathbf{k}$

$\nabla F(1, 1, 1) = 2\mathbf{i} + 2\mathbf{j} \quad \nabla G(1, 1, 1) = \mathbf{i} - \mathbf{k}$

(a) $\nabla F \times \nabla G = \begin{vmatrix} \mathbf{i} & \mathbf{j} & \mathbf{k} \\ 2 & 2 & 0 \\ 1 & 0 & -1 \end{vmatrix} = -2\mathbf{i} + 2\mathbf{j} - 2\mathbf{k} = -2(\mathbf{i} - \mathbf{j} + \mathbf{k})$

Direction numbers: 1, −1, 1

Line: $x - 1 = \dfrac{y - 1}{-1} = z - 1$

(b) $\cos\theta = \dfrac{|\nabla F \cdot \nabla G|}{\|\nabla F\|\|\nabla G\|} = \dfrac{2}{(2\sqrt{2})\sqrt{2}} = \dfrac{1}{2}$

Not orthogonal

32. $F(x, y, z) = x^2 + y^2 - z \quad G(x, y, z) = 4 - y - z$

$\nabla F(x, y, z) = 2x\mathbf{i} + 2y\mathbf{j} - \mathbf{k} \quad \nabla G(x, y, z) = -\mathbf{j} - \mathbf{k}$

$\nabla F(2, -1, 5) = 4\mathbf{i} - 2\mathbf{j} - \mathbf{k} \quad \nabla G(2, -1, 5) = -\mathbf{j} - \mathbf{k}$

(a) $\nabla F \times \nabla G = \begin{vmatrix} \mathbf{i} & \mathbf{j} & \mathbf{k} \\ 4 & -2 & -1 \\ 0 & -1 & -1 \end{vmatrix} = \mathbf{i} + 4\mathbf{j} - 4\mathbf{k}$

Direction numbers: 1, 4, −4. $\dfrac{x - 2}{1} = \dfrac{y + 1}{4} = \dfrac{z - 5}{-4}$

(b) $\cos\theta = \dfrac{|\nabla F \cdot \nabla G|}{\|\nabla F\|\|\nabla G\|} = \dfrac{3}{\sqrt{21}\sqrt{2}} = \dfrac{3}{\sqrt{42}} = \dfrac{\sqrt{42}}{14}$; not orthogonal

33. $F(x, y, z) = x^2 + z^2 - 25 \quad G(x, y, z) = y^2 + z^2 - 25$

$\nabla F = 2x\mathbf{i} + 2z\mathbf{k} \quad \nabla G = 2y\mathbf{j} + 2z\mathbf{k}$

$\nabla F(3, 3, 4) = 6\mathbf{i} + 8\mathbf{k} \quad \nabla G(3, 3, 4) = 6\mathbf{j} + 8\mathbf{k}$

(a) $\nabla F \times \nabla G = \begin{vmatrix} \mathbf{i} & \mathbf{j} & \mathbf{k} \\ 6 & 0 & 8 \\ 0 & 6 & 8 \end{vmatrix} = -48\mathbf{i} - 48\mathbf{j} + 36\mathbf{k} = -12(4\mathbf{i} + 4\mathbf{j} - 3\mathbf{k})$

Direction numbers: 4, 4, −3. $\dfrac{x - 3}{4} = \dfrac{y - 3}{4} = \dfrac{z - 4}{-3}$

(b) $\cos\theta = \dfrac{|\nabla F \cdot \nabla G|}{\|\nabla F\|\|\nabla G\|} = \dfrac{64}{(10)(10)} = \dfrac{16}{25}$; not orthogonal

34. $F(x, y, z) = \sqrt{x^2 + y^2} - z \qquad G(x, y, z) = 5x - 2y + 3z - 22$

$\nabla F(x, y, z) = \dfrac{x}{\sqrt{x^2 + y^2}}\mathbf{i} + \dfrac{y}{\sqrt{x^2 + y^2}}\mathbf{j} - \mathbf{k} \qquad \nabla G(x, y, z) = 5\mathbf{i} - 2\mathbf{j} + 3\mathbf{k}$

$\nabla F(3, 4, 5) = \dfrac{3}{5}\mathbf{i} + \dfrac{4}{5}\mathbf{j} - \mathbf{k} \qquad \nabla G(3, 4, 5) = 5\mathbf{i} - 2\mathbf{j} + 3\mathbf{k}$

(a) $\nabla F \times \nabla G = \begin{vmatrix} \mathbf{i} & \mathbf{j} & \mathbf{k} \\ 3/5 & 4/5 & -1 \\ 5 & -2 & 3 \end{vmatrix} = \dfrac{2}{5}\mathbf{i} - \dfrac{34}{5}\mathbf{j} - \dfrac{26}{5}\mathbf{k}$

Direction numbers: 1, −17, −13

$\dfrac{x - 3}{1} = \dfrac{y - 4}{-17} = \dfrac{z - 5}{-13}$; tangent line

(b) $\cos\theta = \dfrac{|\nabla F \cdot \nabla G|}{\|\nabla F\|\|\nabla G\|} = \dfrac{-(8/5)}{\sqrt{2}\sqrt{38}} = \dfrac{-8}{5\sqrt{76}}$; not orthogonal

35. $F(x, y, z) = x^2 + y^2 + z^2 - 14 \qquad G(x, y, z) = x - y - z$

$\nabla F(x, y, z) = 2x\mathbf{i} + 2y\mathbf{j} + 2z\mathbf{k} \qquad \nabla G(x, y, z) = \mathbf{i} - \mathbf{j} - \mathbf{k}$

$\nabla F(3, 1, 2) = 6\mathbf{i} + 2\mathbf{j} + 4\mathbf{k} \qquad \nabla G(3, 1, 2) = \mathbf{i} - \mathbf{j} - \mathbf{k}$

(a) $\nabla F \times \nabla G = \begin{vmatrix} \mathbf{i} & \mathbf{j} & \mathbf{k} \\ 6 & 2 & 4 \\ 1 & -1 & -1 \end{vmatrix} = 2\mathbf{i} + 10\mathbf{j} - 8\mathbf{k} = 2[\mathbf{i} + 5\mathbf{j} - 4\mathbf{k}]$

Direction numbers: 1, 5, −4

Line: $\dfrac{x - 3}{1} = \dfrac{y - 1}{5} = \dfrac{z - 2}{-4}$

(b) $\cos\theta = \dfrac{|\nabla F \cdot \nabla G|}{\|\nabla F\|\|\nabla G\|} = 0 \Rightarrow$ orthogonal

36. $F(x, y, z) = x^2 + y^2 - z \qquad G(x, y, z) = x + y + 6z - 33$

$\nabla F(x, y, z) = 2x\mathbf{i} + 2y\mathbf{j} - \mathbf{k} \qquad \nabla G(x, y, z) = \mathbf{i} + \mathbf{j} + 6\mathbf{k}$

$\nabla F(1, 2, 5) = 2\mathbf{i} + 4\mathbf{j} - \mathbf{k} \qquad \nabla G(1, 2, 5) = \mathbf{i} + \mathbf{j} + 6\mathbf{k}$

(a) $\nabla F \times \nabla G = \begin{vmatrix} \mathbf{i} & \mathbf{j} & \mathbf{k} \\ 2 & 4 & -1 \\ 1 & 1 & 6 \end{vmatrix} = 25\mathbf{i} - 13\mathbf{j} - 2\mathbf{k}$

Direction numbers: 25, −13, −2. $\dfrac{x - 1}{25} = \dfrac{y - 2}{-13} = \dfrac{z - 5}{-2}$

(b) $\cos\theta = \dfrac{|\nabla F \cdot \nabla G|}{\|\nabla F\|\|\nabla G\|} = 0$; orthogonal

37. $F(x, y, z) = 3x^2 + 2y^2 - z - 15, (2, 2, 5)$

$\nabla F(x, y, z) = 6x\mathbf{i} + 4y\mathbf{j} - \mathbf{k}$

$\nabla F(2, 2, 5) = 12\mathbf{i} + 8\mathbf{j} - \mathbf{k}$

$$\cos\theta = \frac{|\nabla F(2, 2, 5) \cdot \mathbf{k}|}{\|\nabla F(2, 2, 5)\|} = \frac{1}{\sqrt{209}}$$

$$\theta = \arccos\left(\frac{1}{\sqrt{209}}\right) = 86.03°$$

38. $F(x, y, z) = 2xy - z^3, (2, 2, 2)$

$\nabla F = 2y\mathbf{i} + 2x\mathbf{j} - 3z^2\mathbf{k}$

$\nabla F(2, 2, 2) = 4\mathbf{i} + 4\mathbf{j} - 12\mathbf{k}$

$$\cos\theta = \frac{|\nabla F(2, 2, 2) \cdot \mathbf{k}|}{\|\nabla F(2, 2, 2)\|} = \frac{|-12|}{\sqrt{176}} = \frac{3\sqrt{11}}{11}$$

$$\theta = \arccos\left(\frac{3\sqrt{11}}{11}\right) \approx 25.24°$$

39. $F(x, y, z) = x^2 - y^2 + z, (1, 2, 3)$

$\nabla F(x, y, z) = 2x\mathbf{i} - 2y\mathbf{j} + \mathbf{k}$

$\nabla F(1, 2, 3) = 2\mathbf{i} - 4\mathbf{j} + \mathbf{k}$

$$\cos\theta = \frac{|\nabla F(1, 2, 3) \cdot \mathbf{k}|}{\|\nabla F(1, 2, 3)\|} = \frac{1}{\sqrt{21}}$$

$$\theta = \arccos\frac{1}{\sqrt{21}} \approx 77.40°$$

40. $F(x, y, z) = x^2 + y^2 - 5, (2, 1, 3)$

$\nabla F(x, y, z) = 2x\mathbf{i} + 2y\mathbf{j}$

$\nabla F(2, 1, 3) = 4\mathbf{i} + 2\mathbf{j}$

$$\cos\theta = \frac{|\nabla F(2, 1, 3) \cdot \mathbf{k}|}{\|\nabla F(2, 1, 3)\|} = 0$$

$$\theta = \arccos 0 = 90°$$

41. $F(x, y, z) = 3 - x^2 - y^2 + 6y - z$

$\nabla F(x, y, z) = -2x\mathbf{i} + (-2y + 6)\mathbf{j} - \mathbf{k}$

$-2x = 0, x = 0$

$-2y + 6 = 0, y = 3$

$z = 3 - 0^2 - 3^2 + 6(3) = 12$

$(0, 3, 12)$ (vertex of paraboloid)

42. $F(x, y, z) = 3x^2 + 2y^2 - 3x + 4y - z - 5$

$\nabla F(x, y, z) = (6x - 3)\mathbf{i} + (4y + 4)\mathbf{j} - \mathbf{k}$

$6x - 3 = 0, x = \frac{1}{2}$

$4y + 4 = 0, y = -1$

$z = 3\left(\frac{1}{2}\right)^2 + 2(-1)^2 - 3\left(\frac{1}{2}\right) + 4(-1) - 5 = -\frac{31}{4}$

$\left(\frac{1}{2}, -1, -\frac{31}{4}\right)$

43. $F(x, y, z) = x^2 - xy + y^2 - 2x - 2y - z$

$\nabla F(x, y, z) = (2x - y - 2)\mathbf{i} + (-x + 2y - 2)\mathbf{j} - \mathbf{k}$

$2x - y - 2 = 0$

$-x + 2y - 2 = 0$

$y = 2x - 2 \Rightarrow -x + 2(2x - 2) - 2$

$= 3x - 6 = 0 \Rightarrow x = 2$

$y = 2, z = -4$

Point: $(2, 2, -4)$

44. $F(x, y, z) = 4x^2 + 4xy - 2y^2 + 8x - 5y - 4 - z$

$\nabla F(x, y, z) = (8x + 4y + 8)\mathbf{i} + (4x - 4y - 5)\mathbf{j} - \mathbf{k}$

$8x + 4y + 8 = 0$

$4x - 4y - 5 = 0$

Adding, $12x + 3 = 0 \Rightarrow x = -\frac{1}{4} \Rightarrow y = -\frac{3}{2}$, and

$z = -\frac{5}{4}$

Point: $\left(-\frac{1}{4}, -\frac{3}{2}, -\frac{5}{4}\right)$

45. $F(x, y, z) = 5xy - z$

$\nabla F(x, y, z) = 5y\mathbf{i} + 5x\mathbf{j} - \mathbf{k}$

$5y = 0$

$5x = 0$

$x = y = z = 0$

Point: $(0, 0, 0)$

46. $F(x, y, z) = xy + \frac{1}{x} + \frac{1}{y} - z$

$$\nabla F(x, y, z) = \left(y - \frac{1}{x^2}\right)\mathbf{i} + \left(x - \frac{1}{y^2}\right)\mathbf{j} - \mathbf{k}$$

$y = \frac{1}{x^2}$

$x = \frac{1}{y^2} = x^4 \Rightarrow x = 1, y = 1, z = 3$

Point: $(1, 1, 3)$

47. $F(x, y, z) = x^2 + 2y^2 + 3z^2 - 3, (-1, 1, 0)$

$F_x(x, y, z) = 2x \quad F_y(x, y, z) = 4y \quad F_z(x, y, z) = 6z$

$F_x(-1, 1, 0) = -2 \quad F_y(-1, 1, 0) = 4 \quad F_z(-1, 1, 0) = 0$

$$\begin{aligned} -2(x+1) + 4(y-1) + 0(z-0) &= 0 \\ -2x + 4y &= 6 \\ -x + 2y &= 3 \end{aligned}$$

$G(x, y, z) = x^2 + y^2 + z^2 + 6x - 10y + 14, (-1, 1, 0)$

$G_x(x, y, z) = 2x + 6 \quad G_y(x, y, z) = 2y - 10 \quad G_z(x, y, z) = 2z$

$G_x(-1, 1, 0) = 4 \quad G_y(-1, 1, 0) = -8 \quad G_z(-1, 1, 0) = 0$

$$\begin{aligned} 4(x+1) - 8(y-1) + 0(z-0) &= 0 \\ 4x - 8y + 12 &= 0 \\ -x + 2y &= 3 \end{aligned}$$

The tangent planes are the same.

48. $F(x, y, z) = x^2 + y^2 + z^2 - 8x - 12y + 4z + 42, (2, 3, -3)$

$F_x(x, y, z) = 2x - 8 \quad F_y(x, y, z) = 2y - 12 \quad F_z(x, y, z) = 2z + 4$

$F_x(2, 3, -3) = -4 \quad F_y(2, 3, -3) = -6 \quad F_z(2, 3, -3) = -2$

$$\begin{aligned} -4(x-2) - 6(y-3) - 2(z+3) &= 0 \\ -4x - 6y - 2z + 20 &= 0 \\ 2x + 3y + z &= 10 \end{aligned}$$

$G(x, y, z) = x^2 + y^2 + 2z - 7, (2, 3, -3)$

$G_x(x, y, z) = 2x \quad G_y(x, y, z) = 2y \quad G_z(x, y, z) = 2$

$G_x(2, 3, -3) = 4 \quad G_y(2, 3, -3) = 6 \quad G_z(x, y, z) = 2$

$$\begin{aligned} 4(x-2) + 6(y-3) + 2(z+3) &= 0 \\ 4x + 6y + 2z - 20 &= 0 \\ 2x + 3y + z &= 10 \end{aligned}$$

The tangent planes are the same.

49. (a) $F(x, y, z) = 2xy^2 - z,\ F(1, 1, 2) = 2 - 2 = 0$

$G(x, y, z) = 8x^2 - 5y^2 - 8z + 13,\ G(1, 1, 2) = 8 - 5 - 16 + 13 = 0$

So, $(1, 1, 2)$ lies on both surfaces.

(b) $\nabla F = 2y^2\mathbf{i} + 4xy\mathbf{j} - \mathbf{k},\ \nabla F(1, 1, 2) = 2\mathbf{i} + 4\mathbf{j} - \mathbf{k}$

$\nabla G = 16x\mathbf{i} - 10y\mathbf{j} - 8\mathbf{k},\ \nabla G(1, 1, 2) = 16\mathbf{i} - 10\mathbf{j} - 8\mathbf{k}$

$\nabla F \cdot \nabla G = 2(16) + 4(-10) + (-1)(-8) = 0$

The tangent planes are perpendicular at $(1, 1, 2)$.

50. (a) $F(x, y, z) = x^2 + y^2 + z^2 + 2x - 4y - 4z - 12$

$F(1, -2, 1) = 0$

$G(x, y, z) = 4x^2 + y^2 + 16z^2 - 24$

$G(1, -2, 1) = 0$

So, $(1, -2, 1)$ lies on both surfaces.

(b) $\nabla F = (2x + 2)\mathbf{i} + (2y - 4)\mathbf{j} + (2z - 4)\mathbf{k}$

$\nabla F(1, -2, 1) = 4\mathbf{i} - 8\mathbf{j} - 2\mathbf{k}$

$\nabla G = 8x\mathbf{i} + 2y\mathbf{j} + 32z\mathbf{k}$

$\nabla G(1, -2, 1) = 8\mathbf{i} - 4\mathbf{j} + 32\mathbf{k}$

$\nabla F \cdot \nabla G = 32 + 32 - 64 = 0$

The planes are perpendicular at $(1, -2, 1)$.

51. $F(x, y, z) = x^2 + 4y^2 + z^2 - 9$

$\nabla F = 2x\mathbf{i} + 8y\mathbf{j} + 2z\mathbf{k}$

This normal vector is parallel to the line with direction number $-4, 8, -2$.

So, $2x = -4t \Rightarrow x = -2t$

$8y = 8t \Rightarrow y = t$

$2z = -2t \Rightarrow z = -t$

$x^2 + 4y^2 + z^2 - 9 = 4t^2 + 4t^2 + t^2 - 9 = 0 \Rightarrow t = \pm 1$

There are two points on the ellipse where the tangent plane is perpendicular to the line:

$(-2, 1, -1)$ $(t = 1)$

$(2, -1, 1)$ $(t = -1)$

52. $F(x, y, z) = x^2 + 4y^2 - z^2 - 1$

$\nabla F = 2x\mathbf{i} + 8y\mathbf{j} - 2z\mathbf{k}$

The normal to the plane, $\mathbf{n} = \mathbf{i} + 4\mathbf{j} - \mathbf{k}$

must be parallel to ∇F.

So, $2x = t \Rightarrow x = \frac{t}{2}$

$8y = 4t \Rightarrow y = \frac{t}{2}$

$-2z = -t \Rightarrow z = \frac{t}{2}$

$x^2 + 4y^2 - z^2 = \frac{t^2}{4} + t^2 - \frac{t^2}{4} = t^2 = 1 \Rightarrow t = \pm 1.$

Two points: $\left(\frac{1}{2}, \frac{1}{2}, \frac{1}{2}\right)$ $(t = 1)$ and $\left(-\frac{1}{2}, -\frac{1}{2}, -\frac{1}{2}\right)$ $(t = -1)$

53. $F_x(x_0, y_0, z_0)(x - x_0) + F_y(x_0, y_0, z_0)(y - y_0) + F_z(x_0, y_0, z_0)(z - z_0) = 0$

(Theorem 13.13)

54. For a sphere, the common object is the center of the sphere. For a right circular cylinder, the common object is the axis of the cylinder.

55. Answers will vary.

56. (a) At $(4, 0, 0)$, the tangent plane is parallel to the yz-plane.

Equation: $x = 4$

(b) At $(0, -2, 0)$, the tangent plane is parallel to the xz-plane.

Equation: $y = -2$

(c) At $(0, 0, -4)$, the tangent plane is parallel to the xy-plane.

57. $z = f(x, y) = \dfrac{4xy}{(x^2 + 1)(y^2 + 1)}, -2 \le x \le 2, 0 \le y \le 3$

(a) Let $F(x, y, z) = \dfrac{4xy}{(x^2 + 1)(y^2 + 1)} - z$

$$\nabla F(x, y, z) = \frac{4y}{y^2 + 1}\left(\frac{x^2 + 1 - 2x^2}{(x^2 + 1)^2}\right)\mathbf{i} + \frac{4x}{x^2 + 1}\left(\frac{y^2 + 1 - 2y^2}{(y^2 + 1)^2}\right)\mathbf{j} - \mathbf{k} = \frac{4y(1 - x^2)}{(y^2 + 1)(x^2 + 1)^2}\mathbf{i} + \frac{4x(1 - y^2)}{(x^2 + 1)(y^2 + 1)^2}\mathbf{j} - \mathbf{k}$$

$\nabla F(1, 1, 1) = -\mathbf{k}$

Direction numbers: $0, 0, -1$

Line: $x = 1, y = 1, z = 1 - t$

Tangent plane: $0(x - 1) + 0(y - 1) - 1(z - 1) = 0 \Rightarrow z = 1$

(b) $\nabla F\left(-1, 2, -\dfrac{4}{5}\right) = 0\mathbf{i} + \dfrac{-4(-3)}{(2)(5)^2}\mathbf{j} - \mathbf{k} = \dfrac{6}{25}\mathbf{j} - \mathbf{k}$

Line: $x = -1, y = 2 + \dfrac{6}{25}t, z = -\dfrac{4}{5} - t$

Plane: $0(x + 1) + \dfrac{6}{25}(y - 2) - 1\left(z + \dfrac{4}{5}\right) = 0$

$$6y - 12 - 25z - 20 = 0$$

$$6y - 25z - 32 = 0$$

(c)

58. (a) $f(x, y) = \dfrac{\sin y}{x}, -3 \le x \le 3, 0 \le y \le 2\pi$

Let $F(x, y, z) = \dfrac{\sin y}{x} - z$

$$\nabla F(x, y, z) = \frac{-\sin y}{x^2}\mathbf{i} + \frac{\cos y}{x}\mathbf{j} - \mathbf{k}$$

$$\nabla F\left(2, \frac{\pi}{2}, \frac{1}{2}\right) = -\frac{1}{4}\mathbf{i} - \mathbf{k}$$

Direction numbers: $-\dfrac{1}{4}, 0, -1$ or $1, 0, 4$

Line: $x = 2 + t, y = \dfrac{\pi}{2}, z = \dfrac{1}{2} + 4t$

Tangent plane: $1(x - 2) + 0\left(y - \dfrac{\pi}{2}\right) + 4\left(z - \dfrac{1}{2}\right) = 0 \Rightarrow x + 4z - 4 = 0$

(b) $\nabla F\left(-\dfrac{2}{3}, \dfrac{3\pi}{2}, \dfrac{3}{2}\right) = \dfrac{9}{4}\mathbf{i} - \mathbf{k}$

Direction numbers: $\dfrac{9}{4}, 0, -1$ or $9, 0, -4$

Line: $x = -\dfrac{2}{3} + 9t, y = \dfrac{3\pi}{2}, z = \dfrac{3}{2} - 4t$

Tangent plane: $9\left(x + \dfrac{2}{3}\right) + 0\left(y - \dfrac{3\pi}{2}\right) - 4\left(z - \dfrac{3}{2}\right) = 0 \Rightarrow 9x - 4z + 12 = 0$

(c)

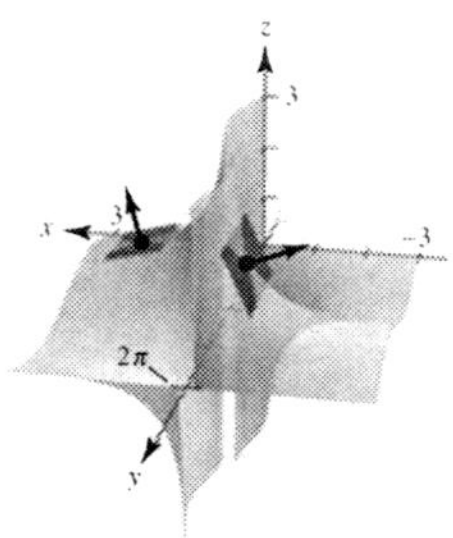

59. $f(x, y) = 6 - x^2 - \dfrac{y^2}{4}, g(x, y) = 2x + y$

(a) $F(x, y, z) = z + x^2 + \dfrac{y^2}{4} - 6 \quad G(x, y, z) = z - 2x - y$

$\nabla F(x, y, z) = 2x\mathbf{i} + \dfrac{1}{2}y\mathbf{j} + \mathbf{k} \quad \nabla G(x, y, z) = -2\mathbf{i} - \mathbf{j} + \mathbf{k}$

$\nabla F(1, 2, 4) = 2\mathbf{i} + \mathbf{j} + \mathbf{k} \quad \nabla G(1, 2, 4) = -2\mathbf{i} - \mathbf{j} + \mathbf{k}$

The cross product of these gradients is parallel to the curve of intersection.

$$\nabla F(1, 2, 4) \times \nabla G(1, 2, 4) = \begin{vmatrix} \mathbf{i} & \mathbf{j} & \mathbf{k} \\ 2 & 1 & 1 \\ -2 & -1 & 1 \end{vmatrix} = 2\mathbf{i} - 4\mathbf{j}$$

Using direction numbers 1, −2, 0, you get $x = 1 + t, y = 2 - 2t, z = 4$.

$$\cos\theta = \frac{\nabla F \cdot \nabla G}{\|\nabla F\|\|\nabla G\|} = \frac{-4 - 1 + 1}{\sqrt{6}\sqrt{6}} = \frac{-4}{6} \Rightarrow \theta \approx 48.2^\circ$$

(b)

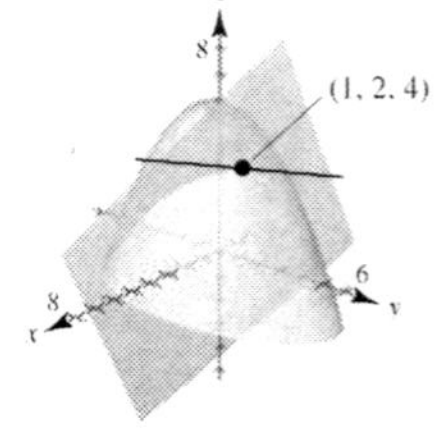

60. (a) $f(x, y) = \sqrt{16 - x^2 - y^2 + 2x - 4y}$

$g(x, y) = \dfrac{\sqrt{2}}{2}\sqrt{1 - 3x^2 + y^2 + 6x + 4y}$

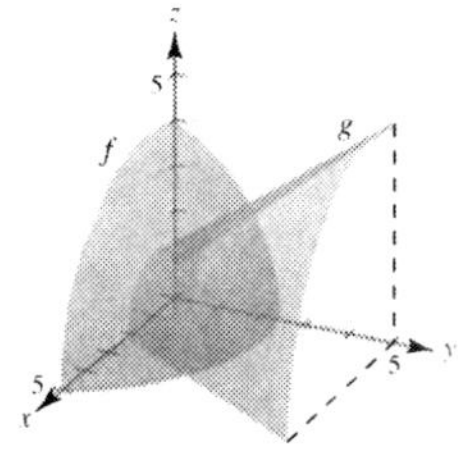

(b)

$$\begin{aligned} f(x, y) &= g(x, y) \\ 16 - x^2 - y^2 + 2x - 4y &= \frac{1}{2}\left(1 - 3x^2 + y^2 + 6x + 4y\right) \\ 32 - 2x^2 - 2y^2 + 4x - 8y &= 1 - 3x^2 + y^2 + 6x + 4y \\ x^2 - 2x + 31 &= 3y^2 + 12y \\ \left(x^2 - 2x + 1\right) + 42 &= 3\left(y^2 + 4y + 4\right) \\ (x - 1)^2 + 42 &= 3(y + 2)^2 \end{aligned}$$

To find points of intersection, let $x = 1$. Then

$$\begin{aligned} 3(y + 2)^2 &= 42 \\ (y + 2)^2 &= 14 \\ y &= -2 \pm \sqrt{14} \end{aligned}$$

$\nabla f\left(1, -2 + \sqrt{14}\right) = -\sqrt{2}\mathbf{j}$, $\nabla g\left(1, -2 + \sqrt{14}\right) = \left(1/\sqrt{2}\right)\mathbf{j}$. The normals to f and g at this point are $-\sqrt{2}\mathbf{j} - \mathbf{k}$ and $\left(-1/\sqrt{2}\right)\mathbf{j} - \mathbf{k}$, which are orthogonal.

Similarly, $\nabla f\left(1, -2 - \sqrt{14}\right) = \sqrt{2}\mathbf{j}$ and $\nabla g\left(1, -2 - \sqrt{14}\right) = \left(-1/\sqrt{2}\right)\mathbf{j}$ and the normals are $\sqrt{2}\mathbf{j} - \mathbf{k}$ and $\left(-1/\sqrt{2}\right)\mathbf{j} - \mathbf{k}$, which are also orthogonal.

(c) No, showing that the surfaces are orthogonal at 2 points does not imply that they are orthogonal at every point of intersection.

61. $F(x, y, z) = \dfrac{x^2}{a^2} + \dfrac{y^2}{b^2} + \dfrac{z^2}{c^2} - 1$

$F_x(x, y, z) = \dfrac{2x}{a^2}$

$F_y(x, y, z) = \dfrac{2y}{b^2}$

$F_z(x, y, z) = \dfrac{2z}{c^2}$

Plane: $\dfrac{2x_0}{a^2}(x - x_0) + \dfrac{2y_0}{b^2}(y - y_0) + \dfrac{2z_0}{c^2}(z - z_0) = 0$

$$\frac{x_0x}{a^2} + \frac{y_0y}{b^2} + \frac{z_0z}{c^2} = \frac{x_0^2}{a^2} + \frac{y_0^2}{b^2} + \frac{z_0^2}{c^2} = 1$$

62. $F(x, y, z) = \dfrac{x^2}{a^2} + \dfrac{y^2}{b^2} - \dfrac{z^2}{c^2} - 1$

$F_x(x, y, z) = \dfrac{2x}{a^2}$

$F_y(x, y, z) = \dfrac{2y}{b^2}$

$F_z(x, y, z) = \dfrac{-2z}{c^2}$

Plane: $\dfrac{2x_0}{a^2}(x - x_0) + \dfrac{2y_0}{b^2}(y - y_0) - \dfrac{2z_0}{c_2}(z - z_0) = 0$

$$\frac{x_0x}{a^2} + \frac{y_0y}{b^2} - \frac{z_0z}{c^2} = \frac{x_0^2}{a^2} + \frac{y_0^2}{b^2} - \frac{z_0^2}{c^2} = 1$$

63. $F(x, y, z) = a^2x^2 + b^2y^2 - z^2$

$F_x(x, y, z) = 2a^2x$

$F_y(x, y, z) = 2b^2y$

$F_z(x, y, z) = -2z$

Plane: $2a^2x_0(x - x_0) + 2b^2y_0(y - y_0) - 2z_0(z - z_0) = 0$

$$a^2x_0x + b^2y_0y - z_0z = a^2x_0^2 + b^2y_0^2 - z_0^2 = 0$$

So, the plane passes through the origin.

64. $$z = xf\left(\frac{y}{x}\right)$$

$$F(x, y, z) = xf\left(\frac{y}{x}\right) - z$$

$$F_x(x, y, z) = f\left(\frac{y}{x}\right) + xf'\left(\frac{y}{x}\right)\left(-\frac{y}{x^2}\right) = f\left(\frac{y}{x}\right) - \frac{y}{x}f'\left(\frac{y}{x}\right)$$

$$F_y(x, y, z) = xf'\left(\frac{y}{x}\right)\left(\frac{1}{x}\right) = f'\left(\frac{y}{x}\right)$$

$$F_x(x, y, z) = -1$$

Tangent plane at (x_0, y_0, z_0):

$$\left[f\left(\frac{y_0}{x_0}\right) - \frac{y_0}{x_0}f'\left(\frac{y_0}{x_0}\right)\right](x - x_0) + f'\left(\frac{y_0}{x_0}\right)(y - y_0) - (z - z_0) = 0$$

$$\left[f\left(\frac{y_0}{x_0}\right) - \frac{y_0}{x_0}f'\left(\frac{y_0}{x_0}\right)\right]x - x_0f\left(\frac{y_0}{x_0}\right) + y_0f'\left(\frac{y_0}{x_0}\right) + yf'\left(\frac{y_0}{x_0}\right) - y_0f'\left(\frac{y_0}{x_0}\right) - z + x_0f\left(\frac{y_0}{x_0}\right) = 0$$

$$\left[f\left(\frac{y_0}{x_0}\right) - \frac{y_0}{x_0}f'\left(\frac{y_0}{x_0}\right)\right]x + f'\left(\frac{y_0}{x_0}\right)y - z = 0$$

So, the plane passes through the origin $(x, y, z) = (0, 0, 0)$.

65. $f(x, y) = e^{x-y}$

$f_x(x, y) = e^{x-y}$, $f_y(x, y) = -e^{x-y}$

$f_{xx}(x, y) = e^{x-y}$, $f_{yy}(x, y) = e^{x-y}$, $f_{xy}(x, y) = -e^{x-y}$

(a) $P_1(x, y) \approx f(0, 0) + f_x(0, 0)x + f_y(0, 0)y = 1 + x - y$

(b) $P_2(x, y) \approx f(0, 0) + f_x(0, 0)x + f_y(0, 0)y + \frac{1}{2}f_{xx}(0, 0)x^2 + f_{xy}(0, 0)xy + \frac{1}{2}f_{yy}(0, 0)y^2 = 1 + x - y + \frac{1}{2}x^2 - xy + \frac{1}{2}y^2$

(c) If $x = 0$, $P_2(0, y) = 1 - y + \frac{1}{2}y^2$. This is the second-degree Taylor polynomial for e^{-y}.

If $y = 0$, $P_2(x, 0) = 1 + x + \frac{1}{2}x^2$. This is the second-degree Taylor polynomial for e^x.

(d)

x	y	$f(x, y)$	$P_1(x, y)$	$P_2(x, y)$
0	0	1	1	1
0	0.1	0.9048	0.9000	0.9050
0.2	0.1	1.1052	1.1000	1.1050
0.2	0.5	0.7408	0.7000	0.7450
1	0.5	1.6487	1.5000	1.6250

(e)

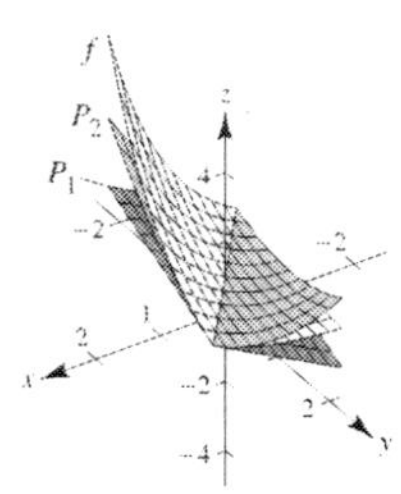

66. $f(x, y) = \cos(x + y)$

$f_x(x, y) = -\sin(x + y),\ f_y(x, y) = -\sin(x + y)$

$f_{xx}(x, y) = -\cos(x + y),\ f_{yy}(x, y) = -\cos(x + y),\ f_{xy}(x, y) = -\cos(x + y)$

(a) $P_1(x, y) \approx f(0, 0) + f_x(0, 0)x + f_y(0, 0)y = 1$

(b) $P_2(x, y) \approx f(0, 0) + f_x(0, 0)x + f_y(0, 0)y + \frac{1}{2}f_{xx}(0, 0)x^2 + f_{xy}(0, 0)xy + \frac{1}{2}f_{yy}(0, 0)y^2$
$= 1 - \frac{1}{2}x^2 - xy - \frac{1}{2}y^2$

(c) If $x = 0$, $P_2(0, y) = 1 - \frac{1}{2}y^2$. This is the second-degree Taylor polynomial for $\cos y$.

If $y = 0$, $P_2(x, 0) = 1 - \frac{1}{2}x^2$. This is the second-degree Taylor polynomial for $\cos x$.

(d)

x	y	$f(x, y)$	$P_1(x, y)$	$P_2(x, y)$
0	0	1	1	1
0	0.1	0.9950	1	0.9950
0.2	0.1	0.9553	1	0.9950
0.2	0.5	0.7648	1	0.7550
1	0.5	0.0707	1	−0.1250

(e)

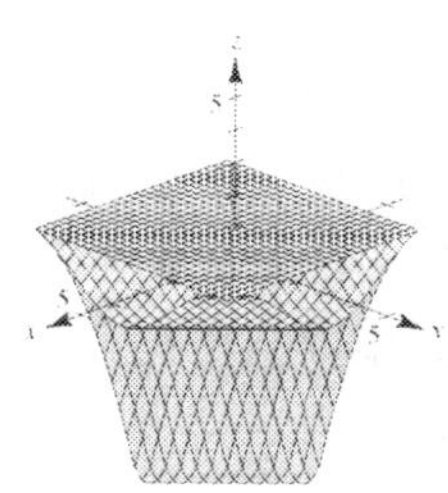

67. Given $z = f(x, y)$, then:

$$F(x, y, z) = f(x, y) - z = 0$$

$$\nabla F(x_0, y_0, z_0) = f_x(x_0, y_0)\mathbf{i} + f_y(x_0, y_0)\mathbf{j} - \mathbf{k}$$

$$\cos\theta = \frac{|\nabla F(x_0, y_0, z_0) \cdot \mathbf{k}|}{\|\nabla F(x_0, y_0, z_0)\|\|\mathbf{k}\|}$$

$$= \frac{|-1|}{\sqrt{[f_x(x_0, y_0)]^2 + [f_y(x_0, y_0)]^2 + (-1)^2}}$$

$$= \frac{1}{\sqrt{[f_x(x_0, y_0)]^2 + [f_y(x_0, y_0)]^2 + 1}}$$

68. Given $w = F(x, y, z)$ where F is differentiable at

(x_0, y_0, z_0) and $\nabla F(x_0, y_0, z_0) \neq \mathbf{0}$,

the level surface of F at (x_0, y_0, z_0) is of the form $F(x, y, z) = C$ for some constant C. Let

$G(x, y, z) = F(x, y, z) - C = 0.$

Then $\nabla G(x_0, y_0, z_0) = \nabla F(x_0, y_0, z_0)$ where $\nabla G(x_0, y_0, z_0)$ is normal to $F(x, y, z) - C = 0$ at (x_0, y_0, z_0). So, $\nabla F(x_0, y_0 z_0)$ is normal to the level surface through (x_0, y_0, z_0).

Section 13.8 Extrema of Functions of Two Variables

1. $g(x, y) = (x - 1)^2 + (y - 3)^2 \ge 0$

Relative minimum: $(1, 3, 0)$

Check: $g_x = 2(x - 1) = 0 \Rightarrow x = 1$

$g_y = 2(y - 3) = 0 \Rightarrow y = 3$

$g_{xx} = 2, g_{yy} = 2, g_{xy} = 0, d = (2)(2) - 0 = 4 > 0$

At critical point $(1, 3)$, $d > 0$ and $g_{xx} > 0 \Rightarrow$ relative minimum at $(1, 3, 0)$.

2. $g(x, y) = 5 - (x - 3)^2 - (y + 2)^2 \le 5$

Relative maximum: $(3, -2, 5)$

Check: $g_x = -2(x - 3) = 0 \Rightarrow x = 3$

$g_y = -2(y + 2) = 0 \Rightarrow y = -2$

$g_{xx} = -2, g_{yy} = -2, g_{xy} = 0$

$d = (-2)(-2) - 0 = 4 > 0$

At critical point $(3, -2)$, $d > 0$ and $g_{xx} < 0 \Rightarrow$ relative maximum at $(3, -2, 5)$.

3. $f(x, y) = \sqrt{x^2 + y^2 + 1} \ge 1$

Relative minimum: $(0, 0, 1)$

Check: $f_x = \dfrac{x}{\sqrt{x^2 + y^2 + 1}} = 0 \Rightarrow x = 0$

$$f_y = \frac{y}{\sqrt{x^2 + y^2 + 1}} = 0 \Rightarrow y = 0$$

$$f_{xx} = \frac{y^2 + 1}{(x^2 + y^2 + 1)^{3/2}}$$

$$f_{yy} = \frac{x^2 + 1}{(x^2 + y^2 + 1)^{3/2}}$$

$$f_{xy} = \frac{-xy}{(x^2 + y^2 + 1)^{3/2}}$$

At the critical point $(0, 0)$, $f_{xx} > 0$ and $f_{xx}f_{yy} - (f_{xy})^2 > 0$.

So, $(0, 0, 1)$ is a relative minimum.

4. $f(x, y) = \sqrt{25 - (x - 2)^2 - y^2} \le 5$

Relative maximum: $(2, 0, 5)$

Check: $f_x = -\dfrac{x - 2}{\sqrt{25 - (x - 2)^2 - y^2}} = 0 \Rightarrow x = 2$

$$f_y = -\frac{y}{\sqrt{25 - (x - 2)^2 - y^2}} = 0 \Rightarrow y = 0$$

$$f_{xx} = -\frac{25 - y^2}{\left[25 - (x - 2)^2 - y^2\right]^{3/2}}$$

$$f_{yy} = -\frac{25 - (x - 2)^2}{\left[25 - (x - 2)^2 - y^2\right]^{3/2}}$$

$$f_{xy} = -\frac{y(x - 2)}{\left[25 - (x - 2)^2 - y^2\right]^{3/2}}$$

At the critical point $(2, 0)$, $f_{xx} < 0$ and $f_{xx}f_{yy} - (f_{xy})^2 > 0$.

So, $(2, 0, 5)$ is a relative maximum.

5. $f(x, y) = x^2 + y^2 + 2x - 6y + 6 = (x + 1)^2 + (y - 3)^2 - 4 \ge -4$

Relative minimum: $(-1, 3, -4)$

Check: $f_x = 2x + 2 = 0 \Rightarrow x = -1$

$f_y = 2y - 6 = 0 \Rightarrow y = 3$

$f_{xx} = 2, f_{yy} = 2, f_{xy} = 0$

At the critical point $(-1, 3)$, $f_{xx} > 0$ and $f_{xx}f_{yy} - (f_{xy})^2 > 0$. So, $(-1, 3, -4)$ is a relative minimum.

6. $f(x, y) = -x^2 - y^2 + 10x + 12y - 64$

$= -(x^2 - 10x + 25) - (y^2 - 12y + 36) + 25 + 36 - 64 = -(x - 5)^2 - (y - 6)^2 - 3 \leq -3$

Relative maximum: $(5, 6, -3)$

Check: $f_x = -2x + 10 = 0 \Rightarrow x = 5$

$f_y = -2y + 12 = 0 \Rightarrow y = 6$

$f_{xx} = -2, f_{yy} = -2, f_{xy} = 0, d = (-2)(-2) - 0 = 4 > 0$

At critical point $(5, 6)$, $d > 0$ and $f_{xx} < 0 \Rightarrow$ relative maximum at $(5, 6, -3)$.

7. $h(x, y) = 80x + 80y - x^2 - y^2$

$\left.\begin{aligned} h_x &= 80 - 2x = 0 \\ h_y &= 80 - 2y = 0 \end{aligned}\right\} x = y = 40$

$h_{xx} = -2, h_{yy} = -2, h_{xy} = 0,$

$d = (-2)(-2) - 0 = 4 > 0$

At the critical point $(40, 40)$, $d > 0$ and

$h_{xx} < 0 \Rightarrow (40, 40, 3200)$ is a relative maximum.

8. $g(x, y) = x^2 - y^2 - x - y$

$\left.\begin{aligned} g_x &= 2x - 1 = 0 \\ g_y &= -2y - 1 = 0 \end{aligned}\right\} \begin{aligned} x &= 1/2 \\ y &= -1/2 \end{aligned}$

$g_{xx} = 2, g_{yy} = -2, g_{xy} = 0, d = 2(-2) - 0 = -4 < 0$

At the critical point $(1/2, -1/2)$, $d < 0$

$\Rightarrow (1/2, -1/2, 0)$ is a saddle point.

9. $g(x, y) = xy$

$\left.\begin{aligned} g_x &= y \\ g_y &= x \end{aligned}\right\} x = 0 \text{ and } y = 0$

$g_{xx} = 0, g_{yy} = 0, g_{xy} = 1$

At the critical point $(0, 0)$, $g_{xx}g_{yy} - (g_{xy})^2 < 0$.

So, $(0, 0, 0)$ is a saddle point.

10. $h(x, y) = x^2 - 3xy - y^2$

$\left.\begin{aligned} h_x &= 2x - 3y = 0 \\ h_y &= -3x - 2y = 0 \end{aligned}\right\}$ Solving simultaneously yields $x = 0$ and $y = 0$.

$h_{xx} = 2, h_{yy} = -2, h_{xy} = -3$

At the critical point $(0, 0)$, $h_{xx}h_{yy} - (h_{xy})^2 < 0$.

So, $(0, 0, 0)$ is a saddle point.

11. $f(x, y) = -3x^2 - 2y^2 + 3x - 4y + 5$

$f_x = -6x + 3 = 0$ when $x = \frac{1}{2}$.

$f_y = -4y - 4 = 0$ when $y = -1$.

$f_{xx} = -6, f_{yy} = -4, f_{xy} = 0$

At the critical point $(\frac{1}{2}, -1)$, $f_{xx} < 0$

and $f_{xx}f_{yy} - (f_{xy})^2 > 0$.

So, $(\frac{1}{2}, -1, \frac{31}{4})$ is a relative maximum.

12. $f(x, y) = 2x^2 + 2xy + y^2 + 2x - 3$

$\left.\begin{aligned} f_x &= 4x + 2y + 2 = 0 \\ f_y &= 2x + 2y = 0 \end{aligned}\right\}$ Solving simultaneously yields $x = -1$ and $y = 1$.

$f_{xx} = 4, f_{yy} = 2, f_{xy} = 2$

At the critical point $(-1, 1)$, $f_{xx} > 0$

and $f_{xx}f_{yy} - (f_{xy})^2 > 0$.

So, $(-1, 1, -4)$ is a relative minimum.

13. $f(x, y) = z = x^2 + xy + \frac{1}{2}y^2 - 2x + y$

$\left.\begin{aligned} f_x &= 2x + y - 2 = 0 \\ f_y &= x + y + 1 = 0 \end{aligned}\right\}$ Solving simultaneously yields $x = 3, y = -4$

$f_{xx} = 2, f_{yy} = 1, f_{xy} = 1, d = 2(1) - 1 = 1 > 0.$

At the critical point $(3, -4)$, $d > 0$

and $f_{xx} > 0 \Rightarrow (3, -4, -5)$ is a relative minimum.

14. $f(x, y) = -5x^2 + 4xy - y^2 + 16x + 10$

$\left.\begin{aligned} f_x &= -10x + 4y + 16 = 0 \\ f_y &= 4x - 2y = 0 \end{aligned}\right\}$ Solving simultaneously yields $x = 8$ and $y = 16$.

$f_{xx} = -10, f_{yy} = -2, f_{xy} = 4$

At the critical point $(8, 16)$, $f_{xx} < 0$

and $f_{xx}f_{yy} - (f_{xy})^2 > 0$.

So, $(8, 16, 74)$ is a relative maximum.

15. $f(x, y) = \sqrt{x^2 + y^2}$

$$\left.\begin{aligned} f_x &= \frac{x}{\sqrt{x^2 + y^2}} = 0 \\ f_y &= \frac{y}{\sqrt{x^2 + y^2}} = 0 \end{aligned}\right\} x = y = 0$$

Because $f(x, y) \geq 0$ for all (x, y) and $f(0, 0) = 0$, $(0, 0, 0)$ is a relative minimum.

16. $h(x, y) = (x^2 + y^2)^{1/3} + 2$

$$\left.\begin{aligned} h_x &= \frac{2x}{3(x^2 + y^2)^{2/3}} = 0 \\ h_y &= \frac{2y}{3(x^2 + y^2)^{2/3}} = 0 \end{aligned}\right\} x = 0,\ y = 0$$

Because $h(x, y) \geq 2$ for all (x, y), $(0, 0, 2)$ is a relative minimum.

17. $f(x, y) = x^2 - xy - y^2 - 3x - y$

$f_x = 2x - y - 3 = 0$

$f_y = -x - 2y - 1 = 0$

Solving simultaneously yields $x = 1$, $y = -1$.

$f_{xx} = 2,\ f_{yy} = -2,\ f_{xy} = -1$

$d = (2)(-2) - (-1)^2 = -5 < 0$

At the critical point $(1, -1)$, $d < 0 \Rightarrow (1, -1, -1)$ is a saddle point.

18. $f(x, y) = 2xy - \frac{1}{2}(x^4 + y^2) + 1$

$$\left.\begin{aligned} f_x &= 2y - 2x^3 \\ f_y &= 2x - 2y^3 \end{aligned}\right\} \text{Solving by substitution yields 3 critical points: } (0, 0), (1, 1), (-1, -1)$$

$f_{xx} = -6x^2,\ f_{yy} = -6y^2,\ f_{xy} = 2$

At $(0, 0)$, $f_{xx}f_{yy} - (f_{xy})^2 < 0 \Rightarrow (0, 0, 1)$ saddle point.

At $(1, 1)$, $f_{xx}f_{yy} - (f_{xy})^2 > 0$ and $f_{xx} < 0 \Rightarrow (1, 1, 2)$ relative maximum.

At $(-1, -1)$, $f_{xx}f_{yy} - (f_{xy})^2 > 0$ and $f_{xx} < 0 \Rightarrow (-1, -1, 2)$ relative maximum.

19. $f(x, y) = e^{-x} \sin y$

$$\left.\begin{aligned} f_x &= -e^{-x} \sin y = 0 \\ f_y &= e^{-x} \cos y = 0 \end{aligned}\right\}$$

Because $e^{-x} > 0$ for all x and $\sin y$ and $\cos y$ are never both zero for a given value of y, there are no critical points.

20. $f(x, y) = \left(\frac{1}{2} - x^2 + y^2\right)e^{1-x^2-y^2}$

$$\left.\begin{aligned} f_x &= (2x^3 - 2xy^2 - 3x)e^{1-x^2-y^2} = 0 \\ f_y &= (2x^2y - 2y^3 + y)e^{1-x^2-y^2} = 0 \end{aligned}\right\} \text{Solving yields the critical points } (0, 0), \left(0, \pm\frac{\sqrt{2}}{2}\right), \left(\pm\frac{\sqrt{6}}{2}, 0\right).$$

$f_{xx} = (-4x^4 + 4x^2y^2 + 12x^2 - 2y^2 - 3)e^{1-x^2-y^2}$

$f_{yy} = (4y^4 - 4x^2y^2 + 2x^2 - 8y^2 + 1)e^{1-x^2-y^2}$

$f_{xy} = (-4x^3y + 4xy^3 + 2xy)e^{1-x^2-y^2}$

At the critical point $(0, 0)$, $f_{xx}f_{yy} - (f_{xy})^2 < 0$. So, $(0, 0, e/2)$ is a saddle point. At the critical points $(0, \pm\sqrt{2}/2)$, $f_{xx} < 0$ and $f_{xx}f_{yy} - (f_{xy})^2 > 0$. So, $(0, \pm\sqrt{2}/2, \sqrt{e})$ are relative maxima. At the critical points $(\pm\sqrt{6}/2, 0)$, $f_{xx} > 0$ and $f_{xx}f_{yy} - (f_{xy})^2 > 0$. So, $(\pm\sqrt{6}/2, 0, -\sqrt{e}/e)$ are relative minima.

21. $z = \dfrac{-4x}{x^2 + y^2 + 1}$

Relative minimum: $(1, 0, -2)$

Relative maximum: $(-1, 0, 2)$

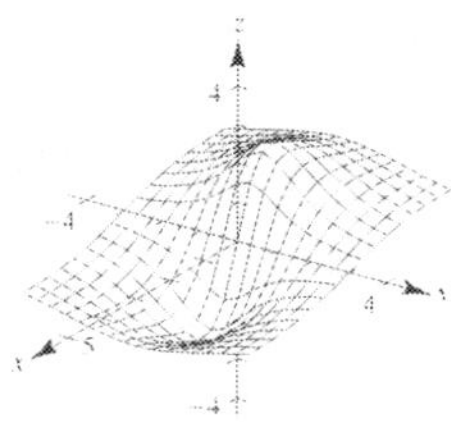

22. $f(x, y) = y^3 - 3yx^2 - 3y^2 - 3x^2 + 1$

Relative maximum: $(0, 0, 1)$

Saddle points:

$(0, 2, -3), (\pm\sqrt{3}, -1, -3)$

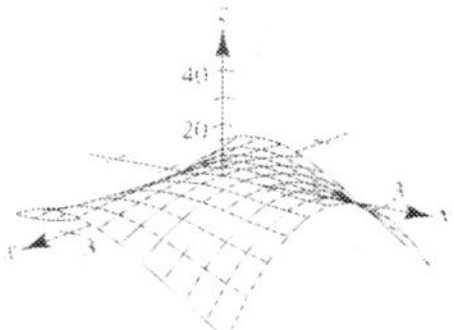

23. $z = (x^2 + 4y^2)e^{1-x^2-y^2}$

Relative minimum: $(0, 0, 0)$

Relative maxima: $(0, \pm 1, 4)$

Saddle points: $(\pm 1, 0, 1)$

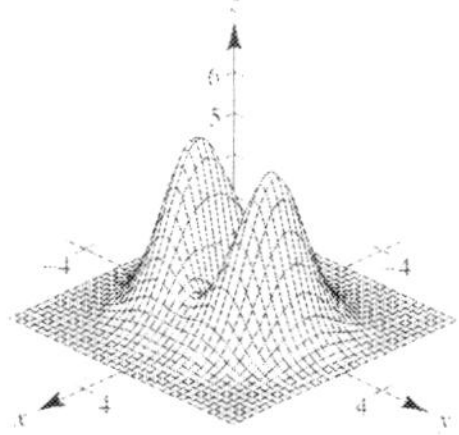

24. $z = e^{xy}$

Saddle point: $(0, 0, 1)$

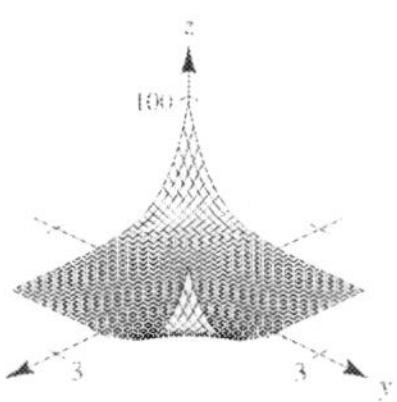

25. $z = \dfrac{(x - y)^4}{x^2 + y^2} \geq 0$. $z = 0$ if $x = y \neq 0$.

Relative minimum at all points $(x, x), x \neq 0$.

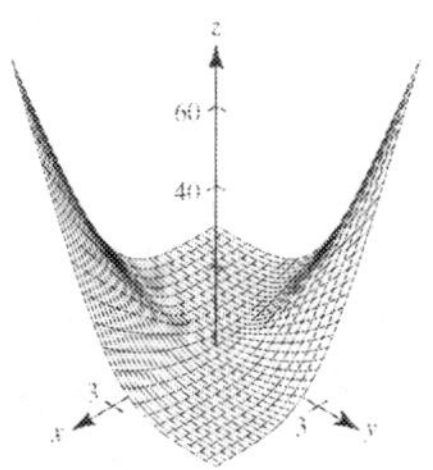

26. $z = \dfrac{(x^2 - y^2)^2}{x^2 + y^2} \geq 0$. $z = 0$ if $x^2 = y^2 \neq 0$.

Relative minima at all points (x, x) and $(x, -x)$, $x \neq 0$.

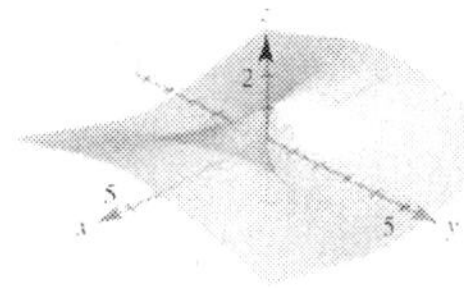

27. $f_{xx}f_{yy} - (f_{xy})^2 = (9)(4) - 6^2 = 0$

Insufficient information.

28. $f_{xx} < 0$ and $f_{xx}f_{yy} - (f_{xy})^2 = (-3)(-8) - 2^2 > 0$

f has a relative maximum at (x_0, y_0)

29. $f_{xx}f_{yy} - (f_{xy})^2 = (-9)(6) - 10^2 < 0$

f has a saddle point at (x_0, y_0).

30. $f_{xx} > 0$ and $f_{xx}f_{yy} - (f_{xy})^2 = (25)(8) - 10^2 > 0$

f has a relative minimum at (x_0, y_0)

31. $d = f_{xx}f_{yy} - f_{xy}^2 = (2)(8) - f_{xy}^2 = 16 - f_{xy}^2 > 0$

$\Rightarrow f_{xy}^2 < 16 \Rightarrow -4 < f_{xy} < 4$

32. $d = f_{xx}f_{yy} - f_{xy}^2 < 0$ if f_{xx} and f_{yy} have opposite signs. So, $(a, b, f(a, b))$ is a saddle point. For example, consider $f(x, y) = x^2 - y^2$ and $(a, b) = (0, 0)$.

33. $f(x, y) = x^3 + y^3$

(a) $\left.\begin{aligned} f_x &= 3x^2 = 0 \\ f_y &= 3y^2 = 0 \end{aligned}\right\} x = y = 0$

Critical point: $(0, 0)$

(b) $f_{xx} = 6x, f_{yy} = 6y, f_{xy} = 0$

At $(0, 0)$, $f_{xx}f_{yy} - (f_{xy})^2 = 0$.

$(0, 0, 0)$ is a saddle point.

(c) Test fails at $(0, 0)$.

(d)

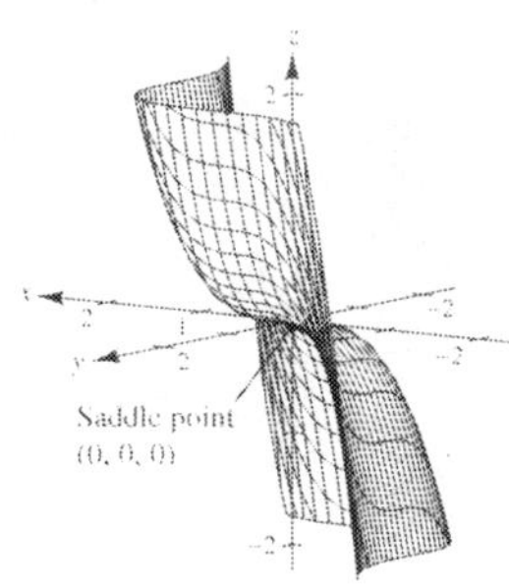

34. $f(x, y) = x^3 + y^3 - 6x^2 + 9y^2 + 12x + 27y + 19$

(a) $\left.\begin{aligned} f_x &= 3x^2 - 12x + 12 = 0 \\ f_y &= 3y^2 + 18y + 27 = 0 \end{aligned}\right\}$ Solving yields $x = 2$ and $y = -3$.

(b) $f_{xx} = 6x - 12,\ f_{yy} = 6y + 18,\ f_{xy} = 0$

At $(2, -3)$, $f_{xx}f_{yy} - (f_{xy})^2 = 0$.

$(2, -3, 0)$ is a saddle point.

(c) Test fails at $(2, -3)$.

(d)

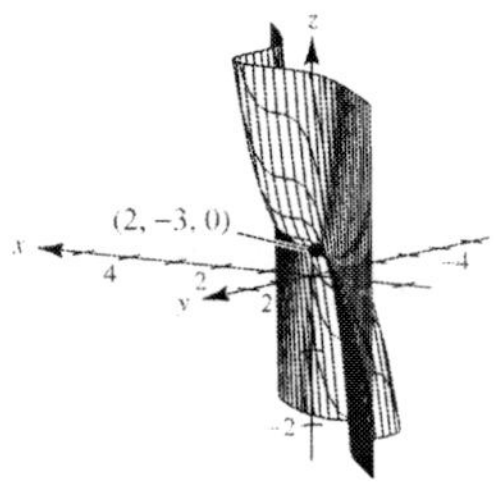

35. $f(x, y) = (x - 1)^2(y + 4)^2 \ge 0$

(a) $\left.\begin{aligned} f_x &= 2(x - 1)(y + 4)^2 = 0 \\ f_y &= 2(x - 1)^2(y + 4) = 0 \end{aligned}\right\}$ critical points: $(1, a)$ and $(b, -4)$

(b) $f_{xx} = 2(y + 4)^2$

$f_{yy} = 2(x - 1)^2$

$f_{xy} = 4(x - 1)(y + 4)$

At both $(1, a)$ and $(b, -4)$, $f_{xx}f_{yy} - (f_{xy})^2 = 0$.

Because $f(x, y) \ge 0$, there are absolute minima at $(1, a, 0)$ and $(b, -4, 0)$.

(c) Test fails at $(1, a)$ and $(b, -4)$.

(d)

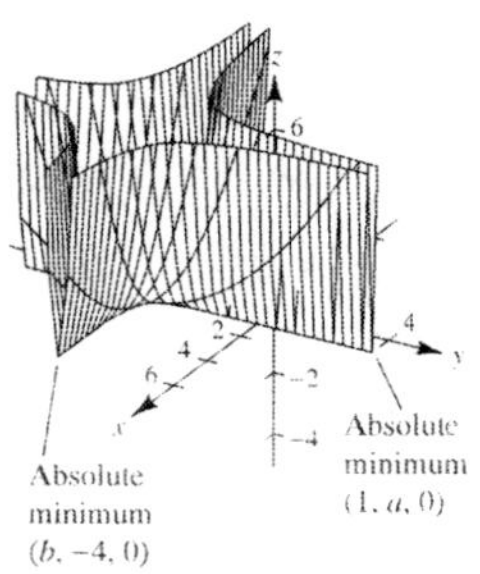

36. $f(x, y) = \sqrt{(x - 1)^2 + (y + 2)^2} \ge 0$

(a) $\left.\begin{aligned} f_x &= \frac{x - 1}{\sqrt{(x - 1)^2 + (y + 2)^2}} = 0 \\ f_y &= \frac{y + 2}{\sqrt{(x - 1)^2 + (y + 2)^2}} = 0 \end{aligned}\right\}$ Solving yields $x = 1$ and $y = -2$.

(b) $f_{xx} = \dfrac{(y + 2)^2}{\left[(x - 1)^2 + (y + 2)^2\right]^{3/2}}$

$f_{yy} = \dfrac{(x - 1)^2}{\left[(x - 1)^2 + (y + 2)^2\right]^{3/2}}$

$f_{xy} = \dfrac{(x - 1)(y + 2)}{\left[(x - 1)^2 + (y + 2)^2\right]^{3/2}}$

At $(1, -2)$, $f_{xx}f_{yy} - (f_{xy})^2$ is undefined.

$(1, -2, 0)$ is an absolute minimum.

(c) Test fails at $(1, -2)$.

(d)

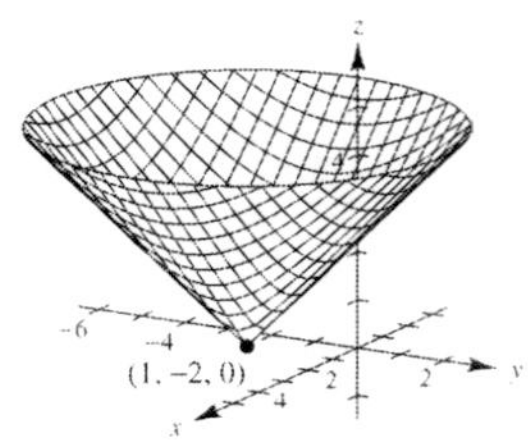

37. $f(x, y) = x^{2/3} + y^{2/3} \ge 0$

(a) $\left.\begin{aligned} f_x &= \frac{2}{3x^{1/3}} \\ f_y &= \frac{2}{3y^{1/3}} \end{aligned}\right\}$ f_x and f_y are undefined at $x = 0$ and $y = 0$. Critical point: $(0, 0)$

(b) $f_{xx} = \dfrac{-2}{9x^{4/3}},\ f_{yy} = \dfrac{-2}{9y^{4/3}},\ f_{xy} = 0$

At $(0, 0)$, $f_{xx}f_{yy} - (f_{xy})^2$ is undefined.

$(0, 0, 0)$ is an absolute minimum.

(c) Test fails at $(0, 0)$.

(d)

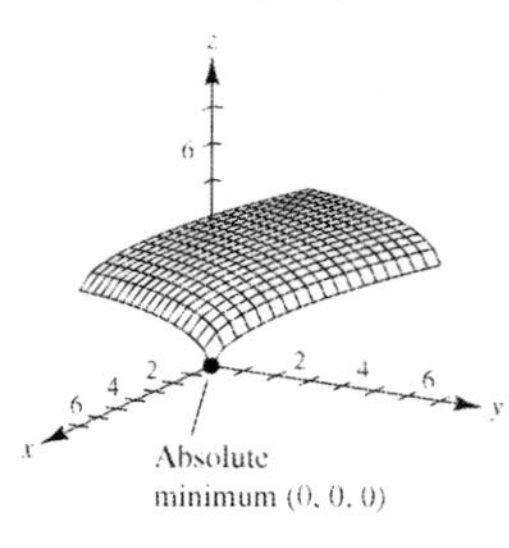

38. $f(x, y) = (x^2 + y^2)^{2/3} \geq 0$

(a) $f_x = \dfrac{4x}{3(x^2 + y^2)^{1/3}}$, $f_y = \dfrac{4y}{3(x^2 + y^2)^{1/3}}$ — f_x and f_y are undefined at $x = 0$, $y = 0$. Critical Point: $(0, 0)$

(b) $f_{xx} = \dfrac{4(x^2 + 3y^2)}{9(x^2 + y^2)^{4/3}}$

$f_{yy} = \dfrac{4(3x^2 + y^2)}{9(x^2 + y^2)^{4/3}}$

$f_{xy} = \dfrac{-8xy}{9(x^2 + y^2)^{4/3}}$

At $(0, 0)$, $f_{xx}f_{yy} - (f_{xy})^2$ is undefined.

$(0, 0, 0)$ is an absolute minimum.

(c) Test fails at $(0, 0)$.

(d)

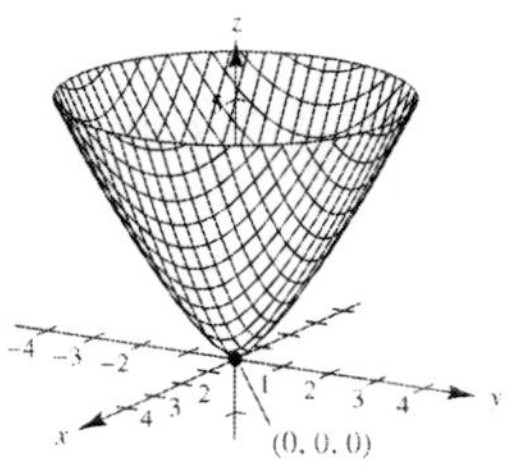

39. $f(x, y, z) = x^2 + (y - 3)^2 + (z + 1)^2 \geq 0$

$f_x = 2x = 0$
$f_y = 2(y - 3) = 0$
$f_z = 2(z + 1) = 0$

Solving yields the critical point $(0, 3, -1)$.

Absolute minimum: 0 at $(0, 3, -1)$

40. $f(x, y, z) = 9 - [x(y - 1)(z + 2)]^2 \leq 9$

The absolute maximum value of f is 9, and realized at all points where $x(y - 1)(z + 2) = 0$.

So, the critical points are of the form $(0, a, b)$, $(c, 1, d)$, $(e, f, -z)$

where a, b, c, d, e, f are real numbers.

41. $f(x, y) = x^2 - 4xy + 5, R = \{(x, y): 1 \le x \le 4, 0 \le y \le 2\}$

$$\left.\begin{aligned} f_x &= 2x - 4y = 0 \\ f_y &= -4x = 0 \end{aligned}\right\} x = y = 0 \quad \text{(not in region R)}$$

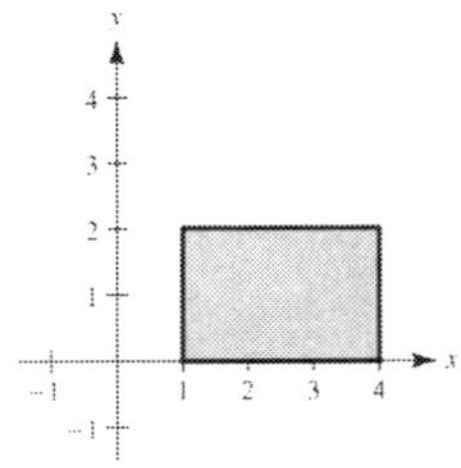

Along $y = 0, 1 \le x \le 4: f = x^2 + 5, f(1, 0) = 6, f(4, 0) = 21.$

Along $y = 2, 1 \le x \le 4: f = x^2 - 8x + 5, f' = 2x - 8 = 0$

$f(1, 2) = -2, f(4, 2) = -11.$

Along $x = 1, 0 \le y \le 2: f = -4y + 6, f(1, 0) = 6, f(1, 2) = -2.$

Along $x = 4, 0 \le y \le 2: f = 21 - 16y, f(4, 0) = 21, f(4, 2) = -11.$

So, the maximum is $(4, 0, 21)$ and the minimum is $(4, 2, -11)$.

42. $f(x, y) = x^2 + xy, R = \{(x, y): |x| \le 2, |y| \le 1\}$

$$\left.\begin{aligned} f_x &= 2x + y = 0 \\ f_y &= x = 0 \end{aligned}\right\} x = y = 0$$

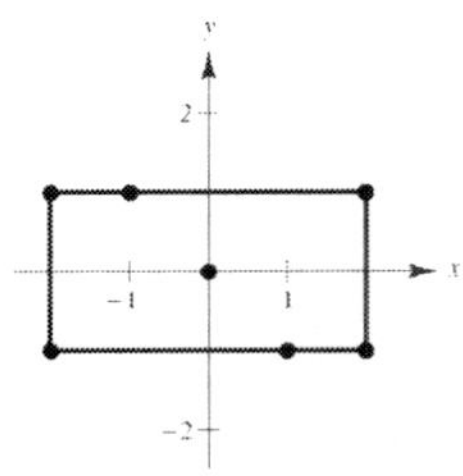

$f(0, 0) = 0$

Along $y = 1, -2 \le x \le 2, f = x^2 + x, f' = 2x + 1 = 0 \Rightarrow x = -\frac{1}{2}.$

Thus, $f(-2, 1) = 2, f(-\frac{1}{2}, 1) = -\frac{1}{4}$ and $f(2, 1) = 6.$

Along $y = -1, -2 \le x \le 2, f = x^2 - x, f' = 2x - 1 = 0 \Rightarrow x = \frac{1}{2}.$

Thus, $f(-2, -1) = 6, f(\frac{1}{2}, -1) = -\frac{1}{4}, f(2, -1) = 2.$

Along $x = 2, -1 \le y \le 1, f = 4 + 2y \Rightarrow f' = 2 \ne 0.$

Along $x = -2, -1 \le y \le 1, f = 4 - 2y \Rightarrow f' = -2 \ne 0.$

So, the maxima are $f(2, 1) = 6$ and $f(-2, -1) = 6$ and the minima are $f(-\frac{1}{2}, 1) = -\frac{1}{4}$ and $f(\frac{1}{2}, -1) = -\frac{1}{4}.$

43. $f(x, y) = 12 - 3x - 2y$ has no critical points. On the line $y = x + 1, 0 \le x \le 1,$

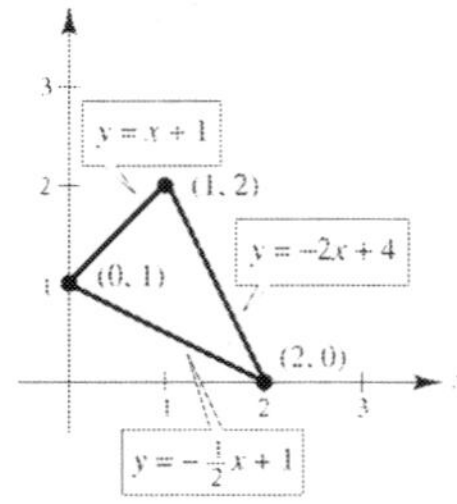

$f(x, y) = f(x) = 12 - 3x - 2(x + 1) = -5x + 10$

and the maximum is 10, the minimum is 5. On the line $y = -2x + 4, 1 \le x \le 2,$

$f(x, y) = f(x) = 12 - 3x - 2(-2x + 4) = x + 4$

and the maximum is 6, the minimum is 5. On the line $y = -\frac{1}{2}x + 1, 0 \le x \le 2,$

$f(x, y) = f(x) = 12 - 3x - 2(-\frac{1}{2}x + 1) = -2x + 10$

and the maximum is 10, the minimum is 6.

Absolute maximum: 10 at $(0, 1)$

Absolute minimum: 5 at $(1, 2)$

44. $f(x, y) = (2x - y)^2$

$f_x = 4(2x - y) = 0 \Rightarrow 2x = y$

$f_y = -2(2x - y) = 0 \Rightarrow 2x = y$

On the line $y = x + 1, 0 \le x \le 1$,

$f(x, y) = f(x) = (2x - (x + 1))^2 = (x - 1)^2$

and the maximum is 1, the minimum is 0. On the line $y = -\frac{1}{2}x + 1, 0 \le x \le 2$,

$f(x, y) = f(x) = (2x - (-\frac{1}{2}x + 1))^2 = (\frac{5}{2}x - 1)^2$

and the maximum is 16, the minimum is 0. On the line $y = -2x + 4, 1 \le x \le 2$,

$f(x, y) = f(x) = (2x - (-2x + 4))^2 = (4x - 4)^2$

and the maximum is 16, the minimum is 0.

Absolute maximum: 16 at $(2, 0)$

Absolute Minimum: 0 at $(1, 2)$ and along the line $y = 2x$.

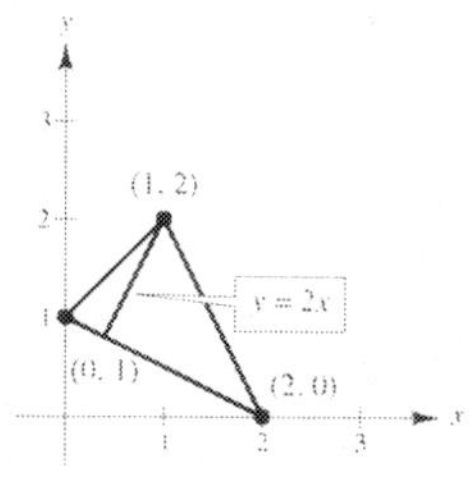

45. $f(x, y) = 3x^2 + 2y^2 - 4y$

$\left.\begin{array}{l} f_x = 6x = 0 \Rightarrow x = 0 \\ f_y = 4y - 4 = 0 \Rightarrow y = 1 \end{array}\right\} f(0, 1) = -2$

On the line $y = 4, -2 \le x \le 2$,

$f(x, y) = f(x) = 3x^2 + 32 - 16 = 3x^2 + 16$

and the maximum is 28, the minimum is 16. On the curve $y = x^2, -2 \le x \le 2$,

$f(x, y) = f(x) = 3x^2 + 2(x^2)^2 - 4x^2 = 2x^4 - x^2 = x^2(2x^2 - 1)$

and the maximum is 28, the minimum is $-\frac{1}{8}$.

Absolute maximum: 28 at $(\pm 2, 4)$

Absolute minimum: -2 at $(0, 1)$

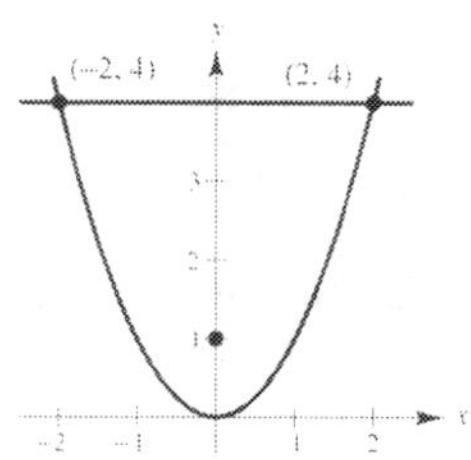

46. $f(x, y) = 2x - 2xy + y^2$

$\left.\begin{array}{l} f_x = 2 - 2y = 0 \Rightarrow y = 1 \\ f_y = 2y - 2x = 0 \Rightarrow y = x \Rightarrow x = 1 \end{array}\right\} f(1, 1) = 1$

On the line $y = 1, -1 \le x \le 1$,

$f(x, y) = f(x) = 2x - 2x + 1 = 1.$

On the curve $y = x^2, -1 \le x \le 1$

$f(x, y) = f(x) = 2x - 2x(x^2) + (x^2)^2 = x^4 - 2x^3 + 2x$

and the maximum is 1, the minimum is $-\frac{11}{16}$.

Absolute maximum: 1 at $(1, 1)$ and on $y = 1$

Absolute minimum: $-\frac{11}{16} = -0.6875$ at $\left(-\frac{1}{2}, \frac{1}{4}\right)$

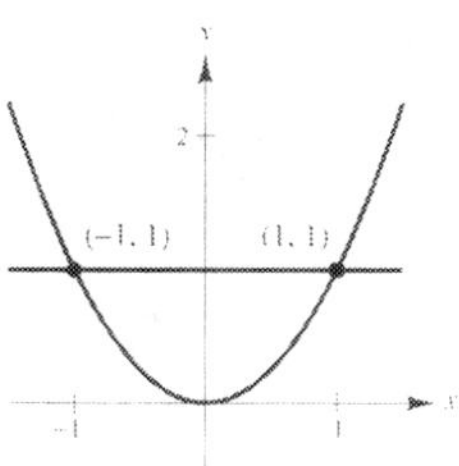

47. $f(x, y) = x^2 + 2xy + y^2, R = \{(x, y): |x| \le 2, |y| \le 1\}$

$$\left.\begin{aligned} f_x &= 2x + 2y = 0 \\ f_y &= 2x + 2y = 0 \end{aligned}\right\} y = -x$$

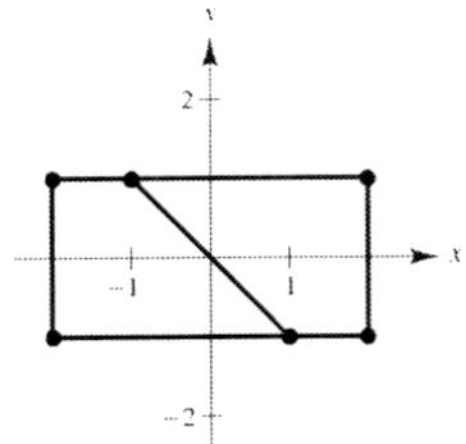

$f(x, -x) = x^2 - 2x^2 + x^2 = 0$

Along $y = 1, -2 \le x \le 2$,

$f = x^2 + 2x + 1, f' = 2x + 2 = 0 \Rightarrow x = -1, f(-2, 1) = 1, f(-1, 1) = 0, f(2, 1) = 9.$

Along $y = -1, -2 \le x \le 2$,

$f = x^2 - 2x + 1, f' = 2x - 2 = 0 \Rightarrow x = 1, f(-2, -1) = 9, f(1, -1) = 0, f(2, -1) = 1.$

Along $x = 2, -1 \le y \le 1, f = 4 + 4y + y^2, f' = 2y + 4 \ne 0.$

Along $x = -2, -1 \le y \le 1, f = 4 - 4y + y^2, f' = 2y - 4 \ne 0.$

So, the maxima are $f(-2, -1) = 9$ and $f(2, 1) = 9$, and the minima are $f(x, -x) = 0, -1 \le x \le 1.$

48. $f(x, y) = \dfrac{4xy}{(x^2 + 1)(y^2 + 1)}, R = \{(x, y): 0 \le x \le 1, 0 \le y \le 1\}$

$$f_x = \frac{4(1 - x^2)y}{(y^2 + 1)(x^2 + 1)^2} = 0 \Rightarrow x = 1 \text{ or } y = 0$$

$$f_y = \frac{4(1 - y^2)x}{(x^2 + 1)(y^2 + 1)^2} \Rightarrow x = 0 \text{ or } y = 1$$

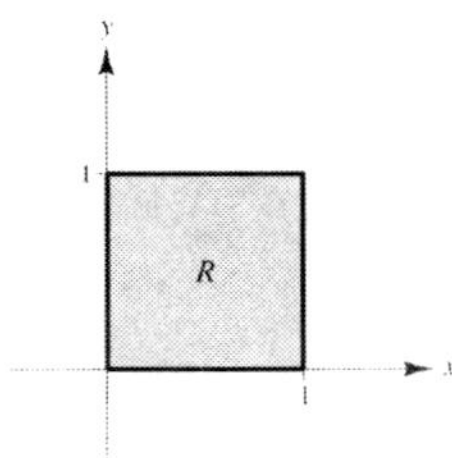

For $x = 0, y = 0$, also, and $f(0, 0) = 0$.

For $x = 1, y = 1, f(1, 1) = 1$.

The absolute maximum is $1 = f(1, 1)$.

The absolute minimum is $0 = f(0, 0)$. (In fact, $f(0, y) = f(x, 0) = 0$.)

49. (a) The function f has a relative minimum at (x_0, y_0) if $f(x, y) \ge f(x_0, y_0)$ for all (x, y) in an open disk containing (x_0, y_0).

(b) The function f has a relative maximum at (x_0, y_0) if $f(x, y) \le f(x_0, y_0)$ for all (x, y) in an open disk containing (x_0, y_0).

(c) The point (x_0, y_0) is a critical point if either

(1) $f_x(x_0, y_0) = 0$ and $f_y(x_0, y_0) = 0$, or

(2) $f_x(x_0, y_0)$ or $f_y(x_0, y_0)$ does not exist.

(d) A critical point is a saddle point if it is neither a relative minimum nor a relative maximum.

50.

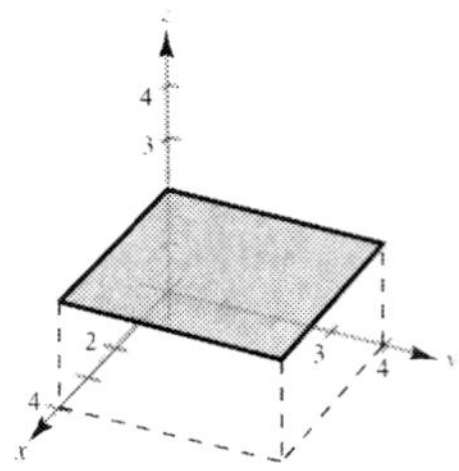

Extrema at all (x, y)

51.

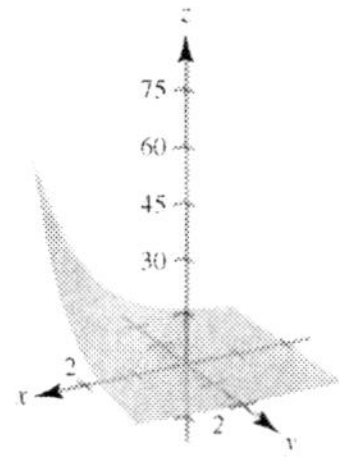

No extrema

52.

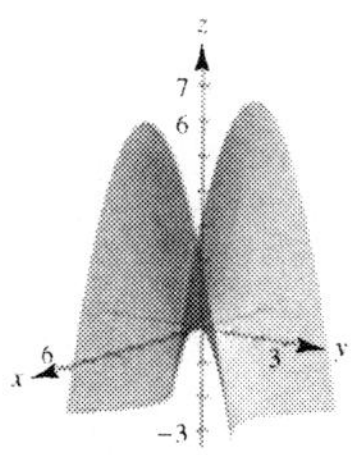

Saddle point

53. $f(x, y) = x^2 - y^2, g(x, y) = x^2 + y^2$

(a) $f_x = 2x = 0, f_y = -2y = 0 \Rightarrow (0, 0)$ is a critical point.

$g_x = 2x = 0, g_y = 2y = 0 \Rightarrow (0, 0)$ is a critical point.

(b) $f_{xx} = 2, f_{yy} = -2, f_{xy} = 0$

$d = 2(-2) - 0 < 0 \Rightarrow (0, 0)$ is a saddle point.

$g_{xx} = 2, g_{yy} = 2, g_{xy} = 0$

$d = 2(2) - 0 > 0 \Rightarrow (0, 0)$ is a relative minimum.

54. A and B are relative extrema.

C and D are saddle points.

55. False.

Let $f(x, y) = 1 - |x| - |y|$.

$(0, 0, 1)$ is a relative maximum, but $f_x(0, 0)$ and $f_y(0, 0)$ do not exist.

56. False. Consider $f(x, y) = x^2 - y^2$.

Then $f_x(0, 0) = f_y(0, 0) = 0$, but $(0, 0, 0)$ is a saddle point.

57. False. Let $f(x, y) = x^2y^2$ (See Example 4 on page 940).

58. False.

Let $f(x, y) = x^4 - 2x^2 + y^2$.

Relative minima: $(\pm 1, 0, -1)$

Saddle point: $(0, 0, 0)$

Section 13.9 Applications of Extrema of Functions of Two Variables

1. A point on the plane is given by $(x, y, z) = (x, y, 3 - x + y)$. The square of the distance from $(0, 0, 0)$ to this point is

$$S = x^2 + y^2 + (3 - x + y)^2.$$
$$S_x = 2x - 2(3 - x + y)$$
$$S_y = 2y + 2(3 - x + y)$$

From the equations $S_x = 0$ and $S_y = 0$ we obtain

$$4x - 2y = 6$$
$$-2x + 4y = -6.$$

Solving simultaneously, we have $x = 1, y = -1, z = 1$.

So, the distance is $\sqrt{1^2 + (-1)^2 + 1^2} = \sqrt{3}$.

2. A point on the plane is given by $(x, y, z) = (x, y, 3 - x + y)$. The square of the distance from $(1, 2, 3)$ to this point is

$$S = (x - 1)^2 + (y - 2)^2 + (3 - x + y - 3)^2$$
$$= (x - 1)^2 + (y - 2)^2 + (y - x)^2.$$

$$S_x = 2(x - 1) - 2(y - x)$$
$$S_y = 2(y - 2) + 2(y - x)$$

From the equation $S_x = 0$ and $S_y = 0$ we obtain

$$4x - 2y = 2$$
$$-2x + 4y = 4.$$

Solving simultaneously, we have $x = 4/3, y = 5/3, z = 10/3$.

So, the distance is

$$\sqrt{\left(\frac{4}{3} - 1\right)^2 + \left(\frac{5}{3} - 2\right)^2 + \left(\frac{5}{3} - \frac{4}{3}\right)^2} = \frac{\sqrt{13}}{3}.$$

3. A point on the surface is given by $(x, y, z) = \left(x, y, \sqrt{1 - 2x - 2y}\right)$. The square of the distance from $(-2, -2, 0)$ to a point on the surface is given by

$$S = (x + 2)^2 + (y + 2)^2 + \left(\sqrt{1 - 2x - 2y} - 0\right)^2 = (x + 2)^2 + (y + 2)^2 + 1 - 2x - 2y.$$

$$S_x = 2(x + 2) - 2$$
$$S_y = 2(y + 2) - 2$$

From the equations $S_x = 0$ and $S_y = 0$, we obtain $\left.\begin{array}{r} 2x + 2 = 0 \\ 2y + 2 = 0 \end{array}\right\} \Rightarrow x = y = -1, z = \sqrt{5}.$

So, the distance is $\sqrt{(-1 + 2)^2 + (-1 + 2)^2 + \left(\sqrt{5}\right)^2} = \sqrt{7}.$

4. A point on the surface is given by $(x, y, z) = \left(x, y, \sqrt{1 - 2x - 2y}\right)$. The square of the distance from $(-4, 1, 0)$ to a point on the surface is given by

$$S = (x + 4)^2 + (y - 1)^2 + (1 - 2x - 2y).$$

$$S_x = 2(x + 4) - 2 = 2x + 6$$

$$S_y = 2(y - 1) - 2 = 2y - 4$$

From the equations $S_x = S_y = 0$, we obtain $x = -3, y = 2$. Hence, $z = \sqrt{3}$.

So the distance is

$$\sqrt{(-3 + 4)^2 + (2 - 1)^2 + \left(\sqrt{3}\right)^2} = \sqrt{5}.$$

5. Let x, y, and z be the numbers. Because $xyz = 27$,

$$z = \frac{27}{xy}.$$

$$S = x + y + z = x + y + \frac{27}{xy}.$$

$$S_x = 1 - \frac{27}{x^2 y} = 0, S_y = 1 - \frac{27}{xy^2} = 0.$$

$$\left.\begin{array}{r} x^2 y = 27 \\ xy^2 = 27 \end{array}\right\} x = y = 3$$

So, $x = y = z = 3$.

6. Because $x + y + z = 32$, $z = 32 - x - y$. So,

$$P = xy^2 z = 32xy^2 - x^2y^2 - xy^3$$
$$P_x = 32y^2 - 2xy^2 - y^3 = y^2(32 - 2x - y) = 0$$
$$P_y = 64xy - 2x^2y - 3xy^2 = y\left(64x - 2x^2 - 3xy\right) = 0.$$

Ignoring the solution $y = 0$ and substituting $y = 32 - 2x$ into $P_y = 0$, we have

$$64x - 2x^2 - 3x(32 - 2x) = 0$$
$$4x(x - 8) = 0.$$

So, $x = 8, y = 16$, and $z = 8$.

7. Let x, y, and z be the numbers and let $S = x^2 + y^2 + z^2$. Because $x + y + z = 30$, we have

$$S = x^2 + y^2 + (30 - x - y)^2$$

$$\left.\begin{array}{l} S_x = 2x + 2(30 - x - y)(-1) = 0 \\ S_y = 2y + 2(30 - x - y)(-1) = 0 \end{array}\right\} \begin{array}{l} 2x + y = 30 \\ x + 2y = 30. \end{array}$$

Solving simultaneously yields $x = 10$, $y = 10$, and $z = 10$.

8. Let x, y, and z be the numbers. Because $xyz = 1$, $z = 1/xy$.

$$S = x^2 + y^2 + z^2 = x^2 + y^2 + \frac{1}{x^2y^2}$$

$$S_x = 2x - \frac{2}{x^3y^2} = 0, S_y = 2y - \frac{2}{x^2y^3} = 0$$

$$\left.\begin{array}{r} x\left(x^3y^2\right) = 1 \\ y\left(x^2y^3\right) = 1 \end{array}\right\} x^4y^2 = x^2y^4 \Rightarrow x = y$$

So, $x = y = z = 1$.

9. The volume is $668.25 = xyz \Rightarrow z = \dfrac{668.25}{xy}$.

$$C = 0.06(2yz + 2xz) + 0.11(xy) = 0.12\left(\frac{668.25}{x} + \frac{668.25}{y}\right) + 0.11(xy)$$

$$C = \frac{80.19}{x} + \frac{80.19}{y} + 0.11(xy)$$

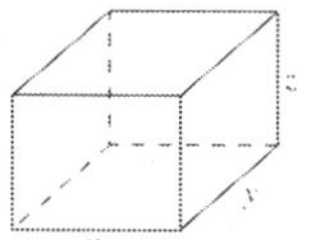

$$C_x = \frac{-80.19}{x^2} + 0.11y = 0$$

$$C_y = \frac{-8.19}{y^2} + 0.11x = 0$$

Solving simultaneously, $x = y = 9$ and $z = 8.25$.

Minimum cost: $\dfrac{80.19}{9} + \dfrac{80.19}{9} + 0.11(xy) = \$26.73$

10. Let x, y, and z be the length, width, and height, respectively. Then $C_0 = 1.5xy + 2yz + 2xz$ and $z = \dfrac{C_0 - 1.5xy}{2(x + y)}$.

The volume is given by

$$V = xyz = \frac{C_0xy - 1.5x^2y^2}{2(x + y)}$$

$$V_x = \frac{y^2(2C_0 - 3x^2 - 6xy)}{4(x + y)^2}$$

$$V_y = \frac{x^2(2C_0 - 3y^2 - 6xy)}{4(x + y)^2}.$$

In solving the system $V_x = 0$ and $V_y = 0$, we note by the symmetry of the equations that $y = x$.

Substituting $y = x$ into $V_x = 0$ yields

$$\frac{x^2(2C_0 - 9x^2)}{16x^2} = 0,\ 2C_0 = 9x^2,\ x = \frac{1}{3}\sqrt{2C_0},\ y = \frac{1}{3}\sqrt{2C_0}, \text{ and } z = \frac{1}{4}\sqrt{2C_0}.$$

11. Let x, y, and z be the length, width, and height, respectively and let V_0 be the given volume.

Then $V_0 = xyz$ and $z = V_0/xy$. The surface area is

$$S = 2xy + 2yz + 2xz = 2\left(xy + \frac{V_0}{x} + \frac{V_0}{y}\right)$$

$$\left.\begin{aligned} S_x &= 2\left(y - \frac{V_0}{x^2}\right) = 0 \\ S_y &= 2\left(x - \frac{V_0}{y^2}\right) = 0 \end{aligned}\right\} \begin{aligned} x^2y - V_0 &= 0 \\ xy^2 - V_0 &= 0. \end{aligned}$$

Solving simultaneously yields $x = \sqrt[3]{V_0}$, $y = \sqrt[3]{V_0}$, and $z = \sqrt[3]{V_0}$.

12. Consider the sphere given by $x^2 + y^2 + z^2 = r^2$ and let a vertex of the rectangular box be $\left(x, y, \sqrt{r^2 - x^2 - y^2}\right)$.

Then the volume is given by

$$V = (2x)(2y)\left(2\sqrt{r^2 - x^2 - y^2}\right) = 8xy\sqrt{r^2 - x^2 - y^2}$$

$$V_x = 8\left(xy\frac{-x}{\sqrt{r^2 - x^2 - y^2}} + y\sqrt{r^2 - x^2 - y^2}\right) = \frac{8y}{\sqrt{r^2 - x^2 - y^2}}\left(r^2 - 2x^2 - y^2\right) = 0$$

$$V_y = 8\left(xy\frac{-y}{\sqrt{r^2 - x^2 - y^2}} + x\sqrt{r^2 - x^2 - y^2}\right) = \frac{8x}{\sqrt{r^2 - x^2 - y^2}}\left(r^2 - x^2 - 2y^2\right) = 0.$$

Solving the system

$$2x^2 + y^2 = r^2$$

$$x^2 + 2y^2 = r^2$$

yields the solution $x = y = z = r/\sqrt{3}$.

13. $R(x_1, x_2) = -5x_1^2 - 8x_2^2 - 2x_1x_2 + 42x_1 + 102x_2$

$$R_{x_1} = -10x_1 - 2x_2 + 42 = 0, 5x_1 + x_2 = 21$$

$$R_{x_2} = -16x_2 - 2x_1 + 102 = 0, x_1 + 8x_2 = 51$$

Solving this system yields $x_1 = 3$ and $x_2 = 6$.

$$R_{x_1x_1} = -10$$

$$R_{x_1x_2} = -2$$

$$R_{x_2x_2} = -16$$

$R_{x_1x_1} < 0$ and $R_{x_1x_1}R_{x_2x_2} - \left(R_{x_1x_2}\right)^2 > 0$

So, revenue is maximized when $x_1 = 3$ and $x_2 = 6$.

14. $P(x_1, x_2) = 15(x_1 + x_2) - C_1 - C_2$

$$= 15x_1 + 15x_2 - \left(0.02x_1^2 + 4x_1 + 500\right) - \left(0.05x_2^2 + 4x_2 + 275\right) = -0.02x_1^2 - 0.05x_2^2 + 11x_1 + 11x_2 - 775$$

$$P_{x_1} = -0.04x_1 + 11 = 0, x_1 = 275$$

$$P_{x_2} = -0.10x_2 + 11 = 0, x_2 = 110$$

$$P_{x_1x_1} = -0.04$$

$$P_{x_1x_2} = 0$$

$$P_{x_2x_2} = -0.10$$

$P_{x_1x_1} < 0$ and $P_{x_1x_1}P_{x_2x_2} - \left(P_{x_1x_2}\right)^2 > 0$

So, profit is maximized when $x_1 = 275$ and $x_2 = 110$.

15. $P(p, q, r) = 2pq + 2pr + 2qr.$

$p + q + r = 1$ implies that $r = 1 - p - q.$

$$P(p, q) = 2pq + 2p(1 - p - q) + 2q(1 - p - q)$$
$$= 2pq + 2p - 2p^2 - 2pq + 2q - 2pq - 2q^2 = -2pq + 2p + 2q - 2p^2 - 2q^2$$

$$\frac{\partial P}{\partial p} = -2q + 2 - 4p; \frac{\partial P}{\partial q} = -2p + 2 - 4q$$

Solving $\frac{\partial P}{\partial p} = \frac{\partial P}{\partial q} = 0$ gives $q + 2p = 1$
$p + 2q = 1$

and so $p = q = \frac{1}{3}$ and $P\left(\frac{1}{3}, \frac{1}{3}\right) = -2\left(\frac{1}{9}\right) + 2\left(\frac{1}{3}\right) + 2\left(\frac{1}{3}\right) - 2\left(\frac{1}{9}\right) - 2\left(\frac{1}{9}\right) = \frac{6}{9} = \frac{2}{3}.$

16. $H = -x \ln x - y \ln y - z \ln z, x + y + z = 1 = -x \ln x - y \ln y - (1 - x - y)\ln(1 - x - y)$

$$H_x = -1 - \ln x + 1 + \ln(1 - x - y) = 0$$
$$H_y = -1 - \ln y + 1 + \ln(1 - x - y) = 0$$

$\ln(1 - x - y) = \ln x = \ln y \Rightarrow x = y.$

So, $\ln(1 - 2x) = \ln x \Rightarrow 1 - 2x = x \Rightarrow x = y = z = \frac{1}{3}.$

$$H = -\tfrac{1}{3}\ln\left(\tfrac{1}{3}\right) - \tfrac{1}{3}\ln\left(\tfrac{1}{3}\right) - \tfrac{1}{3}\ln\left(\tfrac{1}{3}\right) = -\ln\left(\tfrac{1}{3}\right) = \ln 3$$

17. The distance from P to Q is $\sqrt{x^2 + 4}$. The distance from Q to R is $\sqrt{(y - x)^2 + 1}$. The distance from R to S is $10 - y$.

$$C = 3k\sqrt{x^2 + 4} + 2k\sqrt{(y - x)^2 + 1} + k(10 - y)$$

$$C_x = 3k\left(\frac{x}{\sqrt{x^2 + 4}}\right) + 2k\left(\frac{-(y - x)}{\sqrt{(y - x)^2 + 1}}\right) = 0$$

$$C_y = 2k\left(\frac{y - x}{\sqrt{(y - x)^2 + 1}}\right) - k = 0 \Rightarrow \frac{y - x}{\sqrt{(y - x)^2 + 1}} = \frac{1}{2}$$

$$3k\left(\frac{x}{\sqrt{x^2 + 4}}\right) + 2k\left(-\frac{1}{2}\right) = 0$$
$$\frac{x}{\sqrt{x^2 + 4}} = \frac{1}{3}$$
$$3x = \sqrt{x^2 + 4}$$
$$9x^2 = x^2 + 4$$
$$x^2 = \frac{1}{2}$$
$$x = \frac{\sqrt{2}}{2}$$

$$2(y - x) = \sqrt{(y - x)^2 + 1}$$
$$4(y - x)^2 = (y - x)^2 + 1$$
$$(y - x)^2 = \frac{1}{3}$$
$$y = \frac{1}{\sqrt{3}} + \frac{1}{\sqrt{2}} = \frac{2\sqrt{3} + 3\sqrt{2}}{6}$$

So, $x = \frac{\sqrt{2}}{2} \approx 0.707$ km and $y = \frac{2\sqrt{3} + 3\sqrt{2}}{6} \approx 1.284$ km.

18. $A = \frac{1}{2}\left[(30 - 2x) + (30 - 2x) + 2x\cos\theta\right]x\sin\theta = 30x\sin\theta - 2x^2\sin\theta + x^2\sin\theta\cos\theta$

$\frac{\partial A}{\partial x} = 30\sin\theta - 4x\sin\theta + 2x\sin\theta\cos\theta = 0$

$\frac{\partial A}{\partial \theta} = 30x\cos\theta - 2x^2\cos\theta + x^2\left(2\cos^2\theta - 1\right) = 0$

From $\frac{\partial A}{\partial x} = 0$ we have $15 - 2x + x\cos\theta = 0 \Rightarrow \cos\theta = \frac{2x - 15}{x}$.

From $\frac{\partial A}{\partial \theta} = 0$ we obtain $30x\left(\frac{2x-15}{x}\right) - 2x^2\left(\frac{2x-15}{x}\right) + x^2\left(2\left(\frac{2x-15}{x}\right)^2 - 1\right) = 0$

$$30(2x - 15) - 2x(2x - 15) + 2(2x - 15)^2 - x^2 = 0$$
$$3x^2 - 30x = 0$$
$$x = 10.$$

Then $\cos\theta = \frac{1}{2} \Rightarrow \theta = 60°$.

19. Write the equation to be maximized or minimized as a function of two variables. Set the partial derivatives equal to zero (or undefined) to obtain the critical points. Use the Second Partials Test to test for relative extrema using the critical points. Check the boundary points, too.

20. See pages 946 and 947.

21. (a)

x	y	xy	x^2
-2	0	0	4
0	1	0	0
2	3	6	4
$\sum x_i = 0$	$\sum y_i = 4$	$\sum x_i y_i = 6$	$\sum x_i^2 = 8$

$a = \frac{3(6) - 0(4)}{3(8) - 0^2} = \frac{3}{4}, b = \frac{1}{3}\left[4 - \frac{3}{4}(0)\right] = \frac{4}{3}, y = \frac{3}{4}x + \frac{4}{3}$

(b) $S = \left(-\frac{3}{2} + \frac{4}{3} - 0\right)^2 + \left(\frac{4}{3} - 1\right)^2 + \left(\frac{3}{2} + \frac{4}{3} - 3\right)^2 = \frac{1}{6}$

22. (a)

x	y	xy	x^2
-3	0	0	9
-1	1	-1	1
1	1	1	1
3	2	6	9
$\sum x_i = 0$	$\sum y_i = 4$	$\sum x_i y_i = 6$	$\sum x_i^2 = 20$

$a = \frac{4(6) - 0(4)}{4(20) - (0)^2} = \frac{3}{10}, b = \frac{1}{4}\left[4 - \frac{3}{10}(0)\right] = 1, y = \frac{3}{10}x + 1$

(b) $S = \left(\frac{1}{10} - 0\right)^2 + \left(\frac{7}{10} - 1\right)^2 + \left(\frac{13}{10} - 1\right)^2 + \left(\frac{19}{10} - 2\right)^2 = \frac{1}{5}$

23. (a)

x	y	xy	x^2
0	4	0	0
1	3	3	1
1	1	1	1
2	0	0	4
$\sum x_i = 4$	$\sum y_i = 8$	$\sum x_i y_i = 4$	$\sum x_i^2 = 6$

$$a = \frac{4(4) - 4(8)}{4(6) - 4^2} = -2, b = \frac{1}{4}\left[8 + 2(4)\right] = 4, y = -2x + 4$$

(b) $S = (4-4)^2 + (2-3)^2 + (2-1)^2 + (0-0)^2 = 2$

24. (a)

x	y	xy	x^2
3	0	0	9
1	0	0	1
2	0	0	4
3	1	3	9
4	1	4	16
4	2	8	16
5	2	10	25
6	2	12	36
$\sum x_i = 28$	$\sum y_i = 8$	$\sum x_i y_i = 37$	$\sum x_i^2 = 116$

$$a = \frac{8(37) - (28)(8)}{8(116) - (28)^2} = \frac{72}{144} = \frac{1}{2}, b = \frac{1}{8}\left[8 - \frac{1}{2}(28)\right] = -\frac{3}{4}, y = \frac{1}{2}x - \frac{3}{4}$$

(b) $S = \left(\frac{3}{4} - 0\right)^2 + \left(-\frac{1}{4} - 0\right)^2 + \left(\frac{1}{4} - 0\right)^2 + \left(\frac{3}{4} - 1\right)^2 + \left(\frac{5}{4} - 1\right)^2 + \left(\frac{5}{4} - 2\right)^2 + \left(\frac{7}{4} - 2\right)^2 + \left(\frac{9}{4} - 2\right)^2 = \frac{3}{2}$

25. $(0, 0), (1, 1), (3, 4), (4, 2), (5, 5)$

$\sum x_i = 13, \quad \sum y_i = 12,$

$\sum x_i y_i = 46, \quad \sum x_i^2 = 51$

$$a = \frac{5(46) - 13(12)}{5(51) - (13)^2} = \frac{74}{86} = \frac{37}{43}$$

$$b = \frac{1}{5}\left[12 - \frac{37}{43}(13)\right] = \frac{7}{43}$$

$$y = \frac{37}{43}x + \frac{7}{43}$$

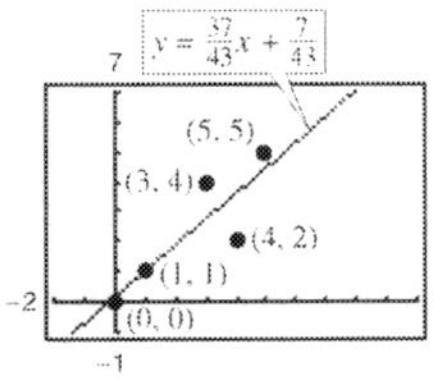

26. $(1, 0), (3, 3), (5, 6)$

$\sum x_i = 9, \quad \sum y_i = 9,$

$\sum x_i y_i = 39, \quad \sum x_i^2 = 35$

$$a = \frac{3(39) - 9(9)}{3(35) - (9)^2} = \frac{36}{24} = \frac{3}{2}$$

$$b = \frac{1}{3}\left[9 - \frac{3}{2}(9)\right] = -\frac{9}{6} = -\frac{3}{2}$$

$$y = \frac{3}{2}x - \frac{3}{2}$$

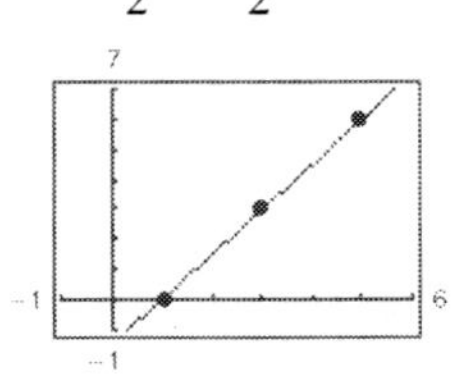

27. $(0, 6), (4, 3), (5, 0), (8, -4), (10, -5)$

$\sum x_i = 27, \quad \sum y_i = 0,$

$\sum x_i y_i = -70, \quad \sum x_i^2 = 205$

$$a = \frac{5(-70) - (27)(0)}{5(205) - (27)^2} = \frac{-350}{296} = -\frac{175}{148}$$

$$b = \frac{1}{5}\left[0 - \left(-\frac{175}{148}\right)(27)\right] = \frac{945}{148}$$

$$y = -\frac{175}{148}x + \frac{945}{148}$$

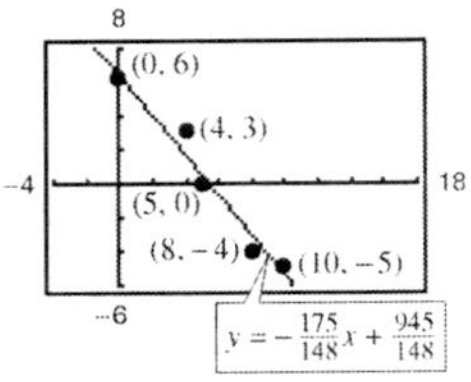

28. $(6, 4), (1, 2), (3, 3), (8, 6), (11, 8), (13, 8); n = 6$

$\sum x_i = 42 \quad \sum y_i = 31$

$\sum x_i y_i = 275 \quad \sum x_i^2 = 400$

$$a = \frac{6(275) - (42)(31)}{6(400) - (42)^2} = \frac{29}{53} \approx 0.5472$$

$$b = \frac{1}{6}\left(31 - \frac{29}{53}42\right) = \frac{425}{318} \approx 1.3365$$

$$y = \frac{29}{53}x + \frac{425}{318}$$

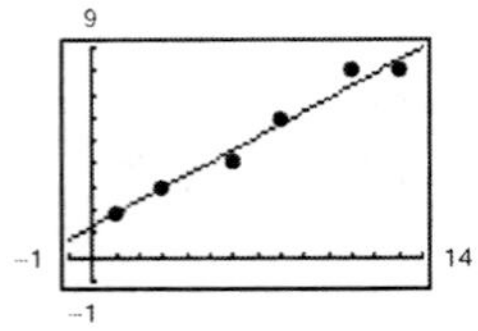

29. (a) Using a graphing utility, $y = 1.6x + 84$.

(b) For each one-year increase in age, the pressure changes by approximately 1.6, the slope of the line.

30. (a) Using a graphing utility, $y = 0.2x - 3$.

(b) When $x = 1300$, $y \approx \$257$ billion.

Answers will vary.

31. $S(a, b, c) = \sum_{i=1}^{n}\left(y_i - ax_i^2 - bx_i - c\right)^2$

$$\frac{\partial S}{\partial a} = \sum_{i=1}^{n} -2x_i^2\left(y_i - ax_i^2 - bx_i - c\right) = 0$$

$$\frac{\partial S}{\partial b} = \sum_{i=1}^{n} -2x_i\left(y_i - ax_i^2 - bx_i - c\right) = 0$$

$$\frac{\partial S}{\partial c} = -2\sum_{i=1}^{n}\left(y_i - ax_i^2 - bx_i - c\right) = 0$$

$$a\sum_{i=1}^{n} x_i^4 + b\sum_{i=1}^{n} x_i^3 + c\sum_{i=1}^{n} x_i^2 = \sum_{i=1}^{n} x_i^2 y_i$$

$$a\sum_{i=1}^{n} x_i^3 + b\sum_{i=1}^{n} x_i^2 + c\sum_{i=1}^{n} x_i = \sum_{i=1}^{n} x_i y_i$$

$$a\sum_{i=1}^{n} x_i^2 + b\sum_{i=1}^{n} x_i + cn = \sum_{i=1}^{n} y_i$$

32. (a) Matches (iv) because the slope in (iv) is approximately 0.22.

(b) Matches (i) because the slope in (i) is approximately −0.35.

(c) Matches (iii) because the slope in (iii) is approximately 0.09.

(d) Matches (ii) because the slope in (ii) is approximately −1.29.

33. $(-2, 0), (-1, 0), (0, 1), (1, 2), (2, 5)$

$\sum x_i = 0$

$\sum y_i = 8$

$\sum x_i^2 = 10$

$\sum x_i^3 = 0$

$\sum x_i^4 = 34$

$\sum x_i y_i = 12$

$\sum x_i^2 y_i = 22$

$34a + 10c = 22, 10b = 12, 10a + 5c = 8$

$a = \frac{3}{7}, b = \frac{6}{5}, c = \frac{26}{35}, y = \frac{3}{7}x^2 + \frac{6}{5}x + \frac{26}{35}$

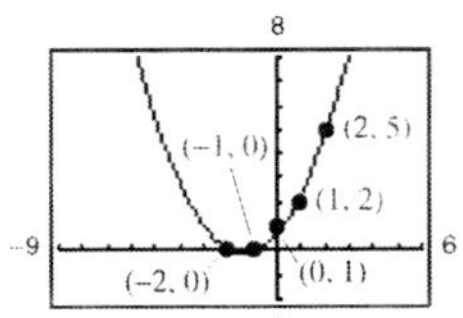

34. $(-4, 5), (-2, 6), (2, 6) (4, 2)$

$\sum x_i = 0$

$\sum y_i = 19$

$\sum x_i^2 = 40$

$\sum x_i^3 = 0$

$\sum x_i^4 = 544$

$\sum x_i y_i = -12$

$\sum x_i^2 y_i = 160$

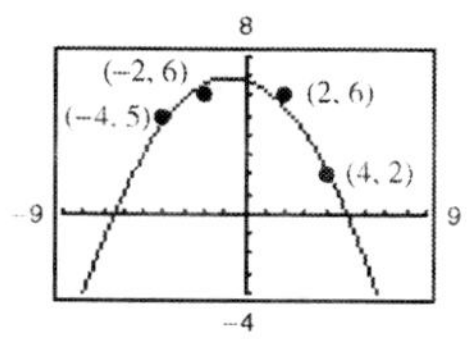

$544a + 40c = 160, 40b = -12, 40a + 4c = 19$

$a = -\frac{5}{24}, b = -\frac{3}{10}, c = \frac{41}{6}, y = -\frac{5}{24}x^2 - \frac{3}{10}x + \frac{41}{6}$

35. $(0, 0), (2, 2), (3, 6), (4, 12)$

$\sum x_i = 9$

$\sum y_i = 20$

$\sum x_i^2 = 29$

$\sum x_i^3 = 99$

$\sum x_i^4 = 353$

$\sum x_i y_i = 70$

$\sum x_i^2 y_i = 254$

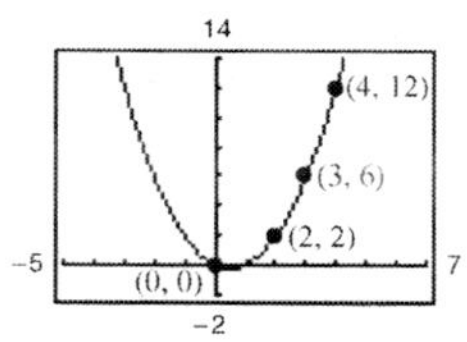

$353a + 99b + 29c = 254$

$99a + 29b + 9c = 70$

$29a + 9b + 4c = 20$

$a = 1, b = -1, c = 0, y = x^2 - x$

36. $(0, 10), (1, 9), (2, 6), (3, 0)$

$\sum x_i = 6$

$\sum y_i = 25$

$\sum x_i^2 = 14$

$\sum x_i^3 = 36$

$\sum x_i^4 = 98$

$\sum x_i y_i = 21$

$\sum x_i^2 y_i = 33$

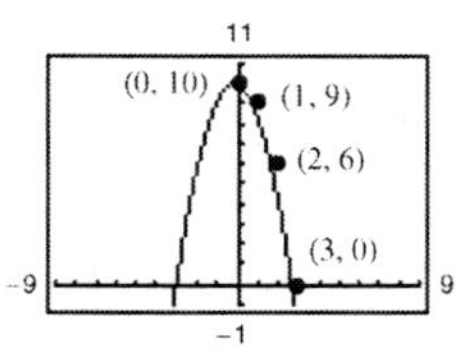

$98a + 36b + 14c = 33$

$36a + 14b + 6c = 21$

$14a + 6b + 4c = 25$

$a = -\frac{5}{4}, b = \frac{9}{20}, c = \frac{199}{20}, y = -\frac{5}{4}x^2 + \frac{9}{20}x + \frac{199}{20}$

37. (a) $(0, 0), (2, 15), (4, 30),$
$(6, 50), (8, 65), (10, 70)$

$\sum x_i = 30$

$\sum y_i = 230$

$\sum x_i^2 = 220$

$\sum x_i^3 = 1800$

$\sum x_i^4 = 15{,}664$

$\sum x_i y_i = 1670$

$\sum x_i^2 y_i = 13{,}500$

$15{,}664a + 1800b + 220c = 13{,}500$

$1800a + 220b + 30c = 1670$

$220a + 30b + 6c = 230$

$y = -\frac{25}{112}x^2 + \frac{541}{56}x - \frac{25}{14} \approx -0.22x^2 + 9.66x - 1.79$

(b)

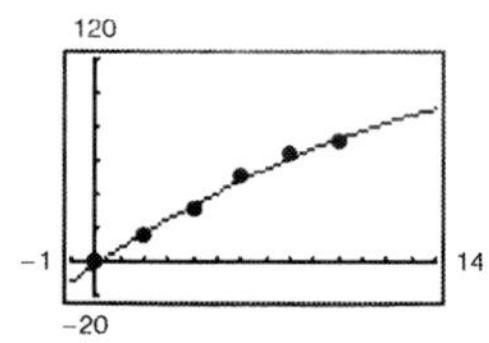

38. (a) Using a graphing utility, $y = 0.08x + 6.1$.

(b) Using a graphing utility,
$y = -0.002x^2 + 0.10x + 6.0$.

(c)

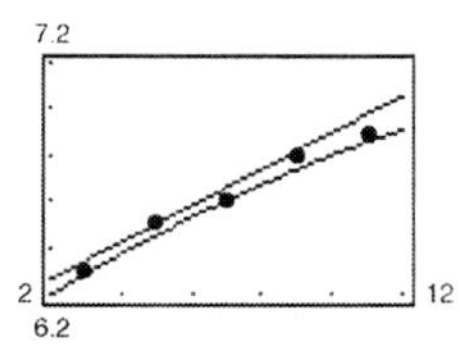

(d) For 2020, $x = 20$,

Linear model:
$y = 0.075(20) + 6.095 \approx 7.6$ billion

Quadratic model:
$y = -0.0018(20)^2 + 0.10(20) + 6.02 \approx 7.3$ billion

The quadratic model is less accurate because of the negative x^2 coefficient

39. (a) $\ln P = -0.1499h + 9.3018$

(b) $\ln P = -0.1499h + 9.3018$

$P = e^{-0.1499h+9.3018} = 10{,}957.7e^{-0.1499h}$

(c)

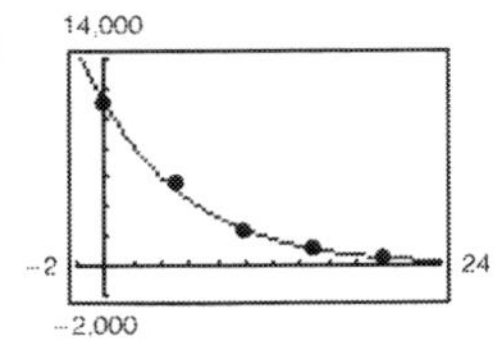

(d) Same answers

40. (a) $\dfrac{1}{y} = ax + b = -0.0074x + 0.445$

$y = \dfrac{1}{-0.0074x + 0.445}$

(b)

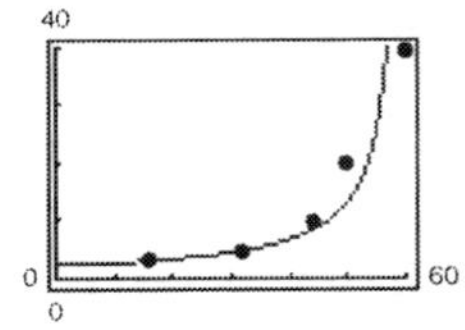

(c) No. For $x = 70$, $y \approx -14$, which is nonsense.

41. $S(a, b) = \sum_{i=1}^{n}(ax_i + b - y_i)^2$

$S_a(a, b) = 2a\sum_{i=1}^{n}x_i^2 + 2b\sum_{i=1}^{n}x_i - 2\sum_{i=1}^{n}x_iy_i$

$S_b(a, b) = 2a\sum_{i=1}^{n}x_i + 2nb - 2\sum_{i=1}^{n}y_i$

$S_{aa}(a, b) = 2\sum_{i=1}^{n}x_i^2$

$S_{bb}(a, b) = 2n$

$S_{ab}(a, b) = 2\sum_{i=1}^{n}x_i$

$S_{aa}(a, b) > 0$ as long as $x_i \neq 0$ for all i. (**Note:** If $x_i = 0$ for all i, then $x = 0$ is the least squares regression line.)

$$d = S_{aa}S_{bb} - S_{ab}^2 = 4n\sum_{i=1}^{n}x_i^2 - 4\left(\sum_{i=1}^{n}x_i\right)^2 = 4\left[n\sum_{i=1}^{n}x_i^2 - \left(\sum_{i=1}^{n}x_i\right)^2\right] \geq 0 \text{ since } n\sum_{i=1}^{n}x_i^2 \geq \left(\sum_{i=1}^{n}x_i\right)^2.$$

As long as $d \neq 0$, the given values for a and b yield a minimum.

Section 13.10 Lagrange Multipliers

1. Maximize $f(x, y) = xy$

Constraint: $x + y = 10$

$\nabla f = \lambda \nabla g$

$y\mathbf{i} + x\mathbf{j} = \lambda(\mathbf{i} + \mathbf{j})$

$$\left.\begin{aligned} y &= \lambda \\ x &= \lambda \\ x + y &= 10 \end{aligned}\right\} \quad x = y = 5$$

$f(5, 5) = 25$

2. Minimize $f(x, y) = 2x + y$

Constraint: $xy = 32$

$\nabla f = \lambda \nabla g$

$2\mathbf{i} + \mathbf{j} = \lambda y\mathbf{i} + \lambda x\mathbf{j}$

$2 = \lambda y \Rightarrow y = 2/\lambda$

$1 = \lambda x \Rightarrow x = 1/\lambda$

$xy = (1/\lambda)(2/\lambda) = 2/\lambda^2 = 32$

$\lambda^2 = 1/16$

$\lambda = 1/4, x = 4, y = 8$

$f(4, 8) = 16$

3. Minimize $f(x, y) = x^2 + y^2$.

Constraint: $x + 2y - 5 = 0$

$\nabla f = \lambda \nabla g$

$2x\mathbf{i} + 2y\mathbf{j} = \lambda(\mathbf{i} + 2\mathbf{j})$

$$\left.\begin{aligned} 2x &= \lambda \\ 2y &= 2\lambda \end{aligned}\right\} \begin{aligned} x &= \lambda/2 \\ y &= \lambda \end{aligned}$$

$x + 2y - 5 = 0$

$\dfrac{\lambda}{2} + 2\lambda = 5 \Rightarrow \lambda = 2, x = 1, y = 2$

$f(1, 2) = 5$

4. Maximize $f(x, y) = x^2 - y^2$.

Constraint: $2y - x^2 = 0$

$\nabla f = \lambda \nabla g$

$2x\mathbf{i} - 2y\mathbf{j} = -2x\lambda\mathbf{i} + 2\lambda\mathbf{j}$

$2x = -2x\lambda \Rightarrow x = 0$ or $\lambda = -1$

If $x = 0$, then $y = 0$ and $f(0, 0) = 0$.

If $\lambda = -1$,

$-2y = 2\lambda = -2 \Rightarrow y = 1 \Rightarrow x^2 = 2 \Rightarrow x = \sqrt{2}$.

$f(\sqrt{2}, 1) = 2 - 1 = 1$, Maximum

5. Maximize $f(x, y) = 2x + 2xy + y$.

Constraint: $2x + y = 100$

$\nabla f = \lambda \nabla g$

$(2 + 2y)\mathbf{i} + (2x + 1)\mathbf{j} = 2\lambda\mathbf{i} + \lambda\mathbf{j}$

$$\left.\begin{aligned} 2 + 2y &= 2\lambda \Rightarrow y = \lambda - 1 \\ 2x + 1 &= \lambda \Rightarrow x = \frac{\lambda - 1}{2} \end{aligned}\right\} y = 2x$$

$2x + y = 100 \Rightarrow 4x = 100$

$x = 25, y = 50$

$f(25, 50) = 2600$

6. Minimize $f(x, y) = 3x + y + 10$.

Constraint: $x^2y = 6$

$\nabla f = \lambda \nabla g$

$3\mathbf{i} + \mathbf{j} = 2xy\lambda\mathbf{i} + x^2\lambda\mathbf{j}$

$$\left.\begin{aligned} 3 &= 2xy\lambda \Rightarrow \lambda = \frac{3}{2xy} \\ 1 &= x^2\lambda \Rightarrow \lambda = \frac{1}{x^2} \end{aligned}\right\} 3x^2 = 2xy \Rightarrow y = \frac{3x}{2} \quad (x \neq 0)$$

$x^2y = 6 \Rightarrow x^2\left(\dfrac{3x}{2}\right) = 6$

$x^3 = 4$

$x = \sqrt[3]{4}, y = \dfrac{3\sqrt[3]{4}}{2}$

$f\left(\sqrt[3]{4}, \dfrac{3\sqrt[3]{4}}{2}\right) = \dfrac{9\sqrt[3]{4} + 20}{2}$

7. Note: $f(x, y) = \sqrt{6 - x^2 - y^2}$ is maximum when $g(x, y)$ is maximum.

Maximize $g(x, y) = 6 - x^2 - y^2$.

Constraint: $x + y - 2 = 0$

$$\left.\begin{aligned} -2x &= \lambda \\ -2y &= \lambda \end{aligned}\right\} x = y$$

$x + y = 2 \Rightarrow x = y = 1$

$f(1, 1) = \sqrt{g(1, 1)} = 2$

8. Note: $f(x, y) = \sqrt{x^2 + y^2}$ is minimum when $g(x, y)$ is minimum.

Minimize $g(x, y) = x^2 + y^2$.

Constraint: $2x + 4y - 15 = 0$

$$\left.\begin{aligned} 2x &= 2\lambda \\ 2y &= 4\lambda \end{aligned}\right\} y = 2x$$

$2x + 4y = 15 \Rightarrow 10x = 15$

$x = \dfrac{3}{2}, y = 3$

$f\left(\dfrac{3}{2}, 3\right) = \sqrt{g\left(\dfrac{3}{2}, 3\right)} = \dfrac{3\sqrt{5}}{2}$

9. Minimize $f(x, y, z) = x^2 + y^2 + z^2$.

Constraint: $x + y + z - 9 = 0$

$$\left.\begin{aligned} 2x &= \lambda \\ 2y &= \lambda \\ 2z &= \lambda \end{aligned}\right\} x = y = z$$

$$x + y + z = 9 \Rightarrow x = y = z = 3$$

$$f(3, 3, 3) = 27$$

10. Maximize $f(x, y, z) = xyz$.

Constraint: $x + y + z - 3 = 0$

$$\left.\begin{aligned} yz &= \lambda \\ xz &= \lambda \\ xy &= \lambda \end{aligned}\right\} yz = xz = xy \Rightarrow x = y = z$$

$$x + y + z = 3 \Rightarrow x = y = z = 1$$

$$f(1, 1, 1) = 1$$

11. Minimize $f(x, y, z) = x^2 + y^2 + z^2$.

Constraint: $x + y + z = 1$

$$\left.\begin{aligned} 2x &= \lambda \\ 2y &= \lambda \\ 2z &= \lambda \end{aligned}\right\} x = y = z$$

$$x + y + z = 1 \Rightarrow x = y = z = \tfrac{1}{3}$$

$$f\left(\tfrac{1}{3}, \tfrac{1}{3}, \tfrac{1}{3}\right) = \tfrac{1}{3}$$

12. Maximize $f(x, y, z) = x + y + z$

Constraint: $x^2 + y^2 + z^2 = 1$

$$\left.\begin{aligned} 1 &= \lambda 2x \\ 1 &= \lambda 2y \\ 1 &= \lambda 2z \end{aligned}\right\} x = y = z = \frac{1}{2\lambda}$$

$$x^2 + y^2 + z^2 = \frac{1}{4\lambda^2} + \frac{1}{4\lambda^2} + \frac{1}{4\lambda^2} = \frac{3}{4\lambda^2} = 1$$

$$\lambda^2 = 3/4 \Rightarrow \lambda = \sqrt{3}/2 \Rightarrow x = y = z = \frac{1}{\sqrt{3}}$$

$$f(x, y, z) = 3/\sqrt{3} = \sqrt{3}$$

13. Maximize or minimize $f(x, y) = x^2 + 3xy + y^2$.

Constraint: $x^2 + y^2 \le 1$

Case 1: On the circle $x^2 + y^2 = 1$

$$\left.\begin{aligned} 2x + 3y &= 2x\lambda \\ 3x + 2y &= 2y\lambda \end{aligned}\right\} x^2 = y^2$$

$$x^2 + y^2 = 1 \Rightarrow x = \pm\frac{\sqrt{2}}{2}, y = \pm\frac{\sqrt{2}}{2}$$

Maxima: $f\left(\pm\frac{\sqrt{2}}{2}, \pm\frac{\sqrt{2}}{2}\right) = \frac{5}{2}$

Minima: $f\left(\pm\frac{\sqrt{2}}{2}, \frac{\sqrt{2}}{2}\right) = -\frac{1}{2}$

Case 2: Inside the circle

$$\left.\begin{aligned} f_x &= 2x + 3y = 0 \\ f_y &= 3x + 2y = 0 \end{aligned}\right\} x = y = 0$$

$$f_{xx} = 2, f_{yy} = 2, f_{xy} = 3, f_{xx}f_{yy} - \left(f_{xy}\right)^2 \le 0$$

Saddle point: $f(0, 0) = 0$

By combining these two cases, we have a maximum of $\frac{5}{2}$ at $\left(\pm\frac{\sqrt{2}}{2}, \pm\frac{\sqrt{2}}{2}\right)$ and a minimum of $-\frac{1}{2}$ at $\left(\pm\frac{\sqrt{2}}{2}, \mp\frac{\sqrt{2}}{2}\right)$.

14. Maximize or minimize $f(x, y) = e^{-xy/4}$.

Constraint: $x^2 + y^2 \le 1$

Case 1: On the circle $x^2 + y^2 = 1$

$$\left.\begin{aligned} -(y/4)e^{-xy/4} &= 2x\lambda \\ -(x/4)e^{-xy/4} &= 2y\lambda \end{aligned}\right\} \Rightarrow x^2 = y^2$$

$$x^2 + y^2 = 1 \Rightarrow x = \pm\frac{\sqrt{2}}{2}$$

Maxima: $f\left(\pm\frac{\sqrt{2}}{2}, \mp\frac{\sqrt{2}}{2}\right) = e^{1/8} \approx 1.1331$

Minima: $f\left(\pm\frac{\sqrt{2}}{2}, \pm\frac{\sqrt{2}}{2}\right) = e^{-1/8} \approx 0.8825$

Case 2: Inside the circle

$$\left.\begin{aligned} f_x &= -(y/4)e^{-xy/4} = 0 \\ f_y &= -(x/4)e^{-xy/4} = 0 \end{aligned}\right\} \Rightarrow x = y = 0$$

$$f_{xx} = \frac{y^2}{16}e^{-xy/4},\ f_{yy} = \frac{x^2}{16}e^{-xy/4},\ f_{xy} = e^{-xy}\left(\frac{1}{16}xy - \frac{1}{4}\right)$$

At $(0, 0)$, $f_{xx}f_{yy} - (f_{xy})^2 < 0$.

Saddle point: $f(0, 0) = 1$

Combining the two cases, we have a maximum of $e^{1/8}$ at $\left(\pm\frac{\sqrt{2}}{2}, \mp\frac{\sqrt{2}}{2}\right)$ and a minimum of $e^{-1/8}$ at $\left(\pm\frac{\sqrt{2}}{2}, \pm\frac{\sqrt{2}}{2}\right)$.

15. Maximize $f(x, y, z) = xyz$.

Constraints: $x + y + z = 32$

$x - y + z = 0$

$\nabla f = \lambda\nabla g + \mu\nabla h$

$yz\mathbf{i} + xz\mathbf{j} + xy\mathbf{k} = \lambda(\mathbf{i} + \mathbf{j} + \mathbf{k}) + \mu(\mathbf{i} - \mathbf{j} + \mathbf{k})$

$$\left.\begin{aligned} yz &= \lambda + \mu \\ xz &= \lambda - \mu \\ xy &= \lambda + \mu \end{aligned}\right\} yz = xy \Rightarrow x = z$$

$$\left.\begin{aligned} x + y + z &= 32 \\ x - y + z &= 0 \end{aligned}\right\} 2x + 2z = 32 \Rightarrow x = z = 8$$

$y = 16$

$f(8, 16, 8) = 1024$

16. Minimize $f(x, y, z) = x^2 + y^2 + z^2$.

Constraints: $x + 2z = 6$

$x + y = 12$

$\nabla f = \lambda\nabla g + \mu\nabla h$

$2x\mathbf{i} + 2y\mathbf{j} + 2z\mathbf{k} = \lambda(\mathbf{i} + 2\mathbf{k}) + \mu(\mathbf{i} + \mathbf{j})$

$$\left.\begin{aligned} 2x &= \lambda + \mu \\ 2y &= \mu \\ 2z &= 2\lambda \end{aligned}\right\} 2x = 2y + z$$

$$x + 2z = 6 \Rightarrow z = \frac{6 - x}{2} = 3 - \frac{x}{2}$$

$x + y = 12 \Rightarrow y = 12 - x$

$$2x = 2(12 - x) + \left(3 - \frac{x}{2}\right) \Rightarrow \frac{9}{2}x = 27 \Rightarrow x = 6$$

$x = 6, z = 0$

$f(6, 6, 0) = 72$

17. Minimize the square of the distance $f(x, y) = (x - 0)^2 + (y - 0)^2 = x^2 + y^2$ subject to the constraint $x + y = 1$.

$$\left.\begin{aligned} 2x &= \lambda \\ 2y &= \lambda \end{aligned}\right\} \quad \left.\begin{aligned} x &= \lambda/2 \\ y &= \lambda/2 \end{aligned}\right\} \Rightarrow x = y$$

$x + y = 1$

$$x = y = \frac{1}{2}$$

The minimum distance is $d = \sqrt{\left(\frac{1}{2}\right)^2 + \left(\frac{1}{2}\right)^2} = \frac{\sqrt{2}}{2}$.

18. Minimize the square of the distance $f(x, y) = x^2 + y^2$ subject to the constraint $2x + 3y = -1$.

$$\left.\begin{aligned} 2x &= 2\lambda \\ 2y &= 3\lambda \end{aligned}\right\} y = \frac{3x}{2}$$

$$2x + 3y = -1 \Rightarrow x = -\frac{2}{13},\ y = -\frac{3}{13}$$

The minimum distance is $d = \sqrt{\left(-\frac{2}{13}\right)^2 + \left(-\frac{3}{13}\right)^2} = \frac{\sqrt{13}}{13}$.

19. Minimize the square of the distance

$f(x, y) = x^2 + (y - 2)^2$

subject to the constraint $x - y = 4$.

$$\left.\begin{aligned} 2x &= \lambda \\ 2(y-2) &= -\lambda \end{aligned}\right\} \begin{aligned} x &= \lambda/2 \\ y &= \frac{4-\lambda}{2} \end{aligned}$$

$$x - y = 4$$

$$\frac{\lambda}{2} - \left(\frac{4-\lambda}{2}\right) = 4$$

$$\lambda = 6$$

$$x = 3, y = -1$$

The minimum distance

is $d = \sqrt{3^2 + (-1-2)^2} = 3\sqrt{2}$.

20. Minimize the square of the distance

$f(x, y) = (x - 1)^2 + y^2$ subject to the constraint

$x + 4y = 3$.

$$\left.\begin{aligned} 2(x-1) &= \lambda \\ 2y &= 4\lambda \end{aligned}\right\} \begin{aligned} x &= \frac{\lambda+2}{2} \\ y &= 2\lambda \end{aligned}$$

$$x + 4y = 3$$

$$\frac{\lambda+2}{2} + 4(2\lambda) = 3$$

$$\lambda + 2 + 16\lambda = 6$$

$$17\lambda = 4$$

$$\lambda = \frac{4}{17}$$

$$x = \frac{19}{17}, y = \frac{8}{17}$$

The minimum distance

is $d = \sqrt{\left(\frac{19}{17}\right)^2 + \left(\frac{8}{17}\right)^2} = \frac{5\sqrt{17}}{17}$.

21. Minimize the square of the distance

$f(x, y) = x^2 + (y - 3)^2$ subject to the constraint

$y - x^2 = 0$.

$$2x = -2x\lambda$$

$$2(y-3) = \lambda$$

$$y = x^2$$

If $x = 0, y = 0$, and $f(0, 0) = 9 \Rightarrow$ distance $= 3$.

If $x \neq 0, \lambda = -1, y = 5/2, x = \pm\sqrt{5/2}$

$$f\left(\pm\sqrt{5/2}, 5/2\right) = 5/2 + \left(\frac{1}{2}\right)^2 = \frac{11}{4} < 3$$

The minimum distance is $d = \frac{\sqrt{11}}{2}$.

22. Minimize the square of the distance

$f(x, y) = (x + 3)^2 + y^2$ subject to the constraint

$y - x^2 = 0$.

$$2(x+3) = -2\lambda x$$

$$2y = \lambda$$

$$y = x^2$$

$$\lambda = 2y = 2x^2$$

$$2(x+3) = -2(2x^3)$$

$$4x^3 + 2x + 6 = 0$$

$$2(x+1)(2x^2 - 2x + 3) = 0 \Rightarrow x = -1, y = 1,$$

The minimum distance is $d = \sqrt{(-1)^2 + (1^2)} = \sqrt{2}$.

23. Minimize the square of the distance $f(x, y) = (x - 4)^2 + (y - 4)^2$ subject to the constraint $x^2 + (y - 1)^2 = 9$.

$$2(x-4) = 2x\lambda$$

$$2(y-4) = 2(y-1)\lambda$$

$$x^2 + (y-1)^2 = 9$$

Solving these equations, you obtain

$x = 12/5, y = 14/5$ and $\lambda = -2/3$.

The minimum distance is $d = \sqrt{\left(\frac{12}{5} - 4\right)^2 + \left(\frac{14}{5} - 4\right)^2} = \sqrt{\frac{64}{25} + \frac{36}{25}} = 2$.

24. Minimize the square of the distance $f(x, y) = x^2 + (y - 10)^2$ subject to the constraint $(x - 4)^2 + y^2 = 4$.

$$\left.\begin{aligned} 2x &= 2(x - 4)\lambda \\ 2(y - 10) &= 2y\lambda \end{aligned}\right\} \frac{x}{x - 4} = \frac{y - 10}{y} \Rightarrow y = -\frac{5}{2}x + 10$$

$$(x - 4)^2 + y^2 = 4 \Rightarrow (x^2 - 8x + 16) + \left(\frac{25}{4}x^2 - 50x + 100\right) = 4$$

$$\frac{29}{4}x^2 - 58x + 112 = 0$$

Using a graphing utility, we obtain $x \approx 3.2572$ and $x \approx 4.7428$ or by the Quadratic Formula,

$$x = \frac{58 \pm \sqrt{58^2 - 4(29/4)(112)}}{2(29/4)} = \frac{58 \pm 2\sqrt{29}}{29/2} = 4 \pm \frac{4\sqrt{29}}{29}.$$

Using the smaller value, we have $x = 4\left(1 - \frac{\sqrt{29}}{29}\right)$ and $y = \frac{10\sqrt{29}}{29} \approx 1.8570$.

The minimum distance is $d = \sqrt{16\left(1 - \frac{\sqrt{29}}{29}\right)^2 + \left(\frac{10\sqrt{29}}{29} - 10\right)^2} \approx 8.77$.

The larger x-value does not yield a minimum.

25. Minimize the square of the distance

$$f(x, y, z) = (x - 2)^2 + (y - 1)^2 + (z - 1)^2$$

subject to the constraint $x + y + z = 1$.

$$\left.\begin{aligned} 2(x - 2) &= \lambda \\ 2(y - 1) &= \lambda \\ 2(z - 1) &= \lambda \end{aligned}\right\} y = z \text{ and } y = x - 1$$

$$x + y + z = 1 \Rightarrow x + 2(x - 1) = 1$$

$$x = 1,\ y = z = 0$$

The minimum distance is

$$d = \sqrt{(1 - 2)^2 + (0 - 1)^2 + (0 - 1)^2} = \sqrt{3}.$$

26. Minimize the square of the distance

$$f(x, y, z) = (x - 4)^2 + y^2 + z^2$$

subject to the constraint $\sqrt{x^2 + y^2} - z = 0$.

$$\left.\begin{aligned} 2(x - 4) &= \frac{x}{\sqrt{x^2 + y^2}}\lambda = \frac{x}{z}\lambda \\ 2y &= \frac{y}{\sqrt{x^2 + y^2}}\lambda = \frac{y}{z}\lambda \\ 2z &= -\lambda \end{aligned}\right\} \begin{aligned} 2(x - 4) &= -2x \\ 2y &= -2y \end{aligned}$$

$$\sqrt{x^2 + y^2} - z = 0,\ x = 2,\ y = 0,\ z = 2$$

The minimum distance is

$$d = \sqrt{(2 - 4)^2 + 0^2 + 2^2} = 2\sqrt{2}.$$

27. Maximize $f(x, y, z) = z$ subject to the constraints

$x^2 + y^2 - z^2 = 0$ and $x + 2z = 4$.

$$\begin{aligned} 0 &= 2x\lambda + \mu \\ 0 &= 2y\lambda \Rightarrow y = 0 \\ 1 &= -2z\lambda + 2\mu \end{aligned}$$

$$x^2 + y^2 - z^2 = 0$$

$$x + 2z = 4 \Rightarrow x = 4 - 2z$$

$$\begin{aligned} (4 - 2z)^2 + 0^2 - z^2 &= 0 \\ 3z^2 - 16z + 16 &= 0 \\ (3z - 4)(z - 4) &= 0 \\ z = \tfrac{4}{3} \text{ or } z &= 4 \end{aligned}$$

The maximum value of f occurs when $z = 4$ at the point of $(-4, 0, 4)$.

28. Maximize $f(x, y, z) = z$ subject to the constraints

$x^2 + y^2 + z^2 = 36$ and $2x + y - z = 2$.

$$\left.\begin{aligned}0 &= 2x\lambda + 2\mu\\ 0 &= 2y\lambda + \mu\\ 1 &= 2z\lambda - \mu\end{aligned}\right\} x = 2y$$

$$x^2 + y^2 + z^2 = 36$$

$$2x + y - z = 2 \Rightarrow z = 2x + y - 2 = 5y - 2$$

$$(2y)^2 + y^2 + (5y - 2)^2 = 36$$
$$30y^2 - 20y - 32 = 0$$
$$15y^2 - 10y - 16 = 0$$
$$y = \frac{5 \pm \sqrt{265}}{15}$$

Choosing the positive value for y we have the point

$$\left(\frac{10 + 2\sqrt{265}}{15}, \frac{5 + \sqrt{265}}{15}, \frac{-1 + \sqrt{265}}{3}\right).$$

29. Optimization problems that have restrictions or constraints on the values that can be used to produce the optimal solution are called contrained optimization problems.

30. See explanation at the bottom of page 953.

31. Minimize $f(x, y, z) = x^2 + y^2 + z^2$.

Constraint: $g(x, y, z) = x - y + z = 3$

$$2x = \lambda \Rightarrow x = \lambda/2$$
$$2y = -\lambda \Rightarrow y = -\lambda/2$$
$$2z = \lambda \Rightarrow z = \lambda/2$$
$$x - y + z = 3$$
$$\frac{\lambda}{2} - \left(-\frac{\lambda}{2}\right) + \frac{\lambda}{2} = 3$$
$$\frac{3\lambda}{2} = 3$$
$$\lambda = 2$$
$$x = 1, y = -1, z = 1$$

Minimum distance $= \sqrt{1^2 + (-1)^2 + 1^2} = \sqrt{3}$

32. Minimize $f(x, y, z) = (x - 1)^2 + (y - 2)^2 + (z - 3)^2$.

Constraint: $g(x, y, z) = x - y + z = 3$

$$2(x - 1) = \lambda \Rightarrow x = \frac{2 + \lambda}{2}$$
$$2(y - 2) = -\lambda \Rightarrow y = \frac{4 - \lambda}{2}$$
$$2(z - 3) = \lambda \Rightarrow z = \frac{6 + \lambda}{2}$$
$$x - y + z = 3$$
$$\frac{2 + \lambda}{2} - \frac{4 - \lambda}{2} + \frac{6 + \lambda}{2} = 3$$
$$3\lambda + 4 = 6$$
$$\lambda = \frac{2}{3}$$
$$x = \frac{4}{3}, y = \frac{5}{3}, z = \frac{10}{3}$$

Minimum distance $= \left(\frac{1}{3}\right)^2 + \left(-\frac{1}{3}\right)^2 + \left(\frac{1}{3}\right)^2 = \frac{\sqrt{3}}{3}$

33. Minimize $f(x, y, z) = x + y + z$.

Constraint: $g(x, y, z) = xyz = 27$

$$\left.\begin{aligned}1 &= \lambda yz \Rightarrow x = \lambda xyz\\ 1 &= \lambda xz \Rightarrow y = \lambda xyz\\ 1 &= \lambda xy \Rightarrow z = \lambda xyz\end{aligned}\right\} \Rightarrow x = y = z$$
$$xyz = 27$$
$$x^3 = 27 \Rightarrow x = y = z = 3$$

34. Maximize $P(x, y, z) = xy^2z$.

Constraint: $g(x, y, z) = x + y + z = 32$

$$y^2z = \lambda$$
$$2xyz = \lambda$$
$$xy^2 = \lambda$$
$$x + y + z = 32$$
$$xy^2 = y^2z \Rightarrow x = z \qquad (y \neq 0)$$
$$2xyz = xy^2 \Rightarrow 2x^2y = xy^2 \Rightarrow 2x = y$$
$$x + 2x + x = 32$$
$$x = 8$$
$$y = 16$$
$$z = 8$$

35. Minimize $f(x, y, z) = 0.06(2yz + 2xz) + 0.11(xy)$.

Constraint: $g(x, y, z) = xyz = 668.25$

$$\begin{aligned} 0.12z + 0.11y &= yz\lambda \\ 0.12z + 0.11x &= xz\lambda \\ 0.12(y + x) &= xy\lambda \\ xyz &= 668.25 \end{aligned}$$

$$0.12xz + 0.11yx = xyz\lambda = 0.12yz + 0.11xy \Rightarrow x = y$$

$$0.12(2x) = x^2\lambda \Rightarrow \lambda = \frac{0.24}{x}$$

$$0.12z + 0.11x = xz\left(\frac{0.24}{x}\right) = 0.24z \Rightarrow z = \frac{0.11x}{0.12} = \frac{11x}{12}$$

$$xyz = x^2\left(\frac{11}{12}x\right) = 668.25 \Rightarrow x = y = 9, z = \frac{33}{4}$$

$$f\left(9, 9, \frac{33}{4}\right) = \$26.73$$

36. Maximize $f(x, y, z) = xyz$ (volume).

Constraint: $g(x, y, z) = 1.5xy + 2xz + 2yz = C$

$$\begin{aligned} yz &= 1.5y\lambda + 2z\lambda \\ xz &= 1.5x\lambda + 2z\lambda \\ xy &= 2x\lambda + 2y\lambda \\ 1.5xy + 2xz + 2yz &= C \end{aligned}$$

$$\begin{aligned} xyz = x[1.5y\lambda + 2z\lambda] &= y[1.5x\lambda + 2z\lambda] \\ 2xz\lambda &= 2yz\lambda \\ x &= y \quad \text{(also by symmetry)} \end{aligned}$$

$$x^2 = 2x\lambda + 2x\lambda \Rightarrow \lambda = x/4.$$

$$xz = 1.5x\left(\frac{x}{4}\right) + 2z\left(\frac{x}{4}\right) \Rightarrow z = \frac{3}{4}x$$

$$1.5x^2 + 2x\left(\frac{3}{4}x\right) + 2x\left(\frac{3}{4}x\right) = C \Rightarrow x^2 = \frac{2}{9}C \Rightarrow x = \frac{\sqrt{2C}}{3},$$

$$y = \frac{\sqrt{2C}}{3}, z = \frac{\sqrt{2C}}{4}$$

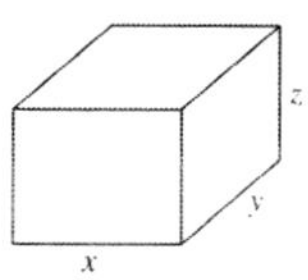

37. Maximize $P(p, q, r) = 2pq + 2pr + 2qr$.

Constraint: $g(p, q, r) = p + q + r = 1$

$$\left.\begin{aligned} 2q + 2r &= \lambda \\ 2p + 2r &= \lambda \\ 2p + 2q &= \lambda \end{aligned}\right\} p = q = r$$

$p + q + r = 3p = 1 \Rightarrow p = \frac{1}{3}$ and

$P\left(\frac{1}{3}, \frac{1}{3}, \frac{1}{3}\right) = 3\left(\frac{2}{9}\right) = \frac{2}{3}$.

38. Maximize $H(x, y, z) = -x \ln x - y \ln y - y \ln z$.

Constraint: $g(x, y, z) = x + y + z = 1$

(a) $$\left.\begin{aligned} -\ln x - 1 &= \lambda \\ -\ln y - 1 &= \lambda \\ -\ln z - 1 &= \lambda \end{aligned}\right\} x = y = z$$

$$x + y + z = 3x = 1 \Rightarrow x = y = z = \frac{1}{3}$$

(b) $H\left(\frac{1}{3}, \frac{1}{3}, \frac{1}{3}\right) = 3\left[-\frac{1}{3}\ln\left(\frac{1}{3}\right)\right] = \ln 3$

39. Maximize $V(x, y, z) = (2x)(2y)(2z) = 8xyz$ subject to the constraint $\dfrac{x^2}{a^2} + \dfrac{y^2}{b^2} + \dfrac{z^2}{c^2} = 1$.

$$\left.\begin{aligned} 8yz &= \frac{2x}{a^2}\lambda \\ 8xz &= \frac{2y}{b^2}\lambda \\ 8xy &= \frac{2z}{c^2}\lambda \end{aligned}\right\} \frac{x^2}{a^2} = \frac{y^2}{b^2} = \frac{z^2}{c^2}$$

$$\frac{x^2}{a^2} + \frac{y^2}{b^2} + \frac{z^2}{c^2} = 1 \Rightarrow \frac{3x^2}{a^2} = 1, \frac{3y^2}{b^2} = 1, \frac{3z^2}{c^2} = 1$$

$$x = \frac{a}{\sqrt{3}}, y = \frac{b}{\sqrt{3}}, z = \frac{c}{\sqrt{3}}$$

So, the dimensions of the box are $\dfrac{2\sqrt{3}a}{3} \times \dfrac{2\sqrt{3}b}{3} \times \dfrac{2\sqrt{3}c}{3}$.

40. (a) $f(1, 2) = 2$

(b) $f(2, 2) = 8$

41. Minimize $C(x, y, z) = 5xy + 3(2xz + 2yz + xy)$ subject to the constraint $xyz = 480$.

$$\left.\begin{aligned} 8y + 6z &= yz\lambda \\ 8x + 6z &= xz\lambda \\ 6x + 6y &= xy\lambda \end{aligned}\right\} x = y, 4y = 3z$$

$$xyz = 480 \Rightarrow \tfrac{4}{3}y^3 = 480$$

$$x = y = \sqrt[3]{360}, z = \tfrac{4}{3}\sqrt[3]{360}$$

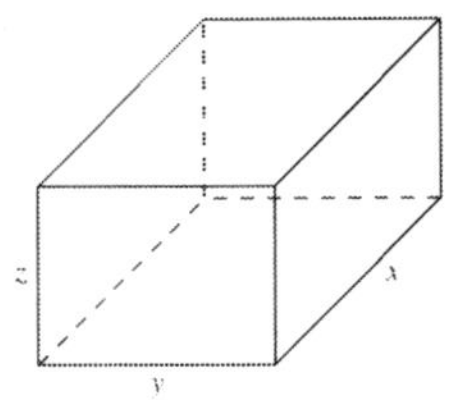

Dimensions: $\sqrt[3]{360} \times \sqrt[3]{360} \times \tfrac{4}{3}\sqrt[3]{360}$ feet.

42. (a) Maximize $P(x, y, z) = xyz$ subject to the constraint $x + y + z = S$.

$$\left.\begin{aligned} yz &= \lambda \\ xz &= \lambda \\ xy &= \lambda \end{aligned}\right\} x = y = z$$

$$x + y + z = S \Rightarrow x = y = z = \frac{S}{3}$$

So, $xyz \le \left(\dfrac{S}{3}\right)\left(\dfrac{S}{3}\right)\left(\dfrac{S}{3}\right)$, $x, y, z > 0$

$$xyz \le \frac{S^3}{27}$$

$$\sqrt[3]{xyz} \le \frac{S}{3}$$

$$\sqrt[3]{xyz} \le \frac{x + y + z}{3}.$$

(b) Maximize $P = x_1x_2x_3\cdots x_n$ subject to the constraint $\displaystyle\sum_{i=1}^{n} x_i = S$.

$$\left.\begin{aligned} x_2x_3\cdots x_n &= \lambda \\ x_1x_3\cdots x_n &= \lambda \\ x_1x_2\cdots x_n &= \lambda \\ &\vdots \\ x_1x_2x_3\cdots x_{n-1} &= \lambda \end{aligned}\right\} x_1 = x_2 = x_3 = \cdots = x_n$$

$$\sum_{i=1}^{n} x_i = S \Rightarrow x_1 = x_2 = x_3 = \cdots = x_n = \frac{S}{n}$$

So,

$$x_1x_2x_3\cdots x_n \le \left(\frac{S}{n}\right)\left(\frac{S}{n}\right)\left(\frac{S}{n}\right)\cdots\left(\frac{S}{n}\right), x_i \ge 0$$

$$x_1x_2x_3\cdots x_n \le \left(\frac{S}{n}\right)^n$$

$$\sqrt[n]{x_1x_2x_3\cdots x_n} \le \frac{S}{n}$$

$$\sqrt[n]{x_1x_2x_3\cdots x_n} \le \frac{x_1 + x_2 + x_3 + \cdots + x_n}{n}.$$

43. Minimize $A(\pi, r) = 2\pi rh + 2\pi r^2$ subject to the constraint $\pi r^2 h = V_0$.

$$\left.\begin{aligned} 2\pi h + 4\pi r &= 2\pi rh\lambda \\ 2\pi r &= \pi r^2\lambda \end{aligned}\right\} h = 2r$$

$$\pi r^2 h = V_0 \Rightarrow 2\pi r^3 = V_0$$

Dimensions: $r = \sqrt[3]{\dfrac{V_0}{2\pi}}$ and $h = 2\sqrt[3]{\dfrac{V_0}{2\pi}}$

44. Maximize $T(x, y, z) = 100 + x^2 + y^2$ subject to the constraints $x^2 + y^2 + z^2 = 50$ and $x - z = 0$.

$$\left.\begin{aligned} 2x &= 2x\lambda + \mu \\ 2y &= 2y\lambda \\ 0 &= 2z\lambda - \mu \end{aligned}\right\}$$

If $y \neq 0$, then $\lambda = 1$ and $\mu = 0, z = 0$.

So, $x = z = 0$ and $y = \sqrt{50}$.

$$T(0, \sqrt{50}, 0) = 100 + 50 = 150$$

If $y = 0$ then $x^2 + z^2 = 2x^2 = 50$ and $x = z = \sqrt{50}/2$.

$$T\left(\frac{\sqrt{50}}{2}, 0, \frac{\sqrt{50}}{2}\right) = 100 + \frac{50}{4} = 112.5$$

So, the maximum temperature is 150.

45. Using the formula $\text{Time} = \dfrac{\text{Distance}}{\text{Rate}}$, minimize

$$T(x, y) = \frac{\sqrt{d_1^2 + x^2}}{v_1} + \frac{\sqrt{d_2^2 + y^2}}{v_2}$$ subject to the constraint $x + y = a$.

$$\left.\begin{aligned} \frac{x}{v_1\sqrt{d_1^2 + x^2}} &= \lambda \\ \frac{y}{v_2\sqrt{d_2^2 + y^2}} &= \lambda \end{aligned}\right\} \frac{x}{v_1\sqrt{d_1^2 + x^2}} = \frac{y}{v_2\sqrt{d_2^2 + y^2}}$$

$$x + y = a$$

Because $\sin\theta_1 = \dfrac{x}{\sqrt{d_1^2 + x^2}}$

and $\sin\theta_2 = \dfrac{y}{\sqrt{d_2^2 + y^2}}$,

we have $\dfrac{x/\sqrt{d_1^2 + x^2}}{v_1} = \dfrac{y/\sqrt{d_2^2 + y^2}}{v_2}$ or

$$\frac{\sin\theta_1}{v_1} = \frac{\sin\theta_2}{v_2}.$$

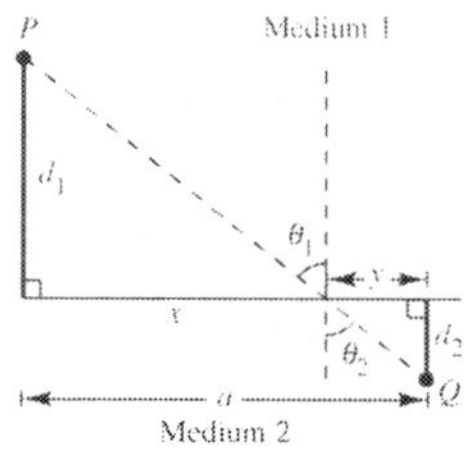

46. Case 1: Minimize $P(l, h) = 2h + l + \left(\dfrac{\pi l}{2}\right)$ subject to the constraint $lh + \left(\dfrac{\pi l^2}{8}\right) = A$.

$$1 + \frac{\pi}{2} = \left(h + \frac{\pi l}{4}\right)\lambda$$

$$2 = l\lambda \Rightarrow \lambda = \frac{2}{l}, 1 + \frac{\pi}{2} = \frac{2h}{l} + \frac{\pi}{2}$$

$$l = 2h$$

Case 2: Minimize $A(l, h) = lh + \left(\dfrac{\pi l^2}{8}\right)$ subject to the constraint $2h + l + \left(\dfrac{\pi l}{2}\right) = P$.

$$h + \frac{\pi l}{4} = \left(1 + \frac{\pi}{2}\right)\lambda$$

$$l = 2\lambda \Rightarrow \lambda = \frac{l}{2}, h + \frac{\pi l}{4} = \frac{l}{2} + \frac{\pi l}{4}$$

$$h = \frac{l}{2} \text{ or } l = 2h$$

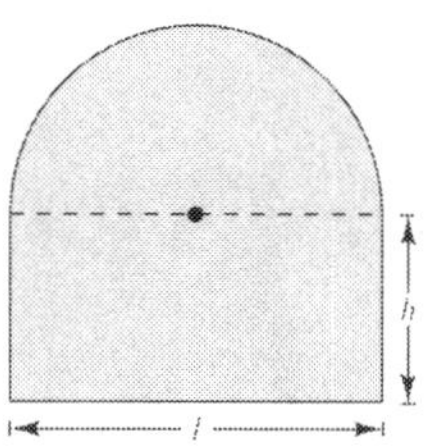

47. Maximize $P(x, y) = 100x^{0.25}y^{0.75}$ subject to the constraint $72x + 60y = 250{,}000$.

$$25x^{-0.75}y^{0.75} = 72\lambda \Rightarrow \left(\frac{y}{x}\right)^{0.75} = \frac{72\lambda}{25}$$

$$75x^{0.25}y^{-0.25} = 60\lambda \Rightarrow \left(\frac{x}{y}\right)^{0.25} = \frac{60\lambda}{75}$$

$$\left(\frac{y}{x}\right)^{0.75}\left(\frac{y}{x}\right)^{0.25} = \left(\frac{72\lambda}{25}\right)\left(\frac{75}{60\lambda}\right)$$

$$\frac{y}{x} = \frac{18}{5}$$

$$y = \frac{18}{5}x$$

$$72x + 60\left(\frac{18}{5}x\right) = 288x = 250{,}000 \Rightarrow x = \frac{15{,}625}{18}$$

$$y = 3125$$

$$P\left(\frac{15625}{18}, 3125\right) \approx 226{,}869$$

48. Maximize $P(x, y) = 100x^{0.4}y^{0.6}$ subject to the constraint $72x + 60y = 250{,}000$.

$$40x^{-0.6}y^{0.6} = 72\lambda \Rightarrow \left(\frac{y}{x}\right)^{0.6} = \frac{72\lambda}{40}$$

$$60x^{0.4}y^{-0.4} = 60\lambda \Rightarrow \left(\frac{x}{y}\right)^{0.4} = \frac{60\lambda}{60} = \lambda$$

$$\left(\frac{y}{x}\right)^{0.6}\left(\frac{y}{x}\right)^{0.4} = \frac{72\lambda}{40}\cdot\frac{1}{\lambda}$$

$$\frac{y}{x} = \frac{9}{5} \Rightarrow y = \frac{9}{5}x$$

$$72x + 60\left(\frac{9}{5}x\right) = 180x = 250{,}000 \Rightarrow x = \frac{125{,}000}{9}$$

$$y = 2500$$

$$P\left(\frac{125{,}000}{9}, 2500\right) \approx 496{,}399$$

49. Minimize $C(x, y) = 72x + 60y$ subject to the constraint $100x^{0.25}y^{0.75} = 50{,}000$.

$$72 = 25x^{-0.75}y^{0.75}\lambda \Rightarrow \left(\frac{y}{x}\right)^{0.75} = \frac{72}{25\lambda}$$

$$60 = 75x^{0.25}y^{-0.25}\lambda \Rightarrow \left(\frac{x}{y}\right)^{0.25} = \frac{60}{75\lambda}$$

$$\left(\frac{y}{x}\right)^{0.75}\left(\frac{y}{x}\right)^{0.25} = \frac{72}{25\lambda}\cdot\frac{75\lambda}{60}$$

$$\frac{y}{x} = \frac{18}{5} \Rightarrow y = \frac{18}{5}x = 3.6x$$

$$100x^{0.25}(3.6x)^{0.75} = 50{,}000$$

$$x = \frac{500}{3.6^{0.75}} \approx 191.3124$$

$$y = 3.6x \approx 688.7247$$

$$C(191.3124, 688.7247) \approx 55{,}097.97$$

50. Minimize $C(x, y) = 72x + 60y$ subject to the constraint $100x^{0.6}y^{0.4} = 50{,}000$.

$$72 = 60x^{-0.4}y^{0.4}\lambda \Rightarrow \left(\frac{y}{x}\right)^{0.4} = \frac{72}{60\lambda}$$

$$60 = 40x^{0.6}y^{-0.6}\lambda \Rightarrow \left(\frac{x}{y}\right)^{0.6} = \frac{60}{40\lambda} = \frac{3}{2\lambda}$$

$$\left(\frac{y}{x}\right)^{0.4}\left(\frac{y}{x}\right)^{0.6} = \frac{72}{60\lambda}\cdot\frac{2\lambda}{3}$$

$$\frac{y}{x} = \frac{4}{5} \Rightarrow y = \frac{4}{5}x$$

$$100x^{0.6}\left(\frac{4}{5}x\right)^{0.4} = 50{,}000$$

$$x = \frac{500}{(4/5)^{0.4}}$$

$$y = \frac{400}{(4/5)^{0.4}}$$

$$C\left(\frac{500}{(4/5)^{0.4}}, \frac{400}{(4/5)^{0.4}}\right) \approx \$65{,}601.72$$

51. Let r = radius of cylinder, and h = height of cylinder = height of cone.

$$S = 2\pi rh + 2\pi r\sqrt{h^2 + r^2} = \text{constant surface area}$$

$$V = \pi r^2 h + \frac{2\pi r^2 h}{3} = \frac{5\pi r^2 h}{3} \text{ volume}$$

We maximize $f(r, h) = r^2 h$ subject to $g(r, h) = rh + r\sqrt{h^2 + r^2} = C$.

$$(C - rh)^2 = r^2(h^2 + r^2)$$

$$C^2 - 2Crh = r^4$$

$$h = \frac{C^2 - r^4}{2Cr}$$

$$f(r, h) = F(r) = r^2\left[\frac{C^2 - r^4}{2Cr}\right] = \frac{Cr}{2} - \frac{r^5}{2C}$$

$$F'(r) = \frac{C}{2} - \frac{5r^4}{2C} = 0$$

$$C^2 = 5r^4$$

$$r^2 = \frac{C}{\sqrt{5}}$$

$$F''(r) = \frac{-10r^3}{C}$$

$$h = \frac{C^2 - r^4}{2Cr} = \frac{C^2 - C^2/5}{2C\left(C^2/5\right)^{1/4}}$$

$$= \frac{(4/5)C}{2\left(C^2/5\right)^{1/4}}$$

$$= \frac{2}{5}\frac{C}{r}$$

$$= \frac{2}{5r}\left(\sqrt{5}r^2\right)$$

$$= \frac{2\sqrt{5}}{5}r$$

So, $\dfrac{h}{r} = \dfrac{2\sqrt{5}}{5}$.

By the Second Derivative Test, this is a maximum.

Review Exercises for Chapter 13

1. $f(x, y) = 3x^2 y$

(a) $f(1, 3) = 3(1)^2(3) = 9$

(b) $f(-1, 1) = 3(-1)^2(1) = 3$

(c) $f(-4, 0) = 3(-4)^2(0) = 0$

(d) $f(x, z) = 3x^2(2) = 6x^2$

2. $f(x, y) = 6 - 4x - 2y^2$

(a) $f(0, 2) = 6 - 4(0) - 2(2)^2 = -2$

(b) $f(5, 0) = 6 - 4(5) - 2(0)^2 = -14$

(c) $f(-1, -2) = 6 - 4(-1) - 2(-2)^2 = 2$

(d) $f(-3, y) = 6 - 4(-3) - 2y^2 = 18 - 2y^2$

3. $f(x, y) = \dfrac{\sqrt{x}}{y}$

The domain is $\{(x, y) : x \geq 0, y \neq 0\}$.

The range is all real numbers.

4. $f(x, y) = \sqrt{36 - x^2 - y^2}$

Domain: $\{(x, y) : x^2 + y^2 \leq 36\}$

Range: $0 \leq z \leq 6$

(The surface is a hemisphere.)

5. $z = 3 - 2x + y$

The level curves are parallel lines of the form $y = 2x - 3 + c$.

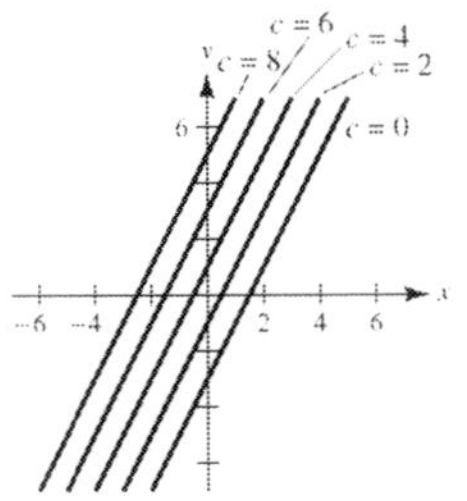

6. $z = 2x^2 + y^2$

The level curves are ellipses of the form $2x^2 + y^2 = c$ (except $2x^2 + y^2 = 0$ is the point (0, 0)).

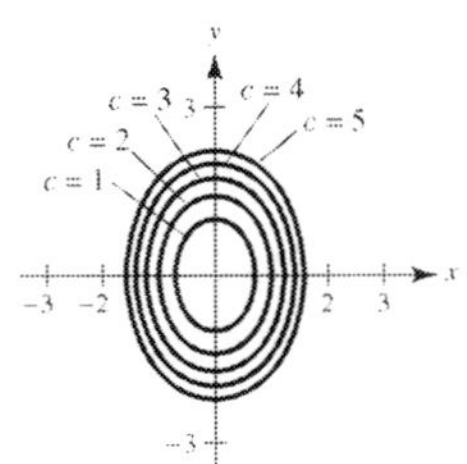

7. $f(x, y) = x^2 + y^2$

(a)

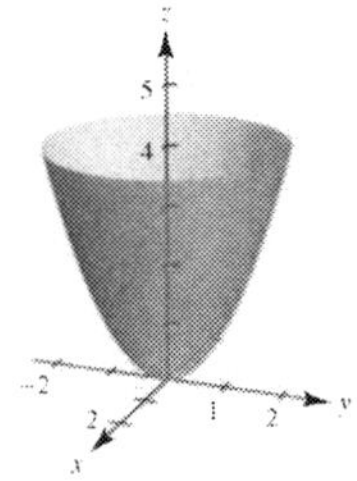

(b) $g(x, y) = f(x, y) + 2$ is a vertical translation of f two units upward.

(c) $g(x, y) = f(x, y - z)$ is a horizontal translation of f two units to the right. The vertex moves from $(0, 0, 0)$ to $(0, 2, 0)$.

(d)

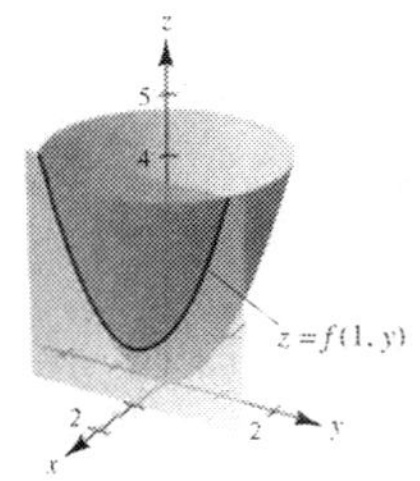

$z = f(1, y)$

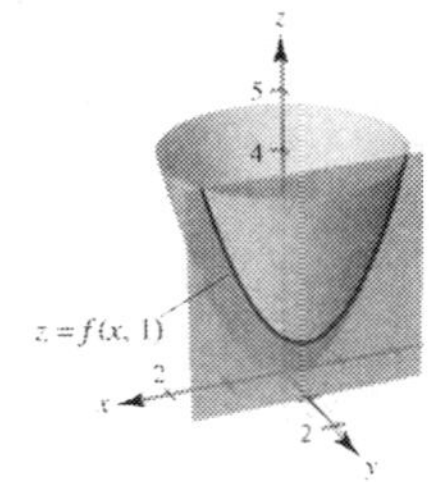

$z = f(x, 1)$

8. $A(r, t) = 2000e^{rt}$

	Number of years			
Rate	5	10	15	20
0.02	2210.34	2442.81	2699.72	2983.65
0.04	2442.81	2983.65	3644.24	4451.08
0.06	2699.72	3644.24	4919.21	6640.23
0.07	2838.14	4027.51	5715.30	8110.40

9. $f(x, y, z) = x^2 - y + z^2 = 2$

$y = x^2 + z^2 - 2$

Elliptic paraboloid

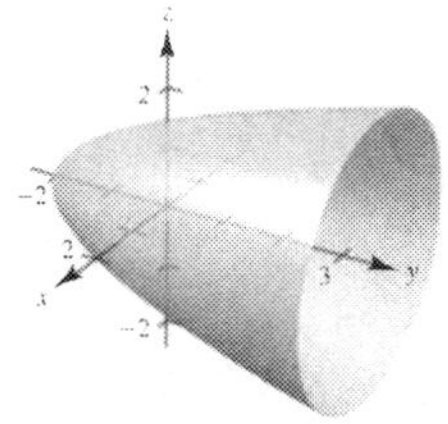

10. $f(x, y, z) = 4x^2 - y^2 + 4z^2 = 0$

Elliptic cone

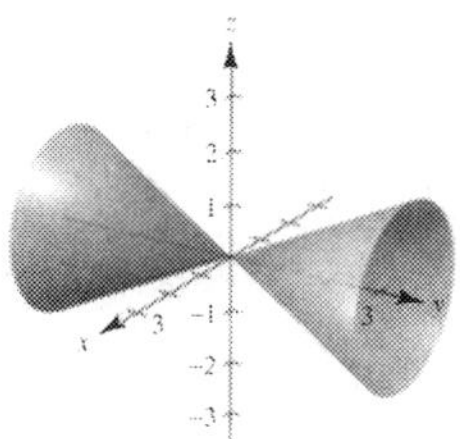

11. $\lim_{(x,y)\to(1,1)} \dfrac{xy}{x^2 + y^2} = \dfrac{1}{2}$

Continuous except at $(0, 0)$.

12. $\lim_{(x,y)\to(1,1)} \dfrac{xy}{x^2 - y^2}$

Does not exist.

Continuous except when $y = \pm x$.

13. $\lim_{(x,y)\to(0,0)} \dfrac{y + xe^{-y^2}}{1 + x^2} = \dfrac{0 + 0}{1 + 0} = 0$

Continuous everywhere.

14. $\lim_{(x,y)\to(0,0)} \dfrac{x^2 y}{x^4 + y^2}$

For $y = x^2$, $\dfrac{x^2 y}{x^4 + y^2} = \dfrac{x^4}{x^4 + x^4} \to \dfrac{1}{2}$.

For $y = 0$, $\dfrac{x^2 y}{x^4 + y^2} = 0$ for $x \neq 0$.

The limit does not exist.

Continuous to all $(x, y) \neq (0, 0)$

15. $f(x, y) = 5x^3 + 7y - 3$

$\dfrac{\partial f}{\partial x} = 15x^2 \qquad \dfrac{\partial f}{\partial y} = 7$

16. $f(x, y) = 4x^2 - 2xy + y^2$

$\dfrac{\partial f}{\partial x} = 8x - 2y$

$\dfrac{\partial f}{\partial y} = -2x + 2y$

17. $f(x, y) = e^x \cos y$

$f_x = e^x \cos y$

$f_y = -e^x \sin y$

18. $f(x, y) = \dfrac{xy}{x + y}$

$f_x = \dfrac{y(x + y) - xy}{(x + y)^2} = \dfrac{y^2}{(x + y)^2}$

$f_y = \dfrac{x^2}{(x + y)^2}$

19. $f(x, y) = y^3 e^{4x}$

$\dfrac{\partial f}{\partial x} = 4y^3 e^{4x}$

$\dfrac{\partial f}{\partial y} = 3y^2 e^{4x}$

20. $z = \ln(x^2 + y^2 + 1)$

$\dfrac{\partial z}{\partial x} = \dfrac{2x}{x^2 + y^2 + 1}$

$\dfrac{\partial z}{\partial y} = \dfrac{2y}{x^2 + y^2 + 1}$

21. $f(x, y, z) = 2xz^2 + 6xyz - 5xy^3$

$\dfrac{\partial f}{\partial x} = 2z^2 + 6yz - 5y^3$

$\dfrac{\partial f}{\partial y} = 6xz - 15xy^2$

$\dfrac{\partial f}{\partial z} = 4xz + 6xy$

22. $w = \sqrt{x^2 - y^2 - z^2}$

$\dfrac{\partial w}{\partial x} = \dfrac{1}{2}(x^2 - y^2 - z^2)^{-1/2}(2x) = \dfrac{x}{\sqrt{x^2 - y^2 - z^2}}$

$\dfrac{\partial w}{\partial y} = \dfrac{-y}{\sqrt{x^2 - y^2 - z^2}}$

$\dfrac{\partial w}{\partial z} = \dfrac{-z}{\sqrt{x^2 - y^2 - z^2}}$

23. $f(x, y) = 3x^2 - xy + 2y^3$

$f_x = 6x - y$

$f_y = -x + 6y^2$

$f_{xx} = 6$

$f_{yy} = 12y$

$f_{xy} = -1$

$f_{yx} = -1$

24. $h(x, y) = \dfrac{x}{x + y}$

$h_x = \dfrac{y}{(x + y)^2}$

$h_y = \dfrac{-x}{(x + y)^2}$

$h_{xx} = \dfrac{-2y}{(x + y)^3}$

$h_{yy} = \dfrac{2x}{(x + y)^3}$

$h_{xy} = \dfrac{(x + y)^2 - 2y(x + y)}{(x + y)^4} = \dfrac{x - y}{(x + y)^3}$

$h_{yx} = \dfrac{-(x + y)^2 + 2y(x + y)}{(x + y)^4} = \dfrac{x - y}{(x + y)^3}$

25. $h(x, y) = x \sin y + y \cos x$

$h_x = \sin y - y \sin x$

$h_y = x \cos y + \cos x$

$h_{xx} = -y \cos x$

$h_{yy} = -x \sin y$

$h_{xy} = \cos y - \sin x$

$h_{yx} = \cos y - \sin x$

26. $g(x, y) = \cos(x - 2y)$

$g_x = -\sin(x - 2y)$

$g_y = 2 \sin(x - 2y)$

$g_{xx} = -\cos(x - 2y)$

$g_{yy} = -4 \cos(x - 2y)$

$g_{xy} = 2 \cos(x - 2y)$

$g_{yx} = 2 \cos(x - 2y)$

27. $z = x^2 \ln(y + 1)$

$\dfrac{\partial z}{\partial x} = 2x \ln(y + 1)$. At $(2, 0, 0)$, $\dfrac{\partial z}{\partial x} = 0$.

Slope in x-direction.

$\dfrac{\partial z}{\partial y} == \dfrac{x^2}{1 + y}$. At $(2, 0, 0)$, $\dfrac{\partial z}{\partial y} = 4$.

Slope in y-direction.

28. $R = 300x_1 + 300x_2 - 5x_1^2 - 10x_1x_2 - 5x_2^2$

(a) $\dfrac{\partial R}{\partial x_1} = 300 - 10x_1 - 10x_2$

At $(x_1, x_2) = (5, 8)$,

$\dfrac{\partial R}{\partial x_1} = 300 - 10(5) - 10(8) = 170.$

(b) $\dfrac{\partial R}{\partial x_2} = 300 - 10x_1 - 10x_2$

At $(x_1, x_2) = (5, 8)$,

$\dfrac{\partial R}{\partial x_2} = 300 - 10(5) - 10(8) = 170.$

29. $z = x \sin xy$

$dz = \dfrac{\partial z}{\partial x}dx + \dfrac{\partial z}{\partial y}dy = (xy \cos xy + \sin xy)dx + (x^2 \cos xy)dy$

30. $z = 5x^4y^3$

$dz = \dfrac{\partial z}{\partial x} dx + \dfrac{\partial z}{\partial y} dy = 20x^3y^3\, dx + 15x^4y^2\, dy$

31. $w = 3xy^2 - 2x^3yz^2$

$$dw = \frac{\partial w}{\partial x} dx + \frac{\partial w}{\partial y} dy + \frac{\partial w}{\partial z} dz$$
$$= (3y^2 - 6x^2yz^2)dx + (6xy - 2x^3z^2)dy - 4x^3yz\, dz$$

32. $w = \dfrac{3x + 4y}{y + 3z}$

$$dw = \frac{\partial w}{\partial x}dx + \frac{\partial w}{\partial y}dy + \frac{\partial w}{\partial z}dz$$

$$= \frac{3}{y + 3z}dx + \frac{3(4z - x)}{(y + 3z)^2}dy + \frac{-3(3x + 4y)}{(y + 3z)^2}dz$$

33. $f(x, y) = 4x + 2y$

(a) $f(2, 1) = 4(2) + 2(1) = 10$

$f(2.1, 1.05) = 4(2.1) + 2(1.05) = 10.5$

$\Delta z = 10.5 - 10 = 0.5$

(b) $dz = 4dx + 2dy$

$= 4(0.1) + 2(0.05) = 0.5$

34. $f(x, y) = 36 - x^2 - y^2$

(a) $f(2, 1) = 36 - 2^2 - 1^2 = 31$

$f(2.1, 1.05) = 36 - (2.1)^2 - (1.05)^2 = 30.4875$

$\Delta z = 30.4875 - 31 = -0.5125$

(b) $dz = -2x\,dx - 2y\,dy$

$= -2(2)(0.1) - 2(1)(0.05) = -0.5$

35. $V = \dfrac{1}{3}\pi r^2 h$

$$dV = \frac{2}{3}\pi rh\,dr + \frac{1}{3}\pi r^2\,dh$$

$$= \frac{2}{3}\pi(2)(5)\left(\pm\frac{1}{8}\right) + \frac{1}{3}\pi(2)^2\left(\pm\frac{1}{8}\right)$$

$$= \pm\frac{5}{6}\pi + \frac{1}{6}\pi = \pm\pi \text{ in.}^3 \quad \text{Propogated error}$$

$$V = \frac{1}{3}\pi(2)^2 5 = \frac{20}{3}\pi \text{ in.}^3$$

$$\text{Relative error} = \frac{dV}{V} = \frac{\pm\pi}{\left(\frac{20}{3}\pi\right)} = \frac{3}{20} = 15\%$$

36. $A = \pi r\sqrt{r^2 + h^2}$

$$dA = \left(\pi\sqrt{r^2 + h^2} + \frac{\pi r^2}{\sqrt{r^2 + h^2}}\right)dr + \frac{\pi rh}{\sqrt{r^2 + h^2}}dh$$

$$= \frac{\pi(2r^2 + h^2)}{\sqrt{r^2 + h^2}}dr + \frac{\pi rh}{\sqrt{r^2 + h^2}}dh = \frac{\pi(8 + 25)}{\sqrt{29}}\left(\pm\frac{1}{8}\right) + \frac{10\pi}{\sqrt{29}}\left(\pm\frac{1}{8}\right) = \pm\frac{43\pi}{8\sqrt{29}} \quad \text{Propogated error}$$

$$A = 2\pi\sqrt{2^2 + 5^2}$$

$$= 2\pi\sqrt{29}$$

$$\text{Relative error} = \frac{dA}{A} = \frac{\pm\frac{43\pi}{8\sqrt{29}}}{2\pi\sqrt{29}} \approx 0.0927 = 9.27\%$$

37. $w = \ln(x^2 + y)$, $x = 2t$, $y = 4 - t$

(a) Chain Rule: $\dfrac{dw}{dt} = \dfrac{\partial w}{\partial x}\dfrac{dx}{dt} + \dfrac{\partial w}{\partial y}\dfrac{dy}{dt}$

$$= \frac{2x}{x^2 + y}(2) + \frac{1}{x^2 + y}(-1)$$

$$= \frac{8t - 1}{4t^2 + 4 - t}$$

(b) Substitution: $w = \ln(x^2 + y) = \ln(4t^2 + 4 - t)$

$$\frac{dw}{dt} = \frac{1}{4t^2 + 4 - t}(8t - 1)$$

38. $w = y^2 - x$, $x = \cos t$, $y = \sin t$

(a) Chain Rule: $\dfrac{dw}{dt} = \dfrac{\partial w}{\partial x}\dfrac{\partial x}{\partial t} + \dfrac{\partial w}{\partial y}\dfrac{\partial y}{\partial t}$

$$= -1(-\sin t) + 2y(\cos t)$$

$$= \sin t + 2(\sin t)\cos t$$

$$= \sin t(1 + 2\cos t)$$

(b) Substitution: $w = \sin^2 t - \cos t$

$$\frac{dw}{dt} = 2\sin t\cos t + \sin t$$

$$= \sin t(1 + 2\cos t)$$

39. $w = \dfrac{xy}{z}, x = 2r + t, y = rt, z = 2r - t$

(a) Chain Rule: $\dfrac{\partial w}{\partial r} = \dfrac{\partial w}{\partial x}\dfrac{\partial x}{\partial r} + \dfrac{\partial w}{\partial y}\dfrac{\partial y}{\partial r} + \dfrac{\partial w}{\partial z}\dfrac{\partial z}{\partial r}$

$$= \frac{y}{z}(2) + \frac{x}{z}(t) - \frac{xy}{z^2}(2)$$

$$= \frac{2rt}{2r - t} + \frac{(2r + t)t}{2r - t} - \frac{2(2r + t)(rt)}{(2r - t)^2}$$

$$= \frac{4r^2t - 4rt^2 - t^3}{(2r - t)^2}$$

$$\frac{\partial w}{\partial t} = \frac{\partial w}{\partial x}\frac{\partial x}{\partial t} + \frac{\partial w}{\partial y}\frac{\partial y}{\partial t} + \frac{\partial w}{\partial z}\frac{\partial z}{\partial t}$$

$$= \frac{y}{z}(1) + \frac{x}{z}(r) = \frac{xy}{z^2}(-1)$$

$$= \frac{4r^2t - rt^2 + 4r^3}{(2r - t)^2}$$

(b) Substitution: $w = \dfrac{xy}{z} = \dfrac{(2r + t)(rt)}{2r - t} = \dfrac{2r^2t + rt^2}{2r - t}$

$$\frac{\partial w}{\partial r} = \frac{4r^2t - 4rt^2 - t^3}{(2r - t)^2}$$

$$\frac{\partial w}{\partial t} = \frac{4r^2t - rt^2 + 4r^3}{(2r - t)^2}$$

40. $w = x^2 + y^2 + z^2, x = r\cos t, y = r\sin t, z = t$

(a) Chain Rule: $\dfrac{\partial w}{\partial r} = \dfrac{\partial w}{\partial x}\dfrac{\partial x}{\partial r} + \dfrac{\partial w}{\partial y}\dfrac{\partial y}{\partial r} + \dfrac{\partial w}{\partial z}\dfrac{\partial z}{\partial r}$

$$= 2x\cos t + 2y\sin t + 2z(0)$$

$$= 2\left(r\cos^2 t + r\sin^2 t\right) = 2r$$

$$\frac{\partial w}{\partial t} = \frac{\partial w}{\partial x}\frac{\partial x}{\partial t} + \frac{\partial w}{\partial y}\frac{\partial y}{\partial t} + \frac{\partial w}{\partial z}\frac{\partial z}{\partial t}$$

$$= 2x(-r\sin t) + 2y(r\cos t) + 2z = 2\left(-r^2\sin t\cos t + r^2\sin t\cos t\right) + 2t = 2t$$

(b) Substitution: $w(r, t) = r^2\cos^2 t + r^2\sin^2 t + t^2 = r^2 + t^2$

$$\frac{\partial w}{\partial r} = 2r$$

$$\frac{\partial w}{\partial t} = 2t$$

41. $x^2 + xy + y^2 + yz + z^2 = 0$

$$2x + y + y\frac{\partial z}{\partial x} + 2z\frac{\partial z}{\partial x} = 0$$

$$\frac{\partial z}{\partial x} = \frac{-2x - y}{y + 2z}$$

$$x + 2y + y\frac{\partial z}{\partial y} + z + 2z\frac{\partial z}{\partial y} = 0$$

$$\frac{\partial z}{\partial y} = \frac{-x - 2y - z}{y + 2z}$$

42. $xz^2 - y\sin z = 0$

$$2xz\frac{\partial z}{\partial x} + z^2 - y\cos z\frac{\partial z}{\partial x} = 0$$

$$\frac{\partial z}{\partial x} = \frac{z^2}{y\cos z - 2xz}$$

$$2xz\frac{\partial z}{\partial y} - y\cos z\frac{\partial z}{\partial y} - \sin z = 0$$

$$\frac{\partial z}{\partial y} = \frac{\sin z}{2xz - y\cos z}$$

43. $f(x, y) = x^2y, P(-5, 5), \mathbf{v} = 3\mathbf{i} - 4\mathbf{j}$

$$\mathbf{u} = \frac{\mathbf{v}}{\|\mathbf{v}\|} = \frac{3}{5}\mathbf{i} - \frac{4}{5}\mathbf{j}$$

$$D_{\mathbf{u}}f(x, y) = \frac{\partial f}{\partial x}\cos\theta + \frac{\partial f}{\partial y}\sin\theta$$

$$= 2xy\cos\theta + x^2\sin\theta$$

$$D_{\mathbf{u}}f(-5, 5) = 2(-5)(5)\left(\frac{3}{5}\right) + (-5)^2\left(-\frac{4}{5}\right)$$

$$= -30 - 20 = -50$$

44. $f(x, y) = \frac{1}{4}y^2 - x^2, P(1, 4), \mathbf{v} = 2\mathbf{i} + \mathbf{j}$

$$\mathbf{u} = \frac{\mathbf{v}}{\|\mathbf{v}\|} = \frac{2}{\sqrt{5}}\mathbf{i} + \frac{1}{\sqrt{5}}\mathbf{j}$$

$$D_{\mathbf{u}}f(x, y) = \frac{\partial f}{\partial x}\cos\theta + \frac{\partial f}{\partial y}\sin\theta$$

$$= -2x\cos\theta + \frac{1}{2}y\sin\theta$$

$$D_{\mathbf{u}}f(1, 4) = -2\left(\frac{2}{\sqrt{5}}\right) + 2\left(\frac{1}{\sqrt{5}}\right) = -\frac{2}{\sqrt{5}} = -\frac{2\sqrt{5}}{5}$$

45.

$$w = y^2 + xz$$

$$\nabla w = z\mathbf{i} + 2y\mathbf{j} + x\mathbf{k}$$

$$\nabla w(1, 2, 2) = 2\mathbf{i} + 4\mathbf{j} + \mathbf{k}$$

$$\mathbf{u} = \tfrac{1}{3}\mathbf{v} = \tfrac{2}{3}\mathbf{i} - \tfrac{1}{3}\mathbf{j} + \tfrac{2}{3}\mathbf{k}$$

$$D_{\mathbf{u}}w(1, 2, 2) = \nabla w(1, 2, 2) \cdot \mathbf{u} = \tfrac{4}{3} - \tfrac{4}{3} + \tfrac{2}{3} = \tfrac{2}{3}$$

46. $w = 5x^2 + 2xy - 3y^2z$

$$\nabla w = (10x + 2y)\mathbf{i} + (2x - 6yz)\mathbf{j} - 3y^2\mathbf{k}$$

$$\nabla w(1, 0, 1) = 10\mathbf{i} + 2\mathbf{j}$$

$$\mathbf{u} = \frac{1}{\sqrt{3}}(\mathbf{i} + \mathbf{j} - \mathbf{k})$$

$$D_{\mathbf{u}}w(1, 0, 1) = \nabla w(1, 0, 1) \cdot \mathbf{u}$$

$$= \frac{10}{\sqrt{3}} + \frac{2}{\sqrt{3}} = \frac{12}{\sqrt{3}} = 4\sqrt{3}$$

47.

$$z = x^2y$$

$$\nabla z = 2xy\mathbf{i} + x^2\mathbf{j}$$

$$\nabla_z(2, 1) = 4\mathbf{i} + 4\mathbf{j}$$

$$\|\nabla z(2, 1)\| = 4\sqrt{2}$$

48.

$$z = e^{-x}\cos y$$

$$\nabla z = -e^{-x}\cos y\mathbf{i} - e^{-x}\sin y\mathbf{j}$$

$$\nabla z\left(0, \frac{\pi}{4}\right) = -\frac{\sqrt{2}}{2}\mathbf{i} - \frac{\sqrt{2}}{2}\mathbf{j} = \left\langle -\frac{\sqrt{2}}{2}, -\frac{\sqrt{2}}{2} \right\rangle$$

$$\left\|\nabla z\left(0, \frac{\pi}{4}\right)\right\| = 1$$

49.

$$z = \frac{y}{x^2 + y^2}$$

$$\nabla z = -\frac{2xy}{(x^2 + y^2)^2}\mathbf{i} + \frac{x^2 - y^2}{(x^2 + y^2)^2}\mathbf{j}$$

$$\nabla z(1, 1) = -\frac{1}{2}\mathbf{i} = \left\langle -\frac{1}{2}, 0 \right\rangle$$

$$\|\nabla z(1, 1)\| = \frac{1}{2}$$

50.

$$z = \frac{x^2}{x - y}$$

$$\nabla z = \frac{x^2 - 2xy}{(x - y)^2}\mathbf{i} + \frac{x^2}{(x - y)^2}\mathbf{j}$$

$$\nabla z(2, 1) = 4\mathbf{j}$$

$$\|\nabla z(2, 1)\| = 4$$

51. $f(x, y) = 9x^2 - 4y^2, c = 65, P(3, 2)$

(a) $\nabla f(x, y) = 18x\mathbf{i} - 8y\mathbf{j}$

$\nabla f(3, 2) = 54\mathbf{i} - 16\mathbf{j}$

(b) Unit normal: $\dfrac{54\mathbf{i} - 16\mathbf{j}}{\|54\mathbf{i} - 16\mathbf{j}\|} = \dfrac{1}{\sqrt{793}}(27\mathbf{i} - 8\mathbf{j})$

(c) Slope $= \dfrac{27}{8}$.

$$y - z = \frac{27}{8}(x - 3)$$

$$y = \frac{27}{8}x - \frac{65}{8} \quad \text{Tangent line}$$

(d)

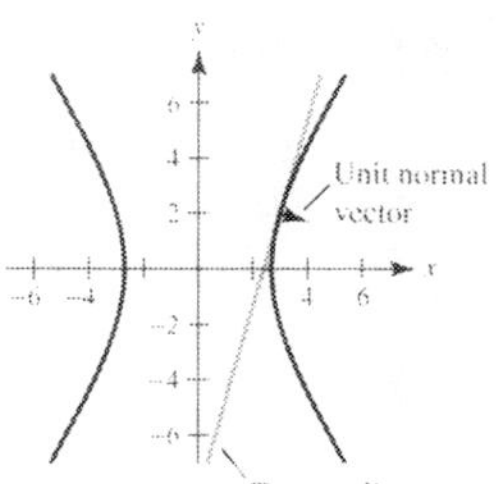

52. $f(x, y) = 4y\sin x - y, c = 3, P\left(\frac{\pi}{2}, 1\right)$

(a) $\nabla f(x, y) = 4y\cos x\mathbf{i} + (4\sin x - 1)\mathbf{j}$

$\nabla f\left(\frac{\pi}{2}, 1\right) = 3\mathbf{j}$

(b) Unit normal vector: $\mathbf{j}$

(c) Tangent line horizontal: $y = 1$

(d)

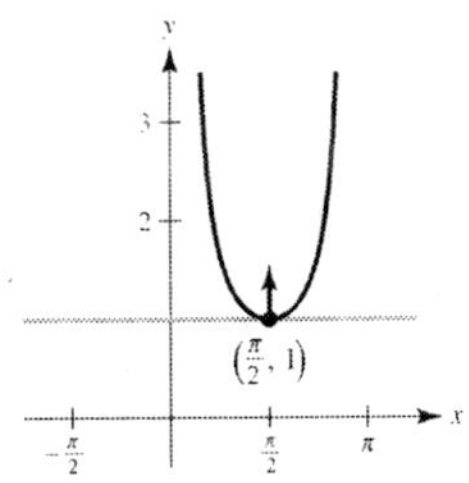

53. $F(x, y, z) = x^2 + y^2 + 2 - z = 0, (1, 3, 12)$

$\nabla F = 2x\mathbf{i} + 2y\mathbf{j} - \mathbf{k}$

$\nabla F(1, 3, 12) = 2\mathbf{i} + 6\mathbf{j} - \mathbf{k}$

Tangent Plane:

$2(x - 1) + 6(y - 3) - (z - 12) = 0$

$2x + 6y - z = 8$

54. $F(x, y, z) = 9x^2 + y^2 + 4z^2 - 25 = 0, (0, -3, 2)$

$\nabla F = 18x\mathbf{i} + 2y\mathbf{j} + 8z\mathbf{k}$

$\nabla F(0, -3, 2) = -6\mathbf{j} + 16\mathbf{k}$

Tangent Plane:

$0(x - 0) - 6(y + 3) + 16(z - 2) = 0$

$-6y + 16z = 50$

$-3y + 8z = 25$

55. $F(x, y, z) = x^2 + y^2 - 4x + 6y + z + 9 = 0$

$\nabla F = (2x - 4)\mathbf{i} + (2y + 6)\mathbf{j} + \mathbf{k}$

$\nabla F(2, -3, 4) = \mathbf{k}$

So, the equation of the tangent plane is

$z - 4 = 0$ or $z = 4$.

56. $F(x, y, z) = y^2 + z^2 - 25 = 0$

$\nabla F = 2y\mathbf{j} + 2z\mathbf{k}$

$\nabla F(2, 3, 4) = 6\mathbf{j} + 8\mathbf{k} = 2(3\mathbf{j} + 4\mathbf{k})$

So, the equation of the tangent plane is

$3(y - 3) + 4(z - 4) = 0$ or $3y + 4z = 25$.

57. $F(x, y, z) = x^2y - z = 0$

$\nabla F = 2xy\mathbf{i} + x^2\mathbf{j} - \mathbf{k}$

$\nabla F(2, 1, 4) = 4\mathbf{i} + 4\mathbf{j} - \mathbf{k}$

So, the equation of the tangent plane is

$4(x - 2) + 4(y - 1) - (z - 4) = 0$ or

$4x + 4y - z = 8$,

and the equation of the normal line is

$x = 4t + 2, y = 4t + 1, z = -t + 4$.

Symmetric equations:

$$\frac{x - 2}{4} = \frac{y - 1}{4} = -\frac{z - 4}{1}$$

58. $F(x, y, z) = x^2 + y^2 + z^2 - 9 = 0$

$\nabla F = 2x\mathbf{i} + 2y\mathbf{j} + 2z\mathbf{k}$

$\nabla F(1, 2, 2) = 2\mathbf{i} + 4\mathbf{j} + 4\mathbf{k} = 2(\mathbf{i} + 2\mathbf{j} + 2\mathbf{k})$

So, the equation of the tangent plane is

$(x - 1) + 2(y - 2) + 2(z - 2) = 0$ or

$x + 2y + 2z = 9$,

and the equation of the normal line is

$$\frac{x - 1}{1} = \frac{y - 2}{2} = \frac{z - 2}{2}.$$

59. $f(x, y, z) = x^2 + y^2 + z^2 - 14$

$\nabla f(x, y, z) = 2x\mathbf{i} + 2y\mathbf{j} + 2z\mathbf{k}$

$\nabla f(2, 1, 3) = 4\mathbf{i} + 2\mathbf{j} + 6\mathbf{k}$ Normal vector to plane.

$$\cos\theta = \frac{|\mathbf{n}\cdot\mathbf{k}|}{\|\mathbf{n}\|} = \frac{6}{\sqrt{56}} = \frac{3\sqrt{14}}{14}$$

$\theta = 36.7°$

60. (a) $f(x, y) = \cos x + \sin y,\ f(0, 0) = 1$

$f_x = -\sin x,\ f_x(0, 0) = 0$

$f_y = \cos y,\ f_y(0, 0) = 1$

$P_1(x, y) = 1 + y$

(b) $f_{xx} = -\cos x,\ f_{xx}(0, 0) = -1$

$f_{yy} = -\sin y,\ f_{yy}(0, 0) = 0$

$f_{xy} = 0,\ f_{xy}(0, 0) = 0$

$P_2(x, y) = 1 + y - \frac{1}{2}x^2$

(c) If $y = 0$, you obtain the 2nd degree Taylor polynomial for $\cos x$.

(d)

x	y	$f(x, y)$	$P_1(x, y)$	$P_2(x, y)$
0	0	1.0	1.0	1.0
0	0.1	1.0998	1.1	1.1
0.2	0.1	1.0799	1.1	1.095
0.5	0.3	1.1731	1.3	1.175
1	0.5	1.0197	1.5	1.0

(e)

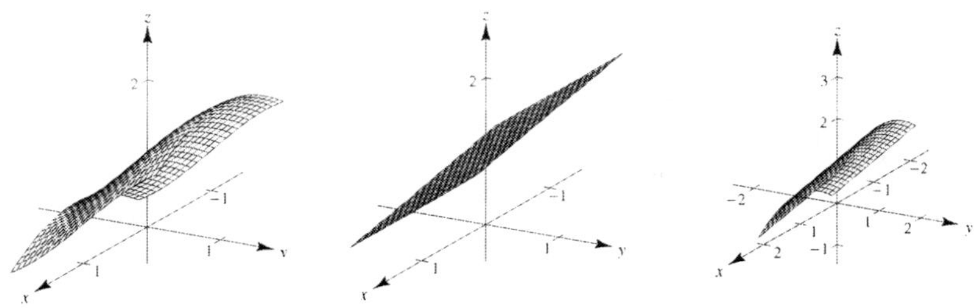

The accuracy lessens as the distance from $(0, 0)$ increases.

61. $f(x, y) = -x^2 - 4y^2 + 8x - 8y - 11$

$f_x = -2x + 8 = 0 \Rightarrow x = 4$

$f_y = -8y - 8 = 0 \Rightarrow y = -1$

$f_{xx} = -2,\ f_{yy} = -8,\ f_{xy} = 0$

$f_{xx}\,f_{yy} - \left(f_{xy}\right)^2 = (-2)(-8) - 0 = 16 > 0$

So, $(4, -1, 9)$ is a relative minimum.

62. $f(x, y) = x^2 - y^2 - 16x - 16y$

$f_x = 2x - 16 = 0 \Rightarrow x = 8$

$f_y = -2y - 16 = 0 \Rightarrow y = -8$

$f_{xx} = 2,\ f_{yy} = -2,\ f_{xy} = 0$

$f_{yy}\,f_{yy} - \left(f_{xy}\right)^2 = 2(-2) - 0 = -4 < 0$

So, $(8, -8, 0)$ is a saddle point.

63. $f(x, y) = 2x^2 + 6xy + 9y^2 + 8x + 14$

$f_x = 4x + 6y + 8 = 0$

$f_y = 6x + 18y = 0,\ x = -3y$

$4(-3y) + 6y = -8 \Rightarrow y = \frac{4}{3},\ x = -4$

$f_{xx} = 4$

$f_{yy} = 18$

$f_{xy} = 6$

$f_{xx}f_{yy} - \left(f_{xy}\right)^2 = 4(18) - (6)^2 = 36 > 0.$

So, $\left(-4, \frac{4}{3}, -2\right)$ is a relative minimum.

64. $f(x, y) = x^2 + 3xy + y^2 - 5x$

$f_x = 2x + 3y - 5 = 0$

$f_y = 3x + 2y = 0 \qquad \Rightarrow y = -\dfrac{3}{2}x$

$2x + 3\left(-\dfrac{3}{2}x\right) = 5$

$4x - 9x = 10$

$x = -2, y = 3$

$f_{xx} = 2, f_{yy} = 2, f_{xy} = 3, d = 4 - 9 < 0$

$\Rightarrow (-2, 3)$ is a saddle point.

65. $f(x, y) = xy + \dfrac{1}{x} + \dfrac{1}{y}$

$f_x = y - \dfrac{1}{x^2} = 0, x^2y = 1$

$f_y = x - \dfrac{1}{y^2} = 0, xy^2 = 1$

So, $x^2y = xy^2$ or $x = y$ and substitution yields the critical point $(1, 1)$.

$f_{xx} = \dfrac{2}{x^3}$

$f_{xy} = 1$

$f_{yy} = \dfrac{2}{y^3}$

At the critical point $(1, 1)$, $f_{xx} = 2 > 0$ and

$f_{xx}f_{yy} - (f_{xy})^2 = 3 > 0.$

So, $(1, 1, 3)$ is a relative minimum.

66. $f(x, y) = -8x^2 + 4xy - y^2 + 12x + 7$

$f_x = -16x + 4y + 12 = 0 \Rightarrow y - 4x = -3$

$f_y = 4x - 2y = 0 \Rightarrow y = 2x$

So, $x = 3/2, y = 3.$

$f_{xx} = -16, f_{yy} = -2, f_{xy} = 4$

$f_{xx}\,f_{yy} - (f_{xy})^2 = (-16)(-2) - 4^2 = 16 > 0$

So, $(3/2, 3, 16)$ is a relative maximum.

67. A point on the plane is given by $(x, y, 4 - x - y)$

The square of the distance from $(2, 1, 4)$ to a point on the plane is

$$S = (x - 2)^2 + (y - 1)^2 + (4 - x - y - 4)^2$$
$$= (x - 2)^2 + (y - 1)^2 + (-x - y)^2.$$

$S_x = 2(x - 2) - 2(-x - y) = 4x + 2y - 4$

$S_y = 2(y - 1) - 2(-x - y) = 2x + 4y - 2$

$S_x = S_y = 0 \Rightarrow \left\{\begin{matrix} 4x + 2y = 4 \\ 2x + 4y = 2 \end{matrix}\right\} \Rightarrow x = 1, y = 0, z = 3$

The distance is $\sqrt{(1 - 2)^2 + (0 - 1)^2 + (-1)^2} = \sqrt{3}$.

68. $xyz = 64 \Rightarrow z = \dfrac{64}{xy}$

$S = x + y + z = x + y + \dfrac{64}{xy}$

$Sx = 1 - \dfrac{64}{x^2y} = 0$

$Sy = 1 - \dfrac{64}{xy^2} = 0$

$$\left.\begin{matrix} \dfrac{64}{x^2y} = 1 \Rightarrow 64 = x^2y \\ \dfrac{64}{xy^2} = 1 \Rightarrow 64 = xy^2 \end{matrix}\right\} \quad x = y = 4$$

So, $x = y = z = 4$.

69. $R = -6x_1^2 - 10x_2^2 - 2x_1x_2 + 32x_1 + 84x_2$

$Rx_1 = -12x_1 - 2x_2 + 32 = 0 \Rightarrow 6x_1 + x_2 = 16$

$Rx_2 = -20x_2 - 2x_1 + 84 = 0 \Rightarrow x_1 + 10x_2 = 42$

Solving this system yields $x_1 = 2$ and $x_2 = 4$.

70. $P = 180(x_1 + x_2) - C_1 - C_2$

$= 180x_1 + 180x_2 - (0.05x_1^2 + 15x_1 + 5400) - (0.03x_2^2 + 15x_2 + 6100)$

$= -0.05x_1^2 - 0.03x_2^2 + 165x_1 + 165x_2 - 11{,}500$

$Px_1 = -0.1x_1 + 165 = 0$

$Px_2 = -0.06x_2 + 165 = 0$

Solving this system yields

$x_1 = 1650$ and

$x_2 = 2750.$

By the Second Derivative Test, this is a maximum.

71. $(0, 4), (1, 5), (3, 6), (6, 8), (8, 10)$

$\sum x_i = 18 \quad \sum y_i = 33$

$\sum x_i y_i = 151 \quad \sum x_i^2 = 110$

$$a = \frac{5(151) - 18(33)}{5(110) - (18)^2} = \frac{161}{226} \approx 0.7124$$

$$b = \frac{1}{5}\left(33 - \frac{161}{226}(18)\right) = \frac{456}{113} \approx 4.0354$$

$$y = \frac{161}{226}x + \frac{456}{113}$$

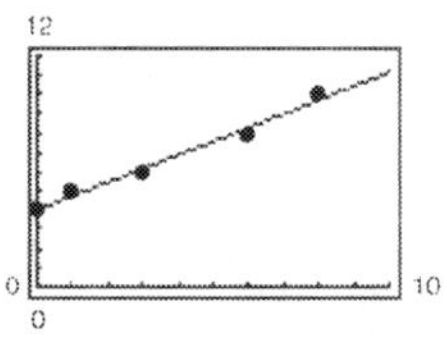

72. $(0, 10), (2, 8), (4, 7), (7, 5), (9, 3), (12, 0)$

$\sum x_i = 34 \quad \sum y_i = 33$

$\sum x_i y_i = 106 \quad \sum x_i^2 = 294$

$$a = \frac{6(106) - 34(33)}{6(294) - (34)^2} = -\frac{243}{304} \approx -0.7993$$

$$b = \frac{1}{6}\left(33 - \left(\frac{-243}{304}\right)(34)\right) = \frac{3049}{304} \approx 10.0296$$

$$y = -\frac{243}{304}x + \frac{3049}{304}$$

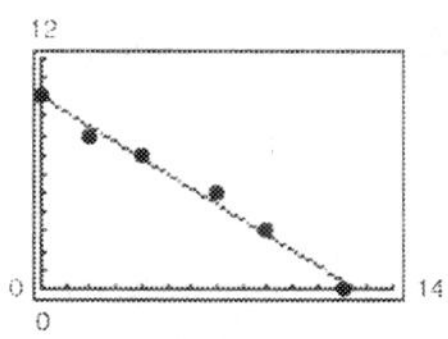

73. $(100, 35), (150, 44), (200, 50), (250, 56)$

(a) Using a graphing utility, you obtain

$y = 0.138x + 22.1.$

(b) If $x = 175$, $y = 0.138(175) + 22.1 = 46.25$ bushels per acre.

74. (a) $y = 2.29t + 2.0$

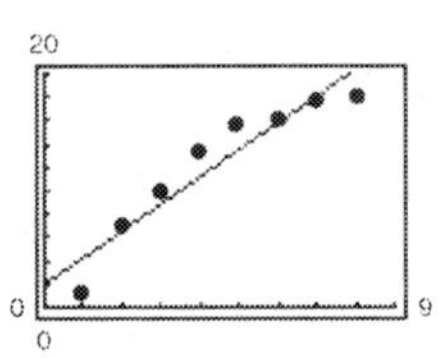

(b)

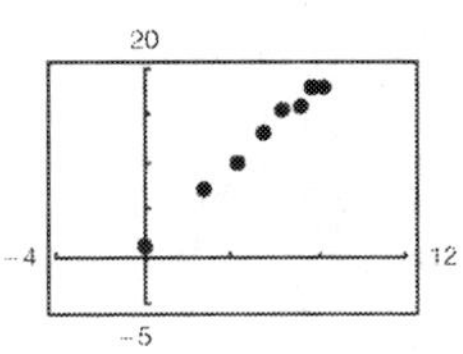

Yes, the data appear linear.

(c) $y = 1.24 + 8.37 \ln t$

(d)

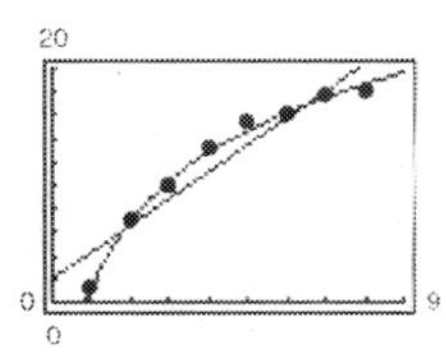

75. Minimize $f(x, y) = x^2 + y^2$

Constraint: $x + y - 8 = 0$

$\nabla f = \lambda \nabla g$

$2x\mathbf{i} + 2y\mathbf{j} = \lambda(\mathbf{i} + \mathbf{j})$

$\left.\begin{array}{l} 2x = \lambda \\ 2y = \lambda \end{array}\right\} \quad x = y$

$x + y - 8 = 2x - 8 = 0 \Rightarrow x = y = 4$

$f(4, 4) = 32$

76. Maximize $f(x, y) = xy$

Constraint: $x + 3y - 6 = 0$

$\nabla f = \lambda \nabla g$

$y\mathbf{i} + x\mathbf{j} = \lambda(\mathbf{i} + 3\mathbf{j})$

$\left.\begin{array}{l} y = \lambda \\ x = 3\lambda \end{array}\right\} \quad x = 3y$

$x + 3y - 6 = 6y - 6 = 0 \Rightarrow y = 1, x = 3$

$f(3, 1) = 3$

77. Maximize $f(x, y) = 2x + 3xy + y$

Constraint: $x + 2y = 29$

$\nabla f = \lambda \nabla g$

$\left.\begin{array}{l} 2 + 3y = \lambda \\ 3x + 1 = 2\lambda \end{array}\right\} \quad 4 + 6y = 3x + 1 \Rightarrow x - 2y = 1$

$\left.\begin{array}{l} x - 2y = 1 \\ x + 2y = 29 \end{array}\right\} \quad x = 15, y = 7$

$f(15, 7) = 2(15) + 3(15)(7) + 7 = 352$

78. Minimize $f(x, y) = x^2 - y^2$

Constraint: $x - 2y + 6 = 0$

$\nabla f = \lambda \nabla g$

$\left.\begin{array}{l} 2x = \lambda \\ -2y = -2\lambda \end{array}\right\} \quad -4x = -2y \Rightarrow y = 2x$

$x - 2y + 6 = x - 4x + 6 = 0 \Rightarrow x = 2, y = 4$

$f(2, 4) = 4 - 16 = -12$

79. Maximize $f(x, y) = 2xy$

Constraint: $2x + y = 12$

$\nabla f = \lambda \nabla g$

$\left.\begin{array}{l} 2y = 2\lambda \\ 2x = \lambda \end{array}\right\} \quad 4x = 2y \Rightarrow y = 2x$

$2x + y = 2x + 2x = 12 \Rightarrow x = 3, y = 6$

$f(3, 6) = 2(3)(6) = 36$

80. Minimize $f(x, y) = 3x^2 - y^2$

Constraint: $2x - 2y + 5 = 0$

$\nabla f = \lambda \nabla g$

$\left.\begin{array}{l} 6x = 2\lambda \\ -2y = -2\lambda \end{array}\right\} \quad 6x = 2y \Rightarrow y = 3x$

$2x - 2y + 5 = 2x - 2(3x) + 5 = 0 \Rightarrow -4x + 5 = 0$

$\Rightarrow x = \frac{5}{4}, y = \frac{15}{4}$

$f\left(\frac{5}{4}, \frac{15}{4}\right) = -\frac{75}{8}$

81. $PQ = \sqrt{x^2 + 4}$,

$QR = \sqrt{y^2 + 1}$,

$RS = z; x + y + z = 10$

$C = 3\sqrt{x^2 + 4} + 2\sqrt{y^2 + 1} + z$

Constraint: $x + y + z = 10$

$\nabla C = \lambda \nabla g$

$\frac{3x}{\sqrt{x^2 + 4}}\mathbf{i} + \frac{2y}{\sqrt{y^2 + 1}}\mathbf{j} + \mathbf{k} = \lambda[\mathbf{i} + \mathbf{j} + \mathbf{k}]$

$3x = \lambda\sqrt{x^2 + 4}$

$2y = \lambda\sqrt{y^2 + 1}$

$1 = \lambda$

$9x^2 = x^2 + 4 \Rightarrow x^2 = \frac{1}{2}$

$4y^2 = y^2 + 1 \Rightarrow y^2 = \frac{1}{3}$

So, $x = \frac{\sqrt{2}}{2} \approx 0.707$ km,

$y = \frac{\sqrt{3}}{3} \approx 0.577$ km,

$z = 10 - \frac{\sqrt{2}}{2} - \frac{\sqrt{3}}{3} \approx 8.716$ km.

Problem Solving for Chapter 13

1. (a) The three sides have lengths 5, 6, and 5.

Thus, $s = \frac{16}{2} = 8$ and $A = \sqrt{8(3)(2)(3)} = 12$.

(b) Let $f(a, b, c) = (\text{area})^2 = s(s - a)(s - b)(s - c)$, subject to the constraint

$a + b + c = $ constant (perimeter).

Using Lagrange multipliers,

$-s(s - b)(s - c) = \lambda$

$-s(s - a)(s - c) = \lambda$

$-s(s - a)(s - b) = \lambda$.

From the first 2 equations

$s - b = s - a \Rightarrow a = b$.

Similarly, $b = c$ and hence $a = b = c$ which is an equilateral triangle.

(c) Let $f(a, b, c) = a + b + c$, subject to $(\text{Area})^2 = s(s - a)(s - b)(s - c)$ constant.

Using Lagrange multipliers,

$1 = -\lambda s(s - b)(s - c)$

$1 = -\lambda s(s - a)(s - c)$

$1 = -\lambda s(s - a)(s - b)$

So, $s - a = s - b \Rightarrow a = b$ and $a = b = c$.

2. $V = \frac{4}{3}\pi r^3 + \pi r^2 h$

Material $= M = 4\pi r^2 + 2\pi rh$

$V = 1000 \Rightarrow h = \dfrac{1000 - (4/3)\pi r^3}{\pi r^2}$

So,
$$M = 4\pi r^2 + 2\pi r\left(\frac{1000 - (4/3)\pi r^3}{\pi r^2}\right)$$
$$= 4\pi r^2 + \frac{2000}{r} - \frac{8}{3}\pi r^2$$
$$\frac{dM}{dr} = 8\pi r - \frac{2000}{r^2} - \frac{16}{3}\pi r = 0$$
$$8\pi r - \frac{16}{3}\pi r = \frac{2000}{r^2}$$
$$r^3\left(\frac{8}{3}\pi\right) = 2000$$
$$r^3 = \frac{750}{\pi} \Rightarrow r = 5\left(\frac{6}{\pi}\right)^{1/3}.$$

Then, $h = \dfrac{1000 - (4/3)\pi(750/\pi)}{\pi r^2} = 0$.

The tank is a sphere of radius $r = 5\left(\dfrac{6}{\pi}\right)^{1/3}$.

3. (a) $F(x, y, z) = xyz - 1 = 0$

$F_x = yz, F_y = xz, F_z = xy$

Tangent plane:

$y_0z_0(x - x_0) + x_0z_0(y - y_0) + x_0y_0(z - z_0) = 0$

$y_0z_0x + x_0z_0y + x_0y_0z = 3x_0y_0z_0 = 3$

(b) $V = \frac{1}{3}(\text{base})(\text{height})$

$= \dfrac{1}{3}\left(\dfrac{1}{2}\dfrac{3}{y_0z_0}\dfrac{3}{x_0z_0}\right)\left(\dfrac{3}{x_0y_0}\right) = \dfrac{9}{2}$

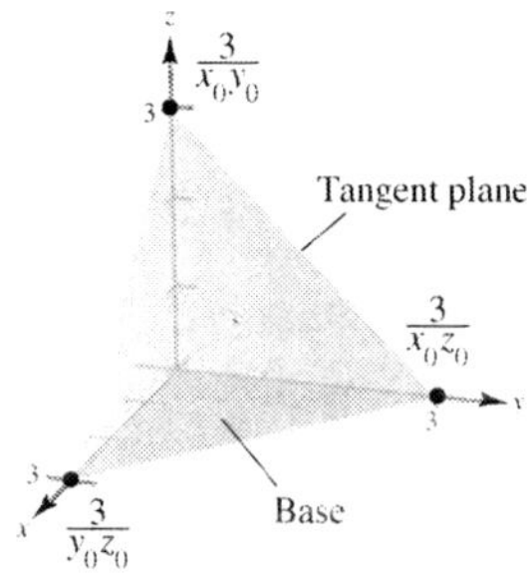

4. (a) As $x \to \pm\infty, f(x) = (x^3 - 1)^{1/3} \to x$ and hence $\lim_{x\to\infty}[f(x) - g(x)] = \lim_{x\to-\infty}[f(x) - g(x)] = 0$.

(b) Let $\left(x_0, (x_0^3 - 1)^{1/3}\right)$ be a point on the graph of f.

The line through this point perpendicular to g is $y = -x + x_0 + \sqrt[3]{x_0^3 - 1}$.

This line intersects g at the point

$\left(\frac{1}{2}\left[x_0 + \sqrt[3]{x_0^3 - 1}\right], \frac{1}{2}\left[x_0 + \sqrt[3]{x_0^3 - 1}\right]\right)$.

The square of the distance between these two points is $h(x_0) = \frac{1}{2}\left(x_0 - \sqrt[3]{x_0^3 - 1}\right)^2$.

h is a maximum for $x_0 = \dfrac{1}{\sqrt[3]{2}}$. So, the point on f farthest from g is $\left(\dfrac{1}{\sqrt[3]{2}}, -\dfrac{1}{\sqrt[3]{2}}\right)$.

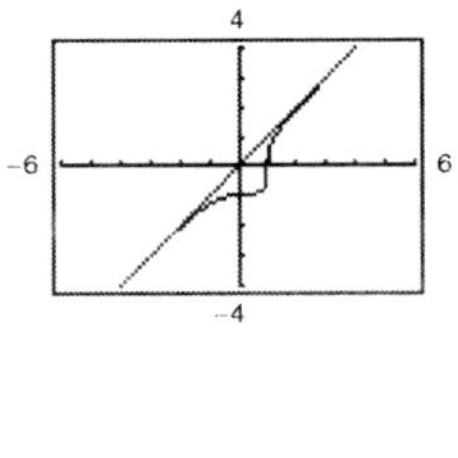

5. (a)

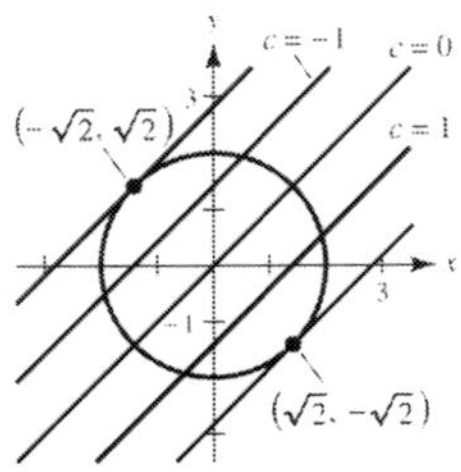

Maximum value of f is $f(\sqrt{2}, -\sqrt{2}) = 2\sqrt{2}$.

Maximize $f(x, y) = x - y$.

Constraint: $g(x, y) = x^2 + y^2 = 4$

$\nabla f = \lambda \nabla g$: $\quad 1 = 2\lambda x$

$-1 = 2\lambda y$

$x^2 + y^2 = 4$

$2\lambda x = -2\lambda y \Rightarrow x = -y$

$2x^2 = 4 \Rightarrow x = \pm\sqrt{2}, y = \mp\sqrt{2}$

$f(\sqrt{2}, -\sqrt{2}) = 2\sqrt{2}, f(-\sqrt{2}, \sqrt{2}) = -2\sqrt{2}$

(b) $f(x, y) = x - y$

Constraint: $x^2 + y^2 = 0 \Rightarrow (x, y) = (0, 0)$

Maximum and minimum values are 0.

Lagrange multipliers does not work:

$$\left.\begin{aligned} 1 &= 2\lambda x \\ -1 &= 2\lambda y \end{aligned}\right\} x = -y = 0, \text{ a contradiction.}$$

Note that $\nabla g(0, 0) = \mathbf{0}$.

6. Heat Loss $= H = k(5xy + xy + 3xz + 3xz + 3yz + 3yz)$

$= k(6xy + 6xz + 6yz)$

$V = xyz = 1000 \Rightarrow z = \dfrac{1000}{xy}$.

Then $H = 6k\left(xy + \dfrac{1000}{y} + \dfrac{1000}{x}\right)$.

Setting $H_x = H_y = 0$, you obtain $x = y = z = 10$.

7. $H = k(5xy + 6xz + 6yz)$

$z = \dfrac{1000}{xy} \Rightarrow H = k\left(5xy + \dfrac{6000}{y} + \dfrac{6000}{x}\right)$.

$H_x = 5y - \dfrac{6000}{x^2} = 0 \Rightarrow 5yx^2 = 6000$

By symmetry, $x = y \Rightarrow x^3 = y^3 = 1200$.

So, $x = y = 2\sqrt[3]{150}$ and $z = \dfrac{5}{3}\sqrt[3]{150}$.

8. (a) $T(x, y) = 2x^2 + y^2 - y + 10 = 10$

$2x^2 + y^2 - y + \dfrac{1}{4} = \dfrac{1}{4}$

$2x^2 + \left(y - \dfrac{1}{2}\right)^2 = \dfrac{1}{4}$

$\dfrac{x^2}{1/8} + \dfrac{(y - (1/2))^2}{1/4} = 1 \qquad$ ellipse

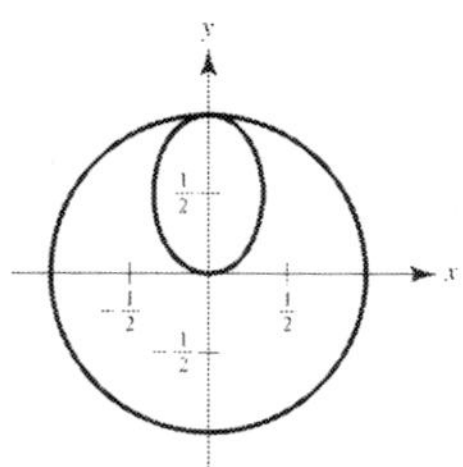

(b) On $x^2 + y^2 = 1, T(x, y) = T(y) = 2(1 - y^2) + y^2 - y + 10 = 12 - y^2 - y$

$T'(y) = -2y - 1 = 0 \Rightarrow y = -\dfrac{1}{2}, x = \pm\dfrac{\sqrt{3}}{2}$.

Inside: $T_x = 4x - 0, T_y = 2y - 1 = 0 \Rightarrow \left(0, \dfrac{1}{2}\right)$

$T\left(0, \dfrac{1}{2}\right) = \dfrac{39}{4}$ minimum

$T\left(\pm\dfrac{\sqrt{3}}{2}, -\dfrac{1}{2}\right) = \dfrac{49}{4}$ maximum

9. (a) $\dfrac{\partial f}{\partial x} = Cax^{a-1}y^{1-a}, \dfrac{\partial f}{\partial y} = C(1-a)x^a y^{-a}$

$$x\frac{\partial f}{\partial x} + y\frac{\partial f}{\partial y} = Cax^a y^{1-a} + C(1-a)x^a y^{1-a}$$
$$= \left[Ca + C(1-a)\right]x^a y^{1-a}$$
$$= Cx^a y^{1-a} = f$$

(b) $f(tx, ty) = C(tx)^a(ty)^{1-a} = Ct^a x^a t^{1-a} y^{1-a} = Cx^a y^{1-a}(t) = tf(x, y)$

10. $x^2 + y^2 = 2x$

$(x-1)^2 + y^2 = 1$ Circle

$\dfrac{x^2}{a^2} + \dfrac{y^2}{b^2} = 1$ Ellipse

The circle and ellipse intersect at (x, y) and $(x, -y)$ for a unique value of x.

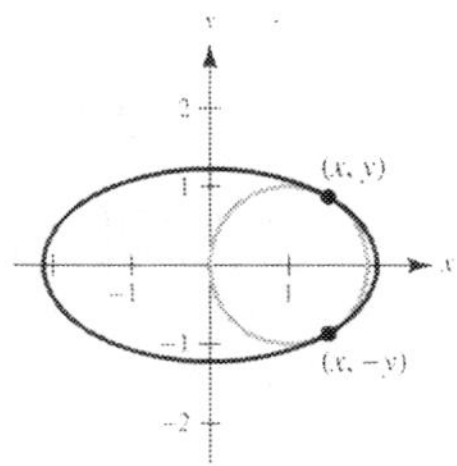

$y^2 = \dfrac{b^2}{a^2}(a^2 - x^2)$ Ellipse

$x^2 + \dfrac{b^2}{a^2}(a^2 - x^2) = 2x$ Circle

$\left(1 - \dfrac{b^2}{a^2}\right)x^2 - 2x + b^2 = 0$ Quadratic

For these to be a unique x-value, the discriminant must be 0.

$$4 - 4\left(1 - \frac{b^2}{a^2}\right)b^2 = 0$$

$$a^2 - a^2b^2 + b^4 = 0$$

We use lagrange multipliers to minimize the area $f(a, b) = \pi ab$ of the ellipse subject to the constraint

$g(a, b) = a^2 - a^2b^2 + b^4 = 0.$

$\nabla f = \lambda \nabla g$

$\langle \pi b, \pi a\rangle = \lambda\langle 2a - 2ab^2, -2a^2b + 4b^3\rangle$

$\pi b = \lambda(2a - 2ab^2)$

$\pi a = \lambda(-2a^2b + 4b^3)$

$$\lambda = \frac{\pi b}{2a - 2ab^2} = \frac{\pi a}{4b^3 - 2a^2b} \Rightarrow 4b^4 - 2a^2b^2 = 2a^2 - 2a^2b^2 \Rightarrow 2b^4 = a^2 \Rightarrow b^2 = \frac{a}{\sqrt{2}}$$

Using the constraint, $a^2 - a^2b^2 + b^4 = 0,\ a^2 - a^2\dfrac{a}{\sqrt{2}} + \dfrac{a^2}{2} = 0$

$$\frac{3}{2} = \frac{a}{\sqrt{2}}$$

$$a = \frac{3}{2}\sqrt{2}, b = \frac{\sqrt{6}}{2}.$$

Ellipse: $\dfrac{x^2}{(9/2)} + \dfrac{y^2}{(3/2)} = 1$

11. (a) $x = 64(\cos 45°)t = 32\sqrt{2}t$

$y = 64(\sin 45°)t - 16t^2 = 32\sqrt{2}t - 16t^2$

(b) $\tan\alpha = \dfrac{y}{x + 50}$

$$\alpha = \arctan\left(\frac{y}{x+50}\right) = \arctan\left(\frac{32\sqrt{2}t - 16t^2}{32\sqrt{2}t + 50}\right)$$

(c) $$\frac{d\alpha}{dt} = \frac{1}{1 + \left(\dfrac{32\sqrt{2}t - 16t^2}{32\sqrt{2}t + 50}\right)^2} \cdot \frac{-64\left(8\sqrt{2}t^2 + 25t - 25\sqrt{2}\right)}{\left(32\sqrt{2}t + 50\right)^2} = \frac{-16\left(8\sqrt{2}t^2 + 25t - 25\sqrt{2}\right)}{64t^4 - 256\sqrt{2}t^3 + 1024t^2 + 800\sqrt{2}t + 625}$$

(d)

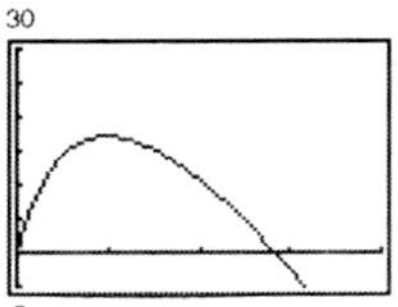

No. The rate of change of α is greatest when the projectile is closest to the camera.

(e) $\dfrac{d\alpha}{dt} = 0$ when

$8\sqrt{2}t^2 + 25t - 25\sqrt{2} = 0$

$$t = \frac{-25 + \sqrt{25^2 - 4\left(8\sqrt{2}\right)\left(-25\sqrt{2}\right)}}{2\left(8\sqrt{2}\right)} \approx 0.98 \text{ second.}$$

No, the projectile is at its maximum height when $dy/dt = 32\sqrt{2} - 32t = 0$ or $t = \sqrt{2} \approx 1.41$ seconds.

12. (a) $d = \sqrt{x^2 + y^2} = \sqrt{\left(32\sqrt{2}t\right)^2 + \left(32\sqrt{2}t - 16t^2\right)^2} = \sqrt{4096t^2 - 1024\sqrt{2}t^3 + 256t^4} = 16t\sqrt{t^2 - 4\sqrt{2}t + 16}$

(b) $$\frac{dd}{dt} = \frac{32\left(t^2 - 3\sqrt{2}t + 8\right)}{\sqrt{t^2 - 4\sqrt{2}t + 16}}$$

(c) When $t = 2$:

$$\frac{dd}{dt} = \frac{32\left(12 - 6\sqrt{2}\right)}{\sqrt{20 - 8\sqrt{2}}} \approx 38.16 \text{ ft/sec}$$

(d) $\dfrac{d^2d}{dt^2} = \dfrac{32\left(t^3 - 6\sqrt{2}t^2 + 36t - 32\sqrt{12}\right)}{\left(t^2 - 4\sqrt{2}t + 16\right)^{3/2}} = 0$ when $t \approx 1.943$ seconds. No. The projectile is at its maximum height when $t = \sqrt{2}$.

13. (a) There is a minimum at $(0, 0, 0)$, maxima at $(0, \pm 1, 2/e)$ and saddle point at $(\pm 1, 0, 1/e)$:

$$f_x = (x^2 + 2y^2)e^{-(x^2+y^2)}(-2x) + (2x)e^{-(x^2+y^2)}$$

$$= e^{-(x^2+y^2)}\left[(x^2 + 2y^2)(-2x) + 2x\right] = e^{-(x^2+y^2)}\left[-2x^3 + 4xy^2 + 2x\right] = 0 \Rightarrow x^3 + 2xy^2 - x = 0$$

$$f_y = (x^2 + 2y^2)e^{-(x^2+y^2)}(-2y) + (4y)e^{-(x^2+y^2)}$$

$$= e^{-(x^2+y^2)}\left[(x^2 + 2y^2)(-2y) + 4y\right] = e^{-(x^2+y^2)}\left[-4y^3 - 2x^2y + 4y\right] = 0 \Rightarrow 2y^3 + x^2y - 2y = 0$$

Solving the two equations $x^3 + 2xy^2 - x = 0$ and $2y^3 + x^2y - 2y = 0$, you obtain the following critical points: $(0, \pm 1)$, $(\pm 1, 0)$, $(0, 0)$. Using the second derivative test, you obtain the results above.

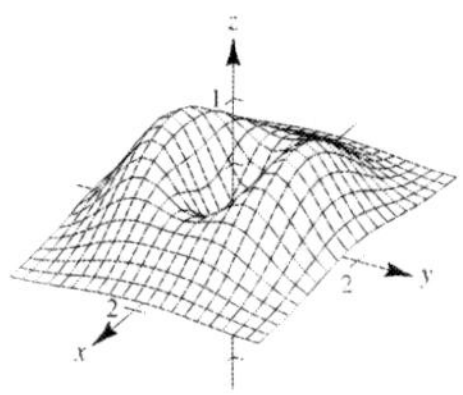

(b) As in part (a), you obtain

$$f_x = e^{-(x^2+y^2)}\left[2x(x^2 - 1 - 2y^2)\right]$$

$$f_y = e^{-(x^2+y^2)}\left[2y(2 + x^2 - 2y^2)\right]$$

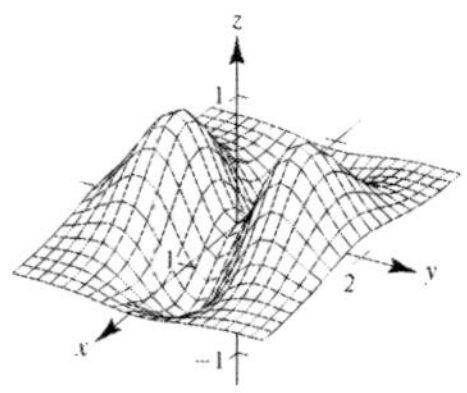

The critical numbers are $(0, 0)$, $(0, \pm 1)$, $(\pm 1, 0)$.

These yield

$(\pm 1, 0, -1/e)$ minima

$(0, \pm 1, 2/e)$ maxima

$(0, 0, 0)$ saddle

(c) In general, for $\alpha > 0$ you obtain

$(0, 0, 0)$ minimum

$(0, \pm 1, \beta/e)$ maxima

$(\pm 1, 0, \alpha/e)$ saddle

For $\alpha < 0$, you obtain

$(\pm 1, 0, \alpha/e)$ minima

$(0, \pm 1, \beta/e)$ maxima

$(0, 0, 0)$ saddle

14. Given that f is a differentiable function such that

$\nabla f(x_0, y_0) = \mathbf{0}$, then $f_x(x_0, y_0) = 0$ and $f_y(x_0, y_0) = 0$.

Therefore, the tangent plane is $-(z - z_0) = 0$ or

$z = z_0 = f(x_0, y_0)$ which is horizontal.

15. (a)

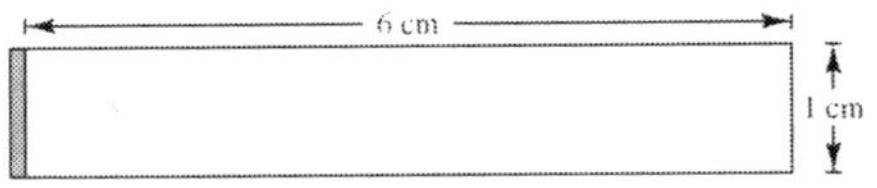

(b)

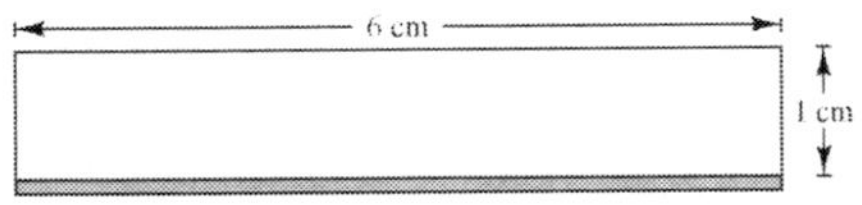

(c) The height has more effect since the shaded region in (b) is larger than the shaded region in (a).

(d) $A = hl \Rightarrow dA = l\,dh + h\,dl$

If $dl = 0.01$ and $dh = 0$, then $dA = 1(0.01) = 0.01$.

If $dh = 0.01$ and $dl = 0$, then $dA = 6(0.01) = 0.06$.

16. Let $g(x, y) = yf\left(\frac{x}{y}\right)$.

$$g_y(x, y) = f\left(\frac{x}{y}\right) + yf'\left(\frac{x}{y}\right)\left(\frac{-x}{y^2}\right) = f\left(\frac{x}{y}\right) - \frac{x}{y}f'\left(\frac{x}{y}\right)$$

$$g_x(x, y) = yf'\left(\frac{x}{y}\right)\left(\frac{1}{y}\right) = f'\left(\frac{x}{y}\right)$$

Tangent plane at (x_0, y_0, z_0) is $f'\left(\frac{x_0}{y_0}\right)(x - x_0) + \left[f\left(\frac{x_0}{y_0}\right) - \frac{x_0}{y_0}f'\left(\frac{x_0}{y_0}\right)\right](y - y_0) - 1\left(z - y_0f\left(\frac{x_0}{y_0}\right)\right) = 0$

$$\Rightarrow f'\left(\frac{x_0}{y_0}\right)x + \left[f\left(\frac{x_0}{y_0}\right) - \frac{x_0}{y_0}f'\left(\frac{x_0}{y_0}\right)\right]y - z = 0.$$

This plane passes through the origin, the common point of intersection.

17. $\frac{\partial u}{\partial t} = \frac{1}{2}\left[-\cos(x - t) + \cos(x + t)\right]$

$\frac{\partial^2 u}{\partial t^2} = \frac{1}{2}\left[-\sin(x - t) - \sin(x + t)\right]$

$\frac{\partial u}{\partial x} = \frac{1}{2}\left[\cos(x - t) + \cos(x + t)\right]$

$\frac{\partial^2 u}{\partial x^2} = \frac{1}{2}\left[-\sin(x - t) - \sin(x + t)\right]$

Then, $\frac{\partial^2 u}{\partial t^2} = \frac{\partial^2 u}{\partial x^2}$.

18. $u(x, t) = \frac{1}{2}\left[f(x - ct) + f(x + ct)\right]$

Let $r = x - ct$ and $s = x + ct$.

Then $u(r, s) = \frac{1}{2}\left[f(r) + f(s)\right]$.

$$\frac{\partial u}{\partial t} = \frac{\partial u}{\partial r}\frac{\partial r}{\partial t} + \frac{\partial u}{\partial s}\frac{\partial s}{\partial t} = \frac{1}{2}\frac{df}{dr}(-c) + \frac{1}{2}\frac{df}{ds}(c)$$

$$\frac{\partial^2 u}{\partial t^2} = \frac{1}{2}\frac{d^2f}{dr^2}(-c)^2 + \frac{1}{2}\frac{d^2f}{ds^2}(c)^2 = \frac{c^2}{2}\left[\frac{d^2f}{dr^2} + \frac{d^2f}{ds^2}\right]$$

$$\frac{\partial u}{\partial x} = \frac{\partial u}{\partial r}\frac{\partial r}{\partial x} + \frac{\partial u}{\partial s}\frac{\partial s}{\partial x} = \frac{1}{2}\frac{df}{dr}(1) + \frac{1}{2}\frac{df}{ds}(1)$$

$$\frac{\partial^2 u}{\partial x^2} = \frac{1}{2}\frac{d^2f}{2dr^2}(1)^2 + \frac{1}{2}\frac{d^2f}{ds^2}(1)^2 = \frac{1}{2}\left[\frac{d^2f}{dr^2} + \frac{d^2f}{ds^2}\right]$$

So, $\frac{\partial^2 u}{\partial t^2} = c^2\frac{\partial^2 u}{\partial x^2}$.

19. $w = f(x, y),\ x = r\cos\theta,\ y = r\sin\theta$

$$\frac{\partial w}{\partial r} = \frac{\partial w}{\partial x}\cos\theta + \frac{\partial w}{\partial y}\sin\theta$$

$$\frac{\partial w}{\partial \theta} = \frac{\partial w}{\partial x}(-r\sin\theta) + \frac{\partial w}{\partial y}(r\cos\theta)$$

(a)

$$r\cos\theta\frac{\partial w}{\partial r} = \frac{\partial w}{\partial x}r\cos^2\theta + \frac{\partial w}{\partial y}r\sin\theta\cos\theta$$

$$-\sin\theta\frac{\partial w}{\partial \theta} = \frac{\partial w}{\partial x}\left(r\sin^2\theta\right) - \frac{\partial w}{\partial y}r\sin\theta\cos\theta$$

$$r\cos\theta\frac{\partial w}{\partial r} - \sin\theta\frac{\partial w}{\partial \theta} = \frac{\partial w}{\partial x}\left(r\cos^2\theta + r\sin^2\theta\right)$$

$$r\frac{\partial w}{\partial x} = \frac{\partial w}{\partial r}(r\cos\theta) - \frac{\partial w}{\partial \theta}\sin\theta$$

$$\frac{\partial w}{\partial x} = \frac{\partial w}{\partial r}\cos\theta - \frac{\partial w}{\partial \theta}\frac{\sin\theta}{r} \quad \text{(First Formula)}$$

$$r\sin\theta\frac{\partial w}{\partial r} = \frac{\partial w}{\partial x}r\sin\theta\cos\theta + \frac{\partial w}{\partial y}r\sin^2\theta$$

$$\cos\theta\frac{\partial w}{\partial \theta} = \frac{\partial w}{\partial x}(-r\sin\theta\cos\theta) + \frac{\partial w}{\partial y}\left(r\cos^2\theta\right)$$

$$r\sin\theta\frac{\partial w}{\partial r} + \cos\theta\frac{\partial w}{\partial \theta} = \frac{\partial w}{\partial y}\left(r\sin^2\theta + r\cos^2\theta\right)$$

$$r\frac{\partial w}{\partial y} = \frac{\partial w}{\partial r}r\sin\theta + \frac{\partial w}{\partial \theta}\cos\theta$$

$$\frac{\partial w}{\partial y} = \frac{\partial w}{\partial r}\sin\theta + \frac{\partial w}{\partial \theta}\frac{\cos\theta}{r} \quad \text{(Second Formula)}$$

(b)

$$\left(\frac{\partial w}{\partial r}\right)^2 + \frac{1}{r^2}\left(\frac{\partial w}{\partial \theta}\right)^2 = \left(\frac{\partial w}{\partial x}\right)^2\cos^2\theta + 2\frac{\partial w}{\partial x}\frac{\partial w}{\partial y}\sin\theta\cos\theta + \left(\frac{\partial w}{\partial y}\right)^2\sin^2\theta + \left(\frac{\partial w}{\partial x}\right)^2\sin^2\theta$$

$$-\ 2\frac{\partial w}{\partial x}\frac{\partial w}{\partial y}\sin\theta\cos\theta + \left(\frac{\partial w}{\partial y}\right)^2\cos^2\theta = \left(\frac{\partial w}{\partial x}\right)^2 + \left(\frac{\partial w}{\partial y}\right)^2$$

20. $w = \arctan\frac{y}{x},\ x = r\cos\theta,\ y = r\sin\theta$

$$= \arctan\left(\frac{r\sin\theta}{r\cos\theta}\right) = \arctan(\tan\theta) = \theta \text{ for } -\frac{\pi}{2} < \theta < \frac{\pi}{2}$$

$$\frac{\partial w}{\partial x} = \frac{-y}{x^2 + y^2},\quad \frac{\partial w}{\partial y} = \frac{x}{x^2 + y^2},\quad \frac{\partial w}{\partial r} = 0,\quad \frac{\partial w}{\partial \theta} = 1$$

$$\left(\frac{\partial w}{\partial x}\right)^2 + \left(\frac{\partial w}{\partial y}\right)^2 = \frac{y^2}{\left(x^2 + y^2\right)^2} + \frac{x^2}{\left(x^2 + y^2\right)^2} = \frac{1}{x^2 + y^2} = \frac{1}{r^2}$$

$$\left(\frac{\partial w}{\partial r}\right)^2 + \left(\frac{1}{r^2}\right)\left(\frac{\partial w}{\partial \theta}\right)^2 = 0 + \frac{1}{r^2}(1) = \frac{1}{r^2}$$

So, $\left(\frac{\partial w}{\partial x}\right)^2 + \left(\frac{\partial w}{\partial y}\right)^2 = \left(\frac{\partial w}{\partial r}\right)^2 + \frac{1}{r^2}\left(\frac{\partial w}{\partial \theta}\right)^2.$

21. $x = r\cos\theta,\ y = r\sin\theta,\ z = z$

$$\frac{\partial u}{\partial \theta} = \frac{\partial u}{\partial x}\frac{\partial x}{\partial \theta} + \frac{\partial u}{\partial y}\frac{\partial y}{\partial \theta} + \frac{\partial u}{\partial z}\frac{\partial z}{\partial \theta} = \frac{\partial u}{\partial x}(-r\sin\theta) + \frac{\partial u}{\partial y}r\cos\theta$$ Similarly,

$$\frac{\partial u}{\partial r} = \frac{\partial u}{\partial x}\cos\theta + \frac{\partial u}{\partial y}\sin\theta.$$

$$\frac{\partial^2 u}{\partial \theta^2} = (-r\sin\theta)\left[\frac{\partial^2 u}{\partial x^2}\frac{\partial x}{\partial \theta} + \frac{\partial^2 u}{\partial x \partial y}\frac{\partial y}{\partial \theta} + \frac{\partial^2 u}{\partial x \partial z}\frac{\partial z}{\partial \theta}\right] - r\frac{\partial u}{\partial x}\cos\theta + (r\cos\theta)\left[\frac{\partial^2 u}{\partial y \partial x}\frac{\partial x}{\partial \theta} + \frac{\partial^2 u}{\partial y^2}\frac{\partial y}{\partial \theta} + \frac{\partial^2 u}{\partial y \partial z}\frac{\partial z}{\partial \theta}\right] - r\frac{\partial u}{\partial y}\sin\theta$$

$$= \frac{\partial^2 u}{\partial x^2}r^2\sin^2\theta + \frac{\partial^2 u}{\partial y^2}r^2\cos^2\theta - 2\frac{\partial^2 u}{\partial x \partial y}r^2\sin\theta\cos\theta - \frac{\partial u}{\partial x}r\cos\theta - \frac{\partial u}{\partial y}r\sin\theta$$

Similarly, $\dfrac{\partial^2 u}{\partial r^2} = \dfrac{\partial^2 u}{\partial x^2}\cos^2\theta + \dfrac{\partial^2 u}{\partial y^2}\sin^2\theta + 2\dfrac{\partial^2 u}{\partial x \partial y}\cos\theta\sin\theta.$

Now observe that

$$\frac{\partial^2 u}{\partial r^2} + \frac{1}{r}\frac{\partial u}{\partial r} + \frac{1}{r^2}\frac{\partial^2 u}{\partial \theta^2} + \frac{\partial^2 u}{\partial z^2} = \left[\frac{\partial^2 u}{\partial x^2}\cos^2\theta + \frac{\partial^2 u}{\partial y^2}\sin^2\theta + 2\frac{\partial^2 u}{\partial x \partial y}\cos\theta\sin\theta\right] + \frac{1}{r}\left[\frac{\partial u}{\partial x}\cos\theta + \frac{\partial u}{\partial y}\sin\theta\right]$$
$$+\left[\frac{\partial^2 u}{\partial x^2}\sin^2\theta + \frac{\partial^2 u}{\partial y^2}\cos^2\theta - 2\frac{\partial^2 u}{\partial x \partial y}\sin\theta\cos\theta - \frac{1}{r}\frac{\partial u}{\partial x}\cos\theta - \frac{1}{r}\frac{\partial u}{\partial y}\sin\theta\right] + \frac{\partial^2 u}{\partial z^2}$$
$$= \frac{\partial^2 u}{\partial x^2} + \frac{\partial^2 u}{\partial y^2} + \frac{\partial^2 u}{\partial z^2}.$$

So, Laplace's equation in cylindrical coordinates, is $\dfrac{\partial^2 u}{\partial r^2} + \dfrac{1}{r}\dfrac{\partial u}{\partial r} + \dfrac{1}{r^2}\dfrac{\partial^2 u}{\partial \theta^2} + \dfrac{\partial^2 u}{\partial z^2} = 0.$

CHAPTER 14
Multiple Integration

CHAPTER 14
Multiple Integration

Section 14.1 Iterated Integrals and Area in the Plane

1. $\int_0^x (x + 2y)\,dy = \left[xy + y^2\right]_0^x = x^2 + x^2 = 2x^2$

2. $\int_x^{x^2} \frac{y}{x}\,dy = \left[\frac{1}{2}\frac{y^2}{x}\right]_x^{x^2} = \frac{1}{2}\left(\frac{x^4}{x} - \frac{x^2}{x}\right) = \frac{x}{2}\left(x^2 - 1\right)$

3. $\int_1^{2y} \frac{y}{x}\,dx = \left[y \ln x\right]_1^{2y}$

$= y \ln 2y - 0 = y \ln 2y, (y > 0)$

4. $\int_0^{\cos y} y\,dx = \left[yx\right]_0^{\cos y} = y \cos y$

5. $\int_0^{\sqrt{4-x^2}} x^2 y\,dy = \left[\frac{1}{2}x^2y^2\right]_0^{\sqrt{4-x^2}} = \frac{4x^2 - x^4}{2}$

6. $\int_{x^3}^{\sqrt{x}} \left(x^2 + 3y^2\right)dy = \left[x^2y + y^3\right]_{x^3}^{\sqrt{x}} = \left(x^2\sqrt{x} + \left(\sqrt{x}\right)^3\right) - \left(x^2x^3 + \left(x^3\right)^3\right) = x^{5/2} + x^{3/2} - x^5 - x^9$

7. $\int_{e^y}^{y} \frac{y \ln x}{x}\,dx = \left[\frac{1}{2}y \ln^2 x\right]_{e^y}^{y} = \frac{1}{2}y\left[\ln^2 y - \ln^2 e^y\right] = \frac{y}{2}\left[(\ln y)^2 - y^2\right], (y > 0)$

8. $\int_{-\sqrt{1-y^2}}^{\sqrt{1-y^2}} \left(x^2 + y^2\right)dx = \left[\frac{1}{3}x^3 + y^2x\right]_{-\sqrt{1-y^2}}^{\sqrt{1-y^2}} = 2\left[\frac{1}{3}\left(1 - y^2\right)^{3/2} + y^2\left(1 - y^2\right)^{1/2}\right] = \frac{2\sqrt{1 - y^2}}{3}\left(1 + 2y^2\right)$

9. $\int_0^{x^3} ye^{-y/x}\,dy = \left[-xye^{-y/x}\right]_0^{x^3} + x\int_0^{x^3} e^{-y/x}\,dy = -x^4e^{-x^2} - \left[x^2e^{-y/x}\right]_0^{x^3} = x^2\left(1 - e^{-x^2} - x^2e^{-x^2}\right)$

$u = y, du = dy, dv = e^{-y/x}\,dy, v = -xe^{-y/x}$

10. $\int_y^{\pi/2} \sin^3 x \cos y\,dx = \int_y^{\pi/2} \left(1 - \cos^2 x\right) \sin x \cos y\,dx$

$= \left[\left(-\cos x + \frac{1}{3}\cos^3 x\right)\cos y\right]_y^{\pi/2} = \left(\cos y - \frac{1}{3}\cos^3 y\right)\cos y$

11. $\int_0^1 \int_0^2 (x + y)\,dy\,dx = \int_0^1 \left[xy + \frac{1}{2}y^2\right]_0^2 dx = \int_0^1 (2x + 2)\,dx = \left[x^2 + 2x\right]_0^1 = 3$

12. $\int_{-1}^1 \int_{-2}^2 \left(x^2 - y^2\right)dy\,dx = \int_{-1}^1 \left[x^2y - \frac{y^3}{3}\right]_{-2}^2 dx = \int_{-1}^1 \left[2x^2 - \frac{8}{3} + 2x^2 - \frac{8}{3}\right]dx$

$= \int_{-1}^1 \left(4x^2 - \frac{16}{3}\right)dx = \left[\frac{4x^3}{3} - \frac{16}{3}x\right]_{-1}^1 = \left(\frac{4}{3} - \frac{16}{3}\right) - \left(-\frac{4}{3} + \frac{16}{3}\right) = -8$

13. $\int_1^2 \int_0^4 \left(x^2 - 2y^2\right)dx\,dy = \int_1^2 \left[\frac{x^3}{3} - 2xy^2\right]_0^4 dy = \int_1^2 \left[\frac{64}{3} - 8y^2\right]dy = \left[\frac{64}{3}y - \frac{8}{3}y^3\right]_1^2 = \left(\frac{128}{3} - \frac{64}{3}\right) - \left(\frac{64}{3} - \frac{8}{3}\right) = \frac{8}{3}$

14. $\int_{-1}^2 \int_1^3 \left(x + y^2\right)dx\,dy = \int_{-1}^2 \left[\frac{x^2}{2} + xy^2\right]_1^3 dy = \int_{-1}^2 \left[\left(\frac{9}{2} + 3y^2\right) - \left(\frac{1}{2} + y^2\right)\right]dy$

$= \int_{-1}^2 \left(4 + 2y^2\right)dy = \left[4y + \frac{2}{3}y^3\right]_{-1}^2 = \left(8 + \frac{16}{3}\right) - \left(-4 - \frac{2}{3}\right) = 18$

15. $\int_0^{\pi/2}\int_0^1 y\cos x\,dy\,dx = \int_0^{\pi/2}\left[\frac{y^2}{2}\cos x\right]_0^1 dx = \int_0^{\pi/2}\frac{1}{2}\cos x\,dx = \left[\frac{1}{2}\sin x\right]_0^{\pi/2} = \frac{1}{2}$

16. $\int_0^{\ln 4}\int_0^{\ln 3} e^{x+y}\,dy\,dx = \int_0^{\ln 4}\left[e^{x+y}\right]_0^{\ln 3} dx$

$= \int_0^{\ln 4}\left[e^{x+\ln 3} - e^x\right]dx = \left[e^{x+\ln 3} - e^x\right]_0^{\ln 4} = \left(e^{\ln 4+\ln 3} - e^{\ln 4}\right) - \left(e^{\ln 3} - 1\right) = (12 - 4) - (3 - 1) = 6$

17. $\int_0^{\pi}\int_0^{\sin x}(1 + \cos x)\,dy\,dx = \int_0^{\pi}\left[(y + y\cos x)\right]_0^{\sin x} dx = \int_0^{\pi}\left[\sin x + \sin x\cos x\right]dx = \left[-\cos x + \frac{1}{2}\sin^2 x\right]_0^{\pi} = 1 + 1 = 2$

18. $\int_1^4\int_1^{\sqrt{x}} 2ye^{-x}\,dy\,dx = \int_1^4\left[y^2e^{-x}\right]_1^{\sqrt{x}} dx = \int_1^4\left(xe^{-x} - e^{-x}\right)dx = \left[-xe^{-x}\right]_1^4 = -4e^{-4} + e^{-1} = \frac{1}{e} - \frac{4}{e^4}$

19. $\int_0^1\int_0^x\sqrt{1 - x^2}\,dy\,dx = \int_0^1\left[y\sqrt{1 - x^2}\right]_0^x dx = \int_0^1 x\sqrt{1 - x^2}\,dx = \left[-\frac{1}{2}\left(\frac{2}{3}\right)\left(1 - x^2\right)^{3/2}\right]_0^1 = \frac{1}{3}$

20. $\int_{-4}^4\int_0^{x^2}\sqrt{64 - x^3}\,dy\,dx = \int_{-4}^4\left[y\sqrt{64 - x^3}\right]_0^{x^2} dx = \int_{-4}^4 x^2\sqrt{64 - x^3}\,dx = \left[-\frac{2}{9}\left(64 - x^3\right)^{3/2}\right]_{-4}^4 = 0 + \frac{2}{9}(128)^{3/2} = \frac{2048}{9}\sqrt{2}$

21. $\int_{-1}^5\int_0^{3y}\left(3 + x^2 + \frac{1}{4}y^2\right)dx\,dy = \int_{-1}^5\left[3x + \frac{x^3}{3} + \frac{1}{4}xy^2\right]_0^{3y} dy$

$= \int_{-1}^5\left[9y + 9y^3 + \frac{3}{4}y^3\right]dy = \int_{-1}^5\left[9y + \frac{39}{4}y^3\right]dy = \left[\frac{9}{2}y^2 + \frac{39}{16}y^4\right]_{-1}^5 = \left(\frac{9}{2}(25) + \frac{39}{16}(625)\right) - \left(\frac{9}{2} + \frac{39}{16}\right) = 1629$

22. $\int_0^2\int_y^{2y}\left(10 + 2x^2 + 2y^2\right)dx\,dy = \int_0^2\left[10x + \frac{2x^3}{3} + 2y^2x\right]_y^{2y} dy = \int_0^2\left[\left(20y + \frac{16}{3}y^3 + 4y^3\right) - \left(10y + \frac{2}{3}y^3 + 2y^3\right)\right]dy$

$= \int_0^2\left[10y + \frac{20}{3}y^3\right]dy = \left[5y^2 + \frac{5y^4}{3}\right]_0^2 = 20 + \frac{80}{3} = \frac{140}{3}$

23. $\int_0^1\int_0^{\sqrt{1-y^2}}(x + y)\,dx\,dy = \int_0^1\left[\frac{1}{2}x^2 + xy\right]_0^{\sqrt{1-y^2}} dy = \int_0^1\left[\frac{1}{2}\left(1 - y^2\right) + y\sqrt{1 - y^2}\right]dy = \left[\frac{1}{2}y - \frac{1}{6}y^3 - \frac{1}{2}\left(\frac{2}{3}\right)\left(1 - y^2\right)^{3/2}\right]_0^1 = \frac{2}{3}$

24. $\int_0^2\int_{3y^2-6y}^{2y-y^2} 3y\,dx\,dy = \int_0^2\left[3xy\right]_{3y^2-6y}^{2y-y^2} dy = 3\int_0^2\left(8y^2 - 4y^3\right)dy = \left[3\left(\frac{8}{3}y^3 - y^4\right)\right]_0^2 = 16$

25. $\int_0^2\int_0^{\sqrt{4-y^2}}\frac{2}{\sqrt{4 - y^2}}\,dx\,dy = \int_0^2\left[\frac{2x}{\sqrt{4 - y^2}}\right]_0^{\sqrt{4-y^2}} dy = \int_0^2 2\,dy = \left[2y\right]_0^2 = 4$

26. $\int_1^3\int_0^y\frac{4}{x^2 + y^2}\,dx\,dy = \int_1^3\left[\frac{4}{y}\arctan\left(\frac{x}{y}\right)\right]_0^y dy = \int_1^3\frac{4}{y}\left(\frac{\pi}{4}\right)dy = \int_1^3\frac{\pi}{y}\,dy = \left[\pi\ln y\right]_1^3 = \pi\ln 3$

27. $\int_0^{\pi/2}\int_0^{2\cos\theta} r\,dr\,d\theta = \int_0^{\pi/2}\left[\frac{r^2}{2}\right]_0^{2\cos\theta} d\theta = \int_0^{\pi/2} 2\cos^2\theta\,d\theta = \left[\theta - \frac{1}{2}\sin 2\theta\right]_0^{\pi/2} = \frac{\pi}{2}$

28. $\int_0^{\pi/4}\int_{\sqrt{3}}^{\sqrt{3}\cos\theta} r\,dr\,d\theta = \int_0^{\pi/4}\left[\frac{r^2}{2}\right]_{\sqrt{3}}^{\sqrt{3}\cos\theta} d\theta = \int_0^{\pi/4}\left(\frac{3\cos^2\theta}{2} - \frac{3}{2}\right)d\theta = \int_0^{\pi/4}\left(\frac{3}{4}(1 + \cos 2\theta) - \frac{3}{2}\right)d\theta$

$= \int_0^{\pi/4}\left(\frac{3}{4}\cos 2\theta - \frac{3}{4}\right)d\theta = \left[\frac{3}{8}\sin 2\theta - \frac{3}{4}\theta\right]_0^{\pi/4} = \frac{3}{8} - \frac{3\pi}{16}$

29. $$\int_0^{\pi/2}\int_0^{\sin\theta} \theta r\,dr\,d\theta = \int_0^{\pi/2}\left[\theta\frac{r^2}{2}\right]_0^{\sin\theta} d\theta = \int_0^{\pi/2}\frac{1}{2}\theta\sin^2\theta\,d\theta$$
$$= \frac{1}{4}\int_0^{\pi/2}(\theta - \theta\cos 2\theta)\,d\theta = \frac{1}{4}\left[\frac{\theta^2}{2} - \left(\frac{1}{4}\cos 2\theta + \frac{\theta}{2}\sin 2\theta\right)\right]_0^{\pi/2} = \frac{\pi^2}{32} + \frac{1}{8}$$

30. $$\int_0^{\pi/4}\int_0^{\cos\theta} 3r^2\sin\theta\,dr\,d\theta = \int_0^{\pi/4}\left[r^3\sin\theta\right]_0^{\cos\theta} d\theta = \int_0^{\pi/4}\cos^3\sin\theta\,d\theta = \left[-\frac{\cos^4\theta}{4}\right]_0^{\pi/4} = -\frac{1}{4}\left[\left(\frac{1}{\sqrt{2}}\right)^4 - 1\right] = \frac{3}{16}$$

31. $$\int_1^{\infty}\int_0^{1/x} y\,dy\,dx = \int_1^{\infty}\left[\frac{y^2}{2}\right]_0^{1/x} dx = \frac{1}{2}\int_1^{\infty}\frac{1}{x^2}\,dx = \left[-\frac{1}{2x}\right]_1^{\infty} = 0 + \frac{1}{2} = \frac{1}{2}$$

32. $$\int_0^3\int_0^{\infty}\frac{x^2}{1+y^2}\,dy\,dx = \int_0^3\left[x^2\arctan y\right]_0^{\infty} dx = \int_0^3 x^2\left(\frac{\pi}{2}\right)dx = \left[\frac{\pi}{2}\cdot\frac{x^3}{3}\right]_0^3 = \frac{9\pi}{2}$$

33. $$\int_1^{\infty}\int_1^{\infty}\frac{1}{xy}\,dx\,dy = \int_1^{\infty}\left[\frac{1}{y}\ln x\right]_1^{\infty} dy = \int_1^{\infty}\left[\frac{1}{y}(\infty) - \frac{1}{y}(0)\right]dy$$

Diverges

34. $$\int_0^{\infty}\int_0^{\infty} xye^{-(x^2+y^2)}\,dx\,dy = \int_0^{\infty}\left[-\tfrac{1}{2}ye^{-(x^2+y^2)}\right]_0^{\infty} dy = \int_0^{\infty}\tfrac{1}{2}ye^{-y^2}\,dy = \left[-\tfrac{1}{4}e^{-y^2}\right]_0^{\infty} = \tfrac{1}{4}$$

35. $$A = \int_0^8\int_0^3 dy\,dx = \int_0^8\left[y\right]_0^3 dx = \int_0^8 3\,dx = \left[3x\right]_0^8 = 24$$
$$A = \int_0^3\int_0^8 dx\,dy = \int_0^3\left[x\right]_0^8 dy = \int_0^3 8\,dy = \left[8y\right]_0^3 = 24$$

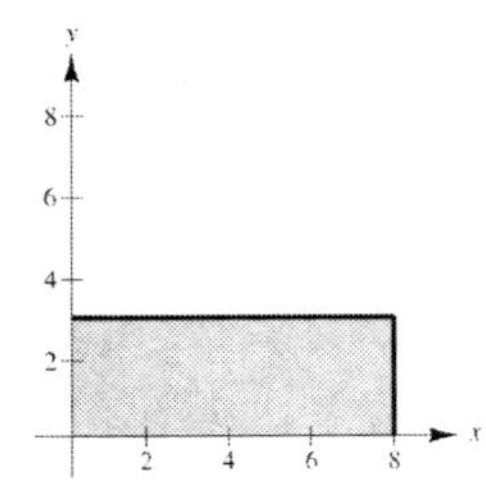

36. $$A = \int_1^2\int_1^3 dy\,dx = \int_1^2\left[y\right]_1^3 dx = \int_1^2 2\,dx = \left[2x\right]_1^2 = 2$$
$$A = \int_1^3\int_1^2 dx\,dy = \int_1^3\left[x\right]_1^2 dy = \int_1^3 dy = \left[y\right]_1^3 = 2$$

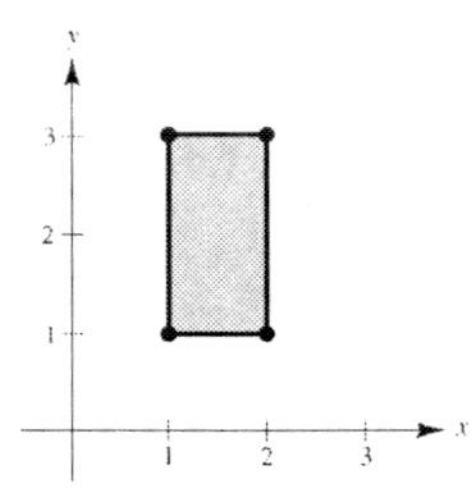

37. $$A = \int_0^2\int_0^{4-x^2} dy\,dx = \int_0^2\left[y\right]_0^{4-x^2} dx = \int_0^2\left(4 - x^2\right)dx = \left[4x - \frac{x^3}{3}\right]_0^2 = \frac{16}{3}$$
$$A = \int_0^4\int_0^{\sqrt{4-y}} dx\,dy = \int_0^4\left[x\right]_0^{\sqrt{4-y}} dy = \int_0^4\sqrt{4-y}\,dy = -\int_0^4(4-y)^{1/2}(-1)\,dy = \left[-\frac{2}{3}(4-y)^{3/2}\right]_0^4 = \frac{2}{3}(8) = \frac{16}{3}$$

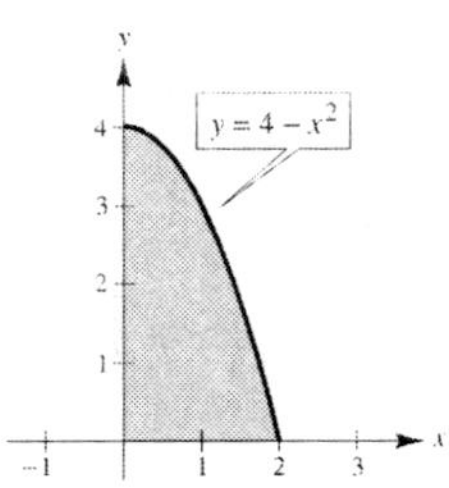

38. $A = \int_2^5 \int_0^{1/\sqrt{x-1}} dy\,dx = \int_2^5 [y]_0^{1/\sqrt{x-1}}\,dx = \int_2^5 \frac{1}{\sqrt{x-1}}\,dx = \left[2\sqrt{x-1}\right]_2^5 = 2$

$A = \int_0^{1/2} \int_2^5 dx\,dy + \int_{1/2}^1 \int_2^{1+(1/y^2)} dx\,dy$

$= \int_0^{1/2} [x]_2^5\,dy + \int_{1/2}^1 [x]_2^{1+(1/y^2)}\,dy = \int_0^{1/2} 3\,dy + \int_{1/2}^1 \left(\frac{1}{y^2} - 1\right)dy = [3y]_0^{1/2} + \left[-\frac{1}{y} - y\right]_{1/2}^1 = 2$

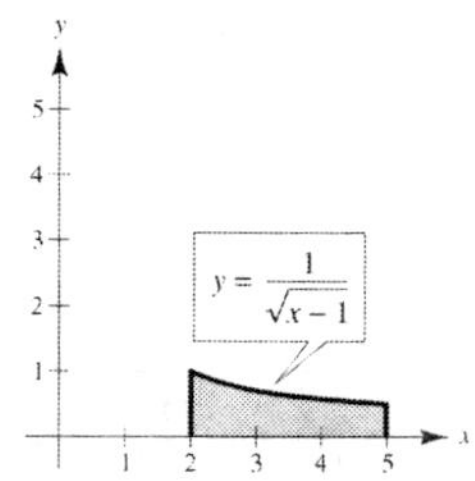

39. $A = \int_0^4 \int_0^{(2-\sqrt{x})^2} dy\,dx = \int_0^4 [y]_0^{(2-\sqrt{x})^2}\,dx = \int_0^4 \left(4 - 4\sqrt{x} + x\right)dx = \left[4x - \frac{8}{3}x\sqrt{x} + \frac{x^2}{2}\right]_0^4 = \frac{8}{3}$

$A = \int_0^4 \int_0^{(2-\sqrt{y})^2} dx\,dy = \frac{8}{3}$

Integration steps are similar to those above.

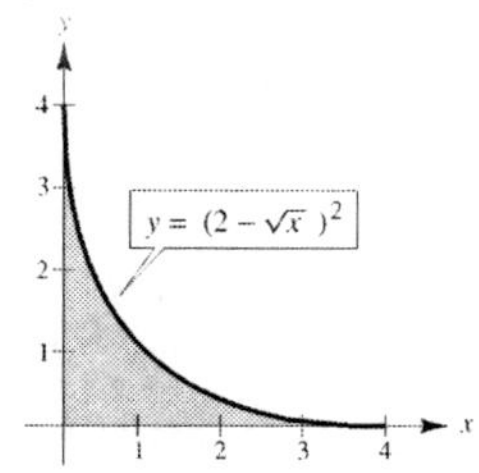

40. $A = \int_0^4 \int_{x^{3/2}}^{2x} dy\,dx$

$= \int_0^4 [y]_{x^{3/2}}^{2x}\,dx = \int_0^4 \left(2x - x^{3/2}\right)dx$

$= \left[x^2 - \frac{2}{5}x^{5/2}\right]_0^4 = 16 - \frac{2}{5}(32) = \frac{16}{5}$

$A = \int_0^8 \int_{y/2}^{y^{2/3}} dx\,dy = \int_0^8 \left(y^{2/3} - \frac{y}{2}\right)dy$

$= \left[\frac{3}{5}y^{5/3} - \frac{y^2}{4}\right]_0^8 = \frac{3}{5}(32) - 16 = \frac{16}{5}$

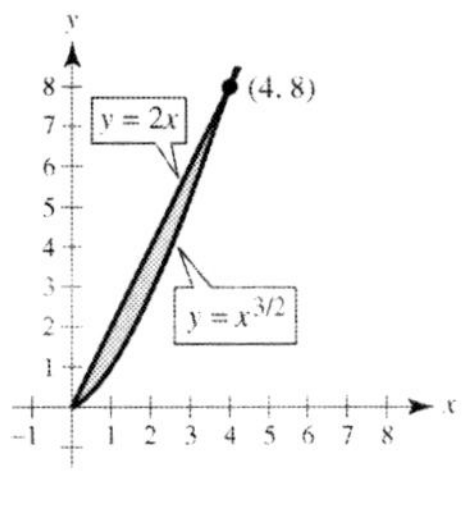

41. $A = \int_0^3 \int_0^{2x/3} dy\,dx + \int_3^5 \int_0^{5-x} dy\,dx$

$= \int_0^3 [y]_0^{2x/3}\,dx + \int_3^5 [y]_0^{5-x}\,dx$

$= \int_0^3 \frac{2x}{3}\,dx + \int_3^5 (5 - x)\,dx$

$= \left[\frac{1}{3}x^2\right]_0^3 + \left[5x - \frac{1}{2}x^2\right]_3^5 = 5$

$A = \int_0^2 \int_{3y/2}^{5-y} dx\,dy$

$= \int_0^2 [x]_{3y/2}^{5-y}\,dy$

$= \int_0^2 \left(5 - y - \frac{3y}{2}\right)dy$

$= \int_0^2 \left(5 - \frac{5y}{2}\right)dy = \left[5y - \frac{5}{4}y^2\right]_0^2 = 5$

42. $\dfrac{A}{4} = \int_0^a \int_0^{(b/a)\sqrt{a^2-x^2}} dy\,dx = \int_0^a \Big[y\Big]_0^{(b/a)\sqrt{a^2-x^2}} dx$

$= \dfrac{b}{a}\int_0^a \sqrt{a^2 - x^2}\,dx = ab\int_0^{\pi/2} \cos^2\theta\,d\theta$

$(x = a\sin\theta,\ dx = a\cos\theta\,d\theta)$

$= \dfrac{ab}{2}\int_0^{\pi/2} (1 + \cos 2\theta)\,d\theta$

$= \left[\dfrac{ab}{2}\left(\theta + \dfrac{1}{2}\sin 2\theta\right)\right]_0^{\pi/2}$

$= \dfrac{\pi ab}{4}$

So, $A = \pi ab$.

$\dfrac{A}{4} = \int_0^b \int_0^{(a/b)\sqrt{b^2-y^2}} dx\,dy = \dfrac{\pi ab}{4}$

So, $A = \pi ab$. Integration steps are similar to those above.

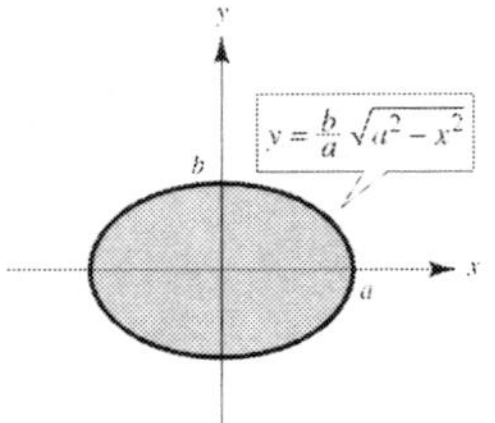

43. $A = \int_{-2}^{1} \int_{x+2}^{4-x^2} dy\,dx$

$= \int_{-2}^{1} \Big[y\Big]_{x+2}^{4-x^2} dx$

$= \int_{-2}^{1} \left(4 - x^2 - x - 2\right) dx$

$= \int_{-2}^{1} \left(2 - x - x^2\right) dx$

$= \left[2x - \frac{1}{2}x^2 - \frac{1}{3}x^3\right]_{-2}^{1} = \frac{9}{2}$

$A = \int_0^3 \int_{-\sqrt{4-y}}^{y-2} dx\,dy + 2\int_3^4 \int_0^{\sqrt{4-y}} dx\,dy$

$= \int_0^3 \Big[x\Big]_{-\sqrt{4-y}}^{y-2} dy + 2\int_3^4 \Big[x\Big]_0^{\sqrt{4-y}} dy$

$= \int_0^3 \left(y - 2 + \sqrt{4 - y}\right) dy + 2\int_3^4 \sqrt{4 - y}\,dy$

$= \left[\frac{1}{2}y^2 - 2y - \frac{2}{3}(4 - y)^{3/2}\right]_0^3 - \left[\frac{4}{3}(4 - y)^{3/2}\right]_3^4 = \frac{9}{2}$

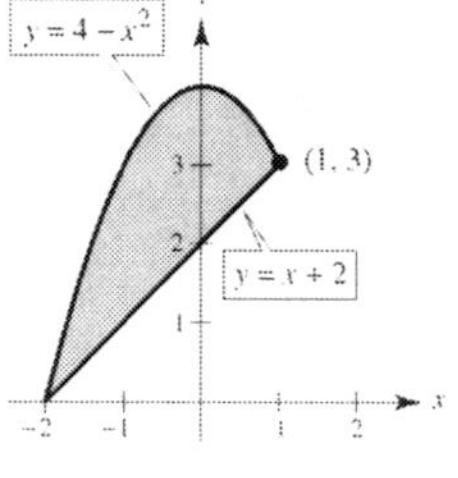

44. $A = \int_0^2 \int_{y/2}^{y} dx\,dy + \int_2^4 \int_{y/2}^{2} dx\,dy$

$= \int_0^2 \frac{y}{2}dy + \int_2^4 \left(2 - \frac{y}{2}\right) dy$

$= \left[\frac{y^2}{4}\right]_0^2 + \left[2y - \frac{y^2}{4}\right]_2^4$

$= 1 + (4 - 3) = 2$

$A = \int_0^2 \int_x^{2x} dy\,dx = \int_0^2 (2x - x)\,dx = \left[\frac{x^2}{2}\right]_0^2 = 2$

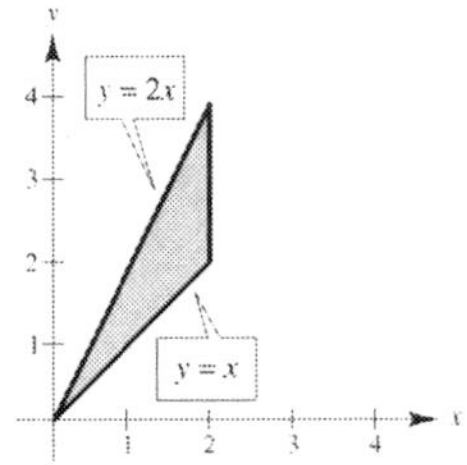

45. $\int_0^4 \int_0^y f(x, y)\,dx\,dy,\ 0 \le x \le y,\ 0 \le y \le 4$

$= \int_0^4 \int_x^4 f(x, y)\,dy\,dx$

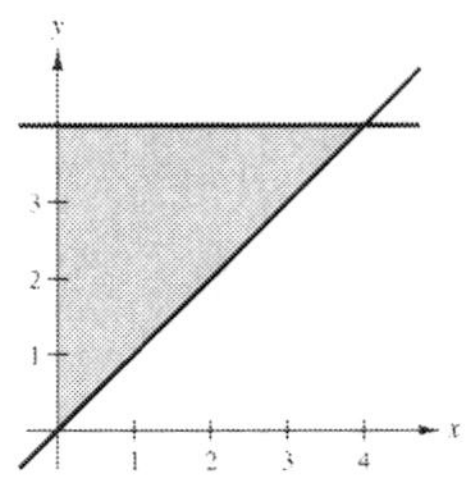

46. $\int_0^4 \int_{\sqrt{y}}^{2} f(x, y)\,dx\,dy,\ \sqrt{y} \le x \le 2,\ 0 \le y \le 4$

$= \int_0^2 \int_0^{x^2} f(x, y)\,dy\,dx$

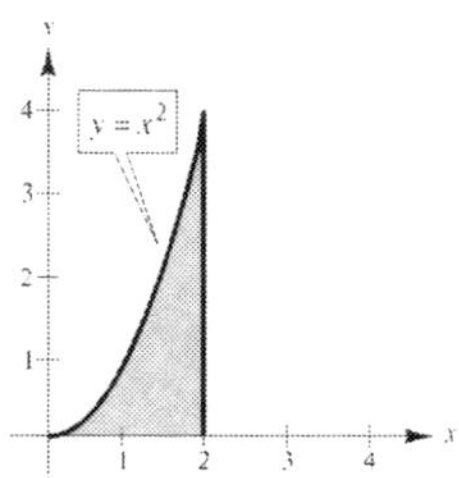

47. $\int_{-2}^{2}\int_{0}^{\sqrt{4-x^2}} f(x, y)\,dy\,dx, 0 \le y \le \sqrt{4-x^2}, -2 \le x \le 2$

$= \int_{0}^{2}\int_{-\sqrt{4-y^2}}^{\sqrt{4-y^2}} dx\,dy$

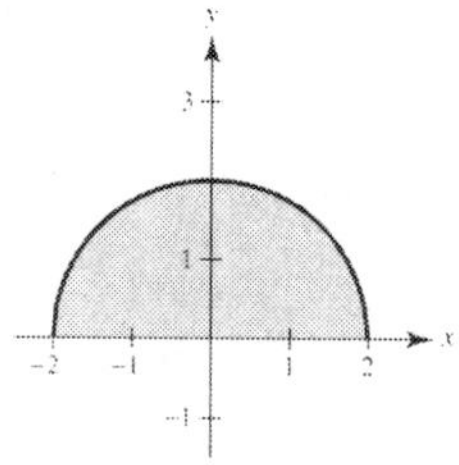

48. $\int_{0}^{2}\int_{0}^{4-x^2} f(x, y)\,dy\,dx, 0 \le y \le 4 - x^2, 0 \le x \le 2$

$= \int_{0}^{4}\int_{0}^{\sqrt{4-y}} f(x, y)\,dx\,dy$

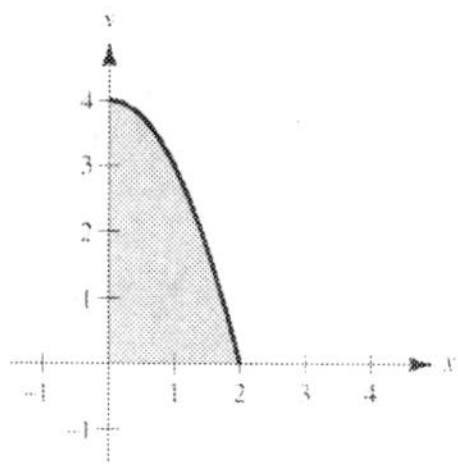

49. $\int_{1}^{10}\int_{0}^{\ln y} f(x, y)\,dx\,dy, 0 \le x \le \ln y, 1 \le y \le 10$

$= \int_{0}^{\ln 10}\int_{e^x}^{10} f(x, y)\,dy\,dx$

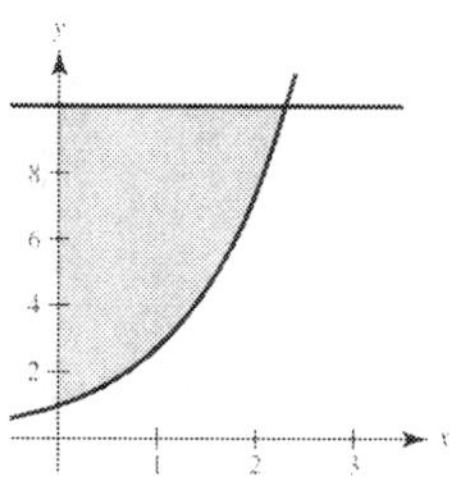

50. $\int_{-1}^{2}\int_{0}^{e^{-x}} f(x, y)\,dy\,dx, 0 \le y \le e^{-x}, -1 \le x \le 2$

$= \int_{0}^{e^{-2}}\int_{-1}^{2} f(x, y)\,dx\,dy + \int_{e^{-2}}^{e}\int_{-1}^{-\ln y} f(x, y)\,dx\,dy$

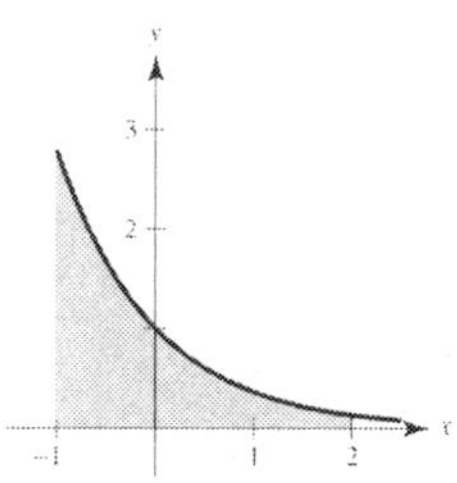

51. $\int_{-1}^{1}\int_{x^2}^{1} f(x, y)\,dy\,dx, x^2 \le y \le 1, -1 \le x \le 1$

$= \int_{0}^{1}\int_{-\sqrt{y}}^{\sqrt{y}} f(x, y)\,dx\,dy$

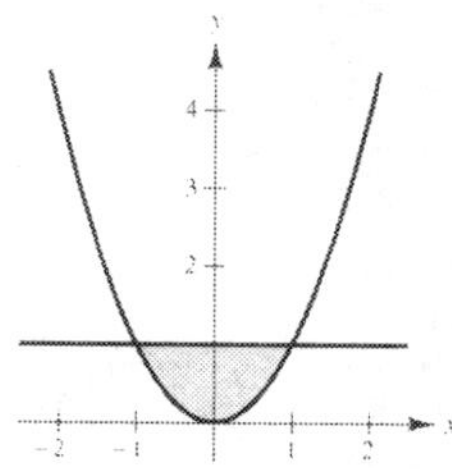

52. $\int_{-\pi/2}^{\pi/2}\int_{0}^{\cos x} f(x, y)\,dy\,dx, 0 \le y \le \cos x, -\frac{\pi}{2} \le x \le \frac{\pi}{2}$

$= \int_{0}^{1}\int_{-\arccos y}^{\arccos y} f(x, y)\,dx\,dy$

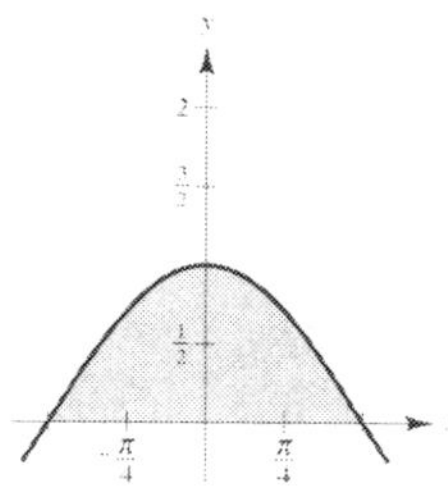

53. $\int_{0}^{1}\int_{0}^{2} dy\,dx = \int_{0}^{2}\int_{0}^{1} dx\,dy = 2$

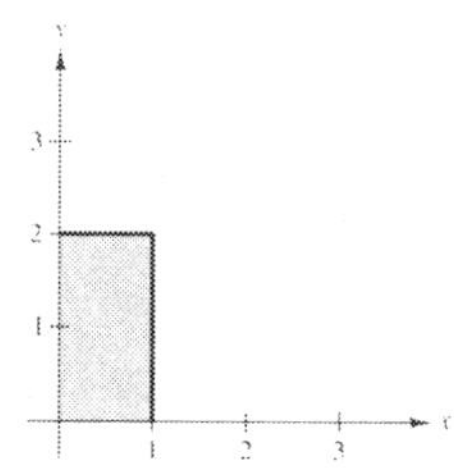

54. $\int_{1}^{2}\int_{2}^{4} dx\,dy = \int_{2}^{4}\int_{1}^{2} dy\,dx = 2$

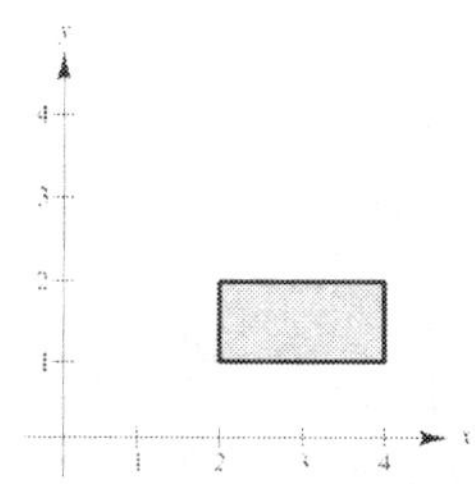

55. $\int_0^1 \int_{-\sqrt{1-y^2}}^{\sqrt{1-y^2}} dx\,dy = \int_{-1}^1 \int_0^{\sqrt{1-x^2}} dy\,dx = \dfrac{\pi}{2}$

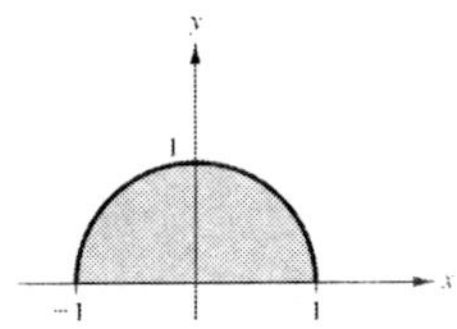

56. $\int_{-2}^2 \int_{-\sqrt{4-x^2}}^{\sqrt{4-x^2}} dy\,dx = \int_{-2}^2 \left(\sqrt{4-x^2} + \sqrt{4-x^2}\right) dx = 4\pi$

$\int_{-2}^2 \int_{-\sqrt{4-y^2}}^{\sqrt{4-y^2}} dx\,dy = 4\pi$

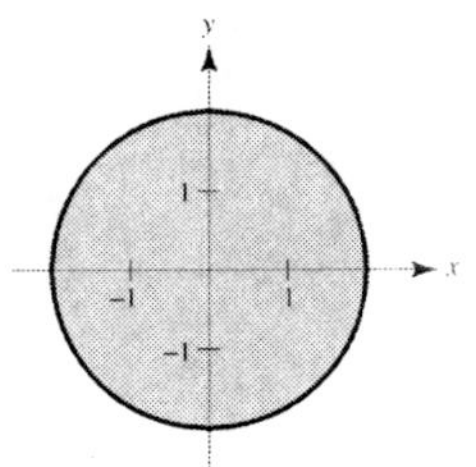

57. $\int_0^2 \int_0^x dy\,dx + \int_2^4 \int_0^{4-x} dy\,dx = \int_0^2 \int_y^{4-y} dx\,dy = 4$

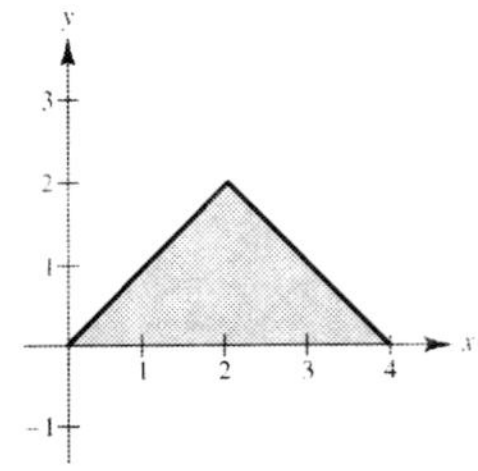

58. $\int_0^4 \int_0^{x/2} dy\,dx + \int_4^6 \int_0^{6-x} dy\,dx = \int_0^4 \dfrac{x}{2}\,dx + \int_4^6 (6 - x)\,dx = 4 + 2 = 6$

$\int_0^2 \int_{2y}^{6-y} dx\,dy = \int_0^2 (6 - 3y)\,dy = \left[6y - \dfrac{3y^2}{2}\right]_0^2 = 6$

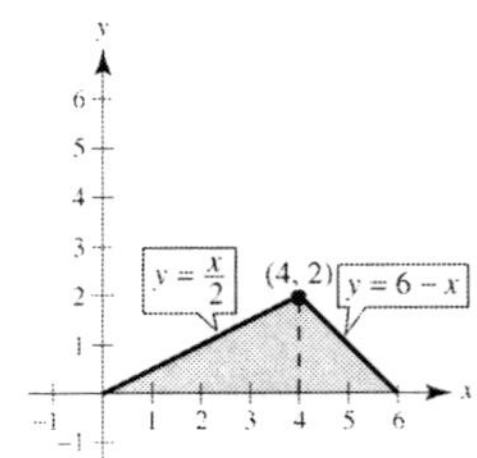

59. $\int_0^2 \int_{x/2}^1 dy\,dx = \int_0^1 \int_0^{2y} dx\,dy = 1$

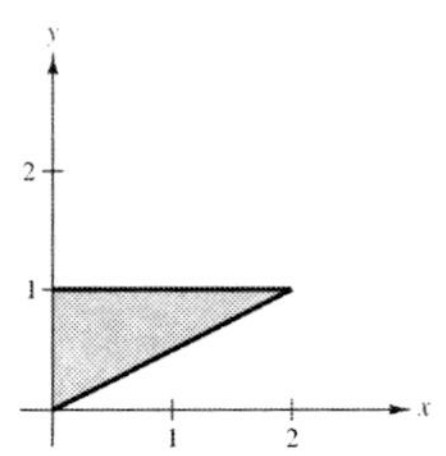

60. $\int_0^9 \int_{\sqrt{x}}^3 dy\,dx = \int_0^9 \left(3 - \sqrt{x}\right) dx$

$= \left[3x - \dfrac{2}{3}x^{3/2}\right]_0^9 = 27 - 18 = 9$

$\int_0^3 \int_0^{y^2} dx\,dy = \int_0^3 y^2\,dy = \left[\dfrac{y^3}{3}\right]_0^3 = 9$

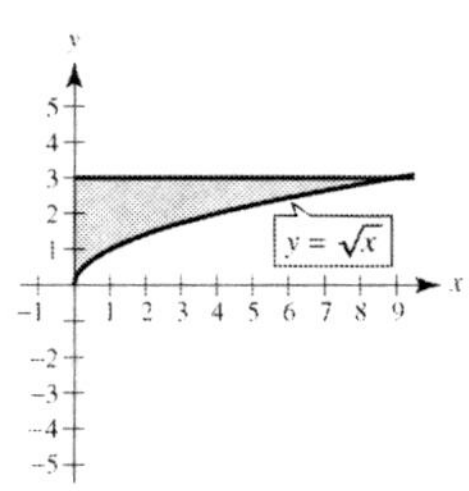

61. $\int_0^1 \int_{y^2}^{\sqrt[3]{y}} dx\,dy = \int_0^1 \int_{x^3}^{\sqrt{x}} dy\,dx = \frac{5}{12}$

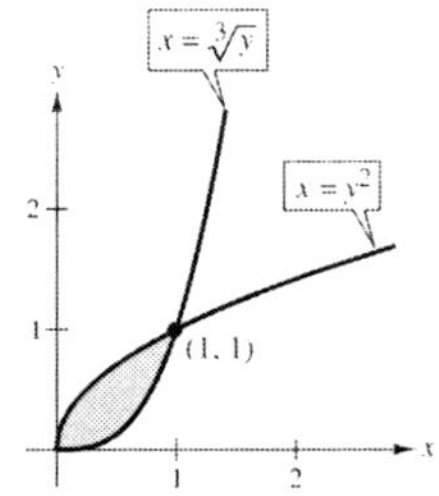

62. $\int_{-2}^2 \int_0^{4-y^2} dx\,dy = \int_0^4 \int_{-\sqrt{4-x}}^{\sqrt{4-x}} dy\,dx = \frac{32}{3}$

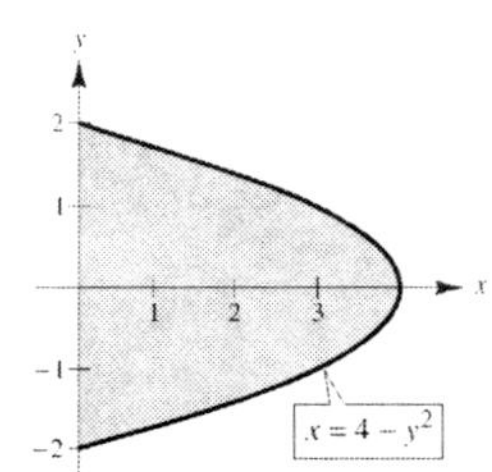

63. The first integral arises using vertical representative rectangles. The second two integrals arise using horizontal representative rectangles.

$$\int_0^5 \int_x^{\sqrt{50-x^2}} x^2y^2\,dy\,dx = \int_0^5 \left[\frac{1}{3}x^2\left(50 - x^2\right)^{3/2} - \frac{1}{3}x^5\right]dx = \frac{15{,}625}{24}\pi$$

$$\int_0^5 \int_0^y x^2y^2\,dx\,dy + \int_5^{5\sqrt{2}} \int_0^{\sqrt{50-y^2}} x^2y^2\,dx\,dy = \int_0^5 \frac{1}{3}y^5\,dy + \int_5^{5\sqrt{2}} \frac{1}{3}\left(50 - y^2\right)^{3/2} y^2\,dy$$

$$= \frac{15{,}625}{18} + \left(\frac{15{,}625}{18}\pi - \frac{15{,}625}{18}\right) = \frac{15{,}625}{24}\pi$$

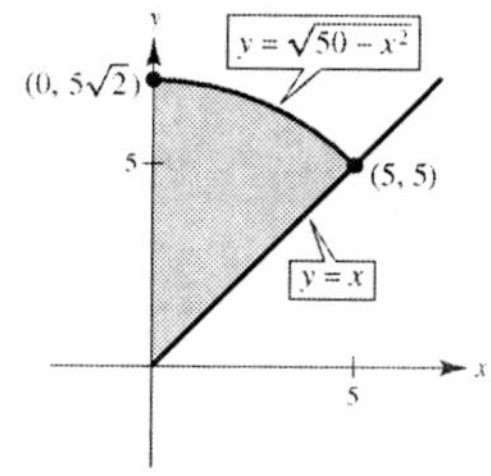

64. (a) $A = \int_0^2 \int_{y^2}^{2y} dx\,dy = \int_0^2 \left[x\right]_{y^2}^{2y} dy = \int_0^2 \left(2y - y^2\right) dy = \left[y^2 - \frac{y^3}{3}\right]_0^2 = 4 - \frac{8}{3} = \frac{4}{3}$

(b) $A = \int_0^4 \int_{x/2}^{\sqrt{x}} dy\,dx = \int_0^4 \left[y\right]_{x/2}^{\sqrt{x}} dx = \int_0^4 \left(\sqrt{x} - \frac{x}{2}\right) dx$

$$= \left[\frac{2}{3}x^{3/2} - \frac{x^2}{4}\right]_0^4 = \frac{16}{3} - 4 = \frac{4}{3}$$

Integrals (a) and (b) are the same.

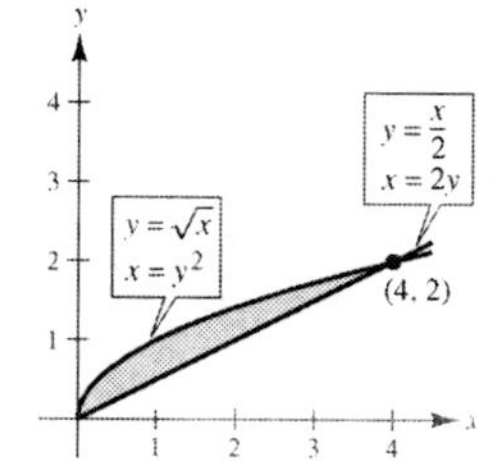

65. $\int_0^2 \int_x^2 x\sqrt{1 + y^3}\,dy\,dx = \int_0^2 \int_0^y x\sqrt{1 + y^3}\,dx\,dy$

$$= \int_0^2 \left[\sqrt{1 + y^3} \cdot \frac{x^2}{2}\right]_0^y dy$$

$$= \frac{1}{2}\int_0^2 \sqrt{1 + y^3}\,y^2\,dy$$

$$= \left[\frac{1}{2} \cdot \frac{1}{3} \cdot \frac{2}{3}\left(1 + y^3\right)^{3/2}\right]_0^2$$

$$= \frac{1}{9}(27) - \frac{1}{9}(1) = \frac{26}{9}$$

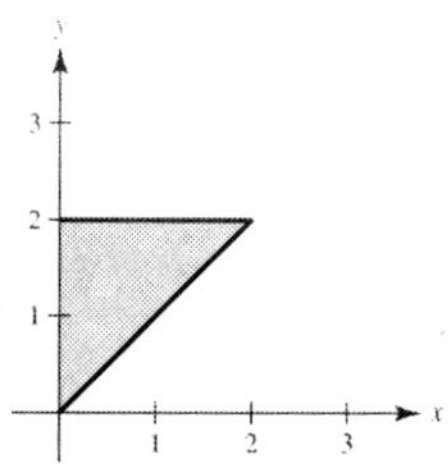

66. $\int_0^4 \int_{\sqrt{x}}^2 \frac{3}{2 + y^3}\,dy\,dx = \int_0^2 \int_0^{y^2} \frac{3}{2 + y^3}\,dx\,dy$

$$= \int_0^2 \left[\frac{3x}{2 + y^3}\right]_0^{y^2} dy$$

$$= \int_0^2 \frac{3y^2}{2 + y^3}\,dy = \left[\ln\left|2 + y^3\right|\right]_0^2$$

$$= \ln 10 - \ln 2 = \ln 5$$

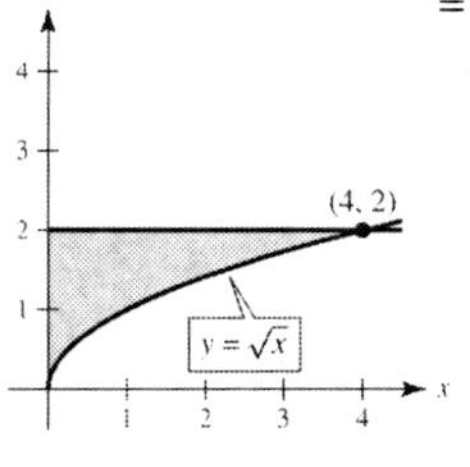

67. $\int_0^1 \int_{2x}^2 4e^{y^2}\,dy\,dx = \int_0^2 \int_0^{y/2} 4e^{y^2}\,dx\,dy$

$$= \int_0^2 \left[4xe^{y^2}\right]_0^{y/2} dy = \int_0^2 2ye^{y^2}\,dy$$

$$= \left[e^{y^2}\right]_0^2 = e^4 - 1$$

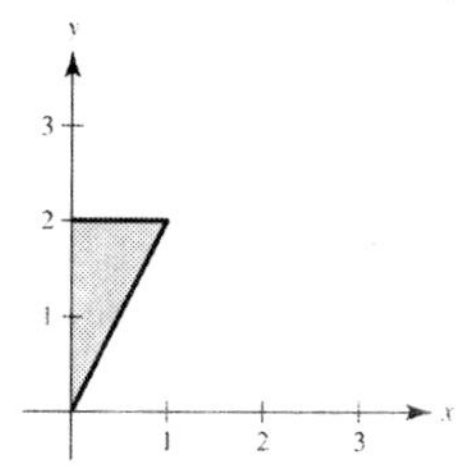

68. $\int_0^2 \int_x^2 e^{-y^2}\,dy\,dx = \int_0^2 \int_0^y e^{-y^2}\,dx\,dy$

$$= \int_0^2 \left[xe^{-y^2}\right]_0^y dy$$

$$= \int_0^2 ye^{-y^2}\,dy$$

$$= \left[-\frac{1}{2}e^{-y^2}\right]_0^2$$

$$= -\frac{1}{2}\left(e^{-4}\right) + \frac{1}{2}e^0$$

$$= \frac{1}{2}\left(1 - \frac{1}{e^4}\right) \approx 0.4908$$

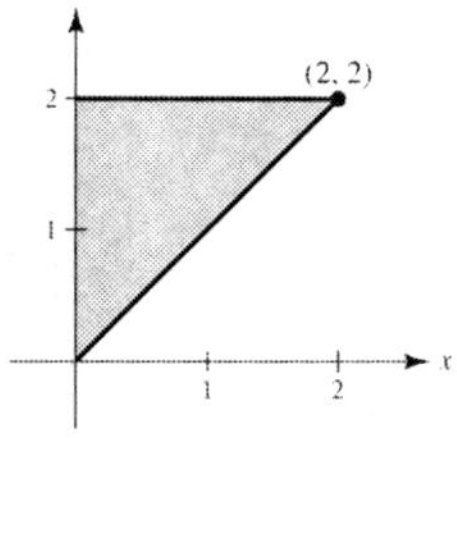

69. $\int_0^1 \int_y^1 \sin(x^2)\,dx\,dy = \int_0^1 \int_0^x \sin(x^2)\,dy\,dx$

$= \int_0^1 \left[y \sin(x^2)\right]_0^x dx$

$= \int_0^1 x \sin(x^2)\,dx$

$= \left[-\frac{1}{2}\cos(x^2)\right]_0^1$

$= -\frac{1}{2}\cos 1 + \frac{1}{2}(1)$

$= \frac{1}{2}(1 - \cos 1) \approx 0.2298$

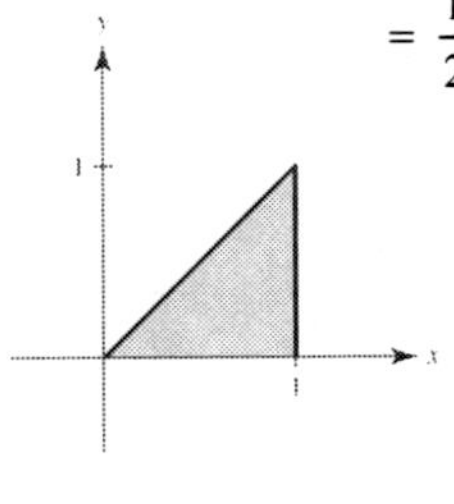

70. $\int_0^2 \int_{y^2}^4 \sqrt{x} \sin x\,dx\,dy = \int_0^4 \int_0^{\sqrt{x}} \sqrt{x} \sin x\,dy\,dx$

$= \int_0^4 \left[y\sqrt{x} \sin x\right]_0^{\sqrt{x}} dx$

$= \int_0^4 x \sin x\,dx$

$= \left[\sin x - x \cos x\right]_0^4$

$= \sin 4 - 4\cos 4 \approx 1.858$

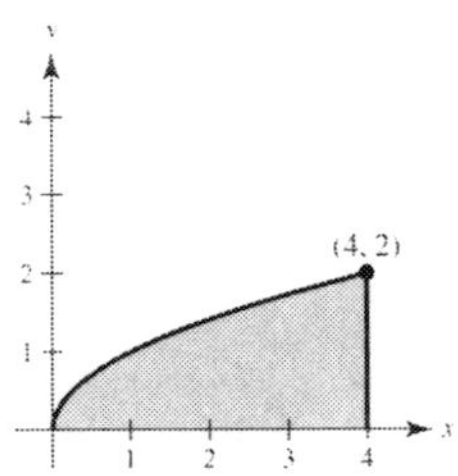

71. $\int_0^2 \int_{x^2}^{2x} \left(x^3 + 3y^2\right) dy\,dx = \frac{1664}{105} \approx 15.848$

72. $\int_0^1 \int_y^{2y} \sin(x + y)\,dx\,dy = \frac{\sin 2}{2} - \frac{\sin 3}{3} \approx 0.408$

73. $\int_0^4 \int_0^y \frac{2}{(x + 1)(y + 1)}\,dx\,dy = (\ln 5)^2 \approx 2.590$

74. $\int_0^a \int_0^{a-x} \left(x^2 + y^2\right) dy\,dx = \frac{a^4}{6}$

75. $\int_0^2 \int_0^{4-x^2} e^{xy}\,dy\,dx \approx 20.5648$

76. $\int_0^2 \int_x^2 \sqrt{16 - x^3 - y^3}\,dy\,dx \approx 6.8520$

77. $\int_0^{2\pi} \int_0^{1+\cos\theta} 6r^2 \cos\theta\,dr\,d\theta = \frac{15\pi}{2}$

78. $\int_0^{\pi/2} \int_0^{1+\sin\theta} 15\,\theta r\,dr\,d\theta = \frac{45\pi^2}{32} + \frac{135}{8} \approx 30.7541$

79. (a) $x = y^3 \Leftrightarrow y = x^{1/3}$

$x = 4\sqrt{2y} \Leftrightarrow x^2 = 32y \Leftrightarrow y = \frac{x^2}{32}$

(b) $\int_0^8 \int_{x^2/32}^{x^{1/3}} \left(x^2 y - xy^2\right) dy\,dx$

(c) Both integrals equal $\frac{67{,}520}{693} \approx 97.43$.

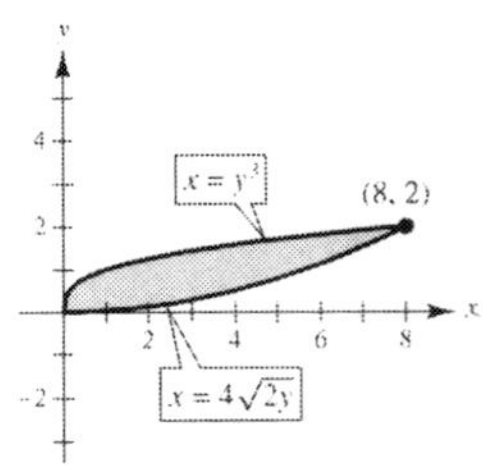

80. (a) $y = \sqrt{4 - x^2} \Leftrightarrow x = \sqrt{4 - y^2}$

$y = 4 - \frac{x^2}{4} \Leftrightarrow x = \sqrt{16 - 4y}$

(b) $\int_0^2 \int_{\sqrt{4-y^2}}^2 \frac{xy}{x^2 + y^2 + 1}\,dx\,dy + \int_2^3 \int_0^2 \frac{xy}{x^2 + y^2 + 1}\,dx\,dy + \int_3^4 \int_0^{\sqrt{16-4y}} \frac{xy}{x^2 + y^2 + 1}\,dx\,dy$

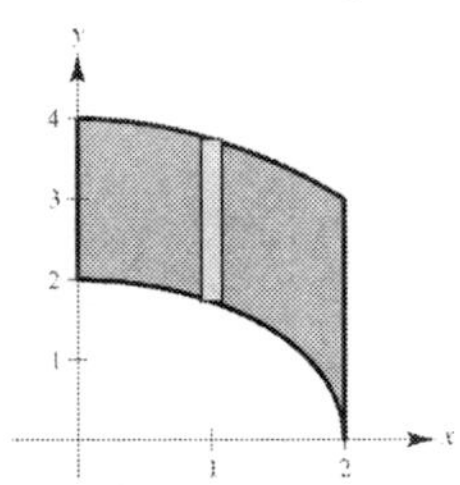

(c) Both orders of integration yield 1.11899.

81. An iterated integral is integration of a function of several variables. Integrate with respect to one variable while holding the other variables constant.

82. A region is vertically simple if it is bounded on the left and right by vertical lines, and bounded on the top and bottom by functions of x. A region is horizontally simple if it is bounded on the top and bottom by horizontal lines, and bounded on the left and right by functions of y.

83. The region is a rectangle.

84. The integrations might be easier. See Exercises 57–60.

85. True

86. False, let $f(x, y) = x$.

Section 14.2 Double Integrals and Volume

For Exercises 1–4, $\Delta x_i = \Delta y_i = 1$ and the midpoints of the squares are

$\left(\frac{1}{2}, \frac{1}{2}\right), \left(\frac{3}{2}, \frac{1}{2}\right), \left(\frac{5}{2}, \frac{1}{2}\right), \left(\frac{7}{2}, \frac{1}{2}\right), \left(\frac{1}{2}, \frac{3}{2}\right), \left(\frac{3}{2}, \frac{3}{2}\right), \left(\frac{5}{2}, \frac{3}{2}\right), \left(\frac{7}{2}, \frac{3}{2}\right).$

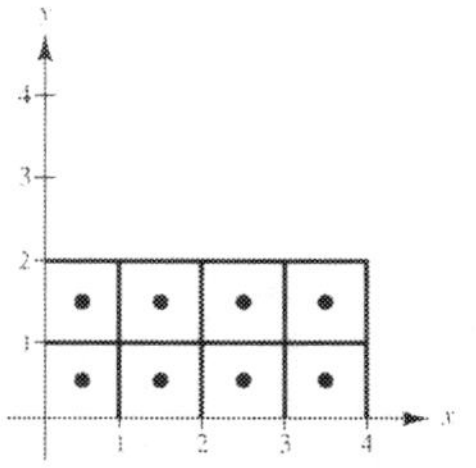

1. $f(x, y) = x + y$

$$\sum_{i=1}^{8} f(x_i, y_i)\,\Delta x_i \Delta y_i = 1 + 2 + 3 + 4 + 2 + 3 + 4 + 5 = 24$$

$$\int_0^4 \int_0^2 (x + y)\,dy\,dx = \int_0^4 \left[xy + \frac{y^2}{2}\right]_0^2 dx = \int_0^4 (2x + 2)\,dx = \left[x^2 + 2x\right]_0^4 = 24$$

2. $f(x, y) = \dfrac{1}{2}x^2 y$

$$\sum_{i=1}^{\infty} f(x_i, y_i)\,\Delta x_i\,\Delta y_i = \frac{1}{16} + \frac{9}{16} + \frac{25}{16} + \frac{49}{16} + \frac{3}{16} + \frac{27}{16} + \frac{75}{16} + \frac{147}{16} = 21$$

$$\int_0^4 \int_0^2 \frac{1}{2}x^2 y\,dy\,dx = \int_0^4 \left[\frac{x^2 y^2}{4}\right]_0^2 dx = \int_0^4 x^2\,dx = \left[\frac{x^3}{3}\right]_0^4 = \frac{64}{3} \approx 21.3$$

3. $f(x, y) = x^2 + y^2$

$$\sum_{i=1}^{8} f(x_i, y_i)\,\Delta x_i\,\Delta y_i = \frac{2}{4} + \frac{10}{4} + \frac{26}{4} + \frac{50}{4} + \frac{10}{4} + \frac{18}{4} + \frac{34}{4} + \frac{58}{4} = 52$$

$$\int_0^4 \int_0^2 \left(x^2 + y^2\right)dy\,dx = \int_0^4 \left[x^2 y + \frac{y^3}{3}\right]_0^2 dx = \int_0^4 \left(2x^2 + \frac{8}{3}\right)dx = \left[\frac{2x^3}{3} + \frac{8x}{3}\right]_0^4 = \frac{160}{3}$$

4. $f(x, y) = \dfrac{1}{(x+1)(y+1)}$

$$\sum_{i=1}^{8} f(x_i, y_i)\,\Delta x_i\,\Delta y_i = \frac{4}{9} + \frac{4}{15} + \frac{4}{21} + \frac{4}{27} + \frac{4}{15} + \frac{4}{25} + \frac{4}{35} + \frac{4}{45} = \frac{7936}{4725} \approx 1.680$$

$$\int_0^4 \int_0^2 \frac{1}{(x+1)(y+1)}\,dy\,dx = \int_0^4 \left[\frac{1}{x+1}\ln(y+1)\right]_0^2 dx$$

$$= \int_0^4 \frac{\ln 3}{x+1}\,dx = \Big[\ln 3 \cdot \ln(x+1)\Big]_0^4 = (\ln 3)(\ln 5) \approx 1.768$$

5. $\displaystyle\int_0^2 \int_0^1 (1 + 2x + 2y)\,dy\,dx = \int_0^2 \Big[y + 2xy + y^2\Big]_0^1 dx = \int_0^2 (2 + 2x)\,dx = \Big[2x + x^2\Big]_0^2 = 8$

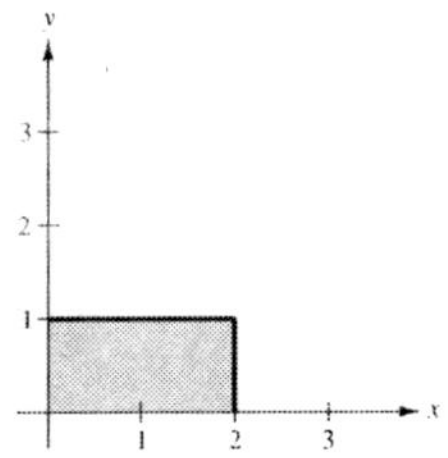

6. $\displaystyle\int_0^{\pi} \int_0^{\pi/2} \sin^2 x \cos^2 y\,dy\,dx = \int_0^{\pi} \left[\frac{1}{2}\sin^2 x\left(y + \frac{1}{2}\sin 2y\right)\right]_0^{\pi/2} dx$

$$= \int_0^{\pi} \frac{1}{2}\sin^2 x\left(\frac{\pi}{2}\right)dx = \frac{\pi}{8}\int_0^{\pi} (1 - \cos 2x)\,dx = \left[\frac{\pi}{8}\left(x - \frac{1}{2}\sin 2x\right)\right]_0^{\pi} = \frac{\pi^2}{8}$$

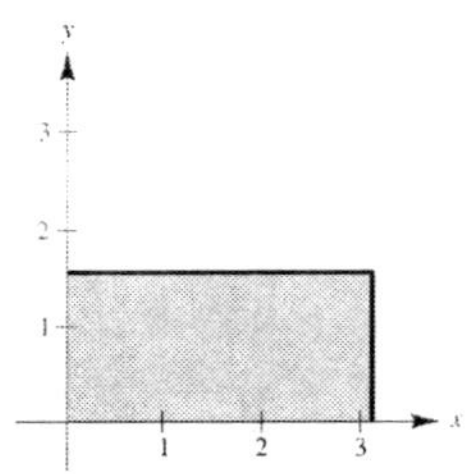

7. $\displaystyle\int_0^6 \int_{y/2}^3 (x + y)\,dx\,dy = \int_0^6 \left[\frac{1}{2}x^2 + xy\right]_{y/2}^3 dy = \int_0^6 \left(\frac{9}{2} + 3y - \frac{5}{8}y^2\right)dy = \left[\frac{9}{2}y + \frac{3}{2}y^2 - \frac{5}{24}y^3\right]_0^6 = 36$

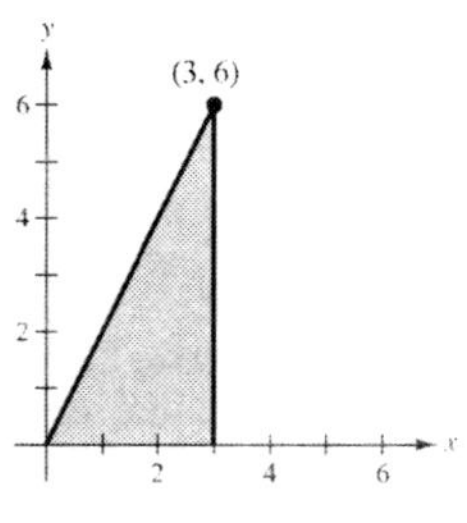

8. $\displaystyle\int_0^4 \int_{(1/2)y}^{\sqrt{y}} x^2y^2\,dx\,dy = \int_0^4 \left[\frac{x^3y^2}{3}\right]_{(1/2)y}^{\sqrt{y}} dy = \int_0^4 \left(\frac{y^{7/2}}{3} - \frac{y^5}{24}\right)dy = \left[\frac{2y^{9/2}}{27} - \frac{y^6}{144}\right]_0^4 = \frac{1024}{27} - \frac{256}{9} = \frac{256}{27}$

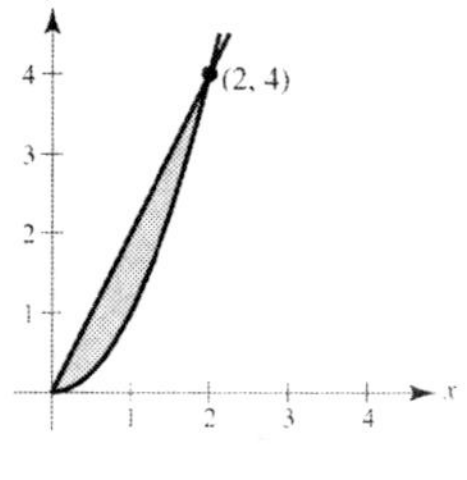

9. $\displaystyle\int_{-a}^{a}\int_{-\sqrt{a^2-x^2}}^{\sqrt{a^2-x^2}}(x+y)\,dy\,dx = \int_{-a}^{a}\left[xy+\frac{1}{2}y^2\right]_{-\sqrt{a^2-x^2}}^{\sqrt{a^2-x^2}}dx = \int_{-a}^{a}2x\sqrt{a^2-x^2}\,dx = \left[-\frac{2}{3}\left(a^2-x^2\right)^{3/2}\right]_{-a}^{a} = 0$

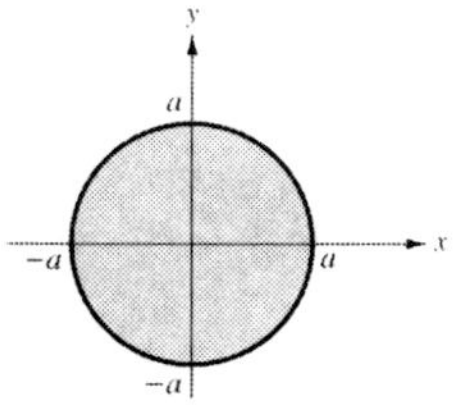

10. $\displaystyle\int_0^1\int_{y-1}^{0}e^{x+y}\,dx\,dy + \int_0^1\int_0^{1-y}e^{x+y}\,dx\,dy = \int_0^1\left[e^{x+y}\right]_{y-1}^{0}dy + \int_0^1\left[e^{x+y}\right]_0^{1-y}dy$

$\displaystyle = \int_0^1\left(e - e^{2y-1}\right)dy = \left[ey - \tfrac{1}{2}e^{2y-1}\right]_0^1 = \tfrac{1}{2}\left(e + e^{-1}\right)$

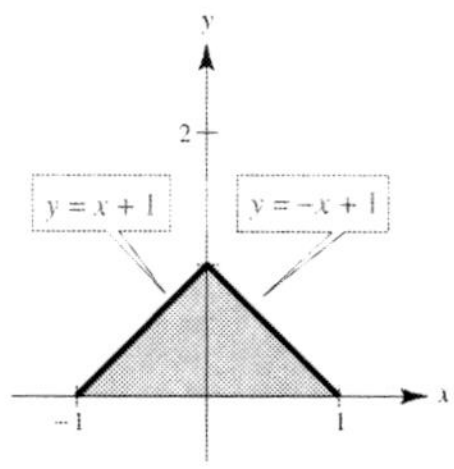

11. $\displaystyle\int_0^5\int_0^3 xy\,dx\,dy = \int_0^3\int_0^5 xy\,dy\,dx$

$\displaystyle = \int_0^3\left[\tfrac{1}{2}xy^2\right]_0^5 dx = \tfrac{25}{2}\int_0^3 x\,dx = \left[\tfrac{25}{4}x^2\right]_0^3 = \tfrac{225}{4}$

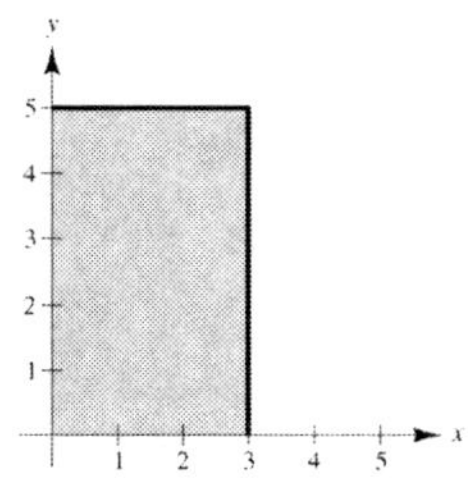

12. $\displaystyle\int_0^{\pi/2}\int_{-\pi}^{\pi}\sin x\sin y\,dx\,dy = \int_{-\pi}^{\pi}\int_0^{\pi/2}\sin x\sin y\,dy\,dx$

$\displaystyle = \int_{-\pi}^{\pi}\left[-\sin x\cos y\right]_0^{\pi/2}dx = \int_{-\pi}^{\pi}\sin x\,dx = 0$

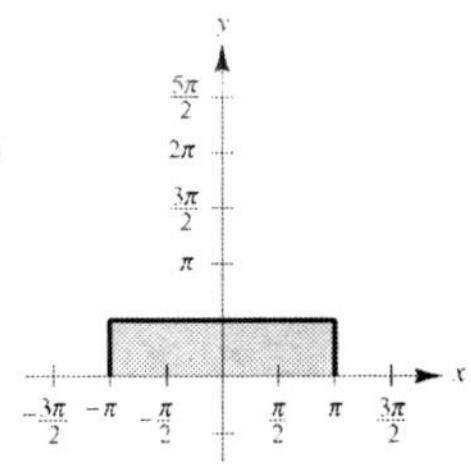

13. $\displaystyle\int_1^2\int_1^y\frac{y}{x^2+y^2}\,dx\,dy + \int_2^4\int_{y/2}^{2}\frac{y}{x^2+y^2}\,dx\,dy = \int_1^2\int_x^{2x}\frac{y}{x^2+y^2}\,dy\,dx$

$\displaystyle = \frac{1}{2}\int_1^2\left[\ln\left(x^2+y^2\right)\right]_x^{2x}dx$

$\displaystyle = \frac{1}{2}\int_1^2\left(\ln 5x^2 - \ln 2x^2\right)dx$

$\displaystyle = \frac{1}{2}\ln\frac{5}{2}\int_1^2 dx = \left[\frac{1}{2}\left(\ln\frac{5}{2}\right)x\right]_1^2 = \frac{1}{2}\ln\frac{5}{2}$

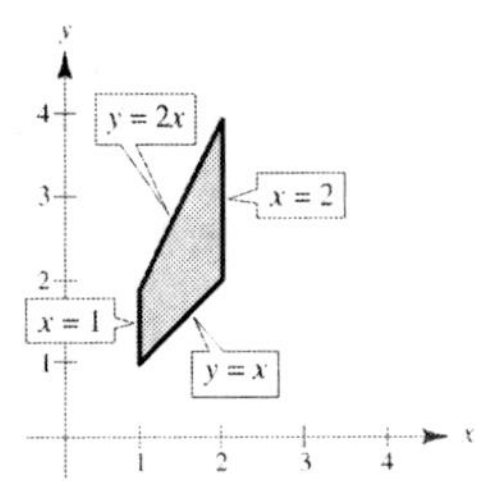

14. $\displaystyle\int_0^4\int_0^{4-x}xe^y\,dy\,dx = \int_0^4\int_0^{4-y}xe^y\,dx\,dy$

For the first integral, you obtain:

$\displaystyle\int_0^4\left[xe^y\right]_0^{4-x}dx = \int_0^4\left(xe^{4-x} - x\right)dx$

$\displaystyle = \left[-e^{4-x}(1+x) - \frac{x^2}{2}\right]_0^4 = (-5-8) + e^4 = e^4 - 13.$

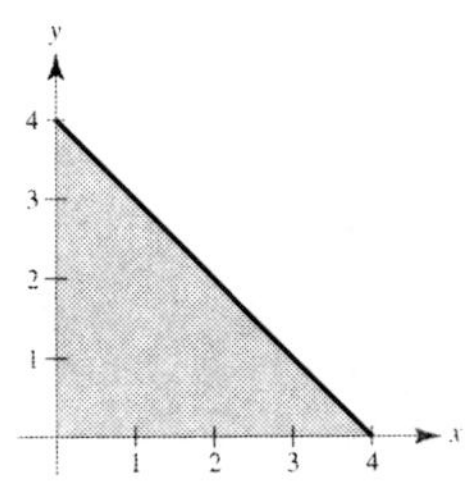

15. $\displaystyle\int_3^4\int_{4-y}^{\sqrt{4-y}} -2y\,dx\,dy = \int_0^1\int_{4-x}^{4-x^2} -2y\,dy\,dx$

$\displaystyle= \int_0^1\left[-y^2\right]_{4-x}^{4-x^2} dx$

$\displaystyle= -\int_0^1\left[\left(4-x^2\right)^2 - \left(4-x\right)^2\right]dx$

$\displaystyle= -\int_0^1\left[16 - 8x^2 + x^4 - \left(16 - 8x + x^2\right)\right]dx$

$\displaystyle= -\left[-3x^3 + \frac{x^5}{5} + 4x^2\right]_0^1 = -\frac{6}{5}$

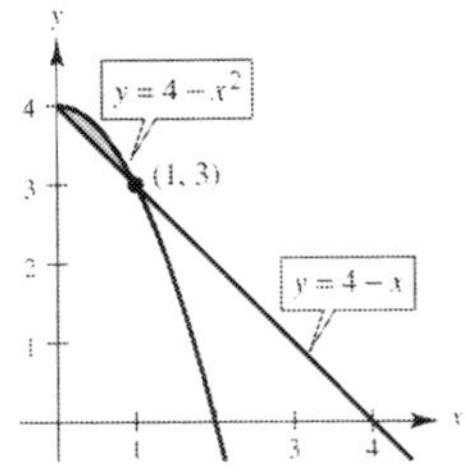

16. $\displaystyle\int_0^2\int_{y^2}^4 \frac{y}{1+x^2}\,dx\,dy = \int_0^4\int_0^{\sqrt{x}} \frac{y}{1+x^2}\,dy\,dx$

$\displaystyle= \frac{1}{2}\int_0^4\left[\frac{y^2}{1+x^2}\right]_0^{\sqrt{x}} dx = \frac{1}{2}\int_0^4 \frac{x}{1+x^2}\,dx = \left[\frac{1}{4}\ln\left(1+x^2\right)\right]_0^4 = \frac{1}{4}\ln(17)$

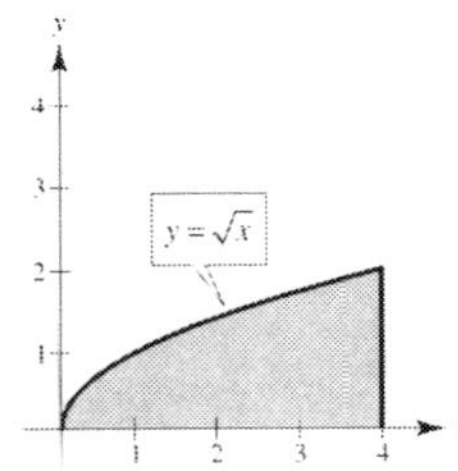

17. $\displaystyle\int_0^4\int_0^{3x/4} x\,dy\,dx + \int_4^5\int_0^{\sqrt{25-x^2}} x\,dy\,dx = \int_0^3\int_{4y/3}^{\sqrt{25-y^2}} x\,dx\,dy$

$\displaystyle= \int_0^3\left[\tfrac{1}{2}x^2\right]_{4y/3}^{\sqrt{25-y^2}} dy$

$\displaystyle= \tfrac{25}{18}\int_0^3\left(9 - y^2\right)dy = \left[\tfrac{25}{18}\left(9y - \tfrac{1}{3}y^3\right)\right]_0^3 = 25$

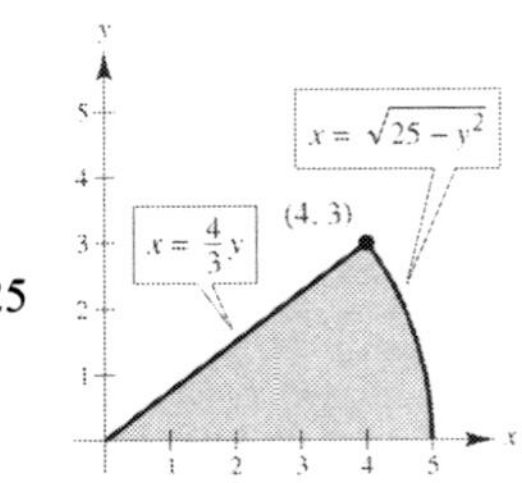

18. $\displaystyle\int_0^2\int_{-\sqrt{4-y^2}}^{\sqrt{4-y^2}}\left(x^2 + y^2\right)dx\,dy$

$\displaystyle= \int_{-2}^2\int_0^{\sqrt{4-x^2}}\left(x^2 + y^2\right)dy\,dx$

$\displaystyle= \int_{-2}^2\left[x^2y + \frac{1}{3}y^3\right]_0^{\sqrt{4-x^2}} dx$

$\displaystyle= \int_{-2}^2\left[x^2\sqrt{4-x^2} + \frac{1}{3}\left(4-x^2\right)^{3/2}\right]dx$

$\displaystyle= \left[-\frac{x}{4}\left(4-x^2\right)^{3/2} + \frac{1}{2}\left(x\sqrt{4-x^2} + 4\arcsin\frac{x}{2}\right) + \frac{1}{12}\left[x\left(4-x^2\right)^{3/2} + 6x\sqrt{4-x^2} + 24\arctan\frac{x}{2}\right]\right]_{-2}^2 = 4\pi$

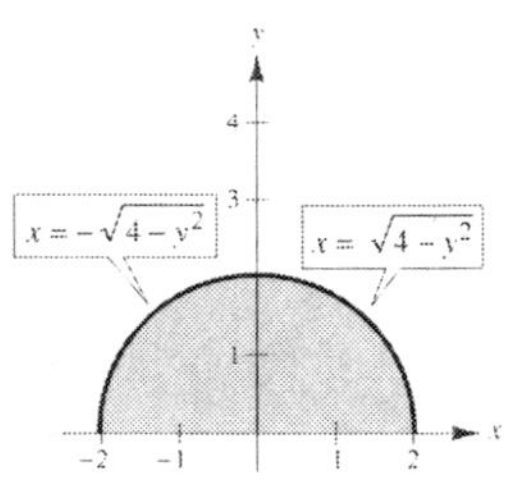

19. $\displaystyle V = \int_0^4\int_0^2 \frac{y}{2}\,dy\,dx = \int_0^4\left[\frac{y^2}{4}\right]_0^2 dx = \int_0^4 dx = 4$

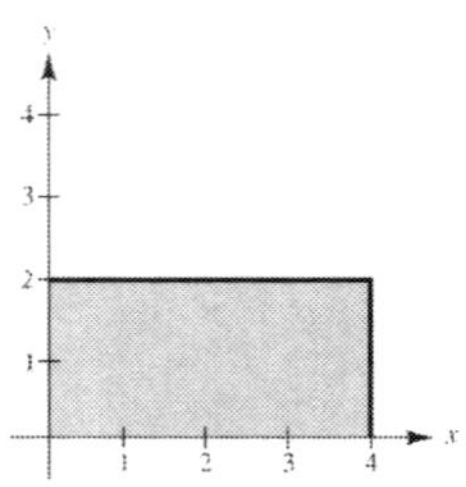

20. $\displaystyle V = \int_0^4\int_0^2 (6 - 2y)\,dy\,dx$

$\displaystyle= \int_0^4\left[6y - y^2\right]_0^2 dx = \int_0^4 8\,dx = 32$

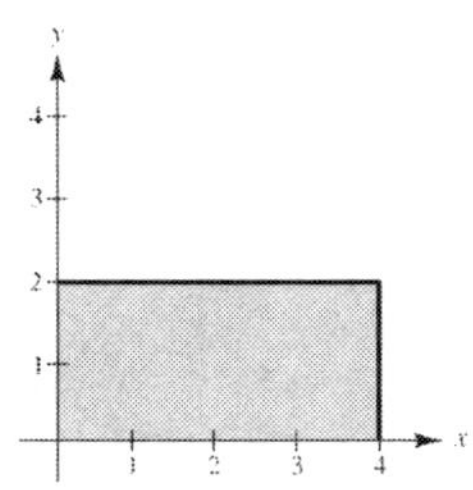

21. $V = \int_0^2 \int_0^y (4 - x - y)\, dx\, dy$

$= \int_0^2 \left[4x - \frac{x^2}{2} - xy \right]_0^y dy$

$= \int_0^2 \left(4y - \frac{y^2}{2} - y^2 \right) dy$

$= \left[2y^2 - \frac{y^3}{6} - \frac{y^3}{3} \right]_0^2$

$= 8 - \frac{8}{6} - \frac{8}{3} = 4$

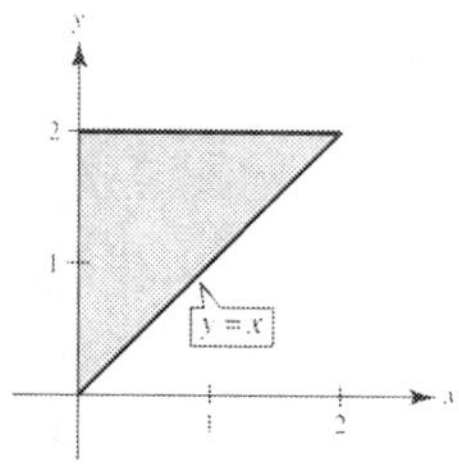

22. $V = \int_0^6 \int_0^{(-2/3)x+4} \left(\frac{12 - 2x - 3y}{4} \right) dy\, dx = \int_0^6 \left[\frac{1}{4}\left(12y - 2xy - \frac{3}{2}y^2 \right) \right]_0^{(-2/3)x+4} dx$

$= \int_0^6 \left(\frac{1}{6}x^2 - 2x + 6 \right) dx = \left[\frac{1}{18}x^3 - x^2 + 6x \right]_0^6 = 12$

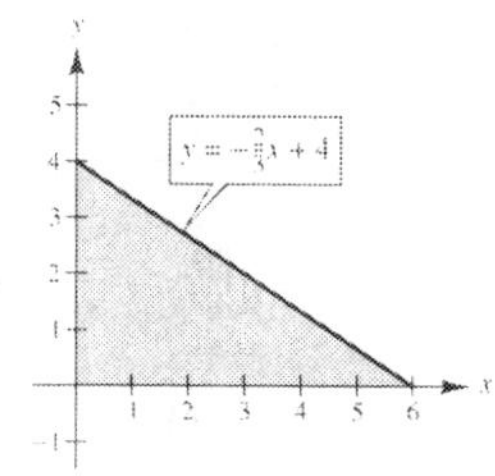

23. $V = \int_0^1 \int_0^y (1 - xy)\, dx\, dy$

$= \int_0^1 \left[x - \frac{x^2 y}{2} \right]_0^y dy = \int_0^1 \left(y - \frac{y^3}{2} \right) dy = \left[\frac{y^2}{2} - \frac{y^4}{8} \right]_0^1 = \frac{3}{8}$

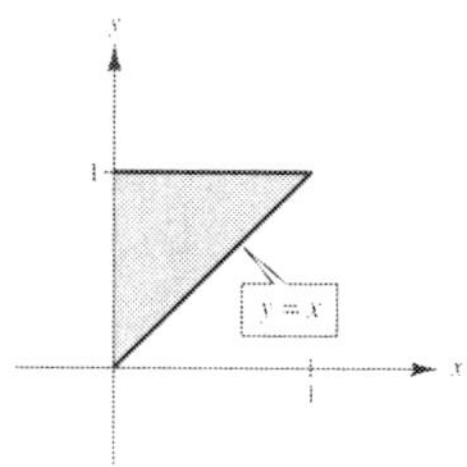

24. $V = \int_0^2 \int_0^y \left(4 - y^2\right) dx\, dy$

$= \int_0^2 \left(4y - y^3\right) dy = \left[2y^2 - \frac{y^4}{4} \right]_0^2 = 4$

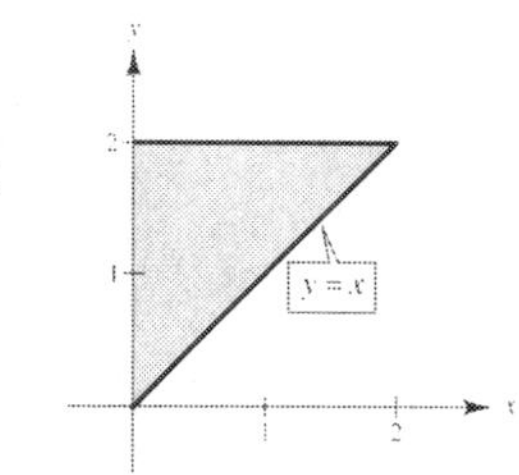

25. $V = \int_0^\infty \int_0^\infty \frac{1}{(x+1)^2(y+1)^2}\, dy\, dx = \int_0^\infty \left[-\frac{1}{(x+1)^2(y+1)} \right]_0^\infty dx = \int_0^\infty \frac{1}{(x+1)^2}\, dx = \left[-\frac{1}{(x+1)} \right]_0^\infty = 1$

26. $V = \int_0^\infty \int_0^\infty e^{-(x+y)/2}\, dy\, dx = \int_0^\infty \left[-2e^{-(x+y)/2} \right]_0^\infty dx = \int_0^\infty 2e^{-x/2}\, dx = \left[-4e^{-x/2} \right]_0^\infty = 4$

27. $V = \int_0^1 \int_0^x xy\, dy\, dx = \int_0^1 \left[\tfrac{1}{2}xy^2 \right]_0^x dx = \tfrac{1}{2}\int_0^1 x^3\, dx = \left[\tfrac{1}{8}x^4 \right]_0^1 = \tfrac{1}{8}$

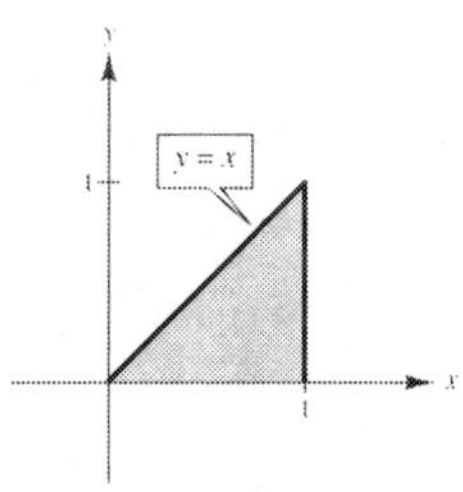

28. $V = \int_0^2 \int_0^4 x^2\, dy\, dx$

$= \int_0^2 \left[x^2 y\right]_0^4 dx = \int_0^2 4x^2\, dx$

$= \left[\frac{4x^3}{3}\right]_0^2 = \frac{32}{3}$

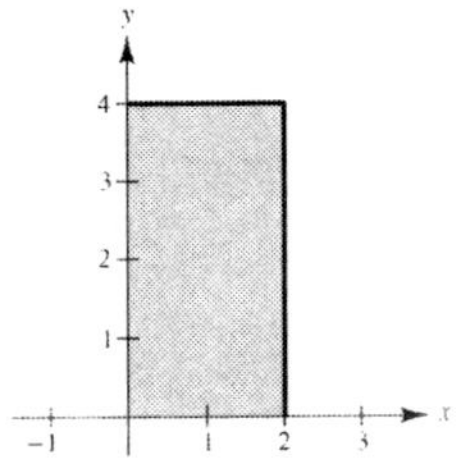

29. Divide the solid into two equal parts.

$V = 2\int_0^1 \int_0^x \sqrt{1 - x^2}\, dy\, dx = 2\int_0^1 \left[y\sqrt{1 - x^2}\right]_0^x dx$

$= 2\int_0^1 x\sqrt{1 - x^2}\, dx = \left[-\frac{2}{3}\left(1 - x^2\right)^{3/2}\right]_0^1 = \frac{2}{3}$

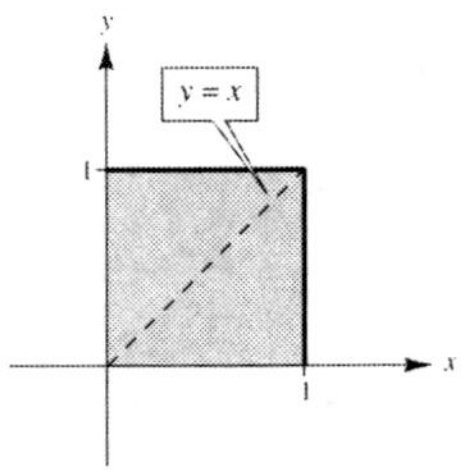

30. $V = \int_0^2 \int_0^{4-x^2} \left(4 - x^2\right) dy\, dx$

$= \int_0^2 \left(4 - x^2\right)\left(4 - x^2\right) dx$

$= \int_0^2 \left(16 - 8x^2 + x^4\right) dx = \left[16x - \frac{8x^3}{3} + \frac{x^5}{5}\right]_0^2 = 32 - \frac{64}{3} + \frac{32}{5} = \frac{256}{15}$

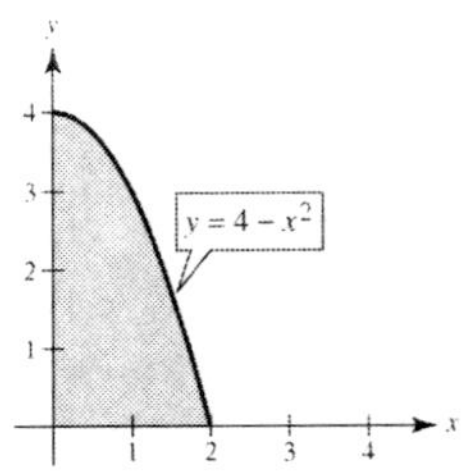

31. $V = \int_0^2 \int_0^{\sqrt{4-x^2}} (x + y)\, dy\, dx = \int_0^2 \left[xy + \frac{1}{2}y^2\right]_0^{\sqrt{4-x^2}} dx$

$= \int_0^2 \left(x\sqrt{4 - x^2} + 2 - \frac{1}{2}x^2\right) dx$

$= \left[-\frac{1}{3}\left(4 - x^2\right)^{3/2} + 2x - \frac{1}{6}x^3\right]_0^2 = \frac{16}{3}$

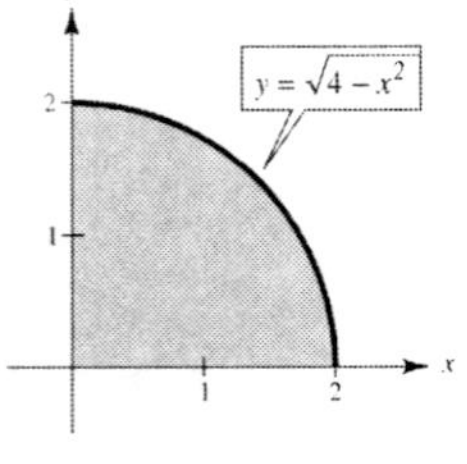

32. $V = \int_0^2 \int_0^{\infty} \frac{1}{1 + y^2}\, dy\, dx = \int_0^2 \left[\arctan y\right]_0^{\infty} dx$

$= \int_0^2 \frac{\pi}{2}\, dx = \left[\frac{\pi x}{2}\right]_0^2 = \pi$

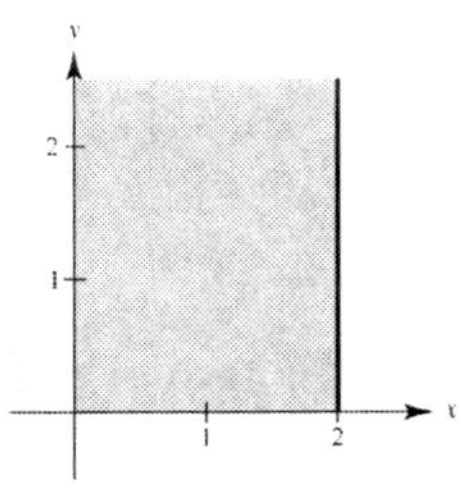

33. $V = 2\int_0^2 \int_0^{\sqrt{1-(x-1)^2}} \left(\left[4 - x^2 - y^2\right] - \left[4 - 2x\right]\right) dy\, dx$

$= 2\int_0^2 \int_0^{\sqrt{1-(x-1)^2}} \left(2x - x^2 - y^2\right) dy\, dx$

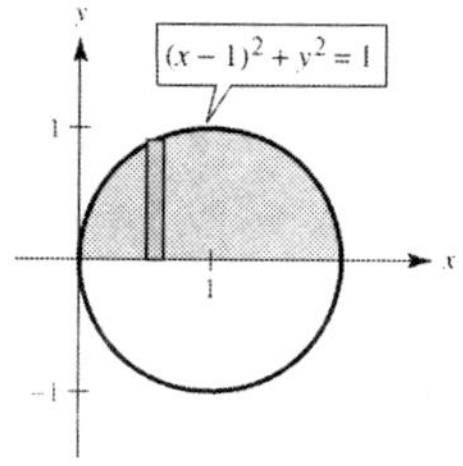

34. $V = 2\int_0^2 \int_0^{\sqrt{1-(x-1)^2}} \left[2x - \left(x^2 + y^2\right)\right] dy\, dx$

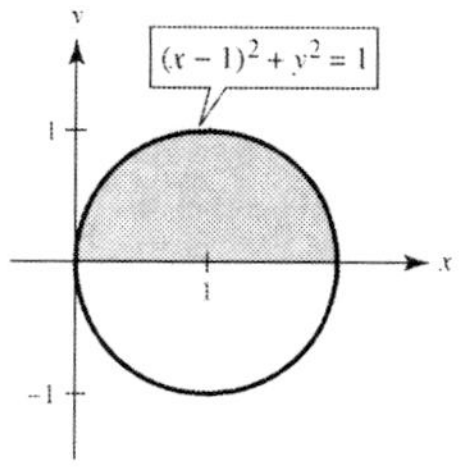

35. $V = 4\int_0^2 \int_0^{\sqrt{4-x^2}} \left(x^2 + y^2\right) dy\, dx$

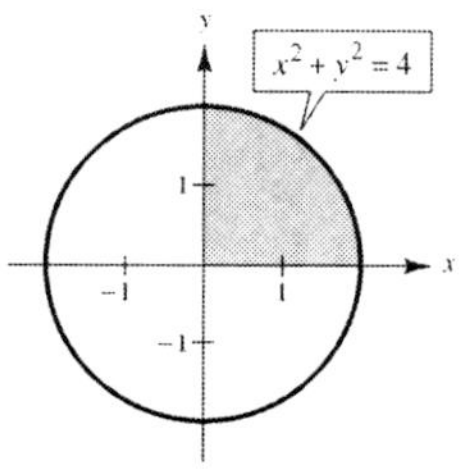

36. $V = \int_0^5 \int_0^{\pi} \sin^2 x\, dx\, dy$

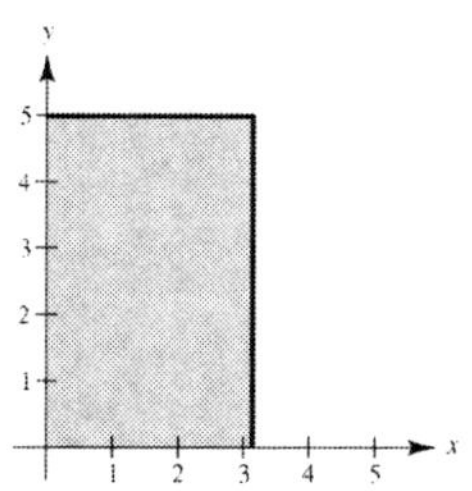

37. $V = \int_0^2 \int_{-\sqrt{2-2(y-1)^2}}^{\sqrt{2-2(y-1)^2}} \left[4y - \left(x^2 + 2y^2\right)\right] dx\, dy$

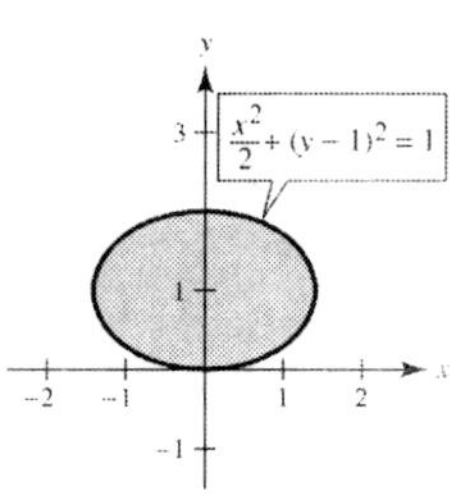

38. $V = \int_{-3}^3 \int_{-\sqrt{9-x^2}}^{\sqrt{9-x^2}} \left(\left[18 - x^2 - y^2\right] - \left[x^2 + y^2\right]\right) dy\, dx$

$= 4\int_0^3 \int_0^{\sqrt{9-x^2}} \left(18 - 2x^2 - 2y^2\right) dy\, dx$

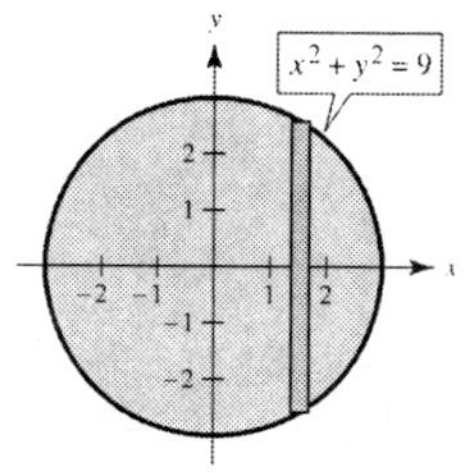

39. $V = 4\int_0^3 \int_0^{\sqrt{9-x^2}} \left(9 - x^2 - y^2\right) dy\, dx = \dfrac{81\pi}{2}$

40. $V = \int_0^9 \int_0^{\sqrt{9-y}} \sqrt{9 - y}\, dx\, dy = \frac{81}{2}$

41. $V = \int_0^2 \int_0^{-0.5x+1} \dfrac{2}{1 + x^2 + y^2}\, dy\, dx \approx 1.2315$

42. $V = \int_0^{16} \int_0^{4-\sqrt{y}} \ln(1 + x + y)\, dx\, dy \approx 38.25$

43. f is a continuous function such that $0 \le f(x, y) \le 1$ over a region R of area 1. Let $f(m, n)$ = the minimum value of f over R and $f(M, N)$ = the maximum value of f over R. Then

$f(m, n)\int_R \int dA \le \int_R \int f(x, y)\, dA \le f(M, N)\int_R \int dA.$

Because $\int_R \int dA = 1$ and

$0 \le f(m, n) \le f(M, N) \le 1$, you have

$0 \le f(m, n)(1) \le \int_R \int f(x, y)\, dA \le f(M, N)(1) \le 1.$

So, $0 \le \int_R \int f(x, y)\, dA \le 1.$

44. $\dfrac{x}{a} + \dfrac{y}{b} + \dfrac{z}{c} = 1$

$$z = c\left(1 - \frac{x}{a} - \frac{y}{b}\right)$$

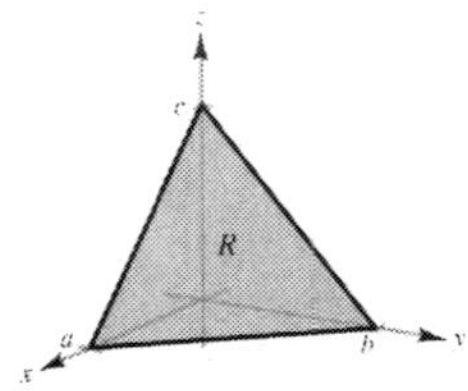

$$V = \int_R \int f(x, y)\, dA = \int_0^a \int_0^{b[1-(x/a)]} c\left(1 - \frac{x}{a} - \frac{y}{b}\right) dy\, dx = c\int_0^a \left[y - \frac{xy}{a} - \frac{y^2}{2b}\right]_0^{b[1-(x/a)]} dx$$

$$= c\int_0^a \left[b\left(1 - \frac{x}{a}\right) - \frac{xb}{a}\left(1 - \frac{x}{a}\right) - \frac{b^2}{2b}\left(1 - \frac{x}{a}\right)^2\right] dx$$

$$= c\left[-\frac{ab}{2}\left(1 - \frac{x}{a}\right)^2 - \frac{x^2 b}{2a} + \frac{x^3 b}{3a^2} + \frac{ab}{6}\left(1 - \frac{x}{a}\right)^3\right]_0^a = c\left[\left(-\frac{ab}{2} + \frac{ab}{3}\right) - \left(-\frac{ab}{2} + \frac{ab}{6}\right)\right] = \frac{abc}{6}$$

45. $\displaystyle\int_0^1 \int_{y/2}^{1/2} e^{-x^2}\, dx\, dy = \int_0^{1/2} \int_0^{2x} e^{-x^2}\, dy\, dx$

$$= \int_0^{1/2} 2xe^{-x^2}\, dx = \left[-e^{-x^2}\right]_0^{1/2} = -e^{-1/4} + 1 = 1 - e^{-1/4} \approx 0.221$$

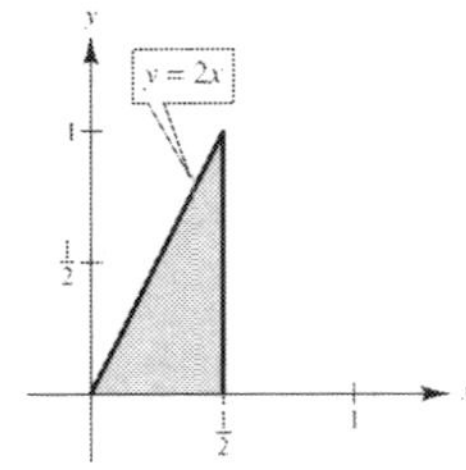

46. $\displaystyle\int_0^{\ln 10} \int_{e^x}^{10} \frac{1}{\ln y}\, dy\, dx = \int_1^{10} \int_0^{\ln y} \frac{1}{\ln y}\, dx\, dy$

$$= \int_1^{10} \left[\frac{x}{\ln y}\right]_0^{\ln y} dy$$

$$= \int_1^{10} dy = \left[y\right]_1^{10} = 9$$

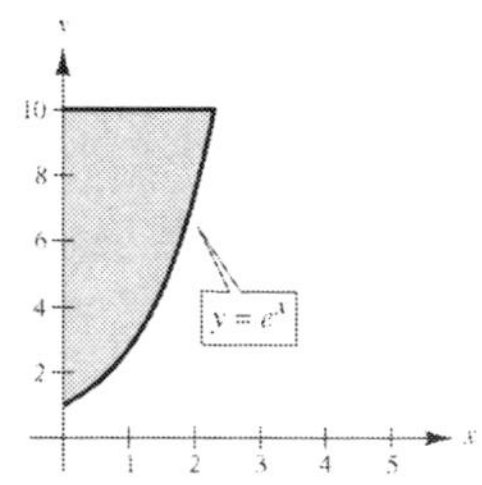

47. $\displaystyle\int_{-2}^2 \int_{-\sqrt{4-x^2}}^{\sqrt{4-x^2}} \sqrt{4 - y^2}\, dy\, dx = \int_{-2}^2 \int_{-\sqrt{4-y^2}}^{\sqrt{4-y^2}} \sqrt{4 - y^2}\, dx\, dy = \int_{-2}^2 \left[x\sqrt{4 - y^2}\right]_{-\sqrt{4-y^2}}^{\sqrt{4-y^2}} dy$

$$= \int_{-2}^2 2\left(4 - y^2\right) dy = \left[8y - \frac{2y^3}{3}\right]_{-2}^2 = \left(16 - \frac{16}{3}\right) - \left(-16 + \frac{16}{3}\right)$$

$$= \frac{64}{3}$$

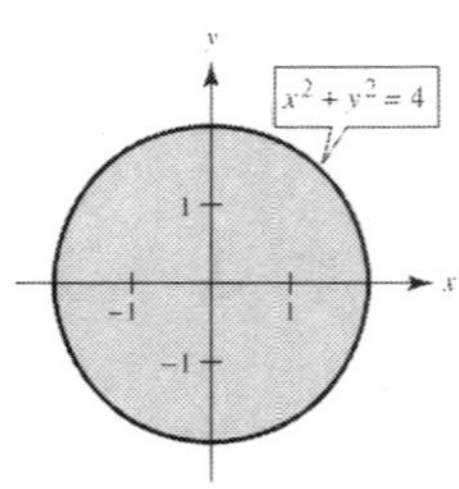

48. $\displaystyle\int_0^3 \int_{y/3}^1 \frac{1}{1 + x^4}\, dx\, dy = \int_0^1 \int_0^{3x} \frac{1}{1 + x^4}\, dy\, dx = \int_0^1 \left[\frac{y}{1 + x^4}\right]_0^{3x} dx$

$$= \int_0^1 \frac{3x}{1 + x^4}\, dx = \left[\frac{3}{2}\arctan\left(x^2\right)\right]_0^1$$

$$= \frac{3}{2}\left(\frac{\pi}{4}\right) = \frac{3\pi}{8}$$

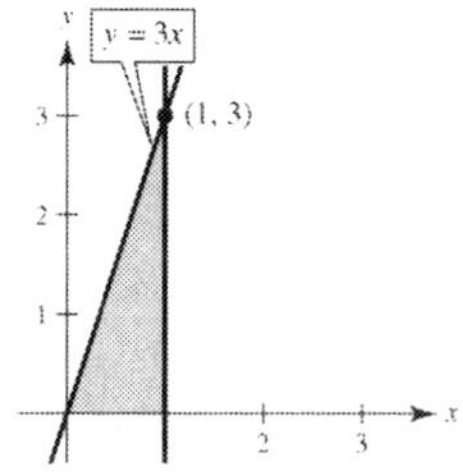

49. $\displaystyle\int_0^1 \int_0^{\arccos y} \sin x\sqrt{1 + \sin^2 x}\, dx\, dy$

$$= \int_0^{\pi/2} \int_0^{\cos x} \sin x\sqrt{1 + \sin^2 x}\, dy\, dx$$

$$= \int_0^{\pi/2} \left(1 + \sin^2 x\right)^{1/2} \sin x \cos x\, dx = \left[\frac{1}{2} \cdot \frac{2}{3}\left(1 + \sin^2 x\right)^{3/2}\right]_0^{\pi/2} = \frac{1}{3}\left[2\sqrt{2} - 1\right]$$

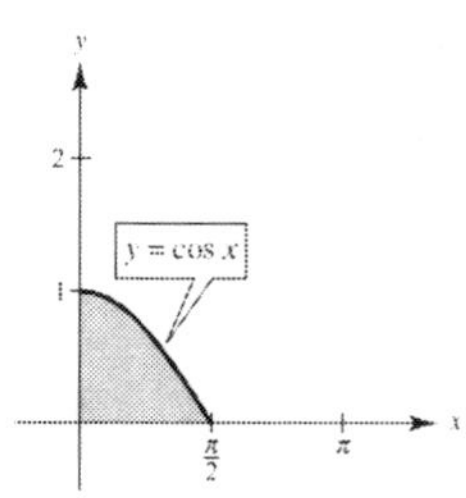

50. $\int_0^2 \int_{(1/2)x^2}^2 \sqrt{y} \cos y \, dy \, dx = \int_0^2 \int_0^{\sqrt{2y}} \sqrt{y} \cos y \, dx \, dy$

$= \int_0^2 \sqrt{2y}\sqrt{y} \cos y \, dy = \sqrt{2}\int_0^2 y \cos y \, dy$

$= \sqrt{2}\big[\cos y + y \sin y\big]_0^2 = \sqrt{2}\big[\cos 2 + 2\sin 2 - 1\big]$

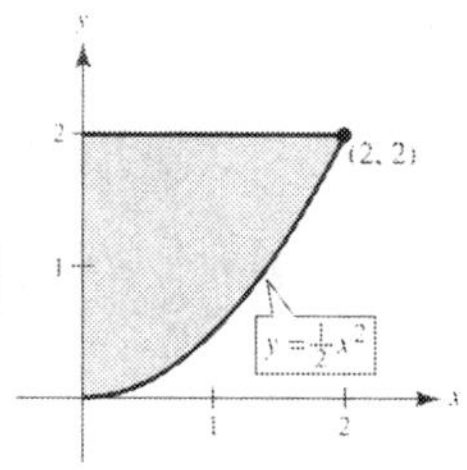

51. Average $= \frac{1}{8}\int_0^4 \int_0^2 x \, dy \, dx = \frac{1}{8}\int_0^4 2x \, dx = \left[\frac{x^2}{8}\right]_0^4 = 2$

52. Average $= \frac{1}{15}\int_0^5 \int_0^3 2xy \, dy \, dx = \frac{1}{15}\int_0^5 \big[xy^2\big]_0^3 \, dx$

$= \frac{1}{15}\int_0^5 9x \, dx = \frac{1}{15}\left[\frac{9x^2}{2}\right]_0^5 = \frac{15}{2}$

53. Average $= \frac{1}{4}\int_0^2 \int_0^2 \left(x^2 + y^2\right) dx \, dy$

$= \frac{1}{4}\int_0^2 \left[\frac{x^3}{3} + xy^2\right]_0^2 dy = \frac{1}{4}\int_0^2 \left(\frac{8}{3} + 2y^2\right) dy$

$= \left[\frac{1}{4}\left(\frac{8}{3}y + \frac{2}{3}y^3\right)\right]_0^2 = \frac{8}{3}$

54. Average $= \frac{1}{(1/2)}\int_0^1 \int_0^x \frac{1}{x+y} \, dy \, dx$

$= 2\int_0^1 \big[\ln|x+y|\big]_0^x \, dx = 2\int_0^1 \big[\ln 2x - \ln x\big] dx$

$= 2\int_0^1 \ln 2 \, dx = 2\big[x \ln 2\big]_0^1 = 2\ln 2$

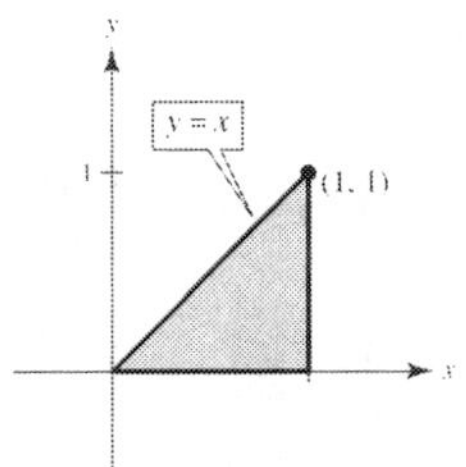

55. Average $= \frac{1}{1/2}\int_0^1 \int_x^1 e^{x+y} \, dy \, dx = 2\int_0^1 e^{x+1} - e^{2x} \, dx$

$= 2\left[e^{x+1} - \frac{1}{2}e^{2x}\right]_0^1 = 2\left[e^2 - \frac{1}{2}e^2 - e + \frac{1}{2}\right]$

$= e^2 - 2e + 1 = (e-1)^2$

56. Average $= \frac{1}{\pi^2}\int_0^\pi \int_0^\pi \sin(x+y) \, dy \, dx$

$= \frac{1}{\pi^2}\int_0^\pi \big[-\cos(x+y)\big]_0^\pi \, dx$

$= \frac{1}{\pi^2}\int_0^\pi \big(-\cos(x+\pi) + \cos x\big) dx$

$= \frac{1}{\pi^2}\int_0^\pi 2\cos x \, dx = \frac{1}{\pi^2}\big[2\sin x\big]_0^\pi = 0$

57. Average $= \frac{1}{1250}\int_{300}^{325}\int_{200}^{250} 100x^{0.6}y^{0.4} \, dx \, dy$

$= \frac{1}{1250}\int_{300}^{325}\left[\left(100y^{0.4}\right)\frac{x^{1.6}}{1.6}\right]_{200}^{250} dy$

$= \frac{128{,}844.1}{1250}\int_{300}^{325} y^{0.4} \, dy$

$= 103.0753\left[\frac{y^{1.4}}{1.4}\right]_{300}^{325} \approx 25{,}645.24$

58. Average $= \frac{1}{8}\int_0^2 \int_0^4 \left(20 - 4x^2 - y^2\right) dy \, dx$

$= \frac{1}{8}\left(\frac{224}{3}\right) = \frac{28}{3}°\text{C}$

59. See the definition on page 976.

60. The value of $\int_R \int f(x, y) \, dA$ would be kB.

61. No, the maximum possible value is $(\text{Area})(6) = 6\pi$.

62. The second is integrable. The first contains $\int \sin y^2 \, dy$ which does not have an elementary antiderivation.

63. $f(x, y) \geq 0$ for all (x, y) and

$\int_{-\infty}^{\infty}\int_{-\infty}^{\infty} f(x, y) \, dA = \int_0^5 \int_0^2 \frac{1}{10} \, dy \, dx$

$= \int_0^5 \frac{1}{5} \, dx = 1$

$P(0 \leq x \leq 2, 1 \leq y \leq 2) = \int_0^2 \int_1^2 \frac{1}{10} \, dy \, dx$

$= \int_0^2 \frac{1}{10} \, dx = \frac{1}{5}.$

64. $f(x, y) \geq 0$ for all (x, y) and

$$\int_{-\infty}^{\infty}\int_{-\infty}^{\infty} f(x, y)\,dA = \int_0^2\int_0^2 \frac{1}{4}xy\,dy\,dx = \int_0^2 \frac{x}{2}\,dx = 1$$

$$P(0 \leq x \leq 1, 1 \leq y \leq 2) = \int_0^1\int_1^2 \frac{1}{4}xy\,dy\,dx = \int_0^1 \frac{3x}{8}\,dx = \frac{3}{16}.$$

65. $f(x, y) \geq 0$ for all (x, y) and

$$\int_{-\infty}^{\infty}\int_{-\infty}^{\infty} f(x, y)\,dA = \int_0^3\int_3^6 \frac{1}{27}(9 - x - y)\,dy\,dx = \int_0^3 \frac{1}{27}\left[9y - xy - \frac{y^2}{2}\right]_3^6 dx = \int_0^3\left(\frac{1}{2} - \frac{1}{9}x\right)dx = \left[\frac{x}{2} - \frac{x^2}{18}\right]_0^3 = 1$$

$$P(0 \leq x \leq 1, 4 \leq y \leq 6) = \int_0^1\int_4^6 \frac{1}{27}(9 - x - y)\,dy\,dx = \int_0^1 \frac{2}{27}(4 - x)\,dx = \frac{7}{27}.$$

66. $f(x, y) \geq 0$ for all (x, y) and

$$\int_{-\infty}^{\infty}\int_{-\infty}^{\infty} f(x, y)\,dA = \int_0^{\infty}\int_0^{\infty} e^{-x-y}\,dy\,dx = \int_0^{\infty} \lim_{b\to\infty}\left[-e^{-x-y}\right]_0^b dx = \int_0^{\infty} e^{-x}\,dx = \lim_{b\to\infty}\left[-e^{-x}\right]_0^b = 1$$

$$P(0 \leq x \leq 1, x \leq y \leq 1) = \int_0^1\int_x^1 e^{-x-y}\,dy\,dx = \int_0^1\left[-e^{-x-y}\right]_x^1 dx = \int_0^1\left(e^{-2x} - e^{-x-1}\right)dx$$

$$= \left[-\frac{1}{2}e^{-2x} + e^{-x-1}\right]_0^1 = \frac{1}{2}e^{-2} - e^{-1} + \frac{1}{2} = \frac{1}{2}\left(e^{-1} - 1\right)^2 \approx 0.1998.$$

67. $$\int_0^4\int_0^4 f(x, y)\,dy\,dx \approx (32 + 31 + 28 + 23) + (31 + 30 + 27 + 22) + (28 + 27 + 24 + 19) + (23 + 22 + 19 + 14)$$
$$= 400$$

Using the corner of the ith square farthest from the origin, you obtain 272.

68. (a) $\iint\limits_R f(x, y)\,dA$ represents the total annual snowfall in Erie County.

(b) $\dfrac{\iint\limits_R f(x, y)\,dA}{\iint\limits_R dA}$ represents the average amount of snowfall at any point (x, y).

69. False

$$V = 8\int_0^1\int_0^{\sqrt{1-y^2}} \sqrt{1 - x^2 - y^2}\,dx\,dy$$

70. True

71. $z = 9 - x^2 - y^2$ is a paraboloid opening downward with vertex $(0, 0, 9)$. The double integral is maximized if $z \geq 0$. That is,

$$R = \{(x, y): x^2 + y^2 \leq 9\}.$$

$$\left[\text{The maximum value is } \iint\limits_R \left(9 - x^2 - y^2\right)dA = \frac{81\pi}{2}.\right]$$

72. $z = x^2 + y^2 - 4$ is a paraboloid opening upward with vertex $(0, 0, -4)$. The double integral is minimized if $z \leq 0$. That is, $R = \{(x, y): x^2 + y^2 \leq 4\}$.

[The minimum value is -8π.]

73. $$\text{Average} = \int_0^1 f(x)\,dx = \int_0^1\int_1^x e^{t^2}\,dt\,dx = -\int_0^1\int_x^1 e^{t^2}\,dt\,dx$$

$$= -\int_0^1\int_0^t e^{t^2}\,dx\,dt = -\int_0^1 te^{t^2}\,dt$$

$$= \left[-\tfrac{1}{2}e^{t^2}\right]_0^1 = -\tfrac{1}{2}(e - 1) = \tfrac{1}{2}(1 - e)$$

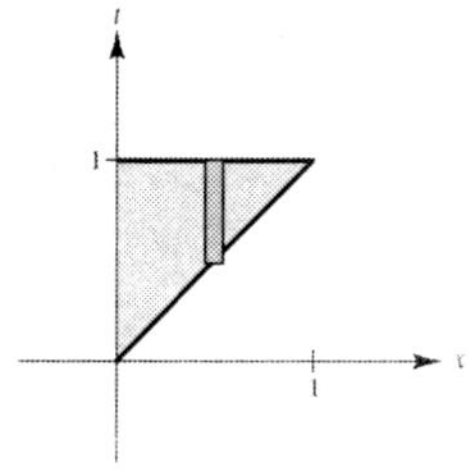

74. $\int_0^3 \int_0^{\sqrt{9-y^2}} \sqrt{9 - x^2 - y^2}\, dx\, dy = \dfrac{9\pi}{2}$

because this double integral represents the portion of the sphere $x^2 + y^2 + z^2 = 9$ in the first octant.

$$V = \frac{1}{8} \cdot \frac{4}{3}\pi(3)^3 = \frac{9\pi}{2}$$

75. Let $I = \int_0^a \int_0^b e^{\max\{b^2x^2,\, a^2y^2\}}\, dy\, dx.$

Divide the rectangle into two parts by the diagonal line $ay = bx$. On lower triangle, $b^2x^2 \geq a^2y^2$ because $y \leq \dfrac{b}{a}x.$

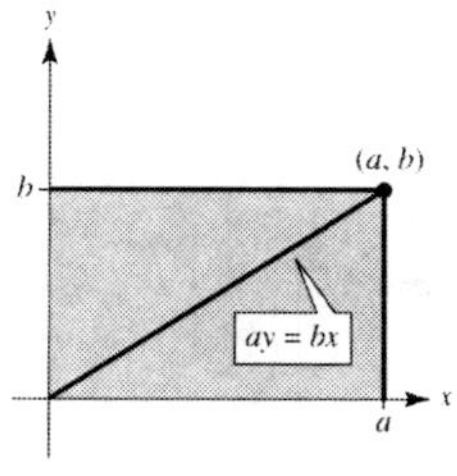

$$I = \int_0^a \int_0^{bx/a} e^{b^2x^2}\, dy\, dx + \int_0^b \int_0^{ay/b} e^{a^2y^2}\, dx\, dy = \int_0^a \frac{bx}{a} e^{b^2x^2}\, dx + \int_0^b \frac{ay}{b} e^{a^2y^2}\, dy$$

$$= \frac{1}{2ab}\Big[e^{b^2x^2}\Big]_0^a + \frac{1}{2ab}\Big[e^{a^2y^2}\Big]_0^b = \frac{1}{2ab}\Big[e^{b^2a^2} - 1 + e^{a^2b^2} - 1\Big] = \frac{e^{a^2b^2} - 1}{ab}$$

76. Assume such a function exists.

$$u(x) = 1 + \lambda \int_x^1 u(y)u(y - x)\, dy;\ \lambda > \frac{1}{2},\ 0 \leq x \leq 1$$

$$\alpha = \int_0^1 u(x)\, dx = \int_0^1 dx + \lambda \int_0^1 \int_x^1 u(y)u(y - x)\, dy\, dx$$

Change the order of integration.

$$\alpha = \int_0^1 u(x)\, dx = 1 + \lambda \int_0^1 \int_0^y u(y)u(y - x)\, dx\, dy = 1 + \lambda \int_0^1 u(y)\left[\int_0^y u(y - x)\, dx\right] dy$$

Hold y fixed and let $z = y - x$, $dz = -dx$.

$$\alpha = 1 + \lambda \int_0^1 u(y)\left[\int_y^0 u(z)(-dz)\right] dy = 1 + \lambda \int_0^1 u(y)\left[\int_0^y u(z)\, dz\right] dy$$

Let $f(y) = \int_0^y u(z)\, dz$. Then $f'(y) = u(y)$, $f(0) = 0$, $f(1) = \alpha$.

$$\alpha = 1 + \lambda \int_0^1 f'(y)f(y)\, dy = 1 + \lambda\left[\frac{f(y)^2}{2}\right]_0^1 = 1 + \lambda\left[\frac{1}{2}f(1)^2 - \frac{1}{2}f(0)^2\right] = 1 + \lambda\frac{1}{2}\alpha^2$$

$\lambda\alpha^2 - 2\alpha + 2 = 0.$

For α to exist, the discriminant of this quadratic must be nonnegative.

$$b^2 - 4ac = 4 - 8\lambda \geq 0 \Rightarrow \lambda \leq \frac{1}{2}$$

But, $\lambda > \dfrac{1}{2}$, a contradiction.

Section 14.3 Change of Variables: Polar Coordinates

1. Rectangular coordinates
2. Polar coordinates
3. Polar coordinates
4. Rectangular coordinates
5. $R = \{(r, \theta): 0 \leq r \leq 8, 0 \leq \theta \leq \pi\}$
6. $R = \{(r, \theta): 0 \leq r \leq 4 \sin\theta, 0 \leq \theta \leq \pi\}$
7. $R = \left\{(r, \theta): 4 \leq r \leq 8, 0 \leq \theta \leq \dfrac{\pi}{2}\right\}$
8. $R = \{(r, \theta): 0 \leq r \leq 4 \cos 3\theta, 0 \leq \theta \leq \pi\}$

9. $\displaystyle\int_0^{\pi}\int_0^{\cos\theta} r\,dr\,d\theta$

$\displaystyle = \int_0^{\pi}\left[\frac{r^2}{2}\right]_0^{\cos\theta} d\theta$

$\displaystyle = \int_0^{\pi}\frac{1}{2}\cos^2\theta\,d\theta$

$\displaystyle = \int_0^{\pi}\frac{1}{4}(1+\cos 2\theta)\,d\theta$

$\displaystyle = \frac{1}{4}\left[\theta + \frac{1}{2}\sin 2\theta\right]_0^{\pi} = \frac{\pi}{4}$

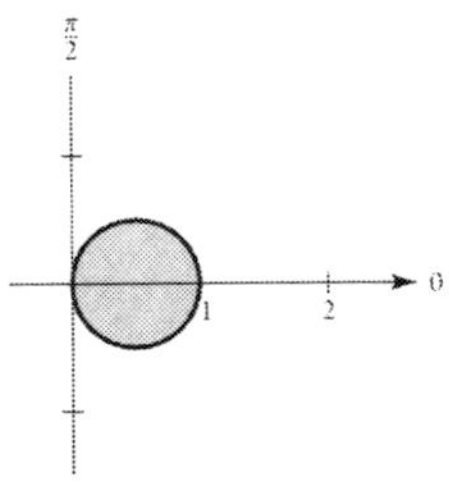

10. $\displaystyle\int_0^{\pi}\int_0^{\sin\theta} r^2\,dr\,d\theta = \int_0^{\pi}\left[\frac{r^3}{3}\right]_0^{\sin\theta} d\theta$

$\displaystyle = \frac{1}{3}\int_0^{\pi}\sin^3\theta\,d\theta = \frac{1}{3}\int_0^{\pi}\left(1-\cos^2\theta\right)\sin\theta\,d\theta$

$\displaystyle = \frac{1}{3}\left[-\cos\theta + \frac{\cos^3\theta}{3}\right]_0^{\pi}$

$\displaystyle = \frac{1}{3}\left[\left(1-\frac{1}{3}\right) - \left(-1+\frac{1}{3}\right)\right]$

$\displaystyle = \frac{4}{9}$

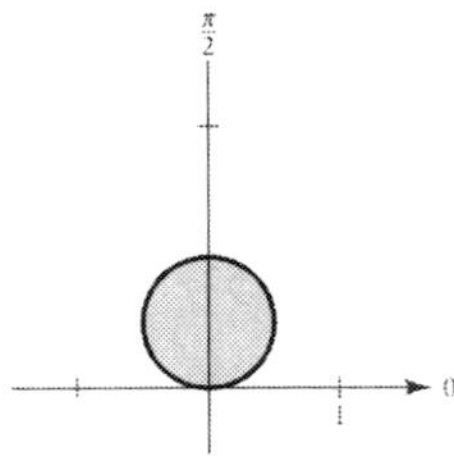

11. $\displaystyle\int_0^{2\pi}\int_0^{6} 3r^2\sin\theta\,dr\,d\theta = \int_0^{2\pi}\left[r^3\sin\theta\right]_0^6 d\theta$

$\displaystyle = \int_0^{2\pi} 216\sin\theta\,d\theta$

$\displaystyle = \left[-216\cos\theta\right]_0^{2\pi} = 0$

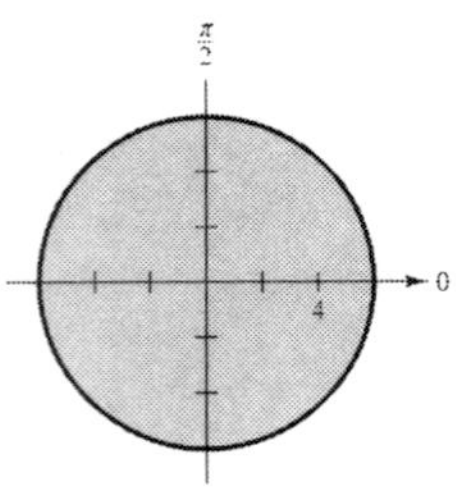

12. $\displaystyle\int_0^{\pi/4}\int_0^{4} r^2\sin\theta\cos\theta\,dr\,d\theta = \int_0^{\pi/4}\left[\frac{r^3}{3}\sin\theta\cos\theta\right]_0^4 d\theta$

$\displaystyle = \left[\left(\frac{64}{3}\right)\frac{\sin^2\theta}{2}\right]_0^{\pi/4} = \frac{16}{3}$

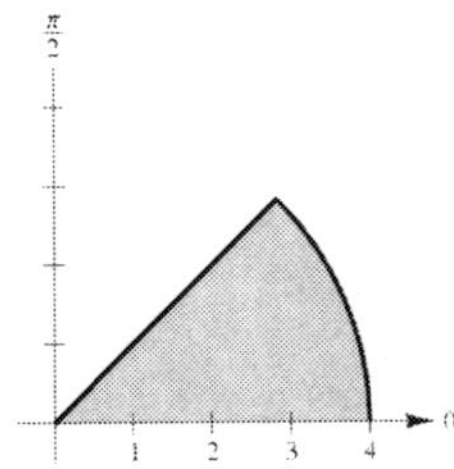

13. $\displaystyle\int_0^{\pi/2}\int_2^{3}\sqrt{9-r^2}\,r\,dr\,d\theta = \int_0^{\pi/2}\left[-\frac{1}{3}\left(9-r^2\right)^{3/2}\right]_2^3 d\theta$

$\displaystyle = \left[\frac{5\sqrt{5}}{3}\theta\right]_0^{\pi/2} = \frac{5\sqrt{5}\pi}{6}$

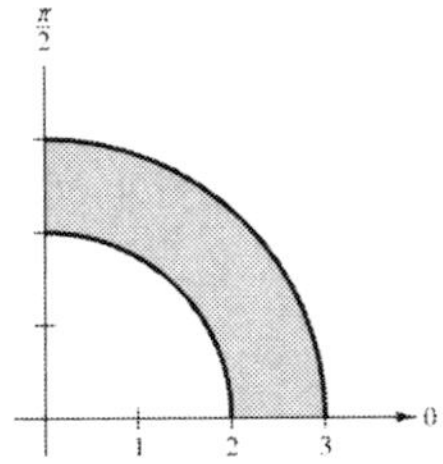

14. $\displaystyle\int_0^{\pi/2}\int_0^{3} re^{-r^2}\,dr\,d\theta = \int_0^{\pi/2}\left[-\frac{1}{2}e^{-r^2}\right]_0^3 d\theta$

$\displaystyle = \left[-\frac{1}{2}\left(e^{-9}-1\right)\theta\right]_0^{\pi/2}$

$\displaystyle = \frac{\pi}{4}\left(1-\frac{1}{e^9}\right)$

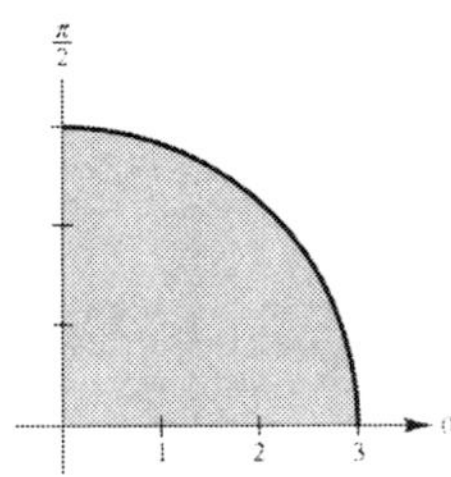

15. $\displaystyle\int_0^{\pi/2}\int_0^{1+\sin\theta}\theta r\,dr\,d\theta = \int_0^{\pi/2}\left[\frac{\theta r^2}{2}\right]_0^{1+\sin\theta} d\theta$

$\displaystyle = \int_0^{\pi/2}\frac{1}{2}\theta(1+\sin\theta)^2\,d\theta$

$\displaystyle = \left[\frac{1}{8}\theta^2 + \sin\theta - \theta\cos\theta + \frac{1}{2}\theta\left(-\frac{1}{2}\cos\theta\cdot\sin\theta + \frac{1}{2}\theta\right) + \frac{1}{8}\sin^2\theta\right]_0^{\pi/2}$

$\displaystyle = \frac{3}{32}\pi^2 + \frac{9}{8}$

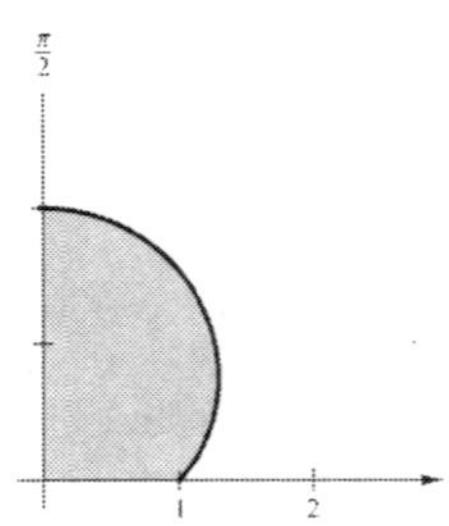

16. $\int_0^{\pi/2}\int_0^{1-\cos\theta}(\sin\theta)r\,dr\,d\theta = \int_0^{\pi/2}\left[(\sin\theta)\frac{r^2}{2}\right]_0^{1-\cos\theta}d\theta = \int_0^{\pi/2}\frac{\sin\theta}{2}(1-\cos\theta)^2\,d\theta = \left[\frac{1}{6}(1-\cos(\theta))^3\right]_0^{\pi/2} = \frac{1}{6}$

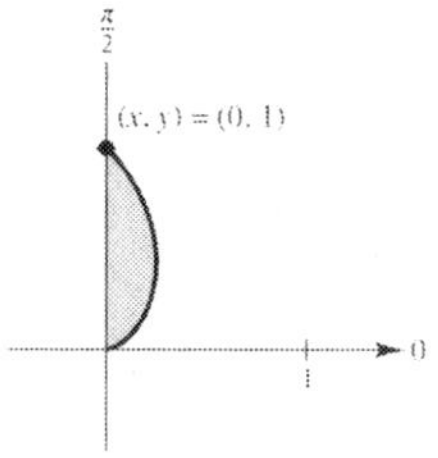

17. $\int_0^a\int_0^{\sqrt{a^2-y^2}}y\,dx\,dy = \int_0^{\pi/2}\int_0^a r^2\sin\theta\,dr\,d\theta = \frac{a^3}{3}\int_0^{\pi/2}\sin\theta\,d\theta = \left[\frac{a^3}{3}(-\cos\theta)\right]_0^{\pi/2} = \frac{a^3}{3}$

18. $\int_0^a\int_0^{\sqrt{a^2-x^2}}x\,dy\,dx = \int_0^{\pi/2}\int_0^a r^2\cos\theta\,dr\,d\theta = \frac{a^3}{3}\int_0^{\pi/2}\cos\theta\,d\theta = \left[\frac{a^3}{3}\sin\theta\right]_0^{\pi/2} = \frac{a^3}{3}$

19. $\int_{-2}^2\int_0^{\sqrt{4-x^2}}\left(x^2+y^2\right)dy\,dx = \int_0^{\pi}\int_0^2 r^2\,r\,dr\,d\theta = \int_0^{\pi}\left[\frac{r^4}{4}\right]_0^2 d\theta = \int_0^{\pi}4\,d\theta = 4\pi$

20. Note that $x - x^2 = -\left(x^2 - x + \frac{1}{4}\right) + \frac{1}{4} = \frac{1}{4} - \left(x - \frac{1}{2}\right)^2$. So $y = \sqrt{x - x^2} \Rightarrow y^2 + \left(x - \frac{1}{2}\right)^2 = \frac{1}{4}$.

$$\int_0^1\int_{-\sqrt{x-x^2}}^{\sqrt{x-x^2}}\left(x^2+y^2\right)dy\,dx = \int_{-\pi/2}^{\pi/2}\int_0^{\cos\theta}r^2\,r\,dr\,d\theta = \int_{-\pi/2}^{\pi/2}\left[\frac{r^4}{4}\right]_0^{\cos\theta}d\theta$$
$$= \frac{1}{4}\int_{-\pi/2}^{\pi/2}\cos^4\theta\,d\theta = \frac{1}{2}\int_0^{\pi}\cos^4\theta\,d\theta$$
$$= \frac{1}{2}\left(\frac{1}{2}\cdot\frac{3}{4}\cdot\frac{\pi}{2}\right) = \frac{3\pi}{32}\quad\text{(Wallis's Formula)}$$

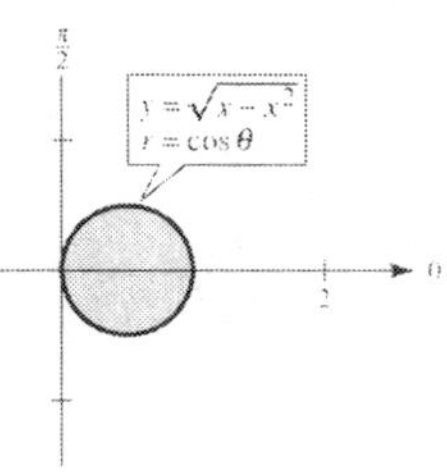

21. $\int_0^3\int_0^{\sqrt{9-x^2}}\left(x^2+y^2\right)^{3/2}dy\,dx = \int_0^{\pi/2}\int_0^3 r^4\,dr\,d\theta = \frac{243}{5}\int_0^{\pi/2}d\theta = \frac{243\pi}{10}$

22. $\int_0^2\int_y^{\sqrt{8-y^2}}\sqrt{x^2+y^2}\,dx\,dy = \int_0^{\pi/4}\int_0^{2\sqrt{2}}r^2\,dr\,d\theta = \int_0^{\pi/4}\frac{\left(2\sqrt{2}\right)^3}{3}d\theta = \left[\frac{\left(2\sqrt{2}\right)^3}{3}\theta\right]_0^{\pi/4} = \frac{\left(2\sqrt{2}\right)^3}{3}\cdot\frac{\pi}{4} = \frac{4\sqrt{2}\pi}{3}$

23. $\int_0^2\int_0^{\sqrt{2x-x^2}}xy\,dy\,dx = \int_0^{\pi/2}\int_0^{2\cos\theta}r^3\cos\theta\sin\theta\,dr\,d\theta = 4\int_0^{\pi/2}\cos^5\theta\sin\theta\,d\theta = \left[-\frac{4\cos^6\theta}{6}\right]_0^{\pi/2} = \frac{2}{3}$

24. $\int_0^4\int_0^{\sqrt{4y-y^2}}x^2\,dx\,dy = \int_0^{\pi/2}\int_0^{4\sin\theta}r^3\cos^2\theta\,dr\,d\theta = \int_0^{\pi/2}64\sin^4\theta\cos^2\theta\,d\theta$

$$= 64\int_0^{\pi/2}\left(\sin^4\theta - \sin^6\theta\right)d\theta = \frac{64}{6}\left[\sin^5\theta\cos\theta - \frac{\sin^3\theta\cos\theta}{4} + \frac{3}{8}(\theta - \sin\theta\cos\theta)\right]_0^{\pi/2} = 2\pi$$

25. $\int_{-1}^1\int_0^{\sqrt{1-x^2}}\cos\left(x^2+y^2\right)dy\,dx = \int_0^{\pi}\int_0^1\cos\left(r^2\right)r\,dr\,d\theta = \int_0^{\pi}\left[\frac{1}{2}\sin\left(r^2\right)\right]_0^1 d\theta = \int_0^{\pi}\frac{1}{2}\sin(1)\,d\theta = \frac{\pi}{2}\sin(1) \approx 1.3218$

26. $\int_0^2 \int_0^{\sqrt{4-x^2}} \sin\sqrt{x^2 + y^2}\, dy\, dx = \int_0^{\pi/2} \int_0^2 \sin(r)\, r\, dr\, d\theta = \int_0^{\pi/2} \left[\sin r - r\cos r\right]_0^2 d\theta$ [Integration by parts]

$= \int_0^{\pi/2} (\sin 2 - 2\cos 2)\, d\theta = \frac{\pi}{2}(\sin 2 - 2\cos 2) \approx 2.7357$

27. $\int_0^2 \int_0^x \sqrt{x^2 + y^2}\, dy\, dx + \int_2^{2\sqrt{2}} \int_0^{\sqrt{8-x^2}} \sqrt{x^2 + y^2}\, dy\, dx = \int_0^{\pi/4} \int_0^{2\sqrt{2}} r^2\, dr\, d\theta$

$= \int_0^{\pi/4} \frac{16\sqrt{2}}{3}\, d\theta = \frac{4\sqrt{2}\pi}{3}$

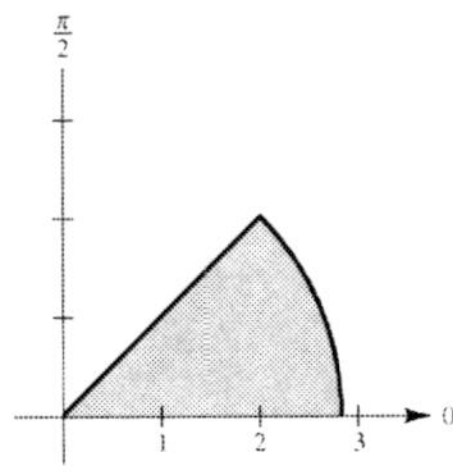

28. $\int_0^{(5\sqrt{2})/2} \int_0^x xy\, dy\, dx + \int_{(5\sqrt{2})/2}^5 \int_0^{\sqrt{25-x^2}} xy\, dy\, dx = \int_0^{\pi/4} \int_0^5 r^3 \sin\theta\cos\theta\, dr\, d\theta$

$= \int_0^{\pi/4} \frac{625}{4} \sin\theta\cos\theta\, d\theta = \left[\frac{625}{8}\sin^2\theta\right]_0^{\pi/4} = \frac{625}{16}$

29. $\int_0^2 \int_0^{\sqrt{4-x^2}} (x + y)\, dy\, dx = \int_0^{\pi/2} \int_0^2 (r\cos\theta + r\sin\theta) r\, dr\, d\theta = \int_0^{\pi/2} \int_0^2 (\cos\theta + \sin\theta) r^2\, dr\, d\theta$

$= \frac{8}{3} \int_0^{\pi/2} (\cos\theta + \sin\theta)\, d\theta = \left[\frac{8}{3}(\sin\theta - \cos\theta)\right]_0^{\pi/2} = \frac{16}{3}$

30. $\int_{-\pi/2}^{\pi/2} \int_0^5 e^{-r^2/2} r\, dr\, d\theta = \int_{-\pi/2}^{\pi/2} \left[-e^{-r^2/2}\right]_0^5 d\theta$

$= \int_{-\pi/2}^{\pi/2} \left(1 - e^{-25/2}\right) d\theta$

$= \left[\left(1 - e^{-25/2}\right)\theta\right]_{-\pi/2}^{\pi/2} = \pi\left(1 - e^{-25/2}\right)$

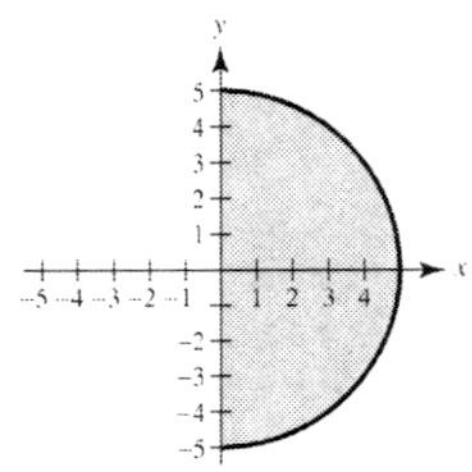

31. $\int_0^{1/\sqrt{2}} \int_{\sqrt{1-y^2}}^{\sqrt{4-y^2}} \arctan\frac{y}{x}\, dx\, dy + \int_{1/\sqrt{2}}^{\sqrt{2}} \int_y^{\sqrt{4-y^2}} \arctan\frac{y}{x}\, dx\, dy$

$= \int_0^{\pi/4} \int_1^2 \theta r\, dr\, d\theta$

$= \int_0^{\pi/4} \frac{3}{2}\theta\, d\theta = \left[\frac{3\theta^2}{4}\right]_0^{\pi/4} = \frac{3\pi^2}{64}$

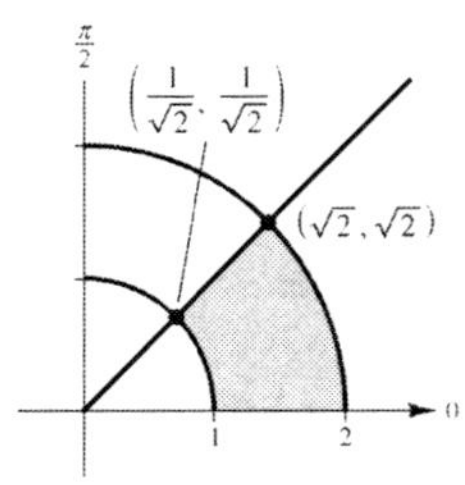

32. $\int_0^3 \int_0^{\sqrt{9-x^2}} \left(9 - x^2 - y^2\right) dy\, dx = \int_0^{\pi/2} \int_0^3 \left(9 - r^2\right) r\, dr\, d\theta$

$= \int_0^{\pi/2} \int_0^3 \left(9r - r^3\right) dr\, d\theta = \int_0^{\pi/2} \left[\frac{9}{2}r^2 - \frac{1}{4}r^4\right]_0^3 d\theta = \frac{81}{4}\int_0^{\pi/2} d\theta = \frac{81\pi}{8}$

33. $V = \int_0^{\pi/2} \int_0^1 (r\cos\theta)(r\sin\theta) r\, dr\, d\theta$

$= \frac{1}{2}\int_0^{\pi/2} \int_0^1 r^3 \sin 2\theta\, dr\, d\theta = \frac{1}{8}\int_0^{\pi/2} \sin 2\theta\, d\theta = \left[-\frac{1}{16}\cos 2\theta\right]_0^{\pi/2} = \frac{1}{8}$

34. $V = 4\int_0^{\pi/2} \int_0^1 \left(r^2 + 3\right) r\, dr\, d\theta = 4\int_0^{\pi/2} \left[\frac{r^4}{4} + \frac{3r^2}{2}\right]_0^1 d\theta = 4\int_0^{\pi/2} \frac{7}{4} d\theta = \frac{7\pi}{2}$

35. $V = \int_0^{2\pi}\int_0^5 r^2\,dr\,d\theta = \int_0^{2\pi}\frac{125}{3}\,d\theta = \frac{250\pi}{3}$

36. $V = \int_R\int \ln\left(x^2 + y^2\right)dA = \int_0^{2\pi}\int_1^2\left(\ln r^2\right)r\,dr\,d\theta = 2\int_0^{2\pi}\int_1^2 r\ln r\,dr\,d\theta$

$$= 2\int_0^{2\pi}\left[\frac{r^2}{4}(-1 + 2\ln r)\right]_1^2 d\theta = 2\int_0^{2\pi}\left(\ln 4 - \frac{3}{4}\right)d\theta = 4\pi\left(\ln 4 - \frac{3}{4}\right)$$

37. $V = 2\int_0^{\pi/2}\int_0^{4\cos\theta}\sqrt{16 - r^2}\,r\,dr\,d\theta = 2\int_0^{\pi/2}\left[-\frac{1}{3}\left(\sqrt{16 - r^2}\right)^3\right]_0^{4\cos\theta} d\theta = -\frac{2}{3}\int_0^{\pi/2}\left(64\sin^3\theta - 64\right)d\theta$

$$= \frac{128}{3}\int_0^{\pi/2}\left[1 - \sin\theta\left(1 - \cos^2\theta\right)\right]d\theta = \frac{128}{3}\left[\theta + \cos\theta - \frac{\cos^3\theta}{3}\right]_0^{\pi/2} = \frac{64}{9}(3\pi - 4)$$

38. $V = \int_0^{2\pi}\int_1^4\sqrt{16 - r^2}\,r\,dr\,d\theta = \int_0^{2\pi}\left[-\frac{1}{3}\left(\sqrt{16 - r^2}\right)^3\right]_1^4 d\theta = \int_0^{2\pi}5\sqrt{15}\,d\theta = 10\sqrt{15}\pi$

39. $V = \int_0^{2\pi}\int_a^4\sqrt{16 - r^2}\,r\,dr\,d\theta = \int_0^{2\pi}\left[-\frac{1}{3}\left(\sqrt{16 - r^2}\right)^3\right]_a^4 d\theta = \frac{1}{3}\left(\sqrt{16 - a^2}\right)^3(2\pi)$

One-half the volume of the hemisphere is $(64\pi)/3$.

$$\frac{2\pi}{3}\left(16 - a^2\right)^{3/2} = \frac{64\pi}{3}$$

$$\left(16 - a^2\right)^{3/2} = 32$$

$$16 - a^2 = 32^{2/3}$$

$$a^2 = 16 - 32^{2/3} = 16 - 8\sqrt[3]{2}$$

$$a = \sqrt{4\left(4 - 2\sqrt[3]{2}\right)} = 2\sqrt{4 - 2\sqrt[3]{2}} \approx 2.4332$$

40. $x^2 + y^2 + z^2 = a^2 \Rightarrow z = \sqrt{a^2 - \left(x^2 + y^2\right)} = \sqrt{a^2 - r^2}$

$V = 8\int_0^{\pi/2}\int_0^a\sqrt{a^2 - r^2}\,r\,dr\,d\theta$ (8 times the volume in the first octant)

$$= 8\int_0^{\pi/2}\left[-\frac{1}{2}\cdot\frac{2}{3}\left(a^2 - r^2\right)^{3/2}\right]_0^a d\theta = 8\int_0^{\pi/2}\frac{a^3}{3}\,d\theta = \left[\frac{8a^3}{3}\theta\right]_0^{\pi/2} = \frac{4\pi a^3}{3}$$

41. $A = \int_0^{\pi}\int_0^{6\cos\theta} r\,dr\,d\theta = \int_0^{\pi}18\cos^2\theta\,d\theta = 9\int_0^{\pi}(1 + \cos 2\theta)\,d\theta = \left[9\left(\theta + \frac{1}{2}\sin 2\theta\right)\right]_0^{\pi} = 9\pi$

42. $A = \int_0^{2\pi}\int_2^4 r\,dr\,d\theta = \int_0^{2\pi}6\,d\theta = 12\pi$

43. $A = \int_0^{2\pi}\int_0^{1+\cos\theta} r\,dr\,d\theta = \frac{1}{2}\int_0^{2\pi}\left(1 + 2\cos\theta + \cos^2\theta\right)d\theta$

$$= \frac{1}{2}\int_0^{2\pi}\left(1 + 2\cos\theta + \frac{1 + \cos 2\theta}{2}\right)d\theta = \frac{1}{2}\left[\theta + 2\sin\theta + \frac{1}{2}\left(\theta + \frac{1}{2}\sin 2\theta\right)\right]_0^{2\pi} = \frac{3\pi}{2}$$

44. $A = \int_0^{2\pi}\int_0^{2+\sin\theta} r\,dr\,d\theta = \frac{1}{2}\int_0^{2\pi}(2+\sin\theta)^2\,d\theta = \frac{1}{2}\int_0^{2\pi}\left(4 + 4\sin\theta + \sin^2\theta\right)d\theta$

$= \frac{1}{2}\int_0^{2\pi}\left(4 + 4\sin\theta + \frac{1-\cos 2\theta}{2}\right)d\theta$

$= \frac{1}{2}\left[4\theta - 4\cos\theta + \frac{1}{2}\theta - \frac{1}{4}\sin 2\theta\right]_0^{2\pi} = \frac{1}{2}[8\pi - 4 + \pi + 4] = \frac{9\pi}{2}$

45. $A = 3\int_0^{\pi/3}\int_0^{2\sin 3\theta} r\,dr\,d\theta = \frac{3}{2}\int_0^{\pi/3} 4\sin^2 3\theta\,d\theta = 3\int_0^{\pi/3}(1-\cos 6\theta)\,d\theta = 3\left[\theta - \frac{1}{6}\sin 6\theta\right]_0^{\pi/3} = \pi$

46. $A = 8\int_0^{\pi/4}\int_0^{3\cos 2\theta} r\,dr\,d\theta = 4\int_0^{\pi/4} 9\cos^2 2\theta\,d\theta = 18\int_0^{\pi/4}(1+\cos 4\theta)\,d\theta = 18\left[\theta + \frac{1}{4}\sin 4\theta\right]_0^{\pi/4} = \frac{9\pi}{2}$

47. $r = 1 = 2\cos\theta \Rightarrow \theta = \pm\frac{\pi}{3}$

$A = 2\int_0^{\pi/3}\int_1^{2\cos\theta} r\,dr\,d\theta = 2\int_0^{\pi/3}\left[\frac{r^2}{2}\right]_1^{2\cos\theta} d\theta = 2\int_0^{\pi/3}\left(2\cos^2\theta - \frac{1}{2}\right)d\theta$

$= 2\int_0^{\pi/3}\left(1 + \cos 2\theta - \frac{1}{2}\right)d\theta = 2\left[\frac{1}{2}\theta + \frac{\sin 2\theta}{2}\right]_0^{\pi/3} = 2\left[\frac{\pi}{6} + \frac{\sqrt{3}}{4}\right] = \frac{\pi}{3} + \frac{\sqrt{3}}{2}$

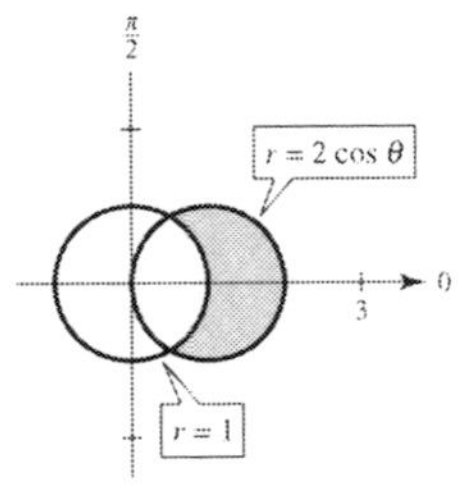

48. $r = 2 + 2\cos\theta = 1 \Rightarrow \cos\theta = -\frac{1}{2} \Rightarrow \theta = \frac{2\pi}{3}, \frac{4\pi}{3}$

$A = 2\int_0^{2\pi/3}\int_1^{2+2\cos\theta} r\,dr\,d\theta = 2\int_0^{2\pi/3}\left[\frac{r^2}{2}\right]_1^{2+2\cos\theta} d\theta$

$= \int_0^{2\pi/3}\left[(2+2\cos\theta)^2 - 1\right]d\theta$

$= \int_0^{2\pi/3}\left[3 + 8\cos\theta + 4\cos^2\theta\right]d\theta = \int_0^{2\pi/3}\left[3 + 8\cos\theta + 2(1+\cos 2\theta)\right]d\theta$

$= \left[5\theta + 8\sin\theta + \sin 2\theta\right]_0^{2\pi/3} = \frac{10\pi}{3} + 4\sqrt{3} - \frac{\sqrt{3}}{2} = \frac{10\pi}{3} + \frac{7\sqrt{3}}{2}$

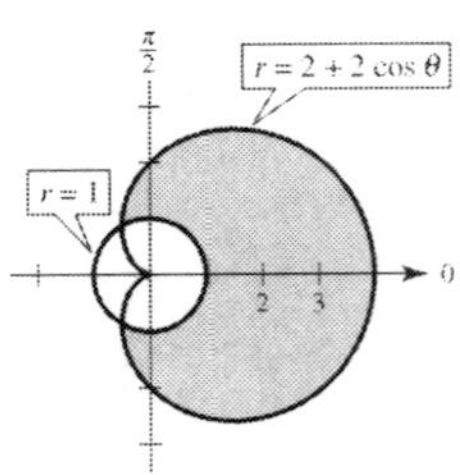

49. $r = 3\cos\theta = 1 + \cos\theta \Rightarrow \cos\theta = \frac{1}{2} \Rightarrow \theta = \pm\frac{\pi}{3}$

$A = 2\int_0^{\pi/3}\int_{1+\cos\theta}^{3\cos\theta} r\,dr\,d\theta = 2\int_0^{\pi/3}\left[\frac{r^2}{2}\right]_{1+\cos\theta}^{3\cos\theta} d\theta = \int_0^{\pi/3}\left[9\cos^2\theta - (1+\cos\theta)^2\right]d\theta$

$= \int_0^{\pi/3}\left(8\cos^2\theta - 2\cos\theta - 1\right)d\theta = \int_0^{\pi/3}\left[4(1+\cos 2\theta) - 2\cos\theta - 1\right]d\theta$

$= \left[3\theta + 2\sin 2\theta - 2\sin\theta\right]_0^{\pi/3} = 3\left(\frac{\pi}{3}\right) + \sqrt{3} - \sqrt{3} = \pi$

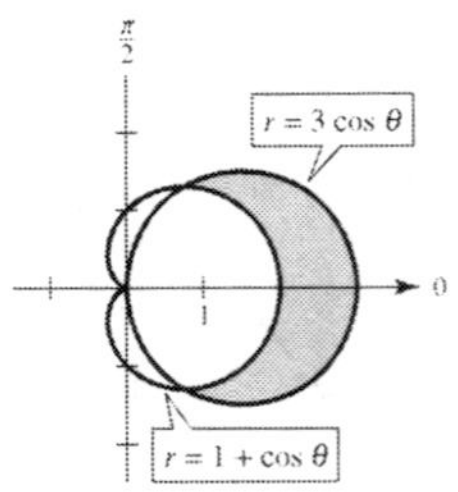

50. $r = 1 + \cos\theta = 3\cos\theta \Rightarrow \cos\theta = \frac{1}{2} \Rightarrow \theta = \pm\frac{\pi}{3}$

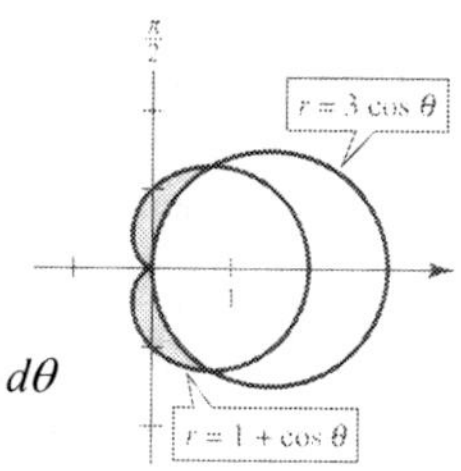

$$\frac{1}{2}A = \int_{\pi/3}^{\pi/2}\int_{3\cos\theta}^{1+\cos\theta} r\,dr\,d\theta + \int_{\pi/2}^{\pi}\int_{0}^{1+\cos\theta} r\,dr\,d\theta$$

$$= \int_{\pi/3}^{\pi/2}\left[\frac{r^2}{2}\right]_{3\cos\theta}^{1+\cos\theta} d\theta + \int_{\pi/2}^{\pi}\left[\frac{r^2}{2}\right]_{0}^{1+\cos\theta} d\theta = \int_{\pi/3}^{\pi/2}\frac{(1+\cos\theta)^2 - 9\cos^2\theta}{2}\,d\theta + \int_{\pi/2}^{\pi}\frac{(1+\cos\theta)^2}{2}\,d\theta$$

$$= \int_{\pi/3}^{\pi/2}\frac{1 + 2\cos\theta - 4(1+\cos 2\theta)}{2}\,d\theta + \int_{\pi/2}^{\pi}\left(\frac{1}{2} + \cos\theta + \frac{1+\cos 2\theta}{4}\right)d\theta$$

$$= \left[-\frac{3}{2}\theta + \sin\theta - \sin 2\theta\right]_{\pi/3}^{\pi/2} + \left[\frac{3}{4}\theta + \sin\theta + \frac{\sin 2\theta}{8}\right]_{\pi/2}^{\pi} = \left(\frac{-3\pi}{4} + 1\right) - \left(\frac{-\pi}{2} + \frac{\sqrt{3}}{2} - \frac{\sqrt{3}}{2}\right) + \left(\frac{3\pi}{4} - \frac{3\pi}{8} - 1\right) = \frac{\pi}{8}$$

So, $A = \frac{\pi}{4}$.

51. $r = 4\sin 3\theta = 2 \Rightarrow \sin 3\theta = \frac{1}{2} \Rightarrow 3\theta = \frac{\pi}{6}, \frac{5\pi}{6} \Rightarrow \theta = \frac{\pi}{18}, \frac{5\pi}{18}$

$$A = 3\int_{\pi/18}^{5\pi/18}\int_{2}^{4\sin 3\theta} r\,dr\,d\theta = 3\int_{\pi/18}^{5\pi/18}\left[\frac{r^2}{2}\right]_{2}^{4\sin 3\theta} d\theta = \frac{3}{2}\int_{\pi/18}^{5\pi/18}\left[(4\sin 3\theta)^2 - 4\right]d\theta$$

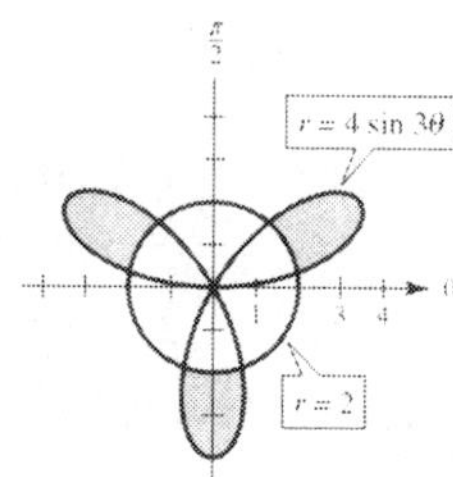

$$= \frac{3}{2}\int_{\pi/18}^{5\pi/18}\left[8(1 - \cos 6\theta) - 4\right]d\theta = \frac{3}{2}\left[4\theta - \frac{4}{3}\sin 6\theta\right]_{\pi/18}^{5\pi/18}$$

$$= \frac{3}{2}\left[\left(\frac{10}{9}\pi - \frac{4}{3}\left(\frac{-\sqrt{3}}{2}\right)\right) - \left(\frac{2\pi}{9} - \frac{4}{3}\left(\frac{\sqrt{3}}{2}\right)\right)\right] = \frac{4}{3}\pi + 2\sqrt{3}$$

52. $r = 2 = 2 - 2\cos\theta \Rightarrow \cos\theta = 0 \Rightarrow \theta = \pm\frac{\pi}{2}$

$$A = 2\int_{0}^{\pi/2}\int_{2-2\cos\theta}^{2} r\,dr\,d\theta$$

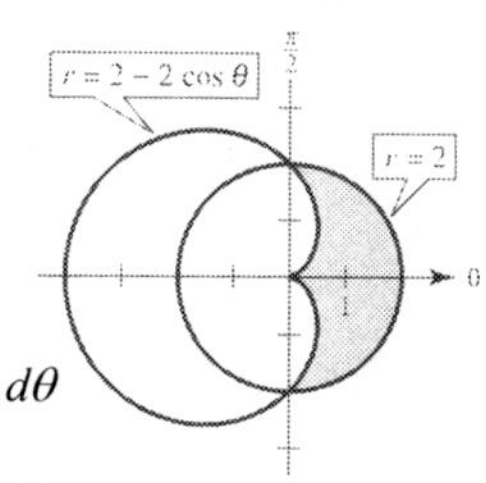

$$= 2\int_{0}^{\pi/2}\left[\frac{r^2}{2}\right]_{2-2\cos\theta}^{2} d\theta$$

$$= \int_{0}^{\pi/2}\left[4 - (2 - 2\cos\theta)^2\right]d\theta$$

$$= \int_{0}^{\pi/2}\left(8\cos\theta - 4\cos^2\theta\right)d\theta$$

$$= \int_{0}^{\pi/2}\left(8\cos\theta - 2(1 + \cos 2\theta)\right)d\theta$$

$$= \left[8\sin\theta - 2\theta - \sin 2\theta\right]_{0}^{\pi/2} = 8 - \pi$$

53. Let R be a region bounded by the graphs of $r = g_1(\theta)$ and $r = g_2(\theta)$, and the lines $\theta = a$ and $\theta = b$.

When using polar coordinates to evaluate a double integral over R, R can be partitioned into small polar sectors.

54. See Theorem 14.3.

55. r-simple regions have fixed bounds for θ.

θ-simple regions have fixed bounds for r.

56. (a) $\int_{-3}^{3}\int_{-\sqrt{9-x^2}}^{\sqrt{9-x^2}} f(x, y)\,dy\,dx$

(b) $\int_{0}^{2\pi}\int_{0}^{3} f(r\cos\theta, r\sin\theta)\,r\,dr\,d\theta$

(c) In general, the integral in part (b) is easier to evaluate. The endpoints of the region of integration are constants.

57. $\int_{-7}^{7}\int_{-\sqrt{49-x^2}}^{\sqrt{49-x^2}} 4000e^{-0.01(x^2+y^2)}\,dy\,dx = \int_{0}^{2\pi}\int_{0}^{7} 4000e^{-0.01r^2}\,r\,dr\,d\theta = \int_{0}^{2\pi}\left[-200{,}000e^{-0.01r^2}\right]_{0}^{7} d\theta$

$$= 2\pi(-200{,}000)\left(e^{-0.49} - 1\right) = 400{,}000\pi\left(1 - e^{-0.49}\right) \approx 486{,}788$$

58. (a) Horizontal or polar representative elements

(b) Polar representative element

(c) Vertical or polar

59. Total volume $= V = \int_0^{2\pi} \int_0^4 25e^{-r^2/4} r\,dr\,d\theta = \int_0^{2\pi} \left[-50e^{-r^2/4}\right]_0^4 d\theta = \int_0^{2\pi} -50\left(e^{-4} - 1\right) d\theta = \left(1 - e^{-4}\right)100\pi \approx 308.40524$

Let c be the radius of the hole that is removed.

$$\frac{1}{10}V = \int_0^{2\pi} \int_0^c 25e^{-r^2/4} r\,dr\,d\theta = \int_0^{2\pi} \left[-50e^{-r^2/4}\right]_0^c d\theta$$

$$= \int_0^{2\pi} -50\left(e^{-c^2/4} - 1\right) d\theta \Rightarrow 30.84052 = 100\pi\left(1 - e^{-c^2/4}\right)$$

$$\Rightarrow e^{-c^2/4} = 0.90183$$

$$-\frac{c^2}{4} = -0.10333$$

$$c^2 = 0.41331$$

$$c = 0.6429$$

$$\Rightarrow \text{diameter} = 2c = 1.2858$$

60. (a) The volume of the subregion determined by the point $(5, \pi/16, 7)$ is base $\times$ height $= (5 \cdot 10 \cdot \pi/8)(7)$.

Adding up the 20 volumes, ending with $(45 \cdot 10 \cdot \pi/8)(12)$, you obtain

$$V \approx 10 \cdot \frac{\pi}{8}\Big[5(7 + 9 + 9 + 5) + 15(8 + 10 + 11 + 8) + 25(10 + 14 + 15 + 11)$$
$$+ 35(12 + 15 + 18 + 16) + 45(9 + 10 + 14 + 12)\Big]$$
$$= \frac{5\pi}{4}[150 + 555 + 1250 + 2135 + 2025] \approx \frac{5\pi}{4}[6115] \approx 24{,}013.5 \text{ ft}^3$$

(b) $(57)(24{,}013.5) = 1{,}368{,}769.5$ pounds

(c) $(7.48)(24{,}103.5) \approx 179{,}621$ gallons

61. $\int_{\pi/4}^{\pi/2} \int_0^5 r\sqrt{1 + r^3} \sin\sqrt{\theta}\,dr\,d\theta \approx 56.051$

[**Note:** This integral equals $\left(\int_{\pi/4}^{\pi/2} \sin\sqrt{\theta}\,d\theta\right)\left(\int_0^5 r\sqrt{1 + r^3}\,dr\right)$.]

62. $\int_0^{\pi/4} \int_0^4 5e^{\sqrt{r\theta}} r\,dr\,d\theta \approx 87.130$

63. False

Let $f(r, \theta) = r - 1$ where R is the circular sector $0 \le r \le 6$ and $0 \le \theta \le \pi$. Then,

$\int_R \int (r - 1)\,dA > 0$ but $r - 1 \not> 0$ for all r.

64. True

65. (a) $I^2 = \int_{-\infty}^{\infty} \int_{-\infty}^{\infty} e^{-\left(x^2 + y^2\right)/2}\,dA = 4\int_0^{\pi/2} \int_0^{\infty} e^{-r^2/2} r\,dr\,d\theta = 4\int_0^{\pi/2} \left[-e^{-r^2/2}\right]_0^{\infty} d\theta = 4\int_0^{\pi/2} d\theta = 2\pi$

(b) So, $I = \sqrt{2\pi}$.

66. (a) Let $u = \sqrt{2}x$, then $\int_{-\infty}^{\infty} e^{-x^2}\,dx = \int_{-\infty}^{\infty} e^{-u^2/2} \frac{1}{\sqrt{2}}\,du = \frac{1}{\sqrt{2}}\left(\sqrt{2\pi}\right) = \sqrt{\pi}$.

(b) Let $u = 2x$, then $\int_{-\infty}^{\infty} e^{-4x^2}\,dx = \int_{-\infty}^{\infty} e^{-u^2} \frac{1}{2}\,du = \frac{1}{2}\sqrt{\pi}$.

67. (a) $\int_2^4 \int_{y/\sqrt{3}}^{y} f\, dx\, dy$

(b) $\int_{2/\sqrt{3}}^{2} \int_{2}^{\sqrt{3}x} f\, dy\, dx + \int_{2}^{4/\sqrt{3}} \int_{x}^{\sqrt{3}x} f\, dy\, dx + \int_{4/\sqrt{3}}^{4} \int_{x}^{4} f\, dy\, dx$

(c) $\int_{\pi/4}^{\pi/3} \int_{2\csc\theta}^{4\csc\theta} fr\, dr\, d\theta$

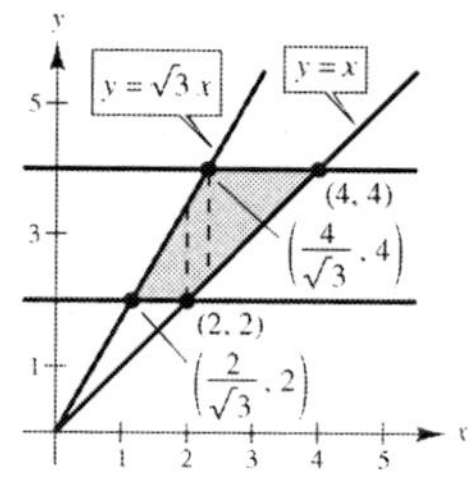

68. (a) $4\int_0^2 \int_2^{2+\sqrt{4-y^2}} f\, dx\, dy$

(b) $4\int_0^2 \int_0^{\sqrt{4-(x-2)^2}} f\, dy\, dx$

(c) $2\int_0^{\pi/2} \int_0^{4\cos\theta} fr\, dr\, d\theta$

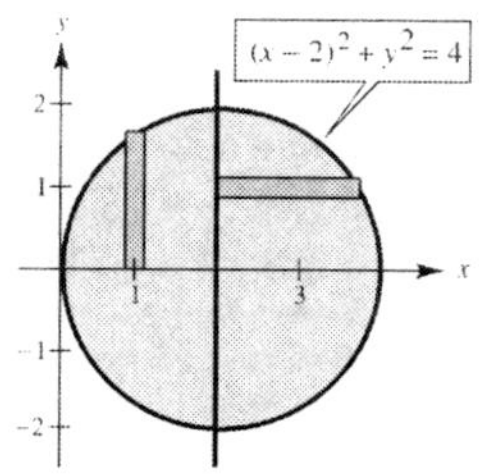

69. $\int_0^{\infty} \int_0^{\infty} ke^{-(x^2+y^2)}\, dy\, dx = \int_0^{\pi/2} \int_0^{\infty} ke^{-r^2}\, r\, dr\, d\theta$

$= \int_0^{\pi/2} \left[-\frac{k}{2}e^{-r^2}\right]_0^{\infty} d\theta$

$= \int_0^{\pi/2} \frac{k}{2}\, d\theta = \frac{k\pi}{4}$

For $f(x, y)$ to be a probability density function,

$\frac{k\pi}{4} = 1$

$k = \frac{4}{\pi}.$

70. $A = \frac{\Delta\theta r_2^2}{2} - \frac{\Delta\theta r_1^2}{2} = \Delta\theta\left(\frac{r_1 + r_2}{2}\right)(r_2 - r_1) = r\Delta r\Delta\theta$

Section 14.4 Center of Mass and Moments of Inertia

1. $m = \int_0^2 \int_0^2 xy\, dy\, dx = \int_0^2 \left[\frac{xy^2}{2}\right]_0^2 dx = \int_0^2 2x\, dx = \left[x^2\right]_0^2 = 4$

2. $m = \int_0^3 \int_0^{9-x^2} xy\, dy\, dx = \int_0^3 \left[\frac{xy^2}{2}\right]_0^{9-x^2} dx = \int_0^3 \frac{x(9-x^2)^2}{2}\, dx = \left[-\frac{1}{4}\frac{(9-x^2)^3}{3}\right]_0^3 = 0 + \frac{1}{4}(243) = \frac{243}{4}$

3. $m = \int_0^{\pi/2} \int_0^1 (r\cos\theta)(r\sin\theta)\, r\, dr\, d\theta = \int_0^{\pi/2} \left[(\cos\theta\sin\theta)\frac{r^4}{4}\right]_0^1 d\theta = \int_0^{\pi/2} \frac{1}{4}\cos\theta\sin\theta\, d\theta = \left[\frac{1}{4}\cdot\frac{\sin^2\theta}{2}\right]_0^{\pi/2} = \frac{1}{8}$

4. $m = \int_0^3 \int_3^{3+\sqrt{9-x^2}} xy\, dy\, dx = \int_0^3 \left[x\frac{y^2}{2}\right]_3^{3+\sqrt{9-x^2}} dx = \int_0^3 \frac{x}{2}\left(\left(3+\sqrt{9-x^2}\right)^2 - 9\right) dx = \frac{1}{2}\int_0^3 \left[6x\sqrt{9-x^2} + 9x - x^3\right] dx$

$= \frac{1}{2}\left[-2(9-x^2)^{3/2} + \frac{9x^2}{2} - \frac{x^4}{4}\right]_0^3 = \frac{1}{2}\left[\frac{81}{2} - \frac{81}{4} + 54\right] = \frac{297}{8}$

5. (a)
$$m = \int_0^a \int_0^a k\,dy\,dx = ka^2$$
$$M_x = \int_0^a \int_0^a ky\,dy\,dx = \int_0^a \frac{ka^2}{2}\,dx = \frac{ka^3}{2}$$
$$M_y = \int_0^a \int_0^a kx\,dy\,dx = \frac{ka^3}{2}$$
$$\bar{x} = \frac{M_y}{m} = \frac{a}{2},\ \bar{y} = \frac{M_x}{m} = \frac{a}{2}$$
$$(\bar{x}, \bar{y}) = \left(\frac{a}{2}, \frac{a}{2}\right) \quad \text{(center of square)}$$

(b)
$$m = \int_0^a \int_0^a ky\,dy\,dx = \frac{1}{2}ka^3$$
$$M_x = \int_0^a \int_0^a ky^2\,dy\,dx = \frac{1}{3}ka^4$$
$$M_y = \int_0^a \int_0^a kyx\,dy\,dx = \frac{1}{4}ka^4$$
$$\bar{x} = \frac{M_y}{m} = \frac{a}{2}, \qquad \bar{y} = \frac{M_x}{m} = \frac{2a}{3}$$
$$(\bar{x}, \bar{y}) = \left(\frac{a}{2}, \frac{2a}{3}\right)$$

(c)
$$m = \int_0^a \int_0^a kx\,dy\,dx = \frac{1}{2}ka^3$$
$$M_x = \int_0^a \int_0^a kxy\,dy\,dx = \frac{1}{4}ka^4$$
$$M_y = \int_0^a \int_0^a kx^2\,dy\,dx = \frac{1}{3}ka^3$$
$$\bar{x} = \frac{M_y}{m} = \frac{2a}{3}, \qquad \bar{y} = \frac{M_x}{m} = \frac{a}{2}$$
$$(\bar{x}, \bar{y}) = \left(\frac{2a}{3}, \frac{a}{2}\right)$$

6. (a)
$$m = \int_0^a \int_0^b kxy\,dy\,dx = \frac{ka^2b^2}{4}$$
$$M_x = \int_0^a \int_0^b kxy^2\,dy\,dx = \frac{ka^2b^3}{6}$$
$$M_y = \int_0^a \int_0^b kx^2y\,dy\,dx = \frac{ka^3b^2}{6}$$
$$\bar{x} = \frac{M_y}{m} = \frac{ka^3b^2/6}{ka^2b^2/4} = \frac{2a}{3},$$
$$\bar{y} = \frac{M_x}{m} = \frac{ka^2b^3/6}{ka^2b^2/4} = \frac{2}{3}b$$
$$(\bar{x}, \bar{y}) = \left(\frac{2a}{3}, \frac{2b}{3}\right)$$

(b)
$$m = \int_0^a \int_0^b k(x^2 + y^2)\,dy\,dx = \frac{kab}{3}(a^2 + b^2)$$
$$M_x = \int_0^a \int_0^b k(x^2y + y^3)\,dy\,dx = \frac{kab^2}{12}(2a^2 + 3b^2)$$
$$M_y = \int_0^a \int_0^b k(x^3 + xy^2)\,dy\,dx = \frac{ka^2b}{12}(3a^2 + 2b^2)$$
$$\bar{x} = \frac{M_y}{m} = \frac{(ka^2b/12)(3a^2 + 2b^2)}{(kab/3)(a^2 + b^2)} = \frac{a(3a^2 + 2b^2)}{4(a^2 + b^2)}$$
$$\bar{y} = \frac{M_x}{m} = \frac{(kab^2/12)(2a^2 + 3b^2)}{(kab/3)(a^2 + b^2)} = \frac{b(2a^2 + 3b^2)}{4(a^2 + b^2)}$$
$$(\bar{x}, \bar{y}) = \left(\frac{a(3a^2 + 2b^2)}{4(a^2 + b^2)}, \frac{b(2a^2 + 3b^2)}{4(a^2 + b^2)}\right)$$

7. (a)
$$m = \int_0^a \int_0^y k\,dx\,dy = \frac{1}{2}ka^2$$
$$M_x = \int_0^a \int_0^y ky\,dx\,dy = \frac{1}{3}ka^3$$
$$M_y = \int_0^a \int_0^y kx\,dx\,dy = \frac{1}{6}ka^3$$
$$\bar{x} = \frac{M_y}{m} = \frac{a}{3} \qquad \bar{y} = \frac{M_x}{m} = \frac{2a}{3}$$
$$(\bar{x}, \bar{y}) = \left(\frac{a}{3}, \frac{2a}{3}\right)$$

(b)
$$m = \int_0^a \int_0^y ky\,dx\,dy = \frac{1}{3}ka^3$$
$$M_x = \int_0^a \int_0^y ky^2\,dx\,dy = \frac{1}{4}ka^4$$
$$M_y = \int_0^a \int_0^y kxy\,dx\,dy = \frac{1}{8}ka^4$$
$$x = \frac{M_y}{m} = \frac{3a}{8}, \qquad \bar{y} = \frac{M_x}{m} = \frac{3a}{4}$$
$$(\bar{x}, \bar{y}) = \left(\frac{3a}{8}, \frac{3a}{4}\right)$$

(c)
$$m = \int_0^a \int_0^y kx\,dx\,dy = \frac{1}{6}ka^3$$
$$M_x = \int_0^a \int_0^y kxy\,dx\,dy = \frac{1}{8}ka^4$$
$$M_y = \int_0^a \int_0^y kx^2\,dx\,dy = \frac{1}{12}ka^4$$
$$\bar{x} = \frac{M_y}{m} = \frac{a}{2} \qquad \bar{y} = \frac{M_x}{m} = \frac{3a}{4}$$
$$(\bar{x}, \bar{y}) = \left(\frac{a}{2}, \frac{3a}{4}\right)$$

8. (a) $m = \int_0^a \int_{y/2}^{a-y/2} k \, dx \, dy = \frac{1}{2} ka^2$

$M_x = \int_0^a \int_{y/2}^{a-y/2} ky \, dx \, dy = \frac{1}{6} ka^3$

$M_y = \int_0^a \int_{y/2}^{a-y/2} kx \, dx \, dy = \frac{1}{4} ka^3$

$\bar{x} = \frac{M_y}{m} = \frac{a}{2}, \bar{y} = \frac{M_x}{m} = \frac{a}{3}, (\bar{x}, \bar{y}) = \left(\frac{a}{2}, \frac{a}{3}\right)$

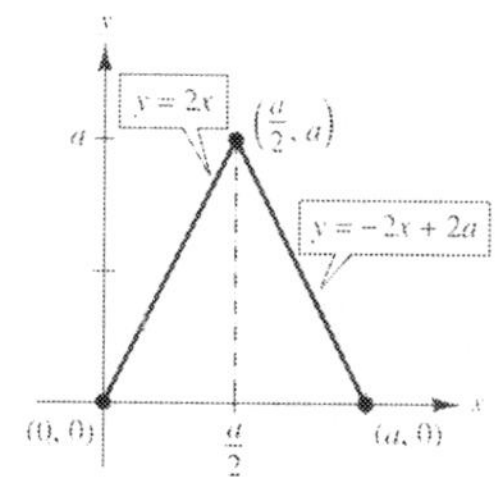

(b) $m = \int_0^a \int_{y/2}^{a-y/2} kxy \, dx \, dy = \frac{1}{12} ka^4$

$M_x = \int_0^a \int_{y/2}^{a-y/2} kxy^2 \, dx \, dy = \frac{1}{24} ka^5$

$M_y = \int_0^a \int_{y/2}^{a-y/2} kx^2y \, dx \, dy = \frac{11}{240} ka^5$

$\bar{x} = \frac{M_y}{m} = \frac{11a}{20}, \bar{y} = \frac{M_x}{m} = \frac{a}{2}, (\bar{x}, \bar{y}) = \left(\frac{11a}{20}, \frac{a}{2}\right)$

9. (a) The x-coordinate changes by 5: $(\bar{x}, \bar{y}) = \left(\frac{a}{2} + 5, \frac{a}{2}\right)$

(b) The x-coordinate changes by 5: $(\bar{x}, \bar{y}) = \left(\frac{a}{2} + 5, \frac{2a}{3}\right)$

(c) $m = \int_5^{a+5} \int_0^a kx \, dy \, dx = \frac{1}{2} ka\left((a + 5)^2 - 25\right)$

$M_x = \int_5^{a+5} \int_0^a kxy \, dy \, dx = \frac{1}{4} ka^2\left((a + 5)^2 - 25\right)$

$M_y = \int_5^{a+5} \int_0^a kx^2 \, dy \, dx = \frac{1}{3} ka\left((a + 5)^3 - 125\right)$

$\bar{x} = \frac{M_y}{m} = \frac{2\left[(a + 5)^3 - 125\right]}{3\left[(a + 5)^2 - 25\right]} = \frac{2(a^2 + 15a + 75)}{3(a + 10)}$

$y = \frac{M_x}{m} = \frac{a}{2}$

$(\bar{x}, \bar{y}) = \left(\frac{2(a^2 + 15a + 75)}{3(a + 10)}, \frac{a}{2}\right)$

10. The x-coordinate changes by c units horizontally and d units vertically. This is not necessarily true for variable densities. See Exercise 9.

11. $m = \int_0^1 \int_0^{\sqrt{x}} ky \, dy \, dx = \frac{1}{4}k$

$M_x = \int_0^1 \int_0^{\sqrt{x}} ky^2 \, dy \, dx = \frac{2}{15}k$

$M_y = \int_0^1 \int_0^{\sqrt{x}} kxy \, dy \, dx = \frac{1}{6}k$

$\bar{x} = \frac{M_y}{m} = \frac{2}{3}$

$\bar{y} = \frac{M_x}{m} = \frac{8}{15}$

$(\bar{x}, \bar{y}) = \left(\frac{2}{3}, \frac{8}{15}\right)$

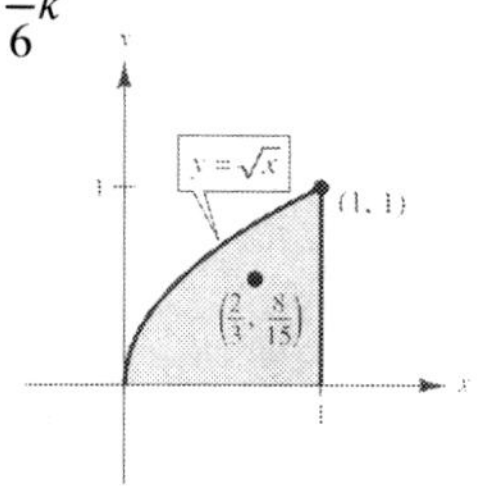

12. $m = \int_0^2 \int_0^{x^2} kxy \, dy \, dx = \frac{16}{3}k$

$M_x = \int_0^2 \int_0^{x^2} kxy^2 \, dy \, dx = \frac{32}{3}k$

$M_y = \int_0^2 \int_0^{x^2} kx^2y \, dy \, dx = \frac{64}{7}k$

$\bar{x} = \frac{M_y}{m} = \frac{12}{7}$

$\bar{y} = \frac{M_x}{m} = 2$

$(\bar{x}, \bar{y}) = \left(\frac{12}{7}, 2\right)$

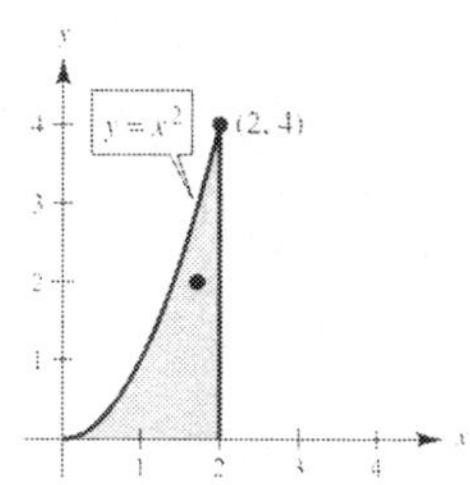

13. $m = \int_1^4 \int_0^{4/x} kx^2 \, dy \, dx = 30k$

$M_x = \int_1^4 \int_0^{4/x} kx^2 y \, dy \, dx = 24k$

$M_y = \int_1^4 \int_0^{4/x} kx^3 \, dy \, dx = 84k$

$\bar{x} = \dfrac{M_y}{m} = \dfrac{84k}{30k} = \dfrac{14}{5}$

$\bar{y} = \dfrac{M_y}{m} = \dfrac{24k}{30k} = \dfrac{4}{5}$

$(\bar{x}, \bar{y}) = \left(\dfrac{14}{5}, \dfrac{4}{5}\right)$

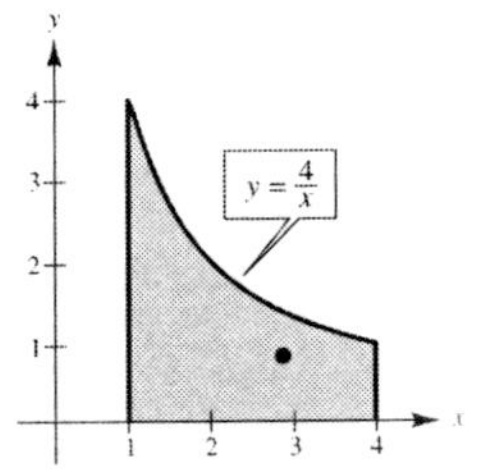

14. $\bar{x} = 0$ by symmetry

$m = \int_{-1}^1 \int_0^{1/(1+x^2)} k \, dy \, dx = \dfrac{k\pi}{2}$

$M_x = \int_{-1}^1 \int_0^{1/(1+x^2)} ky \, dy \, dx = \dfrac{k}{8}(2 + \pi)$

$\bar{y} = \dfrac{M_x}{m} = \dfrac{k}{8}(2 + \pi) \cdot \dfrac{2}{k\pi} = \dfrac{2 + \pi}{4\pi}$

$(\bar{x}, \bar{y}) = \left(0, \dfrac{2 + \pi}{4\pi}\right)$

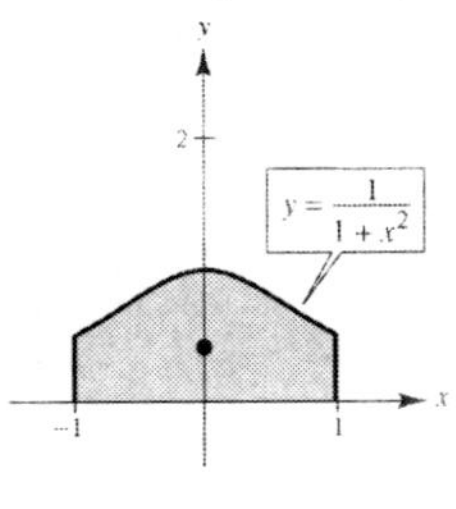

15. (a) $m = \int_0^1 \int_0^{e^x} k \, dy \, dx = k(e - 1)$

$M_x = \int_0^1 \int_0^{e^x} ky \, dy \, dx = \dfrac{1}{4} k(e^2 - 1)$

$M_y = \int_0^1 \int_0^{e^x} kx \, dy \, dx = k$

$\bar{x} = \dfrac{M_y}{m} = \dfrac{1}{e - 1},$

$\bar{y} = \dfrac{M_x}{m} = \dfrac{e^2 - 1}{4(e - 1)} = \dfrac{e + 1}{4},$

$(\bar{x}, \bar{y}) = \left(\dfrac{1}{e - 1}, \dfrac{e + 1}{4}\right)$

(b) $m = \int_0^1 \int_0^{e^x} ky \, dy \, dx = \dfrac{e^2 - 1}{4} k$

$M_x = \int_0^1 \int_0^{e^x} ky^2 \, dy \, dx = \dfrac{e^3 - 1}{9} k$

$M_y = \int_0^1 \int_0^{e^x} kxy \, dy \, dx = \dfrac{e^2 + 1}{8} k$

$\bar{x} = \dfrac{M_y}{m} = \dfrac{e^2 + 1}{2(e^2 - 1)}, \bar{y} = \dfrac{M_x}{m} = \dfrac{4(e^3 - 1)}{9(e^2 - 1)},$

$(\bar{x}, \bar{y}) = \left(\dfrac{e^2 + 1}{2(e^2 - 1)}, \dfrac{4(e^3 - 1)}{9(e^2 - 1)}\right)$

16. (a) $m = \int_0^1 \int_0^{e^{-x}} ky \, dy \, dx = \dfrac{1}{4}(1 - e^{-2})k$

$M_x = \int_0^1 \int_0^{e^{-x}} ky^2 \, dy \, dx = \dfrac{1}{9}(1 - e^{-3})k$

$M_y = \int_0^1 \int_0^{e^{-x}} kxy \, dy \, dx = \dfrac{1}{8}(1 - 3e^{-2})k$

$\bar{x} = \dfrac{M_y}{m} = \dfrac{1 - 3e^{-2}}{2(1 - e^{-2})}$

$\bar{y} = \dfrac{M_x}{m} = \dfrac{4(1 - e^{-3})}{9(1 - e^{-2})}$

$(\bar{x}, \bar{y}) = \left(\dfrac{1 - 3e^{-2}}{2(1 - e^{-2})}, \dfrac{4(1 - e^{-3})}{9(1 - e^{-2})}\right)$

(b) $m = \int_0^1 \int_0^{e^{-x}} ky^2 \, dy \, dx = \dfrac{1}{9}(1 - e^{-3})k$

$M_x = \int_0^1 \int_0^{e^{-x}} ky^3 \, dy \, dx = \dfrac{1}{16}(1 - e^{-4})k$

$M_y = \int_0^1 \int_0^{e^{-x}} kxy^2 \, dy \, dx = \dfrac{1}{27}(1 - 4e^{-3})k$

$\bar{x} = \dfrac{M_y}{m} = \dfrac{1 - 4e^{-3}}{3(1 - e^{-3})}$

$\bar{y} = \dfrac{M_x}{m} = \dfrac{9(1 - e^{-4})}{16(1 - e^{-3})}$

$(\bar{x}, \bar{y}) = \left(\dfrac{1 - 4e^{-3}}{3(1 - e^{-3})}, \dfrac{9(1 - e^{-4})}{16(1 - e^{-3})}\right)$

17. $m = \int_{-2}^{2}\int_{0}^{4-x^2} ky\,dy\,dx = \frac{256}{15}k$

$M_x = \int_{-2}^{2}\int_{0}^{4-x^2} ky^2\,dy\,dx = \frac{4096}{105}k$

$\bar{x} = 0$ (by symmetry)

$\bar{y} = \frac{M_x}{m} = \frac{16}{7}$

$(\bar{x}, \bar{y}) = \left(0, \frac{16}{7}\right)$

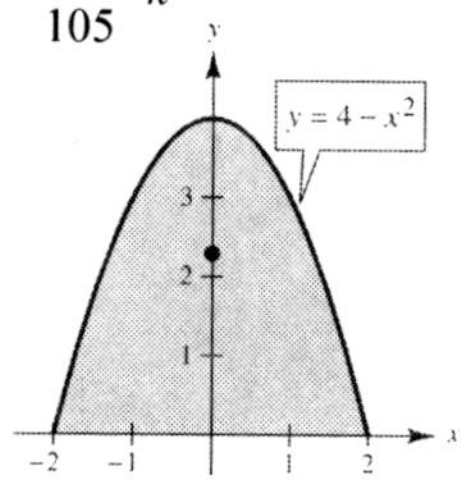

18. $m = \int_{-3}^{3}\int_{0}^{9-y^2} kx\,dx\,dy = \frac{648\,k}{5}$

$M_x = \int_{-3}^{3}\int_{0}^{9-y^2} kxy\,dx\,dy = 0$ (by symmetry)

$M_y = \int_{-3}^{3}\int_{0}^{9-y^2} kx^2\,dx\,dy = \frac{23{,}328\,k}{35}$

$\bar{x} = \frac{M_y}{m} = \frac{36}{7}, \quad \bar{y} = 0$

$(\bar{x}, \bar{y}) = \left(\frac{36}{7}, 0\right)$

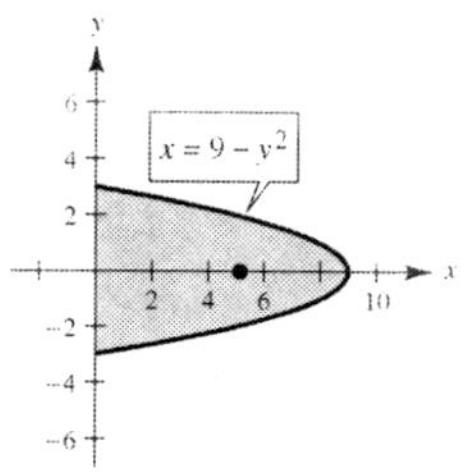

19. $\bar{x} = \frac{L}{2}$ (by symmetry)

$m = \int_{0}^{L}\int_{0}^{\sin(\pi x/2)} k\,dy\,dx = \frac{2kL}{\pi}$

$M_x = \int_{0}^{L}\int_{0}^{\sin(\pi x/L)} ky\,dy\,dx = \frac{kL}{4}$

$\bar{y} = \frac{M_x}{m} = \frac{\pi}{8}$

$(\bar{x}, \bar{y}) = \left(\frac{L}{2}, \frac{\pi}{8}\right)$

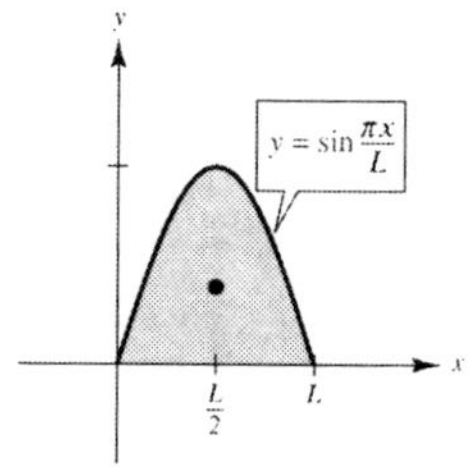

20. $m = \int_{0}^{L/2}\int_{0}^{\cos(\pi x/L)} ky\,dy\,dx = \frac{kL}{8}$

$M_x = \int_{0}^{L/2}\int_{0}^{\cos(\pi y/L)} ky^2\,dy\,dx = \frac{2kL}{9\pi}$

$M_y = \int_{0}^{L/2}\int_{0}^{\cos(\pi x/L)} kxy\,dy\,dx = \frac{L^2k(\pi^2 - 4)}{32\pi^2}$

$\bar{x} = \frac{M_y}{m} = \frac{L(\pi^2 - 4)}{4\pi^2}$

$\bar{y} = \frac{M_x}{m} = \frac{16}{9\pi}$

$(\bar{x}, \bar{y}) = \left(\frac{L(\pi^2 - 4)}{4\pi^2}, \frac{16}{9\pi}\right)$

21. $m = \frac{\pi a^2 k}{8}$

$M_x = \int_R\int ky\,dA = \int_{0}^{\pi/4}\int_{0}^{a} kr^2 \sin\theta\,dr\,d\theta = \frac{ka^3(2 - \sqrt{2})}{6}$

$M_y = \int_R\int kx\,dA = \int_{0}^{\pi/4}\int_{0}^{a} kr^2 \cos\theta\,dr\,d\theta = \frac{ka^3\sqrt{2}}{6}$

$\bar{x} = \frac{M_y}{m} = \frac{ka^3\sqrt{2}}{6} \cdot \frac{8}{\pi a^2 k} = \frac{4a\sqrt{2}}{3\pi}$

$\bar{y} = \frac{M_x}{m} = \frac{ka^3(2 - \sqrt{2})}{6} \cdot \frac{8}{\pi a^2 k} = \frac{4a(2 - \sqrt{2})}{3\pi}$

$(\bar{x}, \bar{y}) = \left(\frac{4a\sqrt{2}}{3\pi}, \frac{4a(2 - \sqrt{2})}{3\pi}\right)$

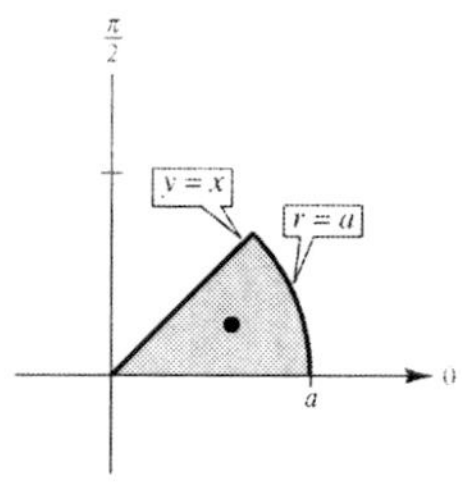

22. $m = \int_{0}^{a}\int_{0}^{\sqrt{a^2 - x^2}} k(x^2 + y^2)\,dy\,dx = \int_{0}^{\pi/2}\int_{0}^{a} kr^3\,dr\,d\theta = \frac{ka^4\pi}{8}$

$M_x = \int_{0}^{a}\int_{0}^{\sqrt{a^2 - x^2}} k(x^2 + y^2)y\,dy\,dx = \int_{0}^{\pi/2}\int_{0}^{a} kr^4 \sin\theta\,dr\,d\theta = \frac{ka^5}{5}$

$M_y = M_x$ by symmetry

$\bar{x} = \bar{y} = \frac{M_y}{m} = \frac{ka^5}{5} \cdot \frac{8}{ka^4\pi} = \frac{8a}{5\pi}$

$(\bar{x}, \bar{y}) = \left(\frac{8a}{5\pi}, \frac{8a}{5\pi}\right)$

23. $m = \int_0^2 \int_0^{e^{-x}} kxy\, dy\, dx = \frac{1 - 5e^{-4}}{8} k$

$M_x = \int_0^2 \int_0^{e^{-x}} kxy^2\, dy\, dx = \frac{1 - 7e^{-6}}{27} k$

$M_y = \int_0^2 \int_0^{e^{-x}} kx^2 y\, dy\, dx = \frac{1 - 13e^{-4}}{8} k$

$\bar{x} = \frac{M_y}{m} = \frac{e^4 - 13}{e^4 - 5}$

$\bar{y} = \frac{M_x}{m} = \frac{8(e^6 - 7)}{27(e^6 - 5e^2)}$

$(\bar{x}, \bar{y}) = \left(\frac{e^4 - 13}{e^4 - 5}, \frac{8(e^6 - 7)}{27(e^6 - 5e^2)}\right)$

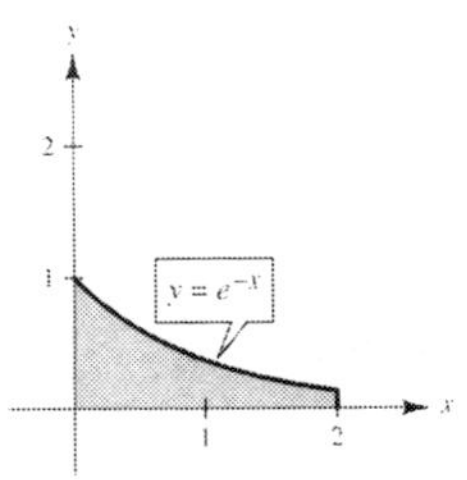

24. $m = \int_1^e \int_0^{\ln x} \frac{k}{x}\, dy\, dx = \frac{k}{2}$

$M_x = \int_1^e \int_0^{\ln x} \frac{k}{x} y\, dy\, dx = \frac{k}{6}$

$M_y = \int_1^e \int_0^{\ln x} \frac{k}{x} x\, dy\, dx = k$

$\bar{x} = \frac{M_y}{m} = \frac{k}{1} \cdot \frac{2}{k} = 2$

$\bar{y} = \frac{M_x}{m} = \frac{k}{6} \cdot \frac{2}{k} = \frac{1}{3}$

$(\bar{x}, \bar{y}) = \left(2, \frac{1}{2}\right)$

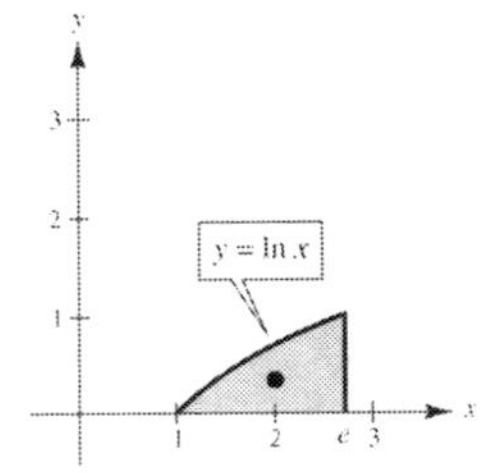

25. $\bar{y} = 0$ by symmetry

$m = \int_R \int k\, dA = \int_{-\pi/6}^{\pi/6} \int_0^{2\cos 3\theta} kr\, dr\, d\theta = \frac{k\pi}{3}$

$M_y = \int_R \int kx\, dA$

$= \int_{-\pi/6}^{\pi/6} \int_0^{2\cos 3\theta} kr^2 \cos\theta\, dr\, d\theta$

$= \frac{27\sqrt{3}}{40} k \approx 1.17k$

$\bar{x} = \frac{M_y}{m} = \frac{81\sqrt{3}}{40\pi} \approx 1.12$

$(\bar{x}, \bar{y}) \approx (1.12, 0)$

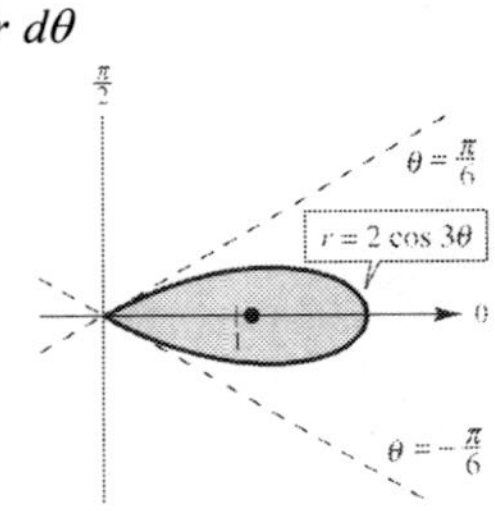

26. $\bar{y} = 0$ by symmetry

$m = \int_R \int k\, dA = \int_0^{2\pi} \int_0^{1+\cos\theta} kr\, dr\, d\theta = \frac{3\pi k}{2}$

$M_y = \int_R \int kx\, dA = \int_0^{2\pi} \int_0^{1+\cos\theta} kr^2 \cos\theta\, dr\, d\theta = \frac{k}{3}\int_0^{2\pi} \cos\theta\left(1 + 3\cos\theta + 3\cos^2\theta + \cos^3\theta\right) d\theta$

$= \frac{k}{3}\int_0^{2\pi} \left[\cos\theta + \frac{3}{2}\left(1 + \cos^2\theta\right) + 3\cos\theta\left(1 - \sin^2\theta\right) + \frac{1}{4}\left(1 + \cos 2\theta\right)^2\right] d\theta = \frac{5k\pi}{4}$

$\bar{x} = \frac{M_y}{m} = \frac{5k\pi}{4} \cdot \frac{2}{3k\pi} = \frac{5}{6}$

$(\bar{x}, \bar{y}) = \left(\frac{5}{6}, 0\right)$

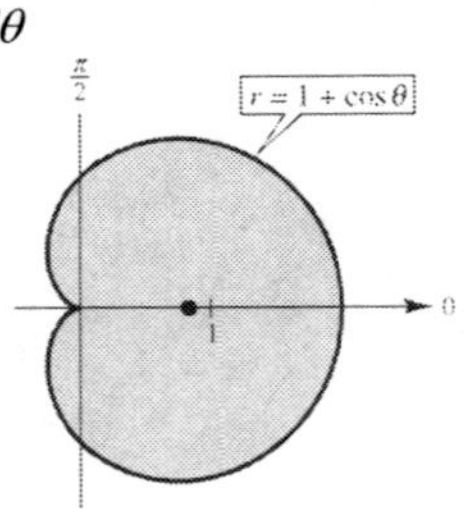

27. $m = bh$

$I_x = \int_0^b \int_0^h y^2\, dy\, dx = \frac{bh^3}{3}$

$I_y = \int_0^b \int_0^h x^2 dy\, dx = \frac{b^3 h}{3}$

$\bar{\bar{x}} = \sqrt{\frac{I_y}{m}} = \sqrt{\frac{b^3 h}{3} \cdot \frac{1}{bh}} = \sqrt{\frac{b^2}{3}} = \frac{b}{\sqrt{3}} = \frac{\sqrt{3}}{3} b$

$\bar{\bar{y}} = \sqrt{\frac{I_x}{m}} = \sqrt{\frac{bh^3}{3} \cdot \frac{1}{bh}} = \sqrt{\frac{h^2}{3}} = \frac{h}{\sqrt{3}} = \frac{\sqrt{3}}{3} h$

28. $m = \int_0^b \int_0^{h-(hx/b)} dy\, dx = \frac{bh}{2}$

$I_x = \int_0^b \int_0^{h-(hx/b)} y^2\, dy\, dx = \frac{bh^3}{12}$

$I_y = \int_0^b \int_0^{h-(hx/b)} x^2\, dy\, dx = \frac{b^3 h}{12}$

$\bar{\bar{x}} = \sqrt{\frac{I_y}{m}} = \sqrt{\frac{b^3 h/12}{bh/2}} = \frac{b}{\sqrt{6}} = \frac{\sqrt{6}}{6} b$

$\bar{\bar{y}} = \sqrt{\frac{I_x}{m}} = \sqrt{\frac{bh^3/12}{bh/2}} = \frac{h}{\sqrt{6}} = \frac{\sqrt{6}}{6} h$

29. $m = \pi a^2$

$$I_x = \int_R \int y^2\, dA = \int_0^{2\pi} \int_0^a r^3 \sin^2\theta\, dr\, d\theta = \frac{a^4\pi}{4}$$

$$I_y = \int_R \int x^2\, dA = \int_0^{2\pi} \int_0^a r^3 \cos^2\theta\, dr\, d\theta = \frac{a^4\pi}{4}$$

$$I_0 = I_x + I_y = \frac{a^4\pi}{4} + \frac{a^4\pi}{4} = \frac{a^4\pi}{2}$$

$$\bar{\bar{x}} = \bar{\bar{y}} = \sqrt{\frac{I_x}{m}} = \sqrt{\frac{a^4\pi}{4} \cdot \frac{1}{\pi a^2}} = \frac{a}{2}$$

30. $m = \dfrac{\pi a^2}{2}$

$$I_x = \int_R \int y^2\, dA = \int_0^{\pi} \int_0^a r^3 \sin^2\theta\, dr\, d\theta = \frac{a^4\pi}{8}$$

$$I_y = \int_R \int x^2\, dA = \int_0^{\pi} \int_0^a r^3 \cos^2\theta\, dr\, d\theta = \frac{a^4\pi}{8}$$

$$I_0 = I_x + I_x = \frac{a^4\pi}{8} + \frac{a^4\pi}{8} = \frac{a^4\pi}{4}$$

$$\bar{\bar{x}} = \bar{\bar{y}} = \sqrt{\frac{I_x}{m}} = \sqrt{\frac{a^4\pi}{8} \cdot \frac{2}{\pi a^2}} = \frac{a}{2}$$

31. $m = \dfrac{\pi a^2}{4}$

$$I_x = \int_R \int y^2\, dA = \int_0^{\pi/2} \int_0^a r^3 \sin^2\theta\, dr\, d\theta = \frac{\pi a^4}{16}$$

$$I_y = \int_R \int x^2\, dA = \int_0^{\pi/2} \int_0^a r^3 \cos^2\theta\, dr\, d\theta = \frac{\pi a^4}{16}$$

$$I_0 = I_x + I_y = \frac{\pi a^4}{16} + \frac{\pi a^4}{16} = \frac{\pi a^4}{8}$$

$$\bar{\bar{x}} = \bar{\bar{y}} = \sqrt{\frac{I_x}{m}} = \sqrt{\frac{\pi a^4}{16} \cdot \frac{4}{\pi a^2}} = \frac{a}{2}$$

32. $m = \pi ab$

$$I_x = 4\int_0^a \int_0^{(b/a)\sqrt{a^2-x^2}} y^2\, dy\, dx = 4\int_0^a \frac{b^3}{3a^3}\left(a^2 - x^2\right)^{3/2} dx = \frac{4b^3}{3a^3}\int_0^a \left[a^2\sqrt{a^2 - x^2} - x^2\sqrt{a^2 - x^2}\right] dx$$

$$= \frac{4b^3}{3a^3}\left[\frac{a^2}{2}\left(x\sqrt{a^2 - x^2} + a^2 \arcsin\frac{x}{a}\right) - \frac{1}{8}\left[x\left(2x^2 - a^2\right)\sqrt{a^2 - x^2} + a^4 \arcsin\frac{x}{a}\right]\right]_0^a = \frac{ab^3\pi}{4}$$

$$I_y = 4\int_0^b \int_0^{(a/b)\sqrt{b^2-y^2}} x^2\, dx\, dy = \frac{a^3b\pi}{4}$$

$$I_0 = I_y + I_x = \frac{a^3b\pi}{4} + \frac{ab^3\pi}{4} = \frac{ab\pi}{4}\left(a^2 + b^2\right)$$

$$\bar{\bar{x}} = \sqrt{\frac{I_y}{m}} = \sqrt{\frac{a^3b\pi}{4} \cdot \frac{1}{\pi ab}} = \frac{a}{2}$$

$$\bar{\bar{y}} = \sqrt{\frac{I_x}{m}} = \sqrt{\frac{ab^3\pi}{4} \cdot \frac{1}{\pi ab}} = \frac{b}{2}$$

33. $\rho = kx$

$$m = k\int_0^2 \int_0^{4-x^2} x\, dy\, dx = 4k$$

$$I_x = k\int_0^2 \int_0^{4-x^2} xy^2\, dy\, dx = \frac{32k}{3}$$

$$I_y = k\int_0^2 \int_0^{4-x^2} x^3\, dy\, dx = \frac{16k}{3}$$

$$I_0 = I_x + I_y = 16k$$

$$\bar{\bar{x}} = \sqrt{\frac{I_y}{m}} = \sqrt{\frac{16k/3}{4k}} = \sqrt{\frac{4}{3}} = \frac{2}{\sqrt{3}} = \frac{2\sqrt{3}}{3}$$

$$\bar{\bar{y}} = \sqrt{\frac{I_x}{m}} = \sqrt{\frac{32k/3}{4k}} = \sqrt{\frac{8}{3}} = \frac{4}{\sqrt{6}} = \frac{2\sqrt{6}}{3}$$

34. $\rho = kxy$

$$m = k\int_0^1 \int_{x^2}^{x} xy\, dy\, dx = \frac{k}{2}\int_0^1 \left(x^3 - x^5\right) dx = \frac{k}{24}$$

$$I_x = k\int_0^1 \int_{x^2}^{x} xy^3\, dy\, dx = \frac{k}{4}\int_0^1 \left(x^5 - x^9\right) dx = \frac{k}{60}$$

$$I_y = k\int_0^1 \int_{x^2}^{x} x^3y\, dy\, dx = \frac{k}{2}\int_0^1 \left(x^5 - x^7\right) dx = \frac{k}{48}$$

$$I_0 = I_x + I_y = \frac{9k}{240} = \frac{3k}{80}$$

$$\bar{\bar{x}} = \sqrt{\frac{I_y}{m}} = \sqrt{\frac{k/48}{k/24}} = \frac{\sqrt{2}}{2}$$

$$\bar{\bar{y}} = \sqrt{\frac{I_x}{m}}\sqrt{\frac{k/60}{k/24}} = \frac{\sqrt{10}}{5}$$

35. $\rho = kxy$

$$m = \int_0^4 \int_0^{\sqrt{x}} kxy\,dy\,dx = \frac{32k}{3}$$

$$I_x = \int_0^4 \int_0^{\sqrt{x}} kxy^3\,dy\,dx = 16k$$

$$I_y = \int_0^4 \int_0^{\sqrt{x}} kx^3y\,dy\,dx = \frac{512k}{5}$$

$$I_0 = I_x + I_y = \frac{592k}{5}$$

$$\bar{\bar{x}} = \sqrt{\frac{I_y}{m}} = \sqrt{\frac{512k}{5} \cdot \frac{3}{32k}} = \sqrt{\frac{48}{5}} = \frac{4\sqrt{15}}{5}$$

$$\bar{\bar{y}} = \sqrt{\frac{I_x}{m}} = \sqrt{\frac{16k}{1} \cdot \frac{3}{32k}} = \sqrt{\frac{3}{2}} = \frac{\sqrt{6}}{2}$$

36. $\rho = kx$

$$m = \int_0^1 \int_{x^2}^{\sqrt{x}} kx\,dy\,dx = \frac{3k}{20}$$

$$I_x = \int_0^1 \int_{x^2}^{\sqrt{x}} kxy^2\,dy\,dx = \frac{3k}{56}$$

$$I_y = \int_0^1 \int_{x^2}^{\sqrt{x}} kx^3\,dy\,dx = \frac{k}{18}$$

$$I_0 = I_x + I_y = \frac{55k}{504}$$

$$\bar{\bar{x}} = \sqrt{\frac{I_y}{m}} = \sqrt{\frac{k}{18} \cdot \frac{20}{3k}} = \frac{\sqrt{30}}{9}$$

$$\bar{\bar{y}} = \sqrt{\frac{I_x}{m}} = \sqrt{\frac{3k}{56} \cdot \frac{20}{3k}} = \frac{\sqrt{70}}{14}$$

37. $I = 2k \int_{-b}^{b} \int_0^{\sqrt{b^2-x^2}} (x-a)^2\,dy\,dx = 2k \int_{-b}^{b} (x-a)^2 \sqrt{b^2-x^2}\,dx$

$$= 2k\left[\int_{-b}^{b} x^2\sqrt{b^2-x^2}\,dx - 2a\int_{-b}^{b} x\sqrt{b^2-x^2}\,dx + a^2\int_{-b}^{b}\sqrt{b^2-x^2}\,dx\right] = 2k\left[\frac{\pi b^4}{8} + 0 + \frac{\pi a^2 b^2}{2}\right] = \frac{k\pi b^2}{4}\left(b^2 + 4a^2\right)$$

38. $I = \int_0^4 \int_0^{\sqrt{x}} kx(x-6)^2\,dy\,dx = \int_0^4 kx\sqrt{x}\left(x^2 - 12x + 36\right)dx = k\left[\frac{2}{9}x^{9/2} - \frac{24}{7}x^{7/2} + \frac{72}{5}x^{5/2}\right]_0^4 = \frac{42{,}752k}{315}$

39. $I = \int_{-a}^{a} \int_0^{\sqrt{a^2-x^2}} ky(y-a)^2\,dy\,dx$

$$= \int_{-a}^{a} k\left[\frac{y^4}{4} - \frac{2ay^3}{3} + \frac{a^2y^2}{2}\right]_0^{\sqrt{a^2-x^2}} dx$$

$$= \int_{-a}^{a} k\left[\frac{1}{4}\left(a^4 - 2a^2x^2 + x^4\right) - \frac{2a}{3}\left(a^2\sqrt{a^2-x^2} - x^2\sqrt{a^2-x^2}\right) + \frac{a^2}{2}\left(a^2 - x^2\right)\right]dx$$

$$= k\left[\frac{1}{4}\left(a^4x - \frac{2a^2x^3}{3} + \frac{x^5}{5}\right) - \frac{2a}{3}\left[\frac{a^2}{2}\left(x\sqrt{a^2-x^2} + a^2\arcsin\frac{x}{a}\right)\right.\right.$$

$$\left.\left.-\frac{1}{8}\left(x\left(2x^2 - a^2\right)\sqrt{a^2-x^2} + a^4\arcsin\frac{x}{a}\right)\right] + \frac{a^2}{2}\left(a^2x - \frac{x^3}{3}\right)\right]_{-a}^{a}$$

$$= 2k\left[\frac{1}{4}\left(a^5 - \frac{2}{3}a^5 + \frac{1}{5}a^5\right) - \frac{2a}{3}\left(\frac{a^4\pi}{4} - \frac{a^4\pi}{16}\right) + \frac{a^2}{2}\left(a^3 - \frac{a^3}{3}\right)\right] = 2k\left(\frac{7a^5}{15} - \frac{a^5\pi}{8}\right) = ka^5\left(\frac{56 - 15\pi}{60}\right)$$

40. $I = \int_{-2}^{2} \int_0^{4-x^2} k(y-2)^2\,dy\,dx = \int_{-2}^{2}\left[\frac{k}{3}(y-1)^3\right]_0^{4-x^2} dx = \int_{-2}^{2} \frac{k}{3}\left[\left(2 - x^2\right) + 8\right]dx$

$$= \frac{k}{3}\int_{-2}^{2}\left(16 - 12x^2 + 6x^4 - x^6\right)dx = \left[\frac{k}{3}\left(16x - 4x^3 + \frac{6}{5}x^5 - \frac{1}{7}x^7\right)\right]_{-2}^{2} = \frac{2k}{3}\left(32 - 32 + \frac{192}{5} - \frac{128}{7}\right) = \frac{1408k}{105}$$

41. $\bar{y} = \frac{L}{2}, A = bL, h = \frac{L}{2}$

$$I_{\bar{y}} = \int_0^b \int_0^L \left(y - \frac{L}{2}\right)^2 dy\,dx$$

$$= \int_0^b \left[\frac{\left[y - (L/2)\right]^3}{3}\right]_0^L dx = \frac{L^3 b}{12}$$

$$y_a = \bar{y} - \frac{I_{\bar{y}}}{hA} = \frac{L}{2} - \frac{L^3 b/12}{(L/2)(bL)} = \frac{L}{3}$$

42. $\bar{y} = \frac{a}{2}, A = ab, h = L - \frac{a}{2}$

$$I_{\bar{y}} = \int_0^b \int_0^a \left(y - \frac{a}{2}\right)^2 dy\,dx = \frac{a^3 b}{12}$$

$$y_a = \frac{a}{2} - \frac{a^3 b/12}{\left[L - (a/2)\right]ab} = \frac{a(3L - 2a)}{3(2L - a)}$$

43. $\bar{y} = \frac{2L}{3}, A = \frac{bL}{2}, h = \frac{L}{3}$

$$I_{\bar{y}} = 2\int_0^{b/2} \int_{2Lx/b}^{L} \left(y - \frac{2L}{3}\right)^2 dy\,dx$$

$$= \frac{2}{3}\int_0^{b/2} \left[\left(y - \frac{2L}{3}\right)^3\right]_{2Lx/b}^{L} dx$$

$$= \frac{2}{3}\int_0^{b/2} \left[\frac{L}{27} - \left(\frac{2Lx}{b} - \frac{2L}{3}\right)^3\right] dx$$

$$= \frac{2}{3}\left[\frac{L^3 x}{27} - \frac{b}{8L}\left(\frac{2Lx}{b} - \frac{2L}{3}\right)^4\right]_0^{b/2} = \frac{L^3 b}{36}$$

$$y_a = \frac{2L}{3} - \frac{L^3 b/36}{L^2 b/6} = \frac{L}{2}$$

44. $\bar{y} = 0, A = \pi a^2, h = L$

$$I_{\bar{y}} = \int_{-a}^{a} \int_{-\sqrt{a^2 - x^2}}^{\sqrt{a^2 - x^2}} y^2\,dy\,dx = \int_0^{2\pi} \int_0^a r^3 \sin^2\theta\,dr\,d\theta$$

$$= \int_0^{2\pi} \frac{a^4}{4} \sin^2\theta\,d\theta = \frac{a^4\pi}{4}$$

$$y_a = -\frac{\left(a^4\pi/4\right)}{L\pi a^2} = -\frac{a^2}{4L}$$

45. Let $\rho(x, y)$ be a continuous density function on the planar lamina R.

The movements of mass with respect to the x- and y-axes are

$$M_x = \int_R \int y\rho(x, y)\,dA \text{ and } M_y = \int_R \int x\rho(x, y)\,dA.$$

If m is the mass of the lamina, then the center of mass is

$$(\bar{x}, \bar{y}) = \left(\frac{M_y}{m}, \frac{M_x}{m}\right).$$

46. $I_x = \int_R \int y^2 \rho(x, y)\,dA$, Moment of inertia about x-axis

$I_y = \int_R \int x^2 \rho(x, y)\,dA$, Moment of inertia about y-axis

47. See the definition on page 999.

48. (a) $\rho(x, y) = ky$

$\bar{y}$ will increase.

(b) $\rho(x, y) = k|2 - x|$

$\bar{y}$ will decrease.

(c) $\rho(x, y) = kxy$

Both $\bar{x}$ and $\bar{y}$ will increase.

(d) $\rho(x, y) = k(4 - x)(4 - y)$

Both $\bar{x}$ and $\bar{y}$ will decrease.

49. Orient the xy-coordinate system so that L is along the y-axis and R is the first quadrant. Then the volume of the solid is

$$V = \int_R \int 2\pi x\,dA$$

$$= 2\pi \int_R \int x\,dA$$

$$= 2\pi \left(\frac{\int_R \int x\,dA}{\int_R \int dA}\right) \int_R \int dA$$

$$= 2\pi \bar{x} A.$$

By our positioning, $\bar{x} = r$. So, $V = 2\pi rA$.

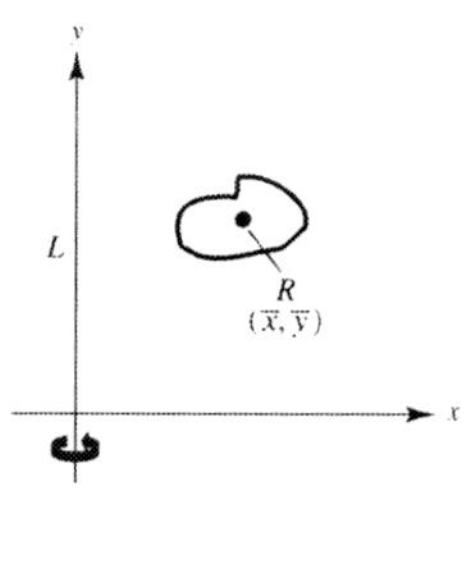

Section 14.5 Surface Area

1. $f(x, y) = 2x + 2y$

$f_x = f_y = 2$

$\sqrt{1 + (f_x)^2 + (f_y)^2} = \sqrt{1 + 4 + 4} = 3$

$S = \int_0^4 \int_0^{4-x} 3\, dy\, dx = 3\int_0^4 (4 - x)\, dx$

$= 3\left[4x - \frac{x^2}{2}\right]_0^4 = 24$

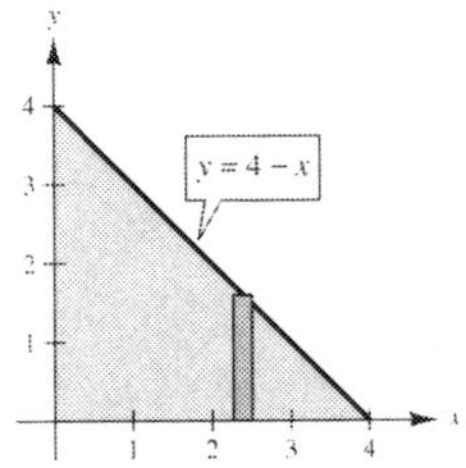

2. $f(x, y) = 15 + 2x - 3y$

$f_x = 2, f_y = -3$

$\sqrt{1 + (f_x)^2 + (f_y)^2} = \sqrt{14}$

$S = \int_0^3 \int_0^3 \sqrt{14}\, dy\, dx = \int_0^3 3\sqrt{14}\, dx = 9\sqrt{14}$

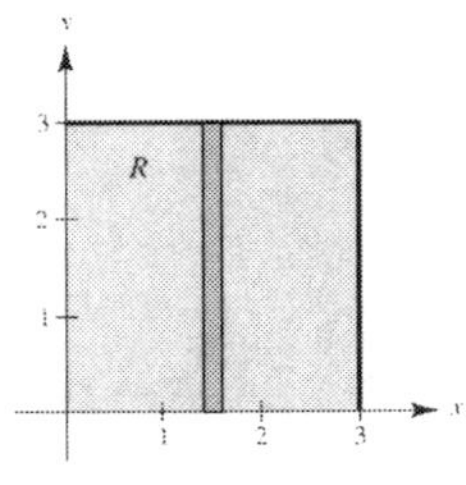

3. $f(x, y) = 7 + 2x + 2y$

$f_x = f_y = 2$

$\sqrt{1 + (f_x)^2 + (f_y)^2} = \sqrt{1 + 4 + 4} = 3$

$S = \int_0^{2\pi} \int_0^2 3\, r\, dr\, d\theta = \int_0^{2\pi} 6\, d\theta = 12\pi$

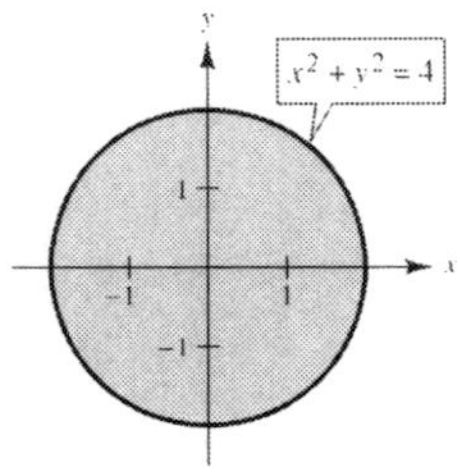

4. $f(x, y) = 12 + 2x - 3y$

$f_x = 2, f_y = -3$

$\sqrt{1 + (f_x)^2 + (f_y)^2} = \sqrt{1 + 4 + 9} = \sqrt{14}$

$S = \int_0^{2\pi} \int_0^3 \sqrt{14}\, r\, dr\, d\theta = \int_0^{2\pi} \frac{9\sqrt{14}}{2}\, d\theta = 9\sqrt{14}\,\pi$

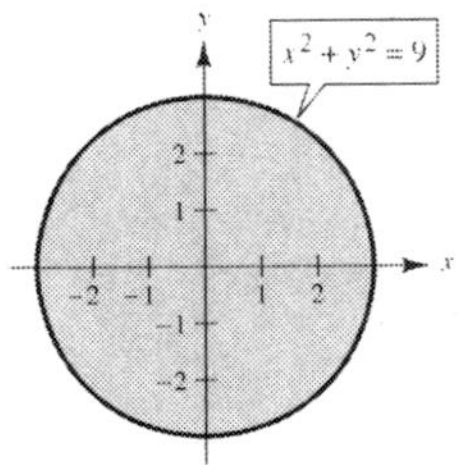

5. $f(x, y) = 9 - x^2$

$f_x = -2x, f_y = 0$

$\sqrt{1 + (f_x)^2 + (f_y)^2} = \sqrt{1 + 4x^2}$

$S = \int_0^2 \int_0^2 \sqrt{1 + 4x^2}\, dy\, dx = 2\int_0^2 \sqrt{1 + 4x^2}\, dx$

$= 2\left[\frac{1}{4}\ln\left(\sqrt{1 + 4x^2} + 2x\right) + \frac{x}{2}\sqrt{1 + 4x^2}\right]_0^2$

$= 2\left[\frac{1}{4}\ln\left(\sqrt{17} + 4\right) + \sqrt{17}\right]$

$= 2\sqrt{17} + \frac{1}{2}\ln\left(4 + \sqrt{17}\right)$

6. $f(x, y) = y^2$

R = square with vertices $(0, 0), (3, 0), (0, 3), (3, 3)$

$f_x = 0, f_y = 2y$

$$\sqrt{1 + (f_x)^2 + (f_y)^2} = \sqrt{1 + 4y^2}$$

$$S = \int_0^3 \int_0^3 \sqrt{1 + 4y^2}\, dx\, dy = \int_0^3 3\sqrt{1 + 4y^2}\, dy$$

$$= \left[\tfrac{3}{4}\left(2y\sqrt{1 + 4y^2} + \ln\left|2y + \sqrt{1 + 4y^2}\right|\right)\right]_0^3$$

$$= \tfrac{3}{4}\left(6\sqrt{37} + \ln\left|6 + \sqrt{37}\right|\right)$$

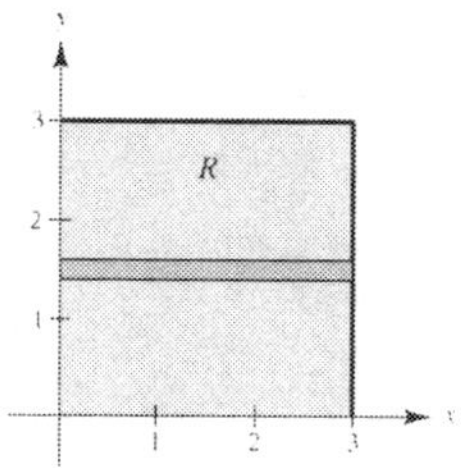

7. $f(x, y) = 3 + x^{3/2}$

$f_x = \dfrac{3}{2}x^{1/2}, f_y = 0$

$$\sqrt{1 + (f_x)^2 + (f_y)^2} = \sqrt{1 + \frac{9}{4}x} = \frac{\sqrt{4 + 9x}}{2}$$

$$S = \int_0^3 \int_0^4 \frac{\sqrt{4 + 9x}}{2}\, dy\, dx = 4\int_0^3 \frac{\sqrt{4 + 9x}}{2}\, dx$$

$$= \left[\frac{4}{27}(4 + 9x)^{3/2}\right]_0^3 = \frac{4}{27}\left(31\sqrt{31} - 8\right)$$

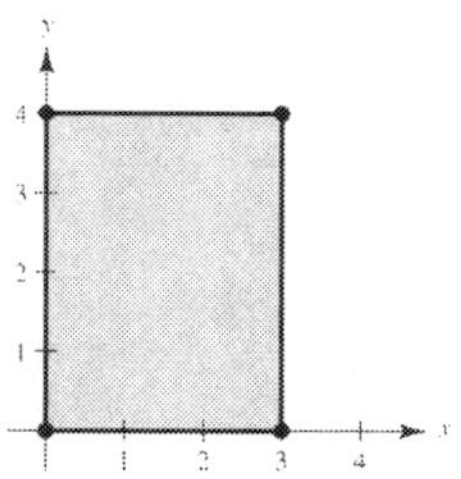

8. $f(x, y) = 2 + \frac{2}{3}y^{3/2}$

$f_x = 0, f_y = y^{1/2}$

$$\sqrt{1 + (f_x)^2 + (f_y)^2} = \sqrt{1 + y}$$

$$S = \int_0^2 \int_0^{2-y} \sqrt{1 + y}\, dx\, dy = \int_0^2 \sqrt{1 + y}(2 - y)\, dy$$

$$= \left[2(1 + y)^{3/2} - \tfrac{2}{5}(1 + y)^{5/2}\right]_0^2$$

$$= 2 \cdot 3^{3/2} - \tfrac{2}{5} \cdot 3^{5/2} - 2 + \tfrac{2}{5} = \tfrac{12}{5}\sqrt{3} - \tfrac{8}{5}$$

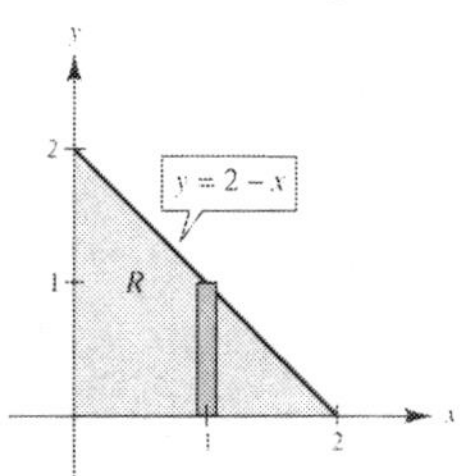

9. $f(x, y) = \ln|\sec x|$

$$R = \left\{(x, y) : 0 \le x \le \frac{\pi}{4}, 0 \le y \le \tan x\right\}$$

$f_x = \tan x, f_y = 0$

$$\sqrt{1 + (f_x)^2 + (f_y)^2} = \sqrt{1 + \tan^2 x} = \sec x$$

$$S = \int_0^{\pi/4} \int_0^{\tan x} \sec x\, dy\, dx$$

$$= \int_0^{\pi/4} \sec x \tan x\, dx$$

$$= \left[\sec x\right]_0^{\pi/4} = \sqrt{2} - 1$$

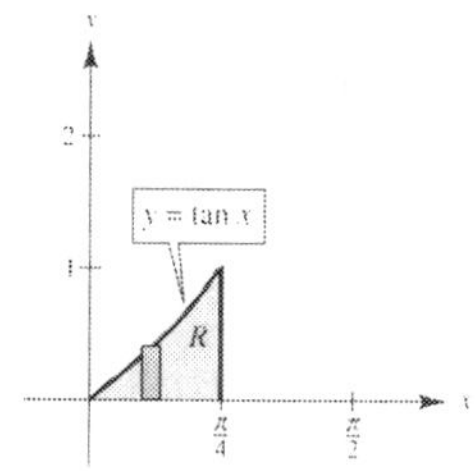

10. $f(x, y) = 13 + x^2 - y^2$

$f_x = 2x, f_y = -2y$

$$\sqrt{1 + (f_x)^2 + (f_y)^2} = \sqrt{1 + 4x^2 + 4y^2}$$

$$S = \int_0^{2\pi} \int_0^{2} \sqrt{1 + 4r^2}\, r\, dr\, d\theta$$

$$= \int_0^{2\pi} \left[\tfrac{1}{12}\left(1 + 4r^2\right)^{3/2}\right]_0^2 d\theta$$

$$= \int_0^{2\pi} \tfrac{1}{12}\left(17^{3/2} - 1\right) d\theta$$

$$= \frac{\pi}{6}\left(17\sqrt{17} - 1\right)$$

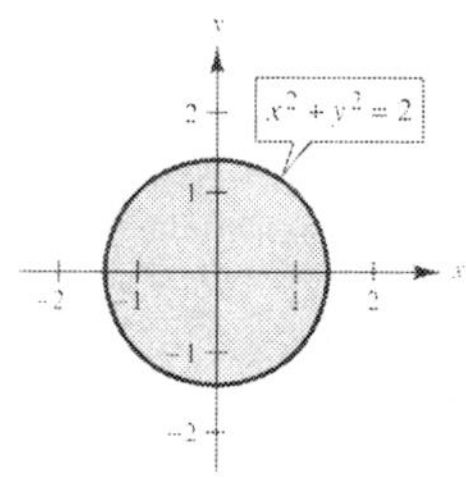

11. $f(x, y) = \sqrt{x^2 + y^2}$

$R = \{(x, y): 0 \le f(x, y) \le 1\}$

$0 \le \sqrt{x^2 + y^2} \le 1, x^2 + y^2 \le 1$

$$f_x = \frac{x}{\sqrt{x^2 + y^2}}, f_y = \frac{y}{\sqrt{x^2 + y^2}}$$

$$\sqrt{1 + (f_x)^2 + (f_y)^2} = \sqrt{1 + \frac{x^2}{x^2 + y^2} + \frac{y^2}{x^2 + y^2}} = \sqrt{2}$$

$$S = \int_{-1}^{1} \int_{-\sqrt{1-x^2}}^{\sqrt{1-x^2}} \sqrt{2}\, dy\, dx = \int_0^{2\pi} \int_0^1 \sqrt{2} r\, dr\, d\theta = \sqrt{2}\pi$$

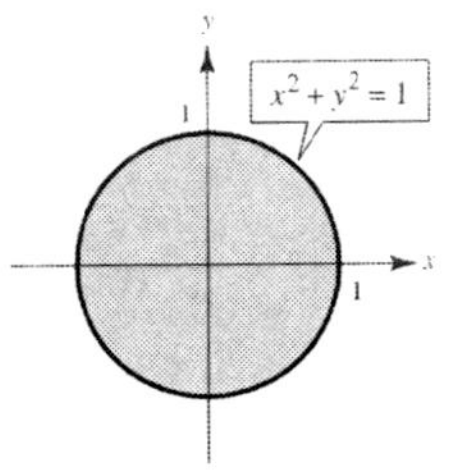

12. $f(x, y) = xy$

$R = \{(x, y): x^2 + y^2 \le 16\}$

$f_x = y, f_y = x$

$$\sqrt{1 + (f_x)^2 + (f_y)^2} = \sqrt{1 + y^2 + x^2}$$

$$S = \int_{-4}^{4} \int_{-\sqrt{16-x^2}}^{\sqrt{16-x^2}} \sqrt{1 + y^2 + x^2}\, dy\, dx$$

$$= \int_0^{2\pi} \int_0^4 \sqrt{1 + r^2}\, r\, dr\, d\theta = \frac{2\pi}{3}\left(17\sqrt{17} - 1\right)$$

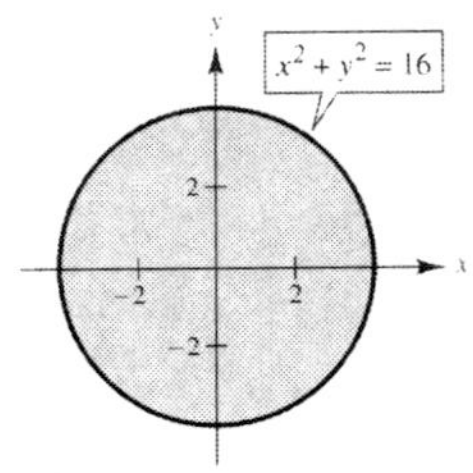

13. $f(x, y) = \sqrt{a^2 - x^2 - y^2}$

$R = \{(x, y): x^2 + y^2 \le b^2, 0 < b < a\}$

$$f_x = \frac{-x}{\sqrt{a^2 - x^2 - y^2}}, f_y = \frac{-y}{\sqrt{a^2 - x^2 - y^2}}$$

$$\sqrt{1 + (f_x)^2 + (f_y)^2} = \sqrt{1 + \frac{x^2}{a^2 - x^2 - y^2} + \frac{y^2}{a^2 - x^2 - y^2}} = \frac{a}{\sqrt{a^2 - x^2 - y^2}}$$

$$S = \int_{-b}^{b} \int_{-\sqrt{b^2-x^2}}^{\sqrt{b^2-x^2}} \frac{a}{\sqrt{a^2 - x^2 - y^2}}\, dy\, dx = \int_0^{2\pi} \int_0^b \frac{a}{\sqrt{a^2 - r^2}}\, r\, dr\, d\theta = 2\pi a\left(a - \sqrt{a^2 - b^2}\right)$$

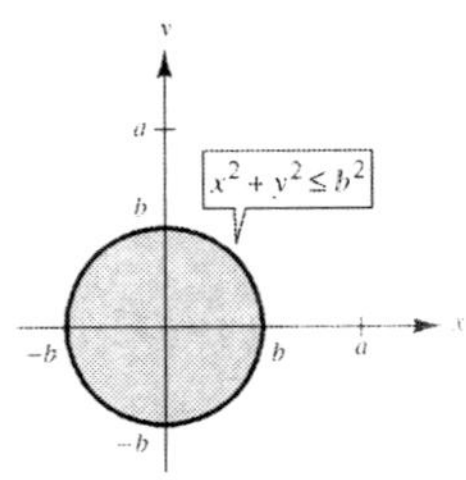

14. See Exercise 13.

$$S = \int_{-a}^{a} \int_{-\sqrt{a^2-x^2}}^{\sqrt{a^2-x^2}} \frac{a}{\sqrt{a^2 - x^2 - y^2}}\, dy\, dx$$

$$= \int_0^{2\pi} \int_0^a \frac{a}{\sqrt{a^2 - r^2}}\, r\, dr\, d\theta = 2\pi a^2$$

15. $z = 24 - 3x - 2y$

$$\sqrt{1 + (f_x)^2 + (f_y)^2} = \sqrt{14}$$

$$S = \int_0^8 \int_0^{-(3/2)x+12} \sqrt{14}\, dy\, dx = 48\sqrt{14}$$

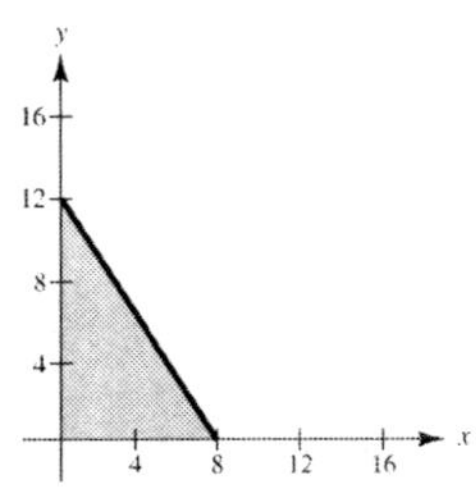

16. $z = 16 - x^2 - y^2$

$$\sqrt{1 + (f_y)^2 + (f_y)^2} = \sqrt{1 + 4x^2 + 4y^2}$$

$$S = \int_0^4 \int_0^{\sqrt{16-x^2}} \sqrt{1 + 4(x^2 + y^2)}\, dy\, dx$$

$$= \int_0^{\pi/2} \int_0^4 \sqrt{1 + 4r^2}\, r\, dr\, d\theta = \frac{\pi}{24}\left(65\sqrt{65} - 1\right)$$

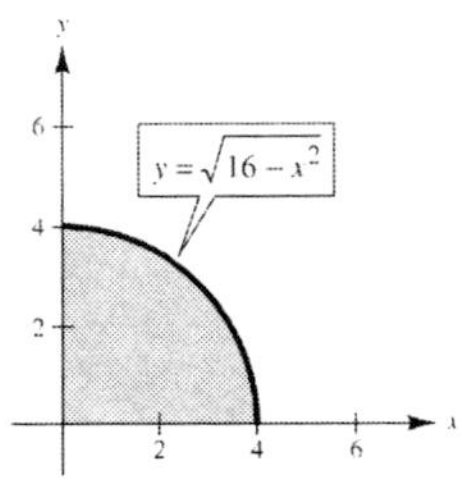

17. $z = \sqrt{25 - x^2 - y^2}$

$$\sqrt{1 + (f_x)^2 + (f_y)^2} = \sqrt{1 + \frac{x^2}{25 - x^2 - y^2} + \frac{y^2}{25 - x^2 - y^2}} = \frac{5}{\sqrt{25 - x^2 - y^2}}$$

$$S = 2\int_{-3}^{3}\int_{-\sqrt{9-x^2}}^{\sqrt{9-x^2}} \frac{5}{\sqrt{25 - (x^2 + y^2)}}\,dy\,dx = 2\int_0^{2\pi}\int_0^3 \frac{5}{\sqrt{25 - r^2}}r\,dr\,d\theta = 20\pi$$

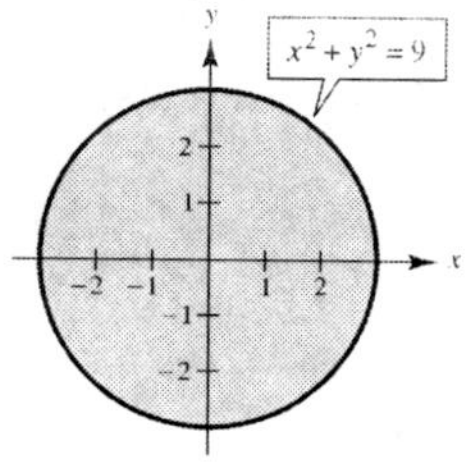

18. $z = 2\sqrt{x^2 + y^2}$

$$\sqrt{1 + (f_x)^2 + (f_y)^2} = \sqrt{1 + \frac{4x^2}{x^2 + y^2} + \frac{4y^2}{x^2 + y^2}} = \sqrt{5}$$

$$S = \int_0^{2\pi}\int_0^2 \sqrt{5}r\,dr\,d\theta = 4\pi\sqrt{5}$$

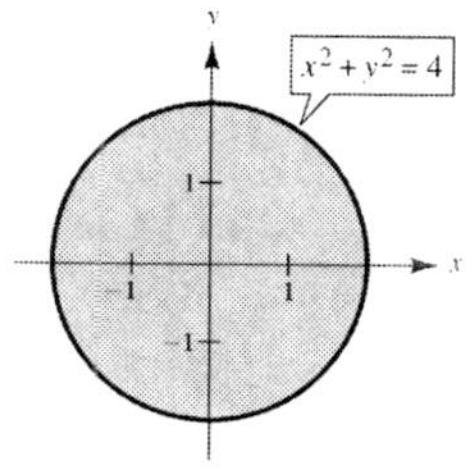

19. $f(x, y) = 2y + x^2$

R = triangle with vertices $(0, 0), (1, 0), (1, 1)$

$$\sqrt{1 + (f_x)^2 + (f_y)^2} = \sqrt{5 + 4x^2}$$

$$S = \int_0^1\int_0^x \sqrt{5 + 4x^2}\,dy\,dx = \frac{1}{12}\left(27 - 5\sqrt{5}\right)$$

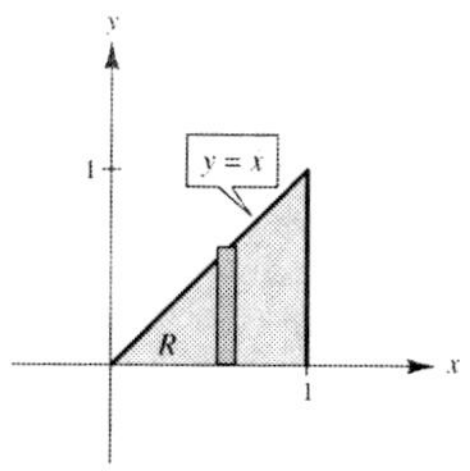

20. $f(x, y) = 2x + y^2$

R = triangle with vertices $(0, 0), (2, 0), (2, 2)$

$$\sqrt{1 + (f_x)^2 + (f_y)^2} = \sqrt{5 + 4y^2}$$

$$S = \int_0^2\int_0^x \sqrt{5 + 4y^2}\,dy\,dx$$

$$= \frac{5}{4}\ln\left(\frac{8\sqrt{21} + 37}{5}\right) + \frac{\sqrt{21}}{4} + \frac{5\sqrt{5}}{12}$$

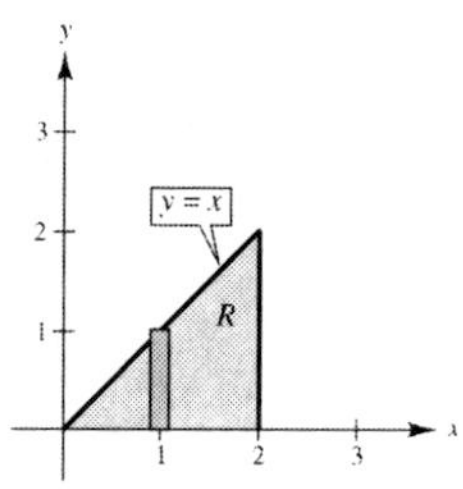

21. $f(x, y) = 9 - x^2 - y^2$

$R = \{(x, y): 0 \le f(x, y)\}$

$0 \le 9 - x^2 - y^2 \Rightarrow x^2 + y^2 \le 9$

$fx = -2x,\ fy = -2y$

$$\sqrt{1 + (f_x)^2 + (f_y)^2} = \sqrt{1 + 4x^2 + 4y^2}$$

$$S = \int_{-3}^{3}\int_{-\sqrt{9-x^2}}^{\sqrt{9-x^2}} \sqrt{1 + 4x^2 + 4y^2}\,dy\,dx$$

$$= \int_0^{2\pi}\int_0^3 \sqrt{1 + 4r^2}\,r\,dr\,d\theta$$

$$= \frac{\pi}{6}\left(37\sqrt{37} - 1\right) \approx 117.3187$$

22. $f(x, y) = x^2 + y^2$

$R = \{(x, y): 0 \le f(x, y) \le 16\}$

$0 \le x^2 + y^2 \le 16$

$f_x = 2x,\ f_y = 2y$

$$\sqrt{1 + (f_x)^2 + (f_y)^2} = \sqrt{1 + 4x^2 + 4y^2}$$

$$S = \int_{-4}^{4}\int_{-\sqrt{16-x^2}}^{\sqrt{16-x^2}} \sqrt{1 + 4x^2 + 4y^2}\,dy\,dx$$

$$= \int_0^{2\pi}\int_0^4 \sqrt{1 + 4r^2}\,dr\,d\theta = \frac{\left(65\sqrt{65} - 1\right)\pi}{6}$$

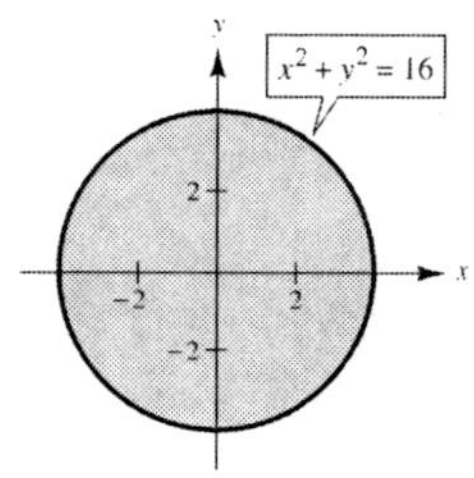

23. $f(x, y) = 4 - x^2 - y^2$

$R = \{(x, y): 0 \le x \le 1, 0 \le y \le 1\}$

$f_x = -2x,\ f_y = -2y$

$\sqrt{1 + (f_x)^2 + (f_y)^2} = \sqrt{1 + 4x^2 + 4y^2}$

$S = \int_0^1 \int_0^1 \sqrt{(1 + 4x^2) + 4y^2}\, dy\, dx \approx 1.8616$

24. $f(x, y) = \frac{2}{3}x^{3/2} + \cos x$

$R = \{(x, y): 0 \le x \le 1, 0 \le y \le 1\}$

$f_x = x^{1/2} - \sin x,\ f_y = 0$

$\sqrt{1 + (f_x)^2 + (f_y)^2} = \sqrt{1 + (\sqrt{x} - \sin x)^2}$

$S = \int_0^1 \int_0^1 \sqrt{1 + (\sqrt{x} - \sin x)^2}\, dy\, dx \approx 1.02185$

25. $f(x, y) = e^{xy}$

$R = \{(x, y): 0 \le x \le 4, 0 \le y \le 10\}$

$f_x = ye^{xy},\ f_y = xe^{xy}$

$\sqrt{1 + (f_x)^2 + (f_y)^2} = \sqrt{1 + y^2e^{2xy} + x^2e^{2xy}} = \sqrt{1 + e^{2xy}(x^2 + y^2)}$

$S = \int_0^4 \int_0^{10} \sqrt{1 + e^{2xy}(x^2 + y^2)}\, dy\, dx$

26. $f(x, y) = x^2 - 3xy - y^2$

$R = \{(x, y): 0 \le x \le 4, 0 \le y \le x\}$

$f_x = 2x - 3y,\ f_y = -3x - 2y = -(3x + 2y)$

$\sqrt{1 + (f_x) + (f_y)^2} = \sqrt{1 + (2x - 3y)^2 + (3x + 2y)^2}$

$= \sqrt{1 + 13(x^2 + y^2)}$

$S = \int_0^4 \int_0^x \sqrt{1 + 13(x^2 + y^2)}\, dy\, dx$

27. $f(x, y) = e^{-x} \sin y$

$f_x = -e^{-x} \sin y,\ f_y = e^{-x} \cos y$

$\sqrt{1 + (f_x^2) + (f_y^2)} = \sqrt{1 + e^{-2x} \sin^2 y + e^{-2x} \cos^2 y} = \sqrt{1 + e^{-2x}}$

$S = \int_{-2}^2 \int_{-\sqrt{4-x^2}}^{\sqrt{4-x^2}} \sqrt{1 + e^{-2x}}\, dy\, dx$

28. $f(x, y) = \cos(x^2 + y^2)$

$R = \left\{(x, y): x^2 + y^2 \le \frac{\pi}{2}\right\}$

$f_x = -2x \sin(x^2 + y^2),\ f_y = -2y \sin(x^2 + y^2)$

$\sqrt{1 + (f_x)^2 + (f_y)^2} = \sqrt{1 + 4x^2 \sin^2(x^2 + y^2) + 4y^2 \sin^2(x^2 + y^2)} = \sqrt{1 + 4[\sin^2(x^2 + y^2)](x^2 + y^2)}$

$S = \int_{-\sqrt{\pi/2}}^{\sqrt{\pi/2}} \int_{-\sqrt{(\pi/2)-x^2}}^{\sqrt{(\pi/2)-x^2}} \sqrt{1 + 4(x^2 + y^2) \sin^2(x^2 + y^2)}\, dy\, dx$

29. See the definition on page 1003.

30. (a) Yes. For example, let R be the square given by

$0 \le x \le 1, 0 \le y \le 1,$

and S the square parallel to R given by

$0 \le x \le 1, 0 \le y \le 1, z = 1.$

(b) Yes. Let R be the region in part (a) and S the surface given by $f(x, y) = xy$.

(c) No.

31. No, the surface area is the same.

$z = f(x, y)$ and $z = f(x, y) + k$

have the same partial derivatives.

32. $f(x, y) = x^2 + y^2$ is a paraboloid opening upward. Using the figure below, you see that the surface areas satisfy:

(b) < (c) < (a)

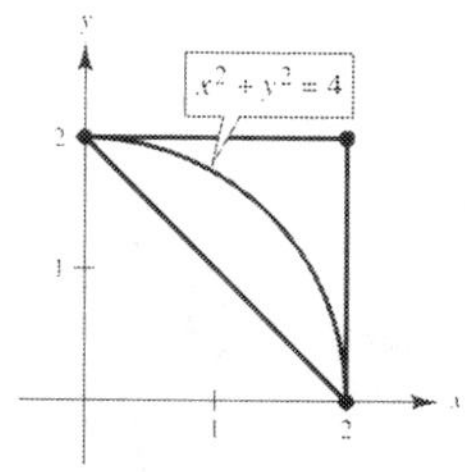

33. (a) $V = \int_R\int f(x, y)$

$= 8\int_R\int \sqrt{625 - x^2 - y^2}\, dA$ where R is the region in the first quadrant

$= 8\int_0^{\pi/2}\int_4^{25} \sqrt{625 - r^2}\, r\, dr\, d\theta = -4\int_0^{\pi/2}\left[\frac{2}{3}\left(625 - r^2\right)^{3/2}\right]_4^{25} d\theta$

$= -\frac{8}{3}\left[0 - 609\sqrt{609}\right]\cdot\frac{\pi}{2} = 812\pi\sqrt{609}$ cm³

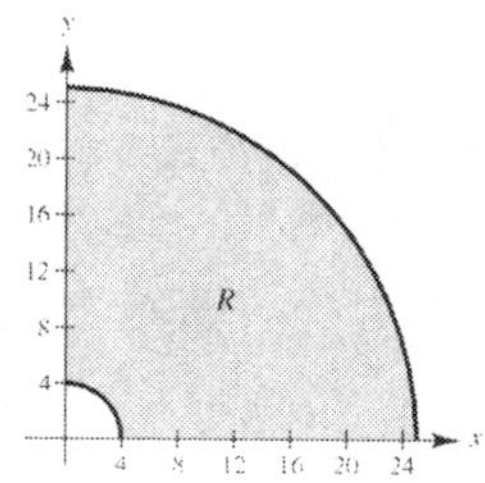

(b) $A = \int_R\int\sqrt{1 + (f_x)^2 + (f_y)^2}\, dA = 8\int_R\int\sqrt{1 + \frac{x^2}{625 - x^2 - y^2} + \frac{y^2}{625 - x^2 - y^2}}\, dA$

$= 8\int_R\int\frac{25}{\sqrt{625 - x^2 - y^2}}\, dA = 8\int_0^{\pi/2}\int_4^{25}\frac{25}{\sqrt{625 - r^2}}\, r\, dr\, d\theta$

$= \lim_{b\to 25^-}\left[-200\sqrt{625 - r^2}\right]_4^b\cdot\frac{\pi}{2} = 100\pi\sqrt{609}$ cm²

34. (a) $z = -\frac{1}{75}y^3 + \frac{4}{25}y^2 - \frac{16}{15}y + 25$

(b) $V \approx 2(50)\int_0^{15}\left(-\frac{1}{75}y^3 + \frac{4}{25}y^2 - \frac{16}{15}y + 25\right)dy = 100(266.25) = 26{,}625$ cubic feet

(c) $f(x, y) = -\frac{1}{75}y^3 + \frac{4}{25}y^2 - \frac{16}{15}y + 25$

$f_x = 0, f_y = -\frac{1}{25}y^2 + \frac{8}{25}y - \frac{16}{15}$

$S = 2\int_0^{50}\int_0^{15}\sqrt{1 + f_y^2 + f_x^2}\, dy\, dx \approx 3087.58$ sq ft

(d) Arc length ≈ 30.8758

Surface area of roof $\approx 2(50)(30.8758) = 3087.58$ sq ft

35. $f(x, y) = \sqrt{1 - x^2}$; $f_x = \dfrac{-x}{\sqrt{1^2 - x^2}}$, $f_y = 0$

$$S = \int_R\int\sqrt{1 + (f_x)^2 + (f_y)^2}\,dA$$
$$= 16\int_0^1\int_0^x \frac{1}{\sqrt{1 - x^2}}\,dy\,dx$$
$$= 16\int_0^1 \frac{x}{\sqrt{1 - x^2}}\,dx = \left[-16(1 - x^2)^{1/2}\right]_0^1 = 16$$

36. $f(x, y) = k\sqrt{x^2 + y^2}$

$$\sqrt{1 + (f_x)^2 + (f_y)^2} = \sqrt{1 + \frac{k^2x^2}{x^2 + y^2} + \frac{k^2y^2}{x^2 + y^2}} = \sqrt{k^2 + 1}$$

$$S = \int_R\int\sqrt{1 + (f_x)^2 + (f_y)^2}\,dA = \int_R\int\sqrt{k^2 + 1}\,dA = \sqrt{k^2 + 1}\int_R\int dA = A\sqrt{k^2 + 1} = \pi r^2\sqrt{k^2 + 1}$$

Section 14.6 Triple Integrals and Applications

1. $$\int_0^3\int_0^2\int_0^1 (x + y + z)\,dx\,dz\,dy = \int_0^3\int_0^2\left[\frac{x^2}{2} + xy + xz\right]_0^1 dz\,dy = \int_0^3\int_0^2\left(\frac{1}{2} + y + z\right)dz\,dy = \int_0^3\left[\frac{1}{2}z + yz + \frac{z^2}{2}\right]_0^2 dy$$
$$= \int_0^3 (1 + 2y + 2)\,dy = \left[3y + y^2\right]_0^3 = 18$$

2. $$\int_{-1}^1\int_{-1}^1\int_{-1}^1 x^2y^2z^2dx\,dy\,dz = \tfrac{1}{3}\int_{-1}^1\int_{-1}^1\left[x^3y^2z^2\right]_{-1}^1 dy\,dz$$
$$= \tfrac{2}{3}\int_{-1}^1\int_{-1}^1 y^2z^2\,dy\,dz = \tfrac{2}{9}\int_{-1}^1\left[y^3z^2\right]_{-1}^1 dz = \tfrac{4}{9}\int_{-1}^1 z^2\,dz = \left[\tfrac{4}{27}z^3\right]_{-1}^1 = \tfrac{8}{27}$$

3. $$\int_0^1\int_0^x\int_0^{xy} x\,dz\,dy\,dx = \int_0^1\int_0^x\left[xz\right]_0^{xy} dy\,dx = \int_0^1\int_0^x x^2y\,dy\,dx = \int_0^1\left[\frac{x^2y^2}{2}\right]_0^x dx = \int_0^1\frac{x^4}{2}\,dx = \left[\frac{x^5}{10}\right]_0^1 = \frac{1}{10}$$

4. $$\int_0^9\int_0^{y/3}\int_0^{\sqrt{y^2-9x^2}} z\,dz\,dx\,dy = \tfrac{1}{2}\int_0^9\int_0^{y/3}\left(y^2 - 9x^2\right)dx\,dy = \tfrac{1}{2}\int_0^9\left[xy^2 - 3x^3\right]_0^{y/3} dy = \tfrac{2}{18}\int_0^9 y^3\,dy = \left[\tfrac{1}{36}y^4\right]_0^9 = \tfrac{729}{4}$$

5. $$\int_1^4\int_0^1\int_0^x 2ze^{-x^2}\,dy\,dx\,dz = \int_1^4\int_0^1\left[\left(2ze^{-x^2}\right)y\right]_0^x dx\,dz = \int_1^4\int_0^1 2zxe^{-x^2}\,dx\,dz$$
$$= \int_1^4\left[-ze^{-x^2}\right]_0^1 dz = \int_1^4 z\left(1 - e^{-1}\right)dz = \left[\left(1 - e^{-1}\right)\frac{z^2}{2}\right]_1^4 = \frac{15}{2}\left(1 - \frac{1}{e}\right)$$

6. $$\int_1^4\int_1^{e^2}\int_0^{1/xz}\ln z\,dy\,dz\,dx = \int_1^4\int_1^{e^2}\left[(\ln z)y\right]_0^{1/xz}dz\,dx = \int_1^4\int_1^{e^2}\frac{\ln z}{xz}\,dz\,dx = \int_1^4\left[\frac{(\ln z)^2}{2x}\right]_1^{e^2} dx = \int_1^4\frac{2}{x}\,dx = \left[2\ln|x|\right]_1^4 = 2\ln 4$$

7. $$\int_0^4\int_0^{\pi/2}\int_0^{1-x} x\cos y\,dz\,dy\,dx = \int_0^4\int_0^{\pi/2}\left[(x\cos y)z\right]_0^{1-x} dy\,dx = \int_0^4\int_0^{\pi/2} x(1 - x)\cos y\,dy\,dx$$
$$= \int_0^4\left[x(1 - x)\sin y\right]_0^{\pi/2} dx = \int_0^4 x(1 - x)\,dx = \left[\frac{x^2}{2} - \frac{x^3}{3}\right]_0^4 = 8 - \frac{64}{3} = -\frac{40}{3}$$

8. $$\int_0^{\pi/2}\int_0^{y/2}\int_0^{1/y}\sin y\,dz\,dx\,dy = \int_0^{\pi/2}\int_0^{y/2}\frac{\sin y}{y}\,dx\,dy = \frac{1}{2}\int_0^{\pi/2}\sin y\,dy = \left[-\frac{1}{2}\cos y\right]_0^{\pi/2} = \frac{1}{2}$$

9. $\int_0^3 \int_{-\sqrt{9-y^2}}^{\sqrt{9-y^2}} \int_0^{y^2} y\,dz\,dx\,dy = \frac{324}{5}$

10. $\int_0^3 \int_0^{2-(2y/3)} \int_0^{6-2y-3z} ze^{-x^2y^2}\,dx\,dz\,dy = \int_0^6 \int_0^{(6-x)/2} \int_0^{(6-x-2y)/3} ze^{-x^2y^2}\,dz\,dy\,dx$

$$= \int_0^6 \int_0^{3-(x/2)} \frac{1}{2}\left(\frac{6-x-2y}{3}\right)^2 e^{-x^2y^2}\,dy\,dx \approx 2.118$$

11. $V = \int_0^5 \int_0^{5-x} \int_0^{5-x-y} dz\,dy\,dx$

12. $V = \int_0^3 \int_0^{2x} \int_0^{9-x^2} dz\,dy\,dx$

13. $V = \int_{-\sqrt{6}}^{\sqrt{6}} \int_{-\sqrt{6-x^2}}^{\sqrt{6-x^2}} \int_0^{6-x^2-y^2} dz\,dy\,dx = \int_{-\sqrt{6}}^{\sqrt{6}} \int_{-\sqrt{6-y^2}}^{\sqrt{6-y^2}} \int_0^{6-x^2-y^2} dz\,dx\,dy$

14. $V = \int_{-4}^{4} \int_{-\sqrt{16-x^2}}^{\sqrt{16-x^2}} \int_0^{\sqrt{16-x^2-y^2}} dz\,dy\,dx$

15. $z = \frac{1}{2}\left(x^2 + y^2\right) \Rightarrow 2z = x^2 + y^2$

$x^2 + y^2 + z^2 = 2z + z^2 = 80 \Rightarrow z^2 + 2z - 80 = 0 \Rightarrow (z-8)(z+10) = 0 \Rightarrow z = 8 \Rightarrow x^2 + y^2 = 2z = 16$

$V = \int_{-4}^{4} \int_{-\sqrt{16-x^2}}^{\sqrt{16-x^2}} \int_{1/2\left(x^2+y^2\right)}^{\sqrt{80-x^2-y^2}} dz\,dy\,dx$

16. $V = \int_{-\sqrt{2}}^{\sqrt{2}} \int_{-\sqrt{\left(4-2x^2\right)/3}}^{\sqrt{\left(4-2x^2\right)/3}} \int_{x^2+3y^2}^{4-x^2} dz\,dy\,dx$

$$z = 4 - x^2 = x^2 + 3y^2$$
$$4 = 2x^2 + 3y^2$$
$$1 = \frac{x^2}{2} + \frac{y^2}{(4/3)} \quad \text{ellipse}$$

17. $V = \int_{-2}^{2} \int_0^{4-y^2} \int_0^x dz\,dx\,dy = \int_{-2}^{2} \int_0^{4-y^2} x\,dx\,dy$

$$= \frac{1}{2}\int_{-2}^{2} \left(4-y^2\right)^2 dy = \int_0^2 \left(16 - 8y^2 + y^4\right) dy = \left[16y - \frac{8}{3}y^3 + \frac{1}{5}y^5\right]_0^2 = \frac{256}{15}$$

18. $V = \int_0^2 \int_0^2 \int_0^{2xy} dz\,dy\,dx = \int_0^2 \int_0^2 2xy\,dy\,dx = \int_0^2 \left[xy^2\right]_0^2 dx = \int_0^2 4x\,dx = 8$

19. $V = 8\int_0^a \int_0^{\sqrt{a^2-x^2}} \int_0^{\sqrt{a^2-x^2-y^2}} dz\,dy\,dx = 8\int_0^a \int_0^{\sqrt{a^2-x^2}} \sqrt{a^2 - x^2 - y^2}\,dy\,dx$

$$= 4\int_0^a \left[y\sqrt{a^2 - x^2 - y^2} + \left(a^2 - x^2\right)\arcsin\left(\frac{y}{\sqrt{a^2 - x^2}}\right)\right]_0^{\sqrt{a^2-x^2}} dx$$

$$= 4\left(\frac{\pi}{2}\right)\int_0^a \left(a^2 - x^2\right) dx = \left[2\pi\left(a^2x - \frac{1}{3}x^3\right)\right]_0^a = \frac{4}{3}\pi a^3$$

20. $V = 4\int_0^6 \int_0^{\sqrt{36-x^2}} \int_0^{36-x^2-y^2} dz\, dy\, dx = 4\int_0^6 \int_0^{\sqrt{36-x^2}} \left(36 - x^2 - y^2\right) dy\, dx = 4\int_0^6 \left[36y - x^2y - \frac{y^3}{3}\right]_0^{\sqrt{36-x^2}} dx$

$$= 4\int_0^6 \left[36\sqrt{36 - x^2} - x^2\sqrt{36 - x^2} - \frac{1}{3}\left(36 - x^2\right)^{3/2}\right] dx$$

$$= 4\left[9x\sqrt{36 - x^2} + 324 \arcsin\left(\frac{x}{6}\right) + \frac{1}{6}x\left(36 - x^2\right)^{3/2}\right]_0^6 = 4(162\pi) = 648\pi$$

21. $V = \int_0^2 \int_0^{4-x^2} \int_0^{4-x^2} dz\, dy\, dx = \int_0^2 \left(4 - x^2\right)^2 dx = \int_0^2 \left(16 - 8x^2 + x^4\right) dx = \left[16x - \frac{8}{3}x^3 + \frac{1}{5}x^5\right]_0^2 = \frac{256}{15}$

22. $V = \int_0^{\sqrt{2}} \int_0^{2-x^2} \int_0^{9-x^3} dz\, dy\, dx$

$$= \int_0^{\sqrt{2}} \int_0^{2-x^2} \left(9 - x^3\right) dy\, dx$$

$$= \int_0^{\sqrt{2}} \left(9 - x^3\right)\left(2 - x^2\right) dx$$

$$= \int_0^{\sqrt{2}} \left(18 - 9x^2 - 2x^3 + x^5\right) dx$$

$$= \left[18x - 3x^3 - \frac{1}{2}x^4 + \frac{x^6}{6}\right]_0^{\sqrt{2}}$$

$$= 18\sqrt{2} - 6\sqrt{2} - 2 + \frac{4}{3} = 12\sqrt{2} - \frac{2}{3}$$

23. $V = \int_0^3 \int_0^2 \int_{2-y}^{4-y^2} dz\, dy\, dx$

$$= \int_0^3 \int_0^2 \left[4 - y^2 - 2 + y\right] dy\, dx$$

$$= \int_0^3 \left[2y - \frac{y^3}{3} + \frac{y^2}{2}\right]_0^2 dx$$

$$= \int_0^3 \left(4 - \frac{8}{3} + 2\right) dx$$

$$= \left[\frac{10}{3}x\right]_0^3 = 10$$

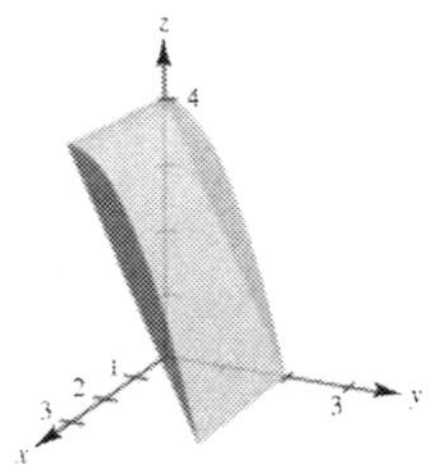

24. The region in the xy-plane is:

$$V = \int_0^2 \int_{x^2}^{x+2} \int_0^x dz\, dy\, dx = \int_0^2 \int_{x^2}^{x+2} x\, dy\, dx$$

$$= \int_0^2 \left[xy\right]_{x^2}^{x+2} dx = \int_0^2 \left(x(x + 2) - x^3\right) dx$$

$$= \left[\frac{x^3}{3} + x^2 - \frac{x^4}{4}\right]_0^2 = \frac{8}{3} + 4 - 4 = \frac{8}{3}$$

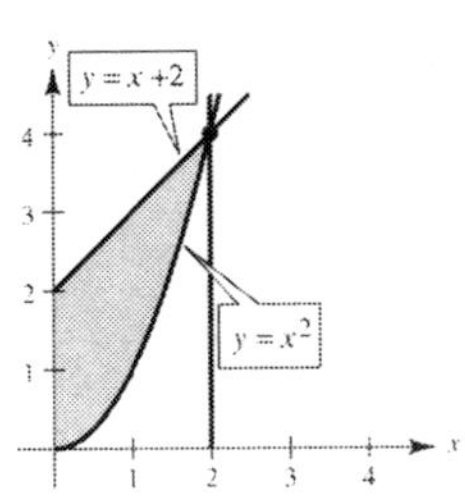

25.

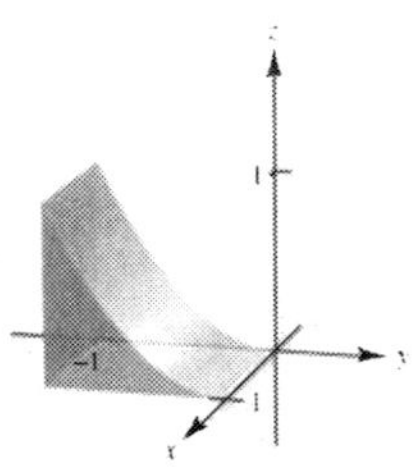

$$\int_0^1 \int_0^1 \int_{-1}^{-\sqrt{z}} dy\, dz\, dx$$

26.

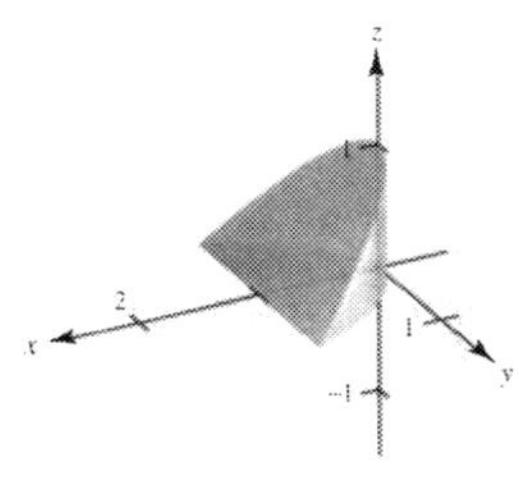

$$\int_{-1}^1 \int_0^{1-y^2} \int_{y^2}^{1-z} dx\, dz\, dy$$

27. Plane: $3x + 6y + 4z = 12$

$$\int_0^3 \int_0^{(12-4z)/3} \int_0^{(12-4z-3x)/6} dy\, dx\, dz$$

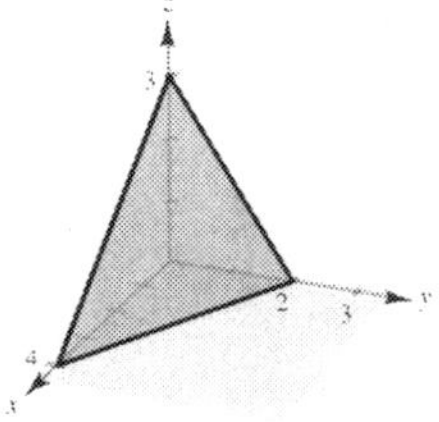

28. Top plane: $x + y + z = 6$

Side cylinder: $x^2 + y^2 = 9$

$$\int_0^3 \int_0^{\sqrt{9-y^2}} \int_0^{6-x-y} dz\, dx\, dy$$

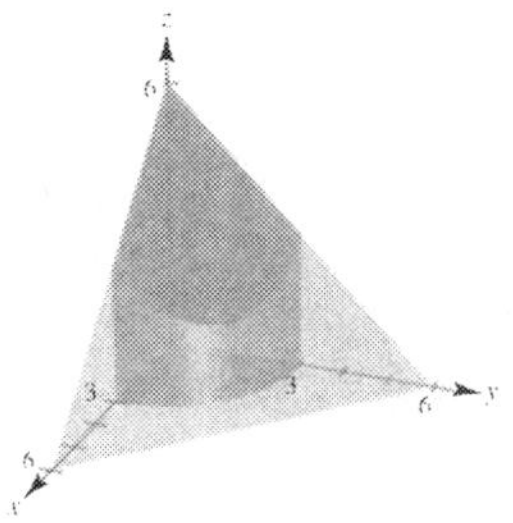

29. Top cylinder: $y^2 + z^2 = 1$

Side plane: $x = y$

$$\int_0^1 \int_0^x \int_0^{\sqrt{1-y^2}} dz\, dy\, dx$$

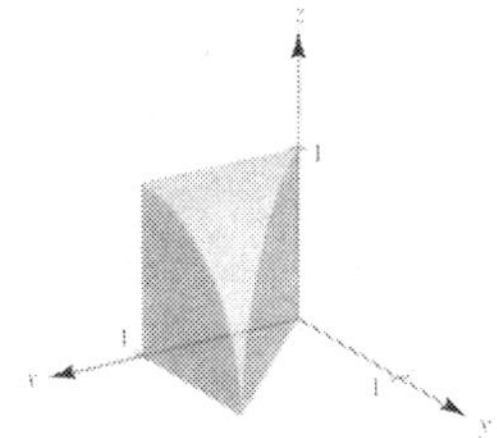

30. Elliptic cone: $4x^2 + z^2 = y^2$

$$\int_0^4 \int_z^4 \int_0^{\sqrt{y^2-z^2}/2} dx\, dy\, dz$$

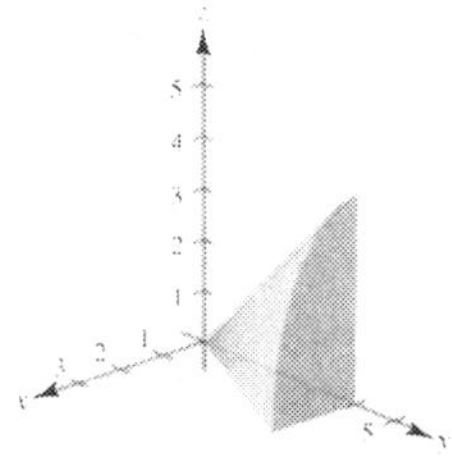

31. $Q = \{(x, y, z): 0 \le x \le 1, 0 \le y \le x, 0 \le z \le 3\}$

$$\iiint_Q xyz\, dV = \int_0^3 \int_0^1 \int_y^1 xyz\, dx\, dy\, dz = \int_0^3 \int_0^1 \int_0^x xyz\, dy\, dx\, dz$$

$$= \int_0^1 \int_0^3 \int_y^1 xyz\, dx\, dz\, dy = \int_0^1 \int_0^3 \int_0^x xyz\, dy\, dz\, dx$$

$$= \int_0^1 \int_y^1 \int_0^3 xyz\, dz\, dx\, dy = \int_0^1 \int_0^x \int_0^3 xyz\, dz\, dy\, dx \left(= \frac{9}{16}\right)$$

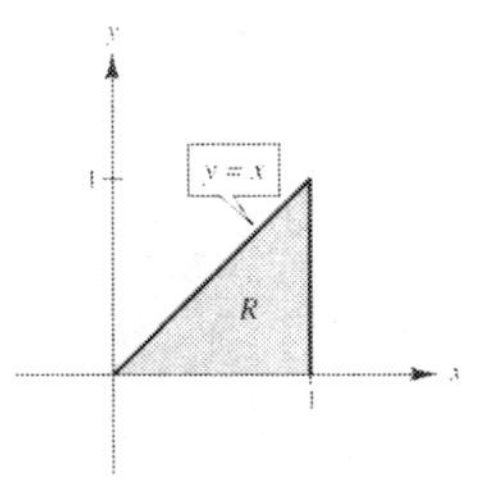

32. $Q = \{(x, y, z): 0 \le x \le 2, x^2 \le y \le 4, 0 \le z \le 2 - x\}$

$$\iiint_Q xyz\, dV = \int_0^2 \int_{x^2}^4 \int_0^{2-x} xyz\, dz\, dy\, dx$$

$$= \int_0^4 \int_0^{\sqrt{y}} \int_0^{2-x} xyz\, dz\, dx\, dy$$

$$= \int_0^2 \int_0^{2-x} \int_{x^2}^4 xyz\, dy\, dz\, dx$$

$$= \int_0^2 \int_0^{2-z} \int_{x^2}^4 xyz\, dy\, dx\, dz$$

$$= \int_0^2 \int_0^{(2-z)^2} \int_0^{\sqrt{y}} xyz\, dx\, dy\, dz + \int_0^2 \int_{(2-z)^2}^4 \int_0^{2-z} xyz\, dx\, dy\, dz$$

$$= \int_0^4 \int_0^{2-\sqrt{y}} \int_0^{\sqrt{y}} xyz\, dx\, dz\, dy + \int_0^4 \int_{2-\sqrt{y}}^2 \int_0^{2-z} dx\, dz\, dy \left(= \frac{104}{21}\right)$$

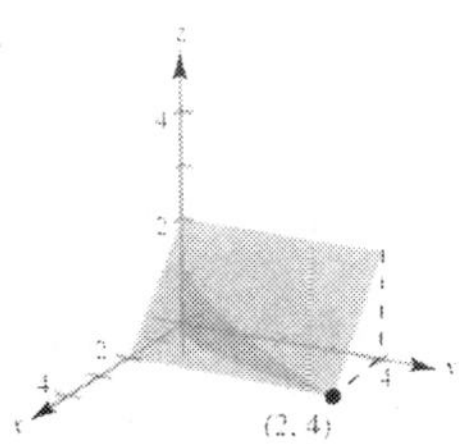

33. $Q = \{(x, y, z): x^2 + y^2 \le 9, 0 \le z \le 4\}$

$$\iiint_Q xyz\,dV = \int_0^4\int_{-3}^3\int_{-\sqrt{9-x^2}}^{\sqrt{9-x^2}} xyz\,dy\,dx\,dz = \int_0^4\int_{-3}^3\int_{-\sqrt{9-y^2}}^{\sqrt{9-y^2}} xyz\,dx\,dy\,dz$$

$$= \int_{-3}^3\int_0^4\int_{-\sqrt{9-y^2}}^{\sqrt{9-y^2}} xyz\,dx\,dz\,dy = \int_{-3}^3\int_{-\sqrt{9-y^2}}^{\sqrt{9-y^2}}\int_0^4 xyz\,dz\,dx\,dy$$

$$= \int_{-3}^3\int_0^4\int_{-\sqrt{9-x^2}}^{\sqrt{9-x^2}} xyz\,dy\,dz\,dx = \int_{-3}^3\int_{-\sqrt{9-x^2}}^{\sqrt{9-x^2}}\int_0^4 xyz\,dz\,dy\,dx\ (= 0)$$

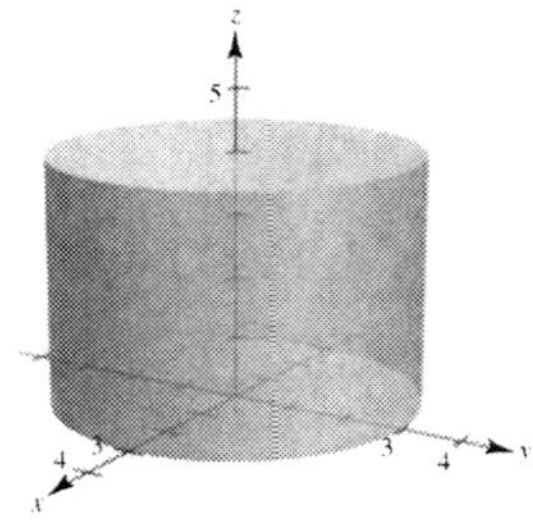

34. $Q = \{(x, y, z): 0 \le x \le 1, y \le 1 - x^2, 0 \le z \le 6\}$

$$\iiint_Q xyz\,dV = \int_0^1\int_0^{1-x^2}\int_0^6 xyz\,dz\,dy\,dx = \int_0^1\int_0^{\sqrt{1-y}}\int_0^6 xyz\,dz\,dx\,dy$$

$$= \int_0^1\int_0^6\int_0^{\sqrt{1-y}} xyz\,dx\,dz\,dy = \int_0^6\int_0^1\int_0^{\sqrt{1-y}} xyz\,dx\,dy\,dz$$

$$= \int_0^1\int_0^6\int_0^{1-x^2} xyz\,dy\,dz\,dx = \int_0^6\int_0^1\int_0^{1-x^2} xyz\,dy\,dx\,dz = \frac{3}{2}$$

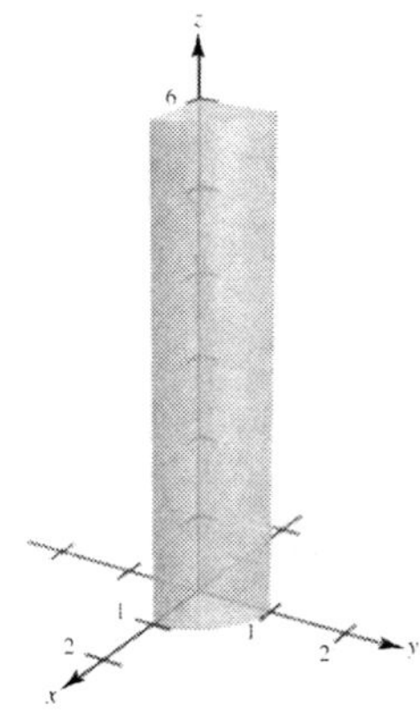

35. $Q = \{(x, y, z): 0 \le y \le 1, 0 \le x \le 1 - y^2, 0 \le z \le 1 - y\}$

$$\int_0^1\int_0^{1-y^2}\int_0^{1-y} dz\,dx\,dy = \int_0^1\int_0^{\sqrt{1-x}}\int_0^{1-y} dz\,dy\,dx$$

$$= \int_0^1\int_0^{2z-z^2}\int_0^{1-z} dy\,dx\,dz + \int_0^1\int_{2z-z^2}^1\int_0^{\sqrt{1-x}} dy\,dx\,dz$$

$$= \int_0^1\int_{1-\sqrt{1-x}}^1\int_0^{1-z} dy\,dz\,dx + \int_0^1\int_0^{1-\sqrt{1-x}}\int_0^{\sqrt{1-x}} dy\,dz\,dx$$

$$= \int_0^1\int_0^{1-y}\int_0^{1-y^2} dx\,dz\,dy = \int_0^1\int_0^{1-z}\int_0^{1-y^2} dx\,dy\,dz = \frac{5}{12}$$

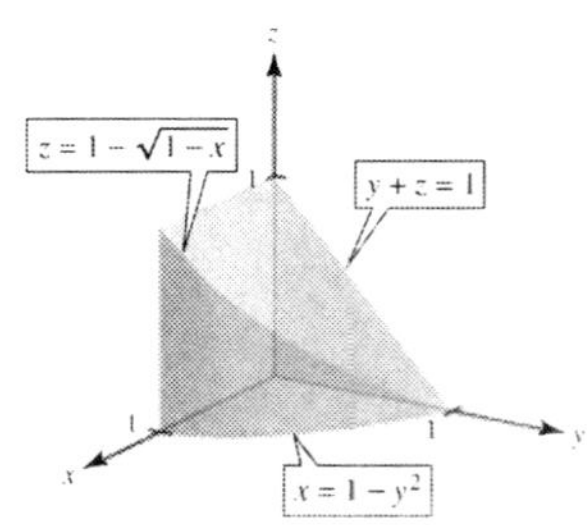

36. $Q = \{(x, y, z): 0 \le x \le 3, 0 \le y \le x, 0 \le z \le 9 - x^2\}$

$$\int_0^3\int_0^x\int_0^{9-x^2} dz\,dy\,dx = \int_0^3\int_y^3\int_0^{9-x^2} dz\,dx\,dy$$

$$= \int_0^3\int_0^{9-x^2}\int_0^x dy\,dz\,dx = \int_0^9\int_0^{\sqrt{9-z}}\int_0^x dy\,dx\,dz$$

$$= \int_0^9\int_0^{\sqrt{9-z}}\int_y^{\sqrt{9-z}} dx\,dy\,dz = \int_0^3\int_0^{9-y^2}\int_y^{\sqrt{9-z}} dx\,dz\,dy = \frac{81}{4}$$

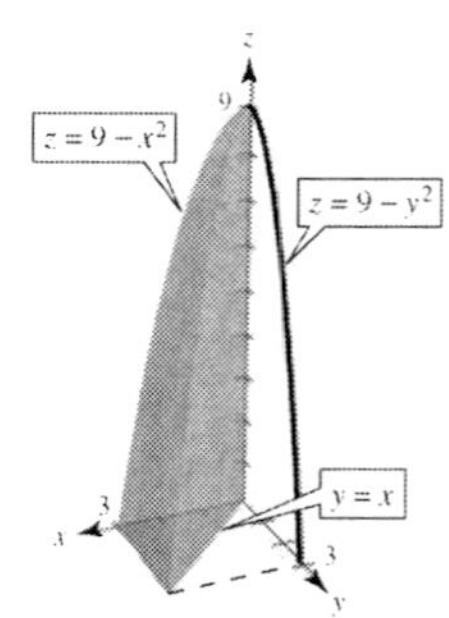

37.
$$m = k\int_0^6\int_0^{4-(2x/3)}\int_0^{2-(y/2)-(x/3)} dz\,dy\,dx = 8k$$

$$M_{yz} = k\int_0^6\int_0^{4-(2x/3)}\int_0^{2-(y/2)-(x/3)} x\,dz\,dy\,dx = 12k$$

$$\bar{x} = \frac{M_{yz}}{m} = \frac{12k}{8k} = \frac{3}{2}$$

38.
$$m = k\int_0^5\int_0^{5-x}\int_0^{1/5(15-3x-3y)} y\,dz\,dy\,dx = \frac{125}{8}k$$

$$M_{xz} = k\int_0^5\int_0^{5-x}\int_0^{1/5(15-3x-3y)} y^2\,dz\,dy\,dx = \frac{125}{4}k$$

$$\bar{y} = \frac{M_{xz}}{m} = 2$$

39. $m = k\int_0^4\int_0^4\int_0^{4-x} x\,dz\,dy\,dx = k\int_0^4\int_0^4 x(4-x)\,dy\,dx$

$= 4k\int_0^4 \left(4x - x^2\right)dx = \frac{128k}{3}$

$M_{xy} = k\int_0^4\int_0^4\int_0^{4-x} xz\,dz\,dy\,dx = k\int_0^4\int_0^4 x\frac{(4-x)^2}{2}\,dy\,dx$

$= 2k\int_0^4 \left(16x - 8x^2 + x^3\right)dx = \frac{128k}{3}$

$\bar{z} = \frac{M_{xy}}{m} = 1$

40. $m = k\int_0^b\int_0^{a[1-(y/b)]}\int_0^{c[1-(y/b)-(x/a)]} dz\,dx\,dy = \frac{kabc}{6}$

$M_{xz} = k\int_0^b\int_0^{a[1-(y/b)]}\int_0^{c[1-(y/b)-(x/a)]} y\,dz\,dx\,dy = \frac{kab^2c}{24}$

$\bar{y} = \frac{M_{xz}}{m} = \frac{kab^2c/24}{kabc/6} = \frac{b}{4}$

41. $m = k\int_0^b\int_0^b\int_0^b xy\,dz\,dy\,dx = \frac{kb^5}{4}$

$M_{yz} = k\int_0^b\int_0^b\int_0^b x^2y\,dz\,dy\,dx = \frac{kb^6}{6}$

$M_{xz} = k\int_0^b\int_0^b\int_0^b xy^2\,dz\,dy\,dx = \frac{kb^6}{6}$

$M_{xy} = k\int_0^b\int_0^b\int_0^b xyz\,dz\,dy\,dx = \frac{kb^6}{8}$

$\bar{x} = \frac{M_{yz}}{m} = \frac{kb^6/6}{kb^5/4} = \frac{2b}{3}$

$\bar{y} = \frac{M_{xz}}{m} = \frac{kb^6/6}{kb^5/4} = \frac{2b}{3}$

$\bar{z} = \frac{M_{xy}}{m} = \frac{kb^6/8}{kb^5/4} = \frac{b}{2}$

42. $m = k\int_0^a\int_0^b\int_0^c z\,dz\,dy\,dx = \frac{kabc^2}{2}$

$M_{xy} = k\int_0^a\int_0^b\int_0^c z^2\,dz\,dy\,dx = \frac{kabc^3}{3}$

$M_{yz} = k\int_0^a\int_0^b\int_0^c xz\,dz\,dy\,dx = \frac{ka^2bc^2}{4}$

$M_{xz} = k\int_0^a\int_0^b\int_0^c yz\,dz\,dy\,dx = \frac{kab^2c^2}{4}$

$\bar{x} = \frac{M_{yz}}{m} = \frac{ka^2bc^2/4}{kabc^2/2} = \frac{a}{2}$

$\bar{y} = \frac{M_{xz}}{m} = \frac{kab^2c^2/4}{kabc^2/2} = \frac{b}{2}$

$\bar{z} = \frac{M_{xy}}{m} = \frac{kabc^3/3}{kabc^2/2} = \frac{2c}{3}$

43. $\bar{x}$ will be greater than 2, whereas $\bar{y}$ and $\bar{z}$ will be unchanged.

44. $\bar{z}$ will be greater than $8/5$, whereas $\bar{x}$ and $\bar{y}$ will be unchanged.

45. $\bar{y}$ will be greater than 0, whereas $\bar{x}$ and $\bar{z}$ will be unchanged.

46. $\bar{x}$, $\bar{y}$ and $\bar{z}$ will all be greater than their original values.

47. $m = \frac{1}{3}k\pi r^2h$

$\bar{x} = \bar{y} = 0$ by symmetry

$M_{xy} = 4k\int_0^r\int_0^{\sqrt{r^2-x^2}}\int_{h\sqrt{x^2+y^2}/r}^{h} z\,dz\,dy\,dx$

$= \frac{2kh^2}{r^2}\int_0^r\int_0^{\sqrt{r^2-x^2}} \left(r^2 - x^2 - y^2\right)dy\,dx$

$= \frac{4kh^2}{3r^2}\int_0^r \left(r^2 - x^2\right)^{3/2}dx = \frac{k\pi r^2h^2}{4}$

$\bar{z} = \frac{M_{xy}}{m} = \frac{k\pi r^2h^2/4}{k\pi r^2h/3} = \frac{3h}{4}$

$(\bar{x}, \bar{y}, \bar{z}) = \left(0, 0, \frac{3h}{4}\right)$

48. $m = 2k\int_0^3\int_0^{\sqrt{9-x^2}}\int_0^y dz\,dy\,dx = 18k$

$M_{yz} = k\int_{-3}^3\int_0^{\sqrt{9-x^2}}\int_0^y x\,dz\,dy\,dx = 0$

$M_{xz} = k\int_{-3}^3\int_0^{\sqrt{9-x^2}}\int_0^y y\,dz\,dy\,dx = \frac{81\pi}{8}k$

$M_{xy} = k\int_{-3}^3\int_0^{\sqrt{9-x^2}}\int_0^y z\,dz\,dy\,dx = \frac{81\pi}{16}k$

$\bar{x} = \frac{M_{yz}}{m} = 0$

$\bar{y} = \frac{M_{xz}}{m} = \frac{9\pi}{16}$

$\bar{z} = \frac{M_{xy}}{m} = \frac{9\pi}{32}$

$(\bar{x}, \bar{y}, \bar{z}) = \left(0, \frac{9\pi}{16}, \frac{9\pi}{32}\right)$

49.
$$m = \frac{128k\pi}{3}$$
$$\bar{x} = \bar{y} = 0 \text{ by symmetry}$$
$$z = \sqrt{4^2 - x^2 - y^2}$$
$$M_{xy} = 4k\int_0^4 \int_0^{\sqrt{4^2-x^2}} \int_0^{\sqrt{4^2-x^2-y^2}} z\,dz\,dy\,dx$$
$$= 2k\int_0^4 \int_0^{\sqrt{4^2-x^2}} \left(4^2 - x^2 - y^2\right) dy\,dx = 2k\int_0^4 \left[16y - x^2y - \frac{1}{3}y^3\right]_0^{\sqrt{4^2-x^2}} dx$$
$$= \frac{4k}{3}\int_0^4 \left(4^2 - x^2\right)^{3/2} dx$$
$$= \frac{1024k}{3}\int_0^{\pi/2} \cos^4\theta\,d\theta \quad (\text{let } x = 4\sin\theta)$$
$$= 64\pi k \quad \text{by Wallis's Formula}$$
$$\bar{z} = \frac{M_{xy}}{m} = \frac{64k\pi}{1}\cdot\frac{3}{128k\pi} = \frac{3}{2}$$
$$(\bar{x}, \bar{y}, \bar{z}) = \left(0, 0, \frac{3}{2}\right)$$

50.
$$\bar{x} = 0$$
$$m = 2k\int_0^2 \int_0^1 \int_0^{1/(y^2+1)} dz\,dy\,dx = 2k\int_0^2 \int_0^1 \frac{1}{y^2+1}\,dy\,dx = 2k\left(\frac{\pi}{4}\right)\int_0^2 dx = k\pi$$
$$M_{xz} = 2k\int_0^2 \int_0^1 \int_0^{1/(y^2+1)} y\,dz\,dy\,dx = 2k\int_0^2 \int_0^1 \frac{y}{y^2+1}\,dy\,dx = k\int_0^2 (\ln 2)\,dx = k\ln 4$$
$$M_{xy} = 2k\int_0^2 \int_0^1 \int_0^{1/(y^2+1)} z\,dz\,dy\,dx$$
$$= k\int_0^2 \int_0^1 \frac{1}{\left(y^2+1\right)^2}\,dy\,dx = k\int_0^2 \left[\frac{y}{2(y^2+1)} + \frac{1}{2}\arctan y\right]_0^1 dx = k\left(\frac{1}{4} + \frac{\pi}{8}\right)\int_0^2 dx = k\left(\frac{1}{2} + \frac{\pi}{4}\right)$$
$$\bar{y} = \frac{M_{xz}}{m} = \frac{k\ln 4}{k\pi} = \frac{\ln 4}{\pi}$$
$$\bar{z} = \frac{M_{xy}}{m} = k\left(\frac{1}{2} + \frac{\pi}{4}\right)\Big/k\pi = \frac{2+\pi}{4\pi}$$
$$(\bar{x}, \bar{y}, \bar{z}) = \left(0, \frac{\ln 4}{\pi}, \frac{2+\pi}{4\pi}\right)$$

51. $f(x, y) = \frac{5}{12}y$

$$m = k\int_0^{20} \int_0^{-(3/5)x+12} \int_0^{(5/12)y} dz\,dy\,dx = 200k$$
$$M_{yz} = k\int_0^{20} \int_0^{-(3/5)x+12} \int_0^{(5/12)y} x\,dz\,dy\,dx = 1000k$$
$$M_{xz} = k\int_0^{20} \int_0^{-(3/5)x+12} \int_0^{(5/12)y} y\,dz\,dy\,dx = 1200k$$
$$M_{xy} = k\int_0^{20} \int_0^{-(3/5)x+12} \int_0^{(5/12)y} z\,dz\,dy\,dx = 250k$$

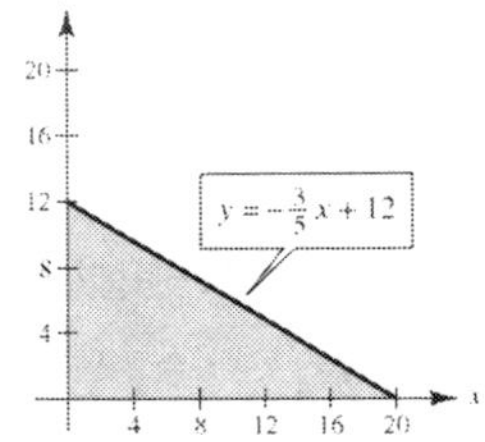

$$\bar{x} = \frac{M_{yz}}{m} = \frac{1000k}{200k} = 5$$
$$\bar{y} = \frac{M_{xz}}{m} = \frac{1200k}{200k} = 6$$
$$\bar{z} = \frac{M_{xy}}{m} = \frac{250k}{200k} = \frac{5}{4}$$
$$(\bar{x}, \bar{y}, \bar{z}) = \left(5, 6, \frac{5}{4}\right)$$

52. $f(x, y) = \frac{1}{15}(60 - 12x - 20y)$

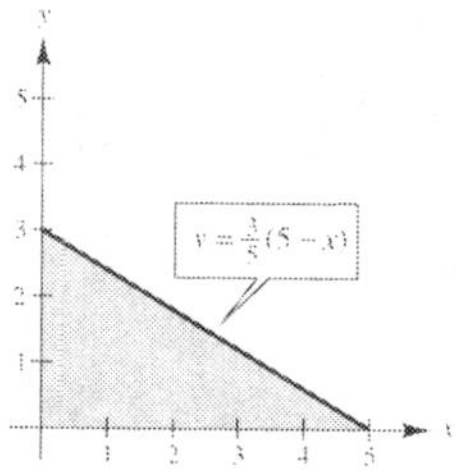

$$m = k\int_0^5 \int_0^{-(3/5)x+3} \int_0^{(1/15)(60-12x-20y)} dz\,dy\,dx = 10k$$

$$M_{yz} = k\int_0^5 \int_0^{-(3/5)x+3} \int_0^{(1/15)(60-12x-20y)} x\,dz\,dy\,dx = \frac{25k}{2}$$

$$M_{xz} = k\int_0^5 \int_0^{-(3/5)x+3} \int_0^{-(1/15)(60-12x-20y)} y\,dz\,dy\,dx = \frac{15k}{2}$$

$$M_{xy} = k\int_0^5 \int_0^{-(3/5)x+3} \int_0^{(1/15)(60-12x-20y)} z\,dz\,dy\,dx = 10k$$

$$\bar{x} = \frac{M_{yz}}{m} = \frac{25k/2}{10k} = \frac{5}{4}$$

$$\bar{y} = \frac{M_{xz}}{m} = \frac{15k/2}{10k} = \frac{3}{4}$$

$$\bar{z} = \frac{M_{xy}}{m} = \frac{10k}{10k} = 1$$

$$(\bar{x}, \bar{y}, \bar{z}) = \left(\frac{5}{4}, \frac{3}{4}, 1\right)$$

53. (a) $I_x = k\int_0^a \int_0^a \int_0^a (y^2 + z^2)\,dx\,dy\,dz = ka\int_0^a \int_0^a (y^2 + z^2)\,dy\,dz$

$$= ka\int_0^a \left[\frac{1}{3}y^3 + z^2 y\right]_0^a dz = ka\int_0^a \left(\frac{1}{3}a^3 + az^2\right) dz = \left[ka\left(\frac{1}{3}a^3 z + \frac{1}{3}az^3\right)\right]_0^a = \frac{2ka^5}{3}$$

$I_x = I_y = I_z = \frac{2ka^5}{3}$ by symmetry

(b) $I_x = k\int_0^a \int_0^a \int_0^a (y^2 + z^2)\,xyz\,dx\,dy\,dz = \frac{ka^2}{2}\int_0^a \int_0^a (y^3 z + yz^3)\,dy\,dz$

$$= \frac{ka^2}{2}\int_0^a \left[\frac{y^4 z}{4} + \frac{y^2 z^3}{2}\right]_0^a dz = \frac{ka^4}{8}\int_0^a (a^2 z + 2z^3)\,dz = \left[\frac{ka^4}{8}\left(\frac{a^2 z^2}{2} + \frac{2z^4}{4}\right)\right]_0^a = \frac{ka^8}{8}$$

$I_x = I_y = I_z = \frac{ka^8}{8}$ by symmetry

54. (a) $I_{xy} = k\int_{-a/2}^{a/2} \int_{-a/2}^{a/2} \int_{-a/2}^{a/2} z^2\,dz\,dy\,dx = \frac{ka^5}{12}$

$I_{xz} = I_{yz} = \frac{ka^5}{12}$ by symmetry

$I_x = I_y = I_z = \frac{ka^5}{12} + \frac{ka^5}{12} = \frac{ka^5}{6}$

(b) $I_{xy} = k\int_{-a/2}^{a/2} \int_{-a/2}^{a/2} \int_{-a/2}^{a/2} z^2(x^2 + y^2)\,dz\,dy\,dx = \frac{a^3 k}{12}\int_{-a/2}^{a/2} \int_{-a/2}^{a/2} (x^2 + y^2)\,dy\,dx = \frac{a^7 k}{72}$

$I_{xy} = k\int_{-a/2}^{a/2} \int_{-a/2}^{a/2} \int_{-a/2}^{a/2} y^2(x^2 + y^2)\,dz\,dy\,dx = ka\int_{-a/2}^{a/2} \int_{-a/2}^{a/2} (x^2 y^2 + y^4)\,dy\,dx = \frac{7ka^7}{360}$

$I_{yz} = I_{xz}$ by symmetry

$I_x = I_{xy} + I_{xz} = \frac{a^7 k}{30}$

$I_y = I_{xy} + I_{yz} = \frac{a^7 k}{30}$

$I_z = I_{yz} + I_{xz} = \frac{7ka^7}{180}$

55. (a) $I_x = k\int_0^4\int_0^4\int_0^{4-x}\left(y^2 + z^2\right)dz\,dy\,dx = k\int_0^4\int_0^4\left[y^2(4-x) + \frac{1}{3}(4-x)^3\right]dy\,dx$

$= k\int_0^4\left[\frac{y^3}{3}(4-x) + \frac{y}{3}(4-x)^3\right]_0^4 dx = k\int_0^4\left[\frac{64}{3}(4-x) + \frac{4}{3}(4-x)^3\right]dx = k\left[-\frac{32}{3}(4-x)^2 - \frac{1}{3}(4-x)^4\right]_0^4 = 256k$

$I_y = k\int_0^4\int_0^4\int_0^{4-x}\left(x^2 + z^2\right)dz\,dy\,dx = k\int_0^4\int_0^4\left[x^2(4-x) + \frac{1}{3}(4-x)^3\right]dy\,dx$

$= 4k\int_0^4\left[4x^2 - x^3 + \frac{1}{3}(4-x)^3\right]dx = 4k\left[\frac{4}{3}x^3 - \frac{1}{4}x^4 - \frac{1}{12}(4-x)^4\right]_0^4 = \frac{512k}{3}$

$I_z = k\int_0^4\int_0^4\int_0^{4-x}\left(x^2 + y^2\right)dz\,dy\,dx = k\int_0^4\int_0^4\left(x^2 + y^2\right)(4-x)\,dy\,dx$

$= k\int_0^4\left[\left(x^2y + \frac{y^3}{3}\right)(4-x)\right]_0^4 dx = k\int_0^4\left(4x^2 + \frac{64}{3}\right)(4-x)\,dx = 256k$

(b) $I_x = k\int_0^4\int_0^4\int_0^{4-x} y\left(y^2 + z^2\right)dz\,dy\,dx = k\int_0^4\int_0^4\left[y^3(4-x) + \frac{1}{3}y(4-x)^3\right]dy\,dx$

$= k\int_0^4\left[\frac{y^4}{4}(4-x) + \frac{y^2}{6}(4-x)^3\right]_0^4 dx = k\int_0^4\left[64(4-x) + \frac{8}{3}(4-x)^3\right]dx = k\left[-32(4-x)^2 - \frac{2}{3}(4-x)^4\right]_0^4 = \frac{2048k}{3}$

$I_y = k\int_0^4\int_0^4\int_0^{4-x} y\left(x^2 + z^2\right)dz\,dy\,dx = k\int_0^4\int_0^4\left[x^2y(4-x) + \frac{1}{3}y(4-x)^3\right]dy\,dx$

$= 8k\int_0^4\left[4x^2 - x^3 + \frac{1}{3}(4-x)^3\right]dx = 8k\left[\frac{4}{3}x^3 - \frac{1}{4}x^4 - \frac{1}{12}(4-x)^4\right]_0^4 = \frac{1024k}{3}$

$I_z = k\int_0^4\int_0^4\int_0^{4-x} y\left(x^2 + y^2\right)dz\,dy\,dx = k\int_0^4\int_0^4\left(x^2y + y^3\right)(4-x)\,dx$

$= k\int_0^4\left[\left(\frac{x^2y^2}{2} + \frac{y^4}{4}\right)(4-x)\right]_0^4 dx = k\int_0^4\left(8x^2 + 64\right)(4-x)\,dx$

$= 8k\int_0^4\left(32 - 8x + 4x^2 - x^3\right)dx = \left[8k\left(32x - 4x^2 + \frac{4}{3}x^3 - \frac{1}{4}x^4\right)\right]_0^4 = \frac{2048k}{3}$

56. (a) $I_{xy} = k\int_0^4\int_0^2\int_0^{4-y^2} z^3\,dz\,dy\,dx = k\int_0^4\int_0^2\frac{1}{4}\left(4-y^2\right)^4 dy\,dx = \frac{k}{4}\int_0^4\int_0^2\left(256 - 256y^2 + 96y^4 - 16y^6 + y^8\right)dy\,dx$

$= \frac{k}{4}\int_0^4\left[256y - \frac{256y^3}{3} + \frac{96y^5}{5} - \frac{16y^7}{7} + \frac{y^9}{9}\right]_0^2 dx = k\int_0^4\frac{16,384}{945}\,dx = \frac{65,536k}{315}$

$I_{xz} = k\int_0^4\int_0^2\int_0^{4-y^2} y^2z\,dz\,dy\,dx = k\int_0^4\int_0^2\frac{1}{2}y^2\left(4-y^2\right)^2 dy\,dx$

$= k\int_0^4\int_0^2\frac{1}{2}\left(16y^2 - 8y^4 + y^6\right)dy\,dx = \frac{k}{2}\int_0^4\left[\frac{16y^3}{3} - \frac{8y^5}{5} + \frac{y^7}{7}\right]_0^2 dx = \frac{k}{2}\int_0^4\frac{1024}{105}\,dx = \frac{2048k}{105}$

$I_{yz} = k\int_0^4\int_0^2\int_0^{4-y^2} x^2z\,dz\,dy\,dx = k\int_0^4\int_0^2\frac{1}{2}x^2\left(4-y^2\right)^2 dy\,dx$

$= k\int_0^4\int_0^2\frac{1}{2}x^2\left(16 - 8y^2 + y^4\right)dy\,dx = \frac{k}{2}\int_0^4\left[x^2\left(16y - \frac{8y^3}{3} + \frac{y^5}{5}\right)\right]_0^2 dx = \frac{k}{2}\int_0^4\frac{256}{15}x^2\,dx = \frac{8192k}{45}$

$I_x = I_{xz} + I_{xy} = \frac{2048k}{9}, I_y = I_{yz} + I_{xy} = \frac{8192k}{21}, I_z = I_{yz} + I_{xz} = \frac{63,488k}{315}$

(b) $$I_{xy} = \int_0^4\int_0^2\int_0^{4-y^2} z^2(4 - z)\,dz\,dy\,dx$$

$$= k\int_0^4\int_0^2\int_0^{4-y^2} 4z^2\,dz\,dy\,dx - k\int_0^4\int_0^2\int_0^{4-y^2} z^3\,dz\,dy\,dx = \frac{32{,}768k}{105} - \frac{65{,}536k}{315} = \frac{32{,}768k}{315}$$

$$I_{xz} = \int_0^4\int_0^2\int_0^{4-y^2} y^2(4 - z)\,dz\,dy\,dx$$

$$= k\int_0^4\int_0^2\int_0^{4-y^2} 4y^2\,dz\,dy\,dx - k\int_0^4\int_0^2\int_0^{4-y^2} y^2z\,dz\,dy\,dx = \frac{1024k}{15} - \frac{2048k}{105} = \frac{1024k}{21}$$

$$I_{yz} = k\int_0^4\int_0^2\int_0^{4-y^2} x^2(4 - z)\,dz\,dy\,dx$$

$$= k\int_0^4\int_0^2\int_0^{4-y^2} 4x^2\,dz\,dy\,dx - k\int_0^4\int_0^2\int_0^{4-y^2} x^2z\,dz\,dy\,dx = \frac{4096k}{9} - \frac{8192k}{45} = \frac{4096k}{15}$$

$$I_x = I_{xz} + I_{xy} = \frac{48{,}128k}{315},\ I_y = I_{yz} + I_{xy} = \frac{118{,}784k}{315},\ I_z = I_{xz} + I_{yz} = \frac{11{,}264k}{35}$$

57. $$I_{xy} = k\int_{-L/2}^{L/2}\int_{-a}^{a}\int_{-\sqrt{a^2-x^2}}^{\sqrt{a^2-x^2}} z^2\,dz\,dx\,dy = k\int_{-L/2}^{L/2}\int_{-a}^{a}\frac{2}{3}(a^2 - x^2)\sqrt{a^2 - x^2}\,dx\,dy$$

$$= \frac{2}{3}\int_{-L/2}^{L/2} k\left[\frac{a^2}{2}\left(x\sqrt{a^2 - x^2} + a^2\arcsin\frac{x}{a}\right) - \frac{1}{8}\left(x(2x^2 - a^2)\sqrt{x^2 - a^2} + a^4\arcsin\frac{x}{a}\right)\right]_{-a}^{a} dy$$

$$= \frac{2k}{3}\int_{-L/2}^{L/2} 2\left(\frac{a^4\pi}{4} - \frac{a^4\pi}{16}\right) dy = \frac{a^4\pi Lk}{4}$$

Because $m = \pi a^2Lk$, $I_{xy} = ma^2/4$.

$$I_{xz} = k\int_{-L/2}^{L/2}\int_{-a}^{a}\int_{-\sqrt{a^2-x^2}}^{\sqrt{a^2-x^2}} y^2\,dz\,dx\,dy = 2k\int_{-L/2}^{L/2}\int_{-a}^{a} y^2\sqrt{a^2 - x^2}\,dx\,dy$$

$$= 2k\int_{-L/2}^{L/2}\left[\frac{y^2}{2}\left(x\sqrt{a^2 - x^2} + a^2\arcsin\frac{x}{a}\right)\right]_{-a}^{a} dy = k\pi a^2\int_{-L/2}^{L/2} y^2\,dy = \frac{2k\pi a^2}{3}\left(\frac{L^3}{8}\right) = \frac{1}{12}mL^2$$

$$I_{yz} = k\int_{-L/2}^{L/2}\int_{-a}^{a}\int_{-\sqrt{a^2-x^2}}^{\sqrt{a^2-x^2}} x^2\,dz\,dx\,dy = 2k\int_{-L/2}^{L/2}\int_{-a}^{a} x^2\sqrt{a^2 - x^2}\,dx\,dy$$

$$= 2k\int_{-L/2}^{L/2}\frac{1}{8}\left[x(2x^2 - a^2)\sqrt{a^2 - x^2} + a^4\arcsin\frac{x}{a}\right]_{-a}^{a} dy = \frac{ka^4\pi}{4}\int_{-L/2}^{L/2} dy = \frac{ka^4\pi L}{4} = \frac{ma^2}{4}$$

$$I_x = I_{xy} + I_{xz} = \frac{ma^2}{4} + \frac{mL^2}{12} = \frac{m}{12}(3a^2 + L^2)$$

$$I_y = I_{xy} + I_{yz} = \frac{ma^2}{4} + \frac{ma^2}{4} = \frac{ma^2}{2}$$

$$I_z = I_{xz} + I_{yz} = \frac{mL^2}{12} + \frac{ma^2}{4} = \frac{m}{12}(3a^2 + L^2)$$

58. $$I_{xy} = \int_{-c/2}^{c/2}\int_{-a/2}^{a/2}\int_{-b/2}^{b/2} z^2\,dz\,dy\,dx = \frac{b^3}{12}\int_{-c/2}^{c/2}\int_{-a/2}^{a/2} dy\,dx = \frac{1}{12}b^2(abc) = \frac{1}{12}mb^2$$

$$I_{xz} = \int_{-c/2}^{c/2}\int_{-a/2}^{a/2}\int_{-b/2}^{b/2} y^2\,dz\,dy\,dx = b\int_{-c/2}^{c/2}\int_{-a/2}^{a/2} y^2\,dy\,dx = \frac{ba^3}{12}\int_{-c/2}^{c/2} dx = \frac{ba^3c}{12} = \frac{1}{12}a^2(abc) = \frac{1}{12}ma^2$$

$$I_{yz} = \int_{-c/2}^{c/2}\int_{-a/2}^{a/2}\int_{-b/2}^{b/2} x^2\,dz\,dy\,dx = ab\int_{-c/2}^{c/2} x^2\,dx = \frac{abc^3}{12} = \frac{1}{12}c^2(abc) = \frac{1}{12}mc^2$$

$$I_x = I_{xy} + I_{xz} = \frac{1}{12}m(a^2 + b^2)$$

$$I_y = I_{xy} + I_{yz} = \frac{1}{12}m(b^2 + c^2)$$

$$I_z = I_{xz} + I_{yz} = \frac{1}{12}m(a^2 + c^2)$$

59. $\int_{-1}^{1}\int_{-1}^{1}\int_{0}^{1-x}\left(x^2 + y^2\right)\sqrt{x^2 + y^2 + z^2}\,dz\,dy\,dx$

60. $\int_{-1}^{1}\int_{-\sqrt{1-x^2}}^{\sqrt{1-x^2}}\int_{0}^{4-x^2-y^2} kx^2\left(x^2 + y^2\right)dz\,dy\,dx$

61. $\rho = kz$

(a) $m = \int_{-2}^{2}\int_{-\sqrt{4-x^2}}^{\sqrt{4-x^2}}\int_{0}^{4-x^2-y^2}(kz)\,dz\,dy\,dx\left(= \dfrac{32k\pi}{3}\right)$

(b) $\bar{x} = \bar{y} = 0$ by symmetry

$$\bar{z} = \frac{M_{xy}}{m} = \frac{1}{m}\int_{-2}^{2}\int_{-\sqrt{4-x^2}}^{\sqrt{4-x^2}}\int_{0}^{4-x^2-y^2} kz^2\,dz\,dy\,dx\ (= 2)$$

(c) $I_z = \int_{-2}^{2}\int_{-\sqrt{4-x^2}}^{\sqrt{4-x^2}}\int_{0}^{4-x^2-y^2}\left(x^2 + y^2\right)kz\,dz\,dy\,dx\left(= \dfrac{32k\pi}{3}\right)$

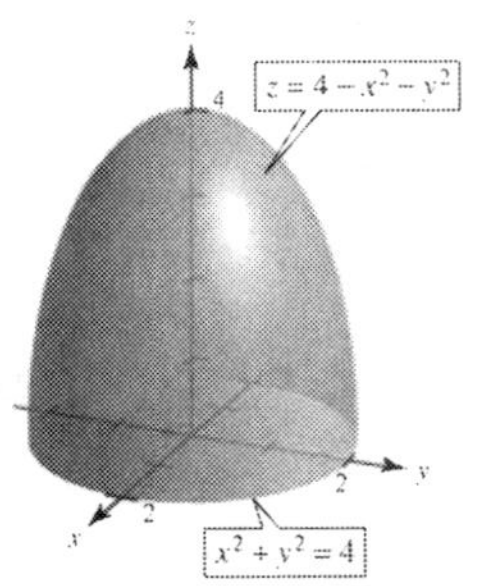

62. $\rho = kxy$

(a) $m = \int_{0}^{5}\int_{0}^{\sqrt{25-x^2}}\int_{0}^{\sqrt{25-x^2-y^2}} kxy\,dz\,dy\,dx\left(= \dfrac{625}{3}k\right)$

(b) $\bar{x} = \dfrac{M_{yz}}{m} = \dfrac{1}{m}\int_{0}^{5}\int_{0}^{\sqrt{25-x^2}}\int_{0}^{\sqrt{25-x^2-y^2}} x(kxy)\,dz\,dy\,dx\left(= \dfrac{25\pi}{32}\right)$

$\bar{y} = \bar{x}$ by symmetry

$$\bar{z} = \frac{M_{xy}}{m} = \frac{1}{m}\int_{0}^{5}\int_{0}^{\sqrt{25-x^2}}\int_{0}^{\sqrt{25-x^2-y^2}} z(kxy)\,dz\,dy\,dx\left(= \frac{25}{16}\right)$$

(c) $I_z = \int_{0}^{5}\int_{0}^{\sqrt{25-x^2}}\int_{0}^{\sqrt{25-x^2-y^2}}\left(x^2 + y^2\right)kxy\,dz\,dy\,dx\left(= \dfrac{62500}{21}k\right)$

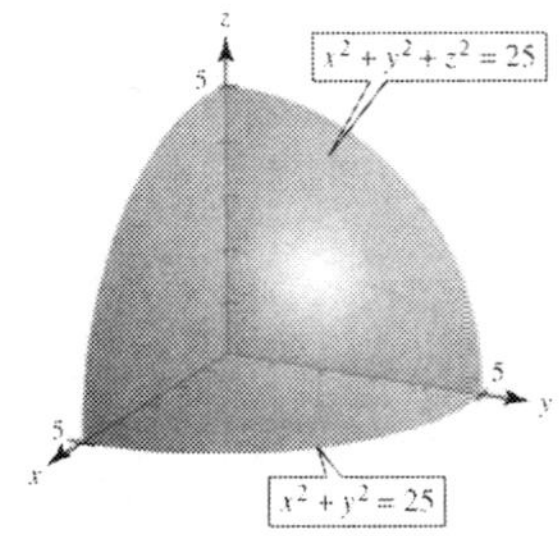

63. $V = 1$ (unit cube)

$$\begin{aligned}\text{Average value} &= \frac{1}{V}\iiint_Q f(x, y, z)\,dV\\ &= \int_{0}^{1}\int_{0}^{1}\int_{0}^{1}\left(z^2 + 4\right)dx\,dy\,dz\\ &= \int_{0}^{1}\int_{0}^{1}\left(z^2 + 4\right)dy\,dz = \int_{0}^{1}\left(z^2 + 4\right)dz\\ &= \left[\frac{z^3}{3} + 4z\right]_0^1 = \frac{1}{3} + 4 = \frac{13}{3}\end{aligned}$$

64. $V = 64$ (cube with sides of length 4)

$$\begin{aligned}\text{Average value} &= \frac{1}{V}\iiint_Q f(x, y, z)\,dV\\ &= \frac{1}{64}\int_{0}^{4}\int_{0}^{4}\int_{0}^{4} xyz\,dx\,dy\,dz\\ &= \frac{1}{64}\int_{0}^{4}\int_{0}^{4} 8yz\,dy\,dz\\ &= \frac{1}{8}\int_{0}^{4} 8z\,dz = \int_{0}^{4} z\,dz = 8\end{aligned}$$

65. $V = \dfrac{1}{3}\text{base} \times \text{height}$

$$= \frac{1}{3}\left(\frac{1}{2}(2)(2)\right)(2) = \frac{4}{3}$$

$f(x, y, z) = x + y + z$

Plane: $x + y + z = 2$

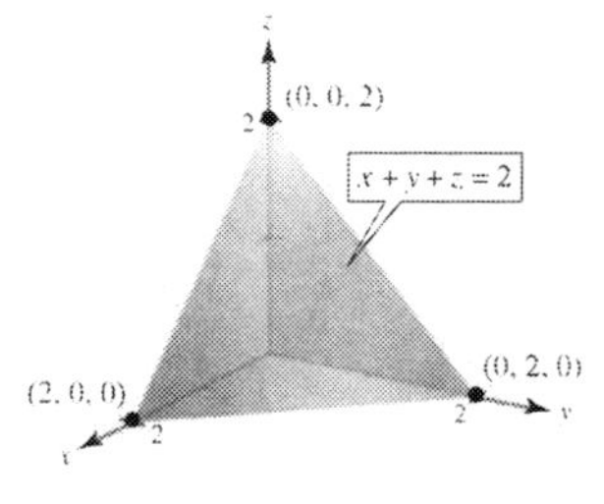

$$\begin{aligned}\text{Average value} &= \frac{1}{V}\iiint_Q f(x, y, z)\,dV\\ &= \frac{3}{4}\int_{0}^{2}\int_{0}^{2-x}\int_{0}^{2-x-y}(x + y + z)\,dz\,dy\,dx = \frac{3}{4}\int_{0}^{2}\int_{0}^{2-x}\frac{1}{2}(2 - x - y)(x + y + 2)\,dy\,dx\\ &= \frac{3}{4}\int_{0}^{2}\frac{1}{6}(x + 4)(x - 2)^2\,dx = \frac{3}{4}(2) = \frac{3}{2}\end{aligned}$$

66. $V = \frac{4}{3}\pi\left(\sqrt{3}\right)^3 = 4\sqrt{3}\pi$

Average value $= \frac{1}{V}\iiint_Q f(x, y, z)\,dV = \frac{1}{4\sqrt{3}\pi}\int_{-\sqrt{3}}^{\sqrt{3}}\int_{-\sqrt{3-x^2}}^{\sqrt{3-x^2}}\int_{-\sqrt{3-x^2-y^2}}^{\sqrt{3-x^2-y^2}}(x + y)\,dz\,dy\,dx = 0$, by symmetry

67. See the definition, pages 1009 and 1010.

See Theorem 14.4.

68. Because the density increases as you move away from the axis of symmetry, the moment of inertia will increase.

69. The region of integration is a cube:

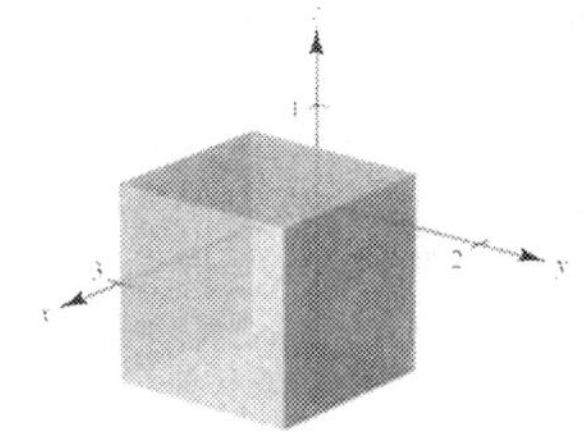

Answer: (b)

70. (a) Solid B has the greater density. Solid B has less volume, but equal weight, than solid A.

(b) Solid B has the greater moment of inertia.

(c) Solid A will reach the bottom first. Solid B has a greater resistance to rotational motion.

71. $1 - 2x^2 - y^2 - 3z^2 \ge 0$

$2x^2 + y^2 + 3z^2 \le 1$

$Q = \{(x, y, z): 2x^2 + y^2 + 3z^2 \le 1\}$ ellipsoid

$$\int_{-1/\sqrt{2}}^{1/\sqrt{2}}\int_{-\sqrt{1-2x^2}}^{\sqrt{1-2x^2}}\int_{-\sqrt{(1-2x^2-y^2)/3}}^{\sqrt{(1-2x^2-y^2)/3}}\left(1 - 2x^2 - y^2 - 3z^2\right)dz\,dy\,dx \approx 0.684$$

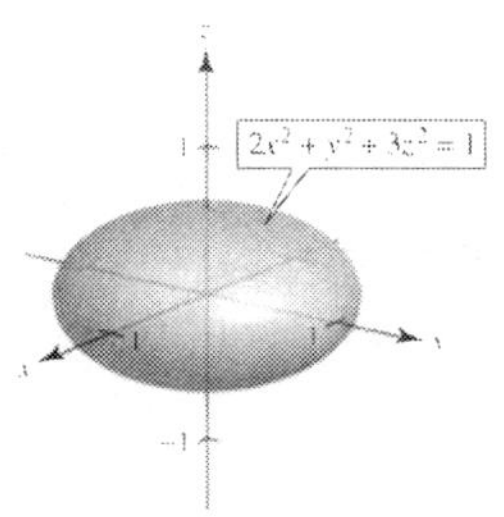

Exact value: $\frac{4\sqrt{6}\pi}{45}$

72. $\frac{14}{15} = \int_0^1\int_0^{3-a-y^2}\int_a^{4-x-y^2} dz\,dx\,dy = \int_0^1\int_0^{3-a-y^2}\left(4 - x - y^2 - a\right)dx\,dy$

$$= \int_0^1\left[\left(4 - y^2 - a\right)x - \frac{x^2}{2}\right]_0^{3-a-y^2} dy = \int_0^1\left[\left(4 - y^2 - a\right)\left(3 - a - y^2\right) - \frac{\left(3 - a - y^2\right)^2}{2}\right]dy = \frac{94}{15} - \frac{11a}{3} + \frac{1}{2}a^2$$

So, $3a^2 - 22a + 32 = 0$

$(a - 2)(3a - 16) = 0$

$a = 2, \frac{16}{3}$.

73. Let $y_k = 1 - x_k$.

$$\frac{\pi}{2n}(x_1 + \cdots + x_n) = \frac{\pi}{2n}(n - y_1 - y_2 - \cdots - y_n) = \frac{\pi}{2} - \frac{\pi}{2n}(y_1 + \cdots + y_n)$$

So,

$$I_1 = \int_0^1\int_0^1\cdots\int_0^1 \cos^2\left\{\frac{\pi}{2n}(x_1 + \cdots + x_n)\right\}dx_1\,dx_2\cdots dx_n$$

$$= \int_1^0\int_1^0\cdots\int_1^0 \sin^2\left\{\frac{\pi}{2n}(y_1 + \cdots + y_n)\right\}(-dy_1)(-dy_2)\cdots(-dy_n) = \int_0^1\int_0^1\cdots\int_0^1 \sin^2\left\{\frac{\pi}{2n}(x_1 + \cdots + x_n)\right\}dx_1\,dx_2\cdots dx_n = I_2$$

$I_1 + I_2 = 1 \Rightarrow I_1 = \frac{1}{2}$.

Finally, $\lim_{n\to\infty} I_1 = \frac{1}{2}$.

Section 14.7 Triple Integrals in Other Coordinates

1. $\int_{-1}^{5}\int_{0}^{\pi/2}\int_{0}^{3} r\cos\theta\,dr\,d\theta\,dz = \int_{-1}^{5}\int_{0}^{\pi/2} \frac{9}{2}\cos\theta\,d\theta\,dz = \int_{-1}^{5}\left[\frac{9}{2}\sin\theta\right]_{0}^{\pi/2} dz = \int_{-1}^{5}\frac{9}{2}\,dz = \left[\frac{9}{2}z\right]_{-1}^{5} = \frac{9}{2}(5-(-1)) = 27$

2. $\int_{0}^{\pi/4}\int_{0}^{6}\int_{0}^{6-r} rz\,dz\,dr\,d\theta = \int_{0}^{\pi/4}\int_{0}^{6}\left[\frac{rz^2}{2}\right]_{0}^{6-r} dr\,d\theta = \int_{0}^{\pi/4}\int_{0}^{6}\frac{1}{2}\left(r^3 - 12r^2 + 36r\right)dr\,d\theta$

$$= \int_{0}^{\pi/4}\frac{1}{2}\left[\frac{r^4}{4} - 4r^3 + 18r^2\right]_{0}^{6} d\theta = \int_{0}^{\pi/4}\frac{1}{2}(108)\,d\theta = 54\left(\frac{\pi}{4}\right) = \frac{27\pi}{2}$$

3. $\int_{0}^{\pi/2}\int_{0}^{2\cos^2\theta}\int_{0}^{4-r^2} r\sin\theta\,dz\,dr\,d\theta = \int_{0}^{\pi/2}\int_{0}^{2\cos^2\theta} r\left(4-r^2\right)\sin\theta\,dr\,d\theta = \int_{0}^{\pi/2}\left[\left(2r^2 - \frac{r^4}{4}\right)\sin\theta\right]_{0}^{2\cos^2\theta} d\theta$

$$= \int_{0}^{\pi/2}\left[8\cos^4\theta - 4\cos^8\theta\right]\sin\theta\,d\theta = \left[-\frac{8\cos^5\theta}{5} + \frac{4\cos^9\theta}{9}\right]_{0}^{\pi/2} = \frac{52}{45}$$

4. $\int_{0}^{\pi/2}\int_{0}^{\pi}\int_{0}^{2} e^{-\rho^3}\rho^2\,d\rho\,d\theta\,d\phi = \int_{0}^{\pi/2}\int_{0}^{\pi}\left[-\frac{1}{3}e^{-\rho^3}\right]_{0}^{2} d\theta\,d\phi = \int_{0}^{\pi/2}\int_{0}^{\pi}\frac{1}{3}\left(1-e^{-8}\right)d\theta\,d\phi = \frac{\pi^2}{6}\left(1-e^{-8}\right)$

5. $\int_{0}^{2\pi}\int_{0}^{\pi/4}\int_{0}^{\cos\phi}\rho^2\sin\phi\,d\rho\,d\phi\,d\theta = \frac{1}{3}\int_{0}^{2\pi}\int_{0}^{\pi/4}\cos^3\phi\sin\phi\,d\phi\,d\theta = -\frac{1}{12}\int_{0}^{2\pi}\left[\cos^4\phi\right]_{0}^{\pi/4} d\theta = \frac{\pi}{8}$

6. $\int_{0}^{\pi/4}\int_{0}^{\pi/4}\int_{0}^{\cos\theta}\rho^2\sin\phi\cos\phi\,d\rho\,d\theta\,d\phi = \frac{1}{3}\int_{0}^{\pi/4}\int_{0}^{\pi/4}\cos^3\theta\sin\phi\cos\phi\,d\theta\,d\phi$

$$= \frac{1}{3}\int_{0}^{\pi/4}\int_{0}^{\pi/4}\sin\phi\cos\phi\left[\cos\theta\left(1-\sin^2\theta\right)\right]d\theta\,d\phi$$

$$= \frac{1}{3}\int_{0}^{\pi/4}\sin\phi\cos\phi\left[\sin\theta - \frac{\sin^3\theta}{3}\right]_{0}^{\pi/4} d\phi$$

$$= \frac{5\sqrt{2}}{36}\int_{0}^{\pi/4}\sin\phi\cos\phi\,d\phi = \left[\frac{5\sqrt{2}}{36}\frac{\sin^2\phi}{2}\right]_{0}^{\pi/4} = \frac{5\sqrt{2}}{144}$$

7. $\int_{0}^{4}\int_{0}^{z}\int_{0}^{\pi/2} re^r\,d\theta\,dr\,dz = \pi\left(e^4+3\right)$

8. $\int_{0}^{\pi/2}\int_{0}^{\pi}\int_{0}^{\sin\theta}(2\cos\phi)\rho^2\,d\rho\,d\theta\,d\phi = \frac{8}{9}$

9. $\int_{0}^{\pi/2}\int_{0}^{3}\int_{0}^{e^{-r^2}} r\,dz\,dr\,d\theta = \int_{0}^{\pi/2}\int_{0}^{3} re^{-r^2}\,dr\,d\theta = \int_{0}^{\pi/2}\left[-\frac{1}{2}e^{-r^2}\right]_{0}^{3} d\theta = \int_{0}^{\pi/2}\frac{1}{2}\left(1-e^{-9}\right)d\theta = \frac{\pi}{4}\left(1-e^{-9}\right)$

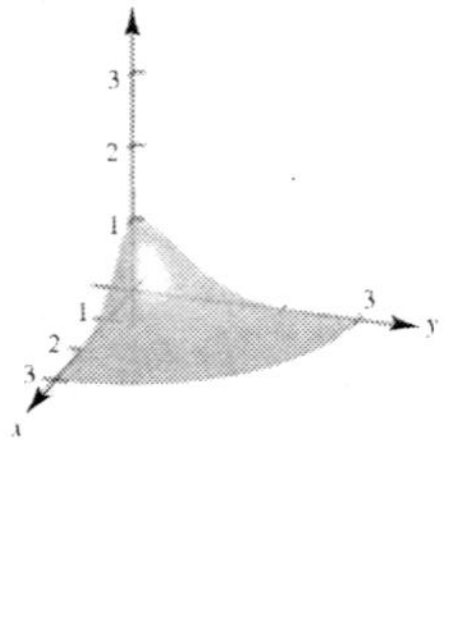

10. $\displaystyle\int_0^{2\pi}\int_0^{\sqrt{5}}\int_2^{5-r^2} r\,dz\,dr\,d\theta = \int_0^{2\pi}\int_0^{\sqrt{5}}\left(5r - r^3\right)dr\,d\theta = \int_0^{2\pi}\left[\frac{5r^2}{2} - \frac{r^4}{4}\right]_0^{\sqrt{5}} d\theta = \int_0^{2\pi}\left(\frac{25}{2} - \frac{25}{4}\right)d\theta = \frac{25}{4}\cdot 2\pi = \frac{25\pi}{2}$

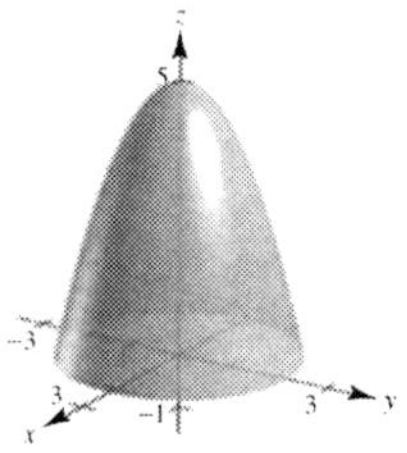

11. $\displaystyle\int_0^{2\pi}\int_{\pi/6}^{\pi/2}\int_0^4 \rho^2 \sin\phi\,d\rho\,d\phi\,d\theta = \frac{64}{3}\int_0^{2\pi}\int_{\pi/6}^{\pi/2}\sin\phi\,d\phi\,d\theta = \frac{64}{3}\int_0^{2\pi}\left[-\cos\phi\right]_{\pi/6}^{\pi/2} d\theta = \frac{32\sqrt{3}}{3}\int_0^{2\pi} d\theta = \frac{64\sqrt{3}\pi}{3}$

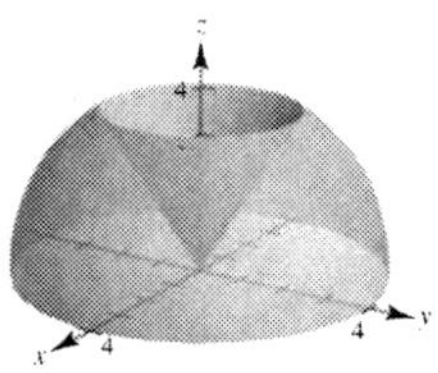

12. $\displaystyle\int_0^{2\pi}\int_0^{\pi}\int_2^5 \rho^2 \sin\phi\,d\rho\,d\phi\,d\theta = \frac{117}{3}\int_0^{2\pi}\int_0^{\pi}\sin\phi\,d\phi\,d\theta = \frac{117}{3}\int_0^{2\pi}\left[-\cos\phi\right]_0^{\pi} d\theta = \frac{468\pi}{3} = 156\pi$

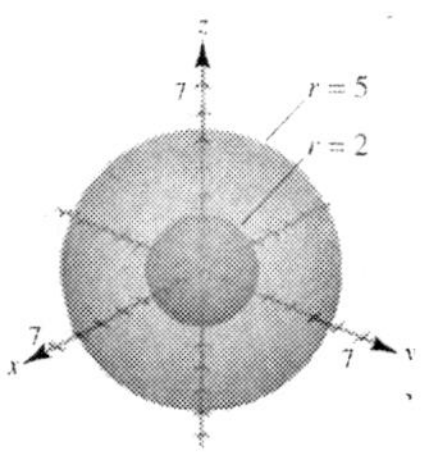

13. $\displaystyle\int_0^{2\pi}\int_0^2\int_{r^2}^4 r^2\cos\theta\,dz\,dr\,d\theta = 0$

$\displaystyle\int_0^{2\pi}\int_0^{\arctan(1/2)}\int_0^{4\sec\phi} \rho^3\sin^2\phi\cos\theta\,d\rho\,d\phi\,d\theta + \int_0^{2\pi}\int_{\arctan(1/2)}^{\pi/2}\int_0^{\cot\phi\csc\phi} \rho^3\sin^2\phi\cos\theta\,d\rho\,d\phi\,d\theta = 0$

14. $\displaystyle\int_0^{\pi/2}\int_0^2\int_0^{\sqrt{16-r^2}} r^2\,dz\,dr\,d\theta = \frac{8\pi^2}{3} - 2\pi\sqrt{3}$

$\displaystyle\int_0^{\pi/2}\int_0^{\pi/6}\int_0^4 \rho^3\sin^2\phi\,d\rho\,d\phi\,d\theta + \int_0^{\pi/2}\int_{\pi/6}^{\pi/2}\int_4^{2\csc\phi} \rho^3\sin^2\phi\,d\rho\,d\phi\,d\theta = \frac{8\pi^2}{3} - 2\pi\sqrt{3}$

15. $\displaystyle\int_0^{2\pi}\int_0^a\int_a^{a+\sqrt{a^2-r^2}} r^2\cos\theta\,dz\,dr\,d\theta = 0$

$\displaystyle\int_0^{\pi/4}\int_0^{2\pi}\int_{a\sec\phi}^{2a\cos\phi} \rho^3\sin^2\phi\cos\theta\,d\rho\,d\theta\,d\phi = 0$

16. $\displaystyle\int_0^{\pi/2}\int_0^3\int_0^{\sqrt{9-r^2}} \sqrt{r^2+z^2}\,r\,dz\,dr\,d\theta = \frac{81\pi}{8}$

$\displaystyle\int_0^{\pi/2}\int_0^{\pi/2}\int_0^3 \rho^3\sin\phi\,d\rho\,d\phi\,d\theta = \frac{81\pi}{8}$

17. $V = 4\int_0^{\pi/2}\int_0^{a\cos\theta}\int_0^{\sqrt{a^2-r^2}} r\,dz\,dr\,d\theta = 4\int_0^{\pi/2}\int_0^{a\cos\theta} r\sqrt{a^2-r^2}\,dr\,d\theta$

$$= \frac{4}{3}a^3\int_0^{\pi/2}\left(1-\sin^3\theta\right)d\theta = \frac{4}{3}a^3\left[\theta + \frac{1}{3}\cos\theta\left(\sin^2\theta+2\right)\right]_0^{\pi/2} = \frac{4}{3}a^3\left(\frac{\pi}{2}-\frac{2}{3}\right) = \frac{2a^3}{9}(3\pi-4)$$

18. $V = \frac{2}{3}\pi(4)^3 + 4\left[\int_0^{\pi/2}\int_0^{2\sqrt{2}}\int_0^{r} r\,dz\,dr\,d\theta + \int_0^{\pi/2}\int_{2\sqrt{2}}^{4}\int_0^{\sqrt{16-r^2}} r\,dz\,dr\,d\theta\right]$

(Volume of lower hemisphere) + 4(Volume in the first octant)

$$V = \frac{128\pi}{3} + 4\left[\int_0^{\pi/2}\int_0^{2\sqrt{2}} r^2\,dr\,d\theta + \int_0^{\pi/2}\int_{2\sqrt{2}}^{4} r\sqrt{16-r^2}\,dr\,d\theta\right]$$

$$= \frac{128\pi}{3} + 4\left[\frac{8\sqrt{2}\pi}{3} + \int_0^{\pi/2}\left[-\frac{1}{3}\left(16-r^2\right)^{3/2}\right]_{2\sqrt{2}}^{4}d\theta\right]$$

$$= \frac{128\pi}{3} + 4\left[\frac{8\sqrt{2}\pi}{3} + \frac{8\sqrt{2}\pi}{3}\right] = \frac{128\pi}{3} + \frac{64\sqrt{2}\pi}{3} = \frac{64\pi}{3}\left(2+\sqrt{2}\right)$$

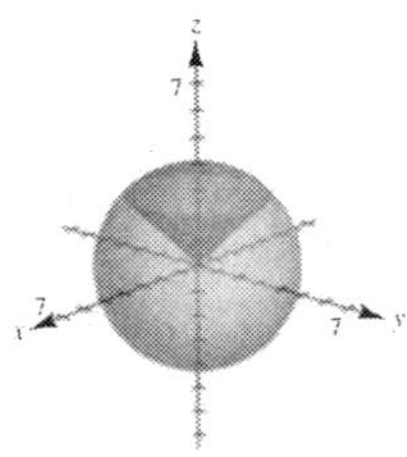

19. In the xy-plane, $2x = 2x^2 + 2y^2 \Rightarrow$

$$0 = x^2 - x + y^2 \Rightarrow \left(x^2 - x + 1/4\right) + y^2 = 1/4$$

$$\Rightarrow (x - 1/2)^2 + y^2 = (1/2)^2$$

In polar coordinates, use $r = \cos\theta$ for this circle.

$$V = \int_0^{\pi}\int_0^{\cos\theta}\int_{2r^2}^{2r\cos\theta} r\,dz\,dr\,d\theta$$

$$= \int_0^{\pi}\int_0^{\cos\theta}\left(2r^2\cos\theta - 2r^3\right)dr\,d\theta$$

$$= \int_0^{\pi}\left[\frac{2r^3}{3}\cos\theta - \frac{r^4}{2}\right]_0^{\cos\theta}d\theta$$

$$= \int_0^{\pi}\left(\frac{2}{3}\cos^4\theta - \frac{\cos^4\theta}{2}\right)d\theta$$

$$= \frac{1}{6}\int_0^{\pi}\cos^4\theta\,d\theta = \frac{\pi}{16}$$

20. $2 - x^2 - y^2 = x^2 + y^2 \Rightarrow x^2 + y^2 = 1$

$$V = \int_0^{2\pi}\int_0^{1}\int_{r^2}^{2-r^2} r\,dz\,dr\,d\theta$$

$$= \int_0^{2\pi}\int_0^{1} r\left(2 - 2r^2\right)dr\,d\theta$$

$$= \int_0^{2\pi}\left[r^2 - \frac{r^4}{2}\right]_0^{1}d\theta = \frac{1}{2}(2\pi) = \pi$$

21. $V = 2\int_0^{\pi}\int_0^{a\cos\theta}\int_0^{\sqrt{a^2-r^2}} r\,dz\,dr\,d\theta$

$$= 2\int_0^{\pi}\int_0^{a\cos\theta} r\sqrt{a^2-r^2}\,dr\,d\theta$$

$$= 2\int_0^{\pi}\left[-\frac{1}{3}\left(a^2-r^2\right)^{3/2}\right]_0^{a\cos\theta}d\theta$$

$$= \frac{2a^3}{3}\int_0^{\pi}\left(1-\sin^3\theta\right)d\theta$$

$$= \frac{2a^3}{3}\left[\theta + \cos\theta - \frac{\cos^3\theta}{3}\right]_0^{\pi} = \frac{2a^3}{9}(3\pi - 4)$$

22. $V = \int_0^{2\pi}\int_0^{\sqrt{2}}\int_r^{\sqrt{4-r^2}} r\,dz\,dr\,d\theta$

$$= \int_0^{2\pi}\int_0^{\sqrt{2}}\left(r\sqrt{4-r^2} - r^2\right)dr\,d\theta$$

$$= \int_0^{2\pi}\left[-\frac{1}{3}\left(4-r^2\right)^{3/2} - \frac{r^3}{3}\right]_0^{\sqrt{2}}d\theta = \frac{8\pi}{3}\left(2-\sqrt{2}\right)$$

23. $m = \int_0^{2\pi}\int_0^{2}\int_0^{9-r\cos\theta-2r\sin\theta}(kr)r\,dz\,dr\,d\theta$

$$= \int_0^{2\pi}\int_0^{2} kr^2(9 - r\cos\theta - 2r\sin\theta)\,dr\,d\theta$$

$$= \int_0^{2\pi} k\left[3r^3 - \frac{r^4}{4}\cos\theta - \frac{r^4}{2}\sin\theta\right]_0^{2}d\theta$$

$$= \int_0^{2\pi} k\left[24 - 4\cos\theta - 8\sin\theta\right]d\theta$$

$$= k\left[24\theta - 4\sin\theta + 8\cos\theta\right]_0^{2\pi}$$

$$= k\left[48\pi + 8 - 8\right] = 48k\pi$$

24. $\int_0^{\pi/2}\int_0^2\int_0^{12e^{-r^2}} k\,r\,dz\,dr\,d\theta = \int_0^{\pi/2}\int_0^2 12ke^{-r^2}\,r\,dr\,d\theta$

$= \int_0^{\pi/2}\left[-6ke^{-r^2}\right]_0^2$

$= \int_0^{\pi/2}\left(-6ke^{-4} + 6k\right)d\theta$

$= 3k\pi\left(1 - e^{-4}\right)$

25. $z = h - \frac{h}{r_0}\sqrt{x^2 + y^2} = \frac{h}{r_0}(r_0 - r)$

$V = 4\int_0^{\pi/2}\int_0^{r_0}\int_0^{h(r_0-r)/r_0} r\,dz\,dr\,d\theta$

$= \frac{4h}{r_0}\int_0^{\pi/2}\int_0^{r_0}\left(r_0r - r^2\right)dr\,d\theta$

$= \frac{4h}{r_0}\int_0^{\pi/2}\frac{r_0^3}{6}\,d\theta$

$= \frac{4h}{r_0}\left(\frac{r_0^3}{6}\right)\left(\frac{\pi}{2}\right) = \frac{1}{3}\pi r_0^2 h$

26. $\bar{x} = \bar{y} = 0$ by symmetry

$m = \frac{1}{3}\pi r_0^2 hk$ from Exercise 25

$M_{xy} = 4k\int_0^{\pi/2}\int_0^{r_0}\int_0^{h(r_0-r)/r_0} zr\,dz\,dr\,d\theta$

$= \frac{2kh^2}{r_0^2}\int_0^{\pi/2}\int_0^{r_0}\left(r_0^2r - 2r_0r^2 + r^3\right)dr\,d\theta$

$= \frac{2kh^2}{r_0^2}\left(\frac{r_0^4}{12}\right)\left(\frac{\pi}{2}\right) = \frac{kr_0^2h^2\pi}{12}$

$\bar{z} = \frac{M_{xy}}{m} = \frac{kr_0^2h^2\pi}{12}\left(\frac{3}{\pi r_0^2hk}\right) = \frac{h}{4}$

27. $\rho = k\sqrt{x^2 + y^2} = kr$

$\bar{x} = \bar{y} = 0$ by symmetry

$m = 4k\int_0^{\pi/2}\int_0^{r_0}\int_0^{h(r_0-r)/r_0} r^2\,dz\,dr\,d\theta = \frac{1}{6}k\pi r_0^3 h$

$M_{xy} = 4k\int_0^{\pi/2}\int_0^{r_0}\int_0^{h(r_0-r)/r_0} r^2\,z\,dz\,dr\,d\theta$

$= \frac{1}{30}k\pi r_0^3\,h^2$

$\bar{z} = \frac{M_{xy}}{m} = \frac{k\pi r_0^3h^2/30}{k\pi r_0^3h/6} = \frac{h}{5}$

$(\bar{x}, \bar{y}, \bar{z}) = \left(0, 0, \frac{h}{5}\right)$

28. $\rho = kz$

$\bar{x} = \bar{y} = 0$ by symmetry

$m = 4k\int_0^{\pi/2}\int_0^{r_0}\int_0^{h(r_0-r)/r_0} zr\,dz\,dr\,d\theta = \frac{1}{12}k\pi r_0^2h^2$

$M_{xy} = 4k\int_0^{\pi/2}\int_0^{r_0}\int_0^{h(r_0-r)/r_0} z^2r\,dz\,dr\,d\theta$

$= \frac{1}{30}k\pi r_0^2h^3$

$\bar{z} = \frac{M_{xy}}{m} = \frac{k\pi r_0^2h^3/30}{k\pi r_0^2h^2/12} = \frac{2h}{5}$

$(\bar{x}, \bar{y}, \bar{z}) = \left(0, 0, \frac{2h}{5}\right)$

29. $I_z = 4k\int_0^{\pi/2}\int_0^{r_0}\int_0^{h(r_0-r)/r_0} r^3\,dz\,dr\,d\theta$

$= \frac{4kh}{r_0}\int_0^{\pi/2}\int_0^{r_0}\left(r_0r^3 - r^4\right)dr\,d\theta$

$= \frac{4kh}{r_0}\left(\frac{r_0^5}{20}\right)\left(\frac{\pi}{2}\right) = \frac{1}{10}k\pi r_0^4h$

Because the mass of the core is $m = kV = k\left(\frac{1}{3}\pi r_0^2h\right)$ from Exercise 25, we have $k = 3m/\pi r_0^2h$. So,

$I_z = \frac{1}{10}k\pi r_0^4h = \frac{1}{10}\left(\frac{3m}{\pi r_0^2h}\right)\pi r_0^4h = \frac{3}{10}mr_0^2.$

30. $I_z = \iiint\limits_Q\left(x^2 + y^2\right)\rho(x, y, z)\,dV$

$= 4k\int_0^{\pi/2}\int_0^{r_0}\int_0^{h(r_0-r)/r_0} r^4\,dz\,dr\,d\theta$

$= 4kh\int_0^{\pi/2}\int_0^{r_0}\frac{r_0 - r}{r_0}r^4\,dr\,d\theta$

$= 4kh\int_0^{\pi/2}\left[\frac{r^5}{5} - \frac{r^6}{6r_0}\right]_0^{r_0}d\theta = 4kh\int_0^{\pi/2}\left[\frac{r_0^5}{5} - \frac{r_0^5}{6}\right]d\theta$

$= 4kh\int_0^{\pi/2}\frac{1}{30}r_0^5\,d\theta = 4kh\frac{1}{30}r_0^5\frac{\pi}{2} = \frac{1}{15}r_0^5\pi kh$

31. $m = k\left(\pi b^2h - \pi a^2h\right) = k\pi h\left(b^2 - a^2\right)$

$I_z = 4k\int_0^{\pi/2}\int_a^b\int_0^h r^3\,dz\,dr\,d\theta$

$= 4kh\int_0^{\pi/2}\int_a^b r^3\,dr\,d\theta = kh\int_0^{\pi/2}\left(b^4 - a^4\right)d\theta$

$= \frac{k\pi\left(b^4 - a^4\right)h}{2} = \frac{k\pi\left(b^2 - a^2\right)\left(b^2 + a^2\right)h}{2}$

$= \frac{1}{2}m\left(a^2 + b^2\right)$

32. $m = k\pi a^2 h$

$$I_z = 2k\int_0^{\pi/2}\int_0^{2a\sin\theta}\int_0^h r^3\,dz\,dr\,d\theta = \frac{3}{2}k\pi a^4 h = \frac{3}{2}ma^2$$

33.
$$\begin{aligned} V &= \int_0^{2\pi}\int_{\pi/4}^{\pi/2}\int_0^3 \rho^2\sin\phi\,d\rho\,d\phi\,d\theta \\ &= \int_0^{2\pi}\int_{\pi/4}^{\pi/2} 9\sin\phi\,d\phi\,d\theta \\ &= \int_0^{2\pi}\Big[-9\cos\phi\Big]_{\pi/4}^{\pi/2}\,d\theta \\ &= \int_0^{2\pi} 9\left(\frac{\sqrt{2}}{2}\right)d\theta = 18\pi\left(\frac{\sqrt{2}}{2}\right) = 9\pi\sqrt{2} \end{aligned}$$

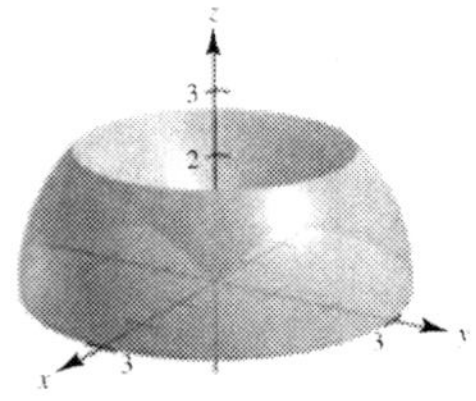

34.
$$\begin{aligned} x^2 + y^2 + z^2 &= z \\ x^2 + y^2 + \left(z^2 - z + \frac{1}{4}\right) &= \frac{1}{4} \\ x^2 + y^2 + \left(z - \frac{1}{2}\right)^2 &= \frac{1}{4} \end{aligned}$$

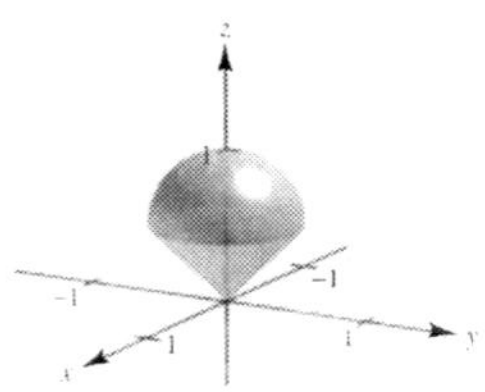

Sphere with center $\left(0, 0, \frac{1}{2}\right)$: $\rho = \cos\phi$

$$V = \int_0^{2\pi}\int_0^{\pi/4}\int_0^{\cos\phi}\rho^2\sin\phi\,d\rho\,d\phi\,d\theta = \int_0^{2\pi}\int_0^{\pi/4}\frac{\cos^3\phi}{3}\sin\phi\,d\phi\,d\theta = \int_0^{2\pi}\left[\frac{-\cos^4\phi}{12}\right]_0^{\pi/4}d\theta = \int_0^{2\pi}\frac{1}{12}\left(1 - \frac{1}{4}\right)d\theta = \frac{\pi}{8}$$

35. $V = \int_0^{2\pi}\int_0^{\pi}\int_0^{4\sin\phi}\rho^2\sin\phi\,d\rho\,d\phi\,d\theta = 16\pi^2$

36.
$$\begin{aligned} V &= 8\int_0^{\pi/4}\int_0^{\pi/2}\int_a^b \rho^2\sin\phi\,d\rho\,d\theta\,d\phi \quad \text{(includes upper and lower cones)} \\ &= \frac{8}{3}\left(b^3 - a^3\right)\int_0^{\pi/4}\int_0^{\pi/2}\sin\phi\,d\theta\,d\phi \\ &= \frac{4\pi}{3}\left(b^3 - a^3\right)\int_0^{\pi/4}\sin\phi\,d\phi \\ &= \left[\frac{4\pi}{3}\left(b^3 - a^3\right)(-\cos\phi)\right]_0^{\pi/4} = \left(1 - \frac{\sqrt{2}}{2}\right)\frac{4\pi}{3}\left(b^3 - a^3\right) = \frac{2\pi}{3}\left(2 - \sqrt{2}\right)\left(b^3 - a^3\right) \end{aligned}$$

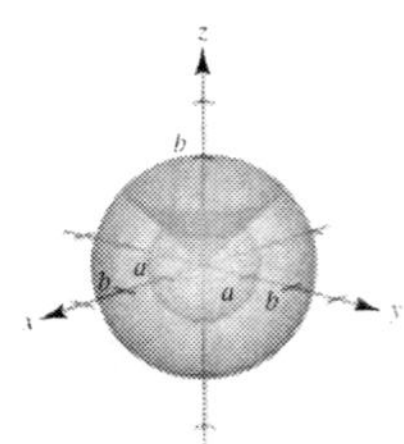

37.
$$\begin{aligned} m &= 8k\int_0^{\pi/2}\int_0^{\pi/2}\int_0^a \rho^3\sin\phi\,d\rho\,d\theta\,d\phi \\ &= 2ka^4\int_0^{\pi/2}\int_0^{\pi/2}\sin\phi\,d\theta\,d\phi \\ &= k\pi a^4\int_0^{\pi/2}\sin\phi\,d\phi = \left[k\pi a^4(-\cos\phi)\right]_0^{\pi/2} = k\pi a^4 \end{aligned}$$

38.
$$\begin{aligned} m &= 8k\int_0^{\pi/2}\int_0^{\pi/2}\int_0^a \rho^3\sin^2\phi\,d\rho\,d\theta\,d\phi \\ &= 2ka^4\int_0^{\pi/2}\int_0^{\pi/2}\sin^2\phi\,d\theta\,d\phi \\ &= k\pi a^4\int_0^{\pi/2}\sin^2\phi\,d\phi \\ &= \left[k\pi a^4\left(\frac{1}{2}\phi - \frac{1}{4}\sin 2\phi\right)\right]_0^{\pi/2} = k\pi a^4\frac{\pi}{4} = \frac{1}{4}k\pi^2 a^4 \end{aligned}$$

39. $m = \frac{2}{3}k\pi r^3$

$\bar{x} = \bar{y} = 0$ by symmetry

$$\begin{aligned} M_{xy} &= 4k\int_0^{\pi/2}\int_0^{\pi/2}\int_0^r \rho^3\cos\phi\sin\phi\,d\rho\,d\theta\,d\phi \\ &= \frac{1}{2}kr^4\int_0^{\pi/2}\int_0^{\pi/2}\sin 2\phi\,d\theta\,d\phi \\ &= \frac{kr^4\pi}{4}\int_0^{\pi/2}\sin 2\phi\,d\phi \\ &= \left[-\frac{1}{8}k\pi r^4\cos 2\phi\right]_0^{\pi/2} = \frac{1}{4}k\pi r^4 \end{aligned}$$

$$\bar{z} = \frac{M_{xy}}{m} = \frac{k\pi r^4/4}{2k\pi r^3/3} = \frac{3r}{8}$$

$$(\bar{x}, \bar{y}, \bar{z}) = \left(0, 0, \frac{3r}{8}\right)$$

40. $\bar{x} = \bar{y} = 0$ by symmetry

$$m = k\left(\frac{2}{3}\pi R^3 - \frac{2}{3}\pi r^3\right) = \frac{2}{3}k\pi\left(R^3 - r^3\right)$$

$$M_{xy} = 4k\int_0^{\pi/2}\int_0^{\pi/2}\int_r^R \rho^3 \cos\phi \sin\phi \, d\rho \, d\theta \, d\phi$$

$$= \frac{1}{2}k\left(R^4 - r^4\right)\int_0^{\pi/2}\int_0^{\pi/2} \sin 2\phi \, d\theta \, d\phi$$

$$= \frac{1}{4}k\pi\left(R^4 - r^4\right)\int_0^{\pi/2} \sin 2\phi \, d\phi$$

$$= \left[-\frac{1}{8}k\pi\left(R^4 - r^4\right)\cos 2\phi\right]_0^{\pi/2} = \frac{1}{4}k\pi\left(R^4 - r^4\right)$$

$$\bar{z} = \frac{M_{xy}}{m} = \frac{k\pi\left(R^4 - r^4\right)/4}{2k\pi\left(R^3 - r^3\right)/3} = \frac{3\left(R^4 - r^4\right)}{8\left(R^3 - r^3\right)}$$

$$(\bar{x}, \bar{y}, \bar{z}) = \left(0, 0, \frac{3\left(R^4 - r^4\right)}{8\left(R^3 - r^3\right)}\right)$$

41. $$I_z = 4k\int_{\pi/4}^{\pi/2}\int_0^{\pi/2}\int_0^{\cos\phi} \rho^4 \sin^3\phi \, d\rho \, d\theta \, d\phi$$

$$= \frac{4}{5}k\int_{\pi/4}^{\pi/2}\int_0^{\pi/2} \cos^5\phi \sin^3\phi \, d\theta \, d\phi$$

$$= \frac{2}{5}k\pi\int_{\pi/4}^{\pi/2} \cos^5\phi\left(1 - \cos^2\phi\right)\sin\phi \, d\phi$$

$$= \left[\frac{2}{5}k\pi\left(-\frac{1}{6}\cos^6\phi + \frac{1}{8}\cos^8\phi\right)\right]_{\pi/4}^{\pi/2} = \frac{k\pi}{192}$$

42. $$I_z = 4k\int_0^{\pi/2}\int_0^{\pi/2}\int_r^R \rho^4 \sin^3\phi \, d\rho \, d\theta \, d\phi$$

$$= \frac{4k}{5}\left(R^5 - r^5\right)\int_0^{\pi/2}\int_0^{\pi/2} \sin^3\phi \, d\theta \, d\phi$$

$$= \frac{2k\pi}{5}\left(R^5 - r^5\right)\int_0^{\pi/2} \sin\phi\left(1 - \cos^2\phi\right) d\phi$$

$$= \left[\frac{2k\pi}{5}\left(R^5 - r^5\right)\left(-\cos\phi + \frac{\cos^3\phi}{3}\right)\right]_0^{\pi/2}$$

$$= \frac{4k\pi}{15}\left(R^5 - r^5\right)$$

43. $x = r\cos\theta$ $\qquad x^2 + y^2 = r^2$

$y = r\sin\theta$ $\qquad \tan\theta = \dfrac{y}{x}$

$z = z$ $\qquad z = z$

44. $x = \rho\sin\phi\cos\theta$ $\qquad \rho^2 = x^2 + y^2 + z^2$

$y = \rho\sin\phi\sin\theta$ $\qquad \tan\theta = \dfrac{y}{x}$

$z = \rho\cos\phi$ $\qquad \cos\phi = \dfrac{z}{\sqrt{x^2 + y^2 + z^2}}$

45. $$\int_{\theta_1}^{\theta_2}\int_{g_1(\theta)}^{g_2(\theta)}\int_{h_1(r\cos\theta, r\sin\theta)}^{h_2(r\cos\theta, r\sin\theta)} f(r\cos\theta, r\sin\theta, z)r \, dz \, dr \, d\theta$$

46. $$\int_{\theta_1}^{\theta_2}\int_{\phi_1}^{\phi_2}\int_{\rho_1}^{\rho_2} f(\rho\sin\phi\cos\theta, \rho\sin\phi\sin\theta, \rho\cos\phi)\rho^2\sin\phi \, d\rho \, d\phi \, d\theta$$

47. (a) $r = r_0$: right circular cylinder about z-axis

$\theta = \theta_0$: plane parallel to z-axis

$z = z_0$: plane parallel to xy-plane

(b) $\rho = \rho_0$: sphere of radius ρ_0

$\theta = \theta_0$: plane parallel to z-axis

$\phi = \phi_0$: cone

48. Spherical coordinates; Triple integrals involving spheres and cones are often easier to evaluate by converting to spherical coordinates.

49. $(x^2 + y^2 + z^2 + 8)^2 \le 36(x^2 + y^2)$

In cylindrical coordinates,

$(r^2 + z^2 + 8)^2 \le 36r^2$

$r^2 + z^2 + 8 \le 6r$

$r^2 - 6r + 9 + z^2 - 1 \le 0$

$(r - 3)^2 + z^2 \le 1.$

This is a torus: rotate $(x - 3)^2 + z^2 = 1$ about the z-axis. By Pappus' Theorem,

$V = 2\pi(3)\pi = 6\pi^2.$

Section 14.8 Change of Variables: Jacobians

1. $x = -\frac{1}{2}(u - v)$

$y = \frac{1}{2}(u + v)$

$$\frac{\partial x}{\partial u}\frac{\partial y}{\partial v} - \frac{\partial y}{\partial u}\frac{\partial x}{\partial v} = \left(-\frac{1}{2}\right)\left(\frac{1}{2}\right) - \left(\frac{1}{2}\right)\left(\frac{1}{2}\right) = -\frac{1}{2}$$

2. $x = au + bv$

$y = cu + dv$

$$\frac{\partial x}{\partial u}\frac{\partial y}{\partial v} - \frac{\partial y}{\partial u}\frac{\partial x}{\partial v} = ad - cb$$

3. $x = u - v^2$

$y = u + v$

$$\frac{\partial x}{\partial u}\frac{\partial y}{\partial v} - \frac{\partial y}{\partial u}\frac{\partial x}{\partial v} = (1)(1) - (1)(-2v) = 1 + 2v$$

4. $x = uv - 2u$

$y = uv$

$$\frac{\partial x}{\partial u}\frac{\partial y}{\partial v} - \frac{\partial y}{\partial u}\frac{\partial x}{\partial v} = (v - 2)u - vu = -2u$$

5. $x = u\cos\theta - v\sin\theta$

$y = u\sin\theta + v\cos\theta$

$$\frac{\partial x}{\partial u}\frac{\partial y}{\partial v} - \frac{\partial y}{\partial u}\frac{\partial x}{\partial v} = \cos^2\theta + \sin^2\theta = 1$$

6. $x = u + a$

$y = v + a$

$$\frac{\partial x}{\partial u}\frac{\partial y}{\partial v} - \frac{\partial y}{\partial u}\frac{\partial x}{\partial v} = (1)(1) - (0)(0) = 1$$

7. $x = e^u \sin v$

$y = e^u \cos v$

$$\frac{\partial x}{\partial u}\frac{\partial y}{\partial v} - \frac{\partial y}{\partial u}\frac{\partial x}{\partial v} = (e^u \sin v)(-e^u \sin v) - (e^u \cos v)(e^u \cos v) = -e^{2u}$$

8. $x = \frac{u}{v}$

$y = u + v$

$$\frac{\partial x}{\partial u}\frac{\partial y}{\partial v} - \frac{\partial y}{\partial u}\frac{\partial x}{\partial v} = \left(\frac{1}{v}\right)(1) - (1)\left(-\frac{u}{v^2}\right) = \frac{1}{v} + \frac{u}{v^2} = \frac{u + v}{v^2}$$

9. $x = 3u + 2v$

$y = 3v$

$v = \frac{y}{3}$

$$u = \frac{x - 2v}{3} = \frac{x - 2(y/3)}{3} = \frac{x}{3} - \frac{2y}{9}$$

(x, y)	(u, v)
(0, 0)	(0, 0)
(3, 0)	(1, 0)
(2, 3)	(0, 1)

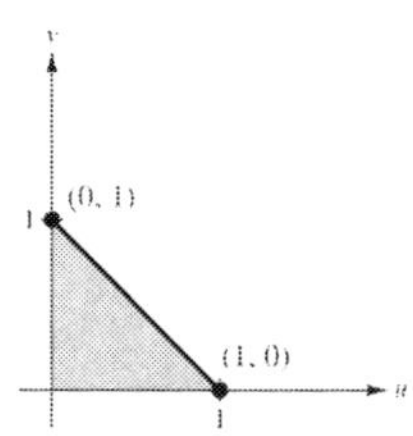

10. $x = \frac{1}{3}(4u - v)$

$y = \frac{1}{3}(u - v)$

$u = x - y$

$v = x - 4y$

(x, y)	(u, v)
$(0, 0)$	$(0, 0)$
$(4, 1)$	$(3, 0)$
$(2, 2)$	$(0, -6)$
$(6, 3)$	$(3, -6)$

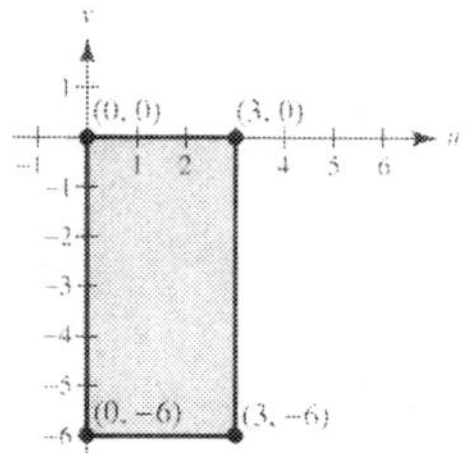

11. $x = \frac{1}{2}(u + v)$

$y = \frac{1}{2}(u - v)$

$u = x + y$

$v = x - y$

(x, y)	(u, v)
$\left(\frac{1}{2}, \frac{1}{2}\right)$	$(1, 0)$
$(0, 1)$	$(1, -1)$
$(1, 2)$	$(3, -1)$
$\left(\frac{3}{2}, \frac{3}{2}\right)$	$(3, 0)$

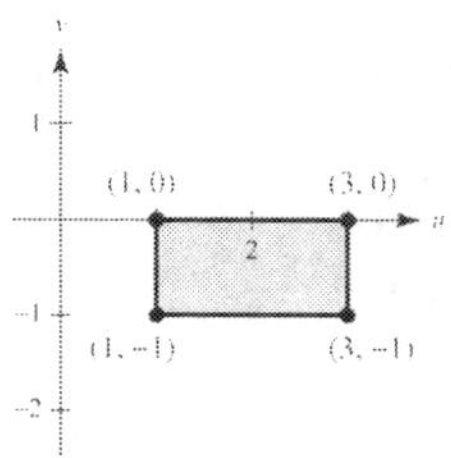

12. $x = \frac{1}{3}(v - u)$

$y = \frac{1}{3}(2v + u)$

$u = y - 2x$

$v = x + y$

(x, y)	(u, v)
$\left(-\frac{1}{3}, \frac{4}{3}\right)$	$(2, 1)$
$\left(\frac{1}{3}, \frac{2}{3}\right)$	$(0, 1)$
$\left(\frac{4}{3}, \frac{8}{3}\right)$	$(0, 4)$
$\left(\frac{2}{3}, \frac{10}{3}\right)$	$(2, 4)$

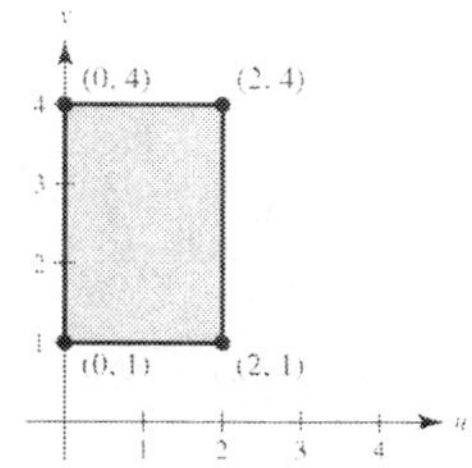

13. $\left.\begin{aligned} x - 2y &= 0 \\ x + y &= 4 \end{aligned}\right\} \Rightarrow \begin{aligned} 3y &= 4 \\ y &= \tfrac{4}{3},\quad x = \tfrac{8}{3} \end{aligned}$

$\left.\begin{aligned} x - 2y &= -4 \\ x + y &= 4 \end{aligned}\right\} \quad \begin{aligned} 3y &= 8 \\ y &= \tfrac{8}{3},\quad x = \tfrac{4}{3} \end{aligned}$

$\left.\begin{aligned} x - 2y &= -4 \\ x + y &= 1 \end{aligned}\right\} \quad \begin{aligned} 3y &= 5 \\ y &= \tfrac{5}{3},\quad x = -\tfrac{2}{3} \end{aligned}$

$\left.\begin{aligned} x - 2y &= 0 \\ x + y &= 1 \end{aligned}\right\} \quad \begin{aligned} 3y &= 1 \\ y &= \tfrac{1}{2},\quad x = \tfrac{2}{3} \end{aligned}$

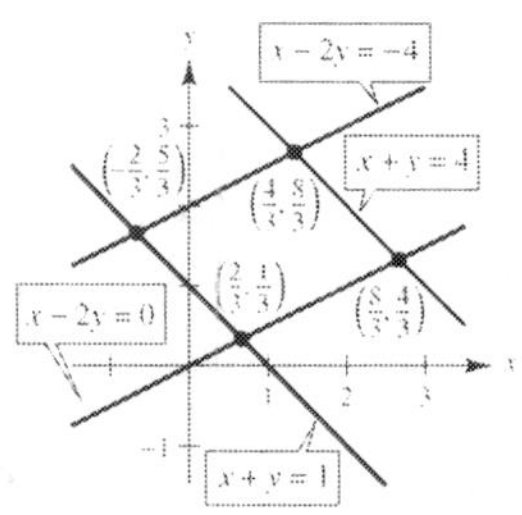

$\left.\begin{aligned} u &= x + y \\ v &= x - 2y \end{aligned}\right\} \quad \begin{aligned} u - v &= 3y \Rightarrow y = \tfrac{1}{3}(u - v) \\ 2u + v &= 3x \Rightarrow x = \tfrac{1}{3}(2u + v) \end{aligned}$

$$\iint_R 3\,xy\,dA = \int_{-2/3}^{2/3}\int_{1-x}^{(x+4)/2} 3\,xy\,dy\,dx + \int_{2/3}^{4/3}\int_{x/2}^{(x+4)/2} 3\,xy\,dy\,dx + \int_{4/3}^{8/3}\int_{x/2}^{4-x} 3\,xy\,dy\,dx = \frac{32}{27} + \frac{164}{27} + \frac{296}{27} = \frac{164}{9}$$

14. $\iint_R (x+y)^2 \sin^2(x-y)\,dA = \int_0^1 \int_{1-x}^{x+1} f(x)\,dy\,dx + \int_1^2 \int_{x-1}^{3-x} f(x)dy\,dx$

$= \left(\dfrac{-\cos^2(1)}{16} - \dfrac{5}{3}\sin(1)\cos(1) + \dfrac{15}{16}\sin^2(1) + \dfrac{17}{16}\right) + \left[\cos^2 1 - \dfrac{8}{3}\sin(1)\cos(1) + 7/3\right]$

$= \dfrac{13}{3} - \dfrac{13}{3}\sin(1)\cos(1) = \dfrac{13}{3} - \dfrac{13}{6}\sin(2) \approx 2.363$

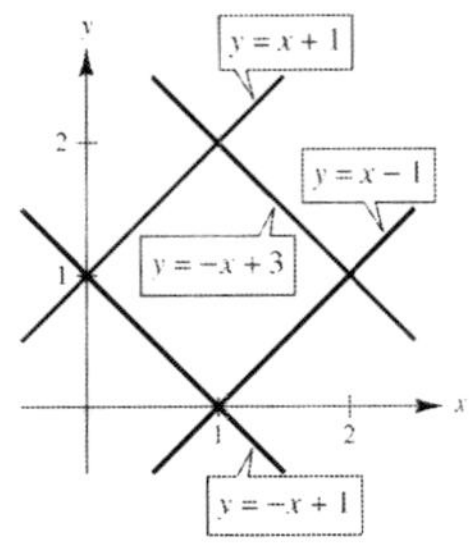

15. $x = \dfrac{1}{2}(u+v)$

$y = \dfrac{1}{2}(u-v)$

$\dfrac{\partial x}{\partial u}\dfrac{\partial y}{\partial v} - \dfrac{\partial y}{\partial u}\dfrac{\partial x}{\partial v} = \left(\dfrac{1}{2}\right)\left(-\dfrac{1}{2}\right) - \left(\dfrac{1}{2}\right)\left(\dfrac{1}{2}\right) = -\dfrac{1}{2}$

$\iint_R 4(x^2+y^2)\,dA = \int_{-1}^1 \int_{-1}^1 4\left[\dfrac{1}{4}(u+v)^2 + \dfrac{1}{4}(u-v)^2\right]\left(\dfrac{1}{2}\right)dv\,du$

$= \int_{-1}^1 \int_{-1}^1 (u^2+v^2)dv\,du = \int_{-1}^1 2\left(u^2 + \dfrac{1}{3}\right)du = \left[2\left(\dfrac{u^3}{3} + \dfrac{u}{3}\right)\right]_{-1}^1 = \dfrac{8}{3}$

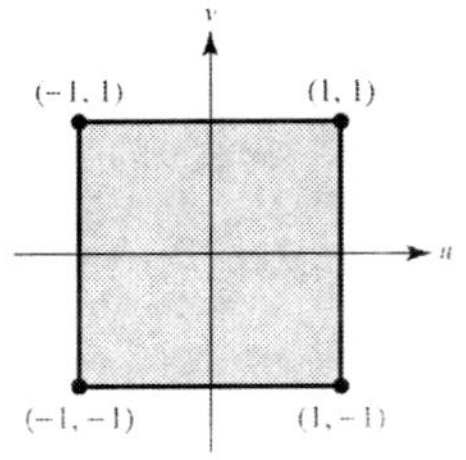

16. $x = \dfrac{1}{2}(u+v),\quad u = x - y$

$y = -\dfrac{1}{2}(u-v),\quad v = x + y$

$\dfrac{\partial x}{\partial u}\dfrac{\partial y}{\partial v} - \dfrac{\partial y}{\partial u}\dfrac{\partial x}{\partial v} = \dfrac{1}{2}\left(\dfrac{1}{2}\right) - \left(-\dfrac{1}{2}\right)\left(\dfrac{1}{2}\right) = \dfrac{1}{2}$

$\int_R\int 60xy\,dA = \int_{-1}^1 \int_1^3 60\left(\dfrac{1}{2}(u+v)\right)\left(-\dfrac{1}{2}(u-v)\right)\left(\dfrac{1}{2}\right)dv\,du$

$= \int_{-1}^1 \int_1^3 -\dfrac{15}{2}(v^2-u^2)dv\,du$

$= \int_{-1}^1 \left[-\dfrac{15}{2}\left(\dfrac{v^3}{3} - u^2v\right)\right]_1^3 du = \int_{-1}^1 \dfrac{15}{2}\left(2u^2 - \dfrac{26}{3}\right)du = \left[\dfrac{15}{2}\left(\dfrac{2}{3}u^3 - \dfrac{26}{3}u\right)\right]_{-1}^1 = 15\left(\dfrac{2}{3} - \dfrac{26}{3}\right) = -120$

(x, y)	(u, v)
(0, 1)	(−1, 1)
(2, 1)	(1, 3)
(1, 2)	(−1, 3)
(1, 0)	(1, 1)

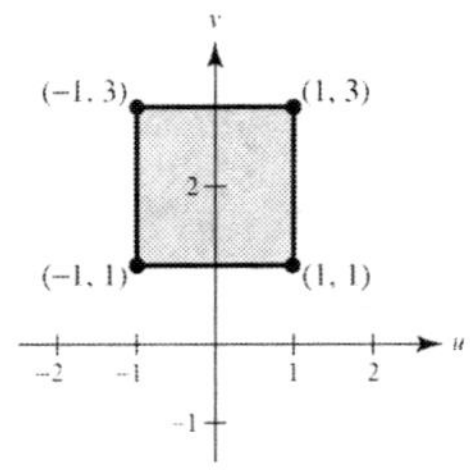

17. $x = u + v$

$y = u$

$\dfrac{\partial x}{\partial u}\dfrac{\partial y}{\partial v} - \dfrac{\partial y}{\partial u}\dfrac{\partial x}{\partial v} = (1)(0) - (1)(1) = -1$

$\int_R\int y(x-y)\,dA = \int_0^3 \int_0^4 uv(1)\,dv\,du$

$= \int_0^3 8u\,du = 36$

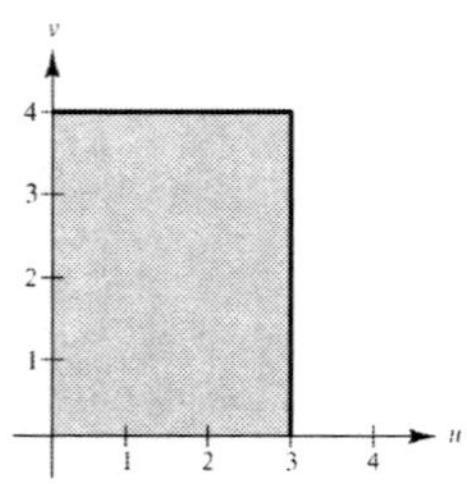

18. $x = \dfrac{1}{2}(u+v)$

$y = \dfrac{1}{2}(u-v)$

$\dfrac{\partial x}{\partial u}\dfrac{\partial y}{\partial v} - \dfrac{\partial y}{\partial u}\dfrac{\partial x}{\partial v} = -\dfrac{1}{2}$

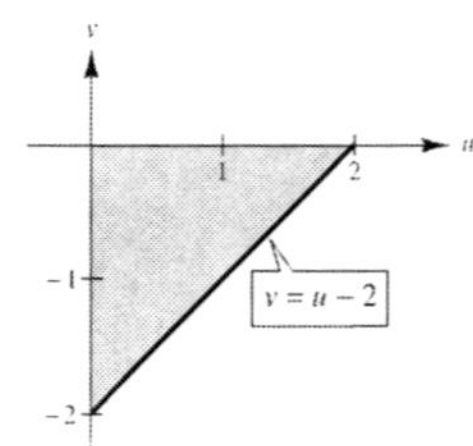

$\int_R\int 4(x+y)e^{x-y}\,dA = \int_0^2 \int_{u-2}^0 4ue^v\left(\dfrac{1}{2}\right)dv\,du$

$= \int_0^2 2u\left(1 - e^{u-2}\right)du$

$= 2\left[\dfrac{u^2}{2} - ue^{u-2} + e^{u-2}\right]_0^2$

$= 2\left(1 - e^{-2}\right)$

19. $\int_R \int e^{-xy/2}\, dA$

R: $y = \frac{x}{4}, y = 2x, y = \frac{1}{x}, y = \frac{4}{x}$

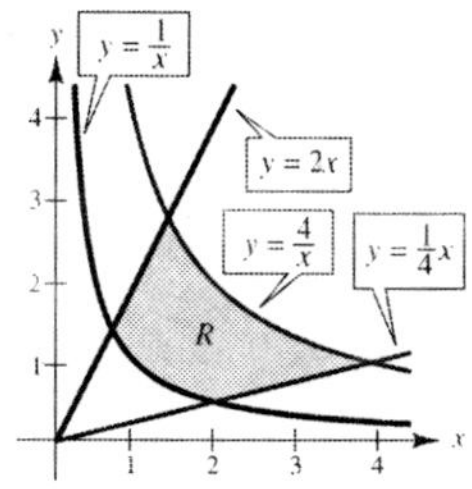

$x = \sqrt{v/u}, y = \sqrt{uv} \Rightarrow u = \frac{y}{x}, v = xy$

$$\frac{\partial(x, y)}{\partial(u, v)} = \begin{vmatrix} \frac{\partial x}{\partial u} & \frac{\partial x}{\partial v} \\ \frac{\partial y}{\partial u} & \frac{\partial y}{\partial v} \end{vmatrix} = \begin{vmatrix} -\frac{1}{2}\frac{v^{1/2}}{u^{3/2}} & \frac{1}{2}\frac{1}{u^{1/2}v^{1/2}} \\ \frac{1}{2}\frac{v^{1/2}}{u^{1/2}} & \frac{1}{2}\frac{u^{1/2}}{v^{1/2}} \end{vmatrix} = -\frac{1}{4}\left(\frac{1}{u} + \frac{1}{u}\right) = -\frac{1}{2u}$$

Transformed Region:

$y = \frac{1}{x} \Rightarrow yx = 1 \Rightarrow v = 1$

$y = \frac{4}{x} \Rightarrow ux = 4 \Rightarrow v = 4$

$y = 2x \Rightarrow \frac{y}{x} = 2 \Rightarrow u = 2$

$y = \frac{x}{4} \Rightarrow \frac{y}{x} = \frac{1}{4} \Rightarrow u = \frac{1}{4}$

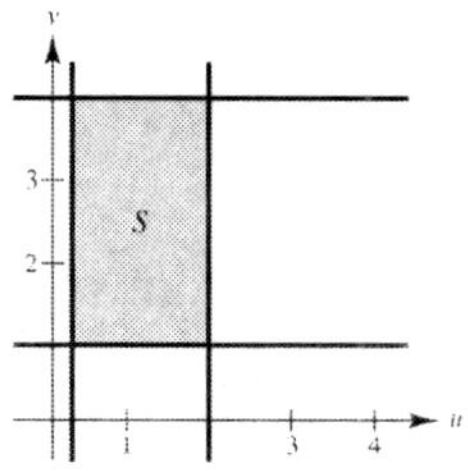

$$\int_R \int e^{-xy/2}\, dA = \int_{1/4}^{2}\int_{1}^{4} e^{-v/2}\left(\frac{1}{2u}\right)dv\, du = -\int_{1/4}^{2}\left[\frac{e^{-v/2}}{u}\right]_1^4 du = -\int_{1/4}^{2}\left(e^{-2} - e^{-1/2}\right)\frac{1}{u}\, du$$

$$= -\left[\left(e^{-2} - e^{-1/2}\right)\ln u\right]_{1/4}^{2} = -\left(e^{-2} - e^{-1/2}\right)\left(\ln 2 - \ln\frac{1}{4}\right) = \left(e^{-1/2} - e^{-2}\right)\ln 8 \approx 0.9798$$

20. $x = \frac{u}{v}$

$y = v$

$\frac{\partial x}{\partial u}\frac{\partial y}{\partial v} - \frac{\partial y}{\partial u}\frac{\partial x}{\partial v} = \frac{1}{v}$

$$\int_R \int y \sin xy\, dA = \int_1^4 \int_1^4 v(\sin u)\frac{1}{v}\, dv\, du = \int_1^4 3 \sin u\, du = \left[-3 \cos u\right]_1^4 = 3(\cos 1 - \cos 4) \approx 3.5818$$

21. $u = x - y = 1, \quad v = x + y = 1$

$u = x - y = -1, \quad v = x + y = 3$

$x = \frac{1}{2}(u + v)$

$y = \frac{1}{2}(v - u)$

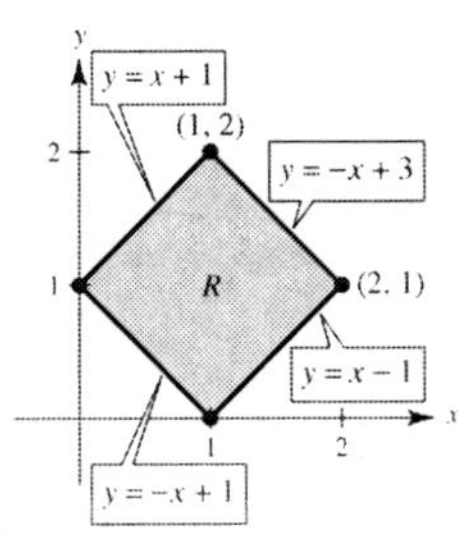

$$\frac{\partial(x, y)}{\partial(u, v)} = \frac{\partial x}{\partial u}\frac{\partial y}{\partial v} - \frac{\partial y}{\partial u}\frac{\partial x}{\partial v} = \frac{1}{2}\left(\frac{1}{2}\right) - \left(-\frac{1}{2}\right)\left(\frac{1}{2}\right) = \frac{1}{2}$$

$$\int_R \int 48 xy\, dA = \int_1^3 \int_{-1}^{1} 48\left(\frac{1}{2}\right)(u + v)\left(\frac{1}{2}\right)(v - u)\left(\frac{1}{2}\right) du\, dv = \int_1^3 \int_{-1}^{1} 6\left(v^2 - u^2\right) du\, dv = 6\int_1^3 \left[uv^2 - \frac{u^3}{3}\right]_{-1}^{1} dv$$

$$= 6\int_1^3 \left(2v^2 - \frac{2}{3}\right) dv = 6\left[\frac{2v^3}{3} - \frac{2}{3}v\right]_1^3 = 6\left[18 - 2 - \frac{2}{3} + \frac{2}{3}\right] = 96$$

22. $u = 2y - x = 0, \quad v = 3x + 2y = 0$

$u = 2y - x = 8, \quad v = 3x + 2y = 16$

$x = \frac{1}{4}(v - u)$

$y = \frac{1}{8}(v + 3u)$

$\frac{\partial(x, y)}{\partial(u, v)} = \left(-\frac{1}{4}\right)\left(\frac{1}{8}\right) - \left(\frac{3}{8}\right)\left(\frac{1}{4}\right) = -\frac{1}{8}$

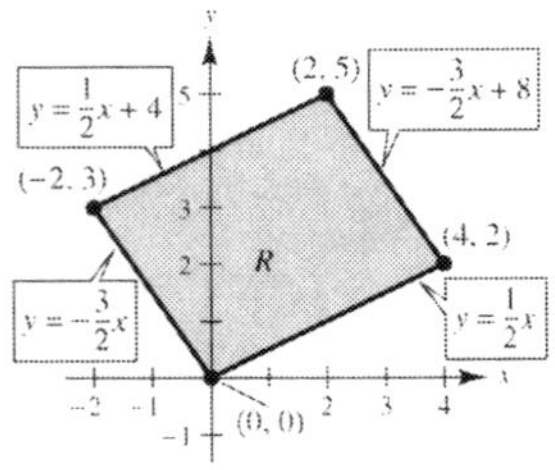

$$\int_R\int (3x + 2y)^2 \sqrt{2y - x}\, dA = \int_0^{16}\int_0^8 v\, u^{1/2}\, du\, dv = \int_0^{16}\left[\frac{2}{3}vu^{3/2}\right]_0^8 dv = \int_0^{16} \frac{32}{3}\sqrt{2}\, v\, dv = \frac{16^2}{2}\left(\frac{32}{3}\sqrt{2}\right) = \frac{4096\sqrt{2}}{3}$$

23. $u = x + y = 4, \quad v = x - y = 0$

$u = x + y = 8, \quad v = x - y = 4$

$x = \frac{1}{2}(u + v) \qquad y = \frac{1}{2}(u - v)$

$\frac{\partial(x, y)}{\partial(u, v)} = -\frac{1}{2}$

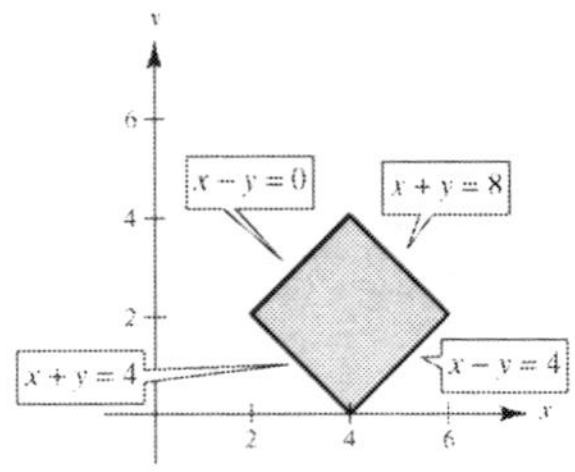

$$\int_R\int (x + y)e^{x-y}\, dA = \int_4^8\int_0^4 ue^v\left(\frac{1}{2}\right) dv\, du = \frac{1}{2}\int_4^8 u\left(e^4 - 1\right)du = \left[\frac{1}{4}u^2\left(e^4 - 1\right)\right]_4^8 = 12\left(e^4 - 1\right)$$

24. $u = x + y = \pi, \quad v = x - y = 0$

$u = x + y = 2\pi, \quad v = x - y = \pi$

$x = \frac{1}{2}(u + v), \qquad y = \frac{1}{2}(u - v)$

$\frac{\partial(x, y)}{\partial(u, v)} = -\frac{1}{2}$

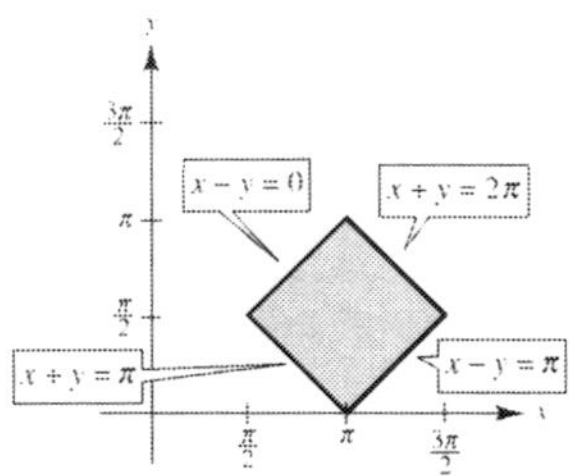

$$\int_R\int (x + y)^2 \sin^2(x - y)\, dA = \int_0^{\pi}\int_{\pi}^{2\pi} u^2 \sin^2 v\left(\frac{1}{2}\right) du\, dv = \int_0^{\pi}\left[\frac{1}{2}\left(\frac{u^3}{3}\right)\frac{1 - \cos 2v}{2}\right]_{\pi}^{2\pi} dy = \left[\frac{7\pi^3}{12}\left(v - \frac{1}{2}\sin 2v\right)\right]_0^{\pi} = \frac{7\pi^4}{12}$$

25. $u = x + 4y = 0, \quad v = x - y = 0$

$u = x + 4y = 5, \quad v = x - y = 5$

$x = \frac{1}{5}(u + 4v), \qquad y = \frac{1}{5}(u - v)$

$\frac{\partial x}{\partial u}\frac{\partial y}{\partial v} - \frac{\partial y}{\partial u}\frac{\partial x}{\partial v} = \left(\frac{1}{5}\right)\left(-\frac{1}{5}\right) - \left(\frac{1}{5}\right)\left(\frac{4}{5}\right) = -\frac{1}{5}$

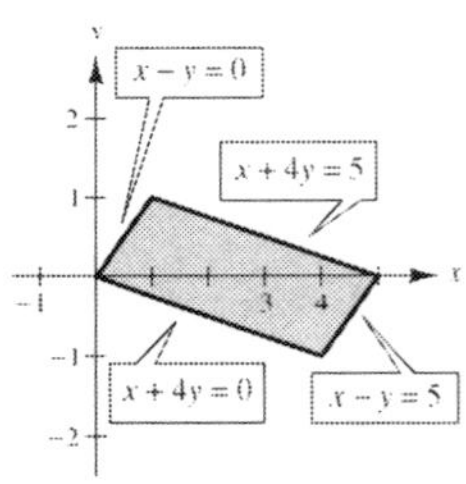

$$\int_R\int \sqrt{(x - y)(x + 4y)}\, dA = \int_0^5\int_0^5 \sqrt{uv}\left(\frac{1}{5}\right) du\, dv = \int_0^5\left[\frac{1}{5}\left(\frac{2}{3}\right)u^{3/2}\sqrt{v}\right]_0^5 dv = \left[\frac{2\sqrt{5}}{3}\left(\frac{2}{3}\right)v^{3/2}\right]_0^5 = \frac{100}{9}$$

26. $u = 3x + 2y = 0, \quad v = 2y - x = 0$

$u = 3x + 2y = 16, \quad v = 2y - x = 8$

$x = \frac{1}{4}(u - v), \qquad y = \frac{1}{8}(u + 3v)$

$\frac{\partial x}{\partial u}\frac{\partial y}{\partial v} - \frac{\partial y}{\partial u}\frac{\partial x}{\partial v} = \frac{1}{4}\left(\frac{3}{8}\right) - \frac{1}{8}\left(-\frac{1}{4}\right) = \frac{1}{8}$

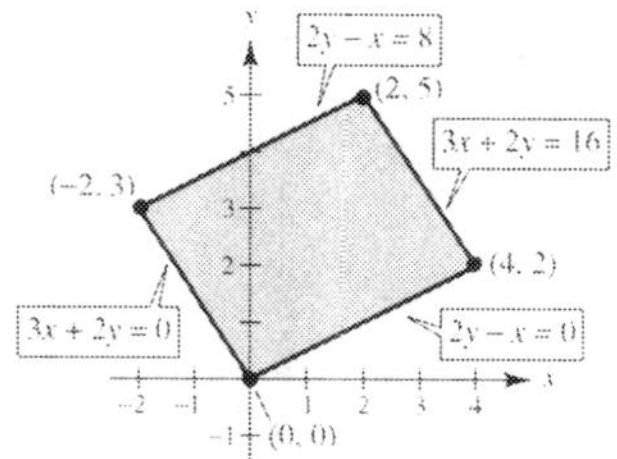

$$\int_R\int(3x + 2y)(2y - x)^{3/2}\,dA = \int_0^8\int_0^{16} u\,v^{3/2}\left(\frac{1}{8}\right)du\,dv = \int_0^8 16v^{3/2}\,dv = \left(\frac{2}{5}\right)16v^{5/2}\Big]_0^8 = \frac{4096}{5}\sqrt{2}$$

27. $u = x + y, v = x - y, x = \frac{1}{2}(u + v), y = \frac{1}{2}(u - v)$

$\frac{\partial x}{\partial u}\frac{\partial y}{\partial v} - \frac{\partial y}{\partial u}\frac{\partial x}{\partial v} = -\frac{1}{2}$

$$\int_R\int\sqrt{x + y}\,dA = \int_0^a\int_{-u}^{u}\sqrt{u}\left(\frac{1}{2}\right)dv\,du = \int_0^a u\sqrt{u}\,du = \left[\frac{2}{5}u^{5/2}\right]_0^a = \frac{2}{5}a^{5/2}$$

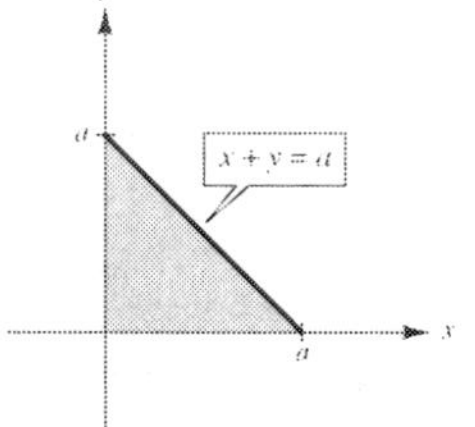

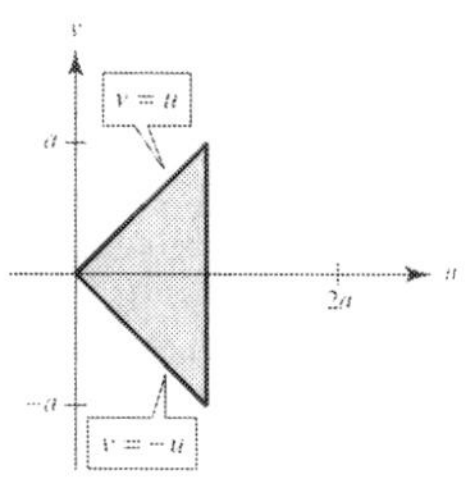

28. $u = x = 1, \quad v = xy = 1$

$u = x = 4, \quad v = xy = 4$

$x = u, \qquad y = \frac{u}{v}$

$\frac{\partial x}{\partial u}\frac{\partial y}{\partial v} - \frac{\partial y}{\partial u}\frac{\partial x}{\partial v} = \frac{1}{u}$

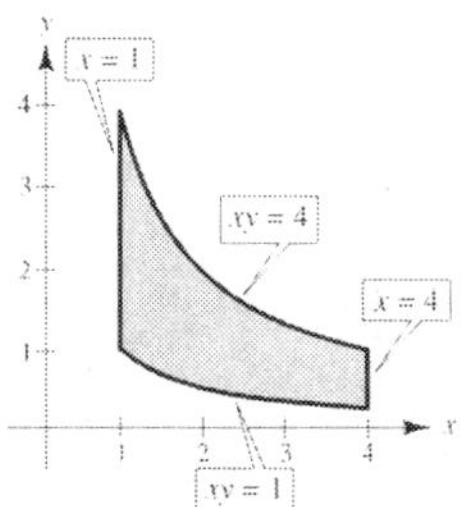

$$\int_R\int\frac{xy}{1 + x^2y^2}\,dA = \int_1^4\int_1^4\frac{v}{1 + v^2}\left(\frac{1}{u}\right)dv\,du = \int_1^4\left[\frac{1}{2}\ln\left(1 + v^2\right)\right]_1^4\frac{1}{u}\,du = \left[\frac{1}{2}[\ln 17 - \ln 2]\ln u\right]_1^4 = \frac{1}{2}\left(\ln\frac{17}{2}\right)(\ln 4)$$

29. $u = 2x - y$

$v = x + y$

$3x = u + v \Rightarrow x = \frac{1}{3}(u + v)$

Then $y = v - x = v - \frac{1}{3}(u + v) = \frac{1}{3}(2v - u)$.

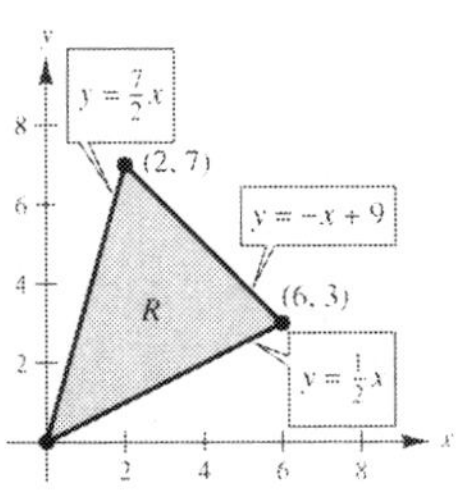

(x, y)	(u, v)
$(0, 0)$	$(0, 0)$
$(6, 3)$	$(9, 9)$
$(2, 7)$	$(-3, 9)$

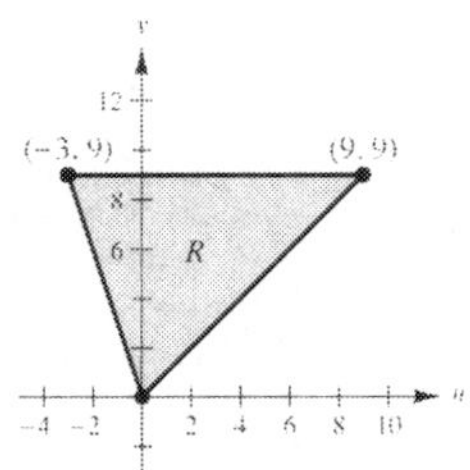

One side is parallel to the u-axis.

30. The transformation in (b) will make the region R into the simpler region S.

(x, y)	(u, v)
$(1, 1)$	$(0, 2)$
$(3, 3)$	$(0, 6)$
$(6, 4)$	$(-2, 6)$
$(4, 2)$	$(-2, 2)$

$$u = y - x \quad \Leftrightarrow \quad x = \frac{1}{2}(v - 3u)$$

$$v = 3y - x \quad \Leftrightarrow \quad y = \frac{1}{2}(v - u)$$

31. $\dfrac{x^2}{a^2} + \dfrac{y^2}{b^2} = 1,\ x = au,\ y = bv$

$$\frac{(au)^2}{a^2} + \frac{(bv)^2}{b^2} = 1$$

$$u^2 + v^2 = 1$$

(a) $\dfrac{x^2}{a^2} + \dfrac{y^2}{b^2} = 1$ $\qquad$ $u^2 + v^2 = 1$

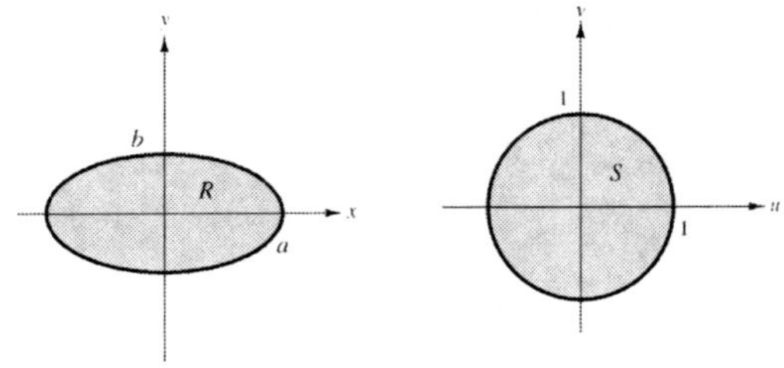

(b) $$\frac{\partial(x, y)}{\partial(u, v)} = \frac{\partial x}{\partial u}\frac{\partial y}{\partial v} - \frac{\partial y}{\partial u}\frac{\partial x}{\partial v} = (a)(b) - (0)(0) = ab$$

(c) $$A = \int_S\int ab\, dS = ab\left(\pi(1)^2\right) = \pi ab$$

32. (a) $f(x, y) = 16 - x^2 - y^2$

$R: \dfrac{x^2}{16} + \dfrac{y^2}{9} \le 1$

$V = \displaystyle\int_R\int f(x, y)\, dA$

Let $x = 4u$ and $y = 3v$.

$$\int_R\int\left(16 - x^2 - y^2\right) dA = \int_{-1}^{1}\int_{-\sqrt{1-u^2}}^{\sqrt{1-u^2}}\left(16 - 16u^2 - 9v^2\right)12\, dv\, du \quad (\text{Let } u = r\cos\theta,\ v = r\sin\theta.)$$

$$= \int_0^{2\pi}\int_0^1\left(16 - 16r^2\cos^2\theta - 9r^2\sin^2\theta\right)12r\, dr\, d\theta$$

$$= 12\int_0^{2\pi}\left[8r^2 - 4r^4\cos^2\theta - \frac{9}{4}r^4\sin^2\theta\right]_0^1 d\theta = 12\int_0^{2\pi}\left[8 - 4\cos^2\theta - \frac{9}{4}\sin^2\theta\right] d\theta$$

$$= 12\int_0^{2\pi}\left[8 - 4\left(\frac{1 + \cos 2\theta}{2}\right) - \frac{9}{4}\left(\frac{1 - \cos 2\theta}{2}\right)\right] d\theta = 12\int_0^{2\pi}\left[\frac{39}{8} - \frac{7}{8}\cos 2\theta\right] d\theta$$

$$= 12\left[\frac{39}{8}\theta - \frac{7}{16}\sin 2\theta\right]_0^{2\pi} = 12\left[\frac{39\pi}{4}\right] = 117\pi$$

(b) $f(x, y) = A\cos\left[\dfrac{\pi}{2}\sqrt{\dfrac{x^2}{a^2} + \dfrac{y^2}{b^2}}\right]$

$R: \dfrac{x^2}{a^2} + \dfrac{y^2}{b^2} \le 1$

Let $x = au$ and $y = bv$.

$$\int_R\int f(x, y)\, dA = \int_{-1}^{1}\int_{-\sqrt{1-u^2}}^{\sqrt{1-u^2}} A\cos\left[\frac{\pi}{2}\sqrt{u^2 + v^2}\right]ab\, dv\, du$$

Let $u = r\cos\theta$, $v = r\sin\theta$.

$$Aab\int_0^{2\pi}\int_0^1\cos\left[\frac{\pi}{2}r\right]r\, dr\, d\theta = Aab\left[\frac{2r}{\pi}\sin\left(\frac{\pi r}{2}\right) + \frac{4}{\pi^2}\cos\left(\frac{\pi r}{2}\right)\right]_0^1(2\pi)$$

$$= 2\pi\, Aab\left[\left(\frac{2}{\pi} + 0\right) - \left(0 + \frac{4}{\pi^2}\right)\right] = \frac{4(\pi - 2)Aab}{\pi}$$

33. Jacobian $= \dfrac{\partial(x, y)}{\partial(u, v)} = \dfrac{\partial x}{\partial u}\dfrac{\partial y}{\partial v} - \dfrac{\partial y}{\partial u}\dfrac{\partial x}{\partial v}$

34. See Theorem 14.5.

35. $x = u(1 - v),\ y = uv(1 - w),\ z = uvw$

$$\frac{\partial(x, y, z)}{\partial(u, v, w)} = \begin{vmatrix} 1-v & -u & 0 \\ v(1-w) & u(1-w) & -uv \\ vw & uw & uv \end{vmatrix} = (1-v)\left[u^2v(1-w) + u^2vw\right] + u\left[uv^2(1-w) + uv^2w\right] = (1-v)\left(u^2v\right) + u\left(uv^2\right) = u^2v$$

36. $x = 4u - v,\ y = 4v - w,\ z = u + w$

$$\frac{\partial(x, y, z)}{\partial(u, v, w)} = \begin{vmatrix} 4 & -1 & 0 \\ 0 & 4 & -1 \\ 1 & 0 & 1 \end{vmatrix} = 17$$

37. $x = \dfrac{1}{2}(u + v),\ y = \dfrac{1}{2}(u - v),\ z = 2uvw$

$$\frac{\partial(x, y, z)}{\partial(u, v, w)} = \begin{vmatrix} 1/2 & 1/2 & 0 \\ 1/2 & -1/2 & 0 \\ 2vw & 2uw & 2uv \end{vmatrix}$$

$$= 2uv\left[-1/4 - 1/4\right] = -uv$$

38. $x = u - v + w,\ y = 2uv,\ z = u + v + w$

$$\frac{\partial(x, y, z)}{\partial(u, v, w)} = \begin{vmatrix} 1 & -1 & 1 \\ 2v & 2u & 0 \\ 1 & 1 & 1 \end{vmatrix}$$

$$= 1(2u) + 1(2v) + 1(2v - 2u) = 4v$$

39. $x = \rho \sin\phi \cos\theta,\ y = \rho \sin\phi \sin\theta,\ z = \rho \cos\phi$

$$\frac{\partial(x, y, z)}{\partial(\rho, \theta, \phi)} = \begin{vmatrix} \sin\phi\cos\theta & -\rho\sin\phi\sin\theta & \rho\cos\phi\cos\theta \\ \sin\phi\sin\theta & \rho\sin\phi\cos\theta & \rho\cos\phi\sin\theta \\ \cos\phi & 0 & -\rho\sin\phi \end{vmatrix}$$

$$= \cos\phi\left[-\rho^2\sin\phi\cos\phi\sin^2\theta - \rho^2\sin\phi\cos\phi\cos^2\theta\right] - \rho\sin\phi\left[\rho\sin^2\phi\cos^2\theta + \rho\sin^2\phi\sin^2\theta\right]$$

$$= \cos\phi\left[-\rho^2\sin\phi\cos\phi\left(\sin^2\theta + \cos^2\theta\right)\right] - \rho\sin\phi\left[\rho\sin^2\phi\left(\cos^2\theta + \sin^2\theta\right)\right]$$

$$= -\rho^2\sin\phi\cos^2\phi - \rho^2\sin^3\phi = -\rho^2\sin\phi\left(\cos^2\phi + \sin^2\phi\right) = -\rho^2\sin\phi$$

40. $x = r\cos\theta,\ y = r\sin\theta,\ z = z$

$$\frac{\partial(x, y, z)}{\partial(r, \theta, z)} = \begin{vmatrix} \cos\theta & -r\sin\theta & 0 \\ \sin\theta & r\cos\theta & 0 \\ 0 & 0 & 1 \end{vmatrix} = 1\left[r\cos^2\theta + r\sin^2\theta\right] = r$$

41. Let $u = \dfrac{x}{3},\ v = y \Rightarrow \dfrac{\partial(x,y)}{\partial(u,v)} = \begin{vmatrix} 3 & 0 \\ 0 & 1 \end{vmatrix} = 3,\ y = \dfrac{x}{2} \Rightarrow v = \dfrac{3u}{2}.$

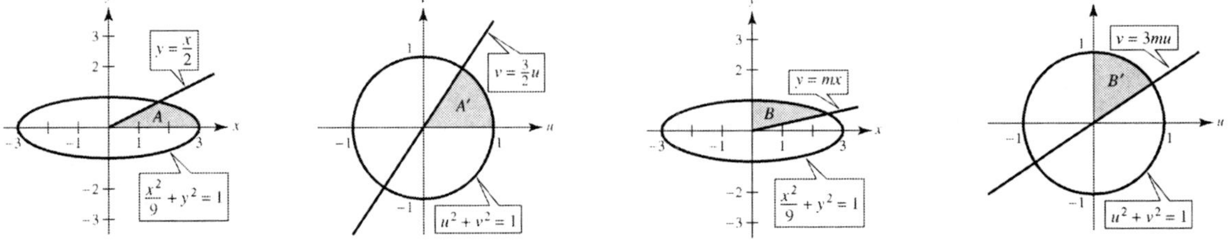

Region A is transformed to region A', and region B is transformed to region B'.

$A' = B' \Rightarrow \dfrac{2}{3} = 3m \Rightarrow m = \dfrac{2}{9}$

Note: You could also calculate the integrals directly.

Review Exercises for Chapter 14

1. $\int_0^{2x} xy^3\,dy = \left[\dfrac{xy^4}{4}\right]_0^{2x} = \dfrac{x(2x)^4}{4} = 4x^5$

2. $\int_y^{2y} \left(x^2 + y^2\right) dx = \left[\dfrac{x^3}{3} + xy^2\right]_y^{2y} = \dfrac{10y^3}{3}$

3. $\int_0^1 \int_0^{1+x} (3x + 2y)\,dy\,dx = \int_0^1 \left[3xy + y^2\right]_0^{1+x} dx$

$= \int_0^1 \left(4x^2 + 5x + 1\right) dx = \left[\frac{4}{3}x^3 + \frac{5}{2}x^2 + x\right]_0^1 = \frac{29}{6}$

4. $\int_0^2 \int_{x^2}^{2x} \left(x^2 + 2y\right) dy\,dx = \int_0^2 \left[x^2y + y^2\right]_{x^2}^{2x} dx$

$= \int_0^2 \left(4x^2 + 2x^3 - 2x^4\right) dx$

$= \left[\frac{4}{3}x^3 + \frac{1}{2}x^4 - \frac{2}{5}x^5\right]_0^2 = \frac{88}{15}$

5. $\int_0^3 \int_0^{\sqrt{9-x^2}} 4x\,dy\,dx = \int_0^3 4x\sqrt{9 - x^2}\,dx$

$= \left[-\frac{4}{3}\left(9 - x^2\right)^{3/2}\right]_0^3 = 36$

6. $\int_0^1 \int_0^{2y} \left(9 + 3x^2 + 3y^2\right) dx\,dy$

$= \int_0^1 \left[9x + x^3 + 3xy^2\right]_0^{2y} dy$

$= \int_0^1 \left(18y + 8y^3 + 6y^3\right) dy$

$= \int_0^1 \left(18y + 14y^3\right) dy$

$= \left[9y^2 + \dfrac{7}{2}y^4\right]_0^1 = \dfrac{25}{2}$

7. $A = \int_0^1 \int_0^{3-3y} dx\,dy = \int_0^1 (3 - 3y)\,dy$

$= \left[3y - \dfrac{3y^2}{2}\right]_0^1$

$= \dfrac{3}{2}$

8. The graphs intersect at $(0, 0)$ and $(4, 8)$.

$A = \int_0^4 \int_{x^2-2x}^{6x-x^2} dy\,dx$

$= \int_0^4 \left(8x - 2x^2\right) dx = \left[4x^2 - \frac{2}{3}x^3\right]_0^4 = \frac{64}{3}$

9. $A = \int_0^4 \int_x^{2x+2} dy\,dx$

$= \int_0^4 (x + 2)\,dx$

$= \left[\dfrac{x^2}{2} + 2x\right]_0^4 = 16$

10. $A = \int_0^2 \int_0^{y^2+1} dx\,dy = \int_0^1 \int_0^2 dy\,dx + \int_1^5 \int_{\sqrt{x-1}}^2 dy\,dx = \frac{14}{3}$

11.

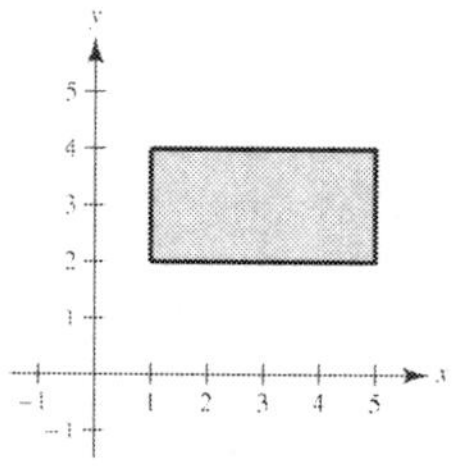

$$\int_1^5 \int_2^4 dy\,dx = \int_1^5 \Big[y\Big]_2^4 dx = \int_1^5 2\,dx = \Big[2x\Big]_1^5 = 8$$

$$\int_2^4 \int_1^5 dx\,dy = \int_2^4 \Big[x\Big]_1^5 dy = \int_2^4 4\,dy = \Big[4x\Big]_2^4 = 8$$

12.

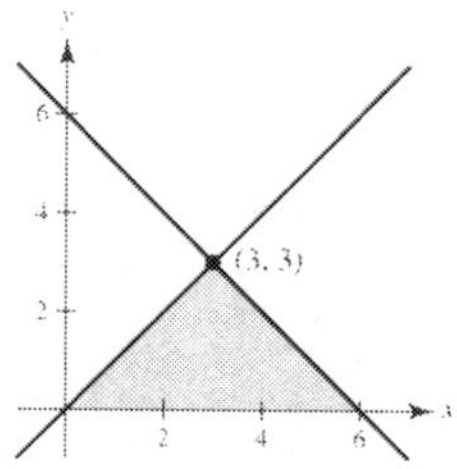

$$\int_0^3 \int_0^x dy\,dx + \int_3^6 \int_0^{6-x} dy\,dx = \int_0^3 \int_y^{6-y} dx\,dy$$

$$\int_0^3 x\,dx + \int_3^6 (6 - x)\,dx = \int_0^3 \big[(6 - y) - y\big]\,dy$$

$$\left[\frac{x^2}{2}\right]_0^3 + \left[6x - \frac{x^2}{2}\right]_3^6 = \left[6y - y^2\right]_0^3$$

$$\frac{9}{2} + \frac{9}{2} = 9$$

13. $\int_0^4 \int_{2x}^8 dy\,dx = \int_0^4 (8 - 2x)\,dx = \left[8x - x^2\right]_0^4 = 16$

$$\int_0^8 \int_0^{y/2} dx\,dy = \int_0^8 \frac{y}{2}\,dy = \left[\frac{y^2}{4}\right]_0^8 = 16$$

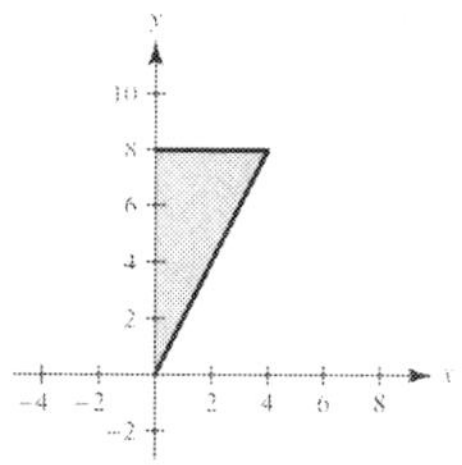

14. $\int_{-3}^3 \int_0^{9-y^2} dx\,dy = \int_{-3}^3 \left(9 - y^2\right) dy = \left[9y - \frac{y^3}{3}\right]_{-3}^3 = (27 - 9) - (-27 + 9) = 36$

$$\int_0^9 \int_{-\sqrt{9-x}}^{\sqrt{9-x}} dy\,dx = \int_0^9 2\sqrt{9 - x}\,dx = \left[-\frac{4}{3}(9 - x)^{3/2}\right]_0^9 = 0 + \frac{4}{3}(9)^{3/2} = 36$$

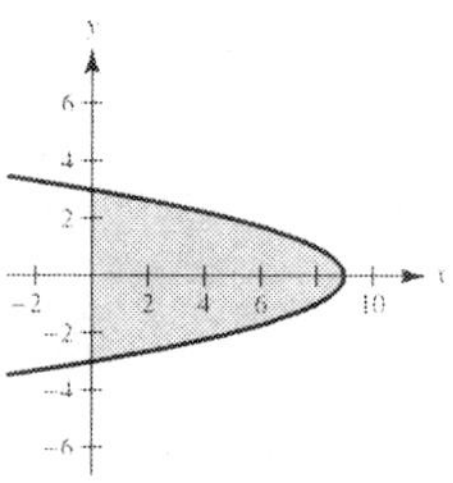

15. $\iint_R 4xy\,dA = \int_0^4 \int_0^2 4xy\,dx\,dy = \int_0^2 \int_0^4 4xy\,dy\,dx$

$$\int_0^4 \int_0^2 4xy\,dx\,dy = \int_0^4 \Big[2x^2y\Big]_0^2 dy$$

$$= \int_0^4 8y\,dy$$

$$= \Big[4y^2\Big]_0^4 = 64$$

16. $\iint_R 6x^2\,dA = \int_0^1 \int_0^{\sqrt{x}} 6x^2\,dy\,dx = \int_0^1 \int_{y^2}^1 6x^2\,dx\,dy$

$$\int_0^1 \int_0^{\sqrt{x}} 6x^2\,dy\,dx = \int_0^1 6x^2\sqrt{x}\,dx$$

$$= \left[\frac{12}{7}x^{7/2}\right]_0^1 = \frac{12}{7}$$

17. $V = \int_0^3 \int_0^2 (5 - x)\, dy\, dx$

$= \int_0^3 (10 - 2x)\, dx$

$= \left[10x - x^2\right]_0^3$

$= 30 - 9 = 21$

18. $V = \int_0^2 \int_0^x 4\, dy\, dx$

$= \int_0^2 4x\, dx$

$= \left[2x^2\right]_0^2 = 8$

19. $V = \int_{-1}^1 \int_{-1}^1 \left(4 - x^2 - y^2\right) dy\, dx$

$= \int_{-1}^1 \left[4y - x^2y - \dfrac{y^3}{3}\right]_{-1}^1 dx$

$= \int_{-1}^1 \left[\left(4 - x^2 - \dfrac{1}{3}\right) - \left(-4 + x^2 - \dfrac{1}{3}\right)\right] dx$

$= \int_{-1}^1 \left[\dfrac{22}{3} - 2x^2\right] dx$

$= \left[\dfrac{22}{3}x - \dfrac{2x^3}{3}\right]_{-1}^1$

$= \dfrac{40}{3}$

Alternate Solution: $V = 4\int_0^1 \int_0^1 \left(4 - x^2 - y^2\right) dy\, dx$

$= 4\int_0^1 \left(4 - x^2 - \dfrac{1}{3}\right) dx$

$= 4\int_0^1 \left(\dfrac{11}{3} - x^2\right) dx = 4\left[\dfrac{11}{3}x - \dfrac{1}{3}x^3\right]_0^1 = \dfrac{40}{3}$

20. $\int_0^2 \int_0^{2-x} (2 - x - y)\, dy\, dx$

$= \int_0^2 \left[2(2 - x) - x(2 - x) - \dfrac{(2 - x)^2}{2}\right] dx$

$= \int_0^2 \left[4 - 2x - 2x + x^2 - \dfrac{1}{2}\left(4 - 4x + x^2\right)\right] dx$

$= \int_0^2 \left(\dfrac{1}{2}x^2 - 2x + 2\right) dx$

$= \left[\dfrac{1}{6}x^3 - x^2 + 2x\right]_0^2$

$= \dfrac{4}{3} - 4 + 4$

$= \dfrac{4}{3}$

21. Area $R = 16$

Average Value $= \dfrac{1}{16}\int_{-2}^2 \int_{-2}^2 \left(16 - x^2 - y^2\right) dy\, dx$

$= \dfrac{1}{16}\int_{-2}^2 \left[16y - x^2y - \dfrac{y^3}{3}\right]_{-2}^2 dx$

$= \dfrac{1}{16}\int_{-2}^2 \left[64 - 4x^2 - \dfrac{16}{3}\right] dx$

$= \dfrac{1}{16}\left[64x - \dfrac{4x^3}{3} - \dfrac{16}{3}x\right]_{-2}^2$

$= \dfrac{1}{16}\left[256 - \dfrac{64}{3} - \dfrac{64}{3}\right] = \dfrac{40}{3}$

22. Area $R = 9$

Average Value $= \dfrac{1}{9}\int_0^3 \int_0^3 \left(2x^2 + y^2\right) dy\, dx = \dfrac{1}{9}\int_0^3 \left[2x^2y + \dfrac{y^3}{3}\right]_0^3 dx = \dfrac{1}{9}\int_0^3 \left(6x^2 + 9\right) dx = \dfrac{1}{9}\left[2x^3 + 9x\right]_0^3 = 9$

23. Area $R = 3(5) = 15$

$$\text{Average temperature} = \frac{1}{15}\int_0^3\int_0^5 \left(40 - 6x^2 - y^2\right) dy\, dx = \frac{1}{15}\int_0^3 \left[40y - 6x^2y - \frac{y^3}{3}\right]_0^5 dx$$

$$= \frac{1}{15}\int_0^3 \left[200 - 30x^2 - \frac{125}{3}\right] dx = \frac{1}{15}\left[200x - 10x^3 - \frac{125x}{3}\right]_0^3 = \frac{1}{15}\left[600 - 270 - 125\right] = 13\frac{2}{3}°\text{C}$$

24. $\text{Average} = \frac{1}{150}\int_{45}^{60}\int_{40}^{50}\left[192x + 576y - x^2 - 5y^2 - 2xy - 5000\right] dx\, dy \approx 13{,}246.67$

25. $$\int_0^h\int_0^x \sqrt{x^2 + y^2}\, dy\, dx = \int_0^{\pi/4}\int_0^{h\sec\theta} r^2\, dr\, d\theta$$

$$= \frac{h^3}{3}\int_0^{\pi/4}\sec^3\theta\, d\theta = \frac{h^3}{6}\Big[\sec\theta\tan\theta + \ln\left|\sec\theta + \tan\theta\right|\Big]_0^{\pi/4} = \frac{h^3}{6}\left[\sqrt{2} + \ln\left(\sqrt{2} + 1\right)\right]$$

26. $$\int_0^4\int_0^{\sqrt{16-y^2}}\left(x^2 + y^2\right) dx\, dy = \int_0^{\pi/2}\int_0^4 r^3\, dr\, d\theta = \int_0^{\pi/2}\left[\frac{r^4}{4}\right]_0^4 d\theta = \int_0^{\pi/2} 64\, d\theta = 32\pi$$

27. $$V = \int_0^{\pi/2}\int_0^3 (r\cos\theta)(r\sin\theta)^2\, r\, dr\, d\theta$$

$$= \int_0^{\pi/2}\int_0^3 \cos\theta\sin^2\theta\, r^4\, dr\, d\theta$$

$$= \int_0^{\pi/2}\cos\theta\sin^2\theta\left[\frac{r^5}{5}\right]_0^3 d\theta$$

$$= \frac{243}{5}\int_0^{\pi/2}\sin^2\theta\cos\theta\, d\theta$$

$$= \frac{243}{5}\left[\frac{\sin^3\theta}{3}\right]_0^{\pi/2} = \frac{81}{5}$$

28. $$V = \int_0^{2\pi}\int_0^4 \sqrt{25 - r^2}\, r\, dr\, d\theta$$

$$= \int_0^{2\pi}\left[\frac{-\left(25 - r^2\right)^{3/2}}{3}\right]_0^4 d\theta$$

$$= \int_0^{2\pi}\left(-\frac{(25 - 16)^{3/2}}{3} + \frac{25^{3/2}}{3}\right) d\theta$$

$$= \int_0^{2\pi}\frac{98}{3}\, d\theta$$

$$= \left[\frac{98}{3}\, d\theta\right]_0^{2\pi} = \frac{196\pi}{3}$$

29. $$A = 2\int_0^{\pi}\int_0^{2+\cos\theta} r\, dr\, d\theta = \int_0^{\pi}(2 + \cos\theta)^2 d\theta = \int_0^{\pi}\left[4 + 4\cos\theta + \frac{1 + \cos 2\theta}{2}\right] d\theta = \left[4\theta + 4\sin\theta + \frac{1}{2}\theta + \frac{\sin 2\theta}{4}\right]_0^{\pi} = \frac{9\pi}{2}$$

30. $$A = 4\int_0^{\pi/2}\int_0^{2\sin 2\theta} r\, dr\, d\theta = 2\int_0^{\pi/2}(2\sin 2\theta)^2 d\theta = 8\int_0^{\pi/2}\frac{1 - \cos 4\theta}{2}\, d\theta = 4\left[\theta - \frac{\sin 4\theta}{4}\right]_0^{\pi/2} = 2\pi$$

31.

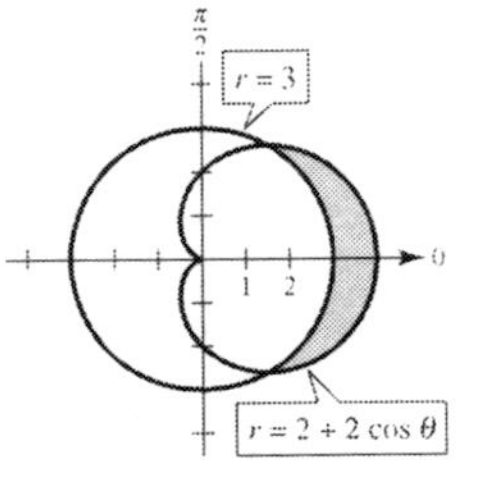

Intersection points: $3 = 2 + 2\cos\theta$

$$\cos\theta = \frac{1}{2}$$

$$\theta = \frac{\pi}{3}, \frac{5\pi}{3}$$

$$A = 2\int_0^{\pi/3}\int_3^{2+2\cos\theta} r\,dr\,d\theta$$

$$= 2\int_0^{\pi/3}\left[\frac{r^2}{2}\right]_3^{2+2\cos\theta} d\theta$$

$$= 2\int_0^{\pi/3}\left[\frac{(2+2\cos\theta)^2}{2} - \frac{9}{2}\right]d\theta$$

$$= \int_0^{\pi/3}\left[4 + 8\cos\theta + 4\cos^2\theta - 9\right]d\theta$$

$$= \int_0^{\pi/3}\left[8\cos\theta + 2(1+\cos 2\theta) - 5\right]d\theta$$

$$= \left[8\sin\theta + \sin 2\theta - 3\theta\right]_0^{\pi/3}$$

$$= \frac{8\sqrt{3}}{2} + \frac{\sqrt{3}}{2} - \pi = \frac{9\sqrt{3}}{2} - \pi$$

32.

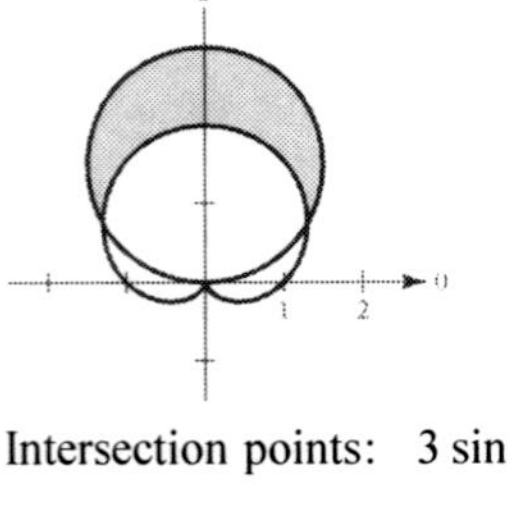

Intersection points: $3\sin\theta = 1 + \sin\theta$

$$\sin\theta = \frac{1}{2} \Rightarrow \theta = \frac{\pi}{6}, \frac{5\pi}{6}$$

$$A = 2\int_{\pi/6}^{\pi/2}\int_{1+\sin\theta}^{3\sin\theta} r\,dr\,d\theta$$

$$= 2\int_{\pi/6}^{\pi/2}\left[\frac{r^2}{2}\right]_{1+\sin\theta}^{3\sin\theta} d\theta$$

$$= \int_{\pi/6}^{\pi/2}\left[(3\sin\theta)^2 - (1+\sin\theta)^2\right]d\theta$$

$$= \int_{\pi/6}^{\pi/2}\left(8\sin^2\theta - 1 - 2\sin\theta\right)d\theta$$

$$= \int_{\pi/6}^{\pi/2}\left[4(1-\cos 2\theta) - 1 - 2\sin\theta\right]d\theta$$

$$= \left[3\theta - 2\sin 2\theta + 2\cos\theta\right]_{\pi/6}^{\pi/2}$$

$$= \frac{3\pi}{2} - \left(\frac{\pi}{2} - \sqrt{3} + \sqrt{3}\right) = \pi$$

33. (a) $\left(x^2 + y^2\right)^2 = 9\left(x^2 - y^2\right)$

$$\left(r^2\right)^2 = 9\left(r^2\cos^2\theta - r^2\sin^2\theta\right)$$

$$r^2 = 9\left(\cos^2\theta - \sin^2\theta\right)$$

$$= 9\cos 2\theta$$

$$r = 3\sqrt{\cos 2\theta}$$

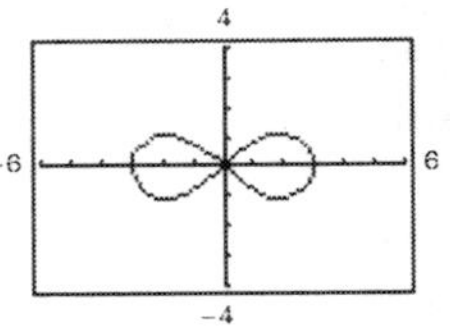

(b) $A = 4\int_0^{\pi/4}\int_0^{3\sqrt{\cos 2\theta}} r\,dr\,d\theta = 9$

(c) $V = 4\int_0^{\pi/4}\int_0^{3\sqrt{\cos 2\theta}} \sqrt{9 - r^2}\,r\,dr\,d\theta \approx 20.392$

34. $\tan\theta = \dfrac{12\sqrt{13}}{8\sqrt{13}} = \dfrac{3}{2} \Rightarrow \theta \approx 0.9828$

The polar region is given by $0 \le r \le 4$ and $0 \le \theta \le 0.9828$. So,

$$\int_0^{\arctan(3/2)}\int_0^4 (r\cos\theta)(r\sin\theta)\,r\,dr\,d\theta = \frac{288}{13}.$$

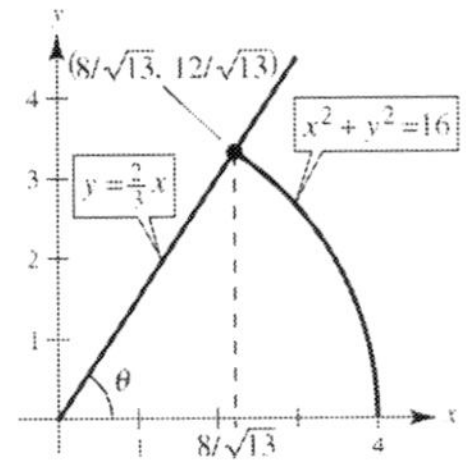

35. $m = \int_0^2\int_0^{x^3} kx\,dy\,dx = \dfrac{32k}{5}$

$$M_x = \int_0^2\int_0^{x^3} kxy\,dy\,dx = 16k$$

$$M_y = \int_0^2\int_0^{x^3} kx^2\,dy\,dx = \frac{32k}{3}$$

$$\bar{x} = \frac{M_y}{m} = \frac{5}{3}$$

$$\bar{y} = \frac{M_x}{m} = \frac{5}{2}$$

$$(\bar{x}, \bar{y}) = \left(\frac{5}{3}, \frac{5}{2}\right)$$

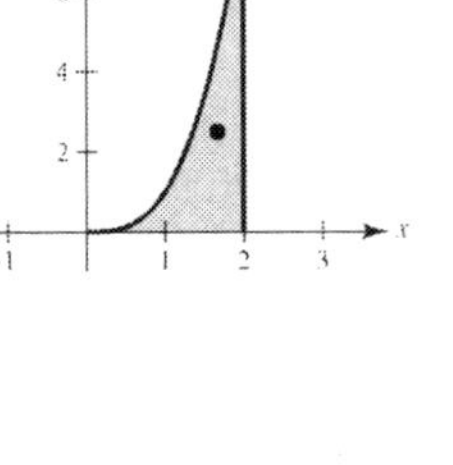

36. $m = \int_1^2 \int_0^{2/x} ky\,dy\,dx = k$

$M_x = \int_1^2 \int_0^{2/x} ky^2\,dy\,dx = k$

$M_y = \int_1^2 \int_0^{2/x} kxy\,dy\,dx = 2k \ln 2$

$\bar{x} = \dfrac{M_y}{m} = 2 \ln 2$

$\bar{y} = \dfrac{M_x}{m} = 1$

$(\bar{x}, \bar{y}) = (2 \ln 2, 1)$

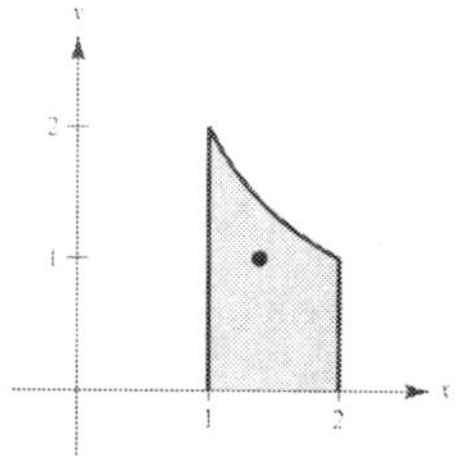

37. $m = k\int_0^1 \int_{2x^3}^{2x} xy\,dy\,dx = \dfrac{k}{4}$

$M_x = k\int_0^1 \int_{2x^3}^{2x} xy^2\,dy\,dx = \dfrac{16k}{55}$

$M_y = k\int_0^1 \int_{2x^3}^{2x} x^2y\,dy\,dx = \dfrac{8k}{45}$

$\bar{x} = \dfrac{M_y}{m} = \dfrac{32}{45}$

$\bar{y} = \dfrac{M_x}{m} = \dfrac{64}{55}$

$(\bar{x}, \bar{y}) = \left(\dfrac{32}{45}, \dfrac{64}{55}\right)$

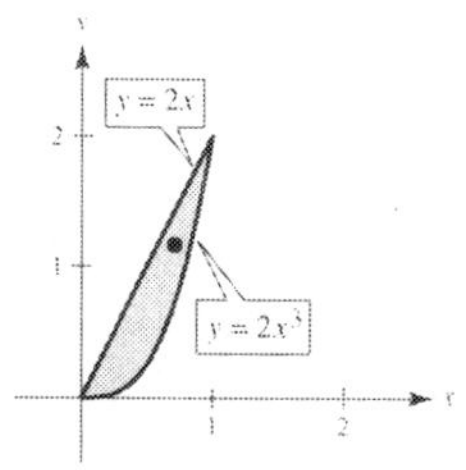

38. $m = \int_0^6 \int_0^{6-x} kx^2\,dy\,dx = 108k$

$M_x = \int_0^6 \int_0^{6-x} kyx^2\,dy\,dx = \dfrac{648k}{5}$

$M_y = \int_0^6 \int_0^{6-x} kx^3\,dy\,dx = \dfrac{1944k}{5}$

$\bar{x} = \dfrac{M_y}{m} = \dfrac{18}{5}$

$\bar{y} = \dfrac{M_x}{m} = \dfrac{6}{5}$

$(\bar{x}, \bar{y}) = \left(\dfrac{18}{5}, \dfrac{6}{5}\right)$

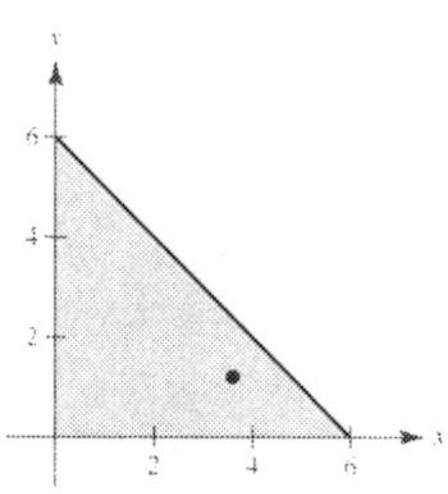

39. $I_x = \int_R \int y^2 \rho(x, y)\,dA = \int_0^a \int_0^b kxy^2\,dy\,dx = \dfrac{1}{6}kb^3a^2$

$I_y = \int_R \int x^2 \rho(x, y)\,dA = \int_0^a \int_0^b kx^3\,dy\,dx = \dfrac{1}{4}kba^4$

$I_0 = I_x + I_y = \dfrac{1}{6}kb^3a^2 + \dfrac{1}{4}kba^4 = \dfrac{ka^2b}{12}\left(2b^2 + 3a^2\right)$

$m = \int_R \int \rho(x, y)\,dA = \int_0^a \int_0^b kx\,dy\,dx = \dfrac{1}{2}kba^2$

$\bar{\bar{x}} = \sqrt{\dfrac{I_y}{m}} = \sqrt{\dfrac{(1/4)kba^4}{(1/2)kba^2}} = \sqrt{\dfrac{a^2}{2}} = \dfrac{a\sqrt{2}}{2}$

$\bar{\bar{y}} = \sqrt{\dfrac{I_x}{m}} = \sqrt{\dfrac{(1/6)kb^3a^2}{(1/2)kba^2}} = \sqrt{\dfrac{b^2}{3}} = \dfrac{b\sqrt{3}}{3}$

40. $I_x = \int_R \int y^2 \rho(x, y)\,dA = \int_0^2 \int_0^{4-x^2} ky^3\,dy\,dx = \dfrac{16{,}384}{315}k$

$I_y = \int_R \int x^2 \rho(x, y)\,dA = \int_0^2 \int_0^{4-x^2} kx^2y\,dy\,dx = \dfrac{512}{105}k$

$I_0 = I_x + I_y = \dfrac{16{,}384k}{315} + \dfrac{512k}{105} = \dfrac{17{,}920}{315}k = \dfrac{512}{9}k$

$m = \int_R \int \rho(x, y)\,dA = \int_0^2 \int_0^{4-x^2} ky\,dy\,dx = \dfrac{128}{15}k$

$\bar{\bar{x}} = \sqrt{\dfrac{I_y}{m}} = \sqrt{\dfrac{512k/105}{128k/15}} = \sqrt{\dfrac{4}{7}} = \dfrac{2\sqrt{7}}{7}$

$\bar{\bar{y}} = \sqrt{\dfrac{I_x}{m}} = \sqrt{\dfrac{16{,}384k/315}{128k/15}} = \sqrt{\dfrac{128}{21}} = \dfrac{8\sqrt{42}}{21}$

41. $f(x, y) = 25 - x^2 - y^2$

$f_x = -2x,\ f_y = -2y$

$S = \int_R \int \sqrt{1 + (f_x)^2 + (f_y)^2}\,dA$

$= \int_R \int \sqrt{1 + 4x^2 + 4y^2}\,dA$

$= 4\int_0^{\pi/2} \int_0^5 \sqrt{1 + 4r^2}\,r\,dr\,d\theta$

$= \dfrac{1}{3}\int_0^{\pi/2} \left[\left(1 + 4r^2\right)^{3/2}\right]_0^5 d\theta$

$= \dfrac{1}{3}\int_0^{\pi/2} \left[(101)^{3/2} - 1\right] d\theta$

$= \dfrac{\pi}{6}\left[101\sqrt{101} - 1\right]$

42. $f(x, y) = 8 + 4x - 5y$

$f_x = 4,\ f_y = -5$

$S = \int_R \int \sqrt{1 + (f_x)^2 + (f_y)^2}\,dA$

$= \int_R \int \sqrt{42}\,dA$

$= \sqrt{42} \iint_R dA$

The area of R is $\pi r^2 = \pi$. Hence, $S = \sqrt{42}\pi$.

43. $f(x, y) = 9 - y^2$

$f_x = 0,\ f_y = -2y$

$$S = \int_R\int\sqrt{1 + (f_x)^2 + (f_y)^2}\,dA$$

$$= \int_0^3\int_{-y}^{y}\sqrt{1 + 4y^2}\,dx\,dy = \int_0^3\left[\sqrt{1 + 4y^2}\,x\right]_{-y}^{y}dy = \int_0^3 2y\sqrt{1 + 4y^2}\,dy = \frac{1}{4}\frac{2}{3}\left(1 + 4y^2\right)^{3/2}\Big]_0^3 = \frac{1}{6}\left[(37)^{3/2} - 1\right]$$

44. $f(x, y) = 4 - x^2,\ f_x = -2x,\ f_y = 0$

$$S = \int_R\int\sqrt{1 + 4x^2}\,dA = \int_{-2}^{0}\int_{-x}^{2}\sqrt{1 + 4x^2}\,dy\,dx + \int_0^2\int_x^2\sqrt{1 + 4x^2}\,dy\,dx$$

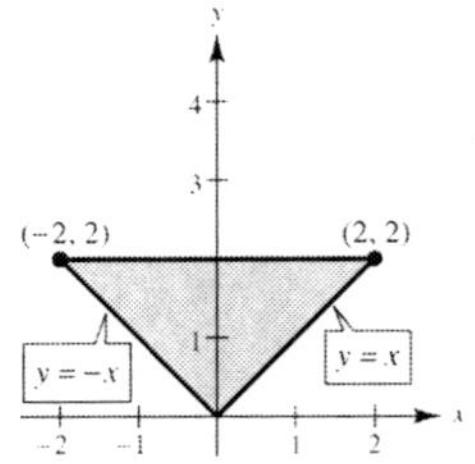

These integrals are equal by symmetry.

$$S = 2\int_0^2\int_x^2\sqrt{1 + 4x^2}\,dy\,dx = \int_0^2\left[2\sqrt{1 + 4x^2} - x\sqrt{1 + 4x^2}\right]dx$$

$$= 2\left[\frac{1}{2}\ln\left(\sqrt{1 + 4x^2} + 2x\right) + x\sqrt{1 + 4x^2} - \frac{1}{12}\left(1 + 4x^2\right)^{3/2}\right]_0^2 = 2\left[\frac{1}{2}\ln\left(\sqrt{17} + 4\right) + 2\sqrt{17} - \frac{17}{12}\sqrt{17} + \frac{1}{12}\right] \approx 7.0717$$

45. (a) $$V = \int_0^{50}\int_0^{\sqrt{50^2 - x^2}}\left(20 + \frac{xy}{100} - \frac{x + y}{5}\right)dy\,dx = \int_0^{50}\left[20\sqrt{50^2 - x^2} + \frac{x}{200}\left(50^2 - x^2\right) - \frac{x}{5}\sqrt{50^2 - x^2} - \frac{50^2 - x^2}{10}\right]dy$$

$$= \left[10\left(x\sqrt{50 - x^2} + 50^2\arcsin\frac{x}{50}\right) + \frac{25}{4}x^2 - \frac{x^4}{800} + \frac{1}{15}\left(50^2 - x^2\right)^{3/2} - 250x + \frac{x^3}{30}\right]_0^{50} \approx 30{,}415.74\text{ ft}^3$$

(b) $z = 20 + \dfrac{xy}{100}$

$$\sqrt{1 + (f_x)^2 + (f_y)^2} = \sqrt{1 + \frac{y^2}{100^2} + \frac{x^2}{100^2}} = \frac{\sqrt{100^2 + x^2 + y^2}}{100}$$

$$S = \frac{1}{100}\int_0^{50}\int_0^{\sqrt{50^2 - x^2}}\sqrt{100^2 + x^2 + y^2}\,dy\,dx = \frac{1}{100}\int_0^{\pi/2}\int_0^{50}\sqrt{100^2 + r^2}\,r\,dr\,d\theta \approx 2081.53\text{ ft}^2$$

46. (a) Graph of $f(x, y) = z = 25\left[1 + e^{-(x^2 + y^2)/1000}\cos^2\left(\dfrac{x^2 + y^2}{1000}\right)\right]$

over region R

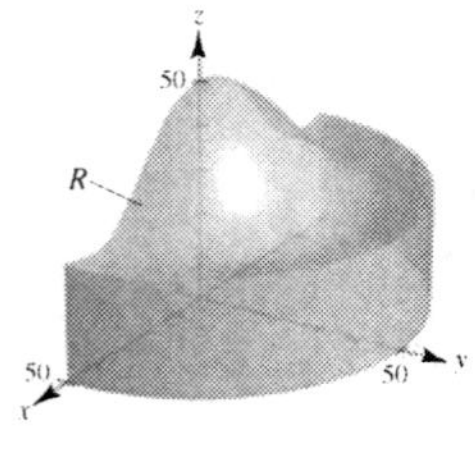

(b) Surface area $= \int_R\int\sqrt{1 + f_x(x, y)^2 + f_y(x, y)^2}\,dA$

Using a symbolic computer program, you obtain surface area $\approx$ 4540 sq. ft.

47. $$\int_0^4 \int_0^1 \int_0^2 (2x + y + 4z)\,dy\,dz\,dx = \int_0^4 \int_0^1 \left[2xy + \frac{y^2}{2} + 4zy\right]_0^2 dz\,dx$$
$$= \int_0^4 \int_0^1 (4x + 2 + 8z)\,dz\,dx$$
$$= \int_0^4 \left[4xz + 2z + 4z^2\right]_0^1 dx$$
$$= \int_0^4 (4x + 2 + 4)\,dx$$
$$= \left[2x^2 + 6x\right]_0^4 = 56$$

48. $$\int_0^2 \int_0^y \int_0^{xy} y\,dz\,dx\,dy = \int_0^2 \int_0^y \left[yz\right]_0^{xy} dx\,dy$$
$$= \int_0^2 \int_0^y xy^2\,dx\,dy = \int_0^2 \left[\frac{x^2y^2}{2}\right]_0^y dy = \int_0^2 \frac{y^4}{2}\,dy = \left[\frac{y^5}{10}\right]_0^2 = \frac{16}{5}$$

49. $$\int_0^a \int_0^b \int_0^c \left(x^2 + y^2 + z^2\right) dx\,dy\,dz = \int_0^a \int_0^b \left(\tfrac{1}{3}c^3 + cy^2 + cz^2\right) dy\,dz$$
$$= \int_0^a \left(\tfrac{1}{3}bc^3 + \tfrac{1}{3}b^3c + bcz^2\right) dz = \tfrac{1}{3}abc^3 + \tfrac{1}{3}ab^3c + \tfrac{1}{3}a^3bc = \tfrac{1}{3}abc\left(a^2 + b^2 + c^2\right)$$

50. $$\int_0^3 \int_{\pi/2}^{\pi} \int_2^5 z \sin x\,dy\,dx\,dz = \int_0^3 \int_{\pi/2}^{\pi} \left[yz \sin x\right]_2^5 dx\,dz$$
$$= \int_0^3 \int_{\pi/2}^{\pi} 3z \sin x\,dx\,dz$$
$$= \int_0^3 \left[-3z \cos x\right]_{\pi/2}^{\pi} dz$$
$$= \int_0^3 3z\,dz$$
$$= \left[\frac{3z^2}{2}\right]_0^3 = \frac{27}{2}$$

51. $$\int_{-1}^1 \int_{-\sqrt{1-x^2}}^{\sqrt{1-x^2}} \int_{-\sqrt{1-x^2-y^2}}^{\sqrt{1-x^2-y^2}} \left(x^2 + y^2\right) dz\,dy\,dx = \int_0^{2\pi} \int_0^1 \int_{-\sqrt{1-r^2}}^{\sqrt{1-r^2}} r^3\,dz\,dr\,d\theta = \frac{8\pi}{15}$$

52. $$\int_0^2 \int_0^{\sqrt{4-x^2}} \int_0^{\sqrt{4-x^2-y^2}} xyz\,dz\,dy\,dx = \tfrac{4}{3}$$

53. $$V = \int_0^3 \int_0^4 \int_0^{xy} dz\,dy\,dx$$
$$= \int_0^3 \int_0^4 xy\,dy\,dx$$
$$= \int_0^3 \left[\frac{xy^2}{2}\right]_0^4 dx$$
$$= \int_0^3 8x\,dx$$
$$= \left[4x^2\right]_0^3 = 36$$

54. $$V = \int_0^3 \int_0^y \int_0^{8-x-y} dz\,dx\,dy$$
$$= \int_0^3 \int_0^y (8 - x - y)\,dx\,dy$$
$$= \int_0^3 \left[8x - \frac{x^2}{2} - xy\right]_0^y dy$$
$$= \int_0^3 \left[8y - \frac{y^2}{2} - y^2\right] dy$$
$$= \left[4y^2 - \frac{y^3}{2}\right]_0^3$$
$$= 36 - \frac{27}{2} = \frac{45}{2}$$

55. $\int_0^1 \int_0^y \int_0^{\sqrt{1-x^2}} dz\,dx\,dy = \int_0^1 \int_x^1 \int_0^{\sqrt{1-x^2}} dz\,dy\,dx$

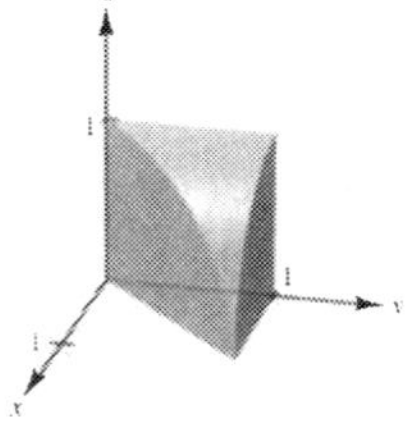

56. $\int_0^6 \int_0^{6-x} \int_0^{6-x-y} dz\,dy\,dx = \int_0^6 \int_0^{6-z} \int_0^{6-z-x} dy\,dx\,dz$

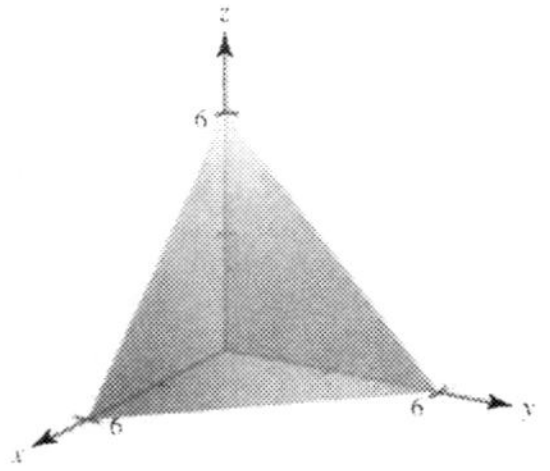

57. $m = \int_0^{10} \int_0^{10-x} \int_0^{10-x-y} k\,dz\,dy\,dx = \dfrac{500}{3}k$

$Myz = \int_0^{10} \int_0^{10-x} \int_0^{10-x-y} kx\,dz\,dy\,dx = \dfrac{1250}{3}k$

$\overline{x} = \dfrac{Myz}{m} = \dfrac{5}{2}$

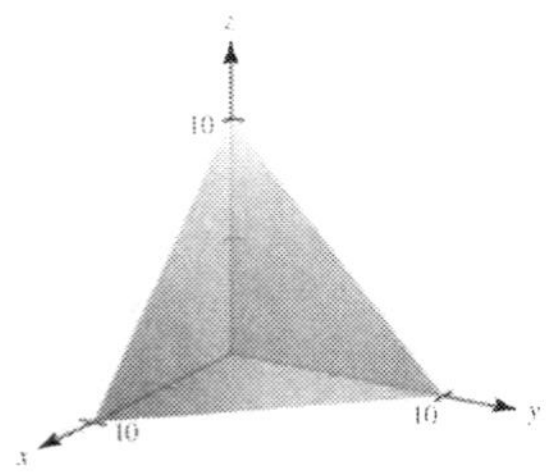

58. $m = \int_0^5 \int_0^5 \int_0^{5-y} kx\,dz\,dy\,dx = \dfrac{625\,k}{4}$

$Mxz = \int_0^5 \int_0^5 \int_0^{5-y} kxy\,dz\,dy\,dx = \dfrac{3125\,k}{12}$

$\overline{y} = \dfrac{Mxz}{m} = \dfrac{5}{3}$

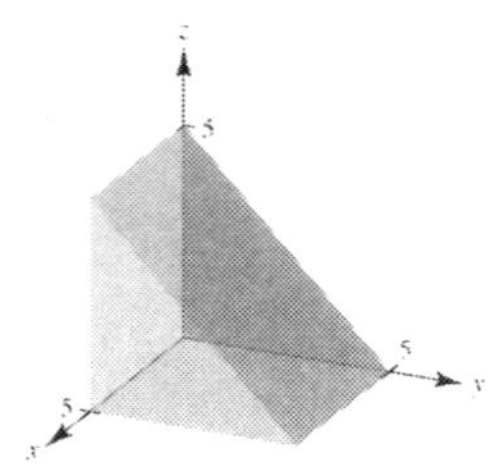

59. $\int_0^3 \int_0^{\pi/3} \int_0^4 r\cos\theta\,dr\,d\theta\,dz$

$= \int_0^3 \int_0^{\pi/3} \left[\dfrac{r^2}{2}\cos\theta\right]_0^4 d\theta\,dz$

$= \int_0^3 \int_0^{\pi/3} 8\cos\theta\,d\theta\,dz$

$= \int_0^3 \left[8\sin\theta\right]_0^{\pi/3} dz$

$= \int_0^3 4\sqrt{3}\,dz = 12\sqrt{3}$

60. $\int_0^{\pi/2} \int_0^3 \int_0^{4-z} z\,dr\,dz\,d\theta = \int_0^{\pi/2} \int_0^3 \left(4z - z^2\right) dz\,d\theta$

$= \int_0^{\pi/2} \left[2z^2 - \dfrac{z^3}{3}\right]_0^3 d\theta$

$= \int_0^{\pi/2} 9\,d\theta = \dfrac{9\pi}{2}$

61. $\int_0^{\pi/2} \int_0^{\pi/2} \int_0^2 \rho^2\,d\rho\,d\theta\,d\phi = \int_0^{\pi/2} \int_0^{\pi/2} \left[\dfrac{\rho^3}{3}\right]_0^2 d\theta\,d\phi$

$= \int_0^{\pi/2} \int_0^{\pi/2} \dfrac{8}{3}\,d\theta\,d\phi$

$= \int_0^{\pi/2} \dfrac{8}{3}\left(\dfrac{\pi}{2}\right) d\theta$

$= \dfrac{8}{3}\left(\dfrac{\pi}{2}\right)\left(\dfrac{\pi}{2}\right) = \dfrac{2}{3}\pi^2$

62. $\int_0^{\pi/4} \int_0^{\pi/4} \int_0^{\cos\phi} \cos\theta\,d\rho\,d\phi\,d\theta$

$= \int_0^{\pi/4} \int_0^{\pi/4} \cos\theta\cos\phi\,d\phi\,d\theta$

$= \int_0^{\pi/4} \left[\cos\theta\sin\phi\right]_0^{\pi/4} d\theta$

$= \int_0^{\pi/4} \dfrac{\sqrt{2}}{2}\cos\theta\,d\theta$

$= \left[\dfrac{\sqrt{2}}{2}\sin\theta\right]_0^{\pi/4} = \dfrac{1}{2}$

63. $\int_0^{\pi} \int_0^2 \int_0^3 \sqrt{z^2 + 4}\,dz\,dr\,d\theta \approx 48.995$

64. $\int_0^{\pi/2} \int_0^{\pi/2} \int_0^{\cos\phi} \rho^2\cos\theta\,d\rho\,d\theta\,d\phi = \dfrac{2}{9}$

65. $z = 8 - x^2 - y^2 = x^2 + y^2$

$$8 = 2(x^2 + y^2)$$

$$x^2 + y^2 = 4$$

$$\begin{aligned}
V &= \int_0^{2\pi}\int_0^2\int_{r^2}^{8-r^2} r\,dz\,dr\,d\theta \\
&= \int_0^{2\pi}\int_0^2 r(8 - r^2 - r^2)\,dr\,d\theta \\
&= \int_0^{2\pi}\int_0^2 (8r - 2r^3)\,dr\,d\theta \\
&= \int_0^{2\pi}\left[4r^2 - \frac{r^4}{2}\right]_0^2 d\theta \\
&= \int_0^{2\pi} 8\,d\theta \\
&= 16\pi
\end{aligned}$$

66. $x^2 + y^2 + z^2 = 36 \Rightarrow \rho = 6$

Intersection of sphere and cone:

$(x^2 + y^2) + z^2 = z^2 + z^2 = 2z^2 = 36 \Rightarrow z^2 = 18 \Rightarrow z = 3\sqrt{2}$

$z = \rho\cos\phi \Rightarrow 3\sqrt{2} = 6\cos\phi \Rightarrow \cos\phi = \dfrac{\sqrt{2}}{2} \Rightarrow \phi = \dfrac{\pi}{4}$

$$\begin{aligned}
V &= \int_0^{2\pi}\int_0^{\pi/4}\int_0^6 \rho^2\sin\phi\,d\rho\,d\phi\,d\theta \\
&= \int_0^{2\pi}\int_0^{\pi/4} 72\sin\phi\,d\phi\,d\theta \\
&= \int_0^{2\pi}\left[-72\cos\phi\right]_0^{\pi/4} d\theta \\
&= \int_0^{2\pi}\left(72 - 72\cdot\frac{\sqrt{2}}{2}\right)d\theta \\
&= 2\pi(72 - 36\sqrt{2}) \\
&= 72\pi(2 - \sqrt{2})
\end{aligned}$$

67. $\dfrac{\partial(x, y)}{\partial(u, v)} = \dfrac{\partial x}{\partial u}\dfrac{\partial y}{\partial v} - \dfrac{\partial y}{\partial u}\dfrac{\partial x}{\partial v} = 1(-3) - 2(3) = -9$

68. $\dfrac{\partial(x, y)}{\partial(u, v)} = \dfrac{\partial x}{\partial u}\dfrac{\partial y}{\partial v} - \dfrac{\partial y}{\partial u}\dfrac{\partial x}{\partial v} = (2u)(-2v) - (2u)(2v) = -8uv$

69.
$$\begin{aligned}
\frac{\partial(x, y)}{\partial(u, v)} &= \frac{\partial x}{\partial u}\frac{\partial y}{\partial v} - \frac{\partial y}{\partial u}\frac{\partial x}{\partial v} \\
&= (\sin\theta)(\sin\theta) - (\cos\theta)(\cos\theta) \\
&= \sin^2\theta - \cos^2\theta
\end{aligned}$$

70.
$$\begin{aligned}
\frac{\partial(x, y)}{\partial(u, v)} &= \frac{\partial x}{\partial u}\frac{\partial y}{\partial v} - \frac{\partial y}{\partial u}\frac{\partial x}{\partial v} \\
&= v\left(\frac{1}{u}\right) - \left(\frac{-v}{u^2}\right)u \\
&= \frac{v}{u} + \frac{v}{u} = \frac{2v}{u}
\end{aligned}$$

71. $\dfrac{\partial(x, y)}{\partial(u, v)} = \dfrac{\partial x}{\partial u}\dfrac{\partial y}{\partial v} - \dfrac{\partial x}{\partial v}\dfrac{\partial y}{\partial u} = \dfrac{1}{2}\left(-\dfrac{1}{2}\right) - \dfrac{1}{2}\left(\dfrac{1}{2}\right) = -\dfrac{1}{2}$

$x = \dfrac{1}{2}(u + v),\ y = \dfrac{1}{2}(u - v) \Rightarrow u = x + y,\ v = x - y$

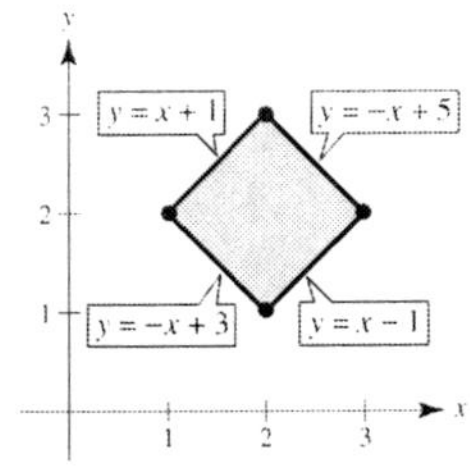

Boundaries in *xy*-plane	Boundaries in *uv*-plane
$x + y = 3$	$u = 3$
$x + y = 5$	$u = 5$
$x - y = -1$	$v = -1$
$x - y = 1$	$v = 1$

$$\int_R\int \ln(x + y)\,dA = \int_3^5\int_{-1}^1 \ln\left(\frac{1}{2}(u + v) + \frac{1}{2}(u - v)\right)\left(\frac{1}{2}\right) dv\,du = \int_3^5\int_{-1}^1 \frac{1}{2}\ln u\,dv\,du = \int_3^5 \ln u\,du = \Big[u\ln u - u\Big]_3^5$$

$$= (5\ln 5 - 5) - (3\ln u - 3) = 5\ln 5 - 3\ln 3 - 2 \approx 2.751$$

72. $\dfrac{\partial(x, y)}{\partial(u, v)} = \dfrac{\partial x}{\partial u}\dfrac{\partial y}{\partial v} - \dfrac{\partial y}{\partial u}\dfrac{\partial x}{\partial v} = \left(\dfrac{1}{4}\right)\left(\dfrac{1}{2}\right) - \left(-\dfrac{1}{2}\right)\left(\dfrac{1}{4}\right) = \dfrac{1}{4}$

$x = \dfrac{1}{4}(u + v),\ y = \dfrac{1}{2}(v - u) \Rightarrow u = 2x - y,\ v = 2x + y$

Boundary in *xy*-plane	Boundary in *uv*-plane
$y + 2x = 2$	$v = 2$
$2x - y = -2$	$u = -2$
$y + 2x = 6$	$v = 6$
$2x - y = 2$	$u = 2$

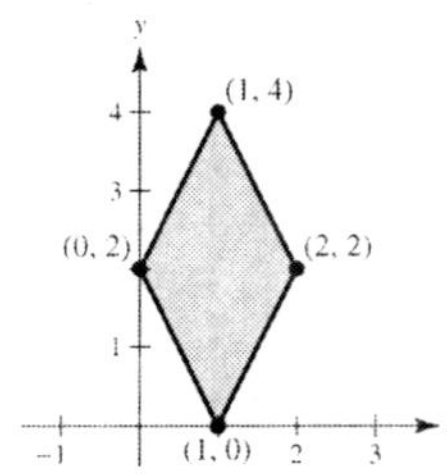

$$\int_R\int 16xy\,dA = \int_{-2}^2\int_2^6 16\left(\frac{1}{4}(u + v)\right)\left(\frac{1}{2}(v - u)\right)\left(\frac{1}{4}\right) dv\,du$$

$$= \int_{-2}^2\int_2^6 \frac{1}{2}\left(v^2 - u^2\right) dv\,du = \frac{1}{2}\int_{-2}^2\left[\frac{v^3}{3} - u^2v\right]_2^6 du$$

$$= \frac{1}{2}\int_{-2}^2\left(72 - 6u^2 - \frac{8}{3} + 2u^2\right) du = \frac{1}{2}\left[72u - \frac{4}{3}u^3 - \frac{8}{3}u\right]_{-2}^2 = 128$$

73. $\dfrac{\partial(x, y)}{\partial(u, v)} = \dfrac{\partial x}{\partial u}\dfrac{\partial y}{\partial v} - \dfrac{\partial y}{\partial u}\dfrac{\partial x}{\partial v} = 1\left(-\dfrac{1}{3}\right) - \dfrac{1}{3}(0) = -\dfrac{1}{3}$

$x = u,\ y = \dfrac{1}{3}(u - v) \Rightarrow u = x,\ v = x - 3y$

Boundary in *xy*-plane	Boundary in *uv*-plane
$x = 1$	$u = 1$
$x = 4$	$u = 4$
$3y - x = 8$	$v = -8$
$3y - x = 2$	$v = -2$

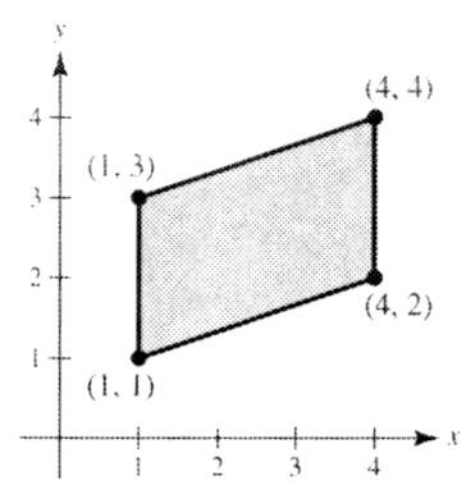

$$\int_R\int\left(xy + x^2\right) dA = \int_1^4\int_{-2}^{-8}\left[u\frac{1}{3}(u - v) + u^2\right]\left(-\frac{1}{3}\right) dv\,du$$

$$= \left(-\frac{1}{3}\right)\int_1^4\int_{-2}^{-8}\left(\frac{4}{3}u^2 - \frac{1}{3}uv\right) dv\,du = \left(-\frac{1}{3}\right)\int_1^4\left[\frac{4}{3}u^2v - \frac{1}{6}uv^2\right]_{-2}^{-8} du$$

$$= \left(-\frac{1}{3}\right)\int_1^4\left(-8u^2 - 10u\right) du = 81$$

74. $\dfrac{\partial(x, y)}{\partial(u, v)} = \dfrac{\partial x}{\partial u}\dfrac{\partial y}{\partial v} - \dfrac{\partial x}{\partial v}\dfrac{\partial y}{\partial u} = 1\left(\dfrac{1}{u}\right) - 0 = \dfrac{1}{u}$

$x = u,\ y = \dfrac{v}{u} \Rightarrow u = x,\ v = xy$

Boundary in xy-plane	Boundary in uv-plane
$x = 1$	$u = 1$
$x = 5$	$u = 5$
$xy = 1$	$v = 1$
$xy = 5$	$v = 5$

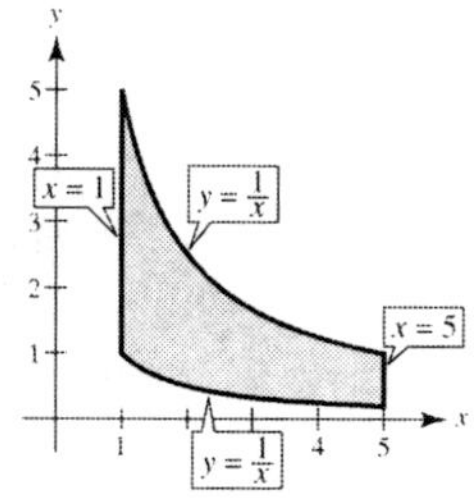

$$\int_R\int \frac{x}{1 + x^2y^2}\,dA = \int_1^5\int_1^5 \frac{u}{1 + u^2(v/u)^2}\left(\frac{1}{u}\right) du\,dv = \int_1^5\int_1^5 \frac{1}{1 + v^2}\,du\,dv = \int_1^5 \frac{4}{1 + v^2}\,dv$$

$$= 4 \arctan v\Big]_1^5 = 4 \arctan 5 - \pi$$

Problem Solving for Chapter 14

1. $V = 16\int_R\int \sqrt{1 - x^2}\,dA$

$$= 16\int_0^{\pi/4}\int_0^1 \sqrt{1 - r^2\cos^2\theta}\; r\,dr\,d\theta = -\frac{16}{3}\int_0^{\pi/4} \frac{1}{\cos^2\theta}\left[\left(1 - \cos^2\theta\right)^{3/2} - 1\right] d\theta$$

$$= -\frac{16}{3}\left[\sec\theta + \cos\theta - \tan\theta\right]_0^{\pi/4} = 8\left(2 - \sqrt{2}\right) \approx 4.6863$$

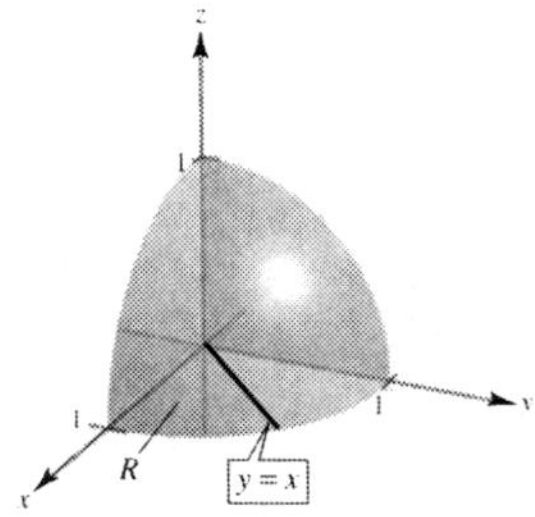

2. $z = \dfrac{1}{c}(d - ax - by)$ Plane

$f_x = -\dfrac{a}{c},\ f_y = -\dfrac{b}{c}$

$$\sqrt{1 + (f_x)^2 + (f_y)^2} = \sqrt{1 + \frac{a^2}{c^2} + \frac{b^2}{c^2}}$$

$$S = \int_R\int \sqrt{1 + \frac{a^2}{c^2} + \frac{b^2}{c^2}}\,dA = \frac{\sqrt{a^2 + b^2 + c^2}}{c}\int_R\int dA = \frac{\sqrt{a^2 + b^2 + c^2}}{c}A(R)$$

3.

Boundary in xy-plane	Boundary in uv-plane
$y = \sqrt{x}$	$u = 1$
$y = \sqrt{2x}$	$u = 2$
$y = \frac{1}{3}x^2$	$v = 3$
$y = \frac{1}{4}x^2$	$v = 4$

$$\frac{\partial(x, y)}{\partial(u, v)} = \begin{vmatrix} \frac{1}{3}\left(\frac{v}{u}\right)^{2/3} & \frac{2}{3}\left(\frac{u}{v}\right)^{1/3} \\ \frac{2}{3}\left(\frac{v}{u}\right)^{1/3} & \frac{1}{3}\left(\frac{u}{v}\right)^{2/3} \end{vmatrix} = -\frac{1}{3}$$

$$A = \int_R\int 1\,dA = \int_S\int 1\left|\frac{\partial(x, y)}{\partial(u, v)}\right| dA = \frac{1}{3}$$

4. $\displaystyle\int_0^1\int_0^1 x^n y^n\,dx\,dy = \int_0^1\left[\frac{x^{n+1}}{n + 1}y^n\right]_0^1 dy$

$$= \int_0^1 \frac{1}{n + 1}y^n\,dy$$

$$= \left[\frac{y^{n+1}}{(n + 1)^2}\right]_0^1 = \frac{1}{(n + 1)^2}$$

$$\lim_{n\to\infty}\int_0^1\int_0^1 x^n y^n\,dx\,dy = \lim_{n\to\infty}\frac{1}{(n + 1)^2} = 0$$

5. (a) $\displaystyle\int \frac{du}{a^2+u^2} = \frac{1}{a}\arctan\frac{u}{a} + c$. Let $a^2 = 2 - u^2,\ u = v$.

Then $\displaystyle\int \frac{1}{(2-u^2)+v^2}\,dv = \frac{1}{\sqrt{2-u^2}}\arctan\frac{v}{\sqrt{2-u^2}} + C.$

(b) $$I_1 = \int_0^{\sqrt{2}/2}\left[\frac{2}{\sqrt{2-u^2}}\arctan\frac{v}{\sqrt{2-u^2}}\right]_{-u}^{u} du$$

$$= \int_0^{\sqrt{2}/2}\frac{2}{\sqrt{2-u^2}}\left(\arctan\frac{u}{\sqrt{2-u^2}} - \arctan\frac{-u}{\sqrt{2-u^2}}\right)du = \int_0^{\sqrt{2}/2}\frac{4}{\sqrt{2-u^2}}\arctan\frac{u}{\sqrt{2-u^2}}\,du$$

Let $u = \sqrt{2}\sin\theta,\ du = \sqrt{2}\cos\theta\,d\theta,\ 2-u^2 = 2 - 2\sin^2\theta = 2\cos^2\theta.$

$$I_1 = 4\int_0^{\pi/6}\frac{1}{\sqrt{2}\cos\theta}\arctan\left(\frac{\sqrt{2}\sin\theta}{\sqrt{2}\cos\theta}\right)\cdot\sqrt{2}\cos\theta\,d\theta = 4\int_0^{\pi/6}\arctan(\tan\theta)d\theta = \frac{4\theta^2}{2}\Bigg]_0^{\pi/6} = 2\left(\frac{\pi}{6}\right)^2 = \frac{\pi^2}{18}$$

(c) $$I_2 = \int_{\sqrt{2}/2}^{\sqrt{2}}\left[\frac{2}{\sqrt{2-u^2}}\arctan\frac{v}{\sqrt{2-u^2}}\right]_{u-\sqrt{2}}^{-u+\sqrt{2}} du$$

$$= \int_{\sqrt{2}/2}^{\sqrt{2}}\frac{2}{\sqrt{2-u^2}}\left[\arctan\left(\frac{-u+\sqrt{2}}{\sqrt{2-u^2}}\right) - \arctan\left(\frac{u-\sqrt{2}}{\sqrt{2-u^2}}\right)\right]du = \int_{\sqrt{2}/2}^{\sqrt{2}}\frac{4}{\sqrt{2-u^2}}\arctan\left(\frac{\sqrt{2}-u}{\sqrt{2-u^2}}\right)du$$

Let $u = \sqrt{2}\sin\theta$.

$$I_2 = 4\int_{\pi/6}^{\pi/2}\frac{1}{\sqrt{2}\cos\theta}\arctan\left(\frac{\sqrt{2}-\sqrt{2}\sin\theta}{\sqrt{2}\cos\theta}\right)\cdot\sqrt{2}\cos\theta\,d\theta = 4\int_{\pi/6}^{\pi/2}\arctan\left(\frac{1-\sin\theta}{\cos\theta}\right)d\theta$$

(d) $$\tan\left(\frac{1}{2}\left(\frac{\pi}{2}-\theta\right)\right) = \sqrt{\frac{1-\cos((\pi/2)-\theta)}{1+\cos((\pi/2)-\theta)}} = \sqrt{\frac{1-\sin\theta}{1+\sin\theta}} = \sqrt{\frac{(1-\sin\theta)^2}{(1+\sin\theta)(1-\sin\theta)}} = \sqrt{\frac{(1-\sin\theta)^2}{\cos^2\theta}} = \frac{1-\sin\theta}{\cos\theta}$$

(e) $$I_2 = 4\int_{\pi/6}^{\pi/2}\arctan\left(\frac{1-\sin\theta}{\cos\theta}\right)d\theta = 4\int_{\pi/6}^{\pi/2}\arctan\left(\tan\left(\frac{1}{2}\left(\frac{\pi}{2}-\theta\right)\right)\right)d\theta = 4\int_{\pi/6}^{\pi/2}\frac{1}{2}\left(\frac{\pi}{2}-\theta\right)d\theta = 2\int_{\pi/6}^{\pi/2}\left(\frac{\pi}{2}-\theta\right)d\theta$$

$$= 2\left[\frac{\pi}{2}\theta - \frac{\theta^2}{2}\right]_{\pi/6}^{\pi/2} = 2\left[\left(\frac{\pi^2}{4}-\frac{\pi^2}{8}\right) - \left(\frac{\pi^2}{12}-\frac{\pi^2}{72}\right)\right] = 2\left[\frac{18-9-6+1}{72}\pi^2\right] = \frac{4}{36}\pi^2 = \frac{\pi^2}{9}$$

(f) $$\frac{1}{1-xy} = 1 + (xy) + (xy)^2 + \cdots \qquad |xy| < 1$$

$$\int_0^1\int_0^1\frac{1}{1-xy}\,dx\,dy = \int_0^1\int_0^1\left[1+(xy)+(xy)^2+\cdots\right]dx\,dy = \int_0^1\int_0^1\sum_{K=0}^{\infty}(xy)^K\,dx\,dy = \sum_{K=0}^{\infty}\int_0^1\frac{x^{K+1}y^K}{K+1}\Bigg|_0^1 dy$$

$$= \sum_{K=0}^{\infty}\int_0^1\frac{y^K}{K+1}\,dy = \sum_{K=0}^{\infty}\frac{y^{K+1}}{(K+1)^2}\Bigg|_0^1 = \sum_{K=0}^{\infty}\frac{1}{(K+1)^2} = \sum_{n=1}^{\infty}\frac{1}{n^2}$$

(g) $u = \dfrac{x + y}{\sqrt{2}}, v = \dfrac{y - x}{\sqrt{2}}$

$$u - v = \frac{2x}{\sqrt{2}} \Rightarrow x = \frac{u - v}{\sqrt{2}}$$

$$u + v = \frac{2y}{\sqrt{2}} \Rightarrow y = \frac{u + v}{\sqrt{2}}$$

$$\frac{\partial(x, y)}{\partial(u, v)} = \begin{vmatrix} 1/\sqrt{2} & -1/\sqrt{2} \\ 1/\sqrt{2} & 1/\sqrt{2} \end{vmatrix} = 1$$

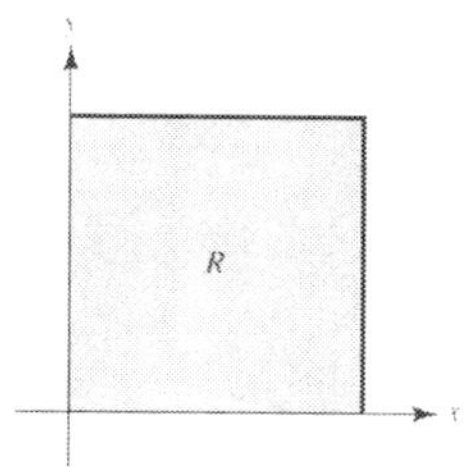

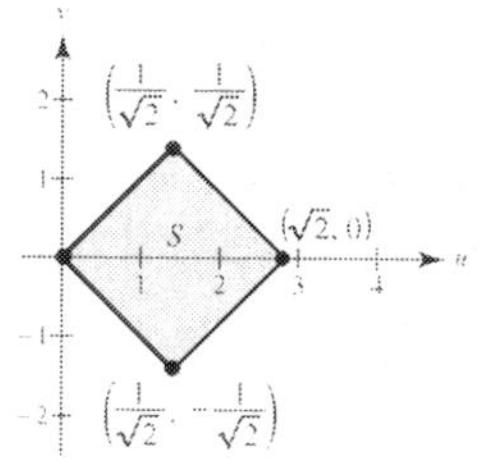

R		S
$(0, 0)$	$\leftrightarrow$	$(0, 0)$
$(1, 0)$	$\leftrightarrow$	$\left(\frac{1}{\sqrt{2}}, -\frac{1}{\sqrt{2}}\right)$
$(0, 1)$	$\leftrightarrow$	$\left(\frac{1}{\sqrt{2}}, \frac{1}{\sqrt{2}}\right)$
$(1, 1)$	$\leftrightarrow$	$\left(\sqrt{2}, 0\right)$

$$\int_0^1 \int_0^1 \frac{1}{1 - xy}\,dx\,dy = \int_0^{\sqrt{2}/2} \int_{-u}^{u} \frac{1}{1 - \frac{u^2}{2} + \frac{v^2}{2}}\,dv\,du + \int_{\sqrt{2}/2}^{\sqrt{2}} \int_{u-\sqrt{2}}^{-u+\sqrt{2}} \frac{1}{1 - \frac{u^2}{2} + \frac{v^2}{2}}\,dv\,du = I_1 + I_2 = \frac{\pi^2}{18} + \frac{\pi^2}{9} = \frac{\pi^2}{6}$$

6. Converting to polar coordinates,

$$\int_0^\infty \int_0^\infty \frac{1}{\left(1 + x^2 + y^2\right)^2}\,dx\,dy = \int_0^\infty \int_0^{\pi/2} \frac{1}{\left(1 + r^2\right)^2} r\,d\theta\,dr$$
$$= \int_0^\infty \frac{r}{\left(1 + r^2\right)^2}\left(\frac{\pi}{2}\right)dr$$
$$= \lim_{t\to\infty} \int_0^t \frac{\pi}{4}\left(1 + r^2\right)^{-2}(2r\,dr)$$
$$= \lim_{t\to\infty} \left[\frac{\pi}{4} \cdot \frac{-1}{1 + r^2}\right]_0^t = \frac{\pi}{4}$$

7. $\displaystyle\int_0^1 \int_0^1 \frac{x - y}{(x + y)^3}\,dx\,dy = -\frac{1}{2}$

$$\int_0^1 \int_0^1 \frac{(x - y)}{(x + y)^3}\,dy\,dx = \frac{1}{2}$$

The results are not the same. Fubini's Theorem is not valid because f is not continuous on the region $0 \le x \le 1, 0 \le y \le 1$.

8. The volume of this spherical block can be determined as follows. One side is length $\Delta\rho$. Another side is $\rho\,\Delta\phi$. Finally, the third side is given by the length of an arc of angle $\Delta\theta$ in a circle of radius $\rho \sin\phi$. Thus:

$$\Delta V \approx (\Delta\rho)(\rho\,\Delta\phi)(\Delta\theta\,\rho\sin\phi)$$
$$= \rho^2 \sin\phi\,\Delta\rho\,\Delta\phi\,\Delta\theta$$

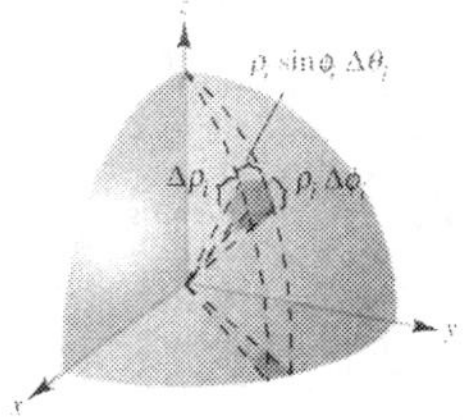

9. From Exercise 65, Section 14.3,

$$\int_{-\infty}^{\infty} e^{-x^2/2}\,dx = \sqrt{2\pi}.$$

So, $\displaystyle\int_0^\infty e^{-x^2/2}\,dx = \frac{\sqrt{2\pi}}{2}$ and $\displaystyle\int_0^\infty e^{-x^2}\,dx = \frac{\sqrt{\pi}}{2}$

$$\int_0^\infty x^2 e^{-x^2}\,dx = \left[-\frac{1}{2}xe^{-x^2}\right]_0^\infty + \frac{1}{2}\int_0^\infty e^{-x^2}\,dx$$
$$= \frac{1}{2} \cdot \frac{\sqrt{\pi}}{2} = \frac{\sqrt{\pi}}{4}$$

10. Let $v = \ln\left(\frac{1}{x}\right)$, $dv = -\frac{dx}{x}$.

$e^v = \frac{1}{x}$, $x = e^{-v}$, $dx = -e^{-v}\,dv$

$$\int_0^1 \sqrt{\ln(1/x)}\,dx = \int_\infty^0 \sqrt{v}\left(-e^{-v}\right)dv = \int_0^\infty \sqrt{v}e^{-v}\,dv$$

Let $u = \sqrt{v}$, $u^2 = v$, $2u\,du = dv$.

$$\int_0^1 \sqrt{\ln(1/x)}\,dx = \int_0^\infty u\,e^{-u^2}(2u\,du)$$

$$= 2\int_0^\infty u^2 e^{-u^2}\,du = 2\left(\frac{\sqrt{\pi}}{4}\right) = \frac{\sqrt{\pi}}{2}$$

(See Problem Solving #9.)

11. $f(x, y) = \begin{cases} ke^{-(x+y)/a} & x \ge 0,\ y \ge 0 \\ 0 & \text{elsewhere} \end{cases}$

$$\int_{-\infty}^{\infty}\int_{-\infty}^{\infty} f(x, y)\,dA = \int_0^\infty \int_0^\infty ke^{-(x+y)/a}\,dx\,dy$$

$$= k\int_0^\infty e^{-x/a}\,dx \cdot \int_0^\infty e^{-y/a}\,dy$$

These two integrals are equal to

$$\int_0^\infty e^{-x/a}\,dx = \lim_{b\to\infty}\left[(-a)e^{-x/a}\right]_0^b = a.$$

So, assuming $a, k > 0$, you obtain

$$1 = ka^2 \text{ or } a = \frac{1}{\sqrt{k}}.$$

12. By the shell method,

$$V = \lim_{b\to\infty}\int_0^b 2\pi x e^{-x^2}\,dx = \lim_{b\to\infty}\left[-\pi e^{-x^2}\right]_0^b = \pi.$$

This same volume is given by

$$\pi = \int_{-\infty}^{\infty}\int_{-\infty}^{\infty} e^{-(x^2+y^2)}\,dy\,dx$$

$$= 4\int_0^\infty \int_0^\infty e^{-(x^2+y^2)}\,dy\,dx$$

$$= 4\int_0^\infty e^{-x^2}\,dx \int_0^\infty e^{-y^2}\,dy$$

$$= 4\left[\int_0^\infty e^{-x^2}\,dx\right]^2.$$

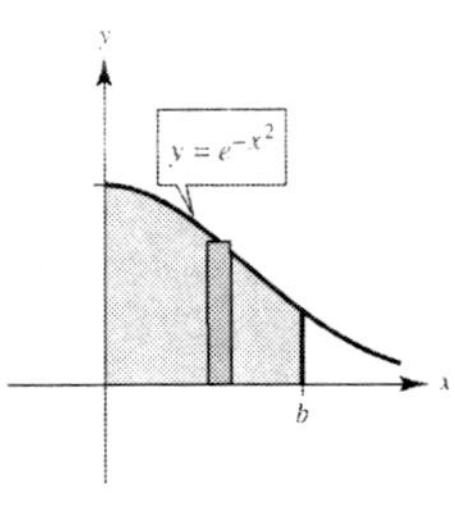

So, $\int_0^\infty e^{-x^2}\,dx = \frac{\sqrt{\pi}}{2}$.

13. Essay

14. $A = l \cdot w = \left(\frac{\Delta x}{\cos\theta}\right)\Delta y = \sec\theta\,\Delta x\,\Delta y$

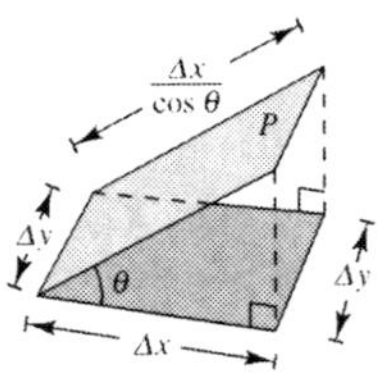

Area in xy-plane: $\Delta x\,\Delta y$

15. The greater the angle between the given plane and the xy-plane, the greater the surface area. So:
$z_2 < z_1 < z_4 < z_3$

16. A: $\int_0^{2\pi}\int_4^5 \left(\frac{r}{16} - \frac{r^2}{160}\right) r\,dr\,d\theta = \frac{1333\pi}{960} \approx 4.36\text{ ft}^3$

$B = \int_0^{2\pi}\int_9^{10}\left(\frac{r}{16} - \frac{r^2}{160}\right) r\,dr\,d\theta = \frac{523\pi}{960} \approx 1.71\text{ ft}^3$

The distribution is not uniform. Less water in region of greater area.
In one hour, the entire lawn receives

$$\int_0^{2\pi}\int_0^{10}\left(\frac{r}{16} - \frac{r^2}{160}\right) r\,dr\,d\theta = \frac{125\pi}{12} \approx 32.72\text{ ft}^3.$$

17. $V = \int_0^3 \int_0^{2x} \int_x^{6-x} dy\,dz\,dx = 18$

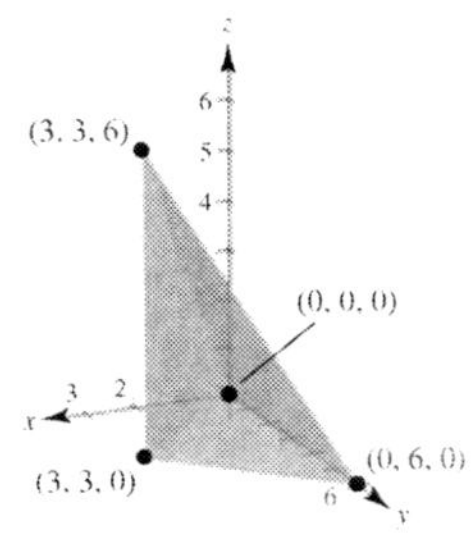

18. (a) $V = \int_0^{2\pi}\int_0^2 \int_2^{\sqrt{8-r^2}} r\,dz\,dr\,d\theta = \frac{8\pi}{3}\left(4\sqrt{2} - 5\right)$

(b) $V = \int_0^{2\pi}\int_0^{\pi/4}\int_{2\sec\phi}^{2\sqrt{2}} \rho^2 \sin\phi\,d\rho\,d\phi\,d\theta$

$= \frac{8\pi}{3}\left(4\sqrt{2} - 5\right)$

CHAPTER 15
Vector Analysis

CHAPTER 15
Vector Analysis

Section 15.1 Vector Fields

1. All vectors are parallel to x-axis.
Matches (d)

2. All vectors are parallel to y-axis.
Matches (c)

3. Vectors are in rotational pattern.
Matches (a)

4. All vectors point outward.
Matches (b)

5. $\mathbf{F}(x, y) = \mathbf{i} + \mathbf{j}$

$\|\mathbf{F}\| = \sqrt{2}$

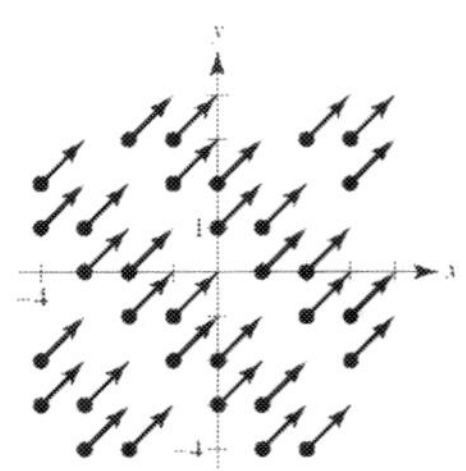

6. $\mathbf{F}(x, y) = y\mathbf{i} - 2x\mathbf{j}$

$\|\mathbf{F}\| = \sqrt{y^2 + 4x^2}$

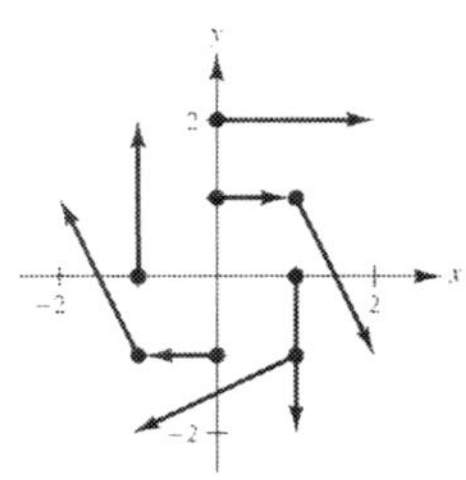

7. $\mathbf{F}(x, y, z) = 3y\mathbf{j}$

$\|\mathbf{F}\| = 3|y| = c$

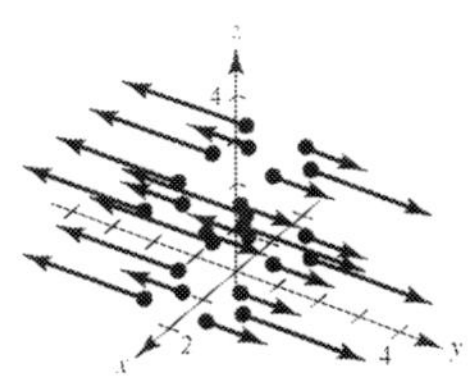

8. $\mathbf{F}(x, y) = y\mathbf{i} + x\mathbf{j}$

$\|\mathbf{F}\| = \sqrt{y^2 + x^2}$

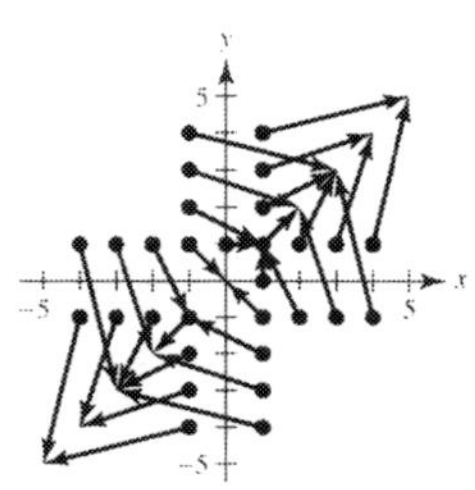

9. $\mathbf{F}(x, y, z) = \mathbf{i} + \mathbf{j} + \mathbf{k}$

$\|\mathbf{F}\| = \sqrt{3}$

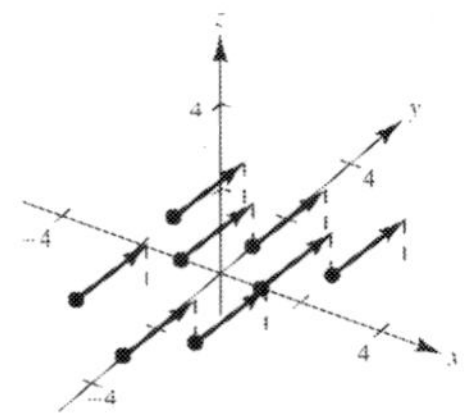

10. $\mathbf{F}(x, y, z) = x\mathbf{i} + y\mathbf{j} + z\mathbf{k}$

$\|\mathbf{F}\| = \sqrt{x^2 + y^2 + z^2} = c$

$x^2 + y^2 + z^2 = c^2$

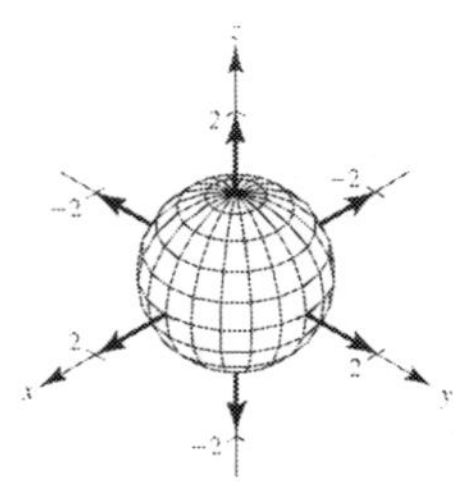

11. $F(x, y) = \frac{1}{8}(2xy\mathbf{i} + y^2\mathbf{j})$

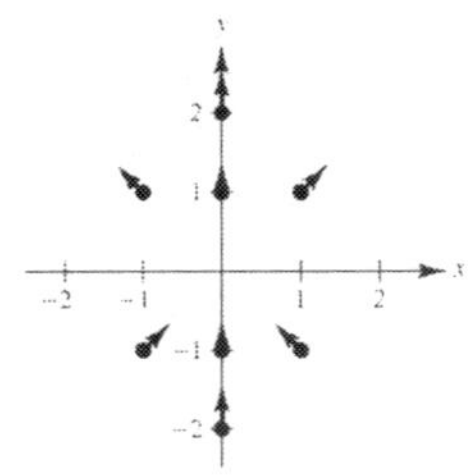

12. $F(x, y) = (2y - x)\,\mathbf{i} + (2y + x)\,\mathbf{j}$

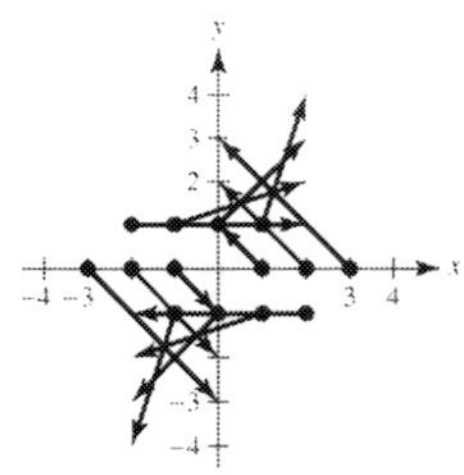

13. $F(x, y, z) = \dfrac{x\mathbf{i} + y\mathbf{j} + z\mathbf{k}}{\sqrt{x^2 + y^2 + z^2}}$

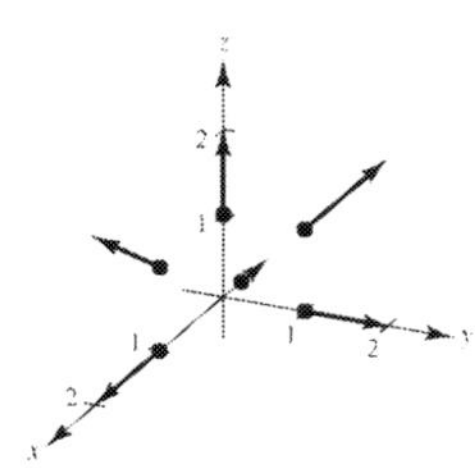

14. $\mathbf{F}(x, y, z) = x\mathbf{i} - y\mathbf{j} + z\mathbf{k}$

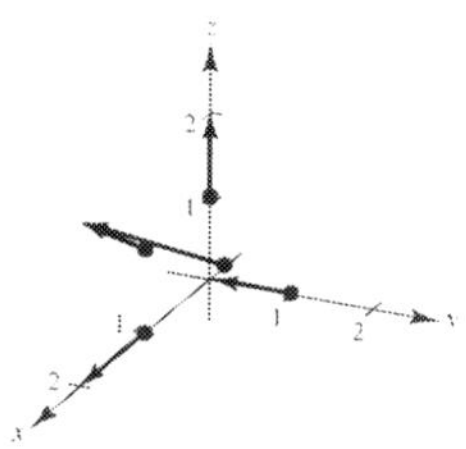

15. $f(x, y) = x^2 + 2y^2$

$f_x(x, y) = 2x$

$f_y(x, y) = 4y$

$\mathbf{F}(x, y) = 2x\mathbf{i} + 4y\mathbf{j}$

Note that $\nabla f = \mathbf{F}$.

16. $f(x, y) = x^2 - \frac{1}{4}y^2$

$f_x(x, y) = 2x$

$f_y(x, y) = -\frac{1}{2}y$

$\mathbf{F}(x, y) = 2x\mathbf{i} - \frac{1}{2}y\mathbf{j}$

17. $g(x, y) = 5x^2 + 3xy + y^2$

$g_x(x, y) = 10x + 3y$

$g_y(x, y) = 3x + 2y$

$\mathbf{G}(x, y) = (10x + 3y)\mathbf{i} + (3x + 2y)\mathbf{j}$

18. $g(x, y) = \sin 3x \cos 4y$

$g_x(x, y) = 3 \cos 3x \cos 4y$

$g_y(x, y) = -4 \sin 3x \sin 4y$

$G(x, y) = 3 \cos 3x \cos 4y\mathbf{i} - 4 \sin 3x \sin 4y\mathbf{j}$

19. $f(x, y, z) = 6xyz$

$f_x(x, y, z) = 6yz$

$f_y(x, y, z) = 6xz$

$f_z(x, y, z) = 6xy$

$\mathbf{F}(x, y, z) = 6yz\mathbf{i} + 6xz\mathbf{j} + 6xy\mathbf{k}$

20. $f(x, y, z) = \sqrt{x^2 + 4y^2 + z^2}$

$$f_x = \frac{x}{\sqrt{x^2 + 4y^2 + z^2}}$$

$$f_y = \frac{4y}{\sqrt{x^2 + 4y^2 + z^2}}$$

$$f_z = \frac{z}{\sqrt{x^2 + 4y^2 + z^2}}$$

$$\mathbf{F}(x, y, z) = \frac{x}{\sqrt{x^2 + 4y^2 + z^2}}\mathbf{i} + \frac{4y}{\sqrt{x^2 + 4y^2 + z^2}}\mathbf{j} + \frac{z}{\sqrt{x^2 + 4y^2 + z^2}}\mathbf{k}$$

21. $g(x, y, z) = z + ye^{x^2}$

$g_x(x, y, z) = 2xye^{x^2}$

$g_y(x, y, z) = e^{x^2}$

$g_z(x, y, z) = 1$

$\mathbf{G}(x, y, z) = 2xye^{x^2}\mathbf{i} + e^{x^2}\mathbf{j} + \mathbf{k}$

22. $g(x, y, z) = \dfrac{y}{z} + \dfrac{z}{x} - \dfrac{xz}{y}$

$$g_x(x, y, z) = -\frac{z}{x^2} - \frac{z}{y}$$

$$g_y(x, y, z) = \frac{1}{z} + \frac{xz}{y^2}$$

$$g_z(x, y, z) = -\frac{y}{z^2} + \frac{1}{x} - \frac{x}{y}$$

$$\mathbf{G}(x, y, z) = \left(-\frac{z}{x^2} - \frac{z}{y}\right)\mathbf{i} + \left(\frac{1}{z} + \frac{xz}{y^2}\right)\mathbf{j} + \left(-\frac{y}{z^2} + \frac{1}{x} - \frac{x}{y}\right)\mathbf{k}$$

23. $h(x, y, z) = xy \ln(x + y)$

$h_x(x, y, z) = y \ln(x + y) + \dfrac{xy}{x + y}$

$h_y(x, y, z) = x \ln(x + y) + \dfrac{xy}{x + y}$

$h_z(x, y, z) = 0$

$\mathbf{H}(x, y, z) = \left[\dfrac{xy}{x + y} + y \ln(x + y)\right]\mathbf{i} + \left[\dfrac{xy}{x + y} + x \ln(x + y)\right]\mathbf{j}$

24. $h(x, y, z) = x \arcsin yz$

$h_x(x, y, z) = \arcsin yz$

$h_y(x, y, z) = \dfrac{xz}{\sqrt{1 - y^2z^2}}$

$h_z(x, y, z) = \dfrac{xy}{\sqrt{1 - y^2z^2}}$

$\mathbf{H}(x, y, z) = (\arcsin yz)\mathbf{i} + \dfrac{xz}{\sqrt{1 - y^2z^2}}\mathbf{j} + \dfrac{xy}{\sqrt{1 - y^2z^2}}\mathbf{k}$

25. $\mathbf{F}(x, y) = xy^2\mathbf{i} + x^2y\mathbf{j}$

$M = xy^2$ and $N = x^2y$ have continuous first partial derivatives.

$\dfrac{\partial N}{\partial x} = 2xy = \dfrac{\partial M}{\partial y} \Rightarrow$ **F** conservative

26. $\mathbf{F}(x, y) = \dfrac{1}{x^2}(y\mathbf{i} - x\mathbf{j}) = \dfrac{y}{x^2}\mathbf{i} - \dfrac{1}{x}\mathbf{j}$

$M = y/x^2$ and $N = -(1/x)$ have continuous first partial derivatives for all $x \neq 0$.

$\dfrac{\partial N}{\partial x} = \dfrac{1}{x^2} = \dfrac{\partial M}{\partial y} \Rightarrow$ **F** is conservative.

27. $\mathbf{F}(x, y) = \sin y\mathbf{i} + x \cos y\mathbf{j}$

$M = \sin y$ and $N = x \cos y$ have continuous first partial derivatives.

$\dfrac{\partial N}{\partial x} = \cos y = \dfrac{\partial M}{\partial y} \Rightarrow$ **F** is conservative.

28. $\mathbf{F}(x, y) = 5y^2(y\mathbf{i} + 3x\mathbf{j})$

$M = 5y^3$, $N = 15xy^2$

$\dfrac{\partial N}{\partial x} = 15y^2 = \dfrac{\partial M}{\partial y} \Rightarrow$ Conservative

29. $\mathbf{F}(x, y) = \dfrac{1}{xy}(y\mathbf{i} - x\mathbf{j}) = \dfrac{1}{x}\mathbf{i} - \dfrac{1}{y}\mathbf{j}$

$M = 1/x$ and $N = -1/y$ have continuous first partial derivatives for all $x, y \neq 0$.

$\dfrac{\partial N}{\partial x} = 0 = \dfrac{\partial M}{\partial y} \Rightarrow$ **F** is conservative.

30. $M = \dfrac{2}{y}e^{2x/y}$, $N = \dfrac{-2x}{y^2}e^{2x/y}$

$\dfrac{\partial N}{\partial x} = \dfrac{-2(y + 2x)}{y^3}e^{2x/y} = \dfrac{\partial M}{\partial y} \Rightarrow$ Conservative

31. $M = \dfrac{1}{\sqrt{x^2 + y^2}}$, $N = \dfrac{1}{\sqrt{x^2 + y^2}}$

$\dfrac{\partial N}{\partial x} = \dfrac{-x}{(x^2 + y^2)^{3/2}} \neq \dfrac{\partial M}{\partial y} = \dfrac{-y}{(x^2 + y^2)^{3/2}}$

$\Rightarrow$ Not conservative

32. $M = \dfrac{y}{\sqrt{1 + xy}}$, $N = \dfrac{x}{\sqrt{1 + xy}}$

$\dfrac{\partial N}{\partial x} = \dfrac{xy + 2}{2(xy + 1)^{3/2}} = \dfrac{\partial M}{\partial y} \Rightarrow$ Conservative

33. $\mathbf{F}(x, y) = y\mathbf{i} + x\mathbf{j}$

$\dfrac{\partial}{\partial y}[y] = 1 = \dfrac{\partial}{\partial x}[x] \Rightarrow$ Conservative

$f_x(x, y) = y$, $f_y(x, y) = x \Rightarrow f(x, y) = xy + k$

34. $\mathbf{F}(x, y) = 3x^2y^2\mathbf{i} + 2x^3y\mathbf{j}$

$$\frac{\partial}{\partial y}\left[3x^2y^2\right] = 6x^2y$$

$$\frac{\partial}{\partial x}\left[2x^3y\right] = 6x^2y$$

Conservative

$$f_x(x, y) = 3x^2y^2$$

$$f_y(x, y) = 2x^3y$$

$$f(x, y) = x^3y^2 + K$$

35. $\mathbf{F}(x, y) = 2xy\mathbf{i} + x^2\mathbf{j}$

$$\frac{\partial}{\partial y}[2xy] = 2x$$

$$\frac{\partial}{\partial x}\left[x^2\right] = 2x$$

Conservative

$f_x(x, y) = 2xy$, $f_y(x, y) = x^2$, $f(x, y) = x^2y + K$

36. $\mathbf{F}(x, y) = xe^{x^2y}(2y\mathbf{i} + x\mathbf{j})$

$$\frac{\partial}{\partial y}\left[2xye^{x^2y}\right] = 2xe^{x^2y} + 2x^3ye^{x^2y}$$

$$\frac{\partial}{\partial x}\left[x^2e^{x^2y}\right] = 2xe^{x^2y} + 2x^3ye^{x^2y}$$

Conservative

$$f_x(x, y) = 2xye^{x^2y}$$

$$f_y(x, y) = x^2e^{x^2y}$$

$$f(x, y) = e^{x^2y} + K$$

37. $\mathbf{F}(x, y) = 15y^3\mathbf{i} - 5xy^2\mathbf{j}$

$$\frac{\partial}{\partial y}\left[15y^3\right] = 45y^2 \neq \frac{\partial}{\partial x}\left[-5xy^2\right] = -5y^2$$

Not conservative

38. $\mathbf{F}(x, y) = \dfrac{1}{y^2}(y\mathbf{i} - 2x\mathbf{j})$

$$= \frac{1}{y}\mathbf{i} - \frac{2x}{y^2}\mathbf{j}$$

$$\frac{\partial}{\partial y}\left[\frac{1}{y}\right] = -\frac{1}{y^2}$$

$$\frac{\partial}{\partial x}\left[-\frac{2x}{y^2}\right] = -\frac{2}{y^2}$$

Not conservative

39. $\mathbf{F}(x, y) = \dfrac{2y}{x}\mathbf{i} - \dfrac{x^2}{y^2}\mathbf{j}$

$$\frac{\partial}{\partial y}\left[\frac{2y}{x}\right] = \frac{2}{x}$$

$$\frac{\partial}{\partial x}\left[-\frac{x^2}{y^2}\right] = -\frac{2x}{y^2}$$

Not conservative

40. $\mathbf{F}(x, y) = \dfrac{x}{x^2 + y^2}\mathbf{i} + \dfrac{y}{x^2 + y^2}\mathbf{j}$

$$\frac{\partial}{\partial y}\left[\frac{x}{x^2 + y^2}\right] = -\frac{2xy}{\left(x^2 + y^2\right)^2}$$

$$\frac{\partial}{\partial x}\left[\frac{y}{x^2 + y^2}\right] = -\frac{2xy}{\left(x^2 + y^2\right)^2}$$

Conservative

$$f_x(x, y) = \frac{x}{x^2 + y^2}$$

$$f_y(x, y) = \frac{y}{x^2 + y^2}$$

$$f(x, y) = \frac{1}{2}\ln\left(x^2 + y^2\right) + K$$

41. $\mathbf{F}(x, y) = e^x(\cos y\mathbf{i} - \sin y\mathbf{j})$

$$\frac{\partial}{\partial y}\left[e^x \cos y\right] = -e^x \sin y$$

$$\frac{\partial}{\partial x}\left[-e^x \sin y\right] = -e^x \sin y$$

Conservative

$$f_x(x, y) = e^x \cos y$$

$$f_y(x, y) = -e^x \sin y$$

$$f(x, y) = e^x \cos y + K$$

42. $\mathbf{F}(x, y) = \dfrac{2x}{(x^2 + y^2)^2}\mathbf{i} + \dfrac{2y}{(x^2 + y^2)^2}\mathbf{j}$

$$\frac{\partial}{\partial y}\left[\frac{2x}{(x^2 + y^2)^2}\right] = -\frac{8xy}{(x^2 + y^2)^3}$$

$$\frac{\partial}{\partial x}\left[\frac{2y}{(x^2 + y^2)^2}\right] = -\frac{8xy}{(x^2 + y^2)^3}$$

Conservative

$$f_x(x, y) = \frac{2x}{(x^2 + y^2)^2}$$

$$f_y(x, y) = \frac{2y}{(x^2 + y^2)^2}$$

$$f(x, y) = -\frac{1}{x^2 + y^2} + K$$

43. $\mathbf{F}(x, y, z) = xyz\,\mathbf{i} + xyz\,\mathbf{j} + xyz\,\mathbf{k},\ (2, 1, 3)$

$$\textbf{curl F} = \begin{vmatrix} \mathbf{i} & \mathbf{j} & \mathbf{k} \\ \dfrac{\partial}{\partial x} & \dfrac{\partial}{\partial y} & \dfrac{\partial}{\partial z} \\ xyz & xyz & xyz \end{vmatrix}$$

$$= (xz - xy)\mathbf{i} - (yz - xy)\mathbf{j} + (yz - xz)\mathbf{k}$$

$$\textbf{curl F}\,(2, 1, 3) = (6 - 2)\mathbf{i} - (3 - 2)\mathbf{j} + (3 - 6)\mathbf{k}$$

$$= 4\mathbf{i} - \mathbf{j} - 3\mathbf{k}$$

44. $\mathbf{F}(x, y, z) = x^2z\mathbf{i} - 2xz\,\mathbf{j} + yz\mathbf{k},\ (2, -1, 3)$

$$\textbf{curl F} = \begin{vmatrix} \mathbf{i} & \mathbf{j} & \mathbf{k} \\ \dfrac{\partial}{\partial x} & \dfrac{\partial}{\partial y} & \dfrac{\partial}{\partial z} \\ x^2z & -2xz & yz \end{vmatrix}$$

$$= (z + 2x)\mathbf{i} - (0 - x^2)\mathbf{j} + (-2z - 0)\mathbf{k}$$

$$= (z + 2x)\mathbf{i} + x^2\mathbf{j} - 2z\mathbf{k}$$

$$\textbf{curl F}\,(2, -1, 3) = 7\mathbf{i} + 4\mathbf{j} - 6\mathbf{k}$$

45. $\mathbf{F}(x, y, z) = e^x \sin y\,\mathbf{i} - e^x \cos y\,\mathbf{j},\ (0, 0, 1)$

$$\textbf{curl F} = \begin{vmatrix} \mathbf{i} & \mathbf{j} & \mathbf{k} \\ \dfrac{\partial}{\partial x} & \dfrac{\partial}{\partial y} & \dfrac{\partial}{\partial z} \\ e^x \sin y & -e^x \cos y & 0 \end{vmatrix} = \left(-e^x \cos y - e^x \cos y\right)\mathbf{k} = -2e^x \cos y\,\mathbf{k}$$

$$\textbf{curl F}\,(0, 0, 1) = -2\mathbf{k}$$

46. $\mathbf{F}(x, y, z) = e^{-xyz}(\mathbf{i} + \mathbf{j} + \mathbf{k}),\ (3, 2, 0)$

$$\textbf{curl F} = \begin{vmatrix} \mathbf{i} & \mathbf{j} & \mathbf{k} \\ \dfrac{\partial}{\partial x} & \dfrac{\partial}{\partial y} & \dfrac{\partial}{\partial z} \\ e^{-xyz} & e^{-xyz} & e^{-xyz} \end{vmatrix} = (-xz + xy)e^{-xyz}\mathbf{i} - (-yz + xy)e^{-xyz}\mathbf{j} + (-yz + xz)e^{-xyz}\mathbf{k}$$

$$\textbf{curl F}\,(3, 2, 0) = 6\mathbf{i} - 6\mathbf{j}$$

47. $\mathbf{F}(x, y, z) = \arctan\left(\dfrac{x}{y}\right)\mathbf{i} + \ln\sqrt{x^2 + y^2}\,\mathbf{j} + \mathbf{k}$

$$\textbf{curl F} = \begin{vmatrix} \mathbf{i} & \mathbf{j} & \mathbf{k} \\ \dfrac{\partial}{\partial x} & \dfrac{\partial}{\partial y} & \dfrac{\partial}{\partial z} \\ \arctan\left(\dfrac{x}{y}\right) & \dfrac{1}{2}\ln(x^2 + y^2) & 1 \end{vmatrix} = \left[\frac{x}{x^2 + y^2} - \frac{(-x/y^2)}{1 + (x/y)^2}\right]\mathbf{k} = \frac{2x}{x^2 + y^2}\mathbf{k}$$

48. $\mathbf{F}(x, y, z) = \frac{yz}{y - z}\mathbf{i} + \frac{xz}{x - z}\mathbf{j} + \frac{xy}{x - y}\mathbf{k}$

$$\textbf{curl F} = \begin{vmatrix} \mathbf{i} & \mathbf{j} & \mathbf{k} \\ \frac{\partial}{\partial x} & \frac{\partial}{\partial y} & \frac{\partial}{\partial z} \\ \frac{yz}{y-z} & \frac{xz}{x-z} & \frac{xy}{x-y} \end{vmatrix} = \left[\frac{x^2}{(x-y)^2} - \frac{x^2}{(x-z)^2}\right]\mathbf{i} - \left[\frac{-y^2}{(x-y)^2} - \frac{y^2}{(y-z)^2}\right]\mathbf{j} + \left[\frac{-z^2}{(x-z)^2} - \frac{-z^2}{(y-z)^2}\right]\mathbf{k}$$

$$= x^2\left[\frac{1}{(x-y)^2} - \frac{1}{(x-z)^2}\right]\mathbf{i} + y^2\left[\frac{1}{(x-y)^2} + \frac{1}{(y-z)^2}\right]\mathbf{j} + z^2\left[\frac{1}{(y-z)^2} - \frac{1}{(x-z)^2}\right]\mathbf{k}$$

49. $\mathbf{F}(x, y, z) = \sin(x - y)\mathbf{i} + \sin(y - z)\mathbf{j} + \sin(z - x)\mathbf{k}$

$$\textbf{curl F} = \begin{vmatrix} \mathbf{i} & \mathbf{j} & \mathbf{k} \\ \frac{\partial}{\partial x} & \frac{\partial}{\partial y} & \frac{\partial}{\partial z} \\ \sin(x-y) & \sin(y-z) & \sin(z-x) \end{vmatrix} = \cos(y-z)\mathbf{i} + \cos(z-x)\mathbf{j} + \cos(x-y)\mathbf{k}$$

50. $\mathbf{F}(x, y, z) = \sqrt{x^2 + y^2 + z^2}(\mathbf{i} + \mathbf{j} + \mathbf{k})$

$$\textbf{curl F} = \begin{vmatrix} \mathbf{i} & \mathbf{j} & \mathbf{k} \\ \frac{\partial}{\partial x} & \frac{\partial}{\partial y} & \frac{\partial}{\partial z} \\ \sqrt{x^2+y^2+z^2} & \sqrt{x^2+y^2+z^2} & \sqrt{x^2+y^2+z^2} \end{vmatrix} = \frac{(y-z)\mathbf{i} + (z-x)\mathbf{j} + (x-y)\mathbf{k}}{\sqrt{x^2+y^2+z^2}}$$

51. $\mathbf{F}(x, y, z) = xy^2z^2\mathbf{i} + x^2yz^2\mathbf{j} + x^2y^2z\mathbf{k}$

$$\textbf{curl F} = \begin{vmatrix} \mathbf{i} & \mathbf{j} & \mathbf{k} \\ \frac{\partial}{\partial x} & \frac{\partial}{\partial y} & \frac{\partial}{\partial z} \\ xy^2z^2 & x^2yz^2 & x^2y^2z \end{vmatrix} = \mathbf{0}$$

Conservative

$f_x(x, y, z) = xy^2z^2$

$f_y(x, y, z) = x^2yz^2$

$f_z(x, y, z) = x^2y^2z$

$f(x, y, z) = \frac{1}{2}x^2y^2z^2 + K$

52. $\mathbf{F}(x, y, z) = y^2z^3\mathbf{i} + 2xyz^3\mathbf{j} + 3xy^2z^2\mathbf{k}$

$$\textbf{curl F} = \begin{vmatrix} \mathbf{i} & \mathbf{j} & \mathbf{k} \\ \frac{\partial}{\partial x} & \frac{\partial}{\partial y} & \frac{\partial}{\partial z} \\ y^2z^3 & 2xyz^3 & 3xy^2z^2 \end{vmatrix} = \mathbf{0}$$

Conservative

$f(x, y, z) = xy^2z^3 + K$

53. $\mathbf{F}(x, y, z) = \sin z\mathbf{i} + \sin x\mathbf{j} + \sin y\mathbf{k}$

$$\textbf{curl F} = \begin{vmatrix} \mathbf{i} & \mathbf{j} & \mathbf{k} \\ \frac{\partial}{\partial x} & \frac{\partial}{\partial y} & \frac{\partial}{\partial z} \\ \sin z & \sin x & \sin y \end{vmatrix}$$

$$= \cos y\mathbf{i} + \cos z\mathbf{j} + \cos x\mathbf{k} \neq \mathbf{0}$$

Not conservative

54. $\mathbf{F}(x, y, z) = ye^z\mathbf{i} + ze^x\mathbf{j} + xe^y\mathbf{k}$

$$\textbf{curl F} = \begin{vmatrix} \mathbf{i} & \mathbf{j} & \mathbf{k} \\ \frac{\partial}{\partial x} & \frac{\partial}{\partial y} & \frac{\partial}{\partial z} \\ ye^z & ze^x & xe^y \end{vmatrix} = (xe^y - e^x)\mathbf{i} - (e^y - ye^z)\mathbf{j} + (ze^x - e^z)\mathbf{k} \neq \mathbf{0}$$

Not conservative

55. $\mathbf{F}(x, y, z) = \frac{z}{y}\mathbf{i} - \frac{xz}{y^2}\mathbf{j} + \frac{x}{y}\mathbf{k}$

$$\text{curl } \mathbf{F} = \begin{vmatrix} \mathbf{i} & \mathbf{j} & \mathbf{k} \\ \frac{\partial}{\partial x} & \frac{\partial}{\partial y} & \frac{\partial}{\partial z} \\ \frac{z}{y} & -\frac{xz}{y^2} & \frac{x}{y} \end{vmatrix} = \left(-\frac{x}{y^2} + \frac{x}{y^2}\right)\mathbf{i} - \left(\frac{1}{y} - \frac{1}{y}\right)\mathbf{j} + \left(-\frac{z}{y^2} + \frac{z}{y^2}\right)\mathbf{k} = \mathbf{0}$$

Conservative

$$f_x(x, y, z) = \frac{z}{y}$$

$$f_y(x, y, z) = -\frac{xz}{y^2}$$

$$f_z(x, y, z) = \frac{x}{y}$$

$$f(x, y, z) = \frac{xz}{y} + K$$

56. $\mathbf{F}(x, y, z) = \frac{x}{x^2 + y^2}\mathbf{i} + \frac{y}{x^2 + y^2}\mathbf{j} + \mathbf{k}$

$$\text{curl } \mathbf{F} = \begin{vmatrix} \mathbf{i} & \mathbf{j} & \mathbf{k} \\ \frac{\partial}{\partial x} & \frac{\partial}{\partial y} & \frac{\partial}{\partial z} \\ \frac{x}{x^2 + y^2} & \frac{y}{x^2 + y^2} & 1 \end{vmatrix} = \mathbf{0}$$

Conservative

$$f_x(x, y, z) = \frac{x}{x^2 + y^2}$$

$$f_y(x, y, z) = \frac{y}{x^2 + y^2}$$

$$f_z(x, y, z) = 1$$

$$f(x, y, z) = \int \frac{x}{x^2 + y^2}\,dx = \frac{1}{2}\ln(x^2 + y^2) + g(y, z) + K_1$$

$$f(x, y, z) = \int \frac{y}{x^2 + y^2}\,dy = \frac{1}{2}\ln(x^2 + y^2) + h(x, z) + K_2$$

$$f(x, y, z) = \int dz = z + p(x, y) + K_3$$

$$f(x, y, z) = \frac{1}{2}\ln(x^2 + y^2) + z + K$$

57. $\mathbf{F}(x, y) = x^2\mathbf{i} + 2y^2\mathbf{j}$

$$\text{div } \mathbf{F}(x, y) = \frac{\partial}{\partial x}(x^2) + \frac{\partial}{\partial y}(2y^2) = 2x + 4y$$

58. $\mathbf{F}(x, y) = xe^x\mathbf{i} + ye^y\mathbf{i}$

$$\text{div } \mathbf{F}(x, y) = \frac{\partial}{\partial x}(xe^x) + \frac{\partial}{\partial y}(ye^y) = xe^x + e^x + ye^y + e^y$$

59. $\mathbf{F}(x, y, z) = \sin x\mathbf{i} + \cos y\mathbf{j} + z^2\mathbf{k}$

$$\text{div } \mathbf{F}(x, y, z) = \frac{\partial}{\partial x}[\sin x] + \frac{\partial}{\partial y}[\cos y] + \frac{\partial}{\partial z}[z^2] = \cos x - \sin y + 2z$$

60. $\mathbf{F}(x, y, z) = \ln(x^2 + y^2)\mathbf{i} + xy\mathbf{j} + \ln(y^2 + z^2)\mathbf{k}$

$$\text{div } \mathbf{F}(x, y, z) = \frac{\partial}{\partial x}\left[\ln(x^2 + y^2)\right] + \frac{\partial}{\partial y}[xy] + \frac{\partial}{\partial z}\left[\ln(y^2 + z^2)\right] = \frac{2x}{x^2 + y^2} + x + \frac{2z}{y^2 + z^2}$$

61. $\mathbf{F}(x, y, z) = xyz\mathbf{i} + xy\mathbf{j} + z\mathbf{k}$

$\text{div } \mathbf{F}(x, y, z) = yz + x + 1$

$\text{div } \mathbf{F}(2, 1, 1) = 1 + 2 + 1 = 4$

62. $\mathbf{F}(x, y, z) = x^2z\mathbf{i} - 2xz\mathbf{j} + yz\mathbf{k}$

$\text{div } \mathbf{F}(x, y, z) = 2xz + y$

$\text{div } \mathbf{F}(2, -1, 3) = 11$

63. $\mathbf{F}(x, y, z) = e^x \sin y\mathbf{i} - e^x \cos y\mathbf{j} + z^2\mathbf{k}$

$\text{div } \mathbf{F}(x, y, z) = e^x \sin y + e^x \sin y + 2z$

$\text{div } \mathbf{F}(3, 0, 0) = 0$

64. $\mathbf{F}(x, y, z) = \ln(xyz)(\mathbf{i} + \mathbf{j} + \mathbf{k})$

$\text{div } \mathbf{F}(x, y, z) = \frac{1}{x} + \frac{1}{y} + \frac{1}{z}$

$\text{div } \mathbf{F}(3, 2, 1) = \frac{1}{3} + \frac{1}{2} + 1 = \frac{11}{6}$

65. See the definition, page 1040. Examples include velocity fields, gravitational fields, and magnetic fields.

66. See the definition of Conservative Vector Field on page 1043. To test for a conservative vector field, see Theorems 15.1 and 15.2.

67. See the definition on page 1046.

68. See the definition on page 1048.

69. $\mathbf{F}(x, y, z) = \mathbf{i} + 3x\mathbf{j} + 2y\mathbf{k}$

$\mathbf{G}(x, y, z) = x\mathbf{i} - y\mathbf{j} + z\mathbf{k}$

$$\mathbf{F} \times \mathbf{G} = \begin{vmatrix} \mathbf{i} & \mathbf{j} & \mathbf{k} \\ 1 & 3x & 2y \\ x & -y & z \end{vmatrix}$$

$$= (3xz + 2y^2)\mathbf{i} - (z - 2xy)\mathbf{j} + (-y - 3x^2)\mathbf{k}$$

$$\mathbf{curl}(\mathbf{F} \times \mathbf{G}) = \begin{vmatrix} \mathbf{i} & \mathbf{j} & \mathbf{k} \\ \frac{\partial}{\partial x} & \frac{\partial}{\partial y} & \frac{\partial}{\partial z} \\ 3xz + 2y^2 & -z + 2xy & -y - 3x^2 \end{vmatrix}$$

$$= (-1 + 1)\mathbf{i} - (-6x - 3x)\mathbf{j} + (2y - 4y)\mathbf{k}$$

$$= 9x\mathbf{j} - 2y\mathbf{k}$$

70. $\mathbf{F}(x, y, z) = x\mathbf{i} - z\mathbf{k}$

$\mathbf{G}(x, y, z) = x^2\mathbf{i} + y\mathbf{j} + z^2\mathbf{k}$

$$\mathbf{F} \times \mathbf{G} = \begin{vmatrix} \mathbf{i} & \mathbf{j} & \mathbf{k} \\ x & 0 & -z \\ x^2 & y & z^2 \end{vmatrix} = yz\mathbf{i} - (xz^2 + x^2z)\mathbf{j} + xy\mathbf{k}$$

$$\mathbf{curl}(\mathbf{F} \times \mathbf{G}) = \begin{vmatrix} \mathbf{i} & \mathbf{j} & \mathbf{k} \\ \frac{\partial}{\partial x} & \frac{\partial}{\partial y} & \frac{\partial}{\partial z} \\ yz & -xz^2 - x^2z & xy \end{vmatrix} = (x + 2xz + x^2)\mathbf{i} - (y - y)\mathbf{j} + (-z^2 - 2xz - z)\mathbf{k} = x(x + 2z + 1)\mathbf{i} - z(z + 2x + 1)\mathbf{k}$$

71. $\mathbf{F}(x, y, z) = xyz\mathbf{i} + y\mathbf{j} + z\mathbf{k}$

$$\mathbf{curl}\ \mathbf{F} = \begin{vmatrix} \mathbf{i} & \mathbf{j} & \mathbf{k} \\ \frac{\partial}{\partial x} & \frac{\partial}{\partial y} & \frac{\partial}{\partial z} \\ xyz & y & z \end{vmatrix} = xy\mathbf{j} - xz\mathbf{k}$$

$$\mathbf{curl}(\mathbf{curl}\ \mathbf{F}) = \begin{vmatrix} \mathbf{i} & \mathbf{j} & \mathbf{k} \\ \frac{\partial}{\partial x} & \frac{\partial}{\partial y} & \frac{\partial}{\partial z} \\ 0 & xy & -xz \end{vmatrix} = z\mathbf{j} + y\mathbf{k}$$

72. $\mathbf{F}(x, y, z) = x^2z\mathbf{i} - 2xz\mathbf{j} + yz\mathbf{k}$

$$\mathbf{curl}\ \mathbf{F} = \begin{vmatrix} \mathbf{i} & \mathbf{j} & \mathbf{k} \\ \frac{\partial}{\partial x} & \frac{\partial}{\partial y} & \frac{\partial}{\partial z} \\ x^2z & -2xz & yz \end{vmatrix} = (z + 2x)\mathbf{i} + x^2\mathbf{j} - 2z\mathbf{k}$$

$$\mathbf{curl}(\mathbf{curl}\ \mathbf{F}) = \begin{vmatrix} \mathbf{i} & \mathbf{j} & \mathbf{k} \\ \frac{\partial}{\partial x} & \frac{\partial}{\partial y} & \frac{\partial}{\partial z} \\ z + 2x & x^2 & -2z \end{vmatrix} = \mathbf{j} + 2x\mathbf{k}$$

73. $\mathbf{F}(x, y, z) = \mathbf{i} + 3x\mathbf{j} + 2y\mathbf{k}$

$\mathbf{G}(x, y, z) = x\mathbf{i} - y\mathbf{j} + z\mathbf{k}$

$$\mathbf{F} \times \mathbf{G} = \begin{vmatrix} \mathbf{i} & \mathbf{j} & \mathbf{k} \\ 1 & 3x & 2y \\ x & -y & z \end{vmatrix} = (3xz + 2y^2)\mathbf{i} - (z - 2xy)\mathbf{j} + (-y - 3x^2)\mathbf{k}$$

$\operatorname{div}(\mathbf{F} \times \mathbf{G}) = 3z + 2x$

74. $\mathbf{F}(x, y, z) = x\mathbf{i} - z\mathbf{k}$

$\mathbf{G}(x, y, z) = x^2\mathbf{i} + y\mathbf{j} + z^2\mathbf{k}$

$$\mathbf{F} \times \mathbf{G} = \begin{vmatrix} \mathbf{i} & \mathbf{j} & \mathbf{k} \\ x & 0 & -z \\ x^2 & y & z^2 \end{vmatrix} = yz\mathbf{i} - (xz^2 + x^2z)\mathbf{j} + xy\mathbf{k}$$

$\operatorname{div}(\mathbf{F} \times \mathbf{G}) = 0$

75. $\mathbf{F}(x, y, z) = xyz\mathbf{i} + y\mathbf{j} + z\mathbf{k}$

$$\mathbf{curl\,F} = \begin{vmatrix} \mathbf{i} & \mathbf{j} & \mathbf{k} \\ \dfrac{\partial}{\partial x} & \dfrac{\partial}{\partial y} & \dfrac{\partial}{\partial z} \\ xyz & y & z \end{vmatrix} = xy\mathbf{j} - xz\mathbf{k}$$

$\operatorname{div}(\mathbf{curl\,F}) = x - x = 0$

76. $\mathbf{F}(x, y, z) = x^2z\mathbf{i} - 2xz\mathbf{j} + yz\mathbf{k}$

$$\mathbf{curl\,F} = \begin{vmatrix} \mathbf{i} & \mathbf{j} & \mathbf{k} \\ \dfrac{\partial}{\partial x} & \dfrac{\partial}{\partial y} & \dfrac{\partial}{\partial z} \\ x^2z & -2xz & yz \end{vmatrix} = (z + 2x)\mathbf{i} + x^2\mathbf{j} - 2z\mathbf{k}$$

$\operatorname{div}(\mathbf{curl\,F}) = 2 - 2 = 0$

77. (a) Let $\mathbf{F} = M\mathbf{i} + N\mathbf{j} + P\mathbf{k}$ and $\mathbf{G} = Q\mathbf{i} + R\mathbf{j} + S\mathbf{k}$ where M, N, P, Q, R, and S have continuous partial derivatives.

$\mathbf{F} + \mathbf{G} = (M + Q)\mathbf{i} + (N + R)\mathbf{j} + (P + S)\mathbf{k}$

$$\mathbf{curl}(\mathbf{F} + \mathbf{G}) = \begin{vmatrix} \mathbf{i} & \mathbf{j} & \mathbf{k} \\ \dfrac{\partial}{\partial x} & \dfrac{\partial}{\partial y} & \dfrac{\partial}{\partial z} \\ M + Q & N + R & P + S \end{vmatrix}$$

$$= \left[\frac{\partial}{\partial y}(P + S) - \frac{\partial}{\partial z}(N + R)\right]\mathbf{i} - \left[\frac{\partial}{\partial x}(P + S) - \frac{\partial}{\partial z}(M + Q)\right]\mathbf{j} + \left[\frac{\partial}{\partial x}(N + R) - \frac{\partial}{\partial y}(M + Q)\right]\mathbf{k}$$

$$= \left(\frac{\partial P}{\partial y} - \frac{\partial N}{\partial z}\right)\mathbf{i} - \left(\frac{\partial P}{\partial x} - \frac{\partial M}{\partial z}\right)\mathbf{j} + \left(\frac{\partial N}{\partial x} - \frac{\partial M}{\partial y}\right)\mathbf{k} + \left(\frac{\partial S}{\partial y} - \frac{\partial R}{\partial z}\right)\mathbf{i} - \left(\frac{\partial S}{\partial x} - \frac{\partial Q}{\partial z}\right)\mathbf{j} + \left(\frac{\partial R}{\partial x} - \frac{\partial Q}{\partial y}\right)\mathbf{k}$$

$$= \mathbf{curl\,F} + \mathbf{curl\,G}$$

(b) Let $f(x, y, z)$ be a scalar function whose second partial derivatives are continuous.

$$\nabla f = \frac{\partial f}{\partial x}\mathbf{i} + \frac{\partial f}{\partial y}\mathbf{j} + \frac{\partial f}{\partial z}\mathbf{k}$$

$$\mathbf{curl}(\nabla f) = \begin{vmatrix} \mathbf{i} & \mathbf{j} & \mathbf{k} \\ \dfrac{\partial}{\partial x} & \dfrac{\partial}{\partial y} & \dfrac{\partial}{\partial z} \\ \dfrac{\partial f}{\partial x} & \dfrac{\partial f}{\partial y} & \dfrac{\partial f}{\partial z} \end{vmatrix} = \left(\frac{\partial^2 f}{\partial y \partial z} - \frac{\partial^2 f}{\partial z \partial y}\right)\mathbf{i} - \left(\frac{\partial^2 f}{\partial x \partial z} - \frac{\partial^2 f}{\partial z \partial x}\right)\mathbf{j} + \left(\frac{\partial^2 f}{\partial x \partial y} - \frac{\partial^2 f}{\partial y \partial x}\right)\mathbf{k} = \mathbf{0}$$

(c) Let $\mathbf{F} = M\mathbf{i} + N\mathbf{j} + P\mathbf{k}$ and $\mathbf{G} = R\mathbf{i} + S\mathbf{j} + T\mathbf{k}$.

$$\operatorname{div}(\mathbf{F} + \mathbf{G}) = \frac{\partial}{\partial x}(M + R) + \frac{\partial}{\partial y}(N + S) + \frac{\partial}{\partial z}(P + T) = \frac{\partial M}{\partial x} + \frac{\partial R}{\partial x} + \frac{\partial N}{\partial y} + \frac{\partial S}{\partial y} + \frac{\partial P}{\partial z} + \frac{\partial T}{\partial z}$$

$$= \left[\frac{\partial M}{\partial x} + \frac{\partial N}{\partial y} + \frac{\partial P}{\partial z}\right] + \left[\frac{\partial R}{\partial x} + \frac{\partial S}{\partial y} + \frac{\partial T}{\partial z}\right]$$

$$= \operatorname{div}\mathbf{F} + \operatorname{div}\mathbf{G}$$

(d) Let $\mathbf{F} = M\mathbf{i} + N\mathbf{j} + P\mathbf{k}$ and $\mathbf{G} = R\mathbf{i} + S\mathbf{j} + T\mathbf{k}$.

$$\mathbf{F} \times \mathbf{G} = \begin{vmatrix} \mathbf{i} & \mathbf{j} & \mathbf{k} \\ M & N & P \\ R & S & T \end{vmatrix} = (NT - PS)\mathbf{i} - (MT - PR)\mathbf{j} + (MS - NR)\mathbf{k}$$

$$\begin{aligned} \operatorname{div}(\mathbf{F} \times \mathbf{G}) &= \frac{\partial}{\partial x}(NT - PS) + \frac{\partial}{\partial y}(PR - MT) + \frac{\partial}{\partial z}(MS - NR) \\ &= N\frac{\partial T}{\partial x} + T\frac{\partial N}{\partial x} - P\frac{\partial S}{\partial x} - S\frac{\partial P}{\partial x} + P\frac{\partial R}{\partial y} + R\frac{\partial P}{\partial y} - M\frac{\partial T}{\partial y} - T\frac{\partial M}{\partial y} + M\frac{\partial S}{\partial z} + S\frac{\partial M}{\partial z} - N\frac{\partial R}{\partial z} - R\frac{\partial N}{\partial z} \\ &= \left[\left(\frac{\partial P}{\partial y} - \frac{\partial N}{\partial z}\right)R + \left(\frac{\partial M}{\partial z} - \frac{\partial P}{\partial x}\right)S + \left(\frac{\partial N}{\partial x} - \frac{\partial M}{\partial y}\right)T\right] - \left[M\left(\frac{\partial T}{\partial y} - \frac{\partial S}{\partial z}\right) + N\left(\frac{\partial R}{\partial z} - \frac{\partial T}{\partial x}\right) + P\left(\frac{\partial S}{\partial x} - \frac{\partial R}{\partial y}\right)\right] \\ &= (\mathbf{curl\,F}) \cdot \mathbf{G} - \mathbf{F} \cdot (\mathbf{curl\,G}) \end{aligned}$$

(e) $\mathbf{F} = M\mathbf{i} + N\mathbf{j} + P\mathbf{k}$

$$\begin{aligned} \nabla \times [\nabla f + (\nabla \times \mathbf{F})] &= \mathbf{curl}(\nabla f + (\nabla \times \mathbf{F})) \\ &= \mathbf{curl}(\nabla f) + \mathbf{curl}(\nabla \times \mathbf{F}) && \text{(Part (a))} \\ &= \mathbf{curl}(\nabla \times \mathbf{F}) && \text{(Part (b))} \\ &= \nabla \times (\nabla \times \mathbf{F}) \end{aligned}$$

(f) Let $\mathbf{F} = M\mathbf{i} + N\mathbf{j} + P\mathbf{k}$.

$$\nabla \times (f\mathbf{F}) = \begin{vmatrix} \mathbf{i} & \mathbf{j} & \mathbf{k} \\ \dfrac{\partial}{\partial x} & \dfrac{\partial}{\partial y} & \dfrac{\partial}{\partial z} \\ fM & fN & fP \end{vmatrix}$$

$$= \left(\frac{\partial f}{\partial y}P + f\frac{\partial P}{\partial y} - \frac{\partial f}{\partial z}N - f\frac{\partial N}{\partial z}\right)\mathbf{i} - \left(\frac{\partial f}{\partial x}P + f\frac{\partial P}{\partial x} - \frac{\partial f}{\partial z}M - f\frac{\partial M}{\partial z}\right)\mathbf{j} + \left(\frac{\partial f}{\partial x}N + f\frac{\partial N}{\partial x} - \frac{\partial f}{\partial y}M - f\frac{\partial M}{\partial y}\right)\mathbf{k}$$

$$= f\left[\left(\frac{\partial P}{\partial y} - \frac{\partial N}{\partial z}\right)\mathbf{i} - \left(\frac{\partial P}{\partial x} - \frac{\partial M}{\partial z}\right)\mathbf{j} + \left(\frac{\partial N}{\partial x} - \frac{\partial M}{\partial y}\right)\mathbf{k}\right] + \begin{vmatrix} \mathbf{i} & \mathbf{j} & \mathbf{k} \\ \dfrac{\partial f}{\partial x} & \dfrac{\partial f}{\partial y} & \dfrac{\partial f}{\partial z} \\ M & N & P \end{vmatrix} = f[\nabla \times \mathbf{F}] + (\nabla f) \times \mathbf{F}$$

(g) Let $\mathbf{F} = M\mathbf{i} + N\mathbf{j} + P\mathbf{k}$, then $f\mathbf{F} = fM\mathbf{i} + fN\mathbf{j} + fP\mathbf{k}$.

$$\begin{aligned} \operatorname{div}(f\mathbf{F}) = \frac{\partial}{\partial x}(fM) + \frac{\partial}{\partial y}(fN) + \frac{\partial}{\partial z}(fP) &= f\frac{\partial M}{\partial x} + M\frac{\partial f}{\partial x} + f\frac{\partial N}{\partial y} + N\frac{\partial f}{\partial y} + f\frac{\partial P}{\partial z} + P\frac{\partial f}{\partial z} \\ &= f\left(\frac{\partial M}{\partial x} + \frac{\partial N}{\partial y} + \frac{\partial N}{\partial z}\right) + \left(\frac{\partial f}{\partial x}M + \frac{\partial f}{\partial y}N + \frac{\partial f}{\partial z}P\right) = f\operatorname{div}\mathbf{F} + \nabla f \cdot \mathbf{F} \end{aligned}$$

(h) Let $\mathbf{F} = M\mathbf{i} + N\mathbf{j} + P\mathbf{k}$.

$$\mathbf{curl\,F} = \left(\frac{\partial P}{\partial y} - \frac{\partial N}{\partial z}\right)\mathbf{i} - \left(\frac{\partial P}{\partial x} - \frac{\partial M}{\partial z}\right)\mathbf{j} + \left(\frac{\partial N}{\partial x} - \frac{\partial M}{\partial y}\right)\mathbf{k}$$

$$\begin{aligned} \operatorname{div}(\mathbf{curl\,F}) &= \frac{\partial}{\partial x}\left[\frac{\partial P}{\partial y} - \frac{\partial N}{\partial z}\right] - \frac{\partial}{\partial y}\left[\frac{\partial P}{\partial x} - \frac{\partial M}{\partial z}\right] + \frac{\partial}{\partial z}\left[\frac{\partial N}{\partial x} - \frac{\partial M}{\partial y}\right] \\ &= \frac{\partial^2 P}{\partial x \partial y} - \frac{\partial^2 N}{\partial x \partial z} - \frac{\partial^2 P}{\partial y \partial x} + \frac{\partial^2 M}{\partial y \partial z} + \frac{\partial^2 N}{\partial z \partial x} - \frac{\partial^2 M}{\partial z \partial y} = 0 \quad \text{(because the mixed partials are equal)} \end{aligned}$$

78. The vectors satisfy $\|\mathbf{F}(x, y)\| = \|\mathbf{G}(x, y)\| = 1$.

The vectors $\mathbf{F}(x, y)$ all point away from the origin. The vectors $\mathbf{G}(x, y)$ all point to the x-axis at a 45° degree angle.

79. True.

$\|\mathbf{F}(x + y)\| = \sqrt{16x^2 + y^4} \to 0$ as $(x, y) \to (0, 0)$.

80. True. If (x, y) is on the positive y-axis, then $x = 0$ and $y > 0$. So,

$\mathbf{F}(x, y) = \mathbf{F}(0, y) = -y^2\mathbf{j}$.

81. False. Curl is defined on vector fields, not scalar fields.

82. False. See Example 7.

83. $\mathbf{F}(x, y) = M(x, y)\mathbf{i} + N(x, y)\mathbf{j} = \dfrac{m}{(x^2 + y^2)^{5/2}}\left[3xy\mathbf{i} + (2y^2 - x^2)\mathbf{j}\right]$

$$M = \frac{3mxy}{(x^2 + y^2)^{5/2}} = 3mxy(x^2 + y^2)^{-5/2}$$

$$\frac{\partial M}{\partial y} = 3mxy\left[-\frac{5}{2}(x^2 + y^2)^{-7/2}(2y)\right] + (x^2 + y^2)^{-5/2}(3mx)$$

$$= 3mx(x^2 + y^2)^{-7/2}\left[-5y^2 + (x^2 + y^2)\right] = \frac{3mx(x^2 - 4y^2)}{(x^2 + y^2)^{7/2}}$$

$$N = \frac{m(2y^2 - x^2)}{(x^2 + y^2)^{5/2}} = m(2y^2 - x^2)(x^2 + y^2)^{-5/2}$$

$$\frac{\partial N}{\partial x} = m(2y^2 - x^2)\left[-\frac{5}{2}(x^2 + y^2)^{-7/2}(2x)\right] + (x^2 + y^2)^{-5/2}(-2mx)$$

$$= mx(x^2 + y^2)^{-7/2}\left[(2y^2 - x^2)(-5) + (x^2 + y^2)(-2)\right]$$

$$= mx(x^2 + y^2)^{-7/2}(3x^2 - 12y^2) = \frac{3mx(x^2 - 4y^2)}{(x^2 + y^2)^{7/2}}$$

So, $\dfrac{\partial N}{\partial x} = \dfrac{\partial M}{\partial y}$ and $\mathbf{F}$ is conservative.

Section 15.2 Line Integrals

1. $\mathbf{r}(t) = \begin{cases} t\mathbf{i} + t\mathbf{j}, & 0 \le t \le 1 \\ (2 - t)\mathbf{i} + \sqrt{2 - t}\,\mathbf{j}, & 1 \le t \le 2 \end{cases}$

2. $\mathbf{r}(t) = \begin{cases} t\mathbf{i} + t^2\mathbf{j}, & 0 \le t \le 2 \\ (4 - t)\mathbf{i} + 4\mathbf{j}, & 2 \le t \le 4 \\ (8 - t)\mathbf{j}, & 4 \le t \le 8 \end{cases}$

3. $\mathbf{r}(t) = \begin{cases} t\mathbf{i}, & 0 \le t \le 3 \\ 3\mathbf{i} + (t - 3)\mathbf{j}, & 3 \le t \le 6 \\ (9 - t)\mathbf{i} + 3\mathbf{j}, & 6 \le t \le 9 \\ (12 - t)\mathbf{j}, & 9 \le t \le 12 \end{cases}$

4. $\mathbf{r}(t) = \begin{cases} t\mathbf{i} + \frac{4}{5}t\mathbf{j}, & 0 \le t \le 5 \\ 5\mathbf{i} + (9 - t)\mathbf{j}, & 5 \le t \le 9 \\ (14 - t)\mathbf{i}, & 9 \le t \le 14 \end{cases}$

5.
$$x^2 + y^2 = 9$$
$$\frac{x^2}{9} + \frac{y^2}{9} = 1$$
$$\cos^2 t + \sin^2 t = 1$$
$$\cos^2 t = \frac{x^2}{9}$$
$$\sin^2 t = \frac{y^2}{9}$$
$$x = 3\cos t$$
$$y = 3\sin t$$
$$\mathbf{r}(t) = 3\cos t\mathbf{i} + 3\sin t\mathbf{j}$$
$$0 \le t \le 2\pi$$

6. $\dfrac{x^2}{16} + \dfrac{y^2}{9} = 1$

$\cos^2 t + \sin^2 t = 1$

$\cos^2 t = \dfrac{x^2}{16}$

$\sin^2 t = \dfrac{y^2}{9}$

$x = 4 \cos t$

$y = 3 \sin t$

$\mathbf{r}(t) = 4 \cos t\mathbf{i} + 3 \sin t\mathbf{j}$

$0 \le t \le 2\pi$

7. $\mathbf{r}(t) = 4t\mathbf{i} + 3t\mathbf{j}, \quad 0 \le t \le 1$

$\mathbf{r}'(t) = 4\mathbf{i} + 3\mathbf{j}$

$$\int_C xy\, ds = \int_0^1 (4t)(3t)\sqrt{4^2 + 3^2}\, dt$$

$$= \int_0^1 60t^2\, dt = \left[20t^3\right]_0^1 = 20$$

8. $\mathbf{r}(t) = t\mathbf{i} + (2 - t)\mathbf{j}, \quad 0 \le t \le 2$

$\mathbf{r}'(t) = \mathbf{i} - \mathbf{j}$

$$\int_C 3(x - y)\, ds = \int_0^2 3(t - (2 - t))\sqrt{1^2 + (-1)^2}\, dt$$

$$= 3\sqrt{2}\int_0^2 (2t - 2)\, dt$$

$$= 3\sqrt{2}\left[t^2 - 2t\right]_0^2 = 0$$

9. $\mathbf{r}(t) = \sin t\mathbf{i} + \cos t\mathbf{j} + 2\mathbf{k}, \quad 0 \le t \le \dfrac{\pi}{2}$

$\mathbf{r}'(t) = \cos t\mathbf{i} - \sin t\mathbf{j}$

$$\int_C \left(x^2 + y^2 + z^2\right) ds = \int_0^{\pi/2} \left(\sin^2 t + \cos^2 t + 4\right)\sqrt{\cos^2 t + \sin^2 t}\, dt = \int_0^{\pi/2} 5dt = \frac{5\pi}{2}$$

10. $\mathbf{r}(t) = 12t\mathbf{i} + 5t\mathbf{j} + 84t\mathbf{k}, \quad 0 \le t \le 1$

$\mathbf{r}'(t) = 12\mathbf{i} + 5\mathbf{j} + 84\mathbf{k}$

$$\int_C 2xyz\, ds = \int_0^1 2(12t)(5t)(84t)\sqrt{(12)^2 + 5^2 + (84)^2}\, dt = \int_0^1 10{,}080\, t^3\, (85)\, dt = 856{,}800\left[\frac{t^4}{4}\right]_0^1 = 214{,}200$$

11. (a) $\mathbf{r}(t) = t\mathbf{i} + t\mathbf{j}, \quad 0 \le t \le 1$

(b) $\mathbf{r}'(t) = \mathbf{i} + \mathbf{j}, \|\mathbf{r}'(t)\| = \sqrt{2}$

$$\int_C \left(x^2 + y^2\right) ds = \int_0^1 \left(t^2 + t^2\right)\sqrt{2}\, dt$$

$$= 2\sqrt{2}\left[\frac{t^3}{3}\right]_0^1 = \frac{2\sqrt{2}}{3}$$

12. (a) $\mathbf{r}(t) = t\mathbf{i} + 2t\mathbf{j}, \quad 0 \le t \le 2$

(b) $\mathbf{r}'(t) = \mathbf{i} + 2\mathbf{j}, \|\mathbf{r}'(t)\| = \sqrt{5}$

$$\int_C \left(x^2 + y^2\right) ds = \int_0^2 \left(t^2 + 4t^2\right)\sqrt{5}\, dt$$

$$= \left[\sqrt{5}\frac{5t^3}{3}\right]_0^2 = \frac{40\sqrt{5}}{3}$$

13. (a) $\mathbf{r}(t) = \cos t\mathbf{i} + \sin t\mathbf{j}, \quad 0 \le t \le \dfrac{\pi}{2}$

(b) $$\int_C \left(x^2 + y^2\right) ds = \int_0^{\pi/2} \left[\cos^2 t + \sin^2 t\right]\sqrt{(-\sin t)^2 + (\cos t)^2}\, dt = \int_0^{\pi/2} dt = \frac{\pi}{2}$$

14. (a) $\mathbf{r}(t) = 2 \cos t\mathbf{i} + 2 \sin t\mathbf{j}, \quad 0 \le t \le \dfrac{\pi}{2}$

(b) $$\int_C \left(x^2 + y^2\right) ds = \int_0^{\pi/2} \left[4 \cos^2 t + 4 \sin^2 t\right]\sqrt{(-2 \sin t)^2 + (2 \cos t)^2}\, dt = \int_0^{\pi/2} 8\, dt = 4\pi$$

15. (a) $\mathbf{r}(t) = t\mathbf{i}, \quad 0 \le t \le 1$

(b) $\mathbf{r}'(t) = \mathbf{i}, \|\mathbf{r}'(t)\| = 1$

$$\int_C \left(x + 4\sqrt{y}\right) ds = \int_0^1 t\, dt = \left[\frac{t^2}{2}\right]_0^1 = \frac{1}{2}$$

16. (a) $\mathbf{r}(t) = t\mathbf{j},\quad 1 \le t \le 9$

(b) $\mathbf{r}'(t) = \mathbf{j}, \|\mathbf{r}'(t)\| = 1$

$$\int_C \left(x + 4\sqrt{y}\right) ds = \int_1^9 4\sqrt{t}\, dt = \left[\tfrac{8}{3}t^{3/2}\right]_1^9 = \tfrac{8}{3}(27 - 1) = \tfrac{208}{3}$$

17. (a) $$\mathbf{r}(t) = \begin{cases} t\mathbf{i}, & 0 \le t \le 1 \\ (2 - t)\mathbf{i} + (t - 1)\mathbf{j}, & 1 \le t \le 2 \\ (3 - t)\mathbf{j}, & 2 \le t \le 3 \end{cases}$$

(b) $$\int_{C_1} \left(x + 4\sqrt{y}\right) ds = \int_0^1 t\, dt = \frac{1}{2}$$

$$\int_{C_2} \left(x + 4\sqrt{y}\right) ds = \int_1^2 \left[(2 - t) + 4\sqrt{t - 1}\right]\sqrt{1 + 1}\, dt = \sqrt{2}\left[2t - \frac{t^2}{2} + \frac{8}{3}(t - 1)^{3/2}\right]_1^2 = \frac{19\sqrt{2}}{6}$$

$$\int_{C_3} \left(x + 4\sqrt{y}\right) ds = \int_2^3 4\sqrt{3 - t}\, dt = \left[-\frac{8}{3}(3 - t)^{3/2}\right]_2^3 = \frac{8}{3}$$

$$\int_C \left(x + 4\sqrt{y}\right) ds = \frac{1}{2} + \frac{19\sqrt{2}}{6} + \frac{8}{3} = \frac{19 + 19\sqrt{2}}{6} = \frac{19\left(1 + \sqrt{2}\right)}{6}$$

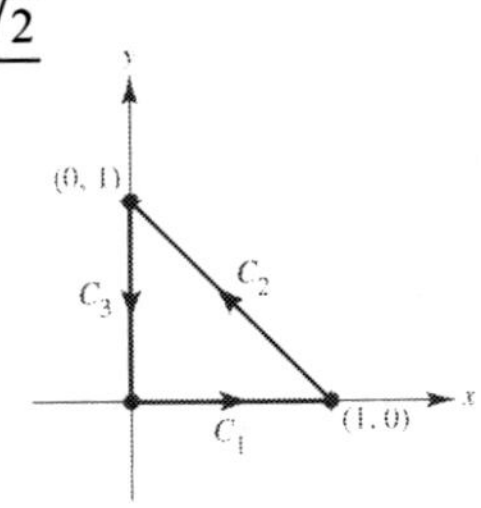

18. (a) $$\mathbf{r}(t) = \begin{cases} t\mathbf{i}, & 0 \le t \le 2 \\ 2\mathbf{i} + (t - 2)\mathbf{j}, & 2 \le t \le 4 \\ (6 - t)\mathbf{i} + 2\mathbf{j}, & 4 \le t \le 6 \\ (8 - t)\mathbf{j}, & 6 \le t \le 8 \end{cases}$$

(b) $$\int_{C_1} \left(x + 4\sqrt{y}\right) ds = \int_0^2 t\, dt = 2$$

$$\int_{C_2} \left(x + 4\sqrt{y}\right) ds = \int_2^4 \left(2 + 4\sqrt{t - 2}\right) dt = 4 + \frac{16\sqrt{2}}{3}$$

$$\int_{C_3} \left(x + 4\sqrt{y}\right) ds = \int_4^6 \left((6 - t) + 4\sqrt{2}\right) dt = 2 + 8\sqrt{2}$$

$$\int_{C_4} \left(x + 4\sqrt{y}\right) ds = \int_6^8 4\sqrt{8 - t}\, dt = \frac{16\sqrt{2}}{3}$$

$$\int_C \left(x + 4\sqrt{y}\right) ds = 2 + 4 + \frac{16\sqrt{2}}{3} + 2 + 8\sqrt{2} + \frac{16\sqrt{2}}{3} = 8 + \frac{56}{3}\sqrt{2}$$

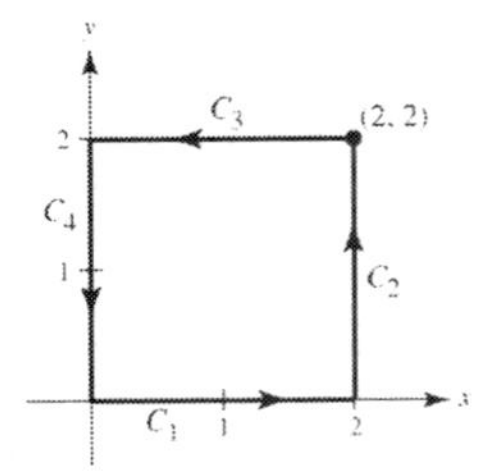

19. (a) $C_1: (0, 0, 0)$ to $(1, 0, 0)$: $\mathbf{r}(t) = t\mathbf{i}, 0 \le t \le 1, \mathbf{r}'(t) = \mathbf{i}, \|\mathbf{r}'(t)\| = 1$

$$\int_{C_1} \left(2x + y^2 - z\right) ds = \int_0^1 2t\, dt = t^2\Big]_0^1 = 1$$

$C_2: (1, 0, 0)$ to $(1, 0, 1)$: $\mathbf{r}(t) = \mathbf{i} + t\mathbf{k}, 0 \le t \le 1, \mathbf{r}'(t) = \mathbf{k}, \|\mathbf{r}'(t)\| = 1$

$$\int_{C_2} \left(2x + y^2 - z\right) ds = \int_0^1 (2 - t)\, dt = \left[2t - \frac{t^2}{2}\right]_0^1 = \frac{3}{2}$$

$C_3: (1, 0, 1)$ to $(1, 1, 1)$: $\mathbf{r}(t) = \mathbf{i} + t\mathbf{j} + \mathbf{k}, 0 \le t \le 1, \mathbf{r}'(t) = \mathbf{j}, \|\mathbf{r}'(t)\| = 1$

$$\int_{C_3} \left(2x + y^2 - z\right) ds = \int_0^1 \left(2 + t^2 - 1\right) dt = \left[t + \frac{t^3}{3}\right]_0^1 = \frac{4}{3}$$

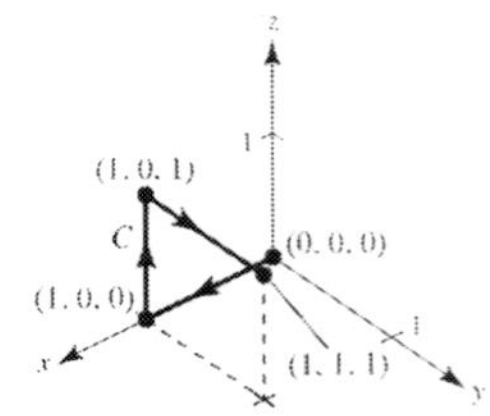

(b) Combining, $\int_C \left(2x + y^2 - z\right) ds = 1 + \dfrac{3}{2} + \dfrac{4}{3} = \dfrac{23}{6}.$

20. (a) $C_1: (0, 0, 0)$ to $(0, 1, 0)$: $\mathbf{r}(t) = t\mathbf{j},\ 0 \le t \le 1,\ \mathbf{r}'(t) = \mathbf{j}, \|\mathbf{r}'(t)\| = 1$

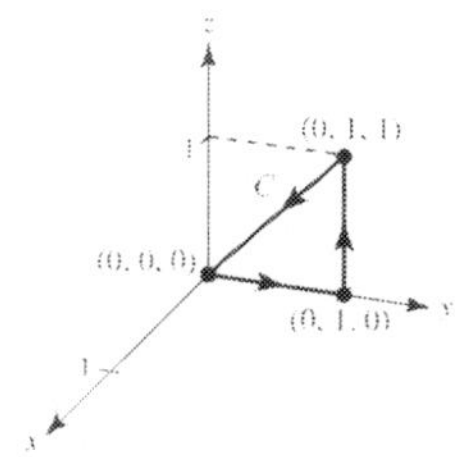

$$\int_{C_1} (2x + y^2 - z)\,ds = \int_0^1 t^2\,dt = \left[\frac{t^3}{3}\right]_0^1 = \frac{1}{3}$$

$C_2: (0, 1, 0)$ to $(0, 1, 1)$: $\mathbf{r}(t) = \mathbf{j} + t\mathbf{k},\ 0 \le t \le 1,\ \mathbf{r}'(t) = \mathbf{k}, \|\mathbf{r}'(t)\| = 1$

$$\int_{C_2} (2x + y^2 - z)\,ds = \int_0^1 (1 - t)\,dt = \left[t - \frac{t^2}{2}\right]_0^1 = \frac{1}{2}$$

$C_3: (0, 1, 1)$ to $(0, 0, 0)$: $\mathbf{r}(t) = (1 - t)\mathbf{j} + (1 - t)\mathbf{k},\ 0 \le t \le 1,\ \mathbf{r}'(t) = -\mathbf{j} - \mathbf{k}, \|\mathbf{r}'(t)\| = \sqrt{2}$

$$\int_{C_3} (2x + y^2 - z)\,ds = \int_0^1 \left[(1 - t)^2 - (1 - t)\right]\sqrt{2}\,dt = \int_0^1 (t^2 - t)\sqrt{2}\,dt = \sqrt{2}\left[\frac{t^3}{3} - \frac{t^2}{2}\right]_0^1 = \frac{-\sqrt{2}}{6}$$

(b) Combining, $\int_C (2x + y^2 - z)\,ds = \frac{1}{3} + \frac{1}{2} - \frac{\sqrt{2}}{6} = \frac{5 - \sqrt{2}}{6}$.

21. $\rho(x, y, z) = \frac{1}{2}(x^2 + y^2 + z^2)$

$\mathbf{r}(t) = 2\cos t\mathbf{i} + 2\sin t\mathbf{j} + t\mathbf{k},\ 0 \le t \le 4\pi$

$\mathbf{r}'(t) = -2\sin t\mathbf{i} + 2\cos t\mathbf{j} + \mathbf{k}$

$\|\mathbf{r}'(t)\| = \sqrt{4\sin^2 t + 4\cos^2 t + 1} = \sqrt{5}$

$$\text{Mass} = \int_C \rho(x, y, z)\,ds$$
$$= \int_0^{4\pi} \frac{1}{2}(4\cos^2 t + 4\sin^2 t + t^2)\sqrt{5}\,dt$$
$$= \frac{\sqrt{5}}{2}\int_0^{4\pi} (4 + t^2)\,dt = \frac{\sqrt{5}}{2}\left[4t + \frac{t^3}{3}\right]_0^{4\pi}$$
$$= \frac{\sqrt{5}}{2}\left[16\pi + \frac{64\pi^3}{3}\right] = \frac{8\pi\sqrt{5}}{3}(4\pi^2 + 3) \approx 795.7$$

22. $\rho(x, y, z) = z$

$\mathbf{r}(t) = 2\cos t\mathbf{i} + 2\sin t\mathbf{j} + t\mathbf{k},\ 0 \le t \le 4\pi$

$\mathbf{r}'(t) = -2\sin t\mathbf{i} + 2\cos t\mathbf{j} + \mathbf{k}$

$\|\mathbf{r}'(t)\| = \sqrt{4\sin^2 t + 4\cos^2 t + 1} = \sqrt{5}$

$$\text{Mass} = \int_C \rho(x, y, z)\,ds$$
$$= \int_0^{4\pi} t\sqrt{5}\,dt = \left[\frac{t^2}{2}\sqrt{5}\right]_0^{4\pi} = 8\pi^2\sqrt{5}$$

23. $\mathbf{r}(t) = \cos t\mathbf{i} + \sin t\mathbf{j},\ \ 0 \le t \le \pi$

$\mathbf{r}'(t) = -\sin t\mathbf{i} + \cos t\mathbf{j},\ \ \|\mathbf{r}'(t)\| = 1$

$$\text{Mass} = \int_C \rho(x, y)\,ds = \int_C (x + y + 2)\,ds$$
$$= \int_0^{\pi} (\cos t + \sin t + 2)\,dt$$
$$= [\sin t - \cos + 2t]_0^{\pi}$$
$$= (1 + 2\pi) - (-1) = 2 + 2\pi$$

24. $\mathbf{r}(t) = t^2\mathbf{i} + 2t\mathbf{j},\ \ 0 \le t \le 1$

$\mathbf{r}'(t) = 2t\mathbf{i} + 2\mathbf{j},\ \ \|\mathbf{r}'(t)\| = \sqrt{4t^2 + 4}$

$$\text{Mass} = \int_C \rho(x, y)\,ds = \int_C \tfrac{3}{4}y\,ds$$
$$= \int_0^1 \tfrac{3}{4}(2t)\sqrt{4t^2 + 4}\,dt$$
$$= \int_0^1 3t(t^2 + 1)^{1/2}\,dt$$
$$= \left[(t^2 + 1)^{3/2}\right]_0^1 = 2\sqrt{2} - 1$$

25. $\mathbf{r}(t) = t^2\mathbf{i} + 2t\mathbf{j} + t\mathbf{k},\ \ 1 \le t \le 3$

$\mathbf{r}'(t) = 2t\mathbf{i} + 2\mathbf{j} + \mathbf{k},\ \ \|\mathbf{r}'(t)\| = \sqrt{4t^2 + 5}$

$$\text{Mass} = \int_C \rho(x, y, z)\,ds = \int_C kz\,ds$$
$$= \int_1^3 kt\sqrt{4t^2 + 5}\,dt$$
$$= \left[\frac{k(4t^2 + 5)^{3/2}}{12}\right]_1^3$$
$$= \frac{k}{12}\left[41\sqrt{41} - 27\right]$$

26. $\mathbf{r}(t) = 2\cos t\mathbf{i} + 2\sin t\mathbf{j} + 3t\mathbf{k},\ \ 0 \le t \le 2\pi$

$\mathbf{r}'(t) = -2\sin t\mathbf{i} + 2\cos t\mathbf{j} + 3\mathbf{k}$

$\|\mathbf{r}'(t)\| = \sqrt{4 + 9} = \sqrt{13}$

$$\text{Mass} = \int_C \rho(x, y, z)\,ds = \int_C (k + z)\,ds$$
$$= \int_0^{2\pi} (k + 3t)\sqrt{13}\,dt$$
$$= \sqrt{13}\left[\left(kt + \frac{3t^2}{2}\right)\right]_0^{2\pi}$$
$$= \sqrt{13}(2\pi k + 6\pi^2)$$

27. $\mathbf{F}(x, y) = x\mathbf{i} + y\mathbf{j}$

$C: \mathbf{r}(t) = t\mathbf{i} + t\mathbf{j},\ 0 \le t \le 1$

$\mathbf{F}(t) = t\mathbf{i} + t\mathbf{j}$

$\mathbf{r}'(t) = \mathbf{i} + \mathbf{j}$

$$\int_C \mathbf{F} \cdot d\mathbf{r} = \int_0^1 (t + t)\,dt = \left[t^2\right]_0^1 = 1$$

28. $\mathbf{F}(x, y) = xy\mathbf{i} + y\mathbf{j}$

$C: \mathbf{r}(t) = 4\cos t\mathbf{i} + 4\sin t\mathbf{j},\ 0 \le t \le \dfrac{\pi}{2}$

$\mathbf{F}(t) = 16\sin t\cos t\mathbf{i} + 4\sin t\mathbf{j}$

$\mathbf{r}'(t) = -4\sin t\mathbf{i} + 4\cos t\mathbf{j}$

$$\int_C \mathbf{F} \cdot d\mathbf{r} = \int_0^{\pi/2} \left(-64\sin^2 t\cos t + 16\sin t\cos t\right) dt = \left[-\frac{64}{3}\sin^3 t + 8\sin^2 t\right]_0^{\pi/2} = -\frac{40}{3}$$

29. $\mathbf{F}(x, y) = 3x\mathbf{i} + 4y\mathbf{j}$

$C: \mathbf{r}(t) = \cos t\mathbf{i} + \sin t\mathbf{j},\ 0 \le t \le \pi/2$

$\mathbf{F}(t) = 3\cos t\mathbf{i} + 4\sin t\mathbf{j}$

$\mathbf{r}'(t) = -\sin t\mathbf{i} + \cos t\mathbf{j}$

$$\int_C \mathbf{F} \cdot d\mathbf{r} = \int_0^{\pi/2} (-3\cos t\sin t + 4\sin t\cos t)\,dt = \left[\frac{\sin^2 t}{2}\right]_0^{\pi/2} = \frac{1}{2}$$

30. $\mathbf{F}(x, y) = 3x\mathbf{i} + 4y\mathbf{j}$

$C: \mathbf{r}(t) = t\mathbf{i} + \sqrt{4 - t^2}\mathbf{j},\ -2 \le t \le 2$

$\mathbf{F}(t) = 3t\mathbf{i} + 4\sqrt{4 - t^2}\mathbf{j}$

$\mathbf{r}'(t) = \mathbf{i} - \dfrac{t}{\sqrt{4 - t^2}}\mathbf{j}$

$$\int_C \mathbf{F} \cdot d\mathbf{r} = \int_{-2}^{2} (3t - 4t)\,dt = \left[-\frac{t^2}{2}\right]_{-2}^{2} = 0$$

31. $\mathbf{F}(x, y, z) = xy\mathbf{i} + xz\mathbf{j} + yz\mathbf{k}$

$C: \mathbf{r}(t) = t\mathbf{i} + t^2\mathbf{j} + 2t\mathbf{k},\ 0 \le t \le 1$

$\mathbf{F}(t) = t^3\mathbf{i} + 2t^2\mathbf{j} + 2t^3\mathbf{k}$

$\mathbf{r}'(t) = \mathbf{i} + 2t\mathbf{j} + 2\mathbf{k}$

$$\int_C \mathbf{F} \cdot d\mathbf{r} = \int_0^1 \left(t^3 + 4t^3 + 4t^3\right) dt = \left[\frac{9t^4}{4}\right]_0^1 = \frac{9}{4}$$

32. $\mathbf{F}(x, y, z) = x^2\mathbf{i} + y^2\mathbf{j} + z^2\mathbf{k}$

$C: \mathbf{r}(t) = 2\sin t\mathbf{i} + 2\cos t\mathbf{j} + \dfrac{1}{2}t^2\mathbf{k},\ 0 \le t \le \pi$

$\mathbf{F}(t) = 4\sin^2 t\mathbf{i} + 4\cos^2 t\mathbf{j} + \dfrac{1}{4}t^4\mathbf{k}$

$\mathbf{r}'(t) = 2\cos t\mathbf{i} - 2\sin t\mathbf{j} + t\mathbf{k}$

$$\int_C \mathbf{F} \cdot d\mathbf{r} = \int_0^{\pi} \left(8\sin^2 t\cos t - 8\cos^2 t\sin t + \frac{1}{4}t^5\right) dt$$

$$= \left[\frac{8}{3}\sin^3 t + \frac{8}{3}\cos^3 t + \frac{t^6}{24}\right]_0^{\pi}$$

$$= -\frac{8}{3} + \frac{\pi^6}{24} - \frac{8}{3} = \frac{\pi^6}{24} - \frac{16}{3}$$

33. $\mathbf{F}(x, y, z) = x^2z\mathbf{i} + 6y\mathbf{j} + yz^2\mathbf{k}$

$\mathbf{r}(t) = t\mathbf{i} + t^2\mathbf{j} + \ln t\mathbf{k},\ 1 \le t \le 3$

$\mathbf{F}(t) = t^2\ln t\mathbf{i} + 6t^2\mathbf{j} + t^2\ln^2 t\mathbf{k}$

$d\mathbf{r} = \left(\mathbf{i} + 2t\mathbf{j} + \dfrac{1}{t}\mathbf{k}\right) dt$

$$\int_C \mathbf{F} \cdot d\mathbf{r} = \int_1^3 \left[t^2\ln t + 12t^3 + t(\ln t)^2\right] dt \approx 249.49$$

34. $\mathbf{F}(x, y, z) = \dfrac{x\mathbf{i} + y\mathbf{j} + z\mathbf{k}}{\sqrt{x^2 + y^2 + z^2}}$

$\mathbf{r}(t) = t\mathbf{i} + t\mathbf{j} + e^t\mathbf{k},\ 0 \le t \le 2$

$\mathbf{F}(t) = \dfrac{t\mathbf{i} + t\mathbf{j} + e^t\mathbf{k}}{\sqrt{2t^2 + e^{2t}}}$

$d\mathbf{r} = \left(\mathbf{i} + \mathbf{j} + e^t\mathbf{k}\right) dt$

$$\int_C \mathbf{F} \cdot d\mathbf{r} = \int_0^2 \frac{1}{\sqrt{2t^2 + e^{2t}}}\left(2t + e^{2t}\right) dt \approx 6.91$$

35. $\mathbf{F}(x, y) = x\mathbf{i} + 2y\mathbf{j}$

$C: \mathbf{r}(t) = t\mathbf{i} + t^3\mathbf{j},\ 0 \le t \le 2$

$\mathbf{r}'(t) = \mathbf{i} + 3t^2\mathbf{j}$

$\mathbf{F}(t) = t\mathbf{i} + 2t^3\mathbf{j}$

$$\text{Work} = \int_C \mathbf{F} \cdot d\mathbf{r} = \int_0^2 \left(t + 6t^5\right) dt = \left[\frac{t^2}{2} + t^6\right]_0^2 = 66$$

36. $\mathbf{F}(x, y) = x^2\mathbf{i} - xy\mathbf{j}$

$C: x = \cos^3 t,\ y = \sin^3 t$ from $(1, 0)$ to $(0, 1)$

$$\mathbf{r}(t) = \cos^3 t\mathbf{i} + \sin^3 t\mathbf{j},\quad 0 \le t \le \frac{\pi}{2}$$

$$\mathbf{r}'(t) = -3\cos^2 t \sin t\mathbf{i} + 3\sin^2 t \cos t\mathbf{j}$$

$$\mathbf{F}(t) = \cos^6 t\mathbf{i} - \cos^3 t \sin^3 t\mathbf{j}$$

$$\mathbf{F}\cdot\mathbf{r}' = -3\cos^8 t \sin t - 3\cos^4 t \sin^5 t = -3\cos^4 t \sin t\left(\cos^4 t + \sin^4 t\right) = -3\cos^4 t \sin t\left[\cos^4 t + \left(1 - \cos^2 t\right)^2\right]$$

$$= -3\cos^4 t \sin t\left(2\cos^4 t - 2\cos^2 t + 1\right) = -6\cos^8 t \sin t + 6\cos^6 t \sin t - 3\cos^4 t \sin t$$

$$\text{Work} = \int_C \mathbf{F}\cdot d\mathbf{r} = \int_0^{\pi/2}\left[-6\cos^8 t \sin t + 6\cos^6 t \sin t - 3\cos^4 t \sin t\right] dt$$

$$= \left[\frac{2\cos^9 t}{3} - \frac{6\cos^7 t}{7} + \frac{3\cos^5 t}{5}\right]_0^{\pi/2} = -\frac{43}{105}$$

37. $\mathbf{F}(x, y) = x\mathbf{i} + y\mathbf{j}$

$$C: \mathbf{r}(t) = \begin{cases} t\mathbf{i} & 0 \le t \le 1 \\ (2 - t)\mathbf{i} + (t - 1)\mathbf{j}, & 1 \le t \le 2 \\ (3 - t)\mathbf{j} & 2 \le t \le 3 \end{cases}$$

On C_1, $\mathbf{F}(t) = t\mathbf{i}$, $\mathbf{r}'(t) = \mathbf{i}$

$$\text{Work} = \int_{C_1} \mathbf{F}\cdot d\mathbf{r} = \int_0^1 t\, dt = \frac{1}{2}$$

On C_2, $\mathbf{F}(t) = (2 - t)\mathbf{i} + (t - 1)\mathbf{j}$, $\mathbf{r}'(t) = -\mathbf{i} + \mathbf{j}$

$$\text{Work} = \int_{C_2} \mathbf{F}\cdot d\mathbf{r} = \int_1^2 \left[(t - 2) + (t - 1)\right] dt$$

$$= \left[t^2 - 3t\right]_1^2$$

$$= (4 - 6) - (1 - 3) = 0$$

On C_3, $\mathbf{F}(t) = (3 - t)\mathbf{j}$, $\mathbf{r}'(t) = -\mathbf{j}$

$$\text{Work} = \int_{C_3} \mathbf{F}\cdot d\mathbf{r} = \int_2^3 (t - 3)\, dt = \left[\frac{t^2}{2} - 3t\right]_2^3$$

$$= \left(\frac{9}{2} - 9\right) - (2 - 6) = -\frac{1}{2}$$

$$\text{Total work} = \frac{1}{2} + 0 - \frac{1}{2} = 0$$

38. $\mathbf{F}(x, y) = -y\mathbf{i} - x\mathbf{j}$

C: counterclockwise along the semicircle $y = \sqrt{4 - x^2}$ from $(2, 0)$ to $(-2, 0)$

$$\mathbf{r}(t) = 2\cos t\mathbf{i} + 2\sin t\mathbf{j},\qquad 0 \le t \le \pi$$

$$\mathbf{r}'(t) = -2\sin t\mathbf{i} + 2\cos t\mathbf{j}$$

$$\mathbf{F}(t) = -2\sin t\mathbf{i} - 2\cos t\mathbf{j}$$

$$\mathbf{F}\cdot\mathbf{r}' = 4\sin^2 t - 4\cos^2 t = -4\cos 2t$$

$$\text{Work} = \int_C \mathbf{F}\cdot d\mathbf{r}$$

$$= -4\int_0^\pi \cos 2t\, dt$$

$$= \left[-2\sin 2t\right]_0^\pi = 0$$

39. $\mathbf{F}(x, y, z) = x\mathbf{i} + y\mathbf{j} - 5z\mathbf{k}$

C: $\mathbf{r}(t) = 2\cos t\mathbf{i} + 2\sin t\mathbf{j} + t\mathbf{k},\quad 0 \le t \le 2\pi$

$$\mathbf{r}'(t) = -2\sin t\mathbf{i} + 2\cos t\mathbf{j} + \mathbf{k}$$

$$\mathbf{F}(t) = 2\cos t\mathbf{i} + 2\sin t\mathbf{j} - 5t\mathbf{k}$$

$$\mathbf{F}\cdot\mathbf{r}' = -5t$$

$$\text{Work} = \int_C \mathbf{F}\cdot d\mathbf{r} = \int_0^{2\pi} -5t\, dt = -10\pi^2$$

40. $\mathbf{F}(x, y, z) = yz\mathbf{i} + xz\mathbf{j} + xy\mathbf{k}$

C: line from $(0, 0, 0)$ to $(5, 3, 2)$

$$\mathbf{r}(t) = 5t\mathbf{i} + 3t\mathbf{j} + 2t\mathbf{k},\quad 0 \le t \le 1$$

$$\mathbf{r}'(t) = 5\mathbf{i} + 3\mathbf{j} + 2\mathbf{k}$$

$$\mathbf{F}(t) = 6t^2\mathbf{i} + 10t^2\mathbf{j} + 15t^2\mathbf{k}$$

$$\mathbf{F}\cdot\mathbf{r}' = 90t^2$$

$$\text{Work} = \int_C \mathbf{F}\cdot d\mathbf{r} = \int_0^1 90t^2\, dt = 30$$

41. Because the vector field determined by **F** points in the general direction of the path C, $\mathbf{F}\cdot\mathbf{T} > 0$ and work will be positive.

42. Because the vector field determined by **F** points for the most part in the opposite direction of the path C, $\mathbf{F}\cdot\mathbf{T} < 0$ and work will be negative.

43. Because the vector field determined by **F** is perpendicular to the path, work will be 0.

44. Because the vector field is perpendicular to the path, work will be 0.

45. $\mathbf{F}(x, y) = x^2\mathbf{i} + xy\mathbf{j}$

(a) $\mathbf{r}_1(t) = 2t\mathbf{i} + (t - 1)\mathbf{j},\quad 1 \le t \le 3$

$\mathbf{r}_1'(t) = 2\mathbf{i} + \mathbf{j}$

$\mathbf{F}(t) = 4t^2\mathbf{i} + 2t(t - 1)\mathbf{j}$

$$\int_{C_1} \mathbf{F}\cdot d\mathbf{r} = \int_1^3 \left(8t^2 + 2t(t - 1)\right) dt = \frac{236}{3}$$

Both paths join $(2, 0)$ and $(6, 2)$. The integrals are negatives of each other because the orientations are different.

(b) $\mathbf{r}_2(t) = 2(3 - t)\mathbf{i} + (2 - t)\mathbf{j},\quad 0 \le t \le 2$

$\mathbf{r}_2'(t) = -2\mathbf{i} - \mathbf{j}$

$\mathbf{F}(t) = 4(3 - t)^2\mathbf{i} + 2(3 - t)(2 - t)\mathbf{j}$

$$\int_{C_2} \mathbf{F}\cdot d\mathbf{r} = \int_0^2 \left[-8(3 - t)^2 - 2(3 - t)(2 - t)\right] dt = -\frac{236}{3}$$

46. $\mathbf{F}(x, y) = x^2y\mathbf{i} + xy^{3/2}\mathbf{j}$

(a) $\mathbf{r}_1(t) = (t + 1)\mathbf{i} + t^2\mathbf{j},\quad 0 \le t \le 2$

$\mathbf{r}_1'(t) = \mathbf{i} + 2t\mathbf{j}$

$\mathbf{F}(t) = (t + 1)^2t^2\mathbf{i} + (t + 1)t^3\mathbf{j}$

$$\int_{C_1} \mathbf{F}\cdot d\mathbf{r} = \int_0^2 \left[(t + 1)^2t^2 + 2t^4(t + 1)\right] dt = \frac{256}{5}$$

(b) $\mathbf{r}_2(t) = (1 + 2\cos t)\mathbf{i} + 4\cos^2 t\mathbf{j},\ 0 \le t \le \frac{\pi}{2}$

$\mathbf{r}_2'(t) = -2\sin t\mathbf{i} - 8\cos t\sin t\mathbf{j}$

$\mathbf{F}(t) = (1 + 2\cos t)^2\left(4\cos^2 t\right)\mathbf{i} + (1 + 2\cos t)\left(8\cos^3 t\right)\mathbf{j}$

$$\int_{C_2} \mathbf{F}\cdot d\mathbf{r} = \int_0^{\pi/2} \left[(1 + 2\cos t)^2\left(4\cos^2 t\right)(-2\sin t) - 8\cos t\sin t(1 + 2\cos t)\left(8\cos^3 t\right) dt = -\frac{256}{5}\right.$$

Both paths join $(1, 0)$ and $(3, 4)$. The integrals are negatives of each other because the orientations are different.

47. $\mathbf{F}(x, y) = y\mathbf{i} - x\mathbf{j}$

$C: \mathbf{r}(t) = t\mathbf{i} - 2t\mathbf{j}$

$\mathbf{r}'(t) = \mathbf{i} - 2\mathbf{j}$

$\mathbf{F}(t) = -2t\mathbf{i} - t\mathbf{j}$

$\mathbf{F}\cdot\mathbf{r}' = -2t + 2t = 0$

So, $\int_C \mathbf{F}\cdot d\mathbf{r} = 0.$

48. $\mathbf{F}(x, y) = -3y\mathbf{i} + x\mathbf{j}$

$C: \mathbf{r}(t) = t\mathbf{i} - t^3\mathbf{j}$

$\mathbf{r}'(t) = \mathbf{i} - 3t^2\mathbf{j}$

$\mathbf{F}(t) = 3t^3\mathbf{i} + t\mathbf{j}$

$\mathbf{F}\cdot\mathbf{r}' = 3t^3 - 3t^3 = 0$

So, $\int_C \mathbf{F}\cdot d\mathbf{r} = 0.$

49. $\mathbf{F}(x, y) = (x^3 - 2x^2)\mathbf{i} + \left(x - \dfrac{y}{2}\right)\mathbf{j}$

$C: \mathbf{r}(t) = t\mathbf{i} + t^2\mathbf{j}$

$\mathbf{r}'(t) = \mathbf{i} + 2t\mathbf{j}$

$\mathbf{F}(t) = (t^3 - 2t^2)\mathbf{i} + \left(t - \dfrac{t^2}{2}\right)\mathbf{j}$

$\mathbf{F} \cdot \mathbf{r}' = (t^3 - 2t^2) + 2t\left(t - \dfrac{t^2}{2}\right) = 0$

So, $\int_C \mathbf{F} \cdot d\mathbf{r} = 0.$

50. $\mathbf{F}(x, y) = x\mathbf{i} + y\mathbf{j}$

$C: \mathbf{r}(t) = 3\sin t\mathbf{i} + 3\cos t\mathbf{j}$

$\mathbf{r}'(t) = 3\cos t\mathbf{i} - 3\sin t\mathbf{j}$

$\mathbf{F}(t) = 3\sin t\mathbf{i} + 3\cos t\mathbf{j}$

$\mathbf{F} \cdot \mathbf{r}' = 9\sin t\cos t - 9\sin t\cos t = 0$

So, $\int_C \mathbf{F} \cdot d\mathbf{r} = 0.$

51. $x = 2t,\ y = 10t,\ 0 \le t \le 1 \Rightarrow y = 5x \text{ or } x = \dfrac{y}{5},\ 0 \le y \le 10$

$$\int_C (x + 3y^2)\,dy = \int_0^{10}\left(\frac{y}{5} + 3y^2\right)dy = \left[\frac{y^2}{10} + y^3\right]_0^{10} = 1010$$

52. $x = 2t,\ y = 10t,\ 0 \le t \le 1 \Rightarrow y = 5x,\ 0 \le x \le 2$

$$\int_C (x + 3y^2)\,dx = \int_0^2 (x + 75x^2)\,dx = \left[\frac{x^2}{2} + 25x^3\right]_0^2 = 202$$

53. $x = 2t,\ y = 10t,\ 0 \le t \le 1 \Rightarrow x = \dfrac{y}{5},\ 0 \le y \le 10,\ dx = \dfrac{1}{5}\,dy$

$$\int_C xy\,dx + y\,dy = \int_0^{10}\left(\frac{y^2}{25} + y\right)dy = \left[\frac{y^3}{75} + \frac{y^2}{2}\right]_0^{10} = \frac{190}{3} \text{ or}$$

$y = 5x,\ dy = 5\,dx,\ 0 \le x \le 2$

$$\int_C xy\,dx + y\,dy = \int_0^2 (5x^2 + 25x)\,dx = \left[\frac{5x^3}{3} + \frac{25x^2}{2}\right]_0^2 = \frac{190}{3}$$

54. $x = 2t,\ y = 10t,\ 0 \le t \le 1 \Rightarrow y = 5x,\ dy = 5\,dx,\ 0 \le x \le 2$

$$\int_C (3y - x)\,dx + y^2\,dy = \int_0^2 (3(5x) - x)\,dx + (5x)^2 5\,dx = \int_0^2 (14x + 125x^2)\,dx$$

$$= \left[7x^2 + \tfrac{125}{3}x^3\right]_0^2 = 28 + \tfrac{125}{3}(8) = \tfrac{1084}{3}$$

55. $\mathbf{r}(t) = t\mathbf{i},\ 0 \le t \le 5$

$x(t) = t,\ y(t) = 0$

$dx = dt,\ dy = 0$

$$\int_C (2x - y)\,dx + (x + 3y)\,dy = \int_0^5 2t\,dt = 25$$

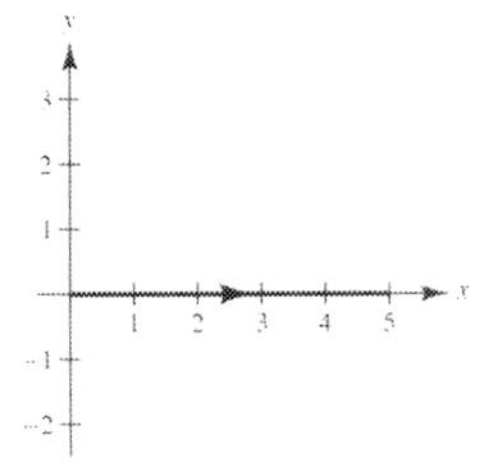

56. $\mathbf{r}(t) = t\mathbf{j},\ 0 \le t \le 2$

$x(t) = 0,\ y(t) = t$

$dx = 0,\ dy = dt$

$$\int_C (2x - y)\,dx + (x + 3y)\,dy = \int_0^2 3t\,dt = \left[\tfrac{3}{2}t^2\right]_0^2 = 6$$

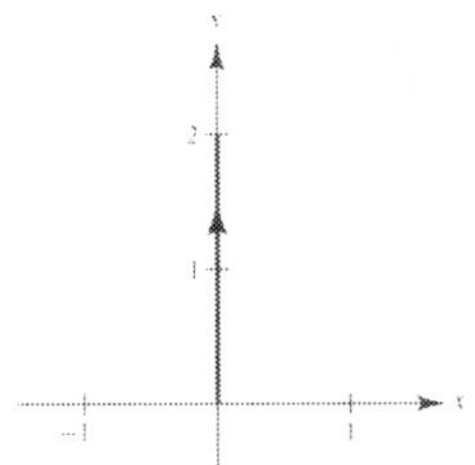

57. $\mathbf{r}(t) = \begin{cases} t\mathbf{i}, & 0 \le t \le 3 \\ 3\mathbf{i} + (t-3)\mathbf{j}, & 3 \le t \le 6 \end{cases}$

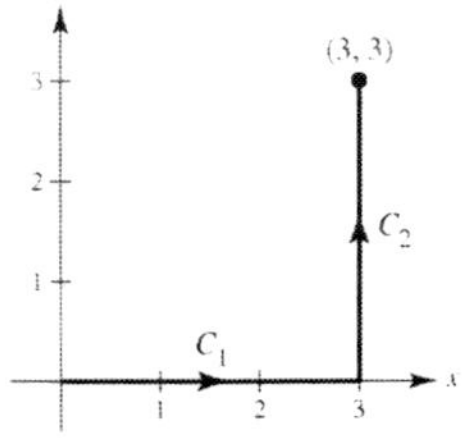

C_1: $x(t) = t,\ y(t) = 0,$

$dx = dt,\ dy = 0$

$$\int_{C_1} (2x - y)\,dx + (x + 3y)\,dy = \int_0^3 2t\,dt = 9$$

C_2: $x(t) = 3,\ y(t) = t - 3$

$dx = 0,\ dy = dt$

$$\int_{C_2} (2x - y)\,dx + (x + 3y)\,dy = \int_3^6 \left[3 + 3(t-3)\right] dt = \left[\frac{3t^2}{2} - 6t\right]_3^6 = \frac{45}{2}$$

$$\int_C (2x - y)\,dx + (x + 3y)\,dy = 9 + \frac{45}{2} = \frac{63}{2}$$

58. $\mathbf{r}(t) = \begin{cases} -t\mathbf{j}, & 0 \le t \le 3 \\ (t-3)\mathbf{i} - 3\mathbf{j}, & 3 \le t \le 5 \end{cases}$

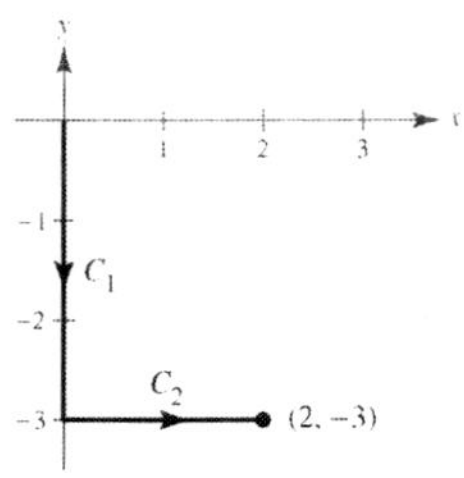

C_1: $x(t) = 0,\ y(t) = -t$

$dx = 0,\ dy = -dt$

$$\int_{C_1} (2x - y)\,dx + (x + 3y)\,dy = \int_0^3 3t\,dt = \tfrac{27}{2}$$

C_2: $x(t) = t - 3,\ y(t) = -3$

$dx = dt,\ dy = 0$

$$\int_{C_2} (2x - y)\,dx + (x + 3y)\,dy = \int_3^5 \left[2(t-3) + 3\right] dt = \left[(t-3)^2 + 3t\right]_3^5 = 10$$

$$\int_C (2x - y)\,dx + (x + 3y)\,dy = \tfrac{27}{2} + 10 = \tfrac{47}{2}$$

59. $x(t) = t,\ y(t) = 1 - t^2,\ 0 \le t \le 1,\ dx = dt,\ dy = -2t\,dt$

$$\int_C (2x - y)\,dx + (x + 3y)\,dy = \int_0^1 \left[\left(2t - 1 + t^2\right) + \left(t + 3 - 3t^2\right)(-2t)\right] dt$$

$$= \int_0^1 \left(6t^3 - t^2 - 4t - 1\right) dt = \left[\frac{3t^4}{2} - \frac{t^3}{3} - 2t^2 - t\right]_0^1 = -\frac{11}{6}$$

60. $x(t) = t,\ y(t) = t^{3/2},\ 0 \le t \le 4,\ dx = dt,\ dy = \frac{3}{2}t^{1/2}\,dt$

$$\int_C (2x - y)\,dx + (x + 3y)\,dy = \int_0^4 \left[\left(2t - t^{3/2}\right) + \left(t + 3t^{3/2}\right)\left(\tfrac{3}{2}t^{1/2}\right)\right] dt$$

$$= \int_0^4 \left(\tfrac{9}{2}t^2 + \tfrac{1}{2}t^{3/2} + 2t\right) dt = \left[\tfrac{3}{2}t^3 + \tfrac{1}{5}t^{5/2} + t^2\right]_0^4 = 96 + \tfrac{1}{5}(32) + 16 = \tfrac{592}{5}$$

61. $x(t) = t,\ y(t) = 2t^2,\ 0 \le t \le 2$

$dx = dt,\ dy = 4t\,dt$

$$\int_C (2x - y)\,dx + (x + 3y)\,dy = \int_0^2 \left(2t - 2t^2\right) dt + \left(t + 6t^2\right)4t\,dt = \int_0^2 \left(24t^3 + 2t^2 + 2t\right) dt = \left[6t^4 + \tfrac{2}{3}t^3 + t^2\right]_0^2 = \tfrac{316}{3}$$

62. $x(t) = 4\sin t,\ y(t) = 3\cos t,\quad 0 \le t \le \dfrac{\pi}{2}$

$dx = 4\cos t\,dt,\ dy = -3\sin t\,dt$

$$\int_C (2x - y)\,dx + (x + 3y)\,dy = \int_0^{\pi/2} (8\sin t - 3\cos t)(4\cos t)\,dt + (4\sin t + 9\cos t)(-3\sin t)\,dt$$

$$= \int_0^{\pi/2} \left(5\sin t\cos t - 12\cos^2 t - 12\sin^2 t\right)dt = \left[\frac{5}{2}\sin^2 t - 12t\right]_0^{\pi/2} = \frac{5}{2} - 6\pi$$

63. $f(x, y) = h$

C: line from $(0, 0)$ to $(3, 4)$

$\mathbf{r} = 3t\mathbf{i} + 4t\mathbf{j},\quad 0 \le t \le 1$

$\mathbf{r}'(t) = 3\mathbf{i} + 4\mathbf{j}$

$\|\mathbf{r}'(t)\| = 5$

Lateral surface area:

$$\int_C f(x, y)\,ds = \int_0^1 5h\,dt = 5h$$

64. $f(x, y) = y$

C: line from $(0, 0)$ to $(4, 4)$

$\mathbf{r}(t) = t\mathbf{i} + t\mathbf{j},\quad 0 \le t \le 4$

$\mathbf{r}'(t) = \mathbf{i} + \mathbf{j}$

$\|\mathbf{r}'(t)\| = \sqrt{2}$

Lateral surface area:

$$\int_C f(x, y)\,ds = \int_0^4 t\left(\sqrt{2}\right)dt = 8\sqrt{2}$$

65. $f(x, y) = xy$

$C: x^2 + y^2 = 1$ from $(1, 0)$ to $(0, 1)$

$\mathbf{r}(t) = \cos t\mathbf{i} + \sin t\mathbf{j},\quad 0 \le t \le \dfrac{\pi}{2}$

$\mathbf{r}'(t) = -\sin t\mathbf{i} + \cos t\mathbf{j}$

$\|\mathbf{r}'(t)\| = 1$

Lateral surface area:

$$\int_C f(x, y)\,ds = \int_0^{\pi/2} \cos t\sin t\,dt = \left[\frac{\sin^2 t}{2}\right]_0^{\pi/2} = \frac{1}{2}$$

66. $f(x, y) = x + y$

$C: x^2 + y^2 = 1$ from $(1, 0)$ to $(0, 1)$

$\mathbf{r}(t) = \cos t\mathbf{i} + \sin t\mathbf{j},\quad 0 \le t \le \dfrac{\pi}{2}$

$\mathbf{r}'(t) = -\sin t\mathbf{i} + \cos t\mathbf{j}$

$\|\mathbf{r}'(t)\| = 1$

Lateral surface area:

$$\int_C f(x, y)\,ds = \int_0^{\pi/2} (\cos t + \sin t)\,dt$$

$$= \left[\sin t - \cos t\right]_0^{\pi/2} = 2$$

67. $f(x, y) = h$

$C: y = 1 - x^2$ from $(1, 0)$ to $(0, 1)$

$\mathbf{r}(t) = (1 - t)\mathbf{i} + \left[1 - (1 - t)^2\right]\mathbf{j},\quad 0 \le t \le 1$

$\mathbf{r}'(t) = -\mathbf{i} + 2(1 - t)\mathbf{j}$

$\|\mathbf{r}'(t)\| = \sqrt{1 + 4(1 - t)^2}$

Lateral surface area:

$$\int_C f(x, y)\,ds = \int_0^1 h\sqrt{1 + 4(1 - t)^2}\,dt = -\frac{h}{4}\left[2(1 - t)\sqrt{1 + 4(1 - t)^2} + \ln\left|2(1 - t) + \sqrt{1 + 4(1 - t)^2}\right|\right]_0^1$$

$$= \frac{h}{4}\left[2\sqrt{5} + \ln\left(2 + \sqrt{5}\right)\right] \approx 1.4789h$$

68. $f(x, y) = y + 1$

$C: y = 1 - x^2$ from $(1, 0)$ to $(0, 1)$

$$\mathbf{r}(t) = (1 - t)\mathbf{i} + \left[1 - (1 - t)^2\right]\mathbf{j},\ 0 \le t \le 1$$

$$\mathbf{r}'(t) = -\mathbf{i} + 2(1 - t)\mathbf{j}$$

$$\|\mathbf{r}'(t)\| = \sqrt{1 + 4(1 - t)^2}$$

Lateral surface area:

$$\int_C f(x, y)\,ds = \int_0^1 \left[2 - (1 - t)^2\right]\sqrt{1 + 4(1 - t)^2}\,dt = 2\int_0^1 \sqrt{1 + 4(1 - t)^2}\,dt - \int_0^1 (1 - t)^2\sqrt{1 + 4(1 - t)^2}\,dt$$

$$= -\frac{1}{2}\left[2(1 - t)\sqrt{1 + 4(1 - t)^2} + \ln\left|2(1 - t) + \sqrt{1 + 4(1 - t)^2}\right|\right]_0^1$$

$$+ \frac{1}{64}\left[2(1 - t)\left[2(4)(1 - t)^2 + 1\right]\sqrt{1 + 4(1 - t)^2} - \ln\left|2(1 - t) + \sqrt{1 + 4(1 - t)^2}\right|\right]_0^1$$

$$= \frac{1}{2}\left[2\sqrt{5} + \ln\left(2 + \sqrt{5}\right)\right] - \frac{1}{64}\left[18\sqrt{5} - \ln\left(2 + \sqrt{5}\right)\right]$$

$$= \frac{23}{32}\sqrt{5} + \frac{33}{64}\ln\left(2 + \sqrt{5}\right) = \frac{1}{64}\left[46\sqrt{5} + 33\ln\left(2 + \sqrt{5}\right)\right] \approx 2.3515$$

69. $f(x, y) = xy$

$C: y = 1 - x^2$ from $(1, 0)$ to $(0, 1)$

You could parameterize the curve C as in Exercises 67 and 68. Alternatively, let $x = \cos t$, then:

$$y = 1 - \cos^2 t = \sin^2 t$$

$$\mathbf{r}(t) = \cos t\mathbf{i} + \sin^2 t\mathbf{j},\ 0 \le t \le \frac{\pi}{2}$$

$$\mathbf{r}'(t) = -\sin t\mathbf{i} + 2\sin t\cos t\mathbf{j}$$

$$\|\mathbf{r}'(t)\| = \sqrt{\sin^2 t + 4\sin^2 t\cos^2 t} = \sin t\sqrt{1 + 4\cos^2 t}$$

Lateral surface area:

$$\int_C f(x, y)\,ds = \int_0^{\pi/2} \cos t\sin^2 t\left(\sin t\sqrt{1 + 4\cos^2 t}\right)dt = \int_0^{\pi/2} \sin^2 t\left[\left(1 + 4\cos^2 t\right)^{1/2}\sin t\cos t\right]dt$$

Let $u = \sin^2 t$ and $dv = \left(1 + 4\cos^2 t\right)^{1/2}\sin t\cos t$, then $du = 2\sin t\cos t\,dt$ and $v = -\frac{1}{12}\left(1 + 4\cos^2 t\right)^{3/2}$.

$$\int_C f(x, y)\,ds = \left[-\frac{1}{12}\sin^2 t\left(1 + 4\cos^2 t\right)^{3/2}\right]_0^{\pi/2} + \frac{1}{6}\int_0^{\pi/2}\left(1 + 4\cos^2 t\right)^{3/2}\sin t\cos t\,dt$$

$$= \left[-\frac{1}{12}\sin^2 t\left(1 + 4\cos^2 t\right)^{3/2} - \frac{1}{120}\left(1 + 4\cos^2 t\right)^{5/2}\right]_0^{\pi/2} = \left(-\frac{1}{12} - \frac{1}{120}\right) + \frac{1}{120}(5)^{5/2} = \frac{1}{120}\left(25\sqrt{5} - 11\right) \approx 0.3742$$

70. $f(x, y) = x^2 - y^2 + 4$

$C: x^2 + y^2 = 4$

$$\mathbf{r}(t) = 2\cos t\mathbf{i} + 2\sin t\mathbf{j},\ 0 \le t \le 2\pi$$

$$\mathbf{r}'(t) = -2\sin t\mathbf{i} + 2\cos t\mathbf{j}$$

$$\|\mathbf{r}'(t)\| = 2$$

Lateral surface area:

$$\int_C f(x, y)\,ds = \int_0^{2\pi}\left(4\cos^2 t - 4\sin^2 t + 4\right)(2)\,dt = 8\int_0^{2\pi}(1 + \cos 2t)\,dt = \left[8\left(t + \frac{1}{2}\sin 2t\right)\right]_0^{2\pi} = 16\pi$$

71. (a) $f(x, y) = 1 + y^2$

$\mathbf{r}(t) = 2\cos t\mathbf{i} + 2\sin t\mathbf{j},\ 0 \le t \le 2\pi$

$\mathbf{r}'(t) = -2\sin t\mathbf{i} + 2\cos t\mathbf{j}$

$\|\mathbf{r}'(t)\| = 2$

$$S = \int_C f(x, y)\,ds = \int_0^{2\pi} \left(1 + 4\sin^2 t\right)(2)\,dt = \Big[2t + 4(t - \sin t\cos t)\Big]_0^{2\pi} = 12\pi \approx 37.70\text{ cm}^2$$

(b) $0.2(12\pi) = \dfrac{12\pi}{5} \approx 7.54\text{ cm}^3$

(c)

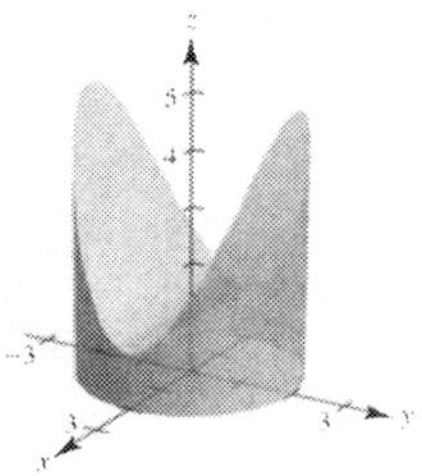

72. $f(x, y) = 20 + \dfrac{1}{4}x$

$C: y = x^{3/2},\ 0 \le x \le 40$

$\mathbf{r}(t) = t\mathbf{i} + t^{3/2}\mathbf{j},\ 0 \le t \le 40$

$\mathbf{r}'(t) = \mathbf{i} + \dfrac{3}{2}t^{1/2}\mathbf{j}$

$\|\mathbf{r}'(t)\| = \sqrt{1 + \left(\dfrac{9}{4}\right)t}$

Lateral surface area: $\displaystyle\int_C f(x, y)\,ds = \int_0^{40} \left(20 + \frac{1}{4}t\right)\sqrt{1 + \left(\frac{9}{4}\right)t}\,dt$

Let $u = \sqrt{1 + \left(\dfrac{9}{4}\right)t}$, then $t = \dfrac{4}{9}\left(u^2 - 1\right)$ and $dt = \dfrac{8}{9}u\,du$.

$$\int_0^{40} \left(20 + \frac{1}{4}t\right)\sqrt{1 + \left(\frac{9}{4}\right)t}\,dt = \int_1^{\sqrt{91}} \left[20 + \frac{1}{9}\left(u^2 - 1\right)\right](u)\left(\frac{8}{9}u\right)du = \frac{8}{81}\int_1^{\sqrt{91}} \left(u^4 + 179u^2\right)du$$

$$= \frac{8}{81}\left[\frac{u^5}{5} + \frac{179u^3}{3}\right]_1^{\sqrt{91}} = \frac{850{,}304\sqrt{91} - 7184}{1215} \approx 6670.12$$

73. $\mathbf{r}(t) = a\cos t\mathbf{i} + a\sin t\mathbf{j},\ 0 \le t \le 2\pi$

$\mathbf{r}'(t) = -a\sin t\mathbf{i} + a\cos t\mathbf{j},\ \|\mathbf{r}'(t)\| = a$

$$I_x = \int_C y^2\rho(x, y)\,ds = \int_0^{2\pi} \left(a^2\sin^2 t\right)(1)a\,dt = a^3\int_0^{2\pi} \sin^2 t\,dt = a^3\pi$$

$$I_y = \int_C x^2\rho(x, y)\,ds = \int_0^{2\pi} \left(a^2\cos^2 t\right)(1)a\,dt = a^3\int_0^{2\pi} \cos^2 t\,dt = a^3\pi$$

74. $\mathbf{r}(t) = a\cos t\mathbf{i} + a\sin t\mathbf{j},\ 0 \le t \le 2\pi$

$\mathbf{r}'(t) = -a\sin t\mathbf{i} + a\cos t\mathbf{j},\ \|\mathbf{r}'(t)\| = a$

$$I_x = \int_C y^2\rho(x, y)\,ds = \int_0^{2\pi} \left(a^2\sin^2 t\right)(\sin t)a^2\,dt = a^4\int_0^{2\pi} \sin^3 t\,dt = 0$$

$$I_y = \int_C x^2\rho(x, y)\,ds = \int_0^{2\pi} \left(a^2\cos^2 t\right)(\sin t)a^2\,dt = a^4\int_0^{2\pi} \cos^2 t\sin t\,dt = 0$$

75. (a) Graph of: $\mathbf{r}(t) = 3\cos t\mathbf{i} + 3\sin t\mathbf{j} + \left(1 + \sin^2 2t\right)\mathbf{k},\ \ 0 \le t \le 2\pi$

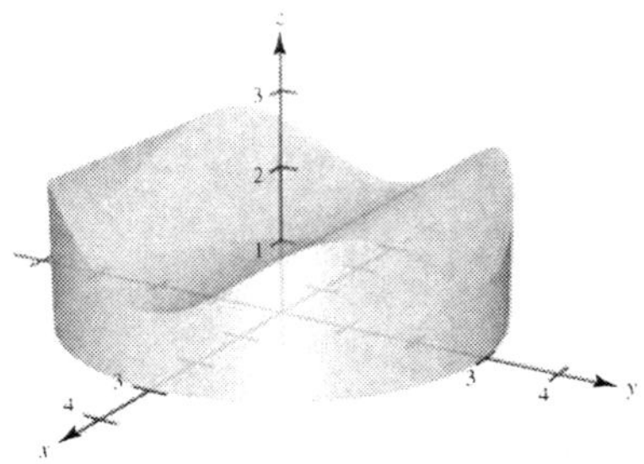

For $y = b$ constant, $3\sin t = b \Rightarrow \sin t = \dfrac{b}{3}$ and

$$1 + \sin^2 2t = 1 + \left(2\sin t\cos t\right)^2$$
$$= 1 + 4\sin^2 t\cos^2 t$$
$$= 1 + 4\sin^2 t\left(1 - \sin^2 t\right) = 1 + \frac{4}{9}b^2\left(1 - \frac{b^2}{9}\right).$$

(b) Consider the portion of the surface in the first quadrant. The curve $z = 1 + \sin^2 2t$ is over the curve $\mathbf{r}_1(t) = 3\cos t\mathbf{i} + 3\sin t\mathbf{j},\ 0 \le t \le \pi/2$. So, the total lateral surface area is

$$4\int_C f(x, y)\,ds = 4\int_0^{\pi/2}\left(1 + \sin^2 2t\right)3\,dt = 12\left(\frac{3\pi}{4}\right) = 9\pi \text{ cm}^2.$$

(c) The cross sections parallel to the xz-plane are rectangles of height $1 + 4\left(y/3\right)^2\left(1 - y^2/9\right)$ and base $2\sqrt{9 - y^2}$. So,

$$\text{Volume} = 2\int_0^3 2\sqrt{9 - y^2}\left(1 + 4\frac{y^2}{9}\left(1 - \frac{y^2}{9}\right)\right)dy = \frac{27\pi}{2} \approx 42.412 \text{ cm}^3.$$

76. $\mathbf{r}(t) = t\mathbf{i} + t^2\mathbf{j},\ \ 0 \le t \le 1$

$\mathbf{r}'(t) = \mathbf{i} + 2t\mathbf{j}$

$$\int_C \mathbf{F}\cdot d\mathbf{r} \approx \frac{1 - 0}{3(4)}\left[5 + 4(4) + 2(4) + 4(6) + 11\right]$$
$$= \frac{16}{3}$$

(x, y)	$(0, 0)$	$\left(\frac{1}{4}, \frac{1}{16}\right)$	$\left(\frac{1}{2}, \frac{1}{4}\right)$	$\left(\frac{3}{4}, \frac{9}{16}\right)$	$(1, 1)$
$\mathbf{F}(x, y)$	$5\mathbf{i}$	$3.5\mathbf{i} + \mathbf{j}$	$2\mathbf{i} + 2\mathbf{j}$	$1.5\mathbf{i} + 3\mathbf{j}$	$\mathbf{i} + 5\mathbf{j}$
$\mathbf{r}'(t)$	$\mathbf{i}$	$\mathbf{i} + 0.5\mathbf{j}$	$\mathbf{i} + \mathbf{j}$	$\mathbf{i} + 1.5\mathbf{j}$	$\mathbf{i} + 2\mathbf{j}$
$\mathbf{F}\cdot\mathbf{r}'$	5	4	4	6	11

77. $\mathbf{r}(t) = 3\sin t\mathbf{i} + 3\cos t\mathbf{j} + \dfrac{10}{2\pi}t\mathbf{k},\ \ 0 \le t \le 2\pi$

$\mathbf{F} = 175\mathbf{k}$

$$d\mathbf{r} = \left(3\cos t\mathbf{i} - 3\sin t\mathbf{j} + \frac{10}{2\pi}\mathbf{k}\right)dt$$

$$\int_C \mathbf{F}\cdot d\mathbf{r} = \int_0^{2\pi}\frac{1750}{2\pi}\,dt = \left[\frac{1750}{2\pi}t\right]_0^{2\pi} = 1750 \text{ ft}\cdot\text{lb}$$

78. $W = \int_C \mathbf{F}\cdot d\mathbf{r} = \int_C M\,dx + N\,dy$

$M = 15\left(4 - x^2y\right) = 60 - 15x^2\left(c - cx^2\right)$

$N = -15xy = -15x\left(c - cx^2\right)$

$dx = dx,\ dy = -2cx\,dx$

$W = \int_{-1}^{1}\left[60 - 15x^2\left(c - cx^2\right) + \left(-15x\left(c - cx^2\right)\right)\left(-2\,cx\right)\right]dx = 120 - 4c + 8c^2$ (parabola)

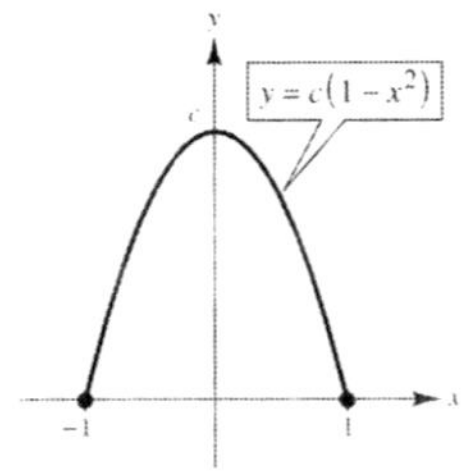

$W' = 16c - 4 = 0 \Rightarrow c = \frac{1}{4}$ yields the minimum work, 119.5. Along the straight line path, $y = 0$, the work is 120.

79. See the definition of Line Integral, page 1052. See Theorem 15.4.

80. See the definition, page 1056.

81. The greater the height of the surface over the curve, the greater the lateral surface area. So, $z_3 < z_1 < z_2 < z_4$.

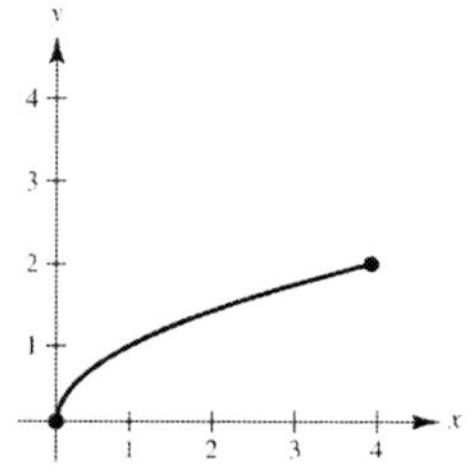

82. (a) Work = 0

(b) Work is negative, because against force field.

(c) Work is positive, because with force field.

83. False

$$\int_C xy\, ds = \sqrt{2}\int_0^1 t^2\, dt$$

84. False, the orientation of C does not affect the form.

$$\int_C f(x, y)\, ds.$$

85. False, the orientations are different.

86. False. For example, see Exercise 32.

87. $\mathbf{F}(x, y) = (y - x)\mathbf{i} + xy\mathbf{j}$

$\mathbf{r}(t) = kt(1 - t)\mathbf{i} + t\mathbf{j}, \quad 0 \le t \le 1$

$\mathbf{r}'(t) = k(1 - 2t)\mathbf{i} + \mathbf{j}$

$$\begin{aligned}\text{Work} = 1 &= \int_C \mathbf{F}\cdot d\mathbf{r}\\ &= \int_0^1 \left[\left(t - kt(1 - t)\right)\mathbf{i} + kt^2(1 - t)\mathbf{j}\right]\cdot\left[k(1 - 2t)\mathbf{i} + \mathbf{j}\right] dt\\ &= \int_0^1 \left[\left(t - kt(1 - t)\right)k(1 - 2t) + kt^2(1 - t)\right] dt\\ &= \int_0^1 \left(-2k^2t^3 - kt^3 - kt^2 + 3k^2t^2 - k^2t + kt\right) dt = \frac{-k}{12}\end{aligned}$$

$k = -12$

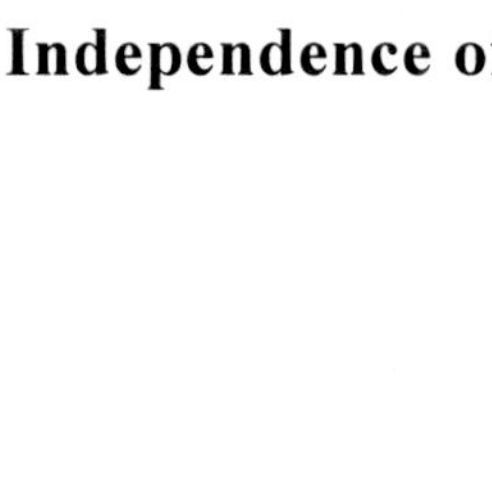

Section 15.3 Conservative Vector Fields and Independence of Path

1. $\mathbf{F}(x, y) = x^2\mathbf{i} + xy\mathbf{j}$

(a) $\mathbf{r}_1(t) = t\mathbf{i} + t^2\mathbf{j}, \quad 0 \le t \le 1$

$\mathbf{r}_1'(t) = \mathbf{i} + 2t\mathbf{j}$

$\mathbf{F}(t) = t^2\mathbf{i} + t^3\mathbf{j}$

$$\int_C \mathbf{F}\cdot d\mathbf{r} = \int_0^1 \left(t^2 + 2t^4\right) dt = \frac{11}{15}$$

(b) $\mathbf{r}_2(\theta) = \sin\theta\mathbf{i} + \sin^2\theta\mathbf{j},\ 0 \le \theta \le \frac{\pi}{2}$

$\mathbf{r}_2'(\theta) = \cos\theta\mathbf{i} + 2\sin\theta\cos\theta\mathbf{j}$

$\mathbf{F}(t) = \sin^2\theta\mathbf{i} + \sin^3\theta\mathbf{j}$

$$\int_C \mathbf{F}\cdot d\mathbf{r} = \int_0^{\pi/2}\left(\sin^2\theta\cos\theta + 2\sin^4\theta\cos\theta\right)d\theta = \left[\frac{\sin^3\theta}{3} + \frac{2\sin^5\theta}{5}\right]_0^{\pi/2} = \frac{11}{15}$$

2. $\mathbf{F}(x, y) = \left(x^2 + y^2\right)\mathbf{i} - x\mathbf{j}$

(a) $\mathbf{r}_1(t) = t\mathbf{i} + \sqrt{t}\mathbf{j},\ 0 \le t \le 4$

$\mathbf{r}_1'(t) = \mathbf{i} + \frac{1}{2\sqrt{t}}\mathbf{j}$

$\mathbf{F}(t) = \left(t^2 + t\right)\mathbf{i} - t\mathbf{j}$

$$\int_C \mathbf{F}\cdot d\mathbf{r} = \int_0^4\left(t^2 + t - \frac{1}{2}\sqrt{t}\right)dt$$

$$= \left[\frac{t^3}{3} + \frac{t^2}{2} - \frac{t^{3/2}}{3}\right]_0^4 = \frac{80}{3}$$

(b) $\mathbf{r}_2(w) = w^2\mathbf{i} + w\mathbf{j},\ 0 \le w \le 2$

$\mathbf{r}_2'(w) = 2w\mathbf{i} + \mathbf{j}$

$\mathbf{F}(w) = \left(w^4 + w^2\right)\mathbf{i} - w^2\mathbf{j}$

$$\int_C \mathbf{F}\cdot d\mathbf{r} = \int_0^2\left[2w\left(w^4 + w^2\right) - w^2\right]dw$$

$$= \left[\frac{w^6}{3} + \frac{w^4}{2} - \frac{w^3}{3}\right]_0^2 = \frac{80}{3}$$

3. $\mathbf{F}(x, y) = y\mathbf{i} - x\mathbf{j}$

(a) $\mathbf{r}_1(\theta) = \sec\theta\mathbf{i} + \tan\theta\mathbf{j},\ 0 \le \theta \le \frac{\pi}{3}$

$\mathbf{r}_1'(\theta) = \sec\theta\tan\theta\mathbf{i} + \sec^2\theta\mathbf{j}$

$\mathbf{F}(\theta) = \tan\theta\mathbf{i} - \sec\theta\mathbf{j}$

$$\int_C \mathbf{F}\cdot d\mathbf{r} = \int_0^{\pi/3}\left(\sec\theta\tan^2\theta - \sec^3\theta\right)d\theta$$

$$= \int_0^{\pi/3}\left[\sec\theta\left(\sec^2\theta - 1\right) - \sec^3\theta\right]d\theta$$

$$= -\int_0^{\pi/3}\sec\theta\, d\theta$$

$$= \Big[-\ln\left|\sec\theta + \tan\theta\right|\Big]_0^{\pi/3}$$

$$= -\ln\left(2 + \sqrt{3}\right) \approx -1.317$$

(b) $\mathbf{r}_2(t) = \sqrt{t + 1}\mathbf{i} + \sqrt{t}\mathbf{j},\ 0 \le t \le 3$

$\mathbf{r}_2'(t) = \frac{1}{2\sqrt{t + 1}}\mathbf{i} + \frac{1}{2\sqrt{t}}\mathbf{j}$

$\mathbf{F}(t) = \sqrt{t}\mathbf{i} - \sqrt{t + 1}\mathbf{j}$

$$\int_C \mathbf{F}\cdot d\mathbf{r} = \int_0^3\left[\frac{\sqrt{t}}{2\sqrt{t + 1}} - \frac{\sqrt{t + 1}}{2\sqrt{t}}\right]dt$$

$$= -\frac{1}{2}\int_0^3\frac{1}{\sqrt{t}\sqrt{t + 1}}\,dt$$

$$= -\frac{1}{2}\int_0^3\frac{1}{\sqrt{t^2 + t + (1/4) - (1/4)}}\,dt$$

$$= -\frac{1}{2}\int_0^3\frac{1}{\sqrt{\left[t + (1/2)\right]^2 - (1/4)}}\,dt$$

$$= \left[-\frac{1}{2}\ln\left|\left(t + \frac{1}{2}\right) + \sqrt{t^2 + t}\right|\right]_0^3$$

$$= -\frac{1}{2}\left[\ln\left(\frac{7}{2} + 2\sqrt{3}\right) - \ln\left(\frac{1}{2}\right)\right]$$

$$= -\frac{1}{2}\ln\left(7 + 4\sqrt{3}\right) \approx -1.317$$

4. $\mathbf{F}(x, y) = y\mathbf{i} + x^2\mathbf{j}$

(a) $\mathbf{r}_1(t) = (2 + t)\mathbf{i} + (3 - t)\mathbf{j}, \quad 0 \le t \le 3$

$\mathbf{r}_1'(t) = \mathbf{i} - \mathbf{j}$

$\mathbf{F}(t) = (3 - t)\mathbf{i} + (2 + t)^2\mathbf{j}$

$$\int_C \mathbf{F} \cdot d\mathbf{r} = \int_0^3 \left[(3 - t) - (2 + t)^2\right] dt = \left[-\frac{(3 - t)^2}{2} - \frac{(2 + t)^3}{3}\right]_0^3 = -\frac{69}{2}$$

(b) $\mathbf{r}_2(w) = (2 + \ln w)\mathbf{i} + (3 - \ln w)\mathbf{j}, \quad 1 \le w \le e^3$

$\mathbf{r}_2'(w) = \dfrac{1}{w}\mathbf{i} - \dfrac{1}{w}\mathbf{j}$

$\mathbf{F}(w) = (3 - \ln w)\mathbf{i} + (2 + \ln w)^2\mathbf{j}$

$$\int_C \mathbf{F} \cdot d\mathbf{r} = \int_1^{e^3} \left[(3 - \ln w)\left(\frac{1}{w}\right) - (2 + \ln w)^2\left(\frac{1}{w}\right)\right] dw = \left[-\frac{(3 - \ln w)^2}{2} - \frac{(2 + \ln w)^3}{3}\right]_1^{e^3} = -\frac{69}{2}$$

5. $\mathbf{F}(x, y) = e^x \sin y\mathbf{i} + e^x \cos y\mathbf{j}$

$\dfrac{\partial N}{\partial x} = e^x \cos y \qquad \dfrac{\partial M}{\partial y} = e^x \cos y$

Because $\dfrac{\partial N}{\partial x} = \dfrac{\partial M}{\partial y}$, $\mathbf{F}$ is conservative.

6. $\mathbf{F}(x, y) = 15x^2y^2\mathbf{i} + 10x^3y\mathbf{j}$

$\dfrac{\partial N}{\partial x} = 30x^2y \qquad \dfrac{\partial M}{\partial y} = 30x^2y$

Because $\dfrac{\partial N}{\partial x} = \dfrac{\partial M}{\partial y}$, $\mathbf{F}$ is conservative.

7. $\mathbf{F}(x, y) = \dfrac{1}{y}\mathbf{i} + \dfrac{x}{y^2}\mathbf{j}$

$\dfrac{\partial N}{\partial x} = \dfrac{1}{y^2} \qquad \dfrac{\partial M}{\partial y} = -\dfrac{1}{y^2}$

Because $\dfrac{\partial N}{\partial x} \ne \dfrac{\partial M}{\partial y}$, $\mathbf{F}$ is not conservative.

8. $\mathbf{F}(x, y, z) = y \ln z\mathbf{i} - x \ln z\mathbf{j} + \dfrac{xy}{z}\mathbf{k}$

curl F $\ne$ **0** so **F** is not conservative.

$$\left(\frac{\partial P}{\partial y} = \frac{x}{z} \ne -\frac{x}{z} = \frac{\partial N}{\partial z}\right)$$

9. $\mathbf{F}(x, y, z) = y^2z\mathbf{i} + 2xyz\mathbf{j} + xy^2\mathbf{k}$

curl F $=$ **0** $\Rightarrow$ **F** is conservative.

10. $\mathbf{F}(x, y, z) = \sin(yz)\mathbf{i} + xz\cos(yz)\mathbf{j} + xy\sin(yz)\mathbf{k}$

curl F $\ne$ **0**, so **F** is not conservative.

11. $\mathbf{F}(x, y) = 2xy\mathbf{i} + x^2\mathbf{j}$

(a) $\mathbf{r}_1(t) = t\mathbf{i} + t^2\mathbf{j}, \quad 0 \le t \le 1$

$\mathbf{r}_1'(t) = \mathbf{i} + 2t\mathbf{j}$

$\mathbf{F}(t) = 2t^3\mathbf{i} + t^2\mathbf{j}$

$$\int_C \mathbf{F} \cdot d\mathbf{r} = \int_0^1 4t^3\, dt = 1$$

(b) $\mathbf{r}_2(t) = t\mathbf{i} + t^3\mathbf{j}, \quad 0 \le t \le 1$

$\mathbf{r}_2'(t) = \mathbf{i} + 3t^2\mathbf{j}$

$\mathbf{F}(t) = 2t^4\mathbf{i} + t^2\mathbf{j}$

$$\int_C \mathbf{F} \cdot d\mathbf{r} = \int_0^1 5t^4\, dt = 1$$

12. $\mathbf{F}(x, y) = ye^{xy}\mathbf{i} + xe^{xy}\mathbf{j}$

(a) $\mathbf{r}_1(t) = t\mathbf{i} - (t - 3)\mathbf{j}, \quad 0 \le t \le 3$

$\mathbf{r}_1'(t) = \mathbf{i} - \mathbf{j}$

$\mathbf{F}(t) = -(t - 3)e^{3t - t^2}\mathbf{i} + te^{3t - t^2}\mathbf{j}$

$$\int_C \mathbf{F} \cdot d\mathbf{r} = \int_0^3 \left[-(t - 3)e^{3t - t^2} - te^{3t - t^2}\right] dt$$

$$= \int_0^3 e^{3t - t^2}(3 - 2t)\, dt$$

$$= \left[e^{3t - t^2}\right]_0^3 = e^0 - e^0 = 0$$

(b) $\mathbf{F}(x, y)$ is conservative because

$$\frac{\partial M}{\partial y} = xye^{xy} + e^{xy} = \frac{\partial N}{\partial x}.$$

The potential function is $f(x, y) = e^{xy} + K$.

By Theorem 15.7, $\int_C \mathbf{F} \cdot d\mathbf{r} = 0$.

13. $\mathbf{F}(x, y) = y\mathbf{i} - x\mathbf{j}$

(a) $\mathbf{r}_1(t) = t\mathbf{i} + t\mathbf{j}, \quad 0 \le t \le 1$

$\mathbf{r}_1'(t) = \mathbf{i} + \mathbf{j}$

$\mathbf{F}(t) = t\mathbf{i} - t\mathbf{j}$

$\int_C \mathbf{F} \cdot d\mathbf{r} = 0$

(b) $\mathbf{r}_2(t) = t\mathbf{i} + t^2\mathbf{j}, \quad 0 \le t \le 1$

$\mathbf{r}_2'(t) = \mathbf{i} + 2t\mathbf{j}$

$\mathbf{F}(t) = t^2\mathbf{i} - t\mathbf{j}$

$\int_C \mathbf{F} \cdot d\mathbf{r} = \int_0^1 -t^2\, dt = -\frac{1}{3}$

(c) $\mathbf{r}_3(t) = t\mathbf{i} + t^3\mathbf{j}, \quad 0 \le t \le 1$

$\mathbf{r}_3'(t) = \mathbf{i} + 3t^2\mathbf{j}$

$\mathbf{F}(t) = t^3\mathbf{i} - t\mathbf{j}$

$\int_C \mathbf{F} \cdot d\mathbf{r} = \int_0^1 -2t^3\, dt = -\frac{1}{2}$

14. $\mathbf{F}(x, y) = xy^2\mathbf{i} + 2x^2y\mathbf{j}$

(a) $\mathbf{r}_1(t) = t\mathbf{i} + \frac{1}{t}\mathbf{j}, \quad 1 \le t \le 3$

$\mathbf{r}_1'(t) = \mathbf{i} - \frac{1}{t^2}\mathbf{j}$

$\mathbf{F}(t) = \frac{1}{t}\mathbf{i} + 2t\mathbf{j}$

$\int_C \mathbf{F} \cdot d\mathbf{r} = \int_1^3 -\frac{1}{t}\, dt = \Big[-\ln|t|\Big]_1^3 = -\ln 3$

(b) $\mathbf{r}_2(t) = (t + 1)\mathbf{i} - \frac{1}{3}(t - 3)\mathbf{j}, \quad 0 \le t \le 2$

$\mathbf{r}_2'(t) = \mathbf{i} - \frac{1}{3}\mathbf{j}$

$\mathbf{F}(t) = \frac{1}{9}(t + 1)(t - 3)^2\mathbf{i} - \frac{2}{3}(t + 1)^2(t - 3)\mathbf{j}$

$$\int_C \mathbf{F} \cdot d\mathbf{r} = \int_0^2 \left[\frac{1}{9}(t + 1)(t - 3)^2 + \frac{2}{9}(t + 1)^2(t - 3)\right] dt$$

$$= \frac{1}{9}\int_0^2 \left(3t^3 - 7t^2 - 7t + 3\right) dt$$

$$= \frac{1}{9}\left[\frac{3t^4}{4} - \frac{7t^3}{3} - \frac{7t^2}{2} + 3t\right]_0^2 = -\frac{44}{27}$$

15. $\int_C y^2\, dx + 2xy\, dy$

Because $\partial M/\partial y = \partial N/\partial x = 2y$, $\mathbf{F}(x, y) = y^2\mathbf{i} + 2xy\mathbf{j}$ is conservative. The potential function is $f(x, y) = xy^2 + k$. So, you can use the Fundamental Theorem of Line Integrals.

(a) $\int_C y^2\, dx + 2xy\, dy = \Big[x^2y\Big]_{(0,0)}^{(4,4)} = 64$

(b) $\int_C y^2\, dx + 2xy\, dy = \Big[x^2y\Big]_{(-1,0)}^{(1,0)} = 0$

(c) and (d) Because C is a closed curve, $\int_C y^2\, dx + 2xy\, dy = 0$.

16. $\int_C (2x - 3y + 1)\, dx - (3x + y - 5)\, dy$

Because $\partial M/\partial y = \partial N/\partial x = -3$, $\mathbf{F}(x, y) = (2x - 3y + 1)\mathbf{i} - (3x + y - 5)\mathbf{j}$ is conservative. The potential function is $f(x, y) = x^2 - 3xy - (y^2/2) + x + 5y + k$.

(a) and (d) Because C is a closed curve, $\int_C (2x - 3y + 1)\, dx - (3x + y - 5)\, dy = 0$.

(b) $\int_C (2x - 3y + 1)\, dx - (3x + y - 5)\, dy = \left[x^2 - 3xy - \frac{y^2}{2} + x + 5y\right]_{(0,-1)}^{(0,1)} = 10$

(c) $\int_C (2x - 3y + 1)\, dx - (3x + y - 5)\, dy = \left[x^2 - 3xy - \frac{y^2}{2} + x + 5y\right]_{(0,1)}^{(2,e^2)} = \frac{1}{2}\left(3 - 2e^2 - e^4\right)$

17. $\int_C 2xy\,dx + (x^2 + y^2)\,dy$

Because $\partial M/\partial y = \partial N/\partial x = 2x$, $\mathbf{F}(x, y) = 2xy\mathbf{i} + (x^2 + y^2)\mathbf{j}$ is conservative.

The potential function is $f(x, y) = x^2y + \dfrac{y^3}{3} + k.$

(a) $\int_C 2xy\,dx + (x^2 + y^2)\,dy = \left[x^2y + \dfrac{y^3}{3}\right]_{(5,0)}^{(0,4)} = \dfrac{64}{3}$

(b) $\int_C 2xy\,dx + (x^2 + y^2)\,dy = \left[x^2y + \dfrac{y^3}{3}\right]_{(2,0)}^{(0,4)} = \dfrac{64}{3}$

18. $\int_C (x^2 + y^2)\,dx + 2xy\,dy$

Because $\partial M/\partial y = \partial N/\partial x = 2y$, $\mathbf{F}(x, y) = (x^2 + y^2)\mathbf{i} + 2xy\mathbf{j}$ is conservative. The potential function is $f(x, y) = (x^3/3) + xy^2 + k.$

(a) $\int_C (x^2 + y^2)\,dx + 2xy\,dy = \left[\dfrac{x^3}{3} + xy^2\right]_{(0,0)}^{(8,4)} = \dfrac{896}{3}$

(b) $\int_C (x^2 + y^2)\,dx + 2xy\,dy = \left[\dfrac{x^3}{3} + xy^2\right]_{(2,0)}^{(0,2)} = -\dfrac{8}{3}$

19. $\mathbf{F}(x, y, z) = yz\mathbf{i} + xz\mathbf{j} + xy\mathbf{k}$

Because $\mathbf{curl\,F} = \mathbf{0}$, $\mathbf{F}(x, y, z)$ is conservative. The potential function is $f(x, y, z) = xyz + k.$

(a) $\mathbf{r}_1(t) = t\mathbf{i} + 2\mathbf{j} + t\mathbf{k}, \quad 0 \le t \le 4$

$\int_C \mathbf{F} \cdot d\mathbf{r} = [xyz]_{(0,2,0)}^{(4,2,4)} = 32$

(b) $\mathbf{r}_2(t) = t^2\mathbf{i} + t\mathbf{j} + t^2\mathbf{k}, \quad 0 \le t \le 2$

$\int_C \mathbf{F} \cdot d\mathbf{r} = [xyz]_{(0,0,0)}^{(4,2,4)} = 32$

20. $\mathbf{F}(x, y, z) = \mathbf{i} + z\mathbf{j} + y\mathbf{k}$

Because $\mathbf{curl\,F} = \mathbf{0}$, $\mathbf{F}(x, y, z)$ is conservative. The potential function is $f(x, y, z) = x + yz + k.$

(a) $\mathbf{r}_1(t) = \cos t\mathbf{i} + \sin t\mathbf{j} + t^2\mathbf{k}, \quad 0 \le t \le \pi$

$\int_C \mathbf{F} \cdot d\mathbf{r} = [x + yz]_{(1,0,0)}^{(-1,0,\pi^2)} = -2$

(b) $\mathbf{r}_2(t) = (1 - 2t)\mathbf{i} + \pi^2 t\mathbf{k}, \quad 0 \le t \le 1$

$\int_C \mathbf{F} \cdot d\mathbf{r} = [x + yz]_{(1,0,0)}^{(-1,0,\pi^2)} = -2$

21. $\mathbf{F}(x, y, z) = (2y + x)\mathbf{i} + (x^2 - z)\mathbf{j} + (2y - 4z)\mathbf{k}$

$\mathbf{F}(x, y, z)$ is not conservative.

(a) $\mathbf{r}_1(t) = t\mathbf{i} + t^2\mathbf{j} + \mathbf{k}, \quad 0 \le t \le 1$

$\mathbf{r}_1'(t) = \mathbf{i} + 2t\mathbf{j}$

$\mathbf{F}(t) = (2t^2 + t)\mathbf{i} + (t^2 - 1)\mathbf{j} + (2t^2 - 4)\mathbf{k}$

$\int_C \mathbf{F} \cdot d\mathbf{r} = \int_0^1 (2t^3 + 2t^2 - t)\,dt = \dfrac{2}{3}$

(b) $\mathbf{r}_2(t) = t\mathbf{i} + t\mathbf{j} + (2t - 1)^2\mathbf{k}, \quad 0 \le t \le 1$

$\mathbf{r}_2'(t) = \mathbf{i} + \mathbf{j} + 4(2t - 1)\mathbf{k}$

$\mathbf{F}(t) = 3t\mathbf{i} + \left[t^2 - (2t - 1)^2\right]\mathbf{j} + \left[2t - 4(2t - 1)^2\right]\mathbf{k}$

$\int_C \mathbf{F} \cdot d\mathbf{r} = \int_0^1 \left[3t + t^2 - (2t - 1)^2 + 8t(2t - 1) - 16(2t - 1)^3\right] dt$

$= \int_0^1 \left[17t^2 - 5t - (2t - 1)^2 - 16(2t - 1)^3\right] dt = \left[\dfrac{17t^3}{3} - \dfrac{5t^2}{2} - \dfrac{(2t - 1)^3}{6} - 2(2t - 1)^4\right]_0^1 = \dfrac{17}{6}$

22. $\mathbf{F}(x, y, z) = -y\mathbf{i} + x\mathbf{j} + 3xz^2\mathbf{k}$

$\mathbf{F}(x, y, z)$ is not conservative.

(a) $\mathbf{r}_1(t) = \cos t\mathbf{i} + \sin t\mathbf{j} + t\mathbf{k}, \quad 0 \le t \le \pi$

$\mathbf{r}_1'(t) = -\sin t\mathbf{i} + \cos t\mathbf{j} + \mathbf{k}$

$\mathbf{F}(t) = -\sin t\mathbf{i} + \cos t\mathbf{j} + 3t^2 \cos t\mathbf{k}$

$$\begin{aligned}\int_C \mathbf{F} \cdot d\mathbf{r} &= \int_0^\pi \left[\sin^2 t + \cos^2 t + 3t^2 \cos t\right] dt \\ &= \int_0^\pi \left[1 + 3t^2 \cos t\right] dt \\ &= \left[t\right]_0^\pi + 3\left[t^2 \sin t\right]_0^\pi - 6\int_0^\pi t \sin t\, dt \\ &= \left[t + 3t^2 \sin t - 6(\sin t - t\cos t)\right]_0^\pi \\ &= -5\pi\end{aligned}$$

(b) $\mathbf{r}_2(t) = (1 - 2t)\mathbf{i} + \pi t\mathbf{k}, \quad 0 \le t \le 1$

$\mathbf{r}_2'(t) = -2\mathbf{i} + \pi\mathbf{k}$

$\mathbf{F}(t) = (1 - 2t)\mathbf{j} + 3\pi^2 t^2(1 - 2t)\mathbf{k}$

$$\begin{aligned}\int_C \mathbf{F} \cdot d\mathbf{r} &= \int_0^1 3\pi^3 t^2(1 - 2t)\, dt \\ &= 3\pi^3 \int_0^1 \left(t^2 - 2t^3\right) dt \\ &= 3\pi^3\left[\frac{t^3}{3} - \frac{t^4}{2}\right]_0^1 = -\frac{\pi^3}{2}\end{aligned}$$

23. $\mathbf{F}(x, y, z) = e^z(y\mathbf{i} + x\mathbf{j} + xy\mathbf{k})$

$\mathbf{F}(x, y, z)$ is conservative. The potential function is $f(x, y, z) = xye^z + k$.

(a) $\mathbf{r}_1(t) = 4\cos t\mathbf{i} + 4\sin t\mathbf{j} + 3\mathbf{k}, \quad 0 \le t \le \pi$

$$\int_C \mathbf{F} \cdot d\mathbf{r} = \left[xye^z\right]_{(4, 0, 3)}^{(-4, 0, 3)} = 0$$

(b) $\mathbf{r}_2(t) = (4 - 8t)\mathbf{i} + 3\mathbf{k}, \quad 0 \le t \le 1$

$$\int_C \mathbf{F} \cdot d\mathbf{r} = \left[xye^z\right]_{(4, 0, 3)}^{(-4, 0, 3)} = 0$$

24. $\mathbf{F}(x, y, z) = y \sin z\mathbf{i} + x \sin z\mathbf{j} + xy \cos x\mathbf{k}$

(a) $\mathbf{r}_1(t) = t^2\mathbf{i} + t^2\mathbf{j}, \; 0 \le t \le 2$

$\mathbf{r}_1'(t) = 2t\mathbf{i} + 2t\mathbf{j}$

$\mathbf{F}(t) = t^4 \cos t^2\mathbf{k}$

$$\int_C \mathbf{F} \cdot d\mathbf{r} = \int_0^2 0\, dt = 0$$

(b) $\mathbf{r}_2(t) = 4t\mathbf{i} + 4t\mathbf{j}, \; 0 \le t \le 1$

$\mathbf{r}_2'(t) = 4\mathbf{i} + 4\mathbf{j}$

$\mathbf{F}(t) = 16t^2 \cos(4t)\mathbf{k}$

$$\int_C \mathbf{F} \cdot d\mathbf{r} = \int_0^1 0\, dt = 0$$

25. $\displaystyle\int_C (3y\mathbf{i} + 3x\mathbf{j}) \cdot d\mathbf{r} = \left[3xy\right]_{(0, 0)}^{(3, 8)} = 72$

26. $\displaystyle\int_C \left[2(x + y)\mathbf{i} + 2(x + y)\mathbf{j}\right] \cdot d\mathbf{r} = \left[(x + y)^2\right]_{(-1, 1)}^{(3, 2)} = 5^2 - 0 = 25$

27. $\displaystyle\int_C \cos x \sin y\, dx + \sin x \cos y\, dy = \left[\sin x \sin y\right]_{(0, -\pi)}^{(3\pi/2, \pi/2)} = -1$

28. $\displaystyle\int_C \frac{y\, dx - x\, dy}{x^2 + y^2} = \left[\arctan\left(\frac{x}{y}\right)\right]_{(1, 1)}^{(2\sqrt{3}, 2)} = \frac{\pi}{3} - \frac{\pi}{4} = \frac{\pi}{12}$

29. $\displaystyle\int_C e^x \sin y\, dx + e^x \cos y\, dy = \left[e^x \sin y\right]_{(0, 0)}^{(2\pi, 0)} = 0$

30. $\displaystyle\int_C \frac{2x}{\left(x^2 + y^2\right)^2} dx + \frac{2y}{\left(x^2 + y^2\right)^2} dy = \left[-\frac{1}{x^2 + y^2}\right]_{(7, 5)}^{(1, 5)} = -\frac{1}{26} + \frac{1}{74} = \frac{-12}{481}$

31. $\displaystyle\int_C (z + 2y)\, dx + (2x - z)\, dy + (x - y)\, dz$

$\mathbf{F}(x, y, z)$ is conservative and the potential function is $f(x, y, z) = xz + 2xy - yz$

(a) $\left[xz + 2xy - yz\right]_{(0, 0, 0)}^{(1, 1, 1)} = 2 - 0 = 2$

(b) $\left[xz + 2xy - yz\right]_{(0, 0, 0)}^{(0, 0, 1)} + \left[xz + 2xy - yz\right]_{(0, 0, 1)}^{(1, 1, 1)} = 0 + 2 = 2$

(c) $\left[xz + 2xy - yz\right]_{(0, 0, 0)}^{(1, 0, 0)} + \left[xz + 2xy - yz\right]_{(1, 0, 0)}^{(1, 1, 0)} + \left[xz + 2xy - yz\right]_{(1, 1, 0)}^{(1, 1, 1)} = 0 + 2 + (2 - 2) = 2$

32. $\int_C zy\,dx + xz\,dy + xy\,dz$

Note: Because $\mathbf{F}(x, y, z) = yz\mathbf{i} + xz\mathbf{j} + xy\mathbf{k}$ is conservative and the potential function is $f(x, y, z) = xyz + k$, the integral is independent of path as illustrated below.

(a) $\left[xyz\right]_{(0,0,0)}^{(1,1,1)} = 1$

(b) $\left[xyz\right]_{(0,0,0)}^{(0,0,1)} + \left[xyz\right]_{(0,0,1)}^{(1,1,1)} = 0 + 1 = 1$

(c) $\left[xyz\right]_{(0,0,0)}^{(1,0,0)} + \left[xyz\right]_{(1,0,0)}^{(1,1,0)} + \left[xyz\right]_{(1,1,0)}^{(1,1,1)} = 0 + 0 + 1 = 1$

33. $\int_C -\sin x\,dx + z\,dy + y\,dz = \left[\cos x + yz\right]_{(0,0,0)}^{(\pi/2,3,4)} = 12 - 1 = 11$

34. $\mathbf{F}(x, y, z)$ is conservative: $f(x, y, z) = 3x^2 - 4yz + 10z^2$

$$\int_C \mathbf{F} \cdot d\mathbf{r} = \left[3x^2 - 4yz + 10z^2\right]_{(0,0,0)}^{(3,4,0)} = 27$$

35. $\mathbf{F}(x, y) = 9x^2y^2\mathbf{i} + \left(6x^3y - 1\right)\mathbf{j}$ is conservative.

$$\text{Work} = \left[3x^3y^2 - y\right]_{(0,0)}^{(5,9)} = 30{,}366$$

36. $\mathbf{F}(x, y)$ is conservative. $f(x, y) = \dfrac{x^2}{y}$

$$\text{Work} = \left[\frac{x^2}{y}\right]_{(-1,1)}^{(3,2)} = \frac{9}{2} - 1 = \frac{7}{2}$$

37. $\mathbf{r}(t) = 2\cos 2\pi t\mathbf{i} + 2\sin 2\pi t\mathbf{j}$

$\mathbf{r}'(t) = -4\pi \sin 2\pi t\mathbf{i} + 4\pi \cos 2\pi t\mathbf{j}$

$\mathbf{a}(t) = -8\pi^2 \cos 2\pi t\mathbf{i} - 8\pi^2 \sin 2\pi t\mathbf{j}$

$\mathbf{F}(t) = m\mathbf{a}(t) = \dfrac{1}{32}\mathbf{a}(t) = -\dfrac{\pi^2}{4}(\cos 2\pi t\mathbf{i} + \sin 2\pi t\mathbf{j})$

$$W = \int_C \mathbf{F} \cdot d\mathbf{r} = \int_C -\frac{\pi^2}{4}(\cos 2\pi t\mathbf{i} + \sin 2\pi t\mathbf{j}) \cdot 4\pi(-\sin 2\pi t\mathbf{i} + \cos 2\pi t\mathbf{j})\,dt = -\pi^3 \int_C 0\,dt = 0$$

38. $\mathbf{F}(x, y, z) = a_1\mathbf{i} + a_2\mathbf{j} + a_3\mathbf{k}$

Because $\mathbf{F}(x, y, z)$ is conservative, the work done in moving a particle along any path from P to Q is

$$f(x, y, z) = \left[a_1x + a_2y + a_3z\right]_{P=(p_1,p_2,p_3)}^{Q=(q_1,q_2,q_3)} = a_1(q_1 - p_1) + a_2(q_2 - p_2) + a_3(q_3 - p_3) = \mathbf{F} \cdot \overrightarrow{PQ}.$$

39. $\mathbf{F} = -175\mathbf{j}$

(a) $\mathbf{r}(t) = t\mathbf{i} + (50 - t)\mathbf{j}, \quad 0 \le t \le 50$

$d\mathbf{r} = (\mathbf{i} - \mathbf{j})\,dt$

$$\int_C \mathbf{F} \cdot d\mathbf{r} = \int_0^{50} 175\,dt = 8750 \text{ ft} \cdot \text{lbs}$$

(b) $\mathbf{r}(t) = t\mathbf{i} + \frac{1}{50}(50 - t)^2\mathbf{j}, \quad 0 \le t \le 50$

$d\mathbf{r} = \mathbf{i} - \frac{1}{25}(50 - t)\mathbf{j}$

$$\int_C \mathbf{F} \cdot d\mathbf{r} = \int_0^{50} (175)\tfrac{1}{25}(50 - t)\,dt$$

$$= 7\left[50t - \frac{t^2}{2}\right]_0^{50} = 8750 \text{ ft} \cdot \text{lbs}$$

40. No. The force field is conservative.

41. See Theorem 15.5.

42. A line integral is independent of path if $\int_C \mathbf{F} \cdot d\mathbf{r}$ does not depend on the curve joining P and Q. See Theorem 15.6.

43. (a) For the circle $\mathbf{r}(t) = a\cos t\mathbf{i} - a\sin t\mathbf{j}$, $0 \le t \le 2\pi$, you have $x^2 + y^2 = a^2$, and

$$\int_C \mathbf{F} \cdot d\mathbf{r} = \int_0^{2\pi} \left(\frac{-a\sin t}{a^2}\mathbf{i} - \frac{a\cos t}{a^2}\mathbf{j} \right) \cdot (-a\sin t\mathbf{i} - a\cos t\mathbf{j})\, dt = \int_0^{2\pi} \left(\sin^2 t + \cos^2 t\right) dt = 2\pi.$$

(b) For this curve, the answer is the same, 2π.

(c) For the opposite overtation, the answer is -2π.

(d) For the curve away from the origin, the answer is 0.

44. (a) The direct path along the line segment joining $(-4, 0)$ to $(3, 4)$ requires less work than the path going from $(-4, 0)$ to $(-4, 4)$ and then to $(3, 4)$.

(b) The closed curve given by the line segments joining $(-4, 0)$, $(-4, 4)$, $(3, 4)$, and $(-4, 0)$ satisfies

$$\int_C \mathbf{F} \cdot d\mathbf{r} \ne 0.$$

45. Conservative. $\int_C \mathbf{F} \cdot d\mathbf{r}$ is independent of path.

46. Not conservative. The value of $\int_C \mathbf{F} \cdot d\mathbf{r}$ from $(-1, 0)$ to $(1, 0)$ is positive if the path is above the x-axis, and negative if the path is below the x-axis.

47. False, it would be true if **F** were conservative.

48. True

49. True

50. False, the requirement is $\partial M/\partial y = \partial N/\partial x$.

51. Let

$$\mathbf{F} = M\mathbf{i} + N\mathbf{j} = \frac{\partial f}{\partial y}\mathbf{i} - \frac{\partial f}{\partial x}\mathbf{j}.$$

Then $\dfrac{\partial M}{\partial y} = \dfrac{\partial}{\partial y}\left(\dfrac{\partial f}{\partial y}\right) = \dfrac{\partial^2 f}{\partial y^2}$ and $\dfrac{\partial N}{\partial x} = \dfrac{\partial}{\partial x}\left(-\dfrac{\partial f}{\partial x}\right) = -\dfrac{\partial^2 f}{\partial x^2}$. Because $\dfrac{\partial^2 f}{\partial x^2} + \dfrac{\partial^2 f}{\partial y^2} = 0$ you have $\dfrac{\partial M}{\partial y} = \dfrac{\partial N}{\partial x}$.

So, **F** is conservative. Therefore, by Theorem 15.7, you have $\displaystyle\int_C \left(\frac{\partial f}{\partial y}\, dx - \frac{\partial f}{\partial x}\, dy\right) = \int_C (M\, dx + N\, dy) = \int_C \mathbf{F} \cdot d\mathbf{r} = 0$

for every closed curve in the plane.

52. Because the sum of the potential and kinetic energies remains constant from point to point, if the kinetic energy is decreasing at a rate of 15 units per minute, then the potential energy is increasing at a rate of 15 units per minute.

53. $\mathbf{F}(x, y) = \dfrac{y}{x^2 + y^2}\mathbf{i} - \dfrac{x}{x^2 + y^2}\mathbf{j}$

(a) $M = \dfrac{y}{x^2 + y^2}$

$$\frac{\partial M}{\partial y} = \frac{(x^2 + y^2)(1) - y(2y)}{(x^2 + y^2)^2} = \frac{x^2 - y^2}{(x^2 + y^2)^2}$$

$$N = -\frac{x}{x^2 + y^2}$$

$$\frac{\partial N}{\partial x} = \frac{(x^2 + y^2)(-1) + x(2x)}{(x^2 + y^2)^2} = \frac{x^2 - y^2}{(x^2 + y^2)^2}$$

So, $\dfrac{\partial N}{\partial x} = \dfrac{\partial M}{\partial y}$.

(b) $\mathbf{r}(t) = \cos t\mathbf{i} + \sin t\mathbf{j}$, $0 \le t \le \pi$

$\mathbf{F} = \sin t\mathbf{i} - \cos t\mathbf{j}$

$d\mathbf{r} = (-\sin t\mathbf{i} + \cos t\mathbf{j})\, dt$

$$\int_C \mathbf{F} \cdot d\mathbf{r} = \int_0^{\pi} \left(-\sin^2 t - \cos^2 t\right) dt = \left[-t\right]_0^{\pi} = -\pi$$

(c) $\mathbf{r}(t) = \cos t\mathbf{i} - \sin t\mathbf{j},\ 0 \le t \le \pi$

$\mathbf{F} = -\sin t\mathbf{i} - \cos t\mathbf{j}$

$d\mathbf{r} = (-\sin t\mathbf{i} - \cos t\mathbf{j})\,dt$

$$\int_C \mathbf{F} \cdot d\mathbf{r} = \int_0^{\pi} (\sin^2 t + \cos^2 t)\,dt = \Big[t\Big]_0^{\pi} = \pi$$

(d) $\mathbf{r}(t) = \cos t\mathbf{i} + \sin t\mathbf{j},\ 0 \le t \le 2\pi$

$\mathbf{F} = \sin t\mathbf{i} - \cos t\mathbf{j}$

$d\mathbf{r} = (-\sin t\mathbf{i} + \cos t\mathbf{j})\,dt$

$$\int_C \mathbf{F} \cdot d\mathbf{r} = \int_0^{2\pi} (-\sin^2 t - \cos^2 t)\,dt = \Big[-t\Big]_0^{2\pi} = -2\pi$$

This does not contradict Theorem 15.7 because **F** is not continuous at $(0, 0)$ in R enclosed by curve C.

(e) $$\nabla\left(\arctan\frac{x}{y}\right) = \frac{1/y}{1 + (x/y)^2}\mathbf{i} + \frac{-x/y^2}{1 + (x/y)^2}\mathbf{j} = \frac{y}{x^2 + y^2}\mathbf{i} - \frac{x}{x^2 + y^2}\mathbf{j} = \mathbf{F}$$

Section 15.4 Green's Theorem

1. $$\mathbf{r}(t) = \begin{cases} t\mathbf{i} + t^2\mathbf{j}, & 0 \le t \le 1 \\ (2 - t)\mathbf{i} + (2 - t)\mathbf{j}, & 1 \le t \le 2 \end{cases}$$

$$\int_C y^2\,dx + x^2\,dy = \int_0^1 \left[t^4(dt) + t^2(2t\,dt)\right] + \int_1^2 \left[(2 - t)^2(-dt) + (2 - t)^2(-dt)\right]$$

$$= \int_0^1 (t^4 + 2t^3)\,dt + \int_1^2 2(2 - t)^2(-dt) = \left[\frac{t^5}{5} + \frac{t^4}{2}\right]_0^1 + \left[\frac{2(2 - t)^3}{3}\right]_1^2 = \frac{7}{10} - \frac{2}{3} = \frac{1}{30}$$

By Green's Theorem,

$$\int_R\int \left(\frac{\partial N}{\partial x} - \frac{\partial M}{\partial y}\right) dA = \int_0^1 \int_{x^2}^{x} (2x - 2y)\,dy\,dx = \int_0^1 \Big[2xy - y^2\Big]_{x^2}^{x} dx$$

$$= \int_0^1 (x^2 - 2x^3 + x^4)\,dx = \left[\frac{x^3}{3} - \frac{x^4}{2} + \frac{x^5}{5}\right]_0^1 = \frac{1}{30}$$

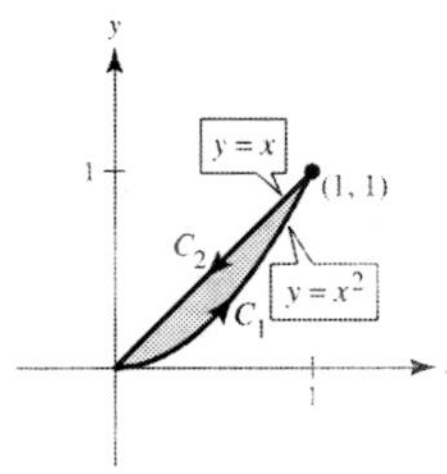

2. $$\mathbf{r}(t) = \begin{cases} t\mathbf{i} + t\mathbf{j}, & 0 \le t \le 1 \\ (2 - t)\mathbf{i} + \sqrt{2 - t}\,\mathbf{j}, & 1 \le t \le 2 \end{cases}$$

$$\int_C y^2\,dx + x^2\,dy = \int_0^1 \left[t^2(dt) + t^2(dt)\right] + \int_1^2 \left[(2 - t)(-dt) + (2 - t)^2\left(-\frac{1}{2\sqrt{2 - t}}\,dt\right)\right]$$

$$= \int_0^1 2t^2\,dt + \int_1^2 \left[(t - 2) - \frac{1}{2}(2 - t)^{3/2}\right] dt$$

$$= \left[\frac{2t^3}{3}\right]_0^1 + \left[\frac{(t - 2)^2}{2} + \frac{(2 - t)^{5/2}}{5}\right]_1^2$$

$$= \frac{2}{3} - \frac{1}{2} - \frac{1}{5} = -\frac{1}{30}$$

By Green's Theorem,

$$\int_R\int \left(\frac{\partial N}{\partial x} - \frac{\partial M}{\partial y}\right) dA = \int_0^1 \int_x^{\sqrt{x}} (2x - 2y)\,dy\,dx = \int_0^1 \Big[2xy - y^2\Big]_x^{\sqrt{x}} dx$$

$$= \int_0^1 \left(2x^{3/2} - x - x^2\right) dx = \left[\frac{4}{5}x^{5/2} - \frac{x^2}{2} - \frac{x^3}{3}\right]_0^1 = -\frac{1}{30}$$

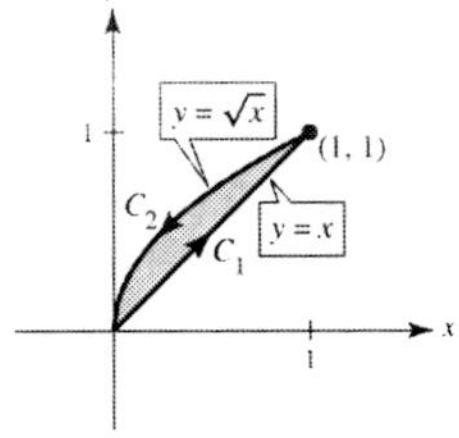

3. $\mathbf{r}(t) = \begin{cases} t\mathbf{i} & 0 \le t \le 1 \\ \mathbf{i} + (t-1)\mathbf{j} & 1 \le t \le 2 \\ (3-t)\mathbf{i} + \mathbf{j} & 2 \le t \le 3 \\ (4-t)\mathbf{j} & 3 \le t \le 4 \end{cases}$

$$\int_C y^2 dx + x^2 dy = \int_0^1 \left[0\,dt + t^2(0)\right] + \int_1^2 \left[(t-1)^2(0) + 1\,dt\right] + \int_2^3 \left[1(-dt) + (3-t)^2(0)\right] + \int_3^4 \left[(4-t)^2(0) + 0(-dt)\right]$$

$$= \int_1^2 dt + \int_2^3 -dt = 1 - 1 = 0$$

By Green's Theorem,

$$\int_R \int \left(\frac{\partial N}{\partial x} - \frac{\partial M}{\partial y}\right) dA = \int_0^1 \int_0^1 (2x - 2y)\,dy\,dx$$

$$= \int_0^1 \left[2xy - y^2\right]_0^1 dx = \int_0^1 (2x - 1)\,dx = \left[x^2 - x\right]_0^1 = 0$$

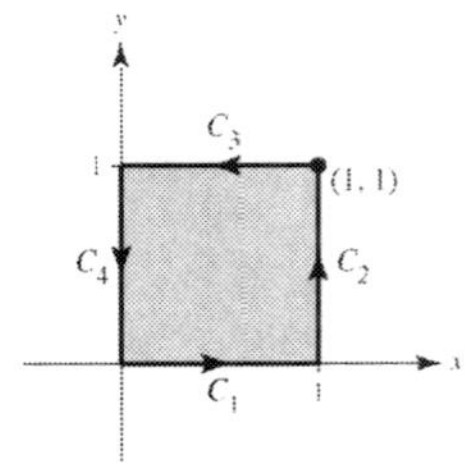

4. $\mathbf{r}(t) = \begin{cases} t\mathbf{i} & 0 \le t \le 3 \\ 3\mathbf{i} + (t-3)\mathbf{j} & 3 \le t \le 7 \\ (10-t)\mathbf{i} + 4\mathbf{j} & 7 \le t \le 10 \\ (14-t)\mathbf{j} & 10 \le t \le 14 \end{cases}$

$$\int_C y^2 dx + x^2 dy = \int_0^3 \left[0(dt) + t^2(0)\right] + \int_3^7 \left[(t-3)^2(0) + 9\,dt\right] + \int_7^{10} \left[16(-dt) + (10-t)^2(0)\right] + \int_{10}^{14} \left[0(-dt) + (14-t)^2(0)\right]$$

$$= \int_3^7 9\,dt + \int_7^{10} -16\,dt = 9(7-3) + (-16)(10-7) = -12$$

By Green's Theorem,

$$\int_R \int \left(\frac{\partial N}{\partial x} - \frac{\partial M}{\partial y}\right) dA = \int_0^3 \int_0^4 (2x - 2y)\,dy\,dx = \int_0^3 \left[2xy - y^2\right]_0^4 dx$$

$$= \int_0^3 \left[8x - 16\right] dx = \left[4x^2 - 16x\right]_0^3 = 36 - 48 = -12$$

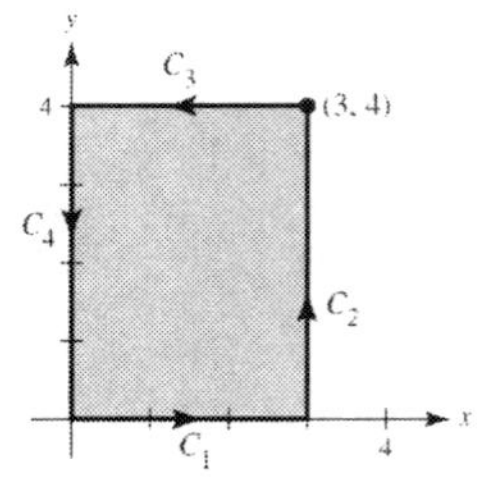

5. $C: x^2 + y^2 = 4$

Let $x = 2\cos t$ and $y = 2\sin t$, $0 \le t \le 2\pi$.

$$\int_C xe^y\,dx + e^x\,dy = \int_0^{2\pi} \left[2\cos t e^{2\sin t}(-2\sin t) + e^{2\cos t}(2\cos t)\right] dt \approx 19.99$$

$$\int_R \int \left(\frac{\partial N}{\partial x} - \frac{\partial M}{\partial y}\right) dA = \int_{-2}^2 \int_{-\sqrt{4-x^2}}^{\sqrt{4-x^2}} \left(e^x - xe^y\right) dy\,dx = \int_{-2}^2 \left[2\sqrt{4-x^2}e^x - xe^{\sqrt{4-x^2}} + xe^{-\sqrt{4-x^2}}\right] dx \approx 19.99$$

6. C: boundary of the region lying between the graphs of $y = x$ and $y = x^3$

$$\int_C xe^y\,dx + e^x\,dy = \int_0^1 \left(xe^{x^3} + 3x^2e^x\right) dx + \int_1^0 \left(xe^x + e^x\right) dx \approx 2.936 - 2.718 \approx 0.22$$

$$\int_R \int \left(\frac{\partial N}{\partial x} - \frac{\partial M}{\partial y}\right) dA = \int_0^1 \int_{x^3}^x \left(e^x - xe^y\right) dy\,dx = \int_0^1 \left(xe^{x^3} - x^3e^x\right) dx \approx 0.22$$

In Exercises 7–10, $\dfrac{\partial N}{\partial x} - \dfrac{\partial M}{\partial y} = 1$.

7. $\int_C (y - x)\,dx + (2x - y)\,dy = \int_0^3 \int_{x^2-2x}^{x} dy\,dx$

$= \int_0^3 \left[x - \left(x^2 - 2x\right)dx\right] = \left[-\dfrac{x^3}{3} + \dfrac{3x^2}{2}\right]_0^3 = -9 + \dfrac{27}{2} = \dfrac{9}{2}$

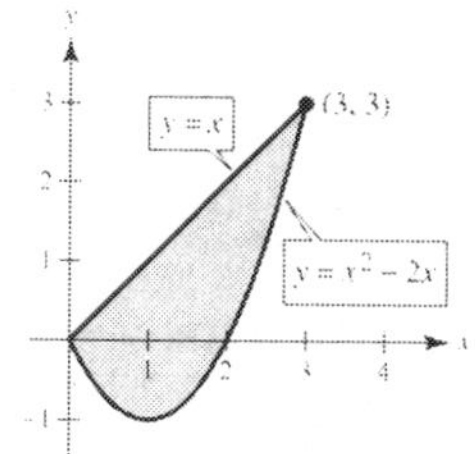

8. Because C is an ellipse with $a = 2$ and $b = 1$, then R is an ellipse of area $\pi ab = 2\pi$. So, Green's Theorem yields

$\int_C (y - x)\,dx + (2x - y)\,dy = \int_R \int 1\,dA = \text{Area of ellipse} = 2\pi.$

9. From the accompanying figure, we see that R is the shaded region. So, Green's Theorem yields

$\int_C (y - x)\,dx + (2x - y)\,dy = \int_R \int 1\,dA = \text{Area of } R = 6(10) - 2(2) = 56.$

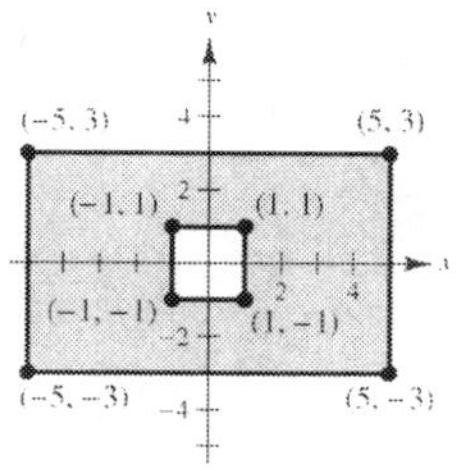

10. R is the shaded region of the accompanying figure.

$\int_C (y - x)\,dx + (2x - y)\,dy = \int_R \int 1\,dA$

$= \text{Area of shaded region}$

$= \frac{1}{2}\pi[25 - 9] = 8\pi$

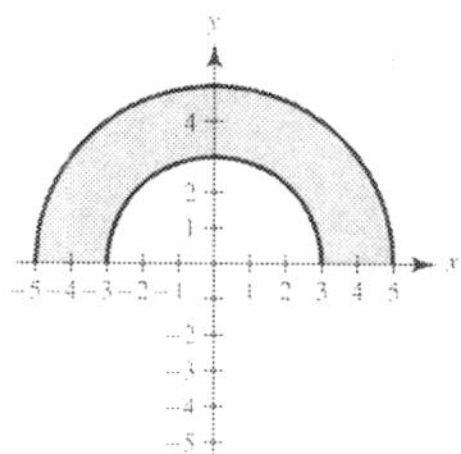

11. $\int_C 2xy\,dx + (x + y)\,dy = \int_R \int \left(\dfrac{\partial N}{\partial x} - \dfrac{\partial M}{\partial y}\right) dA$

$= \int_{-1}^{1} \int_0^{1-x^2} (1 - 2x)\,dy\,dx = \int_{-1}^{1} \left[y - 2xy\right]_0^{1-x^2} dx = \int_{-1}^{1} \left[\left(1 - x^2\right) - 2x\left(1 - x^2\right)\right] dx$

$= \int_{-1}^{1} \left[1 - x^2 - 2x + 2x^3\right] dx = \left[x - \dfrac{x^3}{3} - x^2 + \dfrac{x^4}{2}\right]_{-1}^{1} = \dfrac{1}{6} + \dfrac{7}{6} = \dfrac{4}{3}$

12. The given curves intersect at $(0, 0)$ and $(9, 3)$. So, Green's Theorem yields

$\int_C y^2\,dx + xy\,dy = \int_R \int (y - 2y)\,dA$

$= \int_0^9 \int_0^{\sqrt{x}} -y\,dy\,dx = \int_0^9 \left[\dfrac{-y^2}{2}\right]_0^{\sqrt{x}} dx = \int_0^9 \dfrac{-x}{2}\,dx = \left[\dfrac{-x^2}{4}\right]_0^9 = -\dfrac{81}{4}.$

13. $\int_C \left(x^2 - y^2\right) dx + 2xy\,dy = \int_R \int \left(\dfrac{\partial N}{\partial x} - \dfrac{\partial M}{\partial y}\right) dA = \int_{-4}^{4} \int_{-\sqrt{16-x^2}}^{\sqrt{16-x^2}} (2y + 2y)\,dy\,dx = \int_{-4}^{4} \left[2y^2\right]_{-\sqrt{16-x^2}}^{\sqrt{16-x^2}} dx = 0$

14. In this case, let $y = r\sin\theta$, $x = r\cos\theta$. Then $dA = r\,dr\,d\theta$ and Green's Theorem yields

$$\int_C (x^2 - y^2)\,dx + 2xy\,dy = \int_R\int 4y\,dA = 4\int_0^{2\pi}\int_0^{1+\cos\theta} r\sin\theta r\,dr\,d\theta$$

$$= 4\int_0^{2\pi}\int_0^{1+\cos\theta} r^2\sin\theta\,dr\,d\theta = \frac{4}{3}\int_0^{2\pi}\sin\theta(1+\cos\theta)^3\,d\theta = \left[-\frac{(1+\cos\theta)^4}{3}\right]_0^{2\pi} = 0.$$

15. Because $\dfrac{\partial M}{\partial y} = -2e^x\sin 2y = \dfrac{\partial N}{\partial x}$ you have

$$\int_R\int\left(\frac{\partial N}{\partial x} - \frac{\partial M}{\partial y}\right)dA = 0.$$

16. Because $\dfrac{\partial M}{\partial y} = \dfrac{2x}{x^2+y^2} = \dfrac{\partial N}{\partial x}$,

you have path independence and

$$\int_R\int\left(\frac{\partial N}{\partial x} - \frac{\partial M}{\partial y}\right)dA = 0.$$

17. By Green's Theorem,

$$\int_C \cos y\,dx + (xy - x\sin y)\,dy = \int_R\int(y - \sin y + \sin y)\,dA = \int_0^1\int_x^{\sqrt{x}} y\,dy\,dx = \int_0^1\left[\frac{y^2}{2}\right]_x^{\sqrt{x}}dx$$

$$= \int_0^1\left(\frac{x}{2} - \frac{x^2}{2}\right)dx = \left[\frac{x^2}{4} - \frac{x^3}{6}\right]_0^1 = \frac{1}{4} - \frac{1}{6} = \frac{1}{12}$$

18. By Green's Theorem,

$$\int_C\left(e^{-x^2/2} - y\right)dx + \left(e^{-y^2/2} + x\right)dy = \int_R\int 2\,dA = 2(\text{Area of } R) = 2\left[\pi(6)^2 - \pi(2)(3)\right] = 60\pi.$$

19. By Green's Theorem,

$$\int_C (x - 3y)\,dx + (x + y)\,dy = \int_R\int(1+3)\,dA = 4[\text{Area Large Circle} - \text{Area Small Circle}] = 4[9\pi - \pi] = 32\pi$$

20. By Green's Theorem,

$$\int_C 3x^2e^y\,dx + e^y\,dy = \int_R\int -3x^2e^y\,dA$$

$$= \int_1^2\int_{-2}^2 -3x^2e^y\,dy\,dx + \int_{-1}^1\int_1^2 -3x^2e^y\,dy\,dx$$

$$+ \int_{-2}^{-1}\int_{-2}^2 -3x^2e^y\,dy\,dx + \int_{-1}^1\int_{-2}^{-1} -3x^2e^y\,dy\,dx$$

$$= -7(e^2 - e^{-2}) - 2(e^2 - e) - 7(e^2 - e^{-2}) - 2(e^{-1} - e^{-2})$$

$$= -16e^2 + 16e^{-2} + 2e - 2e^{-1}.$$

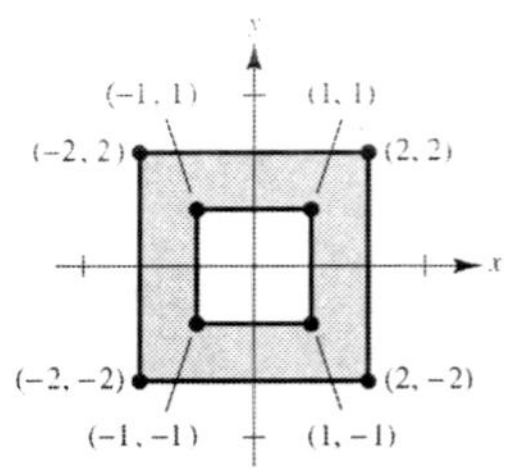

21. $\mathbf{F}(x, y) = xy\mathbf{i} + (x + y)\mathbf{j}$

C: $x^2 + y^2 = 1$

$$\text{Work} = \int_C xy\,dx + (x+y)\,dy = \int_R\int(1 - x)\,dA = \int_0^{2\pi}\int_0^1(1 - r\cos\theta)r\,dr\,d\theta$$

$$= \int_0^{2\pi}\left[\frac{r^2}{2} - \frac{r^2}{2}\cos\theta\right]_0^1 d\theta = \int_0^{2\pi}\frac{1}{2}(1 - \cos\theta)\,d\theta = \left[\frac{1}{2}\theta - \frac{1}{2}\sin\theta\right]_0^{2\pi} = \pi$$

22. $\mathbf{F}(x, y) = (e^x - 3y)\mathbf{i} + (e^y + 6x)\mathbf{j}$

C: $r = 2\cos\theta$

$$\text{Work} = \int_C(e^x - 3y)\,dx + (e^y + 6x)\,dy = \int_R\int 9\,dA = 9\pi$$ because $r = 2\cos\theta$ is a circle with a radius of one.

23. $\mathbf{F}(x, y) = \left(x^{3/2} - 3y\right)\mathbf{i} + \left(6x + 5\sqrt{y}\right)\mathbf{j}$

C: boundary of the triangle with vertices $(0, 0)$, $(5, 0)$, $(0, 5)$

$$\text{Work} = \int_C \left(x^{3/2} - 3y\right) dx + \left(6x + 5\sqrt{y}\right) dy = \int_R \int 9\, dA = 9\left(\tfrac{1}{2}\right)(5)(5) = \tfrac{225}{2}$$

24. $\mathbf{F}(x, y) = \left(3x^2 + y\right)\mathbf{i} + 4xy^2\mathbf{j}$

C: boundary of the region bounded by the graphs of $y = \sqrt{x}$, $y = 0$, $x = 9$

$$\text{Work} = \int_C \left(3x^2 + y\right) dx + 4xy^2\, dy = \int_0^9 \int_0^{\sqrt{x}} \left(4y^2 - 1\right) dy\, dx = \int_0^9 \left(\tfrac{4}{3}x^{3/2} - x^{1/2}\right) dx = \tfrac{558}{5}$$

25. C: let $x = a\cos t$, $y = a\sin t$, $0 \le t \le 2\pi$. By Theorem 15.9, you have

$$A = \frac{1}{2}\int_C x\, dy - y\, dx = \frac{1}{2}\int_0^{2\pi} \left[a\cos t(a\cos t) - a\sin t(-a\sin t)\right] dt = \frac{1}{2}\int_0^{2\pi} a^2\, dt = \left[\frac{a^2}{2}t\right]_0^{2\pi} = \pi a^2.$$

26. From the figure you see that

C_1: $y = \dfrac{3}{2}x$, $dy = \dfrac{3}{2}\, dx$, $0 \le x \le 2$

C_2: $y = -\dfrac{x}{2} + 4$, $dy = -\dfrac{1}{2}\, dx$

C_3: $x = 0$, $dx = 0$.

$$A = \frac{1}{2}\int_0^2 \left(\frac{3}{2}x - \frac{3}{2}x\right) dx + \frac{1}{2}\int_2^0 \left(-\frac{1}{2}x + \frac{x}{2} - 4\right) dx + \frac{1}{2}(0)$$

$$= \frac{1}{2}\int_2^0 (-4)\, dx = 2\int_0^2 dx = 4$$

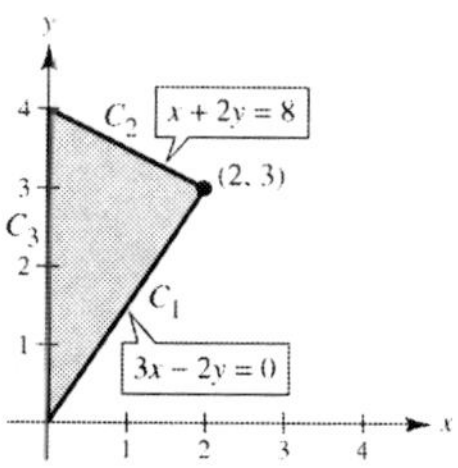

27. C_1: $y = x^2 + 1$, $dy = 2x\, dx$

C_2: $y = 5x - 3$, $dy = 5\, dx$

So, by Theorem 15.9 you have

$$A = \frac{1}{2}\int_1^4 \left(x(2x) - \left(x^2 + 1\right)\right) dx + \frac{1}{2}\int_4^1 \left(x(5) - (5x - 3)\right) dx$$

$$= \frac{1}{2}\left[\frac{x^3}{3} - x\right]_1^4 + \frac{1}{2}\left[3x\right]_4^1 = \frac{1}{2}[18] + \frac{1}{2}[-9] = \frac{9}{2}.$$

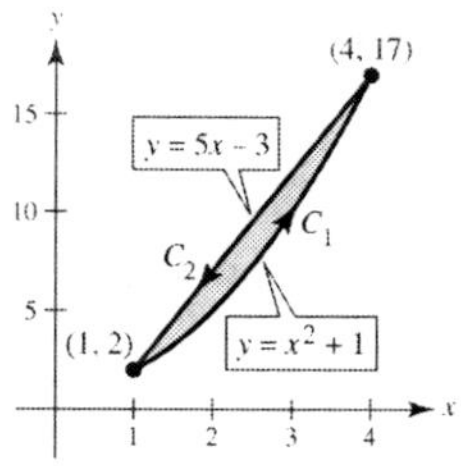

28. Because the loop of the folium is formed on the interval $0 \le t \le \infty$,

$$dx = \frac{3\left(1 - 2t^3\right)}{\left(t^3 + 1\right)^2}\, dt \text{ and } dy = \frac{3\left(2t - t^4\right)}{\left(t^3 + 1\right)^2}\, dt,$$

you have

$$A = \frac{1}{2}\int_0^\infty \left[\left(\frac{3t}{t^3 + 1}\right)\frac{3\left(2t - t^4\right)}{\left(t^3 + 1\right)^2} - \left(\frac{3t^2}{t^3 + 1}\right)\frac{3\left(1 - 2t^3\right)}{\left(t^3 + 1\right)^2}\right] dt$$

$$= \frac{9}{2}\int_0^\infty \frac{t^5 + t^2}{\left(t^3 + 1\right)^3}\, dt = \frac{9}{2}\int_0^\infty \frac{t^2\left(t^3 + 1\right)}{\left(t^3 + 1\right)^3}\, dt = \frac{3}{2}\int_0^\infty 3t^2\left(t^3 + 1\right)^{-2}\, dt = \left[\frac{-3}{2\left(t^3 + 1\right)}\right]_0^\infty = \frac{3}{2}.$$

29. See Theorem 15.8, page 1075.

30. See Theorem 15.9: $A = \frac{1}{2}\int_C x\, dy - y\, dx$.

31. For the moment about the x-axis, $M_x = \int_R \int y \, dA$. Let $N = 0$ and $M = -y^2/2$. By Green's Theorem,

$$M_x = \int_C -\frac{y^2}{2}\,dx = -\frac{1}{2}\int_C y^2\,dx \text{ and } \overline{y} = \frac{M_x}{2A} = -\frac{1}{2A}\int_C y^2\,dx.$$

For the moment about the y-axis, $M_y = \int_R \int x \, dA$. Let $N = x^2/2$ and $M = 0$. By Green's Theorem,

$$M_y = \int_C \frac{x^2}{2}\,dy = \frac{1}{2}\int_C x^2\,dy \text{ and } \overline{x} = \frac{M_y}{2A} = \frac{1}{2A}\int_C x^2\,dy.$$

32. By Theorem 15.9 and the fact that $x = r\cos\theta$, $y = r\sin\theta$, you have

$$A = \frac{1}{2}\int x\,dy - y\,dx = \frac{1}{2}\int (r\cos\theta)(r\cos\theta)\,d\theta - (r\sin\theta)(-r\sin\theta)\,d\theta = \frac{1}{2}\int_C r^2\,d\theta.$$

33. $A = \int_{-2}^{2}\left(4 - x^2\right)dx = \left[4x - \frac{x^3}{3}\right]_{-2}^{2} = \frac{32}{3}$

$$\overline{x} = \frac{1}{2A}\int_{C_1} x^2\,dy + \frac{1}{2A}\int_{C_2} x^2\,dy$$

For C_1, $dy = -2x\,dx$ and for C_2, $dy = 0$. So, $\overline{x} = \frac{1}{2(32/3)}\int_2^{-2} x^2(-2x\,dx) = \left[\frac{3}{64}\left(-\frac{x^4}{2}\right)\right]_2^{-2} = 0.$

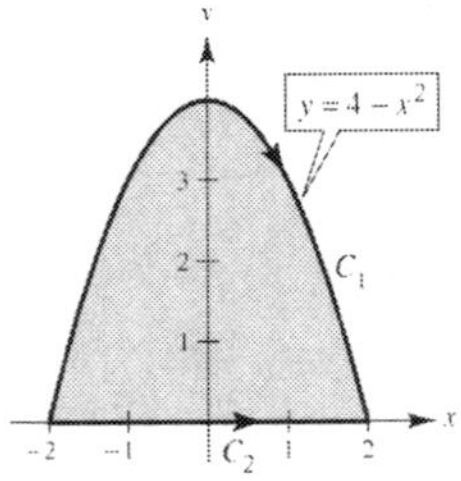

To calculate $\overline{y}$, note that $y = 0$ along C_2. So,

$$\overline{y} = \frac{-1}{2(32/3)}\int_2^{-2}\left(4 - x^2\right)^2 dx = \frac{3}{64}\int_{-2}^{2}\left(16 - 8x^2 + x^4\right)dx = \frac{3}{64}\left[16x - \frac{8x^3}{3} + \frac{x^5}{5}\right]_{-2}^{2} = \frac{8}{5}.$$

$$(\overline{x}, \overline{y}) = \left(0, \frac{8}{5}\right)$$

34. Because $A = \text{area of semicircle} = \frac{\pi a^2}{2}$, you have $\frac{1}{2A} = \frac{1}{\pi a^2}$. Note that $y = 0$ and $dy = 0$ along the boundary $y = 0$.

Let $x = a\cos t$, $y = a\sin t$, $0 \le t \le \pi$, then

$$\overline{x} = \frac{1}{\pi a^2}\int_0^{\pi} a^2\cos^2 t(a\cos t)\,dt = \frac{a}{\pi}\int_0^{\pi}\cos^3 t\,dt = \frac{a}{\pi}\int_0^{\pi}\left(1 - \sin^2 t\right)\cos t\,dt = \frac{a}{\pi}\left[\sin t - \frac{\sin^3 t}{3}\right]_0^{\pi} = 0$$

$$\overline{y} = \frac{-1}{\pi a^2}\int_0^{\pi} a^2\sin^2 t(-a\sin t\,dt) = \frac{a}{\pi}\int_0^{\pi}\sin^3 t\,dt = \frac{a}{\pi}\left[-\cos t + \frac{\cos^3 t}{3}\right]_0^{\pi} = \frac{4a}{3\pi}.$$

$$(\overline{x}, \overline{y}) = \left(0, \frac{4a}{3\pi}\right)$$

35. Because $A = \int_0^1\left(x - x^3\right)dx = \left[\frac{x^2}{2} - \frac{x^4}{4}\right]_0^1 = \frac{1}{4}$, you have $\frac{1}{2A} = 2$. On C_1 you have $y = x^3$, $dy = 3x^2\,dx$ and on C_2 you have $y = x$, $dy = dx$. So,

$$\overline{x} = 2\int_C x^2\,dy = 2\int_{C_1} x^2\left(3x^2\,dx\right) + 2\int_{C_2} x^2\,dx = 6\int_0^1 x^4\,dx + 2\int_1^0 x^2\,dx = \frac{6}{5} - \frac{2}{3} = \frac{8}{15}$$

$$\overline{y} = -2\int_C y^2\,dx = -2\int_0^1 x^6\,dx - 2\int_1^0 x^2\,dx = -\frac{2}{7} + \frac{2}{3} = \frac{8}{21}.$$

$$(\overline{x}, \overline{y}) = \left(\frac{8}{15}, \frac{8}{21}\right)$$

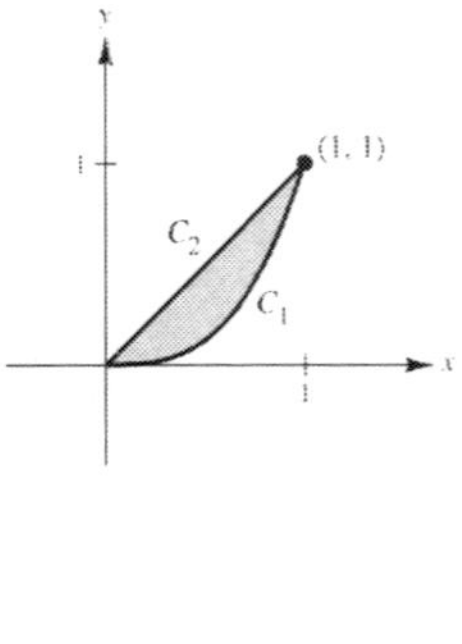

36. Because $A = \frac{1}{2}(2a)(c) = ac$, you have $\frac{1}{2A} = \frac{1}{2ac}$.

C_1: $y = 0, dy = 0$

C_2: $y = \frac{c}{b-a}(x-a), dy = \frac{c}{b-a}\,dx$

C_3: $y = \frac{c}{b+a}(x+a), dy = \frac{c}{b+a}\,dx.$

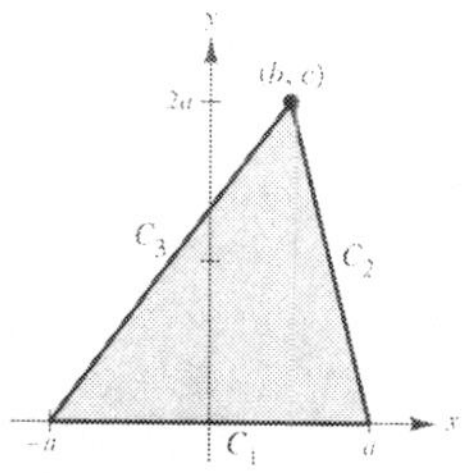

So,

$$\bar{x} = \frac{1}{2ac}\int_C x^2\,dy = \frac{1}{2ac}\left[\int_{-a}^{a} 0 + \int_a^b x^2\frac{c}{b-a}\,dx + \int_b^{-a} x^2\frac{c}{b+a}\,dx\right] = \frac{1}{2ac}\left[0 + \frac{2abc}{3}\right] = \frac{b}{3}$$

$$\bar{y} = \frac{-1}{2ac}\int_C y^2\,dx = \frac{-1}{2ac}\left[0 + \int_a^b\left(\frac{c}{b-a}\right)^2(x-a)^2\,dx + \int_b^{-a}\left(\frac{c}{b+a}\right)^2(x+a)^2\,dx\right] = \frac{-1}{2ac}\left[\frac{c^2(b-a)}{3} - \frac{c^2(b+a)}{3}\right] = \frac{c}{3}.$$

$$(\bar{x}, \bar{y}) = \left(\frac{b}{3}, \frac{c}{3}\right)$$

37. $$A = \frac{1}{2}\int_0^{2\pi} a^2(1-\cos\theta)^2\,d\theta = \frac{a^2}{2}\int_0^{2\pi}\left(1 - 2\cos\theta + \frac{1}{2} + \frac{\cos 2\theta}{2}\right)d\theta = \frac{a^2}{2}\left[\frac{3\theta}{2} - 2\sin\theta + \frac{1}{4}\sin 2\theta\right]_0^{2\pi} = \frac{a^2}{2}(3\pi) = \frac{3\pi a^2}{2}$$

38. $$A = \frac{1}{2}\int_0^{\pi} a^2\cos^2 3\theta\,d\theta = \frac{a^2}{2}\int_0^{\pi}\frac{1+\cos 6\theta}{2}\,d\theta = \frac{a^2}{4}\left[\theta + \frac{\sin 6\theta}{6}\right]_0^{\pi} = \frac{\pi a^2}{4}$$

Note: In this case R is enclosed by $r = a\cos 3\theta$ where $0 \le \theta \le \pi$.

39. In this case the inner loop has domain $\frac{2\pi}{3} \le \theta \le \frac{4\pi}{3}$. So,

$$A = \frac{1}{2}\int_{2\pi/3}^{4\pi/3}\left(1 + 4\cos\theta + 4\cos^2\theta\right)d\theta = \frac{1}{2}\int_{2\pi/3}^{4\pi/3}(3 + 4\cos\theta + 2\cos 2\theta)\,d\theta = \frac{1}{2}\left[3\theta + 4\sin\theta + \sin 2\theta\right]_{2\pi/3}^{4\pi/3} = \pi - \frac{3\sqrt{3}}{2}.$$

40. In this case, $0 \le \theta \le 2\pi$ and you let $u = \frac{\sin\theta}{1+\cos\theta}$, $\cos\theta = \frac{1-u^2}{1+u^2}$, $d\theta = \frac{2\,du}{1+u^2}$.

Now $u \Rightarrow \infty$ as $\theta \Rightarrow \pi$ and you have

$$A = 2\left(\frac{1}{2}\right)\int_0^{\pi}\frac{9}{(2-\cos\theta)^2}\,d\theta = 9\int_0^{\infty}\frac{\frac{2du}{1+u^2}}{4 - 4\left(\frac{1-u^2}{1+u^2}\right) + \frac{(1-u^2)^2}{(1+u^2)^2}} = 18\int_0^{\infty}\frac{1+u^2}{\left(1+3u^2\right)^2}\,du$$

$$= 18\int_0^{\infty}\frac{1/3}{1+3u^2}\,du + 18\int_0^{\infty}\frac{2/3}{\left(1+3u^2\right)^2}\,du = \left[\frac{6}{\sqrt{3}}\arctan\sqrt{3}\,u\right]_0^{\infty} + \frac{12}{\sqrt{3}}\left(\frac{1}{2}\right)\left[\frac{u}{1+3u^2} + \int\frac{\sqrt{3}}{1+3u^2}\,du\right]_0^{\infty}$$

$$= \frac{6}{\sqrt{3}}\left(\frac{\pi}{2}\right) + \frac{6}{\sqrt{3}}\left[\frac{u}{1+3u^2}\right]_0^{\infty} + \left[\frac{6}{\sqrt{3}}\arctan\sqrt{3}u\right]_0^{\infty} = \frac{3\pi}{\sqrt{3}} + 0 + \frac{3\pi}{\sqrt{3}} = 2\sqrt{3}\pi.$$

41. (a) $$\int_{C_1} y^3\,dx + \left(27x - x^3\right)dy = \int_R\int\left[\left(27 - 3x^2\right) - 3y^2\right]dA$$

$$= \int_0^{2\pi}\int_0^1\left(27 - 3r^2\right)r\,dr\,d\theta = \int_0^{2\pi}\left[\frac{27r^2}{2} - \frac{3r^4}{4}\right]_0^1 d\theta = \int_0^{2\pi}\frac{51}{4}\,d\theta = \frac{51}{2}\pi$$

(b) You want to find c such that $\int_0^C (27 - 3r^2)r\,dr\,d\theta$ is a maximum:

$$f(c) = \frac{27c^2}{2} - \frac{3}{4}c^4$$

$$f'(c) = 27c - 3c^2 \Rightarrow c = 3$$

Maximum Value: $\int_0^{2\pi}\int_0^3 (27 - 3r^2)r\,dr\,d\theta = \frac{243\pi}{2}$

42. $\int_C f(x)\,dx + g(y)\,dy = \int_R\int\left[\frac{\partial}{\partial x}g(y) - \frac{\partial}{\partial y}f(x)\right]dA = \int_R\int(0 - 0)\,dA = 0$

43. $\int\limits_C \left(e^{-x^2/2} - y\right)dx + \left(e^{-y^2/2} + x\right)dy$

$$= \iint\limits_R \left(\frac{\partial N}{\partial x} - \frac{\partial M}{\partial y}\right)dA$$

$$= \iint\limits_R (1 - (-1))\,dA$$

$$= 2(\text{area of } R)$$

$$= 2\left(\pi r^2 - \pi ab\right)$$

$$= 2\left(\pi(s^2) - \pi(2)(1)\right)$$

$$= 46\pi$$

44. $\int\limits_C \left(3x^2y + 1\right)dx + \left(x^3 + 4x\right)dy$

$$= \iint\limits_R \left(\frac{\partial N}{\partial x} - \frac{\partial M}{\partial y}\right)dA$$

$$= \iint\limits_R \left[\left(3x^2 + 4\right) - \left(3x^2\right)\right]dA$$

$$= \iint\limits_R 4\,dA$$

$$= 4(\text{area of } R)$$

$$= 4\left(\pi ab - \pi r^2\right)$$

$$= 4\left(\pi(4)(3) - \pi\left(2^2\right)\right) = 32\pi$$

45. $I = \int_C \frac{y\,dx - x\,dy}{x^2 + y^2}$

(a) Let $\mathbf{F} = \frac{y}{x^2 + y^2}\mathbf{i} - \frac{x}{x^2 + y^2}\mathbf{j}$.

$\mathbf{F}$ is conservative because $\frac{\partial N}{\partial x} = \frac{\partial M}{\partial y} = \frac{x^2 - y^2}{\left(x^2 + y^2\right)^2}$.

$\mathbf{F}$ is defined and has continuous first partials everywhere except at the origin. If C is a circle (a closed path) that does not contain the origin, then

$$\int_C \mathbf{F}\cdot d\mathbf{r} = \int_C M\,dx + N\,dy = \int_R\int\left(\frac{\partial N}{\partial x} - \frac{\partial M}{\partial y}\right)dA = 0.$$

(b) Let $\mathbf{r} = a\cos t\mathbf{i} - a\sin t\mathbf{j}$, $0 \le t \le 2\pi$ be a circle C_1 oriented clockwise inside C (see figure). Introduce line segments C_2 and C_3 as illustrated in Example 6 of this section in the text. For the region inside C and outside C_1, Green's Theorem applies. Note that since C_2 and C_3 have opposite orientations, the line integrals over them cancel. So, $C_4 = C_1 + C_2 + C + C_3$ and

$$\int_{C_4}\mathbf{F}\cdot d\mathbf{r} = \int_{C_1}\mathbf{F}\cdot d\mathbf{r} + \int_C \mathbf{F}\cdot d\mathbf{r} = 0.$$

But,

$$\int_{C_1}\mathbf{F}\cdot d\mathbf{r} = \int_0^{2\pi}\left[\frac{(-a\sin t)(-a\sin t)}{a^2\cos^2 t + a^2\sin^2 t} + \frac{(-a\cos t)(-a\cos t)}{a^2\cos^2 t + a^2\sin^2 t}\right]dt$$

$$= \int_0^{2\pi}\left(\sin^2 t + \cos^2 t\right)dt = \left[t\right]_0^{2\pi} = 2\pi.$$

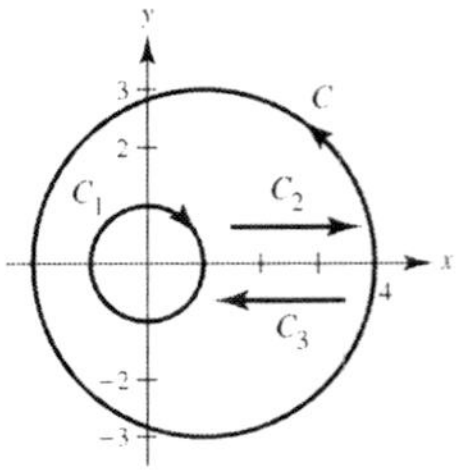

Finally, $\int_C \mathbf{F}\cdot d\mathbf{r} = -\int_{C_1}\mathbf{F}\cdot d\mathbf{r} = -2\pi$.

Note: If C were oriented clockwise, then the answer would have been 2π.

46. (a) $\mathbf{r}(t) = \begin{cases} t\mathbf{i}, & 0 \le t \le 4 \\ 4\mathbf{i} + (t-4)\mathbf{j}, & 4 \le t \le 8 \\ (12-t)\mathbf{i} + (12-t)\mathbf{j}, & 8 \le t \le 12 \end{cases}$

$$\int_C y^2 dx + x^2 dy = \int_0^4 \left[0\,dt + t^2(0)\right] + \int_4^8 \left[(t-4)^2(0) + 16\,dt\right] + \int_8^{12} \left[(12-t)^2(-dt) + (12-t)^2(-dt)\right]$$

$$= 0 + 64 - \frac{128}{3} = \frac{64}{3}$$

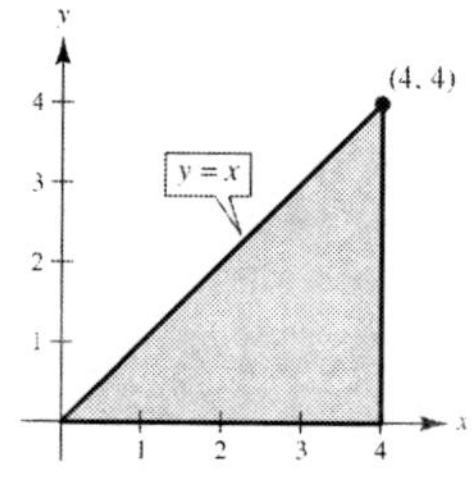

By Green's Theorem, $\int_R \int \left(\frac{\partial N}{\partial x} - \frac{\partial M}{\partial y}\right) dA = \int_0^4 \int_0^x (2x - 2y)\,dy\,dx = \int_0^4 x^2\,dx = \frac{64}{3}.$

(b) $\mathbf{r}(t) = \cos t\mathbf{i} + \sin t\mathbf{j},\ 0 \le t \le 2\pi$

$$\int_C y^2 dx + x^2 dy = \int_0^{2\pi} \left[\sin^2 t(-\sin t\,dt) + \cos^2 t(\cos t\,dt)\right] = \int_0^{2\pi} \left(\cos^3 t - \sin^3 t\right)dt$$

$$= \int_0^{2\pi} \left[\cos t\left(1 - \sin^2 t\right) - \sin t\left(1 - \cos^2 t\right)\right] dt = \left[\sin t - \frac{\sin^3 t}{3} + \cos t - \frac{\cos^3 t}{3}\right]_0^{2\pi} = 0$$

By Green's Theorem,

$$\int_R \int \left(\frac{\partial N}{\partial x} - \frac{\partial M}{\partial y}\right) dA = \int_{-1}^1 \int_{-\sqrt{1-x^2}}^{\sqrt{1-x^2}} (2x - 2y)\,dy\,dx$$

$$= \int_0^{2\pi} \int_0^1 (2r\cos\theta - 2r\sin\theta) r\,dr\,d\theta$$

$$= \frac{2}{3}\int_0^{2\pi} (\cos\theta - \sin\theta)\,d\theta = \frac{2}{3}(0) = 0.$$

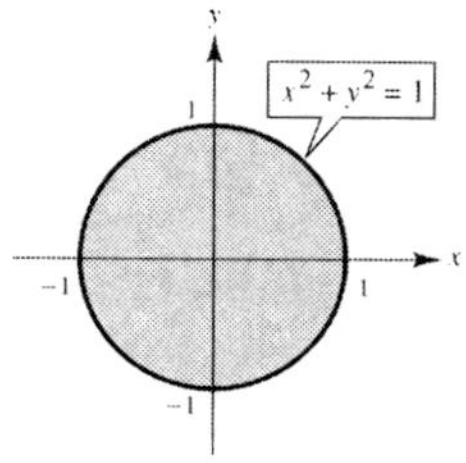

47. (a) Let C be the line segment joining (x_1, y_1) and (x_2, y_2).

$$y = \frac{y_2 - y_1}{x_2 - x_1}(x - x_1) + y_1$$

$$dy = \frac{y_2 - y_1}{x_2 - x_1}\,dx$$

$$\int_C -y\,dx + x\,dy = \int_{x_1}^{x_2} \left[-\frac{y_2 - y_1}{x_2 - x_1}(x - x_1) - y_1 + x\left(\frac{y_2 - y_1}{x_2 - x_1}\right)\right] dx = \int_{x_1}^{x_2} \left[x_1\left(\frac{y_2 - y_1}{x_2 - x_1}\right) - y_1\right] dx$$

$$= \left[\left[x_1\left(\frac{y_2 - y_1}{x_2 - x_1}\right) - y_1\right]x\right]_{x_1}^{x_2} = \left[x_1\left(\frac{y_2 - y_1}{x_2 - x_1}\right) - y_1\right](x_2 - x_1) = x_1(y_2 - y_1) - y_1(x_2 - x_1) = x_1 y_2 - x_2 y_1$$

(b) Let C be the boundary of the region $A = \frac{1}{2}\int_C -y\,dx + x\,dy = \frac{1}{2}\int_R \int (1 - (-1))\,dA = \int_R \int dA.$

So,

$$\int_R \int dA = \frac{1}{2}\left[\int_{C_1} -y\,dx + x\,dy + \int_{C_2} -y\,dx + x\,dy + \cdots + \int_{C_n} -y\,dx + x\,dy\right]$$

where C_1 is the line segment joining (x_1, y_1) and (x_2, y_2), C_2 is the line segment joining (x_2, y_2) and (x_3, y_3), $\cdots$, and C_n is the line segment joining (x_n, y_n) and (x_1, y_1). So,

$$\int_R \int dA = \frac{1}{2}\left[(x_1 y_2 - x_2 y_1) + (x_2 y_3 - x_3 y_2) + \cdots + (x_{n-1} y_n - x_n y_{n-1}) + (x_n y_1 - x_1 y_n)\right].$$

48. (a) Pentagon: $(0, 0), (2, 0), (3, 2), (1, 4), (-1, 1)$

$$A = \frac{1}{2}\left[(0 - 0) + (4 - 0) + (12 - 2) + (1 + 4) + (0 - 0)\right] = \frac{19}{2}$$

(b) Hexagon: $(0, 0), (2, 0), (3, 2), (2, 4), (0, 3), (-1, 1)$

$$A = \frac{1}{2}\left[(0 - 0) + (4 - 0) + (12 - 4) + (6 - 0) + (0 + 3) + (0 - 0)\right] = \frac{21}{2}$$

49. Because $\int_C \mathbf{F} \cdot \mathbf{N}\, ds = \int_R \int \text{div } \mathbf{F}\, dA$, then

$$\int_C fD_{\mathbf{N}}g\, ds = \int_C f\nabla g \cdot \mathbf{N}\, ds = \int_R \int \text{div}(f\nabla g)\, dA = \int_R \int \left(f \text{ div}(\nabla g) + \nabla f \cdot \nabla g\right) dA = \int_R \int \left(f\nabla^2 g + \nabla f \cdot \nabla g\right) dA.$$

50. $\int_C (fD_{\mathbf{N}}g - gD_{\mathbf{N}}f)\, ds = \int_C fD_{\mathbf{N}}g\, ds - \int_C gD_{\mathbf{N}}f\, ds$

$$= \int_R \int \left(f\nabla^2 g + \nabla f \cdot \nabla g\right) dA - \int_R \int \left(g\nabla^2 f + \nabla g \cdot \nabla f\right) dA = \int_R \int \left(f\nabla^2 g - g\nabla^2 f\right) dA$$

51. $\mathbf{F} = M\mathbf{i} + N\mathbf{j}$

$$\frac{\partial N}{\partial x} = \frac{\partial M}{\partial y} \Rightarrow \frac{\partial N}{\partial x} - \frac{\partial M}{\partial y} = 0$$

$$\int_C \mathbf{F} \cdot d\mathbf{r} = \int_C M\, dx + N\, dy = \int_R \int \left(\frac{\partial N}{\partial x} - \frac{\partial M}{\partial y}\right) dA = \int_R \int (0)\, dA = 0$$

52. First, note that the square with vertices $(1, 1), (-1, 1), (-1, -1)$, and $(1, -1)$ has area 4.

We claim that this is the lease possible area. (See figure on the left.)

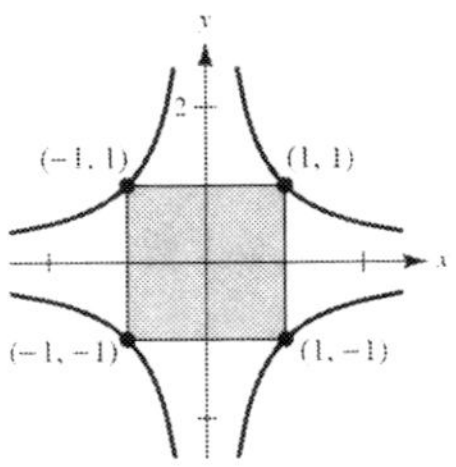

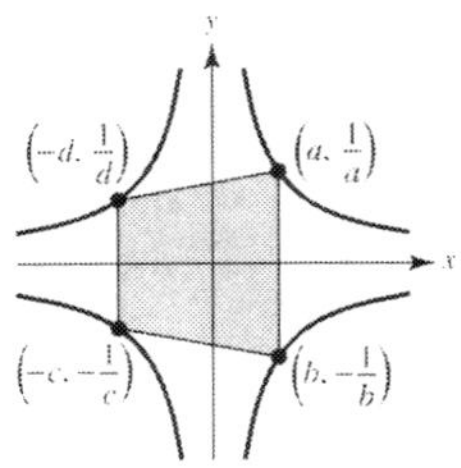

Now, consider a polygon with the indicated vertices (see figure on the right). Its area is a lower bound for any convex set having the same vertices. Using the area formula,

$$A = \frac{1}{2}\left[\left(a\left(\frac{1}{d}\right) - (-d)\left(\frac{1}{a}\right)\right) + \left((-d)\left(-\frac{1}{c}\right) - (-c)\left(\frac{1}{d}\right)\right) + \left((-c)\left(-\frac{1}{b}\right) - b\left(-\frac{1}{c}\right)\right) + \left(b\left(\frac{1}{a}\right) - a\left(-\frac{1}{b}\right)\right)\right.$$

$$= \frac{1}{2}\left[\left(\frac{a}{d} + \frac{d}{a}\right) + \left(\frac{d}{c} + \frac{c}{d}\right) + \left(\frac{c}{b} + \frac{b}{c}\right) + \left(\frac{b}{a} + \frac{a}{b}\right)\right]$$

Each expression inside the parentheses is greater than or equal to 2. For example,

$$(a - d)^2 = a^2 - 2ad + d^2 \geq 0$$

$$a^2 + d^2 \geq 2ad$$

$$\frac{a}{d} + \frac{d}{a} \geq 2.$$

Finally, $A \geq \frac{1}{2}(2 + 2 + 2 + 2) = 4$.

Section 15.5 Parametric Surfaces

1. $\mathbf{r}(u, v) = u\mathbf{i} + v\mathbf{j} + uv\mathbf{k}$

$z = xy$

Matches (e)

2. $\mathbf{r}(u, v) = u \cos v\mathbf{i} + u \sin v\mathbf{j} + u\mathbf{k}$

$x^2 + y^2 = z^2$, cone

Matches (f)

3. $\mathbf{r}(u, v) = u\mathbf{i} + \frac{1}{2}(u + v)\mathbf{j} + v\mathbf{k}$

$2y = x + z$, plane

Matches (b)

4. $\mathbf{r}(u, v) = u\mathbf{i} + \frac{1}{4}v^3\mathbf{j} + v\mathbf{k}$

$4y = z^3$, cylinder

Matches (a)

5. $\mathbf{r}(u, v) = 2 \cos v \cos u\mathbf{i} + 2 \cos v \sin u\mathbf{j} + 2 \sin v\mathbf{k}$

$x^2 + y^2 + z^2 = 4 \cos^2 v \cos^2 u + 4 \cos^2 v \sin^2 u + 4 \sin^2 v = 4 \cos^2 v + 4 \sin^2 v = 4$, sphere

Matches (d)

6. $\mathbf{r}(u, v) = 4 \cos u\mathbf{i} + 4 \sin u\mathbf{j} + v\mathbf{k}$

$x^2 + y^2 = 4$, circular cylinder

Matches (c)

7. $\mathbf{r}(u, v) = u\mathbf{i} + v\mathbf{j} + \dfrac{v}{2}\mathbf{k}$

$y - 2z = 0$

Plane

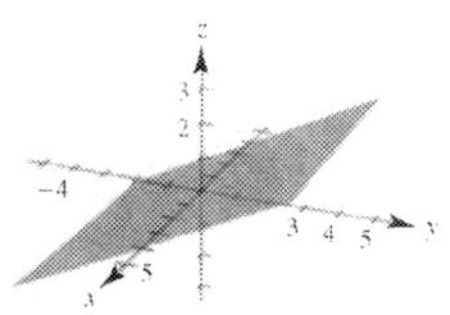

8. $\mathbf{r}(u, v) = 2u \cos v\mathbf{i} + 2u \sin v\mathbf{j} + \frac{1}{2}u^2\mathbf{k}$

$z = \frac{1}{2}u^2,\ x^2 + y^2 = 4u^2 \Rightarrow z = \frac{1}{8}(x^2 + y^2)$

Paraboloid

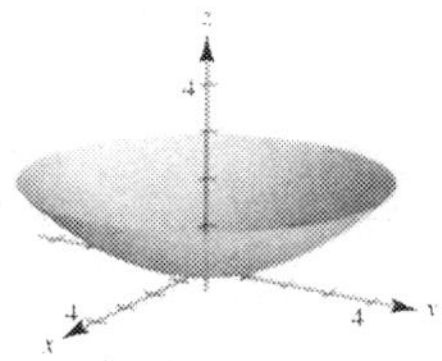

9. $\mathbf{r}(u, v) = 2 \cos u\mathbf{i} + v\mathbf{j} + 2 \sin u\mathbf{k}$

$x^2 + z^2 = 4$

Cylinder

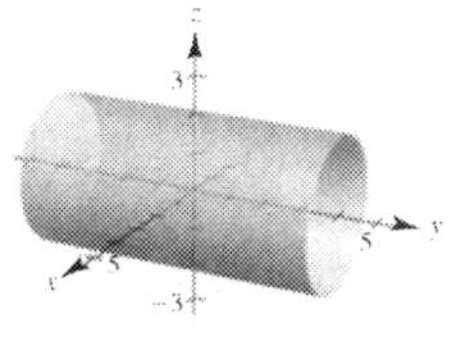

10. $\mathbf{r}(u, v) = 3 \cos v \cos u\mathbf{i} + 3 \cos v \sin u\mathbf{j} + 5 \sin v\mathbf{k}$

$x^2 + y^2 = 9 \cos^2 v \cos^2 u + 9 \cos^2 v \sin^2 u = 9 \cos^2 v$

$$\frac{x^2 + y^2}{9} + \frac{z^2}{25} = \cos^2 v + \sin^2 v = 1$$

$$\frac{x^2}{9} + \frac{y^2}{9} + \frac{z^2}{25} = 1$$

Ellipsoid

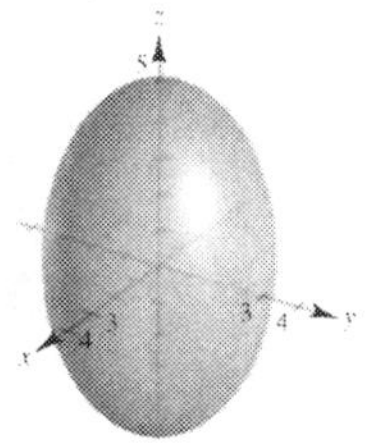

11. $\mathbf{r}(u, v) = 2u \cos v\mathbf{i} + 2u \sin v\mathbf{j} + u^4\mathbf{k}$,

$0 \le u \le 1,\ 0 \le v \le 2\pi$

$$z = \frac{(x^2 + y^2)^2}{16}$$

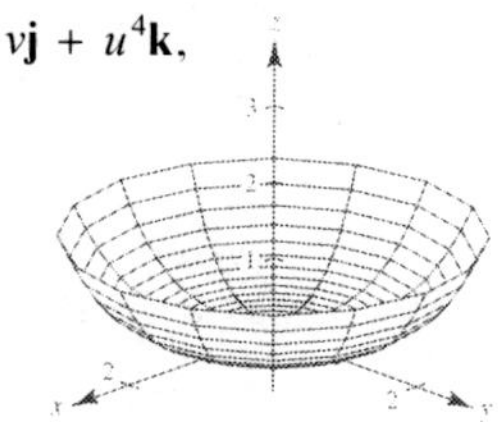

12. $\mathbf{r}(u, v) = 2 \cos v \cos u\mathbf{i} + 4 \cos v \sin u\mathbf{j} + \sin v\mathbf{k}$,

$0 \le u \le 2\pi,\ 0 \le v \le 2\pi$

$$\frac{x^2}{4} + \frac{y^2}{16} + \frac{z^2}{1} = 1$$

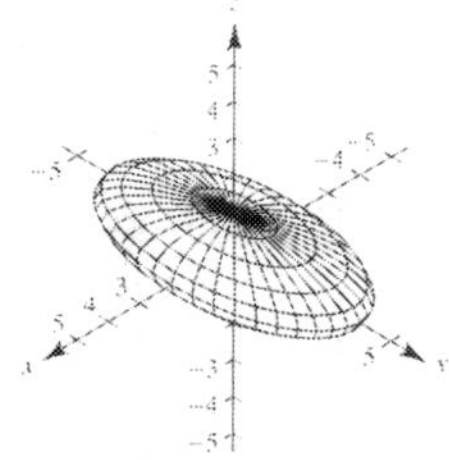

13. $\mathbf{r}(u, v) = 2 \sinh u \cos v\mathbf{i} + \sinh u \sin v\mathbf{j} + \cosh u\mathbf{k}$,

$0 \le u \le 2,\ 0 \le v \le 2\pi$

$$\frac{z^2}{1} - \frac{x^2}{4} - \frac{y^2}{1} = 1$$

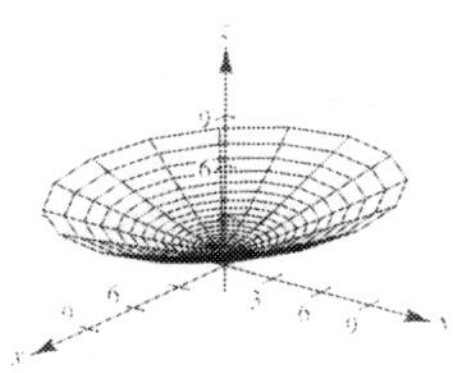

14. $\mathbf{r}(u, v) = 2u \cos v\mathbf{i} + 2u \sin v\mathbf{j} + v\mathbf{k},$

$0 \le u \le 1, \; 0 \le v \le 3\pi$

$\tan z = \dfrac{y}{x}$

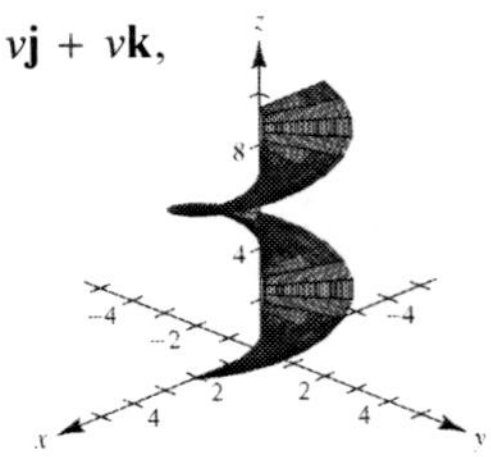

16. $\mathbf{r}(u, v) = \cos^3 u \cos v\mathbf{i} + \sin^3 u \sin v\mathbf{j} + u\mathbf{k},$

$0 \le u \le \dfrac{\pi}{2}, \quad 0 \le v \le 2\pi$

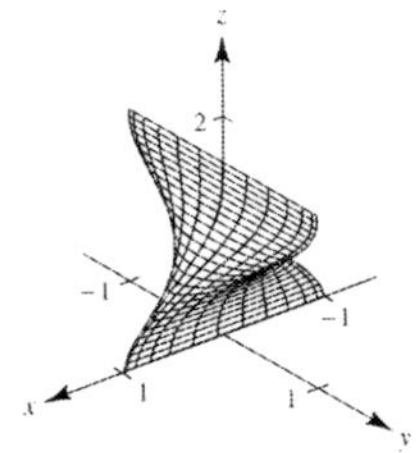

15. $\mathbf{r}(u, v) = (u - \sin u) \cos v\mathbf{i} + (1 - \cos u) \sin v\mathbf{j} + u\mathbf{k},$

$0 \le u \le \pi, \quad 0 \le v \le 2\pi$

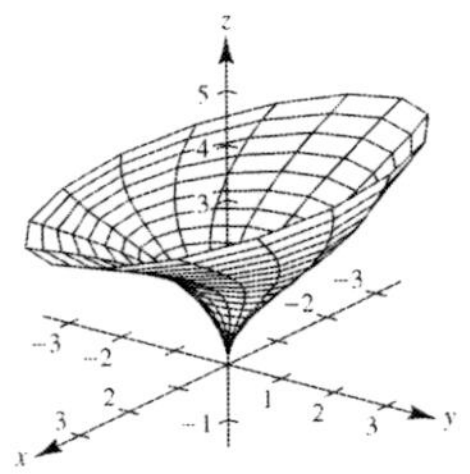

For Exercises 17–20, $\mathbf{r}(u, v) = u \cos v\mathbf{i} + u \sin v\mathbf{j} + u^2\mathbf{k}, \; 0 \le u \le 2, 0 \le v \le 2\pi$.

Eliminating the parameter yields $z = x^2 + y^2, 0 \le z \le 4$.

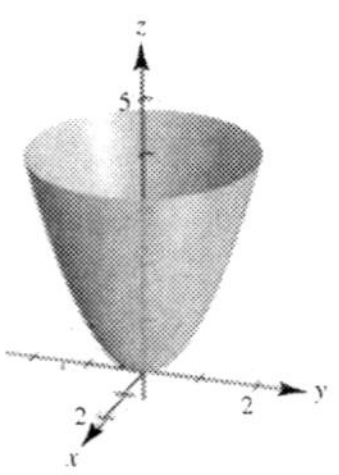

17. $\mathbf{s}(u, v) = u \cos v\mathbf{i} + u \sin v\mathbf{j} - u^2\mathbf{k}, \quad 0 \le u \le 2, \quad 0 \le v \le 2\pi$

$z = -(x^2 + y^2)$

The paraboloid is reflected (inverted) through the xy-plane.

18. $\mathbf{s}(u, v) = u \cos v\mathbf{i} + u^2\mathbf{j} + u \sin v\mathbf{k}, \quad 0 \le u \le 2, \quad 0 \le v \le 2\pi$

$y = x^2 + z^2$

The paraboloid opens along the y-axis instead of the z-axis.

19. $\mathbf{s}(u, v) = u \cos v\mathbf{i} + u \sin v\mathbf{j} + u^2\mathbf{k}, \quad 0 \le u \le 3, \quad 0 \le v \le 2\pi$

The height of the paraboloid is increased from 4 to 9.

20. $\mathbf{s}(u, v) = 4u \cos v\mathbf{i} + 4u \sin v\mathbf{j} + u^2\mathbf{k}, \quad 0 \le u \le 2, \quad 0 \le v \le 2\pi$

$z = \dfrac{x^2 + y^2}{16}$

The paraboloid is "wider." The top is now the circle $x^2 + y^2 = 64$. It was $x^2 + y^2 = 4$.

21. $z = y$

$\mathbf{r}(u, v) = u\mathbf{i} + v\mathbf{j} + v\mathbf{k}$

22. $z = 6 - x - y$

$\mathbf{r}(u, v) = u\mathbf{i} + v\mathbf{j} + (6 - u - v)\mathbf{k}$

23. $y = \sqrt{4x^2 + 9z^2}$

$\mathbf{r}(x, y) = x\mathbf{i} + \sqrt{4x^2 + 9z^2}\,\mathbf{j} + z\mathbf{k}$

or,

$\mathbf{r}(u, v) = \frac{1}{2}u \cos v\mathbf{i} + u\mathbf{j} + \frac{1}{3}u \sin v\mathbf{k},$

$u \ge 0, \quad 0 \le v \le 2\pi$

24. $x = \sqrt{16y^2 + z^2}$

$\mathbf{r}(y, z) = \sqrt{16y^2 + z^2}\,\mathbf{i} + y\mathbf{j} + z\mathbf{k}$

or,

$\mathbf{r}(u, v) = u\mathbf{i} + \frac{1}{4}u \cos v\mathbf{j} + u \sin v\mathbf{k},$

$u \ge 0, \quad 0 \le v \le 2\pi$

25. $x^2 + y^2 = 25$

$\mathbf{r}(u, v) = 5 \cos u\mathbf{i} + 5 \sin u\mathbf{j} + v\mathbf{k}$

26. $4x^2 + y^2 = 16$

$\mathbf{r}(u, v) = 2 \cos u\mathbf{i} + 4 \sin u\mathbf{j} + v\mathbf{k}$

27. $z = x^2$

$\mathbf{r}(u, v) = u\mathbf{i} + v\mathbf{j} + u^2\mathbf{k}$

28. $\dfrac{x^2}{9} + \dfrac{y^2}{4} + \dfrac{z^2}{1} = 1$

$\mathbf{r}(u, v) = 3 \cos v \cos u\mathbf{i} + 2 \cos v \sin u\mathbf{j} + \sin v\mathbf{k}$

29. $z = 4$ inside $x^2 + y^2 = 9$.

$\mathbf{r}(u, v) = v \cos u\mathbf{i} + v \sin u\mathbf{j} + 4\mathbf{k}, \; 0 \le v \le 3$

30. $z = x^2 + y^2$ inside $x^2 + y^2 = 9$.

$\mathbf{r}(u, v) = v \cos u\mathbf{i} + v \sin u\mathbf{j} + v^2\mathbf{k}, \; 0 \le v \le 3$

31. Function: $y = \dfrac{x}{2}, \quad 0 \le x \le 6$

Axis of revolution: x-axis

$x = u, \; y = \dfrac{u}{2} \cos v, \; z = \dfrac{u}{2} \sin v$

$0 \le u \le 6, \; 0 \le v \le 2\pi$

32. Function: $y = \sqrt{x}, \quad 0 \le x \le 4$

Axis of revolution: x-axis

$x = u, \; y = \sqrt{u} \cos v, \; z = \sqrt{u} \sin v$

$0 \le u \le 4, \quad 0 \le v \le 2\pi$

33. Function: $x = \sin z, \quad 0 \le z \le \pi$

Axis of revolution: z-axis

$x = \sin u \cos v, \; y = \sin u \sin v, \; z = u$

$0 \le u \le \pi, \quad 0 \le v \le 2\pi$

34. Function: $z = y^2 + 1, \quad 0 \le y \le 2$

Axis of revolution: y-axis

$x = (u^2 + 1) \cos v, \; y = u, \; z = (u^2 + 1) \sin v$

$0 \le u \le 2, \quad 0 \le v \le 2\pi$

35. $\mathbf{r}(u, v) = (u + v)\mathbf{i} + (u - v)\mathbf{j} + v\mathbf{k}, \; (1, -1, 1)$

$\mathbf{r}_u(u, v) = \mathbf{i} + \mathbf{j}, \; \mathbf{r}_v(u, v) = \mathbf{i} - \mathbf{j} + \mathbf{k}$

At $(1, -1, 1)$, $u = 0$ and $v = 1$.

$\mathbf{r}_u(0, 1) = \mathbf{i} + \mathbf{j}, \; \mathbf{r}_v(0, 1) = \mathbf{i} - \mathbf{j} + \mathbf{k}$

$$\mathbf{N} = \mathbf{r}_u(0, 1) \times \mathbf{r}_v(0, 1) = \begin{vmatrix} \mathbf{i} & \mathbf{j} & \mathbf{k} \\ 1 & 1 & 0 \\ 1 & -1 & 1 \end{vmatrix} = \mathbf{i} - \mathbf{j} - 2\mathbf{k}$$

Tangent plane: $(x - 1) - (y + 1) - 2(z - 1) = 0$

$x - y - 2z = 0$

(The original plane!)

36. $\mathbf{r}(u, v) = u\mathbf{i} + v\mathbf{j} + \sqrt{uv}\mathbf{k}, \; (1, 1, 1)$

$\mathbf{r}_u(u, v) = \mathbf{i} + \dfrac{v}{2\sqrt{uv}}\mathbf{k}, \; \mathbf{r}_v(u, v) = \mathbf{j} + \dfrac{u}{2\sqrt{uv}}\mathbf{k}$

At $(1, 1, 1)$, $u = 1$ and $v = 1$.

$\mathbf{r}_u(1, 1) = \mathbf{i} + \dfrac{1}{2}\mathbf{k}, \; \mathbf{r}_v(1, 1) = \mathbf{j} + \dfrac{1}{2}\mathbf{k}$

$$\mathbf{N} = \mathbf{r}_u(1, 1) \times \mathbf{r}_v(1, 1) = \begin{vmatrix} \mathbf{i} & \mathbf{j} & \mathbf{k} \\ 1 & 0 & \frac{1}{2} \\ 0 & 1 & \frac{1}{2} \end{vmatrix} = -\frac{1}{2}\mathbf{i} - \frac{1}{2}\mathbf{j} + \mathbf{k}$$

Direction numbers: $1, 1, -2$

Tangent plane: $(x - 1) + (y - 1) - 2(z - 1) = 0$

$x + y - 2z = 0$

37. $\mathbf{r}(u, v) = 2u\cos v\mathbf{i} + 3u\sin v\mathbf{j} + u^2\mathbf{k}, \quad (0, 6, 4)$

$\mathbf{r}_u(u, v) = 2\cos v\mathbf{i} + 3\sin v\mathbf{j} + 2u\mathbf{k}$

$\mathbf{r}_v(u, v) = -2u\sin v\mathbf{i} + 3u\cos v\mathbf{j}$

At $(0, 6, 4)$, $u = 2$ and $v = \pi/2$.

$$\mathbf{r}_u\left(2, \frac{\pi}{2}\right) = 3\mathbf{j} + 4\mathbf{k}, \mathbf{r}_v\left(2, \frac{\pi}{2}\right) = -4\mathbf{i}$$

$$\mathbf{N} = \mathbf{r}_u\left(2, \frac{\pi}{2}\right) \times \mathbf{r}_v\left(2, \frac{\pi}{2}\right) = \begin{vmatrix} \mathbf{i} & \mathbf{j} & \mathbf{k} \\ 0 & 3 & 4 \\ -4 & 0 & 0 \end{vmatrix} = -16\mathbf{j} + 12\mathbf{k}$$

Direction numbers: 0, 4, −3

Tangent plane: $4(y - 6) - 3(z - 4) = 0$

$$4y - 3z = 12$$

38. $\mathbf{r}(u, v) = 2u\cosh v\mathbf{i} + 2u\sinh v\mathbf{j} + \frac{1}{2}u^2\mathbf{k},$

$\mathbf{r}_u(u, v) = 2\cosh v\mathbf{i} + 2\sinh v\mathbf{j} + u\mathbf{k}$

$\mathbf{r}_v(u, v) = 2u\sinh v\mathbf{i} + 2u\cosh v\mathbf{j}$

At $(-4, 0, 2)$, $u = -2$ and $v = 0$.

$\mathbf{r}_u(-2, 0) = 2\mathbf{i} - 2\mathbf{k}, \mathbf{r}_v(-2, 0) = -4\mathbf{j}$

$\mathbf{N} = \mathbf{r}_u \times \mathbf{r}_v = -8\mathbf{i} - 8\mathbf{k}$

Direction numbers: 1, 0, 1

Tangent plane: $(x + 4) + (z - 2) = 0$

$$x + z = -2$$

39. $\mathbf{r}(u, v) = 4u\mathbf{i} - v\mathbf{j} + v\mathbf{k}, \quad 0 \le u \le 2, 0 \le v \le 1$

$\mathbf{r}_u(u, v) = 4\mathbf{i}, \mathbf{r}_v(u, v) = -\mathbf{j} + \mathbf{k}$

$$\mathbf{r}_u \times \mathbf{r}_v = \begin{vmatrix} \mathbf{i} & \mathbf{j} & \mathbf{k} \\ 4 & 0 & 0 \\ 0 & -1 & 1 \end{vmatrix} = -4\mathbf{j} - 4\mathbf{k}$$

$\|\mathbf{r}_u \times \mathbf{r}_v\| = \sqrt{16 + 16} = 4\sqrt{2}$

$$A = \int_0^1 \int_0^2 4\sqrt{2}\, du\, dv = 4\sqrt{2}(2)(1) = 8\sqrt{2}$$

40. $\mathbf{r}(u, v) = 2u\cos v\mathbf{i} + 2u\sin v\mathbf{j} + u^2\mathbf{k}, \quad 0 \le u \le 2, 0 \le v \le 2\pi$

$\mathbf{r}_u(u, v) = 2\cos v\mathbf{i} + 2\sin v\mathbf{j} + 2u\mathbf{k}$

$\mathbf{r}_v(u, v) = -2u\sin v\mathbf{i} + 2u\cos v\mathbf{j}$

$$\mathbf{r}_u \times \mathbf{r}_v = \begin{vmatrix} \mathbf{i} & \mathbf{j} & \mathbf{k} \\ 2\cos v & 2\sin v & 2u \\ -2u\sin v & 2u\cos v & 0 \end{vmatrix} = -4u^2\cos v\mathbf{i} - 4u^2\sin v\mathbf{j} + 8u\mathbf{k}$$

$$\|\mathbf{r}_u \times \mathbf{r}_v\| = \sqrt{16u^4\cos^2 v + 16u^4\sin^2 v + 64u^2} = 4u\sqrt{u^2 + 4}$$

$$A = \int_0^{2\pi} \int_0^2 4u\sqrt{u^2 + 4}\, du\, dv = \int_0^{2\pi} \left[\frac{4}{3}(u^2 + 4)^{3/2}\right]_0^2 dv = \int_0^{2\pi} \frac{4}{3}(8\sqrt{8} - 8)\, dv = \frac{4}{3}(16\sqrt{2} - 8)2\pi = \frac{64\pi}{3}(2\sqrt{2} - 1)$$

41. $\mathbf{r}(u, v) = a\cos u\mathbf{i} + a\sin u\mathbf{j} + v\mathbf{k}, \quad 0 \le u \le 2\pi, 0 \le v \le b$

$\mathbf{r}_u(u, v) = -a\sin u\mathbf{i} + a\cos u\mathbf{j}$

$\mathbf{r}_v(u, v) = \mathbf{k}$

$$\mathbf{r}_u \times \mathbf{r}_v = \begin{vmatrix} \mathbf{i} & \mathbf{j} & \mathbf{k} \\ -a\sin u & a\cos u & 0 \\ 0 & 0 & 1 \end{vmatrix} = a\cos u\mathbf{i} + a\sin u\mathbf{j}$$

$\|\mathbf{r}_u \times \mathbf{r}_v\| = a$

$$A = \int_0^b \int_0^{2\pi} a\, du\, dv = 2\pi ab$$

42. $\mathbf{r}(u, v) = a \sin u \cos v\mathbf{i} + a \sin u \sin v\mathbf{j} + a \cos u\mathbf{k}, \quad 0 \le u \le \pi, \quad 0 \le v \le 2\pi$

$\mathbf{r}_u(u, v) = a \cos u \cos v\mathbf{i} + a \cos u \sin v\mathbf{j} - a \sin u\mathbf{k}$

$\mathbf{r}_v(u, v) = -a \sin u \sin v\mathbf{i} + a \sin u \cos v\mathbf{j}$

$$\mathbf{r}_u \times \mathbf{r}_v = \begin{vmatrix} \mathbf{i} & \mathbf{j} & \mathbf{k} \\ a \cos u \cos v & a \cos u \sin v & -a \sin u \\ -a \sin u \sin v & a \sin u \cos v & 0 \end{vmatrix} = a^2 \sin^2 u \cos v\mathbf{i} + a^2 \sin^2 u \sin v\mathbf{j} + a^2 \sin u \cos u\mathbf{k}$$

$\|\mathbf{r}_u \times \mathbf{r}_v\| = a^2 \sin u$

$A = \int_0^{2\pi} \int_0^{\pi} a^2 \sin u \, du \, dv = 4\pi a^2$

43. $\mathbf{r}(u, v) = au \cos v\mathbf{i} + au \sin v\mathbf{j} + u\mathbf{k}, \quad 0 \le u \le b, \quad 0 \le v \le 2\pi$

$\mathbf{r}_u(u, v) = a \cos v\mathbf{i} + a \sin v\mathbf{j} + \mathbf{k}$

$\mathbf{r}_v(u, v) = -au \sin v\mathbf{i} + au \cos v\mathbf{j}$

$$\mathbf{r}_u \times \mathbf{r}_v = \begin{vmatrix} \mathbf{i} & \mathbf{j} & \mathbf{k} \\ a \cos v & a \sin v & 1 \\ -au \sin v & au \cos v & 0 \end{vmatrix} = -au \cos v\mathbf{i} - au \sin v\mathbf{j} + a^2 u\mathbf{k}$$

$\|\mathbf{r}_u \times \mathbf{r}_v\| = au\sqrt{1 + a^2}$

$A = \int_0^{2\pi} \int_0^{b} a\sqrt{1 + a^2} \, u \, du \, dv = \pi ab^2\sqrt{1 + a^2}$

44. $\mathbf{r}(u, v) = (a + b \cos v) \cos u\mathbf{i} + (a + b \cos v) \sin u\mathbf{j} + b \sin v\mathbf{k}, a > b, 0 \le u \le 2\pi, \quad 0 \le v \le 2\pi$

$\mathbf{r}_u(u, v) = -(a + b \cos v) \sin u\mathbf{i} + (a + b \cos v) \cos u\mathbf{j}$

$\mathbf{r}_v(u, v) = -b \sin v \cos u\mathbf{i} - b \sin v \sin u\mathbf{j} + b \cos v\mathbf{k}$

$$\mathbf{r}_u \times \mathbf{r}_v = \begin{vmatrix} \mathbf{i} & \mathbf{j} & \mathbf{k} \\ -(a + b \cos v) \sin u & (a + b \cos v) \cos u & 0 \\ -b \sin v \cos u & -b \sin v \sin u & b \cos v \end{vmatrix}$$

$$= b \cos u \cos v(a + b \cos v)\mathbf{i} + b \sin u \cos v(a + b \cos v)\mathbf{j} + b \sin v(a + b \cos v)\mathbf{k}$$

$\|\mathbf{r}_u \times \mathbf{r}_v\| = b(a + b \cos v)$

$A = \int_0^{2\pi} \int_0^{2\pi} b(a + b \cos v) \, du \, dv = 4\pi^2 ab$

45. $\mathbf{r}(u, v) = \sqrt{u} \cos v\mathbf{i} + \sqrt{u} \sin v\mathbf{j} + u\mathbf{k}, 0 \le u \le 4, \quad 0 \le v \le 2\pi$

$\mathbf{r}_u(u, v) = \dfrac{\cos v}{2\sqrt{u}}\mathbf{i} + \dfrac{\sin v}{2\sqrt{u}}\mathbf{j} + \mathbf{k}$

$\mathbf{r}_v(u, v) = -\sqrt{u} \sin v\mathbf{i} + \sqrt{u} \cos v\mathbf{j}$

$$\mathbf{r}_u \times \mathbf{r}_v = \begin{vmatrix} \mathbf{i} & \mathbf{j} & \mathbf{k} \\ \dfrac{\cos v}{2\sqrt{u}} & \dfrac{\sin v}{2\sqrt{u}} & 1 \\ -\sqrt{u} \sin v & \sqrt{u} \cos v & 0 \end{vmatrix} = -\sqrt{u} \cos v\mathbf{i} - \sqrt{u} \sin v\mathbf{j} + \frac{1}{2}\mathbf{k}$$

$\|\mathbf{r}_u \times \mathbf{r}_v\| = \sqrt{u + \dfrac{1}{4}}$

$A = \int_0^{2\pi} \int_0^{4} \sqrt{u + \dfrac{1}{4}} \, du \, dv = \dfrac{\pi}{6}\left(17\sqrt{17} - 1\right) \approx 36.177$

46. $\mathbf{r}(u, v) = \sin u \cos v\mathbf{i} + u\mathbf{j} + \sin u \sin v\mathbf{k}, 0 \le u \le \pi, \quad 0 \le v \le 2\pi$

$\mathbf{r}_u(u, v) = \cos u \cos v\mathbf{i} + \mathbf{j} + \cos u \sin v\mathbf{k}$

$\mathbf{r}_v(u, v) = -\sin u \sin v\mathbf{i} + \sin u \cos v\mathbf{k}$

$\mathbf{r}_u \times \mathbf{r}_v = \sin u \cos v\mathbf{i} - \cos u \sin u\mathbf{j} + \sin u \sin v\mathbf{k}$

$\|\mathbf{r}_u \times \mathbf{r}_v\| = \sin u\sqrt{1 + \cos^2 u}$

$$A = \int_0^{2\pi}\int_0^{\pi} \sin u\sqrt{1 + \cos^2 u}\, du\, dv = \pi\left[2\sqrt{2} + \ln\left|\frac{\sqrt{2}+1}{\sqrt{2}-1}\right|\right]$$

47. See the definition, page 1084.

48. See the definition, page 1088.

49. Function: $z = x$

Axis of revolution: z-axis

$x = u \cos v,\ y = u \sin v,\ z = u$

$\mathbf{r}(u, v) = u \cos v\mathbf{i} + u \sin v\mathbf{j} + u\mathbf{k}$

$u \le 0, \quad 0 \le v \le 2\pi$

50. (a) From $(-10, 10, 0)$

(b) From $(10, 10, 10)$

(c) From $(0, 10, 0)$

(d) From $(10, 0, 0)$

51. $\mathbf{r}(u, v) = a \sin^3 u \cos^3 v\mathbf{i} + a \sin^3 u \sin^3 v\mathbf{j} + a \cos^3 u\mathbf{k}$

$0 \le u \le \pi, \quad 0 \le v \le 2\pi$

$x = a \sin^3 u \cos^3 v \Rightarrow x^{2/3} = a^{2/3} \sin^2 u \cos^2 v$

$y = a \sin^3 u \sin^3 v \Rightarrow y^{2/3} = a^{2/3} \sin^2 u \sin^2 v$

$z = a \cos^3 u \Rightarrow z^{2/3} = a^{2/3} \cos^2 u$

$$x^{2/3} + y^{2/3} + z^{2/3} = a^{2/3}\left[\sin^2 u \cos^2 v + \sin^2 u \sin^2 v + \cos^2 u\right] = a^{2/3}\left[\sin^2 u + \cos^2 u\right] = a^{2/3}$$

52. Graph of $\mathbf{r}(u, v) = u \cos v\mathbf{i} + u \sin v\mathbf{j} + v\mathbf{k}$

$0 \le u \le \pi, \quad 0 \le v \le \pi$ from

(a) $(10, 0, 0)$

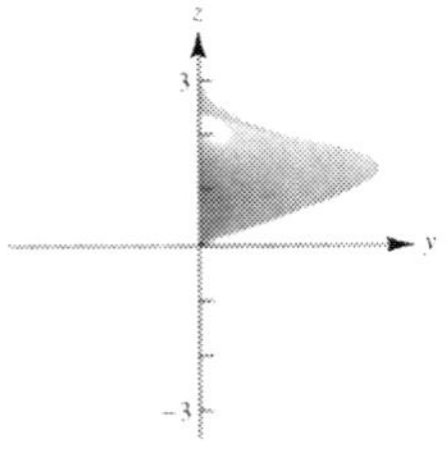

(b) $(0, 0, 10)$

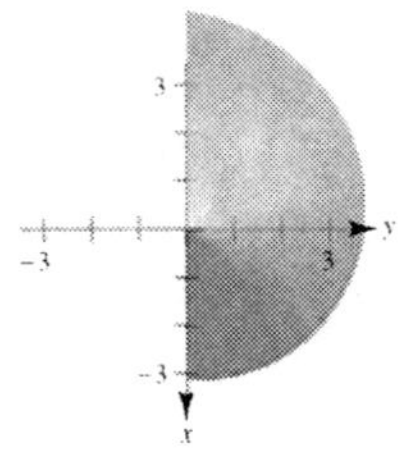

(c) $(10, 10, 10)$

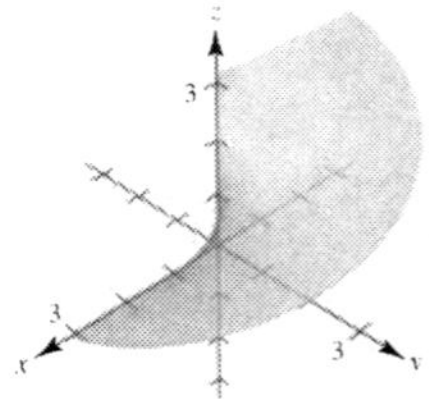

53. (a) $\mathbf{r}(u, v) = (4 + \cos v)\cos u\mathbf{i} + (4 + \cos v)\sin u\mathbf{j} + \sin v\mathbf{k},$
$0 \le u \le 2\pi, 0 \le v \le 2\pi$

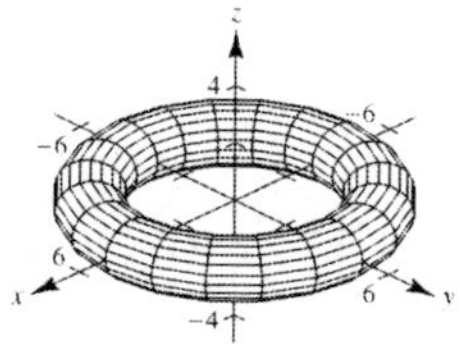

(b) $\mathbf{r}(u, v) = (4 + 2\cos v)\cos u\mathbf{i} + (4 + 2\cos v)\sin u\mathbf{j} + 2\sin v\mathbf{k},$
$0 \le u \le 2\pi, 0 \le v \le 2\pi$

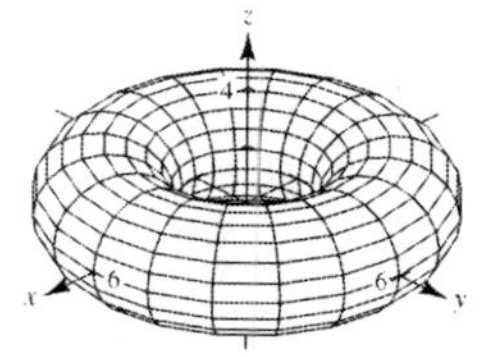

(c) $\mathbf{r}(u, v) = (8 + \cos v)\cos u\mathbf{i} + (8 + \cos v)\sin u\mathbf{j} + \sin v\mathbf{k},$
$0 \le u \le 2\pi, 0 \le v \le 2\pi$

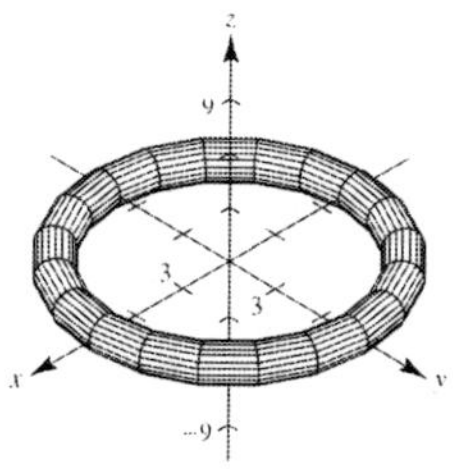

(d) $\mathbf{r}(u, v) = (8 + 3\cos v)\cos u\mathbf{i} + (8 + 3\cos v)\sin u\mathbf{j} + 3\sin v\mathbf{k},$
$0 \le u \le 2\pi, 0 \le v \le 2\pi$

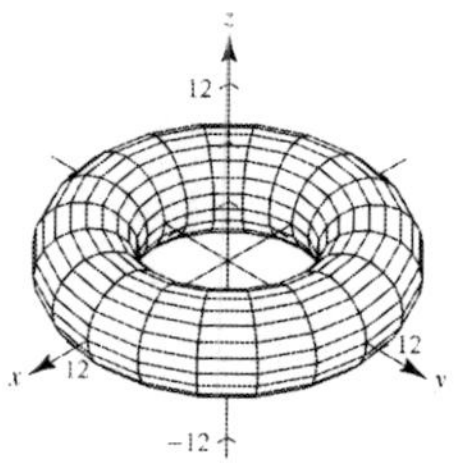

The radius of the generating circle that is revolved about the z-axis is b, and its center is a units from the axis of revolution.

54. $\mathbf{r}(u, v) = 2u\cos v\mathbf{i} + 2u\sin v\mathbf{j} + v\mathbf{k}, \quad 0 \le u \le 1, 0 \le v \le 3\pi$

(a) If $u = 1$:

$\mathbf{r}(1, v) = 2\cos v\mathbf{i} + 2\sin v\mathbf{j} + v\mathbf{k}$

$x^2 + y^2 = 4$

$0 \le z \le 3\pi$

Helix

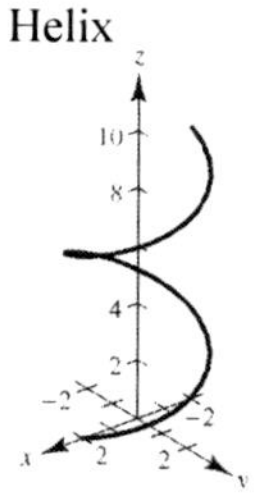

(b) If $v = \dfrac{2\pi}{3}$:

$\mathbf{r}\left(u, \dfrac{2\pi}{3}\right) = -u\mathbf{i} + \sqrt{3}u\mathbf{j} + \dfrac{2\pi}{3}\mathbf{k}$

$y = -\sqrt{3}x$

$z = \dfrac{2\pi}{3}$

Line

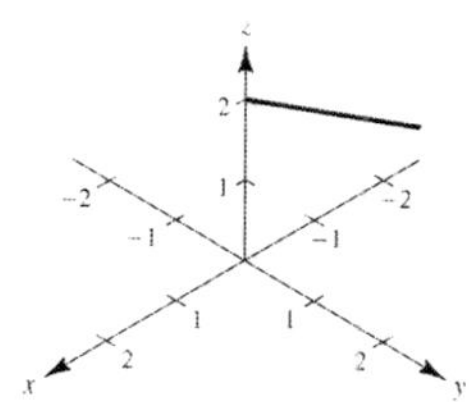

(c) If one parameter is held constant, the result is a **curve** in 3-space.

55. $\mathbf{r}(u, v) = 20 \sin u \cos v\mathbf{i} + 20 \sin u \sin v\mathbf{j} + 20 \cos u\mathbf{k}, 0 \le u \le \pi/3, \quad 0 \le v \le 2\pi$

$\mathbf{r}_u = 20 \cos u \cos v\mathbf{i} + 20 \cos u \sin v\mathbf{j} - 20 \sin u\mathbf{k}$

$\mathbf{r}_v = -20 \sin u \sin v\mathbf{i} + 20 \sin u \cos v\mathbf{j}$

$$\mathbf{r}_u \times \mathbf{r}_v = \begin{vmatrix} \mathbf{i} & \mathbf{j} & \mathbf{k} \\ 20 \cos u \cos v & 20 \cos u \sin v & -20 \sin u \\ -20 \sin u \sin v & 20 \sin u \cos v & 0 \end{vmatrix}$$

$$= 400 \sin^2 u \cos v\mathbf{i} + 400 \sin^2 u \sin v\mathbf{j} + 400\left(\cos u \sin u \cos^2 v + \cos u \sin u \sin^2 v\right)\mathbf{k}$$

$$= 400\left[\sin^2 u \cos v\mathbf{i} + \sin^2 u \sin v\mathbf{j} + \cos u \sin u\mathbf{k}\right]$$

$$\|\mathbf{r}_u \times \mathbf{r}_v\| = 400\sqrt{\sin^4 u \cos^2 v + \sin^4 u \sin^2 v + \cos^2 u \sin^2 u} = 400\sqrt{\sin^4 u + \cos^2 u \sin^2 u} = 400\sqrt{\sin^2 u} = 400 \sin u$$

$$S = \int_S\int dS = \int_0^{2\pi}\int_0^{\pi/3} 400 \sin u \, du \, dv = \int_0^{2\pi}\left[-400 \cos u\right]_0^{\pi/3} dv = \int_0^{2\pi} 200 \, dv = 400\pi \text{ m}^2$$

56. $x^2 + y^2 - z^2 = 1$

Let $x = u \cos v$, $y = u \sin v$, and $z = \sqrt{u^2 - 1}$. Then,

$$\mathbf{r}_u(u, v) = \cos v\mathbf{i} + \sin v\mathbf{j} + \frac{u}{\sqrt{u^2 - 1}}\mathbf{k}$$

$$\mathbf{r}_v(u, v) = -u \sin v\mathbf{i} + u \cos v\mathbf{j}.$$

At $(1, 0, 0)$, $u = 1$ and $v = 0$. $\mathbf{r}_u(1, 0)$ is undefined and $\mathbf{r}_v(1, 0) = \mathbf{j}$. The tangent plane at $(1, 0, 0)$ is $x = 1$.

57. $\mathbf{r}(u, v) = u \cos v\mathbf{i} + u \sin v\mathbf{j} + 2v\mathbf{k}, \quad 0 \le u \le 3, 0 \le v \le 2\pi$

$\mathbf{r}_u(u, v) = \cos v\mathbf{i} + \sin v\mathbf{j}$

$\mathbf{r}_v(u, v) = -u \sin v\mathbf{i} + u \cos v\mathbf{j} + 2\mathbf{k}$

$$\mathbf{r}_u \times \mathbf{r}_v = \begin{vmatrix} \mathbf{i} & \mathbf{j} & \mathbf{k} \\ \cos v & \sin v & 0 \\ -u \sin v & u \cos v & 2 \end{vmatrix} = 2 \sin v\mathbf{i} - 2 \cos v\mathbf{j} + u\mathbf{k}$$

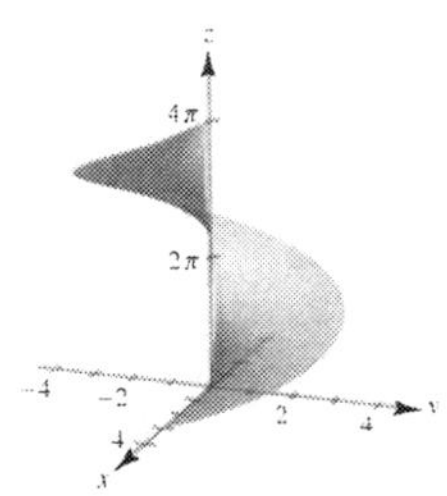

$$\|\mathbf{r}_u \times \mathbf{r}_v\| = \sqrt{4 + u^2}$$

$$A = \int_0^{2\pi}\int_0^3 \sqrt{4 + u^2} \, du \, dv = \pi\left[3\sqrt{13} + 4 \ln\left(\frac{3 + \sqrt{13}}{2}\right)\right]$$

58. $\mathbf{r}(u, v) = u\mathbf{i} + f(u) \cos v\mathbf{j} + f(u) \sin v\mathbf{k}, \quad a \le u \le b, 0 \le v \le 2\pi$

$\mathbf{r}_u(u, v) = \mathbf{i} + f'(u) \cos v\mathbf{j} + f'(u) \sin v\mathbf{k}$

$\mathbf{r}_v(u, v) = -f(u) \sin v\mathbf{j} + f(u) \cos v\mathbf{k}$

$$\mathbf{r}_u \times \mathbf{r}_v = \begin{vmatrix} \mathbf{i} & \mathbf{j} & \mathbf{k} \\ 1 & f'(u) \cos v & f'(u) \sin v \\ 0 & -f(u) \sin v & f(u) \cos v \end{vmatrix} = f(u)f'(u)\mathbf{i} - f(u) \cos v\mathbf{j} - f(u) \sin v\mathbf{k}$$

$$\|\mathbf{r}_u \times \mathbf{r}_v\| = f(u)\sqrt{1 + \left[f'(u)\right]^2}$$

$$A = \int_0^{2\pi}\int_a^b f(u)\sqrt{1 + \left[f'(u)\right]^2} \, du \, dv = 2\pi\int_a^b f(x)\sqrt{1 + \left[f'(x)\right]^2} \, dx \qquad \text{(because } u = x\text{)}$$

59. Answers will vary.

60. Answers will vary.

Section 15.6 Surface Integrals

1. $S: z = 4 - x,\ \ 0 \le x \le 4,\ \ 0 \le y \le 3,\ \ \dfrac{\partial z}{\partial x} = -1,\ \ \dfrac{\partial z}{\partial y} = 0$

$$\int_S\!\!\int (x - 2y + z)\,dS = \int_0^4\!\int_0^3 (x - 2y + 4 - x)\sqrt{1 + (-1)^2 + 0^2}\,dy\,dx = \sqrt{2}\int_0^4\!\int_0^3 (4 - 2y)\,dy\,dx = \sqrt{2}\int_0^4 3\,dx = 12\sqrt{2}$$

2. $S: z = 15 - 2x + 3y,\ \ 0 \le x \le 2,\ \ 0 \le y \le 4,\ \ \dfrac{\partial z}{\partial x} = -2,\ \ \dfrac{\partial z}{\partial y} = 3,\ dS = \sqrt{1 + 4 + 9}\,dy\,dx = \sqrt{14}\,dy\,dx$

$$\int_S\!\!\int (x - 2y + z)\,dS = \int_0^2\!\int_0^4 (x - 2y + 15 - 2x + 3y)\sqrt{14}\,dy\,dx = \sqrt{14}\int_0^2\!\int_0^4 (15 - x + y)\,dy\,dx = 128\sqrt{14}$$

3. $S: z = 2,\ x^2 + y^2 \le 1,\ \dfrac{\partial z}{\partial x} = \dfrac{\partial z}{\partial y} = 0$

$$\int_S\!\!\int (x - 2y + z)\,dS = \int_{-1}^1\!\int_{-\sqrt{1-x^2}}^{\sqrt{1-x^2}} (x - 2y + 2)\sqrt{1 + 0^2 + 0^2}\,dy\,dx = \int_0^{2\pi}\!\int_0^1 (r\cos\theta - 2r\sin\theta + 2)\,r\,dr\,d\theta$$

$$= \int_0^{2\pi}\left[\frac{1}{3}\cos\theta - \frac{2}{3}\sin\theta + 1\right]d\theta = \left[\frac{1}{3}\sin\theta + \frac{2}{3}\cos\theta + \theta\right]_0^{2\pi} = \frac{2}{3} + 2\pi - \frac{2}{3} = 2\pi$$

4. $S: z = \dfrac{2}{3}x^{3/2},\ 0 \le x \le 1,\ 0 \le y \le x,\ \dfrac{\partial z}{\partial x} = x^{1/2},\ \dfrac{\partial z}{\partial y} = 0$

$$\int_S\!\!\int (x - 2y + z)\,dS = \int_0^1\!\int_0^x \left(x - 2y + \frac{2}{3}x^{3/2}\right)\sqrt{1 + \left(x^{1/2}\right)^2 + (0)^2}\,dy\,dx$$

$$= \int_0^1\!\int_0^x \left(x - 2y + \frac{2}{3}x^{3/2}\right)\sqrt{1 + x}\,dy\,dx$$

$$= \frac{2}{3}\int_0^1 x^{5/2}\sqrt{x + 1}\,dx$$

$$= \frac{2}{3}\left[\frac{1}{4}x^{5/2}(1 + x)^{3/2}\right]_0^1 - \frac{5}{12}\int_0^1 x^{3/2}\sqrt{1 + x}\,dx$$

$$= \left[\frac{1}{6}x^{5/2}(1 + x)^{3/2}\right]_0^1 - \frac{5}{12}\left(\frac{1}{3}\right)\left[x^{3/2}(1 + x)^{3/2}\right]_0^1 + \frac{5}{24}\int_0^1 x^{1/2}\sqrt{1 + x}\,dx$$

$$= \frac{\sqrt{2}}{3} - \frac{5\sqrt{2}}{18} + \frac{5}{24}\int_0^1 \sqrt{x + x^2}\,dx$$

$$= \frac{\sqrt{2}}{18} + \frac{5}{24}\int_0^1 \sqrt{\left(x + \frac{1}{2}\right)^2 - \frac{1}{4}}\,dx$$

$$= \frac{\sqrt{2}}{18} + \frac{5}{24}\left(\frac{1}{2}\right)\left[\left(x + \frac{1}{2}\right)\sqrt{x^2 + x} - \frac{1}{4}\ln\left|\left(x + \frac{1}{2}\right) + \sqrt{x^2 + x}\right|\right]_0^1$$

$$= \frac{\sqrt{2}}{18} + \frac{5}{48}\left[\frac{3}{2}\sqrt{2} - \frac{1}{4}\ln\left|\frac{3}{2} + \sqrt{2}\right| + \frac{1}{4}\ln\left|\frac{1}{2}\right|\right]$$

$$= \frac{\sqrt{2}}{18} + \frac{15\sqrt{2}}{96} + \frac{5}{192}\ln\left|\frac{1}{3 + 2\sqrt{2}}\right| = \frac{61\sqrt{2}}{288} - \frac{5}{192}\ln\left|3 + 2\sqrt{2}\right| \approx 0.2536$$

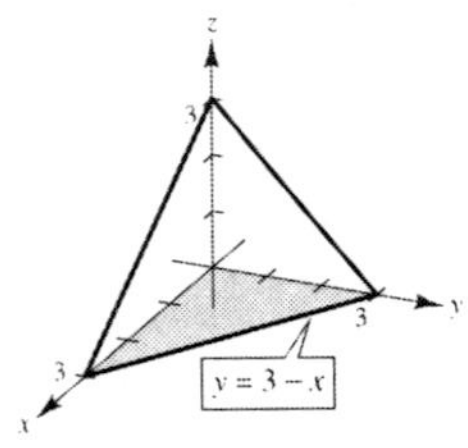

5. $S: z = 3 - x - y$ (first octant), $\dfrac{\partial z}{\partial x} = -1,\ \dfrac{\partial z}{\partial y} = -1$

$$\int_S\int xy\,dS = \int_0^3\int_0^{3-x} xy\sqrt{1 + (-1)^2 + (-1)^2}\,dy\,dx = \sqrt{3}\int_0^3\left[x\frac{y^2}{2}\right]_0^{3-x}$$

$$= \frac{\sqrt{3}}{2}\int_0^3 x(3 - x)^2\,dx = \frac{\sqrt{3}}{2}\left[\frac{x^4}{4} - 2x^3 + \frac{9x^2}{2}\right]_0^3 = \frac{\sqrt{3}}{2}\left[\frac{27}{4}\right] = \frac{27\sqrt{3}}{8}$$

6. $S: z = h,\ 0 \le x \le 2,\ 0 \le y \le \sqrt{4 - x^2},\ \dfrac{\partial z}{\partial x} = \dfrac{\partial z}{\partial y} = 0$

$$\int_S\int xy\,dS = \int_0^2\int_0^{\sqrt{4-x^2}} xy\,dy\,dx = \frac{1}{2}\int_0^2 x\left(4 - x^2\right)dx = \frac{1}{2}\left[2x^2 - \frac{x^4}{4}\right]_0^2 = 2$$

7. $S: z = 9 - x^2,\ 0 \le x \le 2,\ 0 \le y \le x,$

$\dfrac{\partial z}{\partial x} = -2x,\ \dfrac{\partial z}{\partial y} = 0$

$$\int_S\int xy\,dS = \int_0^2\int_y^2 xy\sqrt{1 + 4x^2}\,dx\,dy = \frac{391\sqrt{17} + 1}{240}$$

8. $S: z = \dfrac{1}{2}xy,\ 0 \le x \le 4,\ 0 \le y \le 4,\ \dfrac{\partial z}{\partial x} = \dfrac{1}{2}y,\ \dfrac{\partial z}{\partial y} = \dfrac{1}{2}x$

$$\int_S\int xy\,dS = \int_0^4\int_0^4 xy\sqrt{1 + \frac{y^2}{4} + \frac{x^2}{4}}\,dy\,dx = \frac{3904}{15} - \frac{160\sqrt{5}}{3}$$

9. $S: z = 10 - x^2 - y^2,\ 0 \le x \le 2,\ 0 \le y \le 2$

$$\int_S\int\left(x^2 - 2xy\right)dS = \int_0^2\int_0^2\left(x^2 - 2xy\right)\sqrt{1 + 4x^2 + 4y^2}\,dy\,dx \approx -11.47$$

10. $S: z = \cos x,\ 0 \le x \le \dfrac{\pi}{2},\ 0 \le y \le \dfrac{x}{2}$

$$\int_S\int\left(x^2 - 2xy\right)dS = \int_0^{\pi/2}\int_0^{x/2}\left(x^2 - 2xy\right)\sqrt{1 + \sin^2 x}\,dy\,dx = \int_0^{\pi/2}\frac{x^3}{4}\sqrt{1 + \sin^2 x}\,dx \approx 0.52$$

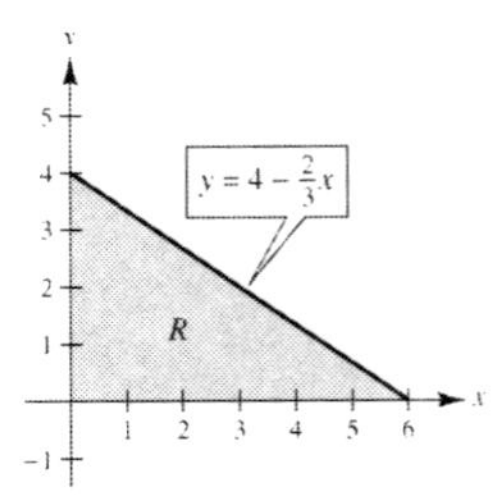

11. $S: 2x + 3y + 6z = 12$ (first octant) $\Rightarrow z = 2 - \frac{1}{3}x - \frac{1}{2}y$

$\rho(x, y, z) = x^2 + y^2$

$$m = \int_R\int\left(x^2 + y^2\right)\sqrt{1 + \left(-\tfrac{1}{3}\right)^2 + \left(-\tfrac{1}{2}\right)^2}\,dA = \frac{7}{6}\int_0^6\int_0^{4-(2x/3)}\left(x^2 + y^2\right)dy\,dx$$

$$= \frac{7}{6}\int_0^6\left[x^2\left(4 - \tfrac{2}{3}x\right) + \tfrac{1}{3}\left(4 - \tfrac{2}{3}x\right)^3\right]dx = \frac{7}{6}\left[\tfrac{4}{3}x^3 - \tfrac{1}{6}x^4 - \tfrac{1}{8}\left(4 - \tfrac{2}{3}x\right)^4\right]_0^6 = \frac{364}{3}$$

12. $S: z = \sqrt{a^2 - x^2 - y^2}$

$\rho(x, y, z) = kz$

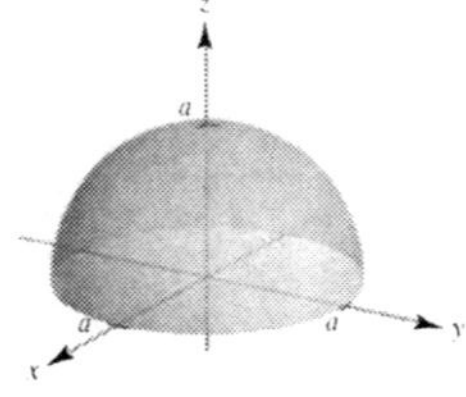

$$m = \int_S\int kz\,dS = \int_R\int k\sqrt{a^2 - x^2 - y^2}\sqrt{1 + \left(\frac{-x}{\sqrt{a^2 - x^2 - y^2}}\right)^2 + \left(\frac{-y}{\sqrt{a^2 - x^2 - y^2}}\right)^2}\,dA$$

$$= \int_R\int k\sqrt{a^2 - x^2 - y^2}\left(\frac{a}{\sqrt{a^2 - x^2 - y^2}}\right)dA = \int_R\int ka\,dA = ka\int_R\int dA = ka\left(2\pi a^2\right) = 2ka^3\pi$$

13. $S: \mathbf{r}(u, v) = u\mathbf{i} + v\mathbf{j} + 2v\mathbf{k}, \quad 0 \le u \le 1, \quad 0 \le v \le 2$

$\mathbf{r}_u = \mathbf{i}, \quad \mathbf{r}_v = \mathbf{j} + 2\mathbf{k}$

$$\mathbf{r}_u \times \mathbf{r}_v = \begin{vmatrix} \mathbf{i} & \mathbf{j} & \mathbf{k} \\ 1 & 0 & 0 \\ 0 & 1 & 2 \end{vmatrix} = -2\mathbf{j} + \mathbf{k}$$

$\|\mathbf{r}_u \times \mathbf{r}_v\| = \sqrt{5}$

$$\int_S\int (y + 5)\, dS = \int_0^2 \int_0^1 (v + 5)\sqrt{5}\, du\, dv = \int_0^2 (v + 5)\sqrt{5}\, dv = \sqrt{5}\left[\frac{v^2}{2} + 5v\right]_0^2 = 12\sqrt{5}$$

14. $\mathbf{r}(u, v) = 2\cos u\mathbf{i} + 2\sin u\mathbf{j} + v\mathbf{k}, \quad 0 \le u \le \pi/2, \quad 0 \le v \le 1$

$\mathbf{r}_u = -2\sin u\mathbf{i} + 2\cos u\mathbf{j}, \qquad \mathbf{r}_v = \mathbf{k}$

$$\mathbf{r}_u \times \mathbf{r}_v = \begin{vmatrix} \mathbf{i} & \mathbf{j} & \mathbf{k} \\ -2\sin u & 2\cos u & 0 \\ 0 & 0 & 1 \end{vmatrix} = 2\cos u\mathbf{i} + 2\sin u\mathbf{j}$$

$\|\mathbf{r}_u \times \mathbf{r}_v\| = \sqrt{4\cos^2 u + 4\sin^2 v} = 2$

$$\int_S\int xy\, dS = \int_0^1 \int_0^{\pi/2} (2\cos u)(2\sin u)2\, du\, dv = 8\int_0^1 \left[\frac{\sin^2 u}{2}\right]_0^{\pi/2} dv = 4$$

15. $S: \mathbf{r}(u, v) = 2\cos u\mathbf{i} + 2\sin u\mathbf{j} + v\mathbf{k}, \quad 0 \le u \le \pi/2, \quad 0 \le v \le 1$

$\mathbf{r}_u = -2\sin u\mathbf{i} + 2\cos u\mathbf{j}, \qquad \mathbf{r}_v = \mathbf{k}$

$$\mathbf{r}_u \times \mathbf{r}_v = \begin{vmatrix} \mathbf{i} & \mathbf{j} & \mathbf{k} \\ -2\sin u & 2\cos u & 0 \\ 0 & 0 & 1 \end{vmatrix} = 2\cos u\mathbf{i} + 2\sin u\mathbf{j}$$

$\|\mathbf{r}_u \times \mathbf{r}_v\| = \sqrt{4\cos^2 u + 4\sin^2 u} = 2$

$$\int_S\int (x + y)\, dS = \int_0^1 \int_0^{\pi/2} (2\cos u + 2\sin u)2\, du\, dv = 4\int_0^1 \big[\sin u - \cos u\big]_0^{\pi/2} dv = 4\int_0^1 2\, dv = 8$$

16. $S: \mathbf{r}(u, v) = 4u\cos v\mathbf{i} + 4u\sin v\mathbf{j} + 3u\mathbf{k}, \quad 0 \le u \le 4, \quad 0 \le v \le \pi$

$\|\mathbf{r}_u \times \mathbf{r}_v\| = \|-12u\cos v\mathbf{i} - 12u\sin v\mathbf{j} + 16u\mathbf{k}\| = 20u$

$$\int_S\int (x + y)\, dS = \int_0^\pi \int_0^4 (4u\cos v + 4u\sin v)20u\, du\, dv = \frac{10{,}240}{3}$$

17. $f(x, y, z) = x^2 + y^2 + z^2$

$S: z = x + y, \quad x^2 + y^2 \le 1, \quad \dfrac{\partial z}{\partial x} = \dfrac{\partial z}{\partial y} = 1$

$$\int_S\int f(x, y, z)\, dS = \int_{-1}^1 \int_{-\sqrt{1-x^2}}^{\sqrt{1-x^2}} \left[x^2 + y^2 + (x + y)^2\right]\sqrt{1 + 1^2 + 1^2}\, dy\, dx$$

$$= \sqrt{3}\int_{-1}^1 \int_{-\sqrt{1-x^2}}^{\sqrt{1-x^2}} \left[2x^2 + 2y^2 + 2xy\right] dy\, dx = \sqrt{3}\int_0^{2\pi} \int_0^1 \left(2r^2 + 2r\cos\theta\, r\sin\theta\right) r\, dr\, d\theta$$

$$= 2\sqrt{3}\int_0^{2\pi} \left[\frac{r^4}{4} + \frac{r^4}{4}\cos\theta\sin\theta\right]_0^1 d\theta = \frac{\sqrt{3}}{2}\int_0^{2\pi} (1 + \cos\theta\sin\theta)\, d\theta = \frac{\sqrt{3}}{2}\left[\theta + \frac{\sin^2\theta}{2}\right]_0^{2\pi} = \sqrt{3}\pi$$

18. $f(x, y, z) = \dfrac{xy}{z}$

$S: z = x^2 + y^2,\ 4 \le x^2 + y^2 \le 16$

$$\int_S\int f(x, y, z)\,dS = \int_S\int \frac{xy}{x^2 + y^2}\sqrt{1 + 4x^2 + 4y^2}\,dy\,dx = \int_0^{2\pi}\int_2^4 \frac{r^2 \sin\theta\cos\theta}{r^2}\sqrt{1 + 4r^2}\,r\,dr\,d\theta$$

$$= \int_0^{2\pi}\int_2^4 r\sqrt{1 + 4r^2}\sin\theta\cos\theta\,dr\,d\theta = \int_0^{2\pi}\left[\frac{1}{12}\left(1 + 4r^2\right)^{3/2}\right]_2^4 \sin\theta\cos\theta\,d\theta$$

$$= \left[\frac{65\sqrt{65} - 17\sqrt{17}}{12}\left(\frac{\sin^2\theta}{2}\right)\right]_0^{2\pi} = 0$$

19. $f(x, y, z) = \sqrt{x^2 + y^2 + z^2}$

$S: z = \sqrt{x^2 + y^2},\ x^2 + y^2 \le 4$

$$\int_S\int f(x, y, z)\,dS = \int_{-2}^{2}\int_{-\sqrt{4-x^2}}^{\sqrt{4-x^2}} \sqrt{x^2 + y^2 + \left(\sqrt{x^2 + y^2}\right)^2}\sqrt{1 + \left(\frac{x}{\sqrt{x^2 + y^2}}\right)^2 + \left(\frac{y}{\sqrt{x^2 + y^2}}\right)^2}\,dy\,dx$$

$$= \sqrt{2}\int_{-2}^{2}\int_{-\sqrt{4-x^2}}^{\sqrt{4-x^2}} \sqrt{x^2 + y^2}\sqrt{\frac{x^2 + y^2 + x^2 + y^2}{x^2 + y^2}}\,dy\,dx$$

$$= 2\int_{-2}^{2}\int_{-\sqrt{4-x^2}}^{\sqrt{4-x^2}} \sqrt{x^2 + y^2}\,dy\,dx = 2\int_0^{2\pi}\int_0^2 r^2\,dr\,d\theta = 2\int_0^{2\pi}\left[\frac{r^3}{3}\right]_0^2 d\theta = \left[\frac{16}{3}\theta\right]_0^{2\pi} = \frac{32\pi}{3}$$

20. $f(x, y, z) = \sqrt{x^2 + y^2 + z^2}$

$S: z = \sqrt{x^2 + y^2},\ (x - 1)^2 + y^2 \le 1$

$$\int_S\int f(x, y, z)\,dS = \int_S\int \sqrt{x^2 + y^2 + \left(\sqrt{x^2 + y^2}\right)^2}\sqrt{1 + \left(\frac{x}{\sqrt{x^2 + y^2}}\right)^2 + \left(\frac{y}{\sqrt{x^2 + y^2}}\right)^2}\,dy\,dx$$

$$= \int_S\int \sqrt{2\left(x^2 + y^2\right)}\sqrt{\frac{2\left(x^2 + y^2\right)}{x^2 + y^2}}\,dy\,dx = 2\int_S\int \sqrt{x^2 + y^2}\,dy\,dx = 2\int_0^{\pi}\int_0^{2\cos\theta} r^2\,dr\,d\theta$$

$$= \frac{16}{3}\int_0^{\pi}\cos^3\theta\,d\theta = \frac{16}{3}\int_0^{\pi}\left(1 - \sin^2\theta\right)\cos\theta\,d\theta = \left[\frac{16}{3}\left(\sin\theta - \frac{\sin^3\theta}{3}\right)\right]_0^{\pi} = 0$$

21. $f(x, y, z) = x^2 + y^2 + z^2$

$S: x^2 + y^2 = 9,\ 0 \le x \le 3,\ 0 \le y \le 3,\ 0 \le z \le 9$

Project the solid onto the yz-plane; $x = \sqrt{9 - y^2},\ 0 \le y \le 3,\ 0 \le z \le 9.$

$$\int_S\int f(x, y, z)\,dS = \int_0^3\int_0^9\left[\left(9 - y^2\right) + y^2 + z^2\right]\sqrt{1 + \left(\frac{-y}{\sqrt{9 - y^2}}\right)^2 + (0)^2}\,dz\,dy$$

$$= \int_0^3\int_0^9\left(9 + z^2\right)\frac{3}{\sqrt{9 - y^2}}\,dz\,dy = \int_0^3\left[\frac{3}{\sqrt{9 - y^2}}\left(9z + \frac{z^3}{3}\right)\right]_0^9 dy$$

$$= 324\int_0^3 \frac{3}{\sqrt{9 - y^2}}\,dy = \left[972\arcsin\left(\frac{y}{3}\right)\right]_0^3 = 972\left(\frac{\pi}{2} - 0\right) = 486\pi$$

22. $f(x, y, z) = x^2 + y^2 + z^2$

$S: x^2 + y^2 = 9,\ 0 \le x \le 3,\ 0 \le z \le x$

Project the solid onto the xz-plane; $y = \sqrt{9 - x^2}$.

$$\int_S\int f(x, y, z)\,dS = \int_0^3\int_0^x \left[x^2 + (9 - x^2) + z^2\right]\sqrt{1 + \left(\frac{-x}{\sqrt{9 - x^2}}\right)^2 + (0)^2}\,dz\,dx$$

$$= \int_0^3\int_0^x (9 + z^2)\frac{3}{\sqrt{9 - x^2}}\,dz\,dx = \int_0^3\left[\frac{3}{\sqrt{9 - x^2}}\left(9z + \frac{z^3}{3}\right)\right]_0^x dx$$

$$= \int_0^3 \frac{3}{\sqrt{9 - x^2}}\left(9x + \frac{x^3}{3}\right)dx = \int_0^3 27x(9 - x^2)^{-1/2}\,dx + \int_0^3 x^3(9 - x^2)^{-1/2}\,dx$$

Let $u = x^2$, $dv = x(9 - x^2)^{-1/2}\,dx$, then $du = 2x\,dx$, $v = -\sqrt{9 - x^2}$.

$$= \left[-27\sqrt{9 - x^2}\right]_0^3 + \left[\left[-x^2\sqrt{9 - x^2}\right]_0^3 + \int_0^3 2x\sqrt{9 - x^2}\,dx\right] = \left[81 - \frac{2}{3}(9 - x^2)^{3/2}\right]_0^3 = 81 + 18 = 99$$

23. $\mathbf{F}(x, y, z) = 3z\mathbf{i} - 4\mathbf{j} + y\mathbf{k}$

$S: z = 1 - x - y$ (first octant)

$G(x, y, z) = x + y + z - 1$

$\nabla G(x, y, z) = \mathbf{i} + \mathbf{j} + \mathbf{k}$

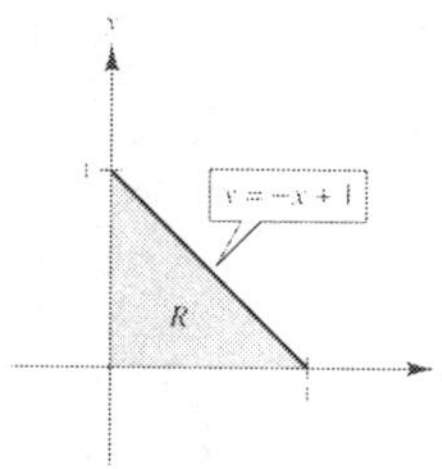

$$\int_S\int \mathbf{F}\cdot\mathbf{N}\,dS = \int_R\int \mathbf{F}\cdot\nabla G\,dA = \int_0^1\int_0^{1-x}(3z - 4 + y)\,dy\,dx$$

$$= \int_0^1\int_0^{1-x}\left[3(1 - x - y) - 4 + y\right]dy\,dx$$

$$= \int_0^1\int_0^{1-x}(-1 - 3x - 2y)\,dy\,dx = \int_0^1\left[-y - 3xy - y^2\right]_0^{1-x} dx$$

$$= -\int_0^1\left[(1 - x) + 3x(1 - x) + (1 - x)^2\right]dx = -\int_0^1 (2 - 2x^2)\,dx = -\tfrac{4}{3}$$

24. $\mathbf{F}(x, y, z) = x\mathbf{i} + y\mathbf{j}$

$S: z = 6 - 3x - 2y$, first octant

$G(x, y, z) = 3x + 2y + z - 6$

$\nabla G(x, y, z) = 3\mathbf{i} + 2\mathbf{j} + \mathbf{k}$

$$\int_S\int \mathbf{F}\cdot\mathbf{N}\,dS = \int_R\int \mathbf{F}\cdot\nabla G\,dA$$

$$= \int_0^2\int_0^{3-\frac{3}{2}x}(3x + 2y)\,dy\,dx$$

$$= \int_0^2\left[3xy + y^2\right]_0^{3-\frac{3}{2}x} dx$$

$$= \int_0^2\left[3x\left(3 - \frac{3}{2}x\right) + \left(3 - \frac{3}{2}x\right)^2\right]dx$$

$$= \int_0^2 \frac{-9}{4}(x^2 - 4)\,dx$$

$$= \frac{-9}{4}\left[\frac{x^3}{3} - 4x\right]_0^2 = \left(\frac{-9}{4}\right)\left(\frac{-16}{3}\right) = 12$$

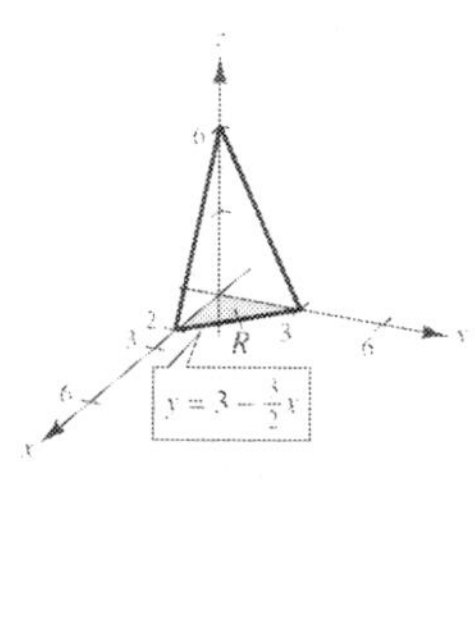

25. $\mathbf{F}(x, y, z) = x\mathbf{i} + y\mathbf{j} + z\mathbf{k}$

$S: z = 1 - x^2 - y^2, \quad z \ge 0$

$G(x, y, z) = x^2 + y^2 + z - 1$

$\nabla G(x, y, z) = 2x\mathbf{i} + 2y\mathbf{j} + \mathbf{k}$

$$\begin{aligned}\int_S\int \mathbf{F}\cdot\mathbf{N}\,dS &= \int_R\int \mathbf{F}\cdot\nabla G\,dA\\ &= \int_R\int\left(2x^2 + 2y^2 + z\right)dA\\ &= \int_R\int\left(2x^2 + 2y^2 + \left(1 - x^2 - y^2\right)\right)dA\\ &= \int_R\int\left(1 + x^2 + y^2\right)dA\\ &= \int_0^{2\pi}\int_0^1\left(r^2 + 1\right)r\,dr\,d\theta\\ &= \int_0^{2\pi}\left[\frac{r^4}{4} + \frac{r^2}{2}\right]_0^1 d\theta = \int_0^{2\pi}\frac{3}{4}\,d\theta = \frac{3\pi}{2}\end{aligned}$$

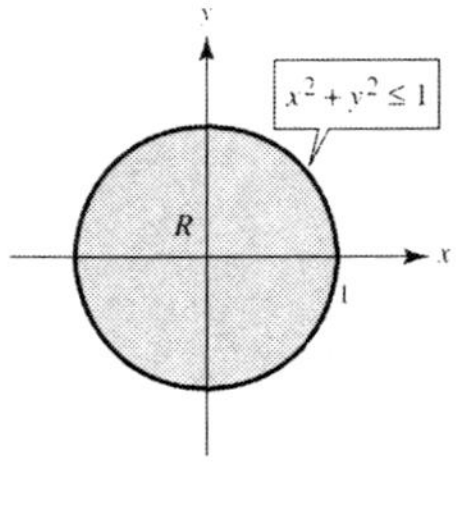

26. $\mathbf{F}(x, y, z) = x\mathbf{i} + y\mathbf{j} + z\mathbf{k}$

$S: x^2 + y^2 + z^2 = 36$ (first octant)

$z = \sqrt{36 - x^2 - y^2}$

$G(x, y, z) = z - \sqrt{36 - x^2 - y^2}$

$\nabla G(x, y, z) = \dfrac{x}{\sqrt{36 - x^2 - y^2}}\mathbf{i} + \dfrac{y}{\sqrt{36 - x^2 - y^2}}\mathbf{j} + \mathbf{k}$

$\mathbf{F}\cdot\nabla G = \dfrac{x^2}{\sqrt{36 - x^2 - y^2}} + \dfrac{y^2}{\sqrt{36 - x^2 - y^2}} + z = \dfrac{36}{\sqrt{36 - x^2 - y^2}}$

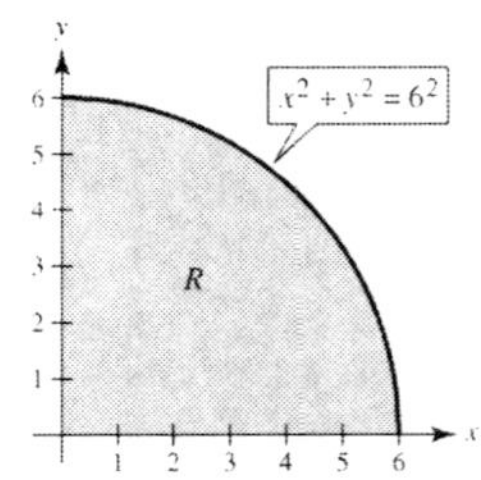

$$\int_S\int \mathbf{F}\cdot\mathbf{N}\,dS = \int_R\int \mathbf{F}\cdot\nabla G\,dA = \int_R\int\frac{36}{\sqrt{36 - x^2 - y^2}}\,dA = \int_0^{\pi/2}\int_0^6\frac{36}{\sqrt{36 - r^2}}\,r\,dr\,d\theta \quad \text{(improper)} = 108\pi$$

27. $\mathbf{F}(x, y, z) = 4\mathbf{i} - 3\mathbf{j} + 5\mathbf{k}$

$S: z = x^2 + y^2, x^2 + y^2 \le 4$

$G(x, y, z) = -x^2 - y^2 + z$

$\nabla G(x, y, z) = -2x\mathbf{i} - 2y\mathbf{j} + \mathbf{k}$

$$\begin{aligned}\int_S\int \mathbf{F}\cdot\mathbf{N}\,dS = \int_R\int \mathbf{F}\cdot\nabla G\,dA &= \int_R\int(-8x + 6y + 5)\,dA\\ &= \int_0^{2\pi}\int_0^2\left[-8r\cos\theta + 6r\sin\theta + 5\right]r\,dr\,d\theta\\ &= \int_0^{2\pi}\left[-\tfrac{8}{3}r^3\cos\theta + 2r^3\sin\theta + \tfrac{5}{2}r^2\right]_0^2 d\theta\\ &= \int_0^{2\pi}\left[-\tfrac{64}{3}\cos\theta + 16\sin\theta + 10\right]d\theta\\ &= \left[-\tfrac{64}{3}\sin\theta - 16\cos\theta + 10\theta\right]_0^{2\pi} = 20\pi\end{aligned}$$

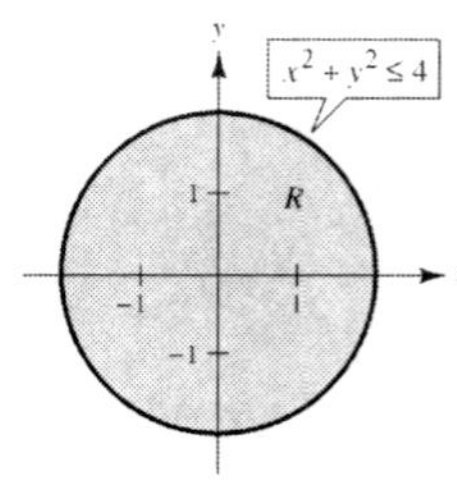

28. $\mathbf{F}(x, y, z) = x\mathbf{i} + y\mathbf{j} - 2z\mathbf{k}$

$S: z = \sqrt{a^2 - x^2 - y^2}$

$G(x, y, z) = z - \sqrt{a^2 - x^2 - y^2}$

$$\nabla G(x, y, z) = \frac{x}{\sqrt{a^2 - x^2 - y^2}}\mathbf{i} + \frac{y}{\sqrt{a^2 - x^2 - y^2}}\mathbf{j} + \mathbf{k}$$

$$\mathbf{F} \cdot \nabla G = \frac{x^2}{\sqrt{a^2 - x^2 - y^2}} + \frac{y^2}{\sqrt{a^2 - x^2 - y^2}} - 2\sqrt{a^2 - x^2 - y^2} = \frac{3x^2 + 3y^2 - 2a^2}{\sqrt{a^2 - x^2 - y^2}}$$

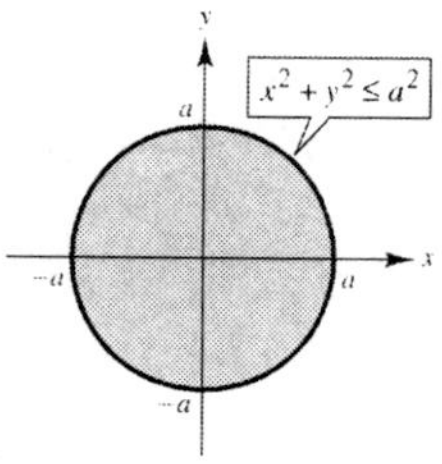

$$\int_S\int \mathbf{F} \cdot \mathbf{N}\, dS = \int_R\int \mathbf{F} \cdot \nabla G\, dA = \int_R\int \frac{3x^2 + 3y^2 - 2a^2}{\sqrt{a^2 - x^2 - y^2}}\, dA = \int_0^{2\pi}\int_0^a \frac{3r^2 - 2a^2}{\sqrt{a^2 - r^2}}\, r\, dr\, d\theta$$

$$= 3\int_0^{2\pi}\int_0^a \frac{r^3}{\sqrt{a^2 - r^2}}\, dr\, d\theta - 2a^2\int_0^{2\pi}\int_0^a \frac{r}{\sqrt{a^2 - r^2}}\, dr\, d\theta$$

$$= 3\left[\int_0^{2\pi}\left[-r^2\sqrt{a^2 - r^2} - \frac{2}{3}\left(a^2 - r^2\right)^{3/2}\right]_0^a d\theta\right] - 2a^2\int_0^{2\pi}\left[-\sqrt{a^2 - r^2}\right]_0^a d\theta$$

$$= 3\int_0^{2\pi}\frac{2}{3}a^3\, d\theta - 2a^2\int_0^{2\pi} a\, d\theta = 0$$

29. $\mathbf{F}(x, y, z) = (x + y)\mathbf{i} + y\mathbf{j} + z\mathbf{k}$

$S: z = 16 - x^2 - y^2, \quad z = 0$

$G(x, y, z) = z + x^2 + y^2 - 16$

$\nabla G(x, y, z) = 2x\mathbf{i} + 2y\mathbf{j} + \mathbf{k}$

$$\mathbf{F} \cdot \nabla G = 2x(x + y) + 2y^2 + z = 2x^2 + 2xy + 2y^2 + 16 - x^2 - y^2 = x^2 + y^2 + 2xy + 16$$

$$\int_S\int \mathbf{F} \cdot \mathbf{N}\, dS = \int_R\int \mathbf{F} \cdot \nabla G\, dA$$

$$= \int_0^{2\pi}\int_0^4 \left(r^2 + 2r^2\cos\theta\sin\theta + 16\right) r\, dr\, d\theta$$

$$= \int_0^{2\pi}\left[\frac{r^4}{4} + \frac{r^4}{2}\cos\theta\sin\theta + 8r^2\right]_0^4 d\theta = \int_0^{2\pi}\left[192 + 128\cos\theta\sin\theta\right] d\theta = \left[192 + 64\sin^2\theta\right]_0^{2\pi} = 384\pi$$

(The flux across the bottom $z = 0$ is 0.)

30. $\mathbf{F}(x, y, z) = 4xy\mathbf{i} + z^2\mathbf{j} + yz\mathbf{k}$

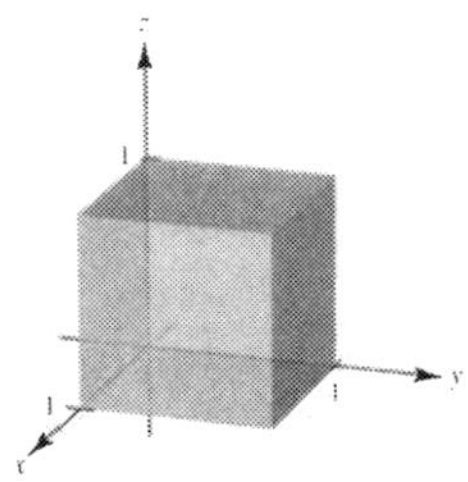

S: unit cube bounded by

$x = 0, x = 1, y = 0, y = 1, z = 0, z = 1$

S_1: The top of the cube

$\mathbf{N} = \mathbf{k}, z = 1$

$$\int_{S_1}\int \mathbf{F}\cdot\mathbf{N}\,dS = \int_0^1\int_0^1 y(1)\,dy\,dx = \frac{1}{2}$$

S_2: The bottom of the cube

$\mathbf{N} = -\mathbf{k}, z = 0$

$$\int_{S_2}\int \mathbf{F}\cdot\mathbf{N}\,dS = \int_0^1\int_0^1 -y(0)\,dy\,dx = 0$$

S_3: The front of the cube

$\mathbf{N} = \mathbf{i}, x = 1$

$$\int_{S_3}\int \mathbf{F}\cdot\mathbf{N}\,dS = \int_0^1\int_0^1 4(1)\,y\,dy\,dz = 2$$

S_4: The back of the cube

$\mathbf{N} = -\mathbf{i}, x = 0$

$$\int_{S_4}\int \mathbf{F}\cdot\mathbf{N}\,dS = \int_0^1\int_0^1 -4(0)\,y\,dy\,dx = 0$$

S_5: The right side of the cube

$\mathbf{N} = \mathbf{j}, y = 1$

$$\int_{S_5}\int \mathbf{F}\cdot\mathbf{N}\,dS = \int_0^1\int_0^1 z^2\,dz\,dx = \frac{1}{3}$$

S_6: The left side of the cube

$\mathbf{N} = -\mathbf{j}, y = 0$

$$\int_{S_6}\int \mathbf{F}\cdot\mathbf{N}\,dS = \int_0^1\int_0^1 -z^2\,dz\,dx = -\frac{1}{3}$$

So,

$$\int_S\int \mathbf{F}\cdot\mathbf{N}\,dS = \frac{1}{2} + 0 + 2 + 0 + \frac{1}{3} - \frac{1}{3} = \frac{5}{2}.$$

31. $\mathbf{E} = yz\mathbf{i} + xz\mathbf{j} + xy\mathbf{k}$

S: $z = \sqrt{1 - x^2 - y^2}$

$$\int_S\int \mathbf{E}\cdot\mathbf{N}\,dS = \int_R\int \mathbf{E}\cdot\left(-g_x(x, y)\mathbf{i} - g_y(x, y)\mathbf{j} + \mathbf{k}\right)dA$$

$$= \int_R\int (yz\mathbf{i} + xz\mathbf{j} + xy\mathbf{k})\cdot\left(\frac{x}{\sqrt{1 - x^2 - y^2}}\mathbf{i} + \frac{y}{\sqrt{1 - x^2 - y^2}}\mathbf{j} + \mathbf{k}\right)dA$$

$$= \int_R\int \left(\frac{2xyz}{\sqrt{1 - x^2 - y^2}} + xy\right)dA = \int_R\int 3xy\,dA = \int_{-1}^1\int_{-\sqrt{1-x^2}}^{\sqrt{1-x^2}} 3xy\,dy\,dx = 0$$

32. $\mathbf{E} = x\mathbf{i} + y\mathbf{j} + 2z\mathbf{k}$

$S: z = \sqrt{1 - x^2 - y^2} = g(x, y)$

$$\int_S\int \mathbf{E}\cdot\mathbf{N}\,dS = \int_R\int \mathbf{E}\cdot\left(-g_x(x, y)\mathbf{i} - g_y(x, y)\mathbf{j} + \mathbf{k}\right)dA$$

$$= \int_R\int (x\mathbf{i} + y\mathbf{j} + 2z\mathbf{k})\cdot\left(\frac{x}{\sqrt{1 - x^2 - y^2}}\mathbf{i} + \frac{y}{\sqrt{1 - x^2 - y^2}}\mathbf{j} + \mathbf{k}\right)dA$$

$$= \int_R\int\left(\frac{x^2}{\sqrt{1 - x^2 - y^2}} + \frac{y^2}{\sqrt{1 - x^2 - y^2}} + 2z\right)dA$$

$$= \int_R\int \frac{x^2 + y^2 + 2\left(1 - x^2 - y^2\right)}{\sqrt{1 - x^2 - y^2}}\,dA = \int_R\int \frac{2 - x^2 - y^2}{\sqrt{1 - x^2 - y^2}}\,dA = \int_0^{2\pi}\int_0^1 \frac{2 - r^2}{\sqrt{1 - r^2}}\,r\,dr\,d\theta = \frac{8\pi}{3}$$

33. $z = \sqrt{x^2 + y^2}, 0 \le z \le a$

$$m = \int_S\int k\,dS = k\int_R\int\sqrt{1 + \left(\frac{x}{\sqrt{x^2 + y^2}}\right)^2 + \left(\frac{y}{\sqrt{x^2 + y^2}}\right)^2}\,dA = k\int_R\int\sqrt{2}\,dA = \sqrt{2}\,k\pi a^2$$

$$I_z = \int_S\int k\left(x^2 + y^2\right)dS = \int_R\int k\left(x^2 + y^2\right)\sqrt{2}\,dA = \sqrt{2}k\int_0^{2\pi}\int_0^a r^3\,dr\,d\theta = \frac{\sqrt{2}ka^4}{4}(2\pi) = \frac{\sqrt{2}k\pi a^4}{2} = \frac{a^2}{2}\left(\sqrt{2}k\pi a^2\right) = \frac{a^2 m}{2}$$

34. $x^2 + y^2 + z^2 = a^2$

$z = \pm\sqrt{a^2 - x^2 - y^2}$

$$m = 2\int_S\int k\,dS = 2k\int_R\int\sqrt{1 + \left(\frac{-x}{\sqrt{a^2 - x^2 - y^2}}\right)^2 + \left(\frac{-y}{\sqrt{a^2 - x^2 - y^2}}\right)^2}\,dA$$

$$= 2k\int_R\int\frac{a}{\sqrt{a^2 - x^2 - y^2}}\,dA = 2ka\int_0^{2\pi}\int_0^a\frac{r}{\sqrt{a^2 - r^2}}\,dr\,d\theta = 2ka\left[-\sqrt{a^2 - r^2}\right]_0^a(2\pi) = 4\pi ka^2$$

$$I_z = 2\int_S\int k\left(x^2 + y^2\right)dS = 2k\int_R\int\left(x^2 + y^2\right)\frac{a}{\sqrt{a^2 - x^2 - y^2}}\,dA = 2ka\int_0^{2\pi}\int_0^a\frac{r^3}{\sqrt{a^2 - r^2}}\,dr\,d\theta$$ (use integration by parts)

Let $u = r^2, dv = r\left(a^2 - r^2\right)^{-1/2}dr, du = 2r\,dr, v = -\sqrt{a^2 - r^2}$.

$$= 2ka\left[-r^2\sqrt{a^2 - r^2} - \frac{2}{3}\left(a^2 - r^2\right)^{3/2}\right]_0^a(2\pi) = 2ka\left(\frac{2}{3}a^3\right)(2\pi) = \frac{2}{3}a^2\left(4\pi ka^2\right) = \frac{2}{3}a^2 m$$

35. $x^2 + y^2 = a^2, 0 \le z \le h$

$\rho(x, y, z) = 1$

$y = \pm\sqrt{a^2 - x^2}$

Project the solid onto the xz-plane.

$$I_z = 4\int_S\int\left(x^2 + y^2\right)(1)\,dS = 4\int_0^h\int_0^a\left[x^2 + \left(a^2 - x^2\right)\right]\sqrt{1 + \left(\frac{-x}{\sqrt{a^2 - x^2}}\right)^2 + (0)^2}\,dx\,dz$$

$$= 4a^3\int_0^h\int_0^a\frac{1}{\sqrt{a^2 - x^2}}\,dx\,dz = 4a^3\int_0^h\left[\arcsin\frac{x}{a}\right]_0^a dz = 4a^3\left(\frac{\pi}{2}\right)(h) = 2\pi a^3 h$$

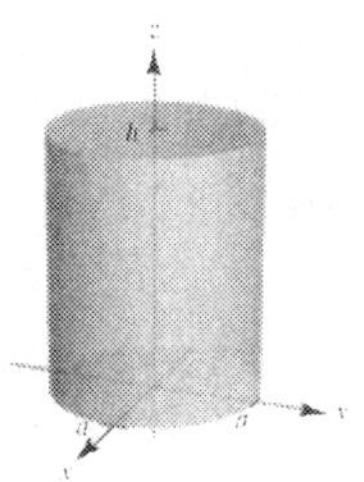

36. $z = x^2 + y^2, 0 \le z \le h$

Project the solid onto the xy-plane.

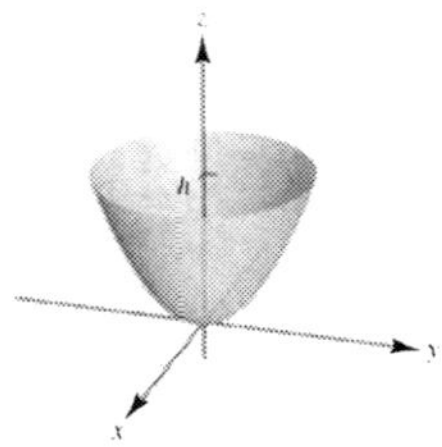

$$I_z = \int_S\int(x^2 + y^2)(1)\,dS = \int_{-\sqrt{h}}^{\sqrt{h}}\int_{-\sqrt{h-x^2}}^{\sqrt{h-x^2}}(x^2 + y^2)\sqrt{1 + 4x^2 + 4y^2}\,dy\,dx$$

$$= \int_0^{2\pi}\int_0^{\sqrt{h}} r^2\sqrt{1 + 4r^2}\,r\,dr\,d\theta = 2\pi\left[\frac{h}{12}(1 + 4h)^{3/2} - \frac{1}{120}(1 + 4h)^{5/2}\right] + \frac{2\pi}{120}$$

$$= \frac{(1 + 4h)^{3/2}\pi}{60}\left[10h - (1 + 4h)\right] + \frac{\pi}{60} = \frac{\pi}{60}\left[(1 + 4h)^{3/2}(6h - 1) + 1\right]$$

37. $S\colon z = 16 - x^2 - y^2, z \ge 0$

$\mathbf{F}(x, y, z) = 0.5z\mathbf{k}$

$$\int_S\int\rho\mathbf{F}\cdot\mathbf{N}\,dS = \int_R\int\rho\mathbf{F}\cdot\left(-g_x(x, y)\mathbf{i} - g_y(x, y)\mathbf{j} + \mathbf{k}\right)dA = \int_R\int 0.5\rho z\mathbf{k}\cdot(2x\mathbf{i} + 2y\mathbf{j} + \mathbf{k})\,dA$$

$$= \int_R\int 0.5\rho z\,dA = \int_R\int 0.5\rho\left(16 - x^2 - y^2\right)dA$$

$$= 0.5\rho\int_0^{2\pi}\int_0^4\left(16 - r^2\right)r\,dr\,d\theta = 0.5\rho\int_0^{2\pi} 64\,d\theta = 64\pi\rho$$

38. $S\colon z = \sqrt{16 - x^2 - y^2}$

$\mathbf{F}(x, y, z) = 0.5z\mathbf{k}$

$$\int_S\int\rho\mathbf{F}\cdot\mathbf{N}\,dS = \int_R\int\rho\mathbf{F}\cdot\left(-g_x(x, y)\mathbf{i} - g_y(x, y)\mathbf{j} + \mathbf{k}\right)dA$$

$$= \int_R\int 0.5\rho z\mathbf{k}\cdot\left[\frac{x}{\sqrt{16 - x^2 - y^2}}\mathbf{i} + \frac{y}{\sqrt{16 - x^2 - y^2}}\mathbf{j} + \mathbf{k}\right]dA$$

$$= \int_R\int 0.5\,\rho z\,dA = \int_R\int 0.5\rho\sqrt{16 - x^2 - y^2}\,dA$$

$$= 0.5\rho\int_0^{2\pi}\int_0^4\sqrt{16 - r^2}\,r\,dr\,d\theta = 0.5\rho\int_0^{2\pi}\frac{64}{3}\,d\theta = \frac{64\pi\rho}{3}$$

39. The surface integral of f over a surface S, where S is given by $z = g(x, y)$, is defined as

$$\int_S\int f(x, y, z)\,dS = \lim_{\|\Delta\|\to 0}\sum_{i=1}^{n} f(x_i, y_i, z_i)\Delta S_i.$$ (page 1112)

See Theorem 15.10, page 1094.

40. A surface is orientable if a unit normal vector N can be defined at every nonboundary point of S in such a way that the normal vectors vary continuously over the surface S.

41. See the definition, page 1100.

See Theorem 15.11, page 1100.

42. Orientable

43. (a)

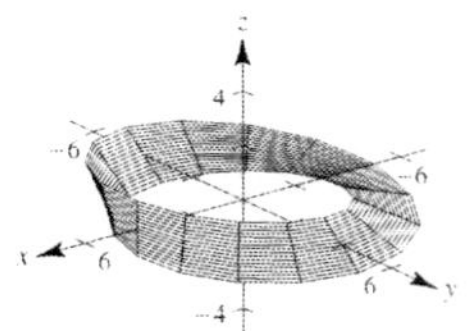

(b) If a normal vector at a point P on the surface is moved around the Möbius strip once, it will point in the opposite direction.

(c) $\mathbf{r}(u, 0) = 4\cos(2u)\mathbf{i} + 4\sin(2u)\mathbf{j}$

This is circle.

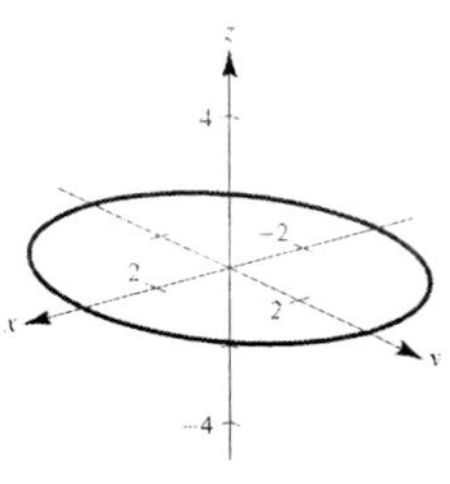

(d) (construction)

(e) You obtain a strip with a double twist and twice as long as the original Möbius strip.

44. (a) $\mathbf{r}_u = \mathbf{i} + \mathbf{j} + 2u\mathbf{k}$

$\mathbf{r}_v = 2v\mathbf{i} - \mathbf{j}$

$$\mathbf{r}_u \times \mathbf{r}_v = \begin{vmatrix} \mathbf{i} & \mathbf{j} & \mathbf{k} \\ 1 & 1 & 2u \\ 2v & -1 & 0 \end{vmatrix} = 2u\mathbf{i} + 4uv\mathbf{j} - (1 + 2v)\mathbf{k}$$

$\mathbf{r}_u \times \mathbf{r}_v$ is a normal vector to the surface.

(b) $\mathbf{F}(u, v) = u^2\mathbf{i} + (u + v^2)\mathbf{j} + (u - v)\mathbf{k}$

$$\mathbf{F} \cdot (\mathbf{r}_u \times \mathbf{r}_v) = 2u^3 + 4uv(u + v^2) - (u - v)(1 + 2v) = 2u^3 + 4u^2v + 4uv^3 + v - u + 2v^2 - 2uv$$

(c) $$\left.\begin{aligned} x &= 3 = u + v^2 \\ y &= 1 = u - v \\ z &= 4 = u^2 \end{aligned}\right\} \quad \begin{aligned} u &= 2\ (u = -2 \text{ not in domain}) \\ v &= 1 \end{aligned}$$

(d) Calculate $\mathbf{F} \cdot \dfrac{\mathbf{r}_u \times \mathbf{r}_v}{\|\mathbf{r}_u \times \mathbf{r}_v\|}$ at P.

$\mathbf{F}(3, 1, 4) = 4\mathbf{i} + 3\mathbf{j} + \mathbf{k}$

$(\mathbf{r}_u \times \mathbf{r}_v)(2, 1) = 4\mathbf{i} + 8\mathbf{j} - 3\mathbf{k}$

$\|\mathbf{r}_u \times \mathbf{r}_v\| = \sqrt{89}$

$$\mathbf{F} \cdot \frac{\mathbf{r}_u \times \mathbf{r}_v}{\|\mathbf{r}_u \times \mathbf{r}_v\|} = \frac{1}{\sqrt{89}}(16 + 24 - 3) = \frac{37}{\sqrt{89}} = \frac{37\sqrt{89}}{89}$$

(e) $$\int_S\!\!\int \mathbf{F} \cdot \mathbf{N}\, dS = \int_R\!\!\int \mathbf{F} \cdot (\mathbf{r}_u \times \mathbf{r}_v)\, dA$$

$$= \int_{-1}^{1}\int_0^2 \left(2u^3 + 4u^2v + 4uv^3 + v - u + 2v^2 - 2uv\right) du\, dv = \int_{-1}^{1}\left(8v^3 + 4v^2 + \frac{26v}{3} + 6\right) dv = \frac{44}{3}$$

Section 15.7 Divergence Theorem

1. Surface Integral: There are six surfaces to the cube, each with $dS = \sqrt{1}\, dA$.

$z = 0,\quad \mathbf{N} = -\mathbf{k},\quad \mathbf{F} \cdot \mathbf{N} = -z^2,\quad \displaystyle\int_{S_1}\!\!\int 0\, dA = 0$

$z = a,\quad \mathbf{N} = \mathbf{k},\quad \mathbf{F} \cdot \mathbf{N} = z^2,\quad \displaystyle\int_{S_2}\!\!\int a^2\, dA = \int_0^a\int_0^a a^2\, dx\, dy = a^4$

$x = 0,\quad \mathbf{N} = -\mathbf{i},\quad \mathbf{F} \cdot \mathbf{N} = -2x,\quad \displaystyle\int_{S_3}\!\!\int 0\, dA = 0$

$x = a,\quad \mathbf{N} = \mathbf{i},\quad \mathbf{F} \cdot \mathbf{N} = 2x,\quad \displaystyle\int_{S_4}\!\!\int 2a\, dy\, dz = \int_0^a\int_0^a 2a\, dy\, dz = 2a^3$

$y = 0,\quad \mathbf{N} = -\mathbf{j},\quad \mathbf{F} \cdot \mathbf{N} = 2y,\quad \displaystyle\int_{S_5}\!\!\int 0\, dA = 0$

$y = a,\quad \mathbf{N} = \mathbf{j},\quad \mathbf{F} \cdot \mathbf{N} = -2y,\quad \displaystyle\int_{S_6}\!\!\int -2a\, dA = \int_0^a\int_0^a -2a\, dz\, dx = -2a^3$

So, $\displaystyle\int_S\!\!\int \mathbf{F} \cdot \mathbf{N}\, dS = a^4 + 2a^3 - 2a^3 = a^4.$

Divergence Theorem: Because div $\mathbf{F} = 2z$, the Divergence Theorem yields

$$\iiint_Q \text{div } \mathbf{F}\, dV = \int_0^a\int_0^a\int_0^a 2z\, dz\, dy\, dx = \int_0^a\int_0^a a^2\, dy\, dx = a^4.$$

2. Surface Integral: There are three surfaces to the cylinder.

Bottom: $z = 0$, $\mathbf{N} = -\mathbf{k}$, $\mathbf{F} \cdot \mathbf{N} = -z^2$

$$\int_{S_1}\int 0\, dS = 0$$

Top: $z = h$, $\mathbf{N} = \mathbf{k}$, $\mathbf{F} \cdot \mathbf{N} = z^2$

$$\int_{S_2}\int h^2\, dS = h^2 \left(\text{Area of circle}\right) = 4\pi h^2$$

Side: $\mathbf{r}(u,v) = 2\cos u\mathbf{i} + 2\sin u\mathbf{j} + v\mathbf{k}, 0 \le u \le 2\pi, 0 \le v \le h$

$\mathbf{r}_u = -2\sin u\mathbf{i} + 2\cos u\mathbf{j}, r_v = \mathbf{k}$

$\mathbf{r}_u \times \mathbf{r}_v = 2\cos u\mathbf{i} + 2\sin u\mathbf{j}$

$\mathbf{F} \cdot (\mathbf{r}_u \times \mathbf{r}_v) = 8\cos^2 u - 8\sin^2 u$

$$\int_{S_3}\int \mathbf{F} \cdot \mathbf{N}\, dS = \int_0^h \int_0^{2\pi} \left(8\cos^2 u - 8\sin^2 u\right) du\, dv = 0$$

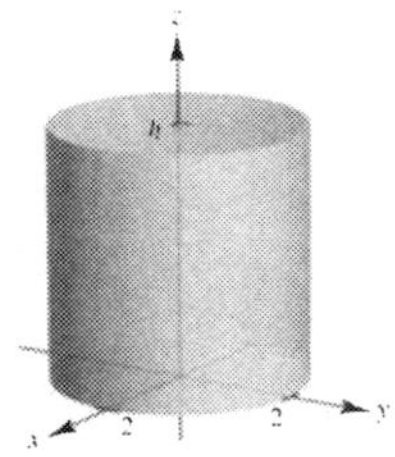

So, $\int_S\int \mathbf{F} \cdot \mathbf{N}\, dS = 0 + 4\pi h^2 + 0 = 4\pi h^2$.

Divergence Theorem: div $\mathbf{F} = 2 - 2 + 2z = 2z$

$$\iiint_Q 2z\, dV = \int_0^{2\pi}\int_0^2\int_0^h 2zr\, dz\, dr\, d\theta = 4\pi h^2.$$

3. Surface Integral: There are four surfaces to this solid.

$z = 0$, $\mathbf{N} = -\mathbf{k}$, $\mathbf{F} \cdot \mathbf{N} = -z$

$$\int_{S_1}\int 0\, dS = 0$$

$y = 0$, $\mathbf{N} = -\mathbf{j}$, $\mathbf{F} \cdot \mathbf{N} = 2y - z$, $dS = dA = dx\, dz$

$$\int_{S_2}\int -z\, dS = \int_0^6\int_0^{6-z} -z\, dx\, dz = \int_0^6 \left(z^2 - 6z\right) dz = -36$$

$x = 0$, $\mathbf{N} = -\mathbf{i}$, $\mathbf{F} \cdot \mathbf{N} = y - 2x$, $dS = dA = dz\, dy$

$$\int_{S_3}\int y\, dS = \int_0^3\int_0^{6-2y} y\, dz\, dy = \int_0^3 \left(6y - 2y^2\right) dy = 9$$

$x + 2y + z = 6$, $\mathbf{N} = \dfrac{\mathbf{i} + 2\mathbf{j} + \mathbf{k}}{\sqrt{6}}$, $\mathbf{F} \cdot \mathbf{N} = \dfrac{2x - 5y + 3z}{\sqrt{6}}$, $dS = \sqrt{6}\, dA$

$$\int_{S_4}\int \left(2x - 5y + 3z\right) dz\, dy = \int_0^3\int_0^{6-2y} (18 - x - 11y)\, dx\, dy = \int_0^3 \left(90 - 90y + 20y^2\right) dy = 45$$

So, $\int_S\int \mathbf{F} \cdot \mathbf{N}\, dS = 0 - 36 + 9 + 45 = 18$.

Divergence Theorem: Because div $\mathbf{F} = 1$, you have

$$\iiint_Q dV = \left(\text{Volume of solid}\right) = \frac{1}{3}\left(\text{Area of base}\right) \times \left(\text{Height}\right) = \frac{1}{3}(9)(6) = 18.$$

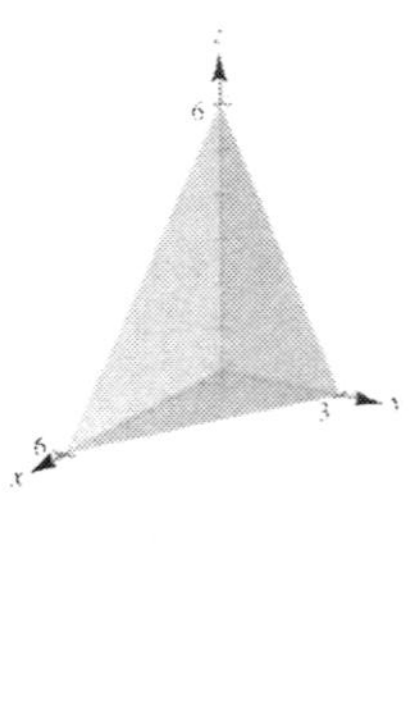

4. $\mathbf{F}(x,y,z) = xy\mathbf{i} + z\mathbf{j} + (x + y)\mathbf{k}$

S: surface bounded by the planes $y = 4$, $z = 4 - x$ and the coordinate planes

Surface Integral: There are five surfaces to this solid.

$z = 0$, $\mathbf{N} = -\mathbf{k}$, $\mathbf{F} \cdot \mathbf{N} = -(x + y)$

$$\int_{S_1}\int -(x + y)\,dS = \int_0^4\int_0^4 -(x + y)\,dy\,dx = -\int_0^4 (4x + 8)\,dx = -64$$

$y = 0$, $\mathbf{N} = -\mathbf{j}$, $\mathbf{F} \cdot \mathbf{N} = -z$

$$\int_{S_2}\int -z\,dS = \int_0^4\int_0^{4-x} -z\,dz\,dx = -\int_0^4 \frac{(4 - x)^2}{2}\,dx = -\frac{32}{3}$$

$y = 4$, $\mathbf{N} = \mathbf{j}$, $\mathbf{F} \cdot \mathbf{N} = z$

$$\int_{S_3}\int z\,dS = \int_0^4\int_0^{4-x} z\,dz\,dx = \int_0^4 \frac{(4 - x)^2}{2}\,dx = \frac{32}{3}$$

$x = 0$, $\mathbf{N} = -\mathbf{i}$, $\mathbf{F} \cdot \mathbf{N} = -xy$

$$\int_{S_4}\int -xy\,dS = \int_0^4\int_0^4 0\,dS = 0$$

$x + z = 4$, $\mathbf{N} = \dfrac{\mathbf{i} + \mathbf{k}}{\sqrt{2}}$, $\mathbf{F} \cdot \mathbf{N} = \dfrac{1}{\sqrt{2}}[xy + x + y]$, $dS = \sqrt{2}\,dA$

$$\int_{S_5}\int \frac{1}{\sqrt{2}}[xy + x + y]\sqrt{2}\,dA = \int_0^4\int_0^4 (xy + x + y)\,dy\,dx = 128$$

So, $\displaystyle\int_S\int \mathbf{F} \cdot \mathbf{N}\,dS = -64 - \frac{32}{3} + \frac{32}{3} + 0 + 128 = 64.$

Divergence Theorem: Because div $\mathbf{F} = y$, you have

$$\iiint_Q \text{div}\,\mathbf{F}\,dV = \int_0^4\int_0^4\int_0^{4-x} y\,dz\,dy\,dx = 64.$$

5. $F(x, y, z) = xz\mathbf{i} + yz\mathbf{j} + 2z^2\mathbf{k}$

Surface Integral: There are two surfaces.

Bottom: $z = 0$, $\mathbf{N} = -\mathbf{k}$, $\mathbf{F} \cdot \mathbf{N} = -2z^2$

$$\int_{S_1}\int \mathbf{F} \cdot \mathbf{N}\,dS = \int_R\int -2z^2\,dA = \iint 0\,dA = 0$$

Side: Outward unit normal is

$$\mathbf{N} = \frac{2x\mathbf{i} + 2y\mathbf{j} + \mathbf{k}}{\sqrt{4x^2 + 4y^2 + 1}}$$

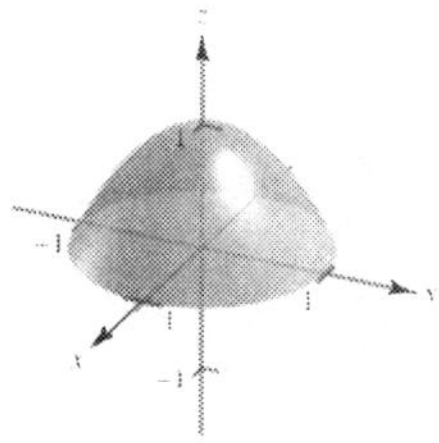

$$\mathbf{F} \cdot \mathbf{N} = \frac{1}{\sqrt{4x^2 + 4y^2 + 1}}\left[2x^2z + 2y^2z + 2z^2\right]$$

$$\int_{S_2}\int \mathbf{F} \cdot \mathbf{N}\,dS = \int_{S_2}\int \left[2(x^2 + y^2)z + 2z^2\right] dA$$

$$= \int_0^{2\pi}\int_0^1 \left[2r^2(1 - r^2) + 2(1 - r^2)^2\right] r\,dr\,d\theta = \int_0^{2\pi}\int_0^1 (2r - 2r^3)\,dr\,d\theta = \int_0^{2\pi} \frac{1}{2}\,d\theta = \pi$$

Divergence Theorem: div $\mathbf{F} = z + z + 4z = 6z$

$$\iiint_Q \text{div}\,\mathbf{F}\,dV = \int_0^{2\pi}\int_0^1\int_0^{1-r^2} 6z\,r\,dz\,dr\,d\theta$$

$$= \int_0^{2\pi}\int_0^1 3(1 - r^2)^2\,r\,dr\,d\theta = \int_0^{2\pi}\int_0^1 (3 - 6r^2 + 3r^4)\,r\,dr\,d\theta = \int_0^{2\pi}\left[\frac{3}{2} - \frac{3}{2} + \frac{1}{2}\right] d\theta = \pi$$

6. $\mathbf{F}(x, y, z) = xy^2\mathbf{i} + yx^2\mathbf{j} + e\mathbf{k}$

S: Surface bounded by $z = \sqrt{x^2 + y^2}$ and $z = 4$

Surface Integral: There are two surfaces.

Top: $z = 4$, $\mathbf{N} = \mathbf{k}$, $\mathbf{F} \cdot \mathbf{N} = e$

$$\int_{S_1}\int \mathbf{F} \cdot \mathbf{N}\, dS = (\text{area circle})\, e = 16\pi e$$

Side: $z = g(x, y) = \sqrt{x^2 + y^2}$, $g_x = \dfrac{x}{\sqrt{x^2 + y^2}}$, $g_y = \dfrac{y}{\sqrt{x^2 + y^2}}$

$$\int_{S_2}\int \mathbf{F} \cdot \mathbf{N}\, dS = \int_{S_2}\int \left[\frac{x^2y^2}{\sqrt{x^2 + y^2}} + \frac{x^2y^2}{\sqrt{x^2 + y^2}} - e\right] dA = \int_0^{2\pi}\int_0^4 \left(\frac{2r^2\cos^2\theta\, r^2\sin^2\theta}{r} - e\right) r\, dr\, d\theta$$

$$= \int_0^{2\pi}\left[\frac{2048\sin^2\theta\cos^2\theta}{5} - 8e\right] d\theta = \left(\frac{512}{5} - 16e\right)\pi$$

So, $\displaystyle\int_S\int \mathbf{F} \cdot \mathbf{N}\, dS = \frac{512}{5}\pi.$

Divergence Theorem: $\text{div}\, \mathbf{F} = y^2 + x^2$

$$\iiint_Q \text{div}\, \mathbf{F}\, dV = \int_0^{2\pi}\int_0^4\int_r^4 (r^2)\, r\, dz\, dr\, d\theta = \int_0^{2\pi}\int_0^4 (4r^3 - r^4)\, dr\, dz = \int_0^{2\pi}\left[r^4 - \frac{r^5}{5}\right]_0^4 dz = \int_0^{2\pi} \frac{256}{5}\, dz = \frac{512\pi}{5}$$

7. Because $\text{div}\, \mathbf{F} = 2x + 2y + 2z$, you have

$$\iiint_Q \text{div}\, \mathbf{F}\, dV = \int_0^a\int_0^a\int_0^a (2x + 2y + 2z)\, dz\, dy\, dx$$

$$= \int_0^a\int_0^a (2ax + 2ay + a^2)\, dy\, dx = \int_0^a (2a^2x + 2a^3)\, dx = \left[a^2x^2 + 2a^3x\right]_0^a = 3a^4.$$

8. Because $\text{div}\, \mathbf{F} = 2xz^2 - 2 + 3xy$, you have

$$\iiint_Q \text{div}\, \mathbf{F}\, dV = \int_0^a\int_0^a\int_0^a (2xz^2 - 2 + 3xy)\, dz\, dy\, dx = \int_0^a\int_0^a \left(\tfrac{2}{3}xa^3 - 2a + 3xya\right) dy\, dx$$

$$= \int_0^a \left(\tfrac{2}{3}xa^4 - 2a^2 + \tfrac{3}{2}xa^3\right) dx = \tfrac{1}{3}a^6 - 2a^3 + \tfrac{3}{4}a^5.$$

9. Because $\text{div}\, \mathbf{F} = 2x - 2x + 2xyz = 2xyz$,

$$\iiint_Q \text{div}\, \mathbf{F}\, dV = \iiint_Q 2xyz\, dV = \int_0^a\int_0^{2\pi}\int_0^{\pi/2} 2(\rho\sin\phi\cos\theta)(\rho\sin\phi\sin\theta)(\rho\cos\phi)\rho^2\sin\phi\, d\phi\, d\theta\, d\rho$$

$$= \int_0^a\int_0^{2\pi}\int_0^{\pi/2} 2\rho^5(\sin\theta\cos\theta)(\sin^3\phi\cos\phi)\, d\phi\, d\theta\, d\rho$$

$$= \int_0^a\int_0^{2\pi} \frac{1}{2}\rho^5 \sin\theta\cos\theta\, d\theta\, d\rho = \int_0^a \left[\left(\frac{\rho^5}{2}\right)\frac{\sin^2\theta}{2}\right]_0^{2\pi} d\rho = 0.$$

10. Because $\text{div}\, \mathbf{F} = y + z - y = z$, you have

$$\iiint_Q \text{div}\, \mathbf{F}\, dV = \int_{-a}^a\int_{-\sqrt{a^2-x^2}}^{\sqrt{a^2-x^2}}\int_0^{\sqrt{a^2-x^2-y^2}} z\, dz\, dy\, dx = \int_0^{2\pi}\int_0^a\int_0^{\sqrt{a^2-r^2}} zr\, dz\, dr\, d\theta$$

$$= \int_0^{2\pi}\int_0^a \left[\frac{a^2r}{2} - \frac{r^3}{2}\right] dr\, d\theta = \int_0^{2\pi}\left[\frac{a^2r^2}{4} - \frac{r^4}{8}\right]_0^a d\theta = \int_0^{2\pi} \frac{a^4}{8}\, d\theta = \frac{\pi a^4}{4}.$$

11. Because div $\mathbf{F} = 3$, you have

$$\iiint_Q 3\,dV = 3\,(\text{Volume of Sphere}) = 3\left[\tfrac{4}{3}\pi\left(3^3\right)\right] = 108\pi.$$

12. Because div $\mathbf{F} = xz$, you have

$$\iiint_Q xz\,dV = \int_0^5\int_{-2}^2\int_{-\sqrt{4-y^2}}^{\sqrt{4-y^2}} xz\,dx\,dy\,dz = \int_0^5\int_{-2}^2\left[\frac{2x^2}{2}\right]_{-\sqrt{4-y^2}}^{\sqrt{4-y^2}} dy\,dz = \int_0^5\int_{-2}^2 0\,dy\,dz = 0.$$

13. Because div $\mathbf{F} = 1 + 2y - 1 = 2y$, you have

$$\iiint_Q 2y\,dV = \int_0^7\int_{-5}^5\int_{-\sqrt{25-y^2}}^{\sqrt{25-y^2}} 2y\,dx\,dy\,dz = \int_0^7\int_{-5}^5 4y\sqrt{25-y^2}\,dy\,dz = \int_0^7\left[\frac{-4}{3}\left(25-y^2\right)^{3/2}\right]_{-5}^5 dz = 0.$$

14. Because div $\mathbf{F} = y^2 + x^2 + e^z$, you have

$$\begin{aligned}\iiint_Q \left(x^2 + y^2 + e^z\right) dV &= \int_0^{16}\int_{-\sqrt{256-x^2}}^{\sqrt{256-x^2}}\int_{(1/2)\sqrt{x^2+y^2}}^{8}\left(x^2 + y^2 + e^z\right) dz\,dy\,dx\\ &= \int_0^{2\pi}\int_0^{16}\int_{r/2}^{8}\left(r^2 + e^z\right) r\,dz\,dr\,d\theta = \int_0^{2\pi}\int_0^{16}\left(8r^3 + re^8 - \tfrac{1}{2}r^4 - re^{r/2}\right) dr\,d\theta\\ &= \int_0^{2\pi}\left(\tfrac{131{,}052}{5} + 100e^8\right) d\theta = \tfrac{262{,}104}{5}\pi + 200e^8\pi.\end{aligned}$$

15. Because div $\mathbf{F} = e^z + e^z + e^z = 3e^z$, you have

$$\iiint_Q 3e^z\,dV = \int_0^6\int_0^4\int_0^{4-y} 3e^z\,dz\,dy\,dx = \int_0^6\int_0^4 3\left[e^{4-y} - 1\right] dy\,dx = \int_0^6 3\left(e^4 - 5\right) dx = 18\left(e^4 - 5\right).$$

16. div $\mathbf{F} = y + 4 + x$. Use spherical coordinates.

$$\begin{aligned}\iiint_Q (y + 4 + x)\,dV &= \int_0^4\int_0^\pi\int_0^{2\pi}(\rho\sin\phi\sin\theta + \rho\sin\phi\cos\theta + 4)\,\rho^2\sin\phi\,d\theta\,d\phi\,d\rho\\ &= \int_0^4\int_0^\pi\int_0^{2\pi}\left(\rho^3\sin^2\phi\sin\theta + \rho^3\sin^2\phi\cos\theta + 4\rho^2\sin\phi\right) d\theta\,d\phi\,d\rho\\ &= \int_0^4\int_0^\pi 8\pi\rho^2\sin\phi\,d\phi\,d\rho = \int_0^4 16\pi\rho^2\,d\rho = \frac{1024\pi}{3}\end{aligned}$$

17. Using the Divergence Theorem, you have

$$\int_S\int \textbf{curl F}\cdot\mathbf{N}\,dS = \iiint_Q \text{div}\,(\textbf{curl F})\,dV$$

$$\textbf{curl F}(x,y,z) = \begin{vmatrix}\mathbf{i} & \mathbf{j} & \mathbf{k}\\ \dfrac{\partial}{\partial x} & \dfrac{\partial}{\partial y} & \dfrac{\partial}{\partial z}\\ 4xy + z^2 & 2x^2 + 6yz & 2xz\end{vmatrix} = -6y\mathbf{i} - (2z - 2z)\mathbf{j} + (4x - 4x)\mathbf{k} = -6y\mathbf{i}$$

$$\text{div}\,(\textbf{curl F}) = 0.$$

So, $\displaystyle\iiint_Q \text{div}\,(\textbf{curl F})\,dV = 0.$

18. Using the Divergence Theorem, you have

$$\int_S\int \mathbf{curl\ F}\cdot\mathbf{N}\,dS = \iiint_Q \text{div}\,(\mathbf{curl\ F})\,dV$$

$$\mathbf{curl\ F}(x, y, z) = \begin{vmatrix} \mathbf{i} & \mathbf{j} & \mathbf{k} \\ \dfrac{\partial}{\partial x} & \dfrac{\partial}{\partial y} & \dfrac{\partial}{\partial z} \\ xy\cos z & yz\sin x & xyz \end{vmatrix} = (xz - y\sin x)\mathbf{i} - (yz + xy\sin z)\mathbf{j} + (yz\cos x - x\cos z)\mathbf{k}.$$

Now, div $\mathbf{curl\ F}(x, y, z) = (z - y\cos x) - (z + x\sin z) + (y\cos x + x\sin z) = 0$. So,

$$\int_S\int \mathbf{curl\ F}\cdot\mathbf{N}\,dS = \iiint_Q \text{div}\,(\mathbf{curl\ F})\,dV = 0.$$

19. See Theorem 15.12.

20. If div $\mathbf{F}(x, y, z) > 0$, then source.

If div $\mathbf{F}(x, y, z) < 0$, then sink.

If div $\mathbf{F}(x, y, z) = 0$, then incompressible.

21. Using the Divergence Theorem, you have $\int_S\int \mathbf{curl\ F}\cdot\mathbf{N}\,dS = \iiint_Q \text{div}\,(\mathbf{curl\ F})\,dV$. Let

$$\mathbf{F}(x, y, z) = M\mathbf{i} + N\mathbf{j} + P\mathbf{k}$$

$$\mathbf{curl\ F} = \left(\frac{\partial P}{\partial y} - \frac{\partial N}{\partial z}\right)\mathbf{i} - \left(\frac{\partial P}{\partial x} - \frac{\partial M}{\partial z}\right)\mathbf{j} + \left(\frac{\partial N}{\partial x} - \frac{\partial M}{\partial y}\right)\mathbf{k}$$

$$\text{div}\,(\mathbf{curl\ F}) = \frac{\partial^2 P}{\partial x\partial y} - \frac{\partial^2 N}{\partial x\partial z} - \frac{\partial^2 P}{\partial y\partial x} + \frac{\partial^2 M}{\partial y\partial z} + \frac{\partial^2 N}{\partial z\partial x} - \frac{\partial^2 M}{\partial z\partial y} = 0.$$

So, $\int_S\int \mathbf{curl\ F}\cdot\mathbf{N}\,dS = \iiint_Q 0\,dV = 0.$

22. At P, the divergence is positive.

23. (a) Using the triple integral to find volume, you need $\mathbf{F}$ so that

$$\text{div}\,\mathbf{F} = \frac{\partial M}{\partial x} + \frac{\partial N}{\partial y} + \frac{\partial P}{\partial z} = 1.$$

So, you could have $\mathbf{F} = x\mathbf{i}$, $\mathbf{F} = y\mathbf{j}$, or $\mathbf{F} = z\mathbf{k}$.

For $dA = dy\,dz$ consider $\mathbf{F} = x\mathbf{i}$, $x = f(y, z)$, then $\mathbf{N} = \dfrac{\mathbf{i} + f_y\mathbf{j} + f_z\mathbf{k}}{\sqrt{1 + f_y^2 + f_z^2}}$ and $dS = \sqrt{1 + f_y^2 + f_z^2}\,dy\,dz$.

For $dA = dz\,dx$ consider $\mathbf{F} = y\mathbf{j}$, $y = f(x, z)$, then $\mathbf{N} = \dfrac{f_x\mathbf{i} + \mathbf{j} + f_z\mathbf{k}}{\sqrt{1 + f_x^2 + f_z^2}}$ and $dS = \sqrt{1 + f_x^2 + f_z^2}\,dz\,dx$.

For $dA = dx\,dy$ consider $\mathbf{F} = z\mathbf{k}$, $z = f(x, y)$, then $\mathbf{N} = \dfrac{f_x\mathbf{i} + f_y\mathbf{j} + \mathbf{k}}{\sqrt{1 + f_x^2 + f_y^2}}$ and $dS = \sqrt{1 + f_x^2 + f_y^2}\,dx\,dy$.

Correspondingly, you then have $\mathbf{V} = \int_S\int \mathbf{F}\cdot\mathbf{N}\,dS = \int_S\int x\,dy\,dz = \int_S\int y\,dz\,dx = \int_S\int z\,dx\,dy.$

(b) $v = \int_0^a\int_0^a x\,dy\,dz = \int_0^a\int_0^a a\,dy\,dz = \int_0^a a^2\,dz = a^3$

Similarly, $\int_0^a\int_0^a y\,dz\,dx = \int_0^a\int_0^a z\,dx\,dy = a^3.$

24. If $\mathbf{F}(x, y, z) = a_1\mathbf{i} + a_2\mathbf{j} + a_3\mathbf{k}$, then $\operatorname{div}\mathbf{F} = 0$.

So,

$$\int_S\int \mathbf{F}\cdot\mathbf{N}\,dS = \iiint_Q \operatorname{div}\mathbf{F}\,dV = \iiint_Q 0\,dV = 0.$$

25. If $\mathbf{F}(x, y, z) = x\mathbf{i} + y\mathbf{j} + z\mathbf{k}$, then $\operatorname{div}\mathbf{F} = 3$.

$$\int_S\int \mathbf{F}\cdot\mathbf{N}\,dS = \iiint_Q \operatorname{div}\mathbf{F}\,dV = \iiint_Q 3\,dV = 3V.$$

26. If $\mathbf{F}(x, y, z) = x\mathbf{i} + y\mathbf{j} + z\mathbf{k}$, then $\operatorname{div}\mathbf{F} = 3$.

$$\frac{1}{\|\mathbf{F}\|}\int_S\int \mathbf{F}\cdot\mathbf{N}\,dS = \frac{1}{\|\mathbf{F}\|}\iiint_Q \operatorname{div}\mathbf{F}\,dV = \frac{1}{\|\mathbf{F}\|}\iiint_Q 3\,dV = \frac{3}{\|\mathbf{F}\|}\iiint_Q dV$$

27. $$\int_S\int f D_{\mathbf{N}}g\,dS = \int_S\int f\nabla g\cdot\mathbf{N}\,dS = \iiint_Q \operatorname{div}(f\nabla g)\,dV = \iiint_Q (f\operatorname{div}\nabla g + \nabla f\cdot\nabla g)\,dV = \iiint_Q (f\nabla^2 g + \nabla f\cdot\nabla g)\,dV$$

28. $$\int_S\int (fD_{\mathbf{N}}g - gD_{\mathbf{N}}f)\,dS = \int_S\int f D_{\mathbf{N}}g\,dS - \int_S\int gD_{\mathbf{N}}f\,dS$$
$$= \iiint_Q (f\nabla^2 g + \nabla f\cdot\nabla g)\,dV - \iiint_Q (g\nabla^2 f + \nabla g\cdot\nabla f)\,dV = \iiint_Q (f\nabla^2 g - g\nabla^2 f)\,dV$$

Section 15.8 Stokes's Theorem

1. $\mathbf{F}(x, y, z) = (2y - z)\mathbf{i} + e^z\mathbf{j} + xyz\mathbf{k}$

$$\mathbf{curl\,F} = \begin{vmatrix} \mathbf{i} & \mathbf{j} & \mathbf{k} \\ \frac{\partial}{\partial x} & \frac{\partial}{\partial y} & \frac{\partial}{\partial z} \\ 2y - z & e^z & xyz \end{vmatrix}$$
$$= (xz - e^z)\mathbf{i} - (yz + 1)\mathbf{j} - 2\mathbf{k}$$

2. $\mathbf{F}(x, y, z) = x\sin y\mathbf{i} - y\cos x\mathbf{j} + yz^2\mathbf{k}$

$$\mathbf{curl\,F} = \begin{vmatrix} \mathbf{i} & \mathbf{j} & \mathbf{k} \\ \frac{\partial}{\partial x} & \frac{\partial}{\partial y} & \frac{\partial}{\partial z} \\ x\sin y & -y\cos x & yz^2 \end{vmatrix}$$
$$= z^2\mathbf{i} + (y\sin x - x\cos y)\mathbf{k}$$

3. $\mathbf{F}(x, y, z) = e^{x^2+y^2}\mathbf{i} + e^{y^2+z^2}\mathbf{j} + xyz\mathbf{k}$

$$\mathbf{curl\,F} = \begin{vmatrix} \mathbf{i} & \mathbf{j} & \mathbf{k} \\ \frac{\partial}{\partial x} & \frac{\partial}{\partial y} & \frac{\partial}{\partial z} \\ e^{x^2+y^2} & e^{y^2+z^2} & xyz \end{vmatrix}$$
$$= \left(xz - 2ze^{y^2+z^2}\right)\mathbf{i} - yz\mathbf{j} - 2ye^{x^2+y^2}\mathbf{k}$$
$$= z\left(x - 2e^{y^2+z^2}\right)\mathbf{i} - yz\mathbf{j} - 2ye^{x^2+y^2}\mathbf{k}$$

4. $\mathbf{F}(x, y, z) = \arcsin y\mathbf{i} + \sqrt{1 - x^2}\mathbf{j} + y^2\mathbf{k}$

$$\mathbf{curl\,F} = \begin{vmatrix} \mathbf{i} & \mathbf{j} & \mathbf{k} \\ \frac{\partial}{\partial x} & \frac{\partial}{\partial y} & \frac{\partial}{\partial z} \\ \arcsin y & \sqrt{1 - x^2} & y^2 \end{vmatrix}$$
$$= 2y\mathbf{i} + \left[\frac{-x}{\sqrt{1 - x^2}} - \frac{1}{\sqrt{1 - y^2}}\right]\mathbf{k}$$
$$= 2y\mathbf{i} - \left[\frac{x}{\sqrt{1 - x^2}} + \frac{1}{\sqrt{1 - y^2}}\right]\mathbf{k}$$

5. $C: x^2 + y^2 = 9, \quad z = 0, dz = 0$

Line Integral:

$\int_C \mathbf{F} \cdot d\mathbf{r} = \int_C -y\,dx + x\,dy$

$x = 3\cos t, dx = -3\sin t\,dt, y = 3\sin t, dy = 3\cos t\,dt$

$$\int_C \mathbf{F} \cdot d\mathbf{r} = \int_0^{2\pi} \left[(-3\sin t)(-3\sin t) + (3\cos t)(3\cos t)\right] dt$$

$$= \int_0^{2\pi} 9\,dt = 18\pi$$

Double Integral: $g(x, y) = 9 - x^2 - y^2, g_x = -2x, g_y = -2y$

curl F $= 2\mathbf{k}$

$\int_S \int \textbf{curl } \mathbf{F} \cdot \mathbf{N}\,dS = \int_R \int 2\,dA = 2(\text{area circle}) = 18\pi$

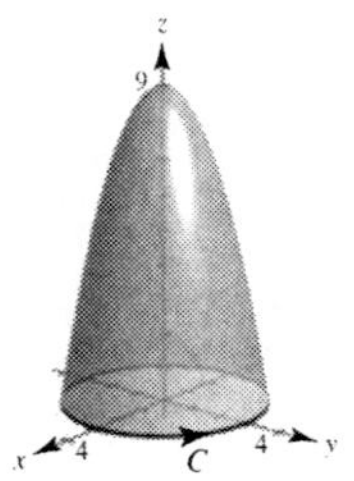

6. In this case, $M = -y + z, N = x - z, P = x - y$ and C is the circle $x^2 + y^2 = 1, z = 0, dz = 0$.

Line Integral: $\int_C \mathbf{F} \cdot d\mathbf{r} = \int_C (-y + z)\,dx + (x - z)\,dy + (x - y)\,dz = \int_C -y\,dx + x\,dy$

Letting $x = \cos t, y = \sin t$, you have $dx = -\sin t\,dt, dy = \cos t\,dt$ and $\int_C -y\,dx + x\,dy = \int_0^{2\pi} (\sin^2 t + \cos^2 t)\,dt = 2\pi$.

Double Integral: Consider $F(x, y, z) = x^2 + y^2 + z^2 - 1$.

Then $\mathbf{N} = \dfrac{\nabla F}{\|\nabla F\|} = \dfrac{2x\mathbf{i} + 2y\mathbf{j} + 2z\mathbf{k}}{2\sqrt{x^2 + y^2 + z^2}} = x\mathbf{i} + y\mathbf{j} + z\mathbf{k}$.

Because $z^2 = 1 - x^2 - y^2, z_x = \dfrac{-2x}{2z} = \dfrac{-x}{z}$, and $z_y = \dfrac{-y}{z}, dS = \sqrt{1 + \dfrac{x^2}{z^2} + \dfrac{y^2}{z^2}}\,dA = \dfrac{1}{z}\,dA$.

Now, because **curl F** $= 2\mathbf{k}$, you have $\int_S \int (\textbf{curl } \mathbf{F}) \cdot \mathbf{N}\,dS = \int_R \int 2z\left(\dfrac{1}{z}\right) dA = \int_R \int 2\,dA = 2(\text{Area of circle of radius } 1) = 2\pi$.

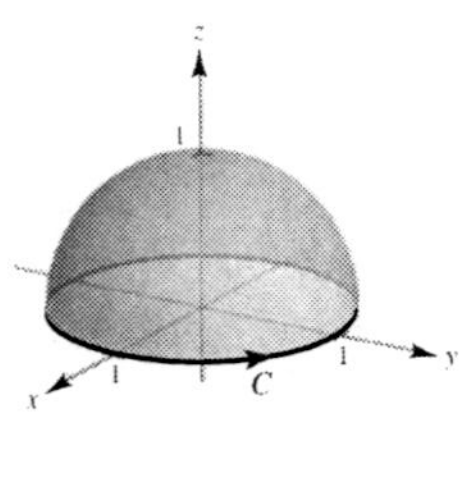

7. Line Integral:

From the figure you see that

C_1: $z = 0, dz = 0$

C_2: $x = 0, dx = 0$

C_3: $y = 0, dy = 0$

$$\int_C \mathbf{F}\cdot d\mathbf{r} = \int_C xyz\,dx + y\,dy + z\,dz = \int_{C_1} y\,dy + \int_{C_2} y\,dy + z dz + \int_{C_3} z\,dz = \int_0^2 y\,dy + \int_2^0 y\,dy + \int_0^{12} z\,dz + \int_{12}^0 z\,dz = 0$$

Double Integral: $\mathbf{curl\,F} = xy\mathbf{j} - xz\mathbf{k}$

Letting $z = 12 - 6x - 6y = g(x, y)$, $g_x = -6 = g_y$.

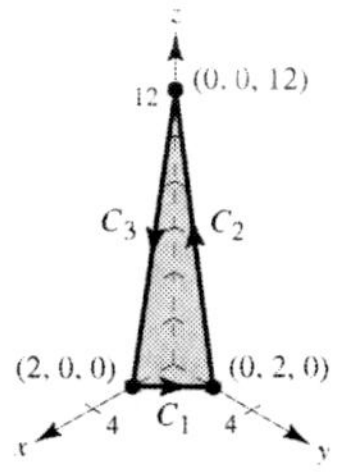

$$\int_S\int(\mathbf{curl\,F})\cdot\mathbf{N}\,dS = \int_R\int(\mathbf{curl\,F})\cdot[6\mathbf{i} + 6\mathbf{j} + \mathbf{k}]\,dA = \int_R\int(6xy - xz)\,dA$$

$$= \int_0^2\int_0^{2-x}\left[6xy - x(12 - 6x - 6y)\right]dy\,dx = \int_0^2\int_0^{2-x}\left(12xy - 12x + 6x^2\right)dy\,dx$$

$$= \int_0^2\left[6xy^2 - 12xy + 6x^2y\right]_0^{2-x}dx = 0$$

8. Line Integral: From the figure you see that

C_1: $y = 0, z = 0, dy = dz = 0$

C_2: $z = y^2, x = 0, dx = 0, dz = 2y\,dy$

C_3: $y = a, z = a^2, dy = dz = 0$

C_4: $z = y^2, x = a, dx = 0, dz = 2y\,dy.$

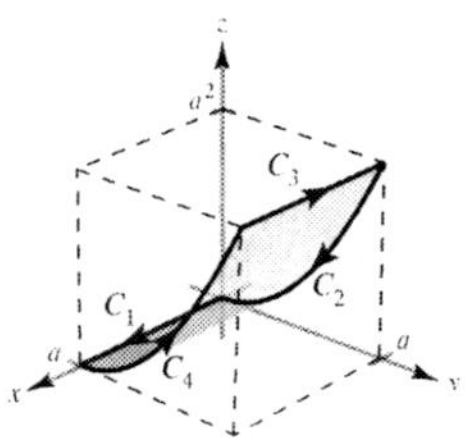

So,

$$\int_C \mathbf{F}\cdot d\mathbf{r} = \int_C z^2\,dx + x^2\,dy + y^2\,dz = \int_{C_1} 0\,dx + \int_{C_2} 2y^3\,dy + \int_{C_3} a^4\,dx + \int_{C_4}\left[a^2\,dy + 2y^3\,dy\right]$$

$$= \int_a^0 2y^3\,dy + \int_a^0 a^4\,dx + \int_0^a\left(a^2 2y^3\right)dy = \left[a^4x\right]_a^0 + \left[a^2y\right]_0^a = -a^5 + a^3 = a^3(1 - a).$$

Double Integral: Because S is given by $-y^2 + z = 0$, you have

$$\mathbf{N} = \frac{2y\mathbf{j} - \mathbf{k}}{\sqrt{1 + 4y^2}} \text{ and } dS = \sqrt{1 + 4y^2}\,dA.$$

Furthermore, $\mathbf{curl\,F} = 2y\mathbf{i} + 2z\mathbf{j} + 2x\mathbf{k}$. So,

$$\int_S\int(\mathbf{curl\,F})\cdot\mathbf{N}\,dS = \int_R\int(4yz - 2x)\,dA = \int_0^a\int_0^a(-4yz + 2x)\,dA = \int_0^a\int_0^a\left(-4y^3 + 2^x\right)dydx$$

$$= \int_0^a\left(-a^4 + 2ax\right)dx = \left[-a^4x + ax^2\right]_0^a = -a^5 + a^3 = a^3\left(1 - a^2\right).$$

9. These three points have equation:

$x + y + z = 2.$

Normal vector: $\mathbf{N} = \mathbf{i} + \mathbf{j} + \mathbf{k}$

$\mathbf{curl\,F} = -3\mathbf{i} - \mathbf{j} - 2\mathbf{k}$

$$\int_S\int \mathbf{curl\,F}\cdot\mathbf{N}\,dS = \int_R\int(-6)\,dA = -6(\text{area of triangle in } xy\text{-plane})$$

$$= -6(2) = -12$$

10. Let $A = (0, 0, 0)$, $B = (1, 1, 1)$, and $C = (0, 0, 2)$. Then $\mathbf{U} = \overline{AB} = \mathbf{i} + \mathbf{j} + \mathbf{k}$, and $\mathbf{V} = \overline{AC} = 2\mathbf{k}$, and

$$\mathbf{N} = \frac{\mathbf{U}\times\mathbf{V}}{\|\mathbf{U}\times\mathbf{V}\|} = \frac{2\mathbf{i} - 2\mathbf{j}}{2\sqrt{2}} = \frac{\mathbf{i} - \mathbf{j}}{\sqrt{2}}.$$

So, $F(x, y, z) = x - y$ and $dS = \sqrt{2}\,dA$. Because $\mathbf{curl\,F} = \dfrac{2x}{x^2 + y^2}\mathbf{k}$, you have $\displaystyle\int_S\int(\mathbf{curl\,F})\cdot\mathbf{N}\,dS = \int_R\int 0\,dS = 0.$

11. $\textbf{curl F} = \begin{vmatrix} \mathbf{i} & \mathbf{j} & \mathbf{k} \\ \dfrac{\partial}{\partial x} & \dfrac{\partial}{\partial y} & \dfrac{\partial}{\partial z} \\ z^2 & 2x & y^2 \end{vmatrix} = 2y\mathbf{i} + 2z\mathbf{j} + 2\mathbf{k}$

$z = G(x, y) = 1 - x^2 - y^2, G_x = -2x, G_y = -2y$

$$\int_S\int \textbf{curl F}\cdot\mathbf{N}\,dS = \int_R\int(2y\mathbf{i} + 2z\mathbf{j} + 2\mathbf{k})\cdot(2x\mathbf{i} + 2y\mathbf{j} + \mathbf{k})\,dA = \int_R\int\left[4xy + 4y(1 - x^2 - y^2) + 2\right]dA$$

$$= \int_{-1}^{1}\int_{-\sqrt{1-x^2}}^{\sqrt{1-x^2}}\left[4xy + 4y - 4x^2y - 4y^3 + 2\right]dy\,dx$$

$$= \int_{-1}^{1} 4\sqrt{1 - x^2}\,dx = 2\left[\arcsin x + x\sqrt{1 - x^2}\right]_{-1}^{1} = 2\pi$$

12. $\mathbf{F}(x, y, z) = 4xz\mathbf{i} + y\mathbf{j} + 4xy\mathbf{k},\quad S\colon 9 - x^2 - y^2, z \le 0$

$\textbf{curl F} = 4x\mathbf{i} + (4x - 4y)\mathbf{j}$

$G(x, y, z) = x^2 + y^2 + z - 9$

$\nabla G(x, y, z) = 2x\mathbf{i} + 2y\mathbf{j} + \mathbf{k}$

$$\int_S\int(\textbf{curl F})\cdot\mathbf{N}\,dS = \int_R\int\left[8x^2 + 2y(4x - 4y)\right]dA = \int_{-3}^{3}\int_{-\sqrt{9-x^2}}^{\sqrt{9-x^2}}\left[8x^2 + 8xy - 8y^2\right]dy\,dx$$

$$= \int_{-3}^{3}\left(16x^2\sqrt{9 - x^2} - \tfrac{16}{3}(9 - x^2)^{3/2}\right)dx = 0$$

13. $\textbf{curl F} = \begin{vmatrix} \mathbf{i} & \mathbf{j} & \mathbf{k} \\ \dfrac{\partial}{\partial x} & \dfrac{\partial}{\partial y} & \dfrac{\partial}{\partial z} \\ z^2 & y & z \end{vmatrix} = 2z\mathbf{j}$

$z = G(x, y) = \sqrt{4 - x^2 - y^2}, G_x = \dfrac{-x}{\sqrt{4 - x^2 - y^2}}, G_y = \dfrac{-y}{\sqrt{4 - x^2 - y^2}}$

$$\int_S\int \textbf{curl F}\cdot\mathbf{N} = \int_R\int(2z\mathbf{j})\cdot\left(\frac{x}{\sqrt{4 - x^2 - y^2}}\mathbf{i} + \frac{y}{\sqrt{4 - x^2 - y^2}}\mathbf{j} + \mathbf{k}\right)dA$$

$$= \int_R\int\frac{2yz}{\sqrt{4 - x^2 - y^2}}\,dA = \int_R\int\frac{2y\sqrt{4 - x^2 - y^2}}{\sqrt{4 - x^2 - y^2}}\,dA = \int_{-2}^{2}\int_{-\sqrt{4-x^2}}^{\sqrt{4-x^2}} 2y\,dy\,dx = 0$$

14. $\mathbf{F}(x, y, z) = x^2\mathbf{i} + z^2\mathbf{j} - xyz\mathbf{k}, \quad S: z = \sqrt{4 - x^2 - y^2}$

$$\mathbf{curl\ F} = \begin{vmatrix} \mathbf{i} & \mathbf{j} & \mathbf{k} \\ \dfrac{\partial}{\partial x} & \dfrac{\partial}{\partial y} & \dfrac{\partial}{\partial z} \\ x^2 & z^2 & -xyz \end{vmatrix} = (-xz - 2z)\mathbf{i} + yz\mathbf{j}$$

$$G(x, y, z) = z - \sqrt{4 - x^2 - y^2}$$

$$\nabla G(x, y, z) = \frac{x}{\sqrt{4 - x^2 - y^2}}\mathbf{i} + \frac{y}{\sqrt{4 - x^2 - y^2}}\mathbf{j} + \mathbf{k}$$

$$\int_S\int(\mathbf{curl\ F}) \cdot \mathbf{N}\, dS = \int_R\int\left[\frac{-z(x + 2)x}{\sqrt{4 - x^2 - y^2}} + \frac{y^2 z}{\sqrt{4 - x^2 - y^2}}\right] dA$$

$$= \int_R\int\left[-x(x + 2) + y^2\right] dA = \int_{-2}^{2}\int_{-\sqrt{4-x^2}}^{\sqrt{4-x^2}}\left(-x^2 - 2x + y^2\right) dy\, dx$$

$$= \int_{-2}^{2}\left[-x^2 y - 2xy + \frac{y^3}{3}\right]_{-\sqrt{4-x^2}}^{\sqrt{4-x^2}} dx$$

$$= \int_{-2}^{2}\left[-2x^2\sqrt{4 - x^2} - 4x\sqrt{4 - x^2} + \frac{2}{3}(4 - x^2)\sqrt{4 - x^2}\right] dx$$

$$= \int_{-2}^{2}\left[-\frac{8}{3}x^2\sqrt{4 - x^2} - 4x\sqrt{4 - x^2} + \frac{8}{3}\sqrt{4 - x^2}\right] dx$$

$$= \left[-\frac{8}{3}\left(\frac{1}{8}\right)\left[x(2x^2 - 4)\sqrt{4 - x^2} + 16\arcsin\frac{x}{2}\right] + \frac{4}{3}(4 - x^2)^{3/2} + \frac{8}{3}\left(\frac{1}{2}\right)\left[x\sqrt{4 - x^2} + 4\arcsin\frac{x}{2}\right]\right]_{-2}^{2}$$

$$= \left[\left(-\frac{1}{3}\right)(8\pi) + \frac{4}{3}(2\pi) + \frac{1}{3}(-8\pi) - \frac{4}{3}(-2\pi)\right] = 0$$

15. $\mathbf{F}(x, y, z) = -\ln\sqrt{x^2 + y^2}\,\mathbf{i} + \arctan\dfrac{x}{y}\mathbf{j} + \mathbf{k}$

$$\mathbf{curl\ F} = \begin{vmatrix} \mathbf{i} & \mathbf{j} & \mathbf{k} \\ \dfrac{\partial}{\partial x} & \dfrac{\partial}{\partial y} & \dfrac{\partial}{\partial z} \\ -1/2\ \ln(x^2 + y^2) & \arctan x/y & 1 \end{vmatrix} = \left[\frac{(1/y)}{1 + (x^2/y^2)} + \frac{y}{x^2 + y^2}\right]\mathbf{k} = \left[\frac{2y}{x^2 + y^2}\right]\mathbf{k}$$

S: $z = 9 - 2x - 3y$ over one petal of $r = 2\sin 2\theta$ in the first octant.

$$G(x, y, z) = 2x + 3y + z - 9$$

$$\nabla G(x, y, z) = 2\mathbf{i} + 3\mathbf{j} + \mathbf{k}$$

$$\int_S\int(\mathbf{curl\ F}) \cdot \mathbf{N}\, dS = \int_R\int\frac{2y}{x^2 + y^2}dA = \int_0^{\pi/2}\int_0^{2\sin 2\theta}\frac{2r\sin\theta}{r^2}r\, dr\, d\theta$$

$$= \int_0^{\pi/2}\int_0^{4\sin\theta\cos\theta} 2\sin\theta\, dr\, d\theta = \int_0^{\pi/2} 8\sin^2\theta\cos\theta\, d\theta = \left[\frac{8\sin^3\theta}{3}\right]_0^{\pi/2} = \frac{8}{3}$$

16. $\mathbf{F}(x, y, z) = yz\mathbf{i} + (2 - 3y)\mathbf{j} + (x^2 + y^2)\mathbf{k}$

$$\textbf{curl F} = \begin{vmatrix} \mathbf{i} & \mathbf{j} & \mathbf{k} \\ \dfrac{\partial}{\partial x} & \dfrac{\partial}{\partial y} & \dfrac{\partial}{\partial z} \\ yz & 2 - 3y & x^2 + y^2 \end{vmatrix} = 2y\mathbf{i} + (y - 2x)\mathbf{j} + z\mathbf{k}$$

S: the first octant portion of $x^2 + z^2 = 16$ over $x^2 + y^2 = 16$

$G(x, y, z) = z - \sqrt{16 - x^2}$

$\nabla G(x, y, z) = \dfrac{x}{\sqrt{16 - x^2}}\mathbf{i} + \mathbf{k}$

$$\int_S\int(\textbf{curl F}) \cdot \mathbf{N}\, dS = \int_R\int\left[\frac{2xy}{\sqrt{16 - x^2}} - z\right] dA = \int_R\int\left[\frac{2xy}{\sqrt{16 - x^2}} - \sqrt{16 - x^2}\right] dA$$

$$= \int_0^4 \int_0^{\sqrt{16-x^2}} \left[\frac{2xy}{\sqrt{16 - x^2}} - \sqrt{16 - x^2}\right] dy\, dx = \int_0^4 \left[\frac{x}{\sqrt{16 - x^2}}y^2 - \sqrt{16 - x^2}\,y\right]_0^{\sqrt{16-x^2}} dx$$

$$= \int_0^4 \left[x\sqrt{16 - x^2} - (16 - x^2)\right] dx = \left[-\frac{1}{3}(16 - x^2)^{3/2} - 16x + \frac{x^3}{3}\right]_0^4 = \left(-64 + \frac{64}{3}\right) - \left(-\frac{64}{3}\right) = -\frac{64}{3}$$

17. $\textbf{curl F} = xy\mathbf{j} - xz\mathbf{k}$

$z = G(x, y) = x^2, G_x = 2x, G_y = 0$

$$\int_S\int \textbf{curl F} \cdot \mathbf{N} = \int_R\int(xy\mathbf{j} - xz\mathbf{k}) \cdot (2x\mathbf{i} - \mathbf{k})\, dA = \int_R\int xz\, dA = \int_0^a\int_0^a x(x^2)\, dy\, dx = \frac{a^5}{4}$$

18. $\mathbf{F}(x, y, z) = xyz\mathbf{i} + y\mathbf{j} + z\mathbf{k}$

$$\textbf{curl F} = \begin{vmatrix} \mathbf{i} & \mathbf{j} & \mathbf{k} \\ \dfrac{\partial}{\partial x} & \dfrac{\partial}{\partial y} & \dfrac{\partial}{\partial z} \\ xyz & y & z \end{vmatrix} = xy\mathbf{j} - xz\mathbf{k}$$

S: the first octant portion of $z = x^2$ over $x^2 + y^2 = a^2$. You have $\mathbf{N} = \dfrac{2x\mathbf{i} - \mathbf{k}}{\sqrt{1 + 4x^2}}$ and $dS = \sqrt{1 + 4x^2}\, dA$.

$$\int_S\int(\textbf{curl F}) \cdot \mathbf{N}\, dS = \int_R\int xz\, dA = \int_R\int x^3\, dA = \int_0^a\int_0^{\sqrt{a^2-x^2}} x^3\, dy\, dx$$

$$= \int_0^a x^3\sqrt{a^2 - x^2}\, dx = \left[-\frac{1}{3}x^2(a^2 - x^2)^{3/2} - \frac{2}{15}(a^2 - x^2)^{5/2}\right]_0^a = \frac{2}{15}a^5$$

19. $\mathbf{F}(x, y, z) = \mathbf{i} + \mathbf{j} - 2\mathbf{k}$

$$\textbf{curl F} = \begin{vmatrix} \mathbf{i} & \mathbf{j} & \mathbf{k} \\ \dfrac{\partial}{\partial x} & \dfrac{\partial}{\partial y} & \dfrac{\partial}{\partial z} \\ 1 & 1 & -2 \end{vmatrix} = \mathbf{0}$$

Letting $\mathbf{N} = \mathbf{k}$, you have $\int_S\int(\textbf{curl F}) \cdot \mathbf{N}\, dS = 0$.

20. $\mathbf{F}(x, y, z) = -z\mathbf{i} + y\mathbf{k}$

S: $x^2 + y^2 = 1$

$$\textbf{curl F} = \begin{vmatrix} \mathbf{i} & \mathbf{j} & \mathbf{k} \\ \dfrac{\partial}{\partial x} & \dfrac{\partial}{\partial y} & \dfrac{\partial}{\partial z} \\ -z & 0 & y \end{vmatrix} = \mathbf{i} - \mathbf{j}$$

Letting $\mathbf{N} = \mathbf{k}$, $\textbf{curl F} \cdot \mathbf{N} = 0$ and $\int_S\int(\textbf{curl F}) \cdot \mathbf{N}\, dS = 0$.

21. See Theorem 15.13.

22. **curl F** measures the rotational tendency. See page 1114.

23. Let $\mathbf{C} = a\mathbf{i} + b\mathbf{j} + c\mathbf{k}$, then $\frac{1}{2}\int_C (\mathbf{C} \times \mathbf{r}) \cdot d\mathbf{r} = \frac{1}{2}\int_S\int \mathbf{curl}(\mathbf{C} \times \mathbf{r}) \cdot \mathbf{N}\, dS = \frac{1}{2}\int_S\int 2\mathbf{C} \cdot \mathbf{N}\, dS = \int_S\int \mathbf{C} \cdot \mathbf{N}\, dS$

because $\mathbf{C} \times \mathbf{r} = \begin{vmatrix} \mathbf{i} & \mathbf{j} & \mathbf{k} \\ a & b & c \\ x & y & z \end{vmatrix} = (bz - cy)\mathbf{i} - (az - cx)\mathbf{j} + (ay - bx)\mathbf{k}$

and $\mathbf{curl}(\mathbf{C} \times \mathbf{r}) = \begin{vmatrix} \mathbf{i} & \mathbf{j} & \mathbf{k} \\ \dfrac{\partial}{\partial x} & \dfrac{\partial}{\partial y} & \dfrac{\partial}{\partial z} \\ bz - cy & cx - az & ay - bx \end{vmatrix} = 2(a\mathbf{i} + b\mathbf{j} + c\mathbf{k}) = 2\mathbf{C}.$

24. Yes. Both S_1 and S_2 are oriented upward and are bounded by the smooth simple closed curve $C = x^2 + y^2 = a^2$. Also, $\mathbf{F}$ is a vector field with continuous partial derivatives. So, by Stokes's Theorem

$$\iint_{S_1} (\mathbf{curl}\ \mathbf{F}) \cdot \mathbf{N}\, dS_1 = \int_C \mathbf{F} \cdot d\mathbf{r} = \iint_{S_2} (\mathbf{curl}\ \mathbf{F}) \cdot \mathbf{N}\, dS_2.$$

25. Let S be the upper portion of the ellipsoid

$x^2 + 4y^2 + z^2 = 4,\ z \geq 0$

Let $C: \mathbf{r}(t) = \langle 2\cos t, \sin t, 0\rangle,\ 0 \leq t \leq 2\pi$, be the boundary of S.

If $\mathbf{F} = \langle M, N, P\rangle$ exists, then

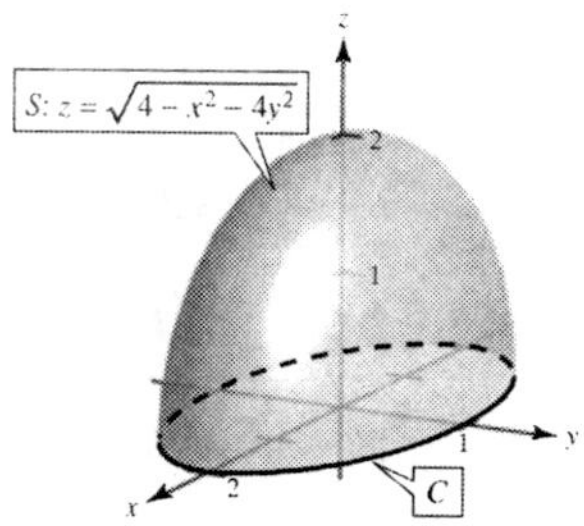

$$\begin{aligned}
0 &= \int_S\int (\mathbf{curl}\ \mathbf{F}) \cdot \mathbf{N}\, dS && \text{(by (i))} \\
&= \int_C \mathbf{F} \cdot d\mathbf{r} && \text{(Stokes's Theorem)} \\
&= \int_C \mathbf{G} \cdot d\mathbf{r} && \text{(by (iii))} \\
&= \int_0^{2\pi} \left\langle \frac{-\sin t}{4}, \frac{2\cos t}{4}, 0\right\rangle \cdot \langle -2\sin t, \cos t, 0\rangle\, dt = \frac{1}{4}\int_0^{2\pi} \left(2\sin^2 t + 2\cos^2 t\right) dt = \pi
\end{aligned}$$

So, there is no such $\mathbf{F}$.

Review Exercises for Chapter 15

1. $\mathbf{F}(x, y, z) = x\mathbf{i} + \mathbf{j} + 2\mathbf{k}$

$\|\mathbf{F}\| = \sqrt{x^2 + 1^2 + 2^2} = \sqrt{x^2 + 5}$

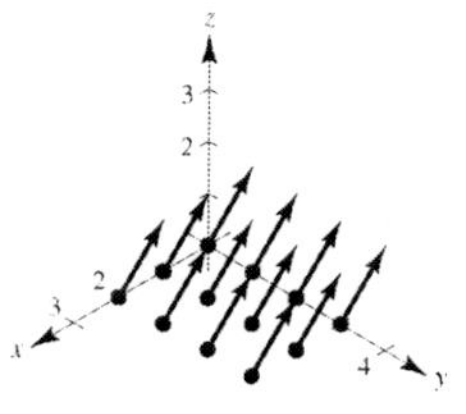

2. $\mathbf{F}(x, y) = \mathbf{i} - 2y\mathbf{j}$

$\|\mathbf{F}\| = \sqrt{1 + 4y^2}$

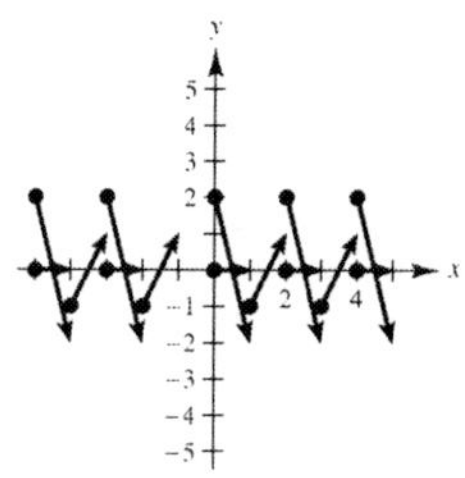

3. $f(x, y, z) = 2x^2 + xy + z^2$

$\mathbf{F}(x, y, z) = \nabla f = (4x + y)\mathbf{i} + x\mathbf{j} + 2z\mathbf{k}$

4. $f(x, y, z) = x^2 e^{yz}$

$\mathbf{F}(x, y, z) = 2xe^{yz}\,\mathbf{i} + x^2 z e^{yz}\,\mathbf{j} + x^2 y e^{yz}\,\mathbf{k}$

$= xe^{yz}(2\mathbf{i} + xz\,\mathbf{j} + xy\,\mathbf{k})$

5. Because $\partial M/\partial y = -1/x^2 = \partial N/\partial x$, **F** is conservative.

From $M = \partial U/\partial x = -y/x^2$ and $N = \partial U/\partial y = 1/x$, partial integration yields $U = (y/x) + h(y)$ and $U = (y/x) + g(x)$ which suggests that $U(x, y) = (y/x) + C$.

6. Because $\partial M/\partial y = -1/y^2 \neq \partial N/\partial x$, **F** is not conservative.

7. Because $\dfrac{\partial M}{\partial y} = 2xy$ and $\dfrac{\partial N}{\partial x} = 2xy$, **F** is conservative.

From $M = \dfrac{\partial U}{\partial x} = xy^2 - x^2$ and

$N = \dfrac{\partial U}{\partial y} = x^2y + y^2$, partial integration yields

$$U = \frac{1}{2}x^2y^2 - \frac{x^3}{3} + h(y)$$

and

$$U = \frac{1}{2}x^2y^2 + \frac{y^3}{3} + g(x).$$

So, $h(y) = y^3/3$ and $g(x) = -x^3/3$. So,

$$U(x, y) = \frac{1}{2}x^2y^2 - \frac{x^3}{3} + \frac{y^3}{3} + C.$$

8. Because $\partial M/\partial y = -6y^2 \sin 2x = \partial N/\partial x$, **F** is conservative. From $M = \partial U/\partial x = -2y^3 \sin 2x$ and $N = \partial U/\partial y = 3y^2(1 + \cos 2x)$, you obtain

$U = y^3 \cos 2x + h(y)$ and

$U = y^3(1 + \cos 2x) + g(x)$ which suggests that

$h(y) = y^3$, $g(x) = C$, and

$U(x, y) = y^3(1 + \cos 2x) + C$.

9. Because $\dfrac{\partial M}{\partial y} = 8xy$ and $\dfrac{\partial N}{\partial x} = 4x$, $\dfrac{\partial M}{\partial y} \neq \dfrac{\partial N}{\partial x}$, so **F** is not conservative.

10. Because

$$\frac{\partial M}{\partial y} = 4x = \frac{\partial N}{\partial x},$$
$$\frac{\partial M}{\partial z} = 2z = \frac{\partial P}{\partial x},$$
$$\frac{\partial N}{\partial z} = 6y \neq \frac{\partial P}{\partial y},$$

F is not conservative.

11. Because

$$\frac{\partial M}{\partial y} = \frac{-1}{y^2z} = \frac{\partial N}{\partial x}, \frac{\partial M}{\partial z} = \frac{-1}{yz^2} = \frac{\partial P}{\partial x},$$
$$\frac{\partial N}{\partial z} = \frac{x}{y^2z^2} = \frac{\partial P}{\partial y},$$

F is conservative. From

$$M = \frac{\partial U}{\partial x} = \frac{1}{yz}, \; N = \frac{\partial U}{\partial y} = \frac{-x}{y^2z}, \; P = \frac{\partial U}{\partial z} = \frac{-x}{yz^2}$$

you obtain

$$U = \frac{x}{yz} + f(y, z), \; U = \frac{x}{yz} + g(x, z),$$
$$U = \frac{x}{yz} + h(x, y) \Rightarrow f(x, y\ z) = \frac{x}{yz} + K.$$

12. Because

$$\frac{\partial M}{\partial y} = \sin z = \frac{\partial N}{\partial x}, \quad \frac{\partial M}{\partial z} = y \cos z \neq \frac{\partial P}{\partial x},$$

F is not conservative.

13. Because $\mathbf{F}(x, y, z) = x^2\,\mathbf{i} + xy^2\,\mathbf{j} + x^2z\,\mathbf{k}$:

(a) $\text{div}\,\mathbf{F} = 2x + 2xy + x^2$

(b) $$\textbf{curl F} = \begin{vmatrix} \mathbf{i} & \mathbf{j} & \mathbf{k} \\ \dfrac{\partial}{\partial x} & \dfrac{\partial}{\partial y} & \dfrac{\partial}{\partial z} \\ x^2 & xy^2 & x^2z \end{vmatrix} = -(2xz)\mathbf{j} + y^2\mathbf{k}$$

14. Because $\mathbf{F}(x, y, z) = y^2\,\mathbf{j} - z^2\,\mathbf{k}$:

(a) $\text{div}\,\mathbf{F} = 2y - 2z$

(b) $$\textbf{curl F} = \begin{vmatrix} \mathbf{i} & \mathbf{j} & \mathbf{k} \\ \dfrac{\partial}{\partial x} & \dfrac{\partial}{\partial y} & \dfrac{\partial}{\partial z} \\ 0 & y^2 & -z^2 \end{vmatrix} = \mathbf{0}$$

15. Because $\mathbf{F} = (\cos y + y \cos x)\mathbf{i} + (\sin x - x \sin y)\mathbf{j} + xyz\mathbf{k}$:

(a) $\text{div } \mathbf{F} = -y \sin x - x \cos y + xy$

(b) $\textbf{curl } \mathbf{F} = xz\mathbf{i} - yz\mathbf{j} + (\cos x - \sin y + \sin y - \cos x)\mathbf{k} = xz\mathbf{i} - yz\mathbf{j}$

16. Because $\mathbf{F} = (3x - y)\mathbf{i} + (y - 2z)\mathbf{j} + (z - 3x)\mathbf{k}$:

(a) $\text{div } \mathbf{F} = 3 + 1 + 1 = 5$

(b) $\textbf{curl } \mathbf{F} = 2\mathbf{i} + 3\mathbf{j} + \mathbf{k}$

17. Because $\mathbf{F} = \arcsin x\mathbf{i} + xy^2\mathbf{j} + yz^2\mathbf{k}$:

(a) $\text{div } \mathbf{F} = \dfrac{1}{\sqrt{1 - x^2}} + 2xy + 2yz$

(b) $\textbf{curl } \mathbf{F} = z^2\mathbf{i} + y^2\mathbf{k}$

18. Because $\mathbf{F} = (x^2 - y)\mathbf{i} - (x + \sin^2 y)\mathbf{j}$:

(a) $\text{div } \mathbf{F} = 2x - 2 \sin y \cos y$

(b) $\textbf{curl } \mathbf{F} = \mathbf{0}$

19. Because $\mathbf{F} = \ln(x^2 + y^2)\mathbf{i} + \ln(x^2 + y^2)\mathbf{j} + z\mathbf{k}$:

(a) $\text{div } \mathbf{F} = \dfrac{2x}{x^2 + y^2} + \dfrac{2y}{x^2 + y^2} + 1 = \dfrac{2x + 2y}{x^2 + y^2} + 1$

(b) $\textbf{curl } \mathbf{F} = \dfrac{2x - 2y}{x^2 + y^2}\mathbf{k}$

20. Because $\mathbf{F} = \dfrac{z}{x}\mathbf{i} + \dfrac{z}{y}\mathbf{j} + z^2\mathbf{k}$:

(a) $\text{div } \mathbf{F} = -\dfrac{z}{x^2} - \dfrac{z}{y^2} + 2z = z\left(2 - \dfrac{1}{x^2} - \dfrac{1}{y^2}\right)$

(b) $\textbf{curl } \mathbf{F} = -\dfrac{1}{y}\mathbf{i} + \dfrac{1}{x}\mathbf{j}$

21. (a) Let $x = 3t,\ y = 4t,\quad 0 \le t \le 1,$

then $ds = \sqrt{9 + 16}\, dt = 5\, dt.$

$$\int_C (x^2 + y^2)\, ds = \int_0^1 (9t^2 + 16t^2)5\, dt = \left[125\frac{t^3}{3}\right]_0^1 = \frac{125}{3}$$

(b) Let $x = \cos t,\ y = \sin t,\quad 0 \le t \le 2\pi,$

then $ds = \sqrt{(-\sin t)^2 + (\cos t)^2} = 1.$

$$\int_C (x^2 + y^2)\, ds = \int_0^{2\pi} dt = 2\pi$$

22. (a) Let $x = 5t,\ y = 4t,\ 0 \le t \le 1,$ then $ds = \sqrt{41}\, dt.$

$$\int_C xy\, ds = \int_0^1 20t^2\sqrt{41}\, dt = \frac{20\sqrt{41}}{3}$$

(b) $C_1\colon x = t,\ y = 0,\ 0 \le t \le 4,\ ds = dt$

$C_2\colon x = 4 - 4t,\ y = 2t,\ 0 \le t \le 1,\ ds = 2\sqrt{5}\, dt$

$C_3\colon x = 0,\ y = 2 - t,\ 0 \le 1 \le 2,\ ds = dt$

So, $\displaystyle\int_C xy\, ds = \int_0^4 0\, dt + \int_0^1 (8t - 8t^2)2\sqrt{5}\, dt + \int_0^2 0\, dt = 16\sqrt{5}\left[\frac{t^2}{2} - \frac{t^3}{3}\right]_0^1 = \frac{8\sqrt{5}}{3}.$

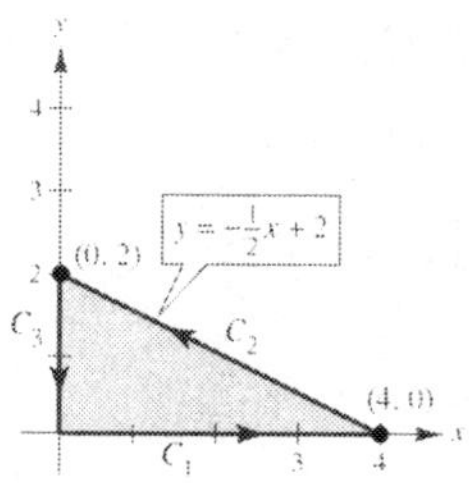

23. $x = 1 - \sin t,\ y = 1 - \cos t,\ 0 \le t \le 2\pi$

$\dfrac{dx}{dt} = -\cos t,\ \dfrac{dy}{dt} = \sin t,\ ds = \sqrt{(-\cos t)^2 + (\sin t)^2}\, dt = dt$

$$\int_C (x^2 + y^2)\, ds = \int_0^{2\pi} \left[(1 - \sin t)^2 + (1 - \cos t)^2\right] dt = \int_0^{2\pi} \left[1 - 2\sin t + \sin^2 t + 1 - 2\cos t + \cos^2 t\right] dt$$

$$= \int_0^{2\pi} [3 - 2\sin t - 2\cos t]\, dt = [3t + 2\cos t - 2\sin t]_0^{2\pi} = 6\pi$$

24. $x = \cos t + t\sin t,\ y = \sin t - t\cos t,\ 0 \le t \le 2\pi,\ \dfrac{dx}{dt} = t\cos t,\ \dfrac{dy}{dt} = t\sin t$

$$\int_C (x^2 + y^2)\, ds = \int_0^{2\pi} \left[(\cos t + t\sin t)^2 + (\sin t - t\cos t)^2\right]\sqrt{t^2\cos^2 t + t^2\sin^2 t}\, dt = \int_0^{2\pi} \left[t^3 + t\right] dt = 2\pi^2(1 + 2\pi^2)$$

25. (a) Let $x = 3t,\ y = -3t, \quad 0 \le t \le 1.$

$$\int_C (2x - y)\,dx + (x + 2y)\,dy = \int_0^1 \left[(6t + 3t)3 + (3t - 6t)(-3)\right] dt = \int_0^1 (27t + 9t)\,dt = 18t^2\Big]_0^1 = 18$$

(b) Let $x = 3\cos t,\ y = 3\sin t,\ dx = -3\sin t\,dt,\ dy = 3\cos t\,dt, 0 \le t \le 2\pi.$

$$\int_C (2x - y)\,dx + (x + 2y)\,dy = \int_0^{2\pi} \left[(6\cos t - 3\sin t)(-3\sin t) + (3\cos t + 6\sin t)(3\cos t)\right] dt = \int_0^{2\pi} 9\,dt = 18\pi$$

26. $x = \cos t + t\sin t,\ y = \sin t - t\sin t, 0 \le t \le \frac{\pi}{2},\ dx = t\cos t\,dt,\ dy = (\cos t - t\cos t - \sin t)\,dt$

$$\int_C (2x - y)\,dx + (x + 3y)\,dy = \int_0^{\pi/2} \left[\sin t\cos t(5t^2 - 6t + 2) + \cos^2 t(t + 1) + \sin^2 t(2t - 3)\right] dt \approx 1.01$$

27. $\int_C (2x + y)\,ds, \mathbf{r}(t) = a\cos^3 t\mathbf{i} + a\sin^3 t\mathbf{j}, 0 \le t \le \frac{\pi}{2}$

$x'(t) = -3a \cdot \cos^2 t\sin t$

$y'(t) = 3a \cdot \sin^2 t\cos t$

$$\int_C (2x + y)\,ds = \int_0^{\pi/2} \left(2(a \cdot \cos^3 t) + a \cdot \sin^3 t\right)\sqrt{x'(t)^2 + y'(t)^2}\,dt = \frac{9a^2}{5}$$

28. $\mathbf{r}(t) = t\mathbf{i} + t^2\mathbf{j} + t^{3/2}\mathbf{k}, 0 \le t \le 4$

$x'(t) = 1, y'(t) = 2t, z'(t) = \frac{3}{2}t^{1/2}$

$$\int_C (x^2 + y^2 + z^2)\,ds = \int_0^4 (t^2 + y^4 + t^3)\sqrt{1 + 4t^2 + \frac{9}{4}t}\,dt \approx 2080.59$$

29. $f(x, y) = 3 + \sin(x + y)$

C: $y = 2x$ from $(0, 0)$ to $(2, 4)$

$\mathbf{r}(t) = t\mathbf{i} + 2t\mathbf{j}, 0 \le t \le 2$

$\mathbf{r}'(t) = \mathbf{i} + 2\mathbf{j}$

$\|\mathbf{r}'(t)\| = \sqrt{5}$

Lateral surface area:

$$\begin{aligned}\int_C f(x, y)\,ds &= \int_0^2 \left[3 + \sin(t + 2t)\right]\sqrt{5}\,dt \\ &= \sqrt{5}\int_0^2 [3 + \sin 3t]\,dt \\ &= \sqrt{5}\left[3t - \frac{1}{3}\cos 3t\right]_0^2 \\ &= \sqrt{5}\left[6 - \frac{1}{3}\cos 6 + \frac{1}{3}\right] \\ &= \frac{\sqrt{5}}{3}(19 - \cos 6) \approx 13.446\end{aligned}$$

30. $f(x, y) = 12 - x - y$

C: $y = x^2$ from $(0, 0)$ to $(2, 4)$

$\mathbf{r}(t) = t\mathbf{i} + t^2\mathbf{j}, 0 \le t \le 2$

$\mathbf{r}'(t) = \mathbf{i} + 2t\mathbf{j}$

$\|\mathbf{r}'(t)\| = \sqrt{1 + 4t^2}$

Lateral surface area:

$$\int_C f(x, y)\,ds = \int_0^2 (12 - t - t^2)\sqrt{1 + 4t^2}\,dt \approx 41.532$$

31. $\mathbf{F}(x, y) = xy\mathbf{i} + 2xy\mathbf{j}$

$\mathbf{r}(t) = t^2\mathbf{i} + t^2\mathbf{j}, \ 0 \le t \le 1$

$\mathbf{r}'(t) = 2t\mathbf{i} + 2t\mathbf{j}$

$$\begin{aligned}\int_C \mathbf{F} \cdot d\mathbf{r} &= \int_0^1 \left[t^2(t^2)(2t) + 2(t^2)(t^2)(2t)\right] dt \\ &= \int_0^1 6t^5\,dt = t^6\Big]_0^1 = 1\end{aligned}$$

32. $d\mathbf{r} = \left[(-4\sin t)\mathbf{i} + 3\cos t\mathbf{j}\right] dt$

$\mathbf{F} = (4\cos t - 3\sin t)\mathbf{i} + (4\cos t + 3\sin t)\mathbf{j}, \ 0 \le t \le 2\pi$

$$\int_C \mathbf{F} \cdot d\mathbf{r} = \int_0^{2x} (12 - 7\sin t\cos t)\,dt = \left[12t - \frac{7\sin^2 t}{2}\right]_0^{2\pi} = 24\pi$$

33. $d\mathbf{r} = \left[(-2\sin t)\mathbf{i} + (2\cos t)\mathbf{j} + \mathbf{k}\right]dt$

$\mathbf{F} = (2\cos t)\mathbf{i} + (2\sin t)\mathbf{j} + t\mathbf{k},\ 0 \le t \le 2\pi$

$\int_C \mathbf{F}\cdot d\mathbf{r} = \int_0^{2\pi} t\,dt = 2\pi^2$

34. $x = 2 - t,\ y = 2 - t,\ z - \sqrt{4t - t^2},\ 0 \le t \le 2$

$$d\mathbf{r} = \left[-\mathbf{i} - \mathbf{j} + \frac{2 - t}{\sqrt{4t - t^2}}\mathbf{k}\right]dt$$

$$\mathbf{F} = \left(4 - 2t - \sqrt{4t - t^2}\right)\mathbf{i} + \left(\sqrt{4t - t^2} - 2 + t\right)\mathbf{j} + 0\mathbf{k}$$

$$\int_C \mathbf{F}\cdot d\mathbf{r} = \int_0^2 (t - 2)\,dt = \left[\frac{t^2}{2} - 2t\right]_0^2 = -2$$

35. $\mathbf{F}(x, y, z) = (y + z)\mathbf{i} + (x + z)\mathbf{j} + (x + y)\mathbf{k}$

Curve of intersection: $x = t,\ y = t,\ z = t^2 + t^2 = 2t^2$

$\mathbf{r}(t) = t\mathbf{i} + t\mathbf{j} + 2t^2\mathbf{k}, \quad 0 \le t \le 2$

$\mathbf{r}'(t) = \mathbf{i} + \mathbf{j} + 4t\mathbf{k}$

$$\int_C \mathbf{F}\cdot d\mathbf{r} = \int_0^2 \left[\left(t + 2t^2\right) + \left(t + 2t^2\right) + (2t)(4t)\right]dt = \int_0^2 \left[12t^2 + 2t\right]dt = \left[4t^3 + t^2\right]_0^2 = 36$$

36. Let $x = 2\sin t,\ y = -2\cos t,\ z = 4\sin^2 t,\ 0 \le t \le \pi.$

$d\mathbf{r} = \left[(2\cos t)\mathbf{i} + (2\sin t)\mathbf{j} + (8\sin t\cos t)\mathbf{k}\right]dt$

$\mathbf{F} = 0\mathbf{i} + 4\mathbf{j} + (2\sin t)\mathbf{k},$

$$\int_C \mathbf{F}\cdot d\mathbf{r} = \int_0^{\pi} \left(8\sin t + 16\sin^2 t\cos t\right)dt = \left[-8\cos t + \tfrac{16}{3}\sin^3 t\right]_0^{\pi} = 16$$

37. For $y = x^2,\ \mathbf{r}_1(t) = t\mathbf{i} + t^2\mathbf{j},\ 0 \le t \le 2$

For $y = 2x,\ \mathbf{r}_2(t) = (2 - t)\mathbf{i} + (4 - 2t)\mathbf{j},\ 0 \le t \le 2$

$$\int_C xy\,dx + \left(x^2 + y^2\right)dy = \int_{C_1} xy\,dx + \left(x^2 + y^2\right)dy + \int_{C_2} xy\,dx + \left(x^2 + y^2\right)dy$$

$$= \tfrac{100}{3} + (-32) = \tfrac{4}{3}$$

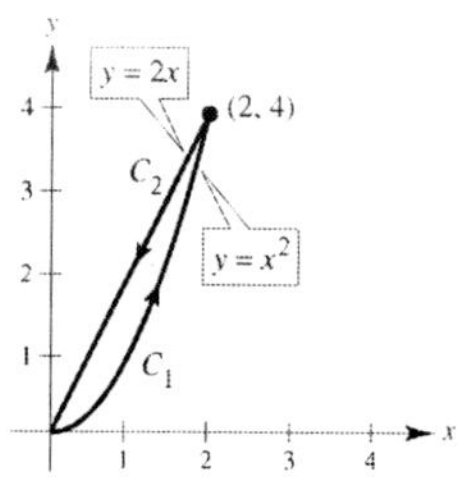

38. $\int_C \mathbf{F}\cdot d\mathbf{r} = \int_C (2x - y)\,dx + (2y - x)\,dy$

$\mathbf{r}(t) = (2\cos t + 2t\sin t)\mathbf{i} + (2\sin t - 2t\cos t)\mathbf{j},\ 0 \le t \le \pi$

$\int_C \mathbf{F}\cdot d\mathbf{r} = 4\pi^2 + 4\pi$

39. $\mathbf{F} = x\mathbf{i} - \sqrt{y}\mathbf{j}$ is conservative.

$$\text{Work} = \left[\tfrac{1}{2}x^2 - \tfrac{2}{3}y^{3/2}\right]_{(0,0)}^{(4,8)} = \tfrac{1}{2}(16) - \left(\tfrac{2}{3}\right)8^{3/2} = \tfrac{8}{3}\left(3 - 4\sqrt{2}\right)$$

40. $\mathbf{r}(t) = 10\sin t\mathbf{i} + 10\cos t\mathbf{j} + \dfrac{2000/5280}{\pi/2}t\mathbf{k} = 10\sin t\mathbf{i} + 10\cos t\mathbf{j} + \dfrac{25}{33\pi}t\mathbf{k},\ 0 \le t \le \dfrac{\pi}{2}$

$\mathbf{F} = 20\mathbf{k}$

$d\mathbf{r} = \left(10\cos t\mathbf{i} - 10\sin t\mathbf{j} + \dfrac{25}{33\pi}\mathbf{k}\right)$

$\int_C \mathbf{F}\cdot d\mathbf{r} = \int_0^{\pi/2} \dfrac{500}{33\pi}\,dt = \dfrac{250}{33}$ mi · ton

41. $\int_c 2xyz\,dx + x^2z\,dy + x^2y\,dz = \Big[x^2yz\Big]_{(0,0,0)}^{(1,3,2)} = 6$

42. $\int_c y\,dx + x\,dy + \dfrac{1}{z}\,dz = \Big[xy + \ln|z|\Big]_{(0,0,1)}^{(4,4,4)} = 16 + \ln 4$

43. (a) $\int_C y^2\,dx + 2xy\,dy = \int_0^1 \left[(1+t)^2(3) + 2(1+3t)(1+t)\right]dt$

$= \int_0^1 3(t^2 + 2t + 1) + 2(3t^2 + 4t + 1)\,dt = \int_0^1 (9t^2 + 14t + 5)\,dt = \Big[3t^2 + 7t^2 + 5t\Big]_0^1 = 15$

(b) $\int_C y^2\,dx + 2xy\,dy = \int_1^4 \left[t(1) + 2(t)\left(\sqrt{t}\right)\dfrac{1}{2\sqrt{t}}\right]dt = \int_1^4 \left[(t + t)\right]dt = \Big[t^2\Big]_1^4 = 15$

(c) $\mathbf{F}(x, y) = y^2\mathbf{i} + 2xy\,\mathbf{j} = \nabla f$ where $f(x, y) = xy^2$.

So, $\int_C \mathbf{F}\cdot d\mathbf{r} = 4(2)^2 - 1(1)^2 = 15.$

44. $x = a(\theta - \sin\theta),\ y = a(1 - \cos\theta),\ 0 \le \theta \le 2\pi$

(a) $A = \dfrac{1}{2}\int_C x\,dy - y\,dx.$

Because these equations orient the curve backwards, you will use

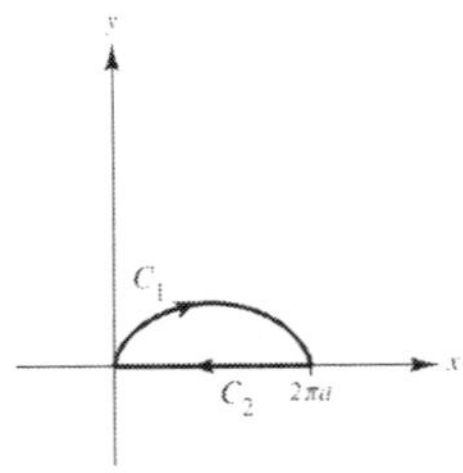

$A = \dfrac{1}{2}\int(y\,dx - x\,dy)$

$= \dfrac{1}{2}\int_0^{2\pi}\left[a^2(1-\cos\theta)(1-\cos\theta) - a^2(\theta - \sin\theta)(\sin\theta)\right]d\theta + \dfrac{1}{2}\int_0^{2\pi a}(0 - 0)\,d\theta$

$= \dfrac{a^2}{2}\int_0^{2\pi}\left[1 - 2\cos\theta + \cos^2\theta - \theta\sin\theta + \sin^2\theta\right]d\theta$

$= \dfrac{a^2}{2}\int_0^{2\pi}(2 - 2\cos\theta - \theta\sin\theta)\,d\theta$

$= \dfrac{a^2}{2}(6\pi) = 3\pi a^2.$

(b) By symmetry, $\bar{x} = \pi a$. From Section 15.4,

$\bar{y} = -\dfrac{1}{2A}\int_C y^2\,dx = \dfrac{1}{2A}\int_0^{2\pi} a^3(1-\cos\theta)^2(1-\cos\theta)\,d\theta = \dfrac{1}{2(3\pi a^2)}a^3(5\pi) = \dfrac{5}{6}a.$

45. $\int_C y\,dx + 2x\,dy = \int_0^1\int_0^1\left(\dfrac{\partial N}{\partial x} - \dfrac{\partial M}{\partial y}\right)dy\,dx$

$= \int_0^1\int_0^1(2 - 1)\,dy\,dx = 1$

46. $\int_C xy\,dx + (x^2 + y^2)\,dy = \int_0^2\int_0^2(2x - x)\,dy\,dx$

$= \int_0^2 2x\,dx = 4$

47. $\int_C xy^2\,dx + x^2y\,dy = \int_R\int\left(\dfrac{\partial N}{\partial x} - \dfrac{\partial M}{\partial y}\right)dA$

$= \int_R\int(2xy - 2xy)\,dA = 0$

48. $\int_C (x^2 - y^2)\,dx + 2xy\,dy = \int_{-a}^{a}\int_{-\sqrt{a^2-x^2}}^{\sqrt{a^2-x^2}} 4y\,dy\,dx$

$= \int_{-a}^{a} 0\,dx = 0$

49. $\int_C xy\,dx + x^2\,dy = \int_R\int\left(\frac{\partial N}{\partial x} - \frac{\partial M}{\partial y}\right) dA$

$$= \int_{-1}^{1}\int_{x^2}^{1}(2x - x)\,dy\,dx$$

$$= \int_{-1}^{1}\left[xy\right]_{x^2}^{1} = dx$$

$$= \int_{-1}^{1}\left(x - x^3\right) dx$$

$$= \left[\frac{x^2}{2} - \frac{x^4}{4}\right]_{-1}^{1} = 0$$

50. $\int_C y^2\,dx + x^{4/3}\,dy = \int_{-1}^{1}\int_{-(1-x^{2/3})^{3/2}}^{(1-x^{2/3})^{3/2}}\left(\frac{4}{3}x^{1/3} - 2y\right) dy\,dx = \int_{-1}^{1}\left[\frac{4}{3}x^{1/3}y - y^2\right]_{-(1-x^{2/3})^{3/2}}^{(1-x^{2/3})^{3/2}} dx$

$$= \int_{-1}^{1}\frac{8}{3}x^{1/3}\left(1 - x^{2/3}\right)^{3/2} dx = \left[-\frac{8}{7}x^{2/3}\left(1 - x^{2/3}\right)^{5/2} - \frac{16}{35}\left(1 - x^{2/3}\right)^{5/2}\right]_{-1}^{1} = 0$$

51. $\mathbf{r}(u, v) = \sec u \cos v\mathbf{i} + (1 + 2\tan u)\sin v\mathbf{j} + 2u\mathbf{k}$

$0 \le u \le \frac{\pi}{3},\ 0 \le v \le 2\pi$

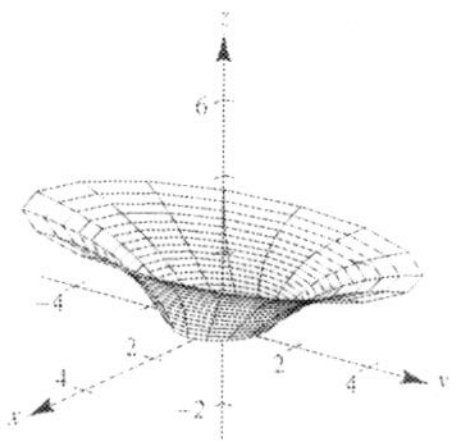

52. $\mathbf{r}(u, v) = e^{-u/4}\cos v\mathbf{i} + e^{-u/4}\sin v\mathbf{j} + \frac{u}{6}\mathbf{k}$

$0 \le u \le 4,\ 0 \le v \le 2\pi$

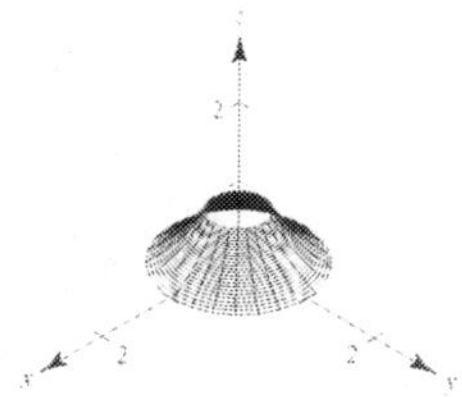

53. (a)

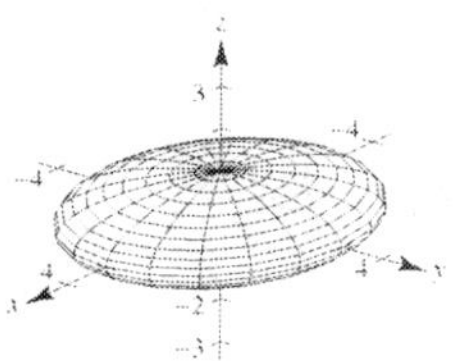

(b)

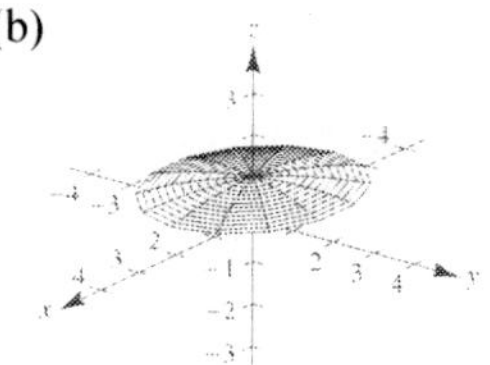

(c)

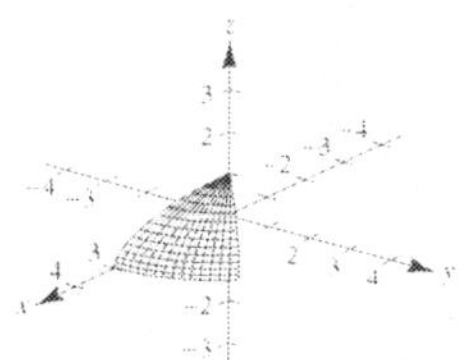

(d)

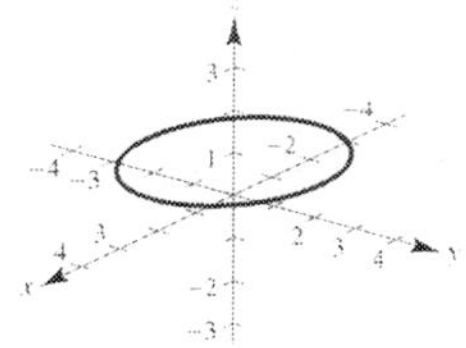

The space curve is a circle: $\mathbf{r}\left(u, \frac{\pi}{4}\right) = \frac{3\sqrt{2}}{2}\cos u\mathbf{i} + \frac{3\sqrt{2}}{2}\sin u\mathbf{j} + \frac{\sqrt{2}}{2}\mathbf{k}$

(e) $\mathbf{r}_u = -3\cos v\sin u\,\mathbf{i} + 3\cos v\cos u\,\mathbf{j}$

$\mathbf{r}_v = -3\sin v\sin u\,\mathbf{i} - 3\sin v\sin u\,\mathbf{j} + \cos v\,\mathbf{k}$

$$\mathbf{r}_u \times \mathbf{r}_v = \begin{vmatrix} \mathbf{i} & \mathbf{j} & \mathbf{k} \\ -3\cos v\sin u & 3\cos v\cos u & 0 \\ -3\sin v\sin u & 3\sin v\sin u & \cos v \end{vmatrix}$$

$$= \left(3\cos^2 v\cos u\right)\mathbf{i} + \left(3\cos^2 v\sin u\right)\mathbf{j} + \left(9\cos v\sin v\sin^2 u + 9\cos v\sin v\cos^2 u\right)\mathbf{k}$$

$$= \left(3\cos^2 v\cos u\right)\mathbf{i} + \left(3\cos^2 v\sin u\right)\mathbf{j} + \left(9\cos v\sin v\right)\mathbf{k}$$

$$\left\|\mathbf{r}_u \times \mathbf{r}_v\right\| = \sqrt{9\cos^4 v\cos^2 u + 9\cos^4 v\sin^2 u + 81\cos^2 v\sin^2 v} = \sqrt{9\cos^4 v + 81\cos^2 v\sin^2 v}$$

Using a Symbolic integration utility, $\int_{\pi/4}^{\pi/2}\int_0^{2\pi}\left\|\mathbf{r}_u \times \mathbf{r}_v\right\| dv\,du \approx 14.44$.

(f) Similarly, $\int_0^{\pi/4}\int_0^{\pi/2}\left\|\mathbf{r}_u \times \mathbf{r}_v\right\| dv\,du \approx 4.27$.

54. S: $\mathbf{r}(u, v) = (u + v)\mathbf{i} + (u - v)\mathbf{j} + \sin v\mathbf{k}, \quad 0 \le u \le 2, 0 \le v \le \pi$

$\mathbf{r}_u(u, v) = \mathbf{i} + \mathbf{j}$

$\mathbf{r}_v(u, v) = \mathbf{i} - \mathbf{j} + \cos v\mathbf{k}$

$$\mathbf{r}_u \times \mathbf{r}_v = \begin{vmatrix} \mathbf{i} & \mathbf{j} & \mathbf{k} \\ 1 & 1 & 0 \\ 1 & -1 & \cos v \end{vmatrix} = \cos v\mathbf{i} - \cos v\mathbf{j} - 2\mathbf{k}$$

$$\left\|\mathbf{r}_u \times \mathbf{r}_v\right\| = \sqrt{2\cos^2 v + 4}$$

$$\int_S\int z\,dS = \int_0^{\pi}\int_0^{2}\sin v\sqrt{2\cos^2 v + 4}\,du\,dv = 2\left[\sqrt{6} + \sqrt{2}\ln\left(\frac{\sqrt{6} + \sqrt{2}}{\sqrt{6} - \sqrt{2}}\right)\right]$$

55. S: $\mathbf{r}(u, v) = u\cos v\mathbf{i} + u\sin v\mathbf{j} + (u - 1)(2 - u)\mathbf{k}, \quad 0 \le u \le 2, 0 \le v \le 2\pi$

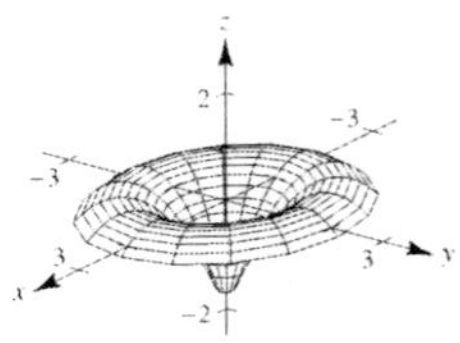

$\mathbf{r}_u(u, v) = \cos v\mathbf{i} + \sin v\mathbf{j} + (3 - 2u)\mathbf{k}$

$\mathbf{r}_v(u, v) = -u\sin v\mathbf{i} + u\cos v\mathbf{j}$

$$\mathbf{r}_u \times \mathbf{r}_v = \begin{vmatrix} \mathbf{i} & \mathbf{j} & \mathbf{k} \\ \cos v & \sin v & 3 - 2u \\ -u\sin v & u\cos v & 0 \end{vmatrix} = (2u - 3)u\cos v\mathbf{i} + (2u - 3)u\sin v\mathbf{j} + u\mathbf{k}$$

$$\left\|\mathbf{r}_u \times \mathbf{r}_v\right\| = u\sqrt{(2u - 3)^2 + 1}$$

$$\int_S\int(x + y)\,dS = \int_0^{2\pi}\int_0^{2}(u\cos v + u\sin v)u\sqrt{(2u - 3)^2 + 1}\,du\,dv = \int_0^{2}\int_0^{2\pi}(\cos v + \sin v)u^2\sqrt{(2u - 3)^2 + 1}\,dv\,du = 0$$

56. (a) $z = a\left(a - \sqrt{x^2 + y^2}\right), 0 \le z \le a^2$

$z = 0 \Rightarrow x^2 + y^2 = a^2$

(b) S: $g(x, y) = z = a^2 - a\sqrt{x^2 + y^2}$

$\rho(x, y) = k\sqrt{x^2 + y^2}$

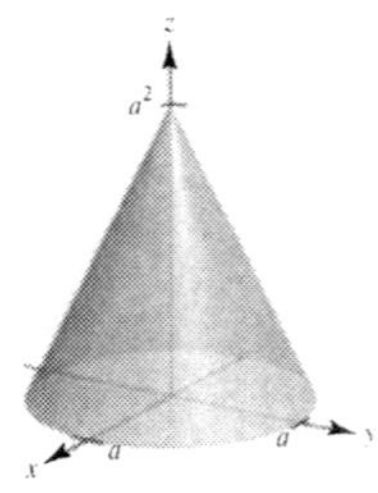

$$m = \int_S\int e(x, y, z)\,dS = \int_R\int k\sqrt{x^2 + y^2}\sqrt{1 + g_x^2 + g_y^2}\,dA$$

$$= k\int_R\int\sqrt{x^2 + y^2}\sqrt{1 + \frac{a^2x^2}{x^2 + y^2} + \frac{a^2y^2}{x^2 + y^2}}\,dA = k\int_R\int\sqrt{a^2 + 1}\left(\sqrt{x^2 + y^2}\right)dA$$

$$= k\sqrt{a^2 + 1}\int_0^{2\pi}\int_0^{a} r^2\,dr\,d\theta = k\sqrt{a^2 + 1}\int_0^{2\pi}\frac{a^3}{3}\,d\theta = \frac{2}{3}k\sqrt{a^2 + 1}\,a^3\pi$$

57. $\mathbf{F}(x, y, z) = x^2\mathbf{i} + xy\mathbf{j} + z\mathbf{k}$

Q: solid region bounded by the coordinates planes and the plane $2x + 3y + 4z = 12$

Surface Integral: There are four surfaces for this solid.

$z = 0,\ \mathbf{N} = -\mathbf{k}, \mathbf{F} \cdot \mathbf{N} = -z, \quad \int_{S_1}\int 0\, dS = 0$

$y = 0,\ \mathbf{N} = -\mathbf{j}, \mathbf{F} \cdot \mathbf{N} = -xy, \quad \int_{S_2}\int 0\, dS = 0$

$x = 0,\ \mathbf{N} = -\mathbf{i}, \mathbf{F} \cdot \mathbf{N} = -x^2, \quad \int_{S_3}\int 0\, dS = 0$

$2x + 3y + 4z = 12,\ \mathbf{N} = \dfrac{2\mathbf{i} + 3\mathbf{j} + 4\mathbf{k}}{\sqrt{29}},\ dS = \sqrt{1 + \left(\dfrac{1}{4}\right) + \left(\dfrac{9}{16}\right)}dA = \dfrac{\sqrt{29}}{4}\, dA$

$$\begin{aligned}
\int_{S_4}\int \mathbf{F} \cdot \mathbf{N}\, dS &= \frac{1}{4}\int_R\int\left(2x^2 + 3xy + 4z\right) dA \\
&= \frac{1}{4}\int_0^6\int_0^{4-(2x/3)}\left(2x^2 + 3xy + 12 - 2x - 3y\right) dy\, dx \\
&= \frac{1}{4}\int_0^6\left[2x^2\left(\frac{12 - 2x}{3}\right) + \frac{3x}{2}\left(\frac{12 - 2x}{3}\right)^2 + 12\left(\frac{12 - 2x}{3}\right) - 2x\left(\frac{12 - 2x}{3}\right) - \frac{3}{2}\left(\frac{12 - 2x}{3}\right)^2\right] dx \\
&= \frac{1}{6}\int_0^6\left(-x^3 + x^2 + 24x + 36\right) dx \\
&= \frac{1}{6}\left[-\frac{x^4}{4} + \frac{x^3}{3} + 12x^2 + 36x\right]_0^6 = 66
\end{aligned}$$

Divergence Theorem: Because $\text{div}\,\mathbf{F} = 2x + x + 1 = 3x + 1$, Divergence Theorem yields

$$\begin{aligned}
\iiint_Q \text{div}\,\mathbf{F}\, dV &= \int_0^6\int_0^{(12-2x)/3}\int_0^{(12-2x-3y)/4}(3x + 1)\, dz\, dy\, dx \\
&= \int_0^6\int_0^{(12-2x)/3}(3x + 1)\left(\frac{12 - 2x - 3y}{4}\right) dy\, dx \\
&= \frac{1}{4}\int_0^6(3x + 1)\left(12y - 2xy - \frac{3}{2}y^2\right)\Bigg]_0^{(12-2x)/3} dx \\
&= \frac{1}{4}\int_0^6(3x + 1)\left[4(12 - 2x) - 2x\left(\frac{12 - 2x}{3}\right) - \frac{3}{2}\left(\frac{12 - 2x}{3}\right)^2\right] dx \\
&= \frac{1}{4}\int_0^6\frac{2}{3}\left(3x^3 - 35x^2 + 96x + 36\right) dx \\
&= \frac{1}{6}\left[\frac{3x^4}{4} - \frac{35x^3}{3} + 48x^2 + 36x\right] = 66.
\end{aligned}$$

58. $\mathbf{F}(x, y, z) = x\mathbf{i} + y\mathbf{j} + z\mathbf{k}$

Q: solid region bounded by the coordinate planes and the plane $2x + 3y + 4z = 12$

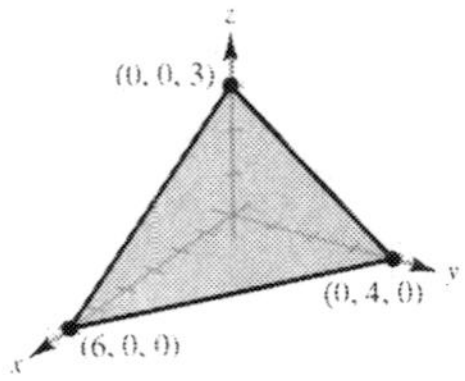

Surface Integral: There are four surfaces for this solid.

$z = 0,\ \mathbf{N} = -\mathbf{k},\ \mathbf{F} \cdot \mathbf{N} = -z,\ \int_{S_1}\int 0\, dS = 0$

$y = 0,\ \mathbf{N} = -\mathbf{j},\ \mathbf{F} \cdot \mathbf{N} = -y,\ \int_{S_2}\int 0\, dS = 0$

$x = 0,\ \mathbf{N} = -\mathbf{i},\ \mathbf{F} \cdot \mathbf{N} = -x,\ \int_{S_3}\int 0\, dS = 0$

$$2x + 3y + 4z = 12,\ \mathbf{N} = \frac{2\mathbf{i} + 3\mathbf{j} + 4\mathbf{k}}{\sqrt{29}},\ dS = \sqrt{1 + \left(\frac{1}{4}\right) + \left(\frac{9}{16}\right)}\, dA = \frac{\sqrt{29}}{4}\, dA$$

$$\int_{S_4}\int \mathbf{N} \cdot \mathbf{F}\, dS = \frac{1}{4}\int_R\int (2x + 3y + 4z)\, dy\, dx = \frac{1}{4}\int_0^6 \int_0^{(12-2x)/3} 12\, dy\, dx = 3\int_0^6 \left(4 - \frac{2x}{3}\right) dx = 3\left[4x - \frac{x^2}{3}\right]_0^6 = 36$$

Triple Integral: Because $\operatorname{div} \mathbf{F} = 3$, the Divergence Theorem yields

$$\iiint_Q \operatorname{div} \mathbf{F}\, dV = \iiint_Q 3\, dV = 3(\text{Volume of solid}) = 3\left[\frac{1}{3}(\text{Area of base})(\text{Height})\right] = \frac{1}{2}(6)(4)(3) = 36.$$

59. $\mathbf{F}(x, y, z) = (\cos y + y \cos x)\mathbf{i} + (\sin x - x \sin y)\mathbf{j} + xyz\mathbf{k}$

S: portion of $z = y^2$ over the square in the xy-plane with vertices $(0, 0), (a, 0), (a, a), (0, a)$

Line Integral: Using the line integral you have:

$C_1: y = 0,\ dy = 0$

$C_2: x = 0,\ dx = 0,\ z = y^2,\ dz = 2y\, dy$

$C_3: y = a,\ dy = 0,\ z = a^2,\ dz = 0$

$C_4: x = a,\ dx = 0,\ z = y^2,\ dz = 2y\, dy$

$$\begin{aligned}\int_C \mathbf{F} \cdot d\mathbf{r} &= \int_C (\cos y + y \cos x)\, dx + (\sin x - x \sin y)\, dy + xyz\, dz \\ &= \int_{C_1} dx + \int_{C_2} 0 + \int_{C_3} (\cos a + a \cos x)\, dx + \int_{C_4} (\sin a - a \sin y)\, dy + ay^3(2y\, dy) \\ &= \int_0^a dx + \int_a^0 (\cos a + a \cos x)\, dx + \int_0^a (\sin a - a \sin y)\, dy + \int_0^a 2ay^4\, dy \\ &= a + \big[x \cos a + a \sin x\big]_a^0 + \big[y \sin a + a \cos y\big]_0^a + \left[2a\frac{y^5}{5}\right]_0^a \\ &= a - a \cos a - a \sin a + a \sin a + a \cos a - a + \frac{2a^6}{5} = \frac{2a^6}{5}\end{aligned}$$

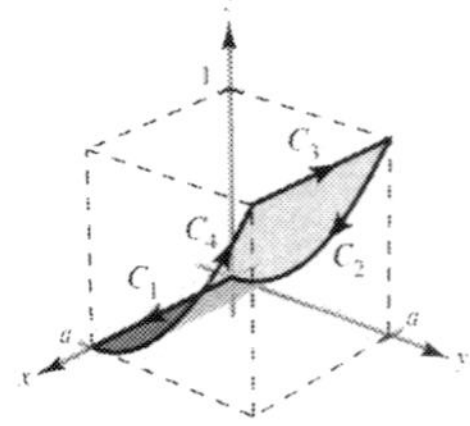

Double Integral: Considering $f(x, y, z) = z - y^2$, you have:

$$\mathbf{N} = \frac{\nabla f}{\|\nabla f\|} = \frac{-2y\mathbf{j} + \mathbf{k}}{\sqrt{1 + 4y^2}},\ dS = \sqrt{1 + 4y^2}\, dA, \text{ and } \mathbf{curl\ F} = xz\mathbf{i} - yz\mathbf{j}.$$

So, $\int_S \int (\mathbf{curl\ F}) \cdot \mathbf{N}\, dS = \int_0^a \int_0^a 2y^2 z\, dy\, dx = \int_0^a \int_0^a 2y^4 z\, dy\, dx = \int_0^a \frac{2a^5}{5}\, dx = \frac{2a^6}{5}.$

60. $\mathbf{F}(x, y, z) = (x - z)\mathbf{i} + (y - z)\mathbf{j} + x^2\mathbf{k}$

S: first octant portion of the plane $3x + y + 2z = 12$

Line Integral:

C_1: $y = 0,\ dy = 0,\ z = \dfrac{12 - 3x}{2},\ dz = -\dfrac{3}{2}\,dx$

C_2: $x = 0,\ dx = 0,\ z = \dfrac{12 - y}{2},\ dz = -\dfrac{1}{2}\,dy$

C_3: $z = 0,\ dz = 0,\ y = 12 - 3x,\ dy = -3\,dx$

$$\int_C \mathbf{F} \cdot d\mathbf{r} = \int_C (x - z)\,dx + (y - z)\,dy + x^2\,dz$$

$$= \int_{C_1} \left[x - \frac{12 - 3x}{2} + x^2\left(-\frac{3}{2}\right)\right] dx + \int_{C_2} \left[y - \frac{12 - y}{2}\right] dy + \int_{C_3} \left[x + (12 - 3x)(-3)\right] dx$$

$$= \int_4^0 \left(-\frac{3}{2}x^2 + \frac{5}{2}x - 6\right) dx + \int_0^{12} \left(\frac{3}{2}y - 6\right) dy + \int_0^4 (10x - 36)\,dx = 8$$

Double Integral: $G(x, y, z) = \dfrac{12 - 3x - y}{2} - z$

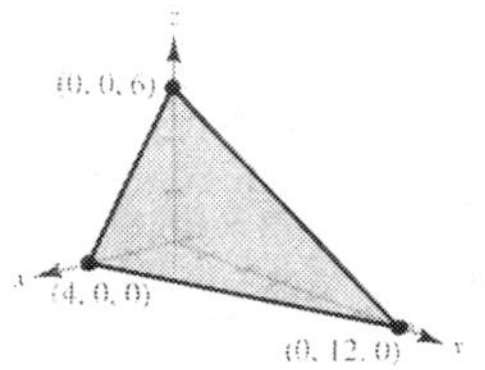

$$\nabla G(x, y, z) = -\frac{3}{2}\mathbf{i} - \frac{1}{2}\mathbf{j} - \mathbf{k}$$

$$\textbf{curl } \mathbf{F} = \mathbf{i} - (2x + 1)\mathbf{j}$$

$$\int_S \int (\textbf{curl } \mathbf{F}) \cdot \mathbf{N}\,dS = \int_0^4 \int_0^{12-3x} (x - 1)\,dy\,dx = \int_0^4 \left(-3x^2 + 15x - 12\right) dx = 8$$

61. If $\textbf{curl}(\mathbf{F}) = x\mathbf{i} + y\mathbf{j} + z\mathbf{k}$, then $\text{div}(\textbf{curl } \mathbf{F}) = 1 + 1 + 1 = 3$, contradicting Theorem 15.3.

Problem Solving for Chapter 15

1. (a) $\nabla T = \dfrac{-25}{\left(x^2 + y^2 + z^2\right)^{3/2}}[x\mathbf{i} + y\mathbf{j} + z\mathbf{k}]$

$\mathbf{N} = x\mathbf{i} + \sqrt{1 - x^2}\,\mathbf{k}$

$dS = \dfrac{1}{\sqrt{1 - x^2}}\,dA$

$$\text{Flux} = \int_S \int -k\nabla T \cdot \mathbf{N}\,dS = 25k \int_R \int \left[\frac{x^2}{\left(x^2 + y^2 + z^2\right)^{3/2}\left(1 - x^2\right)^{1/2}} + \frac{z}{\left(x^2 + y^2 + z^2\right)^{3/2}}\right] dA$$

$$= 25k \int_{-1/2}^{1/2} \int_0^1 \left[\frac{x^2}{\left(x^2 + y^2 + z^2\right)^{3/2}\left(1 - x^2\right)^{1/2}} + \frac{1 - x^2}{\left(x^2 + y^2 + z^2\right)^{3/2}\left(1 - x^2\right)^{1/2}}\right] dy\,dx$$

$$= 25k \int_{-1/2}^{1/2} \int_0^1 \frac{1}{\left(1 + y^2\right)^{3/2}\left(1 - x^2\right)^{1/2}}\,dy\,dx = 25k \int_0^1 \frac{1}{\left(1 + y^2\right)^{3/2}}\,dy \int_{-1/2}^{1/2} \frac{1}{\left(1 - x^2\right)^{1/2}}\,dx = 25k\left(\frac{\sqrt{2}}{2}\right)\left(\frac{\pi}{3}\right) = 25k\frac{\sqrt{2}\pi}{6}$$

(b) $\mathbf{r}(u,v) = \langle \cos u, v, \sin u\rangle$

$\mathbf{r}_u = \langle -\sin u, 0, \cos u\rangle,\ \mathbf{r}_v = \langle 0, 1, 0\rangle$

$\mathbf{r}_u \times \mathbf{r}_v = \langle -\cos u, 0, \sin u\rangle$

$$\nabla T = \frac{-25}{\left(x^2 + y^2 + z^2\right)^{3/2}}[x\mathbf{i} + y\mathbf{j} + z\mathbf{k}] = \frac{-25}{\left(v^2 + 1\right)^{3/2}}[\cos u\mathbf{i} + v\mathbf{j} + \sin u\mathbf{k}]$$

$$\nabla T \cdot (\mathbf{r}_u \times \mathbf{r}_v) = \frac{-25}{\left(v^2 + 1\right)^{3/2}}\left(-\cos^2 u - \sin^2 u\right) = \frac{25}{\left(v^2 + 1\right)^{3/2}}$$

$$\text{Flux} = \int_0^1 \int_{\pi/3}^{2\pi/3} \frac{25}{\left(v^2 + 1\right)^{3/2}}\,du\,dv = 25k\frac{\sqrt{2}\pi}{6}$$

2. (a) $z = \sqrt{1 - x^2 - y^2},\ \dfrac{\partial z}{\partial x} = \dfrac{-x}{\sqrt{1 - x^2 - y^2}},\ \dfrac{\partial z}{\partial y} = \dfrac{-y}{\sqrt{1 - x^2 - y^2}}$

$$dS = \sqrt{\left(\frac{\partial z}{\partial x}\right)^2 + \left(\frac{\partial z}{\partial y}\right)^2 + 1}\,dA = \frac{1}{\sqrt{1 - x^2 - y^2}}$$

$$\nabla T = \frac{-25}{\left(x^2 + y^2 + z^2\right)^{3/2}}(x\mathbf{i} + y\mathbf{j} + z\mathbf{k}) = -25(x\mathbf{i} + y\mathbf{j} + z\mathbf{k})$$

$$\mathbf{N} = \frac{-\dfrac{\partial z}{\partial x}\mathbf{i} + \dfrac{\partial z}{\partial y}\mathbf{j} + \mathbf{k}}{\sqrt{\left(\dfrac{\partial z}{\partial x}\right)^2 + \left(\dfrac{\partial z}{\partial y}\right)^2 + 1}} = \left(\frac{x}{\sqrt{1 - x^2 - y^2}}\mathbf{i} + \frac{y}{\sqrt{1 - x^2 - y^2}}\mathbf{j} + \mathbf{k}\right)\sqrt{1 - x^2 - y^2}$$

$$= x\mathbf{i} + y\mathbf{j} + \sqrt{1 - x^2 - y^2}\,\mathbf{k} = x\mathbf{i} + y\mathbf{j} + z\mathbf{k}$$

$$\text{Flux} = \int_S\int -k\nabla T \cdot \mathbf{N}\,dS = k\int_R\int 25(x\mathbf{i} + y\mathbf{j} + z\mathbf{k}) \cdot (x\mathbf{i} + y\mathbf{j} + z\mathbf{k})\frac{1}{\sqrt{1 - x^2 - y^2}}\,dA$$

$$= k\int_R\int \frac{25}{\sqrt{1 - x^2 - y^2}}\,dA = 25k\int_0^{2\pi}\int_0^1 \frac{1}{\sqrt{1 - r^2}}r\,dr\,d\theta = 50\pi k$$

(b) $\mathbf{r}(u, v) = \langle \sin u \cos v, \sin u \sin v, \cos u\rangle$

$\mathbf{r}_u = \langle \cos u \cos v, \cos u \sin v, -\sin u\rangle$

$\mathbf{r}_v = \langle -\sin u \sin v, \sin u \cos v, 0\rangle$

$\mathbf{r}_u \times \mathbf{r}_v = \left\langle \sin^2 u \cos v, \sin^2 u \sin v, \sin u \cos u \sin^2 v + \sin u \cos u \cos^2 v\right\rangle$

$\|\mathbf{r}_u \times \mathbf{r}_v\| = \sin u$

$$\text{Flux} = 25k\int_0^{2\pi}\int_0^{\pi/2} \sin u\,du\,dv = 50\pi k$$

3. $\mathbf{r}(t) = \langle 3\cos t, 3\sin t, 2t\rangle$

$\mathbf{r}'(t) = \langle -3\sin t, 3\cos t, 2t\rangle, \|\mathbf{r}'(t)\| = \sqrt{13}$

$$I_x = \int_C \left(y^2 + z^2\right)\rho\,ds = \int_0^{2\pi}\left(9\sin^2 t + 4t^2\right)\sqrt{13}\,dt = \frac{1}{3}\sqrt{13}\pi\left(32\pi^2 + 27\right)$$

$$I_y = \int_C \left(x^2 + z^2\right)\rho\,ds = \int_0^{2\pi}\left(9\cos^2 t + 4t^2\right)\sqrt{13}\,dt = \frac{1}{3}\sqrt{13}\pi\left(32\pi^2 + 27\right)$$

$$I_z = \int_C \left(x^2 + y^2\right)\rho\,ds = \int_0^{2\pi}\left(9\cos^2 t + 9\sin^2 t\right)\sqrt{13}\,dt = 18\pi\sqrt{13}$$

4. $\mathbf{r}(t) = \left\langle \frac{t^2}{2}, t, \frac{2\sqrt{2}t^{3/2}}{3} \right\rangle$

$\mathbf{r}'(t) = \left\langle t, 1, \sqrt{2}t^{1/2} \right\rangle, \ \left\| \mathbf{r}'(t) \right\| = t + 1$

$\rho\, ds = \frac{1}{1+t}(t+1)\, dt = 1$

$I_x = \int_C \left(x^2 + z^2\right)\rho\, ds = \int_0^1 \left(\frac{t^4}{4} + \frac{8}{9}t^3\right) dt = \frac{49}{180}$

$I_y = \int_C \left(y^2 + z^2\right)\rho\, ds = \int_0^1 \left(t^2 + \frac{8}{9}t^3\right) dt = \frac{5}{9}$

$I_z = \int_C \left(x^2 + y^2\right)\rho\, ds = \int_0^1 \left(\frac{t^4}{4} + t^2\right) dt = \frac{23}{60}$

$\mathbf{F}(x, y, z) = x\mathbf{i} + y\mathbf{j} + z\mathbf{k}$ **and** $f(x, y, z) = \left\| \mathbf{F}(x, y, z) \right\| = \sqrt{x^2 + y^2 + z^2}$

5. (a) $\ln f = \frac{1}{2}\ln\left(x^2 + y^2 + z^2\right)$

$\nabla(\ln f) = \frac{x}{x^2 + y^2 + z^2}\mathbf{i} + \frac{y}{x^2 + y^2 + z^2}\mathbf{j} + \frac{z}{x^2 + y^2 + z^2}\mathbf{k} = \frac{x\mathbf{i} + y\mathbf{j} + z\mathbf{k}}{x^2 + y^2 + z^2} = \frac{\mathbf{F}}{f^2}$

(b) $\frac{1}{f} = \frac{1}{\sqrt{x^2 + y^2 + z^2}}$

$\nabla\left(\frac{1}{f}\right) = \frac{-x}{\left(x^2 + y^2 + z^2\right)^{3/2}}\mathbf{i} + \frac{-y}{\left(x^2 + y^2 + z^2\right)^{3/2}}\mathbf{j} + \frac{-z}{\left(x^2 + y^2 + z^2\right)^{3/2}}\mathbf{k} = \frac{-(x\mathbf{i} + y\mathbf{j} + z\mathbf{k})}{\left(\sqrt{x^2 + y^2 + z^2}\right)^3} = \frac{\mathbf{F}}{f^3}$

(c) $f^n = \left(\sqrt{x^2 + y^2 + z^2}\right)^n$

$\nabla f^n = n\left(\sqrt{x^2 + y^2 + z^2}\right)^{n-1}\frac{x}{\sqrt{x^2 + y^2 + z^2}}\mathbf{i} + n\left(\sqrt{x^2 + y^2 + z^2}\right)^{n-1}\frac{y}{\sqrt{x^2 + y^2 + z^2}}\mathbf{j}$

$\quad + n\left(\sqrt{x^2 + y^2 + z^2}\right)^{n-1}\frac{z}{\sqrt{x^2 + y^2 + z^2}}\mathbf{k}$

$= n\left(\sqrt{x^2 + y^2 + z^2}\right)^{n-2}(x\mathbf{i} + y\mathbf{j} + z\mathbf{k})$

$= nf^{n-2}\mathbf{F}$

(d) $w = \frac{1}{f} = \frac{1}{\sqrt{x^2 + y^2 + z^2}}$ $\qquad \frac{\partial^2 w}{dx^2} = \frac{2x^2 - y^2 - z^2}{\left(x^2 + y^2 + z^2\right)^{5/2}}$

$\frac{\partial w}{dx} = -\frac{x}{\left(x^2 + y^2 + z^2\right)^{3/2}}$ $\qquad \frac{\partial^2 w}{dy^2} = \frac{2y^2 - x^2 - z^2}{\left(x^2 + y^2 + z^2\right)^{5/2}}$

$\frac{\partial w}{dy} = -\frac{y}{\left(x^2 + y^2 + z^2\right)^{3/2}}$ $\qquad \frac{\partial^2 w}{dz^2} = \frac{2z^2 - x^2 - y^2}{\left(x^2 + y^2 + z^2\right)^{5/2}}$

$\frac{\partial w}{dz} = -\frac{z}{\left(x^2 + y^2 + z^2\right)^{3/2}}$ $\qquad \nabla^2 w = \frac{\partial^2 w}{dx^2} + \frac{\partial^2 w}{dy^2} + \frac{\partial^2 w}{dz^2} = 0$

Therefore $w = \frac{1}{f}$ is harmonic.

6. $\int_C y^n\,dx + x^n\,dy = \int_R\int\left(\dfrac{\partial N}{\partial x} - \dfrac{\partial M}{\partial x}\right)dA$

For the line integral, use the two paths

C_1: $\mathbf{r}_1(x) = x\mathbf{i}, -a \le x \le a$

C_2: $\mathbf{r}_2(x) = x\mathbf{i} + \sqrt{a^2 - x^2}\,\mathbf{j}, x = a$ to $x = -a$

$\int_{C_1} y^n\,dx + x^n\,dy = 0$

$\int_{C_2} y^n\,dx + x^n\,dy = \int_a^{-a}\left[\left(a^2 - x^2\right)^{n/2} + x^n\dfrac{-x}{\sqrt{a^2 - x^2}}\right]dx$

$\int_R\int\left(\dfrac{\partial N}{\partial x} - \dfrac{\partial M}{\partial y}\right)dA = \int_{-a}^{a}\int_0^{\sqrt{a^2-x^2}}\left[nx^{n-1} - ny^{n-1}\right]dy\,dx$

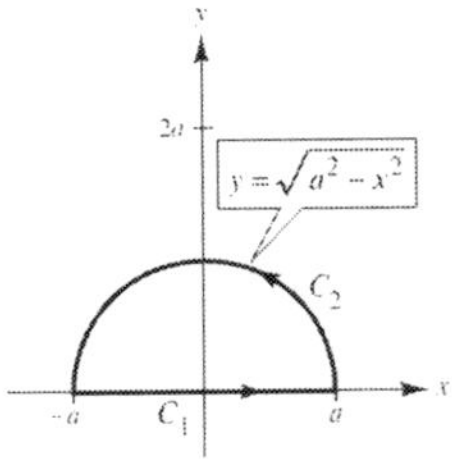

(a) For $n = 1, 3, 5, 7$ both integrals give 0.

(b) For n even, you obtain

$n = 2$: $-\dfrac{4}{3}a^3$ $\quad n = 4$: $-\dfrac{16}{15}a^5$ $\quad n = 6$: $-\dfrac{32}{35}a^7$ $\quad n = 8$: $-\dfrac{256}{315}a^9$

(c) If n is odd then the integral equals 0.

7. $\frac{1}{2}\int_C x\,dy - y\,dx = \frac{1}{2}\int_0^{2\pi}\left[a(\theta - \sin\theta)(a\sin\theta)\,d\theta - a(1 - \cos\theta)(a(1 - \cos\theta))\,d\theta\right]$

$= \frac{1}{2}a^2\int_0^{2\pi}\left[\theta\sin\theta = \sin^2\theta - 1 + 2\cos\theta - \cos^2\theta\right]d\theta = \frac{1}{2}a^2\int_0^{2\pi}(\theta\sin\theta + 2\cos\theta - 2)\,d\theta = -3\pi a^2$

So, the area is $3\pi a^2$.

8. $\frac{1}{2}\int_C x\,dy - y\,dx = 2\int_0^{\pi/2}\left[\frac{1}{2}\sin 2t\cos t - \sin t\cos 2t\right]dt = 2\left(\frac{2}{3}\right)$

So, the area is $\frac{4}{3}$.

9. (a) $\mathbf{r}(t) = t\,\mathbf{j}, 0 \le t \le 1$

$\mathbf{r}'(t) = \mathbf{j}$

$W = \int_C \mathbf{F}\cdot dr = \int_0^1 (t\mathbf{i} + \mathbf{j})\cdot\mathbf{j}\,dt = \int_0^1 dt = 1$

(b) $\mathbf{r}(t) = \left(t - t^2\right)\mathbf{i} + t\mathbf{j}, 0 \le t \le 1$

$\mathbf{r}'(t) = (1 - 2t)\mathbf{i} + \mathbf{j}$

$W = \mathbf{F}\cdot d\mathbf{r} = \int_0^1\left(\left(2t - t^2\right)\mathbf{i} + \left[\left(t - t^2\right)^2 + 1\right]\mathbf{j}\right)\cdot\left((1 - 2t)\mathbf{i} + \mathbf{j}\right)dt$

$= \int_0^1\left[(1 - 2t)\left(2t - t^2\right) + \left(t^4 - 2t^3 + t^2 + 1\right)\right]dt = \int_0^1\left(t^4 - 4t^2 + 2t + 1\right)dt = \dfrac{13}{15}$

(c) $\mathbf{r}(t) = c(t - t^2)\mathbf{i} + t\mathbf{j},\ 0 \le t \le 1$

$\mathbf{r}'(t) = c(1 - 2t)\mathbf{i} + \mathbf{j}$

$$\mathbf{F} \cdot d\mathbf{r} = \left(c(t - t^2) + t\right)\left(c(1 - 2t)\right) + \left(c^2(t - t^2)^2 + 1\right)(1)$$

$$= c^2t^4 - 2c^2t^2 + c^2t - 2ct^2 + ct + 1$$

$$W = \int_C \mathbf{F} \cdot d\mathbf{r} = \frac{1}{30}c^2 - \frac{1}{6}c + 1$$

$$\frac{dW}{dc} = \frac{1}{15}c - \frac{1}{6} = 0 \Rightarrow c = \frac{5}{2}$$

$$\frac{d^2W}{dc^2} = \frac{1}{15} > 0\ c = \frac{5}{2}\text{ minimum.}$$

10. $F(x, y) = 3x^2y^2\mathbf{i} + 2x^3y\mathbf{j}$ is conservative.

$f(x, y) = x^3y^2$ potential function.

Work $= f(2, 4) - f(1, 1) = 8(16) - 1 = 127$

11. $\mathbf{v} \times \mathbf{r} = \langle a_1, a_2, a_3\rangle \times \langle x, y, z\rangle$

$= \langle a_2z - a_3y, -a_1z + a_3x, a_1y - a_2x\rangle$

$\mathbf{curl}(\mathbf{v} \times \mathbf{r}) = \langle 2a_1, 2a_2, 2a_3\rangle = 2\mathbf{v}$

By Stokes's Theorem,

$$\int_C (\mathbf{v} \times \mathbf{r})\,d\mathbf{r} = \int_S \int \mathbf{curl}(\mathbf{v} \times \mathbf{r}) \cdot \mathbf{N}\,dS = \int_S \int 2\mathbf{v} \cdot \mathbf{N}\,dS.$$

12. Area $= \pi ab$

$\mathbf{r}(t) = a\cos t\mathbf{i} + b\sin t\mathbf{j},\ 0 \le t \le 2\pi$

$\mathbf{r}'(t) = -a\sin t\mathbf{i} + b\cos t\mathbf{j}$

$$\mathbf{F} = -\tfrac{1}{2}b\sin t\mathbf{i} + \tfrac{1}{2}a\cos t\mathbf{j}$$

$$\mathbf{F} \cdot d\mathbf{r} = \left[\tfrac{1}{2}ab\sin^2 t + \tfrac{1}{2}ab\cos^2 t\right]dt = \tfrac{1}{2}ab$$

$$W = \int_0^{2\pi} \mathbf{F} \cdot d\mathbf{r} = \tfrac{1}{2}ab(2\pi) = \pi ab$$

Same as area.

13. (a) (i) $\int_C f\nabla g \cdot d\mathbf{r} = \int_S \int \mathbf{curl}[f\nabla g] \cdot \mathbf{N}\,dS$ (Stokes's Theorem)

$$f\nabla g = f\frac{\partial g}{\partial x}\mathbf{i} + f\frac{\partial g}{\partial y}\mathbf{j} + f\frac{\partial g}{\partial z}\mathbf{k}$$

$$\mathbf{curl}(f\nabla g) = \begin{vmatrix} \mathbf{i} & \mathbf{j} & \mathbf{k} \\ \dfrac{\partial}{\partial x} & \dfrac{\partial}{\partial y} & \dfrac{\partial}{\partial z} \\ f(\partial g/\partial x) & f(\partial g/\partial y) & f(\partial g/\partial z) \end{vmatrix}$$

$$= \left[\left[f\left(\frac{\partial^2 g}{\partial y\partial z}\right) + \left(\frac{\partial f}{\partial y}\right)\left(\frac{\partial g}{\partial z}\right)\right] - \left[f\left(\frac{\partial^2 g}{\partial z\partial y}\right) + \left(\frac{\partial f}{\partial z}\right)\left(\frac{\partial g}{\partial y}\right)\right]\right]\mathbf{i}$$

$$-\left[\left[f\left(\frac{\partial^2 g}{\partial x\partial z}\right) + \left(\frac{\partial f}{\partial x}\right)\left(\frac{\partial g}{\partial z}\right)\right] - \left[f\left(\frac{\partial^2 g}{\partial z\partial x}\right) + \left(\frac{\partial f}{\partial z}\right)\left(\frac{\partial g}{\partial x}\right)\right]\right]\mathbf{j}$$

$$+\left[\left[f\left(\frac{\partial^2 g}{\partial x\partial y}\right) + \left(\frac{\partial f}{\partial x}\right)\left(\frac{\partial g}{\partial y}\right)\right] - \left[f\left(\frac{\partial^2 g}{\partial y\partial x}\right) + \left(\frac{\partial f}{\partial y}\right)\left(\frac{\partial g}{\partial x}\right)\right]\right]\mathbf{k}$$

$$= \left[\left(\frac{\partial f}{\partial y}\right)\left(\frac{\partial g}{\partial z}\right) - \left(\frac{\partial f}{\partial z}\right)\left(\frac{\partial g}{\partial y}\right)\right]\mathbf{i} - \left[\left(\frac{\partial f}{\partial x}\right)\left(\frac{\partial g}{\partial z}\right) - \left(\frac{\partial f}{\partial z}\right)\left(\frac{\partial g}{\partial x}\right)\right]\mathbf{j} + \left[\left(\frac{\partial f}{\partial x}\right)\left(\frac{\partial g}{\partial y}\right) - \left(\frac{\partial f}{\partial y}\right)\left(\frac{\partial g}{\partial x}\right)\right]\mathbf{k}$$

$$= \begin{vmatrix} \mathbf{i} & \mathbf{j} & \mathbf{k} \\ \dfrac{\partial f}{\partial x} & \dfrac{\partial f}{\partial y} & \dfrac{\partial f}{\partial z} \\ \dfrac{\partial g}{\partial x} & \dfrac{\partial g}{\partial y} & \dfrac{\partial g}{\partial z} \end{vmatrix} = \nabla f \times \nabla g$$

So, $\int_C f\nabla g \cdot d\mathbf{r} = \int_S \int \mathbf{curl}[f\nabla g] \cdot \mathbf{N}\,dS = \int_S \int [\nabla f \times \nabla g] \cdot \mathbf{N}\,dS.$

(ii) $\int_C (f\nabla f) \cdot d\mathbf{r} = \int_S \int (\nabla f \times \nabla f) \cdot \mathbf{N}\, dS$ (using part a)

$= 0$ because $\nabla f \times \nabla f = \mathbf{0}$.

(iii) $\int_C (f\nabla g + g\nabla f) \cdot d\mathbf{r} = \int_C (f\nabla g) \cdot d\mathbf{r} + \int_C (g\nabla f) \cdot d\mathbf{r}$

$= \int_S \int (\nabla f \times \nabla g) \cdot \mathbf{N}\, dS + \int_S \int (\nabla g \times \nabla f) \cdot \mathbf{N}\, dS$ (using part a)

$= \int_S \int (\nabla f \times \nabla g) \cdot \mathbf{N}\, dS + \int_S \int -(\nabla f \times \nabla g) \cdot \mathbf{N}\, dS = 0$

(b) $f(x, y, z) = xyz,\; g(x, y, z) = z,\; S: z = \sqrt{4 - x^2 - y^2}$

(i) $\nabla g(x, y, z) = \mathbf{k}$

$f(x, y, z)\nabla g(x, y, z) = xyz\mathbf{k}$

$\mathbf{r}(t) = 2\cos t\mathbf{i} + 2\sin t\mathbf{j} + 0\mathbf{k},\; 0 \le t \le 2\pi$

$\int_C \left[f(x, y, z)\nabla g(x, y, z)\right] \cdot d\mathbf{r} = 0$

(ii) $\nabla f(x, y, z) = yz\mathbf{i} + xz\mathbf{j} + xy\mathbf{k}$

$\nabla g(x, y, z) = \mathbf{k}$

$$\nabla f \times \nabla g = \begin{vmatrix} \mathbf{i} & \mathbf{j} & \mathbf{k} \\ yz & xz & xy \\ 0 & 0 & 1 \end{vmatrix} = xz\mathbf{i} - yz\mathbf{j}$$

$$\mathbf{N} = \frac{x}{\sqrt{4 - x^2 - y^2}}\mathbf{i} + \frac{y}{\sqrt{4 - x^2 - y^2}}\mathbf{j} + \mathbf{k}$$

$$dS = \sqrt{1 + \left(\frac{-x}{\sqrt{4 - x^2 - y^2}}\right)^2 + \left(\frac{-y}{\sqrt{4 - x^2 - y^2}}\right)^2}\, dA = \frac{2}{\sqrt{4 - x^2 - y^2}}\, dA$$

$$\int_S \int \left[\nabla f(x, y, z) \times \nabla g(x, y, z)\right] \cdot \mathbf{N}\, dS = \int_S \int \left[\frac{x^2 z}{\sqrt{4 - x^2 - y^2}} - \frac{y^2 z}{\sqrt{4 - x^2 - y^2}}\right] \frac{2}{\sqrt{4 - x^2 - y^2}}\, dA$$

$$= \int_S \int \frac{2(x^2 - y^2)}{\sqrt{4 - x^2 - y^2}}\, dA$$

$$= \int_0^2 \int_0^{2\pi} \frac{2r^2(\cos^2\theta - \sin^2\theta)}{\sqrt{4 - r^2}} r\, d\theta\, dr = \int_0^2 \left[\frac{2r^3}{\sqrt{4 - r^2}}\left(\frac{1}{2}\sin 2\theta\right)\right]_0^{2\pi} dr = 0$$

CHAPTER 16
Additional Topics in Differential Equations

CHAPTER 16
Additional Topics in Differential Equations

Section 16.1 Exact First-Order Equations

1. $(2x + xy^2)\,dx + (3 + x^2y)\,dy = 0$

$$\frac{\partial M}{\partial y} = 2xy$$

$$\frac{\partial N}{\partial x} = 2xy$$

$$\frac{\partial M}{\partial y} = \frac{\partial N}{\partial x} \text{ Exact}$$

2. $(1 - xy)\,dx + (y - xy)\,dy = 0$

$$\frac{\partial M}{\partial y} = -x$$

$$\frac{\partial N}{\partial x} = -y$$

$$\frac{\partial M}{\partial y} \neq \frac{\partial N}{\partial x} \text{ Not exact}$$

3. $x \sin y\,dx + x \cos y\,dy = 0$

$$\frac{\partial M}{\partial y} = x \cos y$$

$$\frac{\partial N}{\partial x} = \cos y$$

$$\frac{\partial M}{\partial y} \neq \frac{\partial N}{\partial x} \text{ Not exact}$$

4. $ye^{xy}\,dx + xe^{xy}\,dy = 0$

$$\frac{\partial M}{\partial y} = e^{xy} + xye^{xy}$$

$$\frac{\partial N}{\partial x} = e^{xy} + xye^{xy}$$

$$\frac{\partial M}{\partial y} = \frac{\partial N}{\partial x} \text{ Exact}$$

5. $(2x - 3y)\,dx + (2y - 3x)\,dy = 0$

$$\frac{\partial M}{\partial y} = -3 = \frac{\partial N}{\partial x} \text{ Exact}$$

$$\begin{aligned} f(x, y) &= \int M(x, y)\,dx \\ &= \int (2x - 3y)\,dx \\ &= x^2 - 3xy + g(y) \end{aligned}$$

$$\begin{aligned} f_y(x, y) &= -3x + g'(y) \\ &= 2y - 3x \Rightarrow g'(y) = 2y \end{aligned}$$

$$\Rightarrow g(y) = y^2 + C_1$$

$$f(x, y) = x^2 - 3xy + y^2 + C_1$$

$$x^2 - 3xy + y^2 = C$$

6. $ye^x\,dx + e^x\,dy = 0$

$$\frac{\partial M}{\partial y} = e^x = \frac{\partial N}{\partial x} \text{ Exact}$$

$$f(x, y) = \int N(x, y)\,dy = \int e^x\,dy = ye^x + g(x)$$

$$f_x(x, y) = ye^x + g'(x) = ye^x \Rightarrow g'(x) = 0$$

$$\Rightarrow g(x) = C_1$$

$$f(x, y) = ye^x + C_1$$

$$ye^x = C$$

7. $(3y^2 + 10xy^2)\,dx + (6xy - 2 + 10x^2y) = 0$

$$\frac{\partial M}{\partial y} = 6y + 20xy = \frac{\partial N}{\partial x} \text{ Exact}$$

$$\begin{aligned} f(x, y) &= \int M(x, y)\,dx = \int (3y^2 + 10xy^2)\,dx \\ &= 3xy^2 + 5x^2y^2 + g(y) \end{aligned}$$

$$\begin{aligned} f_y(x, y) &= 6xy + 10x^2y + g'(y) = 6xy - 2 + 10x^2y \\ &\Rightarrow g'(y) = -2 \Rightarrow g(y) = -2y + C_1 \end{aligned}$$

$$f(x, y) = 3xy^2 + 5x^2y^2 - 2y + C_1$$

$$3xy^2 + 5x^2y^2 - 2y = C$$

8. $2\cos(2x - y)\,dx - \cos(2x - y)\,dy = 0$

$$\frac{\partial M}{\partial y} = 2\sin(2x - y) = \frac{\partial N}{\partial x}$$
$$= 2\sin(2x - y) \text{ Exact}$$

$$f(x, y) = \int M(x, y)\,dx$$
$$= \int 2\cos(2x - y)\,dx = \sin(2x - y) + g(y)$$

$$f_y(x, y) = -\cos(2x - y) + g'(y)$$
$$= -\cos(2x - y) \Rightarrow g'(y) = 0 \Rightarrow g(y) = C_1$$

$$f(x, y) = \sin(2x - y) + C_1$$
$$\sin(2x - y) = C$$

9. $(4x^3 - 6xy^2)\,dx + (4y^3 - 6xy)\,dy = 0$

$$\frac{\partial M}{\partial y} = -12xy$$
$$\frac{\partial N}{\partial x} = -6y$$

Not exact

10. $2y^2e^{xy^2}\,dx + 2xye^{xy^2}\,dy = 0$

$$\frac{\partial M}{\partial y} = 4(xy^3 + y)e^{xy^2}$$
$$\frac{\partial N}{\partial x} = 2(xy^3 + y)e^{xy^2}$$

Not exact

11. $\dfrac{-y}{x^2 + y^2}\,dx + \dfrac{x}{x^2 + y^2}\,dy = 0$

$$\frac{\partial M}{\partial y} = \frac{y^2 - x^2}{(x^2 + y^2)^2} = \frac{\partial N}{\partial x} \text{ Exact}$$

$$f(x, y) = \int M(x, y)\,dx = -\arctan\left(\frac{x}{y}\right) + g(y)$$

$$f_y(x, y) = \frac{x}{x^2 + y^2} + g'(y)$$
$$= \frac{x}{x^2 + y^2} \Rightarrow g'(y) = 0 \Rightarrow g(y) = C_1$$

$$f(x, y) = -\arctan\left(\frac{x}{y}\right) + C_1$$

$$\arctan\left(\frac{x}{y}\right) = C$$

12. $xe^{-(x^2+y^2)}\,dx + ye^{-(x^2+y^2)}\,dy = 0$

$$\frac{\partial M}{\partial y} = -2xye^{-(x^2+y^2)} = \frac{\partial N}{\partial x} \text{ Exact}$$

$$f(x, y) = \int M(x, y)\,dx$$
$$= \int xe^{-(x^2+y^2)}\,dx = -\frac{1}{2}e^{-(x^2+y^2)} + g(y)$$

$$f_y(x, y) = ye^{-(x^2+y^2)} + g'(y) = ye^{-(x^2+y^2)} \Rightarrow g'(y) = 0$$
$$\Rightarrow g(y) = C_1$$

$$f(x, y) = -\frac{1}{2}e^{-(x^2+y^2)} + C_1$$

$$e^{-(x^2+y^2)} = C$$

13. $\left(\dfrac{y}{x - y}\right)^2 dx + \left(\dfrac{x}{x - y}\right)^2 dy = 0$

$$\frac{\partial M}{\partial y} = \frac{2xy}{(x - y)^3}$$
$$\frac{\partial N}{\partial x} = \frac{-2xy}{(x - y)^3}$$

Not exact

14. $ye^y \cos xy\,dx + e^y(x\cos xy + \sin xy)\,dy = 0$

$$\frac{\partial M}{\partial y} = e^y\cos xy + ye^y\cos xy - xye^y\sin xy$$
$$\frac{\partial N}{\partial x} = e^y[\cos xy - xy\sin xy + y\cos xy]$$
$$\frac{\partial M}{\partial y} = \frac{\partial N}{\partial x} \text{ Exact}$$

$$f(x, y) = \int M(x, y)\,dx$$
$$= \int ye^y\cos xy\,dx = e^y\sin xy + g(y)$$

$$f_y(x, y) = e^y\sin xy + xe^y\cos xy + g'(y)$$
$$\Rightarrow g'(y) = 0 \Rightarrow g(y) = C_1$$

$$f(x, y) = e^y\sin xy + C_1$$

$$e^y\sin xy = C$$

15. (a) and (c)

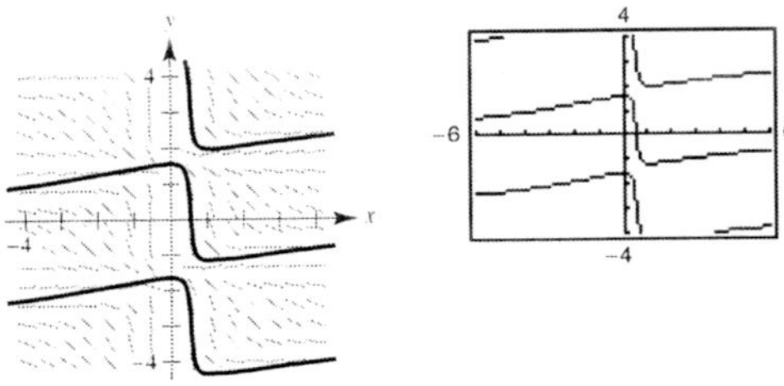

(b) $(2x \tan y + 5)\,dx + (x^2 \sec^2 y)\,dy = 0, \; y\left(\frac{1}{2}\right) = \frac{\pi}{4}$

$$\frac{\partial M}{\partial y} = 2x \sec^2 y = \frac{\partial N}{\partial x} \Rightarrow \text{Exact}$$

$$f(x, y) = \int M(x, y)\,dx = \int (2x \tan y + 5)\,dx$$

$$= x^2 \tan y + 5x + g(y)$$

$$f_y(x, y) = x^2 \sec^2 y + g'(y)$$

$$= x^2 \sec^2 y$$

$$\Rightarrow g'(y) = 0 \Rightarrow g(y) = C$$

$$f(x, y) = x^2 \tan y + 5x = C$$

$$f\left(\frac{1}{2}, \frac{\pi}{4}\right) = \frac{1}{4} + \frac{5}{2} = \frac{11}{4} = C$$

Answer: $x^2 \tan y + 5x = \dfrac{11}{4}$

16. (a) and (c)

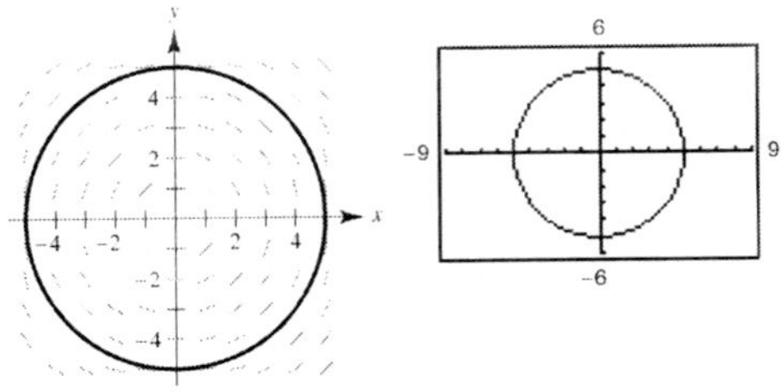

(b) $\dfrac{x}{\sqrt{x^2 + y^2}}\,dx + \dfrac{y}{\sqrt{x^2 + y^2}}\,dy = 0, \; y(4) = 3$

$$\frac{\partial M}{\partial y} = -\frac{xy}{(x^2 + y^2)^{3/2}} = \frac{\partial N}{\partial x} \text{ Exact}$$

$$f(x, y) = \int M(x, y)\,dx = \int \frac{x}{\sqrt{x^2 + y^2}}\,dx$$

$$= \frac{1}{2}\int (x^2 + y^2)^{-1/2}\,2x\,dx$$

$$= \sqrt{x^2 + y^2} + g(y)$$

$$f_y(x, y) = \frac{y}{\sqrt{x^2 + y^2}} + g'(y)$$

$$= \frac{y}{\sqrt{x^2 + y^2}} \Rightarrow g'(y) = 0 \Rightarrow g(y) = C$$

$$f(x, y) = \sqrt{x^2 + y^2} = C$$

$$f(3, 4) = \sqrt{3^2 + 4^2} = \sqrt{25} = 5 = C$$

Solution: $\sqrt{x^2 + y^2} = 5$ or $x^2 + y^2 = 25$

17. $\dfrac{y}{x - 1}\,dx + \left[\ln(x - 1) + 2y\right]dy = 0$

$$\frac{\partial M}{\partial y} = \frac{1}{x - 1} = \frac{\partial N}{\partial x} \text{ Exact}$$

$$f(x, y) = \int M(x, y)\,dx = y \ln(x - 1) + g(y)$$

$$f_y(x, y) = \ln(x - 1) + g'(y)$$

$$\Rightarrow g'(y) = 2y \Rightarrow g(y) = y^2 + C_1$$

$$f(x, y) = y \ln(x - 1) + y^2 + C_1$$

$$y \ln(x - 1) + y^2 = C$$

$$y(2) = 4:\; 4 \ln(2 - 1) + 16 = C \Rightarrow C = 16$$

Solution: $y \ln(x - 1) + y^2 = 16$

18. $\dfrac{x}{x^2 + y^2}\,dx + \dfrac{y}{x^2 + y^2}\,dy = 0$

$\dfrac{\partial M}{\partial y} = \dfrac{-2xy}{\left(x^2 + y^2\right)^2} = \dfrac{\partial N}{\partial x}$ Exact

$f(x, y) = \int M(x, y)\,dx$

$= \int \dfrac{x}{x^2 + y^2}\,dx = \dfrac{1}{2}\ln\left(x^2 + y^2\right) + g(y)$

$f_y(x, y) = \dfrac{y}{x^2 + y^2} + g'(y)$

$\Rightarrow g'(y) = 0 \Rightarrow g(y) = C_1$

$f(x, y) = \dfrac{1}{2}\ln\left(x^2 + y^2\right) + C_1$

$\ln\left(x^2 + y^2\right) = C$

$y(0) = 4$: $\ln(16) = C$

$\ln\left(x^2 + y^2\right) = \ln 16$

Solution: $x^2 + y^2 = 16$

19. $\left(e^{3x}\sin 3y\right)dx + \left(e^{3x}\cos 3y\right)dy = 0$

$\dfrac{\partial M}{\partial y} = 3e^{3x}\cos 3y = \dfrac{\partial N}{\partial x}$ Exact

$f(x, y) = \int M(x, y)\,dx$

$= \int e^{3x}\sin 3y\,dx = \dfrac{1}{3}e^{3x}\sin 3y + g(y)$

$f_y(x, y) = e^{3x}\cos 3y + g'(y)$

$\Rightarrow g'(y) = 0 \Rightarrow g(y) = C_1$

$f(x, y) = \dfrac{1}{3}e^{3x}\sin 3y + C_1$

$e^{3x}\sin 3y = C$

$y(0) = \pi$: $C = 0$

Solution: $e^{3x}\sin 3y = 0$

20. $\left(x^2 + y^2\right)dx + 2xy\,dy = 0$

$\dfrac{\partial M}{\partial y} = 2y = \dfrac{\partial N}{\partial x}$ Exact

$f(x, y) = \int M(x, y)\,dx$

$= \int\left(x^2 + y^2\right)dx = \dfrac{x^3}{3} + xy^2 + g(y)$

$f_y(x, y) = 2xy + g'(y) \Rightarrow g'(y) = 0 \Rightarrow g(y) = C_1$

$f(x, y) = \dfrac{x^3}{3} + xy^2 + C_1$

$\dfrac{x^3}{3} + xy^2 = C$

$y(3) = 1$: $9 + 3 = 12 = C$

$\dfrac{x^3}{3} + xy^2 = 12$

Solution: $x^3 + 3xy^2 = 36$

21. $\left(2xy - 9x^2\right)dx + \left(2y + x^2 + 1\right)dy = 0$

$\dfrac{\partial M}{\partial y} = 2x = \dfrac{\partial N}{\partial x}$ Exact

$f(x, y) = \int M(x, y)\,dx = \int\left(2xy - 9x^2\right)dx$

$= x^2y - 3x^3 + g(y)$

$f_y(x, y) = x^2 + g'(y) = 2y + x^2 + 1$

$\Rightarrow g'(y) = 2y + 1$

$\Rightarrow g(y) = y^2 + y + C_1$

$f(x, y) = x^2y - 3x^2 + y^2 + y + C_1$

$x^2y - 3x^2 + y^2 + y = C$

$y(0) = -3$: $9 - 3 = 6 = C$

Solution: $x^2y - 3x^2 + y^2 + y = 6$

22. $\left(2xy^2 + 4\right)dx + \left(2x^2y - 6\right)dy = 0$

$\dfrac{\partial M}{\partial y} = 4xy = \dfrac{\partial N}{\partial x}$ Exact

$f(x, y) = \int M(x, y)\,dx$

$= \int\left(2xy^2 + 4\right)dx = x^2y^2 + 4x + g(y)$

$f_y(x, y) = 2x^2y + g'(y) = 2x^2y - 6 \Rightarrow g'(y) = -6$

$\Rightarrow g(y) = -6y + C_1$

$f(x, y) = x^2y^2 + 4x - 6y + C_1$

$x^2 + y^2 + 4x - 6y = C$

$y(-1) = 8$: $1 + 64 - 4 - 48 = 13 = C$

Solution:

$x^2 + y^2 + 4x - 6y = 13$

23. $y\,dx - (x + 6y^2)\,dy = 0$

$$\frac{(\partial N/\partial x) - (\partial M/\partial y)}{M} = -\frac{2}{y} = k(y)$$

Integrating factor: $e^{\int k(y)\,dy} = e^{\ln y^{-2}} = \dfrac{1}{y^2}$

Exact equation: $\dfrac{1}{y}\,dx - \left(\dfrac{x}{y^2} + 6\right)dy = 0$

$f(x, y) = \dfrac{x}{y} + g(y)$

$g'(y) = -6$

$g(y) = -6y + C_1$

$\dfrac{x}{y} - 6y = C$

24. $(2x^3 + y)\,dx - x\,dy = 0$

$$\frac{(\partial M/\partial y) - (\partial N/\partial x)}{N} = -\frac{2}{x} = h(x)$$

Integrating factor: $e^{\int h(x)\,dx} = e^{\ln x^{-2}} = \dfrac{1}{x^2}$

Exact equation: $\left(2x + \dfrac{y}{x^2}\right)dx - \dfrac{1}{x}\,dy = 0$

$f(x, y) = x^2 - \dfrac{y}{x} + g(y)$

$g'(y) = 0$

$g(y) = C_1$

$x^2 - \dfrac{y}{x} = C$

25. $(5x^2 - y)\,dx + x\,dy = 0$

$$\frac{(\partial M/\partial y) - (\partial N/\partial x)}{N} = -\frac{2}{x} = h(x)$$

Integrating factor: $e^{\int h(x)\,dx} = e^{\ln x^{-2}} = \dfrac{1}{x^2}$

Exact equation: $\left(5 - \dfrac{y}{x^2}\right)dx + \dfrac{1}{x}\,dy = 0$

$f(x, y) = 5x + \dfrac{y}{x} + g(y)$

$g'(y) = 0$

$g(y) = C_1$

$5x + \dfrac{y}{x} = C$

26. $(5x^2 - y^2)\,dx + 2y\,dy = 0$

$$\frac{(\partial M/\partial y) - (\partial N/\partial x)}{N} = -1 = h(x)$$

Integrating factor: $e^{\int h(x)\,dx} = e^{-x}$

Exact equation: $(5x^2 - y^2)e^{-x}\,dx + 2ye^{-x}\,dy = 0$

$f(x, y) = -5x^2e^{-x} - 10xe^{-x} - 10e^{-x} + y^2e^{-x} + g(y)$

$g'(y) = 0$

$g(y) = C_1$

$y^2e^{-x} - 5x^2e^{-x} - 10xe^{-x} - 10e^{-x} = C$

27. $(x + y)\,dx + (\tan x)\,dy = 0$

$$\frac{(\partial M/\partial y) - (\partial N/\partial x)}{N} = -\tan x = h(x)$$

Integrating factor: $e^{\int h(x)\,dx} = e^{\ln \cos x} = \cos x$

Exact equation: $(x + y)\cos x\,dx + \sin x\,dy = 0$

$f(x, y) = x\sin x + \cos x + y\sin x + g(y)$

$g'(y) = 0$

$g(y) = C_1$

$x\sin x + \cos x + y\sin x = C$

28. $(2x^2y - 1)\,dx + x^3\,dy = 0$

$$\frac{(\partial M/\partial y) - (\partial N/\partial x)}{N} = -\frac{1}{x} = h(x)$$

Integrating factor: $e^{\int h(x)\,dx} = e^{\ln(1/x)} = \dfrac{1}{x}$

Exact equation: $\left(2xy - \dfrac{1}{x}\right)dx + x^2\,dy = 0$

$f(x, y) = x^2y - \ln|x| + g(y)$

$g'(y) = 0$

$g(y) = C_1$

$x^2y - \ln|x| = C$

29. $y^2 dx + (xy - 1)\,dy = 0$

$$\frac{(\partial N/\partial x) - (\partial M/\partial y)}{M} = -\frac{1}{y} = k(y)$$

Integrating factor: $e^{\int k(y)\,dy} = e^{\ln(1/y)} = \dfrac{1}{y}$

Exact equation: $y\,dx + \left(x - \dfrac{1}{y}\right)dy = 0$

$f(x, y) = xy + g(y)$

$g'(y) = -\dfrac{1}{y}$

$g(y) = -\ln|y| + C_1$

$xy - \ln|y| = C$

30. $(x^2 + 2x + y)\,dx + 2\,dy = 0$

$$\frac{(\partial M/\partial y) - (\partial N/\partial x)}{N} = \frac{1}{2} = h(x)$$

Integrating factor: $e^{\int h(x)\,dx} = e^{x/2}$

Exact equation: $(x^2 + 2x + y)e^{x/2}\,dx + 2e^{x/2}\,dy = 0$

$f(x, y) = 2(x^2 - 2x + 4 + y)e^{x/2} + g(y)$

$g'(y) = 0$

$g(y) = C_1$

$(x^2 - 2x + 4 + y)e^{x/2} = C$

31. $2y\,dx + (x - \sin\sqrt{y})\,dy = 0$

$$\frac{(\partial N/\partial x) - (\partial M/\partial y)}{M} = \frac{-1}{2y} = k(y)$$

Integrating factor: $e^{\int k(y)\,dy} = e^{\ln(1/\sqrt{y})} = \dfrac{1}{\sqrt{y}}$

Exact equation: $2\sqrt{y}\,dy + \left(\dfrac{x}{\sqrt{y}} - \dfrac{\sin\sqrt{y}}{\sqrt{y}}\right)dy = 0$

$f(x, y) = 2\sqrt{y}x + g(y)$

$g'(y) = -\dfrac{\sin\sqrt{y}}{\sqrt{y}}$

$g(y) = 2\cos\sqrt{y} + C_1$

$\sqrt{y}\,x + \cos\sqrt{y} = C$

32. $(-2y^3 + 1)\,dx + (3xy^2 + x^3)\,dy = 0$

$$\frac{(\partial M/\partial y) - (\partial N/\partial x)}{N} = \frac{-3}{x} = h(x)$$

Integrating factor: $e^{\int h(x)\,dx} = e^{\ln(1/x^3)} = \dfrac{1}{x^3}$

Exact equation: $\left(\dfrac{-2y^3}{x^3} + \dfrac{1}{x^3}\right)dx + \left(\dfrac{3y^2}{x^2} + 1\right)dy = 0$

$f(x, y) = \dfrac{y^3}{x^2} - \dfrac{1}{2x^2} + g(y)$

$g'(y) = 1$

$g(y) = y + C_1$

$\dfrac{y^3}{x^2} - \dfrac{1}{2x^2} + y = C$

33. $(4x^2y + 2y^2)\,dx + (3x^3 + 4xy)\,dy = 0$

Integrating factor: xy^2

Exact equation:

$(4x^3y^3 + 2xy^4)\,dy + (3x^4y^2 + 4x^2y^3)\,dy = 0$

$f(x, y) = x^4y^3 + x^2y^4 + g(y)$

$g'(y) = 0$

$g(y) = C_1$

$x^4y^3 + x^2y^4 = C$

34. $(3y^2 + 5x^2y)\,dx + (3xy + 2x^3)\,dy = 0$

Integrating factor: x^2y

Exact equation:

$(3x^2y^3 + 5x^4y^2)\,dx + (3x^3y^2 + 2x^5y)\,dy = 0$

$f(x, y) = x^3y^3 + x^5y^2 + g(y)$

$g'(y) = 0$

$g(y) = C_1$

$x^3y^3 + x^5y^2 = C$

35. $\left(-y^5 + x^2y\right) dx + \left(2xy^4 - 2x^3\right) dy = 0$

Integrating factor: $x^{-2}y^{-3}$

Exact equation:

$$\left(-\frac{y^2}{x^2} + \frac{1}{y^2}\right) dx + \left(2\frac{y}{x} - 2\frac{x}{y^3}\right) dy = 0$$

$$f(x, y) = \frac{y^2}{x} + \frac{x}{y^2} + g(y)$$

$$g'(y) = 0$$

$$g(y) = C_1$$

$$\frac{y^2}{x} + \frac{x}{y^2} = C$$

36. $-y^3\,dx + \left(xy^2 - x^2\right) dy = 0$

Integrating factor: $x^{-2}y^{-2}$

Exact equation: $\frac{-y}{x^2}\,dx + \left(\frac{1}{x} - \frac{1}{y^2}\right) dy = 0$

$$f(x, y) = \frac{y}{x} + g(y)$$

$$g'(y) = -\frac{1}{y^2}$$

$$g(y) = \frac{1}{y} + C_1$$

$$\frac{y}{x} + \frac{1}{y} = C$$

37. $y\,dx - x\,dy = 0$

(a) $\frac{1}{x^2}, \frac{y}{x^2}\,dx - \frac{1}{x}\,dy = 0, \frac{\partial M}{\partial y} = \frac{1}{x^2} = \frac{\partial N}{\partial x}$

(b) $\frac{1}{y^2}, \frac{1}{y}\,dx - \frac{x}{y^2}\,dy = 0, \frac{\partial M}{\partial y} = \frac{-1}{y^2} = \frac{\partial N}{\partial x}$

(c) $\frac{1}{xy}, \frac{1}{x}\,dx - \frac{1}{y}\,dy = 0, \frac{\partial M}{\partial y} = 0 = \frac{\partial N}{\partial x}$

(d) $\frac{1}{x^2 + y^2}, \frac{y}{x^2 + y^2}\,dx - \frac{x}{x^2 + y^2}\,dy = 0,$

$$\frac{\partial M}{\partial y} = \frac{x^2 - y^2}{\left(x^2 + y^2\right)^2} = \frac{\partial N}{\partial x}$$

38. $\left(axy^2 + by\right) dx + \left(bx^2y + ax\right) dy = 0$

Exact equation: $\frac{\partial M}{\partial y} = 2axy + b, \frac{\partial N}{\partial x} = 2bxy + a, \frac{\partial M}{\partial y} = \frac{\partial M}{\partial x}$ only if $a = b$.

Integrating factor: $x^m y^n$

$$\left(ax^{m+1}y^{n+2} + bx^m y^{n+1}\right) dx + \left(bx^{m+2}y^{n+1} + ax^{m+1}y^n\right) dy = 0$$

$$\left.\begin{aligned}\frac{\partial M}{\partial y} &= a(n + 2)x^{m+1}y^{n+1} + b(n + 1)x^m y^n \\ \frac{\partial N}{\partial x} &= b(m + 2)x^{m+1}y^{n+1} + a(m + 1)x^m y^n\end{aligned}\right\} \begin{aligned} a(n + 2) &= b(m + 2) \\ b(n + 1) &= a(m + 1)\end{aligned}$$

$$\left.\begin{aligned} an - bm &= 2(b - a) \\ bn - am &= a - b \end{aligned}\right\} \begin{aligned} abn - b^2m &= 2b(b - a) \\ abn - a^2m &= a(a - b) \end{aligned}$$

$$\left(a^2 - b^2\right)m = -(2b + a)(a - b)$$

$$m = -\frac{2b + a}{a + b}$$

$$bn - a\left(-\frac{2b + a}{a + b}\right) = a - b$$

$$bn = \frac{-2ab - a^2 + a^2 - b^2}{a + b} = \frac{-b(2a + b)}{a + b}$$

$$n = -\frac{2a + b}{a + b}$$

39. $\mathbf{F}(x, y) = \dfrac{y}{\sqrt{x^2 + y^2}}\mathbf{i} - \dfrac{x}{\sqrt{x^2 + y^2}}\mathbf{j}$

$$\frac{dy}{dx} = -\frac{x}{y}$$

$$y\,dy + x\,dx = 0$$

$$y^2 + x^2 = C$$

Family of circles

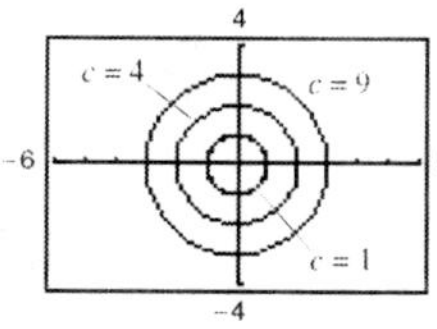

40. $\mathbf{F}(x, y) = \dfrac{x}{\sqrt{x^2 + y^2}}\mathbf{i} - \dfrac{y}{\sqrt{x^2 + y^2}}\mathbf{j}$

$$\frac{dy}{dx} = -\frac{y}{x}$$

$$x\,dy + y\,dx = 0$$

$$xy = C$$

Family of hyperbolas

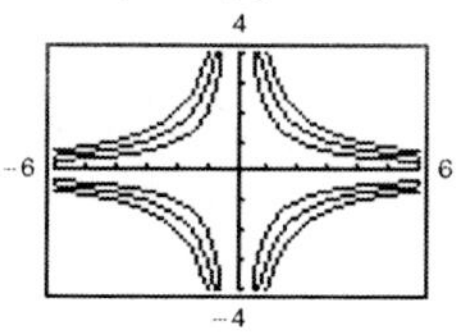

41. $\mathbf{F}(x, y) = 4x^2y\mathbf{i} - \left(2xy^2 + \dfrac{x}{y^2}\right)\mathbf{j}$

$$\frac{dy}{dx} = \frac{-y}{2x} - \frac{1}{4xy^3}$$

$$\frac{8y^3}{2y^4 + 1}\,dy = -\frac{2}{x}\,dx$$

$$\ln\left(2y^4 + 1\right) = \ln\left(\frac{1}{x^2}\right) + \ln C$$

$$2y^4 + 1 = \frac{C}{x^2}$$

$$2x^2y^4 + x^2 = C$$

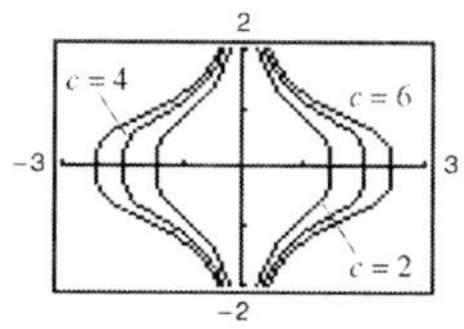

42. $\mathbf{F}(x, y) = \left(1 + x^2\right)\mathbf{i} - 2xy\mathbf{j}$

$$\frac{dy}{dx} = \frac{-2xy}{1 + x^2}$$

$$\frac{1}{y}\,dy = -\frac{2x}{1 + x^2}\,dx$$

$$\ln y = \ln\left(\frac{1}{1 + x^2}\right) + \ln C$$

$$y = \frac{C}{1 + x^2}$$

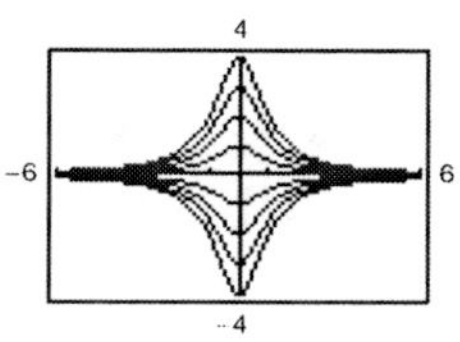

43. $E(x) = \dfrac{20x - y}{2y - 10x} = \dfrac{x}{y}\dfrac{dy}{dx}$

$$\left(20xy - y^2\right)dx + \left(10x^2 - 2xy\right)dy = 0$$

$$\frac{\partial M}{\partial y} = 20x - 2y = \frac{\partial N}{\partial x}$$

$$f(x, y) = 10x^2y - xy^2 + g(y)$$

$$g'(y) = 0$$

$$g(y) = C_1$$

$$10x^2y - xy^2 = K$$

Initial condition: $C(100) = 500,\ 100 \le x,\ K = 25{,}000{,}000$

$$10x^2y - xy^2 = 25{,}000{,}000$$

$$xy^2 - 10x^2y + 25{,}000{,}000 = 0 \text{ Quadratic Formula}$$

$$y = \frac{10x^2 + \sqrt{100x^4 - 4x(25{,}000{,}000)}}{2x} = \frac{5\left(x^2 + \sqrt{x^4 - 1{,}000{,}000x}\right)}{x}$$

44. $\dfrac{dy}{dx} = \dfrac{y - x}{3y - x}$

$(x - y)\,dx + (3y - x)\,dy = 0$

$\dfrac{\partial M}{\partial y} = -1 = \dfrac{\partial N}{\partial x}$

$f(x, y) = \dfrac{x^2}{2} - xy + g(y)$

$g'(y) = 3y$

$g(y) = \dfrac{3y^2}{2} + C_1$

$x^2 - 2xy + 3y^2 = C$

Initial condition: $y(2) = 1, 4 - 4 + 3 = C, C = 3$

Particular solution: $x^2 - 2xy + 3y^2 = 3$

45. $\dfrac{dy}{dx} = \dfrac{-2xy}{x^2 + y^2}$

$2xy\,dx + (x^2 + y^2)\,dy = 0$

$\dfrac{\partial M}{\partial y} = 2x = \dfrac{\partial N}{\partial x}$

$f(x, y) = x^2y + g(y)$

$g'(y) = y^2$

$g(y) = \dfrac{y^3}{3} + C_1$

$3x^2y + y^3 = C$

Initial condition: $y(0) = 2, 8 = C$

Particular solution: $3x^2y + y^3 = 8$

46. From the graph, $\displaystyle\lim_{x \to 100^+} C(x) = 500.$

47. (a) $y(4) \approx 0.5231$

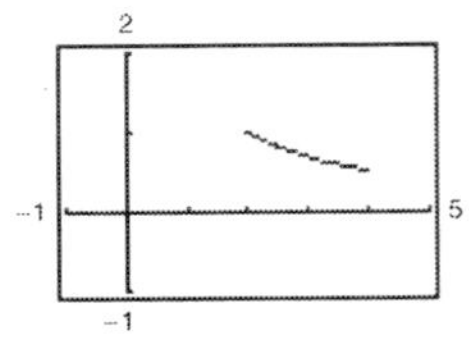

(b) $\dfrac{dy}{dx} = \dfrac{-xy}{x^2 + y^2}$

$xy\,dx + (x^2 + y^2)\,dy = 0$

$\dfrac{1}{M}\left[N_x - M_y\right] = \dfrac{1}{xy}[2x - x] = \dfrac{1}{y}$ function of y alone.

Integrating factor: $e^{\int(1/y)\,dy} = e^{\ln y} = y$

$xy^2\,dx + (x^2y + y^3)\,dy = 0$

$f(x, y) = \int xy^2\,dx = \dfrac{x^2y^2}{2} + g(y)$

$f_y(x, y) = x^2y + g'(y) \Rightarrow g(y) = \dfrac{y^4}{4} + C_1$

$f(x, y) = \dfrac{x^2y^2}{2} + \dfrac{y^4}{4} = C$

Initial condition: $y(2) = 1, \dfrac{4}{2} + \dfrac{1}{4} = \dfrac{9}{4} = C$

Particular solution: $\dfrac{x^2y^2}{2} + \dfrac{y^4}{4} = \dfrac{9}{4}$ or $2x^2y^2 + y^4 = 9.$

For $x = 4$, $32y^2 + y^4 = 9 \Rightarrow y(4) = 0.528$

(c)

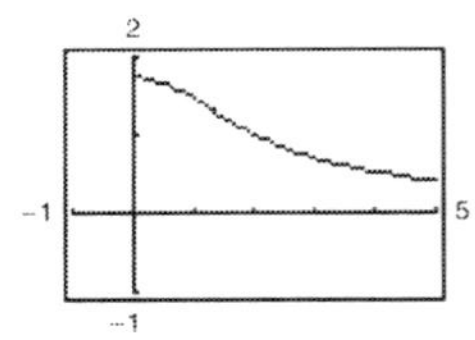

48. (a) $y(5) \approx 6.6980$

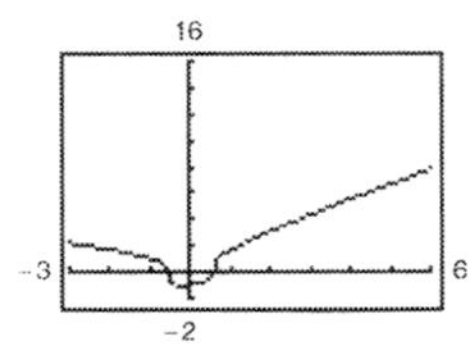

(b) $\dfrac{dy}{dx} = \dfrac{6x + y^2}{y(3y - 2x)}$

$(6x + y^2)\,dx + (2xy - 3y^2)\,dy = 0$

$\dfrac{\partial M}{\partial y} = 2y = \dfrac{\partial N}{\partial x}$ Exact

$f(x, y) = \int(6x + y^2)\,dx = 3x^2 + xy^2 + g(y)$

$f_y = 2xy + g'(y) \Rightarrow g(y) = -y^3 + C_1$

$f(x, y) = 3x^2 + xy^2 - y^3 = C$

Initial condition: $y(0) = 1 \Rightarrow -1 = C$

Particular solution: $3x^2 + xy^2 - y^3 = -1$

For $x = 5$, $75 + 5y^2 - y^3 + 1 = 0 \Rightarrow y = 6.695.$

(c)

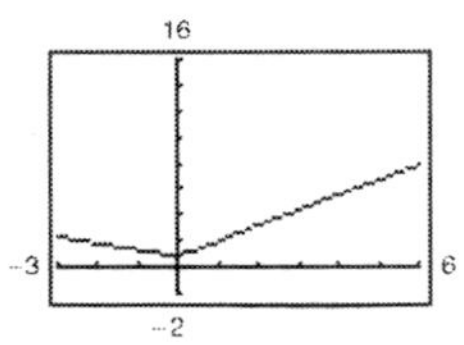

49. (a) $y(4) \approx 0.408$

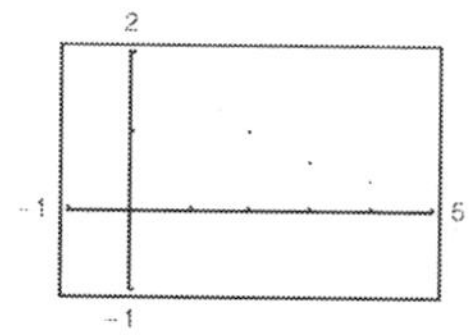

(b) $\frac{dy}{dx} = \frac{-xy}{x^2 + y^2}$

$xy\,dx + (x^2 + y^2)\,dy = 0$

$\frac{1}{M}[N_x - M_y] = \frac{1}{xy}[2x - x]$

$= \frac{1}{y}$ function of y alone.

Integrating factor: $e^{\int 1/y\,dy} = e^{\ln y} = y$

$xy^2\,dx + (x^2y + y^3)\,dy = 0$

$f(x, y) = \int xy^2\,dx = \frac{x^2y^2}{2} + g(y)$

$f_y(x, y) = x^2y + g'(y) \Rightarrow g(y) = \frac{y^4}{4} + C_1$

$f(x, y) = \frac{x^2y^2}{2} + \frac{y^4}{4} = C$

Initial condition: $y(2) = 1, \frac{4}{2} + \frac{1}{4} = \frac{9}{4} = C$

Particular solution: $\frac{x^2y^2}{2} + \frac{y^4}{4} = \frac{9}{4}$ or

$2x^2y^2 + y^4 = 9.$

For $x = 4$, $32y^2 + y^4 = 9 \Rightarrow y(4) = 0.528$

(c)

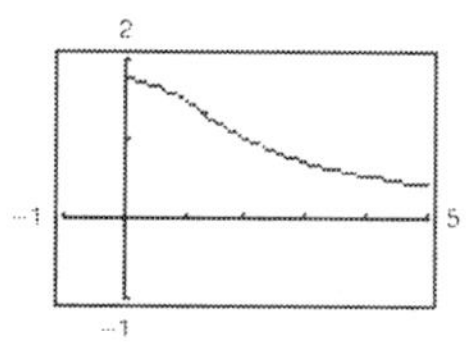

The solution is less accurate. For Exercise 47, Euler's Method gives $y(4) \approx 0.523$, whereas in Exercise 49, you obtain $y(4) \approx 0.408$. The errors are $0.528 - 0.523 = 0.005$ and $0.528 - 0.408 = 0.120$.

50. (a) $y(s) \approx 6.708$

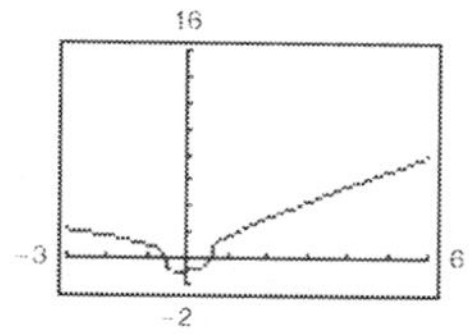

(b) $\frac{dy}{dx} = \frac{6x + y^2}{y(3y - 2x)}$

$(6x + y^2)\,dx + (2xy - 3y^2)\,dy = 0$

$\frac{\partial M}{\partial y} = 2y = \frac{\partial N}{\partial x}$ Exact

$f(x, y) = \int(6x + y^2)\,dx = 3x^2 + xy^2 + g(y)$

$f_y = 2xy + g'(y) \Rightarrow g(y) = -y^3 + C_1$

$f(x, y) = 3x^2 + xy^2 - y^3 = C$

Initial condition: $y(0) = 1 \Rightarrow -1 = C$

Particular solution: $3x^2 + xy^2 - y^3 = -1$

For $x = 5$, $75 + 5y^2 - y^3 + 1 = 0 \Rightarrow y = 6.695.$

(c)

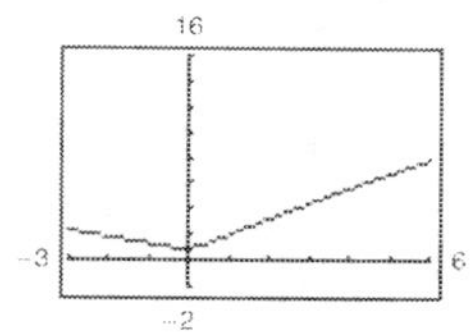

The solution is less accurate. For Exercise 48, Euler's Method gives $y(5) \approx 6.698$, whereas in Exercise 50, you obtain $y(5) \approx 6.708$. The errors are $6.695 - 6.698 = -0.003$ and $6.695 - 6.708 = -0.013$.

51. If M and N have continuous partial derivatives on an open disc R, then $M(x, y)\,dx + N(x, y)\,dy = 0$ is exact if and only if $\frac{\partial M}{\partial y} = \frac{\partial N}{\partial x}$.

52. See Theorem 16.2.

53. False

$\frac{\partial M}{\partial y} = 2x$ and $\frac{\partial N}{\partial x} = -2x$

54. False

$y\,dx + x\,dy = 0$ is exact, but $xy\,dx + x^2dy = 0$ is not exact.

55. True

$\frac{\partial}{\partial y}[f(x) + M] = \frac{\partial M}{\partial y}$ and $\frac{\partial}{\partial x}[g(y) + N] = \frac{\partial N}{\partial x}$

56. True

$$\frac{\partial}{\partial y}[f(x)] = 0 \text{ and } \frac{\partial}{\partial x}[g(y)] = 0$$

57. $M = xy^2 + kx^2y + x^3,\ N = x^3 + x^2y + y^2$

$$\frac{\partial M}{\partial y} = 2xy + kx^2,\ \frac{\partial N}{\partial x} = 3x^2 + 2xy$$

$$\frac{\partial M}{\partial y} = \frac{\partial N}{\partial x} \Rightarrow k = 3$$

58. $M = ye^{2xy} + 2x,\ N = kxe^{2xy} - 2y$

$$\frac{\partial M}{\partial y} = e^{2xy} + 2xye^{2xy},\ \frac{\partial N}{\partial x} = ke^{2xy} + 2kxye^{2xy}$$

$$\frac{\partial M}{\partial y} = \frac{\partial N}{\partial x} \Rightarrow k = 1$$

59. $M = g(y)\sin x,\ N = y^2f(x)$

$$\frac{\partial M}{\partial y} = g'(y)\sin x,\ \frac{\partial N}{\partial x} = y^2f'(x)$$

$$\frac{\partial M}{\partial y} = \frac{\partial N}{\partial x}:\ g'(y)\sin x = f'(x)y^2$$

$$g'(y) = y^2 \Rightarrow g(y) = \frac{y^3}{3} + C_1$$

$$f'(x) = \sin x \Rightarrow f(x) = -\cos x + C_2$$

60. $M = g(y)e^y,\ N = xy$

$$\frac{\partial M}{\partial y} = g'(y)e^y + g(y)e^y,\ \frac{\partial N}{\partial x} = y$$

$$g'(y)e^y + g(y)e^y = y$$

$$[g(y)e^y]' = y$$

$$g(y)e^y = \frac{y^2}{2} + C$$

$$g(y) = e^{-y}\left[\frac{y^2}{2} + C\right]$$

Section 16.2 Second-Order Homogeneous Linear Equations

1.

$$y = C_1e^{-3x} + C_2xe^{-3x}$$

$$y' = -3C_1e^{-3x} + C_2e^{-3x} - 3C_2xe^{-3x}$$

$$y'' = 9C_1e^{-3x} - 6C_2e^{-3x} + 9C_2xe^{-3x}$$

$$y'' + 6y' + 9y = \left(9C_1e^{-3x} - 6C_2e^{-3x} + 9C_2xe^{-3x}\right) + \left(-18C_1e^{-3x} + 6C_2e^{-3x} - 18C_2xe^{-3x}\right) + \left(9C_1e^{-3x} + 9C_2xe^{-3x}\right) = 0$$

y approaches zero as $x \to \infty$.

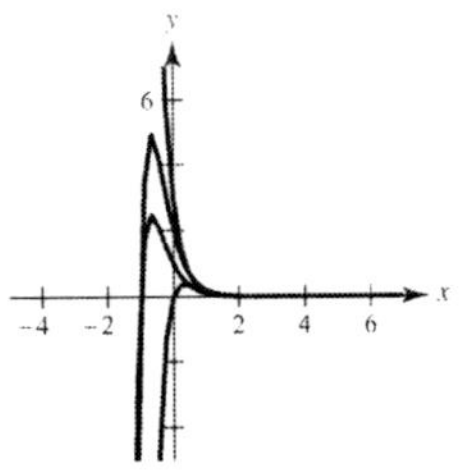

2.

$$y = C_1e^{2x} + C_2e^{-2x}$$

$$y' = 2C_1e^{2x} - 2C_2e^{-2x}$$

$$y'' = 4C_1e^{2x} + 4C_2e^{-2x} = 4y$$

$$y'' - 4y = 4y - 4y = 0$$

The graphs are different combinations of the graphs of e^{2x} and e^{-2x}.

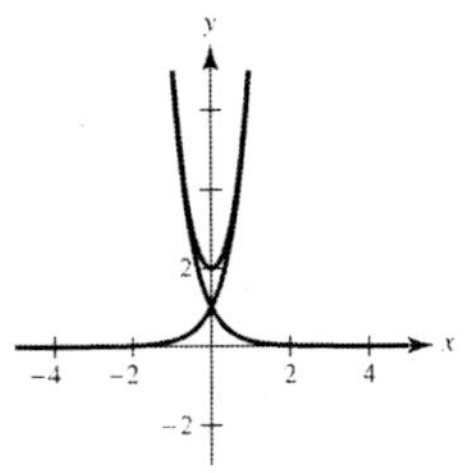

3.

$$y = C_1\cos 2x + C_2\sin 2x$$

$$y' = -2C_1\sin 2x + 2C_2\cos 2x$$

$$y'' = -4C_1\cos 2x - 4C_2\sin 2x = -4y$$

$$y'' + 4y = -4y + 4y = 0$$

The graphs are basically the same shape, with left and right shifts and varying ranges.

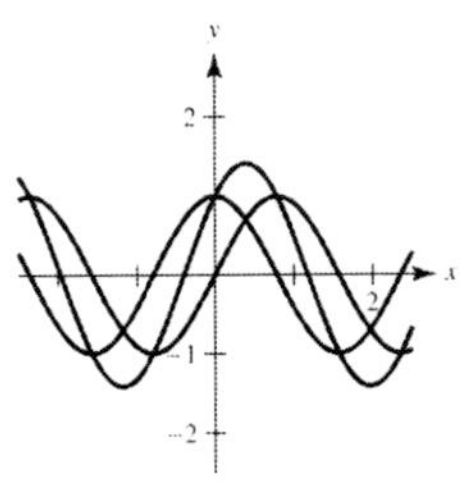

4. $y = C_1e^{-x}\cos 3x + C_2e^{-x}\sin 3x = e^{-x}(C_1\cos 3x + C_2\sin 3x)$

$$y' = -e^{-x}(C_1\cos 3x + C_2\sin 3x) + e^{-x}(-3C_1\sin 3x + 3C_2\cos 3x)$$
$$= e^{-x}[-C_1\cos 3x - C_2\sin 3x - 3C_1\sin 3x + 3C_2\cos 3x]$$

$$y'' = -e^{-x}[-C_1\cos 3x - C_2\sin 3x - 3C_1\sin 3x + 3C_2\cos 3x]$$
$$+ e^{-x}[3C_1\sin 3x - 3C_2\cos 3x - 9C_1\cos 3x - 9C_2\sin 3x]$$

$$y'' + 2y' + 10y = e^{-x}[C_1\cos 3x + C_2\sin 3x + 3C_1\sin 3x - 3C_2\cos 3x$$
$$+ 3C_1\sin 3x - 3C_2\cos 3x - 9C_1\cos 3x - 9C_2\sin 3x]$$
$$+ 2e^{-x}[-C_1\cos 3x - C_2\sin 3x - 3C_1\sin 3x + 3C_2\cos 3x]$$
$$+ 10e^{-x}[C_1\cos 3x + C_2\sin 3x] = 0$$

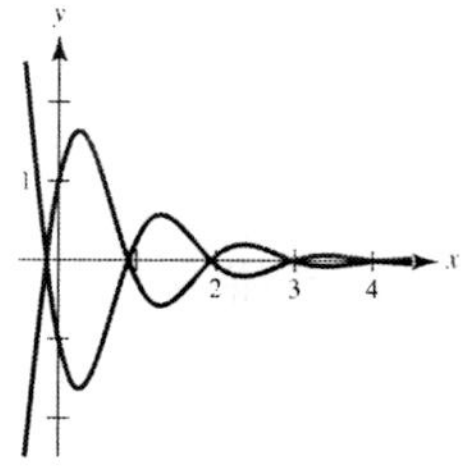

y approaches zero as $x \to \infty$. The graphs are the same only reflected.

5. $y'' - y' = 0$

Characteristic equation: $m^2 - m = 0$

Roots: $m = 0, 1$

$y = C_1 + C_2e^{x}$

6. $y'' + 2y' = 0$

Characteristic equation: $m^2 + 2m = 0$

Roots: $m = 0, -2$

$y = C_1 + C_2e^{-2x}$

7. $y'' - y' - 6y = 0$

Characteristic equation: $m^2 - m - 6 = 0$

Roots: $m = 3, -2$

$y = C_1e^{3x} + C_2e^{-2x}$

8. $y'' + 6y' + 5y = 0$

Characteristic equation: $m^2 + 6m + 5 = 0$

Roots: $m = -1, -5$

$y = C_1e^{-x} + C_2e^{-5x}$

9. $2y'' + 3y' - 2y = 0$

Characteristic equation: $2m^2 + 3m - 2 = 0$

Roots: $m = \frac{1}{2}, -2$

$y = C_1e^{(1/2)x} + C_2e^{-2x}$

10. $16y'' - 16y' + 3y = 0$

Characteristic equation: $16m^2 - 16m + 3 = 0$

Roots: $m = \frac{1}{4}, \frac{3}{4}$

$y = C_1e^{(1/4)x} + C_2e^{(3/4)x}$

11. $y'' + 6y' + 9y = 0$

Characteristic equation: $m^2 + 6m + 9 = 0$

Roots: $m = -3, -3$

$y = C_1e^{-3x} + C_2xe^{-3x}$

12. $y'' - 10y' + 25y = 0$

Characteristic equation: $m^2 - 10m + 25 = 0$

Roots: $m = 5, 5$

$y = C_1e^{5x} + C_2xe^{5x}$

13. $16y'' - 8y' + y = 0$

Characteristic equation: $16m^2 - 8m + 1 = 0$

Roots: $m = \frac{1}{4}, \frac{1}{4}$

$y = C_1e^{(1/4)x} + C_2xe^{(1/4)x}$

14. $9y'' - 12y' + 4y = 0$

Characteristic equation: $9m^2 - 12m + 4 = 0$

Roots: $m = \frac{2}{3}, \frac{2}{3}$

$y = C_1e^{(2/3)x} + C_2xe^{(2/3)x}$

15. $y'' + y = 0$

Characteristic equation: $m^2 + 1 = 0$

Roots: $m = -i, i$

$y = C_1\cos x + C_2\sin x$

16. $y'' + 4y = 0$

Characteristic equation: $m^2 + 4 = 0$

Roots: $m = -2i, 2i$

$y = C_1\cos 2x + C_2\sin 2x$

17. $y'' - 9y = 0$

Characteristic equation: $m^2 - 9 = 0$

Roots: $m = -3, 3$

$y = C_1e^{3x} + C_2e^{-3x}$

18. $y'' - 2y = 0$

Characteristic equation: $m^2 - 2 = 0$

Roots: $m = -\sqrt{2}, \sqrt{2}$

$y = C_1e^{\sqrt{2}x} + C_2^{-\sqrt{2}x}$

19. $y'' - 2y' + 4y = 0$

Characteristic equation: $m^2 - 2m + 4 = 0$

Roots: $m = 1 - \sqrt{3}i, 1 + \sqrt{3}i$

$y = e^x\left(C_1 \cos\sqrt{3}x + C_2 \sin\sqrt{3}x\right)$

20. $y'' - 4y' + 21y = 0$

Characteristic equation: $m^2 - 4m + 21 = 0$

Roots: $m = 2 - \sqrt{17}i, 2 + \sqrt{17}i$

$y = e^{2x}\left(C_1 \cos\sqrt{17}x + C_2 \sin\sqrt{17}x\right)$

21. $y'' - 3y' + y = 0$

Characteristic equation: $m^2 - 3m + 1 = 0$

Roots: $m = \dfrac{3 - \sqrt{5}}{2}, \dfrac{3 + \sqrt{5}}{2}$

$y = C_1e^{\left[(3+\sqrt{5})/2\right]x} + C_2e^{\left[3-\sqrt{5}/2\right]x}$

22. $3y'' + 4y' - y = 0$

Characteristic equation: $3m^2 + 4m - 1 = 0$

Roots: $m = \dfrac{-2 - \sqrt{7}}{3}, \dfrac{-2 + \sqrt{7}}{3}$

$y = C_1e^{\left[(-2+\sqrt{7})/3\right]x} + C_2e^{\left[(-2-\sqrt{7})/3\right]x}$

23. $9y'' - 12y' + 11y = 0$

Characteristic equation: $9m^2 - 12m + 11 = 0$

Roots: $m = \dfrac{2 + \sqrt{7}i}{3}, \dfrac{2 - \sqrt{7}i}{3}$

$$y = e^{(2/3)x}\left[C_1 \cos\left(\frac{\sqrt{7}}{3}x\right) + C_2 \sin\left(\frac{\sqrt{7}}{3}x\right)\right]$$

24. $2y'' - 6y' + 7y = 0$

Characteristic equation: $2m^2 - 6m + 7 = 0$

Roots: $m = \dfrac{3 + \sqrt{5}i}{2}, \dfrac{3 - \sqrt{5}i}{2}$

$$y = e^{(3/2)x}\left[C_1 \cos\left(\frac{\sqrt{5}}{2}x\right) + C_2 \sin\left(\frac{\sqrt{5}}{2}x\right)\right]$$

25. $y^{(4)} - y = 0$

Characteristic equation: $m^4 - 1 = 0$

Roots: $m = -1, 1, -i, i$

$y = C_1e^x + C_2e^{-x} + C_3 \cos x + C_4 \sin x$

26. $y^{(4)} - y'' = 0$

Characteristic equation: $m^4 - m^2 = 0$

Roots: $m = 0, 0, -1, 1$

$y = C_1 + C_2x + C_3e^x + C_4e^{-x}$

27. $y''' - 6y'' + 11y' - 6y = 0$

Characteristic equation: $m^3 - 6m^2 + 11m - 6 = 0$

Roots: $m = 1, 2, 3$

$y = C_1e^x + C_2e^{2x} + C_3e^{3x}$

28. $y''' - y'' - y' + y = 0$

Characteristic equation: $m^3 - m^2 - m + 1 = 0$

Roots: $m = -1, 1, 1$

$y = C_1e^x + C_2xe^x + C_3e^{-x}$

29. $y''' - 3y'' + 7y' - 5y = 0$

Characteristic equation: $m^3 - 3m^2 + 7m - 5 = 0$

Roots: $m = 1, 1 - 2i, 1 + 2i$

$y = C_1e^x + e^x(C_2 \cos 2x + C_3 \sin 2x)$

30. $y''' - 3y'' + 3y' - y = 0$

Characteristic equation: $m^3 - 3m^2 + 3m - 1 = 0$

Roots: $m = 1, 1, 1$

$y = C_1e^x + C_2xe^x + C_3x^2e^x$

31. $y'' + 100y = 0$

$y = C_1 \cos 10x + C_2 \sin 10x$

$y' = -10C_1 \sin 10x + 10C_2 \cos 10x$

(a) $y(0) = 2: 2 = C_1$

$y'(0) = 0: 0 = 10C_2 \Rightarrow C_2 = 0$

Particular solution: $y = 2 \cos 10x$

(b) $y(0) = 0: 0 = C_1$

$y'(0) = 2: 2 = 10C_2 \Rightarrow C_2 = \frac{1}{5}$

Particular solution: $y = \frac{1}{5} \sin 10x$

(c) $y(0) = -1: -1 = C_1$

$y'(0) = 3: 3 = 10C_2 \Rightarrow C_2 = \frac{3}{10}$

Particular solution: $y = -\cos 10x + \frac{3}{10} \sin 10x$

32. $y = C \sin \sqrt{3}t$

$y' = \sqrt{3}C \cos \sqrt{3}t$

$y'' = -3C \sin \sqrt{3}t$

$y'' + \omega y = -3C \sin \sqrt{3}t + \omega \sin \sqrt{3}t$

$= 0 \Rightarrow \omega = 3C$

$y'(0) = -5: -5 = \sqrt{3}C \Rightarrow C = \dfrac{5\sqrt{3}}{3}$ and

$\omega = -5\sqrt{3}$

33. $y'' - y' - 30y = 0,\ y(0) = 1,\ y'(0) = -4$

Characteristic equation: $m^2 - m - 30 = 0$

Roots: $m = 6, -5$

$y = C_1e^{6x} + C_2e^{-5x},\ y' = 6C_1e^{6x} - 5C_2e^{-5x}$

Initial conditions:

$y(0) = 1,\ y'(0) = -4,\ 1 = C_1 + C_2,\ -4 = 6C_1 - 5C_2$

Solving simultaneously: $C_1 = \dfrac{1}{11}, C_2 = \dfrac{10}{11}$

Particular solution: $y = \dfrac{1}{11}\left(e^{6x} + 10e^{-5x}\right)$

34. $y'' - 7y' + 12y = 0,\ y(0) = 3,\ y'(0) = 3$

Characteristic equation: $m^2 - 7m + 12 = 0$

Roots: $m = 3, 4$

$y = C_1e^{3x} + C_2e^{4x},\ y' = 3C_1e^{3x} + 4C_2e^{4x}$

Initial conditions:

$y(0) = 3,\ y'(0) = 3,\ C_1 + C_2 = 3,\ 3C_1 + 4C_2 = 3$

Solving simultaneously: $C_1 = 9, C_2 = -6$

Particular solution: $y = 9e^{3x} - 6e^{4x}$

35. $y'' + 16y = 0,\ y(0) = 0,\ y'(0) = 2$

Characteristic equation: $m^2 + 16 = 0$

Roots: $m = \pm 4i$

$y = C_1 \cos 4x + C_2 \sin 4x$

$y' = -4C_1 \sin 4x + 4C_2 \cos 4x$

Initial conditions: $y(0) = 0 = C_1$

$y'(0) = 2 = 4C_2 \Rightarrow C_2 = \frac{1}{2}$

Particular solution: $y = \frac{1}{2} \sin 4x$

36. $y'' + 2y' + 3y = 0,\ y(0) = 2,\ y'(0) = 1$

Characteristic equation: $m^2 + 2m + 3 = 0$

Roots: $m = \dfrac{-2 \pm \sqrt{4 - 12}}{2} = -1 \pm \sqrt{2}i$

$y = e^{-x}\left(C_1 \cos\sqrt{2}x + C_2 \sin \sqrt{2}x\right)$

$y' = e^{-x}\left(-C_1\sqrt{2} \sin \sqrt{2}x + C_2\sqrt{2} \cos \sqrt{2}x\right) - e^{-x}\left(C_1 \cos \sqrt{2}x + C_2 \sin \sqrt{2}x\right)$

Initial conditions: $y(0) = 2 = C_1$

$y'(0) = 1 = C_2\sqrt{2} - C_1 = C_2\sqrt{2} - 2 \Rightarrow C_2 = \dfrac{3}{\sqrt{2}}$

Particular solution: $y = e^{-x}\left(2 \cos \sqrt{2}x + \dfrac{3}{\sqrt{2}} \sin \sqrt{2}x\right)$

37. $9y'' - 6y' + y = 0,\ y(0) = 2,\ y'(0) = 1$

Characteristic equation: $9m^2 - 6m + 1 = 0$

Roots: $m = \frac{1}{3}, \frac{1}{3}$

$y = C_1e^{(1/3)x} + C_2xe^{(1/3)x}$

$y' = \frac{1}{3}C_1e^{(1/3)x} + \frac{1}{3}C_2xe^{(1/3)x} + C_2e^{(1/3)x}$

Initial conditions: $y(0) = 2,\ y'(0) = 1$

$$\left.\begin{aligned} C_1 &= 2 \\ \tfrac{1}{3}C_1 + C_2 &= 1 \end{aligned}\right\} \Rightarrow C_1 = 2, C_2 = \tfrac{1}{3}$$

Particular solution: $y = 2e^{x/3} + \frac{1}{3}xe^{x/3}$

38. $4y'' + 4y' + y = 0,\ y(0) = 3,\ y'(0) = 1$

Characteristic equation: $4m^2 + 4m + 1 = 0$

$$(2m + 1)^2 = 0$$

Roots: $m = -\frac{1}{2}, -\frac{1}{2}$

$y = C_1e^{(-1/2)x} + C_2xe^{(-1/2)x}$

$y' = -\frac{1}{2}C_1e^{(-1/2)x} - \frac{1}{2}C_2xe^{(-1/2)x} + C_2e^{(-1/2)x}$

Initial conditions:

$y(0) = 3 = C_1$

$y'(0) = 1 = -\frac{1}{2}C_1 + C_2 \Rightarrow C_2 = \frac{5}{2}$

Particular solution: $y = 3e^{x/2} + \frac{5}{2}xe^{-x/2}$

39. $y'' - 4y' + 3y = 0,\ y(0) = 1,\ y(1) = 3$

Characteristic equation: $m^2 - 4m + 3 = 0$

Roots: $m = 1, 3$

$y = C_1e^x + C_2e^{3x}$

$y(0) = 1$: $C_1 + C_2 = 1$

$y(1) = 3$: $C_1e + C_2e^3 = 3$

Solving simultaneously, $C_1 = \dfrac{e^3 - 3}{e^3 - e}, C_2 = \dfrac{3 - e}{e^3 - e}$

Solution: $y = \dfrac{e^3 - 3}{e^3 - e}e^x + \dfrac{3 - e}{e^3 - e}e^{3x}$

40. $4y'' + y = 0,\ y(0) = 2,\ y(\pi) = -5$

Characteristic equation: $4m^2 + 1 = 0$

Roots: $m = \pm\frac{1}{2}i$

$y = C_1 \cos \frac{1}{2}x + C_2 \sin \frac{1}{2}x$

$y(0) = 2$: $C_1 = 2$

$y(\pi) = -5$: $C_2 = -5$

Solution: $y = 2 \cos \frac{1}{2}x - 5 \sin \frac{1}{2}x$

41. $y'' + 9y = 0,\ y(0) = 3,\ y(\pi) = 5$

Characteristic equation: $m^2 + 9m = 0$

Roots: $m = \pm 3i$

$y = C_1 \cos 3x + C_2 \sin 3x$

$y(0) = 3$: $C_1 = 3$

$y(\pi) = 5$: $-C_1 = 5$

No solution

42. $4y'' + 20y' + 21y = 0,\ y(0) = 3,\ y(2) = 0$

Characteristic equation: $4m^2 + 20m + 21 = 0$

Roots: $m = -\dfrac{3}{2}, -\dfrac{7}{2}$

$y = C_1e^{(-3/2)x} + C_2e^{(-7/2)x}$

$y(0) = 3$: $C_1 + C_2 = 3$

$y(2) = 0$: $C_1e^{-3} + C_2e^{-7} = 0 \Rightarrow C_1 + C_2e^{-4} = 0$

Solving simultaneously, $C_1 = \dfrac{-3}{e^4 - 1}, C_2 = \dfrac{3e^4}{e^4 - 1}$

Solution: $y = \dfrac{-3}{e^4 - 1}e^{(-3/2)x} + \dfrac{3e^4}{e^4 - 1}e^{(-7/2)x}$

43. $4y'' - 28y' + 49y = 0,\ y(0) = 2,\ y(1) = -1$

Characteristic equation: $4m^2 - 28m + 49 = 0$

Roots: $m = \dfrac{7}{2}, \dfrac{7}{2}$

$y = C_1e^{(7/2)x} + C_2xe^{(7/2)x}$

$y(0) = 2$: $C_1 = 2$

$y(1) = -1$: $C_1e^{7/2} + C_2e^{7/2} = -1 \Rightarrow C_2 = \dfrac{-1 - 2e^{7/2}}{e^{7/2}}$

Solution: $y = 2e^{(7/2)x} + \left(\dfrac{-1 - 2e^{7/2}}{e^{7/2}}\right)xe^{(7/2)x}$

44. $y'' + 6y' + 45y = 0,\ y(0) = 4,\ y(\pi) = 8$

Characteristic equation: $m^2 + 6m + 45 = 0$

Roots: $m = \dfrac{-6 \pm \sqrt{36 - 180}}{2} = -3 \pm 6i$

$y = C_1e^{-3x}\cos 6x + C_2e^{-3x}\sin 6x$

$y(0) = 4$: $C_1 = 4$

$y(\pi) = 8$: $-C_1e^{-3\pi} = 8$

No solution

45. Answers will vary. See Theorem 16.4.

46. Two functions y_1 and y_2 are linearly independent if the only solution to the equation $C_1y_1 + C_2y_2 = 0$ is the trivial solution $C_1 = C_2 = 0$.

47. The motion of a spring in a shock absorber is damped.

48. (a) y'' is always positive according to the graph (concave upward), but y' is negative when $x < 0$ (decreasing), so $y'' \neq y'$.

(b) y'' is positive for $x > 0$ (concave upward), but $-\frac{1}{2}y' < 0$ for $x > 0$ (increasing). So, $y'' \neq -\frac{1}{2}y'$.

49. $y'' + 9y = 0$

Undamped vibration

Period: $\dfrac{2\pi}{3}$

Matches (b)

50. $y'' + 25y = 0$

Undamped vibration

Period: $\dfrac{2\pi}{5}$

Matches (d)

51. $y'' + 2y' + 10y = 0$

Damped vibration

Matches (c)

52. $y'' + y' + \frac{37}{4}y = 0$

Damped vibration

Matches (a)

53. By Hooke's Law, $F = kx$

$$k = \frac{F}{x} = \frac{32}{2/3} = 48.$$

Also, $F = ma$, and $m = \dfrac{F}{a} = \dfrac{32}{32} = 1.$

So, $y = \dfrac{1}{2}\cos\left(4\sqrt{3}t\right)$

54. By Hooke's Law, $F = kx$

$$k = \frac{F}{x} = \frac{32}{2/3} = 48.$$

Also, $F = ma$, and $m = \dfrac{F}{a} = \dfrac{32}{32} = 1.$

So, $y = -\dfrac{2}{3}\cos\left(4\sqrt{3}t\right)$.

55. $y = C_1\cos\left(\sqrt{k/m}\,t\right) + C_2\sin\left(\sqrt{k/m}\,t\right)$,

$\sqrt{k/m} = \sqrt{48} = 4\sqrt{3}$

Initial conditions: $y(0) = -\dfrac{2}{3},\ y'(0) = \dfrac{1}{2}$

$y = C_1\cos\left(4\sqrt{3}t\right) + C_2\sin\left(4\sqrt{3}t\right)$

$y(0) = C_1 = -\dfrac{2}{3}$

$y'(t) = -4\sqrt{3}\,C_1\sin\left(4\sqrt{3}t\right) + 4\sqrt{3}\,C_2\cos\left(4\sqrt{3}t\right)$

$y'(0) = 4\sqrt{3}\,C_2 = \dfrac{1}{2} \Rightarrow C_2 = \dfrac{1}{8\sqrt{3}} = \dfrac{\sqrt{3}}{24}$

$y(t) = -\dfrac{2}{3}\cos\left(4\sqrt{3}t\right) + \dfrac{\sqrt{3}}{24}\sin\left(4\sqrt{3}t\right)$

56. $y = C_1\cos\left(4\sqrt{3}t\right) + C_2\sin\left(4\sqrt{3}t\right)$

Initial conditions: $y(0) = \dfrac{1}{2},\ y'(0) = -\dfrac{1}{2}$

$y(0) = C_1 = \dfrac{1}{2}$

$y'(t) = -4\sqrt{3}C_1\sin\left(4\sqrt{3}t\right) + 4\sqrt{3}C_2\cos\left(4\sqrt{3}t\right)$

$y'(0) = 4\sqrt{3}C_2 = -\dfrac{1}{2} \Rightarrow C_2 = -\dfrac{1}{8\sqrt{3}}$

$y(t) = \dfrac{1}{2}\cos\left(4\sqrt{3}t\right) - \dfrac{1}{8\sqrt{3}}\sin\left(4\sqrt{3}t\right)$

57. By Hooke's Law, $32 = k(2/3)$, so $k = 48$. Moreover, because the weight w is given by mg, it follows that $m = w/g = 32/32 = 1$. Also, the damping force is given by $(-1/8)(dy/dt)$. So, the differential equation for the oscillations of the weight is

$$m\left(\frac{d^2y}{dt^2}\right) = -\frac{1}{8}\left(\frac{dy}{dt}\right) - 48y$$

$$m\left(\frac{d^2y}{dt^2}\right) + \frac{1}{8}\left(\frac{dy}{dt}\right) + 48y = 0.$$

In this case the characteristic equation is $8m^2 + m + 384 = 0$ with complex roots $m = (-1/16) \pm \left(\sqrt{12{,}287}/16\right)i$.

So, the general solution is $y(t) = e^{-t/16}\left(C_1 \cos\frac{\sqrt{12{,}287}t}{16} + C_2 \sin\frac{\sqrt{12{,}287}t}{16}\right)$.

Using the initial conditions, you have $y(0) = C_1 = \frac{1}{2}$

$$y'(t) = e^{-t/16}\left[\left(-\frac{\sqrt{12{,}287}}{16}C_1 - \frac{C_2}{16}\right)\sin\frac{\sqrt{12{,}287}t}{16} + \left(\frac{\sqrt{12{,}287}}{16}C_2 - \frac{C_1}{16}\right)\cos\frac{\sqrt{12{,}287}t}{16}\right]$$

$$y'(0) = \frac{\sqrt{12{,}287}}{16}C_2 - \frac{C_1}{16} = 0 \Rightarrow C_2 = \frac{\sqrt{12{,}287}}{24{,}574}$$

and the particular solution is

$$y(t) = \frac{e^{-t/16}}{2}\left(\cos\frac{\sqrt{12{,}287}t}{16} + \frac{\sqrt{12{,}287}}{12{,}287}\sin\frac{\sqrt{12{,}287}t}{16}\right).$$

58. By Hooke's Law, $32 = k(2/3)$, so $k = 48$. Also, $m = w/g = 32/32 = 1$. The damping force is given by $(-1/4)(dy/dt)$. So,

$$m\left(\frac{d^2y}{dt^2}\right) = -\frac{1}{4}\left(\frac{dy}{dt}\right) - 48y$$

$$m\left(\frac{d^2y}{dt^2}\right) + \frac{1}{4}\left(\frac{dy}{dt}\right) + 48y = 0.$$

The characteristic equation is $4m^2 + m + 192 = 0$ with complex roots $m = (-1/8) \pm \left(\sqrt{3071}/8\right)i$. So, the general solution is

$$y(t) = e^{-t/8}\left(C_1 \cos\frac{\sqrt{3071}t}{8} + C_2 \sin\frac{\sqrt{3071}t}{8}\right).$$

Using the initial conditions, you have

$$y(0) = C_1 = \frac{1}{2}$$

$$y'(t) = e^{-t/8}\left[\left(-\frac{\sqrt{3071}}{8}C_1 - \frac{C_2}{8}\right)\sin\frac{\sqrt{3071}t}{8} + \left(\frac{\sqrt{3071}C_2}{8} - \frac{C_1}{8}\right)\cos\frac{\sqrt{3071}t}{8}\right]$$

$$y'(0) = \frac{\sqrt{3071}}{8}C_2 - \frac{C_1}{8} = 0 \Rightarrow C_2 = \frac{\sqrt{3071}}{6142}$$

and the particular solution is

$$y(t) = \frac{e^{-t/8}}{2}\left[\cos\frac{\sqrt{3071}t}{8} + \frac{\sqrt{3071}}{3071}\sin\frac{\sqrt{3071}t}{8}\right].$$

59. Because $m = -a/2$ is a double root of the characteristic equation, you have

$$\left(m + \frac{a}{2}\right)^2 = m^2 + am + \frac{a^2}{4} = 0$$

and the differential equation is $y'' + ay' + \left(a^2/4\right)y = 0$. The solution is

$$y = \left(C_1 + C_2x\right)e^{-(a/2)x}$$

$$y' = \left(-\frac{C_1a}{2} + C_2 - \frac{C_2a}{2}x\right)e^{-(a/2)x}$$

$$y'' = \left(\frac{C_1a^2}{4} - aC_2 + \frac{C_2a^2}{4}x\right)e^{-(a/2)x}$$

$$y'' + ay' + \frac{a^2}{4}y = e^{-(a/2)x}\left[\left(\frac{C_1a^2}{4} - C_2a + \frac{C_2a^2}{4}x\right) + \left(-\frac{C_1a^2}{2} + C_2a - \frac{C_2a^2}{2}x\right) + \left(\frac{C_1a^2}{4} + \frac{C_2a^2}{4}x\right)\right] = 0.$$

60. Because $m = \alpha \pm \beta i$ are roots to the characteristic equation, you have

$$\left[m - (\alpha + \beta i)\right]\left[m - (\alpha - \beta i)\right] = m^2 - 2\alpha m + \left(\alpha^2 + \beta^2\right) = 0$$

and the differential equation is $y'' - 2\alpha y' + \left(\alpha^2 + \beta^2\right)y = 0$. (**Note**: $i^2 = -1$.) The solution is

$$y = e^{\alpha x}\left(C_1 \cos \beta x + C_2 \sin \beta x\right)$$

$$y' = e^{\alpha x}\left[\left(C_1\alpha + C_2\beta\right)\cos \beta x + \left(C_2\alpha - C_1\beta\right)\sin \beta x\right.$$

$$y'' = e^{\alpha x}\left[\left(C_1\alpha^2 - C_1\beta^2 + 2C_2\alpha\beta\right)\cos \beta x + \left(C_2\alpha^2 - C_2\beta^2 - 2C_1\alpha\beta\right)\sin \beta x\right]$$

$$-2\alpha y' = e^{\alpha x}\left[\left(-2C_1\alpha^2 - 2C_2\alpha\beta\right)\cos \beta x + \left(-2C_2\alpha^2 + 2C_1\alpha\beta\right)\sin \beta x\right]$$

$$\left(\alpha^2 + \beta^2\right)y = e^{\alpha x}\left[\left(C_1\alpha^2 + C_1\beta^2\right)\cos \beta x + \left(C_2\alpha^2 + C_2\beta^2\right)\sin \beta x\right]$$

So, $y'' - 2\alpha y' + \left(\alpha^2 + \beta^2\right)y = 0$.

61. False. The general solution is $y = C_1e^{3x} + C_2xe^{3x}$.

62. True

63. True

64. False. The solution $y = x^2e^x$ requires that $m = 1$ is a triple root of the characteristic equation. Because the characteristic equation is quadratic, $m = 1$ can be at most a double root.

65. $y_1 = e^{ax},\ y_2 = e^{bx},\ a \neq b$

$$W(y_1, y_2) = \begin{vmatrix} e^{ax} & e^{bx} \\ ae^{ax} & be^{bx} \end{vmatrix}$$

$$= (b - a)e^{ax+bx} \neq 0 \text{ for any value of } x.$$

66. $y_1 = e^{ax},\ y_2 = xe^{ax}$

$$W(y_1, y_2) = \begin{vmatrix} e^{ax} & xe^{ax} \\ ae^{ax} & e^{ax} + axe^{ax} \end{vmatrix}$$

$$= e^{2ax} \neq 0 \text{ for any value of } x.$$

67. $y_1 = e^{ax}\sin bx,\ y_2 = e^{ax}\cos bx,\ b \neq 0$

$$W(y_1, y_2) = \begin{vmatrix} e^{ax}\sin bx & e^{ax}\cos bx \\ ae^{ax}\sin bx + be^{ax}\cos bx & ae^{ax}\cos bx - be^{ax}\sin bx \end{vmatrix}$$

$$= -be^{2ax}\sin^2 bx - be^{2ax}\cos^2 bx = -be^{2ax} \neq 0 \text{ for any value of } x.$$

68. $y_1 = x,\ y_2 = x^2$

$$W(y_1, y_2) = \begin{vmatrix} x & x^2 \\ 1 & 2x \end{vmatrix} = x^2 \neq 0 \text{ for } x \neq 0.$$

Section 16.3 Second-Order Nonhomogeneous Linear Equations

1. $y = 2e^{2x} - 2\cos x$

$y' = 4e^{2x} + 2\sin x$

$y'' = 8e^{2x} + 2\cos x$

$y'' + y = \left(8e^{2x} + 2\cos x\right) + \left(2e^{2x} - \cos x\right) = 10e^{2x}$

2. $y = 2\sin x + \frac{1}{2}x\sin x$

$y' = 2\cos x + \frac{1}{2}x\cos x + \frac{1}{2}\sin x$

$y'' = -2\sin x - \frac{1}{2}x\sin x + \cos x$

$y'' + y = \left(-2\sin x - \frac{1}{2}x\sin x + \cos x\right) + \left(2\sin x + \frac{1}{2}\sin x\right) = \cos x$

3. $y = 3\sin x - \cos x\ln\left|\sec x + \tan x\right|$

$y' = 3\cos x - 1 + \sin x\ln\left|\sec x + \tan x\right|$

$y'' = -3\sin x + \tan x + \cos x\ln\left|\sec x + \tan x\right|$

$y'' + y = \left(-3\sin x + \tan x + \cos x\ln\left|\sec x \tan x\right|\right) + \left(3\sin x - \cos x\ln\left|\sec x + \tan x\right|\right) = \tan x$

4. $y = \left(5 - \ln\left|\sin x\right|\right)\cos x - x\sin x$

$y' = -\left(5 - \ln\left|\sin x\right|\right)\sin x - \cos x\cot x - \sin x - x\cos x = -6\sin x + \sin x\ln\left|\sin x\right| - \cos x\left(\cot x + x\right)$

$y'' = -6\cos x + \cos x + \cos x\ln\left|\sin x\right| - \cos x\left(-\csc^2 x + 1\right) + \sin x\left(\cot x + x\right)$

$= -5\cos x + \cos x\ln\left|\sin x\right| + \csc x\cot x + x\sin x$

$y'' + y = \cos x\left(-5 + \ln\left|\sin x\right|\right) + \csc x\cot x + x\sin x + \left(5 - \ln\left|\sin x\right|\right)\cos x - x\sin x = \csc x\cot x$

5. $y'' + 7y' + 12y = 3x + 1$

$y'' + 7y' + 12y = 0$

$m^2 - 7m + 12 = (m - 3)(m - 4) = 0$ when $m = 3, 4$

$y_h = C_1e^{3x} + C_2e^{4x}$

$y_p = A_0 + A_1x$

$y_p' = A_1,\ y_p'' = 0$

$y_p'' + 7y_p' + 12y_p = 7A_1 + 12\left(A_0 + A_1x\right) = 3x + 1$

$\left.\begin{aligned} 12A_1 &= 3 \\ 7A_1 + 12A_0 &= 1 \end{aligned}\right\} \Rightarrow A_1 = \frac{1}{4},\ A_0 = -\frac{1}{16}$

Solution: $y_p = -\frac{1}{16} + \frac{1}{4}x$

6. $y'' - y' - 6y = 4$

$y'' - y' - 6y = 0$

$m^2 - m - 6 = (m - 3)(m + 2) = 0$ when $m = 3, -2$

$y_h = C_1e^{3x} + C_2e^{-2x}$

$y_p = A,\ y_p' = y_p'' = 0$

$y_p'' - y_p' - 6y_p = -6A = 4 \Rightarrow A = -\frac{2}{3}$

Solution: $y_p = -\frac{2}{3}$

7. $y'' - 8y' + 16y = e^{3x}$

$y'' - 8y' + 16y = 0$

$m^2 - 8m + 16 = (m - 4)^2 = 0$ when $m = 4$

$y_h = C_1e^{4x} + C_2xe^{4x}$

$y_p = Ae^{3x},\ y_p' = 3Ae^{3x},\ y_p'' = 9Ae^{3x}$

$y_p'' - 8y_p' + 16y_p = 9Ae^{3x} - 8\left(3Ae^{3x}\right) + 16\left(Ae^{3x}\right)$

$= e^{3x}$

$9A - 24A + 16A = 1 \Rightarrow A = 1$

Solution: $y_p = e^{3x}$

8. $y'' + y' + 3y = e^{2x}$

$y'' + y' + 3y = 0$

$m^2 + m + 3 = 0 \Rightarrow m = \dfrac{-1 \pm \sqrt{-11}}{2}$

$y_p = Ae^{2x},\ y_p' = 2Ae^{2x},\ y_p'' = 4Ae^{2x}$

$y'' + y' + 3y = 4Ae^{2x} + 2Ae^{2x} + 3\left(Ae^{2x}\right) = e^{2x}$

$4A + 2A + 3A = 1 \Rightarrow A = \dfrac{1}{9}$

Solution: $y_p = \dfrac{1}{9}e^{2x}$

9. $y'' - 2y' - 15y = \sin x$

$y'' - 2y' - 15y = 0$

$m^2 - 2m - 15 = (m - 5)(m + 3) = 0 \Rightarrow m = 5, -3$

$y_p = A \sin x + B \cos x$

$y_p' = A \cos x - B \sin x$

$y_p'' = -A \sin x - B \cos x$

$y_p'' - 2y_p' - 15y_p = (-A \sin x - B \cos x) - 2(A \cos x - B \sin x) - 15(A \sin x + B \cos x) = \sin x$

$(-A + 2B - 15A) \sin x + (-B - 2A - 15B) \cos x = \sin x$

$$\left.\begin{aligned} -16A + 2B &= 1 \\ -2A - 16B &= 0 \end{aligned}\right\} \Rightarrow A = -\frac{4}{65}, B = \frac{1}{130}$$

Solution: $y_p = -\dfrac{4}{65} \sin x + \dfrac{1}{130} \cos x$

10. $y'' + 4y' + 5y = e^x \cos x$

$y'' + 4y' + 5y = 0$

$m^2 + 4m + 5 = 0 \Rightarrow m = \dfrac{-4 \pm \sqrt{-4}}{2} = -2 \pm i$

$y_p = Ae^x \cos x + Be^x \sin x = e^x(A \cos x + B \sin x)$

$y_p' = e^x(A \cos x + B \sin x) + e^x(-A \sin x + B \cos x) = e^x((B - A) \sin x + (A + B) \cos x)$

$y_p'' = e^x((B - A) \sin x + (A + B) \cos x) + e^x((B - A) \cos x - (A + B) \sin x) = e^x(-2A \sin x + 2B \cos x)$

$y_p'' + 4y_p' + 5y_p = e^x(-2A \sin x + 2B \cos x) + 4(e^x((B - A) \sin x + (A + B) \cos x)) + 5(Ae^x \cos x + Be^x \sin x) = e^x \cos x$

$(-2A + 4(B - A) + 5B) \sin x + (2B + 4(A + B) + 5A) \cos x = \cos x$

$$\left.\begin{aligned} -6A + 9B &= 0 \\ 9A + 6B &= 1 \end{aligned}\right\} \Rightarrow A = \frac{3}{39}, B = \frac{2}{39}$$

Solution: $y_p = \dfrac{3}{39} e^x \cos x + \dfrac{2}{39} e^x \sin x$

11. $y'' - 3y' + 2y = 2x$

$y'' - 3y' + 2y = 0$

$m^2 - 3m + 2 = 0$ when $m = 1, 2.$

$y_h = C_1e^x + C_2e^{2x}$

$y_p = A_0 + A_1x$

$y_p' = A_1$

$y_p'' = 0$

$y_p'' - 3y_p' + 2y_p = (2A_0 - 3A_1) + 2A_1x = 2x$

$$\left.\begin{aligned} 2A_0 - 3A_1 &= 0 \\ 2A_1 &= 2 \end{aligned}\right\} A_1 = 1, A_0 = \tfrac{3}{2}$$

$y = C_1e^x + C_2e^{2x} + x + \frac{3}{2}$

12. $y'' - 2y' - 3y = x^2 - 1$

$y'' - 2y' - 3y = 0$

$m^2 - 2m - 3 = 0$ when $m = -1, 3.$

$y_h = C_1e^{-x} + C_2e^{3x}$

$y_p = A_0 + A_1x + a_2x^2$

$y_p' = A_1 + 2A_2x$

$y_p'' = 2A_2$

$y_p'' - 2y_p' - 3y_p = (-3A_2)x^2 + (-3A_1 - 4A_2)x$
$\qquad + (-3A_0 - 2A_1 + 2A_2) = x^2 - 1$

$$\left.\begin{aligned} -3A_2 &= 1 \\ -3A_1 - 4A_2 &= 0 \\ -3A_0 - 2A_1 + 2A_2 &= -1 \end{aligned}\right\} A_0 = -\tfrac{5}{27}, A_1 = \tfrac{4}{9}, A_2 = -\tfrac{1}{3}$$

$y = C_1e^{-x} + C_2e^{3x} - \frac{1}{3}x^2 + \frac{4}{9}x - \frac{5}{27}$

13. $y'' + 2y' = 2e^x$

$y'' + 2y' = 0$

$m^2 + 2m = 0$ when $m = 0, -2$.

$y_h = C_1 + C_2e^{-2x}$

$y_p = Ae^x = y_p' = y_p''$

$y_p'' + 2y_p' = 3Ae^x = 2e^x$ or $A = \frac{2}{3}$

$y = C_1 + C_2e^{-2x} + \frac{2}{3}e^x$

14. $y'' - 9y = 5e^{3x}$

$y'' - 9y = 0$

$m^2 - 9 = 0$ when $m = -3, 3$.

$y_h = C_1e^{-3x} + C_2e^{3x}$

$y_p = Axe^{3x}$

$y_p' = Ae^{3x}(3x + 1)$

$y_p'' = Ae^{3x}(9x + 6)$

$y_p'' - 9y_p = 6Ae^{3x} = 5e^{3x}$ or $A = \frac{5}{6}$

$y = C_1e^{-3x} + \left(C_2 + \frac{5}{6}x\right)e^{3x}$

15. $y'' - 10y' + 25y = 5 + 6e^x$

$y'' - 10y' + 25y = 0$

$m^2 - 10m + 25 = 0$ when $m = 5, 5$.

$y_h = C_1e^{5x} + C_2xe^{5x}$

$y_p = A_0 + A_1e^x$

$y_p' = y_p'' = A_1e^x$

$y_p'' - 10y_p' + 25y_p = 25A_0 + 16A_1e^x = 5 + 6e^x$

or $A_0 = \frac{1}{5}, A_1 = \frac{3}{8}$

$y = (C_1 + C_2x)e^{5x} + \frac{3}{8}e^x + \frac{1}{5}$

16. $16y'' - 8y' + y = 4(x + e^x)$

$16y'' - 8y' + y = 0$

$16m^2 - 8m + 1 = 0$ when $m = \frac{1}{4}, \frac{1}{4}$.

$y_h = (C_1 + C_2x)e^{(1/4)x}$

$y_p = A_0 + A_1x + A_2e^x$

$y_p' = A_1 + A_2e^x$

$y_p'' = A_2e^x$

$16y_p'' - 8y_p' + y_p = (A_0 - 8A_1) + A_1x + 9A_2e^x$

$= 4x + 4e^x$

or $A_2 = \frac{4}{9}, A_1 = 4, A_0 = 32$

$y = (C_1 + C_2x)e^{(1/4)x} + 32 + 4x + \frac{4}{9}e^x$

17. $y'' + 9y = \sin 3x$

$y'' + 9y = 0$

$m^2 + 9 = 0$ when $m = -3i, 3i$.

$y_h = C_1 \cos 3x + C_2 \sin 3x$

$y_p = A_0 \sin 3x + A_1x \sin 3x + A_2 \cos 3x + A_3x \cos 3x$

$y_p'' = (-9A_0 - 6A_3) \sin 3x - 9A_1 x \sin 3x$

$+ (6A_1 - 9A_2) \cos 3x - 9A_3 x \cos 3x$

$y_p'' + 9y_p = -6A_3 \sin 3x + 6A_1 \cos 3x = \sin 3x,$

$A_1 = 0, A_3 = -\frac{1}{6}$

$y = \left(C_1 - \frac{1}{6}x\right) \cos 3x + C_2 \sin 3x$

18. $y''' - 3y' + 2y = 2e^{-2x}$

$y''' - 3y' + 2y = 0$

$m^3 - 3m + 2 = 0$ when $m = 1, 1, -2$.

$y_h = C_1e^x + C_2xe^x + C_3e^{-2x}$

$y_p = A_0e^{-2x} + A_1xe^{-2x}$

$y_p' = (-2A_0 + A_1)e^{-2x} - 2A_1xe^{-2x}$

$y_p'' = (4A_0 - 4A_1)e^{-2x} + 4A_1xe^{-2x}$

$y_p''' = (-8A_0 + 12A_1)e^{-2x} - 8A_1xe^{-2x}$

$y_p''' - 3y_p' + 2y_p = 9A_1e^{-2x} = 2e^{-2x}$ or $A_1 = \frac{2}{9}$

$y = C_1e^x + C_2xe^x + \left(C_3 + \frac{2}{9}x\right)e^{-2x}$

19. $y'' + y = x^3, y(0) = 1, y'(0) = 0$

$y'' + y = 0$

$m^2 + 1 = 0$ when $m = i, -i$.

$y_h = C_1 \cos x + C_2 \sin x$

$y_p = A_0 + A_1x + A_2x^2 + A_3x^3$

$y_p' = A_1 + 2A_2x + 3A_3x^2$

$y_p'' = 2A_2 + 6A_3x$

$y_p'' + y_p = A_3x^3 + A_2x^2 + (A_1 + 6A_3)x + (A_0 + 2A_2)$

$= x^3$

or $A_3 = 1, A_2 = 0, A_1 = -6, A_0 = 0$

$y = C_1 \cos x + C_2 \sin x + x^3 - 6x$

$y' = -C_1 \sin x + C_2 \cos x + 3x^2 - 6$

Initial conditions:

$y(0) = 1, y'(0) = 0, 1 = C_1, 0 = C_2 - 6, C_2 = 6$

Particular solution: $y = \cos x + 6 \sin x + x^3 - 6x$

20. $y'' + 4y = 4,\ y(0) = 1,\ y'(0) = 6$

$y'' + 4y = 0$

$m^2 + 4 = 0$ when $m = 2i, -2i.$

$y_h = C_1 \cos 2x + C_2 \sin 2x$

$y_p = A_0$

$y_p'' = 0$

$y_p'' + 4y_p = 4A_0 = 4$ or $A_0 = 1$

$y = C_1 \cos 2x + C_2 \sin 2x + 1$

$y' = -2C_1 \sin 2x + 2C_2 \cos 2x$

Initial conditions: $y(0) = 1,\ y'(0) = 6,\ 1 = C_1 + 1,$
$C_1 = 0,\ 6 = 2C_2,\ C_2 = 3$

Particular solution: $y = 3 \sin 2x + 1$

21. $y'' + y' = 2 \sin x,\ y(0) = 0,\ y'(0) = -3$

$y'' + y' = 0$

$m^2 + m = 0$ when $m = 0, -1.$

$y_h = C_1 + C_2 e^{-x}$

$y_p = A \cos x + B \sin x$

$y_p' = -A \sin x + B \cos x$

$y_p'' = -A \cos x - B \sin x$

$y_p'' + y_p' = (-A + B)\cos x + (-A - B)\sin x = 2 \sin x$

$$\left.\begin{aligned} -A + B &= 0 \\ -A - B &= 2 \end{aligned}\right\} A = -1,\ B = -1$$

$y = C_1 + C_2 e^{x} - (\cos x + \sin x)$

$y' = -C_2 e^{-x} - (-\sin x + \cos x)$

Initial conditions: $y(0) = 0,\ y'(0) = -3,$
$0 = C_1 + C_2 - 1,\ -3 = -C_2 - 1,$
$C_2 = 2,\ C_1 = -1$

Particular solution: $y = -1 + 2e^{-x} - (\cos x + \sin x)$

22. $y'' + y' - 2y = 3 \cos 2x,\ y(0) = -1,\ y'(0) = 2$

$y'' + y' - 2y = 0$

$m^2 + m - 2 = 0$ when $m = 1, -2.$

$y_h = C_1 e^x + C_2 e^{-2x}$

$y_p = A \cos 2x + B \sin 2x$

$y_p' = -2A \sin 2x + 2B \cos 2x$

$y_p'' = -4A \cos 2x - 4B \sin 2x$

$y_p'' + y_p' - 2y_p = (-6A + 2B)\cos 2x + (-2A - 6B)\sin 2x$
$= 3\cos 2x$

$$\left.\begin{aligned} -6A + 2B &= 3 \\ -2A - 6B &= 0 \end{aligned}\right\} A = -\tfrac{9}{20},\ B = \tfrac{3}{20}$$

$y = C_1 e^x + C_2 e^{-2x} - \frac{9}{20} \cos 2x + \frac{3}{20} \sin 2x$

$y' = C_1 e^x - 2C_2 e^{-2x} + \frac{9}{10} \sin 2x + \frac{3}{10} \cos 2x$

Initial conditions:

$y(0) = -1,\ y'(0) = 2,\ -1 = C_1 + C_2 - \frac{9}{20},$
$2 = C_1 - 2C_2 + \frac{3}{10}$

$$\left.\begin{aligned} C_1 + C_2 &= -\tfrac{11}{20} \\ C_1 - 2C_2 &= \tfrac{17}{10} \end{aligned}\right\} C_1 = \tfrac{1}{5},\ C_2 = -\tfrac{3}{4}$$

Particular solution:

$y = \frac{1}{20}\left(4e^x - 15e^{-2x} - 9 \cos 2x + 3 \sin 2x\right)$

23. $y' - 4y = xe^x - xe^{4x},\ y(0) = \frac{1}{3}$

$y' - 4y = 0$

$m - 4 = 0$ when $m = 4.$

$y_h = Ce^{4x}$

$y_p = (A_0 + A_1 x)e^x + (A_2 x + A_3 x^2)e^{4x}$

$y_p' = (A_0 + A_1 x)e^x + A_1 e^x$
$\quad + 4(A_2 x + A_3 x^2)e^{4x} + (A_2 + 2A_3 x)e^{4x}$

$y_p' - 4y_p = (-3A_0 - 3A_1 x)e^x + A_1 e^x + A_2 e^{4x}$
$\quad + 2A_3 x e^{4x} = xe^x - xe^{4x}$

$A_0 = -\frac{1}{9},\ A_1 = -\frac{1}{3},\ A_2 = 0,\ A_3 = -\frac{1}{2}$

$y = \left(C - \frac{1}{2}x^2\right)e^{4x} - \frac{1}{9}(1 + 3x)e^x$

Initial conditions: $y(0) = \frac{1}{3},\ \frac{1}{3} = C - \frac{1}{9},\ C = \frac{4}{9}$

Particular solution: $y = \left(\frac{4}{9} - \frac{1}{2}x^2\right)e^{4x} - \frac{1}{9}(1 + 3x)e^x$

24. $y' + 2y = \sin x,\ y\left(\frac{\pi}{2}\right) = \frac{2}{5}$

$y' + 2y = 0$

$m + 2 = 0$ when $m = -2$.

$$y_h = Ce^{-2x}$$
$$y_p = A\cos x + B\sin x$$
$$y'_p = -A\sin x + B\cos x$$

$$y'_p + 2y_p = (-A\sin x + B\cos x) + 2(A\cos x + B\sin x)$$
$$= (2B - A)\sin x + (2A + B)\cos x = \sin x$$

$$2B - A = 1, 2A + B = 0 \Rightarrow B = \frac{2}{5}, A = -\frac{1}{5}$$

$$y = y_h + y_p = Ce^{-2x} - \frac{1}{5}\cos x + \frac{2}{5}\sin x$$

Initial conditions: $y\left(\frac{\pi}{2}\right) = \frac{2}{5}, \frac{2}{5} = Ce^{-\pi} + \frac{2}{5}, C = 0$

Particular solution: $y = \frac{2}{5}\sin x - \frac{1}{5}\cos x$

25. $y'' + y = \sec x$

$y'' + y = 0$

$m^2 + 1 = 0$ when $m = -i, i$.

$$y_h = C_1\cos x + C_2\sin x$$
$$y_p = v_1\cos x + v_2\sin x$$

$$v'_1\cos x + v'_2\sin x = 0$$
$$v'_1(-\sin x) + v'_2(\cos x) = \sec x$$

$$v'_1 = \frac{\begin{vmatrix} 0 & \sin x \\ \sec x & \cos x \end{vmatrix}}{\begin{vmatrix} \cos x & \sin x \\ -\sin x & \cos x \end{vmatrix}} = -\tan x$$

$$v_1 = \int -\tan x\,dx = \ln|\cos x|$$

$$v'_2 = \frac{\begin{vmatrix} \cos x & 0 \\ -\sin x & \sec x \end{vmatrix}}{\begin{vmatrix} \cos x & \sin x \\ -\sin x & \cos x \end{vmatrix}} = 1$$

$$v_2 = \int dx = x$$

$$y = (C_1 + \ln|\cos x|)\cos x + (C_2 + x)\sin x$$

26. $y'' + y = \sec x \tan x$

$y'' + y = 0$

$m^2 + 1 = 0$ when $m = -i, i$.

$$y_h = C_1\cos x + C_2\sin x$$
$$y_p = v_1\cos x + v_2\sin x$$

$$v'_1\cos x + v'_2\sin x = 0$$

$$v'_1(-\sin x) + v'_2\cos x = \sec x\tan x$$

$$v'_1 = \frac{\begin{vmatrix} 0 & \sin x \\ \sec x\tan x & \cos x \end{vmatrix}}{\begin{vmatrix} \cos x & \sin x \\ -\sin x & \cos x \end{vmatrix}} = -\tan^2 x$$

$$v_1 = \int -\tan^2 x\,dx = -\int(\sec^2 x - 1)\,dx$$
$$= -\tan x + x$$

$$v'_2 = \frac{\begin{vmatrix} \cos x & 0 \\ -\sin x & \sec x\tan x \end{vmatrix}}{\begin{vmatrix} \cos x & \sin x \\ -\sin x & \cos x \end{vmatrix}} = \tan x$$

$$v_2 = \int \tan x\,dx = -\ln|\cos x| = \ln|\sec x|$$

$$y = y_h + y_p$$
$$= C_1\cos x + C_2\sin x + (x - \tan x)\cos x$$
$$+ \ln|\sec x|\sin x$$
$$= (C_1 + x - \tan x)\cos x + (C_2 + \ln|\sec x|)\sin x$$

27. $y'' + 4y = \csc 2x$

$y'' + 4y = 0$

$m^2 + 4 = 0$ when $m = -2i, 2i$.

$y_h = C_1 \cos 2x + C_2 \sin 2x$

$y_p = v_1 \cos 2x + v_2 \sin 2x = 0$

$v_1' \cos 2x + v_2' \sin 2x = 0$

$v_1'(-2 \sin 2x) + v_2'(2 \cos 2x) = \csc 2x$

$$v_1' = \frac{\begin{vmatrix} 0 & \sin 2x \\ \csc 2x & 2\cos 2x \end{vmatrix}}{\begin{vmatrix} \cos 2x & \sin 2x \\ -2\sin 2x & 2\cos 2x \end{vmatrix}} = -\frac{1}{2}$$

$$v_1 = \int -\frac{1}{2}\,dx = -\frac{1}{2}x$$

$$v_2' = \frac{\begin{vmatrix} \cos 2x & 0 \\ -2\sin 2x & \csc 2x \end{vmatrix}}{\begin{vmatrix} \cos 2x & \sin 2x \\ -2\sin 2x & 2\cos 2x \end{vmatrix}} = \frac{1}{2}\cot 2x$$

$$v_2 = \int \frac{1}{2}\cot 2x\,dx = \frac{1}{4}\ln|\sin 2x|$$

$$y = \left(C_1 - \frac{1}{2}x\right)\cos 2x + \left(C_2 + \frac{1}{4}\ln|\sin 2x|\right)\sin 2x$$

28. $y'' - 4y' + 4y = x^2e^{2x}$

$y'' - 4y' + 4y = 0$

$m^2 - 4m + 4 = 0$ when $m = 2, 2$.

$y_h = (C_1 + C_2x)e^{2x}$

$y_p = (v_1 + v_2x)e^{2x}$

$v_1'e^{2x} + v_2'xe^{2x} = 0$

$v_1'(2e^{2x}) + v_2'(2x + 1)e^{2x} = x^2e^{2x}$

$$v_1' = \frac{\begin{vmatrix} 0 & xe^{2x} \\ x^2e^{2x} & (2x+1)e^{2x} \end{vmatrix}}{\begin{vmatrix} e^{2x} & xe^{2x} \\ 2e^{2x} & (2x+1)e^{2x} \end{vmatrix}} = \frac{-x^3e^{4x}}{e^{4x}} = -x^3$$

$$v_1 = \int -x^3\,dx = -\tfrac{1}{4}x^4$$

$$v_2' = \frac{\begin{vmatrix} e^{2x} & 0 \\ 2e^{2x} & x^2e^{2x} \end{vmatrix}}{e^{4x}} = \frac{x^2e^{4x}}{e^{4x}} = x^2$$

$$v_2 = \int x^2\,dx = \tfrac{1}{3}x^3$$

$$y = \left(C_1 + C_2x + \frac{1}{12}x^4\right)e^{2x}$$

29. $y'' - 2y' + y = e^x \ln x$

$y'' - 2y' + y = 0$

$m^2 - 2m + 1 = 0$ when $m = 1, 1$.

$y_h = (C_1 + C_2x)e^x$

$y_p = (v_1 + v_2x)e^x$

$v_1'e^x + v_2'xe^x = 0$

$v_1'e^x + v_2'(x + 1)e^x = e^x \ln x$

$v_1' = -x \ln x$

$$v_1 = \int -x \ln x\,dx = -\frac{x^2}{2}\ln x + \frac{x^2}{4}$$

$v_2' = \ln x$

$$v_2 = \int \ln x\,dx = x\ln x - x$$

$$y = (C_1 + C_2x)e^x + \frac{x^2e^x}{4}\left(\ln x^2 - 3\right)$$

30. $y'' - 4y' + 4y = \dfrac{e^{2x}}{x}$

$y'' - 4y' + 4y = 0$

$m^2 - 4m + 4 = 0$ when $m = 2, 2$.

$y_h = (C_1 + C_2x)e^{2x}$

$y_p = (v_1 + v_2x)e^{2x}$

$v_1'e^{2x} + v_2'xe^{2x} = 0$

$$v_1e^{2x}(2) + v_2'(2x + 1)e^{2x} = \frac{e^{2x}}{x}$$

$v_1' = -1$

$$v_1 = \int -1\,dx = -x$$

$$v_2' = \frac{1}{x}$$

$$v_2 = \int \frac{1}{x}\,dx = \ln|x|$$

$$y = \left(C_1 + C_2x - x + x\ln|x|\right)e^{2x}$$

31. (a) $y'' - y' - 12y = 0$

$m^2 - m - 12 = (m - 4)(m + 3) = 0 \Rightarrow m = 4, -3$

Let $y_p = Ax^2 + Bx + C$. This is a generalized form of $F(x) = x^2$.

(b) $y'' - y' - 12y = 0$

$m^2 - m - 12 = (m - 4)(m + 3) = 0 \Rightarrow m = 4, -3$

Because $y_h = C_1e^{4x} + C_2e^{-3x}$, let

$y_p = Axe^{4x}$.

32. Answers will vary. See the "Variation of Parameters" box on page 1145.

33. $q'' + 10q' + 25q = 6 \sin 5t, q(0) = 0, q'(0) = 0$

$m^2 + 10m + 25 = 0$ when $m = -5, -5$.

$$q_h = (C_1 + C_2 t)e^{-5t}$$
$$q_p = A \cos 5t + B \sin 5t$$
$$q_p' = -5A \sin 5t + 5B \cos 5t$$
$$q_p'' = -25A \cos 5t - 25B \sin 5t$$

$$q_p'' + 10q_p' + 25q_p = 50B \cos 5t - 50A \sin 5t = 6 \sin 5t,\ A = -\tfrac{3}{25},\ B = 0$$

$$q = (C_1 + C_2 t)e^{-5t} - \tfrac{3}{25} \cos 5t$$

Initial conditions:

$q(0) = 0, q'(0) = 0, C_1 - \frac{3}{25} = 0, -5C_1 + C_2 = 0,$

$C_1 = \frac{3}{25}, C_2 = \frac{3}{5}$

Particular solution: $q = \frac{3}{25}\left(e^{-5t} + 5te^{-5t} - \cos 5t\right)$

34. $q'' + 20q' + 50q = 10 \sin 5t$

$m^2 + 20m + 50 = 0$ when $m = -10 \pm 5\sqrt{2}$.

$$q_h = C_1e^{(-10+5\sqrt{2})t} + C_2e^{(-10-5\sqrt{2})t}$$
$$q_p = A \cos 5t + B \sin 5t$$
$$q_p' = 5B \cos 5t - 5A \sin 5t$$
$$q_p'' = -25A \cos 5t - 25B \sin 5t$$

$$q_p'' + 20q_p' + 50q_p = (25A + 100B) \cos 5t + (25B - 100A) \sin 5t = 10 \sin 5t$$

$$\left.\begin{aligned} 25A + 100B &= 0 \\ 25B - 100A &= 10 \end{aligned}\right\} B = \frac{2}{85},\ A = -\frac{8}{85}$$

$$q = C_1e^{(-10+5\sqrt{2})t} + C_2e^{(-10-5\sqrt{2})t} - \frac{8}{85}\cos 5t + \frac{2}{85}\sin 5t$$

Initial conditions:

$$q(0) = 0, q'(0) = 0, C_1 + C_2 - \frac{8}{85} = 0,$$

$$\left(-10 + 5\sqrt{2}\right)C_1 + \left(-10 - 5\sqrt{2}\right)C_2 + \frac{2}{17} = 0,$$

$$C_1 = \frac{8 + 7\sqrt{2}}{170}, C_2 = \frac{8 - 7\sqrt{2}}{170}$$

Particular solution:

$$q = \frac{8 + 7\sqrt{2}}{170}e^{(-10+5\sqrt{2})t} + \frac{8 - 7\sqrt{2}}{170}e^{(-10-5\sqrt{2})t} - \frac{8}{85}\cos 5t + \frac{2}{85}\sin 5t$$

35. $\frac{24}{32}y'' + 48y = \frac{24}{32}(48\sin 4t),\ y(0) = \frac{1}{4},\ y'(0) = 0$

$\frac{24}{32}m^2 + 48 = 0$ when $m = \pm 8i$.

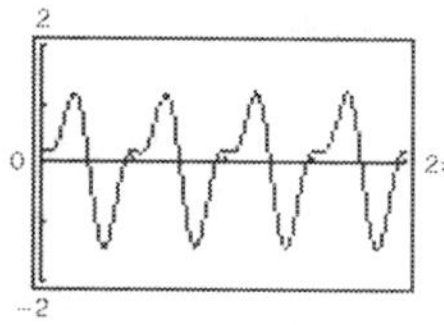

$$\begin{aligned} y_h &= C_1\cos 8t + C_2\sin 8t \\ y_p &= A\sin 4t + B\cos 4t \\ y_p' &= 4A\cos 4t - 4B\sin 4t \\ y_p'' &= -16A\sin 4t - 16B\cos 4t \end{aligned}$$

$$\begin{aligned} \tfrac{24}{32}y_p'' + 48y_p &= 36A\sin 4t + 36B\cos 4t \\ &= \tfrac{24}{32}(48\sin 4t),\ B = 0,\ A = 1 \end{aligned}$$

$y = y_h + y_p = C_1\cos 8t + C_2\sin 8t + \sin 4t$

Initial conditions: $y(0) = \frac{1}{4},\ y'(0) = 0,\ \frac{1}{4} = C_1,$

$0 = 8C_2 + 4 \Rightarrow C_2 = -\frac{1}{2}$

Particular solution: $y = \frac{1}{4}\cos 8t - \frac{1}{2}\sin 8t + \sin 4t$

36. $\frac{2}{32}y'' + 4y = \frac{2}{32}(4\sin 8t),\ y(0) = \frac{1}{4},\ y'(0) = 0$

$\frac{2}{32}m^2 + 4 = 0$

when $m = \pm 8i$.

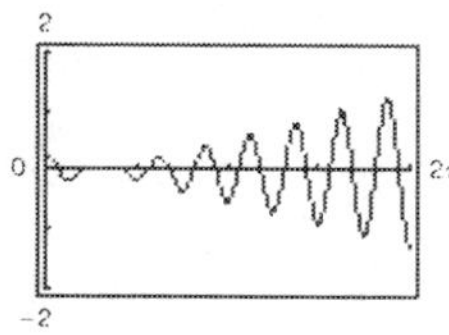

$$\begin{aligned} y_h &= C_1\cos 8t + C_2\sin 8t \\ y_p &= At\sin 8t + Bt\cos 8t \\ y_p'' &= (-64At - 16B)\sin 8t + (16A - 64Bt)\cos 8t \end{aligned}$$

$$\begin{aligned} \tfrac{2}{32}y_p'' + 4y_p &= -B\sin 8t + A\cos 8t \\ &= \tfrac{2}{32}(4\sin 8t),\ A = 0,\ B = -\tfrac{1}{4} \end{aligned}$$

$y = C_1\cos 8t + C_2\sin 8t - \frac{1}{4}t\cos 8t$

Initial conditions:

$y(0) = \frac{1}{4},\ y'(0) = 0,\ \frac{1}{4} = C_1,\ 0 = 8C_2 - \frac{1}{4} \Rightarrow C_2 = \frac{1}{32}$

Particular solution: $y = \frac{1}{4}\cos 8t + \frac{1}{32}\sin 8t - \frac{1}{4}t\cos 8t$

37. $\frac{2}{32}y'' + y' + 4y = \frac{2}{32}(4\sin 8t),\ y(0) = \frac{1}{4},\ y'(0) = -3$

$\frac{1}{16}m^2 + m + 4 = 0$

when $m = -8, -8$.

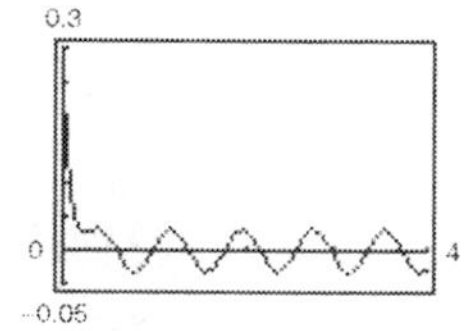

$$\begin{aligned} y_h &= (C_1 + C_2t)e^{-8t} \\ y_p &= A\sin 8t + B\cos 8t \\ y_p' &= 8A\cos 8t - 8B\sin 8t \\ y_p'' &= -64A\sin 8t - 64B\cos 8t \end{aligned}$$

$$\begin{aligned} \tfrac{2}{32}y_p'' + y_p' + 4y_p &= -8B\sin 8t + 8A\cos 8t \\ &= \tfrac{2}{32}(4\sin 8t) - 8B \\ &= \tfrac{1}{4} \Rightarrow B = -\tfrac{1}{32},\ 8A = 0 \Rightarrow A = 0 \end{aligned}$$

Initial conditions:

$y(0) = \frac{1}{4},\ y'(0) = -3,\ \frac{1}{4} = C_1 - \frac{1}{32} \Rightarrow C_1 = \frac{9}{32},$

$-3 = -8C_1 + C_2 \Rightarrow C_2 = -\frac{3}{4}$

Particular solution: $y = \left(\frac{9}{32} - \frac{3}{4}t\right)e^{-8t} - \frac{1}{32}\cos 8t$

38. $\dfrac{4}{32}y'' + \dfrac{1}{2}y' + \dfrac{25}{2}y = 0,$

$y(0) = \dfrac{1}{2},\ y'(0) = -4$

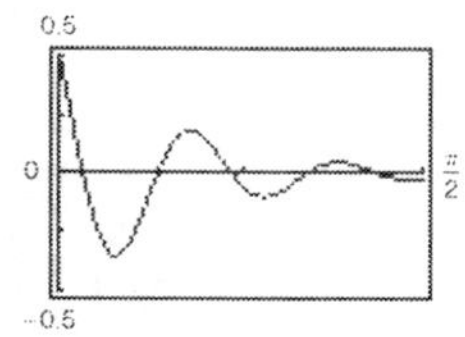

$\dfrac{1}{8}m^2 + \dfrac{1}{2}m + \dfrac{25}{2} = 0$

$m^2 + 4m + 100 = 0$

when $m = -2 \pm 4\sqrt{6}i$.

$y - C_1e^{-2t}\cos\left(4\sqrt{6}t\right) + C_2e^{-2t}\sin\left(4\sqrt{6}t\right)$

Initial conditions:

$y(0) = \dfrac{1}{2},\ y'(0) = -4,\ \dfrac{1}{2} = C_1,\ -4 = -2C_1 + 4\sqrt{6}C_2,$

$C_2 = -\dfrac{3}{4\sqrt{6}} = -\dfrac{\sqrt{6}}{8}$

Particular solution:

$$y = \frac{1}{2}e^{-2t}\cos\left(4\sqrt{6}t\right) - \frac{\sqrt{6}}{8}e^{-2t}\sin\left(4\sqrt{6}t\right)$$

39. In Exercise 35,

$$y_h = \frac{1}{4}\cos 8t - \frac{1}{2}\sin 8t - \frac{\sqrt{5}}{4}\sin\left[8t + \pi + \arctan\left(-\frac{1}{2}\right)\right] = \frac{\sqrt{5}}{4}\sin\left(8t + \pi - \arctan\frac{1}{2}\right) \approx \frac{\sqrt{5}}{4}\sin(8t + 2.6779).$$

40. When $b = 0$, the motion is undamped.

When $b > 0$, the motion is damped.

41. (a) $\frac{4}{32}y'' + \frac{25}{2}y = 0$

$y = C_1 \cos 10x + C_2 \sin 10x$

$y(0) = \frac{1}{2}: \frac{1}{2} = C_1$

$y'(0) = -4: -4 = 10C_2 \Rightarrow C_2 = -\frac{2}{5}$

$y = \frac{1}{2}\cos 10x - \frac{2}{5}\sin 10x$

The motion is undamped.

(b) If $b > 0$, the motion is damped.

(c) If $b > \frac{5}{2}$, the solution to the differential equation is of the form $y = C_1 e^{m_1 x} + C_2 e^{m_2 x}$. There would be no oscillations in this case.

42. (a) $x^2y'' - xy' + y = 4x \ln x$

$y_1 = x$ and $y_2 = x \ln x$

$u_1'x + u_2'x \ln x = 0 \Rightarrow u_1' = -u_2' \ln x$

$u_1' + u_2'(1 + \ln x) = \frac{4}{x}\ln x \Rightarrow u_2' = \frac{4}{x}\ln x$

and $u_1' = -\frac{4}{x}(\ln x)^2$

$u_1 = -\frac{4}{3}(\ln x)^3$ and $u_2 = 2(\ln x)^2$

$y_p = -\frac{4}{3}x(\ln x)^3 + 2x(\ln x)^3 = \frac{2}{3}x(\ln x)^3$

$y = y_h + y_p = C_1x + C_2x \ln x + \frac{2}{3}x(\ln x)^3$

(b) Let $y_p = A\sin(\ln x) + B\cos(\ln x)$.

$y_p' = A\cos(\ln x)\frac{1}{x} - B\sin(\ln x)\frac{1}{x} = \frac{1}{x}\left(A\cos(\ln x) - B\sin(\ln x)\right)$

$y_p'' = \frac{-1}{x^2}\left(A\cos(\ln x) - B\sin(\ln x)\right) + \frac{1}{x}\left(-A\sin(\ln x)\frac{1}{x} - B\cos(\ln x)\frac{1}{x}\right)$

$= \frac{1}{x^2}(B - A)\sin(\ln x) + \frac{1}{x^2}(-A - B)\cos(\ln x)$

$x^2y_p'' + xy' + 4y = (B - A)\sin(\ln x) - (A + B)\cos(\ln x) + \left(A\cos(\ln x) - B\sin(\ln x)\right)$

$+ 4\left(A\sin(\ln x) + B\cos(\ln x)\right) = \sin(\ln x)$

$(B - A - B + 4A)\sin(\ln x) + (-A - B + A + 4B)\cos(\ln x) = \sin(\ln x)$

$3A = 1, 3B = 0 \Rightarrow A = \frac{1}{3}$

So, $y_p = \frac{1}{3}\sin(\ln x)$ and $y = y_h + y_p = C_1\sin(\ln x^2) + C_2\cos(\ln x^2) + \frac{1}{3}\sin(\ln x)$.

43. True. $y_p = -e^{2x}\cos e^{-x}$

$y_p' = e^{2x}\sin e^{-x}(-e^{-x}) - 2e^{2x}\cos e^{-x} = -e^x \sin e^{-x} - 2e^{2x}\cos e^{-x}$

$y_p'' = \left[-e^x\cos e^{-x}(-e^{-x}) - e^x \sin e^{-x}\right] + \left[2e^{2x}\sin e^{-x}(-e^{-x}) - 4e^{2x}\cos e^{-x}\right]$

So,

$y_p'' - 3y_p' + 2y_p = \left[\cos e^{-x} - e^x \sin e^{-x} - 2e^x \sin e^{-x} - 4e^{2x}\cos e^{-x}\right] - 3\left[-e^x\sin e^{-x} - 2e^{2x}\cos e^{-x}\right] - 2e^{2x}\cos e^{-x}$

$= \left[-e^x - 2e^x + 3e^x\right]\sin e^{-x} + \left[1 - 4e^{2x} + 6e^{2x} - 2e^{2x}\right]\cos e^{-x} = \cos e^{-x}$.

44. True.

$y_p = -\frac{1}{8}e^{2x},\ y_p' = -\frac{1}{4}e^{2x},\ y_p'' = -\frac{1}{2}e^{2x}$

$y_p'' - 6y_p' = -\frac{1}{2}e^{2x} - 6\left(-\frac{1}{4}e^{2x}\right) = e^{2x}$

45. $y'' - 2y' + y = 2e^x$

$m^2 - 2m + 1 = 0 \Rightarrow m = 1, 1$

$y_h = C_1e^x + C_2xe^x,\ y_p = x^2e^x$, particular solution

General solution: $f(x) = (C_1 + C_2x)e^x + x^2e^x = \left(C_1 + C_2x + x^2\right)e^x$

$f'(x) = \left(C_2 + 2x + C_1 + C_2x + x^2\right)e^x = \left(x^2 + (C_2 + 2)x + (C_1 + C_2)\right)e^x$

(a) No. If $f(x) > 0$ for all x, then $x^2 + C_2x + C_1 > 0 \Leftrightarrow C_2^2 - 4C_1 < 0$ for all x.

So, let $C_1 = C_2 = 1$. Then $f'(x) = \left(x^2 + 3x + 2\right)e^x$ and $f'\left(-\frac{3}{2}\right) = -\frac{1}{4} < 0$.

(b) Yes. If $f'(x) > 0$ for all x, then

$$\begin{aligned}(C_2 + 2)^2 - 4(C_1 + C_2) &< 0\\ \Rightarrow C_2^2 - 4C_1 + 4 &< 0\\ C_2^2 - 4C_1 &< -4\\ C_2^2 - 4C_1 &< 0\end{aligned}$$

$\Rightarrow f(x) > 0$ for all x.

Section 16.4 Series Solutions of Differential Equations

1. $y' - y = 0$. Letting $y = \sum_{n=0}^{\infty} a_nx^n$:

$$y' - y = \sum_{n=1}^{\infty} na_nx^{n-1} - \sum_{n=0}^{\infty} a_nx^n = \sum_{n=0}^{\infty} (n + 1)a_{n+1}x^n - \sum_{n=0}^{\infty} a_nx^n = 0$$

$$(n + 1)a_{n+1} = a_n$$

$$a_{n+1} = \frac{a_n}{n + 1}$$

$$a_1 = a_0,\ a_2 = \frac{a_1}{2} = \frac{a_0}{2},\ a_3 = \frac{a_2}{3} = \frac{a_0}{1 \cdot 2 \cdot 3},\ \ldots,\ a_n = \frac{a_0}{n!}$$

$$y = \sum_{n=0}^{\infty} \frac{a_0}{n!}x^n = a_0e^x$$

Check: By separation of variables, you have:

$$\int \frac{dy}{y} = \int dx$$

$$\ln y = x + C_1$$

$$y = Ce^x$$

2. $y' - ky = 0$. Letting $y = \sum_{n=0}^{\infty} a_n x^n$:

$$y' - ky = \sum_{n=1}^{\infty} na_n x^{n-1} - k\sum_{n=0}^{\infty} a_n x^n = \sum_{n=0}^{\infty} (n+1)a_{n+1}x^n - \sum_{n=0}^{\infty} ka_n x^n = 0$$

$$(n+1)a_{n+1} = ka_n$$

$$a_{n+1} = \frac{ka_n}{n+1}$$

$$a_1 = ka_0,\ a_2 = \frac{ka_1}{2} = \frac{k^2a_0}{2},\ a_3 = \frac{ka_2}{3} = \frac{k^3a_0}{1\cdot 2\cdot 3}, \ldots, a_n = \frac{k^n}{n!}a_0$$

$$y = \sum_{n=0}^{\infty} \frac{k^n}{n!}a_0x^n = a_0\sum_{n=0}^{\infty} \frac{(kx)^n}{n!} = a_0e^{kx}$$

Check: By separation of variables, you have:

$$\int \frac{dy}{y} = \int k\,dx$$

$$\ln y = kx + C_1$$

$$y = Ce^{kx}$$

3. $y'' - 9y = 0$. Letting $y = \sum_{n=0}^{\infty} a_n x^n$:

$$y'' - 9y = \sum_{n=2}^{\infty} n(n-1)a_n x^{n-2} - 9\sum_{n=0}^{\infty} a_n x^n = \sum_{n=0}^{\infty} (n+2)(n+1)a_{n+2}x^n - \sum_{n=0}^{\infty} 9a_n x^n = 0$$

$$(n+2)(n+1)a_{n+2} = 9a_n$$

$$a_{n+2} = \frac{9a_n}{(n+2)(n+1)}$$

$$a_0 = a_0 \qquad a_1 = a_1$$

$$a_2 = \frac{9a_0}{2} \qquad a_3 = \frac{9a_1}{3\cdot 2}$$

$$a_4 = \frac{9a_2}{4\cdot 3} = \frac{9^2a_0}{4\cdot 3\cdot 2\cdot 1} \qquad a_5 = \frac{9a_3}{5\cdot 4} = \frac{9^2a_1}{5\cdot 4\cdot 3\cdot 2\cdot 1}$$

$$\vdots \qquad \vdots$$

$$a_{2n} = \frac{9^n a_0}{(2n)!} \qquad a_{2n+1} = \frac{9^n a_1}{(2n+1)!}$$

$$y = \sum_{n=0}^{\infty} \frac{9^n a_0}{(2n)!}x^{2n} + \sum_{n=0}^{\infty} \frac{9^n a_1}{(2n+1)!}x^{2n+1} = a_0\sum_{n=0}^{\infty} \frac{(3x)^{2n}}{(2n)!} + \frac{a_1}{3}\sum_{n=0}^{\infty} \frac{(3x)^{2n+1}}{(2n+1)!} = C_0\sum_{n=0}^{\infty} \frac{(3x)^n}{n!} + C_1\sum_{n=0}^{\infty} \frac{(-3x)^n}{n!}$$

$= C_0e^{3x} + C_1e^{-3x}$ where $C_0 + C_1 = a_0$ and $C_0 - C_1 = \frac{a_1}{3}$.

Check: $y'' - 9y = 0$ is a second-order homogeneous linear equation.

$m^2 - 9 = 0 \Rightarrow m_1 = 3$ and $m_2 = -3$

$y = C_1e^{3x} + C_2e^{-3x}$

4. $y = C_0e^{kx} + C_1e^{-kx}$. Follow the solution to Exercise 3 with 9 replaced by k^2.

5. $y'' + 4y = 0$. Letting $y = \sum_{n=0}^{\infty} a_n x^n$:

$$y'' + 4y = \sum_{n=2}^{\infty} n(n-1)a_n x^{n-2} + 4\sum_{n=0}^{\infty} a_n x^n = \sum_{n=0}^{\infty} (n+2)(n+1)a_{n+2}x^n + \sum_{n=0}^{\infty} 4a_n x^n = 0$$

$$(n+2)(n+1)a_{n+2} = -4a_n$$

$$a_{n+2} = \frac{-4a_n}{(n+2)(n+1)}$$

$$a_0 = a_0 \qquad a_1 = a_1$$

$$a_2 = \frac{-4a_0}{2} \qquad a_3 = \frac{-4a_1}{3 \cdot 2}$$

$$a_4 = \frac{-4a_2}{4 \cdot 3} = \frac{(-4)^2 a_0}{4!} \qquad a_5 = \frac{-4a_3}{5 \cdot 4} = \frac{(-4)^2 a_1}{5!}$$

$$\vdots \qquad \vdots$$

$$a_{2n} = \frac{(-1)^n 4^n}{(2n)!}a_0 \qquad a_{2n+1} = \frac{(-1)^n 4^n}{(2n+1)!}a_1$$

$$y = \sum_{n=0}^{\infty} \frac{(-1)^n 4^n a_0}{(2n)!}x^{2n} + \sum_{n=0}^{\infty} \frac{(-1)^n 4^n a_1}{(2n+1)!}x^{2n+1} = a_0\sum_{n=0}^{\infty} \frac{(-1)^n (2x)^{2n}}{(2n)!} + \frac{a_1}{2}\sum_{n=0}^{\infty} \frac{(-1)^n (2x)^{2n+1}}{(2n+1)!} = C_0 \cos 2x + C_1 \sin 2x$$

Check: $y'' + 4y = 0$ is a second-order homogeneous linear equation.

$$m^2 + 4 = 0 \Rightarrow m = \pm 2i$$

$$y = C_1 \cos 2x + C_2 \sin 2x$$

6. $y = C_0 \cos kx + C_1 \sin kx$. Follow the solution to Exercise 5 with 4 replaced by k^2.

7. $y' + 3xy = 0$. Letting $y = \sum_{n=0}^{\infty} a_n x^n$:

$$y' + 3xy = \sum_{n=1}^{\infty} na_n x^{n-1} + \sum_{n=0}^{\infty} 3a_n x^{n+1} = 0$$

$$\sum_{n=-1}^{\infty} (n+2)a_{n+2}x^{n+1} = \sum_{n=0}^{\infty} -3a_n x^{n+1} \Rightarrow a_1 = 0 \text{ and } a_{n+2} = \frac{-3a_n}{n+2}$$

$$a_0 = a_0 \qquad a_1 = 0$$

$$a_2 = -\frac{3a_0}{2} \qquad a_3 = -\frac{3a_1}{3} = 0$$

$$a_4 = -\frac{3}{4}\left(-\frac{3a_0}{2}\right) = \frac{3^2}{2^3}a_0 \qquad a_5 = -\frac{3}{5}\left(-\frac{3a_1}{3}\right) = 0$$

$$a_6 = -\frac{3}{6}\left(-\frac{3^2}{2^3}a_0\right) = -\frac{3^3 a_0}{2^3(3 \cdot 2)} \qquad a_7 = -\frac{3}{7}\left(\frac{3^2 a_1}{3 \cdot 5}\right) = 0$$

$$a_8 = -\frac{3}{8}\left(-\frac{3^3 a_0}{2^3(3 \cdot 2)}\right) = \frac{3^4 a_0}{2^4(4 \cdot 3 \cdot 2)} \qquad a_9 = -\frac{3}{9}\left(-\frac{3^3 a_1}{3 \cdot 5 \cdot 7}\right) = 0$$

$$y = a_0\sum_{n=0}^{\infty} \frac{(-3)^n x^{2n}}{2^n n!}$$

$$\lim_{n\to\infty}\left|\frac{u_{n+1}}{u_n}\right| = \lim_{n\to\infty}\left|\frac{(-3)^{n+1}x^{2n+2}}{2^{n+1}(n+1)!} \cdot \frac{2^n n!}{(-3)^n x^{2n}}\right| = \lim_{n\to\infty} \frac{3x^2}{2(n+1)} = 0$$

The interval of convergence for the solution is $(-\infty, \infty)$.

8. $y' - 2xy = 0$. Letting $y = \sum_{n=0}^{\infty} a_n x^n$:

$$y' - 2xy = \sum_{n=1}^{\infty} na_n x^{n-1} - \sum_{n=0}^{\infty} 2a_n x^{n+1} = 0$$

$$\sum_{n=-1}^{\infty} (n+2)a_{n+2}x^{n+1} = \sum_{n=0}^{\infty} 2a_n x^{n+1} \Rightarrow a_1 = 0 \text{ and}$$

$$a_{n+2} = \frac{2a_n}{n+2}$$

$a_0 = 0$ $\qquad$ $a_1 = 0$

$a_2 = \frac{2a_0}{2} = a_0$ $\qquad$ $a_3 = \frac{2a_1}{3} = 0$

$a_4 = \frac{2}{4}\left(\frac{2a_0}{2}\right) = \frac{2^2 a_0}{2^2 \cdot 2} = \frac{a_0}{2}$ $\qquad$ $a_5 = \frac{2}{5}\left(\frac{2a_1}{3}\right) = 0$

$a_6 = \frac{2}{6}\left(\frac{2^2 a_0}{2^2 \cdot 2}\right) = \frac{2^3 a_0}{2^3 \cdot 3 \cdot 2} = \frac{a_0}{3!}$ $\qquad$ $a_7 = \frac{2}{7}\left(\frac{2^2 a_1}{3 \cdot 5}\right) = 0$

$a_8 = \frac{2}{8}\left(\frac{a_0}{3!}\right) = \frac{a_0}{4!}$ $\qquad$ $a_9 = \frac{2}{9}\left(\frac{2^3 a_1}{3 \cdot 5 \cdot 7}\right) = 0$

$$y = a_0 \sum_{n=0}^{\infty} \frac{x^{2n}}{n!}$$

$$\lim_{n\to\infty}\left|\frac{u_{n+1}}{u_n}\right| = \lim_{n\to\infty}\left|\frac{x^{2n+2}}{(n+1)!}\cdot\frac{n!}{2^n}\right| = \lim_{n\to\infty}\frac{x^2}{n+1} = 0$$

The interval of convergence for the solution is $(-\infty, \infty)$.

9. $y'' - xy' = 0$. Letting $y = \sum_{n=0}^{\infty} a_n x^n$:

$$y'' - xy' = \sum_{n=2}^{\infty} n(n-1)a_n x^{n-2} - x\sum_{n=1}^{\infty} na_n x^{n-1} = 0$$

$$\sum_{n=2}^{\infty} n(n-1)a_n x^{n-2} = \sum_{n=0}^{\infty} na_n x^n$$

$$\sum_{n=0}^{\infty} (n+2)(n+1)a_{n+2}x^n = \sum_{n=0}^{\infty} na_n x^n$$

$$a_{n+2} = \frac{na_n}{(n+2)(n+1)}$$

$a_0 = a_0$ $\qquad$ $a_1 = a_1$

$a_2 = 0$ $\qquad$ $a_3 = \frac{a_1}{3 \cdot 2}$

There are no even powered terms. $a_5 = \frac{3a_3}{5 \cdot 4} = \frac{3a_1}{5!}$

$a_7 = \frac{5a_5}{7 \cdot 6} = \frac{5 \cdot 3a_1}{7!}$

$$y = a_0 + a_1\sum_{n=0}^{\infty}\frac{1\cdot 3\cdot 5\cdot 7\cdots(2n-1)x^{2n+1}}{(2n+1)!} = a_0 + a_1\sum_{n=0}^{\infty}\frac{(2n)!x^{2n+1}}{2^n n!(2n+1)!} = a_0 + a_1\sum_{n=0}^{\infty}\frac{x^{2n+1}}{2^n n!(2n+1)}$$

$$\lim_{n\to\infty}\left|\frac{u_{n+1}}{u_n}\right| = \lim_{n\to\infty}\left|\frac{x^{2n+3}}{2^{n+1}(n+1)!(2n+3)}\cdot\frac{2^n n!(2n+1)}{x^{2n+1}}\right| = \lim_{n\to\infty}\frac{(2n+1)x^2}{2(n+1)(2n+3)} = 0$$

Interval of convergence: $(-\infty, \infty)$

10. $y'' - xy' - y = 0$. Letting $y = \sum_{n=0}^{\infty} a_n x^n$:

$$y'' - xy' - y = \sum_{n=2}^{\infty} n(n-1)a_n x^{n-2} - x\sum_{n=1}^{\infty} na_n x^{n-1} - \sum_{n=0}^{\infty} a_n x^n = 0$$

$$\sum_{n=0}^{\infty}(n+2)(n+1)a_{n+2}x^n = \sum_{n=0}^{\infty}(n+1)a_n x^n$$

$$a_{n+2} = \frac{a_n}{n+2}$$

$$a_0 = a_0 \qquad a_1 = a_1$$

$$a_2 = \frac{a_0}{2} \qquad a_3 = \frac{a_1}{3}$$

$$a_4 = \frac{a_2}{4} = \frac{a_0}{8} = \frac{a_0}{2^2 2!} \qquad a_5 = \frac{a_3}{5} = \frac{a_1}{3 \cdot 5}$$

$$a_6 = \frac{a_4}{6} = \frac{a_0}{2^3 3!} \qquad a_7 = \frac{a_5}{7} = \frac{a_1}{3 \cdot 5 \cdot 7}$$

$$a_8 = \frac{a^6}{8} = \frac{a_0}{2^4 4!} \qquad a_9 = \frac{a_7}{9} = \frac{a_1}{3 \cdot 5 \cdot 7 \cdot 9}$$

$$y = a_0\sum_{n=0}^{\infty}\frac{x^{2n}}{2^n n!} + a_1\sum_{n=0}^{\infty}\frac{x^{2n+1}}{1 \cdot 3 \cdot 5 \cdot 7 \cdots (2n+1)}$$

$$\lim_{n\to\infty}\left|\frac{u_{n+1}}{u_n}\right| = \lim_{n\to\infty}\left|\frac{x^{2n+2}}{2^{n+1}(n+1)!} \cdot \frac{2^n n!}{x^{2n}}\right| = \lim_{n\to\infty}\frac{x^2}{2(n+1)} = 0$$

$$\lim_{n\to\infty}\left|\frac{u_{n+1}}{u_n}\right| = \lim_{n\to\infty}\left|\frac{x^{2n+3}}{1 \cdot 3 \cdot 5 \cdot 7 \cdots (2n+3)} \cdot \frac{1 \cdot 3 \cdot 5 \cdot 7 \cdots (2n+1)}{x^{2n+1}}\right| = \lim_{n\to\infty}\frac{x2}{2n+3} = 0$$

Because the interval of convergence for each series is $(-\infty, \infty)$, the interval of convergence for the solution is $(-\infty, \infty)$.

11. $(x^2 + 4)y'' + y = 0$. Letting $y = \sum_{n=0}^{\infty} a_n x^n$:

$$(x^2+4)y'' + y = \sum_{n=2}^{\infty} n(n-1)a_n x^n + 4\sum_{n=2}^{\infty} n(n-1)a_n x^{n-2} + \sum_{n=0}^{\infty} a_n x^n$$

$$= \sum_{n=0}^{\infty}(n^2 - n + 1)a_n x^n + \sum_{n=0}^{\infty} 4(n+2)(n+1)a_{n+2}x^n = 0$$

$$a_{n+2} = \frac{-(n^2 - n + 1)a_n}{4(n+2)(n+1)}$$

$$a_0 = a_0 \qquad a_1 = a_1$$

$$a_2 = \frac{-a_0}{4(2)(1)} = \frac{-a_0}{8} \qquad a_3 = \frac{-a_1}{4(3)(2)} = \frac{-a_1}{24}$$

$$a_4 = \frac{-3a_2}{4(4)(3)} = \frac{a_0}{128} \qquad a_5 = \frac{-7a_3}{4(5)(4)} = \frac{7a_1}{1920}$$

$$y = a_0\left(1 - \frac{x^2}{8} + \frac{x^4}{128} - \cdots\right) + a_1\left(x - \frac{x^3}{24} + \frac{7x^5}{1920} - \cdots\right)$$

12. $y'' + x^2y = 0$. Letting $y = \sum_{n=0}^{\infty} a_n x^n$:

$$y'' + x^2y = \sum_{n=2}^{\infty} n(n-1)a_n x^{n-2} + \sum_{n=0}^{\infty} a_n x^{n+2} = 0$$

$$\sum_{n=-2}^{\infty} (n+4)(n+3)a_{n+4}x^{n+2} = -\sum_{n=0}^{\infty} a_n x^{n+2}$$

$$a_{n+4} = \frac{-a_n}{(n+4)(n+3)}$$

Also:

$$y = a_0 + a_1x + a_2x^2 + a_3x^3 + \cdots + a_nx^n + \cdots$$

$$y'' = 2a_2 + 3 \cdot 2a_3x + \cdots + n(n-1)a_nx^{n-2} + \cdots$$

$$y'' + x^2y = 2a_2 + 3 \cdot 2a_3x + (a_0 + 4 \cdot 3a_4)x^2 + (a_1 + 5 \cdot 4a_5)x^3 + \cdots = 0$$

$$2a_2 = 0, 6a_3 = 0, 12a_4 + a_0 = 0, 20a_5 + a_1 = 0$$

So, $a_2 = 0$ and $a_3 = 0 \Rightarrow a_6 = 0, a_7 = 0, a_{10} = 0$, and $a_{11} = 0$. Therefore, $a_{4n+2} = 0$ and $a_{4n+3} = 0$.

$$a_0 = a_0 \qquad a_1 = a_1$$

$$a_4 = -\frac{a_0}{4 \cdot 3} \qquad a_5 = -\frac{a_1}{5 \cdot 4}$$

$$a_8 = -\frac{a_4}{8 \cdot 7} = \frac{a_0}{8 \cdot 7 \cdot 4 \cdot 3} \qquad a_9 = -\frac{a_5}{9 \cdot 8} = \frac{a_1}{9 \cdot 8 \cdot 5 \cdot 4}$$

$$a_{12} = -\frac{a_8}{12 \cdot 11} = -\frac{a_0}{12 \cdot 11 \cdot 8 \cdot 7 \cdot 4 \cdot 3} \qquad a_{13} = -\frac{a_9}{13 \cdot 12} = -\frac{a_1}{13 \cdot 12 \cdot 9 \cdot 8 \cdot 5 \cdot 4}$$

$$y'' + x^2y = a_0\left(1 - \frac{x^4}{4 \cdot 3} + \frac{x^8}{8 \cdot 7 \cdot 4 \cdot 3} - \frac{x^{12}}{12 \cdot 11 \cdot 8 \cdot 7 \cdot 4 \cdot 3} + \cdots\right)$$

$$+ a_1\left(x - \frac{x^5}{5 \cdot 4} + \frac{x^7}{9 \cdot 8 \cdot 5 \cdot 4} - \frac{x^9}{13 \cdot 12 \cdot 9 \cdot 8 \cdot 5 \cdot 4} + \cdots\right)$$

13. $y' + (2x - 1)y = 0, y(0) = 2$

$$y' = (1 - 2x)y \qquad y'(0) = 0$$

$$y'' = (1 - 2x)y' - 2y \qquad y''(0) = -2$$

$$y''' = (1 - 2x)y'' - 4y' \qquad y'''(0) = -10$$

$$y^4 = (1 - 2x)y''' - 6y'' \qquad y^{(4)}(0) = 2$$

$$\vdots \qquad \vdots$$

$$y(x) = 2 + \frac{2}{1!}x - \frac{2}{2!}x^2 - \frac{10}{3!}x^3 + \frac{2}{4!}x^4 + \cdots$$

Using the first five terms of the series, $y\left(\frac{1}{2}\right) = \frac{163}{64} \approx 2.547$.

Using Euler's Method with $\Delta x = 0.1$ you have $y' = (1 - 2x)y$.

i	x_i	y_i
0	0	2
1	0.1	2.2
2	0.2	2.376
3	0.3	2.51856
4	0.4	2.61930
5	0.5	2.67169

14. $y' - 2xy = 0,\ y(0) = 1$

$$\begin{aligned} y' &= 2xy & y'(0) &= 0 \\ y'' &= 2(xy' + y) & y''(0) &= 2 \\ y''' &= 2(xy'' + 2y') & y'''(0) &= 0 \\ y^{(4)} &= 2(xy''' + 3y'') & y^{(4)}(0) &= 12 \\ y^{(5)} &= 2\left(xy^{(4)} + 4y'''\right) & y^{(5)}(0) &= 0 \\ y^{(6)} &= 2\left(xy^{(5)} + 5y^{(4)}\right) & y^{(6)}(0) &= 120 \\ &\vdots & &\vdots \end{aligned}$$

$$\begin{aligned} y(x) &= 1 + \frac{2}{2!}x^2 + \frac{12}{4!}x^4 + \frac{120}{6!}x^6 + \cdots \\ &= 1 + x^2 + \frac{1}{2}x^4 + \frac{1}{6}x^6 + \cdots \end{aligned}$$

Using the first four terms of the series, $y(1) = \frac{8}{3} \approx 2.667$.

Using Euler's Method with $\Delta x = 0.1$ you have $y' = 2xy$.

i	x_i	y_i
0	0	1
1	0.1	1
2	0.2	1.02
3	0.3	1.0608
4	0.4	1.1244
5	0.5	1.2144
6	0.6	1.3358
7	0.7	1.4961
8	0.8	1.7056
9	0.9	1.9785
10	1.0	2.3346

So, $y(1) \approx 2.335$.

15. Given a differential equation, assume that the solution is of the form $y = \sum_{n=0}^{\infty} a_n x^n$. Then substitute y and its derivatives into the differential equation. You should then be able to determine the coefficients $a_0, a_1, \ldots$.

16. A recursion formula is a formula for determining the next term of a sequence from one or more of the preceding terms. See Example 1.

17. (a) From Exercise 9, the general solution is

$$y = a_0 + a_1\sum_{n=0}^{\infty} \frac{x^{2n+1}}{2^n n!(2n+1)}.$$

$$y(0) = 0 \Rightarrow a_0 = 0$$

$$y' = a_1\sum_{n=0}^{\infty} \frac{(2n+1)x^{2n}}{2^2 n!(2n+1)} = a_1\sum_{n=0}^{\infty} \frac{x^{2n}}{2^n n!}$$

$$y'(0) = 2 = a_1$$

$$y = 2\sum_{n=0}^{\infty} \frac{x^{2n+1}}{2^n n!(2n+1)}$$

(b) $P_3(x) = 2\left[x + \frac{x^3}{2\cdot 3}\right] = 2x + \frac{x^3}{3}$

$$P_5(x) = 2x + \frac{x^3}{3} + 2\frac{x^5}{4\cdot 2\cdot 5} = 2x + \frac{x^3}{3} + \frac{x^5}{20}$$

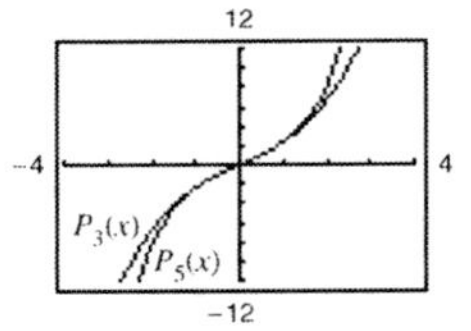

(c) The solution is symmetric about the origin.

18.

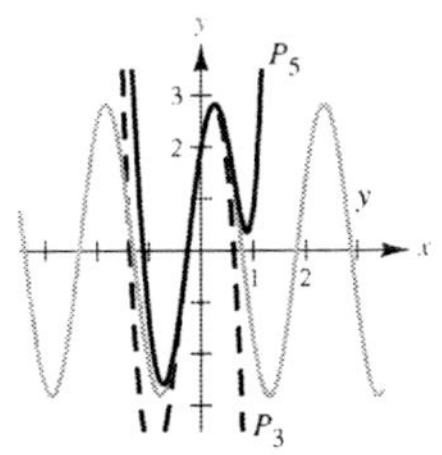

19. $y'' - 2xy = 0,\ y(0) = 1,\ y'(0) = -3$

$y'' = 2xy$	$y''(0) = 0$
$y''' = 2(xy' + y)$	$y'''(0) = 2$
$y^{(4)} = 2(xy'' + 2y')$	$y^{(4)}(0) = -12$
$y^{(5)} = 2(xy''' + 3y'')$	$y^{(5)}(0) = 0$
$y^{(6)} = 2\left(xy^{(4)} + 4y'''\right)$	$y^{(6)}(0) = 16$
$y^{(7)} = 2\left(xy^{(5)} + 5y^{(4)}\right)$	$y^{(7)}(0) = -120$
$\vdots$	$\vdots$

$$y \approx 1 - \frac{3}{1!}x + \frac{2}{3!}x^3 - \frac{12}{4!}x^4 + \frac{16}{6!}x^6 - \frac{120}{7!}x^7$$

Using the first six terms of the series, $y\left(\frac{1}{4}\right) \approx 0.253.$

20. $y'' - 2xy' + y = 0,\ y(0) = 1,\ y'(0) = 2$

$y'' = 2xy' - y$	$y''(0) = -1$
$y''' = 2xy'' + y'$	$y'''(0) = 2$
$y^{(4)} = 2xy''' + 3y''$	$y^{(4)}(0) = 3$
$y^{(5)} = 2xy^{(4)} + 5y'''$	$y^{(5)}(0) = 10$
$y^{(6)} = 2xy^{(5)} + 7y^{(4)}$	$y^{(6)}(0) = -21$
$y^{(7)} = 2xy^{(6)} + 9y^{(5)}$	$y^{(7)}(0) = 90$
$\vdots$	$\vdots$

$$y \approx 1 + \frac{2}{1!}x - \frac{1}{2!}x^2 + \frac{2}{3!}x^3 - \frac{3}{4!}x^4 + \frac{10}{5!}x^5 - \frac{21}{6!}x^6 + \frac{90}{7!}x^7$$

Using the first eight terms of the series, $y\left(\frac{1}{2}\right) \approx 1.911.$

21. $y'' + x^2y' - (\cos x)y = 0,\ y(0) = 3,\ y'(0) = 2$

$y'' = -x^2y' + (\cos x)y \qquad y''(0) = 3$

$y''' = -2x^2y' - x^2y'' - (\sin x)y + (\cos x)y' \qquad y'''(0) = 2$

$$y \approx 3 + \frac{2}{1!}x + \frac{3}{2!}x^2 + \frac{2}{3!}x^3$$

Using the first four terms of the series, $y\left(\frac{1}{3}\right) \approx 3.846.$

22. $y'' + e^xy' - (\sin x)y = 0,\ y(0) = -2,\ y'(0) = 1$

$y'' = -e^xy' + (\sin x)y, \qquad y''(0) = -1$

$y''' = -e^xy' - e^xy'' + (\cos x)y + (\sin x)y'$

$= -e^x(y' + y'') + (\cos x)y + (\sin x)y' \qquad y'''(0) = -(1 - 1) + (-2) = -2$

$$y \approx -2 + \frac{1}{1!}x - \frac{1}{2!}x^2 - \frac{2}{3!}x^3$$

Using the first four terms of the series, $y\left(\frac{1}{5}\right) \approx -1.823.$

23. $f(x) = e^x, f'(x) = e^x, y' - y = 0.$

Assume $y = \sum_{n=0}^{\infty} a_nx^n$, then:

$$y' = \sum_{n=1}^{\infty} na_nx^{n-1}$$

$$\sum_{n=1}^{\infty} na_nx^{n-1} = \sum_{n=0}^{\infty} a_nx^n$$

$$\sum_{n=0}^{\infty} (n+1)a_{n+1}x^n = \sum_{n=0}^{\infty} a_nx^n$$

$$a_{n+1} = \frac{a_n}{n+1},\ n \ge 0$$

$n = 0,\quad a_1 = a_0$

$n = 1,\quad a_2 = \frac{a_1}{2} = \frac{a_0}{2}$

$n = 2,\quad a_3 = \frac{a_2}{3} = \frac{a_0}{2(3)}$

$n = 3,\quad a_4 = \frac{a_3}{4} = \frac{a_0}{2(3)(4)}$

$n = 4,\quad a_5 = \frac{a_4}{5} = \frac{a_0}{2(3)(4)(5)}$

$\vdots$

$$a_{n+1} = \frac{a_0}{(n+1)!} \Rightarrow a_n = \frac{a_0}{n!}$$

$y = a_0\sum_{n=0}^{\infty}\frac{x^n}{n!}$ which converges on $(-\infty, \infty)$. When $a_0 = 1$, you have the Maclaurin Series for $f(x) = e^x$.

24. $f(x) = \cos x, f'(x) = -\sin x, f''(x) = -\cos x,$ $y'' + y = 0.$

Assume $y = \sum_{n=0}^{\infty} a_nx^n$, then:

$$y'' = \sum_{n=2}^{\infty} n(n-1)a_nx^{n-2}$$

$$\sum_{n=2}^{\infty} n(n-1)a_nx^{n-2} + \sum_{n=0}^{\infty} a_nx^{n-2} = 0$$

$$\sum_{n=0}^{\infty} (n+2)(n+1)a_{n+2}x^n = -\sum_{n=0}^{\infty} a_nx^n$$

$$a_{n+2} = -\frac{a_n}{(n+1)(n+2)}, n \ge 0$$

$a_0 = a_0 \qquad a_1 = a_1$

$a_2 = -\frac{a_0}{(1)(2)} \qquad a_3 = -\frac{a_1}{(2)(3)}$

$a_4 = -\frac{a_2}{(3)(4)} = \frac{a_0}{4!} \qquad a_5 = -\frac{a_3}{(4)(5)} = \frac{a_1}{5!}$

$\vdots \qquad \vdots$

$a_{2n} = \frac{(-1)^n a_0}{(2n)!} \qquad a_{2n+1} = \frac{(-1)^n a_1}{(2n+1)!}$

$y = a_0\sum_{n=0}^{\infty}\frac{(-1)^n x^{2n}}{(2n)!} + a_1\sum_{n=0}^{\infty}\frac{(-1)^n x^{2n+1}}{(2n+1)!}$ which converges on $(-\infty, \infty)$

When $a_0 = 1$ and $a_1 = 0$, you have the Maclaurin Series for $f(x) = \cos x$.

25.

$$f(x) = \arctan x$$

$$f'(x) = \frac{1}{1 + x^2}$$

$$f''(x) = \frac{-2x}{\left(1 + x^2\right)^2}$$

$$y'' = \frac{-2x}{1 + x^2}y'$$

$$\left(1 + x^2\right)y'' + 2xy' = 0$$

Assume $y = \sum_{n=0}^{\infty} a_n x^n$, then:

$$y' = \sum_{n=1}^{\infty} na_n x^{n-1}$$

$$y'' = \sum_{n=2}^{\infty} n(n - 1)a_n x^{n-2}$$

$$\left(1 + x^2\right)y'' + 2xy' = \sum_{n=2}^{\infty} n(n - 1)a_n x^{n-2} + \sum_{n=0}^{\infty} n(n - 1)a_n x^n + \sum_{n=0}^{\infty} 2na_n x^n = 0$$

$$\sum_{n=2}^{\infty} n(n - 1)a_n x^{n-2} = -\sum_{n=0}^{\infty} n(n - 1)a_n x^n - \sum_{n=0}^{\infty} 2na_n x^n$$

$$\sum_{n=0}^{\infty} (n + 2)(n + 1)a_{n+2}x^n = -\sum_{n=0}^{\infty} n(n + 1)a_n x^n$$

$$(n + 2)(n + 1)a_{n+2} = -n(n + 1)a_n$$

$$a_{n+2} = -\frac{n}{n + 2}a_n, n \ge 0$$

$n = 0 \Rightarrow a_2 = 0 \Rightarrow$ all the even-powered terms have a coefficient of 0.

$n = 1$, $\quad a_3 = -\frac{1}{3}a_1$

$n = 3$, $\quad a_5 = -\frac{3}{5}a_3 = \frac{1}{5}a_1$

$n = 5$, $\quad a_7 = -\frac{5}{7}a_5 = -\frac{1}{7}a_1$

$n = 7$, $\quad a_9 = -\frac{7}{9}a_7 = \frac{1}{9}a_1$

$\vdots$

$$a_{2n+1} = \frac{(-1)^n a_1}{2n + 1}$$

$y = a_1 \sum_{n=0}^{\infty} \frac{(-1)^n x^{2n+1}}{2n + 1}$ which converges on $(-1, 1)$. When $a_1 = 1$, you have the Maclaurin Series for $f(x) = \arctan x$.

26.
$$f(x) = \arcsin x$$
$$f'(x) = \frac{1}{\sqrt{1-x^2}}$$
$$f''(x) = \frac{x}{\left(1-x^2\right)^{3/2}}$$
$$y'' = \frac{1}{\sqrt{1-x^2}} \cdot \frac{x}{1-x^2} = \frac{x}{1-x^2}y'$$
$$\left(1-x^2\right)y'' - xy' = 0$$

Assume $y = \sum_{n=0}^{\infty} a_n x^n$, then: $\sum_{n=2}^{\infty} a_n n(n-1)x^{n-2} - \sum_{n=0}^{\infty} a_n n(n-1)x^n - \sum_{n=0}^{\infty} a_n nx^n = 0$

$$\sum_{n=0}^{\infty} (n+2)(n+1)a_{n+2}x^n = \sum_{n=0}^{\infty} n^2 a_n x^n$$
$$a_{n+2} = \frac{n^2}{(n+1)(n+2)}a_n,\ n \ge 0$$

$n = 0 \Rightarrow a_2 = 0 \Rightarrow$ all the even-powered terms have a coefficient of 0.

$$a_1 = a_1$$

$n = 1$, $\quad a_3 = \frac{1}{(2)(3)}a_1$

$n = 3$, $\quad a_5 = \frac{9}{(4)(5)}a_3 = \frac{9}{(2)(3)(4)(5)}a_1 = \frac{3}{(2)(4)(5)}a_1$

$n = 5$, $\quad a_7 = \frac{25}{(6)(7)}a_5 = \frac{(9)(25)}{(2)(3)(4)(5)(6)(7)}a_1 = \frac{(3)(5)}{(2)(4)(6)(7)}a_1$

$n = 7$, $\quad a_9 = \frac{49}{(8)(9)}a_7 = \frac{(9)(25)(49)}{(2)(3)(4)(5)(6)(7)(8)(9)}a_1 = \frac{(3)(5)(7)}{(2)(4)(6)(8)}a_1$

$n = 9$, $\quad a_{11} = \frac{81}{(10)(11)}a_9 = \frac{(9)(25)(49)(81)}{(2)(3)(4)(5)(6)(7)(8)(9)(10)(11)}a_1 = \frac{(3)(5)(7)(9)}{(2)(4)(6)(8)(10)(11)}a_1$

$\vdots$

$$a_{2n+1} = \frac{(2n)!}{\left(2^n n!\right)^2(2n+1)}a_1$$

$y = a_1 \sum_{n=0}^{\infty} \frac{(2n)!}{\left(2^n n!\right)^2(2n+1)}x^{2n+1}$ which converges on $(-1, 1)$. When $a_1 = 1$, you have the Maclaurin Series for $f(x) = \arcsin x$.

27. $y'' - xy = 0$. Let $y = \sum_{n=0}^{\infty} a_n x^n$.

$$y'' - xy = \sum_{n=2}^{\infty} n(n-1)a_n x^{n-2} - x\sum_{n=0}^{\infty} a_n x^n = \sum_{n=-1}^{\infty} (n+3)(n+2)a_{n+3}x^{n+1} - \sum_{n=0}^{\infty} a_n x^{n+1} = 0$$

$$2a_2 + \sum_{n=0}^{\infty} \left[(n+3)(n+2)a_{n+3} - a_n\right]x^{n+1} = 0$$

So, $a_2 = 0$ and $a_{n+3} = \dfrac{a_n}{(n+3)(n+2)}$ for $n = 0, 1, 2, \ldots$

The constants a_0 and a_1 are arbitrary.

$a_0 = a_0$ $\qquad$ $a_1 = a_1$

$a_3 = \dfrac{a_0}{3 \cdot 2}$ $\qquad$ $a_4 = \dfrac{a_1}{4 \cdot 3}$

$a_6 = \dfrac{a_3}{6 \cdot 5} = \dfrac{a_0}{6 \cdot 5 \cdot 3 \cdot 2}$ $\qquad$ $a_7 = \dfrac{a_4}{7 \cdot 6} = \dfrac{a_1}{7 \cdot 6 \cdot 4 \cdot 3}$

So, $y = a_0 + a_1 x + \dfrac{a_0}{6}x^3 + \dfrac{a_1}{12}x^4 + \dfrac{a_0}{180}x^6 + \dfrac{a_1}{504}x^7$.

Review Exercises for Chapter 16

1. $\left(y + x^3 + xy^2\right)dx - x\,dy = 0$

$$\frac{\partial M}{\partial y} = 1 + 2xy \neq \frac{\partial N}{\partial x} = -1$$

Not exact

2. $(5x - y)\,dx + (5y - x)\,dy = 0$

$$\frac{\partial M}{\partial y} = -1 = \frac{\partial N}{\partial x}$$

Exact

3. $(10x + 8y + 2)\,dx + (8x + 5y + 2)\,dy = 0$

Exact: $\dfrac{\partial M}{\partial y} = 8 = \dfrac{\partial N}{\partial x}$

$f(x, y) = \int(10x + 8y + 2)\,dx = 5x^2 + 8xy + 2x + g(y)$

$f_y(x, y) = 8x + g'(y) = 8x + 5y + 2$

$g'(y) = 5y + 2$

$g(y) = \dfrac{5}{2}y^2 + 2y + C_1$

$f(x, y) = 5x^2 + 8xy + 2x + \dfrac{5}{2}y^2 + 2y + C_1$

$5x^2 + 8xy + 2x + \dfrac{5}{2}y^2 + 2y = C$

4. $\left(2x - 2y^3 + y\right)dx + \left(x - 6xy^2\right)dy = 0$

Exact: $\dfrac{\partial M}{\partial y} = -6y^2 + 1 = \dfrac{\partial N}{\partial x}$

$$\begin{aligned} f(x, y) &= \int\left(2x - 2y^3 + y\right)dx \\ &= x^2 - 2xy^3 + xy + g(y) \\ f_y(x, y) &= -6xy^2 + x + g'(y) = x - 6xy^2 \\ g'(y) &= 0 \\ g(y) &= C_1 \\ f(x, y) &= x^2 - 2xy^3 + xy + C_1 \end{aligned}$$

$x^2 - 2xy^3 + xy = C$

5. $(x - y - 5)\,dx - (x + 3y - z)\,dy = 0$

$\dfrac{\partial M}{\partial y} = -1 = \dfrac{\partial N}{\partial x}$ Exact

$f(x, y) = \int(x - y - 5)\,dx = \dfrac{x^2}{2} - xy - 5x + g(y)$

$f_y(x, y) = -x + g'(y) = -x - 3y + 2$

$g'(y) = -3y + 2$

$g(y) = \dfrac{-3}{2}y^2 + 2y + C_1$

$\dfrac{x^2}{2} - xy - 5x - \dfrac{3}{2}y^2 + 2y + C_1 = 0$

$x^2 - 2xy - 10x - 3y^2 + 4y = C$

6. $\left(3x^2 - 5xy^2\right)dx + \left(2y^3 - 5xy^2\right)dy = 0$

$$\frac{\partial M}{\partial y} = -10xy \neq \frac{\partial N}{\partial x} - 5y^2$$

Not exact

7. $\frac{x}{y}\,dx - \frac{x}{y^2}\,dy = 0$

$$\frac{\partial M}{\partial y} = \frac{-x}{y^2} \neq \frac{\partial N}{\partial x} = \frac{-1}{y^2}$$

Not exact

8. $y \sin xy\,dx + (x \sin xy + y)\,dy = 0$

$$\frac{\partial M}{\partial y} = xy \cos xy + \sin xy = \frac{\partial N}{\partial x} \text{ Exact}$$

$$f(x, y) = \int y \sin xy\,dx = -\cos xy + g(y)$$

$$f_y(x, y) = x \sin xy + g'(y) = x \sin xy + y$$

$$g'(y) = y \Rightarrow g(y) = \frac{y^2}{2} + C_1$$

$$-\cos xy + \frac{y^2}{2} = C$$

9. (a)

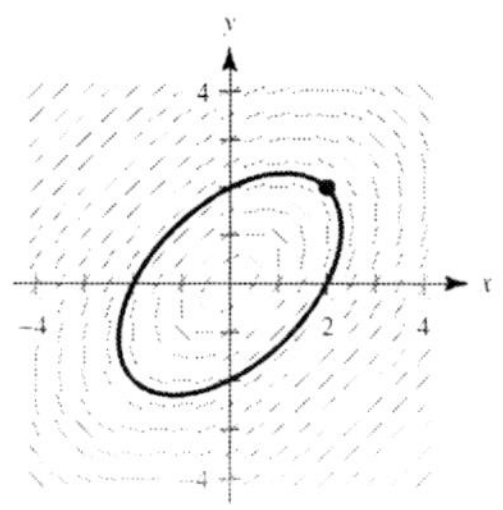

(b) $(2x - y)\,dx + (2y - x)\,dy = 0$

$$\frac{\partial M}{\partial y} = -1 = \frac{\partial N}{\partial x} \text{ Exact}$$

$$f(x, y) = \int (2x - y)\,dx = x^2 - xy + g(y)$$

$$f_y(x, y) = -x + g'(y) = 2y - x$$

$$g'(y) = 2y$$

$$g(y) = y^2 + C_1$$

$$x^2 - xy + y^2 = C$$

$$y(2) = 2 : 4 - 4 + 4 = 4 = C$$

Particular solution: $x^2 - xy + y^2 = 4$

(c)

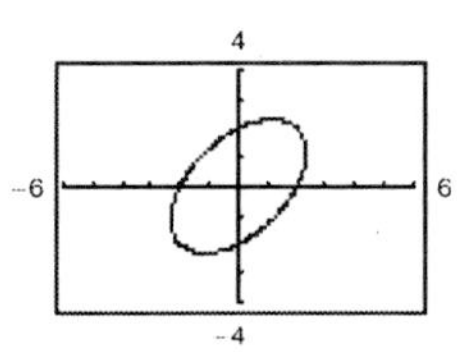

10. (a) and (c)

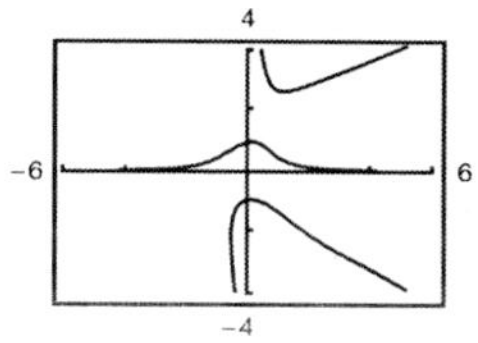

(b) $\left(6xy - y^3\right)dx + \left(4y + 3x^2 - 3xy^2\right)dy = 0$

$$\frac{\partial M}{\partial y} = 6x - 3y^2 = \frac{\partial N}{\partial x} \text{ Exact}$$

$$f(x, y) = \int \left(6xy - y^3\right)dx = 3x^2y - xy^3 + g(y)$$

$$f_y(x, y) = 3x^2 - 3xy^2 + g'(y) = 4y + 3x^2 - 3xy^2$$

$$g'(y) = 4y \Rightarrow g(y) = 2y^2 + C_1$$

$$3x^2y - xy^3 + 2y^2 = C$$

$$y(0) = 1 : 2 = C$$

Particular solution: $3x^2y - xy^3 + 2y^2 = 2$

11. $(2x + y - 3)\,dx + (x - 3y + 1)\,dy = 0$

Exact: $\frac{\partial M}{\partial y} = 1 = \frac{\partial N}{\partial x}$

$$\begin{aligned} f(x, y) &= \int (2x + y - 3)\,dx \\ &= x^2 + xy - 3x + g(y) \\ f_y(x, y) &= x + g'(y) \\ &= x - 3y + 1 \\ g'(y) &= -3y + 1 \\ g(y) &= -\frac{3}{2}y^2 + y + C_1 \\ f(x, y) &= x^2 + xy - 3x \\ &\quad -\frac{3}{2}y^2 + y + C_1 \end{aligned}$$

$$2x^2 + 2xy - 6x - 3y^2 + 2y = C$$

Initial condition:

$$y(2) = 0$$

$$8 + 0 - 12 - 0 + 0 = C \Rightarrow C = -4$$

Particular solution:

$$2x^2 + 2xy - 6x - 3y^2 + 2y = -4$$

12. $3x^2y^2\,dx + \left(2x^3y - 3y^2\right)dy = 0,\ y(1) = 2$

$$\frac{\partial M}{\partial y} = 6x^2y = \frac{\partial N}{\partial x}\ \text{Exact}$$

$$f(x, y) = \int 3x^2y^2\,dx = x^3y^2 + g(y)$$

$$f_y(x, y) = 2x^3y + g'(y) = 2x^3y - 3y^2$$

$$g'(y) = -3y^2$$

$$g(y) = -y^3 + C_1$$

$$x^3y^2 - y^3 = C$$

Initial condition: $y(1) = 2$: $4 - 8 = C$

Particular solution: $x^3y^2 - y^3 = -4$

13. $\left(3x^2 - y^2\right)dx + 2xy\,dy = 0$

$$\frac{(\partial M/\partial y) - (\partial N/\partial x)}{N} = \frac{-2y - 2y}{2xy} = -\frac{2}{x} = h(x)$$

Integrating factor: $e^{\int h(x)\,dx} = e^{\ln x^{-2}} = \dfrac{1}{x^2}$

Exact equation: $\left(3 - \dfrac{y^2}{x^2}\right)dx + \dfrac{2y}{x}\,dy = 0$

$$f(x, y) = \int\left(3 - \frac{y^2}{x^2}\right)dx = 3x + \frac{y^2}{x} + g(y)$$

$$f_y(x, y) = \frac{2y}{x} + g'(y) = \frac{2y}{x}$$

$$g'(y) = 0 \Rightarrow g(y) = C_1$$

$$3x + \frac{y^2}{x} = C$$

14. $2xy\,dx + \left(y^2 - x^2\right)dy = 0$

$$\frac{(\partial N/\partial x) - (\partial M/\partial y)}{M} = \frac{-2x - 2x}{2xy} = -\frac{2}{y} = k(y)$$

Integrating factor: $e^{\int k(y)\,dy} = e^{\ln y^{-2}} = \dfrac{1}{y^2}$

Exact equation: $\dfrac{2x}{y}\,dx + \left(1 - \dfrac{x^2}{y^2}\right)dy = 0$

$$f(x, y) = \int\frac{2x}{y}\,dx = \frac{x^2}{y} + g(y)$$

$$f_y(x, y) = -\frac{x^2}{y^2} + g'(y) = 1 - \frac{x^2}{y^2}$$

$$g'(y) = 1 \Rightarrow g(y) = y + C_1$$

$$\frac{x^2}{y} + y = C$$

15. $dx + \left(3x - e^{-2y}\right)dy = 0$

$$\frac{(\partial N/\partial x) - (\partial M/\partial y)}{M} = \frac{3 - 0}{1} = 3 = k(y)$$

Integrating factor: $e^{\int k(y)\,dy} = e^{3y}$

Exact equation: $e^{3y}\,dx + \left(3xe^{3y} - e^{y}\right)dy = 0$

$$f(x, y) = \int e^{3y}\,dx = xe^{3y} + g(y)$$

$$f_y(x, y) = 3xe^{3y} + g'(y) = 3xe^{3y} - e^{y}$$

$$g'(y) = -e^{y}$$

$$g(y) = -e^{y} + C_1$$

$$xe^{3y} - e^{y} = C$$

16. $\cos y\,dx - \left[2(x - y)\sin y + \cos y\right]dy = 0$

$$\frac{(\partial N/\partial x) - (\partial M/\partial y)}{M} = \frac{-2\sin y + \sin y}{\cos y} = -\tan y = k(y)$$

Integrating factor: $e^{\int k(y)\,dy} = \cos y$

Exact equation:

$$\cos^2 y\,dx - \left[2(x - y)\sin y\cos y + \cos^2 y\right] = 0$$

$$f(x, y) = \int\cos^2 y\,dx = x\cos^2 y + g(y)$$

$$f_y(x, y) = -2x\cos y\sin y + g'(y)$$

$$= -2x\sin y\cos y + 2y\sin y\cos y - \cos^2 y$$

$$g'(y) = 2y\sin y\cos y - \cos^2 y$$

$$\Rightarrow g(y) = -y\cos^2 y + C_1$$

$$x\cos^2 y - y\cos^2 y = C$$

17. $y = C_1e^{2x} + C_2e^{-2x}$

$$y' = 2C_1e^{2x} - 2C_2e^{-2x}$$

$$y'' = 4C_1e^{2x} + 4C_2e^{-2x}$$

$$y'' - 4y = 4C_1e^{2x} + 4C_2e^{-2x} - 4\left(C_1e^{2x} + C_2e^{-2x}\right) = 0$$

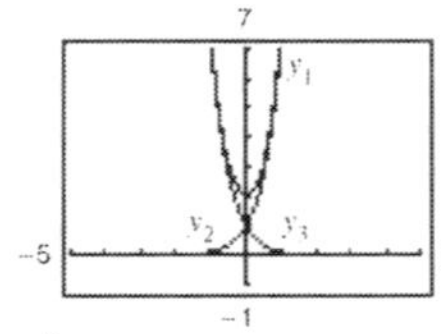

18. $y = C_1\cos 2x + C_2\sin 2x$

$$y' = -2C_1\sin 2x + 2C_2\cos 2x$$

$$y'' = -4C_1\cos 2x - 4C_2\sin 2x$$

$$y'' + 4y = -4C_1\cos 2x - 4C_2\sin 2x + 4(C_1\cos 2x + C_2\sin 2x) = 0$$

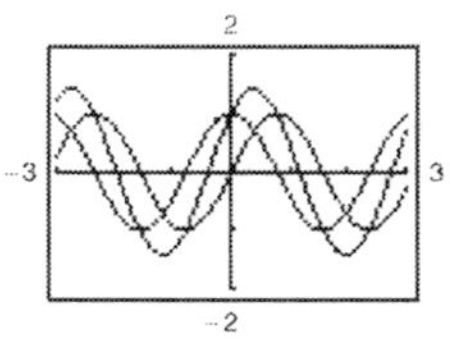

19. $y'' - y' - 2y = 0$

$m^2 - m - 2 = (m-2)(m+1) = 0,\ m = 2, -1$

$$y = C_1e^{2x} + C_2e^{-x}$$
$$y' = 2C_1e^{2x} - C_2e^{-x}$$
$$y(0) = 0 = C_1 + C_2$$
$$y'(0) = 3 = 2C_1 - C_2$$

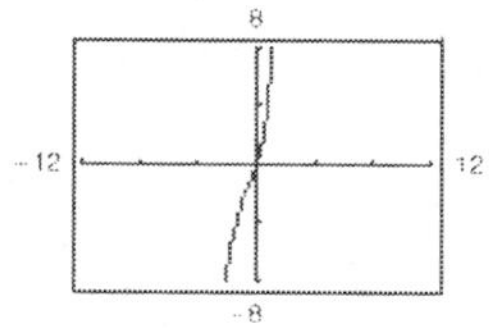

Adding these equations, $3 = 3C_1 \Rightarrow C_1 = 1$ and $C_2 = -1$.

$$y = e^{2x} - e^{-x}$$

20. $y'' + 4y' + 5y = 0$

$m^2 + 4m + 5 = 0 \qquad m = \dfrac{-4 \pm \sqrt{16-20}}{2} = -2 \pm i$

$$y = C_1e^{-2x}\cos x + C_2e^{-2x}\sin x$$
$$y(0) = 2 = C_1 \Rightarrow y = e^{-2x}[2\cos x + C_2 \sin x]$$
$$y' = e^{-2x}[-2\sin x + C_2\cos x] - 2e^{-2x}[2\cos x + C_2\sin x]$$
$$y'(0) = -7 = C_2 - 2(2) \Rightarrow C_2 = -3$$
$$y = 2e^{-2x}\cos x - 3e^{-2x}\sin x$$

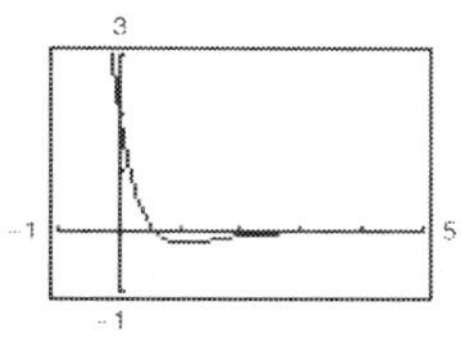

21. $y'' + 2y' - 3y = 0$

$m^2 + 2m - 3 = (m+3)(m-1) = 0 \Rightarrow m = -3, 1$

$$y = C_1e^{-3x} + C_2e^{x}$$
$$y' = -3C_1e^{-3x} + C_2e^{x}$$
$$y(0) = 2 = C_1 + C_2$$
$$y'(0) = 0 = -3C_1 + C_2$$

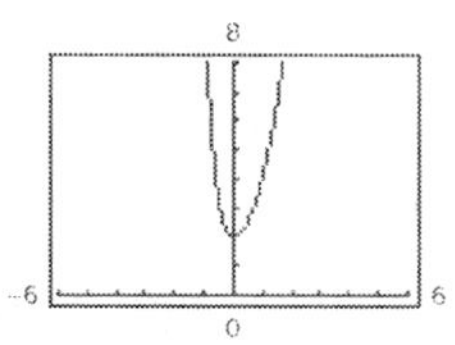

Subtracting these equations, $2 = 4C_1 \Rightarrow C_1 = \frac{1}{2}$ and $C_2 = \frac{3}{2}$.

$$y = \tfrac{3}{2}e^{x} + \tfrac{1}{2}e^{-3x}$$

22. $y'' + 12y' + 36y = 0$

$m^2 + 12m + 36 = (m+6)^2 = 0,\ m = -6, -6$

$$y = C_1e^{-6x} + C_2xe^{-6x}$$
$$y' = -6C_1e^{-6x} + C_2e^{-6x} - 6C_2xe^{-6x}$$
$$y(0) = 2 = C_1$$
$$y'(0) = 1$$
$$= -6(2) + C_2 \Rightarrow C_2 = 13$$

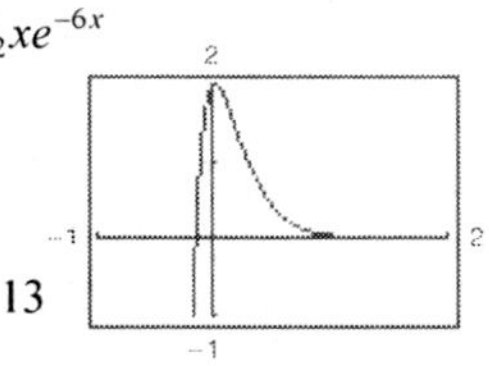

$$y = 2e^{-6x} + 13xe^{-6x}$$

23. $y'' + 2y' + 5y = 0$

$m^2 + 2m + 5 = 0 \Rightarrow m = \dfrac{-2 \pm \sqrt{4-20}}{2} = -1 \pm 2i$

$$y = e^{-x}(C_1\cos 2x + C_2\sin 2x)$$
$$y(1) = 4 = e^{-1}(C_1\cos 2 + C_2\sin 2)$$
$$y(2) = 0 = e^{-2}(C_1\cos 4 + C_2\sin 4)$$

Solving this system, you obtain $C_1 = -9.0496$, $C_2 = 7.8161$.

$$y = e^{-x}(-9.0496\cos 2x + 7.8161\sin 2x)$$

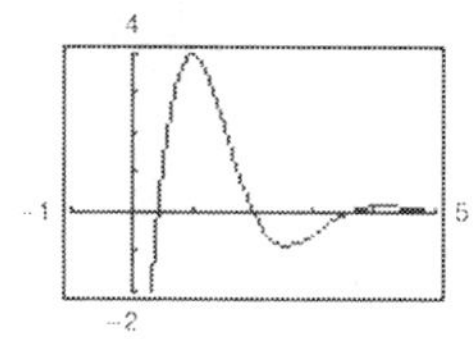

24. $y'' + y = 0$

$m^2 + 1 = 0 \Rightarrow m = \pm i$

$$y = C_1\cos x + C_2\sin x$$
$$y(0) = 2 = C_1$$
$$y\left(\frac{\pi}{2}\right) = 1 = C_2$$
$$y = 2\cos x + \sin x$$

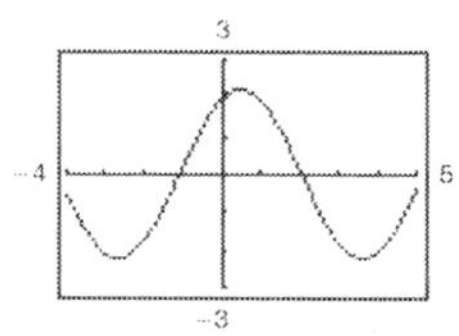

25. No, it is not homogeneous because of the nonzero term $\sin x$.

26. $y'' + 2ky' + ky = 0$

Characteristic equation: $m^2 + 2km + k = 0$

$$m = \frac{-2k \pm \sqrt{4k^2 - 4k}}{2} = -k + \sqrt{k^2 - 1}$$

(a) For $k < -1$ and $k > 1$, $k^2 - 1 > 0$ and there are 2 distinct real roots.

(b) For $k = \pm 1$, $k^2 - 1 = 0$ and the roots are repeated.

(c) For $-1 < k < 1$, the roots are complex.

27. $y'' + y = x^3 + x$

$m^2 + 1 = 0$ when $m = -i, i$.

$$y_h = C_1\cos x + C_2\sin x$$
$$y_p = A_0 + A_1x + A_2x^2 + A_3x^3$$
$$y_p' = A_1 + 2A_2x + 3A_3x^2$$
$$y_p'' = 2A_2 + 6A_3x$$
$$y_p'' + y_p = (A_0 + 2A_2) + (A_1 + 6A_3)x + A_2x^2 + A_3x^3$$
$$= x^3 + x$$
$$A_0 = 0,\ A_1 = -5,\ A_2 = 0,\ A_3 = 1$$
$$y = C_1\cos x + C_2\sin x - 5x + x^3$$

28. $y'' + 2y = e^{2x} + x$

$m^2 + 2 = 0$ when $m = -\sqrt{2}i, \sqrt{2}i.$

$$y_h = C_1 \cos\sqrt{2}x + C_2 \sin\sqrt{2}x$$
$$y_p = Ae^{2x} + B_0 + B_1x$$
$$y_p' = 2Ae^{2x} + B_1$$
$$y_p'' = 4Ae^{2x}$$
$$y_p'' + 2y_p = 6Ae^{2x} + 2B_0 + 2B_1x = e^{2x} + x$$
$$A = \tfrac{1}{6},\ B_0 = 0,\ B_1 = \tfrac{1}{2}$$
$$y = C_1 \cos\sqrt{2}x + C_2 \sin\sqrt{2}x + \tfrac{1}{6}e^{2x} + \tfrac{1}{2}x$$

29. $y'' + y = 2\cos x$

$m^2 + 1 = 0$ when $m = -i, i.$

$$y_h = C_1 \cos x + C_2 \sin x$$
$$y_p = Ax\cos x + Bx\sin x$$
$$y_p' = (Bx + A)\cos x + (B - Ax)\sin x$$
$$y_p'' = (2B - Ax)\cos x + (-Bx - 2A)\sin x$$
$$y_p'' + y_p = 2B\cos x - 2A\sin x = 2\cos x$$
$$A = 0,\ B = 1$$
$$y = C_1\cos x + (C_2 + x)\sin x$$

30. $y'' + 5y' + 4y = x^2 + \sin 2x$

$m^2 + 5m + 4 = 0$ when $m = -1, -4.$

$$y_h = C_1e^{-x} + C_2e^{-4x}$$
$$y_p = A_0 + A_1x + A_2x^2 + B_0\sin 2x + B_1\cos 2x$$
$$y_p' = A_1 + 2A_2x + 2B_0\cos 2x - 2B_1\sin 2x$$
$$y_p'' = 2A_2 - 4B_0\sin 2x - 4B_1\cos 2x$$
$$y_p'' + 5y_p' + 4y_p = (4A_0 + 5A_1 + 2A_2) + (4A_1 + 10A_2)x + 4A_2x^2 - 10B_1\sin 2x + 10B_0\cos 2x = x^2 + \sin 2x$$
$$A_0 = \tfrac{21}{32},\ A_1 = -\tfrac{5}{8},\ A_2 = \tfrac{1}{4},\ B_0 = 0,\ B_1 = -\tfrac{1}{10}$$
$$y = C_1e^{-x} + C_2e^{-4x} + \tfrac{21}{32} - \tfrac{5}{8}x + \tfrac{1}{4}x^2 - \tfrac{1}{10}\cos 2x$$

31. $y'' - 2y' + y = 2xe^x$

$m^2 - 2m + 1 = 0$ when $m = 1, 1.$

$$y_h = (C_1 + C_2x)e^x$$
$$y_p = (v_1 + v_2x)e^x$$
$$v_1'e^x + v_2'xe^x = 0$$
$$v_1'e^x + v_2'(x + 1)e^x = 2xe^x$$
$$v_1' = -2x^2$$
$$v_1 = \int -2x^2\,dx = -\tfrac{2}{3}x^3$$
$$v_2' = 2x$$
$$v_2 = \int 2x\,dx = x^2$$
$$y = \left(C_1 + C_2x + \tfrac{1}{3}x^3\right)e^x$$

32. $y'' + 2y' + y = \dfrac{1}{x^2e^x}$

$m^2 + 2m + 1 = 0$ when $m = -1, -1.$

$$y_h = (C_1 + C_2x)e^{-x}$$
$$y_p = (v_1 + v_2x)e^{-x}$$
$$v_1'e^{-x} + v_2'(xe^{-x}) = 0$$
$$v_1'(-e^{-x}) + v_2'(-x + 1)e^{-x} = \frac{1}{e^xx^2}$$
$$v_1' = -\frac{1}{x}$$
$$v_1 = \int -\frac{1}{x}\,dx = -\ln|x|$$
$$v_2' = \frac{1}{x^2}$$
$$v_2 = \int \frac{1}{x^2}\,dx = -\frac{1}{x}$$
$$y = (C_1 + C_2x - \ln|x| - 1)e^{-x}$$

33. $y'' + y' - 6y = 54,\ y(0) = 2,\ y'(0) = 0$

$m^2 - m - 6 = 0$

$(m - 3)(m + 2) = 0$

$m_1 = 3,\ m_2 = -2$

$y_h = C_1e^{3x} + C_2e^{-2x}$

$y_p = -9$ by inspection

$y = y_h + y_p = C_1e^{3x} + C_2e^{-2x} - 9$

Initial conditions:

$y(0) = 2\colon 2 = C_1 + C_2 - 9 \Rightarrow C_1 + C_2 = 11$

$y'(0) = 0\colon 0 = 3C_1 - 2C_2 \Rightarrow C_1 = \frac{22}{5},\ C_2 = \frac{33}{5}$

$y = \frac{11}{5}\left(2e^{3x} + 3e^{-2x}\right) - 9$

34. $y'' + 25y = e^x,\ y(0) = 0,\ y'(0) = 0$

$y_h = C_1 \cos 5x + C_2 \sin 5x$

$y_p = Ae^x,\ y_p' = y_p'' = Ae^x$

$Ae^x + 25Ae^x = e^x \Rightarrow 26A = 1 \Rightarrow A = \dfrac{1}{26}$

$y = y_h + y_p = C_1 \cos 5x + C_2 \sin 5x + \dfrac{1}{26}e^x$

$y(0) = 0\colon 0 = C_1 + \dfrac{1}{26} \Rightarrow C_1 = -\dfrac{1}{26}$

$y'(0) = 0\colon 0 = 5C_2 + \dfrac{1}{26} \Rightarrow C_2 = -\dfrac{1}{130}$

$y = -\dfrac{1}{26} \cos 5x - \dfrac{1}{130} \sin 5x + \dfrac{1}{26}e^x$

35. $y'' + 4y = \cos x$

$m^2 + 4 = 0 \Rightarrow m = \pm 2i$

$y_h = C_1 \cos 2x + C_2 \sin 2x$

$y_p = A \cos x + B \sin x$

$y_p' = -A \sin x + B \cos x$

$y_p'' = -A \cos x - B \sin x$

$y_p'' + 4y_p = (-A \cos x - B \sin x) + 4(A \cos x + B \sin x) = \cos x$

$3A \cos x + 3B \sin x = \cos x \Rightarrow A = \frac{1}{3}$ and $B = 0$

$y_p = \frac{1}{3} \cos x$

$y = y_h + y_p = C_1 \cos 2x + C_2 \sin 2x + \frac{1}{3} \cos x$

Initial conditions: $y(0) = 6\colon 6 = C_1 + \frac{1}{3} \Rightarrow C_1 = \frac{17}{3}$

$y'(0) = -6\colon -6 = 2C_2 \Rightarrow C_2 = -3$

Particular solution: $y = \frac{17}{3} \cos 2x - 3 \sin 2x + \frac{1}{3} \cos x$

36. $y'' + 3y' = 6x$

$m^2 + 3m = 0 \Rightarrow m_1 = 0$ and $m_2 = -3$

$y_h = C_1 + C_2e^{-3x}$

$y_p = Ax^3 + Bx^2 + Cx + D$

$y_p' = 3Ax^2 + 2Bx + C$

$y_p'' = 6Ax + 2B$

$y_p'' + 3y_p' = (6Ax + 2B) + 3\left(3Ax^2 + 2Bx + C\right) = 9Ax^2 + (6A + 6B)x + (2B + 3C) = 6x,\ A = 0,\ B = 1,$ and $C = -\frac{2}{3}$

$y_p = x^2 - \frac{2}{3}x + D$

$y = y_h + y_p = C_1 + C_2e^{-3x} + x^2 - \frac{2}{3}x + D = C_3 + C_2e^{-3x} + x^2 - \frac{2}{3}x$

Initial conditions: $y(0) = 2\colon 2 = C_3 + C_2$

$y'(0) = \frac{10}{3}\colon \frac{10}{3} = -3C_2 - \frac{2}{3} \Rightarrow C_2 = -\frac{4}{3}$ and $C_3 = \frac{10}{3}$

Particular solution: $y = \frac{10}{3} - \frac{4}{3}e^{-3x} + x^2 - \frac{2}{3}x = \frac{1}{3}\left(10 - 4e^{-3x} + 3x^2 - 2x\right)$

37. $y'' - y' - 2y = 1 + xe^{-x}, y(0) = 1, y'(0) = 3$

$m^2 - m - 2 = (m - 2)(m + 1) = 0 \Rightarrow m = 2, -1$

$$y_h = C_1e^{2x} + C_2e^{-x}$$

$$y_p = A + (Bx + Cx^2)e^{-x}$$

$$y_p' = -(Bx + Cx^2)e^{-x} + (B + 2Cx)e^{-x} = (B + (2C - B)x - Cx^2)e^{-x}$$

$$y_p'' = -(B + (2C - B)x - Cx^2)e^{-x} + (2C - B - 2Cx)e^{-x} = (Cx^2 + (B - 4C)x + 2C - 2B)e^{-x}$$

$$y_p'' - y_p' - 2y_p = (2C - 2B + (-4C + B)x + Cx^2)e^{-x} - (B + (2C - B)x - Cx^2)e^{-x} - 2(A + (Bx + Cx^2)e^{-x})$$

$$= -2A + (-6Cx + 2C - 3B)e^{-x} = 1 + xe^{-x} \Rightarrow A = -\frac{1}{2}, -6C = 1 \text{ and } 2C - 3B = 0.$$

So, $C = -\dfrac{1}{6}$ and $B = -\dfrac{1}{9}$.

$$y = y_h + y_p = C_1e^{2x} + C_2e^{-x} - \frac{1}{2} + \left(-\frac{1}{9}x - \frac{1}{6}x^2\right)e^{-x}$$

Initial conditions: $y(0) = 1 = C_1 + C_2 - \dfrac{1}{2} \Rightarrow C_1 + C_2 = \dfrac{3}{2}$

$y'(0) = 3 = 2C_1 - C_2 - \dfrac{1}{9} \Rightarrow 2C_1 - C_2 = \dfrac{28}{9}$

Adding, $3C_1 = \dfrac{83}{18} \Rightarrow C_1 = \dfrac{83}{54}$.

So, $C_2 = -\dfrac{1}{27}$.

Particular solution: $y = \dfrac{83}{54}e^{2x} - \dfrac{1}{27}e^{-x} - \dfrac{1}{2} - \left(\dfrac{1}{9} + \dfrac{1}{6}x\right)xe^{-x}$

38. $y''' - y'' = 4x^2, y(0) = 1, y'(0) = 1, y''(0) = 1$

$y''' - y'' = 0$

$m^3 - m^2 = 0$ when $m = 0, 0, 1$.

$$y_h = C_1 + C_2x + C_3e^x$$

$$y_p = A_0x^2 + A_1x^3 + A_2x^4$$

$$y_p' = 2A_0x + 3A_1x^2 + 4A_2x^3$$

$$y_p'' = 2A_0 + 6A_1x + 12A_2x^2$$

$$y_p''' = 6A_1 + 24A_2x$$

$$y_p''' - y_p'' = (-2A_0 + 6A_1) + (-6A_1 + 24A_2)x - 12A_2x^2 = 4x^2 \text{ or } A_0 = -4, A_1 = -\tfrac{4}{3}, A_2 = -\tfrac{1}{3}$$

$$y = C_1 + C_2x + C_3e^x - 4x^2 - \tfrac{4}{3}x^3 - \tfrac{1}{3}x^4$$

$$y' = C_2 + C_3e^x - 8x - 4x^2 - \tfrac{4}{3}x^3$$

$$y'' = C_3e^x - 8 - 8x - 4x^2$$

Initial conditions: $y(0) = 1, y'(0) = 1, y''(0) = 1, 1 = C_1 + C_3, 1 = C_2 + C_3, 1 = C_3 - 8, C_1 = -8, C_2 = -8, C_3 = 9$

Particular solution: $y = -8 - 8x - 4x^2 - \tfrac{4}{3}x^3 - \tfrac{1}{3}x^4 + 9e^x$

39. By Hooke's Law, $F = kx$, $k = F/x = 64/(4/3) = 48$. Also, $F = ma$ and $m = F/a = 64/32 = 2$. So,

$$\frac{d^2y}{dt^2} + \left(\frac{48}{2}\right)y = 0$$

$$y = C_2 \cos\left(2\sqrt{6}t\right) + C_2 \sin\left(2\sqrt{6}t\right).$$

Because $y(0) = \frac{1}{2}$ you have $C_1 = \frac{1}{2}$ and $y'(0) = 0$ yields $C_2 = 0$. So, $y = \frac{1}{2}\cos\left(2\sqrt{6}t\right)$.

40. From Exercise 39 you have $k = 48$ and $m = 2$. Also, the damping force is given by $(1/8)(dy/dt)$.

$$2\left(\frac{d^2y}{dt^2}\right) = -\frac{1}{8}\frac{dy}{dt} - 48y$$

$$y'' + \frac{1}{16}y' + 24y = 0$$

$$16y'' + y' + 384y = 0$$

The characteristic equation $16m^2 + m = 384 = 0$ has complex roots

$$m = -\frac{1}{32} \pm \frac{\sqrt{24{,}575}\,i}{32} = -\frac{1}{32} \pm \frac{5\sqrt{983}}{32}i.$$

So, $y(t) = e^{-t/32}\left[C_1 \cos\left(\frac{5\sqrt{983}}{32}t\right) + C_2 \sin\left(\frac{5\sqrt{983}}{32}t\right)\right]$.

Initial conditions:

$$y(0) = \frac{1}{2} \Rightarrow C_1 = \frac{1}{2}$$

$$y'(0) = 0 \Rightarrow \frac{5\sqrt{983}}{32}C_2 - \frac{C_1}{32} = 0 \Rightarrow C_2 = \frac{\sqrt{983}}{9830}$$

Particular solution: $y(t) = e^{-t/32}\left[\frac{1}{2}\cos\left(\frac{5\sqrt{983}}{32}t\right) + \frac{\sqrt{983}}{9830}\sin\left(\frac{5\sqrt{983}}{32}t\right)\right]$

41. (a) (i) $y = \frac{1}{2}\cos 2t + \frac{12\pi}{\pi^2 - 4}\sin 2t + \frac{24}{4 - \pi^2}\sin \pi t$

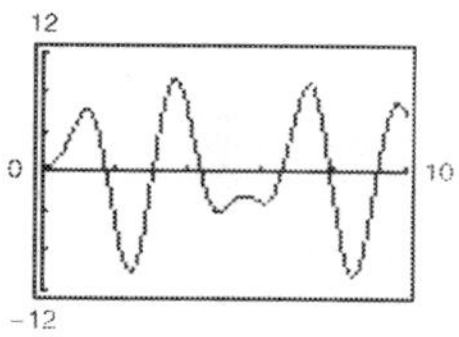

(ii) $y = \frac{1}{2}\left[\left(1 - 6\sqrt{2}t\right)\cos\left(2\sqrt{2}t\right) + 3\sin\left(2\sqrt{2}t\right)\right]$

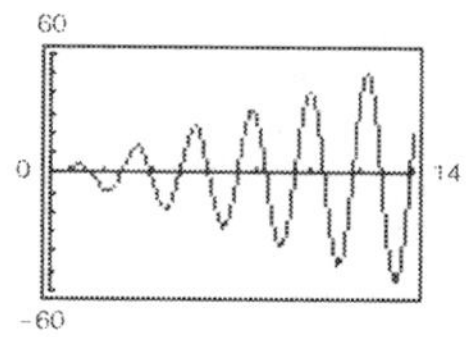

(iii) $y = \frac{e^{-t/5}}{398}\left[199\cos\frac{\sqrt{199}t}{5} + \sqrt{199}\sin\frac{\sqrt{199}t}{5}\right]$

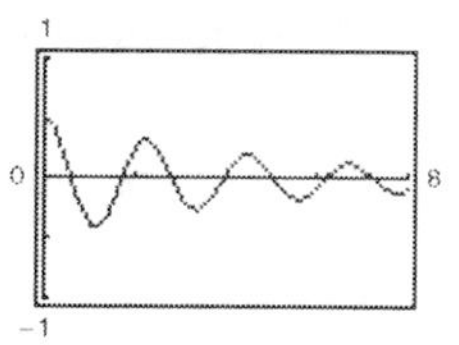

(iv) $y = \frac{1}{2}e^{-2t}(\cos 2t + \sin 2t)$

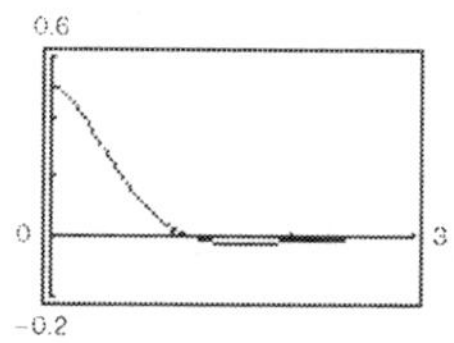

(b) The object comes to rest more quickly. It may not even oscillate, as in part (iv).

(c) It would oscillate more rapidly.

(d) Part (ii). The amplitude becomes increasingly large.

42. $y_p = \frac{1}{4}\cos x,\ y_p' = -\frac{1}{4}\sin x,\ y_p'' = -\frac{1}{4}\cos x$

$$y_p'' + 4y_p' + 5y_p = -\tfrac{1}{4}\cos x + 4\left(-\tfrac{1}{4}\sin x\right) + 5\left(\tfrac{1}{4}\cos x\right)$$
$$= \cos x - \sin x$$

False.

43. (a) $y_p'' = -A\sin x$ and $3y_p = 3A\sin x.$

So, $y_p'' + 3y_p = -A\sin x + 3A\sin x$
$= 2A\sin x = 12\sin x$

(b) $y_p = \dfrac{5}{2}\cos x$

(c) If $y_p = A\cos x + B\sin x$, then $y_p'' = -A\cos x - B\sin x$, and solving for A and B would be more difficult.

44. $y = 5$, because $y' = y'' = 0$ and $6(5) = 30$

45. $(x-4)y' + y = 0$. Letting $y = \sum_{n=0}^{\infty} a_n x^n$:

$$xy' - 4y' + y = \sum_{n=0}^{\infty} na_n x^n - 4\sum_{n=1}^{\infty} na_n x^{n-1} + \sum_{n=0}^{\infty} a_n x^n$$
$$= \sum_{n=0}^{\infty}(n+1)a_n x^n - \sum_{n=1}^{\infty} 4na_n x^{n-1} = \sum_{n=0}^{\infty}(n+1)a_n x^n - \sum_{n=-1}^{\infty} 4(n+1)a_{n+1}x^n = 0$$
$$(n+1)a_n = 4(n+1)a_{n+1}$$
$$a_{n+1} = \frac{1}{4}a_n$$
$$a_0 = a_0,\ a_1 = \frac{1}{4}a_0,\ a_2 = \frac{1}{4}a_1 = \frac{1}{4^2}a_0, \ldots, a_n = \frac{1}{4^n}a_0$$
$$y = a_0\sum_{n=0}^{\infty}\frac{x^n}{4^n}$$

46. $y'' + 3xy' - 3y = 0$. Letting $y = \sum_{n=0}^{\infty} a_n x^n$:

$$y'' + 3xy' - 3y = \sum_{n=2}^{\infty} n(n-1)a_n x^{n-2} + 3x\sum_{n=1}^{\infty} na_n x^{n-1} - 3\sum_{n=0}^{\infty} a_n x^n = 0$$

$$\sum_{n=0}^{\infty} (n+2)(n+1)a_{n+2}x^n = \sum_{n=0}^{\infty} (3 - 3n)a_n x^n$$

$$a_{n+2} = \frac{3(1-n)a_n}{(n+2)(n+1)}$$

$a_0 = a_0$ $\qquad a_1 = a_1$

$a_2 = \dfrac{3}{2 \cdot 1}a_0$ $\qquad a_3 = 0$

There are no odd-powered terms for $n > 1$.

$$a_4 = -\frac{3}{4 \cdot 3}\left(\frac{3}{2 \cdot 1}a_0\right) = -\frac{3(3)a_0}{4!}$$

$$a_6 = -\frac{3(3)}{6 \cdot 5}\left(-\frac{3(3)a_0}{4!}\right) = \frac{3^3(3)a_0}{6!}$$

$$a_8 = -\frac{3(5)}{8 \cdot 7}\left(\frac{3^3(3)a_0}{6!}\right) = -\frac{3^4(5 \cdot 3)a_0}{8!}$$

$$a_{10} = -\frac{3(7)}{10 \cdot 9}\left(-\frac{3^4(5 \cdot 3)a_0}{8!}\right) = \frac{3^5(7 \cdot 5 \cdot 3)a_0}{10!}$$

$$y = a_0 + \frac{3}{2}a_0x^2 + a_0\sum_{n=2}^{\infty} \frac{(-1)^{n+1}3^n[3 \cdot 5 \cdot 7 \cdots (2n-3)]}{(2n)!}x^{2n}$$

47. $y'' + y' - e^x y = 0,\ y(0) = 2,\ y'(0) = 0$

$y'' = -y' + e^x y$ $\qquad y''(0) = 2$

$y''' = -y'' + e^x(y + y')$ $\qquad y'''(0) = -2 + 2 = 0$

$y^{(4)} = -y''' + e^x(y + 2y' + y'')$ $\qquad y^{(4)}(0) = 4$

$y^{(5)} = -y^{(4)} + e^x(y + 3y' + 3y'' + y''')$ $\qquad y^{(5)}(0) = -4 + 8 = 4$

$$y \approx y(0) + y'(0)x + \frac{y''(0)}{2!}x^2 + \frac{y'''(0)}{3!}x^3 + \frac{y^{(4)}(0)}{4!}x^4 + \frac{y^{(5)}(0)}{5!}x^5 = 2 + x^2 + \frac{1}{6}x^4 + \frac{1}{30}x^5$$

Using the first four terms of the series, $y\left(\frac{1}{4}\right) \approx 2.063$.

48. $y'' + xy = 0,\ y(0) = 1,\ y'(0) = 1$

$y'' = -xy$ $\qquad y''(0) = 0$

$y''' = -xy' - y$ $\qquad y'''(0) = -1$

$y^{(4)} = -xy'' - y' - y' = -xy'' - 2y'$ $\qquad y^{(4)}(0) = -2$

$y^{(5)} = -xy''' - y'' - 2y'' = -xy''' - 3y''$ $\qquad y^{(5)}(0) = 0$

$y^{(6)} = -xy^{(4)} - y''' - 3y''' = -xy^{(4)} - 4y'''$ $\qquad y^{(6)}(0) = 4$

$y^{(7)} = -xy^{(5)} - y^{(4)} - 4y^{(4)} = -xy^{(5)} - 5y^{(4)}$ $\qquad y^{(7)}(0) = 10$

$$y \approx y(0) + y'(0)x + \frac{y''(0)}{2!}x^2 + \cdots + \frac{y^{(7)}(0)}{7!}x^7 = 1 + x - \frac{x^3}{6} - \frac{x^4}{12} + \frac{x^6}{180} + \frac{x^7}{504}$$

$y\left(\frac{1}{2}\right) \approx 1.474$

Problem Solving for Chapter 16

1. $\left(3x^2 + kxy^2\right)dx - \left(5x^2y + ky^2\right)dy = 0$

$$\frac{\partial M}{\partial y} = 2kxy$$

$$\frac{\partial N}{\partial x} = -10xy$$

$$\frac{\partial M}{\partial y} = \frac{\partial N}{\partial x} \Rightarrow k = -5$$

$\left(3x^2 - 5xy^2\right)dx - \left(5x^2y - 5y^2\right)dy = 0$ Exact

$$f(x, y) = \int\left(3x^2 - 5xy^2\right)dx = x^3 - \frac{5}{2}x^2y^2 + g(y)$$

$$f_y(x, y) = -5x^2y + g'(y) = -5x^2y + 5y^2$$

$$g'(y) = 5y^2 \Rightarrow g(y) = \frac{5}{3}y^3 + C_1$$

$$x^3 - \frac{5}{2}x^2y^2 + \frac{5}{3}y^3 = C_2$$

$$6x^3 - 15x^2y^2 + 10y^3 = C$$

2. $\left(kx^2 + y^2\right)dx - kxy\,dy = 0$

(a) $\frac{1}{x^2}\left(kx^2 + y^2\right)dx - \frac{1}{x^2}kxy\,dy = 0$

$$\left(k + \frac{y^2}{x^2}\right)dx - \frac{ky}{x}dy = 0$$

$$\frac{\partial M}{\partial y} = \frac{2y}{x^2} = \frac{\partial N}{\partial x} = \frac{ky}{x^2} \Rightarrow k = 2$$

(b) $\left(2 + \frac{y^2}{x^2}\right)dx - \left(\frac{2y}{x}\right)dy = 0$ Exact

$$f(x, y) = \int\left(2 + \frac{y^2}{x^2}\right)dx = 2x - \frac{y^2}{x} + g(y)$$

$$f_y(x, y) = \frac{-2y}{x} + g'(y) = \frac{-2y}{x} \Rightarrow g(y) = C_1$$

$$2x - \frac{y^2}{x} = C$$

3. $y'' - a^2y = 0, y > 0$

$$m^2 - a^2 = (m + a)(m - a) = 0 \Rightarrow m = \pm a$$

$$y = B_1e^{ax} + B_2e^{-ax} = \frac{C_1 + C_2}{2}e^{ax} + \frac{C_1 - C_2}{2}e^{-ax}$$

$$= C_1\left(\frac{e^{ax} + e^{-ax}}{2}\right) + C_2\left(\frac{e^{ax} - e^{-ax}}{2}\right)$$

$$= C_1 \cosh ax + C_2 \sinh ax$$

4. $y'' + \beta^2y = 0$

$$m^2 + \beta^2 = 0 \Rightarrow m = \pm\beta i$$

$$y = C_1 \cos \beta x + C_2 \sin \beta x$$

Let ϕ be given by $\cot \phi = \frac{C_2}{C_1}, 0 \le \phi < 2\pi$.

Then $C_1 \cos \phi = C_2 \sin \phi$.

Let $C = \frac{C_1}{\sin \phi} = \frac{C_2}{\cos \phi}$. Then

$$y = C_1 \cos \beta x + C_2 \sin \beta x$$

$$= C \sin \phi \cos \beta x + C \cos \phi \sin \beta x = C \sin(\beta x + \phi).$$

Note that if $C_1 = 0$, then $\phi = 0$ and

$y = C \sin(\beta x)$. And if $C_2 = 0$, then

$$y = C \sin\left(\beta x + \frac{\pi}{2}\right) = C \cos(\beta x).$$

5. The general solution to $y'' + ay' + by = 0$ is

$$y = B_1e^{(r+s)x} + B_2e^{(r-s)x}.$$

Let $C_1 = B_1 + B_2$ and $C_2 = B_1 - B_2$.

Then $B_1 = \frac{C_1 + C_2}{2}$ and $B_2 = \frac{C_1 - C_2}{2}$.

$$\text{So } y = \left(\frac{C_1 + C_2}{2}\right)e^{(r+s)x} + \left(\frac{C_1 - C_2}{2}\right)e^{(r-s)x}$$

$$= e^{rx}\left[C_1\left(\frac{e^{sx} + e^{-sx}}{2}\right) + C_2\left(\frac{e^{sx} - e^{-sx}}{2}\right)\right]$$

$$= e^{rx}\left[C_1 \cosh sx + C_2 \sinh sx\right].$$

6. The roots of the characteristic equation $m^2 + am + b = 0\ (a, b > 0)$ are

$m = \dfrac{-a \pm \sqrt{a^2 - 4b}}{2}$. You consider three cases:

(i) If the roots are equal,

then $\sqrt{a^2 - 4b} = 0$ and $y = (C_1 + C_2x)e^{\frac{-a}{2}x} \to 0$ as $x \to \infty$.

(ii) If the roots are complex,

$m = \dfrac{-a}{2} \pm \beta i$, then

$y = C_1e^{\frac{-a}{2}x}\cos\beta x + C_2e^{\frac{-a}{2}x}\sin\beta x \to 0$ as $x \to \infty$

(because $\cos\beta x$ and $\sin\beta x$ are bounded).

(iii) If the roots are real and distinct,

then $y = C_1e^{\frac{-a+\sqrt{a^2-4b}}{2}x} + C_2e^{\frac{-a-\sqrt{a^2-4b}}{2}x}$.

The second term clearly tends to 0 as $x \to \infty$.

For the first term, note that

$\sqrt{a^2 - 4b} = a\sqrt{1 - \dfrac{4b}{a^2}} < a$. So

$y = C_1e^{\left(\frac{-a}{2} + \frac{a}{2}\sqrt{1-\frac{4b}{a^2}}\right)x} \to 0$ as $x \to \infty$.

7. $y'' + ay = 0,\ y(0) = y(L) = 0$

(a) If $a = 0$, $y'' = 0 \Rightarrow y = cx + d$. $y(0) = 0 = d$ and $y(L) = 0 = cL \Rightarrow c = 0$. So $y = 0$ is the solution.

(b) If $a < 0$, $y'' + ay = 0$ has characteristic equation $m^2 + a = 0 \Rightarrow m = \pm\sqrt{-a}$.

$$y = C_1e^{\sqrt{-a}\,x} + C_2e^{-\sqrt{-a}\,x}$$

$$y(0) = 0 = C_1 + C_2 \Rightarrow -C_1 = C_2$$

$$\begin{aligned} y(L) = 0 &= C_1e^{\sqrt{-a}\,L} + C_2e^{-\sqrt{-a}\,L} \\ &= C_1e^{\sqrt{-a}\,L} - C_1e^{-\sqrt{-a}\,L} \\ &= 2C_1\left(\frac{e^{\sqrt{-a}\,L} - e^{-\sqrt{-a}\,L}}{2}\right) \\ &= 2C_1\sinh\left(\sqrt{-a}\,L\right) \Rightarrow C_1 = 0 = C_2 \end{aligned}$$

So, $y = 0$ is the only solution.

(c) For $a > 0$:

$$m^2 + a = 0 \Rightarrow m = \pm\sqrt{a}\,i$$

$$y = C_1\cos\left(\sqrt{a}\,x\right) + C_2\sin\left(\sqrt{a}\,x\right).$$

$$y(0) = 0 = C_1$$

$$y = C_2\sin\left(\sqrt{a}\,x\right)$$

$$y(L) = 0 = C_2\sin\left(\sqrt{a}\,L\right)$$

So $\sqrt{a}\,L = n\pi$

$$a = \left(\frac{n\pi}{L}\right)^2, \text{ } n \text{ an integer.}$$

8. $x^2y'' + axy' + by = 0,\ x > 0$

Let $x = e^t$.

(a) $\dfrac{dy}{dx} = \dfrac{dy/dt}{dx/dt} = e^{-t}\dfrac{dy}{dt}$

$$\frac{d^2y}{dx^2} = \frac{d/dt\left[e^{-t}(dy/dt)\right]}{e^t} = e^{-t}\left[e^{-t}\frac{d^2y}{dt^2} - e^{-t}\frac{dy}{dt}\right] = e^{-2t}\left[\frac{d^2y}{dt^2} - \frac{dy}{dt}\right]$$

$$x^2y'' + axy' + by = 0$$

$$e^{2t}\left[e^{-2t}\left(\frac{d^2y}{dt^2} - \frac{dy}{dt}\right)\right] + ae^t\left(e^{-t}\frac{dy}{dt}\right) + by = 0$$

$$\frac{d^2y}{dt^2} + (a - 1)\frac{dy}{dt} + by = 0$$

(b) $x^2y'' + 6xy' + 6y = 0$

Let $x = e^t$. From part (a), you have:

$$\frac{d^2y}{dt^2} + 5\frac{dy}{dt} + 6y = 0$$

$$m^2 + 5m + 6 = 0$$

$$(m + 3)(m + 2) = 0$$

$$m_1 = -3, m_2 = -2$$

$$y = C_1e^{-3t} + C_2e^{-2t} = C_1e^{-3\ln x} + C_2e^{-2\ln x} = C_1e^{\ln(1/x^3)} + C_2e^{\ln(1/x^2)} = \frac{C_1}{x^3} + \frac{C_2}{x^2}.$$

9. $\frac{d^2\theta}{dt^2} + \frac{g}{L}\theta = 0, \frac{g}{L} > 0$

(a) $\theta(t) = C_1 \sin\left(\sqrt{\frac{g}{L}}t\right) + C_2 \cos\left(\sqrt{\frac{g}{L}}t\right)$

Let ϕ be given by $\tan\left(\sqrt{\frac{g}{L}}\phi\right) = -\frac{C_1}{C_2}, -\frac{\pi}{2} < \phi < \frac{\pi}{2}.$

Then $C_2 \sin\left(\sqrt{\frac{g}{L}}\phi\right) = -C_1 \cos\left(\sqrt{\frac{g}{L}}\phi\right).$

Let $A = \frac{C_2}{\cos\left(\sqrt{\frac{g}{L}}\phi\right)} = -\frac{C_1}{\sin\left(\sqrt{\frac{g}{L}}\phi\right)}$

$$\theta(t) = C_1 \sin\left(\sqrt{\frac{g}{L}}t\right) + C_2 \cos\left(\sqrt{\frac{g}{L}}t\right) = -A \sin\left(\sqrt{\frac{g}{L}}\phi\right)\sin\left(\sqrt{\frac{g}{L}}t\right) + A\cos\left(\sqrt{\frac{g}{L}}\phi\right)\cos\left(\sqrt{\frac{g}{L}}t\right) = A\cos\left[\sqrt{\frac{g}{L}}(t + \phi)\right]$$

(b) $\theta(t) = A \cos\left[\sqrt{\frac{g}{L}}(t + \phi)\right], g = 9.8, L = 0.25$

$$\theta(0) = A\cos\left[\sqrt{39.2}\,\phi\right] = 0.1$$

$$\theta'(t) = -A\sqrt{\frac{g}{L}}\sin\left[\sqrt{\frac{g}{4}}(t + \phi)\right]$$

$$\theta'(0) = -A\sqrt{39.2}\sin\left[\sqrt{39.2}\,\phi\right] = 0.5$$

Dividing, $\tan\left[\sqrt{39.2}\,\phi\right] = \frac{-5}{\sqrt{39.2}} \Rightarrow \phi \approx -0.1076 \Rightarrow A \approx 0.128.$

$$\theta(t) = 0.128\cos\left[\sqrt{39.2}(t - 0.108)\right]$$

(c) Period $= \frac{2\pi}{\sqrt{39.2}} \approx 1$ sec

(d) Maximum is 0.128.

(e) $\theta(t) = 0$ at $t \approx 0.359$ sec, and at $t \approx 0.860$ sec.

(f) $\theta'(0.359) \approx -0.801, \theta'(0.860) \approx 0.801$

10. (a) $Ay'' = 2Wx - \frac{1}{2}Wx^2, A > 0$

$$y'' = \frac{2W}{A}x - \frac{W}{2A}x^2 = \frac{W}{2A}(4x - x^2)$$

$$y' = \frac{W}{2A}\left(2x^2 - \frac{x^3}{3}\right) + C_1$$

$$y = \frac{W}{2A}\left(\frac{2x^3}{3} - \frac{x^4}{12}\right) + C_1x + C_2$$

$$y(0) = 0 \Rightarrow C_2 = 0$$

$$y(2) = 0 \Rightarrow \frac{W}{2A}\left(\frac{16}{3} - \frac{16}{12}\right) + 2C_1 = 0$$

$$\frac{2W}{A} = -2C_1 \Rightarrow C_1 = \frac{-W}{A}$$

$$y = \frac{W}{2A}\left(\frac{2x^3}{3} - \frac{x^4}{12} - 2x\right)$$

(b) Using a graphing utility, the maximum deflection is at $x \approx 1.1074$, and the deflection is

$$\frac{W}{2A}(1.43476) \approx 0.7174\frac{W}{A}.$$

11. $y'' + 8y' + 16y = 0, y(0) = 1, y'(0) = 1$

(a) $\lambda = 4, \omega = 4, \lambda^2 - \omega^2 = 0$, critically damped

(b) $m_1 = m_2 = -4$

$$y = (C_1 + C_2t)e^{-4t},$$
$$y' = -4(C_1 + C_2t)e^{-4t} + C_2e^{-4t}$$

$$y(0) = 1 = C_1$$
$$y'(0) = 1 = -4 + C_2 \Rightarrow C_2 = 5$$
$$y = (1 + 5t)e^{-4t}$$

(c)

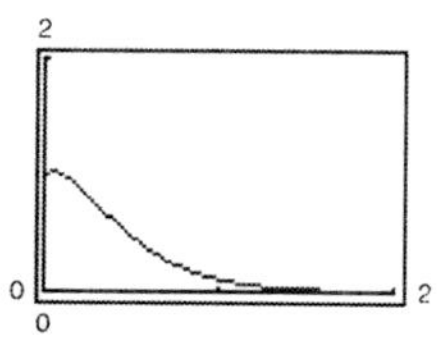

The solution tends to zero quickly.

12. $y'' + 2y' + 26y = 0, y(0) = 1, y'(0) = 4$

(a) $\lambda = 1, \omega = \sqrt{26},$

$\lambda^2 - \omega^2 = -25 < 0$, underdamped

(b) $m_1 = -1 + 5i, m_2 = -1 - 5i$

$$y = C_1e^{-t}\cos(5t) + C_2e^{-t}\sin(5t)$$

$$y(0) = 1 = C_1$$

$$y'(t) = -e^{-t}(C_1\cos 5t + C_2\sin 5t) + e^{-t}(-5C_1\sin 5t + 5C_2\cos 5t)$$

$$y'(0) = 4 = -C_1 + 5C_2 \Rightarrow C_2 = 1$$

$$y = e^{-t}(\cos 5t + \sin 5t)$$

(c)

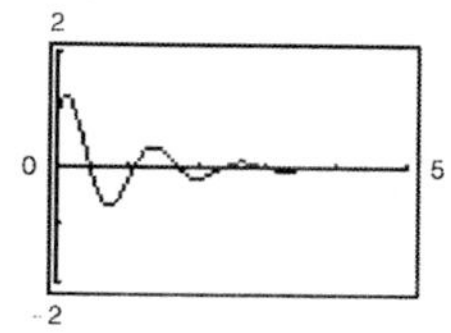

The solution oscillates.

13. $y'' + 20y' + 64y = 0, y(0) = 2, y'(0) = -20$

(a) $\lambda = 10, \omega = 8, \lambda^2 - \omega^2 = 36 > 0$, overdamped

(b) $m_1 = -10 + 6 = -4, m_2 = -10 - 6 = -16$

$$y = C_1e^{-4t} + C_2e^{-16t}$$
$$y(0) = 2 = C_1 + C_2$$
$$y'(t) = -4C_1e^{-4t} - 16C_2e^{-16t}$$
$$y'(0) = -20 = -4C_1 - 16C_2$$

$$\left.\begin{aligned} C_1 + C_2 &= 2 \\ -C_1 - 4C_2 &= -5 \end{aligned}\right\} C_1 = 1, C_2 = 1$$

$$y = e^{-4t} + e^{-16t}$$

(c)

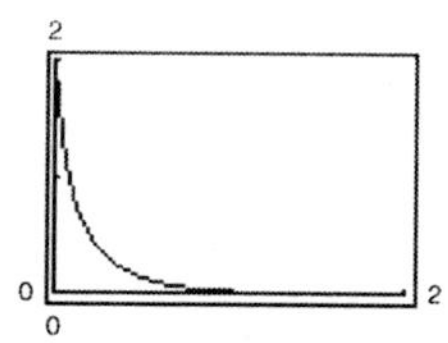

The solution tends to zero quickly.

14. $y'' + 2y' + y = 0, y(0) = 2, y'(0) = -1$

(a) $\lambda = 1, \omega = 1, \lambda^2 - \omega^2 = 0$, critically damped

(b) $m_1 = m_2 = -1$

$$y = (C_1 + C_2t)e^{-t}, y' = -(C_1 + C_2t)e^{-t} + C_2e^{-t}$$

$$y(0) = 2 = C_1$$
$$y'(0) = -1 = -2 + C_2 \Rightarrow C_2 = 1$$
$$y = (2 + t)e^{-t}$$

(c)

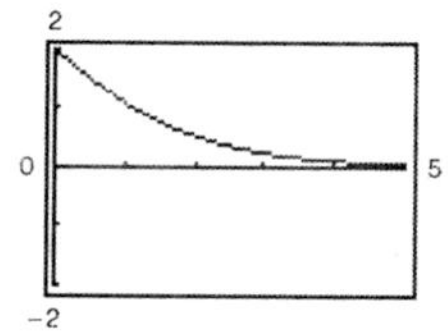

The solution tends to zero quickly.

15. Airy's Equation: $y'' - xy = 0$

$$y'' - xy + y - y = y'' - (x - 1)y - y = 0$$

Let $y = \sum_{n=0}^{\infty} a_n(x-1)^n$, $y' = \sum_{n=1}^{\infty} na_n(x-1)^{n-1}$, $y'' = \sum_{n=2}^{\infty} n(n-1)a_n(x-1)^{n-2}$.

$$y'' - (x - 1)y - y = 0$$

$$\sum_{n=2}^{\infty} n(n-1)a_n(x-1)^{n-2} - (x-1)\sum_{n=0}^{\infty} a_n(x-1)^n - \sum_{n=0}^{\infty} a_n(x-1)^n = 0$$

$$\sum_{n=-1}^{\infty} (n+3)(n+2)a_{n+3}(x-1)^{n+1} - \sum_{n=0}^{\infty} a_n(x-1)^{n+1} - \sum_{n=-1}^{\infty} a_{n+1}(x-1)^{n+1} = 0$$

$$(2a_2 - a_0) + \sum_{n=0}^{\infty}\left[(n+3)(n+2)a_{n+3} - a_n - a_{n+1}\right](x-1)^{n+1} = 0$$

$2a_2 - a_0 = 0 \Rightarrow a_2 = \frac{1}{2}a_0$; a_0, a_1 arbitrary

In general, $a_{n+3} = \dfrac{a_n + a_{n+1}}{(n+3)(n+2)}$.

$$a_3 = \frac{a_0 + a_1}{6}$$

$$a_4 = \frac{a_1 + a_2}{12} = \frac{a_1 + \left(\frac{1}{2}a_0\right)}{12} = \frac{2a_1 + a_0}{24}$$

$$a_5 = \frac{a_2 + a_3}{20} = \frac{\frac{1}{2}a_0 + \frac{a_0 + a_1}{6}}{20} = \frac{4a_0 + a_1}{120}$$

$$a_6 = \frac{a_3 + a_4}{30} = \frac{\left(\frac{a_0 + a_1}{6}\right) + \left(\frac{2a_1 + a_0}{24}\right)}{30} = \frac{5a_0 + 6a_1}{720}$$

$$a_7 = \frac{a_4 + a_5}{42} = \frac{\left(\frac{2a_1 + a_0}{24}\right) + \left(\frac{4a_0 + a_1}{120}\right)}{42} = \frac{9a_0 + 11a_1}{5040}$$

So, the first eight terms are

$$y = a_0 + a_1(x-1) + \frac{a_0}{2}(x-1)^2 + \frac{a_0 + a_1}{6}(x-1)^3 + \frac{2a_1 + a_0}{24}(x-1)^4 + \frac{4a_0 + a_1}{120}(x-1)^5$$
$$+ \frac{5a_0 + 6a_1}{720}(x-1)^6 + \frac{9a_0 + 11a_1}{5040}(x-1)^7.$$

16. (a) $T_{n+1}(x) = 2xT_n(x) - T_{n-1}(x)$

$T_0 = 1, T_1 = x$

$T_2 = 2x(x) - 1 = 2x^2 - 1$

$T_3 = 2x(2x^2 - 1) - x = 4x^3 - 3x$

$T_4 = 2x(4x^3 - 3x) - (2x^2 - 1) = 8x^4 - 8x^2 + 1$

(b) $(1 - x^2)y'' - xy' + k^2y = 0$

Substituting $T_0, \ldots, T_4$ into this equation shows that the polynomials satisfy Chebyshev's equation. For example, for T_4,

$(1 - x^2)[96x^2 - 16] - x[32x^3 - 16x] + 16[8x^4 - 8x^2 + 1] = 0$

(c) $T_5 = 2x(8x^4 - 8x^2 + 1) - (4x^3 - 3x) = 16x^5 - 20x^3 + 5x$

$T_6 = 2x(16x^5 - 20x^3 + 5x) - (8x^4 - 8x^2 + 1) = 32x^6 - 48x^4 + 18x^2 - 1$

$T_7 = 2x(32x^6 - 48x^4 + 18x^2 - 1) - (16x^5 - 20x^3 + 5x) = 64x^7 - 112x^5 + 56x^3 - 7x$

17. $x^2y'' + xy' + x^2y = 0$ Bessell equation of order zero

(a) Let $y = \sum_{n=0}^{\infty} a_n x^n$, $y' = \sum_{n=1}^{\infty} na_n x^{n-1}$, $y'' = \sum_{n=2}^{\infty} n(n-1)a_n x^{n-2}$.

$$x^2y'' + xy' + x^2y = 0$$

$$x^2\sum_{n=2}^{\infty} n(n-1)a_n x^{n-2} + x\sum_{n=1}^{\infty} na_n x^{n-1} + x^2\sum_{n=0}^{\infty} a_n x^n = 0$$

$$\sum_{n=2}^{\infty} n(n-1)a_n x^n + \sum_{n=1}^{\infty} na_n x^n + \sum_{n=0}^{\infty} a_n x^{n+2} = 0$$

$$\sum_{n=0}^{\infty} (n+2)(n+1)a_{n+2}x^{n+2} + \sum_{n=-1}^{\infty} (n+2)a_{n+2}x^{n+2} + \sum_{n=0}^{\infty} a_n x^{n+2} = 0$$

$$a_1x + \sum_{n=0}^{\infty} \left[(n+2)(n+1)a_{n+2} + (n+2)a_{n+2} + a_n\right]x^{n+2} = 0$$

$a_1 = 0$ and $a_{n+2} = \dfrac{-a_n}{(n+2)^2}$.

All odd terms a_i are 0.

$$a_2 = \frac{-a_0}{2^2}$$

$$a_4 = \frac{-a_2}{4^2} = a_0\frac{1}{2^2 \cdot 4^2} = \frac{a_0}{2^4(1 \cdot 2)^2}$$

$$a_6 = \frac{-a_4}{6^2} = -a_0\frac{1}{2^2 \cdot 4^2 \cdot 6^2} = \frac{-a_0}{2^6(3!)^2}$$

$$y = a_0\sum_{n=0}^{\infty} \frac{(-1)^n x^{2n}}{2^{2n}(n!)^2}$$

(b) This is the same function (assuming $a_0 = 1$).

18. (a) Let $y = \sum_{n=0}^{\infty} a_n x^n$, $y' = \sum_{n=1}^{\infty} na_n x^{n-1}$, $y'' = \sum_{n=2}^{\infty} n(n-1)a_n x^{n-2}$

$$x^2y'' + xy' + (x^2 - 1)y = 0$$

$$x^2\sum_{n=2}^{\infty} n(n-1)a_n x^{n-2} + x\sum_{n=1}^{\infty} na_n x^{n-1} + (x^2-1)\sum_{n=0}^{\infty} a_n x^n = 0$$

$$\sum_{n=2}^{\infty} n(n-1)a_n x^n + \sum_{n=1}^{\infty} na_n x^n + \sum_{n=0}^{\infty} a_n x^{n+2} - \sum_{n=0}^{\infty} a_n x^n = 0$$

$$\sum_{n=0}^{\infty} (n+2)(n+1)a_{n+2}x^{n+2} + \sum_{n=-1}^{\infty} (n+2)a_{n+2}x^{n+2} + \sum_{n=0}^{\infty} a_n x^{n+2} - \sum_{n=-2}^{\infty} a_{n+2}x^{n+2} = 0$$

$$-a_0 + (a_1 - a_1)x + \sum_{n=0}^{\infty}\left[(n+2)(n+1)a_{n+2} + (n+2)a_{n+2} + a_n - a_{n+2}\right]x^{n+2} = 0$$

$$a_0 = 0 \text{ and } \left[(n+2)(n+1) + (n+2) - 1\right]a_{n+2} = -a_n \Rightarrow \left[n^2 + 4n + 3\right]a_{n+2} = -a_n \Rightarrow a_{n+2} = \frac{-a_n}{(n+1)(n+3)}$$

All even terms a_i are 0.

$$a_3 = \frac{-a_1}{2 \cdot 4} = \frac{-a_1}{2^3}$$

$$a_5 = \frac{-a_3}{4 \cdot 6} = \frac{-a_1}{2^5\, 3!} = \frac{-2a_1}{2^5 \cdot 2!\, 3!}$$

$$a_7 = \frac{-a_5}{6 \cdot 8} = \frac{-2a_1}{2^7\, 3!\, 4!}$$

$$y = 2a_1\sum_{n=0}^{\infty} \frac{(-1)^n x^{2n+1}}{2^{2n+1} n!(n+1)!}$$

(b) This is the same function (assuming $2a_1 = 1$).

19. (a) Let $y = \sum_{n=0}^{\infty} a_n x^n$, $y' = \sum_{n=1}^{\infty} n a_n x^{n-1}$, $y'' = \sum_{n=2}^{\infty} n(n-1)a_n x^{n-2}$.

$$y'' - 2xy' + 8y = 0$$

$$\sum_{n=2}^{\infty} n(n-1)a_n x^{n-2} - 2x\sum_{n=1}^{\infty} n a_n x^{n-1} + 8\sum_{n=0}^{\infty} a_n x^n = 0$$

$$\sum_{n=0}^{\infty} (n+2)(n+1)a_{n+2}x^n - \sum_{n=0}^{\infty} 2na_n x^n + \sum_{n=0}^{\infty} 8a_n x^n = 0$$

$$\sum_{n=0}^{\infty} \left[(n+2)(n+1)a_{n+2} - 2na_n + 8a_n\right]x^n = 0$$

$$a_{n+2} = \frac{2(n-4)}{(n+2)(n+1)}a_n$$

$$a_4 = 16 = \frac{2(-2)}{4(3)}a_2 = -\frac{1}{3}a_2 \Rightarrow a_2 = -48$$

$$a_2 = -48 = \frac{2(-4)}{2}a_0 = -4a_0 \Rightarrow a_0 = 12$$

$$H_4(x) = 16x^4 - 48x^2 + 12$$

(b) $H_0(x) = \dfrac{(2x)^0}{0!} = 1$

$$H_1(x) = \frac{(2x)^1}{1!} = 2x$$

$$H_2(x) = \sum_{n=0}^{1} \frac{(-1)^n 2!(2x)^{2-2n}}{n!(2-2n)!} = \frac{2(2x)^2}{2!} - \frac{2}{1} = 4x^2 - 2$$

$$H_3(x) = \sum_{n=0}^{1} \frac{(-1)^n 3!(2x)^{3-2n}}{n!(3-2n)!} = \frac{3!(2x)^3}{3!} - \frac{3!(2x)^1}{1} = 8x^3 - 12x$$

$$H_4(x) = \sum_{n=0}^{2} \frac{(-1)^n 4!(2x)^{4-2n}}{n!(4-2n)!} = \frac{4!(2x)^4}{4!} - \frac{4!(2x)^2}{2!} + \frac{4!}{2!} = 16x^4 - 48x^2 + 12$$

20. (a) $xy'' + (1 - x)y' + ky = 0$

Let $y = \sum_{n=0}^{\infty} a_n x^n$, $y' = \sum_{n=1}^{\infty} na_n x^{n-1}$, $y'' = \sum_{n=2}^{\infty} n(n - 1)a_n x^{n-2}$.

$$x\sum_{n=2}^{\infty} n(n - 1)a_n x^{n-2} + (1 - x)\sum_{n=1}^{\infty} na_n x^{n-1} + k\sum_{n=0}^{\infty} a_n x^n = 0$$

$$\sum_{n=2}^{\infty} n(n - 1)a_n x^{n-1} + \sum_{n=1}^{\infty} na_n x^{n-1} - \sum_{n=1}^{\infty} na_n x^n + \sum_{n=0}^{\infty} ka_n x^n = 0$$

$$\sum_{n=1}^{\infty} (n + 1)na_{n+1}x^n + \sum_{n=0}^{\infty} (n + 1)a_{n+1}x^n - \sum_{n=1}^{\infty} na_n x^n + \sum_{n=0}^{\infty} ka_n x^n = 0$$

$$(a_1 + ka_0) + \sum_{n=1}^{\infty} \left[(n + 1)na_{n+1} + (n + 1)a_{n+1} - na_n + ka_n\right]x^n = 0$$

$$a_1 + ka_0 = 0 \qquad \Rightarrow a_1 = -ka_0$$

$$(n + 1)^2 a_{n+1} + (k - n)a_n = 0 \Rightarrow a_{n+1} = \frac{n - k}{(n + 1)^2}a_n$$

Let $a_0 = 1$.

For $k = 0$, $a_1 = a_2 = \cdots = 0 \Rightarrow L_0(x) = 1$.

For $k = 1$, $a_1 = -1$, $a_2 = a_3 = \cdots = 0 \Rightarrow L_1(x) = 1 - x$.

For $k = 2$, $a_1 = -2$, $a_2 = \frac{-1}{2^2}a_1 = \frac{1}{2} \Rightarrow L_2(x) = 1 - 2x + \frac{1}{2}x^2$.

In general, for a given integer $k \geq 0$, $a_{k+1} = a_{k+2} = \cdots = 0$. Furthermore, in the given formula for $L_k(x)$, you can verify that $a_{n+1} = \frac{n - k}{(n + 1)^2}a_n$. Finally, you can see that for $k \geq n$,

$$a_n = \frac{(n - 1) - k}{n^2}a_{n-1} = \frac{(-1)(k - (n - 1))}{n^2}a_{n-1} = \frac{(-1)(k + 1 - n)}{n^2} \cdot \frac{n - 2 - k}{(n - 1)^2}a_{n-2}$$

$$= \frac{(-1)^2(k - (n - 1))(k - (n - 2))}{n^2(n - 1)^2}a_{n-2} = \cdots = \frac{(-1)^n(k - (n - 1))(k - (n - 2))\cdots(k - 0)}{n^2(n - 1)^2\cdots 2^2 \cdot 1^2}a_0$$

$$= \frac{(-1)^2(k - (n - 1))(k - (n - 2))\cdots k(k - n)!}{(n!)^2(k - n)!}a_0 = \frac{(-1)^n k!}{(k - n)!(n!)^2}$$

(b) $$L_0(x) = \sum_{n=0}^{0} \frac{(-1)^n 0!\, x^n}{(0 - n)!(n!)^2} = 1$$

$$L_1(x) = \sum_{n=0}^{1} \frac{(-1)^n 1!\, x^n}{(1 - n)!(n!)^2} = 1 - x$$

$$L_2(x) = \sum_{n=0}^{2} \frac{(-1)^n 2!\, x^n}{(2 - n)!(n!)^2} = 1 - 2x + \frac{x^2}{2}$$

$$L_3(x) = \sum_{n=0}^{3} \frac{(-1)^n 3!\, x^n}{(3 - n)!(n!)^2} = 1 - 3x + \frac{3}{2}x^2 - \frac{x^3}{6}$$

$$L_4(x) = \sum_{n=0}^{4} \frac{(-1)^n 4!\, x^n}{(4 - n)!(n!)^2} = 1 - 4x + 3x^2 - \frac{2}{3}x^3 + \frac{1}{24}x^4$$